CONTENTS—1990 REFRIGERAT

CONTENTS—1989 FUNDAMENTALS

1992 ASHRAE HANDBOOK

Heating, Ventilating, and Air-Conditioning Systems and Equipment

Inch-Pound Edition

American Society of Heating, Refrigerating and Air-Conditioning Engineers, Inc.
1791 Tullie Circle, N.E., Atlanta, GA 30329
404-636-8400

DEDICATED

TO THE ADVANCEMENT OF

THE PROFESSION

AND ITS ALLIED INDUSTRIES

Comments, criticisms, and suggestions regarding the subject matter are invited. Any errors or omissions in the data should be brought to the attention of the Editor. If required, an errata sheet will be issued at approximately the same time as the next Handbook. Notice of any significant errors found after that time will be published in the ASHRAE *Journal*.

ISBN 0-910110-86-7
ISSN 1041-2344

CONTENTS

Contributors

Technical Committees and Task Groups

Preface

AIR-CONDITIONING AND HEATING SYSTEMS

AIR-HANDLING EQUIPMENT

HEATING EQUIPMENT

GENERAL COMPONENTS

UNITARY EQUIPMENT

ADDITIONS AND CORRECTIONS

INDEX

Composite index to the refrigeration-related chapters of the 1988 EQUIPMENT, 1989 FUNDAMENTALS, 1990 REFRIGERATION Systems and Applications, 1991 HVAC Applications, and 1992 HVAC Systems and Equipment volumes.

CONTRIBUTORS

In addition to the Technical Committees, the following individuals contributed significantly to this volume. The appropriate chapter numbers follow each contributor's name.

Arthur B. Sirjord, Jr. (1, 3)
TRA Architects and Engineers

Bryan H. Atkinson (2)
Atkinson Koven Feinberg Engineers

Michael W. Gallagher (2, 3)
Southern California Air
Conditioning Distributors

Howard J. McKew (2)
Balsco, Inc.

Dennis J. Wessel (2)
Bacik Karpinski Associates, Inc.

Kenneth E. Gill (3)
HDR, Inc.

John J. Harmon (3)
H C YU and Associates

Timothy W. O'Connor (3)
Robson & Woese Inc.

Carl N. Lawson (4)
NVT Technologies, Inc.

Bruce Anderson (5)
The Trane Company

Donald M. Eppleheimer (5)
The Trane Company

J. Geoffrey Adair (6)
Adair Industries

Norman A. Buckley (6, 15)
Buckley Associates

Lawrence V. Drake (6)
Teal International Corporation

Herbert P. Becker (7)
Chervenak Keane & Co.

Gerald R. Guinn (7, 34)
University of Alabama

Donald T. McClellan (7, 41)
Columbia Gas of Kentucky

Charles F. Daly (8)

Margaret S. Drake (8)
The Ohio State University

Gary Phetteplace (8, 11)
U.S. Army Cold Regions Laboratory

Joseph A. Pietsch (8)

Frank J. Pucciano (8)
Georgia Power Company

Daniel J. Dempsey (9)
Carrier Corporation

James H. Healy (9, 29)
NHAW

Gregory A. Lynch (9, 29)
Inter-City Products Corporation
(USA)

William A. Ryan (9, 29)
Gas Research Institute

Allen Hanley (10, 14, 39, 42, 43)

Robert A. Bunn (11)
Barge, Waggoner, Sumner and
Cannon
Engineers • Architects • Planners

Gary W. Conkright (11)
Rovanco Corp.

Charles G. McDowell (11)
Alabama Power Company

Vernon P. Meyer (11)
US Army Corps of Engineers

Victor L. Penar (11)
Midwesco

Gordon M. Reistad (11)
Oregon State University

Lewis P. Tomer (11)
Cleveland Thermal Energy Corp.

Joseph J. Wimberly, III (11)
I.C. Thomasson Associates

Albert W. Black (12)
McClure Engineering Associates

William J. Coad (12, 13)
McClure Engineering Associates

J. Gordon Frye (15)
Ohio Power Company

Loukas N. Kalisperis (15)
The Pennsylvania State University

Luis H. Summers (15)
University of Colorado

Herman F. Behls (16)
Sargent & Lundy

John H. Stratton (16)
SMACNA

Carlton Brown (17)
AirConcepts, Inc.

J. Barrie Graham (18)
Graham Consultants

Peter Morris (19)
Honeywell Inc.

Roy T. Otterbein (19)
Albers Technologies

Patricia T. Thomas (19)
Munters Corporation

M. Sohail Aslam (20)
Research Products Corporation

Roland Ares (21, 23, 36)

Lewis G. Harriman, III (22)
Mason-Grant Company

William M. Worek (22)
University of Illinois at Chicago

W. Peter Zheng (22)
University of Illinois at Chicago

Edward Ames (24)
Colmac Coil Manufacturing, Inc.

Carl J. Bauder (25)

Matthew K. Klein (25)
NIOSH

Eugene L. Valerio (25)
AIR FILTER/CONTROL Inc.

Gerhard W. Knutson (26)
SEC Donahue

Charles G. Noll (26)
the SIMCO company, inc.

Leonard J. O'Dell (26)

Thomas A. Butcher (27)
Brookhaven National Laboratory

Richard Krajewski (27)
Brookhaven National Laboratory

Roger McDonald (27)
Brookhaven National Laboratory

Hall Virgil (27, 31)
Carrier Corporation

James B. Rishel (28)
Systecon, Inc.

John I. Woodworth (28, 33)

Lawrence R. Brand (29)
Gas Research Institute

James G. Crawford (29)
The Trane Company

Robert A. Macriss (29)
Phillips Engineering Company

Tony Demases (30)
Lone Star Gas Company

Robert J. Kolodgy (30)
American Gas Association
Laboratories

Jerry E. Lawson (30)
TPI Corporation

James A. Ranfone (30)
American Gas Association

Randy J. Rechgruber (30)
Minnesota Power

Gary T. Satterfield (30)
Hearth Products Association

Darrell D. Paul (31)
Battelle

Allen L. Rutz (31)
Battelle

Richard L. Stone (31)

Gordon Holness (32)
Albert Kahn Associates

Richard A. Hegberg (33, 39, 42, 43)
Hegberg & Associates

Charles J. Cromer (34)
Florida Solar Energy Center

Gerald R. Guinn (34)
University of Alabama

ASHRAE TECHNICAL COMMITTEES AND TASK GROUPS

SECTION 1.0—FUNDAMENTALS AND GENERAL
1.1 Thermodynamics and Psychrometrics
1.2 Instruments and Measurements
1.3 Heat Transfer and Fluid Flow
1.4 Control Theory and Application
1.5 Computer Applications
1.6 Terminology
1.7 Operation and Maintenance Management
1.8 Owning and Operating Costs
1.9 Electrical Systems

SECTION 2.0—ENVIRONMENTAL QUALITY
2.1 Physiology and Human Environment
2.2 Plant and Animal Environment
2.3 Gaseous Air Contaminants and Gas Contaminant Removal Equipment
2.4 Particulate Air Contaminants and Particulate Contaminant Removal Equipment
2.5 Air Flow Around Buildings
2.6 Sound and Vibration Control
TG Global Warming
TG Halocarbon Emissions
TG Safety
TG Seismic Restraint Design

SECTION 3.0—MATERIALS AND PROCESSES
3.1 Refrigerants and Brines
3.2 Refrigerant System Chemistry
3.3 Contaminant Control in Refrigerating Systems
3.4 Lubrication
3.5 Sorption
3.6 Corrosion and Water Treatment
3.7 Fuels and Combustion

SECTION 4.0—LOAD CALCULATIONS AND ENERGY REQUIREMENTS
4.1 Load Calculation Data and Procedures
4.2 Weather Information
4.3 Ventilation Requirements and Infiltration
4.4 Thermal Insulation and Moisture Retarders
4.5 Fenestration
4.6 Building Operation Dynamics
4.7 Energy Calculations
4.8 Energy Resources
4.9 Building Envelope Systems
4.10 Indoor Environmental Modeling
TG Cold Climate Design

SECTION 5.0—VENTILATION AND AIR DISTRIBUTION
5.1 Fans
5.2 Duct Design
5.3 Room Air Distribution
5.4 Industrial Process Air Cleaning (Air Pollution Control)
5.5 Air-to-Air Energy Recovery
5.6 Control of Fire and Smoke
5.7 Evaporative Cooling
5.8 Industrial Ventilation
5.9 Enclosed Vehicular Facilities

SECTION 6.0—HEATING EQUIPMENT, HEATING AND COOLING SYSTEMS AND APPLICATIONS
6.1 Hydronic and Steam Equipment and Systems
6.2 District Heating and Cooling
6.3 Central Forced Air Heating and Cooling Systems
6.4 In-Space Convection Heating
6.5 Radiant Space Heating and Cooling
6.6 Service Water Heating
6.7 Solar Energy Utilization
6.8 Geothermal Energy Utilization
6.9 Thermal Storage

SECTION 7.0—PACKAGED AIR-CONDITIONING AND REFRIGERATION EQUIPMENT
7.1 Residential Refrigerators, Food Freezers and Drinking Water Coolers
7.5 Room Air Conditioners and Dehumidifiers
7.6 Unitary Air Conditioners and Heat Pumps
TG Unitary Combustion-Engine-Driven Heat Pumps

SECTION 8.0—AIR-CONDITIONING AND REFRIGERATION SYSTEM COMPONENTS
8.1 Positive Displacement Compressors
8.2 Centrifugal Machines
8.3 Absorption and Heat Operated Machines
8.4 Air-to-Refrigerant Heat Transfer Equipment
8.5 Liquid-to-Refrigerant Heat Exchangers
8.6 Cooling Towers and Evaporative Condensers
8.7 Humidifying Equipment
8.8 Refrigerant System Controls and Accessories
8.9 Valves
8.10 Pumps and Hydronic Piping
8.11 Electric Motors—Open and Hermetic

SECTION 9.0—AIR-CONDITIONING SYSTEMS AND APPLICATIONS
9.1 Large Building Air-Conditioning Systems
9.2 Industrial Air Conditioning
9.3 Transportation Air Conditioning
9.4 Applied Heat Pump/Heat Recovery Systems
9.5 Cogeneration Systems
9.6 Systems Energy Utilization
9.7 Testing and Balancing
9.8 Large Building Air-Conditioning Applications
9.9 Building Commissioning
TG Clean Spaces
TG Laboratory Systems
TG Tall Buildings

SECTION 10.0—REFRIGERATION SYSTEMS
10.1 Custom Engineered Refrigeration Systems
10.2 Automatic Icemaking Plants and Skating Rinks
10.3 Refrigerant Piping, Controls, and Accessories
10.4 Ultra-Low Temperature Systems and Cryogenics
10.5 Refrigerated Distribution and Storage Facilities
10.6 Transport Refrigeration
10.7 Commercial Food and Beverage Cooling Display and Storage
10.8 Refrigeration Load Calculations

SECTION 11.0—REFRIGERATED FOOD TECHNOLOGY AND PROCESSING
11.2 Foods and Beverages
11.5 Fruits, Vegetables and Other Products
11.9 Thermal Properties of Food

PREFACE

The 1992 ASHRAE Handbook continues to reflect the re-arrangement of topics in the Handbook series. This Handbook includes both systems-related chapters from the 1987 Handbook and HVAC equipment-related chapters from the 1988 Handbook.

The information in this book reflects the work of hundreds of volunteers. Working through the Handbook Committee and technical committees within the Society, these volunteers reviewed all chapters from the previous volumes and made major revisions to over one-third of them.

New or completely revised chapters include all-air systems (Chapter 2), cogeneration systems (Chapter 7), district heating and cooling (Chapter 11), water system design (Chapter 12), and valves (Chapter 43). New or additional descriptions are also included for economizers (Chapter 5), energy recovery (Chapter 7), solid desiccants (Chapter 22), industrial gas cleaning equipment (Chapter 26), boilers (Chapter 28), furnaces (Chapter 29), residential heating (Chapter 30), radiators (Chapter 33), rotary compressors (Chapter 35), and plastic pipes (Chapter 42).

In addition to these changes, this Handbook has been converted from one edition with dual units of measurement to two separate editions—one containing Inch-Pound (I-P) units of measurement and the other, the International System of Units (SI). Now all Handbook volumes are available in either SI or I-P editions.

Instead of being published in this volume, several chapters that cover refrigeration equipment will be revised and included in the 1994 ASHRAE *Handbook—Refrigeration*. These chapters, which are currently in the 1988 ASHRAE *Handbook—Equipment*, include information on absorption equipment (1988 Handbook Chapter 12); air-cycle equipment (Chapter 13); liquid chilling systems (Chapter 17); component balancing (Chapter 18); refrigerant control devices (Chapter 19); factory dehydrating, charging, and testing (Chapter 21); retail food store refrigeration equipment (Chapter 35); food service and general commercial refrigeration (Chapter 36); household refrigerators and freezers (Chapter 37); drinking water coolers (Chapter 38); bottled beverage coolers and vending machines (Chapter 39); and ice makers (Chapter 40).

Additions and corrections for the 1989, 1990, and 1991 Handbooks precede the index. Errors found in this volume will be reported in the 1993 ASHRAE *Handbook—Fundamentals*.

The Handbook Committee welcomes reader input. If you have suggestions or comments on improving a chapter, write or telephone: Handbook Editor, ASHRAE, 1791 Tullie Circle, Atlanta, GA 30329 (Telephone: 404-636-8400).

Robert A. Parsons
Handbook Editor

ASHRAE HANDBOOK COMMITTEE

Byron A. Hamrick, Jr., Chairman

1992 HVAC Systems and Equipment Volume Subcommittee: **Charles J. Procell,** Chairman

John H. Klote   **Thomas W. McDonald**   **Richard F. Sharp**   **Donald H. Spethmann**

ASHRAE HANDBOOK STAFF

Robert A. Parsons, Editor   **Claudia Forman,** Associate Editor

Andrea S. Andersen, Assistant Editor

Ron Baker, Production Manager

Nancy F. Thysell, Christopher R. Hart, and **E. Haven Hawley,** Typography

Lawrence H. Darrow and **Susan M. Boughadou,** Graphics

Frank M. Coda, Publisher
W. Stephen Comstock, Communications and Publications Director

CHAPTER 1

AIR-CONDITIONING SYSTEM SELECTION AND DESIGN

AN air-conditioning system maintains desired environmental conditions within a space. In almost every application, there is a myriad of options available to the designer to satisfy the basic goal. It is in the selection and combination of these options that the engineer must consider all the criteria defined by the functional ambition.

Air-conditioning systems are categorized by how they control cooling in the conditioned area. They are also segregated to accomplish specific purposes by special equipment arrangement. This chapter considers procedures and elements that constitute the system. It also describes and defines the design concepts and characteristics of basic air-conditioning systems. Chapters 2 through 5 describe specific systems and their attributes, based on their terminal cooling medium and their common variations.

SELECTING A SYSTEM

The designer is responsible for considering various systems and recommending the one or two that will perform as desired. It is imperative that the designer and the owner collaborate on identifying and rating the goals of the system design. Some of those criteria are as follows:

1. Performance requirements
2. Capacity requirements
3. Spatial requirements
4. First cost
5. Operating cost
6. Reliability
7. Flexibility
8. Maintainability

Since these factors are interrelated, the owner and the designer must consider how each affects the other. The relative importance of these factors differs with different owners and often changes from one project to another.

SELECTION GOALS

The designer must be aware of and account for specific goals that the owner may require other than merely providing a desired environment. These goals may include:

1. Supporting a process, such as the operation of computer equipment.
2. Promoting an aseptic environment.

___
The preparation of this chapter is assigned to TC 9.1, Large Building Air-Conditioning Systems.

3. Increasing sales.
4. Increasing net rental income.
5. Increasing the salability of a property.

The relative importance of first cost as compared to operating cost, the extent and frequency of maintenance and whether maintenance requires entering the occupied space, how often a system may be expected to fail, how much of the project would be affected by a failure, and how long before the failure can be corrected are typical concerns of owners. Each of these concerns has a different priority, depending on the owner's goals.

The owner can make appropriate value judgments if the designer provides complete information regarding the advantages and disadvantages of each option. Just as the owner does not usually know the relative advantages and disadvantages of different systems, the designer rarely knows all the owner's financial and functional goals. Hence, it is important to involve the owner in selecting the system.

SYSTEM OPTION CONSTRAINTS

The first step in selecting a system is to determine and document constraints dictated by performance, capacity, available space, and other factors important to the project.

Few projects allow detailed quantitative evaluation of all alternatives, and common sense and subjective experience narrow choices to two or three potential systems.

Cooling Loads

Establishing the cooling load often narrows the choice to systems that fit within the available space and are compatible with the building architecture. Chapter 26 of the 1989 ASHRAE *Handbook—Fundamentals* covers how to determine the magnitude and characteristics of the cooling load and how it varies with time and operating conditions. By establishing the capacity requirement, the size of equipment can be estimated. Then, the number of options may be narrowed to those systems that work well on projects of various sizes.

Zoning Requirements

Loads vary over time in different areas due to changes in weather, occupancy, activities, and solar exposure. Each space with a particular exposure requires a different control zone to maintain constant temperature. Some areas with special requirements may need individual control or individual systems, independent of the rest of the building. Variations in indoor conditions, which are acceptable in one space, may be unacceptable in other areas of the same building. The extent of zoning, the degree of control required in each zone, and the space required for individual zones also narrows the system choices.

No matter how efficiently a particular system operates or how economical it may be to install, it cannot be considered if it (1) does not maintain the desired interior environment within an acceptable tolerance under all conditions and occupant activity, and (2) does not physically fit into the building without being objectionable.

Heating and Ventilation

Cooling and humidity control are often the basis of sizing air-conditioning components and subsystems, but the system may also provide other functions, such as heating and ventilation. For example, if the system provides large quantities of outside air for ventilation or replaces air exhausted from the building, only systems that transport large volumes of air need to be considered. In this situation, the ventilation system will require a large air-handling and duct distribution system, thus other means can be discarded.

Effectively delivering heat to an area may be an equally strong factor in system selection. A distribution system that offers high efficiency and comfort for cooling may be a poor choice for heating. This performance compromise may be small for one application in one climate and may be unacceptable in another with more stringent heating requirements.

Architectural Constraints

Air-conditioning systems and the associated distribution systems often occupy a substantial space. Major components may also require special support from the structure. The size and appearance of terminal devices, whether they are diffusers, fan coil units, or radiant panels, have an impact on architectural design because they are visible from the occupied space.

Other factors that limit the selection of a system include (1) acceptable noise levels, (2) space available to house equipment and its location relative to the occupied space, (3) space available for distribution pipes and ducts, and (4) the acceptability of components obtruding into the occupied space, both physically and visually.

NARROWING THE CHOICE

Chapters 2 through 5 each include an evaluation section, which briefly summarizes the positive and negative features of various systems. Comparing the features against a list of design factors and their relative importance will usually identify two or three approaches that closely meet the project criteria. In making subjective choices, it is helpful to keep notes on all systems considered and the reason for eliminating those that are unacceptable.

In most cases, two system selections will evolve: a secondary (or distribution) system delivers heating or cooling to the occupied space from a primary system, which converts energy from fuel or electricity. The two systems are, to a great extent, independent, so that several secondary systems will work with different primary systems. In some cases, however, only one secondary system will work with a primary system.

Once subjective analysis has identified two or three systems—and sometimes only one choice may remain—detailed quantitative evaluations of each system must be made. All systems considered should perform satisfactorily to meet the owner's essential goals. The owner then needs specific data on each system to make an informed choice. Chapter 28 of the 1989 ASHRAE *Handbook—Fundamentals*, outlines how to

estimate annual energy costs. In the 1991 ASHRAE *Handbook—HVAC Applications*, Chapter 35 deals with mechanical maintenance, and Chapter 33 describes life-cycle costing, a method that compares the overall economics of systems.

SELECTION REPORT

As the last step of system selection, a designer should prepare a summary of the design and the selection criteria, a brief outline of the systems deemed inappropriate, and a comparison of the systems selected for detailed study.

If a matrix assessment is made, the designer should have the owner provide input to the analysis. The owner's input can also be applied as weighted multipliers.

There are many grading criteria lists including those in Chapter 33 of the 1991 ASHRAE *Handbook—HVAC Applications*. The examination of total owning and operating costs is an important system selection tool. Other more intangible assessments may include some of the following:

Comfort considerations
    Control options
    Noise
    Ventilation
    Filtration
    Effect of failure

Space considerations
    Floor space
    Plenum space
    Furniture placement
    Maintenance accessibility
    Roofs

The more tangible examinations include:

First cost
    System cost
    Cost to add zones
    Ability to increase capacity
    Contribution to life safety needs
    Air quality control

Operating cost
    Energy cost
    Gas
    Electricity

Maintenance cost
    Economy cycle (free cooling)
    Heat recovery

The system selection report should also address the following inquiries:

1. Does the system fit in the available space, or does it require some architectural modification? Does the system use more floor space than others considered, or does it require construction of additional space for mechanical rooms or shafts?
2. Will the system deliver the desired uniform temperature under varying weather and solar conditions? If compromises are made from the ideal control zoning, how much variation may be expected between spaces?
3. How much will the system cost to own compared to others considered? What is the recovery time of the initial investment, interest on investment, and the future cost of replacement equipment?
4. What are the operating costs—energy costs, maintenance, operating labor, and supplies—of this system compared to others?

5. What reliability can the owner expect compared to other systems? Which component failures might affect the entire building, and which would affect only limited areas? How easily may the system be serviced? How quickly can the system be restored to operation after various equipment failures?
6. Is the system flexible enough to meet changes in the owner's needs? What is required to add a control zone? Can it meet the increased capacity requirements of a space when equipment is added? How will changes in the interior layout and arrangement affect performance?

The system selection report should conclude with a recommended system choice, along with reasons for the choice. The elements of the report should be discussed to be sure the owner's goals have been recognized.

BASIC CENTRAL AIR-CONDITIONING AND DISTRIBUTION SYSTEM

The basic secondary system is an all-air, single-zone, air-conditioning system (Wheeler 1977). It may be designed to supply a constant air volume or a variable air volume for low-, medium-, and high-pressure air distribution. Normally, the equipment is located outside the conditioned area, in a basement, penthouse, or service area. It can, however, be installed within the conditioned area if conditions permit. The equipment can be adjacent to the primary heating and refrigeration equipment or at considerable distance from it by circulating refrigerant, chilled water, hot water, electricity, or steam for energy transfer.

APPLICATIONS

Some central system applications are: (1) spaces with uniform loads, (2) small spaces requiring precision control, (3) multiple systems for large areas, (4) systems for complete environmental control, and (5) a primary source of conditioned air for other subsystems.

Spaces with Uniform Loads

Spaces with uniform loads are generally those with relatively large open areas and small external loads, such as theaters, auditoriums, department stores, and the public spaces of many buildings. Here, the air-conditioning loads are fairly uniform, and adjustment for minor variations can be made by supplying more or less air in the original design and balance of the system.

In office buildings, the interior areas generally meet these criteria as long as local areas of relatively intense and variable heat sources, such as computers, are treated separately. In these applications, nonceiling partitions allow wider diffusion of the conditioned air and equalization of temperatures. These areas usually require year-round cooling, and any isolated spaces with limited occupancy may require special evaluation, as discussed in Chapter 3 of the 1991 ASHRAE *Handbook— HVAC Applications.*

Central systems can also adapt to one-story buildings and to the top floor of single-occupancy spaces, if the exterior walls are part of the main conditioned areas and if the space has a uniform roof load.

In most single-room commercial applications, temperature variations of up to 4°F at the outside walls are usually considered acceptable for tenancy requirements. However, these variations should be carefully determined and limited during design. If people sit or work near the outside walls or if they are isolated by partitions, supplementary heating equipment may be required at the walls, depending on the outdoor design temperature in winter and the thermal characteristics of the wall.

Spaces Requiring Precision Control

These spaces are usually isolated rooms within a larger building and have stringent requirements for cleanliness, humidity, temperature control, and air distribution. Central systems components can be selected and assembled to meet the exact requirements of these areas.

Multiple Systems for Large Buildings

In large buildings such as hangers, factories, large stores, office buildings, and hospitals, practical considerations require installation of multiple central systems. The size of the individual system is usually determined by first cost and operating cost considerations.

Primary Source for Other Systems

Systems for controlling conditions in individual zones are described in Chapters 2 through 5. These tertiary systems move a constant supply of conditioned air for ventilation and control some of the air-conditioning load. This air supply often reduces the amount of conditioned air handled by the central system and, consequently, the space required for ductwork. Ductwork size can be further reduced by moving air at high velocities. However, high-velocity system design must consider the resultant high pressure, sound levels, and energy requirements. Chapters 7 and 32 of the 1989 ASHRAE *Handbook—Fundamentals,* Chapter 3 of the 1988 ASHRAE *Handbook—Equipment,* and Chapter 42 of the 1991 ASHRAE *Handbook— HVAC Applications* present design procedures.

Environmental Control

All-air systems generally provide the necessary outside air supply to dilute the air in controlled spaces in applications requiring contamination control. These applications are usually combinations of supply systems and exhaust systems that circulate the diluting air through the space. Since establishing adequate dilution volumes is closely related to design criteria, occupancy type, air delivery, and scavenging methods, the designer must consider the terminal systems used.

Cleanliness of the air supply also relates directly to the level of contamination control desired. Suitable air filtration should be incorporated in the central system upstream from air-moving and tempering equipment. Some applications, such as hospitals, require downstream filtration to satisfy aseptic requirements.

These systems often incorporate some form of energy recovery. Chapter 25 has information on air cleaners, and Chapters 8, 11, and 12 of the 1989 ASHRAE *Handbook—Fundamentals* include data on physiological factors, contaminants, and odors.

CENTRAL SYSTEM PERFORMANCE

Figure 1 shows a typical draw-through central system that supplies conditioned air to a single-zone or any other system. A blow-through configuration may also be used if space or other conditions dictate. The quantity and quality of this air are fixed by space requirements and determined (Chapter 26 of the 1989 ASHRAE *Handbook—Fundamentals*). Air gains and loses heat by contacting the heat transfer surfaces and by mixing with air of another condition. Some of these mixtures are intentional, as at the outdoor air intake; others are the result of the physical characteristics of a particular component, as when untreated air passes without contacting the fins of a coil (bypass factor).

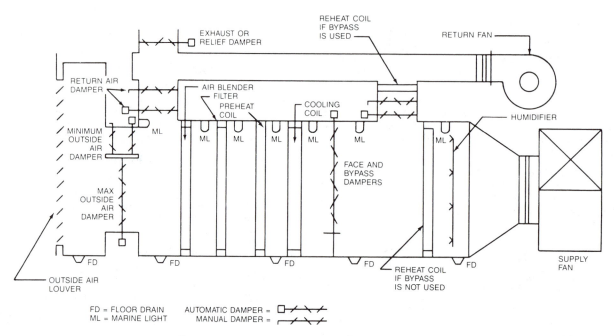

Fig. 1 Equipment Arrangement for Central Systems

All treated and untreated air must be thoroughly mixed for maximum performance of heat transfer surfaces and for uniform temperatures in the airstream. Stratified, parallel paths of treated and untreated air must be avoided, particularly in the vertical plane of systems using double-inlet or multiple-wheel fans. Because these fans do not completely mix the air, different temperatures can occur in branches coming from opposite sides of the supply duct.

LIFE SAFETY

Air-conditioning systems are often used for smoke control during fires. Controlled airflow provides smoke-free areas for occupant evacuation and fire fighter access. Space pressurization techniques create a low-pressure area at the smoke source, surrounding it with high-pressure spaces. The publication, *Design of Smoke Control Systems for Buildings* (ASHRAE 1983), has detailed information.

COMPONENTS

Air-Conditioning Units

To determine the system's air-handling requirement, the designer has to consider the function and physical characteristics of the space to be conditioned and the air volume and thermal exchange capacities required. Then, the various components may be selected and arranged by considering the fundamental requirements of the central system—equipment must be adequate, accessible for easy maintenance, and not too complex in arrangement and control to produce the required conditions.

Further, the designer considers economics in component selection. Both initial cost and operating costs affect design decisions. The designer should not arbitrarily design for a 500 fpm face velocity, which has been common for selection of cooling coils and other components. A 1977 study showed that filter and coil selection at 300 to 400 fpm, with its lower pressure

loss, could produce a substantial payback on constant volume systems. Chapter 33 of the 1991 ASHRAE *Handbook—HVAC Applications* has further energy and life-cycle cost details.

Figure 1 shows a general arrangement of the components of a single-zone, all-air, central system suitable for year-round air conditioning, with close control of temperature and humidity. All these components are seldom used in a comfort application. Although Figure 1 indicates a built-up system, most of the components are available completely assembled by the manufacturer or in sub-assembled sections that can be bolted together in the field.

Factors to be considered when selecting central system components include evaluation of specific design parameters to balance cost, controllability, operating expense, maintenance, noise, and space. The sizing and selection of primary air-handling units substantially affect the results obtained in the conditioned space.

Return Air Fan

A return air fan is optional on small systems but usually essential for the proper operation of large systems. It provides a positive return and exhaust from the conditioned area, particularly when mixing dampers permit cooling with outdoor air in intermediate seasons.

The return air fan ensures that the proper volume of air returns from the conditioned space. It prevents excess pressure when economizer cycles introduce more than the minimum quantity of outside air. It also reduces the static pressure the supply fan has to work against.

A damper controlled by a static pressure regulator may be installed in return air systems to offset stack effect in high-rise buildings.

The supply fan(s) must be carefully matched with the return fan, particularly in variable air volume (VAV) systems. The return air fan should handle a slightly smaller amount of air to account for fixed exhaust systems, such as toilet exhaust, and to ensure a slight positive pressure in the conditioned space. Fan selection and control of VAV systems is specialized (see Chapter 2 for design details).

Automatic Dampers

Opposed blade dampers for the outdoor, return, and relief airstreams provide the highest degree of control. The section titled Mixing Plenums covers the conditions that dictate the use of parallel blade dampers.

Relief Openings

Relief openings in large buildings should be constructed similarly to outdoor air intakes, but they should have motorized or self-acting backdraft dampers to prevent high wind pressures or stack action from causing the airflow to reverse when the automatic dampers are open. The pressure loss through relief openings should be 0.10 in. of water or less. Low leakage dampers, such as those for outdoor intakes, prevent rattling and minimize leakage.

Relief dampers sized for the same air velocity as the maximum outdoor air dampers facilitate control when the air economizer cycle is used. Power relief fans that are interconnected with outside air dampers can exhaust the area, especially when an economizer cycle is part of the system. The relief air opening should be located so that the exhaust air does not short-circuit to the outdoor air intake.

Return Air Dampers

The negative pressure in the outdoor air intake plenum is a function of the resistance or static pressure loss through the outside air louvers, damper, and duct. The positive pressure in the relief air plenum is, likewise, a function of the static pressure loss through the exhaust or relief damper, the exhaust duct between the plenum and outside, and the relief louver. The pressure drop through the return air damper must accommodate the pressure difference between the positive pressure-relief air plenum and the negative pressure outside air plenum. Proper sizing of this damper facilitates both air balancing and mixing. An additional manual damper may be required for proper air balancing.

Outdoor Air Intakes

Resistance through outdoor intakes varies widely, depending on construction. Frequently, architectural considerations dictate the type and style of louver. The designer should ensure that the louvers selected offer a minimum pressure loss, preferably not more than 0.10 in. of water. High-efficiency, low-pressure louvers that effectively limit carryover of rain are available. Flashing installed at outside wall and weep holes or a floor drain will carry away rain and melted snow entering the intake. Cold regions may require a snow baffle to direct fine snow particles to a low-velocity area below the dampers. Outdoor dampers should be low leakage types with special gasketed edges and special end treatment. Separate damper sections for the minimum outdoor air needed for ventilation and the maximum outdoor air needed for economizer cycles are strongly recommended.

Mixing Plenum

If the equipment is alongside outdoor louvers in a wall, the minimum outdoor air damper should be located at the return damper connection. An outside air damper sized for 1500 fpm provides good control. The pressure difference between the relief plenum and outdoor intake plenum must be measured through the return damper section. A higher velocity through the return air damper—high enough to cause this loss at its full open position—facilitates air balance and creates good mixing. To create maximum turbulence and mixing, return air dampers should be set so that any deflection of air is toward the outside air.

Mixing dampers should be placed across the full width of the unit, even though the location of the return duct makes it more convenient to return air through the side. When return dampers are placed at one side, return air passes through one side of the fan, and cold outdoor air passes through the other. If the air return must enter the side, some form of air blender should be used.

While opposed blade dampers offer better control, properly proportioned, parallel blade dampers are more effective than opposed blade dampers for mixing airstreams of different temperatures. If parallel blades are used, each damper should be mounted so that its partially opened blades direct the airstreams toward the other damper for maximum mixing.

Baffles that direct the two airstreams to impinge on each other at right angles and in multiple jets create the turbulence required to mix the air thoroughly. In some instances, unit heaters or propeller fans have been used for mixing regardless of the final type and configuration. Otherwise, the preheat coil will waste heat (if included), or the cooling coil may freeze. Low leakage outdoor air dampers minimize leakage during shutdown.

Coil freezing can be a serious problem with chilled water coils. Full flow circulation of chilled water during freezing weather, or even reduced flow with a small recirculating pump, discourages coil freezing and eliminates stratification. Further, it can provide a source of off-season chilled water in air-water systems. Antifreeze solutions or complete coil draining also prevent coil freezing.

Filter Section

A system's overall performance depends heavily on the filter. Unless the filter is regularly maintained, system resistance increases and airflow is diminished. Accessibility is the primary consideration in filter selection and location. In a built-up system, there should be a minimum of 3 ft between the upstream face of the filter bank and any obstruction. Other requirements for filters can be found in Chapters 25 and 26 and in ASHRAE *Standard* 52.

Good mixing of outdoor and return air is also necessary for good filter performance. A poorly placed outdoor air duct or a bad duct connection to the mixing plenum can cause uneven loading of the filter and poor distribution of air through the coil section. Because of the low resistance of the clean areas, the filter gage may not warn of this condition.

Preheat Coil

The preheat coil should have wide fin spacing, be accessible for easy cleaning, and be protected by filters. If the preheat coil is located in the minimum outdoor airstream rather than in the mixed airstream as shown in Figure 1, it should not heat the air to an exit temperature above 35 to 45°F; preferably, it should become inoperative at outdoor temperatures of 45°F. Inner distributing tube or integral face and bypass coils are preferable with steam. If used, hot water preheat coils should have a constant flow recirculating pump and should be piped for parallel flow so that the coldest air will contact the warmest coil surface first.

Cooling Coil

In this section, sensible and latent heat are removed from the air. In all finned coils, some air passes through without contacting the fins or tubes. The amount of this bypass can vary from 30% for a four-row coil at 700 fpm to less than 2% for an eight-row coil at 300 fpm.

The dew point of the air mixture leaving a four-row coil might satisfy a comfort installation with 25% or less outdoor air, a small internal latent load, and sensible temperature control only. For close control of room conditions for precision work, a deeper coil may be required.

A central station unit that is the primary source of conditioned air for other subsystems, such as in an air-water system, does not need to supply as much air to a space. In this case, the primary air furnishes outdoor air for ventilation and handles space dehumidification and some sensible cooling. This application normally requires deeper coils with more fins and sprays.

Bypass Section

Air is mixed a third time in the bypass section, if used. At reduced internal loads, the room thermostat opens the bypass damper, which permits return air to enter the bypass section. At the same time, the face damper on the cooling section closes and reduces the flow of cooled and dehumidified air. The temperature of the mixture rises, and overcooling is prevented. The amount of heat available depends on the amount of return air. The outdoor air required for ventilation passes through the dehumidifier section, so that this system will control room relative humidity better than any other constant volume system except a dew-point control with reheat system.

Supplemental heat may be required to maintain room temperature in the event of a net heat loss. The relative humidity in the room will increase as the bypass opens under conditions of constant internal latent heat load. Chilled water at a constant flow and temperature will control the dew point. For more precise control, a space humidistat can control chilled water flow and reheat. The ASHRAE Psychrometric Charts in Chapter 6 of the 1989 ASHRAE *Handbook—Fundamentals* show the process.

Because the apparatus has a high air resistance compared to that of the bypass, the total resistance on the supply fan drops as the bypass opens, and the total volume of air delivered to the room increases. A supply fan with a steep and constantly rising pressure curve limits this increased volume; otherwise, a constant volume control can be installed. Without a volume control, this system should only be applied where volume variations do not seriously affect occupant comfort.

Because there is a large pressure drop across the bypass damper, it has an abnormally high leakage when closed. This leakage must be considered in the system design, and the apparatus dew point should be lowered to compensate. For optimum psychrometric performance, a bypass should bypass only return air, as shown in Figure 1.

To maintain a fairly constant supply air volume, the pressure loss through the bypass at full flow must equal the pressure loss through all components from the outdoor air intake to the bypass mixing plenum at full cooling flow. The bypass requires a perforated plate to develop the pressure drop. Return air filters should also be used. Bypass dampers should be sized for a velocity of 2000 to 2500 fpm. Face dampers should match the coil face area, and low-leakage, opposed blade dampers must be used.

Reheat Coil Section

Reheat is strongly discouraged, unless the energy used is recovered from free sources (see ASHRAE *Standard* 90.1-1989). Reheating is limited to laboratory, health care, or similar applications, where it is essential to control temperature and relative humidity accurately.

Heating coils located in the reheat position, as shown in Figure 1, are frequently used for warm-up, although a coil in the preheat position is preferable. The reheat coil may also be located in the bypass duct.

Hot water heating coils provide the highest degree of control. Oversized coils, particularly steam, can stratify the airflow; thus, where cost-effective, inner distributing coils are preferable for steam applications. Electric coils may also be used.

Humidifiers

For comfort installations not requiring close control, moisture can be added to air by mechanical atomizers or point-of-use electric humidifiers. Proper location of this equipment will prevent stratification of moist air in the system.

Steam grid humidifiers with dew-point control usually are used for accurate humidity control. In this application, the heat of evaporation should be replaced by heating the recirculated water rather than by increasing the size of the preheat coil. It is not possible to add moisture to saturated air, even with a steam grid humidifier. Air in a laboratory or other application that requires close humidity control must be reheated after leaving a dehumidifier coil before moisture can be added. This operation increases the amount of air required to cool the space.

The capacity of the humidifying equipment should not exceed the expected peak load by more than 10%. If the humidity is controlled from the room or the return air, a limiting humidistat and fan interlock may be needed in the supply duct to prevent condensation when temperature controls call for cooler air. Humidifiers add some sensible heat that should be accounted for in the psychrometric evaluation.

Supply Air Fan

Either axial flow, centrifugal, or plug fans may be chosen as supply air fans for straight-through flow applications. In factory-fabricated units, more than one centrifugal fan may be tied to the same shaft. If headroom permits, a single-inlet fan should be chosen when air enters at right angles to the flow of air through the equipment. These arrangements permit a direct flow of air from the fan wheel into the supply duct without abrupt change in direction and loss in efficiency. It also permits a more gradual transition from the fan to the duct and increases the static regain in the velocity pressure conversion.

To minimize inlet losses, the distance between the casing walls and the fan inlet should be at least the diameter of the fan wheel. With a single-inlet fan, the length of the transition section should be at least one-half the width or height of the casing, whichever is longer.

If fans blow through the equipment, the air distribution through the downstream components need analyzing, and baffles should be used to ensure uniform air distribution (see Chapter 18).

AIR DISTRIBUTION

Ductwork

Ductwork should deliver conditioned air to an area as directly, quietly, and economically as possible. Structural features of the building generally require some compromise and often limit depth. Chapter 32 of the 1989 ASHRAE *Handbook—Fundamentals* describes ductwork design in detail and gives several methods of sizing duct systems.

Since ductwork is sold on a cost-per-unit-weight basis modified by a fitting complexity multiplier, a design with the lowest weight to handle the required amount of air is the most cost-effective. This, of course, has to accommodate noise, pressure drop, and fan power considerations. Table 1 shows that heat gains and losses are greater in ducts with high aspect ratios

Table 1 Effect of Aspect Ratio on Duct Weight

Aspect Ratio	Standard Size	Perimeter, in.	U.S. Std. Gage Steel	Weight,[a] lb/linear ft
1 to 1	32 by 32	128	22	16.6
2 to 1	44 by 22	132	22	17.0
4 to 1	64 by 16	160	20	22.0
8 to 1	96 by 12	216	18	39.0

[a]Includes 10% for slips, angles, and scrap.

(more perimeter). Also, ducts with large aspect ratios radiate more noise, require greater stiffness, and contribute to greater pressure loss.

It is imperative that the designer coordinate duct design with the structural system. In commercially developed projects, it is common that great pressure is applied to reduce floor-to-floor dimensions. This pressure on the available interstitial space left for ductwork is a major design challenge.

When the duct layout has few outlets, conventional low-velocity designs usually assume a uniform resistance per 100 ft of equivalent length of 0.08 or 0.10 in. of water. On complex systems with long runs and medium and high pressures of 1.5 to 8 in. of water, the self-balancing and energy-saving features of the static pressure regain method of duct sizing should be considered. This method also produces the best results with a VAV duct distribution system.

Room Terminals

In some instances, such as in low-velocity, all-air systems, the air may enter from the supply air ductwork directly into the conditioned space through a grille or diffuser.

In high-velocity air systems, an intermediate device normally controls air volume, reduces duct pressure, or both. Various devices are available, including (1) an air-water induction terminal, which includes a coil or coils in the induced airstream to condition the return air before it mixes with the primary air and enters the space; (2) an all-air induction terminal, which controls the volume of primary air, induces ceiling plenum air, and distributes the mixture through low-velocity ductwork to the space; (3) a fan-powered mixing box, which uses a fan to accomplish the mixing rather than depending on the induction principle; and (4) a VAV box, which varies the amount of air delivered with no induction. This air may be delivered to low-pressure ductwork and then to the space, or the terminal may contain an integral air diffuser.

Insulation

Ductwork that runs outside the conditioned area should be insulated. Acoustical lining is often used for thermal insulation, if its overall heat transfer coefficient is low enough.

Insulation is seldom placed in hung ceilings that serve as return plenums. Exceptions are made when duct runs are so long that the heat pickup requires additional air to offset the temperature rise in the duct or when the duct surface temperature is low enough to cause condensation. The need for insulation should be checked, especially for a ceiling hung below a flat roof, even though the roof load is included in the system capacity.

Ceiling Plenums

The space between a hung ceiling and the floor slab above it is frequently used as a return plenum to reduce sheet metal work. Local and national codes should be consulted before using this approach in new design, since most codes prohibit combustible material in a return air ceiling plenum.

Lobby ceilings with lay-in panels do not work well as return plenums where negative pressures from high-rise elevators or an adjacent factory area may occur. If the plenum leaks to the low-pressure area, the tiles may lift and drop out when the outside door is opened and closed. Return plenums directly below a roof deck have substantially higher return air gain than a ducted return. This has the advantage of reducing the heat gain to or loss from the space.

Controls

Controls should be automatic and simple for best operating and maintenance efficiency. Operations should follow a natural sequence—one controlling thermostat closes a normally open heating valve, opens the outdoor air mixing dampers, modulates the face and bypass dampers, or opens the cooling valve, depending on space need. In certain applications, an enthalpy controller, which compares the heat content of outdoor air to that of return air, may override the temperature controller. This control opens the outdoor air damper when conditions reduce the refrigeration load. In smaller systems, a dry-bulb control saves the cost of the enthalpy control and approaches these savings when an optimum changeover temperature, above the design dew point, is established.

A minimum outdoor air damper with a separate motor, selected for a 1500 fpm velocity, is preferred to one large outdoor air damper with minimum stops. A separate damper simplifies air balancing.

A mixed air temperature control can reduce operating costs and also reduce temperature swings from load variations in the conditioned space. Chapter 41 of the ASHRAE *Handbook—HVAC Applications* shows control diagrams for various arrangements of central system equipment.

Equipment Isolation

Vibration and sound isolation equipment is required for most central system fan installations. Standard mountings of fiberglass, ribbed rubber, neoprene mounts, and springs are available for most fans and prefabricated units.

In some applications, the fans may require concrete inertia blocks in addition to nonenclosed spring mountings. Steel springs require sound-absorbing material inserted between the springs and the foundation. Horizontal discharge fans operating at a high static pressure frequently require thrust arrestors.

Ductwork connections should be made with fireproof fiber cloth sleeves having considerable slack, but without offset between the fan outlet and rigid duct. Misalignment between the duct and fan outlet can cause turbulence, generate noise, and reduce system efficiency. Electrical and piping connections to vibration-isolated equipment should be made with flexible conduit and flexible connections.

Equipment noise transmitted through the ductwork can be reduced by sound-absorbing units, acoustical lining, and other means of attenuation. Sound transmitted through return and relief ducts should not be overlooked. Acoustical lining sufficient to attenuate any objectionable system noise adequately or locally generated noise should be considered. Chapter 42 of the 1991 ASHRAE *Handbook—HVAC Applications,* Chapter 7 of the 1989 ASHRAE *Handbook—Fundamentals,* and ASHRAE *Standard* 68-1986 have further information on sound and vibration control.

The designer must account for seismic restraint requirements established for the seismic zone in which the particular project is located.

Space Heating

Steam is an acceptable medium for central system preheat or reheat coils, while low-temperature hot water provides a simple and more uniform means of perimeter and general space heating. Individual automatic control of each terminal pro-

vides the ideal space comfort. A control system that varies the water temperature inversely with the change in outdoor temperature provides water temperatures that produce acceptable results in some applications. To produce average results, the most satisfactory ratio can be set after the installation is completed and actual operating conditions are ascertained.

Multiple perimeter spaces on one exposure served by a central system may be heated by supplying warm air from the central system. Areas that have heat gain from lights and occupants and no heat loss will require cooling in winter as well as in summer.

In some systems, very little heating of the return and outdoor air is required when the space is occupied. Local codes dictate the amount of outside air required (see ASHRAE *Standard* 62-1989 for recommended optimum outside air ventilation). For example, with return air at 75°F and outside air at 0°F, the temperature of a 25% outdoor/75% return air mixture would be 56°F, which is close to the temperature of the air supplied to cool such a space in summer. A preheat coil installed in the minimum indoor airstream to warm the outdoor air in this instance can produce overheating, unless it is sized as previously recommended. Assuming good mixing, a preheat coil located in the mixed airstream prevents this problem. The outdoor air damper should be kept closed until room temperatures are reached during warm-up. A return air thermostat can terminate the warm-up period.

When a central air-handling unit supplies both perimeter and interior spaces, the supply air must be cool to handle the interior zones. Additional control is needed to heat the perimeter spaces properly. Reheating the air is the simplest solution, but it is not acceptable by most energy codes. An acceptable solution is to vary the volume of air to the perimeter and combine it with a terminal heating coil or a separate perimeter heating system, either a baseboard, overhead air heating system, or a fan-powered mixing box with supplemental heat. The perimeter heating should be individually controlled and integrated with the cooling control.

Resetting the supply water temperature downward when less heat is required generally improves temperature control. For further information, refer to Chapters 12, 24, and 33 in this volume and Chapter 41 in the 1991 ASHRAE *Handbook—HVAC Applications*.

INSTALLATION

Prefabricated Units

Prefabricated units are frequently shipped subassembled for erection in the field. The following installation pointers will help in the maintenance and operation of the equipment:

1. Line up flanges and remove any dents or twists.
2. If gaskets are not provided, caulk or tape joints to prevent leakage.
3. Check each coil and filter section for any bypass leakage.
4. Seal all openings.
5. If the equipment room is not wide enough to permit coil removal, place the coil section opposite a door or a knock-out panel with a steel frame in the wall. Make similar provision for removing the fan shaft and wheels.
6. If space in the equipment room is limited, place piping, filter access, wiring, motor, and drive on the same side with a minimum 4-ft clearance between casing and wall.
7. Keep a minimum clearance of 18 in. on the opposite side for servicing the fan bearing and for painting.
8. Offset all piping branches, and use ground joint unions or flanges to permit coil removal without disturbing the mains or control valves.

9. Be sure all pipe openings in the casing have airtight escutcheons.

Built-Up Units

Figure 2 shows minimum clearances for a large built-up unit operating at medium or high pressure. The unit should be installed as follows:

1. Use insulated single- or double-wall sheet metal for the casing.
2. Install waterproof floor with floor drains as noted or as required.
3. Place a 0.25-in. gasket between all steel flanges of the casing and masonry.
4. Install gasketed access doors that are equipped with wedge-type locks operable from inside and hinged so that either positive or negative plenum pressure keeps them closed to prevent leakage.
5. Install a light switch with a pilot light outside each casing section to control a vaporproof light with a wire guard.
6. Direct the intake louver away from a residential building or quiet area, unless it is baffled for equipment sound.
7. Install a properly drained space between the louver and the outdoor air damper to collect rainwater or fine snow that passes through the louvers. Outside air plenum should be of watertight construction at the bottom for a 2-in. depth.
8. Include a generously sized mixing plenum to permit easy filter maintenance, maximum turbulence, and mixing of outdoor and return air; include room to install baffles to prevent stratification.
9. Install an access door to the mixing plenum large enough to accommodate a standard package of filter media and, in large systems, a hand truck. This section is usually fabricated of sheet metal, but large plenums are often built of masonry and have standard weatherstripped steel doors. Although desirable, it is not necessary to have access between the filter and heating coil, because, as a dry coil, it does not require frequent cleaning.
10. Place material to seal the openings between the coil and the casing on the exit side, so that it is accessible if the heating coil needs to be removed.
11. Install a drain pan for the cooling and dehumidifying section on a concrete pad or structural steel base at least 6 in. above the floor to facilitate connecting the piping to

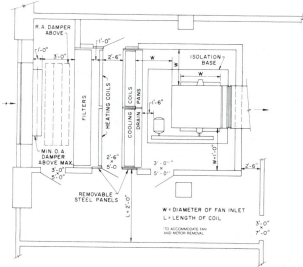

Fig. 2   Recommended Space Requirements and Clearances for Central System Fan Room

the coils and drains. If set on concrete, set the drain pan on heavy-density roofing felt to protect the underside from corrosion and to absorb any irregularities in the concrete surface. If set on steel, insulate the underside with waterproof insulation.

12. Keep the pan level or pitched toward the drain.

13. Place a drain pan under each coil bank stacked more than one high with a separate downspout to the main drain. Baffle the drain pan to prevent airflow from bypassing a dry coil.

14. Trap all drains with a water seal at least 1 in. deeper than the maximum pressure caused by the fan, and pitch 0.25 in./ft to an indirect waste receptacle. Insulate drains if they pass through any space where damage from condensation drip might occur.

15. Use drainage fittings with cleanout plugs at each change of direction.

16. Locate floor drains in the inlet plenum, at the entrance and exit of the cooling coil, and in a humidifier plenum.

17. Place access on the entrance and exit of the cooling and humidifying section to allow for periodic cleaning of the coils (and sprays, if they are used).

18. Install an access door in the fan section large enough for motor and fanwheel removal.

19. Build a masonry fan section, lined with acoustical panels, to lower sound levels in adjacent rooms. This should be a separate partition with an air space between it and the equipment room walls. Heavy-gage sheet metal with a steel framework and external stiffeners, adequately fitted with rigid sound and heat insulation, may be less costly and offer more versatility for internal component arrangement. Acoustical treatment on the fan section ceiling and on the floor under the spring-loaded fan base will reduce sound transmission to other floors.

20. Run piping mains parallel to major ductwork with hangers between the ducts and not through them.

21. Valve drop branches to equipment at the main, and arrange them so that they will not interfere with access doors or the removal of coils and other equipment.

22. Be sure final connections are removable without disturbing control valves or the drop branches.

23. Install all equipment piping after the ductwork so that ductwork dodging pipes will not compromise air handling efficiency.

24. Figure 2 shows the heating and cooling coils in banks split vertically to limit the size of the coils. Control the heating or cooling coils in one bank by one control valve assembly in the main piping.

25. Install a balancing valve in the return branch piping from each individual coil to ensure uniform loading across the face of the coil banks to prevent stratification.

26. Plumb water coils (except hot water preheat coils) to run counterflow (leaving air at entering water).

27. Provide preheat coils with individual control valves and thermostats to minimize freezing in the event of air stratification.

28. Run main electric feeder conduits across the ceiling perpendicular to main ducts and between beams and terminate in readily accessible pull boxes in a clear area of the room.

29. Run subfeeders and branch circuits beneath the floor, stubbing up through waterproof sleeves near the electrical equipment.

30. Run all final connections to moving equipment or vibration and sound isolated structures in a flexible cable with adequate slack.

31. Place the main electric distribution panel and motor control center near the main entrance of the room.

32. Mount the local automatic temperature control panel near the air-conditioning equipment to keep capillary tubing and control connectors as short as possible.

33. Ensure that all electric and temperature control panels are well lighted.

Check Installation and Balancing

After installation:

1. Check for air leaks in the casing and in the sealant around the coils and filter frames by moving a light source along the outside of the joint while observing the darkened interior of the casing. Caulk any leaks.

2. Note points where piping enters the casing to be sure that the escutcheons are tight. Do not rely on pipe insulation to seal these openings because, in time, it will shrink.

3. In prefabricated units, check that all panel fastening holes are filled to prevent whistling.

4. Make a final check for leaks after start-up. Chapter 3 discusses procedures for both air and water systems.

PRIMARY SYSTEMS

The type of central heating and cooling equipment used for air-conditioning systems in large buildings depends chiefly on economic factors, once the total required capacity has been determined. Component choice depends on such factors as the type of fuel available, environmental protection required, structural support, and available space.

Rising energy costs have fostered many designs to recover the internal heat from lights, people, and equipment to reduce the size of the heating plant. Chapter 8 describes several heat-recovery arrangements.

The quest for energy savings has extended to total energy systems, in which on-site power generation has been added to the heating and air-conditioning project. The economics of this function is determined by gas and electric rate differentials and by the ratio of electric to heat demands for the project. In these systems, waste heat from the generators can be input to the cooling equipment, either to drive the turbines of centrifugal compressors or to serve absorption chillers. Chapter 7 presents further details on total energy systems.

Among the largest installations of central mechanical equipment are the central cooling and heating plants serving groups of large buildings. These plants provide higher diversity and generally operate more efficiently and with lower maintenance and labor costs than individual plants.

The economics of these systems require extensive analysis. Boilers, gas and steam turbine-driven centrifugals, and absorption chillers may be installed in combination in one plant. In large buildings with core areas that require cooling while perimeter areas require heating, one of several heat reclaim systems could heat the perimeter to save energy. Chapter 42 gives details of these combinations, and Chapter 11 gives design details of central plants.

Most large buildings, however, have their own central heating/cooling plants in which the choice of equipment depends on the following:

- Required capacity and type of usage
- Costs and types of energy available
- Location of equipment room
- Type of air distribution system
- Owning and operating costs

Many electric utilities impose severe penalties for peak summertime power use, or alternatively, offer incentives for off-peak use. This policy has renewed interest in both water and ice thermal storage systems. The storage capacity paid for by summertime load leveling may also be available for use in winter, making heat reclaim a more viable option. With ice-storage systems, the low-temperature ice water can provide colder air than that available from a conventional system. Use of high water temperature rise and lower temperature air results in less pumping and fan energy and, in some instances, offsets the energy penalty due to the lower temperature required to make ice.

Heating Equipment

Steam and hot water boilers are manufactured for high or low pressure and use coal, oil, electricity, gas, and, sometimes, waste material for fuel. Low-pressure boilers are rated for a working pressure of 15 psig for steam, 160 psig for water, with a maximum temperature limitation of 250°F. Package boilers, with all components and controls assembled as a unit, are available. Electrode or resistance-type electric boilers are available in sizes up to 12,000 kW and larger for either hot water or steam generation. For further information, see Chapter 29.

Where steam or hot water is supplied from a central plant, as on university campuses or in downtown areas of large cities, the service entrance to the building must conform to utility standards. The utility should be contacted at the beginning of the project to determine availability, cost, and the specific requirements of the service.

Fuels

Fuels must be considered when the boiler system is selected to ensure maximum efficiency. Chapter 15 of the 1989 ASHRAE *Handbook—Fundamentals* gives fuel types, properties, and proper combustion factors. Chapter 28 includes information for the design, selection, and operation of automatic fuel-burning equipment.

Refrigeration Equipment

The major types of refrigeration equipment used in large systems are as follows:

- Reciprocating—1/16 to 300 hp multiple units
- Helical rotary—100 to 750 tons
- Centrifugal—100 to 10,000 tons
- Absorption—100 to 2000 tons

Reciprocating, helical rotary, and centrifugal compressors have many types of drives—electric motors, gas and diesel engines, and gas and steam turbines. The compressors may be purchased as part of a refrigeration chiller consisting of compressor, drive, chiller, condenser, and necessary safety and operating controls. Reciprocating and helical rotary compressor units are frequently used in field-assembled systems, with air-cooled or evaporative condensers arranged for remote installation.

Centrifugal compressors are usually included in packaged chillers. Most centrifugal packages are available in limited sizes.

Absorption chillers are water-cooled. They use a lithium bromide/water or water/ammonia cycle and are generally available in the following four configurations: (1) direct fired, (2) indirect fired by low-pressure steam or hot water, (3) indirect fired by high-pressure steam or hot water, and (4) fired by hot exhaust gas. Small, direct-fired chillers are single-effect machines with capacities from 3.5 to 25 tons. Japan produces larger direct-fired, double-effect chillers in the 100 to 2000-ton capacity range.

Low-pressure steam at 15 psig or hot water heats the generator of single-effect absorption chillers with capacities from 50 to 1600 tons. Double-effect machines use higher pressure steam up to 150 psig or hot water at an equivalent temperature. Absorption chillers of this type are available from 350 to 2000 tons.

The absorption chiller is sometimes combined with steam turbine-driven centrifugal compressors in large installations. Steam from the noncondensing turbine is piped to the generator of the absorption machine. When a centrifugal unit is driven by a gas turbine or an engine, an absorption machine generator may be fed with steam or hot water from the jacket. A heat exchanger that will transfer the heat of the exhaust gases to a fluid medium may increase the cycle efficiency. Chapter 29 gives details on absorption air-conditioning and refrigeration equipment.

Cooling Towers

Water is usually cooled by contact with the atmosphere to reject heat from the water-cooled condensers of air-conditioning systems. Either natural draft or mechanical draft cooling towers or spray ponds accomplish the cooling. Of these, the mechanical draft tower, which may be of the forced draft, induced draft, or ejector type, can be designed for most conditions because it does not depend on wind. Air-conditioning systems use towers ranging from small package units of 5 to 500 tons or field-erected towers, with multiple cells in unlimited sizes.

The tower must be winterized if required for cooling at ambient outdoor dry-bulb temperatures below 35°F. Winterizing includes the capability of bypassing water directly into the tower return line (either automatically or manually, depending on the installation) and of heating the tower pan water to a temperature above freezing.

Heat may be added by steam or hot water coils or electric resistance heaters in the tower pan. Also, it is usually necessary to provide an electric heating cable on the condenser water and makeup water pipe and to insulate these heat-traced sections to prevent the pipes from freezing. When it is necessary to operate the cooling tower at or near freezing conditions, the tower fan(s) should be cycled by an immersion thermostat in the basin water, or by a pressure control on the compressor, to minimize freezing of the tower fill.

Where the cooling tower will not operate in freezing weather, provisions for draining the tower and piping are necessary. Draining is the most effective way to prevent tower and piping from freezing.

Careful attention must be given to water treatment to keep maintenance required in the refrigeration machine absorbers and/or condenser to a minimum.

Cooling towers may also cool the building in off-seasons by filtering and directly circulating the condenser water through the chilled water circuit, by cooling the chilled water in a separate heat exchanger, or by using the heat exchangers in the refrigeration equipment to produce thermal cooling. Towers are usually selected in multiples so that they may be run at reduced capacity and shut down for maintenance in cool weather. Chapter 37 includes further design and application details.

Air-Cooled Condensers

Air-cooled condensers pass outdoor air over a dry coil to condense the refrigerant. This results in a higher condensing temperature and, thus, a larger power input at peak condition; however, over 24-h this peak time is short. The air-cooled condenser is popular in small reciprocating systems because of its low maintenance requirements.

Evaporative Condensers

Evaporative condensers pass air over coils sprayed with water, thus taking advantage of adiabatic saturation to lower the condensing temperature. As with the cooling tower, freeze prevention and close control of water treatment are required for success. The lower power consumption of the evaporative versus the air-cooled condenser is gained at the expense of the water used and increased maintenance.

Pumps

Air-conditioning pumps are usually centrifugal pumps. Pumps for large and heavy-duty systems have a horizontal split case with a double-suction impeller for easier maintenance and higher efficiency. End suction pumps, either close coupled or flexible connected, are used for smaller tasks.

Major applications for pumps in the equipment room are as follows:

- Primary and secondary chilled water
- Hot water
- Condenser water
- Condensate pumps
- Boiler feed pumps
- Fuel oil

When the pumps handle hot liquids or have high inlet pressure drops, the required net positive suction head (NPSH) must not exceed the NPSH available at the pump. It is common practice to provide two identical pumps—one a spare, or each sized for 50% of the design flow at the design pressure, operating in parallel to maintain system continuity in case of a pump failure.

Sometimes the chilled water and condenser water system characteristics permit using one spare pump for both systems, with valved connections to the manifolds of each. Chapters 12 and 13 give design recommendations, and Chapter 39 gives information on the selection of pumps.

Piping

Air-conditioning piping systems can be divided into two parts—the piping in the main equipment room and the piping required for air-handling equipment throughout the building. The air-handling system piping follows procedures detailed in Chapters 5, 12, 13, and 14.

The major piping in the main equipment room includes fuel lines, refrigerant piping, and steam and water connections. Chapter 33 of the 1989 ASHRAE *Handbook—Fundamentals* gives details for sizing and installing these pipes. Chapter 10 includes more information on piping for steam systems, and Chapter 13 has information on chilled and dual-temperature water systems.

Instrumentation

All equipment must have adequate gages, thermometers, flow meters, balancing devices, and dampers for effective system performance. In addition, capped thermometer wells, gage cocks, capped duct openings, and volume dampers should be at strategic points for system balancing. Chapter 34 of the 1991 ASHRAE *Handbook—HVAC Applications* indicates the locations and types of fittings required.

Recent advances in electronic and computer technology are changing traditional control system architecture. Pneumatic and electric relay logic is being replaced by microprocessors and software programs to establish control sequences. Output signals may be converted to pneumatic or electric commands to actuate HVAC hardware.

A central control console to monitor the many system points should be considered for any large, complex air-conditioning system. A control panel permits a single operator to monitor and perform functions at any point in the building, to increase occupant comfort and to free maintenance staff for other duties. Chapter 41 of the 1991 ASHRAE *Handbook—HVAC Applications* describes these systems in detail.

SPACE REQUIREMENTS

In the initial phases of building design, the engineer seldom has sufficient information to render the design. Therefore, most experienced engineers have developed rules of thumb to estimate the building space needed.

The air-conditioning system selected, the building configuration, and other variables govern the space required for the mechanical system. The final design is usually a compromise between what the engineer recommends and what the architect can accommodate. Where designers cannot negotiate a space large enough to suit the installation, the judgment of the owner should be called upon.

Although few buildings are identical in design and concept, some basic criteria are applicable to most buildings and help allocate space that approximates the final requirements. These space requirements are often expressed as a percentage of the total building floor area.

Mechanical, Electrical, and Plumbing Facilities

The total mechanical and electrical space requirements range from 4 to 9% of the gross building area, the majority of buildings falling within the 6 to 9% range.

It is desirable to have most of the facilities centrally located to minimize long duct, pipe, and conduit runs and sizes; simplify shaft layouts; and centralize maintenance and operation. A central location also reduces pump and fan motor power, which may reduce building operating costs. But for many reasons, it is often impossible to centrally locate all the mechanical, electrical, and plumbing facilities within the building. In any case, the equipment should be kept together to minimize space requirements, centralize maintenance and operation, and simplify the electrical distribution system.

Equipment rooms generally require a clear ceiling height ranging from 12 to 20 ft, depending on equipment size and the complexity of the ductwork, piping, and conduit.

The main electrical transformer and switchgear rooms should be located as close to the incoming electrical service as practical. If there is an emergency generator, it should be located close to the switchgear room to keep interconnection costs to a minimum. Also, the emergency generator should be near a source of combustion air. Adequate noise control is necessary, and the exhaust gases must be vented to the outdoors.

The main plumbing equipment room usually contains gas and domestic water meters, the domestic hot water system, the fire protection system, and various other elements, such as compressed air, special gases, and vacuum, ejector, and sump pump systems. Some water and gas utilities require remote outdoor meter location.

The heating and air-conditioning equipment room houses the boiler or pressure-reducing station or both; the refrigeration machines, including the chilled water and condensing water pumps, converters for furnishing hot or cold water for air conditioning; control air compressors; vacuum and condensate pumps; and other miscellaneous equipment.

Although the electrical service and structural costs will rise, these may be offset by reduced condenser and chilled water piping, energy consumption, and equipment cost because of lower operating pressure. The boiler plant may be placed on the roof, which eliminates a chimney through the building.

Gas fuel may be more desirable, because oil storage and pumping present added design and operating problems. However, restrictions on gas as a boiler fuel may preclude this option. Heat recovery systems, in conjunction with the refrigeration equipment, can drastically reduce the heating plant size in buildings with large core areas. With well-insulated buildings and electric utility rates structured to encourage a high use factor, even electric resistance heat can be attractive, reducing space requirements even further.

Additional space might be needed for a telephone terminal room, a pneumatic tube equipment room, an incinerator, and similar areas.

Most buildings, especially larger ones, need cooling towers, which often present problems. If the cooling tower is on the ground, it should be at least 100 ft away from the the building for two reasons—to reduce tower noise in the building and to keep intermediate season discharge air from fogging the building's windows. Towers should be kept the same distance from parking lots to avoid staining car finishes with water treatment chemicals. When the tower is on the roof, its vibration and noise must be isolated from the building. Some towers are less noisy than others, and some have attenuation housings to reduce noise levels. These options should be explored before selecting a tower. Provide adequate room for airflow inside screens to prevent recirculation.

The bottom of many towers, especially larger ones, must be set on a steel frame 4 to 5 ft above the roof to allow room for piping and proper tower and roof maintenance. Pumps below the tower should be designed for adequate NPSH. The piping system must be designed to prevent drain-down of the piping system and condensers on condenser water pump shutdown.

Fan Room Requirements

Fan rooms for the substructure block and main floor are usually placed in the lower level of the building—either in the basement or on the second floor. Second-floor mechanical rooms have the advantage of easy outdoor access for makeup air, exhaust, and equipment replacement.

The number of fan rooms on upper floors depend largely on the total area of the floor. Buildings with large floor areas often have multiple fan rooms on each floor. Many high-rise buildings, however, may have one fan room serving 10 to 20 floors—one serving the lower floors, one serving the middle of the building, and one at the roof serving the top quarter of the building.

Life safety is an important factor in fan room location. Chapter 47 discusses principles of fire spread and smoke control.

Interior Shafts

Interior shaft space accommodates return and exhaust air; interior supply air; hot, chilled, and condenser water piping; steam and return piping; electric closets; telephone closets; plumbing piping; and, possibly, pneumatic tubes and conveyor systems.

The shafts must be clear of stairs and elevators on at least two sides to allow maximum headroom when the pipes and ducts come out at the ceiling. In general, duct shafts having an aspect ratio of 2:1 to 4:1 are easier to develop than large square shafts. The rectangular shape also makes it easier to go from the equipment in the fan rooms to the shafts.

The size, number, and location of shafts is important in multistory buildings. Vertical duct distribution systems with little horizontal branch ductwork are desirable because they are usually less costly; easier to balance; create less conflict with pipes, beams, and lights; and enable the architect to design lower floor-to-floor heights.

The number of shafts is a function of building size and shape, but, in larger buildings, it is usually more economical, in cost and space, to have several small shafts rather than one large shaft. Separate supply and exhaust duct shafts may be desired to reduce the number of duct crossovers and to operate the exhaust shaft as a plenum. From 10 to 15% additional shaft space should be allowed for future expansion and modification. The additional space also reduces the initial installation cost.

Equipment Access

Properly designed mechanical equipment rooms must solve the problem of how and where to move large, heavy equipment in, out, and within the building. Equipment replacement and maintenance can be very costly if access is not planned properly.

Because systems vary greatly, it is difficult to estimate space requirements for refrigeration and boiler rooms without making block layouts of the system selected. Block layouts allow the engineer to develop the most efficient arrangement of equipment, with adequate access and serviceability. They further help in preliminary discussions with the owner and architect. Only then can the engineer obtain verification of the estimates and provide a workable and economical design.

REFERENCES

ASHRAE. 1983. Design of smoke control systems for buildings.
Wheeler, A.E. 1977. Air handling unit design for energy conservation. ASHRAE *Journal* (June).

CHAPTER 2

ALL-AIR SYSTEMS

AN all-air system provides complete sensible and latent cooling, preheating, and humidification capacity in the air supplied by the system. No additional cooling or humidification is required at the zone, except in the case of certain industrial systems. Heating may be accomplished by the same airstream, either in the central system or at a particular zone.

In some applications, heating is accomplished by a separate heating system. The term zone implies the provision or the need for separate thermostatic control, while the term room implies a partitioned area that may or may not require separate control.

All-air systems are classified in two basic categories:

- Single-duct systems, which contain the main heating and cooling coils in a series flow air path; a common duct distribution system at a common air temperature feeds all terminal apparatus.
- Dual-duct systems, which contain the main heating and cooling coils in parallel flow or series-parallel flow air paths with either (1) a separate cold and warm air duct distribution system that blends the air at the terminal apparatus (dual-duct systems), or (2) a separate supply air duct to each zone with the supply air blended to the required temperature at the main unit mixing dampers (multizone).

These classifications may be further divided as follows:

Single Duct
    Constant Volume
        Single Zone
        Multiple Zoned Reheat
        Bypass
    Variable Air Volume
        Reheat
        Induction
        Fan Powered
        Dual Conduit
        Variable Diffusers
Dual Duct
    Dual Duct
        Constant Volume
        Variable Volume

Multizone
    Constant Volume
    Variable Volume
    Three-Deck or Texas Multizone

All-air systems may be adapted to many applications for comfort or process work. They are used in buildings that require individual control of multiple zones, such as office buildings; schools and universities; laboratories; hospitals; stores; hotels; and even in ships.

All-air systems are also used virtually exclusively in special applications for close control of temperature and humidity, including clean rooms, computer rooms, hospital operating rooms, research and development facilities, as well as many industrial/manufacturing facilities.

Advantages

All-air systems have the following advantages:

- The central mechanical equipment room location for major equipment allows operation and maintenance to be performed in unoccupied areas and permits the maximum range of choices of filtration equipment, vibration and noise control, and the selection of high quality and durable equipment.

- Complete absence within the conditioned area of piping, electrical equipment, wiring, filters, and vibration and noise-producing equipment reduces potential harm to occupants, furnishings, and processes minimizing service needs.

- These systems have the greatest potential for the use of outside air and "free" cooling systems in place of mechanical refrigeration for cooling.

- Seasonal changeover is simple and readily adaptable to automatic control.

- Gives a wide choice of zoning, flexibility, and humidity control under all operating conditions, with the availability of simultaneous heating and cooling, even during off-season periods.

- Air-to-air and other heat recovery systems may be readily incorporated.

- Permits good design flexibility for optimum air distribution, draft control, and adaptability to varying local requirements.

The preparation of this chapter is assigned to TC 9.1, Large Building Air-Conditioning Systems.

- Well suited to applications requiring unusual exhaust or makeup air quantities (negative or positive pressurization, etc.).
- Adapts well to winter humidification.
- The primary system can be used to introduce outside air required for ventilation without the need for supplemental systems.
- By increasing the air change rate, these systems are able to maintain the closest operating condition of ±0.25°F dry bulb and ±0.5% relative humidity relatively simply. Today there are systems that essentially maintain constant space conditions.

Disadvantages

All-air systems have the following disadvantages:

- They require additional duct clearance, which reduces usable floor space and increases the height of the building.
- Depending on layout, vertical shaft space is needed for distribution requiring larger floor plates.
- The accessibility of terminal devices requires close cooperation between architectural, mechanical, and structural designers.
- Air balancing, particularly on large systems, can be more difficult.
- Perimeter heating systems are not always available for use in providing temporary heat during construction.

Heating and Cooling Calculations

Basic calculations for airflow, temperatures, relative humidity, loads, and psychrometrics are covered in Chapter 6 of the 1989 ASHRAE *Handbook—Fundamentals*. It is important that the designer understand the operation of the various components of a system, their relationship to the psychrometric chart, and their interaction under various operating conditions and system configurations.

PROCESSES

COOLING

Design Considerations

Three basic methods are available for providing cooling:

1. Direct expansion takes advantage of latent heat of a fluid, as shown in the psychrometric diagram in Figure 1.

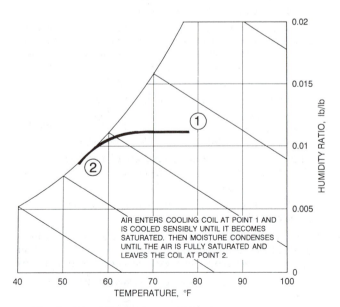

Fig. 1 Direct Expansion or Chilled Water Cooling and Dehumidification

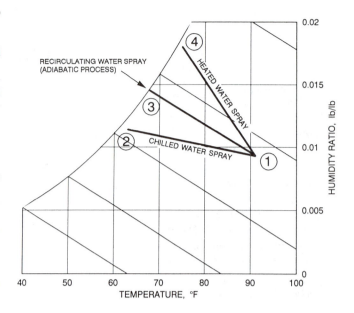

Fig. 2 Water Spray Cooling

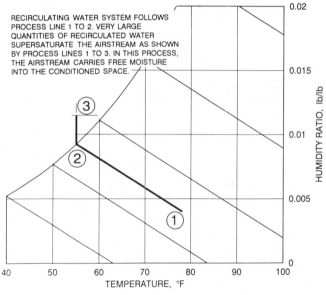

Fig. 3 Water Spray Humidifier

2. Fluid-filled coil, where temperature differences between the fluid and the air exchange energy; same process as in Figure 1 (see section on humidification).
3. Direct spray of water in the airstream (Figure 2). This adiabatic process uses the latent heat of evaporation of water. By spraying chilled water, sensible and latent cooling is also possible. In an evaporative cooler, recirculated water is sprayed or drips onto a filter pad (see section on humidification).
4. The wetted duct or supersaturated system is a variation on direct spray. In this system, free moisture carried by the air evaporates in the conditioned space. This process provides additional cooling and reduces the air needed for a given space load (Figure 3).

Chapter 6 of the 1989 ASHRAE *Handbook—Fundamentals* describes psychrometric processes in more detail.

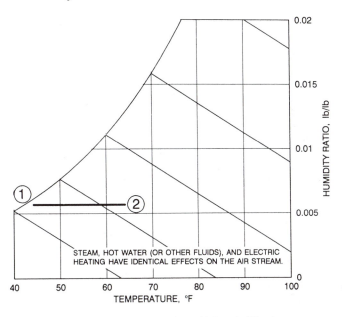

Fig. 4   Steam, Hot-Water, and Electric Heating

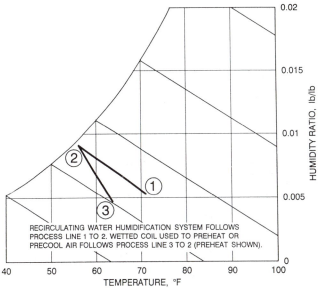

Fig. 5   Compressed Air and Water

HEATING

Design Considerations

Three basic methods of heating are available:

1. Steam, which uses the latent heat of the fluid.
2. Fluid-filled coils, which use temperature differences between the warm fluid and the cooler air (see section on humidification).
3. Electric heating, which also uses temperature differences between the heating coil and the air to exchange energy.

The effect on the airstream for each of these processes is the same and is shown in Figure 4. For basic formulas, refer to Chapter 6 of the 1989 ASHRAE *Handbook—Fundamentals*.

HUMIDIFICATION

Design Considerations

The following methods are available for the humidification of air:

1. Direct spray of recirculated water into the airstream (air washer) reduces the dry-bulb temperature while maintaining an almost constant wet bulb in an adiabatic process [see Figure 2, paths (1) to (3)]. The air may also be cooled, dehumidified, heated, and humidified by changing the temperature of the spray water.

   In one variation, the surface area of water exposed to the air is increased by spraying water onto a cooling/heating coil. The coil surface temperature determines the leaving air conditions. Another method is to spray or distribute water over a porous medium, such as those in evaporative coolers and commercial greenhouses. This method requires careful monitoring of the water condition to keep biological contaminants from the airstream (Figure 5).

2. The use of compressed air to force water through a nozzle into the airstream is essentially a constant dry-bulb process (isothermal). The water must be treated to keep particulates from entering the airstream and contaminating or coating equipment and furnishings. Many types of nozzles are available, each with different characteristics.

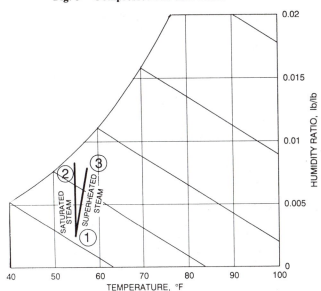

Fig. 6   Steam Humidifier

3. Steam injection is a constant dry-bulb process (Figure 6). However, as the steam injected becomes superheated, the leaving dry-bulb temperature increases. When live steam is injected into the airstream, the boiler water treatment chemical used must be nontoxic to the occupants and, if the air is supplying a laboratory, to the research under way.

DEHUMIDIFICATION

Processes

Moisture will condense on a cooling coil where surface temperature is below the dew-point temperature of the air, which reduces the humidity of the air. In a similar manner, air will also be dehumidified if a fluid with a temperature below the airstream dew point is sprayed into the airstream. The process is identical to that shown in Figure 1.

Chemical dehumidification can be used either by passing air over a solid desiccant or by spraying the air with a solution of the

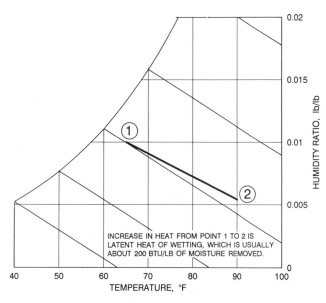

Fig. 7 Chemical Dehumidifier

desiccant and water. Often referred to as latent heat of wetting, both these processes add heat to the air being dehumidified. Usually about 200 Btu/lb of moisture is removed (Figure 7).

It is important to review these systems with the user to ensure that contamination of the space does not occur.

DESIGN

DISTRIBUTION SYSTEMS

Several methods for designing the duct distribution for all-air systems are described in Chapter 32 of the 1989 ASHRAE *Handbook—Fundamentals.*

Depending on their layout, distribution systems can have quite diverse operating characteristics. As a result, if the distribution and terminal devices selected are not compatible, the system will either fail to operate effectively or incur high first, operating, and maintenance costs.

Duct Sizing

All-air duct systems can be designed for high or low velocity. High-velocity systems reduce duct sizes and save space but require higher system pressures. In some low-velocity systems, medium or high fan pressures may be required for balancing or to overcome high pressure drops from terminal devices.

In any variable flow system, changing operating conditions can create rates of airflow in the ducts that are different from those used in the design of the system. Thus, varying airflow in the supply duct requires careful analysis to ensure that the system performs efficiently at all loads. This precaution is particularly true when high-velocity distribution is used.

Return-air ducts are usually sized by the equal friction method. In many applications, ceilings are used as return-air plenums, which means that the return air is collected at a central point. This approach should be reviewed to be sure it meets local code requirements. For example, the National Electric Code requires either conduit or Teflon insulated wire, often referred to as plenum rated cable, to be installed in a return-air plenum ceiling. In addition, return-air plenums often require the space to be divided into smaller areas using fire walls, requiring that fire dampers be installed in ducts and increasing first cost. In research and some

industrial facilities, return ducting must be installed to avoid contamination and growth of biological contaminants within the ceiling cavity.

Corridors should not be used for return air, because they spread smoke and other contaminants. While most codes ban returning air through corridors, they are still used in many older facilities.

Airflow

If net room loads are not determined accurately, a variable volume system having too much airflow will partially throttle at full load and cause excessively noisy throttling at part loads. If a constant volume system is oversized, reheat costs would be excessive. The oversizing of fans, ducts, terminals, control valves, and particularly control dampers on outside or return-air louvers should be avoided.

Air movement below 10 fpm or 2.5 air changes per hour is usually considered uncomfortable, creating a feeling of stuffiness. Minimum air circulation can be maintained by (1) raising supply air temperature on a system basis, which increases space humidity, or with reheat on a zone-by-zone basis; (2) providing auxiliary heat in each room independent of the air system; (3) using individual zone recirculation and blending varying amounts of supply and room air or supply and ceiling plenum air such as with fan-powered VAV boxes, or, if the design permits, at the air-handling unit; and (4) recirculating air with a variable volume induction unit.

ZONING

Exterior

Exterior zones are affected by varying weather conditions—wind, temperature, and sun—and, depending on the geographic area and season, both heating and cooling are required. This variation gives the engineer considerable flexibility in choosing a system and enables the greatest advantages from VAV systems to be realized.

The need for separate perimeter zones/systems is determined by:

• Severity of the heating load, *i.e.,* geographic location.
• Nature and orientation of the building envelope.
• Effects of downdraft at windows and the radiant effect of the cold glass surface, *i.e.,* type of glass, area, height, and U-factor.
• Type of occupancy, *i.e.,* sedentary versus transient.
• Operating costs, *i.e.,* in buildings such as offices and schools that are unoccupied for considerable periods, fan operating costs can be reduced by heating with perimeter radiation during unoccupied periods rather than operation of the main supply fans or local unit fans.

A separate and distinct perimeter heating system can operate with any of the all-air systems. However, its greatest application has been in conjunction with variable air volume (VAV) systems for cooling-only service. The section on VAV systems in this chapter has further details.

Interior

Interior spaces have relatively constant conditions since they are isolated from external influences and usually require cooling throughout the year. The use of VAV systems has limited energy advantages for interior spaces but does provide simple temperature control. Interior spaces with a roof above, however, may require similar treatment to perimeter spaces requiring heat.

Special Conditions

Many situations require particular care in zoning. Laboratories, for example, are usually zoned based on the lab module, hood exhaust, etc. Clean rooms, sterile areas, and manufacturing all require careful analysis to provide proper control.

OUTSIDE AIR REQUIREMENTS

One of the most common complaints regarding buildings, besides the lack of air movement, is the lack of outside air. This problem is especially a concern in VAV systems where outside air quantities are established for peak loads and are then reduced in proportion to the supply air quantity during periods of reduced load. A simple control added to the outside air damper can eliminate this problem and keep the amount of outside air constant, irrespective of the operation of the VAV system. However, the need for preheat must be checked if this type of control is added to the air-handling equipment.

Another problem is that many existing energy codes call for a minimum of 5 cfm per person (about 0.05 cfm/ft^2) of outside air. This amount is far too low for a building in which modern construction materials are used, and many states and building owners are requiring higher outside air quantities to reduce CO_2 levels and other potentially dangerous pollutants. ASHRAE *Standard* 62-1989, for example, recommends increasing the amount of outside air to 20 cfm or more per person.

TEMPERATURE VERSUS AIR QUANTITY

System designers have considerable latitude in selecting supply air temperatures and corresponding air quantities within the limitations of the procedures for determining heating and cooling loads. ASHRAE *Standard* 55-1981 also addresses the effect of these variables on comfort. In establishing the supply air temperature, designers need to consider the initial cost of lower air quantities and low air temperatures (smaller fan and duct systems) against the potential problems of distribution, condensation, air movement, and the presence of increased odors and gaseous or particulate contaminants.

An increasing number of terminal devices are available that use low-temperature air. These devices mix room and primary air to maintain reasonable air movement within the occupied space and to take advantage of reduced air distribution costs. Since the amount of outside air is the same for any system, the percentage in low-temperature systems is high, requiring special care in design to avoid freezing.

Note that the supply of low-temperature air also reduces space humidity. Lower humidities cost more in energy and, if they are too low, may cause respiratory problems.

OTHER CONSIDERATIONS

All-air systems operate with a temperature differential between the supply air and the space, so that any loads imposed by the system that affect the air temperature differential and the required air quantities must be considered. Among these loads are the following:

- *Fan heat gain.* Since fans perform work in the air (they move the air through the system), there is an associated rise in air temperature. The effect of this temperature rise on the overall air quantity supplied to the space depends on the location of the supply fan.

    If the fan is placed after the cooling coil and draws the air through it, the total pressure (static plus velocity pressure) has to be calculated. By placing the fan before the coil (blow-through), only the velocity pressure needs to be accounted for, and the amount of supply air can be reduced. A general rule for this heat gain in medium pressure systems is about 0.5 °F per in. (of water) static pressure.

All fans (supply, return, supplemental) add heat. The effect of these gains can be considerable, particularly in process work. If the fan motor is within the airstream, the inefficiencies in the motor must also be taken into account as heat gain to the air.

- Heat gain or loss in the supply duct system may come from the surroundings. Most energy codes require that the supply air ducting be insulated. It is usually good practice, regardless of the code requirements, to insulate supply duct runs that exceed 75 ft in length.
- Attempting to control humidity levels within a space can affect the quantity of air and become the controlling factor in the selection of supply air quantity. When VAV systems are used, relative humidity control is limited, and if humidity is critical, extreme care must be taken.
- The location of intake and exhaust louvers should be carefully considered; in some jurisdictions, this is governed by codes. Obviously, louvers must be sufficiently separated to avoid short circuiting of air. Furthermore, intake louvers should not be located near a potential source of contaminated air such as a boiler stack or hood exhaust. Relief air should also not interfere with other systems. If heat recovery devices are used, intake and exhaust airstreams may need to run together, such as in air-to-air plate heat exchangers.
- Noise control, both within the occupied spaces and outside near intake or relief louvers, needs to be considered. Some local ordinances may limit external noise produced by these devices.
- Air quality, *i.e.*, the control or reduction of contaminants such as CO_2, formaldehyde from furnishings, and dust, must be reviewed by the engineer. In manufacturing and research and development facilities (and, more recently, in office environments) control of contaminants is particularly important.
- *Heat recovery devices.* Although they are not used commonly in commercial systems, heat recovery devices are used extensively in research and development facilities. Many types are available, and the type of facility will usually determine which is most suitable.

Many countries with extreme climates provide heat exchangers on outside/relief air, even for private homes. This trend is now appearing in larger commercial buildings worldwide.

FIRST, OPERATING, AND MAINTENANCE COSTS

As with all systems, the initial, or first, cost of an air-handling system varies widely depending on geographic location, condition of the local economy, preference of contractor—even for identical systems.

In dual-duct systems, where the amount of material installed for ducting is essentially twice that of a comparable single-duct system, the cost is obviously greater. Systems requiring extensive use of terminal devices will also be comparatively expensive.

Operating costs depend on the system selected and the skill of the designer in selecting and correctly sizing components, efficient duct design, and the effect of the building design and type on the system operation. All-air systems have the greatest potential to minimize operating costs.

Because all-air systems lend themselves to the use of central air-handling equipment separate from occupied spaces, maintenance on major components is more economical. Central air-handling equipment also requires less maintenance when compared to other systems.

The many terminal devices used in all-air systems do, however, require periodic maintenance. Since these devices (including reheat coils) are usually located throughout a facility, maintenance costs for these devices must be considered.

ENERGY

If designed carefully, most systems can keep energy costs to a minimum. In practice, however, systems are usually selected based on first cost or for the performance of a particular task. In general, single-duct systems consume less energy than dual-duct systems, and VAV systems are more energy efficient than constant air volume systems.

Although worthwhile, the savings in fan power for VAV systems are small when compared to the energy savings resulting from the reduced delivery of cooling and heating because of the elimination of overheating or overcooling of spaces and the elimination of simultaneous cooling and heating, as in reheat systems.

Energy waste due to leakage in ductwork and terminal devices can be considerable. The ductwork installed in many commercial buildings can have leakage rates of 20% or more.

Heat recovery devices such as air-to-air plate heat exchangers can save considerable amounts of energy and reduce the required capacity of primary cooling and heating plants by as much as 20% and more under certain circumstances.

The engineer's early involvement in the design of any facility can have a considerable impact on the energy consumed by the building.

SYSTEMS

SINGLE-DUCT SYSTEMS

Constant Volume

While maintaining constant airflow, single-duct constant volume systems change the supply air temperature in response to the space load (Figure 8).

Single-zone systems. The simplest all-air system is a supply unit serving a single-temperature control zone. The unit can be installed either within or remote from the space it serves and may operate with or without distribution ductwork. Ideally, this system responds completely to the space needs, and well-designed systems maintain temperature and humidity closely and efficiently. Single-zone systems can be shut down when not required without affecting the operation of adjacent areas.

A return or relief fan may be needed, depending on the capacity of the system and whether 100% outdoor air is used for cooling at some time during the year. Relief fans can be eliminated if provisions are made to relieve overpressurization by other means, such as gravity dampers.

Zoned reheat. The multiple-zoned reheat system is a modification of the single-zone system. It provides (1) zone or space control for areas of unequal loading; (2) simultaneous heating or cooling of perimeter areas with different exposures; and (3) close tolerance of control for process or comfort applications.

As the word reheat implies, heat is added as a secondary simultaneous process to either preconditioned (cooled, humidified, etc.) primary air or recirculated room air.

Relatively small low-pressure systems have reheat coils in the ductwork at each zone. More sophisticated designs have high-pressure primary distribution ducts to reduce their size and cost, and pressure reduction devices to maintain a constant volume for each reheat zone.

The system uses conditioned air from a central unit generally at a fixed cold air temperature. The temperature can be varied, however, with the proper controls to reduce the amount of reheat required and the associated energy consumption. Care has to be taken to avoid high internal humidity levels when the temperature of air leaving the coil is permitted to rise during the spring and fall.

Bypass. A variation of the constant volume reheat system is the use of a bypass box in lieu of reheat. This system is essentially a constant volume primary system with a VAV secondary system.

The quantity of room supply air is varied to match the space load by dumping excess supply air into the return ceiling plenum or return-air duct, *i.e.*, by bypassing the room. While this system reduces the air volume supplied to the space, the system air volume remains constant. Refrigeration or heating at the air-handling unit is reduced due to the lower return-air temperature.

This system is generally restricted to small systems where a simple method of temperature control is desired, a modest initial cost is desired, and energy conservation is less important.

Variable Volume

A VAV system (as shown in Figure 9) controls temperature within a space by varying the quantity of supply air rather than varying the supply air temperature. A VAV terminal device is used at the zone to vary the quantity of supply air to the space. The supply air temperature is held relatively constant, depending on the season.

Variable air volume systems can be applied to interior or perimeter zones, with common or separate fan systems, common or separate air temperature control, and with or without auxiliary heating devices.

The greatest energy savings associated with VAV systems occur at the perimeter zones, where variations in solar load and outside temperature permit reduction in supply air quantities.

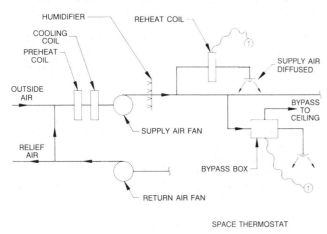

Fig. 8   Constant Volume with Reheat and Bypass Terminal Devices

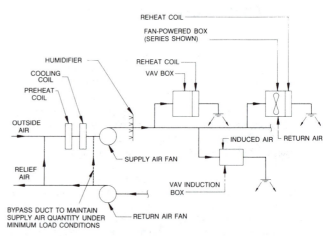

Fig. 9   Variable Air Volume with Reheat, Induction, and Fan-Powered Devices

Humidity control is a potential problem when VAV systems are used. If humidity level is critical, as in certain research and development laboratories, process work, etc., systems may have to be limited to constant volume airflow. Particular care should be taken in areas where the sensible heat ratio is low, such as in conference rooms. In these situations, the VAV box minimum set point is usually limited to about 50%, and reheat is added as necessary to keep humidity levels low during part loads.

Variable air volume terminal devices are available in a number of different configurations, including:

Reheat. This simple VAV system integrates heating at the terminal unit. It is applied to systems requiring full heating and cooling flexibility in interior and exterior zones. The terminal units are set to maintain a predetermined minimum throttling ratio, which is established as the lowest air quantity necessary to (1) offset the heating load; (2) limit the maximum humidity; (3) provide reasonable air movement within the space; and (4) provide required ventilation air.

Variable volume with reheat permits airflow to be reduced as the first step in control; heat is then initiated as the second step. When compared to a constant volume reheat system, this procedure reduces operating cost appreciably, since the amount of primary air to be cooled and secondary air to be heated is reduced.

A feature can be provided to isolate the availability of reheat during the summer, except in situations where low airflow should be avoided or where an increase in humidity causes discomfort, *e.g.*, in conference rooms when the lights are turned off.

Induction. The VAV induction system uses a terminal unit to reduce cooling capacity by simultaneously reducing primary air and inducing room or ceiling air (replaces the reheat coil) to maintain a relatively constant room supply volume. This operation is the reverse of the bypass box described earlier.

The system primary air quantities reduce with load, retaining the savings of VAV, while the air supplied to the space is kept relatively constant to avoid the effect of stagnant air or low air movement.

The terminal device is usually located in the ceiling cavity to take advantage of the heat from lights. This permits the induction box to be used without reheat coils for internal spaces. Provisions must be made for morning warm-up and night heating. Interior spaces that have a roof load must also have heat either separately in the ceiling or at the box.

Fan-powered systems. Fan-powered systems are available in either parallel or series airflow. In parallel flow units, the fan is located outside the primary airstream to allow intermittent fan operation. In series units, the fan is located within the primary airstream and runs continuously when the zone is occupied. Fan-powered systems, both series and parallel, are often selected because they maintain a higher level of air circulation through a room at low loads while still retaining the advantages of VAV systems.

As the cold primary air valve modulates from maximum to minimum (or closed), the unit recirculates more plenum air. When used in a perimeter zone, a hot-water heating coil, electric heater, baseboard heater, or remote radiant heater can be sequenced with the cooling to offset external heat losses. Between heating and cooling operations, a dead band in which the fan recirculates ceiling air only is provided. This operation permits heat from lights to be used to provide space heating for maximum energy savings.

During unoccupied periods, the main supply air-handling unit remains deenergized and individual fan-powered heating zone terminals are cycled to maintain required space temperature, thereby reducing operating costs.

Both systems take advantage of the heat of lights within the ceiling plenum and usually are provided with filters. The constant fan VAV terminal can accommodate minimum (down to zero)

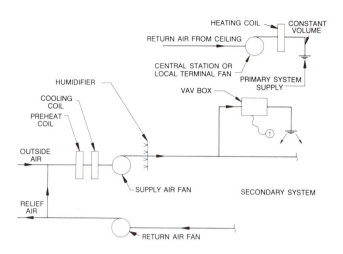

Fig. 10   Dual Conduit System

flow at the primary air inlet while maintaining constant airflow to the space.

Parallel arrangement—Intermittent fan. In this device, primary air is modulated in response to cooling demand and energizes an integral fan at a predetermined reduced primary flow to deliver ceiling air to offset heating demand.

These devices are primarily used in perimeter zones where auxiliary hot-water or electric heating is required. The induction fan operating range normally overlaps the range of the primary air valve. A back-draft damper on the terminal fan prevents conditioned air from escaping into the return-air plenum when the terminal fan is off.

Series arrangement—Constant fan. A constant volume fan-powered box mixes primary air with air from the ceiling space using a continuously operating fan; this provides a relatively constant volume to the space. These devices are used for interior or perimeter zones and are supplied with or without an auxiliary heating coil. They can be used to mix primary and return air to raise the temperature of the air supplied to the space such as with low-temperature air systems.

Dual conduit. The dual-conduit system is designed to provide two air supply paths, one to offset exterior transmission cooling or heating loads, and the other where cooling is required throughout the year (Figure 10).

The first airstream, the primary air, operates as a constant volume system, and the air temperature is varied to offset transmission only (*i.e.*, it is warm in winter and cool in summer). Often, however, the primary air fan is limited to operating only during the peak heating and cooling periods to further save energy. When calculating the heating requirements for this system, the cooling effect of the secondary air must be included, even though the secondary system is operating at minimum flow.

The other airstream, or secondary air, is cool year-round and varies in volume to match the load due to solar heating, lights, power, and occupants. It serves both perimeter and interior spaces.

Variable diffusers. These devices reduce the discharge aperture of the diffuser. This keeps the discharge velocity relatively constant while reducing the conditioned supply airflow. Under these conditions, the induction effect of the diffuser is kept high, and cold air mixes in the space.

These devices are of two basic types—one has a flexible bladder which expands to reduce the aperture, and the other has a diffuser plate that is physically moved. Both devices are typically pressure-dependent, which must be taken into account in the design of the duct distribution system. They are either system powered or pneumatically or electrically driven.

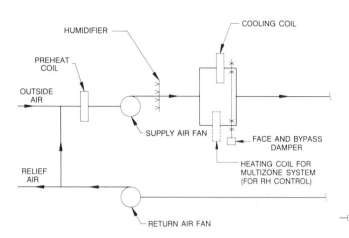

Fig. 11   Single Fan—No Reheat

DUAL-DUCT SYSTEMS

Dual-duct systems condition all the air in a central apparatus and distribute it to the conditioned spaces through two parallel mains or ducts, one duct carrying cold air and the other, warm air. In each conditioned space or zone, a mixing valve mixes the warm and cold air in proper proportions to satisfy the load of the space.

These systems may be designed as constant volume or variable volume and, as with other VAV systems, certain primary air configurations can cause high space relative humidity during the spring and fall.

Dual-duct systems use more energy than single-duct VAV systems but have certain advantages, *e.g.*, no pipes that could leak are required within occupied areas.

Constant volume. These systems are of two types:

Single fan—No reheat. This cycle is similar to a single-duct system, except that it contains a face-and-bypass damper at the cooling coil, which is arranged to bypass a mixture of outdoor and recirculated air as the internal heat load fluctuates in response to a zone thermostat (Figure 11). A problem with this system occurs during periods of high outdoor humidity, when the internal heat load falls, causing the space humidity to rise rapidly (unless reheat is added). It is identical in concept to bypass cooling coils.

This system has limited use in most modern buildings since most occupants demand more consistent temperature and humidity conditions.

Single fan—Reheat. This cycle is similar in effect to a conventional reheat system. The only differences are that reheat is

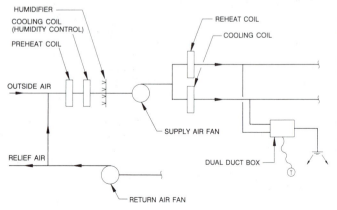

Fig. 12   Dual Duct—Single Fan

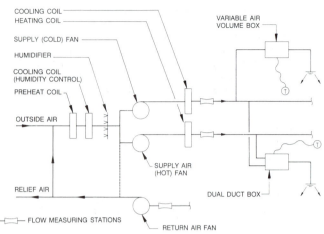

Fig. 13   Variable Air Volume, Dual Duct, Dual Fan

applied at a central point instead of at individual zones (Figure 12).

Variable air volume. Dual-duct variable volume systems blend cold and warm air in various volume combinations. These systems may include single-duct VAV units connected to the cold deck for cooling only of interior spaces.

Single fan. In this system, a single supply fan is sized for the coincident peak of the hot and cold decks. Control of the fan is from two static pressure controllers, one located in the hot deck and the other in the cold deck. The duct requiring the highest pressure governs the fan airflow.

Usually the cold deck is maintained at a fixed temperature, although some central systems permit the temperature to rise during warmer weather to save refrigeration. The temperature of the hot deck is often adjusted higher during periods of low outside temperature and high humidity to increase the flow over the cold deck for dehumidification.

Other systems, particularly in laboratories, use a precooling coil to dehumidify the total airstream or just the outside air to limit humidity within the space.

Return-air quantity can be controlled by either flow-measuring devices within the supply and return duct systems or by static pressure controls which maintain space static pressure.

Dual fan. The volume of each supply fan is controlled independently by the static pressure in its respective duct (Figure 13). The return fan is controlled based on the sum of the hot and cold fan volumes using flow-measuring stations.

Each fan is sized for the anticipated maximum coincident hot or cold volume, not the sum of the instantaneous peaks.

The cold deck can be maintained at a constant temperature either by operating the cooling coil with mechanical refrigeration when minimum fresh air is required or with a cooling economizer when the outside air is below the temperature of the cold deck set point. This operation does not affect the hot deck, which can recover heat from the return air, and the heating coil need only operate when heating requirements cannot be met using return air.

Outdoor air can provide ventilation air via the hot duct system when the outdoor air is warmer than the system return air. However, controls should be used to prohibit the introduction of excessive amounts of outdoor air beyond the required minimum when that air is more humid than the return air.

MULTIZONE SYSTEMS

The multizone system (Figure 11) supplies several zones from a single, centrally located air-handling unit. Different zone requirements are met by mixing cold and warm air through zone dampers at the central air handler in response to zone thermostats. The mixed, conditioned air is distributed throughout the building by a system of single-zone ducts. The return air is handled in a conventional manner.

The multizone system is similar to the dual-duct system. In operation, it has the same potential problem with high humidity levels. This system can provide a smaller building with the advantages of a dual-duct system and uses packaged equipment, which is less expensive. Multizone packaged equipment is usually limited to about 12 zones, while built-up systems can include as many zones as can be physically incorporated in the layout.

A common variation on multizone systems is referred to as the Texas multizone or three-deck multizone. In this system, the hot deck heating coil is removed from the air handler and replaced with an air resistance plate matching the pressure drop of the cooling coil. Individual heating coils are then placed in each perimeter zone duct only. These heating coils are usually located in the equipment room for ease of maintenance.

This system is common in humid climates where the cold deck often produces 48 to 52 °F air for humidity control. Supply air is then mixed with neutral deck (return) air using the air-handling units' zone dampers to maintain zone conditions. Heat is only added if the zone served cannot be maintained by the delivery of return air alone.

SPECIAL SYSTEMS

Primary/Secondary Systems

Some processes use two interconnected all-air systems (Figure 14). In these situations, space gains are very high, a large number of air changes are required, *i.e.*, sterile or clean rooms, or close-tolerance space conditions are needed for process work. The primary systems supply the conditioned outside air requirements for the process either to the secondary system or directly to the space.

The secondary system provides additional cooling and humidification (and heating, if required) to offset space and fan power gains. Normally, the secondary cooling coil is designed to be dry (*i.e.*, sensible cooling only) to reduce the possibility of bacterial growth which can create air quality problems.

Up Air Systems

In this system, air is supplied from the floor rather than from the more conventional overhead system.

Special terminal devices have been developed to use this concept and are installed in the raised floor or as part of the furniture. These devices are used to provide individual control for computer

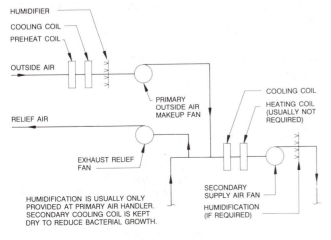

Fig. 14   Primary/Secondary System

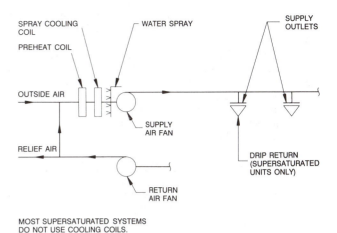

MOST SUPERSATURATED SYSTEMS
DO NOT USE COOLING COILS.

Fig. 15   Supersaturated/Wetted Coil

workstations, and they are becoming more common, particularly in Europe.

Wetted Duct/Supersaturated Systems

Some industries spray water into the airstream at central air-handling units in sufficient quantities to create a controlled carry-over (Figure 15). This supercools the supply air, normally equivalent to an excess oversaturation of about 10 grains per pound of air and allows less air to be distributed for a given space load. These systems are used where high humidities are desirable, such as the textile or tobacco industry, and in climates where adiabatic cooling is sufficient.

Compressed Air and Water Spray Systems

This system is similar to the wetted duct system, except that the water is atomized with compressed air. The nozzles are sometimes located directly within the conditioned space and independent of the air supplied. Depending on the type of nozzle, compressed air and water spray systems can require large and expensive air compressors. Several types of nozzle are available, and the designer should understand their advantages and disadvantages.

Low-Temperature Systems

Since ice storage is used to take advantage of reduced off-hour operating costs, low-temperature systems (where supply air is distributed at approximately 40 °F) are being considered by designers.

These systems allow the designer to use very high temperature differentials and, therefore, the delivery of considerably less air. Since the quantity of air supplied to a zone is so small, special terminal devices are used to maintain minimum airflow for comfort. These devices induce return or room air to be mixed with the cold supply air to increase the air circulation rate within the space.

EQUIPMENT

AIR-HANDLING EQUIPMENT

Air-handling equipment is available as packaged equipment in many configurations using any desired method of cooling, heating, humidification, filtration, etc. In large systems over 50,000 cfm, air-handling equipment is usually custom-designed and fabricated to suit a particular application.

Location

Central mechanical equipment rooms (MER). Usually the type of facility will determine where the air-handling equipment is located. Central fan rooms today are more common in laboratory or industrial facilities, where maintenance is kept isolated from the conditioned space.

Satellite/Local mechanical equipment rooms. Many office buildings locate air-handling equipment at each floor. This not only saves floor space for equipment but minimizes the space required for distribution ductwork and shafts. The reduced size of equipment as a result of duplicity of systems allows the use of less expensive packaged equipment and reduces the need for experienced operating and maintenance personnel.

TERMINAL DEVICES/DIFFUSERS

Many manufacturers provide equipment that is similar, but products vary slightly from one manufacturer to another. A particular product's operation and specifications should be reviewed before making a final selection.

HUMIDIFIERS

Most projects where humidification is required use steam. This is centrally generated as part of the heating plant, where potential contamination from the water treatment of the steam is more easily dealt with and is, therefore, of less concern. Where there is a concern, local generators (*e.g.*, electric) that use treated water are used.

Compressed air and water humidifiers are used to some extent and supersaturated systems are used exclusively for special circumstances, such as industrial processes. Spray-type washers and wetted coils are also more common in industrial facilities.

When using water directly, particularly in recirculating systems, the water must be treated to avoid the accumulation of dust during evaporation and the build-up of bacterial contamination.

DEHUMIDIFIERS

Many dehumidification systems, each with its advantages and disadvantages, are available. In the case of solid desiccants, dusting can be a problem, and for spray types, lithium contamination is of concern.

ODOR CONTROL EQUIPMENT

Odor control is of particular concern today in view of the use of new building materials, energy concerns, and outside air pollution.

Most devices available to control odors and other contaminants use activated carbon or potassium permanganate as a basic filter medium. Other systems use electronic control methods.

FILTERS

Various types and efficiencies of filtration are available. The type used depends on the building being air conditioned. Besides protecting the conditioned space, filters are often applied to the exhaust or relief systems to protect the environment. Filtration can be accomplished by both solid and spray scrubbers.

SPECIAL SYSTEMS

Many systems are available as factory-built packages with the advantage of assembly in a controlled environment. The quality of a packaged system is often better than field-assembled equipment, and it is usually less expensive.

PRIMARY EQUIPMENT

Refrigeration

Either central station or localized equipment, depending on the application, can provide cooling. Most large systems with multiple central air-handling units use a central refrigeration plant. Small, individual air-handling equipment can be supplied from central chilled water generators, can use direct expansion cooling with a central condensing (cooling tower) system, or can be air cooled and totally self-contained.

The decision on whether to provide a central plant versus local equipment is based on similar factors to those for air-handling equipment.

Heating

The same criteria as described for cooling are usually used to determine whether or not a central heating plant or a local one is desirable. Usually, a central, fuel-fired plant is more desirable for heating.

In small facilities, electric heating is a viable option and is often economical, particularly where care has been taken to design energy-efficient systems and buildings.

CONTROLS

Today the trend is toward direct digital control (DDC), and most manufacturers offer either standard or an optional DDC package for equipment including air-handling units, terminal devices, etc. These controls offer a considerable degree of flexibility.

Constant Volume Reheat

Constant volume reheat systems typically use two subsystems for control—one system controls the air-handling unit discharge air conditions, and another maintains the space conditions by controlling the reheat coil.

Variable Air Volume Systems

Air volume can be controlled by duct-mounted terminal units serving a number of air outlets in a control zone or by units integral to each supply air outlet.

Pressure-independent volume regulator units control the flow rate in response to the thermostat's call for heating or cooling. The required flow rate is maintained regardless of fluctuation of the VAV unit inlet or system pressure. These units can be field or factory adjusted for maximum and minimum (or shut off) air settings. They will operate at inlet static pressures as low as 0.2 in. of water for maximum unit design volumes.

Pressure-dependent devices control air volume in response to a maximum volume unit thermostatic (or relative humidity) device, but the flow rate varies with the inlet pressure variation. Generally, airflow oscillates when pressure varies. These thermostatic units do not regulate the flow rate but position the volume-regulating device in response to the thermostat. These units are the least expensive and should only be used where there is no need for maximum or minimum limit control and when the system pressure is stable.

The type of controls available for variable volume units varies with the terminal device. Most use either pneumatic or electric control and may be either self-powered or system air actuated. Self-powered controls position the regulator by using liquid- or wax-filled power elements. System-powered devices use air from the system supplying air to the space to power the operator. Components for both control and regulation are usually contained within the terminal device.

To conserve power and limit noise, especially in larger systems, fan-operating characteristics and system static pressure should be controlled. Many methods are available, including fan speed control, variable inlet vane control, fan bypass, fan discharge damper, and variable pitch fan control.

The location of the pressure-sensing devices depends, to some extent, on the type of variable volume terminal used. Where pressure-dependent units without controllers are used, the system pressure sensor should be near the static pressure midpoint of the duct run to ensure minimum pressure variation in the system. Where pressure-independent units are installed, pressure controllers may be at the end of the duct run with the highest static pressure loss. This sensing point ensures maximum fan power savings while maintaining the minimum required pressure at the last terminal.

As the flow through the various parts of a large system varies, so do the static pressure losses. Some field adjustment is usually required to find the best location for the pressure sensor. In many systems, the initial location is two-thirds to three-fourths of the distance from the supply fan to the end of the main trunk duct. As the pressure at the system control point increases due to the closing of terminal units, the pressure controller signals the fan controller to position the fan volume control, which reduces flow and maintains constant pressure.

Many present-day systems measure flow rather than pressure and, with the development of economical DDC, each terminal box (if necessary) can be monitored and the supply and return air fans modulated to exactly match the demand of the system.

Dual-Duct Systems

As these systems are generally more energy intensive than single-duct systems, their development has been slow. The use of DDC and the ability of this type of control to accurately maintain set points and flow rates can make these systems worthwhile for certain applications. They should be considered as serious alternatives to single-duct systems.

Other Considerations

The personnel operating and maintaining air-conditioning and control systems should be considered. In large research and development or industrial complexes, experienced personnel are available to maintain the system. On small and, sometimes, even large commercial installations, however, office managers are often responsible, and the system must be designed in accordance with their limited capabilities.

Damper selection. The selection of outside, relief, and return-air dampers is critical if the system is to work efficiently. Most dampers are grossly oversized and are, in effect, unable to control. One method of controlling this problem is to provide maximum and minimum dampers.

Valves. Most valves are sized reasonably well, although the same problems can occur as with dampers.

On large hydraulic systems where direct blending is used to maintain secondary water temperatures, the system must be accurately sized for proper control.

Many designers use variable flow for hydronic as well as air systems, and the design must be compatible with the air system to avoid operating problems.

Relief fans. In many applications, relief or exhaust fans can be started in response to a signal from the economizer control or to a space pressure controller. The main unit supply fan must be able to handle the return-air system pressure drop when the relief fan is not running.

AIR-AND-WATER SYSTEMS

AIR-AND-WATER systems condition spaces by distributing air and water sources to terminal units installed in habitable space throughout a building. The air and water are cooled or heated in central mechanical equipment rooms. The air supplied is called primary air; the water supplied is called secondary water. Sometimes a separate electric heating coil is included in lieu of a hot water coil. This chapter is concerned primarily with air-and-water induction units, fan-coil units, and radiant panels as used in air-water systems. Fan-coil units are discussed in Chapter 4, and radiant panels are discussed in Chapter 6.

Air-and-water systems apply primarily to exterior spaces of buildings with high sensible loads and where close control of humidity is not required. They may, however, be applied to interior zones as well. These systems work well in office buildings, hospitals, hotels, schools, apartment houses, research laboratories, and other buildings where their capabilities meet the performance criteria. In most climates, these systems are installed in exterior building spaces and are designed to provide (1) all required space heating and cooling needs and (2) simultaneous heating and cooling in different parts of the building during intermediate seasons.

EXTERIOR ZONE AIR-CONDITIONING LOADS

The air-conditioning load variations in exterior building spaces cause significant variations in space cooling and heating requirements, even when these rooms occupy the same exposure of the building. Accordingly, accurate environmental control in exterior building spaces requires individual control. The following basic loads must be considered.

Internal loads. Heat gain from lights is always a cooling load, and, in most buildings (other than residential), it is relatively constant during the day. Emphasis on energy conservation is causing lights to be turned off when not required, making lighting loads more variable. Heat gain from occupants is also a cooling load and is commonly the only room load that contains a latent component. Heat gain from equipment can vary from zero to maximum and is an important factor in office building design.

External loads. Heat gain from the sun is always a cooling load. It is often the chief cooling load factor and is highly variable. For a given space, solar heat gain will always vary during the day. In addition to normal changes resulting from movement of the sun, solar gain can fluctuate greatly. The magnitude and rate of change of this load component depend on building orientation, glass area, cloud cover, and the building's capacity to store heat. Constantly changing shade patterns from adjacent buildings, trees, or

exterior columns and nonuniform overhangs will cause significant variations in solar load between adjacent offices on the same solar exposure.

Transmission load can be either a heat loss or a heat gain, depending on the outdoor temperature.

Moderate pressurization of the building with ventilation air is normally sufficient to offset summer infiltration provided it is uniformly positive. In winter, however, infiltration can cause a significant heat loss, particularly on the lower floors of high-rise buildings. The magnitude of this component varies with wind and stack effect, as well as with the temperature difference across the outside wall.

The maximum cooling load in a particular space is a function of the magnitude of individual load components and the coincidental times at which they occur. The maximum heating load is a function of transmission and infiltration. Net load variations on a given day during summer or winter can result from variations in occupancy, or in equipment heat gain, and from irregular shadow patterns.

To perform successfully, an air-conditioning system must be capable of satisfying these load variations on a room-by-room basis while fulfilling all other performance criteria, such as humidity control, filtration, air movement, and noise.

SYSTEM DESCRIPTION

An air-and-water system includes central air-conditioning equipment, duct and water distribution systems, and a room terminal. The room terminal may be an induction unit, a fan-coil unit, or a conventional supply air outlet combined with a radiant panel. Larger spaces may be handled for multiple terminals. Generally, the air supply has a constant volume and, as noted earlier, is called primary air to distinguish it from room air or secondary air that has been recirculated. The primary air provides clean outside air for ventilation. In the cooling season, the air is dehumidified sufficiently in the central conditioning unit to achieve comfort humidity conditions throughout the spaces served and to avoid condensation resulting from normal room latent load on the room cooling coil. In winter, moisture can be added centrally, to limit dryness. As the air is dehumidified, it is also cooled to offset a portion of the room sensible loads. The air may be from outdoors, or a mixture of outdoor and return air. A preheater is usually required where freezing temperatures are found. Highly efficient filters will keep the induction unit terminal nozzles clean. Whether or not reheaters are included in the air-handling equipment depends on the type of system, as explained later.

The quantity of primary air supplied to each space is determined by (1) the ventilation requirement and (2) the required sensible cooling capacity at maximum room cooling load.

The preparation of this chapter is assigned to TC 9.1, Large Building Air-Conditioning Systems.

In the ideal air-and-water system design, the secondary cooling coil is always dry; this greatly extends terminal unit life and eliminates odors and the possibility of bacterial growth. The primary air normally controls the space humidity. Therefore, the moisture content of the supply air must be low enough to offset the room latent heat gain and to maintain a room dew point low enough to preclude condensation on the secondary cooling surface.

While some systems operate successfully without a secondary coil drain system, a condensate drain is recommended for all air-and-water systems.

In existing induction systems, an energy conservation technique of raising the chilled water temperature on central air-handling cooling coils can damage the terminal cooling coil by its constantly being used as a dehumidifier. Unlike fan-coil units, the induction unit is not designed or constructed to handle condensation. Therefore, it is critical that an induction terminal operates dry. It is even more critical that the primary air dehumidification be adequate to prevent condensation on the radiant panel.

The water side, in its basic form, consists of a pump and piping to convey water to the heat transfer surface within each conditioned space. The heat exchange coil may be an integral part of the air terminal (as with induction units), a completely separate component within the conditioned space (radiant panel), or a combination (as with fan-coil units). The water can be cooled by direct refrigeration, but it is more commonly cooled by introducing chilled water from the primary cooling system or by heat transfer through a water-to-water exchanger. To distinguish it from the primary chilled water circuit, the water side is usually referred to as the secondary water loop or system.

Air-and-water systems are categorized as two-pipe, three-pipe, or four-pipe systems. They are basically similar in function and include both cooling and heating capabilities for year-round air conditioning. The name is derived from the water distribution system. In two-pipe systems, the water is distributed by one supply and one return pipe for either cold or warm water supply. Two-pipe changeover systems are unsatisfactory during seasonal changeover and should be applied with caution and the owners' full acceptance of their deficiencies. The three-pipe system has a cold water supply, a warm water supply, and a common return. Because the cold and warm water blends, the three-pipe system has excessive energy waste and is not recommended. Existing three-pipe systems are ideal candidates for energy conservation retrofit. The four-pipe system distribution of secondary water has cold water supply, cold water return, warm water supply, and warm water return pipes. As discussed later in this chapter, the arrangements of the secondary water circuits and their control systems differ, depending on the type of secondary water.

AIR-WATER INDUCTION SYSTEMS

Figure 1 shows a basic arrangement for an air-water induction terminal. Centrally conditioned primary air is supplied to the unit plenum at medium to high pressure. The acoustically treated plenum attenuates part of the noise generated in the unit and duct system. A balancing damper adjusts the primary air quantity within limits.

The medium to high-pressure air flows through the induction nozzles and induces secondary air from the room through the secondary coil. This secondary air is either heated or cooled at the coil, depending on the season, the room requirement, or both. Ordinarily, the room coil does no latent cooling, but a drain pan collects condensed moisture from unusual temporary latent loads. Primary and secondary air is mixed and discharged to the room.

A lint screen is normally placed across the face of the secondary coil. Induction units are installed in custom enclosures

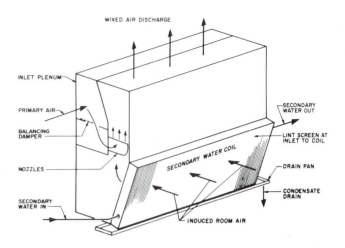

Fig. 1   Air-Water Induction Unit

designed for the particular installation or in standard cabinets provided by the unit manufacturer. These enclosures must permit proper flow of secondary air and discharge of mixed air without imposing excessive pressure losses. They must also allow easy servicing.

Induction units are usually installed under a window at a perimeter wall, although units designed for overhead installation are available. During the heating season, the floor-mounted induction unit can function as a convector during off-hours, with hot water to the coil and without a primary air supply. A number of induction unit configurations are available, including units with low overall height or with larger secondary coil face areas to suit particular space or load needs.

Advantages

1. Individual room temperature control with the capability of adjusting each thermostat for a different temperature at relatively low cost.
2. Separate heating and cooling sources in the primary air and secondary water gives the occupant a choice of heating or cooling.
3. Less space is required for the distribution system when the air supply is reduced by using secondary water for cooling and high velocity air design. The return air duct system is reduced in size and can sometimes be eliminated or combined with the return air system for other building areas, such as the interior spaces.
4. The size of the central air-handling apparatus is smaller than that of other systems, since little air must be conditioned.
5. Dehumidification, filtration, and humidification are performed in a central location remote from conditioned spaces.
6. Ventilation air supply is positive and may accommodate recommended outside air quantities.
7. Space can be heated without operating the air system via the secondary water system. Nighttime fan operation is avoided in an unoccupied building. Emergency power for heating, if required, is much lower than for most all-air systems.
8. System components are long-lasting. Room terminals operated dry have an anticipated life of 15 to 25 years. The piping and ductwork longevity should equal that of the building.
9. Individual induction units do not contain fans, motors, or compressors. Routine service is generally limited to temperature controls, cleaning of lint screens, and infrequent cleaning of the induction nozzles.

Disadvantages

1. Relatively low primary air quantities make the two-pipe changeover design for operating during intermediate seasons more critical than in alternate system types. The four-pipe changeover system overcomes this disadvantage.

2. The two-pipe changeover system has operating complexities not present in other systems. The operator must understand the system cycles and changeover procedures. This disadvantage, and the need for heating at one time of the day and cooling at another, has effectively ruled out the two-pipe changeover system for modern buildings.

3. For most buildings, these systems are limited to perimeter space; separate systems are required for other building areas.

4. Controls tend to be more numerous than for many all-air systems.

5. Secondary airflow can cause the induction or fan-coil unit coils to become dirty enough to affect performance. Lint screens or low efficiency filters used to protect these terminals require frequent in-room maintenance and reduce unit thermal performance.

6. The primary air supply usually is constant with no provision for shutoff. This is a disadvantage in residential applications, where tenants or hotel room guests may prefer to turn off the air conditioning, or where management may desire to do so to reduce operating expense.

7. A low primary chilled water temperature is needed to control space humidity adequately.

8. The system is not applicable to spaces with high exhaust requirements (*e.g.*, research laboratories), unless supplementary ventilation air is provided from other systems.

9. Central dehumidification eliminates condensation on the secondary water heat transfer surface under maximum design latent load. However, abnormal moisture sources (*e.g.*, from open windows or people concentration) can cause condensation that can have annoying or damaging results.

10. Energy consumption for induction systems is higher than for most other systems due to the increased power required by the primary air pressure drop in the terminal units.

11. The initial cost for four-pipe induction systems is greater than that for most all-air systems.

AIR-WATER FAN-COIL SYSTEMS

Fan-coil systems are discussed in Chapter 4.

AIR-WATER RADIANT PANEL SYSTEMS

The advantages and disadvantages of radiant panels are discussed in Chapter 16.

PRIMARY AIR SYSTEMS

Figure 2 illustrates the primary air system for the air-water system. These conventional components are discussed in Chapter 2. Some primary air systems operate with 100% outdoor air at all times. Systems using return air should have provision for operating with 100% outdoor air to reduce operating cost during certain seasons. In some systems, when the quantity of the primary air supplied exceeds the ventilation or exhaust requirements, the excess air is recirculated by a return system common with the interior system. A quality filter is desirable in the central air treatment apparatus. If it is necessary to maintain a given humidity level in cold weather, a humidifier can usually be installed. Steam humidifiers have been used successfully. The water sprays must be operated in conjunction with (1) the pre-

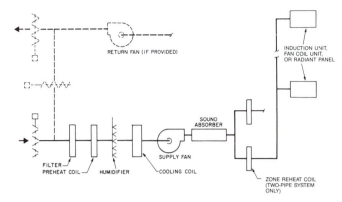

Fig. 2   Primary Air System

heat coil elevating the temperature of incoming air or (2) heaters in the spray water circuit.

The cooling coil is usually selected to provide primary air at a dew point low enough to dehumidify the system totally. Hence, the temperature of the air leaving the cooling coil will be about 50°F or less, and almost completely saturated.

The supply fan should be selected at a point near maximum efficiency to reduce power consumption, heating of the supply air, and noise. Sound absorbers may be required at the fan discharge to attenuate fan noise.

Reheat coils are required in a two-pipe system. Reheat is not required for the primary air supply of four-pipe systems. Many primary air distribution systems for air-water induction units were designed with 8 to 10 in. of water system static pressure. With the energy use restrictions of today, this is no longer economical. Good duct design and elimination of unnecessary restrictions (for example, sound traps) can result in primary systems that operate at 4.5 to 6.0 in. of water. Fan-coil systems can operate at 1.0 to 1.5 in. lower pressure. Careful selection of the primary air cooling coil and induction units for reasonably low air pressure drops is necessary to achieve a medium velocity and medium pressure primary air system. Distribution for fan coil and radiant panel systems may be low velocity or a combination of low and medium velocity systems. The duct system should be designed with the static regain method. Variations in pressure between the first and last terminals should be minimized to limit the pressure drop across balancing dampers.

Room sound characteristics vary depending on unit selection, air system design, and the manufacturer. Units should be selected by considering the unit manufacturer's sound power ratings, the desired maximum room noise level, and the acoustical characteristics of the room. Limits of sound power level can then be specified to obtain acceptable acoustical performance.

PERFORMANCE UNDER VARYING LOAD

Under peak load conditions, the psychrometrics of air-and-water systems are essentially identical for two- and four-pipe systems.

The primary air mixes with secondary air conditioned by the room coil within an induction unit prior to delivery to a room. Mixing also occurs within a fan-coil unit with a direct connected air supply. If the primary air is supplied to the space separately, as in fan-coil systems with independent primary air supplies or as with radiant panel systems, the same effect would occur in the space. The same room conditions result from two physically independent processes as if the air was directly connected to the unit.

During cooling, the primary air system provides a portion of the sensible capacity and all of the dehumidification. The remainder of the sensible capacity is accomplished by the cooling effect of the secondary water circulating through the unit cooling coils or radiant panels. In winter, primary air is provided at a low temperature, and if humidity control is provided, the air is humidified. All room heating is supplied by the secondary water system. All factors that contribute to the cooling load of perimeter space in the summer, with the exception of the transmission, add heat in the winter. The transmission factor becomes negative when the outdoor temperature falls below the room temperature. Its magnitude is directly proportional to the difference between the room and outdoor temperatures.

For all air-water systems, it is important to note that where primary air enters at the terminal unit, the primary air is provided at summer design temperature in winter. For systems where the primary air does not enter at the terminal unit, the primary air should be reset to room temperature in winter. A limited amount of cooling can be accomplished by the primary air operating without supplementary cooling from the secondary coil. As long as internal heat gains are not high, this amount of cooling is usually adequate to satisfy east and west exposures during the fall, winter, and spring, since the solar heat gain is reduced during these seasons. The north exposure is not a significant factor, since the solar gain is very low. For the south, southeast, and southwest exposures, the peak solar heat gain occurs in winter, coincident with lower outdoor temperatures (Figure 3).

In buildings with large areas of glass, the transmitted heat from indoors to the outside, coupled with the normal supply of cool primary air, will not balance internal heat and solar gains until an outdoor temperature well below freezing is reached. Double-glazed windows, with clear or heat-absorbing glass, aggravate this condition because this type of glass permits constant inflow of solar radiation during the winter. However, the insulating effect of the double glass reduces the reverse transmission, and thereby reduces the outdoor temperature to which cooling must be kept available. In buildings with very high internal heat gains from lighting or equipment, the need for cooling from the room coil, as well as from the primary air, can extend well into the winter. In any case, the changeover temperature at which the cooling capacity of the secondary water system is no longer required for a given space is an important calculation.

CHANGEOVER TEMPERATURE

For all systems using outdoor air, there is an outdoor temperature at which secondary cooling is no longer required. The system can accomplish cooling by using outdoor air; at lower tempera-

tures, heating rather than cooling is needed. With all-air systems operating with up to 100% outdoor air, mechanical cooling is seldom required at outdoor temperatures below 55 °F. An important characteristic of air-water systems, however, is that secondary water cooling may continue to be needed, even when the outdoor temperature is considerably less than 50 °F. This cooling may be provided by the mechanical refrigeration unit or by a thermal economizer cycle. Full-flow circulation of the primary air cooling coil at temperatures below 50 °F often provides all necessary cooling while preventing coil freeze-up and reducing preheat requirements. Alternatively, secondary water to condenser water heat exchangers function well. Some systems circulate condenser water directly. This system should be used with caution, recognizing that the vast secondary water system is being operated as an open recirculating system with the potential hazards that may accompany improper water treatment.

The outdoor temperature at which the heat gain to every space can be satisfied by the combination of cold primary air and the transmission loss is termed the changeover temperature. Below this temperature, cooling is no longer required.

The following equation determines the temperature (Carrier 1965).

$$t_{co} = t_r - [q_{is} + q_{es} - 1.10Q_p (t_r - t_p)]/\Delta q_{td} \qquad (2)$$

where

t_{co} = temperature of changeover point, °F
t_r = room temperature at time of changeover, normally taken as 72 °F
t_p = primary air temperature at unit after the system is changed over, normally taken as 56 °F
Q_p = primary air quantity, cfm
q_{is} = internal sensible heat gain, Btu/h
q_{es} = external sensible heat gain, Btu/h
Δq_{td} = heat transmission per degree of temperature difference between room and outdoor air

Since in two-pipe changeover systems the entire system is usually changed from winter to summer operation at the same time, the room with the lowest changeover point should be identified. In northern latitudes, this room usually has a south, southeast, or southwest exposure because the solar heat gains on these exposures reach their maximum during the winter months. However, since changeover is made at comparatively cool outdoor temperatures, substantially increased storage may be available. For buildings of moderate glass area with good room heat storage characteristics, a factor of 0.4 applied to the solar load accounts for the effects of storage and blinds.

If the calculated changeover temperature is below approximately 48 °F, an economizer cycle should operate to allow the refrigeration plant to shut down.

While the factors controlling the changeover temperature of air-water systems are understood by the design engineer, its basic principles are often more difficult for system operators to understand. It is important that the concept and the calculated changeover point be clearly explained in operating instructions prior to operating the system. Some increase from the calculated changeover temperature is normal in actual operation. Also, a range or band of changeover temperatures, rather than a single value is a practical necessity to preclude frequent change in the seasonal cycles and to grant some flexibility in operating procedure. The difficulties associated with operator understanding and the need to change over several times a day in many areas have severely limited the acceptability of the two-pipe changeover system.

REFRIGERATION LOAD

The design refrigeration load is determined by considering the entire portion or block of the building served by the air-and-water system at the same time. Because the load on the secondary water

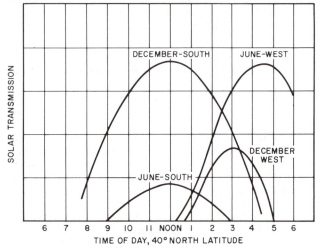

Fig. 3 Solar Radiation Variations with Season

system depends on the simultaneous demand of all spaces, the sum of the individual room or zone peaks is not considered.

The peak load time is influenced by the relative amounts of the east, south, and west exposures; the outdoor wet-bulb temperature; and the period of building occupancy. Where the magnitude of the solar load is about equal for each of the above exposures, the building peak usually occurs in midsummer afternoon when the west solar load and outdoor wet-bulb temperature are at or near concurrent maximums.

The refrigeration load equals the primary air-cooling coil load plus the secondary system heat pickup.

$$q_{re} = q_s + 4.5_1Q_p\ (h_{ea} - h_{la}) - 1.10_2Q_p\ (t_r - t_s) \quad (3)$$

where

q_{re} = refrigeration load, Btu/h
q_s = block room sensible heat for all spaces at time of peak, Btu/h
h_{ea} = enthalpy of primary air upstream of cooling coil at time of peak, Btu/lb
h_{la} = enthalpy of the primary air leaving cooling coil, Btu/lb
Q_p = primary air quantity, cfm
t_r = average room temperature for all exposures at peak time
t_s = average primary air temperature at point of delivery to rooms

Because the latent load is absorbed by the primary air, the resultant room relative humidity can be determined by calculating the block room latent load q_L of all spaces at the time of the peak load and calculating the rise in moisture content of the primary air as:

$$W = (7000\ \text{gr/lb})v_aq_L/[(60\ \text{min/h})h_{fg}Q_p] = q_L/0.68Q_p \quad (4)$$

where

W = moisture content rise per lb of dry air, grains
v_a = air volume = 13.3 ft^3/lb
h_{fg} = latent heat of vaporization = 1050 Btu/lb

Then a psychrometric analysis can be made.

The secondary water system cooling load may be determined by subtracting the primary air cooling coil load from the total refrigeration load.

TWO-PIPE SYSTEMS

Description

Two-pipe systems for induction, fan-coil, or radiant panel systems derive their name from the water distribution circuit, which consists of one supply and one return pipe. Each unit or conditioned space is supplied with secondary water from this distribution system and with conditioned primary air from a central apparatus. The system design and the control of the primary air and secondary water temperatures must be such that all rooms on the same system (or zone, if the system is separated into independently controlled air and water zones) can be satisfied during both heating and cooling seasons. The heating or cooling capacity of any unit at a particular time is the sum of the primary air output plus the secondary water output of that unit.

The primary air quantity is fixed, and the primary air temperature is varied in inverse proportion to outside temperature to provide the necessary amount of heating during summer and intermediate seasons. During winter cycle operation, the primary air is preheated and supplied at approximately 50°F to provide a source of cooling. All room terminals in a given primary air reheater zone must be selected to operate satisfactorily with the common primary air temperature.

The secondary water coil (cooling-heating) within each space is controlled by a space thermostat and can vary from zero to 100% of coil capacity, as required to maintain space temperature. The secondary water is cold in summer and intermediate seasons and warm

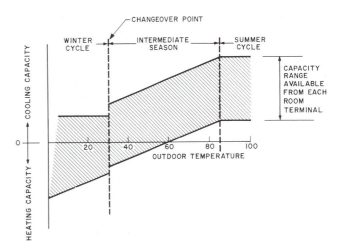

Fig. 4   Capacity Ranges of Air-Water Room Terminal Operating on Two-Pipe System

in winter. All rooms on the same secondary water zone must operate satisfactorily with the same water temperature.

Figure 4 shows the capacity ranges available from a typical air-water two-pipe system. On a hot summer day, loads varying from about 25 to 100% of the design space cooling capacity can be satisfied. On a 50°F intermediate season day, the unit can satisfy a heating requirement by closing off the secondary water coil and using only the output of the warm primary air. A lesser heating or net cooling requirement is satisfied by the cold secondary water coil output, which offsets the warm primary air to obtain cooling. In winter, the unit can provide a small amount of cooling by closing the secondary coil and using only the cold primary air. Smaller cooling loads and all heating requirements are satisfied by using the warm secondary water.

Critical Design Elements

The most critical design elements of a two-pipe system are the calculation of primary air quantities and the final adjustment of the primary air temperature reset schedule. All rooms require a minimum amount of heat during the intermediate season, available from a common primary air supply temperature. The A/T ratio concept to maintain a constant relationship between the primary air quantity and the heating requirements of each space fulfills this requirement. The ratio of primary air to transmission per degree (A/T ratio) determines the primary air temperature and changeover point. This concept is fundamental to proper design and operation of a two-pipe system.

Transmission per degree. The relative heating requirement of every space is determined by calculating the transmission heat flow per degree temperature difference between space temperature and outside temperature (assuming steady-state heat transfer). This is the sum of (1) the glass heat transfer coefficient times glass areas, (2) wall heat transfer coefficient times wall area, and (3) roof heat transfer coefficient times roof area.

Air-to-transmission ratio. The A/T ratio is the ratio of the primary airflow to a given space divided by the transmission per degree of that space:

A/T ratio = Primary air/Transmission per degree

All spaces on a common primary air zone must have approximately the same A/T ratios. The design base A/T ratio establishes the primary air reheat schedule during intermediate seasons. Spaces with A/T ratios higher than the design base A/T ratio will tend to be overcooled during light cooling loads at outdoor temperatures in the 70 to 90°F range, while spaces with A/T ratios lower than the design ratio will lack sufficient heat during the 40 to 60 °F outdoor temperature range when the primary air is warm for heating and the secondary water is cold for cooling.

The minimum primary air quantity that will satisfy the requirements for ventilation, dehumidification, and both summer and winter cooling (as explained in the System Description section) is used to calculate the minimum A/T ratio for each space. If the system will be operated with primary air heating during cold weather, the heating capacity can also be the primary air quantity determinant for two-pipe systems.

The design base A/T ratio is the highest A/T ratio obtained, and the primary airflow to each space is increased, as required, to obtain a uniform A/T ratio in all spaces. An alternate approach is to locate the space with the highest A/T ratio requirement by inspection, establish the design base A/T ratio, and obtain the primary airflow for all other spaces by multiplying this A/T ratio by the transmission per degree of all other spaces.

For each A/T ratio, there is a specific relationship between the outdoor air temperature and the temperature of the primary air that will maintain the room at 72 °F or more during conditions of minimum room cooling load. Figure 5 illustrates this variation based on an assumed minimum room load, equivalent to 10 °F times the transmission per degree. Primary air temperatures over 122 °F at the unit are seldom used. The reheat schedule should be adjusted for hospital rooms or other applications where a higher minimum room temperature is desired, or where there is no minimum cooling load in the space.

Deviation from the A/T ratio is sometimes permissible. A minimum A/T ratio equal to 0.7 of the maximum A/T is suitable, if the building is of massive construction with considerable heat storage effect (Carrier 1965). The heating performance when using warm primary air becomes less satisfactory than that for systems with a uniform A/T ratio. Therefore, systems designed for A/T ratio deviation should be suitable for changeover to warm second-

ary water for heating whenever the outdoor temperature falls below 40 °F. A/T ratios should be more closely maintained on buildings with large glass areas, with curtain wall construction, or on systems with low changeover temperature.

Changeover Temperature Considerations

Transition from summer operations to intermediate season operation is done by gradually raising the primary air temperature as the outdoor temperature falls to keep rooms with small cooling loads from becoming too cold. The secondary water remains cold during both summer and intermediate seasons. Figure 6 illustrates the psychrometrics of summer cycle operation near the changeover temperature.

As the outdoor temperature drops further, the changeover temperature will be reached. The secondary water system can then be changed over to provide hot water for heating.

If the primary airflow is increased to some spaces to elevate the changeover temperature, the A/T ratio for the reheat zone will be affected. Adjustments in the primary air quantities to other spaces on that zone will probably be necessary to establish a reasonably uniform ratio.

System changeover can take several hours and usually temporarily upsets room temperatures. Good design, therefore, includes provision for operating the system with either hot or cold secondary water over a range of 15 to 20 °F below the changeover point. This range makes it possible to operate with warm air and cold secondary water when the outside temperature rises above the daytime changeover temperature. Changeover to hot water is limited to times of extreme or protracted cold weather.

Optional hot or cold water operation below the changeover point is provided by increasing the primary air reheater capacity to provide adequate heat at the colder outside temperatures. Figure 7 shows temperature variations for a system operating with changeover. This indicates the relative temperature of the primary air and secondary water throughout the year and the changeover temperature range. The solid arrows show the temperature variation when changing over from the summer to the winter cycle. The open arrows show the variation when going from the winter to the summer cycle.

Non-Changeover Design

Non-changeover systems should be considered to simplify system operation for buildings with mild winter design climates, or for south exposure zones of buildings with large winter solar

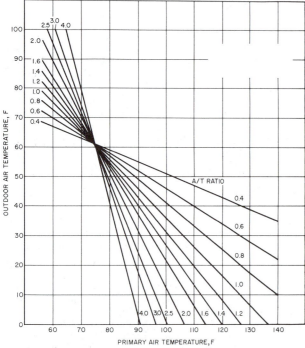

Note: These temperatures are required at the units, and thermostat settings must be adjusted to allow for duct heat gains or losses. Temperatures are based on:
1. Minimum average load in the space equivalent to 10°F, multiplied by the transmission per degree.
2. Preventing the room temperature from dropping below 72°F. These values compensate for radiation and convection effect of the cold outdoor wall.

Fig. 5   Primary Air Temperature versus Outdoor Air Temperature

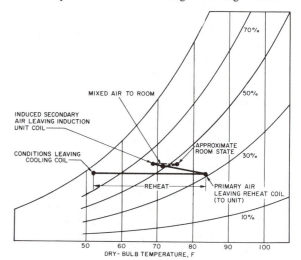

Note: Based on A/T ratio = 1.2/52°F outdoor and reheat schedule on.

Fig. 6   Psychrometric Chart, Two-Pipe System, Off-Season Cooling

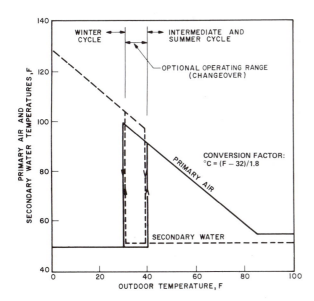

Fig. 7 Typical Changeover System Variations

loads. A non-changeover system operates on an intermediate season cycle throughout the heating season, with cold secondary water to the terminal unit coils and with warm primary air satisfying all the heating requirements. Typical system temperature variations are shown in Figure 8.

Spaces may be heated during unoccupied hours by operating the primary air system with 100% return air. This feature is necessary, since the non-changeover design does not usually include the ability to heat the secondary water. In addition, cold secondary water must be available throughout the winter months. Primary air duct insulation and observance of close A/T ratios for all units are essential for proper heating during cold weather.

Zoning

A properly designed single-zone, two-pipe system can provide good temperature control on all exposures during all seasons of the year. Initial cost or operating cost can be improved by zoning in several ways, such as the following:

- Zoning primary air to permit different A/T ratios on different exposures.
- Zoning primary air to permit solar compensation of primary air temperature.
- Zoning both air and water to permit different changeover temperatures for different exposures.

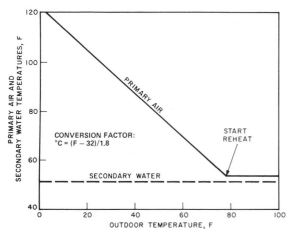

Fig. 8 Typical Non-Changeover System Variations

All spaces on the same primary air zone must have the same A/T ratio. The minimum A/T ratios often are different for spaces on different solar exposures, thus requiring the primary air quantities on some exposures to be increased if they are placed on a common zone with other exposures. The primary air quantity to units serving the north, northeast, or northwest exposures can usually be reduced by using separate primary air zones with different A/T ratios and reheat schedules. Primary air quantity should never be reduced below minimum ventilation requirements.

The peak cooling load for the south exposure occurs during the fall or winter months when outside temperatures are lower. If present or future shading patterns from adjacent buildings or obstructions will not be present, primary air zoning by solar exposure can reduce air quantities and unit coil sizes on the south. Units can be selected for peak capacity with cold primary air instead of reheated primary air. Primary air zoning and solar compensators will save operating cost on all solar exposures by reducing primary air reheat and the secondary water refrigeration penalty.

Separate air and water zoning will save operating cost by permitting north, northeast, or northwest exposures to operate on the winter cycle with warm secondary water at outdoor temperatures as high as 60°F during winter months. Systems with a common secondary water zone must operate with cold secondary water to cool south exposures. Primary airflow can be lower because of separate A/T ratios, resulting in reheat and refrigeration cost savings.

Room Control

During summer, the thermostat must increase the output of the cold secondary coil when the room temperature rises. During winter, the thermostat must decrease the output of the warm secondary coil when the room temperature rises. Changeover from cold water to hot water in the unit coils requires changing the action of the room temperature control system. Room control for non-changeover systems does not require the changeover action, unless it is required to provide gravity heating during shutdown.

Evaluation

Characteristics of two-pipe air-and-water systems are as follows:

- They are usually less expensive to install than four-pipe systems.
- They are less capable of handling widely varying loads or providing a widely varying choice of room temperatures than four-pipe systems.
- They present operational and control changeover problems, increasing the need for competent operating personnel.
- They are more costly to operate than four-pipe systems.

Electric Heat for Two-Pipe Systems

Electric heat can be supplied with a two-pipe air-water system by using a central electric boiler and hot water terminal coils or by individual electric resistance heating coils in the terminal units.

One approach uses small electric resistance terminal heaters for the intermediate season heating requirements and a two-pipe changeover chilled water/hot water system. The electric terminal heater heats when outdoor temperatures are above 40°F, and cooling is available with chilled water in the chilled water/hot water system. System or zone reheating of the primary air is reduced greatly or eliminated entirely. When outdoor temperatures fall below this point, the chilled water/hot water system is switched to hot water, providing greater heating capacity. Changeover is limited to a few times per season, and simultaneous heating/cooling capacity is available, except in extremely cold weather, when little, if any, cooling is needed. If electric resistance terminal heaters are used in this type of system, the heaters should be prevented from operating whenever the secondary water system is operated with hot water.

Another approach is to size electric resistance terminal heaters for the peak winter heating load and to operate the chilled water system as a non-changeover cooling-only system. This system avoids the operating problem of chilled water/hot water system changeover. In fact, this approach functions like a four-pipe system, and, in areas where the electric utility establishes a summer demand charge and has a low unit energy cost for high winter consumption, it may have a lower life cycle cost than hydronic heating with fossil fuel.

THREE-PIPE SYSTEMS

Three-pipe air-and-water systems for induction, fan-coil, and radiant panel systems have three pipes to each terminal unit. These pipes are a cold water supply, a warm water supply, and a common return. These systems are rarely used today because they consume excess energy. For a detailed discussion, refer to Chapter 4 of the 1973 and 1976 volumes of the ASHRAE Handbook.

FOUR-PIPE SYSTEMS

Description

Four-pipe systems have a cold water supply, cold water return, warm water supply, and warm water return. The terminal unit usually has two independent secondary water coils: one served by hot water, the other by cold water. The primary air is cold and remains at the same temperature year-round. During peak cooling and heating, the four-pipe system performs in a manner similar to the two-pipe system with essentially the same operating characteristics. Between seasons, any unit can be operated at any level from maximum cooling to maximum heating, if both cold and warm water are being circulated. Any unit can be operated at or between these extremes without regard to the operation of other units.

All units are selected on the basis of their peak capacity. The A/T ratio design concept for two-pipe systems does not apply to four-pipe systems. There is no need to increase primary air quantities on north or shaded units beyond the amount needed for ventilation and to satisfy cooling loads. The available net cooling is not reduced by heating the primary air. Attention to the changeover point is still important, since cooling of spaces on the south side of the building may continue to require secondary water cooling to supplement the primary air at low outdoor temperatures.

Since the primary air is supplied at a constant cool temperature at all times, it is sometimes feasible for fan-coil or radiant panel systems to extend the interior system supply to the perimeter spaces, eliminating the need for a separate primary air system. The type of terminal unit and the characteristics of the interior system are determining factors.

Zoning

Zoning of primary air or secondary water systems is not required. All terminal units can heat or cool at all times, as long as both hot and cold secondary pumps are operated and sources of heating and cooling are available.

Room Control

The four-pipe terminal usually has two completely separated secondary water coils—one receiving hot water and the other receiving cold water. The coils are operated in sequence by the same thermostat; they are never operated simultaneously. The unit receives either hot water or cold water in varying amounts or else no flow is present, as shown in Figure 9A. Adjustable, dead-band thermostats further reduce operating cost.

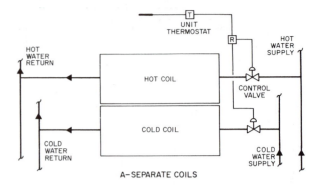

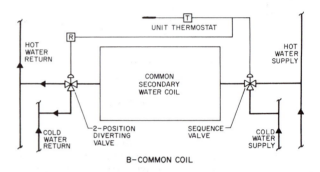

Fig. 9   Four-Pipe System Room Unit Control

Figure 9B illustrates another unit and control configuration. A single secondary water coil at the unit and three-way valves located at the inlet and outlet admit water from either the hot or cold water supply, as required, and divert it to the appropriate return pipe. This arrangement requires a special three-way modulating valve, originally developed for one form of the three-pipe system. It controls the hot or cold water selectively and proportionally but does not mix the streams. The valve at the coil outlet is a two-position valve open to either the hot or cold water return, as required.

When all aspects are considered, the two-coil arrangement provides a superior four-pipe system. The operation of the induction unit controls is the same year-round. Units with secondary air bypass control are not applicable to four-pipe systems.

Evaluation

Compared to the two-pipe system, the four-pipe air-and-water system has the following characteristics:

- It is more flexible and adaptable to widely differing loads, responding quickly to load changes.
- It is simpler to operate.
- It operates without the summer-winter changeover and the primary air reheat schedule.
- Efficiency is greater and operating cost is lower, though initial cost is generally higher.
- The system can be designed with no interconnection of the hot and cold water secondary circuits, and the secondary system can be completely independent of the primary water piping.

SECONDARY WATER DISTRIBUTION

Secondary water system design applies to induction, fan-coil, and radiant panel systems. The secondary water system includes the portion of the water distribution system that circulates water to room terminal air-conditioning units (or radiant coils) when the cooling (or heating) of such water has been accomplished either by extraction from or heat exchange with another source in the primary circuit. In the primary circuit, water is cooled by flow through a chiller or is heated by a heat input source. Primary water is limited to the cooling cycle and is the source of the secondary water cooling.

The term *secondary water* describes a component of the air-water system. The water flow through the unit coil performs secondary cooling when the room air (secondary air) gives up heat to the water. The design of the secondary water system differs for the two- and four-pipe systems.

Secondary water systems are discussed in Chapter 12.

REFERENCE

Carrier Air Conditioning Company. 1965. *Handbook of air conditioning system design.* McGraw-Hill, New York.

BIBLIOGRAPHY

Menacker, R. 1977. Electric induction air-conditioning system. ASHRAE *Transactions* 83(1):664.

McFarlan, A.I. 1967. Three-pipe systems: Concepts and controls. ASHRAE *Journal* 9(8):37.

Ross, D.E. 1974. Variable-air-volume system evaluation for commercial office buildings. ASHRAE *Transactions* 80(1):476.

CHAPTER 4

ALL-WATER SYSTEMS

SYSTEM DESCRIPTION

Convective Radiation Heating

ALL-WATER systems heat and/or cool a space by direct heat transfer between water and circulating air. Hot water systems deliver heat to a space by water that is hotter than the air in contact with the heat transfer surface. Examples of such systems include the following:

• Baseboard radiation
• Freestanding radiators
• Wall or floor radiant
• Bare pipe (racked on wall)
• Other configurations

These types of systems may further be classified as gravity convection systems, where air moves past the transfer surface because of density differences of air caused by heated or cooled surfaces. These systems also lose large amounts of heat by radiation to colder surfaces.

Although these systems provide comfort during the heating season, caution should be exercised in their application for cooling.

Fan Coil Units

The four basic principles of air conditioning listed below also apply to heating.

1. Temperature control
2. Humidity control
3. Air movement
4. Air purity (filtration and outside air makeup)

If all-water systems include cooling as well as heating, they normally move air by forced convection through the conditioned space, filter the circulating air, and introduce outside ventilation air. Terminal units with chilled water coils, heating coils, blowers, replaceable air filters, drain pans for condensate, etc., are designed for these purposes. Terminal units are available in various configurations to fit under windowsills, above furred ceilings, in vertical pilasters built into walls, etc. These units must be properly controlled by thermostats for heating and cooling temperature humidistats control, by humidistats for humidity control, by blower control or other means for regulating air quantity, and by a method for adding ventilation air into the system.

In applying all-water systems, the manufacturer's capacity ratings should be obtained to verify performance under actual operating conditions.

Basic elements of fan-coil units are a finned-tube coil, filter, and fan section (Figure 1). The fan recirculates air continuously from the space through the coil, which contains either hot or chilled water. The unit may contain an additional electric resistance, steam, or hot-water heating coil. The electric heater is often sized for fall and spring to avoid changeover problems in two-pipe systems.

A cleanable or replaceable 35% efficiency filter, located upstream of the fan, prevents clogging the coil with dirt or lint entrained in the recirculated air. It also protects the motor and fan, and reduces the level of airborne contaminants within the conditioned space. The unit is equipped with an insulated drain pan. The fan and motor assembly is arranged for quick removal for servicing. Most manufacturers furnish units—the prototypes of which have been tested and labeled by Underwriters' Laboratories (UL), or Engineering Testing Laboratories (ETL), as required by some codes, with cooling performance that is ARI certified.

Air-conditioning units, with a dampered opening for connection to apertures in the outside wall, are optional. These units are not recommended because wind pressure allows no control over the amount of outside air that is admitted, and caution must be exercised for freeze protection in cold climates. Room fan-coil units for the domestic market are generally available in nominal sizes of 200, 300, 400, 600, 800, and 1200 cfm, often with multispeed, high-efficiency fan motors. Where units do not have individual outside air intakes, means must be provided to introduce pretreated outside air through a duct system engaging each room or space.

The preparation of this chapter is assigned to TC 9.1, Large Building Air-Conditioning Systems.

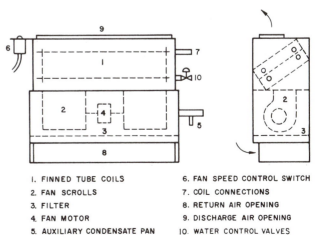

1. FINNED TUBE COILS	6. FAN SPEED CONTROL SWITCH
2. FAN SCROLLS	7. COIL CONNECTIONS
3. FILTER	8. RETURN AIR OPENING
4. FAN MOTOR	9. DISCHARGE AIR OPENING
5. AUXILIARY CONDENSATE PAN	10. WATER CONTROL VALVES

Fig. 1 Typical Fan-Coil Unit

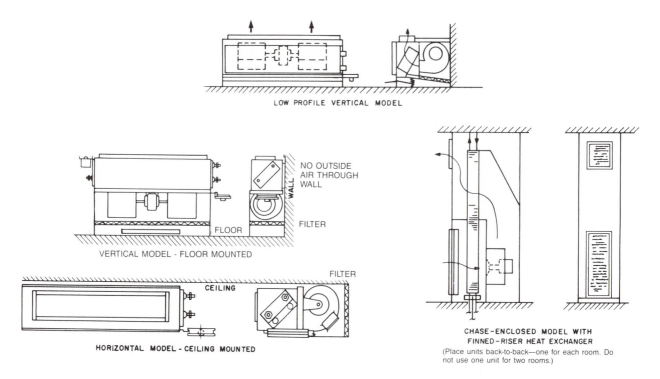

Fig. 2   Typical Fan-Coil Unit Arrangements

Types and Location

Room fan-coil units are available in many configurations. Figure 2 shows several vertical units. Low vertical units are available for use under windows with low sills; however, in some cases, the low silhouette is achieved by compromising such features as filter area, motor serviceability, and cabinet style.

Floor-to-ceiling, chase-enclosed units are available in which the water and condensate drain risers are part of the factory-furnished unit. Supply and return air systems must be isolated from each other to prevent air and sound interchange between rooms.

Horizontal overhead units may be fitted with ductwork on the discharge to supply several outlets. A single unit may serve several rooms, *e.g.*, in an apartment house where individual room control is not essential and a common air return is feasible. High static pressure units with larger fan motors handle the higher pressure drops of units with ductwork.

Central ventilation air may be connected directly to the inlet plenums of horizontal units or introduced directly into the space. If this concept is used, provisions should be made to ensure that this air is pretreated and held at a temperature equal to the room temperature so as not to cause occupant discomfort when the unit is off. One way to prevent air leakage is to provide a spring-loaded motorized damper, which closes off the ventilation air whenever the unit's fan is off. Coil selection must be based on the temperature of the entering mixture of primary and recirculated air, and the air leaving the coil must satisfy the room sensible cooling and heating requirements. Horizontal models conserve floor space and usually cost less, but when located overhead in furred ceilings, they create problems such as condensate collection and disposal, mixing of return air from other rooms, leakage of pans causing damaged ceilings, difficulty of access for filter and component removal, and IAQ concerns.

Vertical models give better results in climates or buildings with high heating requirements. Heating is enhanced by under-window or exterior wall locations. Vertical units can be operated as convectors with the fans turned off during night setback.

Unit Ventilators

Unit ventilators are similar to fan-coil units, except they may serve a multiple fan-coil system with pretreated ventilation air, which eliminates the problems of direct introduction of outside air.

WATER DISTRIBUTION

Chilled and hot water must run to the fan-coil units. The piping arrangement determines quality of performance, ease of operation, and initial cost of the system.

Two-Pipe Changeover

This low initial cost concept supplies either chilled water or hot water through the same piping system. The fan-coil unit has a single coil, and room temperature controls reverse their action, depending on whether hot or cold water is available at the unit coil. This system works well in warm weather when all rooms have a cooling requirement and in cold weather when all rooms have a heating requirement. The two-pipe system *does not have the simultaneous heating or cooling capability* that is required for most projects during intermediate seasons, *i.e.*, when some rooms have a cooling requirement while others have a heating requirement. This problem can be especially troublesome if a single piping zone supplies the entire building. This difficulty may be partly overcome by dividing the piping into zones based on solar exposure. Then each zone may be operated to heat or cool, independent of the others. However, one room may still require cooling while another room on the same solar exposure requires heating—particularly if the building is partially shaded by an adjacent building.

Another difficulty of the two-pipe changeover system is the need for frequent changeover from heating to cooling, thus complicating the operation and increasing energy consumption to the extent that it may become impractical. For example, two-pipe changeover system hydraulics must consider water expansion (and relief) that occurs during the cycling from cooling to heating.

Psychrometric principles also show the loss of dehumidification capability throughout the system during the heating cycle.

Two-pipe changeover with partial electric strip heat. This arrangement provides simultaneous heating/cooling capability in intermediate seasons by using a small electric strip heater in the fan-coil unit. The unit can handle heating requirements in mild weather, typically down to 40°F, while continuing to circulate chilled water to handle any cooling requirements. When the outdoor temperature drops sufficiently to require heating in excess of electric strip heater capacity, the water system must be changed over to hot water.

The designer should consider the disadvantages of the two-pipe system carefully; many installations of this type waste energy and have been unsatisfactory in climates where frequent changeover is required and where interior loads require cooling simultaneously as exterior spaces require heat.

Two-pipe nonchangeover with full electric strip heat. This two-pipe system is not recommended for energy conservation, but it may be practical in areas with a small heating requirement. Any two-pipe system must be carefully analyzed. In any case, they are not recommended for commercial buildings.

Four-Pipe Distribution

The four-pipe concept generally has the highest initial cost, but it provides the best fan-coil system performance, such as (1) all-season availability of heating and cooling at each unit, (2) no summer/winter changeover requirement, (3) simpler operation, and (4) use of any heating fuel, heat recovery, or solar heat. In addition, it can be controlled to maintain a "dead band" between heating and cooling so that there is no possibility of simultaneous heating and cooling.

Central Equipment

Central equipment size is based on the block load of the entire building at the time of building peak load, not on the sum of individual fan-coil unit peak loads. Cooling load should include appropriate diversity factors for lighting and occupant loads. Heating load is based on maintaining the unoccupied building at design temperature, plus an additional allowance for pickup capacity if the building temperature is set back at night.

If water supply temperatures or quantities are to be reset at times other than at peak load, the adjusted settings must be adequate for the most heavily loaded space in the building. An analysis of individual room load variations is required.

If the side exposed to the sun or interior zone loads require chilled water in cold weather, using condenser water with a water-to-water interchanger should be considered. Varying refrigeration loads require the water chiller to operate satisfactorily under all conditions.

The only reason to use central plant equipment in an all-water system is to provide correct amounts of ventilation or makeup air to the various spaces being served by terminal units. The disadvantages of providing direct injection of untreated outside air through apertures in outside walls have previously been mentioned.

Ventilation requires an outside air pretreatment unit complete with sized heating and cooling coils, filters, and fans, to offset the ventilation load. An additional advantage of the ventilation unit is that, if it is sized for the internal latent load, the terminal cooling coil remains dry, which may eliminate the need for piped drains from the condensate pans. A dry system must be carefully designed to be sure no condensate leaks develop. Ventilation units should be sized to maintain the supply air temperature between 70 and 72°F. This neutral temperature removes the outside air

load from the terminal unit, so it can switch from heating to cooling and vice versa without additional internal or external heat loads.

APPLICATIONS

Fan-coil systems are best applied to individual space temperature control. Fan-coil systems also prevent cross-contamination from one room to another. Suitable applications are hotels, motels, apartment buildings, and office buildings. Fan-coil systems are used in a number of hospitals but are less desirable because of the low efficiency filtration and difficulty in maintaining adequate cleanliness in the unit and enclosure.

Where internal sensible loads require heating and/or cooling simultaneously, both mediums must be provided continuously. This practice should only be applied where humidity control is required.

ADVANTAGES

A major advantage of the all-water system is that the delivery system (piping versus duct systems) requires less building space, a smaller or no central fan room, and little duct space. The system has all the benefits of a central water chilling and heating plant, while retaining the ability to shut off local terminals in unused areas. It gives individual room control with little cross-contamination of recirculated air from one space to another. Extra capacity for quick pull down response may be provided. Since this system can heat with low-temperature water, it is particularly suitable for solar or heat recovery refrigeration equipment. For existing building retrofit, it is often easier to install piping and wiring than the large ductwork required for all-air systems.

DISADVANTAGES

All-water systems require much more maintenance than central all-air systems, and this work must be done in occupied areas. Units that operate at low dew points require condensate pans and a drain system that must be cleaned and flushed periodically. Condensate disposal can be difficult and costly. It is also difficult to clean the coil, if necessary. Filters are small, low in efficiency, and require frequent changing to maintain air volume. In some instances, drain systems can be eliminated if dehumidification is positively controlled by a central ventilation air system.

Ventilation is often accomplished by opening windows or by installing outside wall apertures. Ventilation rates are affected by stack effect and wind direction and velocity.

Summer room humidity levels tend to be relatively high, particularly if modulating chilled water control valves are used for room temperature control. Alternatives are two-position control with variable speed fan and the bypass unit variable chilled water temperature control.

VENTILATION

Ventilation air is generally the most difficult factor to control and represents a major load component. To save energy, it should be reduced to a minimum. The designer must select the method that meets local codes, performance requirements, cost constraints, and health requirements.

Central Ventilation Systems

A central outside air pretreatment system, which maintains neutral air at about 70°F, best controls ventilation air with the greatest freedom from problems related to stack effect and infiltration. Ventilation air may then be introduced to the room through

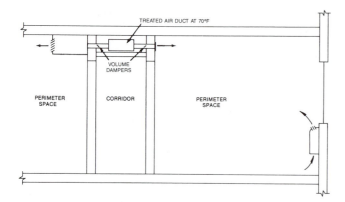

Fig. 3   Ventilation from Separate Duct System

the fan-coil unit, as shown in Figure 3. Any type of fan-coil unit in any location may be used if the ventilation system has separate air outlets.

Ventilation air contributes significantly to the room latent cooling load, so a dehumidifying coil should be installed in the central ventilation system to reduce room humidities during periods of high outside moisture content.

In buildings where fan-coil units serve exterior zones only and a separate all-air system serves interior zones, it is possible to provide exterior zone ventilation air through the interior zone system. This arrangement can provide desirable room humidity control, as well as temperature control of the ventilation air. In addition, ventilation air held in the neutral zone of 70 °F at 50% rh can be introduced into any all-water system without affecting the comfort conditions maintained by the terminal units.

CONTROL AND WIRING

Unit Capacity Controls

Fan-coil unit capacity can be controlled by coil water flow, air bypass, fan speed, or a combination of these. Water flow can be thermostatically controlled by either return air or wall thermostats.

Fan speed control may be automatic or manual. Automatic control is usually on-off with manual speed selection. Units are available with variable speed motors for modulated speed control. Room thermostats are preferred where fan speed control is used. Return air thermostats do not give a reliable index of room temperature when the fan is off.

On-off speed control is poor because (1) alternating shifts in fan noise level are more obvious than the sound of a constantly running fan, and (2) air circulation patterns within the room are noticeably affected.

Wiring

Fan-coil conditioner fans are driven by small motors generally of the shaded pole or capacitor type, with inherent overload protection. Operating wattage of even the largest sizes rarely exceeds 300 W at the high speed setting. Running current rarely exceeds 2.5 A.

Almost all the motors on units sold in the United States are wired for 115 V, single-phase, 60 Hz current, and they provide multiple (usually three) fan speeds and an off position. Other voltages and power characteristics may be encountered, depending on location, and should be investigated before determining the fan motor characteristics.

In planning the wiring circuit, local and national electrical codes must be followed. Wiring methods generally provide separate electrical circuits for the fan-coil units and do not connect them into the lighting circuit.

Separate electrical circuits connected to a central panel allow the building operator to turn off unit fans from a central point during unoccupied hours. While this panel costs more initially, it can lower operating costs in buildings that do not have 24-h occupancy. Use of separate electrical circuits is advantageous in applying a single remote thermostat mounted in a well-exposed perimeter space to operate unit fans.

Unit Selection

Some designers size fan-coil units for nominal cooling at the medium speed setting when a three-speed control switch is provided. This method ensures quieter operation within the space and adds a safety factor, in that capacity can be increased by operating at high speed. Sound power ratings are available from many manufacturers.

Only the internal space heating and cooling loads need to be handled by the terminal fan coil units when outside air is pretreated by a central system to a neutral air temperature of about 70 °F. This pretreatment should reduce the size and cost of the terminal units.

Piping

Even when outside air is pretreated, a condensate removal system should be installed on the terminal units. This precaution ensures that moisture condensed from air from an unexpected open window that bypasses the ventilation system is carried away. Drain pans are an integral feature of all units. Condensate drain lines should be oversized to avoid clogging with dirt and other materials, and provision should be made for periodic cleaning of the condensate drain system. Condensation may occur on the outside of the drain piping, which requires that these pipes be insulated. Many building codes have outlawed systems without condensate drain piping due to the damage they might cause under uncontrollable conditions.

Maintenance

Room fan-coil units are equipped with either cleanable or throwaway filters that should be cleaned or replaced when dirty. Good filter maintenance improves sanitation and full airflow, ensuring full capacity. The frequency of cleaning varies with the application. The presence of lint in apartments, hotels, and hospitals usually requires more filter service in those applications.

Fan-coil unit motors require periodic lubrication. Motor failures are not common, but when they occur, it is possible to replace the entire fan quickly with minimal interruption in the conditioned space. The defective motor can be repaired or replaced. The condensate drain pan and drain system should be cleaned or flushed periodically to prevent overflow and microbiological buildup. Drain pans should be trapped to prevent any gaseous backup.

CHAPTER 5

UNITARY REFRIGERANT-BASED SYSTEMS FOR AIR CONDITIONING

MULTIPLE-packaged unit systems are applied to almost all classes of buildings. They are especially suitable where less demanding performance requirements, low initial cost, and simplified installation are important. These systems have been applied to office buildings, shopping centers, manufacturing plants, hotels, motels, schools, medical facilities, nursing homes, and other multi-occupancy dwellings. They are also suited to air conditioning existing buildings with limited life or income potential. Applications also include specialized facilities requiring high performance levels, such as computer rooms and research laboratories.

SYSTEM CHARACTERISTICS

These systems are characterized by several separate air-conditioning units, each with an integral refrigeration cycle. The components are factory assembled into an integrated package, which includes fans, filters, heating coil, cooling coil, refrigerant compressor(s), refrigerant-side controls, airside controls, and condenser. The equipment is manufactured in various configurations to meet a wide range of applications. Window air conditioners, through-the-wall room air conditioners, unitary air conditioners for indoor and outdoor locations, air source heat pumps, and water source heat pumps are examples. Specialized packages for computer rooms, hospitals, and classrooms are also available.

Components are matched and assembled to achieve specific performance objectives. Although available as large unitary equipment packages, the equipment is available only in preestablished increments of capacity with set performance parameters, such as sensible heat ratio at a given room condition, or cfm of air per ton of refrigeration. These limitations make practical the manufacture of low cost, quality-controlled, factory-tested products. Performance characteristics vary among manufacturers for a particular kind and capacity of unit. All characteristics should be carefully assessed to ensure that the equipment performs as needed for the application. Several trade associations have developed standards by which manufacturers may test and rate their equipment. Chapters 45, 46, and 47 describe the equipment used in multiple-packaged unitary systems and the pertinent industry standards.

Although the equipment can be applied in single units, this chapter covers the application of multiple units to form a complete air-conditioning system for a building. Multiple-packaged unit systems for perimeter spaces are frequently combined with a central all-air or floor-by-floor system. These combinations can provide better humidity control, air purity, and ventilation than

packaged units alone. The all-air system may also serve interior building spaces that could not be conditioned by wall or window-mounted units.

Advantages
- Individual room control is simple and inexpensive.
- Each room has individual air distribution with simple adjustment by the occupant.
- Heating and cooling capability can be provided at all times, independent of the mode of operation of other spaces in the building.
- Individual ventilation air may be provided whenever the conditioner operates.
- Manufacturer matched components have certified ratings and performance data.
- Manufacturer assembly allows improved quality control and reliability.
- Manufacturer instructions and multiple-unit arrangements simplify the installation through repetition of tasks.
- Only one unit conditioner and one zone of temperature control is affected if equipment malfunctions.
- System is readily available.
- One manufacturer is responsible for the final equipment package.
- For improved energy control, equipment serving vacant spaces can be turned off locally or from a central point, without affecting occupied spaces.
- System operation is simple. Trained operators are not required.
- Less mechanical and electrical room space is required than with central systems.
- Initial cost is usually low.
- Equipment can be installed to condition one space at a time as a building is completed, remodeled, or as individual areas are occupied, with favorable initial investment.
- Energy can be metered directly to each tenant.

Disadvantages
- Limited performance options are available because airflow and cooling coil and condenser sizing is fixed.
- Not generally suited for close humidity control, except when using special purpose equipment such as packaged units for computer rooms.
- Energy use may be greater than for central systems, if efficiency of the unitary equipment is less than that of the combined central system components.
- Low cost cooling by outdoor air economizers is not always available.
- Air distribution control may be limited.
- Operating sound levels can be high.

The preparation of this chapter is assigned to TC 9.1, Large Building Air-Conditioning Systems.

- Ventilation capabilities are fixed by equipment design.
- Overall appearance can be unappealing.
- Air filtration options are limited.
- Maintenance may be difficult because of the many pieces of equipment and their location.

WINDOW-MOUNTED AIR CONDITIONERS AND HEAT PUMPS

A window air conditioner (air-cooled room conditioner) is designed to cool/heat individual room spaces. Chapter 45 and ANSI/AHAM *Standard* RAC-1, Room Air Conditioners, describe this equipment.

Design Considerations

Window units may be used as auxiliaries to a central heating or cooling system or to condition selected spaces when the main system is shut down. In such applications, window units usually serve only part of the spaces conditioned by the basic system. Both the basic system and the window units should be sized to cool the space adequately without the other operating.

Window units are furnished with individual electric controls. However, when several units are used in a single space, the controls should be interlocked to prevent simultaneous heating and cooling. In commercial applications (*e.g.*, motels), centrally operated switches can deenergize units in unoccupied rooms.

Window units are used where low initial cost, quick installation, and other operating or performance criteria outweigh the advantages of more sophisticated systems. Room units are also available in through-the-wall sleeve mountings. Sleeve-installed units are popular in low-cost apartments, motels, and homes.

Advantages

- Simple installation.
- No mechanical equipment room space is required.
- Low initial cost.

Disadvantages

- Relatively short equipment life, typically 10 years; window units are built to appliance standards, rather than building equipment standards.
- May have relatively high energy usage.
- Requires outside air; thus, cannot be used for interior rooms.
- Condensate removal can cause dripping on walls, balconies, or sidewalks.
- Temperature control is usually two-position, which causes swings in room temperature.
- Air distribution control is limited.
- Operating sound levels can be high.
- Ventilation capabilities are fixed by equipment design.
- Overall appearance can be unappealing.
- Air filtration options are limited.

THROUGH-THE-WALL MOUNTED AIR CONDITIONERS AND HEAT PUMPS

A through-the-wall air-cooled room air conditioner is designed to cool or heat individual room spaces. Design and manufacturing parameters vary widely. Specifications range from appliance grade through heavy-duty commercial grade. The latter is known as a *packaged terminal air conditioner* (ANSI/ARI *Standard* 310-90). All others are covered in ANSI/AHAM *Standard* RAC/2.

The through-the-wall conditioner system incorporates a complete air-cooled refrigeration and air-handling system in an individual package. Each room is an individual occupant-controlled zone. Cooled or warmed air is discharged in response to thermostatic control to meet room requirements. The section on controls summarizes how controls allow the use of individual room systems during off-schedule hours, yet return all systems automatically to normal schedule use.

Each packaged terminal air conditioner has a self-contained, air-cooled direct expansion cooling system; a heating system (electric, hot water, or steam); and controls. Two general configurations are shown—Figure 1 shows a wall box, an outdoor louver, heater section, cooling chassis, and cabinet enclosure; Figure 2 shows a combination wall sleeve cabinet, plus combination heating and cooling chassis with outdoor louver.

Through-the-wall air conditioner or heat pump systems are installed in buildings requiring many temperature control zones such as (1) office buildings, (2) motels and hotels, (3) apartments and dormitories, (4) schools and other education buildings, and (5) zones of nursing homes or hospitals where air recirculation is allowed.

This system is applicable for renovation of existing buildings, since existing heating systems can still be used. The system lends itself to both low- and high-rise buildings. In buildings where a stack effect is present, this system should be limited to those areas that have dependable ventilation and a tight wall of separation between the interior and exterior.

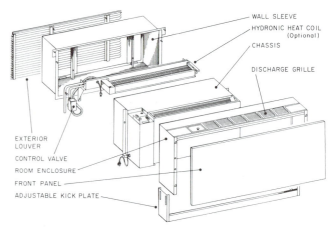

Fig. 1   Packaged Terminal Air Conditioner with Heat Section Separate from Cooling Chassis

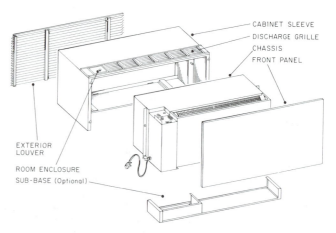

Fig. 2   Packaged Terminal Air Conditioner with Combination Heating and Cooling Chassis

Room air conditioners are often used in parts of buildings primarily conditioned by other systems, especially where spaces to be conditioned are (1) isolated physically from the rest of the building and (2) occupied on a different time schedule, *e.g.*, clergy offices in a church and ticket offices in theaters.

Design Considerations

In choosing through-the-wall equipment adequate to meet the requirement of an application, the system designer should evaluate the following characteristics:

Sound level. The noise levels of these units are not always satisfactory. Check that the noise level of the equipment meets sound level requirements.

Ventilation air. Ventilation air through each terminal may be inadequate with many types of systems, particularly in high-rise structures because of the stack effect. Chapter 23 of the 1989 ASHRAE *Handbook—Fundamentals* explains combined wind and stack effects. Electrically operated outdoor air dampers, which close automatically when the equipment is stopped, reduce heat losses in winter.

Condenser fan operation. Some room air conditioners have only one motor to drive both the evaporator and condenser fans. These units circulate air through the condenser coil whenever the evaporator fan is running, even during the heating season. The annual energy consumption of systems with a single motor is generally higher than those with separate motors, even when energy efficiency ratios (coefficients of performance) are the same for both types of equipment. The year-round continuous flow of air through the condenser increases dirt accumulation on the coil and other components, which increases maintenance costs and reduces equipment life.

Condensate disposal. Because through-the-wall conditioners are seldom installed with drains, they require a positive and reliable means of condensate disposal. Conditioners are available that spray the condensate in a fine mist over the condenser coil. These units can dispose of more condensate than can be developed, without any drip, splash, or spray from the equipment. In heat pump units, provision must be made for disposal of condensate generated from the outside coil during the heating mode.

Cold weather operation. Many air-cooled room conditioners experience evaporator icing and become ineffective when outdoor temperatures fall below about 65 °F. With high lighting levels and high solar radiation experienced in contemporary buildings, be certain that mechanical cooling can be provided at a low enough outdoor ambient temperature.

Life expectancy. Manufacturers project a 10 to 15-year life expectancy for this type of equipment with proper maintenance.

Rainproof louver and wall box. The louver and wall box must stop wind-driven rain from collecting in the wall box and leaking into the building. The wall box should drain to the outside.

Advantages

- Initial cost is generally less than for a central system adapted to heat or cool each room under the control of the room occupants.
- Because no energy is needed to transfer air or chilled water from mechanical equipment rooms, the energy consumption may be lower than central systems. However, this advantage may be offset by better efficiencies of central station equipment.
- Building space is conserved because ductwork and mechanical rooms are not required.
- Installation only requires a hole in the wall for unit mounting and connection to electrical power.
- Generally, the system is well-suited to spaces requiring many zones of individual temperature control.
- Designers can specify electric, hydronic, or steam heat.
- Service can be quickly restored by replacing a defective chassis with a spare one.

Disadvantages

- Humidification, when required, must be provided by separate equipment.
- Packaged terminal air conditioners must be installed on the perimeter of the building.
- Noise levels vary considerably and are not generally suitable for critical applications.
- Routine unit maintenance is required to maintain capacity. Condenser and cooling coils must be cleaned, and filters must be changed regularly.
- Energy use may be greater than for central systems.
- Temperature control is usually two-position, which causes swings in room temperature.
- Air distribution control is limited.
- Equipment life may be relatively short.
- Ventilation capabilities are fixed by equipment design.
- Overall appearance can be unappealing.
- Air filtration options are limited.

Controls

All controls for through-the-wall air conditioners are included as a part of the conditioner. The following control configurations are available:

Thermostat control. Thermostats are either unit mounted or remote wall mounted.

Guest room control for motels and hotels. This has provisions for starting and stopping the equipment from a central point.

Office building and school controls. These controls (for use less than 24 h), through a time clock, start and stop the equipment at a preset time. The conditioners operate normally with the unit thermostat until the preset cutoff time. After this point, each conditioner has its own reset control, which allows the occupant of the conditioned space to reset the conditioner for either cooling or heating, as required.

Master/Slave control. This type of control is used when multiple conditioners are operated by the same thermostat.

Emergency standby control. Standby control allows a conditioner to operate during an emergency, such as a power failure, so that the roomside blowers can operate to provide heating. Units must be specially wired to allow operation on emergency generator circuits.

AIR-TO-AIR HEAT PUMPS

The air-to-air heat pump cycle described in Chapter 46 is available in through-the-wall room air conditioners. Application considerations are quite similar to conventional units without the heat pump cycle, which is used for space heating above 35 to 40 °F outdoor temperature. Electric resistance elements supply heating below this level and during defrost cycles.

The prime advantage of the heat pump cycle is the reduction in annual energy consumption for heating. Savings in heat energy over conventional electric heating range from 10 to 60%, depending on climate.

OUTDOOR UNITARY EQUIPMENT

An outdoor unitary equipment system can be designed to cool or heat an entire building. The complete system includes unitary equipment, a ducted air distribution system, and a temperature control system. The equipment is generally mounted on the roof, but it can also be mounted at grade level.

When a single-unit application is required, the rooftop unit and associated ductwork constitute a central station all-air system, not a multiple unit system. Rooftop units are designed as central station equipment for single-zone, multizone, and variable air volume applications. These systems are described in Chapter 2.

Design Considerations

Location. Centering the rooftop unit over the conditioned space results in reduced fan power, ducting, and wiring. Avoid installation directly above spaces where noise level is critical.

Duct insulation. All outdoor ductwork should be insulated. In addition, the ductwork should be (1) sealed to prevent condensation in the insulation during the heating season and (2) ductwork insulation should be weatherproofed to keep it from getting wet.

Zoning. Use multiple single-zone, not multizone, units where feasible. For large areas such as manufacturing plants, warehouses, gymnasiums, and so forth, single-zone units are less expensive and provide protection against total system failure.

Return fans. Use units with return air fans whenever return air static pressure loss exceeds 0.5 in. of water or the unit introduces a large percentage of outdoor air via an economizer.

Exhaust or relief fans. Units are also available with relief fans for use with an economizer in lieu of continuously running a return fan. Relief fans can be initiated by static pressure control.

Heating. Natural gas, propane, oil, electricity, hot water, steam, and refrigerant gas heating options are available.

Controls. Most operating and safety controls are provided by the equipment manufacturer. Although remote monitoring panels are optional, they are recommended to permit operating personnel to monitor system performance.

Mounting and isolation. Rooftop units are generally mounted using integral support frames or lightweight steel structures. Integral support frames are designed by the manufacturer to connect to the base of the unit. No duct openings are required for supply and return ducts. The completed installation must adequately drain condensed water. Lightweight steel support structures allow the unit to be installed above the roof using separate flashed duct openings. Any condensed water can be drained through the roof drains.

Vibration. Most unitary equipment is available with separate vibration isolation of the rotating equipment; isolation of the entire unit casing is not always required.

Noise. Outdoor noise from unitary equipment should be reduced to a minimum. Sound power levels at all property lines must be evaluated. Airborne noise can be attenuated by silencers in the supply and return air ducts or by acoustically lined ductwork. Avoid installation directly above spaces where noise level is critical.

Special considerations. In a rooftop application, the air handler is outdoors and needs to be weatherproofed against rain, snow, and, in some areas, sand. In cold climates, fuel oil does not atomize and must be warmed to burn properly. Hot water or steam heating coils and piping must be protected against freezing. In some areas, enclosures are needed to maintain units effectively during inclement weather. A permanent safe access to the roof, as well as a roof walkway to protect against roof damage, is essential.

Accessories. Accessories such as economizers, special filters, and humidifiers are available. Factory-installed and wired economizer packages are also available. Other options offered are return and exhaust fans, variable volume controls with hot gas bypass or other form of coil frost protection, smoke and fire detectors, portable external service enclosures, special filters, and microprocessor-based controls with various control options.

Multiple outdoor units are usually the single-zone, constant, or variable volume type. The number of units is determined by the temperature control zoning; each zone has a unit. The zones are determined by the cooling and heating loads for the space served, occupancy, allowable roof loads, flexibility requirements, appearance, duct size limitations, and equipment size availability. Multi-

ple-unit systems have been installed in manufacturing plants, warehouses, schools, shopping centers, office buildings, and department stores. These units also serve core areas of buildings whose perimeter spaces are served by packaged terminal air conditioners. These systems are usually applied to low-rise buildings of one or two floors, but have been used for conditioning multistory buildings, as well.

Advantages

- Equipment location allows for shorter duct runs, reduced duct space requirements, lower initial cost, and ease of service access.
- Installation is simplified.
- Valuable building space for mechanical equipment is conserved.
- Suitable for floor-by-floor control in low-rise office buildings.

Disadvantages

- Maintenance or servicing of outdoor units is difficult.
- Frequent removal of panels for access may destroy the weatherproofing of the unit, causing electrical component failures, rusting, and water leakage.
- Rusting of casings is a potential problem. Many manufacturers prevent rusting by using vinyl coating and other protective measures.
- Equipment life is reduced by outdoor installation.

INDOOR UNITARY EQUIPMENT SYSTEMS

Description

Unitary equipment for indoor locations is designed to cool and heat entire buildings. The complete system consists of an indoor unit with either a water-cooled condenser, integral air-cooled condenser, or remote air-cooled condenser. Air distribution and temperature control complete the system. Equipment is generally installed in service areas adjacent to the conditioned space. When a single unit is required, the indoor unit and its related ductwork constitute a central all-air system, as described in Chapter 2. Figure 3 shows an indoor unit with some commonly used components.

Typical components of a self-contained air conditioner include compressors, condensers, evaporator coils, economizer coils, heating coils, filters, valves, and controls. To complete the system, a building has cooling towers and condenser water pumps for water-cooled units and remote or integral condensers for air-cooled units. Either system has independent air distribution ductwork for each packaged unit.

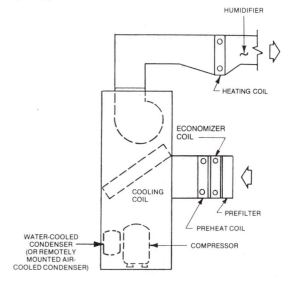

Fig. 3 Unitary Package with Accessories

Air-cooled systems are common to single-story or low-rise buildings, industrial applications, or where condenser water is not readily available. Water-cooled systems are typically used in low- to high-rise buildings or in industrial buildings.

There are typically two classifications of this system:

- Unitary or vertical self-contained systems, in which unit sizes generally range from 36 to 240 mbh cooling.
- Commercial self-contained systems, in which unit sizes generally range from 240 to 1500 mbh cooling.

Design Considerations

Building characteristics that favor self-contained systems include:

- Multiple floors
- Renovation work
- Historic structures
- Multiple tenants per floor with variable schedules
- Condenser water required for computer or other equipment cooling
- Common return air paths

Unitary self-contained systems (Figure 4) allow refrigeration and air-handling equipment to be placed in close proximity to the air-conditioning load. This equipment configuration provides ample air distribution to the air-conditioned space with a minimum amount of ductwork and fan power. Heat rejection through a water-cooled condenser or remote air-cooled condenser allows the final heat rejector to be located distant from the air-conditioned space.

Multiple unit sizes and fan discharge arrangements allow proper application of equipment to load and space requirements. The modest space requirements of unitary self-contained equipment make it an excellent choice for renovation work or for providing air-conditioning to previously nonair-conditioned historic structures. Control is usually two-step cool and two-step or modulating heat. Variable air volume (VAV) operation is possible with a supply air bypass.

The close proximity of a 1000 to 8000 cfm air handler to the air-conditioned space requires special attention to unit inlet and outlet airflow and to building acoustics around the unit. Ducting ventilation air to the unit and removing condensate from the cooling coil should also be considered. Separate outdoor air systems are commonly applied in conjunction with unitary self-contained systems.

Commercial self-contained systems (Figure 5) evolved to fill the requirement for central air distribution, refrigeration, and system control on a floor-by-floor basis. In commercial self-contained systems, each floor is totally independent of other floors, which permits the comfort cooling requirements of each floor or tenant to be scheduled individually. Commercial self-contained systems are well-suited to office environments with variable occupancy schedules.

A floor-by-floor arrangement can offer reduced fan power consumption since the air handlers are located close to the air-conditioned space. Large vertical duct shafts and fire dampers are eliminated. Electrical wiring, condenser water piping, and condensate removal are centrally located. Commercial self-contained systems generally reduce mechanical room area. Equipment is generally located in the building interior near elevators and other service areas and does not interfere with the building perimeter.

This equipment integrates refrigeration, air-handling, and systems controls into a factory package, thus eliminating many field integration problems. Commercial self-contained units are available as constant volume equipment for use in atriums, public areas, and industrial applications. Unit options include a fan modulation device and control for VAV systems. When applied with VAV terminals, commercial self-contained systems provide excellent comfort and individual zone control.

Units that employ an airside economizer must be located near an outside wall or outdoor air shaft. Commercial self-contained units do not include return air fans. A separate relief fan and discharge damper with space pressure controls must be incorporated if airside economizers are used. Many commercial self-contained designs incorporate an integral waterside economizer coil and controls, thus allowing an interior equipment location. In this configuration, the outdoor air shaft is reduced to meet only ventilation and space pressurization requirements.

Many commercial self-contained units include a factory-engineered acoustical discharge plenum, which facilitates smooth supply air discharge from the equipment room. This allows lower fan power and lower sound power levels. This discharge arrangement also reduces the size of the equipment room.

UNITARY SELF-CONTAINED SYSTEMS

Multiple-unit systems generally use single-zone unitary air conditioners with a unit for each zone (Figure 6). Zoning is determined by (1) cooling and heating loads, (2) occupancy consider-

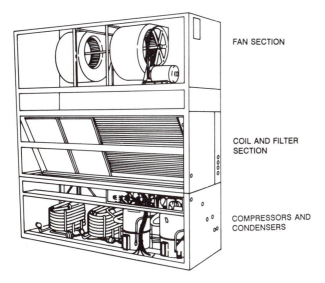

Fig. 4   Vertical Self-Contained Unit

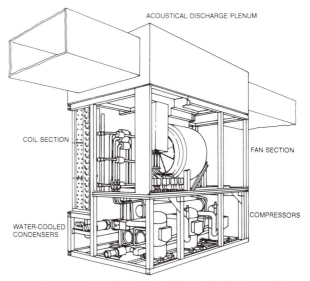

Fig. 5   Commercial Self-Contained Unit with Discharge Plenum

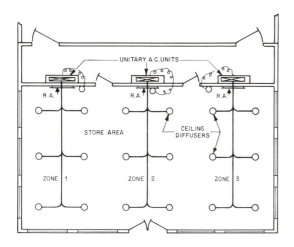

Fig. 6   Multiple-Packaged Units

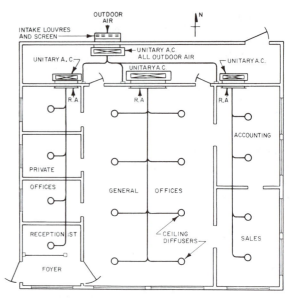

**Fig. 8   Multiple-Packaged Units with Separate Outdoor
Air Makeup Unit**

ations, (3) flexibility requirements, (4) appearance considerations, and (5) equipment and duct space availability. Multiple-unit systems are popular for office buildings, manufacturing plants, shopping centers, department stores, and apartment buildings. Unitary self-contained units are excellent for renovation.

Systems under 20 tons are typically of the constant volume (CV) type. Variable air volume distribution is accomplished with a bypass damper that allows excess supply air to bypass to the return air duct.

The bypass damper assures constant airflow across the direct expansion cooling coil and can be controlled by supply duct pressure.

Core systems. Unitary systems can be used throughout a building or to supplement perimeter area packaged terminal air-conditioning systems (Figure 7). Since core areas frequently have little or no heat loss, unitary equipment with water-cooled condensers can be applied, with water-source heat pumps serving the perimeter.

Separate unit for outdoor air. In this multiple-unit system, one unit preconditions outside air for a group of units (Figure 8). This all-outdoor air unit prevents hot, humid air from entering the conditioned space under periods of light load. The outdoor unit should have sufficient capacity to cool the required ventilation air from outdoor design conditions to interior design dew point. Zone units are then sized to handle only the internal load for their particular area.

Economizer cycle. Energy use can be reduced in many locales by cooling with outdoor air in lieu of mechanical refrigeration when outdoor temperature permits. Units must be located close to an outside wall or outside air duct shafts. Where this is not possible, it may be practical to add an economizer cooling coil

adjacent to the preheat coil (Figure 3). Cold water is obtained by cooling the condenser water through a winterized cooling tower. Chapter 37 has further details.

Specialized system. Special purpose unitary equipment is frequently used to cool, dehumidify, humidify, and reheat to maintain close control of space temperature and humidity in computer areas. Chapter 17 of the 1991 ASHRAE *Handbook—HVAC Applications* includes more information about this system and its design.

Acoustics

Since self-contained units are typically located on each floor near building occupants, they can have a significant impact on acoustics and productivity. Improved air distribution is often accomplished with a plenum located on top of the self-contained unit. This plenum facilitates multiple duct discharges that can reduce the amount of airflow over an occupied space adjacent to the equipment room. Reducing the airflow in any one direction can reduce the sound power from the breakout path of the discharge duct.

In addition to the airflow breakout path, the designer must study the unit's radiated sound power when selecting equipment room wall and door construction. Locating noncritical work spaces such as restrooms, elevators, or storage areas around the equipment room helps reduce the impact of air-conditioning sound in the occupied space.

Advantages

- Installation is simple. Equipment is readily available and its size allows easy handling.
- Relocation of units to other spaces or buildings is practical, if necessary.
- Units are available with complete, self-contained control systems including variable volume control, economizer cycle, night setback, and morning warm-up.
- Easy access to equipment facilitates routine maintenance.

Disadvantages

- Fans may have limited static pressure ratings.
- Integral air-cooled units must be located along outside walls.
- Multiple units and equipment closets or rooms may occupy rentable floor space.
- Close proximity to building occupants may create acoustical problems.

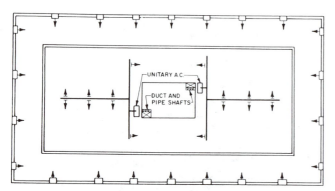

**Fig. 7   Multiroom, Multistory Office Building with Unitary
Core and Through-the-Wall Perimeter Air Conditioners**

COMMERCIAL SELF-CONTAINED SYSTEMS

Commercial self-contained systems provide central air distribution, refrigeration, and system control on a floor-by-floor basis. Typical components of a commercial self-contained air conditioner include compressors, condensers, evaporator coils, economizer coils, heating coils, filters, valves, and controls (Figure 5). To complete the system, a building needs cooling towers and condenser water pumps.

Commercial self-contained units can serve either VAV or CV systems. These units contain one or two fans—either forward curved (FC), backward inclined (BI), or airfoil (AF)—inside the cabinet. The fans are commonly in a draw-through arrangement. Along with the fan are the appropriate drive and motor selected to rotate the fan at the design airflow and static pressure.

The size and diversity of the zones served often dictate which system is optimal. For comfort applications, VAV self-contained units coupled with VAV air terminal boxes are popular for their energy savings, individual zone control, and acoustic benefits. Constant volume self-contained units have a low installation cost and are often used in noncomfort or industrial air-conditioning applications or in single-zone comfort applications.

Unit airflow is reduced in response to VAV boxes closing. Several common methods that modulate delivered airflow to match system requirements include inlet guide vanes, fan speed control, inlet/discharge dampers, and multiple fan motors.

Appropriate outside air and exhaust fans and dampers work in conjunction with the self-contained unit. Their operation must be coordinated with the unit operation to maintain design air exchanges and building pressurization.

ACOUSTICS

Since self-contained units are typically located near occupied space, their performance can significantly affect tenant comfort. Units of less than 15 tons are often placed inside a closet with a discharge grille penetrating the common wall to the occupied space. Larger units have their own equipment room and duct system. Three common sound paths to consider include:

- Fan inlet and compressor sound radiating through the unit casing to enter the space through the separating wall.
- Fan discharge sound is airborne through the supply duct and enters the space through duct breakout and diffusers.
- Airborne fan inlet sound finds its way to the space through the return air system.

Discharge air transition off the self-contained unit is often accomplished with a plenum located on top of the unit. This plenum facilitates multiple duct discharges that reduce the amount of airflow over a single occupied space adjacent to the equipment room (Figure 5). Reducing the airflow in one direction reduces the sound that breaks out from the discharge duct. Several feet of internally lined round duct immediately off the discharge plenum significantly reduces noise levels in adjacent areas.

In addition to the airflow breakout path, the system designer must study unit radiated sound power when selecting equipment room wall and door construction. A unit's airside inlet typically has the highest radiated sound component. The inlet should be located away from the critically occupied space and return air ducts to reduce the impact of this sound path.

Selecting a fan's operating point near its peak efficiency point helps in the design of quiet systems. Fans are typically dominant in the first three octave bands, and selections at high fan static pressures or near the fan's "surge" region should be avoided.

Units may be isolated from the structure with neoprene pads or spring isolators. Manufacturers often isolate the fan and compressors internally, which generally reduces external isolation requirements.

ECONOMIZERS

Self-contained units can have multiple compressors as a means of providing refrigeration capacity control. For VAV systems, compressors are turned on or off or unloaded to maintain discharge air temperature. As system airflow is decreased, the leaving air temperature from the unit can often be reset upward so that a minimum ventilation rate can be maintained. Resetting the discharge air temperature is a way of demand limiting the unit, thus saving energy.

An attractive option for reducing energy costs with self-contained systems is the waterside or airside economizer. While climate often dictates which economizer is selected, either one can provide system advantages.

The waterside economizer consists of a water coil located within the self-contained unit upstream of the direct expansion cooling coil. All economizer control valves, piping between the economizer coil and the condenser, and economizer control wiring can be factory installed (Figure 9).

The waterside economizer takes advantage of low cooling tower water temperature to either precool the entering air, assist mechanical cooling or, if cooling water is cold enough, to provide total system cooling. If the economizer is unable to maintain the supply air set point for VAV units or zone set point for CV units, factory-mounted controls integrate economizer and compressor operation to meet system cooling requirements.

Cooling water flow is controlled by two valves, one at the economizer coil inlet (A) and one in the bypass loop to the condenser (B). Two control methods are common—constant water flow and variable water flow.

Standard modulating control allows constant condenser water flow during unit operation. The two control valves are factory wired for complementary control, where one valve is driven open while the other is driven closed. This keeps water flow through the unit relatively constant.

Energy-saving modulating control allows variable condenser water flow during unit operation. The valve in the bypass loop (B)

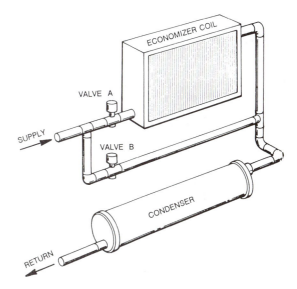

Fig. 9   Waterside Economizer and Valves

is an on-off valve and is closed when the economizer is enabled. Water flow through the economizer coil is modulated by valve (A), thus allowing variable cooling water flow. As the cooling load increases, valve (A) opens, increasing water flow through the economizer coil. If the economizer is unable to satisfy system cooling requirements, factory-mounted controls integrate economizer and compressor operation. In this operating mode, valve (A) is fully open. When the self-contained unit is not in the cooling mode, both valves are closed. Reducing or eliminating cooling water flow reduces pumping energy.

Advantages of Waterside Economizers

- Reduces compressor energy by precooling entering air.
- Can often completely satisfy the building load with entering condenser water temperatures less than 55 °F.
- Building humidification may not be required if return air contains sufficient humidity to satisfy winter requirement.
- No external wall penetration required for exhaust or outdoor air ducts.
- Mechanical equipment rooms can be centrally located in a building.
- Controls are less complex than airside, because they often reside inside the packaged unit.
- Coil can be mechanically cleaned.
- Net usable floor area is increased by eliminating large fresh air and relief air ducts.

Disadvantages of Waterside Economizers

- Increased tower water treatment cost.
- Airside pressure drop may be increased.
- Condenser water pump may see slightly higher pressure.
- Cooling tower must be designed for winter operation.

The airside economizer is usually a field-installed accessory that includes an outdoor air damper, relief damper, return air damper, filters, actuator, and linkage. Economizer controls are usually a factory-installed option. The airside economizer takes advantage of cool outdoor air to either assist mechanical cooling or, if the outdoor air is cool enough, provide total system cooling.

Self-contained units do not include return air fans. It is necessary to include a variable volume relief fan when airside economizers are employed. The relief fan volume is generally controlled with discharge dampers in response to building space pressure. The relief fan is off and discharge dampers are closed when the airside economizer is inactive.

Advantages of Airside Economizers

- Substantially reduces compressor, cooling tower, and condenser water pump energy requirements.
- Has a lower airside pressure drop than a waterside economizer.
- Reduces tower makeup water and related water treatment.

Disadvantages of Airside Economizers

- Humidification may be required during winter operation.
- Equipment room is generally placed along the building's exterior wall.
- Installed cost is generally higher than that for a waterside economizer due to exhaust system requirements.

CONTROLS

Self-contained units typically have built-in capacity controls for refrigeration, economizers, and fans. While units under 15 tons tend to have basic on-off/auto controls, many larger systems have sophisticated microprocessors that monitor and take action based on local or remote programming. These controls provide for stand-alone operation, or they can be tied to a building automation system.

A building automation system allows for more sophisticated unit control by time of day scheduling, optimal start/stop, duty cycling, demand limiting, custom programming, etc. This control can keep the units operating at their peak efficiency by alerting the operator to conditions that could cause substandard unit performance.

Self-contained units may control capacity with multiple compressors. For VAV systems, compressors are turned on or off or unloaded to maintain discharge air temperature. As system airflow decreases, the temperature of air leaving the unit is often reset upward so that a minimum ventilation rate can be maintained. Resetting the discharge air temperature limits the unit's demand, thus saving energy.

The unit's control panel can sequence the modulating valves and dampers of an economizer. A waterside economizer is located upstream of the evaporator coil, and when condenser water temperature is lower than entering air temperature to the unit, water flow can be through the economizer coil to meet building load. If the coil alone cannot meet design requirements, but the entering condenser water temperature remains cool enough to provide some useful precooling, the control panel can keep the economizer coil active as stages of compressors are activated. When entering condenser water exceeds entering air temperature to the unit, the coil is valved off, and water is circulated through the unit's condensers only.

Typically, in an airside economizer an enthalpy switch or dry-bulb temperature energizes the unit to bring in outside air as the first stage of cooling. An outside air damper modulates the flow to meet a design temperature, and when outside air can no longer provide enough cooling, compressors are energized.

A temperature input to the control panel, either from a discharge air sensor or a zone sensor, provides information for integrated economizer and compressor control. Supply air temperature reset is commonly applied to VAV systems.

In addition to capacity controls, units have safety features for the refrigerant-side, airside, and electrical systems. Refrigeration protection controls typically consist of high and low refrigerant pressure sensors and temperature sensors wired into a common control panel. The controller then cycles compressors on and off or introduces hot gas bypass to meet system requirements.

Constant volume units typically have high-pressure cut-out controls, which protect the unit and ductwork from high static pressure. Variable air volume units typically have some type of static pressure probe inserted in the discharge duct several feet downstream of the unit. As VAV boxes close, the control modulates airflow to meet the set point, which is determined by calculating the static pressure required to deliver design airflow to a zone farthest from the unit.

Advantages

- Commercial self-contained systems are well-suited for office environments with variable occupancy schedules.
- The floor-by-floor arrangement can offer reduced fan power consumption.
- Large vertical duct shafts and fire dampers are eliminated.
- Electrical wiring, condenser water piping, and condensate removal are centrally located.
- Commercial self-contained systems generally reduce mechanical room area.
- Equipment is generally located in the building interior near elevators and other service areas and does not interfere with the building perimeter.

- An integral waterside economizer coil and controls allows an interior equipment location and eliminates large outdoor air and exhaust ducts and relief fans.
- An acoustical discharge plenum allows lower fan power and lower sound power levels.
- Cost of operating cooling equipment after normal building hours is reduced by operating only units in occupied areas.

Disadvantages

- Units must be located near an outside wall or outdoor air shaft to incorporate an airside economizer.

- A separate relief fan and discharge damper with space pressure controls must be incorporated if airside economizers are used.
- Close proximity to building occupants requires careful analysis of space acoustics.
- Filter options may be limited.

REFERENCES

Trane Company. 1984. *Self-contained VAV system design*. Form AM-SYS9 (984).

Trane Company. 1988. *Commercial self-contained systems*. Form ICS-AM3 (888).

PANEL HEATING AND COOLING

RADIANT panel systems use controlled temperature surfaces in the floor, walls or ceiling; the temperature is maintained by circulating water, air, or electric resistance. Radiant panel systems may be combined with a central station air system of one-zone, constant temperature, constant volume design, or with dual-duct, reheat, multizone or variable volume systems. A controlled temperature surface is referred to as a radiant panel if 50% or more of the heat transfer is by radiation to other surfaces seen by the panel. This chapter is concerned with surfaces whose temperatures are controlled and are the primary source of heating and cooling within the conditioned space.

High-temperature surface radiant panels over about 300°F energized by gas, electricity, or high-temperature water are discussed in Chapter 15.

RADIANT ENERGY TRANSFER PRINCIPLES

Radiant energy (1) is transmitted at the speed of light; (2) travels in straight lines and can be reflected; and (3) elevates the temperature of solid objects by absorption, but does not heat the air through which it travels.

Radiant energy exchanges continuously between all bodies in a building environment. The rate at which radiant energy is transferred depends on the following factors:

- Temperature (of the emitting surface and receiver)
- Emissivity (of the radiating surface)
- Reflectivity, absorptivity, and transmissivity (of the receiver)
- The view factor between the emitting surface and receiver (viewing angle of the occupant to the radiant source)

A critical factor is the structure of the body surface. In general, rough surfaces have low reflectivity and high emissivity/absorptivity characteristics. Conversely, smooth or polished surfaces have high reflectivity and low absorptivity/emissivity characteristics.

One example of radiant heating is the feeling of warmth when standing in the sun's rays on a cool, sunny day. Some of the rays come directly from the sun and include the entire electromagnetic spectrum. Other rays from the sun impinge on surrounding objects and are absorbed or reflected. This generates a secondary source of energy, creating rays that are a combination of the wavelength produced by the temperature of the objects and the wavelength of the reflected rays. If a cloud passes in front of the sun, there is an instant sensation of cold. This sensation is caused by the decrease in radiant energy received from the sun, although there is little, if any, change in the surrounding ambient air temperature.

Thermal comfort, as defined in ANSI/ASHRAE *Standard* 55-1981, is "that condition of mind which expresses satisfaction with the thermal environment." No system is completely satisfactory unless the three main factors controlling heat transfer from the human body (radiation, convection, and evaporation) result in thermal neutrality.

Designers sometimes think that a radiant heat transfer system is desirable only for certain types of buildings and only in some climates. However, maintaining correct conditions for human comfort by radiant heat transfer is possible for even the most severe climatic conditions (Buckley 1989).

Panel heating and cooling systems provide an acceptable thermal environment by controlling surface temperature within an occupied space, thus affecting the radiant heat transfer. With a properly designed system, a person should not be aware that the environment is being heated or cooled. The mean radiant temperature (MRT) has a strong influence on body comfort. When the temperature of the surfaces comprising the building deviates excessively from the ambient temperature (particularly outside walls with large amounts of glass), convective systems sometimes have difficulty in counteracting the discomfort caused by cold or hot surfaces. Heating and cooling panels neutralize these deficiencies and minimize radiation losses or gains by the body.

Most building materials have surfaces with relatively high emissivity factors and, therefore, absorb, reradiate, and reflect radiant heat from active panels. Warm ceiling panels are effective because their radiant heat is absorbed and reflected by the irradiated surfaces and not transmitted through the construction. Glass is also opaque to the wavelengths emitted by active panels, and therefore, transmits little of the long-wave radiant heat to the outside. This is significant because all surfaces within the room tend to assume temperatures that result in an acceptable thermal comfort condition within the space.

GENERAL EVALUATION

Principal advantages of radiant panel systems are:
- Comfort levels can be better than those of other conditioning systems because radiant loads are treated directly and air motion in the space is at normal ventilation levels.
- Mechanical equipment is not needed at the outside walls, simplifying the wall, floor, and structural systems.
- All pumps, fans, filters, and so forth are centrally located, simplifying maintenance and operation.
- No space is required within the air-conditioned room for mechanical equipment. This feature is especially valuable in hospital patient rooms and other applications where space is at a premium, where maximum cleanliness is essential, or where dictated by legal requirements.
- Cooling and heating can be simultaneous, without central zoning or seasonal changeover, when four-pipe systems are used.
- Supply air quantities usually do not exceed those required for ventilation and dehumidification.

The preparation of this chapter is assigned to TC 6.5, Radiant Space Heating and Cooling.

- Draperies and curtains can be installed at the outside wall without interfering with the heating and cooling system.
- The modular panel provides flexibility to meet changes in partitioning.
- A 100% outdoor air system may be installed with less severe penalties in terms of refrigeration load because of reduced air quantities.
- A common central air system can serve both the interior and perimeter zones.
- Wet surface cooling coils are eliminated from the occupied space, reducing the potential for septic contamination.
- The panel system can use the automatic sprinkler system piping (see NFPA 13, Chapter 5, Sections 5-6). The maximum water temperature must not fuse the heads.
- Radiant cooling and minimum supply air quantities provide a draft-free environment.
- Noise associated with fan coil or induction units is eliminated.

Disadvantages are similar to those listed in Chapter 3 of the 1989 ASHRAE *Handbook—Fundamentals*.

HEAT TRANSFER BY PANEL SURFACES

A heated or cooled panel transfers heat to or from a room by convection and radiation.

Radiation Transfer

The basic equation for a multisurface enclosure with gray, diffuse isothermal surfaces is derived by radiosity formulation methods (Chapter 3 of the 1989 ASHRAE *Handbook—Fundamentals*). This equation may be written as:

$$q_r = J_p - \sum_{j=1}^{N} F_{pj} J_i \qquad (1)$$

where

q_r = net radiation heat transferred by panel surface, Btu/h·ft^2
J_p = total radiosity that leaves panel surface, Btu/h·ft^2
J_i = total radiosity from all other surfaces in room, Btu/h·ft^2
F_{pj} = radiation angle factor between panel surface and another surface in room (dimensionless)

Equation (1) is applicable to simple and complex enclosures with varying surface temperatures and emittances. The net radiation transferred by the panels can be found by determining unknown J_i if the number of surfaces is small. More complex enclosures require computer calculations.

Radiation angle factors can be evaluated using charts in Chapter 3 of the 1989 ASHRAE *Handbook—Fundamentals*. Fanger (1972) shows room-related angle factors or they may also be developed from algorithms in ASHRAE *Energy Calculations* I (1976).

Several methods have been developed to simplify Equation (1) by reducing a multisurface enclosure to a two-surface approximation. In the MRT method, the radiant interchange in a room is modeled by assuming that each surface radiates to a fictitious surface that has an area emissivity, and temperature giving about the same heat transfer from the surfaces as the real multisurface case (Walton 1980). In addition, angle factors do not need to be determined in the evaluation of a two-surface enclosure. The MRT equation may be written as:

$$q_r 2 = \sigma F_e \left[\frac{T_p^4}{100} - \frac{T_r^4}{100} \right] \qquad (2)$$

where

q_r = net radiation heat transferred by panel surface, Btu/h·ft^2
T_p = average temperature of heated (cooled) panel, °R
σ = Stefan-Boltzman constant, 0.1714 × 10^{-8} Btu/h·ft^2·°R

The temperature of the fictitious surface is given by an area emissivity weighted average of all surfaces other than the panel.

$$T_r = \sum_{j \neq p}^{n} A_j \epsilon_j T_j \Big/ \sum_{j \neq p}^{n} A_j \epsilon_j \qquad (3)$$

where

T_r = temperature of fictitious surface, °R
A_j = area of surfaces other than panel
ϵ_j = surface emissivities other than panel (dimensionless)

The radiation interchange factor or Hottel equation for two-surface radiation heat exchange is given by:

$$F_c = \frac{1}{1/F_{p-r} + [(1/\epsilon_p) - 1] + A_p/A_r [(1/\epsilon_f) - 1]} \qquad (4)$$

where

F_c = radiation interchange factor (dimensionless)
F_{p-r} = radiation angle factor from panel to fictitious surface (1.0 for flat panel)
A_p, A_r = area of panel surface and fictitious surface, respectively
ϵ_p, ϵ_r = emittance of panel surface and fictitious surface, respectively (dimensionless)

When the emissivities of an enclosure are closely equal and surfaces exposed to the panel are marginally unheated (uncooled), then Equation (3) becomes the area-weighted average unheated (or uncooled) surface temperature (AUST) exposed to the panels.

In practice, the emissivity of nonmetallic or painted metal nonreflecting surfaces is about 0.9. When this emissivity is used in Equation (4), the combined factor is about 0.87 for most rooms. Substituting this value in Equation (2), the constant becomes about 0.15, and the equation for heating can be rewritten:

$$q_r = 0.15 \left[\left(\frac{t_p + 460}{100} \right)^4 - \left(\frac{\text{AUST} + 460}{100} \right)^4 \right] \qquad (5A)$$

or for cooling:

$$q_r = 0.15 \left[\left(\frac{\text{AUST} + 460}{100} \right)^4 - \left(\frac{t_p + 460}{100} \right)^4 \right] \qquad (5B)$$

where

q_r = heat transferred by the panel to or from room surfaces by radiation, Btu/h·ft^2
t_p = average panel surface temperature, °F
AUST = area-weighted average temperature of unheated surfaces in room, °F

The actual radiation transfer in a room may be somewhat different from that given by Equations (5A) or (5B) because of non-uniform temperatures, irregular room surfaces, variations in emissivity of materials, and so forth. However, the equation is accurate to within 10% when used in conventional heating and cooling calculations. Min *et al.* (1956) showed that the value of the constant of Equation (5A) and (5B) was 0.152 in the test room. The design information in this chapter is based on that constant value.

Stemman (1989) found the accuracy of Equations (5A), (5B), and (2) was improved by adding back geometric angle factors to these equations. These factors allow the accuracy of the MRT method to approach that of Equation (1) and still maintain a first order temperature solution.

Radiation exchange calculated from Equation (5A) is given in Figure 1. The values apply to ceiling, floor, or wall panel output. Radiation removed by a cooling panel for a range of normally encountered temperatures and as calculated from Equation (5B) is given in Figure 2.

In many specific instances where normal multistory commercial construction and fluorescent lighting are used, the room

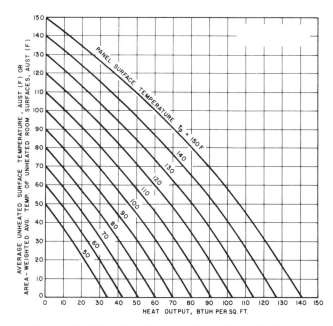

**Fig. 1   Radiation Heat Transfer from Heated Ceiling,
Floor, or Wall Panel**

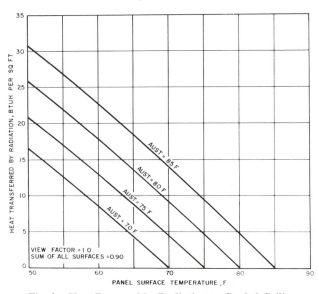

**Fig. 2   Heat Removed by Radiation to Cooled Ceiling
or Wall Panel**

temperature at the 5-ft level will closely approach the average uncooled surface temperatures (AUST). In structures where the main heat gain is through the walls or where incandescent lighting is used, the wall surface temperatures tend to rise considerably above the room air temperature.

Convection Transfer

The convection coefficient q_c is defined as the heat transferred by convection in Btu/h·ft^2·°F difference between air and panel temperatures. Heat transfer convection values are not easily established. Convection in panel systems is usually natural; that is, air motion is generated by the warming or cooling of the boundary layer of air. In practice, however, many factors such as a room's configuration interfere with or affect natural convection. Infiltration, the movement of persons, and mechanical ventilating systems can introduce some forced convection that will disturb the natural process.

Parmelee and Huebscher (1947) included the effect of forced convection on heat transfer from panels as an increment to be added to the natural convection coefficient. However, increased heat transfer from forced convection should not be used, because the increments are unpredictable in pattern and performance, and forced convection does not significantly increase the total capacity of the panel system.

Convection in a panel system is a function of the panel surface temperature and the temperature of the airstream layer directly below the panel. The most consistent measurements are obtained when the air layer temperature is measured close to the region where the fully developed stream begins, usually 2 to 2.5 in. below the panels.

Min *et al.* (1956) determined natural convection coefficients 5 ft above the floor in the center of a 12 ft by 24 ft room. Equations (6) to (11), derived from this research, can be used to calculate heat transfer from panels by natural convection.

Natural convection from a heated ceiling

$$q_c = 0.041 \, (t_p - t_a)^{1.25}/D_e^{0.25} \tag{6}$$

Natural convection from a heated floor or cooled ceiling

$$q_c = 0.39 \, (t_p - t_a)^{1.31}/D_e^{0.08} \tag{7}$$

Natural convection from a heated or cooled wall panel

$$q_c = 0.29 \, (t_p - t_a)^{1.32}/H^{0.05} \tag{8}$$

where

q_c = heat transfer by natural convection, Btu/h·ft^2
t_p = temperature of panel surface, °F
t_a = temperature of air, °F
D_e = equivalent diameter of panel (4 area/perimeter), ft
H = height of wall panel, ft

Schutrum and Humphreys (1954) measured panel performance in furnished test rooms that did not have uniform temperature surfaces and found no variations large enough to be significant in heating practice. Schutrum and Vouris (1954) established that the effect of room size was also usually insignificant. The convection equations can therefore be simplified to:

Natural convection from a heated ceiling

$$q_c = 0.021 \, (t_p - t_a)^{1.25} \tag{9}$$

Natural convection from a heated floor or cooled ceiling

$$q_c = 0.32 \, (t_p - t_a)^{1.31} \tag{10}$$

Natural convection from a heated or cooled wall panel

$$q_c = 0.26 \, (t_p - t_a)^{1.32} \tag{11}$$

Figure 3 shows heat output by natural convection from floor, wall, and ceiling heating panels as calculated from Equations (9) and (10).

Figure 4 shows heat removed by natural convection by cooled ceiling panels as calculated by Equation (10) and data from Wilkes and Peterson (1938) for specific panel sizes. An additional curve illustrates the effect of forced convection on the latter data. Similar adjustment of the ASHRAE data is inappropriate, but the effects would be much the same. For preliminary design, use 1 Btu/h·ft^2·°F between room design and panel temperature.

Combined Heat Transfer (Radiation and Convection)

The combined heat transfer from a panel to a room can be determined by adding the radiant heat transfer from Figure 1 or 2 to the convective heat transfer from Figure 3 or 4, respectively. Figures 1 and 2 require calculating the AUST in the room. In calculating the AUST, the surface temperature of the interior walls is assumed to be the same as the room air temperature. The inside surface temperatures of outside walls and exposed floors or ceil-

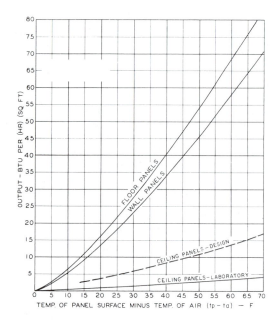

Fig. 3   **Heat Output by Natural Convection from Floor and Ceiling Panels [Equations (10) and (11)]**

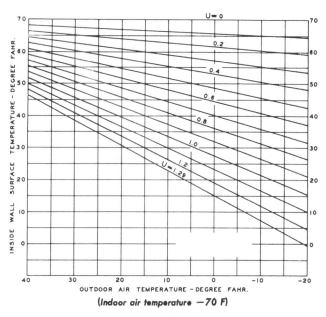

Fig. 5   **Relation of Inside Surface Temperature to Overall Coefficient of Heat Transfer**

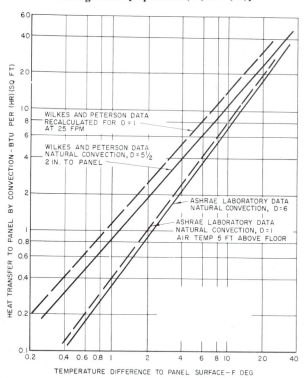

Fig. 4   **Heat Removed by Natural Convection to Ceiling Cooling Panels [Equation (10)]**

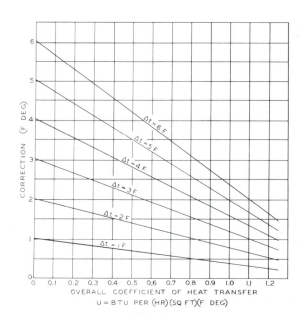

Fig. 6   **Inside Wall Surface Temperature Correction for Air Temperatures Other than 70°F**

ings can be obtained from Figure 5 for a 70°F room temperature. Figure 6 gives corrections for other temperatures. Steinman *et al.* (1989) noted that these temperatures may not be appropriate for enclosures with large glass areas or a high percentage of outside wall and/or ceiling surface area. These cold surfaces have a lower AUST, which increases the radiant heat transfer.

The combined heat transfer for ceiling and floor panels used to heat rooms in which the air temperature is 70 to 76°F can be read directly from Figures 7 and 8, respectively. These diagrams apply to rooms in which the AUST does not differ greatly from

room air temperatures. Tests by Schutrum *et al.* (1953a, 1953b) and simulation by Kalisperis (1985) based on a CAD program developed by Kalisperis and Summers (1985) show that the temperatures are almost equal.

The performance of radiant transfer equipment, in its various forms and applications to a particular space, may differ substantially from the outputs given in Figures 7 and 8. Refer to manufacturer's data for the specific system selected.

Figure 9 shows the combined radiation and convection transfer for cooling, as given in Figures 2 and 4. The data in Figure 9 do not include heat gains from sun, lights, people, or equipment: manufacturer's data includes these heat gains.

In suspended ceiling panel systems, heat transfers from the ceiling panel to the floor slab above (heating) and vice versa (cooling).

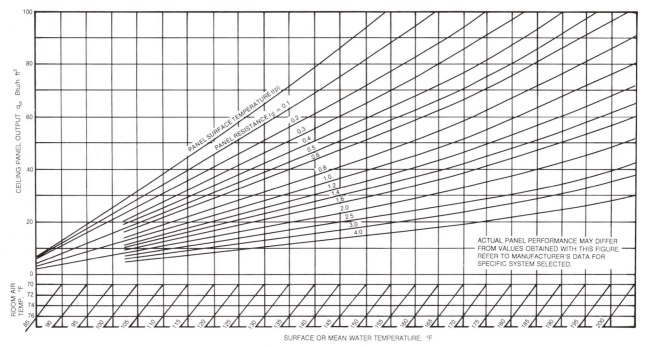

Fig. 7 Ceiling Panel Design Graph Showing Panel Surface Temperature and Mean Water Temperature versus Output Downward

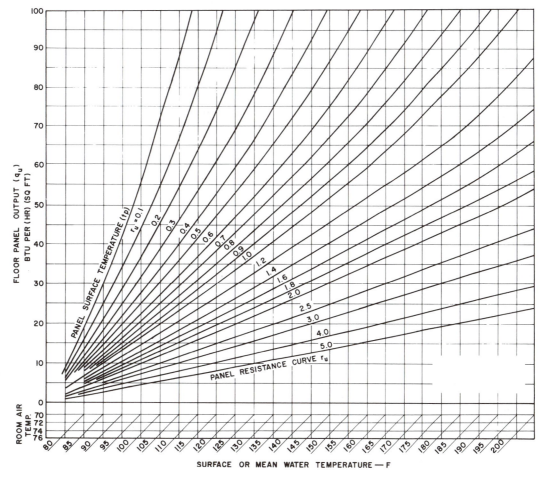

Fig. 8 Floor Panel Design Graph Showing Panel Surface Temperature and Mean Water Temperature versus Output Upward

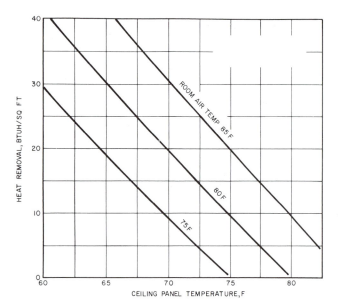

Fig. 9   Cooled Ceiling Panel Performance in Uniform Environment with No Infiltration and No Internal Heat Sources

The ceiling panel surface temperature is affected because of heat transfer to or from the panel and the slab by radiation and, to a much smaller extent, by convection. The radiation component can be approximated using Figure 1. The convection component can be estimated from Figure 3 or 4. In this case, the temperature difference is that between the top of the ceiling panel and the mid-space of the ceiling. The temperature of the ceiling space should be determined by testing, since it varies with different panel systems. However, much of this heat transfer is nullified when insulation is placed over the ceiling panel, which, for perforated metal panels, also provides acoustical control.

If lighting fixtures are recessed into the suspended ceiling space, radiation from the top of the fixtures raises the overhead slab temperature and transfers heat to the space by convection. This energy is absorbed at the top of the cooled ceiling panels by radiation, as in Figure 2, and by convection, generally in accordance with Equation (6). The amount the top of the panel absorbs depends on the system. Most manufacturers have information available. Similarly, panels installed under a roof absorb additional heat, depending on configuration and insulation.

GENERAL DESIGN CONSIDERATIONS

Radiant panel systems are similar to other air-water systems in the arrangement of the system components. Room thermal conditions are maintained primarily by direct transfer of radiant energy, rather than by convection heating and cooling. The room heating and cooling loads are calculated in the conventional manner. Manufacturers' ratings generally are for total performance and can be applied directly to the calculated room load.

Because the mean radiant temperature (MRT) within a panel heated space increases as the heating load increases, the controlled air temperature during this increase may be lowered without affecting comfort. In ordinary structures with normal infiltration loads, the required reduction in air temperature is small, enabling a conventional room thermostat to be used.

In panel heating systems, lowered night temperature can produce less satisfactory results with heavy panels such as concrete floors. These panels cannot respond to a quick increase or decrease in heating demand within the relatively short time required, resulting in a very slow reduction of the space temperature at night and a correspondingly slow pickup in the morning. Light panels, such as plaster or metal ceilings and walls, may respond to changes in demand quickly enough for satisfactory results from lowered night temperatures. Tests on a metal ceiling panel demonstrated the speed of response to be comparable to that of convection systems. However, very little fuel savings can be expected even with light panels unless the lowered temperature is maintained for long periods. If reduced nonoccupancy temperatures are employed, some means of providing a higher-than-normal rate of heat input for rapid warm-up is necessary; for example, fast-acting radiant ceiling panels (Berglund 1982).

Metal radiant heating panels, hydronic and electric, are applied to building perimeter spaces for heating in much the same way as finned-tube convectors. Metal panels are usually installed in the ceiling and are integrated into the ceiling design. The layout and arrangement of panels usually considers architectural design.

Partitions may be erected to the face of hydronic panels but not to the active heating portion of electric panels because of possible element overheating and burnout. Electric panels are often sized to fit the building module with a small removable filler or dummy panel at the window mullion to accommodate future partitions. Hydronic panels can run continuously.

By cutting and fitting, field modification of hydronic panels is possible; however, modification should be kept to a minimum to keep installation costs down. Electric panels cannot be modified in the field.

Panel Thermal Resistance

The thermal resistance to heat flow may vary considerably among panel systems, depending on the type of bond between the water tube and the panel material. Corrosion between lightly touching surfaces, the method of maintaining contact, and other factors may change the bond with time. The actual thermal resistance of any proposed system should be verified by testing whenever practical. Tables 1 through 5 show some typical values for thermal resistance factors for various types of floor and ceiling panels.

The thermal resistance factors shown are for ferrous and nonferrous pipe and tube. Recently, the use of nonmetallic, plastic tubing has increased. Long lengths are available in coils up to 1000 ft, which simplify installation. Resistance and performance data should be obtained from the manufacturer.

Table 1   Thermal Resistance of Metal Ceiling Panels

Type of Panel	Spacing, in.	Thermal Resistance,[a] ft²·°F·h/Btu
STEEL PIPE — PAN EDGE HELD AGAINST PIPE BY SPRING CLIP / ALUMINUM PAN 0.032 IN. THICK	3	—
	6	0.31
	12	0.61
COPPER TUBE BONDED TO 0.040 IN. ALUMINUM SHEET	4	0.071
	8	0.15
COPPER TUBE PRESSED INTO ALUMINUM EXTRUSION, 0.050 IN. T	5	0.071

[a]Manufacturer's data.

Table 2 Thermal Resistance of Bare Concrete Floor Panels (Heating)

Panel Construction	Spacing in.	Heat Flow Ratio, q_u/q_d or q_u/q_{de} 1 Panel Thermal Resistance, ft²·°F·h/Btu Up	Down	3 Up	Down	5 Up	Down	10 Up	Down
4-in. Concrete Slab with 2-in. Cover									
0.5-in nonferrous tube	9	0.57	0.52	0.46	0.84	0.43	1.17	0.42	1.97
	12	0.73	0.68	0.58	1.16	0.54	1.65	0.51	2.86
0.5-in. ferrous or 0.75 in. nonferrous tube	9	0.49	0.42	0.41	0.66	0.39	0.90	0.38	1.80
	12	0.63	0.55	0.50	0.93	0.48	1.30	0.46	2.35
6-in. Concrete Slab with 2-in. Cover									
0.5-in. nonferrous tube	9	0.59	0.70	0.47	1.05	0.45	1.39	0.43	2.25
	12	0.78	0.90	0.60	1.40	0.56	1.97	0.54	3.21
0.75-in. nonferrous pipe	9	0.51	0.61	0.43	0.87	0.41	1.13	0.40	1.78
	12	0.68	0.78	0.54	1.23	0.51	1.63	0.49	2.61
0.75-in. ferrous pipe	9	0.47	0.55	0.40	0.77	0.39	0.98	0.38	1.50
	12	0.63	0.71	0.50	1.07	0.48	1.44	0.46	2.36
1-in. nonferrous tube or ferrous pipe	12	0.59	0.66	0.48	0.98	0.46	1.30	0.45	2.11
	15	0.73	0.83	0.57	1.21	0.54	1.73	0.51	2.74

Tube diameters are nominal.

q_u = upward heat flow from panel
q_d = downward heat flow from panel
q_{de} = apportioned downward and edgewise heat flow from panel

Table 3 Thermal Resistance of Concrete Ceiling Panels (Heating)

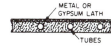

Panel Construction[a]	Spacing in.	Heat Flow Ratio,[b] q_u/q_d 0 Panel Thermal Resistance, ft²·°F·h/Btu Up	Down	0.5 Up	Down	1.0 Up	Down
6-in. Concrete Slab with 1-in. Cover							
0.5-in. nonferrous tube	9	3.6	0.30	0.9	0.35	0.7	0.45
	12	5.1	0.35	1.1	0.45	0.9	0.55
0.5-in. ferrous or 0.75-in. nonferrous tube	9	2.6	0.25	0.7	0.30	0.6	0.35
	12	4.0	0.30	0.9	0.40	0.8	0.50
0.75-in. ferrous pipe or 0.75-in. nonferrous tube	9	2.1	0.20	0.6	0.25	0.6	0.30
	12	3.3	0.30	0.8	0.35	0.7	0.40
	15	4.5	0.35	1.0	0.45	0.8	0.55
1-in. ferrous pipe	9	1.6	0.20	0.5	0.25	0.5	0.25
	12	2.6	0.25	0.7	0.30	0.9	0.40
	15	3.6	0.30	0.9	0.40	0.7	0.45
8-in. Concrete Slab with 1-in. Cover							
0.5-in. nonferrous tube	9	3.6	0.30	1.0	0.35	0.8	0.40
	12	5.2	0.35	1.2	0.45	1.0	0.55
0.5-in. ferrous pipe or 0.75-in. nonferrous tube	9	2.9	0.25	0.9	0.30	0.8	0.35
	12	4.0	0.30	1.1	0.40	0.9	0.45
0.75-in. ferrous pipe or 1-in. nonferrous tube	9	2.2	0.20	0.8	0.30	0.7	0.30
	12	3.3	0.30	1.0	0.35	0.8	0.40
	15	4.3	0.35	1.1	0.40	0.9	0.50
1-in. ferrous pipe	9	1.7	0.20	0.7	0.25	0.7	0.25
	12	2.7	0.25	0.9	0.30	0.8	0.35
	15	3.7	0.30	1.0	0.40	0.9	0.45

q_u = upward heat flow from panel
q_d = downward heat flow from panel

[a]Tube diameters are nominal.
[b]Any ceiling panel also acts as a floor panel to the extent of its upward heat flow. If the upward heat flow is high and the space above is occupied, check floor surface temperature for possible foot discomfort. Also check effect on heating requirements of the space above. Do not supply the major portion of the upper room's heating requirements by the upward heat flow of a ceiling panel below.

Table 4 Thermal Resistance of Plaster Ceiling Panels (Heating or Cooling)

Panel Construction

Standard gypsum plaster (three coats) with 3/8-in. nominal nonferrous tube or 1/2-in. nominal ferrous pipe above metal lath tied at 8-in. intervals with good tube imbedment, or 3/8-in. nominal nonferrous tube below metal or gypsum lath.

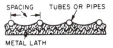

Spacing, in.	Thermal Resistance to Downward Heat Flow r_d, ft²·°F·h/Btu Heat Flow Ratio q_u/q_d 0	0.2	0.4	0.6	0.8	1.0
4.5	0.30	0.34	0.38	0.42	0.46	0.50
6	0.45	0.51	0.57	0.63	0.69	0.75
9	0.75	0.85	0.95	1.05	1.15	1.25
16	1.15	1.29	1.43	1.57	1.71	1.85

Notes:
1. Any ceiling panel also acts as a floor panel to the extent of its upward heat flow. If the upward heat flow is high and the space above is occupied, check floor surface temperature for possible discomfort (see Schutrum *et al.* 1953). Also check effect on heating requirements of the space above. Do not supply the major portion of the upper room's heating requirements by the upward heat flow of a ceiling panel below.
2. Recommended maximum inlet water temperature t_{max} = 140°F.

Table 5 Thermal Resistance of Floor Coverings

Description	Resistance r_{uc}, ft²·°F·h/Btu
Bare concrete, no covering	0.00
Asphalt tile	0.05
Rubber tile	0.05
Light carpet	0.6
Light carpet with rubber pad	1.0
Light carpet with light pad	1.4
Light carpet with heavy pad	1.7
Heavy carpet	0.8
Heavy carpet with rubber pad	1.2
Heavy carpet with light pad	1.6
Heavy carpet with heavy pad	1.9

Effect of Floor Coverings

Floor coverings can have a pronounced effect on the performance of a floor heating panel system. The added thermal resistance of the floor covering reduces upward heat flow and increases the heat flow to the underside of the panel. To maintain a given upward heat flow after a floor covering has been added, the temperature of the heating medium must be increased. Data on the thermal resistance of common floor coverings are given in Table 5.

Where covered and bare floor panels exist in the same system, it may be possible to maintain a high enough water temperature to satisfy the covered panels and balance the system by throttling the flow to the bare slabs. In some instances, however, the increased water temperature required when carpeting is applied over floor panels makes it impossible to balance floor panel systems in which only some rooms have carpeting, unless the piping is arranged to permit zoning using more than one water temperature.

Panel Heat Losses

Heat transferred from the upper surface of ceiling panels, the back surface of wall panels, the underside of floor panels or the

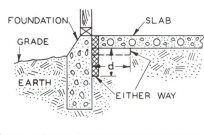

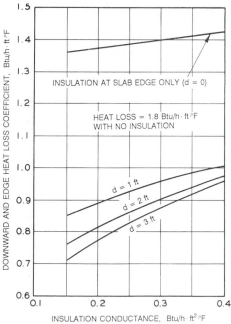

Fig. 10 Downward and Edgewise Heat Loss Coefficient for Concrete Floor Slabs on Grade

edges of any panel is considered a panel heat loss. Panel heat losses are part of the building heat loss if the heat is transferred outside of the building. If the heat is transferred to another heated space, the panel loss is a source of heat for the space instead. In either case, the magnitude of panel loss should be determined.

Panel heat loss to space outside the room should be kept to a reasonable amount by insulation. For example, a floor panel may overheat the basement below, and a ceiling panel may cause the temperature of a floor surface above it to be too high for comfort unless it is properly insulated.

The heat loss from most panels can be calculated by using the coefficients given in Chapter 22 of the 1989 ASHRAE *Handbook—Fundamentals*. These coefficients should not be used to determine the downward heat loss from panels built on grade because the heat flow from them is not uniform (Sartain and Harris 1956; ASHAE 1956, 1957). The heat loss from panels built on grade can be estimated from Figure 10 or from data in Chapter 25 of the 1989 ASHRAE *Handbook—Fundamentals*.

PANEL HEATING AND COOLING SYSTEMS

The most common forms of panels applied in panel heating systems are as follows:

1. Metal ceiling panels
2. Embedded piping in ceilings, walls, or floors
3. Electric ceiling panels
4. Electrically heated ceilings or floors
5. Air-heated floors

Residential heating applications usually consist of pipe coils or electric elements embedded in masonry floors or plaster ceilings. This construction is suitable where loads are stable and solar effects are minimized by building design. However, in buildings where glass areas are large and load changes occur faster, the slow response, lag, and override effect of masonry panels are unsatisfactory. Light metal panel ceiling systems respond quickly to load changes (Berglund *et al.* 1982).

Radiant panels are often located in the ceiling because it is exposed to all other surfaces and objects in the room. It is not likely to be covered, as are the floors, and higher surface temperatures can be used. Also, its smaller mass enables it to respond more quickly to load changes.

Warm air and electric heating elements are two design concepts used in systems influenced by local factors. The warm air system has a special cavity construction where air is supplied to a cavity behind or under the panel surface. The air leaves the cavity through a normal diffuser arrangement and is supplied to the room. Generally, these systems are used as floor radiant panels in schools and in floors subject to extreme cold, such as in an overhang. Cold outdoor temperatures and heating medium temperatures must be analyzed with regard to potential damage to the building construction. Electric heating elements embedded in the floor or ceiling construction and unitized electric ceiling panels are used in various applications to provide both full heating and spot heating of the space.

HYDRONIC PANEL SYSTEMS

Design Considerations

Hydronic radiant panels can be used with two- and four-pipe distribution systems. Figure 11 shows the arrangement of a typical system. It is common to design for a 20°F temperature drop for heating across a given grid and a 5°F rise for cooling, but larger temperature differentials may be used, if applicable.

Panel design requires determining panel area, panel type, supply water temperature, water flow rate, and panel arrangement. Panel performance is directly related to room conditions. Air-side design also must be established. Heating and cooling loads may be calculated by procedures covered in Chapters 22 through 27 in the 1989 ASHRAE *Handbook—Fundamentals*. The procedure is as follows:

1. Determine room design dry-bulb temperature, relative humidity, and dew point.
2. Calculate room sensible and latent heat gains.
3. Select mean water temperature for cooling.
4. Establish minimum supply air quantity.
5. Calculate latent cooling available from the air.
6. Calculate sensible cooling available from the air.
7. Determine panel cooling load.
8. Determine panel area for cooling.
9. Designate room design dry-bulb temperature for heating.
10. Calculate room heat loss.
11. Select mean water temperature for heating.
12. Determine surface temperatures of unheated surfaces. Refer to Figures 5 and 6 to find surface temperatures of exterior walls and exposed floors and ceilings. Interior walls are assumed to have surface temperatures equal to the room air temperature.
13. Determine AUST of surfaces in room.
14. Determine surface temperature of heated radiant surface. Refer to Figure 7 if the AUST does not greatly differ from room air temperatures. Refer to Figures 1 and 3 otherwise.
15. Determine panel area for heating. Refer to Figure 7 if the AUST does not vary greatly from room air temperature. Re-

fer to manufacturer's data for panel surface temperatures higher than those given in Figure 7.

16. Design the panel arrangement.

17. Thermal comfort requirements should be checked in the following steps (see Chapter 8 of the 1989 ASHRAE *Handbook—Fundamentals* and NRB 1981).

    a. Determine occupant's clothing value (clo value) and metabolic rate (MET) (see Tables 1D and 4A, Chapter 8 of the 1989 ASHRAE *Handbook—Fundamentals*).

    b. Determine the optimum operative temperature at the coldest point in the room (see Figure 15, Chapter 8 of the 1989 ASHRAE *Handbook—Fundamentals* or NRB Figure 2.3.22 for other values).

    c. Determine the mean radiant temperature (MRT) at the coldest point in the room (see Fanger 1972).

    d. From the definition of operative temperature, establish the optimum room design temperature at the coldest point in the room. If the optimum room design temperature varies greatly from the designated room design temperature, designate a new temperature.

    e. Determine the MRT at the hottest point in the room.

    f. Calculate the operative temperature at the hottest point in the room.

    g. Compare the operative temperatures at the hottest and coldest points in the room. For light activity and normal clothing, the acceptable operative temperature range is 68 to 75 °F (see NRB 1981 for other ranges). If the range is not acceptable, the heating system must be modified.

    h. Calculate radiant temperature asymmetry (NRB 1981). Acceptable ranges are < 18 °F for windows and < 9 °F for warm ceilings.

18. Determine water flow rate and pressure drop.

The application, design, and installation of panel systems have certain requirements and techniques such as the following:

1. As with any hydronic system, look closely at the piping system design. Piping should be designed to ensure that water of the proper temperature and in sufficient quantity is available to every grid or coil at all times. Reverse-return systems should be considered to minimize balancing problems.

2. Individual panels can be connected for parallel flow using headers, or for sinuous or serpentine flow. To avoid flow irregularities within a header-type grid, the water channel or lateral length should be greater than the header length. If the laterals in a header grid are forced to run in a short direction, this problem can be solved by using a combination series-parallel arrangement. Serpentine flow will ensure a more even panel surface temperature throughout the heating or cooling zone.

3. Noises from entrained air, high-velocity, or high-pressure drop devices or from pump and pipe vibrations must be avoided. Water velocities should be high enough to prevent separated air from accumulating and causing air binding. Where possible, avoid automatic air venting devices over ceilings of occupied spaces.

4. Design piping systems to accept thermal expansion adequately. Do not allow forces from piping expansion to be transmitted to panels. Thermal expansion of the ceiling panels must be considered.

5. In circulating water systems, plastic, steel, and copper pipe or tube are used widely in ceiling, wall, or floor panel construction. Where coils are embedded in concrete or plaster, no threaded joints should be used for either pipe coils or mains. Steel pipe should be the all-welded type. Copper tubing should be soft-drawn coils. Fittings and connections should be minimized. Changes in direction should be made by bending. Solder-joint fittings for copper tube should be used with a medium temperature solder of 95% tin, 5% antimony, or capillary brazing alloys. All piping should be subjected to a hydrostatic test of at least three times the working pressure. Maintain adequate pressure in embedded piping while pouring concrete.

6. Placing the thermostat on a side wall where it can see the outside wall and the warm panel should be considered. The normal thermostat cover reacts to the warm panel, and the radiant effect of the panel on the cover tends to alter the control point so that the thermostat controls 2 to 3 °F lower when the outdoor temperature is a minimum and the panel temperature is a maximum. Experience indicates that radiantly heated rooms are more comfortable under these conditions than when the thermostat is located on a back wall.

7. If throttling valve control is used, either the end of the main should have a fixed bypass, or the last one or two rooms on the mains should have a bypass valve to maintain water flow in the main. Thus, when a throttling valve modulates, there will be a rapid response.

8. When selecting heating design temperatures for a ceiling panel surface, the design parameters are:

    a. Excessively high temperatures over the occupied zone will cause the occupant to experience a "hot head effect."

    b. Temperatures that are too low can result in an oversized, uneconomical panel and a feeling of coolness at the outside wall.

    c. Locate ceiling panels adjacent to perimeter walls and/or areas of maximum load.

    d. With normal ceiling heights of 8 to 9 ft, panels less than 3 ft wide at the outside wall can be designed for 235 °F surface temperature. If panels extend beyond 3 ft into the room, the panel surface temperature should be limited the values as given in Figure 16. The surface temperature of concrete or plaster panels is limited by construction.

9. Floor panels are limited to surface temperatures of less than 85 °F for comfort reasons.

10. When the panel chilled water system is started, the circulating water temperature should be maintained at room temperature until the air system is completely balanced, the dehumidification equipment is operating properly, and building humidity is at design value.

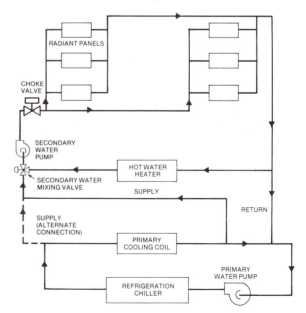

Fig. 11   Primary/Secondary Water Distribution System with Mixing Control

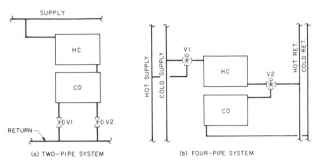

Fig. 12 Split Panel Piping Arrangement for Two-Pipe and Four-Pipe Systems

11. When the panel area for cooling is greater than the area required for heating, a two-panel arrangement (Figure 12) can be used. Panel HC (heating and cooling) is supplied with hot or chilled water year-round. When chilled water is used, the controls function activate panel CO (cooling only), and both panels are used for cooling.

12. To prevent condensation on the room side of cooling panels, the panel water supply temperature should be maintained at least 1°F above the room design dew-point temperature. This minimum difference is recommended to allow for the normal drift of temperature controls for the water and air systems, and also to provide a factor of safety for temporary increase in space humidity.

13. Selection of summer design room dew point below 50°F generally is not economical.

14. The most frequently applied method of dehumidification utilizes cooling coils. If the main cooling coil is six rows or more, the dew point of the air leaving will approach the temperature of the water leaving. The cooling water leaving the dehumidifier can then be used for the panel water circuit.

15. Several chemical dehumidification methods are available to control latent and sensible loads separately. In one application, cooling tower water is used to remove heat from the chemical drying process, and additional sensible cooling is necessary to cool the dehumidified air to the required system supply air temperature.

16. When chemical dehumidification is used, hygroscopic chemical-type dew-point controllers are required at the central apparatus and at various zones to monitor dehumidification.

17. When cooled ceiling panels are used with a variable air volume (VAV) system, the air supply rate should be near maximum volume to assure adequate dehumidification before the cooling ceiling panels are activated.

Other factors to consider when using panel systems are:

1. Evaluate the panel system to take full advantage in optimizing the physical building design.

2. Select recessed lighting fixtures, air diffusers, hung ceilings, and other ceiling devices to provide the maximum ceiling area possible for use as radiant panels.

3. The air-side design must be able to maintain humidity levels at or below design conditions at all times to eliminate any possibility of condensation on the panels. This becomes more critical if space dry- and wet-bulb temperatures are allowed to drift as an energy conservation measure, or if duty cycling of the fans is used.

4. Do not place cooling panels in or adjacent to high humidity areas.

5. Anticipate thermal expansion of the ceiling and other devices in or adjacent to the ceiling.

6. Design operable windows to discourage unauthorized opening.

HYDRONIC METAL CEILING PANELS

Metal ceiling panels can be integrated into a system that heats and cools. In such a system, a source of dehumidified ventilation air is required in summer, so the system is classed as an air-water system. Also, various amounts of forced air are supplied year-round. When metal panels are applied for heating only, a ventilation system may be required, depending on local codes.

The ceiling panel systems are an outgrowth of the perforated metal, suspended acoustical ceilings. These radiant ceiling systems are usually designed into buildings where the suspended acoustical ceiling can be combined with panel heating and cooling. The panels can be designed as small units to fit the building module, which provides extensive flexibility for zoning and control; or the panels can be arranged as large continuous areas for maximum economy. Some ceiling installations require active panels to cover only a portion of the room and compatible matching acoustical panels for the remaining ceiling area.

Three types of metal ceiling systems are available. The first consists of light aluminum panels, usually 12 in. by 24 in., attached in the field to 0.5-in. galvanized pipe coils. Figure 13 illustrates a metal ceiling panel system that uses 0.5-in. pipe laterals, on either 6-, 12-, or 24-in. centers, hydraulically connected in a sinuous or parallel flow welded system. Aluminum ceiling panels are clipped to these pipe laterals and act as a heating panel when warm water is flowing, or as a cooling panel when chilled water is flowing.

The second type of panel consists of a copper coil metallurgically bonded to the aluminum face sheet to form a modular panel. Modular panels are available in sizes up to about 36 in. by 60 in. and are held in position by various types of ceiling suspension systems, most typically, a standard suspended T-bar 24 in. by 48 in. exposed grid system. Figure 14 illustrates a metal panel ceiling system using copper tubing, bonded to an aluminum panel.

Metal ceiling panels can be perforated so that the ceiling becomes sound absorbent when acoustical material is installed on the back of the panels. The acoustical blanket is also required for thermal reasons, so that the reverse loss or upward flow of heat from the metal ceiling panels is minimized.

The third type of panel is an aluminum extrusion face sheet with a copper tube mechanically fastened into a channel housing on the back of the face sheet. Extruded panels can be manufactured in almost any shape and size. Extruded aluminum panels are often used as long, narrow panels at the outside wall and are independent of the ceiling system. Panels 15 or 20 in. wide usually satisfy the heating requirements of a typical office building. Lengths up to 20 ft are available. Figure 15 illustrates metal panels using a copper tube pressed into an aluminum extrusion.

Performance data for extruded aluminum panels vary with extrusion surface configuration, copper tube/aluminum contact, and test procedures used. Hydronic ceiling panels have a low ther-

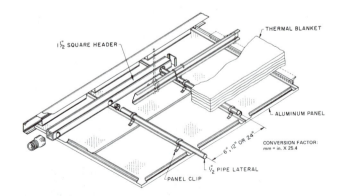

Fig. 13 Metal Ceiling Panels Attached to Pipe Laterals

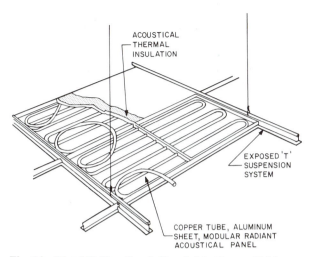

Fig. 14   Metal Ceiling Panels Bonded to Copper Tubing

mal resistance and respond quickly to changes in space conditions. Table 1 shows thermal resistance values for various ceiling constructions.

Metal radiant ceiling panels can be used with any of the all-air cooling systems described in Chapter 2. Chapters 23 through 25 of the 1989 ASHRAE *Handbook—Fundamentals* describe how to calculate heating loads. Double glazing and heavy insulation in outside walls has reduced transmission heat losses. As a result, infiltration and reheat have become of greater concern. Additional design considerations are as follows:

1. Perimeter radiant heating panels, not extending more than 3 ft into the room, may operate at higher temperatures, as described under "Design Considerations" in the Heating and Cooling Systems section.
2. Hydronic panels operate efficiently at low temperature and are suitable for condenser water heat reclaim systems.
3. Locate ceiling panels adjacent to the outside wall and as close as possible to the areas of maximum load. The panel area

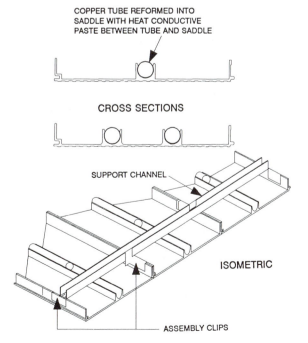

Fig. 15   Extruded Aluminum Panels with Integral Copper Tube

within 3 ft of the outside wall should have a heating capacity equal to or greater than 50% of the wall transmission load.
4. Ceiling system designs based on passing return air through perforated modular panels into the plenum space above the ceiling are not recommended, because much of the panel heat transfer is lost to the return air system.
5. When selecting heating design temperatures for a ceiling panel surface or mean water temperature, the design parameters are:

   a. Excessively high temperatures over the occupied zone will cause the occupant to experience a "hot head effect."
   b. Temperatures that are too low can result in an oversized, uneconomical panel and a feeling of coolness at the outside wall.
   c. Give the technique in item 3 priority.
   d. With normal ceiling heights of 8 to 9 ft, panels less than 3 ft wide at the outside wall can be designed for 235 °F surface temperature. If panels extend beyond 3 ft into the room, the panel surface temperature should be limited to the values as given in Figure 16.
6. Allow sufficient space above the ceiling for installation and connection of the piping that forms the radiant panel ceiling.

Metal radiant acoustic panels provide heating, cooling, sound absorption, insulation, and unrestricted access to the plenum space. They are easily maintained, can be repainted to look new, and have a life expectancy in excess of 30 years. The system is quiet, comfortable, draft-free, easy to control, and responds quickly. The system is a basic air-and-water system. First costs are competitive with other systems, and a life cycle cost analysis often shows that the long life of the equipment makes it the lowest cost in the long run. The system has been used in hospitals, schools, office buildings, colleges, airports, and exposition facilities.

Metal radiant panels can also be integrated into the ceiling design to provide a narrow band of radiant heating around the perimeter of the building. The radiant system offers advantages over baseboard or overhead air in appearance, comfort, operating efficiency and cost, maintenance, and product life.

DISTRIBUTION AND LAYOUT

Chapter 3 and chapters covering hydronic systems apply to radiant panels. Layout and design of metal radiant ceiling panels for heating and cooling begins early in the job. The type of ceiling chosen influences the radiant design and conversely, thermal considerations may dictate what ceiling type to be used. Heating panels should be located adjacent to the outside wall. Cooling panels may be positioned to suit other elements in the ceiling. In applications with normal ceiling heights, heating panels that

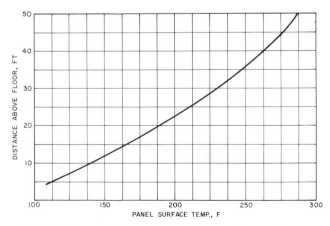

Fig. 16   Suggested Design Ceiling Surface Temperatures at Various Ceiling Heights

Piping Embedded in Ceilings

exceed 160°F should not be located over the occupied area. In hospital applications, valves should be located in the corridor outside patient rooms.

One of the following types of construction is generally used.

1. Pipe or tube is embedded in the lower portion of a concrete slab, generally within 1 in. of its lower surface. If plaster is to be applied to the concrete, the piping may be placed directly on the wood forms. If the slab is to be used without plaster finish, the piping should be installed not less than 0.75 in. above the undersurface of the slab. Figure 17 shows this method of construction. The minimum coverage must comply with local building code requirements.
2. Pipe or tube is embedded in a metal lath and plaster ceiling. If the lath is suspended to form a hung ceiling, the lath and heating coils are securely wired to the supporting members so that the lath is below, but in good contact with, the coils. Plaster is then applied to the metal lath, carefully embedding the coil as shown in Figure 18.
3. Smaller diameter copper or plastic tube is attached to the underside of wire lath or gypsum lath. Plaster is then applied to the lath to embed the tube, as shown in Figure 19.
4. Other forms of ceiling construction are composition board, wood paneling, etc., with warm water piping, tube, or channels built into the panel sections.

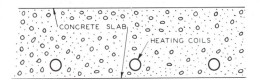

Fig. 17   Coils in Structural Concrete Slab

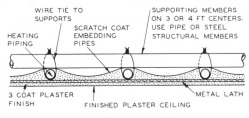

Fig. 18   Coils in Plaster above Lath

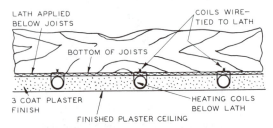

Fig. 19   Coils in Plaster below Lath

Coils are usually the sinuous type, although some header or grid-type coils have been used in ceilings. Coils may be plastic, ferrous, or nonferrous pipe or tube, with coil pipes spaced from 4.5 to 9 in. on centers, depending on the required output, pipe or tube size, and other factors.

Where plastering is applied to pipe coils, a standard three-coat gypsum plastering specification is followed, with a minimum of 0.38 in. of cover below the tubes when they are installed below the lath. Generally, the surface temperature of plaster panels should not exceed 120°F. This can be accomplished by limiting the water temperature in the pipes or tubes in contact with the plaster to a maximum temperature of 140°F. Insulation should be placed above the coils to reduce *reverse loss*, the difference between heat supplied to the coil and net useful output to the heated room.

To protect the plaster installation and to ensure proper air drying, heat must not be applied to the panels for two weeks after all plastering work has been completed. When the system is started for the first time, the water supplied to the panels should not be higher than 20°F above the prevailing room temperature at that time and not in excess of 90°F. Water should be circulated at this temperature for about two days, then increased at a rate of about 5°F per day to 140°F.

During the air drying and preliminary warm-up periods, there should be adequate ventilation to carry moisture from the panels. No paint or paper should be applied to the panels before these periods have been completed or while the panels are being operated. After paint and paper have been applied, an additional shorter warm-up period, similar to first-time starting, is also recommended.

Hydronic Wall Panels

Although embedded piping in walls is not as universally used as floor and ceiling panels, they can be constructed by any of the methods outlined for ceilings.

Hydronic Floor Panels

Interest has developed in radiant floor heating with the introduction of polybutylene tubing and new design techniques. The systems are energy-efficient and use low water temperatures available from solar collector systems.

Embedded Piping in Concrete Slab

Plastic, ferrous, and nonferrous pipe and tube are used in floor slabs that rest on grade. The coils are constructed as sinuous-continuous pipe coils or arranged as header coils with the pipes spaced from 6 to 18 in. on centers. The coils are generally installed with 1.5 to 4 in. of cover above them. Insulation is recommended to reduce the perimeter and reverse losses. Figure 20 shows the application of pipe coils in slabs resting on grade. Coils should be embedded completely and should not rest on an interface. Any supports used for positioning the heating coils should be nonabsorbent and inorganic. Reinforcing steel, angle iron, pieces of pipe

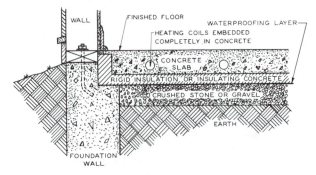

Fig. 20   Coils in Floor Slab on Grade

or stone, or concrete mounds can be used. No wood, brick, concrete block, or similar materials should support coils. A waterproofing layer is desirable to protect insulation and piping.

Where coils are embedded in structural load-supporting slabs above grade, construction codes may affect their position. Otherwise, the coil piping is installed as described for slabs resting on grade.

The warm-up and start-up period for concrete panels are similar to those outlined for plaster panels.

Embedded systems may fail sometime during their life. Adequate valves and properly labeled drawings will help to isolate the point of failure.

Suspended Floor Piping

Piping may be applied on or under suspended wood floors using several methods of construction. Piping may be attached to the surface of the floor and be embedded in a layer of concrete or gypsum, mounted in or below the subfloor, or attached directly to the underside of the subfloor using metal panels to improve heat transfer from the piping. Whichever method is used for optimum floor output and comfort, it is important that the heat be evenly distributed throughout the floor. Pipes are generally spaced on 4- to 12-in. spacing. Wide spacing under tile or bare floors can cause uneven surface temperatures.

Figure 21 illustrates construction with piping embedded in concrete or gypsum. The thickness of the embedding material is generally 1 to 2 in. when applied to a wood subfloor. Gypsum products specifically designed for floor heating can generally be installed 1 to 1.5 in. thick because they are more flexible and crack resistant than concrete. When concrete is used, it should be of structural quality to reduce cracking due to movement of the wood frame or shrinkage. Embedding material must provide a hard, flat, smooth surface that can accommodate a variety of floor coverings.

As illustrated in Figure 22, tubing may also be installed in the subfloor. The tubing is installed on top of the rafters between the subflooring members. Heat transfer can be improved by the addition of metal heat transfer plates which spread the heat beneath the finished flooring. This construction is illustrated in the figure.

A third construction option is to attach the tube to the underside of the subfloor with or without metal heat transfer plates. The construction is illustrated in Figure 23.

Transfer from the hot-water tube to the surface of the floor is the important consideration in all cases. The floor surface temperature affects the actual heat transfer to the space. Any hin-

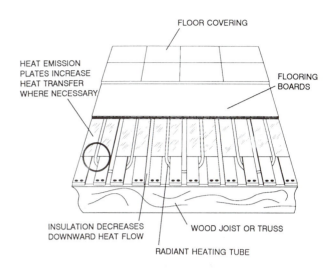

Fig. 22   Pipe in Subfloor

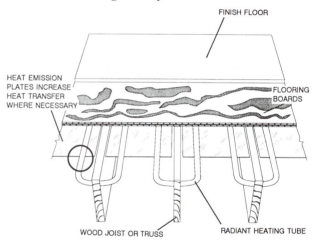

Fig. 23   Pipe under Subfloor

drance between the heated water tube and the floor surface will affect the efficiency of the system. As pointed out in the Panel Thermal Resistance section, the method that most effectively transfers and spreads the heat evenly through the subfloor with the least resistance produces the best results.

ELECTRICALLY HEATED SYSTEMS

Several different forms of electric resistance units are available for heating interior room surfaces. These include: (1) electric heating cables that may be embedded in concrete or plaster or laminated in drywall ceiling construction; (2) prefabricated electric heating panels to be attached to room surfaces; and (3) electrically heated fabrics or other materials for application to, or incorporation into, finished room surfaces.

Electrically Heated Ceilings

Prefabricated electric ceiling panels. A variety of prefabricated electric heating panels are available for either supplemental or full-room heating. These panels are available in sizes from 2 ft by 4 ft to 6 ft by 12 ft. They are constructed from a variety of materials such as gypsum board, glass, steel fiberglass, or vinyl. Different panels have rated inputs varying from 10 to 95 W/ft^2 for 120, 208, 240, and 277 V service. Maximum operating temperatures vary from about 100 to about 300 °F, depending on watt density. National

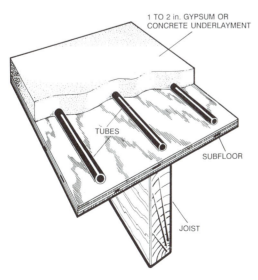

Fig. 21   Embedded Pipe

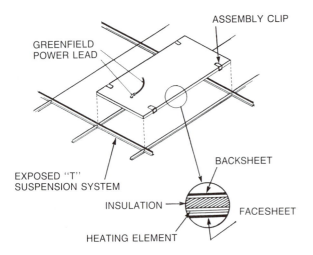

Fig. 24   Electric Heating Panel

and local codes should be followed when placing partitions, lights, and air grilles adjacent to or near electric panels.

Panel heating elements may be embedded conductors, laminated conductive coatings, or printed circuits. Nonheating leads are connected and furnished as part of the panel. Some panels can be cut to fit available space; others must be installed as received. Panels may be either flush or surface mounted and, in some cases, are finished as part of the ceiling. Rigid panels that are about 1 in. thick and weigh from 6 lb to about 25 lb for a 2 ft by 4 ft panel are usually made for lay-in ceilings (see Figure 24). Panels may also be (1) surface-mounted on gypsum board and wood ceilings, or

(2) recessed between ceiling joists. Panels range in size from 4 ft wide to 8 ft long. The maximum output is 95 W/ft^2.

Electrical cables embedded in ceilings. Electric heating cables for embedded or laminated ceiling panels are factory-assembled units furnished in standard lengths of 75 to 1800 ft. These cable lengths cannot be altered in the field. The cable assemblies are normally rated at 2.75 W per linear foot and are supplied in capacities from 200 to 5000 W in roughly 200-W increments. Standard cable assemblies are available for 120, 208, and 240 V. Each cable unit is supplied with 7-ft nonheating leads for connection at the thermostat or junction box.

Electric cables for panel heating have electrically insulated coverings resistant to medium temperature, water absorption, aging effects, and chemical action with plaster, cement, or ceiling lath material. This insulation is normally a polyvinylchloride (PVC) covering which may have a nylon jacket. The outside diameter of the insulation covering is usually about 0.12 in.

For plastered ceiling panels, the heating cable may be stapled to gypsum board, plaster lath, or similar fire-resistant materials with rust-resistant staples (Figure 25). With metal lath or other conducting surfaces, a coat of plaster (brown or scratch coat) is applied to completely cover the metal lath or conducting surface before the cable is attached. After fastening on the lath and applying the first plaster coat, each cable is tested for continuity of circuit and for insulation resistance of at least 100,000 ohms measured to ground.

The entire ceiling surface is finished with a covering of thermally noninsulating sand plaster about 0.50 to 0.75 in. thick or other approved noninsulating material applied according to manufacturer's specifications. The plaster is applied parallel to the heating cable rather than across the runs. While new plaster is drying, the system should not be energized, and the range and

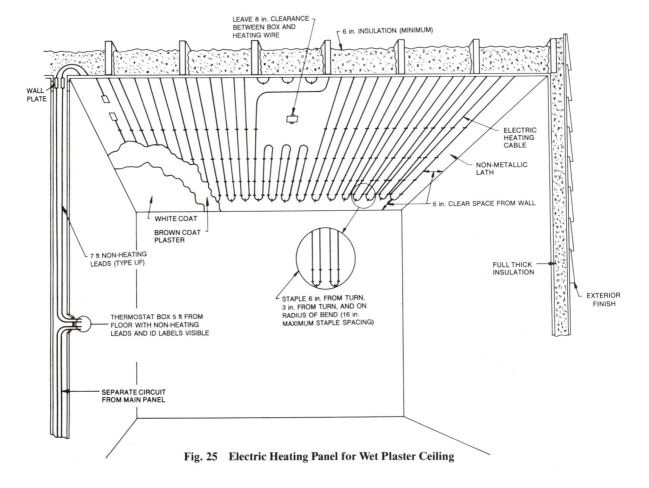

Fig. 25   Electric Heating Panel for Wet Plaster Ceiling

rate of temperature change should be kept low by other heat sources or by ventilation until the plaster is thoroughly cured. Vermiculite or other insulating plaster causes cables to overheat and is contrary to code provisions.

For laminated drywall ceiling panels, the heating cable is placed between two layers of gypsum board, plasterboard, or other thermally noninsulating fire-resistant ceiling lath. The cable is stapled directly to the first (or upper) lath, and the two layers are held apart by the thickness of the heating cable. It is essential that the space between the two layers of lath *completely* filled with a noninsulating plaster or similar material. This fill holds the cable firmly in place and improves heat transfer between the cable and the finished ceiling. Failure to fill the space completely between the two layers of plasterboard may allow the cable to overheat in the resulting voids and cause cable failure. The plaster fill should be applied according to manufacturer's specifications.

Electric heating cables are ordinarily installed with a 6-in. nonheating border around the periphery of the ceiling. An 8-in. clearance must be provided between heating cables and the edges of the outlet or junction boxes used for surface-mounted lighting fixtures. A 2-in. clearance must be provided from recessed lighting fixtures, trim, and ventilating or other openings in the ceiling.

Heating cables or panels must be installed only in ceiling areas that are not covered by partitions, cabinets, or other obstructions. However, it is permissible for a single run of isolated embedded cable to pass over a partition.

The *National Electric Code* requires that all general power and light wiring be run above the thermal insulation or at least 2 in. above the heated ceiling surface, or that the wiring be derated.

In drywall ceiling construction, the heating cable is always installed with the cable runs parallel to the joist. A 2.5-in. clearance between adjacent cable runs must be left centered under each joist for nailing. Cable runs that cross over the joist must be kept to a minimum. Where possible, these crossings should be in a straight line at one end of the room.

For cable having a watt density of 2.75 W/ft, the minimum permissible spacing is 1.5 in. between adjacent runs. Some manufacturers recommend a minimum spacing of 2 in. for drywall construction.

The spacing between adjacent runs of heating cable can be determined using Equation (12):

$$s = 12 A_n /C \qquad (12)$$

where

s = cable spacing, in.
A_n = net panel heated area, ft^2
C = length of cable, ft

Net panel area A_n in Equation (12) is the net ceiling area available after deducting the area covered by the nonheating border, lighting fixtures, cabinets, and other ceiling obstructions. For simplicity, Equation (12) contains a slight safety factor, and small lighting fixtures are usually ignored in determining net ceiling area.

The 2.5-in. clearance required under each joist for nailing in drywall applications occupies one-fourth of the ceiling area, if the joists are 16 in. on center. Therefore, for drywall construction, the net area A_n must be multiplied by 0.75. Many installations have a spacing of 1.5 in. for the first 2 ft from the cold wall. Remaining cable is then spread over the balance of the ceiling.

Electrically Heated Wall Panels

Cable embedded in walls similar to ceiling construction is occasionally found in Europe. Because of possible damage from nails driven for hanging pictures or from building alterations, most codes in the United States prohibit such panels. Some of the prefabricated panels described in the preceding section are also used for wall panel heating.

Electrically Heated Floors

Electric heating cable assemblies, such as those used for ceiling panels, are sometimes used for concrete floor heating systems. Since the possibility of cable damage during installation is greater for concrete floor slabs than for ceiling panels, these assemblies must be carefully installed. After the cable has been placed, all unnecessary traffic should be eliminated until the concrete covering has been placed and hardened.

Preformed mats are sometimes used for electric floor slab heating systems. These mats usually consist of PVC-insulated heating cable woven in, or attached to, metallic or glass fiber mesh. Such mats are available as prefabricated assemblies in many sizes from 2 to 100 ft^2 and with various watt densities ranging from 15 to 25 W/ft^2. When used with a thermally treated cavity beneath the floor, a heat storage system is provided, which may be controlled for off-peak heating.

Mineral-insulated (MI) heating cable is another effective method of slab heating. MI cable is a small-diameter, highly durable, flexible heating cable composed of solid electric-resistance heating wire or wires surrounded by tightly compressed magnesium oxide electrical insulation and enclosed by a metal sheath. MI cable is available in stock assemblies in a variety of standard voltages, watt densities, and lengths. A cable assembly consists of the specified length of heating cable, waterproof hot-cold junctions, 7-ft cold sections, UL-approved end fittings, and connection leads. Several standard MI cable constructions are available, such as single conductor, twin conductor, and double cable. Custom-designed MI heating cable assemblies can be ordered for specific installations.

Other outer-covering materials that are sometimes specified for electric floor heating cable include (1) silicone rubber, (2) lead, and (3) tetrafluoroethylene (Teflon).

For a given floor heating cable assembly, the required cable spacing is determined from Equation (12). In general, cable watt density and spacing should be such that floor panel watt density is not greater than 15 W/ft^2. Higher watt densities (up to 25 W/ft^2) are often specified for the 2-ft border next to cold walls. Check with the latest issue of the *National Electric Code* and other applicable codes to obtain information on maximum panel watt density and other required criteria and parameters.

Floor heating cable installation. When PVC-jacketed electric heating cable is used for floor heating, the concrete slab is laid in two pourings. The first pour should be at least 3 in. thick and, where practical, should be insulating concrete to reduce downward heat loss. For a proper bond between the layers, the finish slab should be placed within 24 h of the first pour, with a bonding grout applied. The finish layer should be at least 1.5 in. and not more than 2 in. thick. This top layer must not be insulating concrete. At least 1 in. of perimeter insulation should be installed as shown in Figure 26.

The cable is installed on top of the first pour of concrete not closer than 2 in. from adjoining walls and partitions. Methods of fastening the cable to the concrete include:

1. Staple the cable to wood nailing strips fixed in the surface of the rough slab. The predetermined cable spacing is maintained by daubs of cement, plaster of paris, or tape.
2. In light or uncured concrete, staple the cable directly to the slab using hand-operated or powered stapling machines.
3. Nail special anchor devices to the first slab to hold the cable in position while the top layer is being poured.

Preformed mats can be embedded in the concrete in a continuous pour. The mats are positioned in the area between expansion and/or construction joints and electrically connected to a junction box. The slab is poured to within 1.5 to 2 in. of the finished level. The surface is rough screeded and the mats placed in position. The final cap is applied immediately. Since the first pour

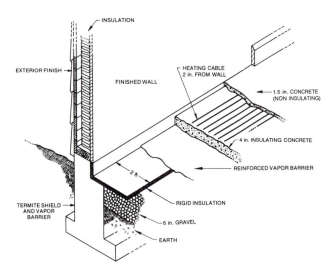

Fig. 26 Electric Heating Cable in Concrete Slab

has not set, there is no adhesion problem between the first and second pour, and a monolithic slab results. A variety of contours can be developed by using heater wire attached to glass fiber mats. Allow for circumvention of obstructions in the slab.

MI electric heating cable can be installed in concrete slab using either one or two pours. For single-pour applications, the cable is fastened to the top of the reinforcing steel before the pour is started. For two-layer applications, the cable is laid on top of the bottom structural slab and embedded in the finish layer. Proper spacing between adjacent cable runs is maintained by using prepunched copper spacer strips nailed to the lower slab.

Air-Heated Floors

Several methods have been devised to warm interior room surfaces by circulating heated air through passages in the floor. In some cases, the heated air is recirculated in a closed system. In others, all or a part of the air is passed through the room on its way back to the furnace to provide supplementary heating and ventilation. Figure 27 indicates one common type of construction. Compliance with applicable building codes is important.

CONTROLS

Automatic controls for panel heating may differ from those for convective heating because of the thermal inertial characteristics of the panel and the increase in the mean radiant temperature within the space under increasing loads. However, lightweight

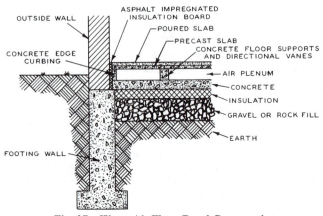

Fig. 27 Warm Air Floor Panel Construction

systems using thin metal panels or thin underlay with low thermal heat capacity may be successfully controlled with conventional control technology using indoor sensors. Many of the control principles for hot water heating systems described in Chapter 12 and Chapter 14 also apply to panel heating. Because radiant panels do not depend on air-side equipment to distribute energy, many control methods have been used successfully; however, a control interface between heating and cooling should be installed to prevent simultaneous heating and cooling.

High mass panels such as concrete radiant slabs require a control approach different from low mass panels. Because of thermal inertia, significant time is required to bring such massive panels from one operating point to another, say from vacation setback to standard operating conditions. This will result in long periods of discomfort from low temperature, then possibly periods of uncomfortable and wasteful overshoot. Careful economic analysis may reveal that a nighttime setback strategy is not warranted.

Once a slab is at operating conditions, the control strategy should endeavor to supply the slab with heat at the rate that heat is being lost from the space (MacCluer *et al.* 1989). For hydronic slabs with constant circulator flow rate, this equates to modulating the difference between the outgoing water and the returning water temperatures, accomplished via mixing valves, fuel modulation, or, for constant thermal power sources, via pulse-width modulation ("bang-bang" or "on-off" control). Slabs with embedded electric resistance cable can be controlled by pulse-width modulators such as the common round thermostat with anticipator or its solid state equivalent.

A related approach, outdoor reset control, has enjoyed wide acceptance. An outdoor reset control measures the outdoor air temperature, calculates the supply water temperature required for steady operation, and operates a mixing valve or boiler to achieve that supply water temperature. If the heating load of the controlled space is primarily a function of the outdoor air temperature, or indoor temperature measurement of the controlled space is impractical, then outdoor reset control alone is an acceptable control strategy. When other factors such as solar or internal gains are also significant, then indoor temperature feedback should be added to the outdoor reset.

In all radiant panel applications, precautions must be taken to prevent excessive temperatures. A manual boiler bypass or other means of reducing the water temperature may be necessary to prevent new panels from drying out too rapidly.

Cooling Controls

Controlling the panel water circuit temperature by mixing, heat exchange, or using the water leaving the dehumidifier is typical. Other considerations are listed in the General Design Conditions section. It is imperative to dry out the building space before starting the panel water system, particularly after extended down periods, such as weekends. Such delayed starting action can be controlled manually or by a device.

Panel cooling systems require the following basic areas of temperature control: (1) exterior zones, (2) areas under exposed roofs to compensate for transmission and solar loads, and (3) control of each typical interior zone to compensate for internal loads. For optimum results, each exterior corner zone and similarly loaded face zone should be treated as a separate subzone. Panel cooling systems may also be zoned to control temperature in individual exterior offices, particularly in applications where there is a high lighting load or for corner rooms with large glass areas on both walls.

The temperature control of the interior air and panel water supply should not be functions of the outdoor weather. The normal thermostat drift is usually adequate compensation for the slightly lower temperatures desirable during winter weather. This drift should be limited to result in a room temperature change of not more than 1.5 °F. Control of the interior zones is best accomplished by devices that reflect the actual presence of the internal load elements. Frequently, time clocks and current-sensing devices are used on lighting feeders.

Because air quantities are generally small, constant volume supply air systems should be used. With the apparatus arranged to supply air at an appropriate dew point at all times, comfortable indoor conditions can be maintained throughout the year with a panel cooling system. As with all systems, to prevent condensation on window surfaces, the supply air dew point should be reduced during extremely cold weather according to the type of glazing installed.

Electric Heating Slab Controls

For comfort heating applications, the surface temperature of a floor slab is held to a maximum of 80 to 85 °F. Therefore, when the slab is the primary heating system, thermostatic controls sensing air temperature should not be used to control temperature; instead, the heating system should be wired in series with a slab-sensing thermostat. The remote sensing thermostat in the slab acts as a limit switch to control maximum surface temperatures allowed on the slab. The ambient sensing thermostat controls the comfort level. For supplementary slab heating, as in kindergarten floors, a remote sensing thermostat in the slab is commonly used to tune in the desired comfort level. Indoor-outdoor thermostats are used to vary the floor temperature inversely with the outdoor temperature. If the heat loss of the building is calculated for 70 to 0 °F and the floor temperature range is held from 70 to 85 °F with a remote sensing thermostat, the ratio of outdoor temperature to slab temperature is 70:15, or approximately 5:1. This means that a 5 °F drop in outdoor temperature requires a 1 °F increase in the slab temperature. An ambient sensing thermostat is used to vary the ratio between outdoor and slab temperatures. A time clock is used to control each heating zone if off-peak slab heating is desirable.

REFERENCES

ASHAE. 1956. Thermal design of warm water ceiling panels. ASHAE *Transactions* 62:71.

ASHAE. 1957. Thermal design of warm water concrete floor panels. ASHAE *Transactions* 63:239.

ASHRAE. 1981. Thermal environmental conditions for human occupancy. ASHRAE *Standard* 55-1981.

Berglund, L., R. Rascati, and M.L. Markel. 1982. Radiant heating and control for comfort during transient conditions. ASHRAE *Transactions* (88):765-75.

Buckley, N.A. 1989. Application of radiant heating saves energy. ASHRAE *Journal* 31(9):17-26.

Fanger, P.O. 1972. Thermal comfort analysis and application in environmental engineering. McGraw Hill, Inc., New York.

Hogan, R.E., Jr., and B. Blackwell. 1986. Comparison of numerical model with ASHRAE designed procedure for warm-water concrete floor-heating panels. ASHRAE *Transactions* 92(1B):589-601.

Kalisperis, L.N. 1985. Design patterns for mean radiant temperature prediction. Department of Architectural Engineers, Pennsylvania State University, University Park, PA.

Kalisperis, L.N. and L.H. Summers. 1985. MRT33GRAPH—A CAD Program for the design evaluation of thermal comfort conditions. Tenth National Passive Solar Conference, Raleigh, NC.

MacCluer, C.R., M. Miklavcic, and Y. Chait. 1989. The temperature stability of a radiant slab-on-grade. ASHRAE *Transactions* 95(1): 1001-1009.

Min, T.C., *et al.* 1956. Natural convection and radiation in a panel hated room. ASHAE Research Report No. 1576. ASHAE *Transactions* 62:337.

NRB. 1981. Indoor climate. Technical Report No. 41, The Nordic Committee on Building Regulations, Stockholm, Sweden.

Parmelee, G.V. and R.G. Huebscher. 1947. Forced convection, heat transfer from flat surfaces. ASHVE *Transactions* 53:245.

Sartain, E.L. and W.S. Harris. 1956. Performance of covered hot water floor panels, Part I—Thermal characteristics. ASHAE *Transactions* 62:55.

Schutrum, L.F. and C.M. Humphreys. 1954. Effects of non-uniformity and furnishings on panel heating performance. ASHVE *Transactions* 60:121.

Schutrum, L.F., G.V. Parmelee, and C.M. Humphreys. 1953a. Heat exchangers in a ceiling panel heated room. ASHVE *Transactions* 59:197.

Schutrum, L.F., G.V. Parmelee, and C.M. Humphreys. 1953b. Heat exchangers in a floor panel heated room. ASHVE *Transactions* 59:495.

Schutrum, L.F. and J.D. Vouris. 1954. Effects of room size and non-uniformity of panel temperature on panel performance. ASHVE *Transactions* 60:455.

Steinman, M., L.N. Kalisperis, and L.H. Summers. 1989. The MRT-correction method—An improved method for radiant heat exchange. ASHRAE *Transactions* 96(1).

Walton, G.N. 1980. A new algorithm for radiant interchange in room loads calculations. ASHRAE *Transactions* 86(2).

Wilkes, G.B. and C.M.F. Peterson. 1938. Radiation and convection from surfaces in various positions. ASHVE *Transactions* 44:513.

COGENERATION SYSTEMS

COGENERATION is the sequential use of energy from a single primary source such as oil, coal, natural gas, or biomass fuels to produce two useful energy forms—heat and power. By capturing and applying heat that would otherwise be rejected, cogeneration systems can operate at efficiencies greater than those achieved when heat and power are produced in separate or distinct processes.

Cogeneration can be a topping or bottoming cycle. If the fuel generates power first, and then the resulting thermal energy is recovered and productively used, it is called a topping cycle. If the power is generated last, from the thermal energy left over after the higher level thermal energy has been used to satisfy thermal loads, it is called a bottoming cycle.

Isolated cogeneration systems, whose electrical output is used on-site to satisfy all site power requirements, are referred to as total energy systems. A cogeneration system that is actively tied (paralleled) to the utility grid can, on a contractual basis or on a tariff basis, exchange power with the public utility. This lessens the need for redundant on-site generating capacity and allows operation at maximum thermal efficiency when satisfying the facility's thermal load requires more electric power than the facility needs.

System feasibility and design depend on the magnitude, duration, and coincidence of electrical and thermal loads, as well as the selection of prime movers and waste heat recovery. Integrating the design of the project's electrical and thermal requirements with the cogeneration plant is required for optimum economic benefit. The basic components of the cogeneration plant are: (1) prime movers, (2) generators, (3) waste heat recovery systems, (4) control systems, (5) electrical and thermal transmission and distribution systems, and (6) connections to building mechanical and electrical services.

This chapter covers prime movers, generators, heat recovery, and control systems. Thermal transmission and distribution systems are discussed in Chapter 11.

PRIME MOVERS

The basic component in a cogeneration system is the prime mover, which converts fuel or thermal energy to shaft energy. The conversion devices normally used are reciprocating internal combustion engines, combustion gas turbines, expansion turbines, and steam boiler-turbine combinations. Engine and turbine drives are discussed in Chapter 41.

Reciprocating engines are the most common type of prime mover used in cogeneration plants. These engines are available in sizes up to 27,000 brake horsepower (bhp) and use all types of liquid and gaseous fuels. Internal combustion engines that use the diesel (compression ignition) cycle can be fueled by a wide range of petroleum products, although No. 2 diesel oil is most commonly used. Diesel cycle engines can also be fired with gaseous fuel in combination with liquid fuel. Under these conditions, the liquid acts as the ignition agent and is called pilot oil.

Spark ignition Otto cycle engines are produced in sizes up to 18,000 bhp and can use natural gas, liquefied petroleum gas (LPG), and other gaseous and volatile liquid fuels. Engines usually operate in the range of 360 to 1200 rpm, with some up to 1800 rpm. The specific operating speed selected depends on the size of machine, characteristics of manufacturers, the generator, and the desired length of time between complete engine overhauls. Engines are usually selected to provide a minimum of 15,000 to 30,000 operating hours between minor overhauls, and up to 50,000 operating hours between major overhauls; the higher operating hours correspond to the larger lower speed engines.

The engine components include a starter system; fuel input system; fuel air mixing cycle; ignition system; combustion chamber; exhaust gas collection and removal system; lubrication system; and power transmission gear with pistons, connecting rods, crankshafts, and flywheel. Some engines also increase power output using turbochargers (combustion air compressors driven by exhaust gas turbines) that increase the amount of air delivered to the combustion chamber. The engines are generally from 20 to 40% efficient in converting fuel to shaft energy. Figure 1 is a typical fuel curve for a gas-

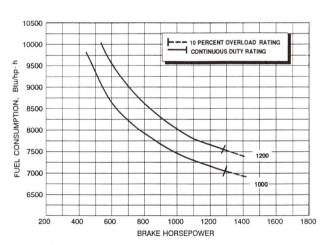

Fig. 1 Typical Fuel Consumption Curve for Gas-Fired Internal Combustion Engine

The preparation of this chapter is assigned to TC 9.5, Cogeneration Systems.

fired internal combustion engine, based on the lower heating value of the fuel. The lower heating value is determined by deducting the energy necessary to keep the water produced in the combustion process in the vapor state since this latent heat does no work on the piston.

Combustion gas turbines, although originally used for aircraft propulsion, have been developed for stationary use and are gaining acceptance as prime movers. Turbines are available in sizes from 50 to 173,000 bhp and can burn a wide range of liquid and gaseous fuels. Turbines, and some dual-fuel engines, are capable of shifting from one fuel to another without loss of service. Gas turbines consist of an air compressor section to boost combustion air pressure, a combination fuel-air mixing and combustion chamber (combustor), and an expansion turbine section that extracts energy from the combustion gases. In addition to these components, some turbines use regenerators that are turbine discharge gas-to-combustion inlet air-heat exchangers to increase machine efficiency. Most turbines are the single-shaft type, *i.e.*, air compressor and turbine on a common shaft. However, split-shaft machines that use one turbine stage on the same shaft as the compressor and a separate power turbine driving the output shaft are available. Present turbines rotate at speeds varying from 3600 to 60,000 rpm and often need speed reduction gearboxes to obtain shaft speeds suitable for generators. Turbines are completely rotary in motion and relatively vibration-free. This feature, coupled with their low mass and high power output, provides an advantage over reciprocating engines with regard to space and foundation requirements. However, in smaller sizes, the combustion gas turbine has a fuel-to-power efficiency of 12 to 20%, which is lower than efficiencies obtainable from reciprocating engines. Larger turbines can approach a fuel-to-power efficiency well above 30%. Expansion turbines are gas turbines that use compressed gases available from gas-producing processes or decompress natural gas from high-pressure pipelines.

There are two types of steam turbines: the condensing turbine, which operates at exhaust pressures below atmospheric and requires vacuum-type condensers and pumps; and the noncondensing or back-pressure type, which operates at exhaust pressures of atmospheric or above to achieve any desired pressure. If the entire exhaust steam from a base-loaded back-pressure turbine is fully used, the maximum efficiency of a steam turbine cogeneration cycle is achieved. However, if the facility's thermal/electric load requirements cannot absorb the fully loaded outputs of the turbine, the profile that is lower must be tracked, and the reduced power output or steam flow must be accepted. The annual efficiency can still be high if the machine operates at significant combined loads for substantial periods.

Straight steam condensing turbines offer no opportunity for topping cycles, but are not unusual in bottoming cycles since the waste steam from process can be most efficiently used by full-condensing turbines, when there is no use for low-pressure steam.

Either back-pressure or extraction turbines may be used as extraction turbines. The steam in an extraction turbine expands part of the way through the turbine until the pressure and temperature level required by the external thermal load are attained. The remaining steam continues through the low-pressure turbine stages; however, it is easier to adjust for noncoincident electrical and thermal loads. Because steam cycles operate at pressures exceeding those allowed by ASME and local codes for unattended operation, their use in cogeneration plants is limited to large systems where attendants are required for other reasons or the labor burden of operating personnel does not seriously affect overall economics. Figure 2 shows the performance of a 1500-kW condensing turbine,

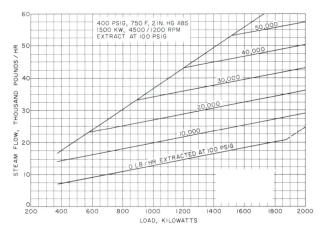

Fig. 2 Effect of Extraction Rate on Total Steam Requirement for Condensing Turbine

indicating the effect of various extraction rates on total steam requirements illustrated as follows:

At zero extraction and 1500 kW, 17,500 lb/h is required or a water rate of 11.67 lb/kWh. When 45,000 lb/h are extracted, only 4000 lb/h more (49,000 to 45,000) is required at the throttle to develop the same 1500 kW, chargeable to the generation of electric power. The portion of the heat rate chargeable to the power is represented by the sum of the enthalpy of this 2.7 lb/kWh at throttle conditions and the difference in enthalpy of the extracted portion of steam between its throttle and extraction conditions. In effect, as the extraction rate increases, the overall efficiency increases. However at "full" extraction, a significant flow of "cooling" steam must pass through the final turbine stages for condensing, and at this condition, a simple back-pressure turbine would be more efficient, if all of the exhaust steam could be used.

The selection of the type and size of a prime mover can be driven by the facility's thermal or electric load profile. The choice depends on: (1) the rate of heat/power output of the prime mover and its ability to match those of the facility loads; (2) the desire to parallel with the public utility or be totally independent of it; (3) the desire to sell excess power to the utility or not; or (4) the desire to maximize machine utilization by tracking the low thermal loads and base loading the prime movers at a power output level absorbed on-site or sold to the utility at all times.

No matter which basis is used to choose the prime mover, the degree of use of the available heat determines the overall system efficiency; this is a critical factor in the economic feasibility. Therefore, each prime mover's heat/power characteristics over its range of operating loads must be analyzed to make the best choice. Maximum efficiency may not be as important as maximizing the total economic value of the cogeneration system output.

Full-condensing steam turbines have the maximum plant shaft efficiency (power output as a percentage of input fuel to the boiler) ranging from 20 to 36%, but without any useful thermal output. Therefore, with overall plant efficiencies no better than their shaft efficiencies, they are unsuitable as topping cogeneration cycles.

At maximum extraction, the heat/power ratio of extraction turbines is relatively high, making it difficult to match facility loads, except for those with very high base thermal loads, if reasonable annual efficiencies are to be achieved. As extraction rates decrease, plant efficiency approaches that of a condensing turbine, but can never reach it. Thus, the 11.67 lb/kWh illustrated in Figure 2, at 1500 kW and zero extraction, represents a steam-to-electric efficiency of 28.6%, but a fuel-to-electric efficiency of 24.3%, with a boiler plant efficiency of 85%, developed as follows:

Isentropic Δh from 400 psig steam
$- 2$ in. Hg (abs.) condensing
$= 1022$ Btu/lb output

$$\text{Actual } \Delta h = \frac{3413 \text{ Btu/kWh}}{\text{Actual steam rate, lb/kWh}}$$
$$= \frac{3413}{11.67} = 292 \text{ Btu/lb}$$

Plant shaft efficiency $= 100 \times 292 \times 0.85/1022$
$= 24.3\%$ (fuel input to electric output)

GENERATORS

Generators are available in a wide range of sizes, speeds, types, electrical characteristics, and control options. Many engine manufacturers offer package arrangements in which the generator and prime mover come as a unit on the same skid.

Criteria influencing the selection of alternating current (ac) generators for cogeneration systems are: (1) machine efficiency in converting mechanical input into electrical output at various loads; (2) electrical load requirements, including frequency, power factor, voltage, and harmonic distortion; (3) phase balance capabilities; (4) equipment cost; and (5) motor-starting current requirements. Industrial generator manufacturers have detailed information available on their equipment.

Generator speed is a direct function of the number of poles and output frequency. For 60-Hz output, the speed can range from 3600 rpm for a two-pole machine to 900 rpm for an eight-pole machine. A wide latitude exists in matching generator speed to prime mover speed without reducing the efficiency of either unit. This range in speed and resultant frequency suggests that electrical equipment, whose operation is improved at a special frequency, might be accommodated.

Induction generators are similar to induction motors in construction and control requirements. The generator draws exciting current from the plant electrical system and produces power when driven above its synchronous speed. In the typical induction generator, full output occurs at 5% above synchronous speed. Special precautions are required to prevent the induction generator from being isolated from the electrical system while connected to power factor correction capacitors to prevent large transient overvoltage in the isolated circuit.

Generator efficiency is a nonlinear function of the load and is maximum usually at or near the rated load (see Figure 3). The rated load estimate should include a safety factor to cover such transient conditions as short-term peaking and equipment start-up. Industrial generators are designed to handle a steady-state overload of 20 to 25% for several hours of continuous operation. If sustained overloads are possible, the generator ventilation system must be able to relieve the temperature rise of the windings. The prime mover must be able to accommodate the overload. Proper phase balance is extremely important. Driving three-phase motors from the three-phase generator presents the best phase balance, assuming that the power factor requirements have been met.

Driving single-phase motors and building lighting or building distribution systems may cause an unbalanced distribution of the single-phase loads, leading to harmonic distortion, overheating, and electrical imbalance of the generator. Practically, maximum phase imbalance can be held within 5 to 10% by proper distribution system loading. Voltage is regulated by using static convertors or rotating dc generators to excite the

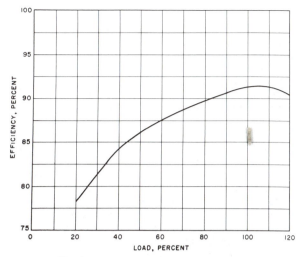

Fig. 3   Typical Generator Efficiency

alternator. Voltage regulation should be within 0.5% from full load to no load during static conditions. Good electronic three-phase voltage sensing is necessary to control the system response to load changes and the excitation of paralleled alternators to ensure reactive load division.

The system power factor is reflected to the generator and should be no less than 0.8 for generator efficiency. To fall within this limit, the planned electrical load may have to be adjusted so that the combined leading power factor substantially offsets the combined lagging power factor. Although more expensive, individual power factor correction with properly sized capacitors is preferred to total power factor correction on the bus. On-site generators can correct some power factor problems, and for those cogeneration systems interconnected to the grid, they improve the power factor seen by the grid.

Generation systems transfer considerable heat to the equipment space. It is necessary to remove this heat to provide acceptable working conditions and to protect the electrical systems from overloading because of high ambient conditions. Heat can be removed by outdoor air ventilation systems that include dampers and fans and thermostatic controls regulated to prevent overheating or excessively low temperatures in extreme weather.

GENERAL HEAT RECOVERY

Cogeneration provides an opportunity to use the fuel energy that the prime mover does not convert into shaft energy. If the heat cannot be used effectively, the plant efficiency is limited to that of the prime mover. However, if the site heat energy requirements can be met effectively by the normally wasted heat at the level it is available from the prime mover, this salvaged heat will reduce the normal fuel requirements of the site and increase overall plant efficiency.

The prime mover furnishes two kinds of energy: (1) mechanical energy from the shaft, and (2) unused heat energy that remains after the fuel or steam has acted on the shaft. Shaft loads (generators, centrifugal chillers, compressors, and process equipment) require a given amount of rotating mechanical energy. Once the prime mover is selected to provide the required shaft output, it has a fixed relationship to heat availability and system efficiency, depending on the prime mover fuels versus heat balance curves. The ability to use the prime mover waste heat determines overall system efficiency and is one of the critical factors in economic feasibility.

Steam turbine exhaust or extracted steam, reduced in both pressure and temperature from throttle conditions as it generates the prime mover shaft power, may be fed to heat exchangers, absorption chillers, steam turbine-driven centrifugal chillers; or pumps, or other equipment. Its value as recovered energy is the same as that of the fuel otherwise needed to generate the same steam flow at the condition leaving the prime mover in an independent boiler plant cycle.

In the gas turbine cycle, the average fuel-to-electrical efficiency is approximately 12 to 30%, with the remainder of the fuel energy discharged in the exhaust, through radiation, or through internal coolants in large turbines. A minimum stack exhaust temperature of about 300°F is required to prevent condensation. Because of the lower fuel-to-power efficiency, the quantity of heat recoverable per unit of power is greater for the gas turbine than for a reciprocating engine. This heat is generally available at the higher temperature of the exhaust gas. The net result is an overall first law thermal efficiency of approximately 60% and higher. Since gas turbine exhaust contains a large percentage of excess air, it is possible to install afterburners or boost burners in the exhaust, thereby creating a supplementary boiler system. This system can provide additional steam or level out the steam production during reduced turbine loads. Boost burning can increase cycle efficiency to a maximum of 93%. Finally, absorption chillers capable of direct operation off of turbine exhausts are available, offering another design option with coefficients of performance of 1.14 or more.

The conventional method of controlling steam or hot water production in a waste heat recovery system is to bypass a portion of the exhaust gases around the boiler tubes and out of the exhaust stack through a gas bypass valve assembly (Figure 4).

In all reciprocating internal combustion engines, except small air-cooled units, heat can be reclaimed from the lubricating system, jacket cooling system, and the exhaust. These engines require extensive cooling of the machine to remove excessive heat not conducted into the power train during combustion and the heat resulting from friction. Coolant fluids and lubricating oil are circulated to remove this engine heat. Some engines are also constructed to permit the coolant to reach temperatures of up to 250°F at above atmospheric conditions, for flashing into low-pressure steam (15 psig) after the coolant has left the engine jacket (ebullient cooling).

Waste heat in the form of hot water or low-pressure steam is recovered from the engine jacket manifolds and exhaust, as shown in Figures 5 through 10. Additional heat can be recovered from the lubrication system. Provisions similar to those used with gas turbines are necessary if supplemental heat is required, except that boost firing of an engine exhaust is rarely done because it contains insufficient oxygen. If electric elements are used for supplemental heat, the additional electrical load is reflected back to the prime mover, which reacts

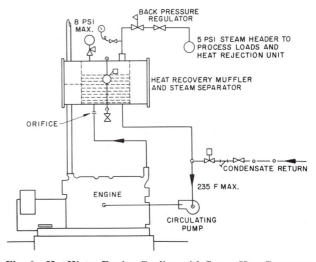

Fig. 5   Hot Water Heat Recovery System

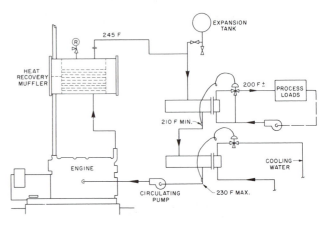

Fig. 6   Hot Water Engine Cooling with Steam Heat Recovery (Forced Circulation)

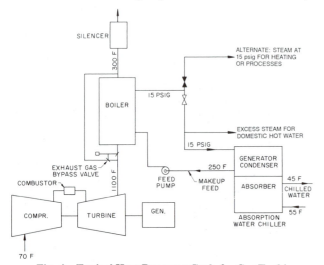

Fig. 4   Typical Heat Recovery Cycle for Gas Turbine

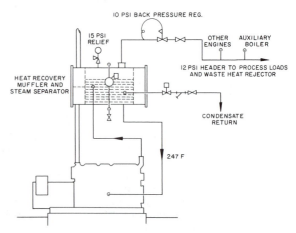

Fig. 7   Engine Cooling with Gravity Circulation and Steam Heat Recovery

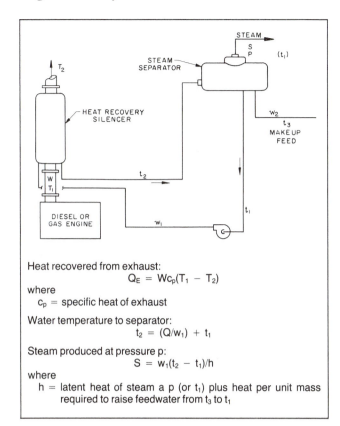

Heat recovered from exhaust:
$$Q_E = Wc_p(T_1 - T_2)$$
where
c_p = specific heat of exhaust

Water temperature to separator:
$$t_2 = (Q/w_1) + t_1$$

Steam produced at pressure p:
$$S = w_1(t_2 - t_1)/h$$
where
h = latent heat of steam a p (or t_1) plus heat per unit mass required to raise feedwater from t_3 to t_1

Fig. 8   Exhaust Heat Recovery System Using Separate Steam Separator

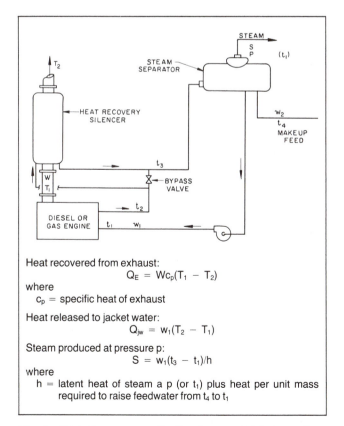

Heat recovered from exhaust:
$$Q_E = Wc_p(T_1 - T_2)$$
where
c_p = specific heat of exhaust

Heat released to jacket water:
$$Q_{jw} = w_1(T_2 - T_1)$$

Steam produced at pressure p:
$$S = w_1(t_3 - t_1)/h$$
where
h = latent heat of steam a p (or t_1) plus heat per unit mass required to raise feedwater from t_4 to t_1

Fig. 9   High-Temperature Cooling System with Exhaust Heat Recovery Using Separate Steam Separator

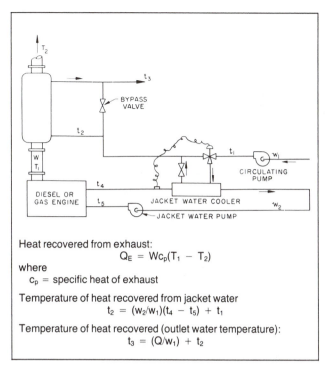

Heat recovered from exhaust:
$$Q_E = Wc_p(T_1 - T_2)$$
where
c_p = specific heat of exhaust

Temperature of heat recovered from jacket water
$$t_2 = (w_2/w_1)(t_4 - t_5) + t_1$$

Temperature of heat recovered (outlet water temperature):
$$t_3 = (Q/w_1) + t_2$$

Fig. 10   Combined Exhaust and Jacket Water Heat Recovery System

accordingly by making additional waste heat, creating a feedback effect, and stabilizing system operation.

The approximate distribution of input fuel energy under selective control of the thermal demand for an engine operating at rated load is as follows:

Shaft power	33%
Convection and radiation	7%
Rejected in jacket water	30%
Rejected in exhaust	30%

These amounts vary with engine load and design.

A four-cycle engine heat balance for naturally aspirated (Figures 11 and 12) and turbocharged gas engines (Figure 13) show typical heat distribution. The exhaust gas temperature for these engines is about 1200°F at full load and 1000°F at 60% load.

Two-cycle and lower speed (900 rpm and below) engines operate at lower exhaust gas temperatures, particularly at light loads, because the scavenger air volumes remain high through the entire range of capacity. High-volume, lower temperature exhaust gas offers less efficient exhaust gas heat recovery possibilities. The exhaust gas temperature is approximately 700°F at full load and drops below 500°F at low loads. Low-speed engines generally cannot be ebulliently cooled because their cooling systems often operate at 170°F and below.

Cogeneration systems paralleled with the utility grid can operate at peak efficiency under conditions such as the following: (1) if the electric generator can be sized to meet the valley of the thermal load profile, and operates in a "base" electric load fashion (100% full load) at all times, purchasing the balance of the site's electric needs from the utility; (2) if the electric generators are sized for 100% of the sites' electric demand and the heat recovered can be fully utilized at that condition, with thermal demands in excess of the recovered heat provided by supplementary means, and excess power sold to the utility.

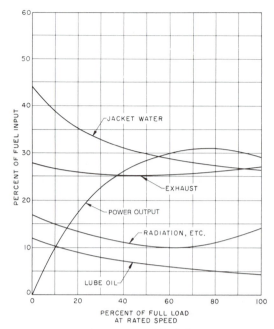

Fig. 11   Heat Balance for Typical Naturally Aspirated Engine with Hot Exhaust Manifold

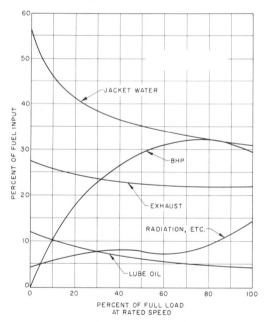

Fig. 12   Heat Balance for Typical Naturally Aspirated Engine with Water-Cooled Exhaust Manifold

PACKAGED SYSTEMS

Packaged cogeneration systems are available in sizes from 6 to 650 kW and larger. Some of these units are designed for specific applications. General-type designs are available with short lead times.

LUBRICATING SYSTEM HEAT RECOVERY

All engine designs use the lubricating system to remove some heat from the machine. Some configurations only cool the piston skirt with oil; other designs remove more of the

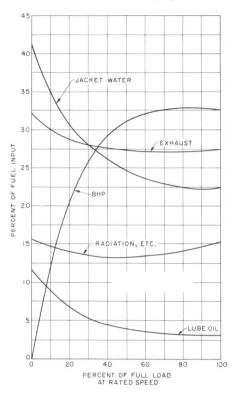

Fig. 13   Heat Balance for Typical Turbocharged Engine

engine heat with the lubricating system. The operating temperature of the engine may be significant in determining the proportion of engine heat removed by the lube oil. Radiator-cooled units generally use the same fluid to cool the engine water jacket and the lube oil, thus the temperature difference between the oil and the jacket coolant is not significant. If the oil temperature rises in one area (such as around the piston skirts), the heat may be transferred to other engine oil passages and then removed by the jacket coolant.

When the engine jacket water temperatures are much higher than the lubricant temperatures, the reverse process occurs and the oil removes heat from the engine oil passages. Heat is dissipated to lube oil in a four-cycle engine with a high-temperature (225 to 250°F) jacket water coolant at a rate of about 7 or 8 Btu/min·bhp; oil heat is rejected in the same engine at 3 to 4 Btu/min·bhp. However, this engine uses more moderate (180°F) coolant temperatures for both lube oil and engine jacket. Determining the lube oil cooling effect is necessary to design lube oil heat exchangers and coolant systems.

High quality lubricating oils are generally required for operation at temperatures between 160 and 200°F, with longer oil life expected at lower temperatures. Condensation may occur in the crankcase if the oil is too cool, thereby reducing its useful life. Each lubricant, engine, and application condition has individual characteristics, and only periodic laboratory analysis of lube oil samples can establish optimum lubricant service periods. Typically, a shorter life at higher temperatures is one of the limitations. Between 5 and 10% of the total fuel input produces heat that must be extracted from the lube oil; this may warrant using oil coolant at temperatures high enough to permit economic use in a process such as domestic water heating. Avoid copper piping and oil side surfaces in oil coolers and heat exchangers to reduce the possibility of oil breakdown caused by contact with copper.

Lube oil heat exchanger design should be able to maintain oil temperatures at 190°F, with the highest coolant temperature consistent with the economics of salvage heat value. Engine manufacturers usually size their oil cooler heat transfer surface on the basis of 130°F entering coolant water, and without provision for additional lubricant heat gains that occur with high engine operating temperatures. The costs of obtaining a reliable supply of lower temperature cooling water must be compared with costs of increasing the size of the oil cooling heat exchanger and operating at a cooling water temperature of 165°F. In applications where engine jacket coolant temperatures are above 220°F and where there is use for heat at a level of 155 to 165°F, the lube oil heat can be salvaged profitably.

The oil coolant should not foul the oil cooling heat exchanger. Do not use raw water unless it is free of silt, calcium carbonates, sand, and other contaminants. A good solution is a closed circuit, treated water system using an air-cooled transfer coil with freeze protected coolant. A domestic water heater can be installed on the closed circuit to act as a reserve heat exchanger and to salvage some useful heat when needed. Inlet air temperatures are not as critical with diesel engines as with turbocharged natural gas engines, and the aftercooler water on diesel engines can be run in series with the lube oil cooler (Figure 14).

Turbochargers on natural gas engines require medium fuel gas pressure (12 to 20 psi) and rather low aftercooler water temperatures (90°F or less) for high compression ratios and best fuel economy. Aftercooler water at 90°F is a premium coolant in many applications, because the usual sources are raw domestic water and evaporative cooling systems, such as cooling towers. Aftercooler water at temperatures as high as 135°F can be used, although the engine is somewhat derated. Using a domestic water solution may be expensive because (1) the coolant is continuously needed while the engine is running, and (2) the available heat exchanger designs require using a large amount of water even though the load is less than 200 Btu/h·bhp. A cooling tower can be used, but will require winter freeze protection and water quality control, both of which increase initial costs. If a cooling tower is used, the lube oil coolant load can be included in the tower load when there is no use for salvage lube oil heat.

The turbocharger increases engine capacity and extends the optimum fuel consumption curve, because the usual limitation on larger gas-fueled engines is the volume of combustion air and fuel that can be inserted in the available combustion chamber. Many stationary reciprocating engines were developed for use with diesel fuel. The energy ratings with the liquid fuel are higher than with naturally aspirated gaseous fuel. Turbocharging on diesel service permits more air pressure to

be applied in the cylinder so that larger quantities of the separately injected fuel can be burned efficiently.

In the gaseous fuel system, the fuel must always have a pressure high enough to enter the carburetor and mix with the boosted pressure of the combustion air. Since the gas and air mixture will ignite at a specific temperature-pressure relationship, the lower the inlet air temperature is, the higher the compression ratio can be before spontaneous combustion (preignition) occurs.

JACKET HEAT RECOVERY

Engine jacket cooling passages for reciprocating engines, including the water cooling circuits in the block, heads, and exhaust manifolds, must remove about 30% of the heat input to the engine. If the machine operates at temperatures above 180°F coolant temperature, condensation of combustion products should produce no effect. Some engines have modified gasket and seal designs enabling satisfactory operation up to 250°F at 30 psig. To avoid thermal stress, the maximum temperature rise through the engine jacket should not exceed 15°F. Flow rates must be kept within the engine manufacturers' design limitations to avoid erosion from excessive flow or inadequate distribution within the engine from low flow rates.

Engine-mounted, water-circulating pumps driven from an auxiliary shaft can be modified with proper shaft seals and bearings to give good service life at the elevated temperatures. Configurations that have a circulating pump for each engine increase reliability because the remaining engines can operate if one engine pump assembly fails. An alternative design uses an electric drive pump assembly to circulate water to several engines and has a standby pump assembly in reserve.

Forced circulation hot water, 250°F at 30 psig, can be used for many process loads, including hot water heating systems for comfort and process, absorption refrigeration chillers, and domestic hot water heating. System-wide distribution of engine jacket coolant must be limited to lessen the risk of leaks or other failures in downstream equipment that would prevent engine cooling. One solution is to confine each engine circuit to its individual engine, using a heat exchanger to transfer the salvaged heat to another circuit that might serve several engines and an extensive distribution system. An additional heat exchanger is needed in each engine circuit to remove heat whenever the waste heat recovery circuit does not extract all the heat produced (Figure 5). The reliability of this approach is high, but because the using circuit must be at a lower temperature than the engine operating level, it requires larger heat exchangers, piping and pumps, or a sacrifice of system efficiency that might result from these lower temperatures.

Controls that include low-temperature limits prevent excessive heat loads from seriously reducing the operating temperature of the engines. Engine castings may crack if they are subjected to sudden cooling when a large demand for heating from a large heat circuit is turned on and temporarily overloads the system. A heat storage tank is an excellent buffer for such occasions, since it can provide heat at a very high rate for short periods and protects the machinery serving the heat loads. The heat level can be controlled with supplementary heat input, such as an auxiliary boiler, to ensure adequate capacity in the system for any loads.

Another method of protecting the engine heat recovery system from instantaneous load is by flashing the recovered heat into steam and using the steam as the distribution fluid (Figure 6). By using a back-pressure regulator, the steam pressure on the engine can be kept uniform, and the auxiliary steam boilers can supplement the distribution steam header.

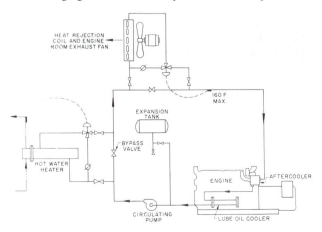

Fig. 14   Typical Lube Oil and Aftercooler System

The same engine coolant system using forced circulation water at 235°F entering and 250°F leaving the engine can produce steam at 235°F, 8 psig. The engine cooling circuit is the same as in other high-temperature water systems. This is accomplished by maintaining a static head above the circulating pump inlet adequate to avoid cavitation at the pump, and restricting the flow rate at the entrance to the steam flash chamber. This causes a change in pressure and prevents flash steam, which can cause severe damage within the engine.

A limited steam distribution system that distributes steam through nearby heat exchangers can salvage heat without contaminating the engine cooling system from downstream piping systems. The salvage heat temperature is kept high enough for most low-pressure steam loads. Returning condensate must be treated to prevent engine oxidation. The flash tank system can accumulate the sediment from treatment chemicals.

Some engines can use natural convection, ebullient cooling with water circulated by gravity flow into the bottom of the engine. There, it is heated by the engine to form steam bubbles (Figure 7). This heat lowers the density of the fluid, which then rises to a separating chamber, or flash tank, where the steam is released and the water is recirculated to the engine. Maintaining the flash tank at a static pressure slightly above the highest part of the engine causes a rapid flow of coolant to circulate through the engine, so only a small temperature difference is maintained between inlet and outlet. A constant back pressure must be maintained at the steam outlet of the flash chamber to prevent sudden lowering of the operating pressure. If the operating pressure changes rapidly, the steam bubbles in the engine could quickly expand and interfere with heat transfer, permitting overheating at the most critical points in the engine. This method of engine cooling is suitable for operation at up to 15-psi steam, 250°F, using components built to meet Section VIII of the ASME *Boiler and Pressure Vessel Code*, Unfired Pressure Vessels.

The engine coolant passages must be designed for gravity circulation and must eliminate the formation of freely bubbling steam. It is important to consider how coolant is allowed to flow out of the engine heads and exhaust manifolds. Each coolant passage must vent upward to encourage gravity flow and the free release of steam toward the steam separating chamber. Engine heat removal is most satisfactory when these circuits are used because fluid temperatures are uniform throughout the machine. The free convection cooling system depends less on mechanical accessories and is more readily arranged for completely independent assembly on each engine with its own coolant system replacing standard engine cooling systems.

Any of the systems used to generate steam from an engine have inherent design hazards from both the engine and system standpoint. It is preferable, when the downstream facility thermal profiles and system arrangements permit, to employ the conventional hot water heat recovery shown in Figure 5.

When a facility has neither adequate low-temperature water nor low-pressure steam demand to support a small cogeneration system, but does have medium-pressure steam loads up to 125 psig, cycles are available that divert a variable portion of the shaft power into a steam vapor compressor. Modules up to 650 kW can be controlled to provide more electric power and less medium-pressure steam or vice versa, for flexible tracking of either load.

EXHAUST GAS RECOVERY

Almost all of the heat transferred to the engine jacket cooling system can be reclaimed in a standard jacket cooling process or in combination with exhaust gas heat recovery.

However, only part of the exhaust heat can be salvaged due to the practical limitations of heat transfer equipment and the need to prevent flue gas condensation. Energy balances are often based on standard air at 60°F; however, exhaust temperatures cannot easily be reduced to this level. A recommended minimum exhaust temperature of 250°F has been established by the Diesel Engine Manufacturers Association (DEMA). Many heat recovery boiler designs are based on a minimum exhaust temperature of 300°F to avoid water vapor condensation and acid formation in the exhaust piping. Final exhaust temperature at part load is important on generator sets that operate at part load most of the time. Depending on the initial exhaust temperature, approximately 50 to 60% of the available exhaust heat can be recovered. A complete heat recovery system, which includes jacket water, lube oil, and exhaust, can increase the overall thermal efficiency from 30% for the engine generator only to approximately 75%. Exhaust heat recovery equipment is available in the same categories as standard firetube and water-tube boilers. Because engine exhaust must be muffled to reduce ambient noise levels, most recovery units also act as silencers. Figure 15 illustrates a typical noise curve. Figure 16 shows typical attenuation curves for various types of silencers. Chapter 42 of the 1991 ASHRAE *Handbook— HVAC Applications* lists acceptable noise level criteria for various applications.

Other heat recovery silencers and boilers include coil-type hot water heaters with integral silencers; water-tube boilers

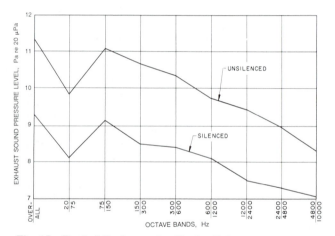

Fig. 15   Typical Reciprocating Engine Exhaust Noise Curve

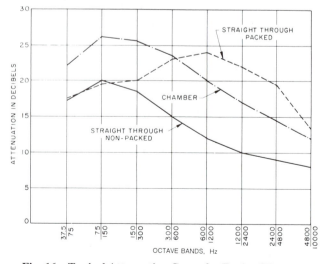

Fig. 16   Typical Attenuation Curve for Engine Silencers

with steam separators for gas turbines; and engine exhausts and steam separators for high-temperature cooling of engine jackets. Recovery boiler design should facilitate inspection and cleaning of the exhaust gas and water sides of the heat transfer surface. Diesel engine units should have a means of soot removal. Soot deposits can quickly reduce the heat exchanger effectiveness (Figure 17). These recovery boilers can also serve other requirements of the heat recovery system, such as surge tanks, steam separators, and fluid level regulators.

In many applications for heat recovery equipment, the demand for heat requires some method of automatic control. In vertical recovery boilers, control can be achieved by varying the water level in the boiler. Figure 18 shows a control system using an air-operated pressure controller with diaphragm or bellows control valves. When steam production begins to exceed the using system demand, the feed control valve begins to close, throttling the feedwater supply. Concurrently, the dump valve begins to open, and the valves reach an equilibrium position that maintains a level in the boiler to match the steam demand. This system can be fitted with an overriding exhaust temperature controller that regulates the boiler output to maintain a preset minimum exhaust temperature at the outlet. This type of automatic control is limited to vertical boilers, since the ASME *Boiler and Pressure Vessel Code* does not permit horizontal boilers to be controlled by varying the water level. With the latter type, a control condenser, radiator, or thermal storage can be used to absorb excess steam production.

In hot water units, a temperature-controlled bypass valve can be used to divert the water or the exhaust gas to achieve automatic modulation with heat load demand (Figure 10). If the water is diverted, precautions must be taken to limit the temperature rise of the lower flow of water in the recovery device, which could otherwise cause steaming. The heat recovery equipment should not adversely affect the primary function of the engine to produce work. Therefore, the design of waste heat recovery boilers should begin by determining the back pressure imposed on the engine exhaust gas flow. Limiting back pressures vary widely with the make of engine, but the general range is 6 in. of water. The next step is to calculate the heat transfer area that gives the most economical heat recovery without reducing the final exhaust temperature below 300°F.

Heat recovery silencers are designed to adapt to all engines; efficient heat recovery depends on the initial exhaust temperature. Most designs can be modified by adding or deleting heat transfer surface to suit the initial exhaust conditions and to maximize heat recovery down to a minimum temperature of 300°F. Figure 19 illustrates the effect of lowering the exhaust temperature below 300°F. This curve is based on a specific heat recovery silencer design with an initial exhaust temperature of 1000°F. Lowering the final temperature from 300 to 200°F increases heat recovery 14%, with a 38% surface increase. Similarly, a reduction from 300 to 100°F increases heat recovery 29% but requires a 130% surface increase. Therefore, the cost of the heat transfer surface is a factor that must be considered when setting the final temperature.

The other factor to consider is the problem of water vapor condensation and acid formation when the exhaust gas temperature passes through the dew point. This point varies with fuel and atmospheric conditions, and is usually in the range of 125 to 150°F. This gives an adequate margin of safety for the 250°F minimum temperature recommended by DEMA. Also, it allows for other conditions that could cause condensation, such as an uninsulated boiler shell or other cold surface in the exhaust system, or part loads on an engine.

Limited data have been published on the effect of water vapor in exhaust gas. The quantity varies with the fuel type and the intake air humidity. The standard combustion equation for methane fuel, with the correct amount of air for complete combustion, gives a relationship of 2.25 lb of water vapor in the exhaust for every pound of fuel burned. Similarly, for diesel fuel, the ratio is 1.38 lb of water vapor per pound of fuel. The condensates formed at low exhaust temperatures can

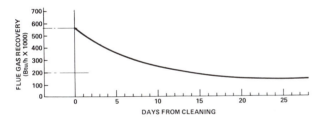

Fig. 17 Thermal Energy Recovered from 600 kW Diesel Engine, Flue Gas Heat Recovery Unit

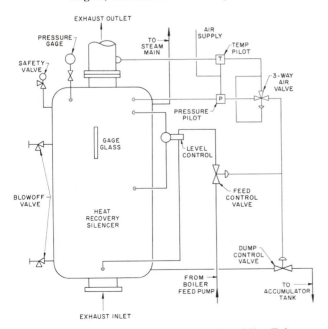

Fig. 18 Automatic Boiler System with Overriding Exhaust Temperature Control

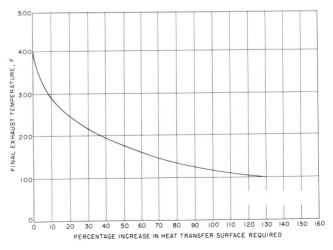

Fig. 19 Effect of Lowering Exhaust Temperature Below 300°F

be highly acidic, such as sulfuric acid from diesel fuels and carbonic acid from natural gas fuels, both of which can cause severe corrosion in the exhaust stack as well as in colder sections of the recovery device.

Several heat recovery equations and rules can be applied quantitatively to determine the feasibility of a heat recovery application. If engine exhaust flow and temperature data are available, and maximum recovery to 300°F final exhaust temperature is desired, the basic equation is:

$$Q = m_e(c_p)_e(t_e - t_f)$$

where

Q = heat recovered, Btu/h
m_e = exhaust flow, lb/h
t_e = exhaust temperature, °F
t_f = final exhaust temperature, °F
$(c_p)_e$ = specific heat of exhaust gas, 0.25 Btu/lb·°F

This equation applies to both steam and hot water units. To estimate the quantity of steam obtainable, divide the total heat recovered Q by the latent heat of steam at the desired pressure. The latent heat value should include an allowance for the temperature of the feedwater return to the boiler. The basic equation is:

$$Q = m_s(h_s - h_f)$$

where

m_s = mass flow rate (steam)
h_s = enthalpy of steam
h_f = enthalpy of feedwater

Similarly, the quantity of hot water can be determined by:

$$Q = m_w(c_p)_w(t_o - t_i)$$

where

$(c_p)_w$ = specific heat of water, 1.0 Btu/lb·°F
t_o = temperature of water out, °F
t_i = temperature of input water, °F
m_w = mass flow rate of water, lb/h

If the shaft power is known but engine data is not available, the heat available from the exhaust can be estimated at about 1000 Btu/h per horsepower output or 1 lb/h steam per horsepower output. The exhaust recovery equation is also applicable to gas turbines, although the quantity of flow will be much greater. The estimating figure for gas turbine boilers is 8 to 10 lb/h of steam per horsepower output. These factors are reasonably accurate for steam pressures in the range of 15 to 150 psig.

The normal procedure is to design and fabricate heat recovery boilers to the ASME *Code* (Section VIII, Unfired Pressure Vessels) for the working pressure required. Since the temperature levels in most exhaust systems are not excessive, it is common to use flange- or firebox-quality steels for the pressure parts and low carbon steels for the nonpressure components. Wrought iron or copper can be used for extended fin surfaces to improve heat transfer capacities. In special applications such as sewage gas engines, where exhaust products are highly corrosive, wrought iron or special steels are used to improve corrosion resistance.

When waste heat is to be converted and used to produce chilled water, several alternatives exist. The conventional method is to use hot water (>200°F) or low-pressure steam (<15 psig) in single-stage absorption chillers. These single-stage absorbers have a COP of 0.6 or less. At 12,000 Btu/ton, 18,000 Btu/h of recovered heat can produce about 1 ton of cooling.

If the direct exhaust, two-stage absorption chiller is used, the formula for cooling produced from recoverable heat is:

$$Q = mc_p(t_1 - t_2)(1.14)(0.97)/12000$$

where

Q = cooling produced, tons
m = mass flow of exhaust gas, lb/h
c_p = specific heat of gas = 0.268 Btu/lb·°F
t_1 = exhaust temperature in, °F
t_2 = exhaust temperature out = 375°F
1.14 = coefficient of performance
0.97 = connecting duct system efficiency
12000 = Btu/ton

For internal combustion engines, jacket water heat at 180 to 210°F may be added to the recovered heat of the engine exhaust to produce chilled water in a single-stage absorption machine.

$$Q = 0.6m_w(c_p)_w(t_1 - t_2)/12000$$

where

Q = heat recovered, tons
m = mass flow of water, lb/h
$(c_p)_w$ = specific heat of water, 1.0 Btu/lb·°F
t_1 = water temperature out of engine
t_2 = water temperature returned to engine, typically $t_1 - 15 = t_2$
0.6 = COP
12000 = Btu/ton

Heat may be recovered from engines and gas turbines as high-pressure steam, depending on exhaust temperature. Steam pressures greater than 15 psig up to 200 psig are not uncommon.

When steam is produced at pressures over 43 psig, two-stage absorption chillers can also be considered. The COP of a two-stage absorption chiller is 1.21 or greater. The steam input required is 9.3 to 9.7 lb/ton. This compares to 18 lb/ton for the single-stage absorption machine using 15 psig steam.

Local air pollution control authorities may require engine exhaust gases to be conditioned or may even require special engines or fuels. Turbines may need water injection systems or selective catalyst reduction systems (SCR). Reciprocating engines may be required to have catalytic converters or SCR.

GENERAL INSTALLATION PROBLEMS

The circulating fluid systems must be kept clean because the internal coolant passages of the engine are not readily accessible for service. The installation of piping, heat exchangers, valves, and accessories must include provisions for internal cleaning of these circuits before they are placed in service. Coolant fluids must be noncorrosive and free from salts, minerals, or chemical additives that can deposit on hot engine surfaces or form sludge in relatively inactive fluid passages. Generally, engines cannot be drained and flushed effectively without major disassembly, making any chemical treatment of the coolant fluid that produces sediment or sludge undesirable.

An initial step toward maintaining clean coolant surfaces is to limit fresh water makeup. The coolant system should be tight and leak-free. Softened water or mineral-free water is effective for initial fill and makeup. Forced circulation hot water systems may require only minor corrosion-inhibiting additives to ensure long, trouble-free service; this feature is one of the major assets of hot water heat recovery systems.

The ebullient cooling method presents some special hazards for water treatment. Evaporation takes place in the engine, and the concentration of dissolved minerals increases at this critical location. This problem can be minimized by using gravity or forced circulation to encourage high flow rates and by reducing minerals added to the system to a negligible amount. Elevating the flash tank or using pumps to maintain a high static pressure also reduces the tendency for engine fouling. The water treatment must be adequate to prevent corrosion from free oxygen brought in with returned condensate and makeup water in the flash tank. Additional treatment should control corrosion in the heat recovery muffler, the steam separator, and the condensate system. If the cost of pumping the water and the discount in utilization temperature can be absorbed in the system design, a hot water system is preferred. Most available engines are designed for forced circulation of the coolant.

Where several engines are used in one process, independent coolant systems for each machine will avoid a complete plant shutdown from a common coolant system component failure. Such independence can be a disadvantage in that the unused engines are not maintained at operating temperature, as they are when all units are in a common circulating system. If the idle machine temperature drops below the dew point of the combustion products, corrosive condensate may form in the exhaust gas passages each time the idle machine is started.

When substantial water volume and machinery mass must be heated up to operating temperature, the condensate volume is quite significant and must be drained. Some contaminants will get into the lube oil and reduce the service life. If the machinery gets very cold, it may be difficult to start. Units that are started and stopped frequently require an off-cycle heating sequence to lessen the exposure to corrosion. Forced lubrication by an auxiliary, externally powered lubricating oil pump is beneficial to the engine when the unit is off. Continuously bathing the engine parts in oil and maintaining engine temperatures near their normal operating temperatures improves engine life and lowers engine maintenance costs. This practice should be evaluated against the pumping losses and the radiation heat losses to the engine room environment.

Water level control of separate steam-producing engine cooling systems is needed to prevent backflow through the steam nozzle of idle units. Using a back-pressure regulating valve, a steam check valve, or an equalizing line can prevent this problem. Analysis of the installation costs, available product designs, and operating experience indicates that forced circulation hot water heat recovery systems are best suited to installations of about 1000 kW or smaller total capacity as well as where future expansion is limited by other design considerations. In projects of much larger capacity, the fluid pumping systems become unwieldy, and future capacity increase becomes difficult. Larger projects are better served with free convection, ebullient-cooled packages for each engine, with a low-pressure steam header connecting various sources of heat input to utilization loads. Larger projects can have special water conditioning and testing apparatus. The added complexity of coolant quality control in the steam systems can be justified by the greater reliability of the independent engine coolant systems.

CONTROL SYSTEMS

Controls for cogeneration systems are required for (1) system output, (2) safety, (3) prime mover automation, and (4) waste heat recovery and disposal.

Building requirements determine the level of automation. Every system must have controls to regulate the output energy and to protect the equipment. Independent power plants require constantly available control energy to actuate cranking motors, fuel valves, circuit breakers, alarms, and emergency lighting. Battery systems are generally used for these functions, since stored energy is available if the generating system malfunctions. Automatic equalizer battery chargers maintain the energy level with minimum battery maintenance.

Generators operating in parallel with utility system grids have different control requirements than those that operate isolated from the utility grid. A system that operates in parallel and provides emergency standby power if a utility system source is lost must also be able to operate in the same control mode as the system that normally operates isolated from the electric utility grid.

Control requirements for systems that provide electricity and heat for equipment and electronic processors differ, depending on the number of energy sources and the type of operation relative to the electric utility grid. Isolated systems generally use more than one prime mover during normal operation to allow for load following and redundancy.

Table 1 shows the control functions required for systems operating isolated from the utility grid and systems operating in parallel with the grid with single and multiple prime movers. Frequency and voltage are directly controlled in a single-engine isolated system. The power is determined by the load characteristics and is met by automatic adjustment of the throttle. Reactive power is also determined by the load and is automatically met by the exciter in conjunction with voltage control.

The heat output to the primary process is determined by the engine load. It must be balanced with actual requirements by supplementation or excessive heat rejection through peripheral devices such as cooling towers. Similarly, if more than one level of heat is required, controls are needed to (1) reduce it from a higher level when it is available at the primary level, (2) supplement it if it is not available, or (3) provide for its rejection when availability exceeds the requirements.

In an isolated plant with more than one prime mover, controls must be added to balance the power output of the prime movers and to balance the reactive power flow between the generators. Synchronizing equipment must be added to parallel the second and any additional generators with the first. Generally, an isolated system requires that the prime movers supply the needed electrical output with the heat availability controlled by the electrical output requirements. Any imbalance in heat requirement results in burning supplemental fuels or wasting surplus recoverable heat through the heat rejection system.

Supplemental firing and heat loss can be minimized during parallel operation of the generators and the electric utility system grid and by adjusting the prime mover throttle for the required amount of heat. This is referred to as *thermal load following*. The electric generation then depends on heat requirements; imbalances between the heat and the electrical load are carried by the electric utility system, either through absorption of excess generation or through the delivery of supplemental electrical energy to the electrical system. In parallel operation, both frequency and voltage are determined by the utility service. The power output is determined by the throttle setting, which responds to the system heat requirements. Only the reactive power flow is independently controlled by the generator controls. When additional generators are added to the system, there must be a means for controlling the power division between multiple prime movers and to continue to divide and control the reactive power flow. All units require synchronizing equipment.

Additional electric utility interface requires safety on the electric grid and the ability to meet the operating problems of the electric grid and the generating system. Additional control functions depend on the desired operating method during loss

Table 1　Generator Control Functions

	Isolated		Parallel	
	One Engine	Two or More Engines	One Engine	Two or More Engines
Frequency	Yes	Yes	No	No
Voltage	Yes	Yes	No	No
Power	No (Load following)	Yes (Division of load)	No	Yes (Division of load)
Reactive kVAR	No (Load following)	Yes (Division of kVAR)	Yes	Yes
Heat t_1	Supplement only	Supplement only	Load following	Load following
Heat $t_2 - t_x$	Reduce from t_1 or supplement	Reduce from t_1 or supplement	Reduce from t_1 and load following	Reduce from t_1 and load following
Cooling	Remove excess heat (tower, fan, etc.)	Remove excess heat (tower, fan, etc.)	Normally no (Emergency yes)	Normally no (Emergency yes)
Synchronizing	No	Yes	Yes	Yes
Black start	Yes	Yes (one engine)	Emergency use	Emergency use (one engine)

of interconnection. For example, the throttle setting on a single generator operating in parallel with the utility is determined by the heat recovery requirements and its exciter current, which is set by the reactive power flow through the interconnection. When the interconnection with the utility is lost, the generator control system must detect that loss, assume voltage and frequency control, and immediately disconnect the intertie to prevent an unsynchronized reconnection. With the throttle control now determined by the electric load, the heat produced may not match the requirements for supplemental or discharge heat from the system. When the utility source is reestablished, the system must be manually or automatically synchronized and the control functions restored to normal operation.

Loss of the utility source may be sensed by one or more of the following factors: overfrequency, underfrequency, overcurrent, overvoltage, undervoltage, or any combination of these factors. The most severe condition occurs when the generator is delivering all electrical requirements of the system up to the point of disconnection, whether it is on the electric utility system or at the plant switchgear. Under such conditions, the generator tends to operate until the load changes. At this time, it either speeds up or slows down, allowing the over- or underfrequency device to sense loss of source and to reprogram the generator controls to isolated system operation. The interconnection is normally disconnected during such a change and automatically prevented from reclosing to the electric system until the electric source is reestablished and stabilized and the generator is brought back to synchronous speed.

The start-stop control may include manual or automatic activation of the engine fuel supply, engine cranking cycle, and establishment of the engine heat removal circuits. Stop circuits always shut off the fuel supply and, for spark-ignited engines, the ignition system is generally grounded as a precaution against incomplete fuel valve closing.

The prime mover must be protected from malfunction by alarms that warn against unusual conditions and by safety shutdown under unsafe conditions. The control system must protect against failure of (1) speed control (overspeed); (2) lubrication (low oil pressure, high oil temperature); (3) heat removal (high coolant temperature or lack of coolant flow); and (4) combustion process (fuel, ignition).

The generator system must be protected from overload, overheating, short-circuit faults, and reverse power. The minimum protection is a properly sized circuit breaker with a shunt-trip coil for immediate automatic disconnect in the event of low voltage, overload, or reverse power. The voltage regulation control must prevent overvoltage. Circuit breakers should be of the air type and operate within 5 Hz.

Generally, a simple control system is adequate where labor is available to make minor adjustments and to oversee system operation. Fully automatic, completely unattended systems have the same advantages as automatic temperature control systems and are gaining acceptance.

Fully automated generator controls should be considered for office buildings, hotels, motels, apartments, large shopping centers, some schools, and manufacturing plants, where there is the need for closely controlled frequency and voltage. The control system must be highly reliable, regardless of load change or malfunction, and must protect the system equipment from electrical transients and malfunctions. To do this, each generator must be controlled. Generators that operate in parallel require interconnecting controls. The complete system must be integrated to the building use, give properly sequenced operation, and provide overall protection. In addition to those controls required for a single prime mover installation, the following further controls are required for multiple generator installations:

1. Simultaneous regulation of fuel or steam flow to each prime mover to maintain required shaft output.
2. Load division and frequency regulation of generators by a signal to the fuel controls.
3. System voltage regulation and reactive load division by maintaining the generator output at the required level.
4. Automatic starting and stopping of each unit for protection in the event of a malfunction.
5. Load demand and unit sequencing by determining when a unit should be added or taken off the bus as a function of total load.
6. Automatic paralleling of the oncoming unit to the bus after it has been started and reaches synchronous speed.
7. Safety protection for prime movers, generators, and waste heat recovery equipment in the event of overload or abnor-

mal operation, including a means of load dumping (automatic removal of building electrical loads) in case the prime mover overloads or fails.

When the system is operated in isolation of the utility grid, the engine speed control must maintain frequency within close tolerances, both at steady-state and transient conditions. Generators operating in parallel require a speed control to determine frequency and to balance real load between operating units. To obtain precise control, an electronic governing system is generally used to throttle the engine; the engine responds quickly and is capable of responding to control functions as follows:

1. Engine speed sensing and controlling to operating frequency, *i.e.*, 60 Hz.
2. Synchronizing the oncoming generating unit to the bus so that it can be paralleled.
3. Sensing real load rather than current to divide real load between parallel units. (This must be done through the speed control to the throttle.)
4. Proper throttle operation during start-up to ensure engine starting and prevent overspeed as the engine approaches rated speed.
5. Time reference control is desirable to maintain clock accuracy to within 60 s per month.

Voltage must be held to close tolerances by the voltage regulator from no load to full load. A tolerance of 0.5% is realistic for steady-state conditions from no load to full load. The voltage regulator must allow the system to respond to load changes with minimum transient voltage variations. During parallel operation, the reactive load must be divided through the voltage regulator to maintain equal excitation of the alternators connected to the bus. True reactive load sensing is of prime importance to good reactive load division; current sensing is not adequate. An electronic voltage control responds rapidly and, if all three phases are sensed, better voltage regulation is obtained even if the loads are unbalanced on the phases. The alternator construction of a well-designed voltage regulator dictates the voltage transient variation.

Engine sizing can be influenced by the control system's accuracy in dividing real load. If one engine lags another in carrying its share of the load, the capacity that it lags is never used. Therefore, if the load-sharing tolerance is small, the engines can be sized more closely to the power requirements. A load-sharing tolerance of less than 5% of unit rating is necessary to use the engine capacity to good advantage.

This is also true for reactive load sharing and alternator sizing. If reactive load sharing is not close, a circulating current results between the alternators. The circulating current uses up alternator capacity, which is determined by the heat generated by the alternator current. The heat generated, and therefore the alternator capacity, is proportional to the square of the current. Therefore, it is advisable to install a precise control system; the added cost is justified by the possible installation of smaller engines and alternators.

When starting a unit, the following must be performed in sequence: (1) initiate engine crank; (2) open fuel valve and throttle; (3) when engine starts, terminate cranking; (4) when desired speed is reached, eliminate overshoot; (5) if the engine refuses to start in a given period of time, register a malfunction and start the next unit in sequence; and (6) on reaching the desired speed, synchronize the started engine alternator to the bus to be paralleled without causing severe mechanical and electrical transients. A time period should be allowed for synchronization; if it does not occur during the allotted time, a malfunction should register and the next unit in sequence should start.

Once a unit is on the line, real and reactive load division should be effective immediately. The throttle of the unit coming on the line must be advanced so that it will accept its share of the load. The unit or units that were carrying all the load will have their throttle(s) retarded so that they will give up load to the oncoming unit until all units share equally, or in proportion to the unit size if they are sized differently. During the load-sharing process, control frequency must be maintained and, once paralleled, the load-sharing control must correct continuously to maintain the load balance. Reactive load sharing is similar but is done by the voltage regulator varying the excitation of the alternator exciter field.

Problems that increase installation cost can arise during installation. These can be minimized by carefully checking out the cogeneration system, particularly the engine alternator units. Typical problems arising during installation include (1) engine actuators placed too near exhaust pipes, causing excessive wear of actuator bearings and loose parts in the engine actuator; (2) improper wiring; (3) sloppy linkage between engine actuator and carburetor; and (4) problems arising out of failure to coordinate electrical design with the utility. Sensors used to send a malfunction signal to the control system can also cause trouble when not properly installed. It is practical to use two sensors, one for alarm indication of an abnormal condition and the other for shutdown malfunction.

ENERGY USER INTERCONNECTIONS

A cogeneration system provides both power and thermal energy. Depending on the design and operating decisions, the users can intertie with the electric utility with energy delivered to the utility grid. For the cogeneration system to operate thermally and be economically efficient, the energy delivered by the system must be at levels required by the energy users.

The selection of prime movers is partially based on user thermal requirements. Internal combustion reciprocating engines have the lowest ratio of thermal-to-electric energy with low-temperature heat at a maximum of 93°C and 120°C available from lube oil and jacket water, respectively. This heat can be used to supply thermal energy to applications requiring low-temperature heat or where the low-temperature heat can be used effectively for preheat. In addition, for maximum heat recovery, the building thermal application must take sufficient energy from the recovered heat stream to lower its temperature to that required to cool the prime mover effectively. A supplementary means for rejecting prime mover heat may be required if the thermal load does not provide adequate cooling or as a backup to thermal load loss.

Gas turbines can provide higher quantities and better qualities of heat per unit of power, while extraction steam turbines can provide even greater flexibility in both the quantity and quality (temperature and pressure) of heat delivered. Externally fired prime mover cycles, such as the Stirling cycle, and gas turbines with steam injection are also flexible in quantity and quality of thermal energy.

In selecting prime movers and evaluating cogeneration economics, consider the amount of thermal energy available, its quality, and the degree to which this recovered energy can satisfy user requirements. Also, consider the degree to which the user's energy systems provide necessary cooling to the cogeneration facility.

Electrical energy can be delivered to the utility grid, directly to the user, or both. Generators for on-site power plants can deliver electrical energy at levels equal in quality to those provided by the electric utility for voltage regulation, frequency control, harmonic content, reliability, and phase balance. They are also capable of satisfying stringent requirements imposed by user computer applications, medical equipment, high-frequency equipment, and emergency power supplies. The cogenerator's electrical interface should be designed according to user or utility electrical characteristics.

PREVENTATIVE MAINTENANCE

One of the most important provisions for healthy and continuous plant operation is implementation of a comprehensive preventative maintenance program. This should include written schedules of daily wipedown and observation of equipment, weekly and periodic inspection for replacement of degradable components, engine oil analysis, and maintenance of proper water treatment. Immediate access to repair services may be accomplished by subcontract or with in-house plant personnel. An inventory of critical parts should be maintained on site.

In multiengine plants, or where any battery of more than one unit is serving a given distribution system, it is undesirable to operate the units with the equal life approach; *i.e.*, each unit in the same state of wear and component deterioration. If there is only one standby unit and another unit suffers a major failure or shutdown, all units now needed to carry 100% load would be prone to additional failure while the first failed unit is undergoing repair. The preferred operation is to keep one unit in continuous reserve, with the shortest possible running hours between overhauls or major repair, and to schedule operation of all others for unequal running hours. Thus any two units would have a minimum statistical chance of a simultaneous failure. All units, however, should be exercised for several hours in any week.

COGENERATION SYSTEM PLANNING

Planning Considerations

The planning for a cogeneration system is considerably more involved than that for an HVAC system. HVAC systems must be sized to meet peak loads, while cogeneration systems need not. HVAC systems do not have to be coordinated and integrated with other energy systems as extensively as do cogeneration systems. A much more comprehensive energy analysis combined with economic analysis must be used to select the size of a cogeneration system that maximizes the efficiency and economic return on investment. To identify and screen potential system candidates, a simplified, but accurate, performance analysis that considers the dynamics of the facility electrical and thermal loads, as well as the size and fuel consumption of the prime mover, must be conducted. The need for accurate analysis is especially important for commercial and institutional cogeneration applications because of the large changes in magnitude with time and the noncoincidental nature of the power and thermal loads. A facility containing a generator sized and operated to meet the thermal demand may occasionally be required to purchase supplemental power and sometimes may produce power in excess of facility demand. Even in the early planning stages of a project, it is necessary to provide reasonably accurate estimates of (1) the fuel consumed (if it is a topping cycle); (2) the amount of supplemental electricity that must be purchased; (3) the amount of supplemental boiler fuel (if any) that must be purchased; (4) the amount of excess power available for sale; (5) the electrical capacity required of the utility to provide supplemental and standby power; and (6) the electrical capacity represented by the excess power if the utility is offering capacity credits. Obtaining estimates of the previously mentioned performance

values for multiple time-varying loads is a difficult task, further complicated by utility rate structures which may be based on time-of-day or time-of-year purchase and sale of power. Data collection intervals must be short enough to give the desired levels of accuracy.

Sizing and Operating Options

Selecting a cogeneration system requires the evaluation of a large number of factors. Aside from the selection of the type of prime mover, the most important step in evaluating or planning a cogeneration system involves determining the cogeneration plant capacity and operating options. The possibilities for sizing are:

- Peak electrical demand
- Minimum (or baseload) electric demand
- Peak thermal demand
- Minimum (or baseload) thermal demand
- Maximum economic return with available funds
- Marginal cost/benefit ratio

Plant operating options are:

- Operate at rated output
- Thermally dispatched (tracks facility thermal load)
- Electrically dispatched (tracks facility electric load)
- Maximum economic return
- Utility dispatched
- Peak shaving

The significance of each of these is explained in the following discussion.

Load Duration Curve Method of Analysis

A basic method for analyzing the performance of HVAC systems is the bin method. The basic tool for sizing and evaluating performance of power systems is the *load duration curve*. This curve contains the same information as bins, but the load data are arranged in a slightly different manner. The load duration curve is a plot of hourly averaged instantaneous load data over a period, rearranged to indicate the frequency, or hours-per-period, at which the load is at, or below, the stated value. The load duration curve is constructed by rearranging the hourly averaged load values of the facility into a descending order sort. The sorting of large volumes of load data can be easily accomplished with desktop computers and electronic spreadsheets or databases. The load duration curve produces a visually intuitive tool for sizing cogeneration systems and for accurately estimating system performance.

Figure 20 is an example of a hypothetical steam load profile for a plant operating with two shifts each weekday and one shift each on the weekend days; no steam is consumed during nonworking hours for this example.

The data provides little information for thermally sizing a generator, except to indicate that the peak demand for steam is about 46,000 lb/h and the minimum demand is about 13,000 lb/h.

Figure 21 is a load duration curve of the steam load data in Figure 20. As previously stated, it was obtained by a descending order sort of Figure 20 data. Mathematically, the load duration curve is the frequency at which the load will be equal to or exceed a given value, and is one minus the integral of the frequency distribution function for a random variable. Since the frequency distribution is a continuous representation of a histogram, the load duration curve is simply another arrangement of the familiar bin data.

In the frequency domain, or load duration curve form, the baseload and peakload can be identified readily. Note that the practical baseload at the "knee" of the curve is about 21,000

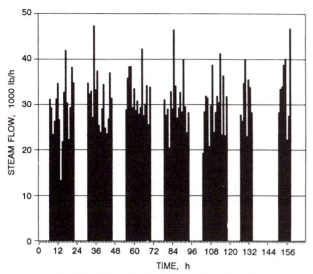

Fig. 20   Hypothetical Steam Load Profile

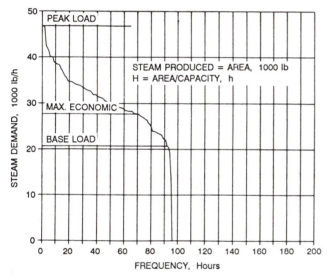

Fig. 21   Load Duration Curve

lb/h rather than the 13,000 lb/h absolute minimum identified on the load profile curve.

The cogenerator sized at the baseload provides the greatest efficiency and best use of capital investment. However, it may not offer the shortest payback because of the high value of electrical power. An appropriately sized cogeneration plant might be sized somewhat larger than baseload to achieve minimum payback through increased electrical savings. Obviously the determination of this economically sized cogeneration plant must be done with a combination of load analysis and economic analysis. For illustration purposes, the plant sized for maximum economy is arbitrarily assumed to be 28,000 lb/h. Cogeneration systems sized this way produce high equipment utilization, which depends on the utility to serve the peak load.

The load duration curve allows the designer to provide estimates of the total amount of steam generated within the interval. Since the total amount of steam is the area under the curve, the calculation may be performed by using formulae for rectangles and triangles or an appropriate curve analysis program. Thus, for a thermally tracked plant sized at the baseload, the hours of operation are about 93 h per week. Therefore,

based on the area under the rectangle, the total steam produced is:

$$93 \times 21,000 = 1.953 \times 10^6 \text{ lb/week}$$

Typically, a boiler would be used to meet the supplemental thermal requirements of the facility. This may be roughly estimated using the area of a triangle:

$$\text{Boiler steam required} = 93(46,000 - 21,000)/2$$
$$= 1.16 \times 10^6 \text{ lb/week}$$

$$\text{Steam produced} = 91 \times 21,000 + 70(28,000 - 21,000) +$$
$$(28,000 - 21,000)(93 - 70)/2$$
$$= 2.524 \times 10^6 \text{ lb/week}$$
$$\text{EFLH} = 2.524 \times 10^6/28,000 = 90.1 \text{ h}$$
$$\text{Electricity produced} = (\text{EFLH})Q_{FL}$$

where Q_{FL} is the electrical capacity at rated output.

Estimating the fuel consumption and the electrical energy output of a system sized for the peak load is not as simple as for the baseload design, as changes in the performance of the prime mover at part load must be considered. In this case, average value estimates of the performance at part load must be used for preliminary studies.

In some cases, an installation has only one prime mover; however, several smaller units operating in parallel provide increased reliability and performance through less part-load operation. Figure 22 illustrates the use of three prime movers rated at 10,000 lb/h each. For this example, generators #1 and #2 operate fully loaded, and #3 operates between full load capacity and 50% capacity while tracking the facility thermal demand. Since operation at less than 50% load is inefficient, further reduction in total output must be achieved by part-load operation of generators #1 and #2.

Offsetting the advantages of multiple units are higher specific investment and maintenance costs, the control complexity, and the usually lower efficiency of smaller units.

Facilities such as hospitals often seek to reduce their utility costs by using existing standby generators to share the electrical peak (peak shaving). Such an operation is not strictly cogeneration because heat recovery is rarely justified. Figure 23 illustrates a hypothetical electrical load duration curve with frequency as a percent of total hours in the year (8760). A generator rated at 1100 kW, for example, would reduce the

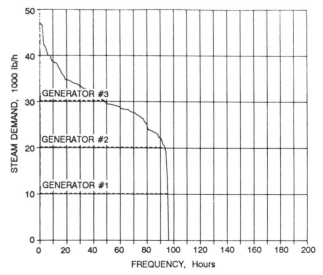

Fig. 22   Load Duration Curve Illustrating Multiple Generators

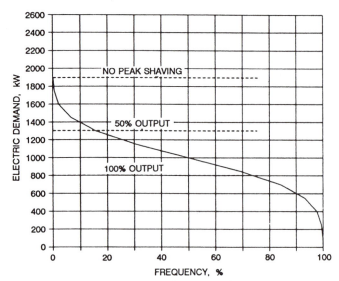

Fig. 23   Illustration of Peaking Generator

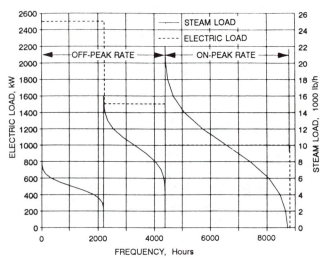

Fig. 24   Example of Two-Dimensional Load Duration Curve

peak demand by 550 kW if it is operated between 50 and 100% rated output. Clearly, there would be some electrical energy savings as well as a reduction in demand. However, this savings is idealistic because it can only be obtained if the operators or the control system can anticipate in advance when facility demand will exceed 1400 kW in sufficient time to bring the generator up to operating condition.

This must be done without anticipating too early in advance the need for the peak shaver, thereby wasting fuel in the process. Also, many utilities include ratchet clauses in their rate schedules. As a result, if the peak shaving generator is inoperative for any reason when the facility monthly peak occurs, the ratchet is set for a year hence, and the demand savings potential of the peaking generator will not be realized until a year later. Even though an existing standby generator may seem to offer "free" peak shaving capacity, careful planning and operation is required to realize its full potential. Conversely, continuous duty/standby systems give the benefits of heat recovery while satisfying the standby requirements of the facility. During emergencies, the generator load is switched from its normal nonessential load to the emergency load. Authorities have accepted and approved the premise that a spinning generator in a hospital application is more reliable than one that needs a cold start in emergencies.

Two-Dimensional Load Duration Curve

A two-dimensional load duration curve becomes necessary when the designer must consider simultaneous steam and electrical load variations. Such a situation occurs, for example, when the facility electric demand drops below the output of a steam-tracking generator and the excess power capacity cannot be exported. During periods of excess generator capacity, it is throttled to curtail electrical output to that of facility demand. In other words, it is now operating in an electrical tracking mode.

To develop the two-dimensional load duration curve method for simultaneous loads, either the electrical demand or the steam demand must be broken into discrete periods defined by the number of hours-per-year when the electric or steam demand is within a certain range of values, or bin.

Duration curves for the remaining load are then created for each period as before, using coincident values. Figure 24 illustrates such a representation for three bins. The electric load values indicated on the figure represent the center value

of each bin. In general, a large number of periods gives a more accurate representation of facility loads. Also, the greater the load fluctuation, the greater the number of periods required for accurate representation. Note that the total number of hours for all periods add up to 8760, the number of hours in a year.

Another situation that requires a two-dimensional load duration curve is when the facility buys or sells power whose price depends on the time-of-day or time-of-year. Many electric rate structures contain explicit time periods of the purchase or sale of power. In some cases, there is only a summer-winter distinction. Other rate schedules may have several periods to reflect time-of-use or time-of-sale rates. The two-dimensional analysis is required to consider these rate schedule periods when defining the load bins, because the greatest annual savings operating schedule is influenced by energy prices as well as energy demands. Using Figure 24 as an illustration of a summer peaking utility, the first two bins might coincide with the winter-fall-spring off-peak rate and the third bin the on-peak rate. Obviously, the analysis process becomes burdensome when there are large load swings and several rate periods requiring a large number of periods. For this reason, the two-dimensional load duration method of analysis is implemented by computers.

Analysis by Simulations

The load duration curve is a convenient, intuitive graphical tool for preliminary sizing and analysis of cogeneration systems. It lacks the capability, however, for detailed analysis of cogeneration systems.

A commercial or institutional facility can have as many as four different loads that must be considered simultaneously. These are cooling, noncooling electrical, steam or high-temperature hot water for space heating, and low-temperature domestic hot water. These loads are never in balance at any instant, which complicates sizing of equipment, establishing operating modes, and determining the quantity of heat rejected from the cogenerator that is usefully applied to the facility loads.

Further complicating the evaluation of commercial cogeneration systems is the fact that prime movers rarely operate at full rated load; therefore, part-load operating characteristics such as fuel consumption, exhaust mass flow, exhaust temperature, and heat rejected from the jacket and intercooler of internal combustion (IC) engines must be considered. If the prime mover is a combustion gas turbine (CGT), then the effects of ambient temperature on full-load capacity and part-load fuel consumption and exhaust characteristics should be considered.

In addition to the prime movers, a commercial or institutional cogeneration system may include absorption chillers that use jacket heat and exhaust heat to produce chilled water at two different COPs. Some commercial cogeneration systems use thermal energy storage (TES) to store hot water, thereby reducing the heat that must be dumped during periods of low demand. Internal combustion engines coupled with steam compressors and steam injected gas turbines (STIGs) have been produced to allow some variability of the output heat-to-power ratio from a single package.

Computer programs that analyze cogeneration systems are available. Many of these programs emphasize the financial aspects of cogeneration systems with elaborate rate structures, energy price forecasts, and economic models. However, equipment part-load performance, load schedules, and other technical characteristics that greatly affect system economics are modeled only superficially in many programs. Also, many computer codes contain built-in equipment data and perform system selection and sizing automatically, thereby excluding the designer from important design decisions that could affect system viability. While these programs have their place, they should only be used with care by those who know the program limitations.

The primary consideration in selecting a computer program for analyzing commercial cogeneration technical feasibility is the ability to handle multiple time-varying loads.

The four methods of modeling the thermal and electrical loads are as follows:

- Hourly average values for a complete year
- Monthly average values
- A truncated year consisting of hourly average values for one or more typical (usually working and nonworking) days of each month
- Bin methods

For the greatest accuracy, a cogeneration simulation using hourly averaged values for load representation would be used; however, the 8760 values for each load type can make the management of load data a formidable task. Guinn (1987) describes a public domain simulation that can consider a full year of multiple-load data.

The data needed to perform a monthly averaged load representation can often be obtained from utility billings. This model should only be used as an initial analysis; best accuracy is obtained when thermal and electric profiles are relatively consistent.

The truncated year model is a compromise between the accuracy offered by the full hourly model and the minimal data handling requirements of the monthly average model. It involves the development of hourly values for each load over a typical average day of each month. Usually, two typical days are considered, a working day and a nonworking day. Thus, instead of 8760 values to represent a load over a year, only 576 values are required. This type of load model is often used in cogeneration computer programs. Pedreyera (1988) and Somasundarum (1986) describe programs that use this method of load modeling.

Bin methods are based on the frequency distribution, or histogram, of load values. The method determines the number of hours per year the load was between ranges or values, or bins. This method of representing weather data is widely used to perform building energy analysis. It is a convenient way to condense a large database into a smaller set of values but is no more accurate than the time resolution of the original set. Furthermore, bin methods become unacceptably cumbersome for cogeneration analysis if more than two loads must be considered.

GLOSSARY

Avoided cost The incremental cost for the electric utility to generate or purchase electricity that is avoided through the purchase of power from a cogeneration facility.

Backup power Electric energy available from or to an electric utility during an unscheduled outage to replace energy ordinarily generated by the facility or the utility. Frequently referred to as standby power.

Baseload The minimum electric or thermal load generated or supplied continuously over a period of time.

Bottoming cycle A cogeneration facility in which the energy input to the system is first applied to another thermal energy process; the reject heat that emerges from the process is then used for power production.

Capability The maximum load that a generating unit, generating station, or other electrical apparatus can carry under specified conditions for a given period of time without exceeding approved limits of temperature and stress.

Capacity The load for which a generating unit, generating station, or other electrical apparatus is rated.

Capacity credits The value included in the utility's rate for purchasing energy, based on the savings accrued through the reduction or postponement of new generation capacity that results from purchasing power from cogenerators.

Capacity factor The ratio of the actual annual plant electricity output to the rated plant output.

Central cooling The same as central heating except that cooling (heat removal) is supplied instead of heating; usually a chilled water distribution system and return system for air conditioning.

Central heating Supply of thermal energy from a central plant to multiple points of end use, usually by steam or hot water, for space and/or service water heating; central heating may be large scale as in plants serving university campuses, medical centers, military installations, or in central buildings systems serving multiple zones; also district heating plants.

Cogeneration The sequential production of electrical or mechanical energy and useful thermal energy from a single energy stream. To qualify in the United States as a cogeneration facility, certain energy efficiency standards must be met.

Coproduction The conversion of energy from a fuel (possibly including solid or other wastes) into shaft power (which may be used to generate electricity) and a second or additional useful form. The process generally entails a series topping and bottoming arrangement of conversion to shaft power and either process or space heating. Cogeneration is a form of coproduction.

Demand The rate at which electric energy is delivered at a given instant or averaged over any designated time period.

Annual demand The greatest of all demands that occur during a prescribed demand interval billing cycle in a calendar year.

Billing demand The demand on which customer billing is based, as specified in a rate schedule or contract. It can be based on the contract year, a contract minimum, or a previous maximum and is not necessarily based upon the actual measured demand of the billing period.

Coincident demand The sum of two or more demands occurring in the same demand interval.

Instantaneous peak demand The maximum demand at the instant of greatest load.

Demand charge The specified charge for electrical capacity on the basis of the billing demand.

Energy charge That portion of the billed charge for electric service based on the electric energy (kilowatt-hours) supplied, as contrasted with the demand charge.

FERC efficiency Electrical output plus one-half the thermal heat utilized divided by the energy input (LHV).

Grid The system of interconnected transmission lines, substations, and generating plants of one or more utilities.

Grid interconnection The intertie of a cogeneration plant to an electric utility system to allow electricity flow in either direction.

Harmonics Waveforms whose frequencies are multiples of the fundamental (60 Hz or 50 Hz) wave. The combination of harmonics and fundamental wave causes a nonsinusoidal, periodic wave. Harmonics in power systems are the result of nonlinear effects. Typically, harmonics are associated with rectifiers and inverters, arc furnaces, arc welders, and transformer magnetizing current. There are voltage and current harmonics.

Heat rate A measure of generating station thermal efficiency, generally expressed in Btu per net kilowatt-hour.

Heating value The energy content in a fuel that is available as useful heat. The high heating value (HHV) includes the energy transmitted to water formed during combustion, whereas the low heating value (LHV) deducts this energy since it does no work on the piston.

Interruptible power Electric energy supplied by an electric utility subject to interruption by the electric utility under specified conditions.

Load factor The ratio of the average load supplied or required during a designated period to the peak or maximum load occurring in that period.

Maintenance power Electric energy supplied by an electric utility during scheduled outages of the cogenerator.

Off-peak Time periods when power demands are below average; for electric utilities, generally nights and weekends; for gas utilities, summer months.

Power factor The ratio of real power (kW) to apparent power (kVA) for any given load and time; generally expressed as a decimal.

Shaft efficiency Shaft energy output of a prime mover/energy input to it. For the fuel-fired prime mover, it is prime mover efficiency. For a steam turbine, it can be the thermal value of the steam, or the fuel value to produce the steam.

Selective energy systems A form of cogeneration in which part, but not all, of the site's electrical needs are met with on-site generation, with additional electricity purchased from a utility as needed.

Supplemental thermal The heat required when recovered engine heat is insufficient to meet thermal demands.

Supplementary firing The injection of fuel into an exhaust gas stream to raise its energy content (heat).

Supplementary power Electric energy supplied by an electric utility in addition to the energy the facility generates.

Thermal capacity The maximum amount of heat that a system can produce.

Thermal efficiency Electrical output divided by the energy input.

Total efficiency (First Law Efficiency) Electrical output plus thermal heat utilized divided by the energy input in consistent units.

Topping cycle A cogeneration facility in which the energy input to the facility is first used to produce useful power, with the reject heat from power production then used for other purposes.

Total energy systems A form of cogeneration in which all electrical and thermal energy needs are met by on-site systems; a total energy system can be completely isolated or switched over to a normally disconnected electrical utility system for backup.

Voltage flicker Term used to describe a significant fluctuation of voltage.

Wheeling The use of the transmission facilities of one system to transmit power for another system.

BIBLIOGRAPHY

Bos, P.G. and S.A. Davis. 1985. *Economic screening guidebook for cogeneration in buildings.* Gas Research Institute.

Bullard, C.W. and S.J. Pien. 1987. Optimal sizing of cogeneration systems. ASHRAE *Transactions* 92(2):321-32.

Guinn, G.R. 1987. Analysis of cogeneration systems using a public domain simulation. ASHRAE *Transactions* 93(2).

Heery Energy Consultants, Inc. 1988. *Georgia cogeneration handbook.* The Governor's Office of Energy Resources, Atlanta, GA.

Internal Combustion Engine Institute. 1962. *Engine Installation Manual.* National Engine Use Council, Chicago, IL.

National Engine Use Council. 1967. NEUC Engine Criteria, Chicago, IL.

Orlando, J.A. 1986. *Cogeneration feasibility analysis.* American Gas Association, Arlington, VA.

Orlando, J.A. 1987. *Natural gas prime movers.* American Gas Association, Arlington, VA.

Pedreyra, D.C. 1988. A microcomputer version of a large mainframe program for use in cogeneration analysis. ASHRAE *Transactions* 94(1).

Somasundarum, S. 1986. An analysis of a cogeneration system at T.W.U. *The Cogeneration Journal* 1(4):16-26.

Waukesha Engine Co. 1986. *Cogeneration handbook.* Waukesha Engine Co., Waukesha, WI.

CHAPTER 8

APPLIED HEAT PUMP AND HEAT RECOVERY SYSTEMS

APPLIED HEAT PUMPS

A HEAT pump is a device that extracts heat from one substance and transfers it to another portion of the same substance or to a second substance at a higher temperature. In a physical sense, all refrigeration equipment, including air conditioners and chillers with refrigeration cycles, are heat pumps. In engineering, however, the term "heat pump" is reserved for equipment that heats for beneficial purposes, rather than that which removes heat for cooling only. Dual-mode heat pumps provide separate heating and cooling, while heat reclaim heat pumps provide heating or simultaneous heating and cooling. An applied heat pump requires engineering for the specific application as opposed to the direct use of a manufacturer-designed unitary product. Applied systems include built-up (field- or custom-assembled from components) and industrial heat pumps. Almost all modern heat pumps use a vapor compression (modified Rankine) cycle. Any of the other refrigeration cycles discussed in Chapter 1 of the 1989 ASHRAE *Handbook—Fundamentals* are also suitable, and their use is being researched and developed. While most modern heat pumps are powered by electric motors, limited use is made of engine and turbine drives and absorption equipment. Applied heat pump systems are most commonly used for heating and cooling buildings, but they are gaining popularity for efficient domestic and service water, pool, and process heating.

Modern central heat pumps having capacities from 50 thousand Btu/h to 150 million Btu/h operate in many facilities. Some of these machines are capable of output water temperatures of 220°F and up to 60 psig steam.

Compressors in large central systems vary from one or more reciprocating or screw types to staged centrifugal types. A single or central system is often used throughout the building, but in some instances, the total capacity is divided among several separate heat pump systems to facilitate zoning. Ground, well water, solar, air, and internal building heat are used as heat sources. Compression can be single- or multistage. Frequently, heating and cooling are supplied simultaneously to separate zones.

Decentralized systems with water loop heat pumps are commonly used. These systems employ multiple water-source heat pumps connected by a water loop. They can also include heat rejectors (cooling towers and dry coolers), supplementary heaters

(boilers and steam heat exchangers), loop reclaim heat pumps, solar collection devices, and thermal storage. Initial costs are relatively low, and building reconfiguration is flexible.

Community and district heating systems based on both centralized and distributed heat pump systems are feasible. A few such systems are in use.

TYPES OF HEAT PUMPS

Several types of applied heat pumps (both open and closed cycle) are available; some reverse their cycle for both heating and cooling in HVAC systems, while others are for heating only in HVAC and industrial process applications. The basic types follow:

Closed vapor compression cycle (Figure 1). Using a conventional, separate refrigerant cycle, this is the most common type for both HVAC and industrial process use. It may use a compound, multistage, or cascade refrigeration cycle.

Mechanical vapor recompression (MVR) cycle with heat exchanger (Figure 2). Process vapor is compressed to a temperature and pressure sufficient for reuse directly in a process. Energy consumption is minimal, since temperature levels are optimum for the process. Typical applications for this cycle include evaporators (concentrators) and distillation columns.

Open vapor recompression cycle (Figure 3). A typical application for this cycle is in an industrial plant with a series of steam pressure levels and an excess of steam at a lower-than-desired pressure level. The heat is pumped to a higher pressure by compressing the lower pressure steam.

Waste heat driven Rankine cycle (Figure 4). This cycle is useful where large quantities of heat are wasted and where energy costs are high. The heat pump portion of the cycle may be either open or closed, but the Rankine cycle is usually closed.

Heat pumps are classified by (1) heat source and sink, (2) heating and cooling distribution fluid, (3) thermodynamic cycle, (4) building structure, (5) size and configuration, and (6) limitation of the source and the sink. Table 1 shows the more common types of closed cycle heat pumps for heating and cooling service.

The **air-to-air** heat pump is the most common type and is particularly suitable for factory-built unitary heat pumps. It is widely used in residential and commercial applications (see Chapter 47). The first diagram in Table 1 is typical of the refrigeration circuit used. A few installations have been made in which the forced convection indoor heat transfer surface has been replaced by a radiant panel.

The preparation of this chapter is assigned to TC 9.4, Applied Heat Pump/Heat Recovery Systems.

Table 1   Common Heat Pump Types

Heat Source and Sink	Distribution Fluid	Thermal Cycle	Diagram
Air	Air	Refrigerant Changeover	
Air	Air	Air Changeover	
Water	Air	Refrigerant Changeover	
Air	Water		
Earth	Air	Refrigerant Changeover	
Water	Water	Water Changeover	

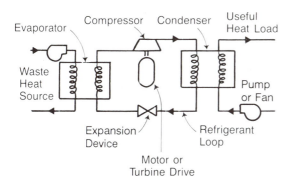

Fig. 1 **Closed Vapor Compression Cycle**

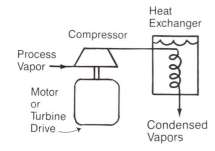

Fig. 2 **Vapor Recompression Cycle with Heat Exchanger**

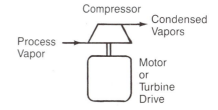

Fig. 3 **Open Vapor Recompression Concept**

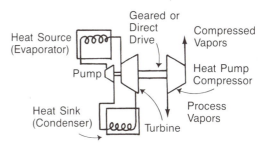

Fig. 4 **Waste Heat Driven Rankine Cycle**

In air-to-air heat pump systems, as shown in the second diagram of Table 1, the air circuits can be interchanged by motor-driven or manually operated dampers to obtain either heated or cooled air for the conditioned space. In this system, one heat exchanger coil is always the evaporator and the other is always the condenser. The conditioned air passes over the evaporator during the cooling cycle, and the outdoor air passes over the condenser. The positioning of the dampers causes the change from cooling to heating.

A **water-to-air** heat pump relies on water as the heat source and sink, and uses air to transmit heat to or from the conditioned space.

Air-to-water heat pumps without changeover are commonly called heat pump water heaters. With changeover controls, they may be found in large buildings where zone control is necessary. They are also used to produce hot or cold water in industrial applications.

Earth-to-air heat pumps can use the direct expansion of the refrigerant in a buried coil, as presented in Table 1, or they can be indirect, described under the earth-to-water type.

A **water-to-water** heat pump uses water as the heat source and sink for both cooling and heating. Heating-cooling changeover can be done in the refrigerant circuit, but it is often more convenient to perform the switching in the water circuits, as shown in Table 1. Although the diagram shows direct admittance of the water source to the evaporator, in some cases, it may be necessary to apply the water source indirectly through a heat exchanger (or double-wall evaporator) to avoid contaminating the water source into the normally treated closed chilled water system. Another method is to use a closed circuit condenser water system.

An **earth-to-water** heat pump may be like the earth-to-air type shown but may have a refrigerant-water heat exchanger like the water-to-air type shown in Table 1. Ground (earth) coupled heat pumps use the earth as a heat source or sink and are coupled to the buried heat exchanger by a secondary fluid circuit. A water or antifreeze solution is pumped through the horizontal or vertical pipes embedded in the earth. Soil, moisture, composition, density, and uniformity close to the surrounding field areas affect the success of this method of heat exchange. The materials of con-

struction for the pipe and the corrosiveness of the local soil and underground water affect the heat transfer and the service life. A variation of this cycle could transfer heat from the evaporator plus the heat of compression to a water-cooled condenser. The condenser heat would then be available for heating air, domestic hot water, or other items. A further variation could use refrigerant in direct expansion (DX), flooded, or recirculation evaporator circuits for the earth pipe coils or for an aboveground exchanger for air or other fluids.

Internal source, including water-loop, heat pumps use the high internal cooling load generated in modern buildings either directly or with storage.

Solar-assisted heat pumps rely on stored low-temperature solar heat as the heat source. Solar heat pumps may resemble water-to-air, or other types, depending on the form of solar heat collector and type of heating and cooling distribution system.

Waste source heat pump installations have been made that use sanitary sewage waste heat or laundry waste heat as a heat source. The waste fluid can be introduced directly into the heat pump evaporator after waste filtration, or it can be taken from a storage tank, depending on the application. An intermediate loop may also be used for heat transfer between the evaporator and the waste heat source.

Refrigerant-to-water heat pumps condense a refrigerant by the cascade principle shown on page 4.15 of the 1990 ASHRAE *Handbook—Refrigeration.* Cascading pumps the heat to a higher level, where it is rejected to water or another liquid. This type of heat pump can also serve as a condensing unit to cool almost any fluid or process. More than one heat source can be used to offset those times when insufficient heat is available from the primary source.

HEAT SOURCES AND SINKS

Table 2 shows the principal media as heat sources and heat sinks. The choice for an application is influenced primarily by geographic location, climatic conditions, initial cost, availability, and type of structure. Table 2 presents the various factors to be considered for each type. A discussion of design and selection factors for each source and sink follows.

Air

Outdoor air is a universal heat-source, heat-sink medium for the heat pump. Extended surface, forced convection heat transfer

Table 2 Heat Pump Sources and Sinks

Source or Sink	Examples	Suitability		Availability		Cost		Temperature		Common Practice	
		Heat Source	Heat Sink	Location Relative to Need	Coincidence with Need	Installed	Operation and Maintenance	Level	Variation	Use	Limitations
AIR											
outdoor	ambient air	good, but performance and capacity fall when very cold	good, but performance and capacity fall when very hot	universal	continuous	low	low	variable	generally extreme	most common, many standard products	defrosting and supplemental heat usually required
exhaust	building ventilation	excellent	fair	excellent if planned for in building design	excellent	low to moderate	low unless exhaust is dirt or grease laden	excellent	very low	emerging as conservation measure	insufficient for typical loads
WATER											
well	groundwater, well often shared with potable water source	excellent	excellent	poor to excellent, practical depth varies by location	continuous	low if existing well used or shallow wells suitable, can be high otherwise	low, but periodic maintenance required	generally excellent, varies by location	extremely stable	common	water disposal and required permits may limit, double wall exchangers may be required, may foul or scale
surface	lakes, rivers, ocean	excellent with large water bodies or high flow rates	excellent with large water bodies or high flow rates	limited, depends on proximity	usually continuous	depends on proximity and water quality	depends on proximity and water quality	usually satisfactory	depends on source	available, particularly for fresh water	often regulated or prohibited; may clog, foul, or scale
tap (city)	municipal water supply	excellent	excellent	excellent	continuous	low	low unless water use or disposal is costly	excellent	usually very low	excellent	use or disposal may be regulated or prohibited, may corrode or scale
condensing	cooling towers, refrigeration systems	excellent	poor to good	varies	varies with cooling loads	usually low	moderate	favorable as heat source	depends on source	available	suitable only if heating need is coincident with heat rejection
closed loops	building water-loop heat pump systems	good, loop may need supplemental heat	favorable, loop heat rejection may be needed	excellent if designed as such	as needed	low	moderate	as designed	as designed	very common	high cost for small buildings
waste	raw or treated sewage, gray water	fair to excellent	fair, varies with source	varies	varies, may be adequate	depends on proximity, high for raw sewage	varies, high for raw sewage	good	usually low	uncommon, practical only in large systems	usually regulated; may clog, foul, scale, or corrode
GROUND											
ground-coupled	horizontal or vertical buried pipe loops	good if ground is wet, otherwise poor	fair to good if ground is wet, otherwise poor	depends on soil suitability	continuous	high	low	usually good	low, particularly for vertical systems	available, increasing	high initial costs
direct	refrigerant circulated in ground	varies with soil conditions	varies with soil conditions	varies with soil conditions	continuous	high	favorable	varies by design	generally low	extremely limited	leaks, very expensive; large refrigerant quantities
SOLAR											
direct or heated water	solar collectors and panels	fair	poor, usually unacceptable	universal	highly intermittent, night use requires storage	extremely high	moderate to high	varies	extreme	very limited	supplemental source or storage required
INDUSTRIAL											
process heat or exhaust	distillation, molding, refining, washing	fair to excellent	varies, often impractical	varies	varies	varies	generally low	varies	varies	varies	often impractical unless heat need is near rejected source

coils transfer the heat between the air and the refrigerant. Typically, these surfaces are 50 to 100% larger than the corresponding surface on the indoor side of heat pumps that use air as the distributive medium. The volume of outdoor air handled is also greater in about the same proportions. The temperature difference during the heating operation between the outdoor air and the evaporating refrigerant is generally from 10 to 25 °F. The performance of air heating and cooling coils is presented in more detail in Chapters 21 and 24.

When selecting or designing an air-source heat pump, two factors in particular must be considered: (1) the temperature variation in a given locality, and (2) frost formation.

As the outdoor temperature decreases, the heating capacity of an air-source heat pump decreases. Selecting equipment for a given outdoor heating design temperature is therefore more critical than for a fuel-fired system. Consequently, the equipment must be sized for as low a balance point as is practical for heating without having excessive and unnecessary cooling capacity during the summer. A procedure for finding this balance point, which is defined as the outdoor temperature at which capacity matches heating requirements, is discussed in Chapter 47.

When the surface temperature of an outdoor air coil is 32 °F or lower, frost may form on it. If allowed to accumulate, this can inhibit heat transfer. Therefore, the outdoor coil must be defrosted periodically. The number of defrosting operations are influenced by the climate, air-coil design, and the hours of operation. Experience shows that little defrosting is generally required below 17 °F and below 60% rh. This can be confirmed by psychrometric analysis using the principles given in Chapter 21. However, under very humid conditions when small suspended water droplets are present in the air, the rate of frost deposit can be about three times as great as predicted from psychrometric theory. Then, a heat pump may require defrosting after as little as 20 min of operation. The effect of this condition on loss of available heating capacity should be taken into account in applying an air-source heat pump.

The early application of air-source heat pumps followed commercial refrigeration practice and used relatively wide fin spacing, i.e., 4 to 5 fin/in., on the theory that this would minimize the frequency of defrosting. However, experience has proven that effective hot gas defrosting permits much closer fin spacing and reduces the size and bulk of the system. In current practice, fin spacings of 8 to 13 per in. are widely used.

In many institutional and commercial buildings, some air must be continuously exhausted year-round. This exhaust air can be used as a heat source, although supplemental heat is generally added.

The high humidity in indoor swimming pools causes condensation on ceiling structural members, walls, windows, and floors and causes discomfort to spectators. Traditionally, outside air and dehumidification coils with reheat from a boiler that also heats the pool water are used. This application is ideal for air-to-air and air-to-water heat pumps because energy costs can be reduced. Suitable materials must be chosen to resist corrosion from chlorine and high humidity.

Water

Water can be a satisfactory heat source, subject to the considerations listed in Table 2. City water is seldom used because of cost and municipal restrictions. Geothermal water is an innovative heat source in some areas. Well water is particularly attractive because of its relatively high and nearly constant temperature. The water temperature is often a function of source depth, but it is generally about 50 °F in northern areas and 60 °F or higher in the south. Frequently, sufficient water is available from wells where the water is reinjected into the aquifer. The use is nonconsumptive and, with proper system design, only the temperature changes. The water quality should be analyzed, and the possible effect of scale formation and corrosion should be minimized. In some instances, it may be necessary to separate the well fluid from the system equipment with an additional heat exchanger. Special consideration must also be given to filtering and settling ponds for specific fluids. Other considerations are the costs of drilling, piping, pumping, and means for disposal of used water. Information on well water availability, temperature, and chemical and physical analysis is available from the U.S. Geological Survey offices located in many major cities.

Interest in geothermal energy has led to the identification of many geothermal energy sources with temperatures compatible with heat pump systems. They range from the typical shallow well temperatures mentioned above to temperatures where direct use for heating is possible. When using temperatures higher than typical groundwater temperatures, special attention should be given to designing the system for the proper fluid temperature drop. This is necessary, since increasing the temperature drop decreases the fluid flow requirements for a specified heating rate. When the available temperature is high enough for direct use, but the available water resource flow is insufficient to satisfy the heating load, the heat pump can be used to meet the load by achieving a greater drop in the resource temperature than the temperature necessary for direct use. Geothermal system designs are presented in more detail in Chapter 29 of the 1991 ASHRAE *Handbook—HVAC Applications*.

Surface or stream water may be used, but under reduced winter temperatures, the temperature drop across the evaporator may need to be limited to prevent freezeup.

In industrial applications, waste process water (e.g., spent warm water in laundries, plant effluent, and warm condenser water) may be a source for heat pump operations.

Use of water during cooling operations follows the conventional practice for water-cooled condensers.

Water-refrigerant heat exchangers are generally direct-expansion or flooded water coolers, usually of the shell-and-coil or shell-and-tube type. In small capacity refrigerant changeover systems, they are connected as refrigerant condensers during the heating cycle and as refrigerant evaporators during the cooling cycle. In applied heat pumps, the water is usually reversed instead of the refrigerant and the heat pump typically is used for heating only.

Earth

Earth as a heat source and sink, by heat transfer through buried coils, has recently been used extensively in many parts of the United States and Canada. Soil composition varies widely from wet clay to sandy soil and has a predominant effect on thermal properties and expected overall performance. The heat transfer process in soil depends on transient heat flow. Thermal diffusivity (α), the ratio of thermal conductivity to the product of unit density and specific heat $(\alpha = k/pc)$, is a dominant factor and is difficult to determine without local soils data. The soil moisture content also influences thermal conductivity.

Earth coils are generally one of two basic types. The first is single, or multiple, serpentine heat exchanger pipes buried 3 to 6 ft apart in a horizontal plane at a depth of 3 to 6 ft below grade. Pipes may be buried deeper, but excavation costs and temperature must be considered. The second type of coil is a vertical concentric tube or U-tube heat exchanger. A vertical coil may consist of one long or several shorter exchangers. The ASHRAE *Design/Data Manual* for closed loop ground coupled heat pumps provides information on earth coil design.

Solar

The use of solar energy as a heat source, either on a primary basis or in combination with other sources, has attracted interest but has not emerged as an economical system. Air, surface water, shallow groundwater, and shallow ground-source systems all use solar energy indirectly. The principal advantage of using solar energy directly as a heat pump heat source is that, when available, it provides heat at a higher temperature level than other sources, resulting in an increase in coefficient of performance. Compared to a solar heating system without a heat pump, the collector efficiency and capacity are increased because of the lower collector temperature required.

Research and development in the solar-source heat pump field has been concerned with two basic systems—direct and indirect. The direct system places refrigerant evaporator tubes in a solar collector, usually a flat-plate type. Research shows that a collector without glass cover plates can also extract heat from the outdoor air. The same surface may then serve as a condenser using outdoor air as a heat sink for cooling. The refrigeration circuit resembles the earth-to-air heat pump shown in Table 1.

An indirect system circulates either water or air through the solar collector. When air is used, the first system shown in Table 1 for an air-to-air heat pump may be selected; the collector is added in such a way that (1) the collector can serve as an outdoor-air preheater, (2) the outdoor-air loop can be closed so that all source heat is derived from the sun, or (3) the collector can be disconnected from the outdoor air serving as the source or sink. When water is circulated through the collector, the heat pump circuit may be either water-to-air or water-to-water, as illustrated in Table 1.

All heat pump systems using solar energy directly as the only heat source require either an alternate heating system or a means of storing heat during periods of insufficient solar radiation. Heat storage is discussed in the section Auxiliaries and in Chapter 30 of the 1991 ASHRAE *Handbook—HVAC Applications*.

HEAT PUMP COMPONENTS

For the most part, the components and practices associated with heat pumps evolved from work with low-temperature refrigeration. This section outlines the major components used and points out characteristics or special considerations that apply to heat pumps in combined room heating and air-conditioning applications or higher temperature industrial applications.

Compressors

Centrifugal compressors. Most centrifugal applications in heat pumps have been limited to water-to-water or refrigerant-to-water heat pump systems, heat transfer systems, storage systems, and hydronically cascaded systems (see the section Operating Cycles). With these applications, the centrifugal compressor enables heat pumps to enter the field of industrial plants, as well as large multistory buildings; many installations with double-bundle condensers have been made. The transfer cycles permit low pressure ratios, and many single- and two-stage units with various refrigerants are operational with high COPs. Centrifugal compressor characteristics do not always meet the needs of air-source heat pumps. High pressure ratios, or high lifts, associated with low gas volume at low load conditions cause the centrifugal compressor to surge.

Reciprocating compressors. These compressors are used more than any other type on systems in the range of 0.5 to 100 tons. A brief description of this compressor is given in Chapter 35. For the most economical concept applied to combined heating and cooling, it is general practice to select the compressor for the heating duty and a separate, high-efficiency chiller for the remaining cooling duty. When heating, the capacity of a particular compressor is the sum of the evaporator capacity (*i.e.*, heat-source coil capacity) and the heat equivalent of the compressor work minus heat

losses outside the conditioned space. Some machines are designed for heating only and others for dual heating and cooling.

A compressor used for comfort cooling has a medium clearance volume (ratio of gas volume remaining in cylinder after compression to the total swept volume is about 0.05). For an air-source heat pump, a compressor with a small clearance volume ratio, *e.g.*, 0.025, is more suitable for low-temperature operation and provides, for example, about 15% greater refrigerating capacity at an evaporator temperature of 0°F and a condenser temperature of 110°F. However, this compressor has somewhat more power demand under maximum cooling load conditions than does one of medium clearance.

More total heat capacity can be obtained at low outdoor temperatures by deliberately oversizing the compressor. When this is done, some capacity reduction for operation at higher outdoor temperatures can be provided by multi- or variable-speed drives, cylinder cutouts, or other methods.

The disadvantage of this arrangement is that the greater number of operating hours that occur at the higher suction temperatures must be served with the compressor in the unloaded condition, which generally causes lower efficiency and higher annual operating cost. The additional initial cost of the oversized compressor must be economically justified by the gain in heating capacity. One method proposed for increasing the heating output at low temperatures uses *staged compression,* in which one compressor may compress from −29°F saturated suction temperature (SST) to 40°F saturated discharge temperature (SDT), and a second compressor compresses the vapor from 40°F SST to 120°F SDT. In this arrangement, any two compressors may be interconnected in parallel, with both pumping from about 45°F SST to 120°F at the normal cooling rate point. Then at some predetermined outdoor temperature on heating, they are reconnected in series and compress in two successive stages.

Figure 5 shows the performance of a pair of compressors for units of both medium and low clearance volume. At low suction temperatures, the reconnection in series adds some capacity. However, the motor for this case must be selected for the maximum loading conditions for summer operation, even though the low stage compressor has a greatly reduced power requirement under the heating condition.

Rotary vane and scroll compressors. These compressors can be used for the low stage multistage plant; they have a high capacity but are generally limited to lower pressure ratios. They also have limited means for capacity reduction.

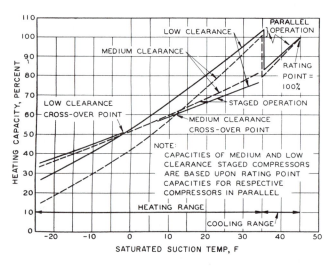

Fig. 5 Comparison of Parallel and Staged Operation for Air-Source Heat Pumps

Screw compressors. Screw compressors offer high pressure ratios at low to high capacities. Capacity control is usually provided by such means as variable porting or valving. Generally, large oil separators are required, since currently available compressors use oil injection. They are less susceptible to damage from liquid spillover and have fewer parts than do reciprocating compressors. They also simplify capacity modulation.

Heat Transfer Components

Refrigerant-to-air and refrigerant-to-water heat exchangers, as previously described in the section Heat Sources and Sinks, are similar to heat exchangers currently used in air-conditioning refrigeration. A refrigerant subcooler coil can be paired with an indoor air coil or used to preheat ventilation air on systems with a ventilation air supply. A substantial gain in capacity and coefficient of performance can result.

VESSELS AND PIPING COMPONENTS

Refrigeration Components

Refrigerant piping, receivers, expansion devices, and refrigeration accessories in heat pumps are usually the same as components found in other types of refrigeration and air-conditioning systems.

A **reversing valve** changes the system from the cooling to the heating mode. This changeover requires the use of a valve(s) in the refrigerant circuit, except where the change occurs in fluid circuits external to the refrigerant circuit (see Table 1). Reversing valves are usually pilot-operated by solenoid valves, which admit head and suction pressures to move the operating elements.

Expansion devices for controlling the refrigerant flow are normally thermostatic expansion valves, as described in Chapter 19 of the 1988 ASHRAE *Handbook—Equipment*. The control bulb must be located carefully. If the circuiting is arranged so that the refrigerant line on which the control bulb is placed can become the compressor discharge line, the resulting pressure developed in the valve power element may be excessive, requiring a special control charge or pressure-limiting element. When a thermostatic expansion valve is applied to an outdoor air coil, it is desirable to have a special cross-charge to limit the superheat at low temperatures for better use of the coil. When an expansion valve is attached to a coil operated as a condenser, a bypass with a check valve is normally provided, as indicated in Table 1.

On an air-source heat pump that operates over a wide range of evaporating temperatures, a capillary tube passes refrigerant at an excessive rate at low back pressures, causing liquid floodback to the compressor. In some cases, suction line accumulators or charge-control devices are added to minimize this effect.

A **refrigerant receiver**, which is commonly used as a storage place for liquid refrigerant, is particularly useful in a heat pump to take care of the unequal refrigerant requirements of heating and cooling. The receiver is usually omitted on heat pumps used for heating only.

Defrost Control (Air Source)

A variety of defrosting control schemes sense the need for defrosting air-source heat pumps and initiate (and terminate) the defrost cycle. The cycle is usually initiated by demand rather than a timer. Termination of the cycle may be timer-controlled.

The defrost cycle can also be terminated either by a control sensing the coil pressure or a thermostat that measures the temperature of the liquid refrigerant in the outdoor coil. When the temperature (or corresponding saturation pressure) of the liquid leaving the outdoor coil rises to about 70°F, the completion of defrosting is ensured.

Another means of starting the defrost cycle is with a pressure control that reacts to the air pressure drop across the coil. Under conditions of frost accumulation, the airflow is reduced and the increased pressure drop across the coil initiates the defrost cycle. Again, the preferred method of terminating the defrost cycle is by using a refrigerant temperature control that measures the temperature of the liquid refrigerant in the coil.

A third method for defrosting involves a temperature differential control and two temperature-sensing elements. One element is responsive to outdoor-air temperature and the other to the temperature of the refrigerant in the coil. As frost accumulates, the differential between the outdoor temperature and the refrigerant temperature increases, initiating a defrost cycle. The system is restored to operation when the refrigerant temperature in the coil reaches a specified temperature, indicating that defrosting has been completed. When the outdoor air temperature decreases, the differential between outdoor air temperature and refrigerant temperature decreases, initiating the defrost cycle with greater frost buildup unless compensation is provided.

Auxiliaries

Supplementary heaters may be included either within the heat pump package or external in the system. They are frequently controlled to temper the air during defrost operation on units using reverse-cycle defrost. They are required for start-up where the heat source and sink are the same (*e.g.*, heating process water using process effluent as the source). Usually, these heaters do not require sizing for the total heat load.

Thermal storage in a heat pump system can improve its performance characteristics and is essential where the heat source and heating loads are not coincident. All materials have thermal storage properties to a greater or lesser degree. In buildings, the structural materials are almost always either absorbing heat from or delivering heat to the interior space. This effect is more pronounced in the cooling operation where greater air temperature variation is tolerated. Storage tends to reduce the rate of temperature change and helps to reduce the peak equipment requirements. In this sense, every heating and cooling system can be said to involve heat storage.

Many attempts have been made, particularly in recent years, to increase the heat storage effect by using special heat storage materials as part of the heating or cooling system. The result has been a reduction in the size of the heating or cooling equipment necessary to take care of peak demands.

In the case of the heat pump, a provision for heat storage not only reduces the size of the heat pump necessary for a given load, but it also provides a more desirable electrical load by shifting part of the load to the time of day when the cost of power is least. The off-peak electric hot water heater is a common example of such a heat storage application.

BUILDING DESIGN CONSIDERATIONS

Modern buildings use double-glazed windows and better insulation, as well as more internal areas with considerably higher internal heat gain from lights, people, computers, and other heat-dissipating equipment. Systems for such buildings are normally required, particularly during mild weather, to supply heating to the exterior zones and cooling to the interior zones simultaneously. In many instances, these systems may have to change daily or even hourly between the heating and cooling cycle to maintain the required indoor conditions.

The heat pump is well-qualified for such applications and frequently shows a considerable saving in operating cost over other systems because of its inherent ability to transfer heat from an interior zone, where cooling may be desired, to the exterior, where heating may be needed.

OPERATING CYCLES

Figures 6, 7, and 8 show three of the several possible operating cycles for simultaneously providing heating and cooling. Figure 6 illustrates one method of using water as the heat source or sink and as the heating and cooling medium. The compressor, evaporator, condenser, refrigerant piping, and accessories are essentially standard and are available as a factory-packaged water-to-water heat pump.

The cycle is flexible, and the heating or cooling medium is instantly available at all times. Heating can be provided exclusively to the zone conditioners by closing valves 2 and 3 and opening valves 1 and 4. With the valves in these positions, the water is divided into two separate circuits. The warm water circuit consists of the condenser (where the heat is supplied by the high-temperature refrigerant), valve 1, zone conditioners, and a circulating pump. The cold water circuit consists of the evaporator (where heat is taken from the water by the low-temperature refrigerant), valve 4, exchanger (where heat is taken from the source water), valve 1, and a circulating pump. The refrigerating compressor operates to maintain the desired leaving water temperature from the condenser.

Similarly, cooling can be exclusively obtained in the cycle of Figure 6 by opening valves 2 and 3 and closing valves 1 and 4. With this arrangement, the cold water circuit consists of the evaporator (where heat is removed from the water by the low-temperature refrigerant), valve 2, zone conditioners, and a circulating pump. The warm water circuit consists of the condenser (which receives the heat from the refrigerant), valve 3, exchanger (where heat is rejected to the source water), valve 2, and a circulating pump. The refrigerating compressor operates to maintain the desired water temperature leaving the evaporator.

During the intermediate season, simultaneous heating and cooling can be provided by the cycle shown in Figure 6. Valves 3 and 4 are modulated when valves 1 and 2 are open. Valve 3 is adjusted to maintain 85 to 140 °F water in the condenser circuit and valve 4 to maintain 45 to 50 °F in the evaporator circuit. The excess heating or cooling effect is wasted to the exchanger, which passes it on to the source water.

The source or sink water, if of suitable quality, can be supplied directly to the condenser and evaporator instead of by an exchanger, as illustrated by Figure 6, eliminating one heat transfer surface and its performance penalty.

Figure 7A shows an air-source, air-to-water, single-stage heat pump working with a four-pipe system for simultaneous heating and cooling. The refrigerant cycle of this package is factory preassembled. However, the outside summer condenser or winter evaporator must be field connected to the compressor unit. In such heat pumps, which are available in packages to 150 tons of cooling, the only operational reversal takes place in the outdoor condenser-evaporator. The shell-and-tube evaporator and condenser remain as cooler and heater throughout, respectively. Most

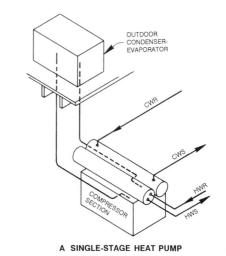

A SINGLE-STAGE HEAT PUMP

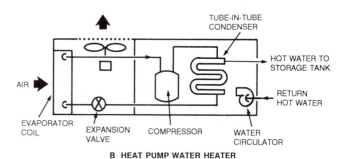

B HEAT PUMP WATER HEATER
(Adapted from McQuay, Inc.)

Fig. 7 Air-to-Water Heat Pumps

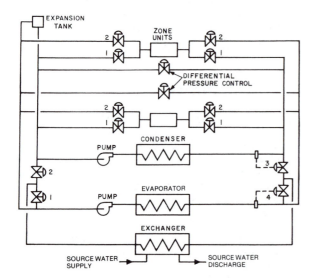

Fig. 6 Water-to-Water Heat Pump Cycle

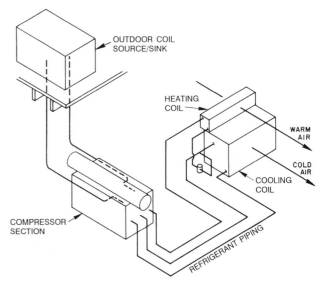

Fig. 8 Air-to-Air Heat Pump

commercial units of this type use the flooded principle. Figure 7B shows a typical air-to-water heat pump water heater schematic.

The air-to-air heat pump (Figure 8) is more efficient than the air-to-water system, since the conditioned air is heated or cooled directly by the refrigerant without the addition of another media. No shell-and-tube heat exchangers are required, which reduces initial costs. All previously described heat pumps are also available without the simultaneous feature. Careful evaluation of each application permits the designer to select the best system for each particular case.

APPLIED HEAT RECOVERY

Many commercial structures require simultaneous heating and cooling for prolonged periods during occupancy. Frequently, the mechanical cooling equipment and the heating equipment must be operated jointly to provide the necessary comfort. Many cycles have been developed to permit the transfer of heat from areas of heat surplus in the building to areas requiring heat.

The details of specific systems are limited only by economics and the designer's imagination. Optimum use of countercurrent heat exchange surfaces and innovative circuiting of chillers and condensers to reduce pumping pressure, optimum circuiting of storage tanks, and outside air exchange all improve efficiency and enhance processes.

Conceptually, the designer must ask (1) whether the total heat energy entering and/or generated within the facility is fully used before it is rejected from the facility, and (2) if it is not, what percentage of the total heat input is wasted, and what portion of this waste could be recovered to defer or reduce HVAC or water heating requirements cost effectively?

Comprehensive designs require (1) knowledge of the building, processes, operating patterns, available energy sources, and an economic analysis; (2) designers with an understanding of the material contained in this volume; and (3) specific performance data on primary HVAC equipment (many components are customized to meet specific job requirements).

While first cost of equipment is important, payback to the owner in operating and maintenance cost savings should be a primary concern. Another concern is the volume and floor space required by air economizer units. An analysis of internal loads and available temperatures and appropriate scheduling of equipment operation during all seasons can help keep space requirements to a minimum.

DEFINITIONS

Internal heat is total passive heat generated within the conditioned space. It includes heat generated by lighting, computers, business machines, occupants, and mechanical and electrical equipment such as fans, pumps, compressors, and transformers. Normally, this heat must be removed to control the environment.

Internal process heat is heat from industrial activities and sources such as waste water, boiler flue gas, coolants, exhaust air, and some waste materials. This heat is normally wasted unless equipment is included to extract it for further use.

External heat is heat generated from sources outside the conditioned area. This heat from gas, oil, steam, electricity, or solar sources supplements internal heat and internal process heat sources. Recovered heat can reduce the demand for external heat.

Waste heat is heat rejected from the building (or process) because its temperature is too low for economical recovery or direct use.

Recovered (or reclaimed) heat comes from internal heat sources. It is used for space heating, domestic or service water heating, air reheat in air conditioning, or other similar requirements. Recovered heat may be stored for later use.

Stored heat from external or recovered heat sources is held in reserve for later use.

Usable temperature is the temperature or range of temperatures at which heat energy can be absorbed, rejected, or stored for use within the system.

Breakeven temperature (t_{be}) is the outdoor temperature at which the total heat losses from conditioned spaces equal the internally generated heat gains.

Changeover temperature (t_{co}) is the outdoor temperature that the designer selects as the changeover point from net cooling to net heating by the air-conditioning system.

Balanced heat recovery occurs when internal heat gain equals recovered heat and no external heat is introduced to the conditioned space. Maintaining balance may require raising the temperature of the recovered heat. The following section provides more details.

BALANCED HEAT RECOVERY IN BUILDINGS

In an ideal balanced system, all components work year-round to recover all the internal heat before adding external heat. Any excess heat is either stored, rejected, or both. The section on Heat Storage includes more details.

When the outdoor temperature drops significantly, or when, on nights and weekends, the building is shut down, internal heat gain may be insufficient to meet space conditioning requirements. Then, a balanced system provides heat from storage or an external source. When internal heat is again generated, the external heat is automatically reduced to maintain proper temperature in the space. Some time occurs before equilibrium is reached.

The size of the equipment and the external heat source can be reduced in a storage balanced system. Regardless of the system, a heat balance analysis establishes the merits of balanced heat recovery at various outdoor temperatures.

Outdoor air less than 55 to 65 °F may be used to cool building spaces in an air economizer cycle. When considering this method of cooling, the space required by ducts, air shafts, and fans, as well as the increased filtering requirements to remove contaminants and the hazard of possible freezeups of dampers and coils must be weighed against alternatives such as the use of deep row coils with antifreeze fluids and efficient heat exchange. Innovative use of heat pump principles may give considerable energy savings and more satisfactory human comfort than an air economizer. In any case, hot and cold air should never be mixed (if avoidable) to control zone temperatures because it wastes energy.

HEAT REDISTRIBUTION

Many building projects, especially those with computers or large interior areas, generate more heat than can be used for most of the year. Operating cost is kept to a minimum when the system changes over from net heating to net cooling at the breakeven outdoor temperature at which external heat losses equal internal heat loads. If heat is unnecessarily rejected or added to the space, the changeover temperature will vary from the natural breakeven temperature, and operating costs will increase. Heating costs can be reduced or eliminated if excess heat is stored for later distribution.

The various means of using internal heat provide additional criteria for selecting both equipment and the external heat source. The optimum selection is often discovered after the heat balance and its effect on operation is analyzed. Fuel or electric power costs may not be the dominant factor regarding operating costs until a proper changeover temperature is chosen. Equipment that substantially lowers the changeover temperature can cause higher operating costs than selections with higher fuel or power costs.

HEAT BALANCE CONCEPT

The ideal heat balance concept in an overall building project or a single space requires that one of the following takes place on demand:

- Heat must be removed.
- Heat must be added.
- Heat generated must exactly balance the heat required, in which case heat should be neither added nor removed.

In small air-conditioning projects serving only one space, either cooling or heating satisfies thermostat demands. If humidity control is not required, the problem is simple. Assuming both heating and cooling are available, the automatic controls will respond to the thermostat to supply either. A system should not heat and cool the same space simultaneously.

Multiroom buildings commonly require heating in some rooms and cooling in others. Optimum design considers the building as a whole and transfers excess internal heat from one area to another, as required, without introducing external heat that would require waste heat disposal at the same time. The heat balance concept is violated when this occurs.

Humidity control must also be considered. Any system should add or remove only enough heat to maintain the desired temperature, plus any heat required for humidity control. Large percentages of outdoor air with high wet-bulb temperatures, as well as certain types of humidity control, may require reheat, which could upset the desirable balance. Usually, humidity control can be obtained without upsetting the balance. When reheat is unavoidable, internally transferred heat from heat recovery should always be used to the extent available before using an external heat source such as a boiler. However, the effect of the added reheat must be analyzed, since it affects the heat balance and may have to be treated as a variable internal load.

When a building requires heat and the refrigeration plant is not in use, dehumidification is not usually required and the outdoor air is dry enough to compensate for any internal moisture gains. (This should be carefully reviewed for each design.)

HEAT BALANCE STUDIES

The following analytical examples illustrate situations that can occur in nonrecovery and unbalanced heat recovery situations. Figure 9 graphically shows the major components that comprise the total air-conditioning load of a building. Values above the zero line are cooling loads, and values below the zero line are heating loads. On an individual basis, the ventilation and conduction loads cross the zero line, which indicates that these loads can be a heating or a cooling load, depending on outdoor temperature. The solar load and internal loads are always a cooling load and are, therefore, above the zero line.

Figure 10 combines all the loads shown in Figure 9. The graph is obtained by plotting the conduction load of a building at various outdoor temperatures, and then adding or subtracting the other loads at each temperature. The project load lines with and without solar effect cross the zero line at 16 and 30°F, respectively. These are the outdoor temperatures for the plotted conditions when the naturally created internal load exactly balances the losses.

As plotted, this heat balance diagram includes only the building loads with no allowance for additional external heat from a

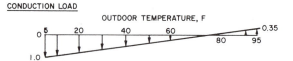

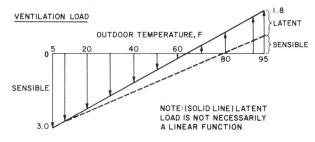

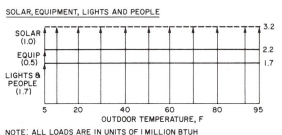

NOTE: ALL LOADS ARE IN UNITS OF 1 MILLION BTUH

Fig. 9 Major Load Components

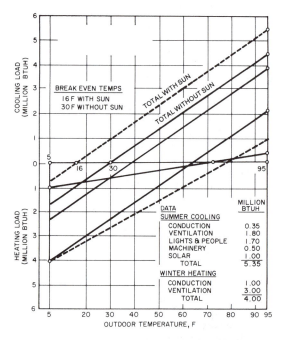

**Fig. 10 Composite Plot of Loads in Figure 1
(Adjust for Internal Motor Heat)**

boiler or other source. If external heat is necessary because of system design, the diagram should include the additional heat.

Figure 11 illustrates what happens when heat recovery is eliminated. Here, it is assumed that at a temperature of 70°F, heat from an external source is added to balance conduction through the skin of the building in increasing amounts down to the minimum outdoor temperature winter design condition. Figure 11 also adds the heat required for the outdoor air intake. The outdoor air compris-

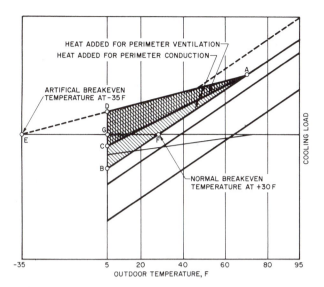

Fig. 11   Non-Heat Recovery System

ing part or all of the supply air must be heated from outdoor temperature to room temperature. Only the temperature range above the room temperature is effective for heating to balance the perimeter conduction loss.

These loads are plotted at the minimum outdoor winter design temperature, resulting in a new line passing through points A, D, and E. This line crosses the zero line at −35 °F, which becomes the artificially created breakeven temperature rather than 30 °F, when not allowing for solar effect. When the sun shines, the added solar heat at the minimum design temperature would further drop the −35 °F breakeven temperature.

Such a design adds more heat than the overall project requires and does not use a balanced heat recovery system to use the available internal heat. This problem is most evident during mild weather on systems not designed to take full advantage of internally generated heat on a year-round basis.

Two examples of situations that can be shown in a heat balance study are as follows:

1. As the outdoor air wet-bulb temperature drops, the total heat of the air falls. If a mixture of outdoor and recirculated air is cooled to 55 °F in the summer and the same dry-bulb temperature is supplied by an economizer cycle for interior space cooling in the winter, there will be an entirely different result. As the outdoor wet-bulb temperature drops below 55 °F, each unit volume of air introduced does more cooling. To make matters more difficult, this increased cooling is latent cooling, which requires the addition of latent heat to prevent too low a relative humidity, yet this air is intended to cool. The extent of this added external heat for free cooling is shown to be very large when plotted on a heat balance analysis at 0 °F outside temperature.

Figure 11 is typical for many current non-heat recovery systems. There may be a need for cooling, even at the minimum design temperature, but the need to add external heat for humidification can be eliminated by using available internal heat. When this asset is thrown away and external heat is added, operation is inefficient.

Some systems recover heat from exhaust air to heat the incoming air. When a system operates below its natural breakeven temperature (t_{be}) such as 30 or 16 °F (shown in Figure 10), the heat recovered from exhaust air is useful and beneficial. This assumes that only the available internal heat is used and that no supplementary heat is added at or above the t_{be}. Above the t_{be}, the internal heat is sufficient and any recovered air would

become excessive heat to be removed by more outdoor air or refrigeration.

If heat is added to a central system to create an artificial t_{be} of −35 °F as in Figure 11, any recovered heat above −35 °F requires an equivalent amount of heat removal elsewhere. If the project were in an area with a minimum design temperature of 0 °F, heat recovery from exhaust air could be a liability at all times for the conditions stipulated in Figure 10. This does not mean that the value of heat recovered from exhaust air should be forgotten. The emphasis should be on recovering heat from exhaust air rather than on adding heat from external sources.

Proprietary equipment with counterflow chiller and condenser circuits, as well as efficient exchange with exhaust or ambient air can often reduce operating costs without the large air volumes required by air economizers.

2. A heat balance shows that insulation, double glazing, and so forth can be extremely valuable on some projects. However, these practices may be undesirable in certain regions during the heating season, when excess heat must usually be removed from large buildings. For instance, for minimum winter design temperatures of approximately 35 to 40 °F, it is improbable that the interior core of a large office building will ever reach its breakeven temperature. The temperature lag for shutdown periods, such as nights and weekends, at minimum design conditions could never economically justify the added cost of double windows. Therefore, double windows merely require the amount of heat saved to be removed elsewhere.

These are only two of many factors that can be analyzed. A temperature-frequency scale superimposed on a properly constructed heat balance diagram can be another valuable analytical tool. It is possible to further superimpose a kilowatt-hour or steam consumption cost and integrate it with a time frequency scale to predict approximate operating cost with different types of power or fuel. Many computer programs with weather tapes are available for this analysis.

GENERAL APPLICATIONS

System Considerations

A properly applied heat reclaim system automatically responds to the balanced heat recovery concept. An example is a reciprocating water chiller with a hot gas diverting valve and both a water-cooled and an air-cooled condenser. Hot gas from the compressor is rejected to the water-cooled condenser. This hot water provides internal heat as long as it is needed. At a predetermined temperature, the hot gas is diverted to the air-cooled condenser, rejecting excess heat from the total building system. Larger projects with centrifugal compressors use double condenser chiller units, which are available from many manufacturers. For typical buildings, chillers normally provide hot water for space heating at 105 to 110 °F.

Many buildings that run chillers all or most of the year are reclaiming some of the condenser heat to provide domestic hot water (see Figure 12).

Designers should include a source of external heat for backup. The control system should ensure that backup heat is not injected unless all internal heat has been used. For example, if electric backup coils are in series with hot water coils fed from a hot water storage tank, they may automatically start when the system restarts after the building temperature has dropped to a "night low limit" setting. An adjustable time delay in the control circuit gives the stored hot water time to warm the building before energizing the electric heat.

This type of heat reclaim system is readily adaptable to smaller projects using a reciprocating chiller with numerous air terminal units or a common multizone air handler. The multizone air handler

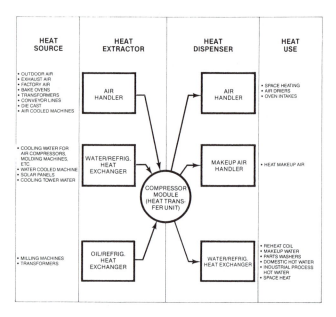

Fig. 12 Industrial Heat Recovery Schematic

should have individual zone duct heating coils and controls arranged to prevent simultaneous heating and cooling in the same zone.

Properly applied heat reclaim systems not only meet all space heating needs, but they also provide hot water required for public showers, food service facilities, and reheat in conjunction with dehumidification cycles.

Heat reclaim chillers or heat pumps should not be used with air-handling systems that have modulating damper economizer control. This free cooling concept may result in a higher annual operating cost than a minimum fresh air system with a heat reclaim chiller. Careful study will show if the economizer cycle violates the heat balance concept.

Heat reclaim chillers or heat pumps are available in many sizes and configurations. Combinations include (1) centrifugal, reciprocating, and screw compressors; (2) single- and double-bundle condensers; (3) cascade design for higher temperatures (up to 220°F); and (4) air- or water-cooled or both.

The designer can make the best selection after doing both a heating and cooling load calculation and a preliminary economic analysis and after understanding the building, processes, operating patterns, and available energy sources.

The application of heat reclaim chillers or heat pumps ranges from simple systems with few control modes to complex systems having many control modes and incorporating two-, three-, or four-pipe circulating systems. Certain system concepts using double-bundle condensers and single-condenser bundles coupled with exterior closed circuit coolers have been patented. These patents may impose some constraints on design considerations, but royalties or other arrangements may be acceptable. Potential infringements should be checked early in the planning stage.

A successful heat recovery design depends on the performance of the total system, not just the chiller or heat pump. A careful and thorough analysis is often time-consuming and requires more design time than a nonrecovery system. The balanced heat recovery concept should guide all phases of planning and design, and the effects of economic compromise should be studied. There may be little difference between the initial cost (installed cost) of a heat recovery system and a nonrecovery system, especially in larger projects. Also, in view of energy costs, life cycle analysis usually shows dramatic savings when using balanced heat recovery. Many project budgets, however, will not fund all measures required to optimize heat reclaim.

MULTIPLE BUILDINGS

A multiple building complex is particularly suited to heat recovery. Variations in occupancy and functions provide an abundance of heat sources and uses. Applying the balanced heat concept to a large multibuilding complex can result in large energy savings. Each building captures its own total heat by interchange. Heat rejected from one building could possibly heat adjacent buildings.

INDUSTRIAL HEAT RECOVERY

Many opportunities for heat recovery exist in industrial facilities. Heat exchangers can be effective energy devices in individual plants. Chapter 44 covers air-to-air recovery equipment, including guidance for economic evaluation.

Proximity of the heat source to the heat use is the most significant constraint when using heat recovery equipment. The cost of transporting air over great distances can be prohibitive.

In many instances, a piping system coupled with a refrigeration compressor (heat transfer unit) can replace an expensive duct system. Figure 12 illustrates this concept and is intended to stimulate ideas. The primary compressor module can be reciprocating, centrifugal, or screw. The transfer fluid can be water, glycol, or refrigerant. The final selection and arrangement depends on several factors, including temperature, distances, total load, and configuration of the heat sources. For specific industrial processes, the application manuals and documented case histories available from manufacturers of recovery refrigeration machines should be consulted.

A variety of heat exchangers are available for heat recovery, especially for industrial applications. Both parallel and counterflow exchangers may be used, although counterflow is most commonly used for heat recovery. Tube-in-shell exchangers installed so the fluid most likely to cause fouling passes through the tubes are also common. Removable tube heads facilitate mechanical cleaning.

Appropriate materials that resist corrosion and fouling from the fluids should be specified for use. The supplier needs accurate information on fluid properties, flow rates, fouling factors, and pressure drops to furnish reliable and cost-effective exchangers. In some cases, intermediate heat exchangers are needed between a process fluid and the water or brine that passes through the chillers and condensers.

For good operating economy, plate-type heat exchangers permit very close temperature approaches in the range of 2 to 5°F. These exchangers are made with parallel corrugated titanium, stainless steel, or other appropriate material separated by gaskets and then clamped together. Because the plates can be easily separated for cleaning and inspection, they are commonly used in the food and dairy industry.

Spiral plate heat exchangers have the advantage of enabling close temperature approaches in a compact unit. Thus, they require less space than other exchangers.

Another type of heat exchanger consists of trombone-shaped horizontal coils stacked in vertical rows. The coils are cooled by water or other fluids flowing over them. This exchanger requires more space than others; it has the advantage of easy access for cleaning and inspection during continuous operation.

HEAT STORAGE

Heat can be stored for later use in liquids and phase-change materials. The demand for stored heat varies with the heating load profile versus the rejected heat available from various sources. The amount of rejected heat seldom matches the requirement; there is either a simultaneous excess or deficiency. Heat storage is discussed in Chapter 30 of the 1991 ASHRAE *Handbook—HVAC Applications.*

The designer must provide a cost-effective means of (1) accepting rejected heat, (2) storing it for a satisfactory time with minimal thermal losses, and (3) delivering it where needed at the proper temperature, consistent with the requirements for the flow and heat transfer capabilities of the devices demanding the heat. Supplemental heat sources are used to increase the amount of heat and/or boost the temperature of heat from storage. Suitable baffles and piping orientation reduce the need to boost the temperature.

ARI *Standard* 410-87 lists entering fluid temperatures of 120 to 250 °F for air heating coils. Traditionally, 180 °F fluid with a 10 to 20 °F range has been used in coils for cost-effective air units, although special applications have been found for up to a 50 °F range. Heating coils usually cannot be used for fluids below 110 °F, although entering fluids down to 105 °F temperature level have been used successfully.

Standard centrifugal units modified for heat recovery can supply 105 °F fluid, whereas special centrifugal units can routinely supply up to 120 °F fluid. Positive displacement compressor R-22 systems can provide up to 125 °F fluid; R-12 heat pump systems can provide up to 160 °F fluid for heating purposes with full heat rejection. Desuperheaters can provide less heat at higher temperatures. Temperatures of 220 °F can be reached by other halocarbon refrigerants. Beyond these temperatures, booster heaters are generally required to avoid exceeding the practical limits on compression equipment and lubricants.

Sensible Heat Storage

Sensible heat storage alternatives include (1) tank storage, (2) aquifer thermal energy storage (ATES), (3) cavern/mine storage, (4) earth storage, (5) lake and pond storage, and (6) rock bed storage.

Heat is stored sensibly by changing the temperature of the heat storage material (usually a liquid and/or a solid). The material's heat content changes without the material changing phase. In the cases of tank, lake, and pond storage, the storage material also serves as the heat transfer fluid. For other storage types, heat is also stored in the containment or matrix materials.

Size and economics dictate which storage is best suited to a given heat recovery application. Storage size is determined by energy availability and demand profiles, storage losses, and storage temperature. During initial design phases, a storage efficiency is assumed, and preliminary sizing and costs are established. Storage alternatives for the project can then be compared.

Mixing the storage fluid reduces the quality of the stored heat because the difference between the usable temperature and storage temperature is increased. As this difference increases, so does the energy required by the heat recovery heat pump to boost the temperature to a *usable temperature* high enough to heat the space effectively. Mixing is caused by thermal diffusion, buoyancy, and the inertia effect of the jets into or out of the store. In tank storage, thermal stratification of the stored fluid can be promoted by carefully designing the supply/return jets, by placing diaphragms or baffles between the liquid layers, or by segmenting the tank into compartments.

Stratification in heated stored fluids is assisted by the buoyancy differential to prevent temperature blending. Stratification is enhanced by tank height-to-diameter ratios above four and by using baffles and diffusers with proper nozzle orientations.

Heat storage system designers have a wide choice of fluid and phase-change materials, with each satisfying a specific need. The bibliography includes references that give design parameters for various heat storage media. The temperatures available for consideration depend on the cooling and heating equipment chosen

for the specific installation. The basic function of satisfying peak demand requirements at optimum owning and operating costs continues to justify the need for heat storage systems.

WATER LOOP HEAT PUMP SYSTEMS

Another cycle that combines load transfer characteristics with water-to-air heat pump units is shown in Figure 13. Each module, or space, has one or more water-to-air heat pumps. The units in both the building core and perimeter areas are connected hydronically with a common two-pipe system. Each unit cools conventionally, supplying air to the individual module and rejecting the heat removed to the two-pipe system through its integral condenser. The total heat gathered by the two-pipe system is expelled through a common heat rejection device. This device often includes a closed circuit evaporative cooler with an integral spray pump. If and when some of the modules, particularly on the northern side, require heat, the individual units switch (by means of four-way refrigerant valves) into the heating cycle. The units derive their heat source from the two-pipe water loop, basically obtaining heat from a relatively high source; *i.e.,* the condenser water of the other units. When only heating is required, all units are in the heating cycle and, consequently, an external heat input source is needed to provide 100% heating capability. The heat of compression should be considered in heat source sizing. The water loop is usually 60 to 90 °F in temperature and, therefore, seldom requires piping insulation.

A water-to-water heat pump can be added in the closed water loop before the heat rejection device for further heat reclaim. This heat pump reuses the heat and provides domestic hot water or elevates water temperatures in a storage tank to be bled back into the loop.

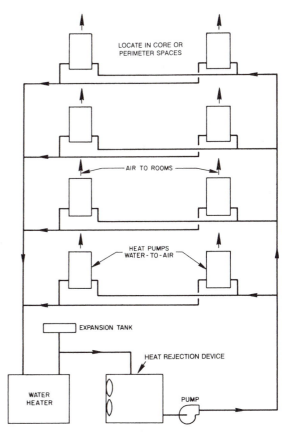

**Fig. 13   Heat Recovery System Using Water-to-Air
Heat Pumps in a Closed Loop**

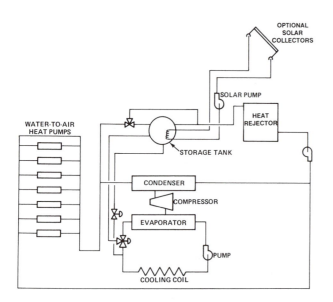

**Fig. 14 Closed Loop Solar-Assisted Heat Pump System
with Thermal Storage**

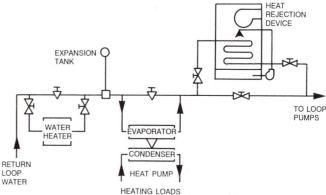

Fig. 15 Secondary Heat Recovery from WLHP System
(Adapted from *Marketing the Industrial Heat Pump*,
Edison Electric Institute, 1989)

Some aspects of these cycles may be proprietary and are covered by patents, and should not be used without appropriate investigation. Illustrations and their description are not meant to imply any endorsement of the cycles by ASHRAE but are used to illustrate the variations made in practical applications.

Any number of water-to-air heat pumps may be installed in such a system. Water circulates through each unit via the closed loop.

The water circuit should include two pumps—one is 100% standby and a means for adding and rejecting heat to and from the loop. Each heat pump can either heat or cool and maintain temperature in each zone.

Units on the heating mode extract heat from the circulated water, while units on the cooling mode reject heat to the water. Thus, the system recovers and redistributes heat, where needed. Unlike air-source heat pumps, heat available for this system does not depend on outdoor temperature. The water loop conveys rejected heat, but a secondary source, typically a boiler, is usually provided.

Another version of the water loop heat pump (WLHP) system uses a coil buried in the earth as a heat source and heat sink. This earth coupled system does not normally need the boiler and cooling tower incorporated in conventional WLHP systems to keep the circulating water within acceptable temperature limits. However, earth coupled heat pumps must operate at lower entering water (or antifreeze solution) temperatures. Some applications may require a heat pump tolerant of supply water temperatures of 25 to 110°F.

Figure 14 illustrates a solar-assisted system with a storage tank and solar collectors. The storage tank in the condenser circuit can store the excess heat during occupied hours and provide heat to the loop during unoccupied hours. During this process of storing and draining heat from the tank, solar heat can also be added within the limitations of the temperatures of the condenser system. The solar collectors are more effective and efficient under these circumstances because of the temperature ranges involved. The system may eliminate the need for a supplementary water heater in the WLHP system.

Figure 15 illustrates a system for extracting excess heat from a WLHP system. A secondary heat pump is used to divert excess heat to some need external to the WLHP piping system.

Many facilities require large cooling loads (*e.g.*, for interior zones, lights, people, business machines, computers, switchgear, and production machinery) that result in net loop heat rejection during all or most of the year, particularly during occupied hours. This rejection of waste heat often occurs while other heating loads in the facility (ventilation air, reheat, domestic hot water heating) are using external purchased energy to supply heat.

By including a secondary water-to-water heat pump in the system design, additional balanced heat recovery can be economically achieved. The secondary heat pump can effectively be used to reclaim this otherwise rejected heat, amplify its temperature, and use it to serve other heating loads, thus minimizing the use of expensive purchased energy.

Design Considerations

Water loop heat pump systems have been successfully applied in many types of multiroom buildings. A popular application has been in office buildings, where heat gains from the interior can be redistributed to the perimeter during the winter. Other applications include hotels and motels, schools, apartment buildings, nursing homes, manufacturing facilities, and hospitals. Operating costs for this system are most favorable in applications where both heating and cooling are required.

Unit types. Chapter 47 describes various types and styles of units and the control options available.

Zoning. The WLHP system offers excellent zoning capability. Since equipment can be placed in interior areas, the system can accommodate the future relocation of partitions with minimum duct changes. Some systems use heat pumps for perimeter zones and the top floor, with cooling-only units serving interior zones; all units are connected into the same water circuit.

Heat recovery and heat storage. This system lends itself well to heat storage. Installations that cool most of the day in winter and heat at night (such as a school) make excellent use of heat storage. The water may be stored in a large storage tank in the closed loop circuit ahead of the boiler. In this application, the loop temperature is allowed to build up to 90°F during the day. The stored water at 90°F can be used during unoccupied hours to maintain heat in the building, with the loop temperature allowed to drop to 60°F. The water heater (or boiler) would not be used until the loop had dropped the entire 30°F. The storage tank operates as a flywheel to prolong the period of operation where neither heat makeup nor heat rejection is required.

Concealed units. Equipment in the ceiling spaces must have access for maintenance and servicing filters, control panels, compressors, and so forth. The condensate drain lines also require space between the drain connection on the unit and the top of the ceiling.

Ventilation. Outdoor air for ventilation may be (1) ducted from a ventilation supply system to the units or (2) drawn in directly through a damper into the individual units. To operate satisfactorily, the air entering the water-source heat pumps should be above 60 °F. In cold climates, the ventilation air must be preheated. Stack effect, wind, and balancing difficulties can greatly vary the quantity of ventilation air entering directly through individual units.

Secondary heat source. The secondary heat source for heat makeup may be electric, gas, oil, solar, and/or waste heat. Normally, a water heater or boiler is used; however, electric resistance heat in the individual heat pumps, with suitable changeover controls, may also be considered. The changeover control may be an aquastat set to switch from heat pump to resistance heaters when the loop water reaches the minimum 60 °F. When the loop temperature reaches 70 °F, the resistance heat is cut out and the heat pump is again operated for heating.

An electric water heater is readily controlled to provide 60 °F outlet water and can be used directly in the loop. With a gas or oil-fired combustion boiler, a heat exchanger may be used to transfer heat to the loop or, depending on the type of boiler used, a modulating valve may blend water from the boiler into the loop.

Solar or ground energy can supply part or all of the secondary heat. Water or antifreeze solution circulated through collectors can add heat to the system directly or indirectly via a secondary heat exchanger.

Heater capacity. If used, the heater for supplementary heat input must be carefully selected. Total requirements of all heat pumps in the heating mode must be considered.

Night setback. A building with night setback may need a supplementary heater boiler sized for the installed capacity, not the building heat loss, since the morning warm-up cycle may require every heat pump to operate at full heating capacity until the building is up to temperature. In this case, the boiler should be sized for 75% of the total heating capacity of all-water source heat pumps installed in the building, plus the heat rejector heat loss. Night setback must not allow the temperature in any space to fall below 55 °F to ensure proper restarting of the units. Newer electronic controls with ramped functions may alleviate this potential problem.

Heat rejector selection. A closed loop circuit requires a heat rejector that is either a heat exchanger (loop water to cooling tower water) or a closed circuit evaporative cooler or a ground coil. The heat rejector is selected in accordance with manufacturer's selection curves, using the following parameters:

Water flow rates. Manufacturers' recommendations on water flow rates vary between 2 and 3 gpm/ton of installed cooling capacity. The lower flow rates are generally preferred in regions having a relatively low summer outdoor design wet-bulb temperature. In more humid climates, a higher flow rate will allow a higher water temperature to be supplied from the heat rejector to the heat pumps, without a corresponding increase in temperature leaving the heat pumps. Thus, cooling tower or evaporative cooler size and cost is minimized without penalizing performance of the heat pumps.

Water temperature range. Range (the difference between the leaving and entering water temperatures at the heat rejector) is affected by heat pump EER, water flow rate, and diversity. It is typically between 10 and 15 °F.

Approach. Approach is the difference between the water temperature leaving the cooler and the wet-bulb temperature of the outside air. The maximum water temperature expected in the loop supply is a function of the design wet-bulb temperature.

Diversity. Diversity is the maximum instantaneous cooling load of the building divided by the installed cooling capacity. Diversity times the average range of the heat pumps is the applied range of the total system (the rise through all units in the system and the drop through the heat rejector). Hence:

$$D = Q_m/Q_i$$

where

Q_m = maximum instantaneous cooling load
Q_i = total installed cooling capacity
D = diversity

$$R_s = DR_p$$

where

R_s = range of system
R_p = average range of heat pumps

The average leaving water temperature of the heat pumps is the entering water temperature of the heat rejector. The leaving water temperature of the heat rejector is the entering water temperature of the heat pumps.

Winterization. On buildings with some potential year-round cooling (*i.e.*, office buildings), all loop water may be continuously pumped through the heat rejector. This control procedure reduces the danger of freezing. In addition, a stand-by pump in the system starts automatically in case the main loop pump fails. However, it is important to winterize the heat rejector to minimize the heat loss.

In northern climates, the most important winterization step for evaporative coolers is to install a discharge air plenum with positive closure, motorized, ice-proof dampers. The entire casing that houses the tube bundle and the discharge plenum may be insulated. The sump, if outside the heated space, should be equipped with electric heaters. The heat pump equipment manufacturer's instructions will help in the selection and control of the heat rejector.

If sections of the water circuit are to be exposed to freezing temperatures, the addition of an antifreeze solution should be considered. In a serpentine pipe circuit having no automatic valves that might totally isolate individual components, ethylene glycol concentrations between 10 and 15% provide protection against bursting pipes, with a minimal effect on system performance (Trelease 1978).

An open cooling tower with a separate heat exchanger is a practical alternative to the closed water cooler (Figure 16). An additional pump is required to circulate the tower water through the heat exchanger. In such an installation where no tubes are exposed to the atmosphere, it may not be necessary to provide freeze protection on the tower. The sump may be indoors or, if outdoors, may be heated to keep the water from freezing. This arrangement allows the use of small, remotely located towers. Temperature control necessary for tower operation is maintained by a sensor in the water loop system controlling operation of the tower fan(s).

The combination of an open tower, heat exchanger, and tower pump frequently is lower in first cost than an evaporative cooler. In addition, operating costs are lower because no heat is lost from the loop in the winter and, frequently, less power is required for the cooling tower fans.

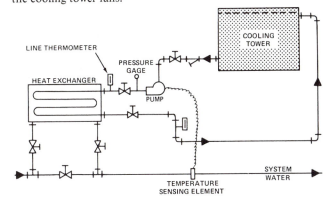

Fig. 16   Cooling Tower with Heat Exchanger

Ductwork layout. Often, a WLHP system has ceiling-concealed units, and the ceiling area is used as the return plenum. Troffered light fixtures are a popular means of returning air to the ceiling plenum.

The air supply from the heat pumps should be designed for quiet operation. Heat pumps connected to ductwork require external static pressure. The heat pump manufacturers' recommendations should be consulted for the maximum and minimum external static pressure allowable with each piece of equipment.

Piping layout. A reverse return piping system should be used wherever possible with the WLHP system. This is particularly true where all units are essentially the same capacity. Balancing is then minimized except for each of the system branches. If a direct return system is used, balancing the water flow is required at each individual heat pump. The entire system flow may circulate through the boiler and heat rejector in series. Water makeup should be at the constant pressure point of the entire loop water system. Piping system design is similar to the secondary water distribution of air-and-water systems.

A clean piping system is vital to successful performance of the water-source heat pump system. The pipe should be clean when installed, kept clean during construction, and thoroughly cleaned and flushed on completion. Startup water filters in the system bypass (pump discharge to suction) should be included on large, extensive systems.

Advantages of a WLHP System

1. Affords opportunity for energy conservation by recovering heat from interior zones and/or waste heat and by storing excess heat from daytime cooling for nighttime heating.
2. Allows recovery of solar energy at a relatively low fluid temperature where solar collector efficiency is likely to be greater.
3. The building does not require wall openings to reject heat from air-cooled condensers.
4. Provides environmental control in small occupied zones during nights or weekends without the need to start a large central refrigeration machine.
5. Units are not exposed to outdoor weather, which allows installation on the coast and in other corrosive atmospheres.
6. Units have a longer expected life than air-cooled heat pumps.
7. Noise levels can be lower than air-cooled equipment because condenser fans are eliminated and the compression ratio is lower.
8. Two-pipe fan coil systems are potentially convertible to this system.
9. Should a unit fail, the entire system is not shut down. However, loss of pumping capability, heat rejection, or secondary heating could affect the entire system.
10. Energy for the heat pumps can be metered directly to each tenant. However, this metering would not include energy consumed by the central pump, heat rejector, or boiler.
11. Total life cycle cost of this system frequently compares favorably to central systems when considering relative installed cost, operating costs, and system life.

Disadvantages

1. Space is required for boiler, heat exchangers, pumps, and heat rejector.
2. Initial cost is higher than for most other multiple-packaged unit systems.
3. Reduced airflow can cause the heat pump to cycle cutout. Good filter maintenance is imperative.
4. Piping loop must be very clean; special attention is required.

Controls

The closed loop heat pump system has simpler controls than those for other totally central systems. There are only two control points: one to add heat when the water temperature drops to 60°F and the other to reject heat when the water temperature rises to 90°F.

The manufacturer includes thermostatic controls for the individual heat pumps. The boiler controls should be checked to be sure that outlet water is controlled at 60°F, since controls normally supplied with boilers are in a much higher range. Some heat pump manufacturers provide a control and alarm panel that sounds an alarm when the loop water temperature exceeds recommended limits or if water flow stops completely.

An evaporative cooler should be controlled by increasing or decreasing heat rejection capacity in response to the loop water temperature leaving the cooler. A reset schedule that operates the system at lower water temperatures (to take advantage of lower outdoor wet-bulb temperatures) can save energy when heat from the loop storage is not likely to be used.

System abnormal condition alarms should operate as follows:

1. On a fall in loop temperature to 50°F, sound alarm horn. Open heat pump control circuits at 45°F.
2. On a rise in loop temperature to 105°F, sound alarm horn. Open heat pump control circuits at 115°F.
3. On sensing insufficient system water flow, flow switch sounds the alarm horn and opens heat pump control circuits.

An outside ambient control should be provided to prevent operation of the sump pump at freezing temperatures.

Optional system control arrangements include:

1. Night set-back control
2. Automatic unit start-stop, with after-hour restart as a tenant option
3. Warm-up cycle
4. Pump alternator control
5. Central no-flow control to deenergize heat pumps.

Waste Heat Recovery

In many large buildings, internal heat gains require year-round chiller operation. This internal heat is often wasted through a cooling tower. Figure 17 illustrates a heat pump installed in the water line from the chiller's condenser before rejection at the cooling tower. This arrangement uses the otherwise wasted heat to provide heat at the higher temperatures required for space heating, reheat, and domestic water heating.

Prudent design may dictate cascade systems with chillers in parallel or series. Manufacturers can assist with custom components to meet a wide range of load and temperature requirements.

The double-bundle condenser working with a reciprocating or centrifugal compressor is most often used in this application. Figure 18 shows the basic configuration of this system, which

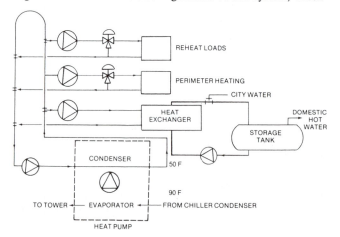

Fig. 17 Typical Waste Heat Recovery System

makes heat available in the range of 100 to 130 °F. The warm water is supplied as a secondary function of the heat pump and represents recovered heat.

Figure 19 shows a similar cycle, except that a storage tank has been added, enabling the system to store heat during occupied hours by raising the temperature of the water in the tank. During unoccupied hours, water from the tank is gradually fed to the evaporator providing load for the compressor and condenser that heats the building during off hours.

Figure 20 is another transfer system capable of generating 130 to 140 °F or warmer water whenever there is a cooling load by cascading two compressors hydronically. In this configuration, one chiller can be considered as a chiller only and the second unit as a heating only heat pump.

Industrial Process Heat Pumps

Heat recovery in industrial plants offers numerous opportunities for applied heat pumps. Factory-packaged machines can produce condenser water temperatures of up to 220 °F on special order, and steam can also be delivered.

Heat pumps are generally used in industrial applications for other than room heating and air conditioning. These applied heat pump systems are specifically engineered for the intended applications. Many such applications are found in the process industries. The operating principles and equipment are the same as discussed above, but the economic trade-offs are different from room heating and air conditioning and must be specifically investigated for each case. Guidelines should be reconsidered in view of the operating economics and technical issues involved for industrial process applications.

Two major classes of industrial heat pump systems can be distinguished: closed and open systems.

Closed cycle systems use a suitable working fluid, usually a refrigerant in a sealed system. They can use either the absorption or the vapor compression principle, depending on the temperature levels, and process economics. Heat is transferred to and from the system through heat exchangers similar to refrigeration system heat exchangers. Closed cycle heat pump systems are often classed with industrial refrigeration systems with operation occurring at higher temperatures. Refrigerant and oil choices must be consistent with available component and material restraints and limits, as well as mutual compatibilities at the expected operating temperatures. In addition, the viscosity and foaming characteristics of the oil and refrigerant mixtures must be consistent with the lubrication requirements at the specific mechanical load imposed on the equipment. Proper oil return and heat transfer at the evaporators and condensers must be considered (see Figure 21).

Open cycle systems use the process fluid to raise the temperature of the available heat energy by vapor compression. The most important class of applications is steam process fluid. Compression can be provided with a mechanical compressor or by a thermocompression ejector driven by the required quantity of high-pressure steam. A distinction must be made between systems

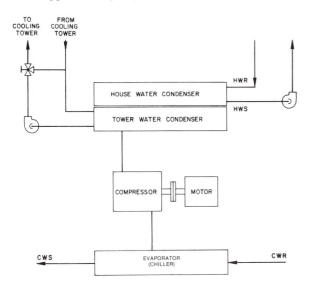

Fig. 18   Heat Transfer Heat Pump with Double-Bundle Condenser

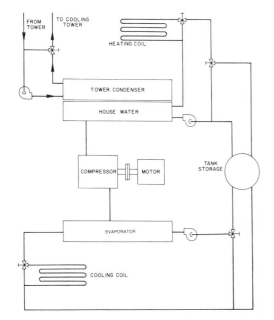

Fig. 19   Heat Transfer System with Storage Tank

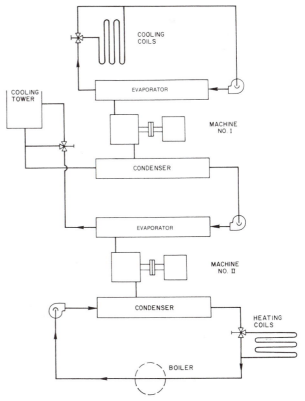

Fig. 20   Multistage (Cascade) Heat Transfer System

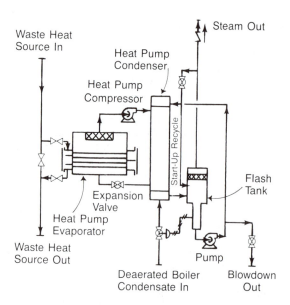

Fig. 21   Water-to-Steam Heat Pump

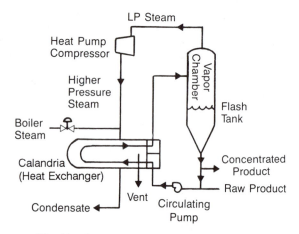

Fig. 22   Open Cycle Single-Effect Evaporator

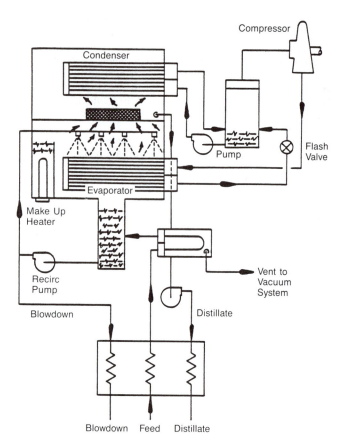

Fig. 23   Closed Cycle Vapor Compression

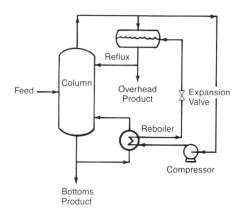

Fig. 24   Mechanical Vapor Compression for Distillation

compressing relatively clean, process-quality steam as opposed to contaminated steam from evaporation processes.

Boiler-generated process steam usually conforms to cleanliness standards that ensure corrosion-free operation of the compression equipment. However, boiler treatment chemicals often prevent the use of process steam in direct contact with food or potable water. An open cycle heat pump compressing process steam can provide exceptional steam energy management flexibility to a process with variable steam demand, if the heat pump controls are integrated properly with the process controls.

Application of open cycle heat pumps to evaporation is exceptionally important (see Figure 22). The most frequent applications are in the dairy, food, paper, and chemical industries. The water vapor from the evaporator is compressed to a higher pressure and temperature and is condensed at the heat transfer surface of the evaporator liberating its heat of condensation. Very high efficiencies are possible. Depending on the temperature differential required for heat transfer in the evaporator (as low as 4 °F), the compression work required is a small fraction of the heat energy recovered. A separate condenser for waste steam is eliminated, and cooling water consumption is greatly reduced. However, the vapors from the evaporator may contain contaminants from the materials being concentrated, and fouling, corrosion, or erosion of the compression system can take place. Adequate separators,

baffles, and flow path must be provided with proper attention given to the materials and equipment selected. Coefficients of performance in excess of 25 are common. When process vapors are incompatible with compressor materials, or when low-temperature evaporation is desired, a closed cycle evaporator, such as the one shown in Figure 23, can be used.

Vapor compositions can be different from liquid compositions and boiling point rise results. Mechanical vapor compression heat pumps can be applied to distillation columns (Figure 24). The overhead vapors are compressed to a higher pressure and temperature and then condensed in the reboiler. This eliminates the need for boiler steam in the reboiler and reduces overall energy consumption. After the pressure of condensed vapor as liquid from the reboiler is reduced through the expansion valve, some of the liquid at a lower temperature is returned to the column as reflux, and the balance forms the overhead product, both as liquid and vapor. Vapor is recycled as necessary. Distillation processes attractive for this application include ethanol-water, methanol-water, and mixed hydrocarbon splitters.

An added benefit is obtained when capacity control is provided using a variable-speed driver such as an engine, gas or steam turbine, or electric motor with variable-speed provisions. Additional capacity for emergencies is also available by temporarily overspeeding the driver, while optimal efficiency is retained by sizing for nominal operating conditions. Proper integration of heat pump controls and process system controls is essential.

EQUIPMENT SELECTION

The heat source and heat sink surfaces, heating and cooling surfaces, heat pumps, and accessory equipment are sized and selected in accordance with current refrigeration and air-conditioning practices. The capacity of the system must meet the design heating and cooling requirements.

Supplemental Heating

Heating needs may exceed the capacity available from equipment selected for the cooling load, particularly if outdoor air is used as the heat source. When this occurs, supplemental heating or additional compressor capacity should be considered. The additional compressor capacity or the supplemental heat is generally used only in the severest winter weather and, consequently, has a low usage factor. Both possibilities must be evaluated to determine the most economical selection.

When supplemental heaters are used, the elements should always be located in the air or water circuit downstream from the heat pump condenser. This permits the heat pump system to operate at a lower condensing temperature, increasing the system heating capacity and improving the coefficient of performance. The controls should sequence the heaters so that they are energized after all heat pump compressors are fully loaded. An outdoor thermostat is recommended to limit or prevent energizing of heater elements during mild weather when they are not needed. Where 100% supplemental heat is provided for emergency operation, it may be desirable to keep one or more stages of the heaters deenergized whenever the compressor is running. In this way, the cost of electrical service to the building is reduced to meet the maximum coincidental demand.

A flow switch should be used to prevent operation of the heating elements and the heat pump, unless there is air or water flow.

Heating and Cooling Load Estimates

Realistic winter outdoor design temperatures should be selected to obtain authentic heating requirements and prevent installation of oversized supplemental heating equipment.

The winter design temperatures given in Table 1 of Chapter 24 in the 1989 ASHRAE *Handbook—Fundamentals* should be used, unless experience and more authentic data for the particular locality are available. The local utility company may have such data.

Conservative heating requirements and lower nightly indoor temperatures are not recommended, because the resulting additional heating capacity would unnecessarily increase both the initial and operating cost of the system. Instead, consider operating cycles using auxiliary equipment to provide higher operating efficiencies and load reduction, including the following:

- Using the maximum, practical amount of insulation with proper vapor barrier in sidewalls, ceilings, and floors of a structure. Windows and doors should at least be weatherstripped and properly caulked; in most areas of the country, storm windows and doors (or the equivalent) can be justified.

- Use of available, dependable heat sources within the structures, such as lights, motors, heat-producing machinery, and people.

- Automatically closing the outdoor air intakes during nights, weekends, and other unoccupied periods, so that the heat loss will be limited to transmission and infiltration.

- Proper selection of the ventilation air. Because the ventilation air load is an appreciable part of the total, reasonable ventilation quantities, without sacrificing a healthy environment, should be selected. Minimum quantities of outdoor air may even be unacceptable, because of objectionable, irritable, and toxic odors in the atmosphere due to a high percentage of hydrogen sulfide, smog, and other chemical effluences from industrial processes. Physical and chemical odor removal units, such as activated charcoal in combination with a conventional or electric air filter, are effective in purifying and removing odor-causing substances (see Chapter 12 in the 1989 ASHRAE *Handbook—Fundamentals* and Chapter 7 in the 1990 ASHRAE *Handbook—Refrigeration*). The quantity of ventilation air can often be reduced considerably, resulting in savings in both initial and operating cost.

For buildings with highly variable occupancies, the ventilation air quantity might be reduced at lower outdoor temperatures because the building is occupied by fewer people. This is particularly true in retail stores where fewer people shop during extremely cold weather. However, code requirements should be checked for restrictions on ventilating requirements before this principle is applied.

Reciprocating or Centrifugal Compressors

While most air-source installations for combined heating and cooling incorporate single-stage reciprocating compressors, the use of multistage compressors has increased. The two-stage compression system is particularly applicable in the extreme northern climate, where the frequent occurrence of low outdoor temperatures makes the single-stage system impractical. In two-stage systems, the compressors can be used in parallel during the cooling cycle and in series during the heating cycle. The main advantages of staging are the increased capacity per unit of compressor displacement and increased compressor efficiency, which results in proportionally smaller (or perhaps eliminates) auxiliaries needed to provide a given heating effect at the lower outdoor temperatures. Whether these improvements in efficiency will outweigh the disadvantages of the higher initial cost of staging depends on the requirements of a particular installation.

Centrifugal compressors for larger capacity applications are primarily used on water-source heat pumps but also are readily applicable to air-source systems, particularly those using the reclaiming cycle (Figure 18). Multistaging of the centrifugal machines, similar to that described for the reciprocating units, can be prescribed using a brine when low-temperature outdoor air is used as the heat source. Centrifugal assemblies normally use water as the heat transfer fluid on both the high and low sides of the refrigerating machine. When the unit is exposed to temperatures below 30 °F, an antifreeze liquid or brine with a suitable viscosity, corrosiveness, and initial cost is required.

Compressor Floodback Protection

On the more recent air source-sink (central plant-type) installations, a suction line separator, similar to that shown in Figure 25, has been used successfully. This separator, in combination with a liquid-gas heat exchanger, is a protection against migration and harmful liquid floodbacks to the compressor. The solenoid valve should be wired to open when the compressor is operating, and the hand valve should then be adjusted to provide an acceptable bleed rate into the suction line.

Liquid Subcooling Coils

A refrigerant subcooling coil can be added to the heat pump cycle, as illustrated by Figure 26, to preheat the ventilation air during the heating cycle and, at the same time, lower the temperature of the liquid refrigerant. Depending on the refrigerant circulation rate and the quantity and temperature of ventilation air, the heating capacity can be increased as much as 15 to 20%, as indicated by Figure 27.

Similarly, a source water coil can be incorporated in the water-to-water systems (Figure 6) to either preheat or precool the ventilation air, depending on the operating cycle, and thereby improve the performance. Means must be provided to prevent freezing of the water within the coil when exposed to ventilation air temperatures below 32°F.

Water Temperatures for Heating

A realistic warm water temperature should be selected for the design condition on all systems using water as the heat transfer fluid. Designing for water temperatures higher than actually required results in not only higher initial cost, but higher operating cost as well. When the same air coil is used for both heating and cooling, a coil selected for the summer cooling design load is generally sufficient to use heating water temperatures from 95 to 110°F.

When a building contains areas that must be heated but not cooled, it may be necessary to select a slightly deeper coil to satisfy the heating load with relatively low-temperature water. The addi-

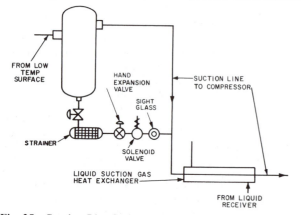

Fig. 25 Suction Line Separator for Protection against Liquid

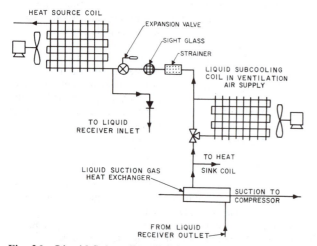

Fig. 26 Liquid Subcooling Coil in Ventilation Air Supply to Increase Heating Effect and Improve Coefficient of Performance on Heating Cycle

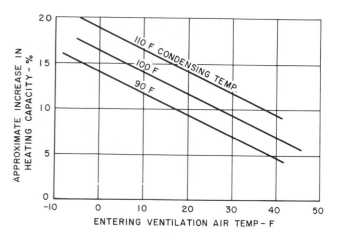

Fig. 27 Anticipated Increase in Heating Capacity Resulting from Use of Liquid Subcooling Coil

tional coil cost is relatively small and easily justified in comparison with the operating cost in raising the warm water temperature schedule of the entire building to meet requirements of several small areas for a few hours per year. Another alternative for these critical areas is to use direct electric heat, either as a supplement or as the total heating source.

The objection that low heating water temperatures result in drafty conditioned spaces is a misconception. Where a building is air conditioned both in summer and winter, the air quantities circulated are based on the design summer cooling load. Consequently, if sufficient heat is provided to spaces to offset any winter losses, the air temperature entering the space will be the same, regardless of the water temperature supplied to the coil. In fact, the lower water temperatures are more advantageous from the control standpoint, since the throttling water valve, bypass water valve, or face-and-bypass dampers need not throttle through such an extreme range.

Defrosting of Heat Source Coils

Frost accumulates rather heavily on outdoor heat source coils when the temperature reaches approximately 40°F, but lessens somewhat with the simultaneous decrease in outdoor temperature and moisture content. Methods for defrost are discussed in the section Defrost Control (Air Source).

Another method of defrosting is by spraying heated water over an outdoor coil. The water can be heated by the refrigerant or by auxiliaries.

Draining of Heat Source and Sink Coils

Direct expansion indoor and outdoor heat transfer surfaces serve as condensers and evaporators. Both surfaces, therefore, must have proper refrigerant distribution headers to serve as evaporators and have suitable refrigerant drainage while serving as condensers.

In a flooded system, the normal float control to maintain the required refrigerant level is used.

CONTROLS

The control system should provide flexible and effective heat pump operation. Capacity modulation, a convenient method of switching from heating to cooling, and automatic defrosting of the air-source heat surfaces should be provided.

The control system should prevent energizing the supplemental electric heater until (1) the heat pump system is unable to satisfy the heating requirement when operating at full capacity and (2) the source water or outdoor air is below a predetermined outdoor temperature. In this way, the coincident energy demand will be minimized with a resulting lower average energy cost. In some installations, it may be necessary to energize the supplemental heat during the defrosting cycle to minimize any cooling effect within the structure, but this normally will not exceed the maximum coincident demand during the heating cycle.

On air-to-water and water-to-water systems, an outdoor reset control of the hot water temperature may be desirable for improved economy. During milder weather, the warm water temperature is rescheduled to a lower level, since less heat dissipation is required; this permits the compressor to operate at a lower condensing temperature.

The following controls may be used to change from heating to cooling.

- **Conditioned space thermostat** on residences and small commercial applications.
- **Outdoor air thermostat** (with provision made for manual overriding for variable solar and internal load conditions) can reverse the cycle on larger installations, where it may be difficult to find a location in the conditioned space, which reflects the desired operating cycle for the total building.
- **Manual changeover.**
- **Sensing device,** which responds to the greater load requirement, heating or cooling, is generally applied on simultaneous heating and cooling systems.
- **Dedicated microcomputer** to automate changeover and perform all the other control functions needed, as well as to simultaneously monitor the performance of the system. This may be a stand-alone device or incorporated as part of a larger building automation system.

On the heat pump system, it is important that space thermostats are interlocked with ventilation dampers so that both operate on the same cycle. During the heating cycle, the fresh air damper should be positioned for minimum ventilation air, with the space thermostat calling for increased ventilation air only if the conditioned space becomes too warm. Fan and/or pump interlocks are generally provided to prevent the heat pump system from operating if the use of the accessory equipment is not available. On commercial and industrial installations, some form of heat pressure control is required on the condenser at outdoor air temperatures from 60 to 0°F.

Industrial process controls may have combinations of electrical and pneumatic controls with proportional auto reset and batch modifications to prevent temperature drifts.

HEAT RECLAIMING CYCLE

The heat pump adapts readily to installations requiring simultaneous heating and cooling by transferring heat from the core of a structure (*e.g.*, where cooling is desired) to the perimeter, where heating is needed. This type of application frequently results in a considerable cost saving compared with other designs, because of the resulting relatively high operating efficiency of the equipment under these conditions.

An indication of the magnitude of the recoverable heat in a modern multistory office building is shown in Figures 29 and 30. Figures 28 and 29 show the gross heat loss and the internal heat gain of the exterior and interior zones during the day. Figure 30 shows the result of transferring this internal excessive heat gain to

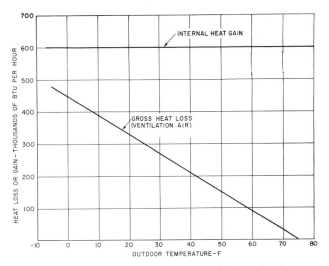

Fig. 29 Heat Loss and Heat Gain at Various Outdoor Temperatures for Interior Zone of Typical Multistory Office Building

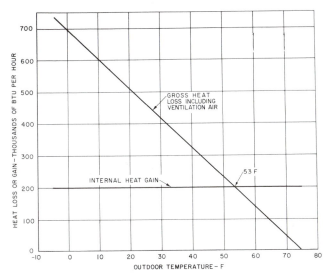

Fig. 28 Heat Loss and Heat Gain at Various Outdoor Temperatures for Exterior Zone of Typical Multistory Office Building

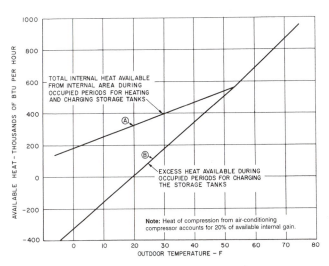

Fig. 30 Quantity of Internal Heat Available to External Areas and Storage Tank at Various Outdoor Temperatures during Occupied Periods

the exterior zone by a heat pump. Curve A of Figure 30 is the total internal heat available, and curve B is excess heat (above that needed by the exterior zone), which can be kept in a storage tank (using a cycle similar to that shown by Figure 19). During the occupied periods, no outside heat source or supplemental heat is needed at outdoor temperatures of 19°F or above for this hypothetical building.

It must be understood that relative performances are indicated to illustrate general cases. Each building is likely to have unique parameters that depart from these illustrations.

BIBLIOGRAPHY

Arnold, H.C., G. Grossman, and Perez H. Blanco. Investigations of advanced industrial heat pumps and chillers for low grade heat recovery. Oak Ridge National Laboratory, TN.

ASHRAE. 1981. Industrial heat pump systems. ASHRAE Seminar, J.M. Calm, Chairman, Chicago, January 28.

ASHRAE. 1984. Survey of thermal energy installations in the United States and Canada. ASHRAE, Inc., Atlanta.

Bard Manufacturing Company. Air source heat pump operation and components. Manual No. 2100-059.

Bard Manufacturing Company. Dual fuel add-on heat pump guide for operational cost savings for regions 2 through 5. Manual No. 2100-070 (Region 2), No. 2100-071 (Region 3), No. 2100-072 (Region 4), No. 2100-073 (Region 5).

Bard Manufacturing Company. Heat pump sizing (air-to-air and water-to-air). Manual No. 2100-057.

Bard Manufacturing Company. Operation and troubleshooting water-to-air heat pumps. Manual No. 2100-054.

Bard Manufacturing Company. Water systems and well pump and pipe sizing for water source heat pumps. Manual 2100-078.

Becker, F.E. and A.I. Zakak. 1985. Recovering energy by mechanical vapor recompression. *Chemical Engineering Progress* (July):45-49.

Beesley, A.H. and R.D. Rhinesmith. 1980. Energy conservation by vapor compression evaporation. *Chemical Engineering Progress* (August): 37-41.

Berntsson, T., *et al.* 1980. The use of the ground as a heat source for heat pumps in urban areas. An overall review of technical-economic aspects (in English/Swedish). Doc. D39: Swed. Counc. Bldg. Res., Stockholm. Available from Svensk Byggtjanst, Box 7853, S-103 99 Stockholm.

Bierwirth, H.C. 1982. Packaged heat pump primer. *Heating-Piping* (July).

Boland, D. and J.C. Hill. Heat pumps application in an energy integrated process. ICI Petrochemicals Ltd., United Kingdom.

Booth, B. 1982. Industrial and commercial applications—Heat pumps. Electricity Council, London. Paper presented at Electrotechnologies in Industry Conference, May 25-27, Montreal.

Bose, J.E., J.D. Parks, and F.C. McQuiston. 1985. Design/data manual for closed-loop ground-coupled heat pump systems. ASHRAE.

Braham, G.D. Heat recovery technology for swimming pools by the application of heat pumps. Electricity Council, London.

Bryson, S. 1980. Warm up with your computer. *Data Processing* 22 (October):38.

Bschorr, O. 1979. Heat pump for air heating and air drying (in German). *Brennst.-Waerme-Kraft* 31:483-85.

Burton, D.C. and D.W. Chaudoir. 1980. Industrial heat pumps for the manufacture of power alcohol (ethanol). Paper presented at Third World Energy Engineering Conference, October, Atlanta, GA.

Carrier Corporation. 1977. Engineering guide for reciprocating chiller heat reclaim systems. Catalog No. 592-026.

Carrier Corporation. 1982. Water source heat pump system design guide for commercial systems. Catalog No. 592-045.

Cerina, A., D. Portoso, and A. Reali. 1981. Tests on the industrial application of the heat pump in the testing room for engines of an Italian firm (in Italian). *Cond. Aria, IT.* 25:390-95.

CHP Company. 1985. Selection criteria, closed circuit evaporative water cooler for use with California heat pump units and system. Form CHP-H-070-18009.

Clark, E.C. 1984. Chemical heat pumps drive to upgrade waste heat. *Chemical Engineering*, February 20, 50-51.

Curis, O. and J.D. Laine. *Gas motors driving heat pumps in the malting industry.* Cie Electro-Mechanique, France.

Curl, R.S. 1978. Cutting waste to reduce energy costs. *Heating-Piping* 50 (August):73-81.

Davis, N. 1982. Waste heat recovery the simple way. *Process Engineering* 63 (January):51.

Dizier, M. 1982. Recompression de vapeur dans une sucerie. Paper presented at Electrotechnologies in Industry Conference, May 25-27, Montreal.

Duane, B.P. 1982. Vapour recompression—A case study. Paper presented at Electrotechnologies in Industry Conference, May 25-27, Montreal.

Edison Electric Institute. *Application handbook—Electric heat pumps in waste heat recovery systems.* Washington, D.C.

Ekroth, I.A. 1979. Thermodynamic evaluation of heat pumps working with high temperatures. Proceedings of the Inter-Society Energy Conversion Engineering Conference 2:1713-19.

Forwalter, J. 1979. Waste heat from refrigeration system provides energy for 140°F hot water. *Food Processing* (December).

Freedman, G.M. 1981. Strategy criterion for transition of a central system from heat recovery to conventional heating and refrigerating. ASHRAE *Journal* 23 (December):27-30.

Garcia, P. 1981. Use of the heat pump in wine-making (in French). *Rev. Prat. Froid Cond. Air,* FR 36, No. 504:43-46.

Gilbert, J.S. 1983. Heat pump strategies and payoffs. Paper presented at the Industrial Energy Conservation Technology Conference, April, Houston, TX.

Gilbert, J.S. 1983. Industrial heat pumps...Concepts, energy savings, and cost guidelines. *Plant Engineering* (November):74-77.

Gilbert, J.S. and R.C. Niess. 1987. Marketing the industrial heat pump. Pub. No. 07-89-21. Edison Electric Institute, Washington, D.C.

Guarino, L.J. 1982. Application of the refrigeration cycle for distilling liquids compared with conventional evaporators. ASHRAE *Journal* (February):32-36.

Hagstedt, B., P. Helland, and K. Rosberg. Evaluation and measurement of existing large refrigeration and heat pump installations operated by the Swedish Cooperative Association (in Swedish). Available from Byggdok, Halsingegatan 49, S-113 31 Stockholm.

Heiburg, O. 1978. Heat pump installation with gas engine drive-design and operational experience (German). *Klima Kaelte Ing.* 6, No. 7-8:261-64.

Hoffman, D. 1981. Low temperature evaporation plants. *Chemical Engineering Progress* (October):59-62.

Holland, F.A., F.A. Watson, and S. Devotta. 1982. *Thermodynamic design data for heat pump systems.* Pergamon Press Ltd., Oxford, England.

Hospital heat recovery system saves a bundle. 1980. *Commercial Remodeling* (August):50, 51.

Hoyos, G.H. and J.D. Muzzy. 1980. Use low-grade waste heat for refrigeration. *Chemical Engineering* 87(May):140.

Hughes, C.H. and D.K. Emmermann. 1981. VTE/VC for sea water. *Chemical Engineering Progress* (July):72, 73.

Kearney, D.W. 1981. Industrial applications of vapor compression heat pumps. International Gas Research Conference, 1123-33.

King, R.J. 1984. Mechanical vapor recompression crystallizers. *Chemical Engineering Progress* (July):63-69.

Koebbeman, W.F. 1981. High COP Rankine-driven heat pump for the generation of process steam. Mechanical Technology Incorporated, Latham, NY.

Koebbeman, W.F. 1982. Industrial applications of a Rankine-powered heat pump for the generation of process steam. Paper presented at the 1st International Symposium on the Industrial Application of Heat Pumps, March 24-26, Coventry, United Kingdom.

Krueger, W. 1979. Optimierung Der Waermeruechegewinnung Mit Kompressions-Waermepumpen (Application of heat pumps in heat recovery loops) (in German). *Waerme,* 85, No. 2:27-30.

Leatherman, H.R. 1983. Cost effective mechanical vapor recompression. *Chemical Engineering Progress* (January):40-42.

Legendre, E. and F. Bonduelle. 1977. Application of the heat pump for concentration by evaporation and for drying in agricultural foodstuffs industries. Comite Francais d'Electrothermie Versailles Symposium, Paper II, 4 April 21-22.

Lieberman, L. 1981. Heat recovery system heats/cools huge complex. *Specified Engineering* 45 (March):77-78.

Limberg, G.E. *Brayton-cycle heat pump for solvent recovery.* AiResearch Mfg. Co., Torrance, CA.

Lloyd, A.S. 1983. Heat pump water heating systems. *Heating/Piping/Air Conditioning* (May).

Manning, E. 1980. Design plant-wide heat recovery. *Hydrocarbon Process* 59 (November):245-47.

McFarlan, A.I. 1983. Late developments in the field of heat recovery. SAIRAC *Journal of Heating, Air Conditioning & Refrigeration* 15(7):14.

McGuigan, D. 1981. *Heat pumps.* Garden Way Publishing, Charlotte, VT.

McQuay-Perfex Inc. Templifier industrial heat pumps. *Bulletins* TP, TP-AL, TPB-Al. Staunton, VA.

Merrill, R.F. 1979. Calculating heat-exchanger capacity for a heat recovery system. *Plastics Engineering* 33 (August):30-32.

Michel, J.W., R.N. Lyon, and F.C. Chen. 1978. Low-temperature heat utilization program. ORNL-5513, 218-44, September 30.

Miller, B. 1981. Tuning in to the fact of life: Heat from cooling water. *Plastics World* 30 (September):60-72.

Mueller, J.H. 1980. Energy savings of heat recovery equipment. *Iron & Steel Engineering* 57 (June):64-66.

Nakanishi, T., *et al.* 1981. Industrial high-temperature heat pump. *Hitachi Zosen Tech Rev.* 42(1):7-13.

Newbert, G.J. 1982. Industrial heat pump demonstrations. ETSU, Harwell, United Kingdom.

Niess, R.C. 1978. How to recycle low-grade process heat. *Electrical World.*

Niess, R.C. 1978. Putting the squeeze on BTU's. Proceedings of the 4th Annual Heat Pump Technology Conference, 4-1/6, April.

Niess, R.C. 1979. The case for high temperature heat pumps. Proceedings of the 5th Annual Heat Pump Technology Conference, April.

Niess, R.C. 1979. Utilization of geothermal energy with an emphasis on heat pumps. Paper presented at the Symposium on Geothermal Energy and Its Direct Use in the Eastern U.S., April 5-7.

Niess, R.C. 1980. High temperature heat pumps can accelerate the use of geothermal energy. ASHRAE *Transactions* 86(1):755-62.

Niess, R.C. 1980. High-temperature heat pumps make waste heat profitable. *Electrical Energy Management* (April).

Niess, R.C. 1980. Industrial heat pump applications for waste heat recovery. Proceedings of the EPRI/RWE Heat Pump Conference at Dusseldorf, West Germany, June.

Niess, R.C. 1981. Effluents and energy economics. *Water/Engineering and Management* (August).

Niess, R.C. 1981. High temperature heat pump applications. Proceedings of the 4th World Energy Engineering Congress, Atlanta, GA, October.

Niess, R.C. 1982. Geothermal heat pump systems are competing today. *Geothermal Resources Council Bulletin* (December).

Niess, R.C. 1982. Simultaneous heating and cooling through heat recovery. Specifying Engineer (August):96-99.

Niess, R.C. and A. Weinstein. 1982. Cutting industrial solar system costs in half. Paper presented at the Industrial Energy Conservation Technology Conference, Houston, TX, April.

Norelli, P. 1980. Industrial heat pumps. *Plant Engineering* (August/September).

North, C.D.R. Large gas engine driven heat pumps. GEA Airchangers Ltd., United Kingdom.

Ober, A. 1980. Chilled water supply and distribution for large air conditioning installations (in German). *Heiz. Luft. Haustech.,* DE. 31(12):452-58.

Oklahoma State University. 1979. Proceedings of the 4th Annual Heat Pump Technology Conference—A University Extension Program of the OSU Division of Engineering, Technology and Architecture, Oklahoma State University, Stillwater, OK, April 9-10.

Oklahoma State University. 1980. Proceedings of the 5th Annual Heat Pump Technology Conference—A University Extension Program of the OSU Division of Engineering, Technology and Architecture, Oklahoma State University, Stillwater, OK, April 14-15.

Oliver, T.N. 1982. Process drying with a dehumidifying heat pump. Westair Ltd., United Kingdom.

Palnchet, R.J., *et al.* 1978. Method picks best heat recovery schemes. *Oil and Gas Journal* 76 (November):51-54.

Parker, J.D. 1975. Heat pump can recycle industrial heat. *Electric World* (May):69-80.

Phetteplace, G.E. and H.T. Ueda. 1989. Primary effluent as a heat source for heat pumps. ASHRAE *Transactions* 95(1).

Podhoreski, A. 1979. Improving the efficiency of refrigeration systems. Paper presented at Ontario Hydro Seminar on Energy Management in the Food and Beverage Industry, October 10 and 17. Processing beer with waste heat. 1980. *Heating/Piping/Air Conditioning* (August).

Process steam generation: A look at an alternative energy source. 1983. *Energy Management Technology* (May/June):24-27.

Reay, D.A. 1980. Industrial applications of heat pumps. *Chemical and Mechanical Engineers* 27, No. 4:86-93, No. 5:73-78.

Reistad, G.M. and P. Means. 1980. Heat pumps for geothermal applications: Availability and performance. Report DOE/ID/12020-TI.

Reynaud, J.F. 1982. Development of heat pumps in industrial processes in France. Paper presented at the Electrotechnologies in Industry Conference, Montreal, May 25-27.

Robb, G.A. 1978. Electric heat pumping and co-generation economic considerations by regions. Paper presented at 64th Annual Meeting of Canadian Pulp and Paper Association, Montreal.

Rogers, J.T. 1975. Industrial use of low grade heat in Canada. Energy Research Group, Carleton University, Report ERG 75-4.

Ross, J.L. 1983. Rooftop VAV vs water source heat pumps. *Heating/Piping/Air Conditioning* 55 (May).

Smith, I.E. and C.O.B. Carey. Thermal transformers for up-grading industrial waste heat. Cranfield Institute of Technology, United Kingdom.

Solar heat pumps: Solution for cold climates. 1979. *Electric Comfort Conditioning News* (June):16-17.

Stamm, R.H. 1983. Energy pump. *Heating/Piping/Air Conditioning* (March).

Strack, J.T. 1982. Heat pumping in industry—A state-of-the-art study. Ontario Hydro Research Division, Toronto, Report to Canadian Electrical Association, Contract 130 U Soa, September.

Stucki, A. 1979. Heat pump cuts oil use by 40,000 gallons/year. *Contractors Electrical Equipment* (November).

Svedinger, B., *et al.* 1981. Heat from earth, rock and water (in English). Summary S2, 1981 Report T1: Swed. Counc. Bldg. Res., Stockholm.

Swearingen, J.S. and J.E. Ferguson. 1983. Optimized power recovery from waste heat. *Chemical Engineering Progress* (August):66-70.

Tabb, E.S. and D.W. Kearney. An overview of the industrial heat pump applications assessment and technology development programs of the Gas Research Institute. Gas Research Institute and Insights West, Inc.

Trane Co. 1981. Water source heat pump system design. *Applications Engineering Manual,* AM-SYS7.

Trelease, S.W. 1978. Closed loop reverse cycle air conditioning. *Building Operating Management* (January):50.

Trelease, S.W. 1980. Water source heat pump evaluation. *Heating/Piping/Air Conditioning* (October).

Truedsson, G.R. 1980. Industrial waste heat recovery: A case in point. *Energy Engineering* 77 (February):14-16.

U.S. Patent 4,344,828. 1982. Energy efficient distillation apparatus. Inventor: J.D. Melton. Assignee: None. Issued August 17.

U.S. Patent 4,345,971. 1982. Distillation employing heat pump. Inventor: W.K.R. Watson. Assignee: None. Issued August 24.

U.S. Patent 4,390,396. 1983. Apparatus for the distillation of vaporizable liquids. Inventor: H. Koblenzer. Assignee: Langbein-Phanhauser Werke AG. Issued June 28.

U.S. Patent 4,461,675. Energy efficient process for vaporizing a liquid and condensing the vapors thereof. Inventors: H.F. Osterman, G.C. Nylen. Assignee: Allied Corp.

Vacuum-evaporator system cuts energy costs. *Chemical Engineering,* 43 and 45, August 23, 1983.

Villadsen, V. 1983. Oil in refrigeration plants. *Papers Referaat,* Frigair '83.

Villaume, M. 1981. Use of the heat pump in apartment buildings (in French). *Rev. Gen. Froid,* FR, 71, No. 4:217-29.

Waste heat recovery. 1980. *Mechanical Engineering* 102 (August):51.

Waste heat recovery: Special report. 1979. *Chemical Engineering Progress* 75 (December):25-42.

Weinstein, A., S. Sero, and R.C. Niess. 1981. Solar assisted domestic hot water heat pump system. Paper presented at Annual Meeting of American Section of International Solar Energy Society, Philadelphia, PA, May 28.

Weinstein, A. and G.J. van Zuiden. 1979. Reducing solar costs with solar-assisted templifier. Proceedings of International Solar Energy Society.

Whitehead, E.R. and R.D. Roley, Jr. 1976. The heat pump—A proven device for heat recovery systems. ASHRAE *Journal* 18 (May):31.

Wise, J.L. 1982. Energy pumps—Closing the energy cycle. TAPPI Proceedings, 211-16.

Woods, D. and R.F. Ellis. 1983. Waste heat recovery system. *Food Processing* (August).

Yundt, B. 1984. Troubleshooting VC evaporators. *Chemical Engineering* (December):46-55.

CHAPTER 9

AIR DISTRIBUTION DESIGN FOR SMALL HEATING AND COOLING SYSTEMS

THIS chapter describes the design of small forced-air heating and cooling systems, covers system component selection, and explains their importance. Residential and certain small commercial systems may be designed using this procedure, but large commercial systems are beyond the scope of this chapter.

SYSTEM COMPONENTS

Forced-air systems are heating and/or cooling systems that use motor-driven blowers to distribute heated, cooled, and otherwise treated air for the comfort of individuals within confined spaces. A typical residential or small commercial system includes: (1) a heating and/or cooling unit, (2) accessory equipment, (3) supply and return ductwork, (4) supply and return grilles, and (5) controls. These components are described briefly in the following sections and are illustrated in Figure 1.

Heating and Cooling Units

The three main types of forced-air heating and cooling devices are: (1) furnaces, (2) air conditioners, and (3) heat pumps.

Furnaces are the basic component of most forced-air heating systems. They are augmented with an air-conditioning coil when cooling is included in the system, and they are manufactured to use specific fuels such as oil, natural gas, or liquefied petroleum gas. The fuel used dictates installation requirements and safety considerations (see Chapter 29).

The most common air-conditioning system uses a split-system configuration with a furnace. In this application, the air-conditioning evaporator coil is installed on the discharge air side of the furnace. The compressor and condensing coil are located outside the structure, and refrigerant lines connect the outdoor and indoor units.

Self-contained air conditioners are another type of forced-air system. These units contain all necessary air-conditioning components, including circulating air blowers, and may or may not include fuel-fired heat exchangers or electric heating elements.

The heat pump cools and heats using the refrigeration cycle. It is available in split-system and packaged (self-contained) configurations. Generally, the air source heat pump requires supplemental heating; therefore, electric heating elements are usually included with the heat pump as part of the forced-air system. Heat pumps are also combined with fossil fuel furnaces to take advantage of their high efficiency at mild temperatures as a means to minimize heating costs.

Accessory Equipment

Forced-air systems may be equipped to humidify and dehumidify the environment, remove contaminants from recirculated air,

and provide for circulation of outside air in an economizer operation.

Humidifiers. Several types of humidifiers are available, including self-contained steam, atomizing, evaporative, and heated pan. Chapter 20 describes the units in detail. The Air Conditioning and Refrigeration Institute (ARI) rates various humidifiers in ARI *Standard* 610-89.

Humidifiers must match the heating unit. Discharge air temperatures on heating systems vary, and some humidifiers do not provide their own heat source for humidification. These humidifiers should be applied with caution to heat pumps and other low air temperature rise heating units.

Structures with complete vapor barriers (walls, ceilings, and floors) normally require no supplemental moisture during the heating season, since internally generated moisture maintains an acceptable relative humidity of 20 to 60%.

Electronic air cleaners. These units attract oppositely charged particles, fine dust, smoke, and other particulates to collecting plates in the air cleaner. While these plates remove finer particles, larger particles are trapped in an ordinary throwaway or permanent filter. A nearly constant pressure drop and efficiency can be expected, unless the cleaner becomes severely loaded with dust.

Custom accessories. The many variations in solar, off-peak storage, and other custom systems are not covered specifically in this chapter. However, their components may be classified as duct system accessories.

Economizer control. This device monitors outdoor temperature and humidity and automatically shuts down the air-conditioning unit when a preset outdoor condition is met. Damper motors open outdoor return air dampers, letting outside air enter the system to provide comfort cooling. When outdoor air conditions are no longer acceptable, the outdoor air dampers close and the air-conditioning unit comes back on.

All these accessories affect system airflow and pressure requirements. Losses must be taken into account when selecting the heating and cooling equipment and sizing ductwork.

Ducts

Most systems require ducts composed of manufactured components or on-site constructed air passages. A properly designed duct system is critical to the successful performance of the central equipment and accessories.

In small commercial applications and residential work, ductwork design depends on the air-moving characteristics of the blower included with the selected equipment. It is important to recognize this difference between small commercial or residential systems and large commercial and industrial systems. The designer of smaller systems must determine resistances to the movement of air and adjust the ductwork size to limit the static pressure against which the blower operates. Manufacturers publish static

The preparation of this chapter is assigned to TC 6.3, Central Forced Air Heating and Cooling Systems.

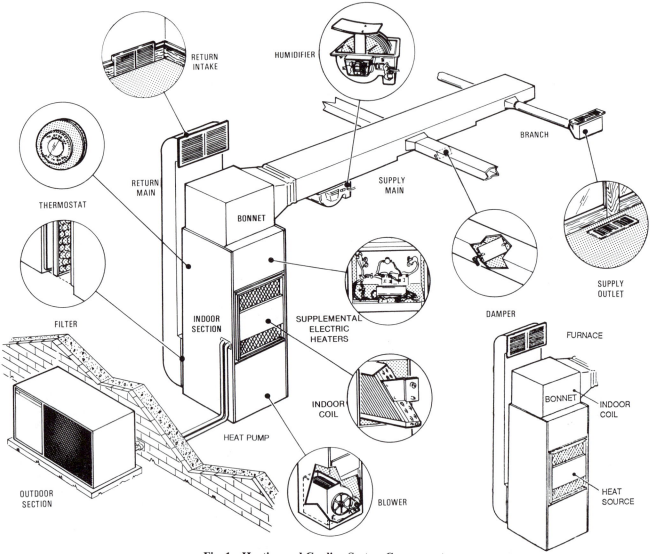

Fig. 1   Heating and Cooling System Components

pressure versus flow rate information for the blower so the designer can determine the maximum static pressure against which the blower will operate while delivering the proper volume of air (see Chapter 19).

Supply Outlets and Return Grilles

Supply air should be directed to the sources of greatest heat loss and/or heat gain to offset their effects. Registers and grilles for the supply and return systems should accommodate all aspects of the supply air distribution patterns such as throw, spread, and drop. The outlet and return grille velocities must be held within recommended limits, as noise generated at the grille is of equal or greater importance than duct noise.

Controls

Forced-air heating and/or cooling systems may be adequately controlled in several ways. Simple on/off cycling of central equipment is frequently adequate to maintain comfort. Spaces with large variations in load requirements may require zone control systems or multiple units with separate ductwork. Systems with minimal variations in load requirements in the space may function adequately with one central wall thermostat or a return air thermostat. Residential conditioning systems of 60,000 Btu/h capacity or less are typically operated with one central thermostat.

Forced-air system control may require several devices, depending on the sophistication of the system and the accessories used (see Chapter 41 of the 1991 ASHRAE *Handbook—HVAC Applications*). Energy conservation has increased the importance of control systems, so now methods that were considered too expensive for small systems may be cost-effective.

Temperature control, the primary consideration in forced-air systems, may be accomplished by a single-stage thermostat. When properly located with correctly adjusted heat anticipators, this device accurately controls temperature. Multistage thermostats are required on many systems (*e.g.*, a heat pump with auxiliary heating) and may improve temperature regulation. Outdoor thermostats, in series with indoor control, can stage heating increments adequately.

Indoor thermostats may incorporate many control capabilities within one device, including continuous or automatic fan control and automatic or manual changeover between heating and cooling. Where more than one system conditions a common space, manual control is preferred to prevent simultaneous heating and cooling.

Thermostats with programmable temperature controls allow the occupant to vary the temperature set point for different periods. These devices save substantial energy by applying automatic night setback and/or daytime temperature setback for all systems when appropriately matched.

Two-speed fan control may be desirable for fossil-fueled systems but should not be applied to heat pump systems unless recommended by the manufacturer. Humidistats should be specified for humidifier control.

When applying unusual controls, manufacturer recommendations should always be observed to guard against equipment damage or misuse.

SYSTEM DESIGN

The size and performance characteristics of system components are interrelated, and the overall system design should proceed in the organized manner described. For example, furnace selection depends on heat gain and loss; however, the sizing is also affected by ductwork location (attic, basement, and so forth), night setback, and humidifier usage. Here is a recommended procedure:

1. Estimate heating and cooling loads.
2. Determine preliminary ductwork location.
3. Determine heating and cooling unit location.
4. Select accessory equipment. Frequently, accessory equipment is not provided with initial construction; however, the system may be designed to add these components later.
5. Select control components.
6. Determine maximum airflow (cooling or heating) for each supply and return location.
7. Determine airflow at reduced heating and cooling loads (two-speed and variable-speed fan).
8. Select heating/cooling equipment.
9. Select control system.
10. Finalize duct design and size system.
11. Select supply and return grilles.

This procedure requires certain preliminary information such as location, weather conditions, and architectural considerations. The following sections cover the preliminary considerations and discuss how to follow this recommended procedure.

Locating Outlets, Ducts, and Equipment

The characteristics of a residence determine the location and the type of forced-air system that can be installed. The presence or absence of particular areas within a residence has a direct influence on equipment and ductwork location. The structure's size, room or area use, and air distribution system determine how many central systems will be needed to maintain comfort temperatures in all areas.

A full basement is an ideal location to install equipment and ductwork. If a residence has a crawl space, the ductwork and equipment can be located there, or the equipment can be placed in a closet or utility room. The equipment's enclosure must meet all fire and safety code requirements; adequate service clearance must also be provided. In a home built on a concrete slab, equipment should be located in the conditioned space (for systems that do not require combustion air), an unconditioned closet, an attached garage, the attic space, or outdoors. The ductwork normally is located in a furred-out space, in the slab, or in the attic.

Duct construction must conform to local code requirements which often reference NFPA *Standard* 90B, Installation of Warm Air Heating and Air Conditioning Systems, published by the National Fire Protection Association, or Installation Standards for Residential Heating and Air-Conditioning Systems, published by the Sheet Metal and Air Conditioning Contractors National Association, Inc.

Weather conditions should be considered when locating equipment and ductwork. Packaged outdoor units for houses in severely cold climates must be installed according to manufacturer recommendations. Most houses in cold climates have basements, making them well suited for indoor furnaces and split-system air conditioners or heat pumps. In mild and moderate climates, the ductwork is frequently in the attic or crawl space. Ductwork located outdoors, in attics, crawl spaces, and basements must be insulated, as outlined in Chapter 21 of the 1989 ASHRAE *Handbook—Fundamentals*.

Although the principles of air distribution discussed in Chapter 31 of the 1989 ASHRAE *Handbook—Fundamentals* apply in forced-air system design, simplified methods of selecting outlet size and location are generally used.

Supply outlets fall into four general groups, defined by their air discharge patterns: (1) horizontal high, (2) vertical nonspreading, (3) vertical spreading, and (4) horizontal low. Straub and Chen (1957) and Wright *et al.* (1963) describe these types and their performance characteristics under controlled laboratory and actual residence conditions. Table 1 lists the general characteristics of supply outlets. It includes the performance of various outlet types for cooling as well as heating, since one of the advantages of forced-air systems is that they may be used for both heating and cooling. However, as indicated in Table 1, no single outlet type is best for both heating and cooling.

Table 1   General Characteristics of Supply Outlets

Group	Outlet Type	Outlet Flow Pattern	Most Effective Application		Selection Criteria (see Figure 2)
1	Ceiling and high sidewall	Horizontal	Cooling	*Ceiling outlets*	
				Full circle or widespread type	Select for throw equal to distance from outlet to nearest wall at design flow rate and pressure limitations.
				Narrow spread type	Select for throw equal to 0.75 to 1.2 times distance from outlet to nearest wall at design flow rate and pressure limitations.
				Two adjacent ceiling outlets	Select each so that throw is about 0.5 distance between them at design flow rate and pressure limits.
				High sidewall outlets	Select for throw equal to 0.75 to 1.2 times distance to nearest wall at design flow rate and pressure limits. If pressure drop is excessive, use several smaller outlets rather than one large one to reduce pressure drop.
2	Floor diffusers, baseboard, and low sidewall	Vertical, nonspreading	Cooling and heating	For cooling only	Select for 6 to 8 ft throw at design flow rate and pressure limitations.
3	Floor diffusers, baseboard, and low sidewall	Vertical, spreading	Heating and cooling	For heating only	Select for 4 to 6 ft throw at design flow rate and pressure limitations.
4	Baseboard and low sidewall	Horizontal	Heating only		Limit face velocity to 300 ft/min.

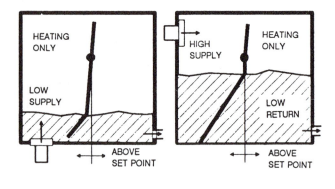

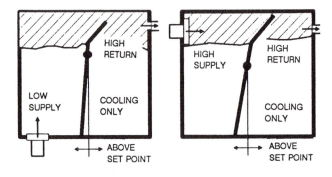

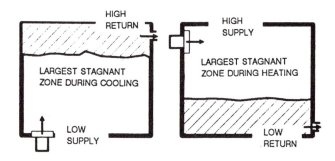

Fig. 2 Preferred Return Locations for Various Supply Outlet Positions

The best outlets for heating are located near the floor at outside walls and provide a vertical spreading air jet, preferably under windows, to blanket cold areas and counteract cold drafts. Called perimeter heating, this arrangement mixes the warm supply air with both the cool air from the area of high heat loss and the cold air from infiltration, preventing drafts.

The best outlet types for cooling are located in the ceiling and have a horizontal air discharge pattern. For year-round systems, supply outlets are located to satisfy the more critical load.

Figure 2 illustrates preferred return locations for different supply outlet positions and system functions. These return locations are based on the presence of the stagnant layer in a room, which is beyond the influence of the supply outlet and thus experiences little air motion (*e.g.*, cigarette smoke "hanging" in a spot in a room is evidence of a stagnant region). The stagnant layer degrades room comfort.

The stagnant layer develops near the floor during heating and near the ceiling during cooling. Returns help remove air from this region if the return face is placed within the stagnant zone. Thus, for heating, returns should be placed low; for cooling, returns should be placed high.

In a year-round heating and cooling application, a compromise must be made by placing returns where the largest stagnant zone will develop. With low supply outlets, the largest stagnant zone will develop during cooling, so returns should be placed high or opposite the supply locations. Conversely, high supply outlets will not perform as well during heating; therefore, returns should be placed low to be of maximum benefit.

Determining Heating and Cooling Loads

Design heating loads can be calculated by following the procedures outlined in Chapter 25 of the 1989 ASHRAE *Handbook—Fundamentals*. Design cooling loads can be calculated by the procedures presented in Chapters 26 and 27 of the 1989 ASHRAE *Handbook—Fundamentals*.

When calculating design loads, heat losses or gains from the air distribution system must be included in the total load for each room. Losses from ducts can be calculated by following the procedures in Chapter 32 of the 1989 ASHRAE *Handbook—Fundamentals*.

Frequently, local codes may require outdoor air ventilation in residential applications. This ventilation load is in addition to the building load.

Selecting Equipment

A furnace heating output should match or slightly exceed the estimated design load. A 40% limit on oversizing has been recommended by the Air Conditioning Contractors of America for fossil fuel furnaces. This limit minimizes venting problems associated with oversized equipment and improves part-load performance. Note that the calculated load must include duct losses, humidification load, night setback recovery load, as well as building conduction and infiltration heat losses. Chapter 29 has detailed information on how to size and select a furnace.

To help conserve energy, manufacturers have added features to improve furnace efficiency. Electric ignition has replaced the standing pilot; vent dampers and more efficient motors are also available. Furnaces with fan-assisted combustion systems (FACS) and condensing furnaces also improve efficiency. Two-stage heating and cooling, variable-speed heat pump systems, and two-speed and variable-speed blowers are also available.

A system designed to both heat and cool, and one that cycles the cooling equipment on and off by sensing dry-bulb temperature alone, should be sized to match the design heat gain as closely as possible. Oversizing under this control strategy could lead to higher than desired indoor humidity levels. Chapter 26 of the 1989 ASHRAE *Handbook—Fundamentals* recommends that cooling units not be oversized by more than 25% of the sensible load. Other sources suggest limiting oversizing to 15% of the sensible load if it is not an air source heat pump application.

Airflow Requirements

After the equipment is selected and prior to duct design, the following decisions must be made:

- Determine the air quantities required for each room or space during heating and cooling based on each room's heat loss or heat gain. The air quantity selected should be the greater of the heating or cooling requirement.
- Determine the number of supply outlets needed for each space to supply the selected air quantity, giving detailed consideration to discharge velocity, spread, throw, terminal velocity, occupancy patterns, location of heat gain and heat loss sources, and register or diffuser design.
- Determine the return system type (multiple or central), the availability of space for the grilles, the filtering system, maximum velocity limitations for sound, efficient filtration velocity, and space use limitations.

The required airflow rate, along with the static pressure limitation of the blower, are the parameters around which the duct system is designed. The total airflow selected must be proportioned to the room or space load requirements. The heat loss or gain for each space determines the proportion of the total airflow supplied to each space.

The static pressure drop in supply registers should be limited to about 0.03 in. of water. The required pressure drop must be deducted from the static pressure available from the system for duct design.

The flow rate delivered by a single supply outlet should be determined by considering (1) the space limitations on the number of registers that can be installed, (2) the pressure drop for the register considered at the flow rate selected, (3) the adequacy of the air delivery patterns for offsetting heat losses or gains, and (4) the space use patterns.

Manufacturer's specifications provide blower airflow rates for each blower speed and external static pressure combination. Determining static pressure available for duct design should include the possibility of adding accessories in the future, *e.g.*, electronic air cleaners or humidifiers. Therefore, it is recommended that the highest available fan speed not be used for design.

For systems that heat only, determine the blower rate from the manufacturer's data. The temperature rise of air passing through the heat exchanger of a fossil fuel fired furnace must be within the manufacturer's recommended range (usually from 40 to 80 °F). The possible later addition of a cooling system should also be considered by selecting a blower that operates in the midrange of the fan speed and settings.

For cooling only, or for heating and cooling systems, the design flow rate can be estimated by the following equation:

$$Q = q_s/(1.1 \, \Delta t)$$

where

Q = flow rate, cfm
q_s = sensible load, Btu/h
Δt = dry-bulb temperature difference between air entering and leaving equipment, °F

For preliminary design, an approximate Δt is listed as follows:

Sensible Heat Ratio (SHR)	Δt
0.75 to 0.79	21
0.80 to 0.85	19
0.85 to 0.90	17

SHR = Calculated sensible load/Calculated total load

For example, if a calculation indicates the sensible load is 23,000 Btu/h and the latent load is 4900 Btu/h, the SHR is calculated as follows:

$$SHR = 23,000/(23,000 + 4900) = 0.82$$

and

$$Q = 23,000/(1.1 \times 19) = 1100 \text{ cfm}$$

This value is the estimated design flow rate. The exact design flow rate can only be determined after the cooling unit is selected. The unit that is ultimately selected should supply an airflow that is in the range of the estimated flow rate, and it must also have adequate sensible and latent cooling capacity when operating at design conditions.

Duct Design Recommendations

Because of the competitive nature of the residential construction business and the practical necessity of using less sophisticated design techniques for these systems, the modified equal-friction duct loss calculation method is recommended. This method is

Table 2 Recommended Division of Duct Pressure Loss

System Characteristics	Supply, %	Return, %
A Single return at blower	90	10
B Single return at or near equipment	80	20
C Single return with appreciable return duct run	70	30
D Multiple return with moderate return duct system	60	40
E Multiple return with extensive return duct system	50	50

satisfactory as long as the designer understands its strengths and weaknesses. It should be applied only to those systems requiring less than a normal air-moving capacity of 2250 cfm or a 60,000 Btu/h cooling requirement.

With the equal friction method, the pressure available for supply and return duct losses is found by deducting coil, filter, grille, and accessory losses from the manufacturer's specified blower pressure. The remaining pressure is divided between the supply and return system as shown in Table 2.

Using duct calculators and the friction chart simplifies the calculations, but pressures must be in proper units. Branch ducts are sized to balance the available pressure without exceeding recommended maximum velocities. Low resistance duct runs may require dampers.

The velocity reduction method provides similar results but without the need to check on resulting duct velocities. Hand calculators and computer programs simplify the number of manual calculations required.

The ductwork distributes the total air supplied by the system to spaces according to the space heating and/or cooling requirements. The return air systems may be single, multiple, or any combination that will return air to the equipment within design static pressure and with satisfactory air movement patterns.

Following are some general rules in duct design:

- Keep main ducts as straight as possible.
- Streamline transitions.
- Design elbows with an inside radius of at least one-third the duct width. If this inside radius is not possible, include turning vanes.
- Make ducts tight and sealed to limit air loss.
- Insulate and/or line ducts, where necessary, to conserve energy and limit noise.
- Locate branch duct takeoffs at least 4 ft downstream from a fan or transition, if possible.
- Isolate the air-moving equipment from the duct system using flexible connectors to isolate noise.

Large air distribution systems are designed to meet specific noise criteria (NC) decibel levels. Quiet, small systems are usually achieved by limiting airflow velocities in mains and branches to the following:

Main ducts	700 to 900 fpm
Branch ducts	600 fpm
Branch risers	500 fpm

Considerable difference may exist between the cooling and heating flow rate requirements. Since many systems cannot be rebalanced seasonally, a compromise must be made in the duct design to accommodate the most critical need. For example, a kitchen may require 165 cfm for cooling but only 65 cfm for heating. Since the kitchen may be used heavily during design cooling periods, the cooling flow rate should be used. Normally, the maximum design flow rate should be used, since register dampers do allow some optional reduction in airflows.

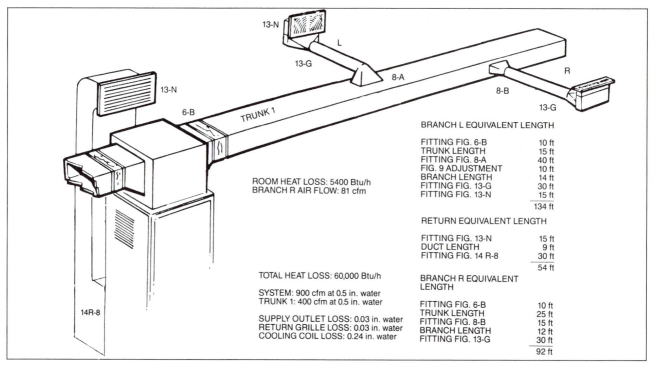

Fig. 3   Equal Friction Duct Design Example

Duct Design Procedures—Equal Friction

Refer to Figure 3 for numerical example.

1. Determine the heating and cooling load to be supplied by each outlet using the previously outlined ASHRAE or ACCA procedure. Include duct losses or gains.
2. Make a simple diagram of the supply and return duct systems—at least one outlet in a room or area for each 8000 Btu/h loss or 4000 Btu/h sensible heat gain, whichever is greater.
3. Label all fittings and transitions to show equivalent lengths on the drawing. (Refer to Figures 6 to 15 for approximate equivalent lengths of fittings. The letter code in the figures identifies the type of fitting shown on the drawings.)
4. Show measured lengths of ductwork on the drawing.
5. Determine the total effective length of each branch supply. (Begin at the air handler and add all equivalent lengths and measured lengths to each outlet: Effective length = Equivalent length + Measured length.)
   Example: Branch R in Figure 3 is 92 ft.
6. Proportion the total airflow rate to each supply outlet for both heating and cooling.

   Example: Supply outlet flow rate =

   $$\frac{\text{Outlet heat loss (gain)}}{\text{Total heat loss (gain)}} \times \text{Total system flow rate}$$

   Branch R in Figure 3:

   $$(5400/60{,}000) \times 900 = 81 \text{ cfm}$$

7. The flow rate required for each supply outlet is equal to the heating or cooling flow rate requirement (normally the larger of the two rates).
8. Label the supply outlet flow rate requirement on the drawing for each outlet.
9. Determine the total external static pressure available from the unit at the selected airflow rate.

10. Subtract the supply and return register pressure, external coil pressure, filter pressure, and box plenum (if used) from the available static pressure to determine the static pressure available for the duct design.

    Example: Figure 3

Equipment	0.50 in. water
Cooling coil	− 0.24 in. water
Supply outlet	− 0.03 in. water
Return grille	− 0.03 in. water
Available static pressure	0.20 in. water

11. Proportion the available static pressure between the supply and return systems.

    Example:

Supply (75%)	0.15 in. water
Return (25%)	0.05 in. water
Total	0.20 in. water

Sizing the Branch Supply Air System

12. Use the supply static pressure available to calculate each branch design static pressure for 100 ft of equivalent length.

    Example: Branch R design static pressure rate =

    $$\frac{100 \text{ (Supply static pressure available)}}{\text{Effective length of each branch supply}}$$

    $$100 \times 0.15/92 = 0.163 \text{ in. water/100 ft (Branch R)}$$

13. Enter the friction chart (Chapter 32 of the 1989 ASHRAE *Handbook—Fundamentals*) at the branch design static pressure (0.163) opposite the flow rate for each supply, and read the round duct size (5 in.) and velocity (600 fpm).

14. If velocity exceeds maximum recommended values, increase the size and specify runout damper.
15. Convert the round duct to rectangular, where needed.

Sizing the Supply Trunk System

16. Determine the branch supply with the longest total effective length and, from this, determine the static pressure to size the supply trunk duct system.

    Example: Supply trunk design static pressure rate =

    $$\frac{100 \text{ (Total supply static pressure available)}}{\text{Longest effective length of branch duct supplies}}$$

    $0.15 \times 100/134$ (Branch L) = 0.112 in. water/100 ft

17. Total the heating airflow rate and the cooling airflow rate for each trunk duct section. Select the larger of the two flow rates for each section of duct between runouts or groups of runouts.
18. Design each supply trunk duct section by entering the friction chart at the supply trunk static pressure (0.112) and sizing each trunk section for the appropriate air volume handled by that section of duct. (Trunk 1 – 400 cfm; duct size = 9.5 in. at 800 fpm.)

    Trunks should be checked for size after each runout and reduced, as required, to maintain velocity above branch duct design velocity.

19. Convert round duct size to rectangular, where needed.

Sizing the Return Air System

20. Select the number of return air openings to be used.
21. Determine the volume of air that will be returned by each of the return air openings.
22. From Step 11, select the return air static pressure.
23. Determine the static pressure available per 100 ft effective length for each return run, and design the same as for the supply trunk system.

    Example: Return trunk design static pressure =

    $$\frac{100 \text{ (Total return static pressure available)}}{\text{Longest effective length of return duct runs}}$$

    $0.05 \times 100/54 = 0.093$ in. water/100 ft

24. The trunk design for the return air is the same as the trunk design for the supply system.

    Example: Return duct in Figure 3 is as follows—enter friction chart at 900 cfm and 0.093 in. of water. Duct size is 13.1 in. The velocity, however, is too high at 950 fpm, so use a larger duct.

Sizing Using Velocity Reduction

The velocity reduction method assigns duct velocities throughout the system in decreasing order from furnace to furthest branch. The duct sizes needed to achieve these assigned values are determined using the friction chart, which also provides the pressure drop in each run of duct. This method is fast, and noise control is relatively assured.

Example:

Mains sized for 800 fpm velocity
Branches sized for 400 fpm velocity

Total airflow: 1200 cfm

Step 1: Determine equivalent length of fittings (listed in Figure 4).

Step 2: Measure actual length of duct in mains and branches (listed in Figure 4).

Step 3: Determine duct size and pressure loss using friction chart.

Step 4: Compute duct static loss in longest run (Figure 5).

The loss through the supply outlet and return grille must be added to the duct loss to determine the design static pressure at 1200 cfm.

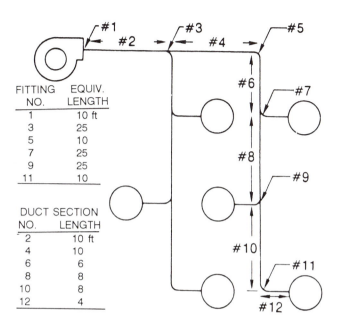

FITTING NO.	EQUIV. LENGTH
1	10 ft
3	25
5	10
7	25
9	25
11	10

DUCT SECTION NO.	LENGTH
2	10 ft
4	10
6	6
8	8
10	8
12	4

Fig. 4   Velocity Reduction Duct Design Example

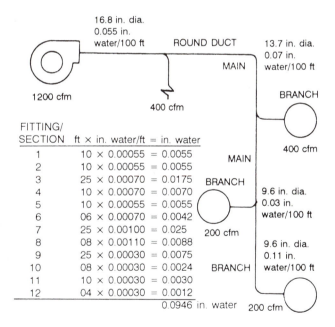

FITTING/ SECTION	ft × in. water/ft	= in. water
1	10 × 0.00055	= 0.0055
2	10 × 0.00055	= 0.0055
3	25 × 0.00070	= 0.0175
4	10 × 0.00070	= 0.0070
5	10 × 0.00055	= 0.0055
6	06 × 0.00070	= 0.0042
7	25 × 0.00100	= 0.025
8	08 × 0.00110	= 0.0088
9	25 × 0.00030	= 0.0075
10	08 × 0.00030	= 0.0024
11	10 × 0.00030	= 0.0030
12	04 × 0.00030	= 0.0012
		0.0946 in. water

Fig. 5   Example Duct Pressure Losses

Zone Control for Small Systems

In residential applications, some complaints of rooms that are too cold or too hot are related to the limitations of the system. No matter how carefully a single-zone system is designed, problems will occur if the control strategy is unable to accommodate the various load conditions that are simultaneously occurring throughout the house at any given time of day and/or during any season.

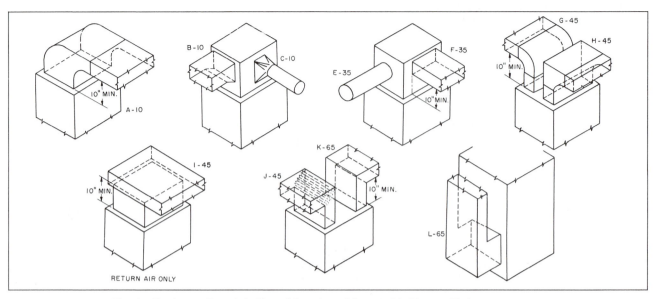

Fig. 6   Equivalent Length in Feet of Supply and Return Air Plenum Fittings (ACCA 1984)

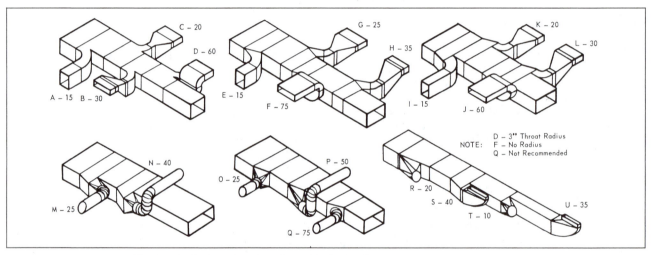

Fig. 7   Equivalent Length in Feet of Reducing Trunk Duct Fittings (ACCA 1984)

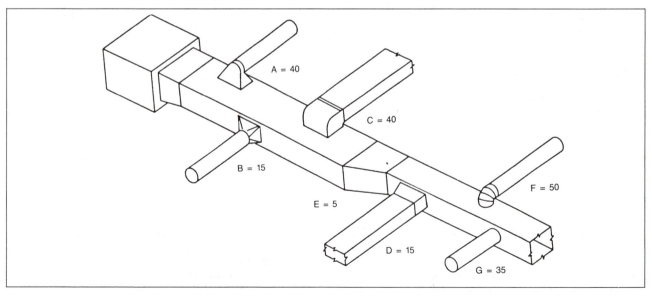

Fig. 8   Equivalent Length in Feet of Extended Plenum Fittings (ACCA 1984)

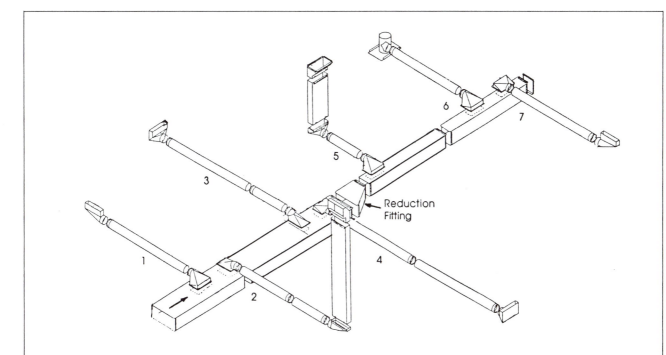

The pressure loss (expressed in equivalent length) of a fitting in a branch takeoff of an extended plenum depends on the location of the branch in the plenum. Takeoff branches nearest the furnace impose a higher loss, and those downstream, a lower loss. The equivalent lengths listed in Figure 8 are for the lowest possible loss—that is, the values are based on the assumption that the takeoff is the last (or only) branch in the plenum.

To correct these values, add velocity factor increments of 10 ft to each upstream branch, depending on the number of branches remaining downstream. For example, if two branches are downstream from a takeoff, add 2 × 10 or 20 ft of equivalent length to the takeoff loss listed.

In addition, consider a long extended plenum that has a reduction in width as two (or more) separate plenums—one before the reduction and one after. The loss for each top takeoff fitting in this example duct system is shown in the following table.

Branch Number	Takeoff Loss, ft	Downstream Branches	Velocity Factor, ft	Design Equiv. Length, ft
Before plenum reduction				
1	40	3	30	70
2	40	2	20	60
3	40	1	10	50
4	40	0	0	40
After plenum reduction				
5	40	2	20	60
6	40	1	10	50
7	40	0	0	40

Fig. 9   Example Duct System and Equivalent Length and Extended Plenum Fittings

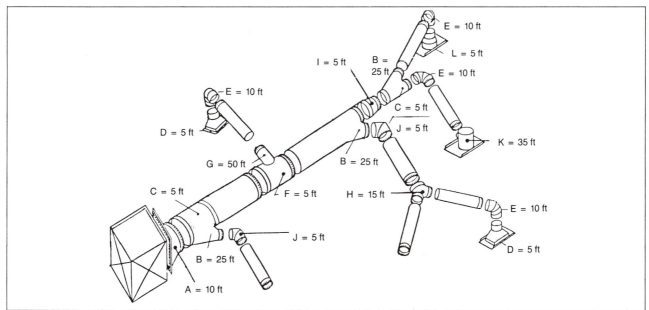

Fig. 10   Equivalent Length of Round Supply System Fittings

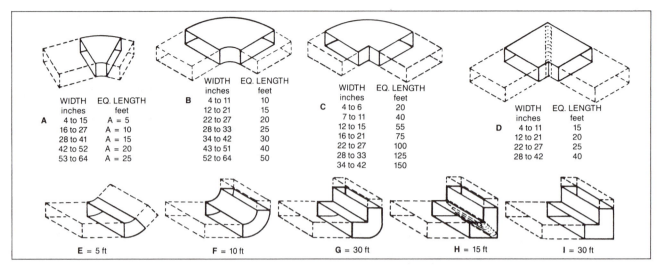

Fig. 11   Equivalent Length in Feet of Angles and Elbows for Trunk Ducts (ACCA 1984)

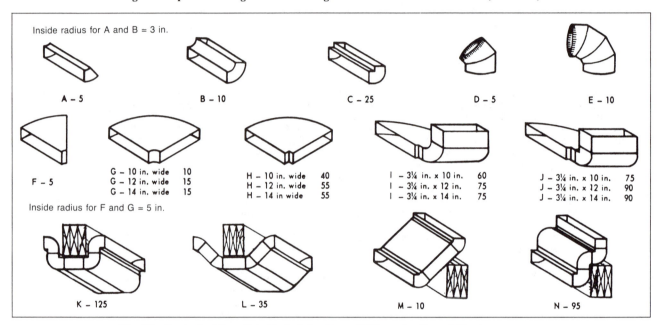

Fig. 12   Equivalent Length in Feet of Angles and Elbows for Branch Ducts (ACCA 1984)

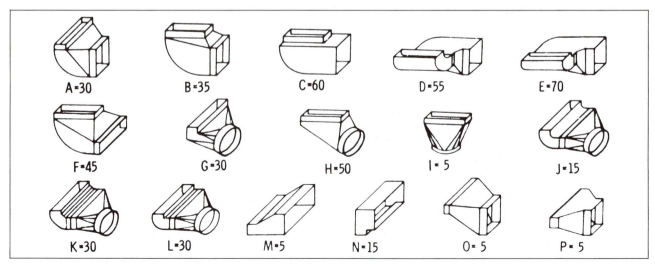

Fig. 13   Equivalent Length in Feet of Boot Fittings (ACCA 1984)

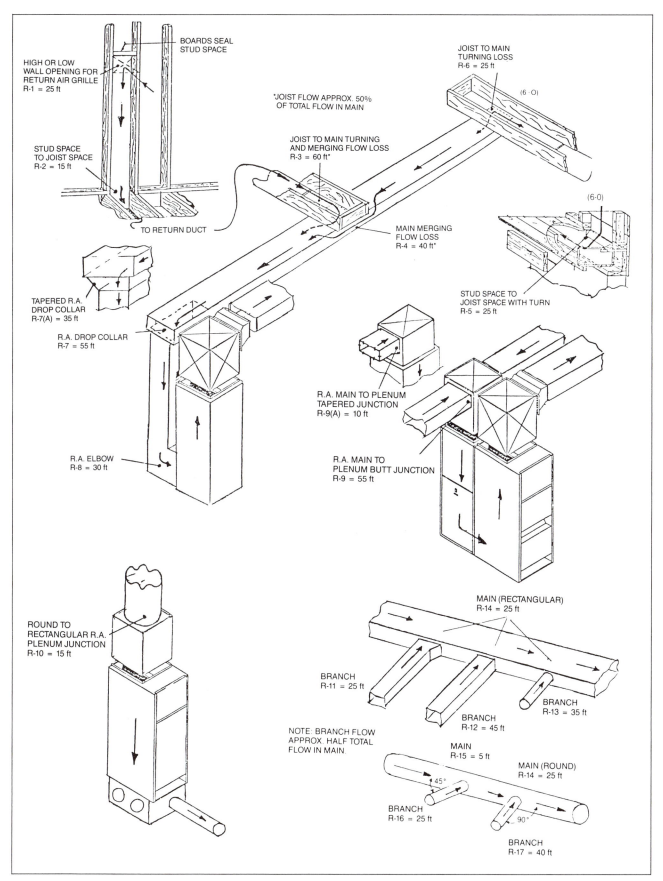

Fig. 14   Equivalent Length of Special Return Fittings

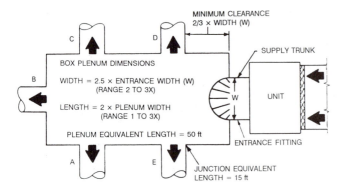

Fig. 15   Dimensions for Efficient Box Plenum

A single-zone control strategy works as long as the various rooms are open to each other. In this case, room-to-room temperature differences are minimized by the convection currents between the rooms. For small rooms, an open door is adequate. For large rooms, an opening that is equal to or greater than 25% of the partition has been proposed in ACCA *Technical Bulletin* 58 as being necessary to ensure adequate air interchange for a single-zone control strategy.

When rooms are isolated from each other, temperature differences cannot be moderated by convection currents, and the conditions in the room with the thermostat may not be representative of the conditions in the other rooms. In this situation, comfort can be improved by continuous blower operation, but this strategy may not completely solve the problem.

Zone control is required when the conditions at the thermostat are not representative of all the rooms. This situation will almost certainly occur if:

- House has more than one level
- One or more rooms are used for entertaining large groups
- One or more rooms have large glass areas
- House has an indoor swimming pool and/or hot tub
- House has a solarium or atrium

In addition, zoning may be required when several rooms are isolated from each other and from the location of the thermostat. This situation is likely to occur when:

- House spreads out in many directions (wings)
- Some rooms are distinctly isolated from the rest of the house
- Envelope only has one or two exposures
- House has a room or rooms in a finished basement
- House has a room or rooms in a finished attic
- House has one or more rooms with slab or exposed floor

Zone control can be achieved by installing:

- Discrete heating/cooling duct systems for each zone requiring control
- Automatic zone damper systems in a single heating/cooling duct system

The airflow delivered to each room must be able to offset the peak room load during cooling. The peak room load can be determined using Table 57 in Chapter 26 of the 1989 ASHRAE *Handbook—Fundamentals*. Then the following equation using the same supply air temperature difference used to size equipment can be used:

$$Q_r = q_s/(1.1 \, \Delta t)$$

where

Q_r = room flow rate, cfm
q_s = sensible load, from peak room load calculations, Btu/h
Δt = temperature difference of entering and leaving air, °F

The design flow rate for any zone is equal to the sum of the peak room flow rates assigned to a zone.

Duct Sizing for Zone Damper Systems

The following guidelines have been proposed in ACCA *Technical Bulletin* 58 to size the various duct runs.

1. Use the design blower airflow value to size a plenum or a main trunk that feeds the zone trunks. Size plenum and main trunk ducts at 800 fpm.
2. Use the zone airflow values (those based on the sum of the peak room loads) to size the zone trunk ducts. Size all zone trunks at 800 fpm.
3. Use the peak room airflow values (those based on the peak room loads) to size the branch ducts or runouts. Size all branch runouts at a friction rate of 0.10 in. of water per 100 ft.
4. Size return ducts for 600 fpm air velocity.

Box Plenum Systems Using Flexible Duct

In some climates, an overhead duct system with a box plenum feeding a series of individual, flexible duct, branch runouts is popular. The pressure drop through a flexible duct is higher than through a rigid sheet metal duct, however. Recognizing this larger loss is important when designing box plenum/flexible duct systems.

The design of the box plenum is critical to avoid excessive pressure losses and to minimize unstable air rotation within the plenum, which can change direction between blower cycles. This in turn may change the air delivery through individual branch takeoffs. Unstable rotation can be avoided by entering the box plenum from the side and by using a special splitter entrance fitting.

Gilman *et al.* (1951) proposed box plenum dimensions and entrance fitting designs to minimize unstable conditions as summarized in Figures 15 and 16.

For residential systems with less than 2250 cfm capacity, the pressure loss through the box plenum is approximately 0.05 in. of water. This loss should be deducted from the available static pressure to determine the static pressure available for duct runouts. In terms of equivalent length, add approximately 50 equivalent feet to the measured branch runs.

Embedded Loop Duct Systems

In cold climates, floor slab construction requires that the floor and perimeter of the slab be heated to provide comfort and prevent condensation.

The temperature drop (or rise) in the supply air is significant, and special design tables must be used to account for the different supply air temperatures at distant registers. In loop duct systems, duct heat losses may cause a large temperature drop. This requires the feed ducts to be placed at critical points in the loop. ACCA *Manual* 4 (1961) describes a procedure for sizing small embedded loop systems.

A second aspect of loop systems is installation. The building site must be well drained and the surrounding grade sloped away from the structure. A vapor retarder must be installed under the slab. The bottom of the embedded duct must not be lower than the finished grade. Because a concrete slab loses heat from its edges outward through the foundation walls and downward through the earth, the edge must be properly insulated.

A typical loop duct is buried in the slab 2 to 18 in. from the outside edge and about 2.5 in. beneath the slab surface. If galvanized sheet metal is used for the duct system, it must be coated on the outside to comply with federal specifications (SS-A-701). Other special materials used for ducts must be installed according to the manufacturer's instructions. In addition, care must be taken when the slab is poured to not puncture the vapor retarder or to crush or dislodge the ducts.

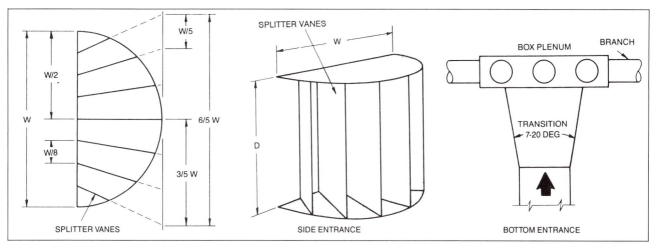

Fig. 16 Entrance Fittings to Eliminate Unstable Airflow in Box Plenum

SELECTING SUPPLY AND RETURN GRILLES AND REGISTERS

Grilles and registers are selected from a manufacturer's catalog with appropriate engineering data after the duct design is completed. Rule-of-thumb selection should be avoided. Carefully determine the suitability of the register or grille selected for each location according to its performance specifications for the quantity of air to be delivered and the discharge velocity from the duct system.

Generally, in small commercial and residential applications, the selection and application of registers and grilles is important, especially since system size and air-handling capacity are small in energy-efficient structures. Proper selection ensures the satisfactory delivery of heating and/or cooling. Cooling systems should not be oversized to attain air recirculation.

Table 1 summarizes selection criteria for common types of supply outlets. Pressure loss is usually limited to 0.03 in. of water or less at the required flow rate. Face velocities range from 400 to 750 fpm at the required airflow rate.

Return grilles are usually sized to provide a face velocity of 400 to 600 fpm, or 2.7 to 4.1 cfm per square inch of grille-free area. Some central return grilles are designed to hold an air filter. This design allows the air to be filtered close to the occupied area and also allows easy access for filter maintenance. Easy access is important when the furnace is located in a remote area such as a crawl space or attic. The velocity of the air through a filter grille should not exceed 300 fpm, which means that the volume of air should not exceed 2.1 cfm per square inch of free filter area.

COMMERCIAL SYSTEMS

The duct design procedure described in this chapter can be applied to small commercial systems using residential equipment, provided the commercial application does not include moisture sources that lead to large latent loads.

In commercial applications that do not require low noise levels, air duct velocities may be increased, reducing duct size. Long throws from supply outlets are also required for large areas, and higher velocities may be required for that reason.

Commercial systems with significant variation in airflow rates for cooling, heating, and large internal loads (*e.g.*, kitchens and theaters) should be designed in accordance with Chapters 31 and 32 of the 1989 ASHRAE *Handbook—Fundamentals* and Chapters 17, 18, and 19 of this volume.

REFERENCES

ACCA. 1964. Installation technique for perimeter heating and cooling. ACCA *Manual* 4, 10th ed. Air Conditioning Contractors of America, Washington, D.C.

ACCA. 1977. Equipment selection and system design procedures for commercial summer and winter air conditioning (using unitary equipment). ACCA *Manual* Q. Air Conditioning Contractors of America, Washington, D.C.

ACCA. 1984. Equipment selection and system design procedures. ACCA *Manual* D, 2nd ed. Air Conditioning Contractors of America, Washington, D.C.

ARI. 1982. Central system humidifiers. ARI *Standard* 610-82. Air-Conditioning and Refrigeration Institute, Arlington, VA.

ARI. Updated annually. Directory of certified unitary air conditioners, unitary air-source heat pumps, and sound rated outdoor unitary equipment. Air Conditioning and Refrigeration Institute, Arlington, VA.

GAMA. Consumer's directory of certified efficiency ratings for residential heating and water heating equipment. Gas Appliance Manufacturers Association, Arlington, VA.

Gilman, S.F., R.J. Martin, and S. Konzo. 1951. Investigation of the pressure characteristics and air distribution in box-type plenums for air conditioning duct systems. University of Illinois Engineering Experiment Station *Bullentin* No. 393 (July), Champagne-Urbana, IL.

NFPA. 1984. Installation of warm air heating and air conditioning systems. NFPA No. 90B-89. National Fire Protection Association, Quincy, MA.

NHAW. 1986. Selection of supply outlets and return inlets. North American Heating and Air Conditioning Wholesalers Association, Columbus, OH.

Rutowski, H. and J.H. Healy. 1990. Selecting residential air-cooled cooling equipment based on sensible and latent performance. ASHRAE *Transactions* 96:2.

SMACNA. 1988. Installation standards for residential heating and air conditioning systems. Sheet Metal and Air Conditioning Contractors National Association, Inc., Marrifield, VA.

Straub, H.E. and M. Chen. 1957. Distribution of air within a room for year-round air conditioning—Part II. University of Illinois Engineering Experimental Station Bulletin No. 442 (March), Champagne-Urbana, IL.

Wright, J.R., D.R. Bahnfleth, and E.G. Brown. 1963. Comparative performance of year-round systems used in air conditioning research no. 2. University of Illinois Engineering Experimental Station Bulletin No. 465 (January), Champagne-Urbana, IL.

STEAM SYSTEMS

A STEAM system uses the vapor phase of water to supply heat or kinetic energy through a piping system. As a source of heat, steam can heat a conditioned space with suitable terminal heat transfer equipment such as fan coil units, unit heaters, radiators, and convectors (fin tube or cast iron), or through a heat exchanger that supplies hot water or some other heat transfer medium to the terminal units. In addition, steam is commonly used in heat exchangers (shell-and-tube, plate, or coil types) to heat domestic hot water and supply heat for industrial and commercial processes such as laundry and kitchen equipment. Steam is also used as a heat source for certain cooling processes such as single- and two-stage absorption refrigeration machines.

STEAM SYSTEM ADVANTAGES

Steam offers the following advantages:

- Steam flows through the system unaided by external energy sources such as pumps.
- Because of its low density, steam can be used in tall buildings where water systems create excessive pressure.
- Terminal units can be added to or removed from the system without making basic changes to the system design.
- Steam components can be repaired or replaced by closing the steam supply without the difficulties associated with draining and refilling a water system.
- Steam is pressure-temperature dependent; therefore, the system temperature can be controlled by varying either steam pressure or temperature.
- Steam can be distributed throughout a heating system with little change in temperature.

In view of the above advantages, steam is applicable to the following facilities:

- Where heat is required for processes and comfort heating, such as in industrial plants, hospitals, restaurants, dry-cleaning plants, laundries, and commercial buildings.
- Where the heating medium must travel great distances, such as in facilities with scattered building locations, or where the building height would result in excessive pressures in water systems.
- Where intermittent changes in heat load occur.

The preparation of this chapter is assigned to TC 6.1, Hydronic and Steam Equipment and Systems.

FUNDAMENTALS

Steam is the vapor phase of water, which is generated by adding more heat than required to maintain its liquid phase at a given pressure, causing the liquid to change to vapor without any further increase in temperature. Table 1 illustrates the pressure-temperature relationship and various other properties of steam.

Temperature is the thermal state of both liquid and vapor at any given pressure. The figures shown are for dry saturated steam. The vapor temperature can be raised by adding more heat, resulting in superheated steam, which is used (1) where higher temperatures are required, (2) in large distribution systems to compensate for heat losses and to ensure that steam is delivered at the desired saturated pressure and temperature, and (3) to ensure that the steam is dry and contains no entrained liquid that could damage some turbine-driven equipment.

Enthalpy of the liquid h_f (sensible heat) is the amount of heat in Btu required to raise the temperature of a pound of water from 32°F to the boiling point at the pressure indicated.

Enthalpy of the evaporation h_{fg} (latent heat of vaporization) is the amount of heat required to change a pound of boiling water at a given pressure to a pound of steam at the same pressure. This same amount of heat is released when the vapor is condensed back to a liquid.

Enthalpy of the steam h_g (total heat) is the combined enthalpy of liquid and vapor and represents the total heat above 32°F in the steam.

Specific volume, the reciprocal of density, is the volume of unit mass and indicates the volumetric space that 1 lb of steam or water occupies.

An understanding of the above helps explain some of the following unique properties and advantages of steam:

1. Most of the heat content of steam is stored as latent heat, which permits large quantities of heat to be transmitted efficiently with little change in temperature. Since the temperature of saturated steam is pressure-dependent, a negligible temperature reduction occurs from a reduction in pressure caused by pipe friction losses as steam flows through the system, regardless of the insulation efficiency, as long as the boiler maintains the initial pressure and the steam traps remove the condensate. Conversely, in a hydronic system, inadequate insulation can significantly reduce fluid temperature.

Table 1 Properties of Saturated Steam

Pressure, psi		Saturation Temperature, °F	Specific Volume, ft³/lb		Enthalpy, Btu/lb		
Gage	Absolute		Liquid v_f	Steam v_g	Liquid h_f	Evaporation h_{fg}	Steam h_g
25 in. of Hg	2.4	134	.0163	146.4	101	1018	1119
9.56 vacuum	10	193	.0166	38.4	161	982	1143
0	14.7	212	.0167	26.8	180	970	1150
2	16.7	218	.0168	23.8	187	966	1153
5	19.7	227	.0168	20.4	195	961	1156
15	29.7	250	.0170	13.9	218	946	1164
50	64.7	298	.0174	6.7	267	912	1179
100	114.7	338	.0179	3.9	309	881	1190
150	164.7	366	.0182	2.8	339	857	1196
200	214.7	388	.0185	2.1	362	837	1179

Values are rounded off or approximated to illustrate the various properties discussed in the text. For calculation and design, use values of thermodynamic properties of water shown in Chapter 6 of the 1989 ASHRAE *Handbook — Fundamentals* or a similar table.

2. Steam, as all fluids, flows from areas of high pressure to areas of low pressure and is able to move throughout a system without an external energy source. Heat dissipation causes the vapor to condense, creating a reduction in pressure caused by the dramatic change in specific volume (1600:1 at atmospheric pressure).

3. As steam gives up its latent heat at the terminal equipment, the condensate that forms is initially at the same pressure and temperature as the steam. When this condensate is discharged to a lower pressure (as when a steam trap passes condensate to the return system), the condensate contains more heat than necessary to maintain the liquid phase at the lower pressure; this excess heat causes some of the liquid to vaporize or "flash" to steam at the lower pressure. The amount of liquid that flashes to steam can be calculated by Equation (1).

$$\% \text{ Flash Steam} = \frac{100(h_{f1} - h_{f2})}{h_{fg2}} \qquad (1)$$

where

h_{f1} = enthalpy of liquid at pressure p_1

h_{f2} = enthalpy of liquid at pressure p_2

h_{fg2} = latent heat of vaporization at pressure p_2

Flash steam contains significant and useful heat energy that can be recovered and used (see Heat Recovery). This re-evaporation of condensate can be controlled (minimized) by subcooling the condensate within the terminal equipment before it discharges into the return piping. The amount of subcooling should not be so large as to cause a significant loss of heat transfer (condensing) surface.

EFFECTS OF WATER, AIR, AND GASES

The enthalpies shown in Table 1 are for dry saturated steam. Most systems operate near these theoretically available values, but the presence of water and gases can affect enthalpy, as well as have other adverse operating effects.

Dry saturated steam is pure vapor without entrained water droplets. However, some amount of water usually carries over as condensate forms because of heat losses in the distribution system. Steam quality describes the amount of water present and can be determined by calorimeter tests.

While steam quality might not have a significant effect on the heat transfer capabilities of the terminal equipment, the backing up or presence of condensate can be significant because the enthalpy of condensate h_f is negligible compared with the enthalpy of evaporation h_{fg}. If condensate does not drain properly from pipes and coils, the rapidly flowing steam

can push a slug of condensate through the system. This can cause water hammer and result in objectionable noise and damage to piping and system components.

The presence of air also reduces steam temperature. Air reduces heat transfer because it migrates to and insulates heat transfer surfaces. Further, oxygen in the system causes pitting of iron and steel surfaces. Carbon dioxide (CO_2) traveling with steam dissolves in condensate, forming carbonic acid, which is extremely corrosive to pipes and heat transfer equipment.

The combined adverse effects of water, air, and CO_2 necessitate their prompt and efficient removal.

HEAT TRANSFER

The quantity of steam that must be supplied to a heat exchanger to transfer a specific amount of heat is a function of (1) the steam temperature and quality, (2) the character and entering and leaving temperature of the medium to be heated, and (3) the heat exchanger design. For a more detailed discussion of heat transfer, see Chapter 3 of the 1989 ASHRAE *Handbook—Fundamentals*.

BASIC STEAM SYSTEM DESIGN

Because of the various codes and regulations governing the design and operation of boilers, pressure vessels, and systems, steam systems are classified according to operating pressure. Low-pressure systems operate up to 15 psig, and high-pressure systems operate over 15 psig. There are many subclassifications within these broad classifications, especially for heating systems such as one- and two-pipe, gravity, vacuum, or variable vacuum return systems. However, these subclassifications relate to the distribution system or temperature-control method. Regardless of classification, all steam systems include a source of steam, a distribution system, and terminal equipment, where steam is used as the source of power or heat.

STEAM SOURCE

Steam can be generated directly by boilers using oil, gas, coal, wood, or waste as a fuel source or solar, nuclear, or electrical energy as a heat source, and indirectly by recovering heat from processes or equipment such as gas turbines and diesel or gas engines. The cogeneration of electricity and steam should always be considered for facilities that have year-round steam requirements. Where steam is used as a power source (such as in turbine-driven equipment), the exhaust steam may be used in heat transfer equipment for process and space heating.

Steam can be provided by a facility's own boiler or cogeneration plant or can be purchased from a central utility serving a city or specific geographic area. This distinction can be very important. A facility with its own boiler plant usually has a closed loop system and requires the condensate to be as hot as possible when it returns to the boiler. Conversely, condensate return pumps require a few degrees of subcooling to prevent cavitation or flashing of condensate to vapor at the suction eye of pump impellers. The degree of subcooling will vary, depending on the hydraulic design or characteristics of the pump in use. (See Chapter 39, Net Positive Suction Head section.)

Central utilities often do not take back condensate, so it is discharged by the using facility and results in an open loop system. If a utility does take back condensate, it rarely gives credit for its heat content. Except for losses to the atmosphere, which for the most part can be eliminated, every pound of steam eventually becomes a pound of condensate. If condensate is returned at 180°F and a heat recovery system reduces this temperature to 80°F, 100,000 Btu are recovered from every 1000 lb of steam purchased. The heat remaining in the condensate represents 10 to 15% of the heat purchased from the utility. Using this heat effectively can reduce steam and heating costs by 10% or more (see Heat Recovery).

Boilers

Fired and waste heat boilers are usually constructed and labeled according to the ASME Boiler and Pressure Vessel Code since pressures normally exceed 15 psig. Details on design, construction, and application can be found in Chapter 28. Boiler selection is based on the combined loads, including heating processes and equipment that use steam, hot water generation, piping losses, and pickup allowance.

The Hydronics Institute standards are used to test and rate most low-pressure heating boilers that have net and gross ratings. In smaller systems, selection is based on a net rating. Larger system selection is made on a gross load basis. The occurrence and nature of the load components, with respect to the total load, determines the number of boilers used in an installation.

Heat Recovery and Waste Heat Boilers

Steam can be generated by waste heat, such as exhaust from fuel-fired engines and turbines. Figure 1 schematically shows a typical exhaust boiler and heat recovery system used for diesel engines. A portion of the water used to cool the engine block is diverted as preheated makeup water to the exhaust heat boiler to obtain maximum heat and energy efficiency. Where the quantity of steam generated by the waste heat boiler is not steady or ample enough to satisfy the facility's steam requirements, a conventional boiler must generate supplemental steam.

Heat Exchangers

Heat exchangers are used in most steam systems. Steam-to-water heat exchangers (sometimes called converters or storage tanks with steam heating elements) are used to heat domestic hot water and to supply the terminal equipment in hot-water heating systems. These heat exchangers are usually the shell-and-tube type, where the steam is admitted to the shell and the water is heated as it circulates through the tubes. Condensate coolers (water-to-water) are sometimes used to subcool the condensate while reclaiming the heat energy.

Water-to-steam heat exchangers (steam generators) are used in high-temperature water (HTW) systems to provide process steam. Such heat exchangers generally consist of a U-tube bundle, through which the HTW circulates, installed in a tank or pressure vessel.

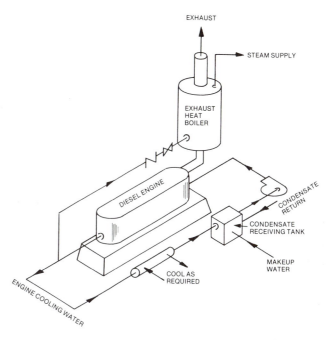

Fig. 1   Exhaust Heat Boiler

All heat exchangers should be constructed and labeled according to the applicable ASME Boiler and Pressure Vessel Code.

BOILER CONNECTIONS

Many jurisdictions require double-wall construction in shell-and-tube heat exchangers between steam and potable water. Recommended boiler connections for pumped and gravity return systems are shown in Figure 2; check local codes for specific legal requirements.

Supply Piping

Small boilers usually have one steam outlet connection sized to reduce steam velocity to minimize carryover of water into supply lines. Large boilers can have several outlets that minimize boiler water entrainment.

Figure 2 shows piping connections to the steam header. Although some engineers prefer to use an enlarged steam header for additional storage space, there usually is not a sudden demand for steam, except during the warmup period, so an oversized header is a disadvantage. The boiler header can be the same size as the boiler connection or the pipe used on the steam main. The horizontal runouts from the boiler (or boilers) to the header should be sized by calculating the heaviest load that will be placed on the boiler. The runouts should be sized on the same basis as the building mains. Any change in size after the vertical uptakes should be made by reducing elbows.

Return Piping

Cast-iron boilers have return tappings on both sides, while steel boilers may have one or two return tappings. Where two tappings are provided, both should be used to effect proper circulation through the boiler. Condensate in boilers can be returned by a pump or a gravity return system. Return connections shown for a multiple-boiler gravity return installation may not always maintain correct water levels in all boilers. Extra controls or accessories may be required.

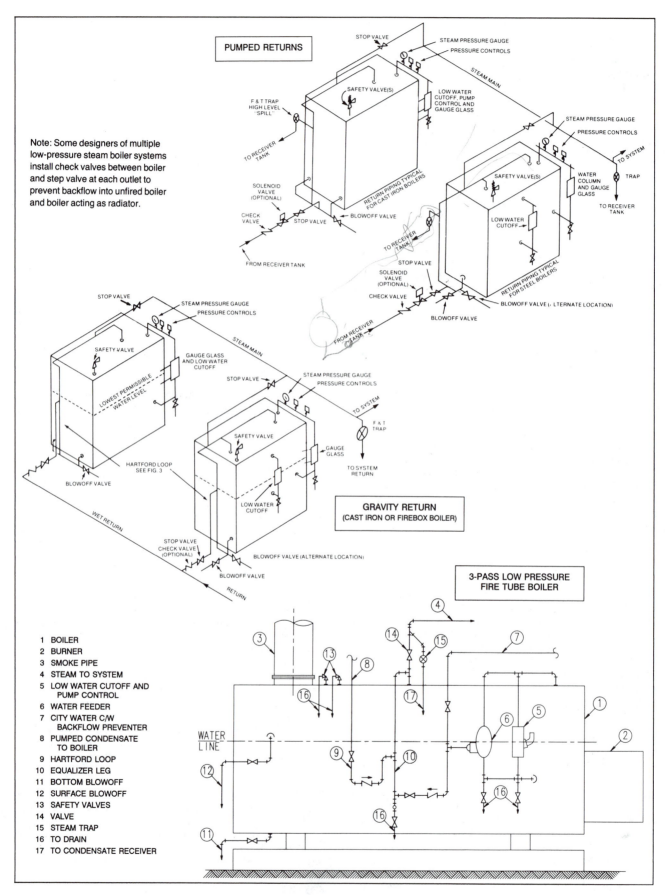

Note: Some designers of multiple low-pressure steam boiler systems install check valves between boiler and step valve at each outlet to prevent backflow into unfired boiler and boiler acting as radiator.

1 BOILER
2 BURNER
3 SMOKE PIPE
4 STEAM TO SYSTEM
5 LOW WATER CUTOFF AND
   PUMP CONTROL
6 WATER FEEDER
7 CITY WATER C/W
   BACKFLOW PREVENTER
8 PUMPED CONDENSATE
   TO BOILER
9 HARTFORD LOOP
10 EQUALIZER LEG
11 BOTTOM BLOWOFF
12 SURFACE BLOWOFF
13 SAFETY VALVES
14 VALVE
15 STEAM TRAP
16 TO DRAIN
17 TO CONDENSATE RECEIVER

Fig. 2   Boiler Connections

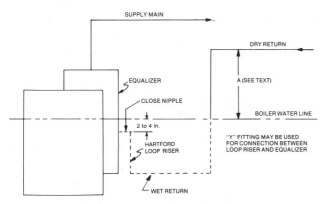

Fig. 3   Boiler with Gravity Return

Recommended return piping connections for systems using gravity return are detailed in Figure 3. Dimension A must be a minimum of 28 in. for each 1 psig maintained at the boiler to provide the pressure required to return the condensate to the boiler. To provide a reasonable safety factor, make Dimension A a minimum of 14 in. for small systems with piping sized for pressure drop of 1/8 psi, and a minimum of 28 in. for larger systems with piping sized for a pressure drop of 0.5 psi. The Hartford Loop protects against a low water condition, which can occur if a leak develops in the wet return portion of the piping system. The Hartford Loop takes the place of a check valve on the wet return; however, certain local codes require check valves. Because of hydraulic pressure limitations, gravity return systems are only suitable for systems operating at a boiler pressure between 0.5 to 1 psig. However, since these systems have minimum mechanical equipment and low initial installed cost, they are appropriate for many small systems. Kremers (1982) and Stamper and Koral (1979) provide additional design information on piping for gravity return systems.

Recommended piping connections for steam boilers with pump-returned condensate are shown in Figure 2. Common practice provides an individual condensate or boiler feedwater pump for each boiler. Pump operation is controlled by the boiler water level control on each boiler. However, one pump may be connected to supply the water to each boiler from a single manifold by using feedwater control valves regulated by the individual boiler water level controllers. When such systems are used, the condensate return pump runs continuously to pressurize the return header.

Return piping should be sized based on total load. The line between the pump and boiler should be sized for a very small pressure drop and the maximum pump discharge flow rate.

DESIGN STEAM PRESSURE

One of the most important decisions in the design of a steam system is the selection of the generating, distribution, and utilization pressures. Considering investment cost, energy efficiency, and control stability, the pressure should be held to the minimum values above atmospheric pressures that accomplish the required heating task, unless detailed economic analysis indicates advantages in higher pressure generation and distribution.

The first step in selecting pressures is to analyze the load requirements. Space heating and domestic water heating can best be served, directly or indirectly, with low-pressure steam less than 15 psig or 250°F. Other systems that can be served with low-pressure steam include absorption units (10 psig), cooking, warming, dishwashers, and snow melting heat exchangers. Thus, from the standpoint of load requirements, high-pressure steam (above 15 psig) is required only for loads

such as dryers, presses, molding dies, power drives, and other processing, manufacturing, and power requirements. The load requirement establishes the pressure requirement.

When the source is close to the load(s), the generation pressure should be high enough to provide the (1) load design pressure, (2) friction losses between the generator and the load, and (3) control range. Losses are caused by flow through the piping, fittings, control valves, and strainers. If the generator(s) is located remotely from the loads, there could be some economic advantage in distributing the steam at a higher pressure to reduce the pipe sizes. When this is considered, the economic analysis should include the additional investment and operating costs associated with a higher pressure generation system. When an increase in the generating pressure requires a change from below to above 15 psi, the generating system equipment changes from low-pressure class to high-pressure class and there are significant increases in both investment and operating cost.

Where steam is provided from a nonfired device or prime mover such as a diesel engine cooling jacket, the source device can have an inherent pressure limitation.

STEAM SYSTEM PIPING

The piping system distributes the steam, returns the condensate, and removes air and noncondensable gases. In steam heating systems, it is important that the piping system distribute steam, not only at full design load, but at partial loads and excess loads that can occur on system warmup. The usual average winter steam demand is less than half the demand at the lowest outdoor design temperature. However, when the system is warming up, the load on the steam mains and returns can exceed the maximum operating load for the coldest design day, even in moderate weather. This load comes from raising the temperature of the piping to the steam temperature and the building to the indoor design temperature. Supply and return piping should be sized according to Chapter 33 of the 1989 ASHRAE *Handbook — Fundamentals*.

Supply Piping Design Considerations

1. Size pipe according to Chapter 33 of the 1989 ASHRAE *Handbook — Fundamentals*, taking into consideration pressure drop and steam velocity.
2. Pitch piping uniformly down in the direction of steam flow at 0.25 in. in 10 ft. If pipe cannot be pitched down in the direction of the steam flow, refer to Chapter 33 of the 1989 ASHRAE *Handbook — Fundamentals* for rules on pipe sizing and pitch.
3. Insulate piping well to avoid unnecessary heat loss (see Chapters 20 and 22 of the 1989 ASHRAE *Handbook — Fundamentals*).
4. Condensate from unavoidable heat loss in the distribution system must be removed promptly to eliminate water hammer and degrading steam quality and heat transfer capability. Install drip legs at all low points and natural drainage points in the system, such as at the ends of mains, bottoms of risers, and ahead of the pressure regulators, control valves, isolation valves, pipe bends, and expansion joints. On straight horizontal runs with no natural drainage points, space drip legs at intervals not exceeding 300 ft when the pipe is pitched down in the direction of the steam flow and a maximum of 150 ft when the pipe is pitched up so that condensate flow is opposite of steam flow. Reduce these distances by about half in systems that are warmed up automatically rather than in a system where valves are opened manually to remove air and excess condensate that forms during warmup conditions.

5. Where horizontal piping must be reduced in size, use eccentric reducers that permit the continuance of uniform pitch along the bottom of piping (in downward pitched systems). Avoid concentric reducers on horizontal piping, since they can cause water hammer.

6. Take off all branch lines from the steam mains the top, preferably at a 45° angle, although vertical 90° connections are acceptable.

7. Where the length of a branch takeoff is less than 10 ft, the branch line can be pitched back 0.5 in. per 10 ft, providing drip legs as described in (4) above.

8. Size drip legs properly to separate and collect the condensate. Drip legs at vertical risers should be full and extend beyond the rise, as shown in Figure 4. Drip legs at other locations should be the same diameter as the main. In steam mains 6 in. and over, this can be reduced to half the diameter of the main, but to not less than 4 in. Where warmup is supervised, the length of the collecting leg is not critical. However, the recommended length is one and a half times the pipe diameter and not less than 8 in. For automatic warmup, collecting legs should always be the same size as the main and should be at least 28 in. long to provide the hydraulic heat for the pressure differential necessary for the trap to discharge before a positive pressure is built up in the steam main.

9. Condensate should flow by gravity from the trap to the return piping system. Where the steam trap is located below the return line, the condensate must be lifted. In systems operating above 40 psi, the trap discharge can usually be piped directly to the return system (Figure 5). However, back pressure at the trap discharge (return line pressure plus hydraulic pressure created by height of lift) must not exceed steam main pressure, and the trap must be sized after considering back pressure. In systems operating under 40 psi, drip legs must be installed on equipment where the temperature is regulated by modulating the steam control valves and where the back pressure at the trap exceeds or is close to system pressure. The trap discharge should flow by gravity to a vented condensate receiver from which it is pumped to the overhead return.

10. Strainers installed before the pressure-reducing and control valves are a natural water collection point. Since water carryover can erode the valve seat, install a trap at the strainer blowdown connection (Figure 6).

Terminal Equipment Piping Design Considerations

1. Size piping the same as the supply and return connections of the terminal equipment.

2. Keep equipment and piping accessible for inspection and maintenance of the steam traps and control valves.

3. Minimize strain caused by expansion and contraction with pipe bends, loops, or three elbow swings to take advantage of piping flexibility, or with expansion joints or flexible pipe connectors.

4. In multiple-coil applications, separately trap each coil for proper drainage (Figure 7). Piping two or more coils to a common header served by a single trap can cause condensate backup, improper heat transfer, and inadequate temperature control.

5. Terminal equipment, where temperature is regulated by a modulating steam control valve, requires special consideration. Refer to the section Condensate Removal from Temperature-Regulated Equipment.

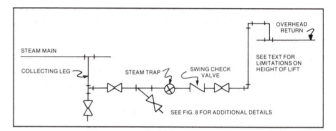

Fig. 5   Trap Discharging to Overhead Return

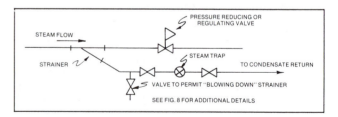

Fig. 6   Trapping Strainers

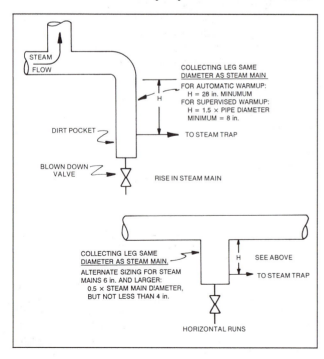

Fig. 4   Method of Dripping Steam Mains

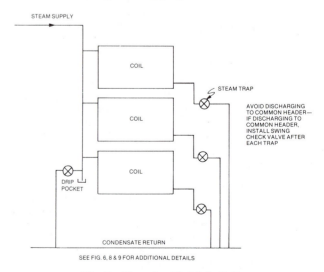

Fig. 7   Trapping Multiple Coils

Return Piping Design Considerations

1. Flow in the return line is two-phase, consisting of steam and condensate. See Chapter 33 of the 1989 ASHRAE *Handbook — Fundamentals* for sizing considerations.
2. Pitch return lines downward in the direction of the condensate flow at 0.5 in. per 10 ft to ensure prompt condensate removal.
3. Insulate the return line well, especially where the condensate is returned to the boiler or the condensate enthalpy is recovered.
4. Where possible and practical, use heat recovery systems to recover the condensate enthalpy. See the section on heat recovery.
5. Equip dirt pockets of the drip legs and strainer blowdowns with valves to remove dirt and scale.
6. Install steam traps close to drip legs and make them accessible for inspection and repair. Servicing is simplified by making the pipe sizes and configuration identical for a given type and size of trap. The piping arrangement shown in Figure 8 facilitates inspection and maintenance of steam traps.
7. When elevating condensate to an overhead return, consider the pressure at the trap inlet and the fact that it requires approximately 1 psi to elevate condensate 2 ft. See (9) of Supply Piping Design Considerations for complete discussion.

CONDENSATE REMOVAL FROM TEMPERATURE-REGULATED EQUIPMENT

When air, water, or another product is heated, the temperature or heat transfer rate can be regulated by a modulating steam pressure control valve. Since pressure and temperature do not vary at the same rate as load, the steam trap capacity, which is determined by the pressure differential between the trap inlet and outlet, may be adequate at full load, but not at some lesser load.

Analysis shows that steam pressure must be reduced dramatically to achieve a slight lowering of temperature. In most applications, this can result in subatmospheric pressure in the coil, while as much as 75% of full condensate load has to be handled by the steam trap. This is especially important for coils exposed to outside air, since subatmospheric conditions can occur in the coil at outside temperatures below 32°F and the coil will freeze if the condensate is not removed.

There are detailed methods for determining condensate load under various operating conditions (Kremers 1982). However, in most cases, this load need not be calculated if the coils are piped as shown in Figure 9 and this procedure is followed:

1. Place the steam trap 1 to 3 ft below the bottom of the steam coil to provide a pressure of approximately 0.5 to 1.5 psig.

Locating the trap at less than 12 in. minimum usually results in improper drainage and operating difficulties.

2. Install vacuum breakers between the coil and trap inlet to ensure that the pressure can drain the coil when it is atmospheric or subatmospheric. The vacuum breaker should respond to a differential pressure of no greater than 3 in. of water. For atmospheric returns, the vacuum breaker should be opened to the atmosphere, and the return system must be designed to ensure no pressurization of the return line. In vacuum return systems, the vacuum breaker should be piped to the return line.
3. Discharge from the trap must flow by gravity, without any lifts in the piping, to the return system, which must be vented properly to the atmosphere to eliminate any back pressure that could prevent the trap from draining the coil. Where the return main is overhead, the trap discharge should flow by gravity to a vented receiver, from which it is then pumped to the overhead return.
4. Design traps to operate at maximum pressure at the control valve inlet and size them to handle the full condensate load at a pressure differential equal to the hydraulic pressure between the trap and coil. Since the actual condensate load can vary from the theoretical design load because of the safety factors used in coil selection and the fact that condensate does not always form at a uniform steady rate, size steam traps according to the following:
   - 0 to 15 psig at the control valve inlet: size the trap for twice the full condensate load (coil condensing rate at maximum design conditions) at 0.5 psi pressure differential.
   - 16 to 30 psig at the control valve inlet: size the trap for twice the full condensate load at 2 psi pressure differential.
   - 31 psig and over at the control valve inlet: size the trap to handle triple the full condensate load at a pressure differential equal to half the control valve inlet pressure.

STEAM TRAPS

Steam traps are an essential part of all steam systems, except one-pipe steam heating systems. Traps discharge condensate, which forms as steam gives up some of its heat, and direct the air and noncondensable gases to a point of removal. Condensate forms in steam mains and distribution piping because of unavoidable heat losses through less-than-perfect insulation,

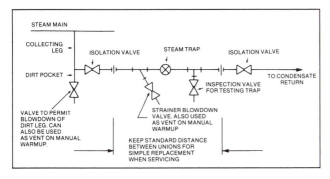

Fig. 8   Recommended Steam Trap Piping

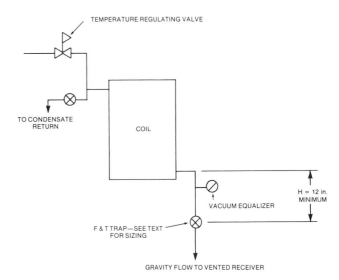

Fig. 9   Trapping Temperature-Regulated Coils

as well as in terminal equipment such as radiators, convectors, fan coil units, and heat exchangers, where steam gives up heat during normal operation. Condensate must always be removed from the system as soon as it accumulates for the following reasons:

- Although condensate contains some valuable heat, using this heat by holding the condensate in the terminal equipment reduces the heat transfer surface. It also causes other operating problems because it retains air, which further reduces heat transfer, and noncondensable gases such as CO_2, which cause corrosion. As discussed in Basic Steam System Design, recovering condensate heat is usually only desirable when the condensate is not returned to the boiler. Methods for this are discussed in the Heat Recovery section.
- Steam moves rapidly in mains and supply piping, so when condensate accumulates to the point where the steam can push a slug of it, serious damage can occur from the resulting water hammer.

Ideally, the steam trap should remove all condensate promptly, along with air and noncondensable gases that might be in the system, with little or no loss of live steam. A steam trap is an automatic valve that can distinguish between steam and condensate or other fluids. Traps are classified as follows:

- *Thermostatic traps* react to the difference in temperature between steam and condensate.
- *Mechanical traps* operate by the difference in density between steam and condensate.
- *Kinetic traps* rely on the difference in flow characteristics of steam and condensate.

The following points apply to all steam traps:

- No single type of steam trap is best suited to all applications, and most systems require more than one type of trap.

- Steam traps, regardless of type, should be carefully sized for the application and condensate load to be handled, since both undersizing and oversizing can cause serious problems. Undersizing can result in undesirable condensate backup and excessive cycling that can cause premature failure. Oversizing might appear to solve this problem and make selection much easier because fewer different sizes are required, but if the trap fails, excessive steam can be lost.

Thermostatic Traps

In thermostatic traps, a bellows or bimetallic element operates a valve that opens in the presence of condensate and closes in the presence of steam (Figure 10). Because condensate is initially at the same temperature as the steam from which it was condensed, the thermostatic element must be designed and calibrated to open at a temperature below the steam temperature; otherwise, the trap would blow live steam continuously. Therefore, the condensate must be subcooled by allowing it to back up in the trap and a portion of the upstream drip leg piping, both of which are left uninsulated. Some thermostatic traps operate with a continuous water leg behind the trap so there is no steam loss; however, this prohibits the discharge of air and noncondensable gases, and can cause excessive condensate to back up into the mains or terminal equipment, thereby causing operating problems. Devices that operate without significant backup can lose steam before the trap closes.

Although both bellows and bimetallic traps are temperature sensitive, their operations are significantly different. The *bellows' thermostatic trap* has a lower boiling point than water. When the trap is cold, the element is contracted and the discharge port is open. As hot condensate enters the trap, it causes the bellows' contents to boil and vaporize before the condensate temperature rises to steam temperature. Because

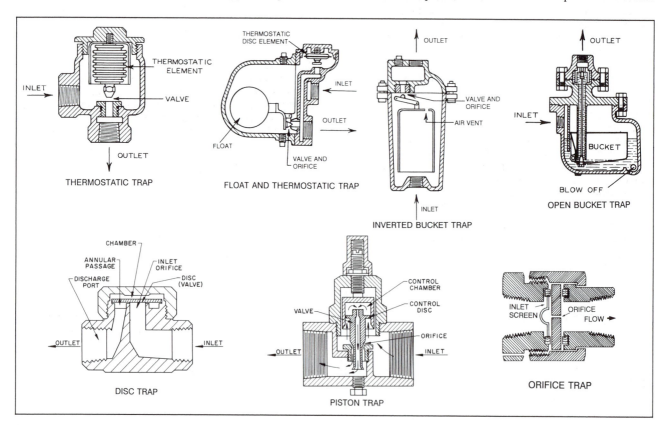

Fig. 10 Thermostatic Traps

the bellows' contents boil at a lower temperature than water, the vapor pressure inside the bellows element is greater than the steam pressure outside, causing the element to expand and close the discharge port.

Assuming the contained liquid has a pressure-temperature relationship similar to that of water, the balance of forces acting on the bellows' element remains relatively constant, no matter how the steam pressure varies. Therefore, this is a balanced pressure device that can be used at any pressure within the operating range of the device. However, this device should not be used where superheated steam is present, since the temperature is no longer in step with the pressure and damage or rupture of the bellows' element can occur.

Bellows' thermostatic traps are best suited for steady light loads on low-pressure service. They are most widely used in radiators and convectors in HVAC applications.

The *bimetallic thermostatic trap* has an element made from metals with different expansion coefficients. Heat causes the element to change shape, permitting the valve port to open or close. Because a bimetallic element responds only to temperature, most traps have the valve on the outlet so that steam pressure is trying to open the valve. Therefore, by properly designing the bimetallic element, the trap can operate on a pressure-temperature curve approaching the steam saturation curve, but not as closely as with a balanced pressure bellows' element.

Unlike the bellows' thermostatic trap, bimetallic thermostatic traps are not adversely affected by superheated steam or subject to damage by water hammer, so they can be readily used for high-pressure applications. They are best suited for steam tracers, jacketed piping, and heat transfer equipment, where some condensate backup is tolerable. If they are used on steam main drip legs, the element should not back up condensate.

Mechanical Traps

Mechanical traps are buoyancy operated, depending on the difference in density between steam and condensate. The *float and thermostatic trap,* commonly called the F&T trap (Figure 10), is actually a combination of two types of traps in a single trap body: (1) a bellows thermostatic element operating on temperature difference, which provides automatic venting and (2) a float portion, which is buoyancy operated. Float traps without automatic venting should not be used for steam systems.

On start-up, the float valve is closed and the thermostatic element is open for rapid air venting, permitting the system or equipment to rapidly fill with steam. When steam enters the trap body, the thermostatic element closes and, as condensate enters, the float rises and the condensate discharges. The float regulates the valve opening so it continuously discharges the condensate at the rate at which it reaches the trap.

The F&T trap has large venting capabilities, continuously discharges condensate without backup, handles intermittent loads very well, and can operate at extremely low pressure differentials. Float and thermostatic traps are suited for use with temperature-regulated steam coils. They also are well suited for steam main and riser drip legs on low-pressure steam-heating systems. Although F&T traps are available for pressure to 250 psig or higher, they are susceptible to water hammer, so other traps are usually a better choice for high-pressure applications.

Bucket traps operate on buoyancy, but they use a bucket that is either open at the top or inverted instead of a closed float. Initially, the bucket in an *open bucket trap* is empty and rests on the bottom of the trap body with the discharge vented, and, as condensate enters the trap, the bucket floats up and

closes the discharge port. Additional condensate overflows into the bucket causing it to sink and open the discharge port, allowing steam pressure to force the condensate out of the bucket. At the same time, it seals the bottom of the discharge tube, prohibiting air passage. Therefore, to prevent air binding, this device has an automatic air vent, as does the F&T trap.

Inverted bucket traps eliminate the size and venting problems associated with open bucket traps. Steam or air entering the submerged inverted bucket causes it to float and close the discharge port. As more condensate enters the trap, it forces air and steam out of the vent on top of the inverted bucket into the trap body where the steam condenses by cooling. When the mass of the bucket exceeds the buoyancy effect, the bucket drops, opening the discharge port, and steam pressure forces the condensate out, and the cycle repeats.

Unlike most cycling-type traps, the inverted bucket trap continuously vents air and noncondensable gases. Although it discharges condensate intermittently, there is no condensate backup in a properly sized trap. Inverted bucket traps are made for all pressure ranges and are well suited for steam main drip legs and most HVAC applications. Although they can be used for temperature-regulated steam coils, the F&T trap is usually better because it has the high venting capability desirable for such applications.

Kinetic Traps

Numerous devices operate on the difference between the flow characteristics of steam and condensate and on the fact that condensate discharging to a lower pressure contains more heat than necessary to maintain its liquid phase. This excess heat causes some of the condensate to flash to steam at the lower pressure.

Thermodynamic traps are simple devices with only one moving part. When air or condensate enters the trap on start-up, it lifts the disc off its seat and is discharged. When steam or hot condensate (some of which "flashes" to steam upon exposure to a lower pressure) enters the trap, the increased velocity of this vapor flow decreases the pressure on the underside of the disc and increases the pressure above the disc, causing it to snap shut. Pressure is then equalized above and below the disc, but since the area exposed to pressure is greater above than below it, the disc remains shut until the pressure above is reduced by condensing or bleeding, permitting the disc to snap open and repeat the cycle. This device does not cycle open and shut as a function of condensate load; it is a time cycle device that opens and shuts at fixed intervals as a function of how fast the steam above the disc condenses. Since disc traps require a significant pressure differential to operate properly, they are not well suited for low-pressure systems or for systems with significant back pressure. They are best suited for high-pressure systems and are widely applied to steam main drip legs.

Impulse traps or *piston traps* continuously pass a small amount of steam or condensate through a "bleed" orifice, changing the pressure positions within the piston. When live steam or very hot condensate that flashes to steam enters the control chamber, the increased pressure closes the piston valve port. When cooler condensate enters, the pressure decreases, permitting the valve port to open. Most impulse traps cycle open and shut intermittently, but some modulate to a position that passes condensate continuously.

Impulse traps can be used for the same applications as disc traps; however, because they have a small "bleed" orifice and close piston tolerances, they can stick or clog if dirt is present in the system.

Orifice traps have no moving parts. All other traps have discharge ports or orifices, but in the traps described above, this opening is oversized, and some type of closing mechanism controls the flow of condensate and prevents the loss of live steam.

The orifice trap has no such closing mechanism, and the flow of steam and condensate is controlled by two-phase flow through an orifice. A simple explanation of this theory is that an orifice of any size has a much greater capacity for condensate than it does for steam because of the significant differences in their densities and because "flashing" condensate tends to choke the orifice. An orifice is selected larger than required for the actual condensate load; therefore, it continuously passes all condensate along with the air and noncondensable gases, plus a small controlled amount of steam. The steam loss is usually comparable to that of most cycling-type traps.

Orifice traps must be sized more carefully than cycling-type traps. On light condensate loads, the orifice size is small and, like impulse traps, tends to clog. Orifice traps are suitable for all system pressures and can operate against any back pressure. They are best suited for steady pressure and load conditions such as steam main drip legs.

PRESSURE-REDUCING VALVES

Where steam is supplied at pressures higher than required, one or more pressure-reducing valves (pressure regulators) are required. The pressure-reducing valve reduces pressure to a safe point and regulates pressure to that required by the equipment. The district heating industry refers to valves according to their functional use. There are two classes of service: (1) where the steam must be shut off completely (dead-end valves) to prevent buildup of pressure on the low-pressure side during no load (single-seated valves should be used) and (2) where the low-pressure lines condense enough steam to prevent buildup of pressure from valve leakage (double-seated valves can be used). Valves available for either service are direct-operated, or spring, weight, air-loaded, or pilot-controlled, using either the flowing steam or auxiliary air or water as the operating medium. The direct-operated, double-seated valve is less affected by varying inlet steam pressure than the direct-operated, single-seated valve. Pilot-controlled valves, either single- or double-seated, tend to eliminate the effect of variable inlet pressures.

Installation

Pressure-reducing valves should be readily accessible for inspection and repair. There should be a bypass around each reducing valve equal to the area of the reducing-valve seat ring. The globe valve in a bypass line should have plug disc construction and must have an absolutely tight shutoff. A steam pressure gage, graduated up to the initial pressure, should be installed on the low-pressure side. The gage should be ahead of the shutoff valve because the reducing valve can be adjusted with the shutoff valve closed. A similar gage should be installed downstream from the shutoff valve for use during manual operation. Typical service connections are shown in Figure 11 for low-pressure service and Figure 12 for high-pressure service. In the smaller sizes, the standby pressure-regulating valve can be removed, a filler installed, and the inlet stop valves used for manual pressure regulation until repairs are made.

Strainers should be installed on the inlet of the primary pressure-reducing valve and before the second-stage reduction if there is considerable piping between the two stages. If a two-stage reduction is made, it is advisable to install a pressure gage immediately before the reducing valve of the second-

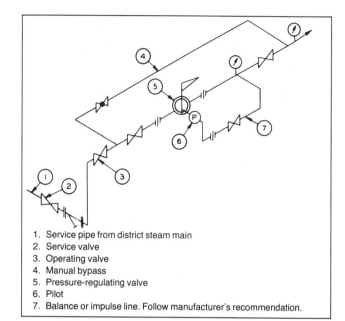

1. Service pipe from district steam main
2. Service valve
3. Operating valve
4. Manual bypass
5. Pressure-regulating valve
6. Pilot
7. Balance or impulse line. Follow manufacturer's recommendation.

Fig. 11 Pressure-Reducing Valve Connections—Low Pressure

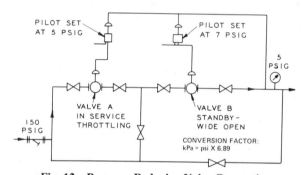

Fig. 12 Pressure-Reducing Valve Connections

stage reduction to set and check the operation of the first valve. A drip trap should be installed between the two reducing valves.

Where pressure-reducing valves are used, one or more relief devices or safety valves must be provided, and the equipment on the low-pressure side must meet the requirements for the full initial pressure. The relief or safety devices are adjoining or as close as possible to the reducing valve. The combined relieving capacity must be adequate to avoid exceeding the design pressure of the low-pressure system if the reducing valve does not open. Local codes in most areas dictate the safety relief valve installation requirements.

Safety valves should be set at least 5 psi higher than the reduced pressure when the reduced pressure is under 35 psig and at least 10 psi higher than the reduced pressure, if the reduced pressure is above 35 psig or the first-stage reduction of a double reduction. The outlet from relief valves should not be piped to a location where the discharge can jeopardize persons or property or is a violation of local codes.

Figure 13 shows a typical service installation, with a separate line to various heating zones and process equipment. If the initial pressure is below 50 psig, the first-stage pressure-reducing valve can be omitted. In Assembly A (Figure 13), the single-stage pressure-reducing valve is also the pressure-regulating valve.

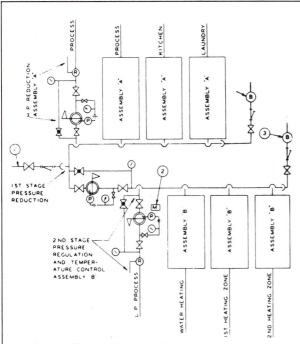

1. Service pipe from district steam main
2. Motorized pilot on pressure-regulating valve
3. Bucket trap on header drip line
Note: Where process equipment is individually controlled, temperature control on header may be omitted. All fittings are American Standard Class 250 cast iron or 300 steel in both reduction and regulation assemblies.

Fig. 13   Steam Supply

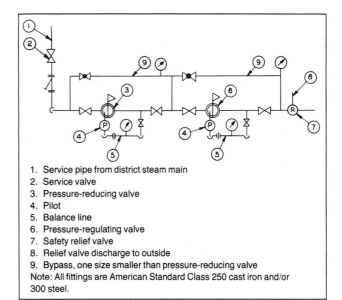

1. Service pipe from district steam main
2. Service valve
3. Pressure-reducing valve
4. Pilot
5. Balance line
6. Pressure-regulating valve
7. Safety relief valve
8. Relief valve discharge to outside
9. Bypass, one size smaller than pressure-reducing valve
Note: All fittings are American Standard Class 250 cast iron and/or 300 steel.

Fig. 14   Two-Stage Pressure-Regulating Valve
(Used where high-pressure steam is supplied for low-pressure requirements.)

When making a two-stage reduction, *e.g.*, 150 to 50 psig and then 50 to 2 psig, allow for the expansion of steam on the low-pressure side of each reducing valve by increasing the pipe area to about double the area of a pipe the size of the reducing valve. This also allows steam to flow at a more uniform velocity. It is recommended that the valves be separated by a distance of at least 20 ft to reduce excessive hunting action of the first valve.

Figure 14 shows a typical double-reduction installation where the pressure in the district steam main is higher than can safely be applied to building heating systems. The first or pressure-reducing valve effects the initial pressure reduction. The pressure-regulating valve regulates the steam to the desired final pressure.

Pilot-controlled or air-loaded direct-operated reducing valves can be used without limitation for all reduced-pressure settings for all heating, process, laundry, and hot-water services. Spring-loaded direct-operated valves can be used for reduced pressures up to 50 psig, providing they can pass the required steam flow without excessive deviation in reduced pressure.

Weight-loaded valves are used for reduced pressures below 15 psig and for moderate steam flows.

Pressure-equalizing or impulse lines must be connected to serve the type of valve selected. With direct-operated diaphragm valves having rubber-like diaphragms, the impulse line should be connected into the bottom of the reduced-pressure steam line to allow for maximum condensate on the diaphragm and in the impulse or equalizing line. If it is connected to the top of the steam line, a condensate accumulator should be used to reduce variations in the pressure of condensate on the diaphragm. Equalizing or impulse lines for pilot-

controlled and direct-operated reducing valves using metal diaphragms should be connected into the expanded outlet piping approximately 2 to 4 ft from the reducing valve and pitched away from the reducing valve to prevent condensate accumulation. Pressure impulse lines for externally pilot-controlled reducing valves using compressed air or fresh water should be installed according to manufacturer's recommendations.

Valve Size Selection

Pressure-regulating valves should be sized to supply the maximum steam requirements of the heating system or equipment. Consideration should be given to rangeability, speed of load changes, and regulation accuracy required to meet system needs, especially with temperature control systems using intermittent steam flow to heat the building.

The reducing valve should be selected carefully. The manufacturer should be consulted. Piping to and from the reducing valve should be adequate to pass the desired amount of steam at the desired maximum velocity. A common error is to make the size of the reducing valve the same as the service or outlet pipe size; this makes the reducing valve oversized and causes wiredrawing or erosion of the valve and seat because of the high-velocity flow caused by the small lift of the valve.

On installations where the steam requirements are large and variable, wiredrawing and cycling control can occur during mild weather or during reduced demand periods. To overcome this condition, two reducing valves are installed in parallel, with the sizes selected on a 70 and 30% proportion of maximum flow. For example, if 10,000 lb of steam per hour is required, the size of one valve is based on 7000 lb of steam per hour, and the other is based on 3000 lb of steam per hour. During mild weather (spring and fall), the larger valve is set for slightly lower reduced pressure than the smaller one and remains closed as long as the smaller one can meet the demand. During the remainder of the heating season, the valve settings are reversed to keep the smaller one closed, except when the larger one is unable to meet the demand.

TERMINAL EQUIPMENT

A variety of terminal units are used in steam-heating systems. All are suited for use on low-pressure systems. Terminal units used on high-pressure systems have heavier construction, but are otherwise similar to those on low-pressure systems.

Terminal units are usually classified as follows:

1. *Natural convection units* transfer some heat by radiation. The equipment includes cast-iron radiators, finned-tube convectors, and cabinet and baseboard units with convection-type elements (see Chapter 33).
2. *Forced-convection units* employ a forced air movement to increase heat transfer and distribute the heated air within the space. The devices include unit heaters, unit ventilators, induction units, fan-coil units, the heating coils of central air-conditioning units, and many process heat exchangers. When such units are used for both heating and cooling, there is a steam coil for heating and a separate chilled water or refrigerant coil for cooling. See Chapters 1 through 4, 21, 24, and 32.
3. *Radiant panel systems* transfer some heat by convection. Because of the low-temperature and high-vacuum requirements, this type of unit is rarely used on steam systems.

Selection of Terminal Equipment

The primary consideration in selecting terminal equipment is comfortable heat distribution. The following briefly describes suitable applications for specific types of steam terminal units.

Natural Convection Units

Radiators, convectors, and convection-type cabinet units and baseboard convectors are commonly used for (1) facilities that require heating only, rather than both heating and cooling, and (2) in conjunction with central air-conditioning systems as a source of perimeter heating or for localized heating in spaces such as corridors, entrances, halls, toilets, kitchens, and storage areas.

Forced-Convection Units

Forced-convection units can be used for the same types of applications as natural-convection units but are primarily used for facilities that require both heating and cooling, as well as spaces that require localized heating.

Unit heaters are often used as the primary source of heat in factories, warehouses, and garages and as supplemental freeze protection for loading ramps, entrances, equipment rooms, and fresh air plenums.

Unit ventilators are forced-convection units with dampers that introduce controlled amounts of outside air. They are used in spaces with ventilation requirements not met by other system components.

Cabinet heaters are often used in entranceways and vestibules that can have high intermittent heat loads.

Induction units are similar to fan-coil units, but the air is supplied by a central air system rather than individual fans in each unit. Induction units are most commonly used as perimeter heating for facilities with central systems.

Fan-coil units are designed for heating and cooling. On water systems, a single coil can be supplied with hot water for heating and chilled water for cooling. On steam systems, single- or dual-coil units can be used. Single-coil units require steam-to-water heat exchangers to provide hot water to meet heating requirements. Dual-coil units have a steam coil for heating and a separate coil for cooling. The cooling media for dual-coil units can be chilled water from a central chiller or refrigerant provided by a self-contained compressor. Single-coil units require the entire system to be either in a heating or cooling mode and do not permit simultaneous heating and cooling to satisfy individual space requirements. Dual-coil units can eliminate this problem.

Central air-handling units are used for most larger facilities. These units have a fan, a heating and cooling coil, and a compressor if chilled water is not available for cooling. Multizone units are arranged so that each air outlet has separate controls for individual space heating or cooling requirements. Large central systems distribute either warm or cold conditioned air through duct systems and employ separate terminal equipment such as mixing boxes, reheat coils, or variable volume controls to control the temperature to satisfy each space. Central air-handling systems can be factory assembled, field erected, or built at the job site from individual components.

CONVECTION STEAM-HEATING SYSTEMS

Any system that uses steam as the heat transfer medium can be considered a steam-heating system; however, the term "steam-heating system" is most commonly applied to convection systems using radiators or convectors as terminal equipment. Other types of steam-heating systems use forced convection, in which a fan or air-handling system is used with a convector or steam coil such as unit heaters, fan coil units, and central air-conditioning and heating systems.

Convection-type steam-heating systems are used in facilities that have a heating requirement only, such as factories, warehouses, and apartment buildings. They are often used in conjunction with central air-conditioning systems to heat the perimeter of the building. Also, steam is commonly used with incremental units that are designed for cooling and heating and have a self-contained air-conditioning compressor.

Steam-heating systems are classified as one-pipe or two-pipe systems, according to the piping arrangement that supplies steam to and returns condensate from the terminal equipment. These systems can be further subdivided by (1) the method of condensate return (gravity flow or mechanical flow by means of condensate pump or vacuum pump) and (2) by the piping arrangement (upfeed or downfeed and parallel or counterflow for one-pipe systems).

One-Pipe Steam-Heating Systems

The one-pipe system has a single pipe through which steam flows to and condensate is returned from the terminal equipment (Figure 15). These systems are designed as gravity return, although a condensate pump can be used where there is insufficient height above the boiler water level to develop enough pressure to return condensate directly to the boiler.

A one-pipe system with gravity return does not have steam traps; instead it has air vents at each terminal unit and at the ends of all supply mains to vent the air so the system can fill with steam. There must be an air vent at each terminal unit

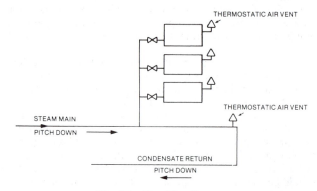

Fig. 15   One-Pipe System

and steam traps at the ends of each supply main in systems with a condensate pump.

The one-pipe system with gravity return has low initial cost and is simple to install, since it requires a minimum of mechanical equipment and piping. One-pipe systems are most commonly used in small facilities such as small apartment buildings and office buildings. In larger facilities, the larger pipe sizes required for two-phase flow, problems of distributing steam quickly and evenly throughout the system, inability to zone the system, and difficulty in controlling the temperature make the one-pipe system less desirable than two-pipe systems.

The heat input to the system is controlled by cycling the steam on and off. In the past, temperature control in individual spaces has been a problem. Many systems have adjustable vents at each terminal unit to help balance the system, but these are seldom effective. A practical approach is to use a self-contained thermostatic valve in series with the air vent (as explained in the Temperature Control for Steam Heating Systems section) that provides individual thermostatic control for each space.

Many designers do not favor one-pipe systems because of their distribution and control problems. However, when a self-contained thermostatic valve is used to eliminate the problems, one-pipe systems can be considered for small facilities, where initial cost and simple installation and operation are prime factors.

Most one-pipe gravity return systems are in facilities that have their own boiler, and, since returning condensate must overcome boiler pressure, these systems usually operate from a fraction of a psi to a maximum of 5 psi. The boiler hookup is critical and the Hartford Loop (described in the section on Boiler Connections) is used to avoid problems that can occur with boiler low-water condition. Stamper and Koral (1979) and Hoffman give piping design information for one-pipe systems.

Two-Pipe Steam-Heating Systems

The two-pipe system uses separate pipes to deliver the steam and return the condensate from each terminal unit (Figure 16). Thermostatic traps are installed at the outlet of each terminal unit to keep the steam in the unit until it gives up its latent heat, at which time the trap cycles open to pass the condensate and permits more steam to enter the radiator. If orifices are installed at the inlet to each terminal unit, as discussed in the Steam Distribution and Temperature Control sections, and if the system pressure is precisely regulated so only the amount of steam is delivered to each unit that it is capable of condens-

ing, the steam traps can be omitted. However, omitting steam traps is generally not recommended for initial design.

Two-pipe systems can have either gravity or mechanical returns; however, gravity returns are restricted to use in small systems and are generally outmoded. In larger systems that require higher steam pressures to distribute steam, some mechanical means, such as a condensate pump or vacuum pump, must return condensate to the boiler. A vacuum return system is used on larger systems and has the following advantages:

1. The system fills quickly with steam. The steam in a gravity return system must push the air out of the system, resulting in delayed heat-up and condensate return that can cause low-water problems. A vacuum return system can eliminate these problems.
2. The steam supply pressure can be lower, resulting in more efficient operation.

Variable vacuum or subatmospheric systems are a variation of the vacuum return system in which a controllable vacuum is maintained in both the supply and return sides. This permits using the lowest possible system temperature and prompt steam distribution to the terminal units. The primary purpose of variable vacuum systems is to control temperature as discussed in the Temperature Control section.

Unlike one-pipe systems, two-pipe systems can be simply zoned where piping is arranged to supply heat to individual sections of the building that have similar heating requirements. Heat is supplied to meet the requirements of each section without overheating other sections. The heat also can be varied according to hours of use, type of occupancy, sun load, and similar factors.

STEAM DISTRIBUTION

Steam supply piping should be sized so that the pressure drops in all branches of the same supply main are nearly uniform. Return piping should be sized for the same pressure drop as supply piping for quick and even steam distribution. Since it is impossible to size piping so that pressure drops are exactly the same, the steam flows first to those units that can be reached with the least resistance, resulting in uneven heating. Units farthest from the source of steam will heat last, while other spaces are overheated. This problem is most evident when the system is filling with steam. It can be severe on systems in which temperature is controlled by cycling the steam on and off. The problem can be alleviated or eliminated by balancing valves or inlet orifices.

Balancing valves are installed at the unit inlet and contain an adjustable valve port to control the amount of steam delivered. The main problem with such devices is that they are seldom calibrated accurately, and variations in orifice size that are as small as 0.003 in² can make a significant difference.

Inlet orifices are thin brass or copper plates installed in the unit inlet valve or pipe unions (Figure 17). Inlet orifices can

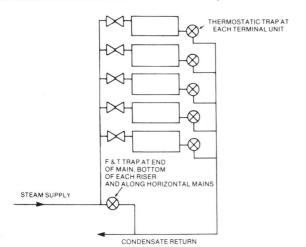

THERMOSTATIC TRAP AT EACH TERMINAL UNIT

F & T TRAP AT END OF MAIN, BOTTOM OF EACH RISER AND ALONG HORIZONTAL MAINS

STEAM SUPPLY

CONDENSATE RETURN

Fig. 16   Two-Pipe System

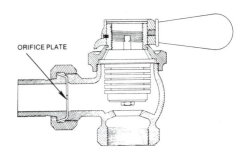

ORIFICE PLATE

Fig. 17   Inlet Orifice

solve distribution problems, since they can be drilled for appropriate size and changed easily to compensate for unusual conditions. Properly sized inlet orifices can compensate for oversized heating units, reduce energy waste and system control problems caused by excessive steam loss from defective steam traps, and provide a means of temperature control and balancing within individual zones. Manufacturers of orifice plates provide data for calculating the required sizes for most systems. Also, Sanford and Sprenger (1931) developed data for sizing orifices for low-pressure, gravity return systems.

If orifices are installed in valves or pipe unions that are conveniently accessible, minor rebalancing among individual zones can be accomplished through replacement of individual orifices.

TEMPERATURE CONTROL

All heating systems require some means of temperature control to achieve desired comfort conditions and operating efficiencies. In convection-type steam-heating systems, the temperature and resulting heat output of the terminal units must be increased or decreased. This can be done by (1) permitting the steam to enter the heating unit intermittently, (2) varying the steam temperature delivered to each unit, and (3) varying the amount of steam delivered to each unit.

There are two types of controls: those that control the temperature or heat input to the entire system and those that control the temperature of individual spaces or zones. Often, both types are used together for maximum control and operating efficiency.

Intermittent flow controls, commonly called heat timers, control the temperature of the system by cycling the steam on and off during a certain portion (or fraction) of each hour as a function of the outdoor and/or indoor temperature. Most of these devices provide for night setback, and newer "computer-style" models optimize morning startup as a function of outdoor and indoor temperature and make anticipatory adjustments. Used independently, these devices do not permit varying heat input to various parts of the building or spaces, so they should be used with zone control valves or individual thermostatic valves for maximum energy efficiency.

Zone control valves control the temperature of spaces with similar heating requirements as a function of outdoor and/or indoor temperature, as well as permit duty cycling. These intermittent flow control devices operate full open or full closed. They are controlled by an indoor thermostat and are used in conjunction with a heat timer, variable vacuum, or pressure differential control that controls heat input to the system.

Individual thermostatic valves installed at each terminal unit can provide the proper amount of heat to satisfy the individual requirements of each space and eliminate overheating. Valves can be actuated electrically, pneumatically, or by a self-contained mechanism that has a wax- or liquid-filled sensing element requiring no external power source.

Individual thermostatic valves can be and are often used as the only means of temperature control. However, these systems are always "on," resulting in inefficiency and no central control of heat input to the system or to the individual zones. Electric and pneumatic operators allow centralized control to be built into the system. However, it is desirable to have a supplemental system control with self-contained thermostatic valves in the form of zone control valves, or a system that controls the heat input to the entire system, relying on self-contained valves as a local high-temperature shutoff only.

Self-contained thermostatic valves can also be used to control temperature on one-pipe systems that cycle on and off when installed in series with the unit air vent. On initial startup, the air vent is open, and inherent system distribution problems still exist. However, on subsequent on-cycles where the room temperature satisfies the thermostatic element, the vent valve remains closed, preventing the unit from filling with steam.

Variable vacuum or subatmospheric systems control the temperature of the system by varying the steam temperature through pressure control. These systems differ from the regular vacuum return system in that the vacuum is maintained in both supply and return lines. Such systems usually operate at a pressure range from 2 psig to 25 in. Hg vacuum. Inlet orifices are installed at the terminal equipment for proper steam distribution during mild weather.

Since design-day conditions are seldom encountered, the system usually operates at a substantial vacuum. Because the specific volume of steam increases as the pressure decreases, it takes less steam to fill the system and operating efficiency results. The variable vacuum system can be used in conjunction with zone control valves or individual self-contained valves for increased operating efficiency.

Pressure differential control (two-pipe orifice) systems provide centralized temperature control with any system that has properly sized inlet orifices at each heating unit. This method operates on the principle that flow through an orifice is a function of the pressure differential across the orifice plate. The pressure range is selected to fill each heating unit on the coldest design day; on warmer days, the supply pressure is lowered so that heating units are only partially filled with steam, thereby reducing their heat output. An advantage of this method is that it virtually eliminates all heating unit steam trap losses because the heating units are only partially filled with steam on all but the coldest design days.

The required system pressure differential can be achieved manually with throttling valves or with an automatic pressure differential controller. Table 2 shows the required pressure

Table 2   Pressure Differential Temperature Control

Outdoor Temperature, °F	Required Pressure Differential[a], in. Hg.
0	6.0
10	4.5
20	3.2
30	2.1
40	1.2
50	0.5
60	0.1

[a]To maintain 70°F indoors for 0°F outdoor design.

differentials to maintain 70°F indoors for 0°F outdoor design, with orifices selected according to Table 2. Required pressure differential curves can be established for any combination of supply and return line pressures by sizing orifices to deliver the proper amount of steam for the coldest design day, calculating the amount of steam required for warmer days using Equation (2) and then selecting the pressure differential that will provide this flow rate.

$$Q_r = Q_d \frac{(t_i)_{design} - t_o}{(t_i)_{design} - (t_o)_{design}} \qquad (2)$$

where

Q_r = required flow rate
Q_d = design day flow rate
$(t_i)_{design}$ = design indoor temperature
$(t_o)_{design}$ = design outdoor temperature
t_o = outside temperature

Pressure differential systems can be used with zone control valves or individual self-contained thermostatic valves to achieve increased operating efficiency.

HEAT RECOVERY

Two methods are generally employed to recover heat from condensate: (1) the enthalpy of the liquid condensate (sensible heat) can be used to vaporize or "flash" some of the liquid to steam at a lower pressure or (2) it can be used directly in a heat exchanger to heat air, fluid, or a process.

The particular methods used vary with the type of system. As explained in the Basic Steam System Design section, facilities that purchase steam from a utility generally do not have to return condensate and, therefore, can recover heat to the maximum extent possible. On the other hand, facilities with their own boiler generally want the condensate to return to the boiler as hot as possible, limiting heat recovery since any heat removed from condensate has to be returned to the boiler to generate steam again.

Flash Steam

Flash steam is an effective use for the enthalpy of the liquid condensate. It can be used in any facility that has a requirement for steam at different pressures, regardless of whether steam is purchased or generated by a facility's own boiler. Flash steam can be used in any heat exchange device to heat air, water, or other liquids or directly in processes with lower pressure steam requirements.

Equation (1) may be used to calculate the amount of flash steam generated, and Figure 18 provides a graph for calculating the amount of flash steam as a function of system pressures.

Although flash steam can be generated directly by discharging high-pressure condensate to a lower pressure system, most designers prefer a flash tank to control flashing. Flash tanks can be mounted either vertically or horizontally, but the vertical arrangement shown in Figure 19 is preferred because it provides better separation of steam and water, resulting in the highest possible steam quality.

The most important dimension in the design of vertical flash tanks is the internal diameter, which must be large enough to

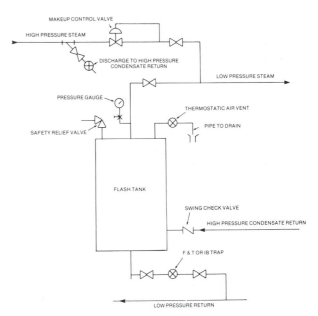

Fig. 19   Vertical Flash Tank

ensure low upward velocity of flash to minimize water carry-over. If the upward velocity is low enough, the height of the tank is not important, but it is good practice to use a minimum height of 2 to 3 ft. The chart in Figure 20 can be used to determine the internal diameter and is based on a steam velocity of 10 ft/s, which is the maximum velocity in most systems.

Installation is important for proper flash tank operation. Condensate lines should pitch towards the flash tank. If more than one condensate line discharges to the tank, each line should be equipped with a swing check valve to prevent backflow. Condensate lines and the flash tank should be well insulated to prevent any unnecessary heat loss. A thermostatic air vent should be installed at the top of the tank to vent any air that accumulates. The tank should be trapped at the bottom with an inverted bucket or float and thermostatic trap sized to triple the condensate load.

The demand load must always be greater than the amount of flash steam available or the low-pressure system can be

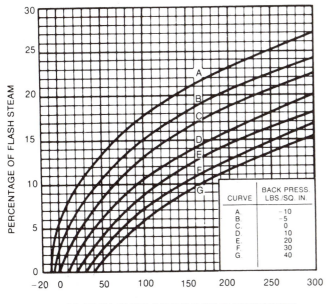

PSI FROM WHICH CONDENSATE IS DISCHARGED

CURVE	BACK PRESS. LBS./SQ. IN.
A.	−10
B.	−5
C.	0
D.	10
E.	20
F.	30
G.	40

Fig. 18   Flash Steam

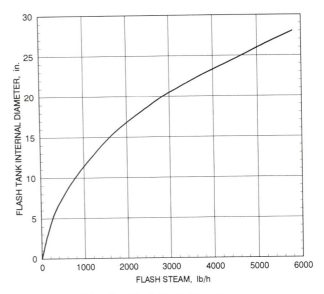

Fig. 20   Flash Tank Diameters

overpressurized. A safety relief valve should always be installed at the top of the flash tank to preclude such a condition.

Since the flash steam available is generally less than the demand for low-pressure steam, a makeup valve ensures that the low-pressure system maintains design pressure.

Flash tanks are considered pressure vessels and must be constructed in accordance with ASME and local codes.

Direct Heat Recovery

Direct heat recovery that uses the enthalpy of the liquid in some type of heat exchange device is appropriate when condensate is not returned to a facility's own boiler; any lowering of condensate temperature below 212°F requires reheating at the boiler to regenerate steam.

The enthalpy of the condensate can be used in fan-coil units, unit heaters, or convectors to heat spaces where temperature control is not critical such as garages, ramps, loading docks, and entrance halls or in a shell-and-tube heat exchanger to heat water or other fluids.

In most HVAC applications, the enthalpy of the liquid condensate may be used most effectively and efficiently to heat domestic hot water with a shell-and-tube or plate-type heat exchanger, commonly called an economizer. Many existing economizers do not use the enthalpy of the condensate effectively because they are designed only to preheat makeup water flowing directly through the heat exchanger at the time of water usage. Hot water use seldom coincides with condensate load, and most of the enthalpy is wasted in these preheat economizer systems.

Heat recovery can only be effective with a storage-type hot water heater. This can be a shell-and-tube heat exchanger with a condensate coil for heat recovery and a steam or electric coil when the condensate enthalpy cannot satisfy the demand. Another option is a storage-type heat exchanger incorporating only a coil for condensate with a supplemental heat exchanger to satisfy peak loads. Note that many areas require a double-wall heat exchanger between steam and the potable water.

Chapter 44 in the 1991 ASHRAE *Handbook — HVAC Applications* provides useful information on determining hot water loads for various facilities, but in general the following provides optimum heat recovery.

1. Install the greatest storage capacity in the available space. Although all systems must have supplemental heaters for peak load conditions, with ample storage capacity these heaters may seldom function if all the necessary heat is provided by the enthalpy of the condensate.

2. For maximum recovery, permit stored water to heat to 180°F or higher, using a mixing valve to temper the water to the proper delivery temperature.

COMBINED STEAM AND WATER SYSTEMS

Combined steam and water systems are often used to take advantage of the unique properties of steam, described in the Advantages of Steam Systems and Fundamentals sections.

Combined systems are used where a facility must generate steam to satisfy the heating requirements of certain processes or equipment. They are usually used where steam is available from a utility and economic considerations of local codes preclude the facility from operating its own boiler plant. There are two types of combined steam and water systems: (1) where steam is used directly as a heating medium; the terminal equipment must have two separate coils, one for heating with steam and one for cooling with chilled water and (2) where steam is used indirectly and is piped to heat exchangers that generate the hot water for use at terminal equipment; the exchanger for terminal equipment may use either one coil or separate coils for heating and cooling.

Combined steam and water systems may be two-, three-, or four-pipe systems. Chapters 3 and 4 have further descriptions.

COMMISSIONING

After design and construction of a system, care should be taken to ensure correct performance of the system and that building personnel understand the operating and maintenance procedure required to maintain operating efficiency.

REFERENCES

ANSI/ASME. 1989. Power piping. ANSI/ASME *Standard* B31.1-89. American Society of Mechanical Engineers, New York.

Hoffman steam heating systems design manual. *Bulletin* No. TES-181, Hoffman Specialty ITT Fluid Handling Division, Indianapolis, IN.

Kremers, J.A. 1982. Modulating steam pressure in coils compound steam trap selection procedures. *Armstrong Trap Magazine* 50(1). Armstrong Machine Works, Three Rivers, MI.

Sanford, S.S. and C.B. Springer. 1931. Flow of steam through orifices. *ASHVE Transactions* 37:371-94.

Stamper, E. and R.L. Koral. 1979. *Handbook of air conditioning, heating and ventilating,* 3rd ed. Industrial Press, New York.

CHAPTER 11

DISTRICT HEATING AND COOLING

A DISTRICT heating and cooling (DHC) system distributes thermal energy from a central source to residential, commercial, and/or industrial consumers for use in space heating, cooling, water heating, and/or process heating. The energy is transferred by steam or hot or chilled water lines. Thus, consumers obtain thermal energy from a distribution medium rather than generate it on-site at each facility.

Components. District heating and cooling systems consist of three primary components: the central plant, the distribution network, and the user systems (Figure 1).

The central source or production plant may be any type of boiler, refuse incinerator, geothermal source, solar energy, or thermal energy developed as a byproduct of electrical generation. The latter approach, called cogeneration, has a high energy utilization efficiency.

Chilled water can be produced by an absorption refrigeration machine, electric-driven centrifugal chiller, gas/steam turbine or engine-driven centrifugal chiller, or a combination of mechanically driven systems and thermal energy-driven absorption systems.

The second component is the piping network that conveys the energy. The piping for these systems usually consists of a combination of preinsulated and field-insulated pipe in both concrete tunnel and direct burial applications.

These networks require substantial permitting and coordinating with nonusers of the system. Because they are expensive, it is important to maximize their use.

The third component is the consumer system, which includes in-building equipment. When steam is supplied, (1) it may be used directly for heating; (2) it may be directed through a pressure-reducing station for use in low-pressure (0 to 15 psig) steam space heating, service water heating, and absorption cooling; or (3) it may be passed through a steam-to-water heat exchanger, which transfers energy from one fluid to another.

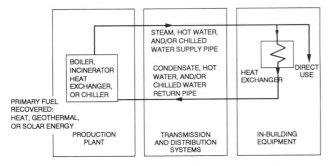

Fig. 1   Major Components of District Heating System

The preparation of this chapter is assigned to TC 6.2, District Heating and Cooling.

Applicability. District heating and cooling systems are best used in markets where (1) the thermal load density is high and (2) the annual load factor is high. A high load density is important to cover the capital investment for the transmission and distribution system, which usually constitutes most of the capital cost for the overall system, often ranging from 50 to 75% of the total. The annual load factor is important because the total system is capital intensive. These factors make district heating and cooling systems most attractive in serving (1) industrial complexes, (2) densely populated urban areas, and (3) high-density building clusters with high thermal loads. Low-density residential areas have usually not been attractive markets for district heating, although there have been some successful applications. District heating is best suited to areas with a high building and population density in relatively cold climates. District cooling applies in most areas that have appreciable concentrations of cooling loads, usually associated with tall buildings.

Emissions from central plants are also easier to control than those of individual plants. A central plant that burns high sulfur coal can economically remove noxious sulfur emissions, where individual combustors could not. Similarly, the thermal energy from municipal wastes can provide an environmentally sound system.

ECONOMIC BENEFITS

The following economic benefits should be considered when a district heating and cooling system is considered.

Building Mechanical Rooms

Operating personnel. One of the primary advantages to a building owner is that operating personnel for the HVAC system can be reduced or eliminated. Most municipal codes require operating engineers to be on site when high-pressure boilers are in operation. Some older systems require trained operating personnel to be in the boiler/mechanical room at all times. When thermal energy is brought into the building as a utility, the need for skilled help is eliminated. Depending on the sophistication of the building HVAC control system, there may be opportunity to reduce or minimize operating personnel.

Insurance. Both property and liability insurance costs are significantly reduced with the elimination of a boiler in the mechanical room since risk of a fire or accident is reduced.

Usable space. Usable space in the basement area increases when a boiler and/or chiller and related equipment are no longer necessary. The noise associated with such in-building equipment is also eliminated. Although this space usually cannot be converted into prime office space, it does provide the opportunity for increased storage or other use.

Equipment maintenance. With less mechanical equipment, there is proportionately less equipment maintenance, resulting in less expense and a reduced maintenance staff.

Higher thermal efficiency. A larger central plant can achieve higher thermal and emission efficiencies than can several smaller units. When strict regulations have to be met, additional pollution control equipment is also more economical for larger plants.

Partial load performance of central plants may be more efficient than that of many isolated small systems, because the larger plant can operate one or more capacity modules as the combined load requires and can modulate output. Central plants generally have efficient base load units and less costly peaking equipment for use in extreme loads or emergencies.

Wider range of available fuels. Smaller heating plants are usually designed for one fuel type. Generally, the fuel is limited to gas or oil. Central DHC plants can operate on less expensive coal or refuse. Larger facilities can often be designed for more than one fuel, *i.e.,* coal and oil.

Life cycle cost analysis. In this analysis, the cash outlays and income stream are projected over the model period. See Chapter 33 of the 1991 ASHRAE *Handbook—HVAC Applications* for detailed information on life cycle cost analysis.

Energy source economics. If an existing facility is the energy source, the available temperature and pressure of the thermal fluid is predetermined. If exhaust steam from an existing electrical generating turbine is used to provide thermal energy, the conditions of the bypass determine the maximum operating pressure and temperature of the DHC system. A trade-off analysis must be conducted to determine what percentage of the energy will be diverted for thermal generation and what percentage will be used for electrical generation. Based on the marginal value of energy, the determination of the operating conditions is critical to the economic analysis.

If a new central plant is being considered, a decision for cogeneration or thermal energy only must be made. An example of a cogeneration system is a diesel or natural gas engine-driven generator with heat recovery equipment. The engine drives a generator to produce electricity, and heat is recovered from the exhaust, cooling, and lubrication systems. Another system is one of several variable extraction steam turbine designs for cogeneration. These turbine systems combine the thermal and electrical output to maximize use of available energy.

The selection of temperature and pressure is crucial because it can dramatically affect the economic feasibility of a DHC design. If the temperature and/or pressure level chosen is too low, a potential customer base can be eliminated from the system. On the other hand, if there is no demand for absorption chillers or high-temperature industrial processes, a low-temperature system usually provides the lowest delivered energy cost.

The availability and location of fuel sources must also be considered in optimizing the economic design of a DHC system. For example, a natural gas boiler might not be feasible where abundant sources of natural gas are not available.

Initial capital investment. The initial capital investment for a DHC system is usually the major economic driving force. Normally, the initial capital investment includes the four components of (1) concept planning, (2) design, (3) construction, and (4) consumer interconnect.

Concept planning. Although the construction cost usually accounts for most of the initial capital investment, neglect in any of the other three areas could mean the difference between an economic success or failure.

In concept planning, three areas are generally reviewed. First, the technical feasibility of a DHC system must be consid-

ered. Conversion of an existing heat source, for example, usually requires the services of an experienced power plant or district heating and cooling engineering firm.

Secondly, the financial feasibility must be considered. For example, a municipal or governmental body must consider availability of general obligation (GO) or industrial revenue (IR) bond financing. Alternative energy choices for potential customers must be reviewed because consumers are often asked to sign long-term contracts in order to justify a DHC system.

Lastly, political feasibility must be considered, particularly if a municipality or governmental body is considering a DHC installation. Historically, successful DHC systems have had the political backing and support of the community.

Design. The distribution system accounts for a significant portion of the initial investment. Its design depends on the heat transfer medium chosen, its operating temperature and pressure, and the routing. Failure to consider these key variables will result in higher-than-planned installation costs. An analysis must be done to optimize the insulating properties of the system. The methods for these determinations are discussed in the Distribution section of this chapter.

Construction. The construction costs of the central plant and distribution system depend on the quality of the concept planning and design. Field changes usually increase the final cost and delay start-up. Even a small delay in start-up can adversely affect both economics and consumer confidence.

Lead time needed to obtain equipment generally determines the time required to build a DHC system. In some cases, lead time on major components in the central plant can be over a year.

Installation time of the distribution system depends on the routing interference with existing utilities. A distribution system running in a new industrial park is simpler and requires less time to install than a system being installed in an established business district. All these factors should be considered in the economic analysis.

Consumer interconnect. Interconnect costs are usually borne by the consumer. High interconnect costs may favor an in-building plant instead of a DHC system. For example, if an existing building is equipped for steam service, the interconnect to a hot water DHC system may be so costly that it is not economical, even though the cost of energy is cheaper. A payback period for the consumer typically has to be less than 3 years before a major retrofit project for the interconnect will be approved.

CENTRAL PLANT

The components of the central production plant vary depending on services performed, type of fuel source used, environmental concerns, and, to some extent, the source of equipment. Assuming the plant is to provide both heating and cooling, a boiler generates steam or hot water, and a chiller provides chilled water. The boiler may be fired by coal, oil, gas, electric or nuclear power, or other locally available fuels. In some cases, firing may be augmented by solid waste incineration with heat recovery. Chilled water may be produced by an absorption refrigeration machine or by a chiller driven by electric power, a gas/steam turbine, or an engine.

Fire-tube and water-tube boilers are available for gas/oil firing. If coal is used, either package-type coal-fired boilers in small sizes (less than 20,000 to 25,000 lb/h) or field-erected boilers in larger sizes are available. Coal-firing underfeed stokers are available up to a 30,000 to 35,000 lb/h capacity; travelling grate and spreader stokers are available up to a

160,000 lb/h capacity in single-boiler installations. Larger coal-fired boilers are typically multiple installations of the three types of stokers previously described, or larger, pulverized fired or fluidized bed boilers. Generally, the complexity of fluidized bed or pulverized firing does not lend itself to central plant operation but to utility operations where electricity generation is the primary function.

Other factors to consider include:

• Plan for expansion and future growth when deemed appropriate
• Cogeneration when energy and electric energy use are coincident
• Availability of solid waste, which may provide an economic opportunity for a solid waste fueled plant

Instrumentation systems can be either electronic or pneumatic. Pneumatic instrumentation is known for its ruggedness and dependability. Electronic instrumentation systems, including programmable controllers for boiler and chiller plant operation, offer the flexibility of combining control systems with data acquisition systems. This combination provides for improved efficiency, better energy management, and reduced operating staff for the central heating/cooling plant.

Numerous pieces of auxiliary support equipment related to the boiler and chiller operations are not unique to the production plant of a DHC system and are found in similar installations. Some components of a DHC system deserve special consideration due to their critical nature and potential impact on operations.

Boiler feedwater treatment has a direct bearing on equipment life. Condensate receivers, filters, polishers, and chemical feed equipment must be accessible for proper management, maintenance, and operation. Depending on the temperature, pressure, and quality of the heating medium, water treatment systems may require softeners, alkalizers, and/or demineralizers for higher operating temperatures and pressure systems.

Equipment and layout of a central heating and cooling plant should reflect the refinements required for proper plant operation and maintenance. The plant should have an adequate service area for equipment and a sufficient number of electrical power outlets and floor drains. Equipment should be placed on housekeeping pads for better housekeeping.

Environmental equipment, including electrostatic precipitators, baghouses, and scrubbers, is required to meet emission standards for coal-fired or solid waste-fired operations. Proper controls are critical to their operation, and they should be designed and located for easy access by maintenance personnel.

A baghouse gas filter provides good service if gas flow and temperature are properly maintained. Because baghouses were designed for continuous on-line use, they are less suited for cycle operation. Heating and cooling significantly reduces the useful life of the bags due to acidic condensation. The use of an economizer to preheat boiler feedwater and help control flue gas temperature may enhance baghouse operation. Contaminants generated by plant operation and maintenance, such as washdown of floors and equipment, may need to be contained.

Multiple air-conditioning loads interconnected with a central chilled water system provide some economic advantages and energy conservation opportunities. The size of air-conditioning loads served, as well as the diversity between load profiles and their distance from the chilling plant are principal factors in determining the feasibility of central plants. The distribution system pipe capacity is sensitive to the operating temperature difference between the supply and return lines and the diversity of the connected load.

An economic evaluation of piping and pumping costs versus chiller power requirements can establish the most suitable supply water temperature. It is often more efficient to use isolated auxiliary equipment for special process requirements and to allow the central plant supply water temperature to float upward for higher chiller efficiency at decreased system loads. However, the designer must investigate the effects of higher chilled water supply temperatures on chilled water secondary system distribution flows and air-side systems performance (humidity control) before applying this to individual central chilled water plants.

Thermal storage and DHC. Chilled or hot water storage allows sizing of chiller or heating capacity to meet average rather than peak loads. In the large capacities common in DHC facilities, water storage may cost less than the avoided chiller or heating plant. Distributed storage (located at one or more points on the distribution system) can improve the central plant capacity and the distribution piping capacity (see Chapter 47 of the 1991 ASHRAE *Handbook—HVAC Applications*).

Design considerations. Primary water distribution systems are designed either as constant flow (variable return temperature) or variable flow (constant return temperature) systems. The design decision between constant or variable volume flow affects (1) selection and arrangement of chiller(s); (2) design of the primary distribution system; and (3) methods of building load connection to the primary system.

The constant flow primary distribution system is generally applied to smaller systems where simplicity of design and operation are important and where primary distribution pumping costs are insignificant. Chillers are usually arranged in series.

The flow volume through a full-load primary distribution system depends on the type of constant flow system used. One technique connects the building and its terminals across the primary distribution system, and the central plant circulating pump pumps chilled water through wild-flowing or three-way, valve-controlled air-side terminal units. Balancing problems may occur because many separate flow circuits are interconnected (Figure 2).

Constant flow primary distribution is also applied to in-building circuits with separate pumps. This arrangement isolates the flow balance problem between buildings. In this case, the flow through the primary distribution system can be significantly lower than the sum of the flows needed by the terminals if the secondary in-building system supply temperature is higher than the primary system supply temperature (Figure 3). The return water temperature in the primary system is always between the primary supply water temperature and the effect of all connected secondary returns.

In this design, chillers arranged in parallel have decreased entering water temperatures at part load; thus several machines may need to run simultaneously, each at a reduced load, to produce the required chilled water flow. In this case, chillers in series are better because constant flow can be maintained through the chilled water plant at all times, with only the chiller machines required for producing chilled water energized. Carefully analyze constant flow systems when considering multiple chillers in a parallel arrangement, as the auxiliary burdens of condenser water pumps, tower fans, and central plant circulating pumps are a significant part of the total energy input.

The variable water flow rate can improve energy use and expand the capacity of the primary system piping by using diversity. The design temperature rises only during peak load conditions of the total system. Design each secondary in-building circuit for two-way throttling valve control so that leaving water temperatures are as high as the process can tolerate. Constant flow, three-way control valves should not be used in a variable flow system.

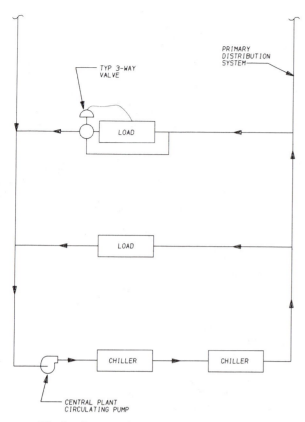

Fig. 2   Constant Flow Primary Distribution

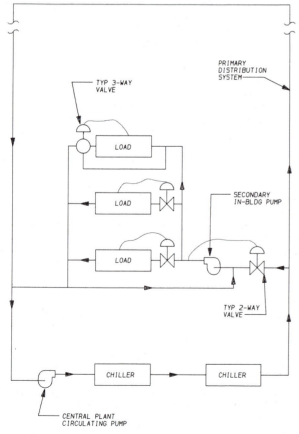

Fig. 3   Constant Flow Primary Distribution Secondary Pumping

To maintain high return temperatures at part load, the primary distribution system flow rate must track the load imposed on the central chilled water plant. Multiple parallel pumps or variable speed pumps can reduce flow and pressure and lower pumping energy at part load. Select refrigeration equipment that ensures high efficiency at part-load conditions. Control the flow at each terminal device to ensure that variable flow system objectives are met. Flow-throttling valves provide the continuous high return temperature needed to correlate the system load change to a system flow change.

Secondary systems in each building are usually two-pipe, with individual in-building pumping. In some cases, the pressure of the primary distribution system may cause flow through the in- building system without secondary in-building pumping. Primary distribution system pumps can provide total building system pumping if (1) the primary distribution system pressure drops are minimal and (2) the primary distribution system is relatively short-coupled (3000 ft or less). To implement this pumping method, the total flow through the system must be pumped at the pressure required by the building with the largest pressure requirement. If the designer has control over the design of each secondary in-building system, this pumping method can be achieved in a reasonable manner. However, if the primary system operates at an artificially high pressure, the gains of not having individual in-building pumps may be negated.

The maximum differential pressure that can be applied to the secondary circuit control valves is a function of the shut-off rating of the control valves and the controllability of the control loop. If the pressure drop across the control valve at low flow rates is greater than five times the design pressure drop at maximum flow rate, controllability is compromised. Typically, commercial grade automatic temperature control valves have lower shut-off ratings and higher minimum controllable flow rates than do industrial quality control valves. All chilled water control valves must be normally closed. Then, when any of the secondary in-building systems are deenergized, the valves close and will not bypass chilled water to the return system.

When buildings are circulated separately, isolating primary/secondary piping and pumping techniques are used, thereby ensuring that two-way control valves are subjected only to the differential pressure established by the in-building pump. Figures 4a through 4e show five types of primary/secondary connections using in-building pumping schemes.

Where the supply water temperature need not be as low as the water from the primary distribution system, temperature-actuated diversity control valves can be applied (Figure 4e). These valves allow chilled water from the primary distribution system to mix with the secondary in-building return water to deliver the required in-building supply water temperature. Flow sensor controllers in the crossover bridge balance primary flow to secondary building flow demand (Figures 4d and 4e).

Control excessive differential pressures in the primary loop by sizing the distribution system for a low total pressure drop. Unless the system is extremely long (over 10,000 ft equivalent length), size the distribution system total pressure drop in the range of a 20 to 50-ft total pressure drop.

When secondary in-building pumps are used, all series interconnections between the primary system pump and the in-building pumps must be removed. A series connection can cause the primary system return to operate at a higher pressure than the distribution system supply and disrupt normal flow patterns. Series operation usually occurs during improper use of three-way mixing valves in the primary to secondary connection.

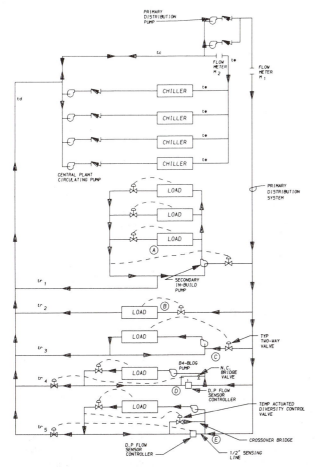

A. Variable flow primary/variable flow secondary system with in-building pump. Constant return water temperature variable secondary supply water temperature. Secondary system could be constant flow if three-way load valves are used.
B. Chilled water circulated through building system with differential pressure created by secondary distribution pump.
C. In-building pump with building secondary water supply temperature controlled by building load. Constant flow secondary; variable flow primary.
D. In-building pump with two-way valve control of building loads. Differential pressure flow sensor/controller to modulate return water control valve to match primary flow from distribution system and secondary building flow equally.
E. Blending of primary and secondary water temperatures with crossover bridge control valve controlled by either building supply or return water temperature. Allows for higher building water temperatures than primary chilled water temperatures. Building load controlled by two-way valves. Differential pressure flow sensor controller to modulate return water control valve to match primary flow from distribution stem and secondary building flow equally.

Fig. 4 Variable Flow Primary/Secondary Systems

Usually, a positive pressure must be maintained at the highest point on the system at all times. This height determines the static pressure at which the expansion tank will operate. Excessively tall buildings that do not isolate the in-building systems from the distribution systems can impose unacceptable static pressures. To prevent excessive operating pressures in chilled water distribution systems, heat exchangers have been used to isolate the in-building system from the distribution system. To ensure reasonable temperature differentials between supply and return temperatures, flow control must be controlled on the primary distribution system side of the heat exchanger.

In high-rise buildings, all piping, valves, coils, and other equipment may be required to withstand high pressures. Where system static pressure exceeds safe or economical op-

erating pressure, either the heat exchanger method or pressure sustaining valves in the return line with check valves in the supply line may be used to minimize pressures. However, the pressure sustaining/check valve arrangement may overpressurize the entire primary distribution system if a malfunction of either valve occurs.

Expansion tanks, relief valves, and water makeup. The expansion tank is usually located in the central chilling plant building. To control system pressure, either air or nitrogen is introduced to the air space in the expansion tank. To function properly, the expansion tank is the single point of the system where no pressure change occurs. Multiple, air-filled tanks cause erratic and possibly harmful movement of air through the piping system. Although diaphragm expansion tanks eliminate air movement, the possibility of hydraulic surge should be considered. The addition of several pumps in auxiliary circuits inside buildings served by the primary system can complicate the pressure relationship throughout the secondary in-building systems. On large chilled water systems, makeup for water losses is generally accomplished by a makeup water pump. The pump is typically controlled from level switches on the expansion tank or from desired suction pressure.

A conventional water meter on the makeup line can show water loss in a closed system. This meter also provides necessary data for water treatment. The fill valve should be controlled to open or close and not modulate to a very low flow, so that the water meter can detect all makeup.

Guidelines for plant design and operation include:

- Design for a reasonable temperature differential of 12 to 14°F. A 12 to 16°F maximum temperature differential with a 10 to 12°F minimum temperature differential can be achieved with this design.
- Select six-row, 12 to 14 fins-per-inch coils as the minimum size coil applied to central station air-handling units to provide adequate performance. With this type of coil, the return water temperature rise can be expected to be in the 12 to 16°F range at full load. To maintain a reasonable temperature differential at design conditions, size fan coil units for an entering water temperature several degrees above the main chilled water plant supply temperature. This will require that temperature-actuated diversity control valves be applied to the primary distribution system crossover bridge (Figure 4e).
- Limit the use of constant flow systems to relatively small central chilled water plants. Chillers should be in series.
- Larger central chilled water plants can benefit from primary/secondary pumping with constant flow in central plant and variable flow in primary distribution systems. Size the primary distribution system for a low overall total pressure loss. Short-coupled distribution systems (3000 ft or less) can be sized for a total pressure loss of 20 to 40 ft. With this maximum differential between any point in the system, size the primary distribution pumps to provide the necessary pressure to circulate chilled water through the secondary in-building systems, eliminating the need for in-building pumping systems. This decreases the complexity of operating central chilled water systems.
- All two-way valves must have proper close-off ratings and a design pressure drop of at least 20% of the maximum design pressure drop for controllability. Commercial quality automatic temperature control valves generally have low shutoff ratings, but industrial valves can achieve higher ratings.
- Variable speed pumping saves energy and should be considered for distribution system pumping.
- The lower practical limit for chilled water supply temperatures is 40°F. Carefully analyze temperatures below 40°F, although systems with thermal energy storage may operate advantageously at lower temperatures.

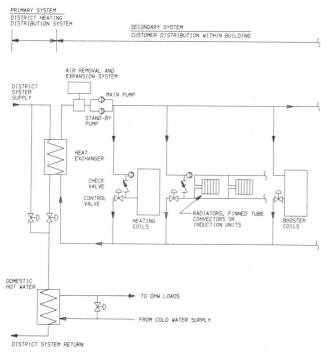

Fig. 5   Basic Heating System Schematic

Cogeneration. Analyze each plant's thermal cycle for cogeneration opportunities. Cogeneration can be achieved by:

- Using noncondensing turbines for pressure-reducing valves
- Base loading steam boilers and using condensing turbines to condense surplus steam (if fuel costs permit)
- Using gas- or oil-fired combustion turbines or engines as the prime driver, with heat recovered in a steam-generating heat recovery boiler

DISTRICT HEATING DESIGN

Hot water district heating systems supply hot water at a temperature that varies according to the outside temperature. By reducing the water temperature when the heating demand decreases, energy losses are reduced and energy efficiency is increased. Annual heat losses in efficient, medium temperature hot water distribution systems are about 5 to 10%.

In general, systems reflect the following design temperatures: at an outdoor temperature of −20°F, the district heat supply temperature is 250°F; at 45°F ambient, the supply water temperature is 170°F. The supply temperature commonly ranges from 170 to 250°F. The low end of the range is often the temperature required to meet summertime domestic hot water needs. However, summer temperatures may be higher if there are process heating loads or other higher temperature loads.

Building systems may be connected directly or indirectly to the district heating distribution system. With direct connection, the district heating water is distributed within the building to directly provide heat to terminal equipment such as radiators and unit heaters.

The building heat exchanger station is divided into primary and secondary sides, which are separated by one or more heat exchangers. The primary side of the heat exchanger is connected directly to the district heating distribution system. Figure 5 shows an example of a basic building system schematic including heat exchangers and secondary systems.

Extraction of energy in the heat exchanger is governed by a control valve that is sensitive to the building heat demand.

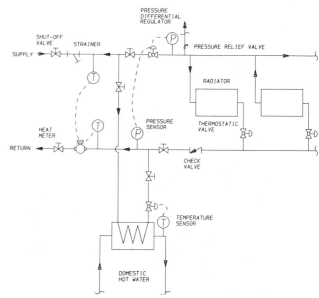

**Fig. 6   Direct Connection of Building System to
District Hot Water**

Benefits include potential use of low-grade, low-cost heating sources and low-cost piping materials and designs. However, these savings must be balanced against higher distribution costs due to a lower primary temperature differential and, therefore, larger pipe sizes and/or higher pumping costs.

Generally, maintaining a high temperature differential between supply and return lines is most cost effective, because this allows for smaller pipe sizes in the primary distribution system. These savings must be weighed against higher building conversion costs that may result from the need for a low primary return temperature.

The secondary supply temperature must be controlled in a manner similar to the primary supply temperature. To ensure a low primary return temperature, the secondary system should have a low return temperature. This design helps lower costs through reductions in pipe sizes, smaller pumps, less insulation, and smaller pump motors. A combined return temperature on the secondary side of 130°F or lower is reasonable, although this temperature could be difficult to achieve in smaller buildings with only baseboard or radiator space heating.

To keep the return temperature low, water flow through heating equipment should be controlled according to the heating demand in the space.

Direct connection. Direct connection of terminal heating equipment (such as heating coils, radiators, and unit heaters) to the primary distribution system increases the possibility of damage or contamination that could affect the entire system. However, a direct connection is often more economical than the indirect method. Because no heat exchangers, pumps, or water treatment systems are required in the customer installation, investment costs are reduced and a lower district heating return temperature is possible. Figure 6 shows the simplest form of direct connection, which includes a pressure differential regulator, a thermostatic valve on each heating unit, a pressure relief valve, and a check valve.

In a direct system, the main distribution system pressure must meet local building codes to protect the customer's installation and the reliability of the district heating system. To minimize noise and control problems, install a constant pressure differential control valve in the buildings. Pay special attention to potential noise problems at the thermostatic valves. These valves must correspond to the design pressure differential.

If the temperature in the main system is higher than that required in the building systems, a larger temperature differential between supply and return may occur, which reduces the required pipe size. The desired supply temperature within the building is then attained by mixing return water with supply water. This approach requires a centralized temperature control system with a temperature controller connected to an outdoor temperature sensor.

Depending on the size and design of the main system, elevation differences, and types of customers and building systems, additional safety equipment such as automatic shut-off valves on both supply and return lines may be required.

Interface with Hot Water District Heating

The following building system design elements are important regardless of the type of building heating system.

Heat exchangers. Many district heating utilities select and/or approve heat exchangers to be used by customers for indirect connection. Reliability is increased if two heat exchangers (each designed for about two-thirds of the load) are installed for space heating. One heat exchanger can carry the full load for 80 to 85% of the heating season, but if both exchangers are operating continually, the return temperature can be reduced.

Several types of heat exchangers are designed specifically for interface with hot water district heating systems. Plate heat exchangers generally have a cost advantage and require less space because they require one-third to one-half the surface of shell- and-tube units for the same operating conditions. With plate exchangers, the approach temperature difference between the primary return and the secondary return is generally about 4°F, compared to about 12°F for most shell-and-tube exchangers.

Plate heat exchanger gasket material must be selected specifically for district heating applications, because gasket replacement is costly. Some manufacturers produce an all-brazed heat exchanger that does not require gaskets. These units are compact and require only minimal maintenance.

Pipe strainers should be installed with all plate exchangers to protect them from grit and debris.

Temperature controls. Usually, only one control valve is installed, but, if the single valve required is 2.5 in. or more, use two valves connected in parallel and operating in sequence. For best control, size the two valves to handle one-third and two-thirds of total capacity. When redundancy requirements are high or equal loads are desired for all exchangers, consider installing separate control equipment for each heat exchanger.

Electronic control valves should remain in a fixed position when a power failure occurs and should be manually operable. Pneumatic control valves should close upon loss of air pressure. A manual override on the control valves allows the operator the option of controlling flow.

Locate temperature sensors by the exchangers being controlled rather than in the common pipe. Improperly located sensors will cause one control valve to open and others to close, resulting in unequal loads in the exchangers.

The heat exchanger control valves must match the building loads. Oversizing reduces valve life and causes valve "hunting". Select control valves that have high rangeability; low leakage; and proportional plus integrating control for close adjustment, balancing, temperature accuracy, and response time. Control valves should also have enough power to open and close under the maximum pressure differential in the system. The control valve should have a pressure drop through the valve equal to at least one-half the pressure drop between the primary supply and return lines. The relationship between valve travel and heat output should be linear, with an equal percentage characteristic.

Control valves are normally installed in the return line because the lower temperature in the line reduces the risk of cavitation and increases valve life.

Primary control valves are the most important element in the interface with the district heating system, and proper valve adjustment and calibration will save energy. Select high quality, industrial grade control valves for more precise control, longer service life, and minimum maintenance.

PIPING FOR THERMAL DISTRIBUTION SYSTEMS

General Considerations

Choice of temperature and medium. Central heat distribution systems may use either hot water or steam as a heat-conveying medium. Regardless of the medium used, the temperature and pressures used for heating should be only as high as to satisfy the consumer requirements. Higher temperatures and pressures require additional engineering and planning to avoid higher heat losses, higher leakage rates, and lower safety and comfort levels for operators and maintenance personnel. Higher temperatures may also require higher pressure ratings for piping and fittings and may preclude the use of desirable materials such as polyurethane foam insulation and nonmetallic conduits.

Hot water systems are divided into three temperature classes. High-temperature systems supply temperatures over 350°F; medium temperature systems supply temperatures in the range of 250 to 350°F; and low-temperature systems supply temperatures of 250°F or lower. For chilled water systems, the range of feasible temperatures is limited. Supply temperatures are normally around 40 to 45°F, with return temperatures in the 50 to 60°F range.

For hot water systems, design for a high temperature drop at the consumer end. This reduces both the flow rate through the system and the pumping power required, and results in lower return temperatures, causing lower return line heat loss and lower condensing temperatures in cogeneration power plants.

In many instances, existing equipment and processes require the use of steam. Low-pressure steam systems operate below 15 psig or 250°F; high-pressure steam systems operate above this level (see the section Consumer Interconnect for further information). The common attributes and relative merits of each medium are discussed below.

Heat capacity. Steam relies primarily on the latent heat capacity of water rather than on sensible heat. The net heat content for saturated steam at 100 psig (338°F) condensed and cooled to 180°F is approximately 1040 Btu/lb$_m$. Hot water cooled from 350 to 250°F has a net heat effect of 103 Btu/lb$_m$, or only about 10% as much as that of steam. Thus, a hot water system must circulate about ten times more mass than a steam system of similar heat capacity.

Pipe sizes. Despite the fact that less steam is required for a given heat load and flow velocities are greater, steam usually requires a larger pipe size for the supply line due to its lower density (Aamot *et al.* 1978). This is compensated for by a much smaller condensate return pipe. Therefore, piping costs for steam and condensate as opposed to hot water supply and return are often comparable.

Return system. Condensate return systems require more maintenance than hot water return systems. Corrosion of piping and other components, particularly in areas where feedwater is high in bicarbonates, is a problem. Nonmetallic piping has been used successfully in some applications, such as systems with pumped returns, where it has been possible to isolate the nonmetallic piping from live steam.

Similar concerns are associated with condensate drainage systems (steam traps, condensate pumps, and receiver tanks) for steam supply lines. Condensate collection and return should be carefully considered when designing a steam system. Although similar problems with water treatment occur in hot water systems, they present less of a concern because makeup rates are lower.

Pressure requirements. Flowing steam and hot water both incur pressure losses. Hot water systems may use intermediate booster pumps to increase the pressure at points between the plant and the consumer. Due to the higher density of water, pressure variations caused by elevation differences within a hot water system are much greater than for steam systems. This can adversely affect the economics of a hot water system by requiring the use of a higher pressure class of piping and/or booster pumps.

Piping system layout. Generally, aboveground systems are acceptable only where they are naturally hidden from view or can be hidden by landscaping. The best application for shallow trench systems is where the covers can serve as sidewalks. Direct burial, shallow concrete trench, and poured envelope systems should not be used where future access would interfere with existing facilities (*e.g.,* cutting of pavement). Tunnels that provide walk-through or crawl-through access can be buried in nearly any location without causing future problems. Regardless of the type of system construction, it is usually advantageous to route piping through the basements of buildings, but only after liability issues are addressed.

In laying out the main supply and return piping, redundancy of supply should be considered. If pipe loops are provided for redundancy, flow rates under all possible failure modes must be addressed when sizing the piping.

Manholes provide access to the system at critical points such as: (1) high or low points on the system profile; (2) major branches with valves; (3) condensate drainage points on steam lines; and (4) locations with mechanical expansion devices.

To facilitate leak location and repair, manholes should be spaced no farther than 500 ft apart (in the absence of other requirements).

Piping materials and standards. All piping, fittings, and accessories should be in accordance with ANSI *Standard* B31.1 or with local requirements. For steam and hot water, pipe should conform to either ASTM A53 seamless or ERW, Grade B or ASTM A106 seamless, Grade B. Pipe wall thickness is determined by the maximum operating temperature and pressure. In the United States, most piping for steam and hot water is Schedule 40 for 10 in. NPS and below, and standard weight for 12 in. NPS and above. Many European systems use piping with a wall thickness similar to Schedule 10 on low-temperature water systems. Due to reduced piping material, such systems are not only less expensive, but they also develop reduced expansion forces and thus require simpler methods of expansion compensation. Welding pipes with thinner walls requires extra care and may require additional inspection. Where steel piping is used for condensate return lines, Schedule 80 is recommended to handle the heavy corrosion normally experienced. Fiberglass reinforced plastic (FRP) pipe should be lined with a corrosion-resistant liner at least 0.02 in. thick and must be rated for the temperature and pressure requirements of the system. For chilled water systems, a variety of materials, such as ductile iron, PVC, and FRP, have been used in addition to steel. As with any piping material, adequate temperature and pressure ratings for the intended service should be specified.

TYPES OF SYSTEM CONSTRUCTION

Distribution system construction falls into two categories: field fabricated and partially factory prefabricated. In field-fabricated systems, the pipes, insulation, pipe supports, insulation jackets, and protective enclosures are fabricated from commercially available material. Field-fabricated systems must be designed in detail, and all materials must be specified. In factory prefabricated systems, the pipe, insulation, and protective jacket are assembled in a factory and shipped in lengths up to 40 ft. Elbows, tees, loops, bends, and straight lengths are factory preassembled. Much of the component design work is done by the factory that assembles the prefabricated sections; however, the fieldwork must be designed and specified. All systems require field construction such as trenching, splicing of prefabricated sections, connection to buildings, connection to distribution systems, construction of valve vaults, and some electrical work.

Only the aboveground system and the walk-through tunnel can be inspected and maintained easily. The shallow concrete trench system requires lifting of concrete lids, and other underground systems require excavation for repairs. Leak detection is easy in the aboveground system, the walk-through tunnel system, and the concrete surface trench system. In other underground systems, leak detection is more difficult and often involves excavation of portions of a suspected branch. All systems are likely to encounter problems on wet sites.

Field-fabricated distribution systems include aboveground distribution systems and several underground systems, including walk-through tunnels, shallow concrete trenches, deep-bury small tunnels, and underground systems that use poured insulation to form an envelope around the carrier pipe.

The *aboveground system* consists of a distribution pipe, insulation surrounding the pipe, and a protective jacket that surrounds that insulation and which may have an integral vapor retarder. When the distribution system carries chilled water or any other cold medium, a vapor retarder is required if open cell insulations are used. In heating applications, the vapor retarder is not needed; however, a watertight jacket is required to keep storm water out of the insulation. The jacket material can be aluminum, stainless steel, plastic, a multilayered fabric and organic cement lay-up, or a combination. Plastics and organic cements exposed to sunshine must be resistant to ultraviolet light. The structural supports for this system are typically wood, steel, or concrete columns. Pipe expansion and contraction is taken up in loops, elbows, and bends. Mechanical expansion joints may be used, but their maintenance must be considered. Welded joints have a higher level of integrity, but joining systems may have lower installation costs.

The aboveground system usually has the lowest first cost and is the easiest to inspect and maintain; therefore, it has the lowest life cycle cost. It is frequently the system that all other systems are compared to when making a decision about which type of system to use. Drawbacks include poor aesthetics and the hazard of being struck by vehicles and equipment; also, the working medium can freeze in winter if circulation is stopped and heat is not added.

Factory prefabricated systems (Figures 7 through 10) consist of a pipe that carries the working medium, insulation around that pipe, and a protective jacket surrounding the insulation. This pipe is also called a carrier pipe, service pipe, or pressure pipe, and is designated as either air gap or nonair gap. Systems with an air gap between the insulation and the casing are usually used when the working medium is above 250°F. The perimeter jacket is called a casing, shell, or conduit. The term conduit is often used to denote the entire factory prefabricated system. These systems are designed so that the air space is continuous throughout the line from one manhole to the next. If water enters the conduit, it can be drained from the system. Once the water is drained and the required repairs made, the insulation can be dried in place by blowing air through the conduit and thus restoring it to its original thermal efficiency. Underground systems will be flooded several times during their design life, even on dry sites; therefore, a reliable water removal system should be used in the valve vaults and for systems that have an air space between the insulation and casing. Two schemes are used to ensure satisfactory insulation performance throughout the life of the system. One scheme uses insulation that can survive flooding. An annular air space around the insulation allows air to pass through the system to dry the insulation.

Some closed cell, high-temperature insulations essentially absorb no water, and little if any drying is required. Any of these systems must withstand earth loads and expansion and contraction forces due to temperature changes.

The second scheme used to retain insulation efficiency is to enclose the insulation in a waterproof envelope. Some manufacturers form the insulation envelope in each prefabricated section; others use watertight casing field joints to extend the envelope to the next termination point. The first cost of these

District distribution systems, in particular, require careful design and construction to ensure the lowest maintenance and longest life. Proper slope, water drainage, flawless pipe and casing welds, proper supports, and anchors are essential for a long system life with low maintenance. The designer must clearly delineate on the drawings and in the specification those parts of the system the conduit manufacturer must design. Unless a leak detection system is installed, these systems essentially cannot be inspected between vaults. If a problem develops in the system, excavation is required to repair it. Locating the excavation point may be difficult, even though some commercially available leak detection systems may narrow down the excavation area.

types of systems is more than that in the aboveground system but less than that in the walk-through tunnel system.

Metal conduits are coated to protect against corrosion. In addition to the outside coating, cathodic protection of the conduit metal casing is used to extend the casing life in all but the most noncorrosive soils. Corrosion on the inside of a steel casing may develop but is of minimum consequence if water is kept out of the air gap.

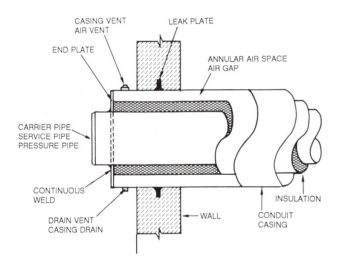

Fig. 7   Conduit System Components

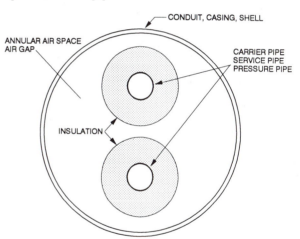

**Fig. 9   Conduit System with Two Carrier Pipes and
Annular Air Space**

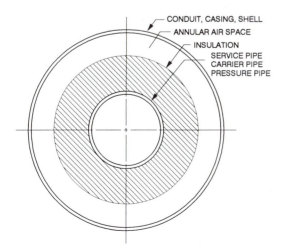

**Fig. 8   Conduit System with Annular Air Space and
Single Carrier Pipe**

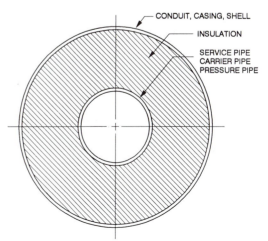

**Fig. 10   Conduit System with Single Carrier Pipe and
No Air Space**

Products with nonmetallic casings are available, but, during design, carefully consider moisture presence and thermal expansion to avoid deterioration of the plastic. Not all plastic casings can tolerate the range of temperatures that can vary from the groundwater temperature during wet site conditions to the carrier pipe working medium temperature during extremely dry site conditions. A site survey of installed systems is recommended to establish the overall performance and system life of a particular system or product with special attention given to the soil type and groundwater conditions.

The *walk-through tunnel system* (Figure 11) consists of a field-erected tunnel that is large enough for a person to walk through after the distribution pipes are in place. The top cover of the tunnel is covered with earth. The preferred construction material is reinforced concrete. Masonry units and metal preformed sections have been used with less success due to groundwater leakage and metal corrosion problems. Distribution pipes are supported from the tunnel wall or floor with conventional pipe supports. Walk-through tunnels generally cannot accommodate thermal expansion movement with pipe loops and, therefore, expansion joints are required, as described in Chapter 42. Because groundwater will penetrate the tunnel top and walls, a water drainage system must be provided. To ease inspection and maintenance, electric lights and electric service outlets may be provided. Tunnels usually have the highest first cost of all systems; however, they can have the lowest life cycle cost because (1) maintenance is easy, (2) construction errors can be corrected easily, and (3) they can be placed in areas where right-of-way is near the ground surface. In some cases, water or sewer utilities may be included in the tunnel, which further improves the economics and is an alternative in cold regions where it is necessary to protect water and sewer systems from freezing.

The *shallow concrete surface trench system* (Figure 12) is only partially buried. The floor is usually about 3 ft below the surface grade and is only wide enough for the carrier pipes, insulation, clearance to allow for pipe movement, and possibly enough room to stand on the floor. Generally, the trench is at least as wide as it is deep. The top is constructed of precast or cast-in-place reinforced concrete that protrudes slightly above grade to serve as a sidewalk. The floor and walls are usually cast-in-place reinforced concrete. Precast concrete floor and wall sections have not been successful because of the large number of oblique joints and nonstandard sections required to follow the surface topography and to slope the floor to drain. Because this system handles the storm and groundwater entering the system, the floor should slope toward a drainage point.

Pipes should be supported from cross beams attached to the side walls, because floor-mounted pipe supports tend to corrode and interfere with water drainage. This mounting system allows the distribution pipes to be assembled before lowering them on the pipe supports. These supports should be galvanized or oversized to mitigate deterioration due to corrosion. The carrier pipes, pipe supports, expansion loops and bends or expansion joints, and the insulation jacket are similar to those used in aboveground systems, with the exception of the pipe insulation. Insulation is covered with a metal or plastic jacket to protect it from abuse in the valve vaults and from storm water that enters at the top cover joints. Provide inspection ports of about 12 in. diameter in the top covers at key locations so that the system may be inspected without removing the top covers. All replaceable elements such as valves, condensate pumps, steam traps, strainers, sump pumps, and meters should be located in valve vaults. Remember that aesthetics are sometimes a consideration in the design of trench systems. The first cost and life cycle cost of this system can be among the lowest for underground systems.

Deep-bury trench systems (Figure 13) are used on sites where the groundwater elevation is typically lower than the system elevation. These systems are only large enough to contain the distribution system piping, the pipe insulation, and the pipe support system. The system is covered with earth and is essentially not maintainable between valve vaults without excavation; therefore, select materials that will last for the intended life of the system and ensure that the groundwater drainage system will function reliably. Depending on the watertightness

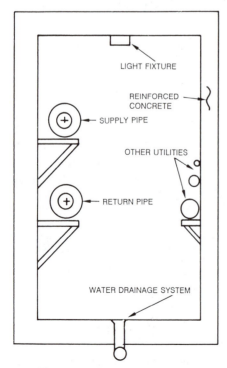

Fig. 11   Walk-Through Tunnel

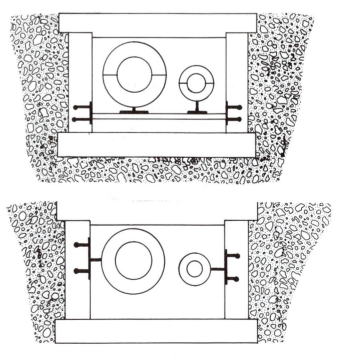

Fig. 12   Shallow Concrete Trench

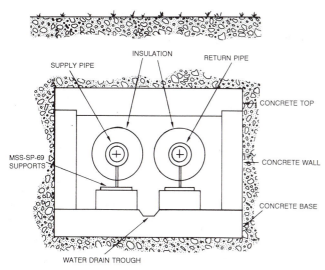

Fig. 13   Deep Buried Trench/Tunnel

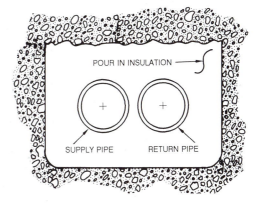

Fig. 14   Poured Envelope System

of the construction and the capacity and reliability of the internal drainage system, this system can tolerate some groundwater flooding.

The distribution pipe insulation must withstand stringent boiling tests and still show less than 5% degradation in thermal efficiency. The shallow concrete surface trench system is covered with earth and sloped independent of the earth topography. Construction of other versions of this system is typically started in an excavated trench by pouring a cast-in-place concrete base that slopes so that groundwater can drain to the valve vaults. The slope must also be compatible with the pipe slope requirements of the distribution system. The concrete base has provisions for supports for the distribution pipes, the groundwater drainage system, and the mating surface for the top cover. If the top cover is to rest on cast-in-place concrete walls, tie the walls to the base using construction techniques such as reinforcing steel protruding upward. Install the pipe supports, the distribution pipes, and the pipe insulation before installing the top cover. The cast-in-place concrete bottom contains the groundwater drainage system, which may be a trough formed into the concrete bottom or a perforated drainage pipe either cast into the concrete bottom or located slightly below the concrete base. Typically, the cover for the system is either cast-in-place concrete or preformed concrete or half-round clay tile sections. The top cover mates to the bottom as tightly as possible to limit the entry of groundwater.

Poured envelope systems (Figure 14) usually encase the distribution system pipes in an envelope of insulating material, and the insulation envelope is covered with a thick layer of earth as required to match existing topography. This system is used on sites where the groundwater is below the system elevation. Like other underground systems, the design must accommodate groundwater flood conditions. The insulation material must (1) serve as a structural member to support the distribution pipes and earth loads; (2) prevent groundwater from entering the interior of the envelope; (3) allow the distribution pipes to expand and contract axially as the pipes change temperature; and (4) at elbows, expansion loops, and bends, the insulation must allow cavities to form to allow lateral movement of the pipes, or it must flow around the pipe without significant distortion while still retaining the required structural load-carrying capacity. Special attention must be given to corrosion of metal parts and water penetration at anchors and structural supports that penetrate the insulation envelope. When the distribution pipes are hot, the heat loss tends to

drive moisture out of the insulation as steam vapor; however, when distribution pipes circulate a cooling medium, vapor condenses and is retained in the insulation, greatly reducing the thermal efficiency of the insulation. A groundwater drainage system may be required, depending on the insulation material selected and the volume of the groundwater; however, if such a drainage system is needed, this is not the proper system for the site conditions.

Construction of this system is started by excavating a trench with a bottom slope that matches the desired slope of the distribution piping. The width of the bottom of the trench is the same as the insulation envelope. The distribution piping is assembled in the trench and supported by the anchors and blocks that should be removed when the insulation is poured in place. The trench bottom and sides, wood, or sheets of plastic can be used as the form to hold the insulation. The insulation envelope is then covered with earth. Typical insulation materials include (1) insulating concrete, (2) hydrocarbon powder, and (3) hydrophobic powder.

Insulating concrete (Figure 15) is a low-mass aggregate with additives in a matrix of portland cement. This insulation material has some structural properties to resist earth loads and to support distribution piping. Insulating concrete may inter-

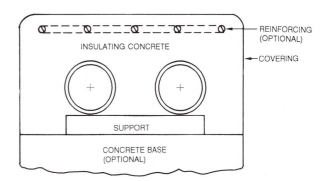

Fig. 15   Insulated Concrete, Poured Envelope System

fere with proper operation of rubber ring couplings; distribution pipe joints should be welded and a parting agent placed around the pipes so that they can freely expand and contract axially.

Steel plate anchors are embedded in a thickened and reinforced base slab and welded to the distribution pipe. The design of anchor points, loops, and bends is similar to other welded systems. Voids built into the corners of elbows, loops, and bends allow for pipe movement. Internal drains and vents remove construction water and other unexpected water, and also act as leak detectors.

Hydrocarbon powders (Figure 16) are thermoplastic materials composed of low-mass aggregate with asphalt binders. This insulation is poured around the pipe and its supports, which are fastened to a concrete base. A continuous, waterproof membrane helps keep water out of this system. Weld distribution pipe joints because the insulation material interferes with rubber ring joints. Special care must be given to the design for lateral pipe movement at elbows, loops, and bends, because this material hardens as the number of expansion cycles increases and limits the amount of pipe movement that can be tolerated.

Hydrophobic powders are treated to make them water repellent. As a result, water will not dampen the powder, which may prevent water from entering the insulation envelope. Hydrophobic powders are installed in the same way as hydrocarbon powder, except that compaction of the insulation material is required only for chilled water distribution systems and cryogenic applications.

THERMAL CONSIDERATIONS

Thermal Design Conditions

Three thermal design conditions must be met to ensure satisfactory system performance:

1. The "normal" condition used for the life cycle cost analysis determines appropriate insulation thickness. Average values for the temperatures, burial depth, and thermal properties of the materials are used for design. If the thermal properties of the insulating material are expected to degrade over the useful life of the system, appropriate allowances should be made in the analysis.
2. Maximum heat transfer rate determines the load on the central plant due to the distribution system. It also determines the temperature drop (or increase in the case of chilled water distribution) that determines the delivered temperature to the consumer. For this calculation, the thermal conductivity of each component must be taken at its maximum value, and the temperatures must be assumed

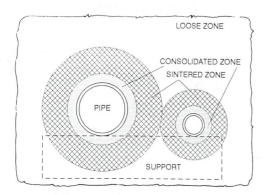

Fig. 16   Insulated Hydrocarbon Poured Envelope System

to take on their extreme values that would result in the greatest temperature difference between the carrier medium and the soil or air. The burial depth will normally be at its lowest value for this calculation.
3. During operation, none of the thermal capabilities of the materials (or any other materials within the area influenced thermally by the system) must exceed design conditions. To satisfy this objective, each of the components and the surrounding environment must be examined to determine if thermally induced damage is possible. A heat transfer analysis may be necessary in some cases. The conditions of these analyses must be chosen to represent the worse case scenario from the perspective of the component being examined.

For example, in assessing the suitability of a coating material for a metallic conduit, assume that the thermal insulation is saturated, that the soil moisture is at its lowest probable level, and that the burial depth is maximum. These conditions, combined with the highest anticipated pipe and soil temperatures, give the highest conduit surface temperature to which the coating could be exposed.

Heat transfer in buried systems is influenced by the thermal conductivity of the soil and by the depth of burial, particularly when the insulation has low thermal resistance. Soil thermal conductivity changes significantly with moisture content; for example, Bottorf (1951) indicates that the soil thermal conductivity ranges from 0.083 Btu/h·ft·°F during dry soil conditions to 1.25 Btu/h·ft·°F during wet soil conditions.

Thermal Properties

Uncertainty in heat transfer calculations for thermal distribution systems results from the uncertainty in the thermal properties of the materials involved. Generally, the designer must rely on manufacturers' specifications and handbook data to obtain proper values.

Insulation provides primary thermal resistance against heat loss or gain in thermal distribution systems. Thermal properties and other characteristics of insulations normally used in thermal distribution systems are given in Table 1.

If an analysis of the soil is available or can be done, the thermal conductivity of the soil can be estimated from Table 12, Chapter 22 of the 1989 ASHRAE *Handbook—Fundamentals*. The thermal conductivity factors in Table 2 may be used as an approximation where detailed information on the soil is not known. Because dry soil is rare in the United States, a low moisture content should be assumed only where it can be validated. Values of 0.8 to 1 Btu/h·ft·°F are commonly used where soil moisture content is unknown. Because moisture will migrate towards a chilled pipe, use a thermal conductivity value of 1.25 Btu/h·ft·°F for chilled water systems in the absence of any site-specific soil data. For steady-state analyses, only the thermal conductivity of the soil is required. If a transient analysis is required, the specific heat and density are also required. Kersten (1949), Farouki (1981), Lunardini (1981), and others give this data for some of the more common soil types.

METHODS OF HEAT TRANSFER ANALYSIS

Because heat transfer in a piping system is not related to the load factor, it can be a large part of the total load. The most important factors affecting heat transfer are the difference between earth and fluid temperatures and the thermal insulation. For example, the extremes might be a 6-in., insulated, 400°F water line in 40°F earth with 100 to 200 Btu/h·ft loss,

Table 1 Comparison of Commonly Used Insulations in Underground Piping Systems

Item	Calcium Silicate	Urethane Foam	Cellular Glass	Preformed Glass Fiber	Loose Glass Fiber	Insulating Concrete	Hydrocarbon Powder	Hydrophobic Powder
Thermal conductivity, Btu/h·ft·°F								
at 100°F	0.028	0.013	0.033	0.022	0.027	0.050 to 0.333	0.050 to 0.067	0.044 to 0.067
at 200°F	0.031	0.014	0.039	0.025	—	—	—	0.050 to 0.068
at 300°F	0.034	—	0.046	0.028	—	—	—	0.055 to 0.071
at 400°F	0.038	—	0.053	—	—	—	—	—
Density, lb/ft³	10 to 14	1.5 to 4	9 to 10	3 to 7	2 to 11	20 to 30	45 to 55	35 to 65
Maximum temperature, °F	1200	260	800	370 to 500	1000	1800	300 to 450	500
Compressive strength,								
psi	75 to 165	20 to 50	100	5	0	100 to 150	70 to 85	70 to 85
at compression	5%	5%	—	—	—	—	5%	5%
Moisture absorption	Great	Slight	Nil	Great	Great	Moderate	Moderate	Nil
Effect on k factor	Large	Slight	Slight	Large	Large	Slight	Large	Slight
Resistance to boiling	Good	Poor	Fair	Good	Good	Good	Poor	Poor
Recovery-drying	Good	Poor	Good	Good	Good	Good	Poor	Poor
Resistance to abrasion	Fair	Fair	Fair	Poor	Nil	Good	Poor	Good
Resistance to vibration	Fair	Good	Poor	Good	Poor	Fair	Fair	Good
Stability: shrink, shock	Good	Good	Fair	Excellent	Excellent	Good	Good	Good
Combustible	No	Yes	No	Yes	No	No	Yes	Yes

1. The descriptive terms in this table are only approximations, since any insulation and the conditions under which it is used will vary.
2. The thermal conductivities for these materials and for urethane foam used in the trench are for dry material. Since foam may absorb groundwater, and water vapor may eventually enter the interstices between powder particles, the ultimate thermal conductivity values are difficult to estimate.

Table 2 Soil Conductivites

Soil Moisture Content, mass	Thermal Conductivity, Btu/h · ft · °F		
	Sand	Silt	Clay
Low, <4%	0.17	0.08	0.08
Medium, 4 to 20%	1.08	0.75	0.58
High, >20%	1.25	1.25	1.25

and a 6-in., uninsulated, 55°F chilled water return in 60°F earth with 10 Btu/h·ft gain. The former requires analysis to determine the required insulation and its effect on the total heating system; the latter suggests analysis and insulation needs might be minimal. Other factors that affect heat transfer are: (1) depth of burial, related to the earth temperature and soil thermal resistance; (2) soil conductivity, related to soil moisture content and density; and (3) distance between adjacent pipes.

To compute transient heat gains or losses in underground piping systems, numerical methods that approximate any physical problem and include such factors as the effect of temperature on thermal properties must be used. For further information on numerical methods, refer to Albert and Phetteplace (1983), Rao (1982), and Minkowycz *et al.* (1988).

For most designs, numerical analyses may not be warranted, except where the potential exists to thermally damage something adjacent to the distribution system. Also, complex geometries probably require numerical analysis.

Steady-state calculations are appropriate for determining the annual heat loss/gain from a buried system if average annual earth temperatures are used. Steady-state calculations may also be appropriate for worst-case analyses of thermal effects on materials. Steady-state calculations for a one-pipe system can be done without a computer, but it becomes increasingly difficult for a two-, three-, or four-pipe system, unless a computer is used.

The following steady-state methods of analysis use resistance formulations developed by Phetteplace and Meyer (1990) that simplify the calculations needed to determine temperatures within the system. Each type of resistance is given a unique subscript and is defined only when introduced. In each case, the resistances are on a unit length basis so that heat flows per unit length result directly when the temperature difference is divided by the resistance. *Note:* For consistency and simplic-ity, all thermal conductivities are given in Btu/h·ft·°F with all dimensions in feet (not the more traditional Btu·in/h·ft²·°F).

Single Uninsulated Buried Pipe

For this case, an approximation for the soil thermal resistance has been used extensively. This approximation is sufficiently accurate (within 1%) for the specified range of the following pipe diameter and burial depth given below. Both the actual resistance and the approximate resistance are presented, along with the depth/radius criteria for each.

$$R_s = \frac{\ln[(d/r_o) + \{(d/r_o)^2 - 1\}^{1/2}]}{2\pi k_s} \quad \text{for } d/r_o > 1 \quad (1)$$

$$R_s = \frac{\ln[2d/r_o]}{2\pi k_s} \quad \text{for } d/r_o > 4 \quad (2)$$

where:

R_s = thermal resistance of soil, h·ft·°F/Btu
k_s = thermal conductivity of soil, Btu/h·ft·°F
d = burial depth to centerline of pipe, ft
r_o = outer radius of pipe or conduit, ft

Include the thermal resistance of the pipe if it is significant when compared to the soil resistance. The thermal resistance of a pipe or any concentric circular region is given by:

$$R_p = \frac{\ln[r_o/r_i]}{2\pi k_p} \quad (3)$$

where:

R_p = thermal resistance of pipe wall, h·ft·°F/Btu
k_p = thermal conductivity of pipe, Btu/h·ft·°F
r_i = inner radius of pipe, ft

Example 1. Consider an uninsulated, 3 in. Schedule 40 PVC chilled water supply line carrying 45°F water. Assume the pipe is buried 3 ft deep in soil with a thermal conductivity of 1 Btu/h·ft·°F and no other pipes or thermal anomalies are within close proximity. Assume the average annual soil temperature is 60°F.

r_i = 1.54 = 0.128 ft
r_o = 1.75 in. = 0.146 ft
d = 3 ft
k_s = 1 Btu/h·ft·°F
k_p = 0.10 Btu/h·ft·°F

Solution: Calculate the thermal resistance of the pipe using Equation (3):

$$R_p = 0.21 \ \text{h} \cdot \text{ft} \cdot {}^\circ\text{F/Btu}$$

Calculate the thermal resistance of the soil using Equation (2). [*Note*: $d/r_o = 21$ is greater than 4; thus Equation (2) may be used in lieu of Equation (1).]

$$R_s = 0.59 \ \text{h} \cdot \text{ft} \cdot {}^\circ\text{F/Btu}$$

Calculate the rate of heat transfer by dividing the overall temperature difference by the total thermal resistance:

$$q = \frac{t_f - t_s}{R_t} = \frac{(45 - 60)}{0.80 \ \text{h} \cdot \text{ft} \cdot {}^\circ\text{F/Btu}} = -19 \ \text{Btu/h} \cdot \text{ft}$$

where:

R_t = total thermal resistance, *i.e.*, the sum of all thermal resistances for pure series heat flow, $\text{h} \cdot \text{ft} \cdot {}^\circ\text{F/Btu}$
t_f = fluid temperature, °F
t_s = average annual soil temperature, °F
q = heat loss or gain per unit length of system, $\text{Btu/h} \cdot \text{ft}$

The negative result indicates a heat gain rather than a loss. Note that the thermal resistance of the fluid/pipe interface has been neglected, which is a reasonable assumption because such resistances tend to be very small for flowing fluids. Also note that in this case, the thermal resistance of the pipe comprises a significant portion of the total thermal resistance. This results from the relatively low thermal conductivity of PVC compared to other piping materials and the fact that no other major thermal resistances exist in the system to overshadow it. If any significant amount of insulation were included in the system, its thermal resistance would dominate, and it might be possible to neglect that of the piping material.

Single Buried Insulated Pipe

Equation (3) can be used to calculate the thermal resistance of any concentric circular region of material such as a pipe or insulation layer.

Example 2. Consider the effect of adding 1 in. of urethane foam insulation and a 1/8 in. thick PVC jacket to the chilled water line in the above example. Calculate the thermal resistance of the insulation layer from Equation (3) as follows:

$$R_i = \frac{\ln[0.229/0.146]}{2\pi \times 0.0125} = 5.75 \ \text{h} \cdot \text{ft} \cdot {}^\circ\text{F/Btu}$$

For the PVC jacket material, use Equation (3) again:

$$R_j = \frac{\ln[0.240/0.229]}{2\pi \times 0.10} = 0.07 \ \text{h} \cdot \text{ft} \cdot {}^\circ\text{F/Btu}$$

The thermal resistance of the soil as calculated by Equation (2) decreases slightly to $R_s = 0.51 \ \text{h} \cdot \text{ft} \cdot {}^\circ\text{F/Btu}$ because of the increase in the outer radius of the piping system. The total thermal resistance is now:

$$R_t = R_p + R_i + R_j + R_s = 0.21 + 5.75 + 0.07 + 0.51$$
$$= 6.54 \ \text{h} \cdot \text{ft} \cdot {}^\circ\text{F/Btu}$$

The heat gain by the chilled water pipe is reduced to about 2 $\text{Btu/h} \cdot \text{ft}$. In this case, the thermal resistance of the piping material and the jacket material could be neglected with a resultant error of <5%. Considering that the uncertainties in the material properties are likely greater than 5%, it is usually appropriate to neglect minor resistances such as those of piping and jacket materials if insulation is present.

Single Buried Pipe in Conduit with Air Space

Systems with air spaces may be treated by adding an appropriate resistance for the air space. For simplicity, assume a heat

transfer coefficient of 3 $\text{Btu/h} \cdot \text{ft}^2 \cdot {}^\circ\text{F}$ (based on the outer surface are of the insulation) applies in most cases. The resistance due to this heat transfer coefficient is then:

$$R_a = 1/(3 \times 2\pi \times r_{oi}) = 0.053/r_{oi} \tag{4}$$

where:

r_{oi} = outer radius of insulation, ft
R_a = resistance of air space, $\text{h} \cdot \text{ft} \cdot {}^\circ\text{F/Btu}$

A more precise value for the resistance of an air space can be developed with empirical relations available for convection in enclosures such as those given by Grober *et al.* (1961). The effect of radiation within the annulus should also be considered when high temperatures are expected within the air space. For the treatment of radiation, refer to Siegel and Howell (1981).

Example 3. Consider a 6 in. nominal diameter (6.625 in. outer diameter) high-temperature water line operating at 375°F. Assume the pipe is insulated with 2.5 in. of mineral wool with a thermal conductivity $k_i = 0.026 \ \text{Btu/h} \cdot \text{ft} \cdot {}^\circ\text{F}$ at 200°F and $k_i = 0.030 \ \text{Btu/h} \cdot \text{ft} \cdot {}^\circ\text{F}$ at 300°F.

The pipe will be encased in a steel conduit with a concentric air gap of 1 in. The steel conduit will be 0.125 in. thick and will have a corrosion-resistant coating approximately 0.125 in. thick. The pipe will be buried 4 ft deep to pipe centerline in soil with an average annual temperature of 60°F. The soil thermal conductivity is assumed to be 1 $\text{Btu/h} \cdot \text{ft} \cdot {}^\circ\text{F}$. The thermal resistances of the pipe, conduit, and conduit coating will be neglected.

Solution: Calculate the thermal resistance of the pipe insulation. To do so, assume a mean temperature of the insulation of 250°F to establish its thermal conductivity, which is equivalent to assuming the insulation outer surface temperature is 125°F. Interpolating the data listed previously, the insulation thermal conductivity $k_i = 0.028 \ \text{Btu/h} \cdot \text{ft} \cdot {}^\circ\text{F}$. Then calculate insulation thermal resistance using Equation (3) as:

$$R_i = \frac{\ln(0.484/0.276)}{2\pi \times 0.028} = 3.19 \ \text{h} \cdot \text{ft} \cdot {}^\circ\text{F/Btu}$$

Calculate the thermal resistance of the air space using Equation (4):

$$R_a = 0.053/0.484 = 0.11 \ \text{h} \cdot \text{ft} \, {}^\circ\text{F/Btu}$$

Calculate the thermal resistance of the soil from Equation (2):

$$R_s = \frac{\ln[8.0/0.589]}{2\pi \times 1} = 0.42 \ \text{h} \cdot \text{ft} \cdot {}^\circ\text{F/Btu}$$

The total thermal resistance is:

$$R_t = R_i + R_a + R_s = 3.19 + 0.11 + 0.42 = 3.72 \ \text{h} \cdot \text{ft} \cdot {}^\circ\text{F/Btu}$$

The first estimate of the heat loss is then:

$$q = (375 - 60)/3.72 = 84.7 \ \text{Btu/h} \cdot \text{ft}$$

Now repeat the calculation procedure with an improved estimate of the mean insulation temperature obtained and calculate the outer surface temperature as:

$$t_{io} = t_{po} - (q \times R_i) = 375 - (84.7 \times 3.19) = 105°F$$

where t_{po} = outer surface temperature of pipe, °F.

The new estimate of the mean insulation temperature is 240°F, which is close to the original estimate. Thus the insulation thermal conductivity changes only slightly, and the resulting thermal resistance is $R_i = 3.24 \ \text{h} \cdot \text{ft} \cdot {}^\circ\text{F/Btu}$. The other thermal resistances in the system remain unchanged, and the heat loss becomes $q = 83.8 \ \text{Btu/h} \cdot \text{ft}$. The insulation surface temperature is now approximately $t_{io} = 104 \ °F$, and no further calculations are needed.

Two Pipes Buried in Common Conduit with Air Space

Make the same assumption as in the previous section regarding heat transfer with the air space. For convenience, add some of the thermal resistances as follows:

$$R_1 = R_{p1} + R_{i1} + R_{a1} \tag{5}$$

$$R_2 = R_{p2} + R_{i2} + R_{a2} \tag{6}$$

The subscripts 1 and 2 differentiate between the two pipes within the conduit. The combined heat loss is then given by:

$$q = \frac{[(t_{f1} - t_s)/R_1] + [t_{f2} - t_s)/R_2]}{1 + (R_{cs}/R_1) + (R_{cs}/R_2)} \quad (7)$$

where R_{cs} = total thermal resistance of conduit shell and soil, h·ft·°F/Btu.

Once the combined heat flow is determined, calculate the bulk temperature within the air space from:

$$t_a = t_s + qR_{cs} \quad (8)$$

Then calculate the insulation outer surface temperature from:

$$t_{i1} = t_a + (t_{f1} - t_a)(R_{a1}/R_1) \quad (9)$$

$$t_{i2} = t_a + (t_{f2} - t_a)(R_{a2}/R_2) \quad (10)$$

The heat flows from each pipe are given by:

$$q_1 = (t_{f1} - t_a)/R_1 \quad (11)$$

$$q_2 = (t_{f2} - t_a)/R_2 \quad (12)$$

When the insulation thermal conductivity is a function of its temperature, as is usually the case, an iterative calculation is required as illustrated in the following example.

Example 4. A pair of 4 in. NPS medium temperature hot water supply and return lines run in a common 21-in. outside diameter conduit. Assume that the supply temperature is 325°F and the return temperature is 225°F. The supply pipe is insulated with 2.5 in. of mineral wool insulation and the return pipe has 2 in. of mineral wool insulation. This insulation has the same thermal properties as those given in the example for a single buried pipe in conduit with air space (Example 3). The pipe is buried 4 ft to centerline in soil with a thermal conductivity of 1 Btu/h·ft·°F. Assume the thermal resistance of the pipe, the conduit, and the conduit coating are negligible. As a first estimate, assume the bulk air temperature within the conduit is 100°F. In addition, use this temperature as a first estimate of the insulation surface temperatures to obtain the mean insulation temperatures and subsequent insulation thermal conductivities.

By interpolation, estimate the insulation thermal conductivities from the data given in Example 3 are 0.0265 for the supply pipe and 0.0245 for the return pipe. Calculate the thermal resistances using Equations (5) and (6):

$$R_1 = R_{i1} + R_{a1} = \ln(0.396/0.188)/(2\pi \times 0.0265) + 0.053/0.396$$
$$= 4.48 + 0.13 = 4.61 \text{ h·ft·°F/Btu}$$

$$R_2 = R_{i2} + R_{a2} = \ln(0.354/0.188)/2\pi \times 0.0245] + 0.053/0.354$$
$$= 4.11 + 0.15 = 4.26 \text{ h·ft·°F/Btu}$$

$$R_{cs} = R_s = \ln(8/0.875)/2\pi \times 1 = 0.352 \text{ h·ft·°F/Btu}$$

Calculate the first estimate of the combined heat flow from Equation (7) as:

$$q = \frac{[(325 - 60)/4.61] + [(225 - 60)/4.26]}{1 + (0.352/4.61) + (0.352/4.26)} = 83.0 \text{ Btu/h·ft}$$

Estimate the bulk air temperature within the conduit with Equation (8):

$$t_a = 60 + (83.0 \times 0.352) = 89.2°F$$

Then revise estimates of the insulation surface temperatures with Equations (9) and (10):

$$t_{i1} = 89.2 + (325 - 89.2)(0.13/4.61) = 95.8°F$$

$$t_{i2} = 89.2 + (225 - 89.2)(0.15/4.26) = 94.0°F$$

These insulation surface temperatures are close enough to the original estimate of 100°F that further iterations are not warranted. If the

individual supply and return heat losses are desired, calculate them using Equations (11) and (12).

Two Buried Pipes or Conduits

This case may be formulated in terms of the thermal resistances used for a single buried pipe or conduit and some correction factors. The correction factors needed are:

$$\theta_1 = (t_{p2} - t_s)/(t_{p1} - t_s) \quad (13)$$

$$\theta_2 = 1/\theta_1 = (t_{p1} - t_s)/(t_{p2} - t_s) \quad (14)$$

$$P_1 = \frac{1}{2\pi k_s} \ln\left(\frac{(d_1 + d_2)^2 + a^2}{(d_1 - d_2)^2 + a^2}\right)^{0.5} \quad (15)$$

$$P_2 = \frac{1}{2\pi k_s} \ln\left(\frac{(d_2 + d_1)^2 + a^2}{(d_2 - d_1)^2 + a^2}\right)^{0.5} \quad (16)$$

where a = horizontal separation distance between centerline of two pipes, ft.

And the thermal resistance for each pipe or conduit is given by:

$$R_{e1} = \frac{R_{t1} - (P_1^2/R_{t2})}{1 - (P_1\theta_1/R_{t2})} \quad (17)$$

$$R_{e2} = \frac{R_{t2} - (P_2^2/R_{t1})}{1 - (P_2\theta_2/R_{t1})} \quad (18)$$

where:

θ = temperature correction factor, dimensionless
P = geometric/material correction factor, h·ft·°F/Btu
R_e = effective thermal resistance of one pipe/conduit in two-pipe system, h·ft·°/Btu
R_t = total thermal resistance of one pipe/conduit if buried separately, h·ft·°F/Btu

Heat flow from each pipe is then calculated from:

$$q_1 = (t_{p1} - t_s)/R_{e1} \quad (19)$$

$$q_2 = (t_{p2} - t_s)/R_{e2} \quad (20)$$

Example 5. Consider buried supply and return lines for a low-temperature hot water system. The carrier pipes are 4 in. NPS (4.5 in. outer diameter) with 1.5 in. of urethane foam insulation. The insulation is protected by a 0.25 in. thick PVC jacket. The thermal conductivity of the insulation is 0.013 Btu/h·ft·°F and is assumed constant with respect to temperature. The pipes are buried 4 ft deep to the centerline in soil with a thermal conductivity of 1.0 Btu/h·ft·°F and a mean annual temperature of 60°F. The horizontal distance between the pipe centerlines is 2 ft. The supply water is at 250°F and the return water is at 150°F.

Neglect the thermal resistances of the carrier pipes and the PVC jacket.

First, calculate the resistances as if the pipes were independent of each other from Equations (2) and (3):

$$R_{i1} = R_{i2} = \frac{\ln(0.313/0.188)}{2\pi \times 0.013} = 6.25 \text{ h·ft·°F/Btu}$$

$$R_{s1} = R_{s2} = \frac{\ln(8.0/0.333)}{2\pi \times 1.0} = 0.51 \text{ h·ft·°F/Btu}$$

$$R_{t1} = R_{t2} = 6.25 + 0.51 = 6.76 \text{ h·ft·°F/Btu}$$

And the correction factors:

$$P_1 = P_2 = \frac{1}{2\pi \times 1} \ln\left(\frac{(4 + 4)^2 + 2^2}{(4 - 4)^2 + 2^2}\right)^{0.5} = 0.225 \text{ h·ft·°F/Btu}$$

$$\theta_1 = (150 - 60)/(250 - 60) = 0.474$$

$$\theta_2 = 1/\theta_1 = 2.11$$

Calculate the effective total thermal resistances as:

$$R_{e1} = \frac{6.76 - (0.225^2/6.76)}{1 - (0.225 \times 0.474/6.76)} = 6.87 \ \text{h} \cdot \text{ft} \cdot {}^\circ\text{F/Btu}$$

$$R_{e2} = \frac{6.76 - (0.225^2/6.76)}{1 - (0.225 \times 2.11/6.76)} = 7.32 \ \text{h} \cdot \text{ft} \cdot {}^\circ\text{F/Btu}$$

The heat flows are then:

$$q_1 = (250 - 60)/6.87 = 27.7 \ \text{Btu/h} \cdot \text{ft}$$

$$q_2 = (150 - 60)/7.32 = 12.3 \ \text{Btu/h} \cdot \text{ft}$$

$$q_t = 27.7 + 12.3 = 40.0 \ \text{Btu/h} \cdot \text{ft}$$

Note that when the resistances and geometry for the two pipes are identical, the total heat flow from the two pipes is the same if the temperature corrections are used or if they are set to unity. The individual losses will vary somewhat, however. These equations may also be used with air space systems. When the thermal conductivity of the pipe insulation is a function of temperature, iterative calculations must be run.

Pipes in Buried Trenches or Tunnels

Buried rectangular trenches or tunnels require several assumptions to obtain approximate solutions for the heat transfer. First, assume that the air within the tunnel or trench is uniform in temperature and that the same is true for the inside surface of the trench/tunnel walls. Field measurements on operating shallow trenches (Phetteplace et al. 1991) indicate maximum spatial air temperature variations of about 10°F. Air temperature variations of this magnitude within a tunnel or trench will not cause significant errors for systems with normal operating temperatures when using the following calculation methods.

Unless numerical methods are used, an approximation must be made for the resistance of a rectangular region such as the walls of a trench or tunnel. One procedure is to assume linear heat flow through the trench or tunnel walls, which yields the following resistance for the walls (Phetteplace et al. 1981):

$$R_w = x_w/[2k_w(a + b)] \tag{21}$$

where:

R_w = thermal resistance of trench/tunnel walls, $\text{h} \cdot \text{ft} \cdot {}^\circ\text{F/Btu}$
x_w = thickness of trench/tunnel walls, ft
a = width of trench/tunnel inside, ft
b = height of trench/tunnel inside, ft
k_w = thermal conductivity of trench/tunnel wall material, $\text{Btu/h} \cdot \text{ft} \cdot {}^\circ\text{F}$

As an alternative to Equation (21), the thickness of the trench/tunnel walls may be included in the soil burial depth. This approximation is only acceptable when the thermal conductivity of the trench/tunnel wall material is similar to that of the soil.

The thermal resistance of the soil surrounding the buried trench/tunnel is calculated using the following equation (Rohsenow and Hartnett 1973):

$$R_{ts} = \frac{\ln[3.5d/(b_o^{0.75} \ a_o^{0.25})]}{k_s[(a_o/2b_o) + 5.7]} \quad a_o > b_o \tag{22}$$

where:

R_{ts} = thermal resistance of soil surrounding trench/tunnel, $\text{h} \cdot \text{ft} \cdot {}^\circ\text{F/Btu}$
a_o = width of trench/tunnel outside, ft
b_o = height of trench/tunnel outside, ft
d = burial depth of trench to centerline, ft

Equations (21) and (22) can be combined with the equations already presented to calculate approximations for the heat flows and temperatures for trenches/tunnels. As with the conduits described in earlier examples, the heat transfer processes within the air space inside the trench/tunnel are too complex to warrant a complete treatment for design purposes. The thermal resistance of this air space may be approximated by several methods. For example, Equation (4) may be used to calculate an approximate resistance for the air space.

Alternately, heat transfer coefficients could be calculated using Equations (4) and (5), Chapter 22 of the 1989 ASHRAE *Handbook—Fundamentals*. Thermal resistances at the pipe insulation/air interface would be calculated from these heat transfer coefficients as done in the section Pipes in Air. If the thermal resistance of the air/trench wall interface is also included, use Equation (23):

$$R_{aw} = 1/[2h_t(a + b)] \tag{23}$$

where:

R_{aw} = thermal resistance of air/trench wall interface, $\text{h} \cdot \text{ft} \cdot {}^\circ\text{F/Btu}$
h_t = total heat transfer coefficient at air/trench wall interface, $\text{Btu/h} \cdot \text{ft}^2 \cdot {}^\circ\text{F}$

The total heat loss from the trench/tunnel is calculated from the following relationship:

$$q = \frac{[(t_{p1} - t_s)/R_1] + [(t_{p2} - t_s)/R_2]}{1 + (R_{ss}/R_1) + (R_{ss}/R_2)} \tag{24}$$

where:

R_1, R_2 = thermal resistances of two-pipe/insulation systems within trench/tunnel, $\text{h} \cdot \text{ft} \cdot {}^\circ\text{F/Btu}$
R_{ss} = total thermal resistance on soil side of air within trench/tunnel, $\text{h} \cdot \text{ft} \cdot {}^\circ\text{F/Btu}$

Once the total heat loss has been found, the air temperature within the trench/tunnel may be found as:

$$t_{ta} = t_s + qR_{ss} \tag{25}$$

where T_{ta} = air temperature within trench/tunnel, °F.

The individual heat flows for the two pipes within the trench/tunnel are then:

$$q_1 = (t_{p1} - t_{ta})/R_1 \tag{26}$$

$$q_2 = (t_{p2} - t_{ta})/R_2 \tag{27}$$

If the thermal conductivity of the pipe insulation is a function of temperature, assume an air temperature for the air space before starting calculations. Iterate the calculations if the air temperature calculated with Equation (25) differs significantly from the initial assumption.

Example 6. The walls of a buried trench are 6 in. thick and the trench is 3 ft wide and 2 ft tall. The trench is constructed of concrete, with a thermal conductivity of $k_w = 1 \ \text{Btu/h} \cdot \text{ft} \cdot {}^\circ\text{F}$. The soil surrounding the trench also has a thermal conductivity of $k_s = 1 \ \text{Btu/h} \cdot \text{ft} \cdot {}^\circ\text{F}$. The centerline of the trench is 4 ft below grade, and the soil temperature is assumed be 60°F. The trench contains supply and return lines for a medium temperature water system with the physical and operating parameters identical to those of Example 4.

Solution: Assuming the air temperature within the trench is 100°F, the thermal resistances for the pipe/insulations systems are identical to those of Example 4, or:

$$R_1 = 4.61 \ \text{h} \cdot \text{ft} \cdot {}^\circ\text{F/Btu}$$

$$R_2 = 4.26 \ \text{h} \cdot \text{ft} \cdot {}^\circ\text{F/Btu}$$

Solution: The thermal resistance of the soil surrounding the trench is given by Equation (22):

$$R_{ts} = \frac{\ln[14/(3^{0.75} \times 4^{0.25})]}{1[(4/6) + 5.7]} = 0.231 \ \text{h} \cdot \text{ft} \cdot {}^\circ\text{F/Btu}$$

The thermal resistance of the trench walls is calculated using Equation (21):

$$R_w = 0.5/[2(3 + 2)] = 0.050 \ \text{h} \cdot \text{ft} \cdot {}^\circ\text{F/Btu}$$

If the thermal resistance of the air/trench wall is neglected, the total thermal resistance on the soil side of the air space is:

$$R_{ss} = R_w + R_{ts} = 0.050 + 0.231 = 0.281 \ \text{h} \cdot \text{ft} \cdot {}^\circ\text{F/Btu}$$

Find a first estimate of the total heat loss using Equation (24):

$$q = \frac{[(325 - 60)/4.61] + [(225 - 60)/4.26]}{1 + [0.281/4.61] + [0.281/4.26]} = 85.4 \ \text{Btu/h} \cdot \text{ft}$$

The first estimate of the air temperature within the trench is given by Equation (25):

$$t_{ta} = 60 + (85.4 \times 0.281) = 84.0{}^\circ\text{F}$$

Refined estimates of the pipe insulation surface temperatures are then calculated using Equations (9) and (10):

$$t_{i1} = 84.0 + [(325 - 84.0)(0.13/4.61)] = 90.8{}^\circ\text{F}$$

$$t_{i2} = 84.0 + [(225 - 84.0)(0.15/4.26)] = 89.0{}^\circ\text{F}$$

From these estimates, calculate the revised mean insulation temperatures to find the resultant resistance values. Repeat the calculation procedure until satisfactory agreement between successive estimates of the trench air temperature are obtained. Calculate the individual heat flows from the pipes with Equations (26) and (27).

If the thermal resistance of the trench walls are added to the soil thermal resistance, the thermal resistance on the soil side of the air space is:

$$R_{ss} = \frac{\ln[14/(2^{0.75} \times 3^{0.25})]}{1[(3/4) + 5.7]} = 0.286 \ \text{h} \cdot \text{ft} \cdot {}^\circ\text{F/Btu}$$

The result is less than 2% higher than the resistance previously calculated by treating the trench walls and soil separately. In the event that the soil and trench wall material have significantly different thermal conductivities, this simpler calculation will not yield as favorable results and should not be used.

Pipes in Shallow Trenches

The cover of a shallow trench is exposed to the environment. Thermal calculations for such trenches require the following assumptions: (1) the interior air temperature is uniform as discussed in the previous section on trenches/tunnels, and (2) the soil and the trench wall material have the same thermal conductivity. This assumption will yield reasonable results if the thermal conductivity of the trench material is used, since most of the heat flows directly through the cover. The thermal resistance of the trench walls and surrounding soil are usually a small portion of the total thermal resistance and thus the heat losses are not usually highly dependent on this thermal resistance. Using these assumptions, Equations (22) and (24) through (27) may be used for shallow trench systems.

Example 7. Consider a shallow trench having the same physical parameters and operating conditions as the buried trench in the previous section. However, the top of the trench is at grade level. Calculate the thermal resistance of the shallow trench using Equation (22) as:

$$R_{ts} = R_{ss} = \frac{\ln[(3.5 \times 1.5)/(2^{0.75} \times 3^{0.25})]}{1[(3/4) + 5.7]} = 0.134 \ \text{h} \cdot \text{ft} \cdot {}^\circ\text{F/Btu}$$

Use thermal resistances for the pipe/insulation systems from the previous example, and use Equation (24) to calculate q as:

$$q = \frac{[(325 - 60)/4.61] + [(225 - 60)/4.26]}{1 + [0.134/4.61] + [0.134/4.26]} = 90.7 \ \text{Btu/h} \cdot \text{ft}$$

From this, calculate the first estimate of the air temperature, using Equation (25):

$$t_{ta} = 60 + (90.7 \times 0.134) = 72.2{}^\circ\text{F}$$

Then refine estimates of the pipe insulation surface temperatures using Equations (9) and (10):

$$t_{i1} = 72.2 + [(325 - 72.2)(0.13/4.61)] = 79.3{}^\circ\text{F}$$

$$t_{i2} = 72.2 + [(225 - 72.2)(0.15/4.26)] = 77.6{}^\circ\text{F}$$

From these, calculate the revised mean insulation temperatures to find resultant resistance values. Repeat the calculation procedure until satisfactory agreement between successive estimates of the trench air temperatures are obtained. If the individual heat flows from the pipes are desired, calculate them using Equations (26) and (27).

Another method for calculating the heat losses in a shallow trench assumes an interior air temperature and treats the pipes as pipes in air (see the following section). Interior air temperatures in the range of 70 to 120°F have been observed in a temperate climate (Phetteplace *et al.* 1991).

Buried Pipes with Other Geometries

Other geometries have been used for buried thermal utilities that are not specifically addressed by the previous cases. In some instances, the equations presented previously may be used to approximate the system. For instance, the soil thermal resistance for a buried system with a half-round clay tile on a concrete base could be approximated as a circular system using Equation (1) or (2). In this case, the outer radius r_o is taken as that of a cylinder with the same circumference as the outer perimeter of the clay tile system. The remainder of the resistances and subsequent calculations would be similar to those for a buried trench/tunnel. The accuracy of such calculations will vary inversely with the portion that the thermal resistance in question comprises of the total thermal resistance. In most instances, the thermal resistance of the pipe insulation overshadows other resistances, and the errors induced by approximations in the other resistances are acceptable for design calculations.

Pipes in Air

Pipes surrounded by gases may transfer heat via conduction, convection, and/or thermal radiation. Heat transfer modes depend mainly on the surface temperatures and geometry of the system being considered. For air, conduction is usually dominated by the other modes and thus may be neglected. Equations for convective and radiative heat transfer coefficients are given in Equations (4) and (5) in Chapter 22 of the 1989 ASHRAE *Handbook—Fundamentals*. The thermal resistances for cylindrical systems can be found from these heat transfer coefficients using the equations that follow. Generally, the piping systems used for thermal distribution have sufficiently low surface temperatures to preclude any significant heat transfer by thermal radiation.

Example 8. Consider a 6-in. nominal (6.625 in. outside diameter), high-temperature hot water pipe operates in air at 375°F with 2.5 in. of mineral wool insulation. The surrounding air annually averages 60°F. The average annual wind speed is 4 mph. The insulation is covered with an aluminum jacket with an emissivity of 0.26. The thickness and thermal resistance of the jacket material are negligible. As the heat transfer coefficient at the outer surface of the insulation is a function of the temperature there, initial estimates of this temperature must be made. This temperature estimate is also required to determine an estimated mean insulation temperature. Assuming that the insulation surface is at 100°F as a first estimate, the mean insulation temperature is calculated as 238°F. Using the properties of mineral wool given in Example 3, the insulation thermal conductivity is k_i = 0.0275 Btu/h·ft·°F. Using Equation (3), the thermal resistance of the insulation is R_i = 3.25 h·ft·°F/Btu. Equation (4), Chapter 22 of the 1989 *ASHRAE Handbook—Fundamentals* gives the forced convective heat transfer coefficient at the surface of the insulation as:

$$
\begin{aligned}
h_{cv} &= 1.016 \, (1/d)^{0.2}(2/[t_{oi} + t_a])^{0.181} \, (t_{oi} - t_a)^{0.266} \, (1 + 1.277V)^{0.5} \\
&= 1.016 \, (1/11.625)^{0.2} \, (2/160)^{0.181} \, (40)^{0.266} \, (6.11)^{0.5} \\
&= 1.86 \, \text{Btu/h·ft}^2\text{·°F}
\end{aligned} \tag{28}
$$

where:

d = outer diameter of surface, in.
t_a = ambient air temperature, °F
V = wind speed, mph

The radiative heat transfer coefficient must be added to this convective heat transfer coefficient. Determine the radiative heat transfer coefficient from Equation (5), Chapter 22 of the 1989 *ASHRAE Handbook—Fundamentals* as:

$$
\begin{aligned}
h_{rad} &= 1.713 \times 10^{-9}\epsilon \, [(t_{oi} + 459.7)^4 - (t_a + 459.7)^4]/(t_{oi} - t_a) \\
&= 1.713 \times 10^{-9} \, (0.26)[(559.7)^4 - (519.7)^4]/(100 - 60) \\
&= 0.28 \, \text{Btu/h·ft}^2\text{·°F}
\end{aligned} \tag{29}
$$

where ϵ = emissivity of surface, dimensionless.

Add the convective and radiative coefficients to obtain a total surface heat transfer coefficient h_t of 2.14 Btu/h·ft²·°F. The equivalent thermal resistance of this heat transfer coefficient is calculated from the following:

$$
\begin{aligned}
R_{surf} &= 1/(2\pi r_{oi}h_t) \\
&= 1/(2\pi \times 0.484 \times 2.14) = 0.15 \, \text{h·ft·°F/Btu}
\end{aligned} \tag{30}
$$

With this, the total thermal resistance of the system becomes R_t = 3.40 h·ft·°F/Btu and the first estimate of the heat loss is q = 92.6 Btu/h·ft.

An improved estimate of the insulation surface temperature is t_{oi} = 375 − (92.6 × 3.25) = 74°F. From this, a new mean insulation temperature, insulation thermal resistance, and surface resistance can be calculated. The heat loss is then 90.0 Btu/h·ft and the insulation surface temperature is calculated as 77°F. These results are close enough to the previous results so that further iterations are not warranted.

Note that the contribution of thermal radiation to the heat transfer could have been omitted with negligible effect on the results. In fact, the entire surface resistance could have been neglected and the resulting heat loss would have increased by only about 4%.

In the above example, the convective heat transfer was forced. In cases with no wind, where the convection is free rather than forced, the radiative heat transfer is more significant, as is the total thermal resistance of the surface. However, in instances where the piping is well insulated, the thermal resistance of the insulation dominates and minor resistances can often be neglected with little resultant error. By neglecting resistances, a conservative result is obtained, *i.e.*, the heat transfer is overpredicted.

Economic Thickness for Pipe Insulation

A life cycle cost analysis may be run to determine the economic thickness of pipe insulation. Because the insulation thickness affects other parameters in some systems, each insulation thickness must be considered as a separate system. For example, a conduit system or one with a jacket around the insulation requires a larger conduit or jacket for greater insu-

lation thicknesses. The cost of the extra conduit or jacket material may exceed that of the additional insulation and is therefore usually included in the analysis. It is usually not necessary to include excavation, installation, and backfill costs in the analysis.

The life cycle cost of a system is the sum of the initial capital cost and the present worth of the subsequent cost of heat lost or gained over the life of the system. The initial capital cost needs only to include those costs that are affected by insulation thickness. Use Equation (31) to calculate the life cycle cost as:

$$
\text{LCC} = \text{CC} + (qt_uC_h\text{PWF}) \tag{31}
$$

where:

LCC = present worth of life cycle costs associated with pipe insulation thickness, $/ft
CC = capital costs associated with pipe insulation thickness, $/ft
q = annual average rate of heat loss, Btu/h·ft
t_u = utilization time for system each year, h
C_h = cost of heat lost from system, $/Btu
PWF = present worth factor for future annual heat loss costs, dimensionless

If heat costs are expected to escalate, the present worth factor may be multiplied by an appropriate escalation factor and the result in place of the present worth factor in Equation (31).

Example 9. Consider a steel conduit system with an air space. The insulation is mineral wool with thermal conductivity as given in Examples 3 and 4. The carrier pipe is 4 in. NPS and operates at 350°F for the entire year (8760 h). The conduit is buried 4 ft to the centerline in soil with a thermal conductivity of 1 Btu/h·ft·°F and an annual average temperature of 60°F. Neglect the thermal resistance of the conduit and carrier pipe. The useful lifetime of the system is assumed to be 20 years and the interest rate is taken as 10%. The present worth factor is the reciprocal of the capital recovery factor and is found from Equation (32).

$$
\begin{aligned}
\text{CRF} &= i(1 + i)^n/[(1 + i)^n - 1] \\
&= 0.10(1 + 0.10)^{20}/[(1 + 0.10)^{20} - 1] \\
&= 0.11746
\end{aligned} \tag{32}
$$

The value of heat lost from the system is assumed to be $10 per million Btu. Table 3 summarizes the heat loss and cost data for several available insulation thicknesses.

Table 3 Heat Loss and Cost Data

Insulation thickness, in.	1.5	2.0	2.5	3.0	3.5	4.0
Insulation outer radius, ft	0.313	0.354	0.396	0.438	0.479	0.521
Insulation thermal conductivity, Btu/h·ft·°F	0.027	0.027	0.027	0.027	0.027	0.027
R_i, Equation (3), h·ft·°F/Btu	3.00	3.73	4.39	4.99	5.52	6.00
R_a, Equation (4), h·ft·°F/Btu	0.17	0.15	0.13	0.12	0.11	0.10
Conduit outer radius, ft	0.448	0.448	0.531	0.531	0.583	0.667
R_s, Equation (2), h·ft·°F/Btu	0.46	0.46	0.43	0.43	0.42	0.40
R_t, Total thermal resistance, h·ft·°F/Btu	3.63	4.34	4.95	5.54	6.04	6.50
q, Heat loss rate, Btu/h·ft	79.9	66.8	58.6	52.3	48.0	44.6
Conduit system cost, $/ft	23.00	24.50	28.25	30.00	33.00	40.00
Life cycle cost, Equation (31), $/ft	82.59	74.32	71.95	69.00	68.80	73.27

Table 3 indicates that 3.5 in. of insulation yields the lowest life cycle cost for the system considered in the example. Note that the results depend highly on the economic parameters used. Thus, take care when determining these parameters.

EXPANSION PROVISIONS

All piping systems move due to thermal expansion and/or contraction. Field conditions determine the method used to absorb movement. Where the pipeline routing changes direction frequently, use these turns to build flexibility in the system. When the distance between the changes in direction is long, position expansion loops at key locations. If loops are required, obtain a right of way. This method of absorbing pipe movement is called *natural flexibility*. If field conditions do not allow the use of expansion loops or changes in direction, then use mechanical methods such as expansion joints or ball joints. In addition to field conditions, consider items such as material and installation cost, life of the system, and pressure losses when selecting the method of absorbing the pipe movement.

When the pipe movement absorption method is selected, other criteria must be considered, such as design and location of pipe supports and anchors.

In order to analyze each method of absorption, consider each routing shown in Figure 17 with the following data:

Pipe size = 8 in. Schedule 40, black steel GR B-A-53 ERW
Operation pressure = 300 psig
Operation temperature = 400°F
Ambient temperature = 40°F

For Routing A, two different types of supports are required. The dotted line indicates that an expansion support is required for length A. The expansion support allows pipe movement in longitudinal and lateral directions. For the length indicated between the anchor and the dotted line, standard supports, which allow pipe movement in the longitudinal direction only, are used.

The length of A is calculated using the formulas indicated in the ASME *Code for Pressure Piping* B 31.1 and a finite element program. Note that as the A dimension increases, the thermal stress of the system decreases. For simple configurations as indicated in Routing A, charts are also available for determining the A dimension. Clearance for pipe movement must be considered for both a direct buried system or a concrete trench. Maximum lateral movement occurs at the elbow, with no lateral movement at the first standard support. The amount of lateral movement at any given point in the A dimension can be approximated using the proportional formula:

$$X/A = X_1/A_1$$

where:

X = total amount of movement
A = distance between elbow and first standard support
X_1 = amount of movement at point A
A_1 = any given point between elbow and standard support

The forces at the anchors are calculated by the finite element method. For this configuration, the force shown at each anchor

is 1274 lbs. Anchor reaction forces for this type of system are much less than those for a system that contains expansion joints.

For Routing B, two different supports are required. The first type allows only longitudinal pipe movement. This support can consist of a roller or sliding support. The second type is a guide or alignment support. When using an expansion joint, the pipe movement must be directed into the expansion joint. The first pipe guide is located 4 pipe diameters from the expansion joint; the second guide is 14 pipe diameters from the first guide. Depending on the pressure of service lines, additional guides may be required. If an offset is located between the expansion joint and the anchor, position guides before and after the offset. The guides and pipeline must be able to withstand the moment due to the forces generated by the expansion joint.

Expansion joint anchors must withstand forces due to thermal expansion (spring rate) and the operating pressure of the system. The spring rate force equals the force required to compress the expansion joint. Calculate force due to pressure by multiplying the effective cross-sectional area of the expansion joint by the operating pressure of the system. In most cases, the force due to pressure is much greater than the spring rate. Because expansion joints must be accessible for maintenance, position one anchor within four pipe diameters from the joint, which is located inside a manhole or building. By placing the anchor adjacent to the expansion joint, an expansion guide can be omitted. For the example shown, the forces (end loads) at the anchor are approximately 24,000 lbs due to pressure and 4400 lbs due to spring rate, for a total force of 28,400 lbs.

Space the supports so that the pipes do not sag between them and cause a low point in the system. Since a pipeline is at a slope minimum of 1 in. in 40 ft, this spacing is important. For a direct burial system, supports are usually located on 10-ft centers.

Cold springing is used whenever thermal expansion compensation is used. With natural flexibility (*i.e.*, Routing A), cold springing minimizes the clearance required for pipe movement only. The pipes are sprung 50% of the total amount of movement in the direction toward the anchor. Note that cold springing is not considered in calculating the stresses of the piping system. When expansion joints are used (*i.e.*, Routing B), cold springing positions the expansion joint in an extended position. However, the amount of extension is not equal to 50%, and the manufacturer should be contacted for the proper amount.

HYDRAULIC CONSIDERATIONS

Objectives of Hydraulic Design

Although the distribution of a thermal utility, such as hot water, encompasses many of the aspects of domestic tap water distribution, many dissimilarities exist and thus the design should not be approached in the same manner. Thermal utilities must supply sufficient energy at the appropriate temperature and pressure to meet consumer needs. Within the constraints imposed by the consumers end use and equipment, the required thermal energy can be delivered with various combinations of temperature and pressure. Computer-aided design methods are available for thermal piping networks (Reisman 1985, Rasmussen 1987, and COWIconsult 1985). The use of such methods allows the rapid evaluation of many alternate designs.

General steam system design can be found in Chapter 10, as well as in IDHA (1983). For water systems, consult Chapter 12 and IDHA (1983)

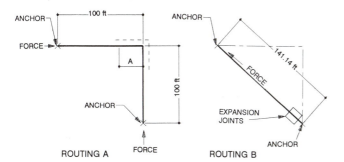

Fig. 17   Expansion Compensation Example

Water Hammer

Water hammer is used to describe several phenomena that occur in fluid flow. Although these phenomena differ in nature, they all result in stresses in the piping system that are higher than normally encountered. Water hammer can have disastrous effects on thermal utility systems, bursting pipes and fittings and threatening life and property.

In steam systems, water hammer is caused primarily by condensate collecting in the bottom of the steam piping. Steam flowing at velocities ten times greater than normal water flow will pick up a slug of condensate and accelerate it to a high velocity. The slug of condensate subsequently collides with the piping system wall at a point where flow changes direction. To prevent this type of water hammer, condensate must not be allowed to collect in steam pipes by using proper steam pipe pitch and adequate condensate collection and return facilities. Water hammer also occurs in steam systems due to rapid condensation of steam during system warm-up. Rapid condensation decreases specific volume and pressure, which precipitates pressure shock waves. This form of water hammer is prevented by controlled warm-up of the piping system.

Valves should be opened slowly and in stages during warm-up of the piping system. Provide large steam valves with smaller bypass valves to facilitate slow warm-up.

Water hammer in hot and chilled water distribution systems is caused by sudden changes in flow velocity and the subsequent pressure shock waves. The two primary causes are pump failure and sudden valve closures. Methods of analysis can be found in Stephenson (1981) and Fox (1977). Preventative measures include operational procedures and special piping fixtures such as surge columns.

Pressure Losses

Friction pressure losses occur between the inner wall of a pipe and a flowing fluid within the pipe due to shear stresses at the interface. In steam systems, these pressure losses are compensated for with increased steam pressure at the point of steam generation. In water systems, pumps are used to increase pressure at either the plant or intermediate points in the distribution system. The calculation of pressure losses is discussed in Chapter 2 of the 1989 ASHRAE *Handbook—Fundamentals*.

Pipe Sizing

The appropriate pipe size for each portion of a thermal distribution system should be determined from an economic study of the life cycle cost for construction and operation. However, in practice this is seldom done due to the effort involved. Instead, criteria that have evolved from practice are frequently used for design. These criteria normally take the form of constraints on the maximum flow velocity or pressure drop. Chapter 33 of the 1989 ASHRAE *Handbook—Fundamentals* provides velocity and pressure drop constraints. Noise generated by excessive flow velocities is usually not a concern for thermal utility distribution systems outside of buildings. For steam systems, maximum flow velocities of 200 to 250 ft/s are recommended (IDHA 1983). For water systems, Europeans use the criteria that pressure losses should be limited to 0.44 psi per 100 ft of pipe (Bohm 1988). Recent studies indicate that higher levels of pressure loss may be acceptable (Stewart 1986) and warranted from an economic standpoint (Bohm 1986, Koskelainen 1980, Phetteplace 1989).

When establishing design flows for thermal distribution systems, the diversity of consumer demands should be considered; *i.e.*, each consumer's maximum demand will not occur at the same time. Thus the heat supply and main distribution piping may be sized for a maximum load that is somewhat less than the sum of the individual consumer's maximum demands. For steam systems, Geiringer (1963) suggests diversity factors of 0.80 for space heating and 0.65 for domestic hot water heating and process loads. Geiringer also suggests that these factors may be reduced by approximately 10% for high-temperature water systems. Werner (1984) conducted a study of the heat load on six operating low- temperature hot water systems in Sweden and found diversity factors ranging from 0.57 to 0.79, with the average being 0.685.

Network Calculations

Calculating the flow rates and pressures in a piping network with branches, loops, pumps, and heat exchangers can be difficult without the aid of a computer. Methods have been developed primarily for domestic water distribution systems (Jeppson 1976, Stephenson 1981). These may apply to thermal distribution systems with appropriate modifications. Computer-aided design methods usually incorporate methods for hydraulic analysis as well as calculating heat losses and delivered water temperature at each consumer. Calculations are usually carried out in an iterative fashion, starting with constant supply and return temperatures throughout the network. After initial estimates of the design flow rates and heat losses are determined, refined estimates of the actual supply temperature at each consumer are computed. Flow rates at each consumer are then adjusted to ensure that the load is met with the reduced supply temperature and the calculations are repeated.

Condensate Drainage and Return

Condensate forms in operating steam lines as a result of heat loss. When the operating temperature of a steam system is increased, condensate will also form as steam warms the piping system material. At system start-up, these loads usually exceed any operating heat loss loads; thus special provisions should be provided.

To drain the condensate, steam piping should slope toward a collection point called a drip station. Drip stations are located within manholes or buildings where they are accessible for maintenance. Steam piping should slope toward the drip station at least 1 in. in 40 ft. If possible, the steam pipe should slope in the same direction as steam flow. If it is not possible to slope the steam pipe in the direction of steam flow, increase the pipe size to at least one size greater than that which would normally be used. This will reduce the flow velocity of the steam and provide better condensate drainage against the steam flow. Drip stations should be spaced no further than 500 ft apart in the absence of other requirements.

Drip stations consist of a short piece of pipe (called a *drip leg*) positioned vertically on the bottom of the steam pipe as well as a steam trap and appurtenant piping. The drip leg should be the same diameter as the steam pipe. The length of the drip leg should provide a volume equal to 50% of the condensate load from system start-up for steam pipes of 4 in. diameter and larger and 25% of the start-up condensate load for smaller steam pipes (IDHA 1983). Steam traps should be sized to meet the normal load from operational heat losses only. Start-up loads should be accommodated by manual operation of the bypass valve.

Steam traps are used to separate the condensate and noncondensable gases from the steam. For drip stations on steam distribution piping, use inverted bucket- or bimetallic-type thermostatic traps. Take care to ensure that drip leg capacity is adequate when thermostatic traps are used, because they will always back up some condensate.

If it is to be returned, condensate leaving the steam trap flows into the condensate return system. Where steam pressure is sufficiently high, it may be used to force the condensate through the condensate return system. On low-pressure steam systems or systems where a large pressure exists between the drip stations and the ultimate destination of the condensate, condensate receivers and pumps must be provided.

Schedule 80 steel piping is acceptable for condensate lines. Fiberglass reinforced (FRP) piping is also a good choice for the condensate return system because it does not corrode. When nonmetallic piping is used, ensure that it is not subjected to temperature/pressure conditions that will exceed its capability. Because steam traps have the potential of failing in an open position, nonmetallic piping must be protected from live steam where its temperature/pressure would exceed the limitations of the piping. Nonmetallic piping should not be located so close to steam pipes that heat losses from the steam pipes could overheat it. Sizing of condensate return piping may be found in Chapter 33 of the 1989 ASHRAE *Handbook—Fundamentals*.

Valve Vaults and Entry Pits

Valve vaults allow a user to isolate problems to a small portion of the system rather than analyze the entire system. This feature is important if the underground distribution system cannot be maintained between valve vaults without excavation. The correct number of valve vaults is that which affords lowest life cycle cost of owning the system and that dictated by design parameters. Valve vaults provide a space in which to put valves, steam traps, carrier pipe drains, carrier pipe vents, casing vents and drains, condensate pumping units, condensate cooling devices, flash tanks, expansion joints, system ground water drains, electrical leak detection equipment and wiring, electrical isolation couplings, branch line isolation valves, carrier pipe isolation valves, and flow meters. Valve vaults allow for elevation changes in the distribution system piping while maintaining an acceptable slope on the system; they also allow the designer to better match the topography and avoid unreasonable and expensive burial depths.

Ponding Water

The most significant problem with valve vaults is that water ponds in them. When the heat and chilled water distribution system share the same valve vault, plastic chilled water lines often fail because ponded water heats the plastic to failure. Note that water will gather in the valve vault irrespective of climate; therefore, design to eliminate the water for the entire life of the underground distribution system. The most successful water removal systems are those that drain to sanitary or storm drainage systems; this technique is successful because it is affected very little by corrosion and it has no moving parts to fail. Backwater valves are recommended in case the drainage system backs up.

Duplex pumps with lead-lag controllers and a failure annunciation system are used when storm drains and sanitary drains are not accessible. Because pumps have a history of frequent failures, duplex pumps help eliminate short cycling and provide standby pumping capacity. Steam ejector pumps can be used only if the distribution system is never shut down, because the carrier pipe insulation can be severely damaged during even short outages. Use a labeled, lockable, dedicated electrical service for electric pumps. The circuit label should indicate what the circuit is used for; it should also warn of the damage that will occur if the circuit is deenergized. Electrical components in valve vaults have experienced accelerated corrosion in the high heat and high humidity of closed, unventilated vaults. The pump that works well at 50°F often performs poorly at 200°F and 100% rh. To resolve this problem, one approach specifies components that have demonstrated high reliability at 200°F and 100% rh with a damp-proof electrical service. The pump should have a corrosion-resistant shaft (when immersed in water) and impeller and have demonstrated 200,000 cycles of successful operation, including the electrical switching components, at the referenced temperature and humidity. The pump must also pass foreign matter; therefore, the requirement to pass a 3/8-in. ball should be specified.

Another method drains the valve vaults into a separate sump adjacent to the valve vault. Then the pumps are placed in this sump where it is cool and the environment is more nearly that of the typical sump pump environment. Combining methods may be necessary if maximum reliability is desired or future maintenance is questionable. The pump can discharge to either the sanitary or storm drainage system, or to a splash block near the valve vault. Water pumped to a splash block has a tendency to enter the valve vault; this will not be a significant problem if the valve vault construction joints have been sealed properly. Exercise extreme caution if the bottom of the valve vault has French drains. These drains work backward during high groundwater conditions, allowing groundwater to enter the valve vault and flood the insulation in the distribution system. Adequate ventilation of the valve vault is also important.

Crowding of Components

The valve vault must be laid out in three dimensions, considering standing room for the worker, wrench swings, the size of valve operators, deviation in the size of appurtenances between manufacturers, and all other deviations that the specifications allow with respect to any item placed in the valve vault. To achieve desired results, the valve vault layout must be shown to scale on the contract drawings.

High Humidity

High humidity occurs in valve vaults when they have no working ventilation system. A gravity ventilation system is often provided where cool air enters the valve vault and sinks to the bottom. At the bottom of the vault, the air is warmed and then travels to the top of the valve vault, where it exits. Some designers have used a closed top with a pipe that is elbowed down for the exiting air. However, the elbowed-down vent hood tends to trap the exiting air and prevent the ventilation system from working. Open structural grate tops are the most successful covers for ventilation purposes. Open grates allow rain to enter the vault; however, the techniques previously mentioned for the prevention of ponding water are sufficient to handle the rainwater. Open grates with sump basins have worked well in extremely cold climates and in warm climates. Some vaults have screened sides to allow free ventilation. In this design, the solid vault sides extend slightly above grade; then, a screened window is placed in the wall in at least two sides. The overall above-grade height may be only 18 in.

High Temperatures

The temperature rises in the valve vault when no systematic way is provided to remove heat losses from the distribution system. The ventilation rate is usually not sufficient to transport heat from the vault. Part of this heat transfers to the earth; however, an equilibrium temperature is reached that may be higher than desired. Ventilation techniques discussed in the High Humidity section can resolve the problem of high temperature if the heat loss from the distribution system is normal. Typical problems that greatly increase the amount of heat released include:

- A leak in the distribution system from a carrier pipe, leaking gaskets, leaking packings, or leaks from appurtenances

- Insulation that has deteriorated due to flooding or abuse
- Standing water in a vault that touches the distribution pipe
- Steam vented to the vault from partial flooding between valve vaults
- Vents from flash tanks
- Insulation that was removed during routine maintenance and not replaced

To prevent heat release in a new system, design a workable ventilation system. On existing valve vaults, ensure that the valve vault is ventilated properly, correct all leaks, and replace all insulation that was damaged or left off. Install commercially available insulation jackets that can easily be removed and reinstalled from fittings and valves. If flooding occurs between valve vaults, portions of the distribution system may have be to excavated and repaired or replaced. Vents that exhaust steam may have to be routed aboveground if the ventilation technique used will not handle the quantity of steam exhausted.

Deep Burial

When a valve vault is buried too deeply, the structure is exposed to groundwater pressures, entry and exit often become a safety problem, construction becomes more difficult, and the cost of the vault is greatly increased. Valve vault spacing should be no more than 500 ft (NAS 1975). If greater spacing is desired, use an accurate life cycle analysis to determine spacing. The most common way to limit burial depth is to place the valve vaults closer together. Steps in the distribution system slope are made in the valve vault; *i.e.*, the carrier pipes come into the valve vault at one elevation and leave at a different elevation. If the slope of the distribution system is changed to more nearly match the earth topography, the valve vaults will be shallower; however, this method is restricted by the allowable range of slope of the carrier pipes. In most systems, the slope between valve vaults can be reversed in a valve vault, but not out in the system. The minimum allowable slope for the carrier pipes is 1 in. in 40 ft. If the entire distribution system is buried too deeply, the designer must determine the minimum allowable burial depth of the system and survey the topography of the distribution system to determine where the minimum depth of burial will occur. All system elevations must be adjusted with respect to the minimum burial depth location.

Freezing Conditions

Failure of distribution systems due to water freezing in components is common. The designer must consider the coldest temperature that may occur at a site and not the 97.5 or 99% condition used in building design. Drain legs or vent legs that allow water to stagnate are usually the cause of system failure. Insulate appurtenances that contain water that can stagnate. Insulation should be on all items that can freeze and must be kept in good condition. Electrical heat tape and pipe-type heat tracing can be used under insulation. Circulation may be required in the chilled water system or it may have to be drained if not used in winter.

Safety and Access

Some of the working fluids used in underground distribution systems can cause severe bodily injury and death if accidentally released in a confined space such as a valve vault. The shallow

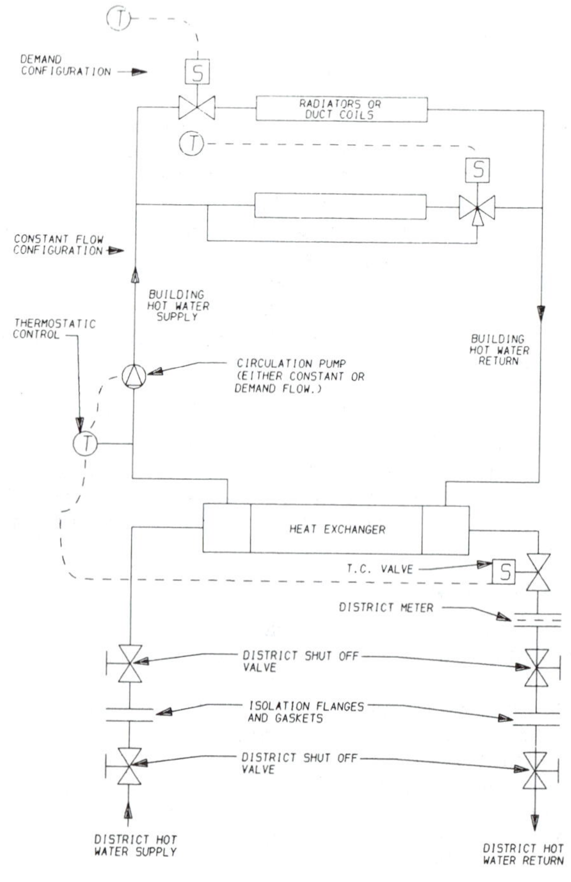

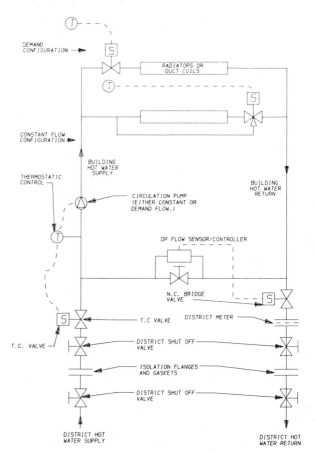

Fig. 18   District/Building Direct Interconnect Hot Water System

Fig. 19   District/Building Direct Interconnect Hot Water System

valve vault with large openings is desirable because it allows personnel to escape quickly in an emergency. The layout of the pipes and appurtenances must allow easy access for maintenance without requiring maintenance personnel to crawl underneath or between other pipes. The goal of the designer is to keep clear work spaces for maintenance personnel so that they can work efficiently and, if necessary, exit quickly. Engineering drawings must show pipe insulation thickness; otherwise they will give a false impression of the available space.

The location and type of ladder is important for safety and ease of egress. It is best to lay out the ladder and access openings when laying out the valve vault pipes and appurtenances as a method of exercising control over safety and ease of access. Ladder steps, when cast in the concrete valve vault walls, may corrode. Corrosion is most common in steel rungs; cast iron or prefabricated, OSHA-approved, galvanized steel ladders that sit on the valve vault floor and are anchored near the top to hold it into position are best. If the design uses lockable access doors, the locks must be operable from inside or have some "keyed-open" device that allows workers to take the key while working in the valve vault.

Valve Vault Construction

The most successful valve vaults are those constructed of cast-in-place reinforced concrete. These vaults conform to the earth excavation profile and show little movement when backfilled properly. Leakproof connections can be made with mating tunnels and conduit casings even though they may enter or leave at oblique angles. In contrast, prefabricated valve vaults may settle and move after construction is complete. Penetrations for prefabricated valve vaults, as well as the angles of

entry and exit, are difficult to locate exactly. As a result, much of the work associated with penetrations must be done in the field, which greatly lowers the quality and greatly increases the chances of a groundwater leak.

Construction deficiencies that go unnoticed in the buildings will destroy a heating and cooling distribution system; therefore, the designer must clearly pass on to the contractor that a valve bears no relation to a sanitary manhole. What is sufficient design for a sanitary manhole will assure premature failure of a heating and cooling distribution system.

CONSUMER INTERCONNECTS

General considerations. Chapter 33 of the 1989 ASHRAE *Handbook—Fundamentals* outlines procedures for sizing piping. All systems in the building must be electrically isolated from the distribution system to prevent electrolytic corrosion on both systems. An isolating flange set is used for this purpose.

Hot water system interconnects. Hot water system interconnects can be divided into systems that isolate the customer building system from the district system and systems that flow district hot water through the building. Figure 18 illustrates a typical interconnection using a heat exchanger between the district hot water system and the customer's system. It shows the radiator configurations typically used in both constant flow and demand flow systems. Figure 19 shows a typical direct connection between the district hot water system and the building system. It includes the typical configurations for both the demand flow and constant flow systems and the additional check valve and piping required for the constant flow system.

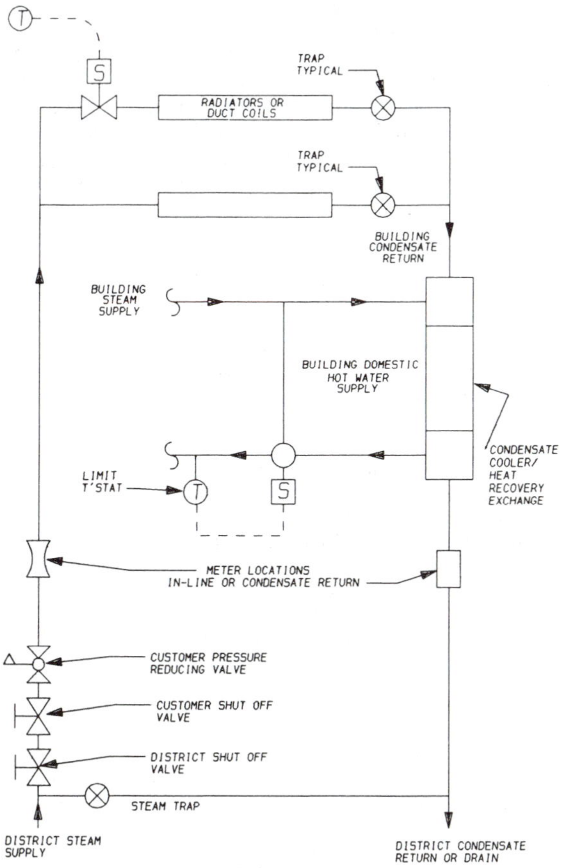

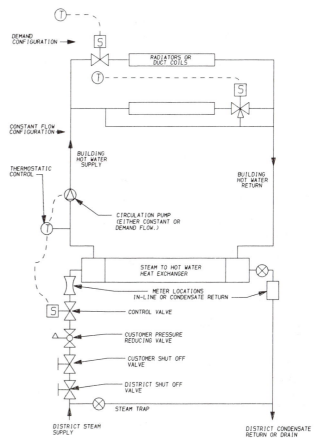

Fig. 20   District/Building Interconnect with Heat Exchange Steam System

Fig. 21   District/Building Interconnect with Heat Exchange Steam System

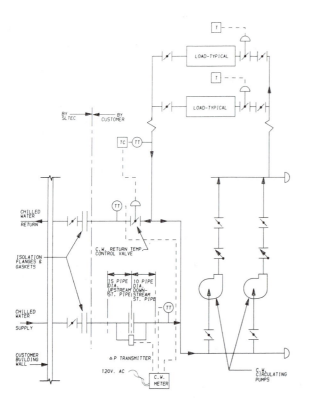

Fig. 22 Typical Chilled Water Piping and Metering Diagram

Line sizing on these systems must be done using the same design considerations used for the main feed lines of an in-building power plant system. In general, demand flow systems permit better energy transfer efficiency and smaller line sizing for a given energy transfer requirement.

Steam system interconnects. Most district heating systems supply saturated steam at pressures between 5 and 250 psig to customers' facilities, although higher pressures are sometimes used. The steam is pretreated to maintain a neutral pH, and the condensate is either cooled and removed by the building drainage system or returned back to the central plant for recycling. Many buildings run the condensate through a heat exchanger to heat the domestic hot water supply of the building prior to returning it to the central plant or to the building drains.

Interconnection between the district system and the building system is simple when the building extracts heat energy from steam at the radiators or the VAV box heat exchanger coils. A typical installation is shown in Figure 20. Chapter 10 gives more information on the building distribution piping, valving, traps, and other system requirements.

Some buildings use a hot water circulation system and extract heating energy from the district steam at a central (or zoned) heat exchanger. Figure 21 shows a typical example of this type of system interconnection.

Chilled Water System Interconnects

District chilled water systems can operate either as constant flow or demand-type systems. Demand systems can interconnect with either building demand or constant flow systems. Demand systems are best if dehumidification is required along with the cooling.

Figure 22 illustrates a typical configuration for a demand flow building system interconnection. The typical constant flow system found in older buildings can be converted to a demand-

type system by blocking off the bypass line around the air handler heat exchanger coil. This effectively converts a 3-way bypass-type valve to a 2-way modulating shutoff valve.

Peak demand requirements must be determined for the building at maximum design conditions. These conditions usually include direct bright sunlight on the building, 95 to 100°F dry-bulb temperature, and 73 to 78°F wet-bulb temperature occurring at peak conditions.

The designer must consider the effect of return water temperature control. As the load decreases, the supply water temperature tends to rise and, with a low load condition, the supply water temperature rises. As a result, no process or critical humidity control systems should be connected to a system where return water temperature control is used to achieve acceptable temperature differentials between the chilled water supply and return temperatures. Other techniques, such as separately pumping each chilled water coil, should be used where constant supply water temperatures are necessary year round.

METERING

Hot water and chilled water system metering are both accomplished by measuring the flow rate of the energy transfer medium and the temperature differential between the supply and return lines. Thermal transducers, resistance bulbs, or liquid expansion capillaries are usually used to measure the temperature of the water in the supply and return lines.

The water flow rate measurement can be done with either propeller or displacement meters. Displacement meters are more accurate than propeller meters, but are also larger. They can handle flow ranges from less than 2% up to 100% of the maximum rated flow with ± 1.0% accuracy. On some models, the water flow through the meter supplies the energy required to power the meter. Turbine-type meters require the smallest physical space for a given maximum flow rate. However, they require at least 10 diameters of straight pipe upstream and at least 6 diameters of straight pipe downstream of the turbine to ensure accurate readings. See IDHA (1969) and IDHA (1983) for further information on metering.

REFERENCES

Bohm, B. 1986. On the optimal temperature level in new district heating networks. *Fernwarme International* 15(5):301-306.

Bohm, B. 1988. Energy-economy of Danish district heating systems: A technical and economic analysis. Laboratory of Heating and Air Conditioning, Technical University of Denmark, Lyngby, Denmark.

Bottorf, J.D. 1951. Summary of thermal conductivity as a function of moisture content. Purdue University Thesis, West Lafayette, IN.

COWIconsult. 1985. Computerized planning and design of district heating networks. COWIconsult Consulting Engineers and Planners AS, Virum, Denmark.

Fox, J.A. 1977. *Hydraulic analysis of unsteady flow in pipe networks.* John Wiley and Sons, New York.

Geiringer, P.L. 1963. *High temperature water heating: Its theory and practice for district and space heating applications.* John Wiley and Sons, New York.

IDHA. 1969. Code for steam metering. International District Heating and Cooling Association (formerly International District Heating Association), Washington, D.C.

IDHA. 1983. *District heating handbook.* International District Heating and Cooling Association (formerly International District Heating Association), Washington, D.C.

Kersten, M.S. 1949. Laboratory research for the determination of the thermal properties of soils. ACFEL Technical Report 23.

Koskelainen, L. 1980. Optimal dimensioning of district heating networks. *Fernwarme International* 9(4):84-90.

Lunardini, V.J. 1981. *Heat transfer in cold climates.* Van Nostrand Reinhold, New York.

National Academy of Sciences. 1975. National Research Council Building Research Advisory Board (BRAB), Federal Construction Council, Technical Report No. 66.

Phetteplace, G.E. 1989. Simulation of district heating systems for piping design. International Symposium on District Heat Simulation, Reykjavik, Iceland.

Phetteplace, G. and V. Meyer. 1990. Piping for thermal distribution systems. CRREL Internal Report 1059, U.S. Army Cold Regions Research and Engineering Laboratory, Hanover, NH.

Phetteplace, G.E., D. Carbee, and M. Kryska. 1991. Field measurement of heat losses from three types of heat distribution systems. CRREL SR 9-19, U.S. Army Cold Regions Research and Engineering Laboratory, Hanover, NH.

Rasmussen, C.H. and J.E. Lund. 1987. Computer aided design of district heating systems. *District Heating Research and Technological Development in Denmark,* Danish Ministry of Energy, Copenhagen, Denmark.

Reisman, A.W. 1985. *District heating handbook: A handbook of district heating and cooling models.* International District Heating and Cooling Association, Washington, D.C.

Rohsenow, W.M. and J.P. Hartnett. 1973. *Handbook of heat transfer.* McGraw-Hill, Inc., New York.

Stephenson, D. 1981. *Pipeline design for water engineers.* Elsevier Scientific Publishing Co., New York.

Stewart, W.E. 1986. Water flow rate limitations. ASHRAE Research Project 450, Energy Research Laboratory, University of Missouri-Columbia/Kansas City, MO.

Werner, S.E. 1984. *The heat load in district heating systems.* Chalmers University of Technology, Goteborg, Sweden.

BIBLIOGRAPHY

Bowling, T. 1990. A technology assessment of potential telemetry technologies for district heating. IEA (International Energy Agency), Published by NOVEM, Sittard, The Netherlands.

Holtse, C. and P. Randlov, eds. 1989. District heating and cooling R&D project review. IEA (International Energy Agency), Published by NOVEM, Sittard, The Netherlands.

Morck, O. and T. Pedersen, eds. 1989. Advanced district heating production technologies. IEA (International Energy Agency), Published by NOVEM, Sittard, The Netherlands.

Sleimen, A.H. *et al.* 1990. Guidelines for converting building heating systems for hot water district heating. IEA (International Energy Agency), Published by NOVEM, Sittard, The Netherlands.

Ulseth, R., ed. 1990. Heat meters, report of research activities. IEA (International Energy Agency), Published by NOVEM, Sittard, The Netherlands.

HYDRONIC HEATING AND COOLING SYSTEM DESIGN

WATER systems that convey heat to or from a conditioned space or process with hot or chilled water are frequently called hydronic systems. The water flows through piping that connects a boiler, water heater, or chiller to suitable terminal heat transfer units located at the space or process.

Water systems can be classified by (1) temperature, (2) flow generation, (3) pressurization, (4) piping arrangement, and (5) pumping arrangement. Water systems can be either once-through or recirculating systems. This chapter describes forced recirculating systems.

Hydronic heating systems classified by flow generation are: (1) gravity systems, which use the difference in density between the supply and return water columns of a circuit or system to circulate water; and (2) forced systems, in which a pump, usually driven by an electric motor, maintains the flow. Gravity systems are seldom used today. (See the ASHVE *Heating, Ventilating and Air Conditioning Guide* issued prior to 1957 for detailed information on gravity systems.)

TEMPERATURE CLASSIFICATIONS

Water system classifications by operating temperature are:

Low-temperature water system (LTW). This hydronic heating system operates within the pressure and temperature limits of the ASME Boiler Code for low-pressure boilers. The maximum allowable working pressure for low-pressure boilers is 160 psig with a maximum temperature limitation of 250 °F. The usual maximum working pressure for boilers for LTW systems is 30 psi, although boilers specifically designed, tested, and stamped for higher pressures are frequently used. Steam-to-water or water-to-water heat exchangers are also used for heating low-temperature water.

Medium-temperature water system (MTW). This hydronic heating system operates at temperatures between 250 °F and 350 °F, with pressures not exceeding 160 psi. The usual design supply temperature is approximately 250 to 325 °F, with a usual pressure rating for boilers and equipment of 150 psi.

High-temperature water system (HTW). These hydronic heating systems operate at temperatures over 350 °F and usual pres-

sures of about 300 psi. The maximum design supply water temperature is usually about 400 °F, with a pressure rating for boilers and equipment of about 300 psi. The pressure-temperature rating of each component must be checked against the system's design characteristics.

Chilled water system (CW). A hydronic cooling system normally operates with a design supply water temperature of 40 to 55 °F, usually 44 or 45 °F, and within a pressure range of 120 psi. Antifreeze or brine solutions may be used for applications (usually process applications) that require temperatures below 40 °F or for coil freeze protection. Well water systems can use supply temperatures of 60 °F or higher.

Dual-temperature water system (DTW). This hydronic combination heating and cooling system circulates hot and/or chilled water through common piping and terminal heat transfer apparatus. These systems operate within the pressure and temperature limits of LTW systems, with usual winter design supply water temperatures of about 100 to 150 °F and summer supply water temperatures of 40 to 45 °F.

Low-temperature water systems are used in buildings ranging in size from small, single dwellings to very large and complex structures. Terminal heat transfer units include convectors, cast iron radiators, baseboard and commercial finned-tube units, fan-coil units, unit heaters, unit ventilators, central station air-handling units, radiant panels, and snow melting panels. A large storage tank may be included in the system to store energy to use when such heat input devices as the boiler or a solar energy collector are not supplying energy.

This chapter covers the principles and procedures for designing and selecting piping and components for low-temperature water, chilled water, and dual-temperature water systems.

PRINCIPLES

The design of effective and economical water systems is affected by complex relationships between the various system components; the design water temperature, flow rate, piping layout, pump selection, terminal unit selection, and control method are all interrelated. The size and complexity of the system determine the importance of these relationships in affecting the total system

---

The preparation of this chapter is assigned to TC 6.1, Hydronic and Steam Equipment and Systems.

operating success. Present hot water heating system design practice originated in residential heating applications, where a 20 °F temperature drop Δt was used to determine flow rate. Besides producing satisfactory operation and economy in small systems, this Δt enabled simple calculations because 1 gpm conveys 10,000 Btu/h. However, almost universal use of hydronic systems for both heating and cooling of large buildings and building complexes has rendered this simplified concept obsolete.

Closed Water Systems

Since most hot and chilled water systems are closed, this chapter addresses only these systems. The fundamental difference between a closed and an open water system is the interface of the water with a compressible gas (such as air) or an elastic surface (such as a diaphragm). A closed water system is defined as one with no more than one point of interface with a compressible gas or surface. This definition is fundamental to understanding the hydraulic dynamics of these systems. Earlier literature referred to a system with an open or vented expansion tank as an "open system." Such a system is actually a closed system. The atmospheric interface of the tank in that system simply establishes the system pressure.

An open system, on the other hand, has more than one such interface. As an example, a cooling tower system has at least two such points of interface: (1) the tower basin and (2) the discharge pipe or nozzles entering the tower. The difference between the hydraulics of the two systems becomes increasingly more evident as one analyzes the two systems, but one of the major differences is that certain hydraulic characteristics of open systems cannot occur in closed systems; *e.g.*, in a closed system (1) flow cannot be motivated by static head differences, (2) pumps do not provide static lift, and (3) the entire piping system is always filled with water.

Basic System

Figure 1 shows the fundamental components of a closed hydronic system. Actual systems generally have additional components such as valves, vents, regulators, etc., but they are not essential to the basic principles underlying the concept of the system.

These fundamental components are:

- Source system
- Load system
- Pump system
- Distribution system
- Expansion chamber

Conceptually, a hydronic system could operate with only these five components.

The components are subdivided into two groups—thermal components and hydraulic components. The thermal components consist of (1) the source, (2) the load, and (3) the expansion chamber. The hydraulic components consist of (1) the distribution sys-

tem, (2) the pump, and (3) the expansion chamber. Thus, the expansion chamber is the only component that serves both a thermal and a hydraulic function.

THERMAL COMPONENTS

LOADS

The load is the point where heat flows out of or into the system from the space or process; it is the independent variable to which the remainder of the system must respond. Outward heat flow characterizes a heating system and inward heat flow characterizes a cooling system. The quantity of heating or cooling is calculated by one of the following means.

Sensible heating or cooling. The quantity of heat entering or leaving the airstream is expressed by:

$$q = 60 \, Q_a \rho_a c_p \Delta t \qquad (1)$$

where:

q = heat transfer rate, Btu/h
Q_a = airflow rate, ft^3/min
ρ_a = density of air, lb/ft^3
c_p = specific heat of air, Btu/lb · °F
Δt = temperature increase or decrease of air, °F

For standard air with a density of 0.075 lb/ft^3 and a specific heat of 0.24 Btu/lb · °F, this equation becomes:

$$q = 1.08 \, Q_a \Delta t \qquad (2)$$

The heat exchanger or coil must then transfer this heat to the water. The quantity of sensible heat transferred to the heated or cooled medium in a specific heat exchanger is a function of the surface area; the mean temperature difference between the water and medium; and the overall heat transfer coefficient, which is a function of the fluid velocities, properties of the medium, geometry of the heat transfer surfaces, and other factors. It may be expressed by:

$$q = UA(\text{LMTD}) \qquad (3)$$

where:

q = heat transfer rate, Btu/h
U = overall coefficient of heat transfer, Btu/h·ft^2 · °F
A = surface area, ft^2
LMTD = logarithmic mean temperature difference, heated medium to water, °F

Cooling and dehumidification. The quantity of heat removed from the cooled medium when both sensible cooling and dehumidification are present is expressed by:

$$q_t = W\Delta h \qquad (4)$$

where:

q_t = total heat transfer rate, Btu/h
W = mass flow rate of cooled medium, lb/h
Δh = enthalpy difference between entering and leaving conditions of cooled medium, Btu/lb

Expressed for a cooling coil, this equation becomes:

$$q_t = 60 \, Q_a \rho_a \Delta h \qquad (5)$$

where:

Q_a = airflow rate, ft^3/min
ρ_a = density of air, lb/ft^3

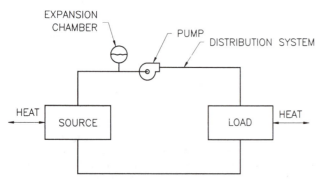

Fig. 1   Hydronic System—Fundamental Components

which, for standard air with a density of 0.075 lb/ft^3, reduces to

$$q_t = 4.5 \, Q_a \Delta h \tag{6}$$

Heat transferred to or from water. The quantity of heat transferred to or from the water is a function of the flow rate, the specific heat, and the temperature drop or rise of the water as it passes through the heat exchanger. The heat transferred to or from the water is expressed by:

$$q_w = \dot{m} c_p \Delta t \tag{7}$$

where:

q_w = heat transfer rate from water, Btu/h
$\dot{m}$ = mass flow of water, lb/h
c_p = specific heat of water, Btu/lb·°F
Δt = temperature increase or decrease across unit, °F

With water systems, it is common to express the flow rate in gallons per minute, in which case Equation (7) becomes:

$$q_w = 8.02 \, \rho_w c_p Q_w \Delta t \tag{8}$$

where:

Q_w = water flow rate, gal/min
ρ_w = density of water, lb/ft^3

For standard conditions in which the density is 62.4 lb/ft^3 and the specific heat is 1 Btu/lb·°F, the equation becomes:

$$q_w = 500 \, Q_w \Delta t \tag{9}$$

Equation (8) or (9) can be used to express the heat transfer across a single load or source device, or any quantity of such devices connected across a piping system. In the design or diagnosis of a system, the load side may be balanced with the source side by these equations.

Heat carrying capacity of piping. Equations (8) and (9) are also used to express the heat carrying capacity of the piping or distribution system, or of any portion thereof. In this regard, the temperature differential Δt, sometimes called the temperature range, is established or identified; for any flow rate through the piping, the q_w is called the *heat carrying capacity*.

Load systems can be any system in which heat is conveyed from the water for heating or to the water for cooling the space or process. Most load systems are basically a water-to-air finned-coil heat exchanger or a water-to-water exchanger. The specific configuration is usually used to describe the load device. The most common configurations include:

Heating load devices

    Preheat coils in central units
    Heating coils in central units
    Zone or central reheat coils
    Finned-tube radiation
    Baseboard radiation
    Convectors
    Unit heaters
    Fan-coil units
    Water-to-water heat exchangers
    Radiant heating panels
    Snow melting panels

Cooling load devices

    Coils in central units
    Fan-coil units
    Induction unit coils
    Radiant cooling panels
    Water-to-water heat exchangers

SOURCE

The source is the point where heat is added to (heating) or removed from (cooling) the system. Ideally, the amount of energy entering or leaving the sources equals the amount entering or leaving through the load system(s). Under steady-state conditions, the load energy and source energy are equal and opposite. Also, when properly measured or calculated, temperature differentials and flow rates across the source and loads are all equal. Equations (8) and (9) are used to express the source capacities as well as the load capacities.

Any device that can be used to heat or cool water under controlled conditions can be used as a source device. The most common source devices for heating and cooling systems are:

Heating source devices

    Hot water generator or boiler
    Steam-to-water heat exchanger
    Water-to-water heat exchanger
    Solar heating panels
    Heat recovery or salvage heat device, such as the water jacket of an internal combustion engine
    Exhaust gas heat exchanger
    Incinerator heat exchanger
    Heat pump condenser
    Air-to-water heat exchanger

Cooling source devices

    Electric compression chiller
    Thermal absorption chiller
    Heat pump evaporator
    Air-to-water heat exchanger
    Water-to-water heat exchanger

The two primary considerations in selecting a source device are the design capacity and the part-load capability, sometimes called the turndown ratio. The turndown ratio, expressed in percent of design capacity, is:

$$\text{Turndown Ratio} = 100 \, \frac{\text{Minimum Capacity}}{\text{Design Capacity}} \tag{10}$$

The reciprocal of the turndown ratio is sometimes used (for example, a turndown ratio of 25% may also be expressed as a turndown ratio of 4).

The turndown ratio has a significant effect on the successful performance of a system, and lack of consideration for this capability of the source system has been responsible for many systems that either do not function properly, or do so at the expense of excess energy consumption. The turndown ratio has a significant impact on the ultimate system design selection.

System temperatures. Design temperatures and temperature ranges are selected by consideration of the performance requirements and the economics of the components. As an example, for a cooling system that must maintain 50% rh at 75°F, the dewpoint temperature is 55°F, which sets the maximum return water temperature at something near 55°F (60°F maximum); on the other hand, the lowest practical temperature for refrigeration, considering the freezing point and economics, is about 40°F. This temperature spread then sets constraints for a chilled water system. For a heating system, the maximum hot water temperature is normally established by the ASME Low-Pressure Code as 250°F, and with space temperature requirements of little over 75°F, the actual operating supply temperatures and the temperature ranges are set by the design of the load devices. Most economic considerations relating to the distribution and pumping systems favor the use of the maximum possible temperature range Δt.

EXPANSION CHAMBER

The expansion chamber (also called an expansion or compression tank) serves both a thermal function and a hydraulic function. In its thermal function the tank provides a space into which the noncompressible liquid can expand or from which it can contract as the liquid undergoes volumetric changes with changes in temperature. To allow for this expansion or contraction, the expansion tank provides an interface point between the system fluid and a compressible gas. *Note*: By definition, a closed system can have only one such interface; thus a system designed to function as a closed system can have only one expansion chamber.

Expansion tanks are of three basic configurations: (1) an open tank *i.e.*, a tank open to the atmosphere; (2) a closed tank, which contains a captured volume of compressed air and water, with an air-water interface (sometimes called a plain steel tank); and (3) a diaphragm tank, in which a flexible membrane is inserted between the air and the water.

In the open tank and the plain steel tank, gases can enter the system water through the interface and can adversely affect the system performance. Thus, current design practice normally specifies diaphragm tanks.

Sizing the tank is the primary thermal consideration in incorporating a tank into a system. However, prior to sizing the tank, the control or elimination of air must be considered. The amount of air which will be absorbed and can be held in solution with the water is expressed by Henry's Equation:

$$x = p/H \qquad (11)$$

where:

x = solubility of air in water (% by volume)
p = absolute pressure, lb/in^2 absolute
H = Henry's Constant

Henry's Constant, however is constant only for a given temperature (Figure 2). Combining the data of Figure 2 with Equation (11) results in the solubility diagram of Figure 3. With that diagram, the solubility can be determined if the temperature and pressure are known.

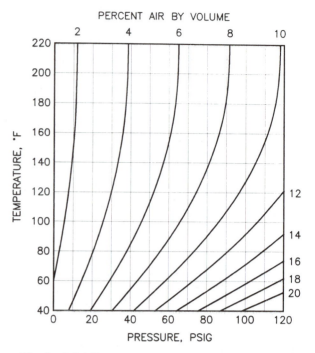

Fig. 3   Solubility versus Temperature and Pressure for Air/Water Solutions

If the water is not at its maximum solubility, it will, at the interface surface between air and water, absorb air until the point of maximum solubility has been reached. Once absorbed, the air will move throughout the body of water either by mass migration or by molecular diffusion until the water is uniformly saturated. If the air/water solution changes to a state that reduces solubility, the excess air will be released as a gas. As an example of this principle, if the air/water interface is at a high pressure point, the water will absorb air to its limit of solubility at that point; if at another point in the system the pressure is reduced, some of the dissolved air will be released.

In the design of systems with open or plain steel expansion tanks, it is common practice to utilize the tank as the major air control or release point in the system.

Equations for sizing the three common configurations of expansion tanks follow.

For closed tanks with air/water interface:

$$V_t = V_s \frac{[(v_2/v_1) - 1] - 3\alpha\Delta t}{(P_a/P_1) - (P_a/P_2)} \qquad (12)$$

For diaphragm tanks:

$$V_t = V_s \frac{[(v_2/v_1) - 1] - 3\alpha\Delta t}{1 - (P_1/P_2)} \qquad (13)$$

For open tanks with air/water interface:

$$V_t = 2\{V_s [(v_2/v_1) - 1] - 3\alpha\Delta t\} \qquad (14)$$

where:

V_t = volume of expansion tank, gal
V_s = volume of water in system, gal
t_1 = lower temperature, °F
t_2 = higher temperature, °F
P_a = atmospheric pressure (absolute), psia
P_1 = absolute pressure at lower temperature, psia
P_2 = absolute pressure at higher temperature, psia

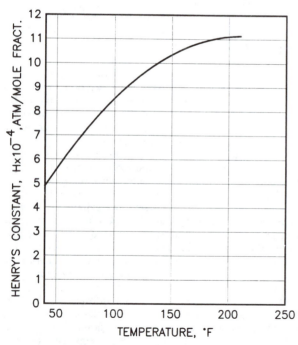

Fig. 2   Henry's Constant versus Temperature for Air and Water

v_1 = specific volume of water at lower temperature, ft^3/lb
v_2 = specific volume of water at higher temperature, ft^3/lb
α = linear coefficient of thermal expansion, in/in·°F
α = 6.5 × 10^{-6} in/in·°F for steel
α = 9.5 × 10^{-6} in/in·°F for copper
Δt = $(t_2 - t_1)$, °F

As an example, usually the lower temperature for a heating system is normal ambient temperature at fill conditions (*e.g.*, 50°F) and the higher temperature is the operating supply water temperature for the system. For a chilled water system, the lower temperature is the design chilled water supply temperature and the higher temperature is ambient temperature (*e.g.*, 95°F). For a hot-chilled system, the lower temperature is the chilled water design supply temperature and the higher temperature is the heating water design supply temperature.

The specific volume and saturation pressure of water at various temperatures are listed in Table 2, Chapter 6 of the 1989 ASHRAE *Handbook—Fundamentals*.

At the tank connection point, the pressure in closed tank systems increases as the water temperature increases. Pressures at the expansion tank are generally set by the following parameters.

- The lower pressure is usually selected to hold a positive pressure at the highest point in the system (usually about 10 psig).
- The higher pressure is normally set by the maximum pressure allowable at the location of the safety relief valve(s) without opening them.

Other considerations are to ensure (1) that the pressure at no point in the system will ever drop below the saturation pressure at the operating system temperature, and (2) that all pumps have adequate net positive suction head (NPSH) available to prevent cavitation.

Example 1. Size an expansion tank for a heating water system that will be operated at a design temperature range of 180 to 220°F. The minimum pressure at the tank is 10 psig (24.7 psia) and the maximum pressure is 25 psig (39.7 psia). (Atmospheric pressure is 14.7 psia.) The volume of water is 3000 gal. The piping is steel.

1. Calculate the required size for a closed tank with an air/water interface.

Solution: Lower temperature t_1, use 40°F
Higher temperature, t_2 = 220°F

From Table 2, Chapter 6 of the 1989 ASHRAE *Handbook—Fundamentals*:

v_1 (at 40°F) = 0.01602 ft^3/lb
v_2 (at 220°F) = 0.01677 ft^3/lb

Using Equation (12):

$$V_t = 3000 \times \frac{[(0.01677/0.01602) - 1] - 3(6.5 \times 10^{-6})(220 - 40)}{(14.7/24.7) - (14.7/39.7)}$$

V_t = 578 gal

2. If a diaphragm tank were to be used in lieu of the plain steel tank, what tank size in gallons would be required?

Solution: Using Equation (13),

$$V_t = 3000 \times \frac{[(0.01677/0.01602) - 1] - 3(6.5 \times 10^{-6})(220 - 40)}{1 - (24.7/39.7)}$$

V_t = 344 gal

HYDRAULIC COMPONENTS

DISTRIBUTION SYSTEM

The distribution system is the piping system connecting the various other components of the system. The primary considerations in designing this system are (1) sizing the piping to handle the heat ing or cooling capacity required and (2) arranging the piping to ensure flow in the quantities required at design conditions and at all other loads.

The flow requirement of the pipe is determined by Equation (8) or (9). After establishing Δt based on the thermal requirements, either of these equations (as applicable) can be used to determine the flow rate. First cost economics and energy consumption make it advisable to design for the greatest practical Δt, since the flow rate is inversely proportional to the Δt: *i.e.*, if the Δt doubles, the flow rate is reduced by half.

The three related variables in sizing the pipe are the flow rate, pipe size, and pressure drop. The primary consideration in selecting a design pressure drop is the relationship between the economics of first cost and energy costs.

Once the distribution system is designed, the pressure loss at design flow is calculated by the methods discussed in Chapter 33 of the 1989 ASHRAE *Handbook—Fundamentals*. The relationship between flow rate and pressure loss can be expressed by:

$$Q = C_v \sqrt{\Delta p} \qquad (15)$$

where:

Q = system flow rate, gal/min
Δp = pressure drop in system, lb/in^2
C_v = system constant (sometimes called valve coefficient)

Equation (15) may be modified as shown in Equation (16):

$$Q = C_s \sqrt{\Delta h} \qquad (16)$$

where:

Δh = system head loss, feet of fluid
C_s = system constant (C_s = 0.67 C_v for water with a density of 62.4 lb/ft^3)

Equations (15) and (16) are the system constant form of the Darcy-Weisbach equation. If the flow rate and head loss are known for a system, the system constant C_s can be calculated. From this calculation, the pressure loss can be determined at any other flow rate. Equation (16) can be graphed as a system curve (Figure 4).

The system curve changes if anything occurs that changes the flow-pressure drop characteristics. Examples of this are a strainer that starts to block or a control valve closing, either of which increases the head loss at any given flow rate, thus changing the system curve from curve A to curve B in Figure 4.

PUMP OR PUMPING SYSTEM

Centrifugal pumps are most commonly used in hydronic systems (see Chapter 39). Circulating pumps used in water systems can vary in size from small in-line circulators delivering 5 gpm at 6 or 7 ft head to base-mounted or vertical, pipe mounted pumps han-

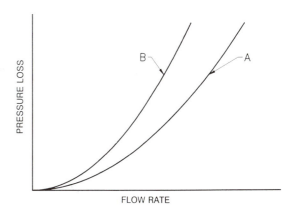

Fig. 4 Typical System Curves for Closed System

dling hundreds or thousands of gpm with pressures limited only by the characteristics of the system. Pump operating characteristics must be carefully matched to system operating requirements.

Pump Curves and Water Temperature

Performance characteristics of centrifugal pumps are described by pump curves, which plot flow versus head or pressure as well as efficiency and power information. The point at which a pump operates is described by the point at which the pump curve intersects the system curve (Figure 5).

A complete piping system follows the same water flow pressure-drop relationships as any component of the system [see Equation (16)]. Thus, the pressure required for any proposed flow rate through the system may be determined and a system curve constructed. A pump may be selected by using the calculated system pressure at the design flow rate as the base point value.

Figure 6 illustrates how a shift of the system curve to the right affects system flow rate. This shift can be caused by incorrectly calculating the system pressure drop by using arbitrary safety factors or overstated pressure drop charts. Variable system flow caused by control valve operation or improperly balanced systems (subcircuits having substantially lower pressure drops than the longest circuit) can also cause a shift to the right.

As described in Chapter 39, pumps for closed loop piping systems should have a flat pressure characteristic and should operate slightly to the left of the peak efficiency point on their curves. This characteristic permits the system curve to shift to the right without causing undesirable pump operation, overloading, or reduction in available pressure across circuits with large pressure drops.

Many dual-temperature systems are designed so that the chillers are bypassed during the winter months. The chiller pressure drop, which may be quite high, is thus eliminated from the system pressure drop, and the pump shift to the right may be quite large. For such systems, system curve analysis should be used to check winter operating points.

Operating points may be highly variable, depending on (1) load conditions, (2) the types of control valves used, and (3) the piping circuitry and heat transfer elements. The best selection generally will be:

1. For design flow rates, and made using pressure drop charts that illustrate actual closed loop hydronic system piping pressure drops.
2. To the left of the maximum efficiency point of the pump curve to allow shifts to the right caused by system circuit unbalance, direct-return circuitry applications, and modulating three-way valve applications.
3. A pump with a flat curve to compensate for unbalanced circuitry and to provide a minimum pressure differential increase across two-way control valves.

Parallel Pumping

When pumps are applied in parallel, each pump operates at the same head and provides its share of the system flow at that pressure (Figure 7). Generally, pumps of equal size are used, and the parallel pump curve is established by doubling the flow of the single pump curve.

Plotting a system curve across the parallel pump curve shows the operating points for both single and parallel pump operation (Figure 8). Note that single pump operation does not yield 50% flow. The system curve crosses the single pump curve considerably to the right of its operating point when both pumps are running. This factor leads to two important concerns: (1) the pumps must be powered to prevent overloading during single-pump operation and (2) a single-pump can provide standby service of up to 80% of design flow; the actual amount depending on the specific pump curve and system curve.

Series Pumping

When operated in series, each pump operates at the same flow rate and provides its share of the total pressure at that flow (Figure 9). A system curve plotted across the series pump curve shows the operating points for both single and series pump operation (Figure 10). Note that the single pump can provide up to 80% flow for standby and at a lower power requirement.

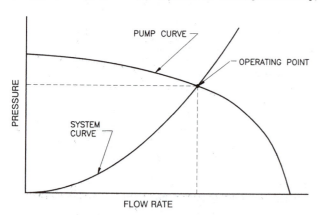

Fig. 5 Pump Curve and System Curve

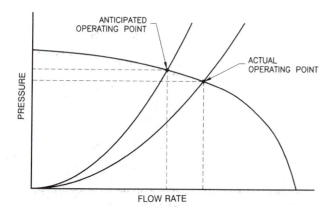

Fig. 6 Shift of System Curve Due to Circuit Unbalance

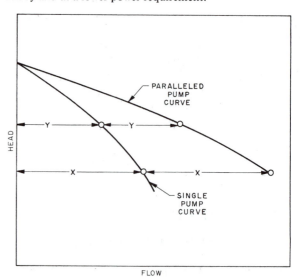

Fig. 7 Pump Curve for Parallel Operation

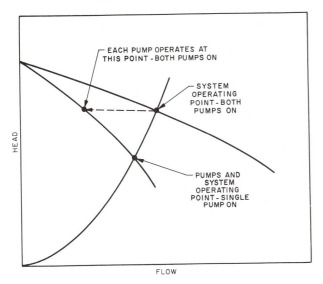

Fig. 8   Operating Conditions for Parallel Pump Installation

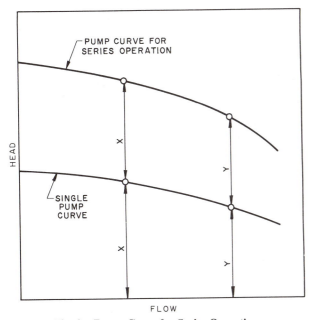

Fig. 9   Pump Curve for Series Operation

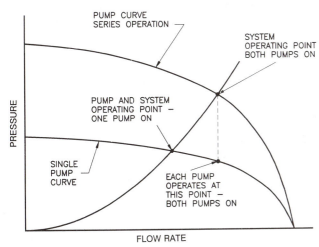

Fig. 10   Operating Conditions for Series Pumps Installation

Series pump installations are often used in heating and cooling systems so that both pumps operate during the cooling season to provide maximum flow and head while only a single pump operates during the heating season. Note that both parallel and series pump applications require that the actual pump operating points be used to accurately determine the actual pumping points. Adding artificial safety factor head, using improper pressure drop charts, or incorrectly calculating pressure drops may lead to an unwise selection.

Care must be taken in designing systems with multiple pumps to ensure that if pumps are ever operated in either parallel or series, such operation is fully understood and considered by the designer. Pumps performing unexpectedly in series or parallel have been the cause of performance problems in hydronic systems. Typical problems resulting from pumps of unequal size functioning in parallel and series are:

Parallel. Pumps of unequal pressures may result in one pump creating a pressure across the other pump in excess of its cutoff pressure, causing flow through the second pump to reduce significantly or to cease. This phenomenon can cause flow problems or pump damage.

Series. Pumps of different flow capacity in series can result in the pump of greater capacity overflowing the pump of lesser capacity, which could cause damaging cavitation in the smaller pump and could actually cause a pressure drop rather than a pressure rise across that pump.

Standby Pump Provision

If total flow standby capacity is required, a properly valved equal-capacity standby pump is installed to operate when the normal pump is inoperable. Also, a single standby may be provided for several similarly sized pumps. Parallel or series pump installation can provide up to 80% standby, which is often sufficient.

Compound Pumping

In larger systems, compound pumping, also known as primary-secondary pumping, is often employed to provide system advantages that would not be available with a single pumping system. The concept of compound pumping is illustrated in Figure 11.

In Figure 11, Pump No. 1 can be referred to as the source pump or the primary pump and Pump No. 2 the load or secondary pump. The short section of pipe between A and B is called the common pipe, because it is common to both the source and load circuits. Other terms used for the common pipe are the decoupling line and the neutral bridge. In the design of compound systems, the common pipe should be kept as short and as large a diameter as practical to minimize the pressure loss between those two points. There should never be a valve or check valve in the common pipe. If these conditions are met and the pressure loss in the common

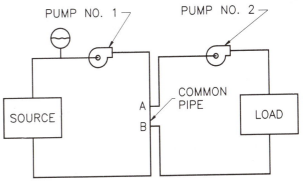

Fig. 11   Compound Pumping
(Primary-Secondary Pumping)

pipe can be assumed to be zero, then neither pump will affect the other. Then, except for the system static pressure at any given point, the circuits can be designed and analyzed and will function dynamically independently of one another.

In Figure 11, if Pump No. 1 has the same flow capacity in its circuit as Pump No. 2 has in its circuit, all of the flow entering Point A from Pump No. 1 will leave in the branch supplying Pump No. 2, and no water will flow in the common pipe. Also at this condition, the water entering the load will be at the same temperature as that leaving the source.

If the flow capacity of Pump No. 1 exceeds that of Pump No. 2, some water will flow downward in the common pipe. Under this condition, Tee A is a diverting tee, and Tee B becomes a mixing tee. Again, the temperature of the fluid entering the load is the same as that leaving the source. However, because of the mixing taking place at Point B, the temperature of the water returning to the source is between the source supply temperature and the load return temperature.

On the other hand, if the flow capacity of Pump No. 1 is less than that of Pump No. 2, then Point A becomes a mixing point since some water must recirculate upward in the common pipe from Point B. Under this condition, the temperature of the water entering the load is between the supply water temperature from the source and the return water temperature from the load.

As an example, if Pump No. 1 circulates 25 gpm of water leaving the source at 200°F, and Pump No. 2 circulates 50 gpm of water leaving the load at 100°F, then the water temperature entering the load would be:

$$t_{load} = 200 - (25/50)(200 - 100) = 150°F$$

Some advantages of compound circuits are:

1. They enable the designer to achieve different water temperatures and temperature ranges in different elements of the system.
2. They decouple the circuits hydraulically, thereby making the control, operation, and analysis of large systems much less complex. Also, hydraulic decoupling prevents unwanted series or parallel operations.
3. Circuits can be designed for different flow characteristics. For example, a chilled water load system can be designed with two-way valves for better control and energy conservation while the source system operates at constant flow to protect the chiller from freezing.

EXPANSION TANK

As a hydraulic device, the expansion tank serves as the reference pressure point in the system, analogous to a ground in an electrical system. Where the tank connects to the piping, the pressure equals the pressure of the air in the tank plus or minus any fluid pressure due to the elevation difference between the tank liquid surface and the pipe (Figure 12).

As previously stated, a closed system should have but one expansion chamber. The presence of more than one chamber or

of excessive amounts of undissolved air in a piping system can cause the closed system to behave in unexpected (but understandable) ways, which can cause extensive damage from shock waves or water hammer.

With a single chamber on a system, assuming isothermal conditions for the air, the air pressure can change only as a result of displacement by the water. The only thing that can cause the water to move into or out of the tank (assuming no water is being added to or removed from the system) is expansion or shrinkage of the water in the system. Thus, in sizing the tank, thermal expansion is related to the pressure extremes of the air in the tank [Equations (12), (13), and (14)].

The point of connection of the tank should be based on the pressure requirements of the system, remembering that the pressure at the tank connection will not change as the pump is turned on or off. For example, consider a system containing an expansion tank at 30 psig pressure and a pump with a pump head of 23.1 ft (10 psig). Figure 13 shows alternative locations for connecting the expansion tank; in either case, with the pump off, the pressure will be 30 psig on both the pump suction and discharge. With the tank on the pump suction side, when the pump is turned on, the pressure increases on the discharge side with the tank on the pump suction by an amount equal to the pump pressure (Figure 13a). With the tank connected on the discharge side of the pump, the pressure decreases on the suction side by the same amount (Figure 13b).

Other considerations relating to the tank connection are:

- A tank open to the atmosphere must be located above the highest point in the system.
- A tank with an air/water interface is generally used with an air control system that continually revents the air into the tank. For this reason, it should be connected at a point where air can best be released.
- Within reason, the lower the pressure in a tank, the smaller the tank [see Equations (12) and (13)]. Thus, in a vertical system, the higher the tank is placed, the smaller it can be.

PRACTICES

Piping Circuits

Common practice in the design of hydronic systems employs numerous configurations of piping circuits. In addition to simple preference by the design engineer, the method of arranging the circuiting can be dictated by such factors as the shape or configuration of the building, economics of installation, energy economics, nature of the load, part load capabilities or requirements, and others.

Each piping system is a network; the more extensive the network, the more complex it is to understand, analyze, or control. Thus, a major design objective is to maintain the highest degree of simplicity.

Load distribution circuits are of four general types:

- Full series
- Diverting series
- Parallel direct return
- Parallel reverse return

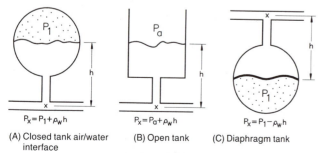

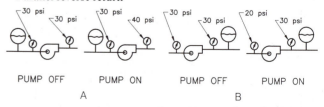

(A) Closed tank air/water interface $P_x = P_1 + \rho_w h$

(B) Open tank $P_x = P_a + \rho_w h$

(C) Diaphragm tank $P_x = P_1 - \rho_w h$

Fig. 12   Tank Pressure Related to "System" Pressure

Fig. 13   Effect of Expansion Tank Location with Respect to Pump Pressure

Series circuit. A simple series circuit is as shown in Figure 14. Series loads generally have the advantage of lower piping costs, and higher temperature drops that result in smaller pipe sizes and lower energy consumption. A disadvantage is that the different circuits cannot be controlled separately. Simple series circuits are generally limited to residential and small commercial standing radiation systems. Figure 15 shows a typical layout of such a system with two zones for residential or small commercial heating.

Diverting series. The simplest diverting series circuit diverts some of the flow from the main piping circuit through a special diverting tee to a load device (usually standing radiation) that has a low pressure drop. This system is generally limited to heating systems in residential or small commercial applications.

Figure 16 illustrates a typical one-pipe diverting tee circuit. For each terminal unit, a supply and a return tee are installed on the main. One of the two tees is a special diverting tee, which creates a pressure drop in the main flow to divert part of it to the unit. One (return) diverting tee is usually sufficient for upfeed (units above the main) systems. Two special fittings (supply and return tees) are usually required to overcome thermal pressure in downfeed units.

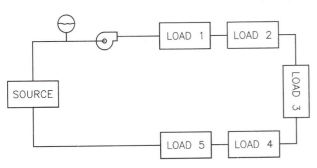

Fig. 14   Flow Diagram of Simple Series Circuit

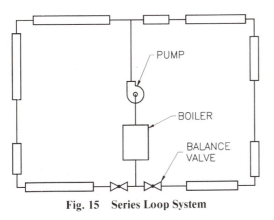

Fig. 15   Series Loop System

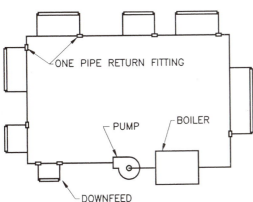

Fig. 16   One-Pipe Diverting Tee System

Special tees are proprietary; consult manufacturer's literature for flow rates and pressure drop data on these devices. Unit selection can be only approximate without these data.

One-pipe diverting series circuits allow manual or automatic control of flow to individual heating units. On-off rather than flow modulation control is preferred because of the relatively low pressure drop allowable through the control valve in the diverted flow circuit. This system is likely to cost more than the series loop because extra branch pipe and fittings, including special tees, are required. Each unit usually requires a manual air vent because of the low water velocity through the unit. The length and load imposed on a one-pipe circuit are usually small because of these limitations.

Since only a fraction of the main flow is diverted in a one-pipe circuit, the flow rate and pressure drop as unit water flow is controlled are less variable than in some other circuits. When two or more one-pipe circuits are connected to the same two-pipe mains, the circuit flow may need to be mechanically balanced. After balancing, sufficient flow must be maintained in each one-pipe circuit to ensure adequate flow diversion to the loads.

When coupled with compound pumping systems, series circuits can be applied to multiple control zones on larger commercial or institutional systems (Figure 17). Note that in the series circuit with compound pumping, the load pumps need not be equal in capacity to the system pump. If, for example, Load Pump LP-1 circulates less flow Q_{LP1} than system pump SP-1 Q_{SP1}, the temperature difference across Load 1 would be greater than the circuit temperature difference between A and B, (*i.e.*, water would flow in the common pipe from A to B). If, on the other hand, the load pump LP-2 is equal in flow capacity to the system pump SP-1, the temperature differential across Load 2 and the system from C to D would be equal and no water would flow in the common pipe. If Q_{LP3} exceeds Q_{SP1}, mixing occurs at Point E and, in a heating system, the temperature entering pump LP-3 would be lower than that available from the system leaving load connection D.

Thus, it can be seen that series utilization circuits employing compound or load pumps offer many design options. It might also be noted that each of the loads shown in Figure 16 could be a complete piping circuit or network.

Parallel piping. These networks are most commonly used in hydronic systems because these circuits allow the same temperature water to be available to all loads. The two types of parallel networks are direct return and reverse return (Figure 18).

In the direct-return system, the length of supply and return piping through the subcircuits is unequal, which may cause unbalanced flow rates and require careful balancing to provide each subcircuit with design flow. Ideally, the reverse-return system provides nearly equal total lengths for all terminal circuits.

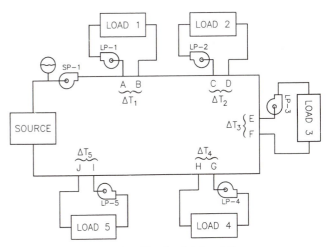

Fig. 17   Series Circuit with Load Pumps

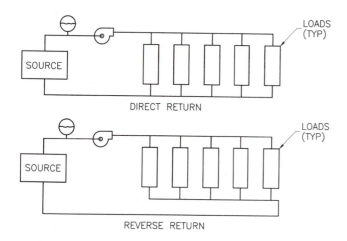

Fig. 18   Direct- and Reverse-Return Two-Pipe Systems

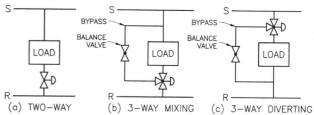

Fig. 19   Load Control Valves

Direct-return piping has been successfully applied where the designer has guarded against major flow unbalance by:

1. Providing for pressure drops in the subcircuits or terminals that are significant percentages of the total, usually establishing close subcircuit pressure drops at higher values than the far subcircuits.
2. Minimizing distribution piping pressure drop. (In the limit, if the distribution piping loss is zero and the loads are of equal flow resistance, the system is inherently balanced.)
3. Including balancing devices and some means of measuring flow at each terminal or branch circuit.
4. Using control valves with a high head loss at the terminals.

CAPACITY CONTROL OF LOAD SYSTEM

The two alternatives for controlling the capacity of hydronic systems are on-off control and variable capacity or modulating control. The on-off option is generally limited to smaller systems such as residential or small commercial and individual components of larger systems. In smaller systems where the entire building is a single zone, control is accomplished by cycling the source device (the boiler or chiller) on and off. Usually a space thermostat allows the chiller or boiler to run, then a water temperature thermostat (aquastat) controls the capacity of the chiller(s) or boiler(s) as a function of either supply or return water temperature. The pump can be either cycled with the load device (usually the case in a residential heating system) or left running (usually done in small commercial chilled water systems).

In these single-zone applications, the piping design requires no special consideration for control. Where multiple zones of control are required, the various load devices are controlled first; then the source system capacity is controlled to follow the capacity requirement of the loads.

Control valves are commonly used to control loads. These valves control the capacity of each load by varying the amount of water flow through the load device when load pumps are not used. Control valves for hydronic systems are straight through (or two-way) valves and three-way valves (Figure 19). The effect of either valve is to vary the amount of water flowing through the load device.

When using a two-way valve (Figure 19a), as the valve strokes from full open to full closed, the quantity of water flowing through the load gradually decreases from design flow to no flow. When using a three-way mixing valve (Figure 19b) in one position, the valve is open from Port A to AB, with Port B closed off. In that

position, all the flow is through the load. As the valve moves from the A-AB position to the B-AB position, some of the flow bypasses the load flowing through the bypass line, causing decreasing flow through the load. At the end of the stroke, Port A is closed, and all of the fluid will flow from B to AB with no flow through the load. Thus, the three-way valve has the same effect on the load as the two-way valve—as the load reduces, the quantity of water flowing through the load decreases.

The effect on load control with the three-way diverting valve (Figure 19c) is the same as with the mixing valve in a closed system—the flow is either directed through the load or through the bypass in proportion to the load. Because of the dynamics of valve operation, diverting valves are more complex in their design and are thus more expensive than mixing valves; since they accomplish the same function as the simpler mixing valve, they are seldom used in closed hydronic systems.

The decision on whether to use two-way or three-way valves is influenced by several factors:

1. Insofar as the load control is concerned, a two-way valve and a three-way valve perform identically the same function—they both vary the flow through the load as the load changes.
2. The fundamental difference between the two-way valve and the three-way valve is as the source or distribution system sees the load: the two-way valve provides a variable flow load response and the three-way valve provides a constant flow load response.

Referring to Equation (9), the load q is proportional to the product of Q and Δt. Ideally, as the load changes, Q changes, while Δt remains fixed. However, as the system sees it, as the load changes with the two-way valve, Q varies and Δt is fixed, whereas with a three-way valve, Δt varies and Q is fixed. This principle is illustrated in Figure 20. An understanding of this concept is fundamental to the design or analysis of hydronic systems.

The flow characteristics of two-way and three-way valve ports need to be carefully understood and are described in Chapter 41, Automatic Control, of the 1991 ASHRAE *Handbook—HVAC Applications*. The equal percentage characteristic is recommended for proportional control of the load flow for two-way and three-way valves; the bypass flow port of three-way valves should have the linear characteristic to maintain a uniform flow during part-load operation.

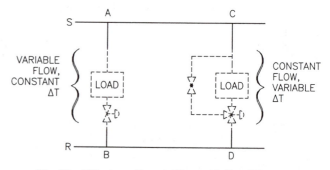

Fig. 20   Effects on System Flow with Two-Way and Three-Way Valves

SIZING CONTROL VALVES

For stable control, the pressure drop in the control valve at the full open position should be no less than one-half the pressure drop in the branch. For example, in Figure 20, the pressure drop at full open position for the two-way valve should equal one-half the pressure drop from A to B, and for the three-way valve, the full open pressure drop should be half that from C to D. The bypass balancing valve in the three-way valve circuit should be set to equal that in the coil (load).

Control valves may be sized on the basis of the valve coefficient C_v, where the C_v in the full open position is:

$$Q \sqrt{\Delta p}$$

where:

C_v = valve coefficient
Q = flow rate, gpm
Δp = pressure drop, lb/in^2

If a system is to be designed with multiple zones of control such that load response is to be by constant flow through the load and variable Δt, control cannot be achieved by valve control alone. To accomplish this, a load pump is required.

Several control arrangements of load pump and control valve configurations are shown in Figure 21. Note that in all three configurations the common pipe has no restriction or check valve. In all configurations there is no difference in control as seen by the load. However, the basic differences in control are:

1. With the two-way valve configuration, the distribution system sees a variable flow, constant Δt, whereas with both three-way configurations, the distribution system sees a constant flow, variable Δt.
2. Configuration b differs from c in that the pressure required through the three-way valve in Figure 21b is provided by the load pump, while in Figure 21c it is provided by the distribution pump(s).

LOW-TEMPERATURE HEATING SYSTEMS

These systems are used for heating of spaces or processes directly as with standing radiation and process heat exchangers or indirectly through air-handling unit coils for preheating, reheating, or in hot water unit heaters. These systems are generally designed with supply water temperatures from 180 to 240°F and temperature ranges from 20 to 100°F.

Historically, in the United States, hot water heating systems were designed for a 200°F supply water temperature and a 20°F temperature drop. This practice evolved from earlier gravity system designs and provides convenient design relationships for heat transfer coefficients related to coil tubing and finned tube radiation and for calculations (one gallon per minute conveys 10,000 Btu/h at 20°F Δt). Since many terminal devices still require these

flow rates, it is important to recognize this relationship in selecting devices and designing systems.

However, the greater the temperature range (and related lower flow rate) that can be applied, the less costly it is to install and operate. A lower flow rate requires smaller and less expensive piping, less secondary building space, and smaller pumps. Also, smaller pumps require less energy, so that operating costs are lower.

Nonresidential Heating Systems

Three approaches are possible to enhance the economics of large heating systems: (1) higher supply temperatures, (2) primary-secondary pumping, and (3) terminal equipment designed for smaller flow rates. All three may be used, either singly or in combination.

Using higher supply water temperatures achieves higher temperature drops and smaller flow rates. Terminal units with a reduced heating surface can be used. These smaller terminals are not necessarily less expensive, however, because their required operating temperatures and pressures may increase manufacturing costs and the problems of pressurization, corrosion, expansion, and control. System components may not increase in cost uniformly with temperature, but may increase in steps conforming to the three major temperature classifications. Within each classification, the most economical design uses the highest temperature within that classification.

Primary-secondary or compound pumping reduces the size and cost of the distribution system and also may use larger flows and lower temperatures in the terminal or secondary circuits. A primary pump circulates water in the primary distribution system while one or more secondary pumps circulate the terminal circuits. The connection between primary and secondary circuits provides complete hydraulic isolation of both circuits and permits a controlled interchange of water between the two. Thus, a high supply water temperature can be used in the primary circuit at a low flow rate and high temperature drop, while a lower temperature and conventional temperature drops can be used in the secondary circuit(s).

For example, a system could be designed with primary-secondary pumping in which the supply temperature from the boiler was 240°F, the supply temperature in the secondary was 200°F, and the return temperature was 180°F. This design results in a conventional 20°F Δt in the secondary zones, but permits the primary circuit to be sized on the basis of a 60°F drop. This primary-secondary pumping arrangement is most advantageous with terminal units such as convectors and finned radiation, which are generally unsuited for small flow rate design.

Many types of terminal heat transfer units are being designed to use smaller flow rates with temperature drops up to 100°F in low-temperature systems and up to 150°F in medium-temperature systems. Fan apparatus, the heat transfer surface used for air heating in fan systems, and water-to-water heat exchangers are most adaptable to such design.

Another technique is to put certain loads in series utilizing a combination of control valves and compound pumping (Figure 22).

In the system illustrated, the capacity of the boiler or heat exchanger is 2×10^6 Btu/h and each of the four loads is 0.5×10^6. Under design conditions, the system is designed for an 80°F water temperature drop and the loads each provide 20°F of the total Δt. The loads in these systems, as well as the smaller or simpler systems in residential or commercial applications, can be connected in a direct return or a reverse return piping system. The different features of each load include:

1. The domestic hot water heat exchanger has a two-way valve, and is thus arranged for variable flow (while the main distribution circuit provides constant flow for the boiler circuit).

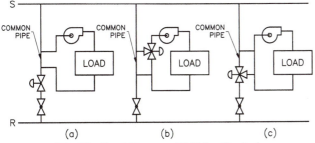

Fig. 21   Load Pumps with Valve Control

2. The finned-tube radiation circuit is a 20°F Δt circuit with the design entering water temperature reduced to and controlled at 200°F.
3. The reheat coil circuit takes a 100°F temperature drop for a very low flow rate.
4. The preheat coil circuit provides constant flow through the coil to keep it from freezing.

When water-to-air heating coil type loads in low-temperature hot water heating systems are valve controlled (flow varies), they have a heating characteristic of flow versus capacity as shown in Figure 23 for 20°F Δt and 60°F Δt temperature drops. For a 20°F Δt coil, 50% flow provides approximately 90% capacity; valve control will tend to be unstable. For this reason, proportional temperature control is required and equal percentage characteristic two-way valves should be selected such that 10% flow is achieved with 50% valve lift. This combination of the valve characteristic and the heat transfer characteristic of the coil makes the control linear with respect to the control signal. This type of control can only be obtained with equal percentage two-way valves and can

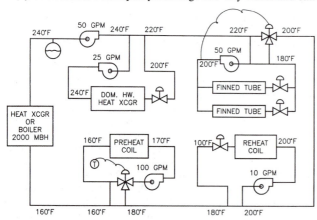

Fig. 22 Hydronic Heating System Example Series Connected Loads

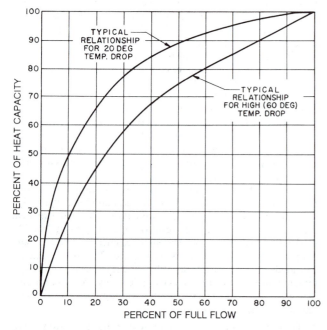

Fig. 23 Heat Emission versus Flow Characteristic of Typical Hot Water Heating Coil

be further enhanced if piped with a secondary pump arrangement as shown in Figure 21a. See Chapter 41, Automatic Control, 1991 ASHRAE *Handbook—HVAC Applications* for further information on automatic controls.

CHILLED WATER SYSTEMS

Designers have less latitude in selecting supply water temperatures for cooling applications because there is only a narrow range of water temperatures low enough to provide adequate dehumidification and high enough to avoid chiller freeze-up. Circulated water quantities can be reduced by selecting proper air quantities and heat transfer surface at the terminals. Terminals suited for a 12°F rise, rather than an 8°F rise, reduce circulated water quantity and pump power by one-third and increase chiller efficiency.

A proposed system should be economically evaluated to achieve the desired balance between installation cost and operating cost. Table 1 shows the effect of coil circuiting and chilled water temperature on the water flow and temperature rise. The coil rows, fin spacing, air-side performance, and cost are identical for all selections. Morabito (1960) showed how such changes affect the overall system. Considering the investment cost of piping and insulation versus the operating cost of refrigeration and pumping motors, higher temperature rises, *i.e.*, about 1.0 to 1.5 gpm/ton at 16 to 24°F temperature rise can be applied on chilled water systems with long distribution piping runs; larger flow rates should be used only where reasonable in close-coupled systems.

For the most economical design, the minimum flow rate to each terminal heat exchanger is calculated. For example, if one terminal can be designed for an 18°F rise, another for 14°F, and others for 12°F, the highest rise to each terminal should be used, rather than designing the system for an overall temperature rise based on the least capabilities.

The control system selected also influences the design water quantity. For systems using multiple terminal units, diversity factors can be applied to water quantities before sizing pump and piping mains if exposure or use prevents the unit design loads from occurring simultaneously and if water flow control using two-way valves is used. If air-side control, *e.g.*, face-and-bypass or fan cycling, or three-way valves on the water side are used, diversity should not be a consideration in pump and piping design, although it should be considered in the chiller selection.

A primary consideration with chilled water system design is the control of the source systems at reduced loads. The constraints on the temperature parameters are (1) a water freezing temperature of 32°F, (2) economics of the refrigeration system in generating chilled water, and (3) the dew-point temperature of the air at nominal indoor comfort conditions (55°F dew point at 75°F

Table 1 Chilled Water Coil Performance

Coil Circuiting	Chilled Water Inlet Temp., °F	Coil Pressure Drop, psi	Chilled Water Flow, gpm/ton	Chilled Water Temp. Rise, °F
Full[a]	45	1.0	2.2	10.9
Half[b]	45	5.5	1.7	14.9
Full[a]	40	0.5	1.4	17.1
Half[b]	40	2.5	1.1	21.8

Table based on cooling air from 81°F dry bulb, 67°F wet bulb to 58°F dry bulb, 56°F wet bulb.

[a]Full circuiting (also called single circuit). In this circuit, water at the inlet temperature flows simultaneously through all tubes in a plane transverse to airflow; it then flows simultaneously through all tubes, in unison, in successive planes (*i.e.*, rows) of the coils.

[b]Half circuiting. Tube connections are arranged so there are half as many circuits as there are tubes in each plane, or row, thereby using higher water velocities through the tubes. This circuiting is used with small water quantities.

and 50% rh). These parameters have led to the common practice of designing for a supply chilled water temperature of 44 to 45 °F and a return water temperature between 55 and 64 °F.

Historically, most chilled water systems have utilized three-way control valves to achieve constant water flow through the chillers. However, as systems have become larger, as designers have turned to multiple chillers for reliability and controllability, and as energy economics have become an increasing concern, the use of two-way valves and source pumps for the chillers has greatly increased.

A typical configuration of a small chilled water system utilizing two parallel chillers and loads with three-way valves is illustrated in Figure 24. Note that the flow is essentially constant. A simple energy balance [Equation (9)] dictates that at one-half of design load, with a constant flow rate, the water temperature differential drops to one-half of design. At this load, if one of the chillers is turned off, the return water circulating through this off chiller would mix with the supply water. This mixing raises the temperature of the supply chilled water and can cause a loss of control if the designer does not consider this operating mode.

A typical configuration of a large chilled water system with multiple chillers and loads and compound piping is shown in Figure 25. This system provides variable flow, constant supply temperature chilled water, multiple chillers, more stable two-way control valves, and the advantage of adding chilled water storage with little additional complexity.

One design issue illustrated in Figure 25 is the placement of the common pipe for the chillers. With the common pipe as shown, the chillers will unload from right to left. With the common pipe in the alternate location, the chillers will unload equally in proportion to their capacity (*i.e.*, equal percentage).

DUAL-TEMPERATURE SYSTEMS

Dual-temperature systems are used when the same load devices and distribution systems are used for both heating and cooling; *e.g.*, fan coil units and central station air-handling unit coils. In the design of dual-temperature systems, the cooling cycle design usually dictates the requirements of the load heat exchangers and distribution systems. Dual-temperature systems are basically of three different configurations, each requiring different design techniques:

1. Two-pipe systems
2. Four-pipe common load systems
3. Four-pipe independent load systems

Two-Pipe Systems

In a two-pipe system, the load devices and the distribution system circulate chilled water when cooling is required and hot water when heating is required (Figure 26). Design considerations for these systems include:

- Loads must all require cooling or heating coincidentally; *i.e.*, if cooling is required for some loads and heating for other loads at a given time, this type system should not be used.
- When designing the system, the flow and temperature requirements for both the cooling and heating media must be calculated first. Then (1) design the load and distribution system for the most stringent, and (2) design for the water temperatures and temperature differential as dictated for the other mode.
- Design the changeover procedure such that the chiller evaporator is not exposed to damaging high water temperatures and the boiler is not subjected to damaging low water temperatures. To accommodate these limiting requirements, the changeover of a system from one mode to the other requires considerable time. If rapid load swings are anticipated, a two-pipe system should not be selected, although it is the least costly of the three options.

Four-Pipe Common Load Systems

In the four-pipe common load system, load devices are used for both heating and cooling as in the two-pipe system. The four-pipe common load system differs from the two-pipe system in that both heating and cooling are available to each load device, and the changeover from one mode to the other takes place at each individual load device, or grouping of load devices, rather than at the source. Figure 27 is a flow diagram of a four-pipe common load system, with multiple loads and a single boiler and chiller. The fundamental difference between the four-pipe common load system and the two-pipe system is that some of the load systems can be in the cooling mode while others are in the heating mode.

Although many of these systems have been installed, many such installations have not performed successfully due to problems in implementing the design concepts.

One problem that must be addressed is the expansion tank connection(s). Many four-pipe systems were designed with two expansion tanks—one for the cooling circuit and one for the heating

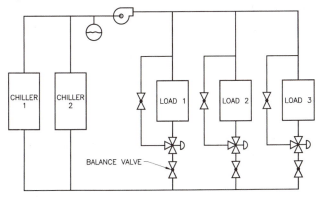

Fig. 24   Constant Flow Chilled Water System

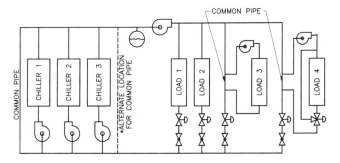

Fig. 25   Variable Flow Chilled Water System

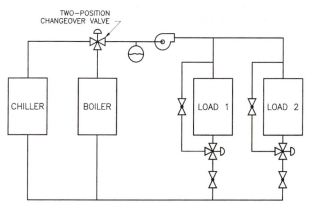

Fig. 26   Simplified Diagram of Two-Pipe System

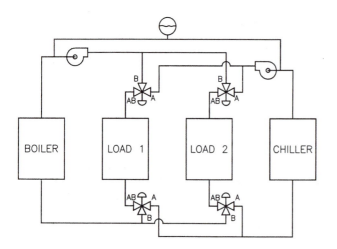

Fig. 27 Four-Pipe Common Load System

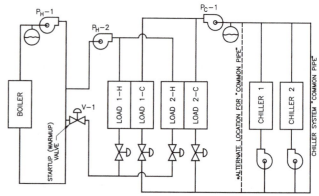

Fig. 28 Four-Pipe Independent Load System

circuit. However, with multiple loads, these circuits become hydraulically interconnected, thus creating a system with two expansion chambers. The preferred method of handling the expansion tank connection sets the point of reference pressure equal in both circuits (Figure 27).

Another potential problem is the mixing of hot and chilled water. At each load connection, two three-way valves are required—a mixing valve on the inlet and a diverting valve on the outlet. These valves operate in unison in just two positions—opening either Port B to AB or Port A to AB. If, for example, the valve on the outlet does not seat tightly and Load 1 is indexed to cooling and Load 2 is indexed to heating, return heating water from Load 2 will flow into the chilled water circuit, and return chilled water from Load 1 will flow into the heating water circuit. The probability of this occurring increases as the number of loads increases, because the number of control valves increases.

Another disadvantage of this system is that the loads have no individual capacity control as far as the water system is concerned. That is, each valve must be positioned to either full heating or full cooling with no control in between.

Because of these disadvantages, four-pipe common load systems should be limited to those applications in which independent load circuits do not apply (*i.e.*, radiant ceiling panels or induction unit coils).

Four Pipe Independent Load Systems

The four-pipe independent load system is preferred for those hydronic applications in which some of the loads are in the heating mode while others are in the cooling mode. Control is simpler and more reliable than for the common load systems, and in many applications, the four-pipe independent load system is less costly to install. Also, the flow through the individual loads can be modulated, providing both the control capability for variable capacity and the opportunity for variable flow in either or both circuits.

A simplified example of a four-pipe independent load system with two loads, one boiler, and two chillers is shown in Figure 28. Note that both hydronic circuits are independent, so each can be designed essentially with disregard for the other system. Although both circuits in the figure are shown as variable flow distribution systems, they could be constant flow (three-way valves) or one variable flow and one constant flow. Generally, the control modulates the two load valves in sequence with a dead band at the control midpoint.

This type of system offers additional flexibility when some selective loads are arranged for heating only or cooling only, such

as unit heaters or preheat coils. Then, central station systems can be designed for humidity control with reheat by configuration at the coil locations and with proper control sequences.

OTHER DESIGN CONSIDERATIONS

Makeup and Fill Water Systems

Generally, a hydronic system is filled with water through a valved connection to a domestic water source, with a service valve, a backflow preventer, and a pressure gage. (The domestic water source pressure must exceed the system fill pressure.)

Since the expansion chamber is the reference pressure point in the system, the water makeup point is usually located at or near the expansion chamber.

Many designers prefer to install automatic makeup valves, which consist of a pressure regulating valve in the makeup line. However, the quantity of water being made up must be monitored to avoid scaling and oxygen corrosion in the system.

Safety Relief Valves

Safety relief valves should be installed at any point at which pressures can be expected to exceed the safe limits of the system components. The causes of excessive pressures include:

- Overpressurization from fill system
- Pressure increases due to thermal expansion
- Pressure surges caused by momentum changes (shock or water hammer)

Overpressurization from the fill system could occur by an accident in filling the system or the failure of an automatic fill regulator. To prevent this, a safety relief valve is usually installed at the fill location. Figure 29 shows a typical piping configuration for a system with a plain steel or air/water interface expansion tank. Note that no valves are installed between the hydronic system piping and the safety relief valve. This is a mandatory design requirement if the valve in this location is to also serve as a protection against pressure increases due to thermal expansion.

The water in a hydronic system changes in volume as the water temperature changes. To allow for these volumetric changes, the expansion chamber is installed in the system. However, if any part of the system is configured such that (1) it can be isolated from the expansion tank and (2) its temperature can increase while it is isolated, then overpressure relief should be provided.

The relationship between pressure change due to temperature change and the temperature change in a piping system is expressed by Equation (17).

$$\Delta p = \frac{(\beta - 3\alpha)\Delta t}{(5/4)(D/E\Delta r) + \gamma} \qquad (17)$$

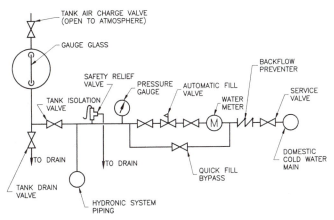

Fig. 29   Typical Makeup Water and Expansion Tank Piping Configuration for Plain Steel Expansion Tank

where:

Δp = pressure increase, lb/in²
β = volumetric expansion of water, 1/°F
α = linear coefficient of expansion for piping material, 1/°F
Δt = water temperature increase, °F
D = pipe diameter, in.
E = modulus of elasticity of piping material, lb/in²
γ = volumetric compressibility of water, in²/lb
Δr = thickness of pipe wall, in.

Figure 30 shows a solution to Equation (17), demonstrating the pressure increase caused by any given temperature increase, calculated for 1- and 10-in. steel piping. If the temperature in a chilled water system with piping spanning sizes between 1 and 10 in. were to increase by 15 °F, the pressure would increase between 340 and 420 psi, depending on the average pipe size in the system.

Safety relief should be provided to protect boilers, heat exchangers, cooling coils, chillers, and the entire system when the expansion tank is isolated for air charging or other service. As a minimum, the ASME Boiler Code requires that a dedicated safety relief valve be installed on each boiler and that isolating or service valves be provided on the supply and return connections to each boiler.

Potential forces caused by shock waves or water hammer should also be considered in design. Chapter 33 of the 1989 ASHRAE *Handbook—Fundamentals* discusses the causes of shock forces and the methodology for calculating the magnitude of these forces.

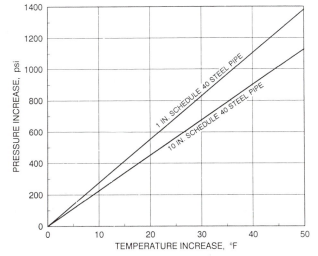

Fig. 30   Pressure Increase as Function of Temperature Increase Resulting from Thermal Expansion

Air Elimination

If air and other gases are not eliminated from the flow circuit, they may cause binding in the terminal heat transfer elements, corrosion, noise, reduced pumping capacity, and loss of hydraulic stability (see Principles section). A closed tank without a diaphragm can be installed at the point of the lowest solubility of air in water. When a diaphragm tank is used, air in the system can be removed by an air separator and air elimination valve installed at the point of lowest solubility. Manual vents should be installed at high points to remove all trapped air during initial operation and to ensure that the system is tight. Shutoff valves should be installed on any automatic air removal device to permit servicing without draining the system.

Drain and Shutoff

All low points should have drains. Separate shutoff and draining of individual equipment and circuits should be possible so that the entire system does not have to be drained to service a particular item. Whenever a device or section of the system is isolated and the water in that section or device could increase in temperature following isolation, overpressure safety relief protection must be provided.

Balance Fittings

Balance fittings or valves and a means of measuring flow quantity should be applied as needed to permit balancing of individual terminals and subcircuits.

Pitch

Piping need not pitch but can run level, providing flow velocities exceeding 1.5 ft/s are maintained or a diaphragm tank is used.

Strainers

Strainers should be used where necessary to protect system elements. Strainers in the pump suction need to be checked carefully to avoid cavitation. Large separating chambers can serve as main air venting points and dirt strainers ahead of pumps. Automatic control valves or other devices operating with small clearances require protection from pipe scale, gravel, and welding slag, which may readily pass through the pump and its protective separator. Individual fine mesh strainers may therefore be required ahead of each control valve.

Thermometers

Thermometers or thermometer wells should be installed to assist the system operator in routine operation and troubleshooting. Permanent thermometers, with the correct scale range and separate sockets, should be used at all points where temperature readings are regularly needed. Thermometer wells should be installed where readings will be needed only during start-up and infrequent troubleshooting. If a central monitoring system is provided, a calibration well should be installed adjacent to each sensing point in insulated piping systems.

Flexible Connectors and Expansion Compensation

Flexible connectors are sometimes installed at pumps and machinery to reduce pipe stress. See Chapter 42 of the 1991 ASHRAE *Handbook—HVAC Applications* for vibration isolation information. Expansion, flexibility, and hanger and support information is in Chapter 42 of this volume.

Gage Cocks

Gage cocks or quick-disconnect test ports should be installed at points requiring pressure readings. Gages permanently installed in the system will deteriorate because of vibration and pulsation and will, therefore, be unreliable. It is good practice to install gage cocks and provide the operator with several quality gages for troubleshooting.

Insulation

Insulation should be applied to minimize pipe thermal loss and to prevent condensation during chilled water operation (see Chapter 20 of the 1989 ASHRAE *Handbook—Fundamentals*). On chilled water systems, special rigid metal sleeves or shields should be installed at all hanger and support points, and all valves should be provided with extended bonnets to allow for the full insulation thickness without interference with the valve operators.

Condensate Drains

Condensate drains from dehumidifying coils should be trapped and piped to an open-sight plumbing drain. Traps should be deep enough to overcome the air pressure differential (between drain inlet and room), which ordinarily will not exceed 2 in. of water. Pipe should be noncorrosive and insulated to prevent moisture condensation. Depending on the quantity and temperature of condensate, plumbing drain lines may require insulation to prevent sweating.

Common Pipe

In compound pumping systems (primary-secondary pumping), the common pipe is used to dynamically decouple the two pumping circuits. Ideally, there is no pressure drop in this section of piping; however, in actual systems, it is recommended that this section of piping be a minimum of 10 diameters in length to reduce the likelihood of unwanted mixing resulting from velocity (kinetic) energy or turbulence.

DESIGN PROCEDURES

Preliminary Equipment Layout

Determining flows in mains and laterals. Regardless of the method used to determine the flow through each item of terminal equipment, the desired result should be listed in terms of mass flow (in pounds per hour) on the preliminary plans or in a schedule of flow rates for the piping system. (In the design of small systems, or in chilled water systems, the determination may be made in terms of gallons per minute).

In an equipment schedule or on the plans, starting from the most remote terminal and working toward the pump, progressively list the cumulative flow in each of the mains and branch circuits in the distribution system.

Preliminary pipe sizing. For each portion of the piping circuit, select a tentative pipe size from the unified flow chart (Figure 1, Chapter 33 of the 1989 ASHRAE *Handbook—Fundamentals*), using a value of pipe friction loss ranging from 0.75 to 4 ft per 100 ft (approximately 0.1 to 0.5 in/ft).

Residential piping size is often based on pump preselection, using pipe sizing tables, which are available from the Hydronics Institute or from manufacturers.

Preliminary pressure drop. Using the preliminary pipe sizing indicated above, determine the pressure drop through each portion of the piping. Determine the total pressure drop in several of the longest circuits to determine the maximum pressure drop through the piping, including the terminals and control valves, that must be available in the form of pump pressure.

Preliminary pump selection. The preliminary selection should be based on the pump's ability to fulfill the determined capacity requirements. It should be selected at a point left-of-center on the pump curve and should not overload the motor. Because pressure

drop in a flow system varies as the square of the flow rate, the flow variation between the nearest size of stock pump and an exact point selection will be relatively minor.

Final Pipe Sizing and Pressure Drop Determination

Final piping layout. Examine the overall piping layout to determine if pipe sizes in some areas need to be readjusted. Several principal circuits should have approximately equal pressure drops so that excessive pressures are not needed to serve a small portion of the building.

Consider both the initial costs of the pump and piping system and the pump's operating cost when determining final system friction loss. Generally, lower heads and larger piping are more economical when longer amortization periods are considered, especially in larger systems. However, in small systems such as in residences, it may be most economical to select the pump first and design the piping system to meet the available pressure. In all cases, adjust the piping system design and pump selection until the optimum design is found.

When the final piping layout has been established, determine the friction loss for each section of the piping system from the pressure drop charts (Chapter 33 of the 1989 ASHRAE *Handbook—Fundamentals*) for the mass flow rate in each portion of the piping system.

After calculating the friction loss at design flow for all sections of the piping system and all fittings, terminal units, and control valves, summarize them for several of the longest piping circuits to determine the precise pressure against which the pump must operate at design flow.

Final pump selection. After completing the final pressure drop calculations, select the final pump by plotting a system curve and pump curve and selecting the pump or pump assembly that operates closest to the actual calculated design point.

Freeze Prevention

All circulating water systems require precautions to prevent freezing, particularly in makeup air applications in temperate climates where coils are exposed to 100% outdoor air at below-freezing temperatures, where undrained chilled water coils are in the winter airstream, or where piping passes through unheated spaces. Freezing will not occur as long as flow is maintained and the water is at least warm. Unfortunately, during extremely cold weather or in the event of a power failure, water flow and temperature cannot be guaranteed. Additionally, continuous pumping can be energy intensive and cause system wear. Precautions to avoid flow stoppage or damage from freezing are:

1. All load devices that are subjected to outdoor air temperatures (such as preheat coils) should be designed for constant flow, variable Δt control.
2. Cooling coils with valve control that are dormant in winter months should have their coil valves position to the full-open position at those times.
3. If intermittent pump operation is used as an economy measure, use an automatic override to operate both chilled water and heating water pumps in below-freezing weather.
4. Pump starters should automatically restart after power failure, (*i.e.,* maintain contact control).
5. Select nonoverloading pumps.
6. Instruct operating personnel never to shut down pumps in subfreezing weather.
7. Do not use aquastats, which can stop a pump, in boiler circuits.
8. Avoid sluggish circulation, which may cause air binding or dirt deposit. Properly balance and clean systems. Provide proper air control or means to eliminate air.
9. Freezestats should be phase change capillaries wound in a serpentine pattern across the leaving face of the upstream coil.

In fan equipment handling outdoor air, take precautions to avoid stratification of air entering the coil. The best methods for proper mixing of indoor and outdoor air are:

1. Select dampers for pressure drops adequate to provide stable control of mixing, preferably with dampers installed several equivalent diameters upstream of the air-handling unit.
2. Design intake and approach duct systems to promote natural mixing.
3. Select coils with circuiting to allow parallel flow of air and water.

Freeze-up may still occur with any of these precautions. If an antifreeze solution is not used, water should circulate at all times. Valve-controlled elements should have low-limit thermostats and sensing elements should be located to ensure accurate air temperature readings. Primary-secondary pumping of coils with three-way valve injection (as in Figures 21b and 21c) is advantageous. Use outdoor reset of water temperature wherever possible.

ANTIFREEZE SOLUTIONS

In systems in danger of freeze-up, water solutions of ethylene glycol and propylene glycol are commonly used. Freeze protection may be needed (1) in snow melting applications (see Chapter 45 of the 1991 ASHRAE *Handbook—HVAC Applications*); (2) in systems subjected to 100% outdoor air where the methods outlined above may not provide absolute antifreeze protection; (3) in isolated parts or zones of a heating system where intermittent operation or long runs of exposed piping increase the danger of freezing; or (4) in process cooling applications requiring temperatures below 40 °F. While using ethylene glycol or propylene glycol is comparatively expensive and tends to create corrosion problems unless suitable inhibitors are used, it may be the only practical solution in many cases.

Solutions of triethylene glycol, as well as certain other heat transfer fluids, may also be used. However, ethylene glycol and propylene glycol are the most common substances used in hydronic systems since they are relatively inexpensive and provide the most effective heat transfer.

Heat Transfer and Flow

Figures 13 through 18 in Chapter 18 of the 1989 ASHRAE *Handbook—Fundamentals* show specific heat, specific gravity, and viscosity of various aqueous solutions of ethylene glycol and propylene glycol. Figure 10 of that chapter indicates the freezing points for the two solutions.

System flow rate is affected by relative density and specific heat according to the following equation:

$$q_w = 500Q(\rho/\rho_w)c_p\Delta t \qquad (18)$$

where:

q_w = total heat transfer rate, Btu/h
Q = water or solution flow rate, gal/min
ρ = fluid density
ρ_w = density of water at 60 °F
c_p = specific heat, Btu/lb·°F
Δt = temperature increase or decrease, °F

Effect on Heat Source or Chiller

Generally, ethylene glycol solutions should not be used directly in a boiler because of the danger of chemical corrosion caused by glycol breakdown on direct heating surfaces. However, properly inhibited glycol solutions can be used in low-temperature water systems directly in the heating boiler if proper operations can be ensured. Automobile antifreeze solutions are not recommended because the silicate inhibitor can cause fouling, pump seal wear, fluid gelation, and reduced heat transfer. The area or zone requiring the antifreeze protection can be isolated in a separate zone with a heat exchanger or converter. Glycol solutions are used directly in water chillers in many cases.

Glycol solutions affect the output of a heat exchanger by changing the film coefficient of the surface contacting the solution. This change in film coefficient is primarily caused by viscosity changes. Figure 31 illustrates typical changes in output for two types of heat exchangers, Curve A for a steam-to-liquid converter and Curve B for a refrigerant-to-liquid chiller. The curves are plotted for one set of operating conditions only and reflect the change in ethylene glycol concentration as the only variable. Propylene glycol has similar effects on heat exchanger output.

Since many other variables, such as liquid velocity, steam or refrigerant loading, temperature difference, and unit construction affect the overall coefficient of a heat exchanger, designers should consult manufacturers' ratings when selecting such equipment. The curves only indicate the magnitude of these output changes.

Effect on Terminal Units

Since the effect of glycol on the capacity of terminal units may vary widely with temperature, the manufacturer's rating data should be consulted when selecting heating or cooling units in glycol systems.

Effect on Pump Performance

Centrifugal pump characteristics are affected to some degree by glycol solutions because of viscosity changes. Figure 32 shows these effects on pump capacity, head, and efficiency. Figures 15 and 16 in Chapter 18 of the 1989 ASHRAE *Handbook—Fundamentals* plot the viscosity of ethylene glycol, propylene glycol, and water. Centrifugal pump performance is normally cataloged for water at 60 to 80 °F. Hence, absolute viscosity effects below 1.1 centipoises can safely be ignored as far as pump performance is concerned. In intermittently operated systems, such as snow melting applications, viscosity effects at start-up may decrease flow enough to slow pickup.

Effect on Piping Pressure Loss

The friction loss in piping also varies with viscosity changes. Figure 33 gives correction factors for various ethylene glycol and

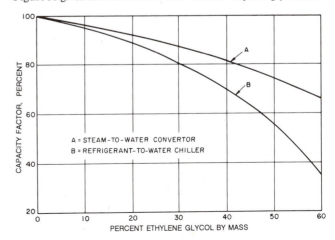

Fig. 31   Example of Effect of Aqueous Ethylene Glycol Solutions on Heat Exchanger Output

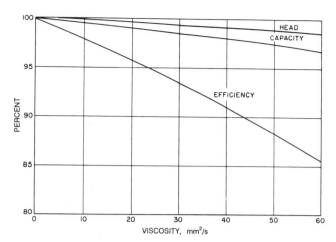

Fig. 32   Effect of Viscosity on Pump Characteristics

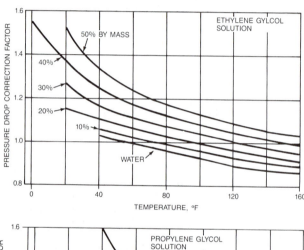

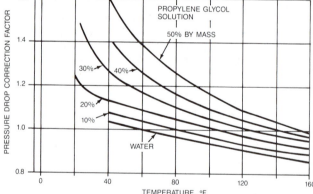

Fig. 33   Presure Drop Correction for Glycol Solutions

propylene glycol solutions. These factors are applied to the calculated pressure loss for water, as determined previously. Ethylene glycol and propylene glycol solutions need no correction above 160°F.

Installation and Maintenance

Since glycol solutions are comparatively expensive, the smallest possible concentrations to produce the desired antifreeze properties should be used. The total water content of the system should be calculated carefully to determine the required amount of glycol. The solution can be mixed outside the system in drums or barrels and then pumped in. Air vents should be watched during filling to prevent loss of solution. The system and the cold water supply should not be permanently connected, so automatic fill valves are usually not used.

Ethylene glycol and propylene glycol normally include an inhibitor to help prevent corrosion. Solutions should be checked each year using a suitable refractometer to determine glycol concentration. Certain precautions regarding the use of inhibited ethylene glycol solutions should be taken to extend their service life and to preserve equipment:

1. Before installing the glycol solution, thoroughly clean and flush the system.

2. Use waters that are soft and are low in chloride and sulfate ions to prepare the solution whenever possible.

3. Limit the maximum operating temperature to 250°F in a closed hydronic system. In a heat exchanger, limit glycol film temperatures to 300 to 350°F (steam pressures 120 psi or less) to prevent deterioration of the solution.

4. Check the concentration of inhibitor periodically, following procedures recommended by the glycol manufacturer.

REFERENCES

Carlson, G.F. 1981. The design influence of air on hydronic systems. ASHRAE *Transactions* 87(1):1293-1300.

Coad, W.J. 1985. Variable flow in hydronic systems for improved stability, simplicity and energy economics. ASHRAE *Transactions* 91.

Hull, R.F. 1981. Effect of air on hydraulic performance of the HVAC system. ASHRAE *Transactions* 87(1):1301-25.

Lockhart, H.A. and G.F. Carlson. 1953. Compression tank selection for hot water heating systems. ASHVE *Journal* (April).

Morabito, B.P. 1960. How higher cooling coil differentials affect system economics. ASHRAE *Journal* 2(8):60.

Pierce, J.D. 1963. Application of fin tube radiation to modern hot water heating systems. ASHRAE *Journal* 5(2):72.

Pompei, F. 1981. Air in hydronic systems: How Henry's law tells us what happens. ASHRAE *Transactions* 87(1):1326-42.

CHAPTER 13

CONDENSER WATER SYSTEMS

CONDENSER water systems for refrigerant compressors are classified as cooling tower systems, or as once-through systems, such as city water, well water, or pond or lake water systems. They are usually open systems, in which there are at least two points of interface between the system water and the atmosphere, and require a different approach to hydraulic design, pump selection, and sizing than do closed heating and cooling systems. Exceptions occur when closed-circuit towers or plate-type heat exchangers are used. Some heat conservation systems rely on a split condenser heating system that includes a two-section condenser. Heat from one section of the condenser is used for heating in closed-circuit systems and is occasionally interconnected with chilled water systems. The other section of the condenser serves as a heat-reject circuit, which is an open system connected to a cooling tower.

In selecting a pump for a condenser water system, consideration must be given to the static head and the system friction loss. The pump inlet must have an adequate net positive suction head. In addition, continuous contact with air in an open system introduces oxygen impurities and concentrates minerals that can cause scale and corrosion on a continuing basis. Fouling factors and an increased pressure drop caused by aging of the piping must be included in the condenser system design (see Chapter 36). The amount of water flow required depends on the refrigeration unit used and on the available temperature of the condenser water.

Cooling tower water is available at a temperature several degrees above the design wet-bulb temperature, depending on tower performance. In city, lake, river, or well water systems, the maximum water temperature that occurs during the operating season must be used for equipment selection.

With manufacturers' performance data, the required flow rate through a condenser may be determined for various condensing temperatures and capacities.

Once-Through Systems

Figure 1 shows a water-cooled condenser using city, well, or river water. The return is run higher than the condenser so that the condenser is always full of water. Water flow through the condenser is modulated by a control valve in the supply line, usually actuated from condenser head pressure to maintain a constant condensing temperature with load variations. City water systems should always include approved backflow prevention devices and open sight drains. When more than one condenser is used on the same circuit, individual control valves are used.

Piping should be sized according to the principles outlined in Chapter 33 of the 1989 ASHRAE *Handbook—Fundamentals*, with velocities of 5 to 10 fps for flow rates. A pump is not required where city water is used. Pumps may be necessary for well or river water, in which case the procedure for pump sizing is the same as for cooling tower systems.

Cooling Tower Systems

Figure 2 illustrates a typical cooling tower system for a refrigerant condenser. Water flows to the pump from the tower basin and

is discharged under pressure to the condenser and then back to the tower. Since it is desirable to maintain condenser water temperature above a predetermined minimum, water is diverted through a control valve to maintain minimum supply temperature. Piping from the tower sump to the pump requires some cautions. Sump

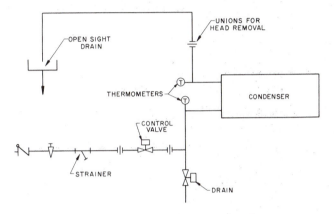

Fig. 1   Water-Cooled Condenser Connections for City Water

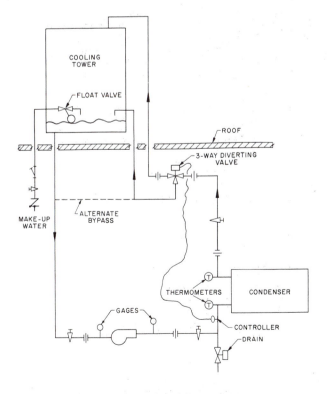

Fig. 2   Cooling Tower Piping Systems

The preparation of this chapter is assigned to TC 6.1, Hydronic and Steam Equipment and Systems.

13.1

level should be above the top of the pump casing for positive prime, and piping pressure drop should be minimized. All piping must pitch up either to the tower or the pump suction to eliminate air pockets. If used, suction strainers should be equipped with inlet and outlet gages to indicate when cleaning is required. Vortexing in the tower basin is prevented by piping connections at the tower according to the manufacturer's specifications and by limiting flow to the maximum allowed by the sump design. Pump performance is enhanced by a straight section of suction pipe five times the diameter in length or by using a suction diffuser.

The elements of required pump head are illustrated in Figure 3. Since there is an equal head of water between the level in the tower sump or interior reservoir and the pump on both the suction and discharge sides, these heads cancel each other and can be disregarded. The elements of pump head are: (1) static head from tower sump or interior reservoir level to the tower header, (2) friction loss in suction and discharge piping, (3) pressure loss in the condenser, (4) control valves, and (5) strainer and tower nozzles, if used. These elements added in feet of water determine the required pump total dynamic head.

Normally, piping is sized to water velocities between 5 and 12 fps. Friction factors for 15-year-old pipe are commonly used. The manufacturers' data contain pressure drops for the condenser, cooling tower, control valves, and strainers. If condensers are installed in parallel, only the one with the highest pressure drop is counted. Combination flow measuring and balancing valves can be used to equalize pressure drops.

If multiple cooling towers are to be connected, the piping should be designed so that the pressure loss from the tower to the pump suction is *exactly* equal for each tower. Additionally, large equalizing lines or a common reservoir can also be used to assure the same water level in each tower.

Evaporation in a cooling tower causes a concentration of dissolved solids in the circulating water. This concentration is limited by discharging a portion of the water as overflow or blowdown. Also, the drift loss that results from droplets of water drifting out of the tower helps to maintain a limited concentration.

Makeup water is required to replace the water lost by drift, evaporation, and blowdown. Automatic float valves are usually installed to maintain a constant water level.

Depending on the chemistry of the water available for the cooling tower, water treatment may be necessary to prevent scaling, cor-rosion, or biological fouling of the condenser and circulating system. On large systems, fixed continuous feeding chemical treatment systems are frequently installed in which chemicals, including acids for pH control, must be diluted and blended and then pumped into the condenser water system. Corrosion-resistant materials may be required for surfaces that come in contact with these chemicals. For further information on water treatment, refer to Chapter 43 of the 1991 ASHRAE *Handbook—HVAC Applications.*

Special precautions are required if cooling towers are to operate in subfreezing weather. Ice formation can hinder the performance of the tower by obstructing airflow, and periodic manual shutdown of the tower fan may be necessary for deicing. Operating periods in cold weather can be longer if the tower fan is thermostatically controlled with the bypass, so that gravity airflow is achieved and larger water quantities are circulated through the tower. The control for the diverting valve that bypasses the tower should have a wide throttling range from 12 to 15°F to avoid cycling, since periodic flow into the tower can cause rapid icing in subfreezing weather.

Exposed piping must be protected or drained when a tower operates intermittently during cold weather. The most satisfactory arrangement is to provide an indoor receiving tank into which the cold water basin drains by gravity. The makeup line, overflow, and pump suction are then connected to the reservoir tank rather than to the tower basin. This leaves only the return water line exposed to the weather; it can be drained by a small bleed line connected to the tower return water line (see Figure 4).

Tower basin heaters or heat exchangers connected across the supply and return line to the tower can also be selected, although their cost can be high. Steam or electric basin heaters are most commonly used. An arrangement incorporating an indoor heater is also shown in Figure 4.

Cooling towers that are unused in cold-weather months for considerable lengths of time should be drained. Using glycol solutions in cooling towers is not recommended.

City water is occasionally piped to condenser cooling systems for wintertime use. Whenever such connections are made, they should be connected as discussed in the section, Once-Through Systems.

For automatic control information, refer to Chapter 41 of the 1991 ASHRAE *Handbook—HVAC Applications.*

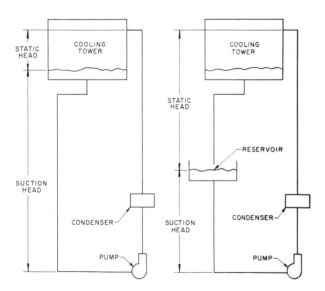

Fig. 3   Schematic Piping Layout Showing Static and Suction Head

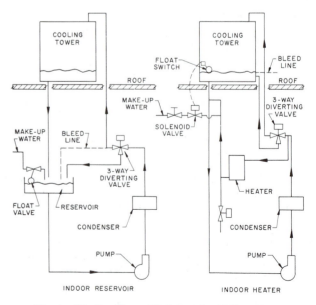

Fig. 4   Cooling Tower Piping to Avoid Freezeup

CHAPTER 14

MEDIUM AND HIGH-TEMPERATURE WATER HEATING SYSTEMS

HIGH-temperature water systems are classified as those operating with supply water temperatures above 350°F and designed to a pressure rating of 300 psig. Medium temperature water systems have operating temperatures below 350°F and permit design to a pressure rating of 125 to 150 psig. The usual practical temperature limit is about 450°F because of pressure limitations on pipe fittings, equipment, and accessories. The rapid rate of pressure rise that occurs as the temperature rises above 450°F increases system cost since higher pressure rated components are required (see Figure 1). The design principles for both medium temperature and high-temperature systems are basically the same. In this chapter, HTW refers to both systems.

This chapter presents the general principles and practices that apply to HTW and distinguishes them from low-temperature water systems operating below 250°F. Refer to Chapter 12 for basic design considerations applicable to all hot-water systems.

SYSTEM CHARACTERISTICS

The following characteristics distinguish HTW systems from steam distribution or low-temperature water systems:

- The system is a completely closed circuit with supply and return mains maintained under pressure. There are no losses from flashing, and heat that is not used in the terminal heat transfer equipment is returned to the HTW generator. Tight systems have minimal corrosion.
- Mechanical equipment that does not control the performance of individual terminal units is concentrated at the central station.
- Piping can slope up or down or run at a variety of elevations to suit the terrain and the architectural and structural requirements without provision for trapping at each low point. This may reduce the amount of excavation required and eliminate drip points and return pumps required with steam.
- Greater temperature drops are used and less water is circulated than in low-temperature water systems.
- The pressure in any part of the system must always be well above the pressure corresponding to the temperature at saturation in the system to prevent flashing of the water into steam.
- Terminal units requiring different water temperatures can be served at their required temperatures by: regulating the flow of water, modulating the water supply temperature, placing some units in series, and using heat exchangers or other methods.

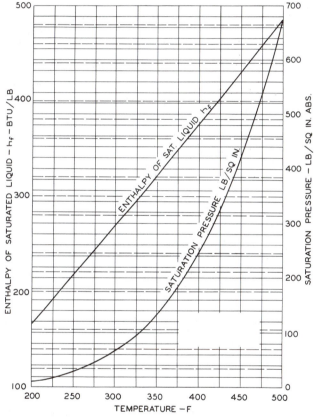

Fig. 1   Relation of Saturation Pressure and Enthalpy to Water Temperature

- The high heat content of the water in the high-temperature water circuit acts as a thermal flywheel, evening out fluctuations in the load. The heat storage capacity can be further increased by adding heat storage tanks or by increasing the temperature in the return mains during periods of light load.
- The high heat content of the heat carrier makes high-temperature water unsuitable for two-pipe dual temperature (hot and chilled water) applications and for intermittent operation if rapid start-up and shutdown are desired, unless the system is designed for minimum water volume and is operated with rapid response controls.
- Higher engineering skills are required to design a HTW system that is simple, yet safer and more convenient to operate than a comparable steam or low-temperature water system.

The preparation of this chapter is assigned to TC 6.1, Hydronic and Steam Equipment and Systems.

• HTW system design requires careful attention to basic laws of chemistry and physics as these systems are less forgiving than standard hydronic systems.

BASIC SYSTEM

High-temperature water systems are similar to conventional forced hot-water heating systems. They require a heat source, which can be a direct-fired HTW generator, a steam boiler, or an open or closed heat exchanger, to heat the water. The expansion of the heated water is usually taken up in an expansion vessel, which simultaneously pressurizes the system. Heat transport depends on circulating pumps. The distribution system is closed, comprising supply and return pipes under the same basic pressure. Heat emission at the terminal unit is indirect by heat transfer through heat transfer surfaces. The basic system is shown in Figure 2.

The principal differences from low-temperature water systems are the higher pressure used, the consequently heavier equipment, the generally smaller pipe sizes, and the manner in which water pressure is maintained.

Most systems are either (1) a saturated steam cushion system, in which the high-temperature water develops its own pressure or (2) a gas- or pump-pressurized system, in which the pressure is imposed externally.

HTW generators and all auxiliaries (such as water makeup and feed equipment, pressure tanks, and circulating pumps) are usually located in a central station. Cascade HTW generators sometimes use an existing steam distribution system and are installed remote from the central plant.

DESIGN CONSIDERATIONS

Selection of the system pressure, supply temperature, temperature drop, type of HTW generator, and the pressurization method are the most important initial design considerations. Some determining factors are:

• Type of load (space heating and/or process). Load fluctuations during a 24-h period and a 1-year period. Process loads might require water at a given minimum supply temperature continuously, while space heating can permit temperature modulation as a function of outdoor temperature or other climatic influences.
• Terminal unit temperature requirements.
• Distance between heating plant and space or process requiring heat.
• Quantity and pressure of steam used for power equipment in the central plant.
• Elevation variations within the system and the effect of basic pressure distribution.

Usually, distribution piping is the major investment in an HTW system. A distribution system with the widest tempera-

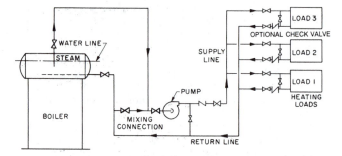

Fig. 2   Elements of High-Temperature Water System

ture spread (ΔT) between supply and return will have the lowest initial and operating costs. Economical designs have a ΔT of 150°F or higher.

The requirements of terminal equipment or user systems determine the system selected. For example, if the users are 10 psig steam generators, the return temperatures would be 250°F. A 300 psig rated system operated at 400°F would be selected to serve the load. In another example, where the primary system serves predominantly 140 to 180°F hot-water heating systems, an HTW system that operates at 325°F could be selected. The supply temperature is reduced by blending with 140°F return water to the desired 180°F hot-water supply temperature in a direct-connected hot-water secondary system. This highly economical design has a 140°F return temperature in the primary water system and a ΔT of 185°F.

Because the danger of water hammer is always present if flashing occurs when the pressure drops to the point when pressurized hot water flashes to steam, the primary HTW system should be designed with steel valves and fittings of 150 psi. The secondary water, which operates below 212°F and is not subject to flashing and water hammer, can be designed for 125 psi and standard HVAC equipment.

Theoretically, water temperatures up to about 350°F can be provided using equipment suitable for 125 psi. But in practice, unless push-pull pumping is used, maximum water temperatures will be limited by the system design, pump pressures, and elevation characteristics to values between 300 and 325°F.

Many systems designed for self-generated steam pressurization have a steam drum through which the entire flow is taken, and which also serves as an expansion vessel. A circulating pump in the supply line takes water from the tank. The temperature of the water from the steam drum cannot exceed the steam temperature in the drum that corresponds to its pressure at saturation. The point of maximum pressure is at the discharge of the circulating pump. If, for example, this pressure is to be maintained below 125 psig, the pressure in the drum that corresponds to the water temperature cannot exceed 125 psig minus the sum of the pump pressure and the pressure that is caused by the difference in elevation between the drum and the circulating pump.

Most systems are designed for inert gas pressurization. In most of these systems, the pressurizing tank is connected to the system by a single balance line on the suction side of the circulating pump. The circulating pump is located at the inlet side of the HTW generator. There is no flow through the pressurizing tank, and a reduced temperature will normally establish itself inside. A special characteristic of the gas-pressurized systems is the apparatus that creates and maintains gas pressure inside the tank.

In designing and operating an HTW system, it is important to maintain a pressure that always exceeds the vapor pressure of the water, even if the system is not operating. This may require limiting the water temperature and thereby the vapor pressure, or increasing the imposed pressure.

Elevation and the pressures required to prevent water from flashing into steam in the supply system can also limit the maximum water temperature that may be used and must therefore be studied in evaluating the temperature-pressure relationships and method of pressurizing the system.

The properties of water that govern design are as follows:

• Temperature versus pressure at saturation (Figure 1)
• Density or specific volume versus temperature
• Enthalpy or sensible heat versus temperature
• Viscosity versus temperature
• Type and amount of pressurization

The relationships among temperature, pressure, specific volume, and enthalpy are all available in steam tables. Some properties of water are summarized in Table 1 and Figure 3.

Direct-Fired High-Temperature Water Generators

In direct-fired HTW generators, the central stations are comparable to steam boiler plants operating within the same pressure range. The generators should be selected for size and type in keeping with the load and design pressures, as well as the circulation requirements peculiar to high-temperature water. In some systems, both steam for power or processing and high-temperature water are supplied from the same boiler; in others, steam is produced in the boilers and used for generating high-temperature water; and in many others, the burning fuel directly heats the water.

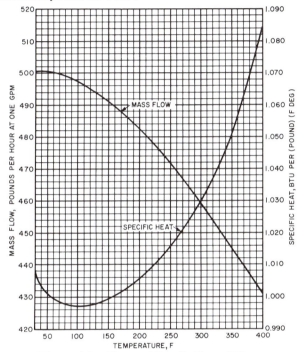

Fig. 3   Mass Flow and Specific Heat of Water

The HTW generators can be the water-tube or fire-tube type, and can be equipped with any conventional fuel firing apparatus. Water-tube generators can have either forced circulation, gravity circulation, or a combination of both. The recirculating pumps of forced circulation generators must operate continuously while the generator is being fired. Steam boilers relying on natural circulation might require internal baffling when used for HTW generation. In scotch marine boilers, thermal shock may occur, caused by a sudden drop in the temperature of the return water or when the ΔT exceeds 40°F. Forced-circulation HTW generators are usually the once-through type and rely solely on pumps to achieve circulation. Depending on the design, internal orifices in the various circuits might be required to regulate the water flow rates in proportion to the heat absorption rates. Circulation must be maintained at all times while the generator is being fired, and the flow rate must never drop below a minimum indicated by the manufacturer.

Where gravity circulation steam boilers are used for HTW generation, the steam drum usually serves as an expansion vessel. In steam pressurized forced-circulation HTW generators, a separate vessel is commonly used for maintaining the steam pressure cushion and for expansion. A separate vessel is always used when the system is cushioned by an inert gas or auxiliary steam. Proper internal circulation is essential in all types of boilers to prevent tube failures from overheating or unequal expansion.

In early HTW systems, the generator is a steam boiler with an integral steam drum used to absorb the expansion of the water level and for pressurization. A dip pipe removes water below the water line (Figure 4) (Applegate 1958). This dip pipe should be installed so that it picks up water at or near the saturation temperature, without too many steam bubbles. If a pipe breaks somewhere in the system, the boiler must not empty to a point where heating surfaces are bared and the danger of a boiler explosion results. The same precautions must be taken with the return pipe. If the return pipe is connected in the lower part of the boiler, a check valve should be placed in the connecting line to the boiler to preclude the danger of emptying the boiler.

When two or more such boilers supply a common system, the same steam pressure and water level must be maintained

Table 1   Properties of Water—212 to 400°F

Temperature, °F	Absolute Pressure, psia[a]	Density, lb/ft³	Specific Heat, Btu/lb · °F	Total Heat above 32°F		Dynamic Viscosity, Centipoise
				Btu/lb[a]	Btu/ft³	
212	14.70	59.81	1.007	180.07	10,770	0.2838
220	17.19	59.63	1.009	188.13	11,216	0.2712
230	20.78	59.38	1.010	198.23	11,770	0.2567
240	24.97	59.10	1.012	208.34	12,313	0.2436
250	29.83	58.82	1.015	218.48	12,851	0.2317
260	35.43	58.51	1.017	228.64	13,378	0.2207
270	41.86	58.24	1.020	238.84	13,910	0.2107
280	49.20	57.94	1.022	249.06	14,430	0.2015
290	57.56	57.64	1.025	259.31	14,947	0.1930
300	67.01	57.31	1.032	269.59	15,450	0.1852
310	77.68	56.98	1.035	279.92	15,950	0.1779
320	89.66	56.66	1.040	290.28	16,437	0.1712
330	103.06	56.31	1.042	300.68	16,931	0.1649
340	118.01	55.96	1.047	311.13	17,409	0.1591
350	134.63	55.59	1.052	321.63	17,879	0.1536
360	153.04	55.22	1.057	332.18	18,343	0.1484
370	173.37	54.85	1.062	342.79	18,802	0.1436
380	195.77	54.47	1.070	353.45	19,252	0.1391
390	220.37	54.05	1.077	364.17	19,681	0.1349
400	247.31	53.65	1.085	374.97	20,117	0.1308

[a]Reprinted by permission from *Thermodynamic Properties of Steam*, J. H. Keenan and F. G. Keyes, John Wiley and Sons, Inc., 1936 edition.

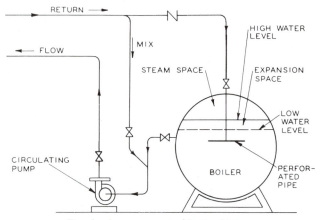

Fig. 4　Arrangement of Boiler Piping

in each. Water and steam balance pipes are usually installed between the drums (Figure 5). These should be liberally sized. The following table shows recommended sizes.

Boiler Rating, million Btu/h	Balance Pipe Dia, in.
2.5	3
5	3.5
10	4
15	5
20	6
30	8

A difference of only 0.25 psi in the system pressure between two boilers operated at 100 psig would cause a difference of 9 in. in the water level. The situation is further aggravated because an upset is not self-balancing. Rather, when too high a heat release in one of the boilers has caused the pressure to rise and the water level to fall in this boiler, the decrease in the flow of colder return water into it causes a further pressure rise, while the opposite happens in the other boiler. It is therefore important that the firing rates match the flow through each boiler at all times. Modern practice is to use either flooded HTW heaters with a single external pressurized expansion drum common to all the generators, or the combination of steam boilers and a direct contact (cascade) heater.

Expansion and Pressurization

In addition to the information in Chapter 12, the following factors should be considered:

• The connection point of the expansion tank used for pressurization greatly affects the pressure distribution throughout the system and the avoidance of HTW flashing.

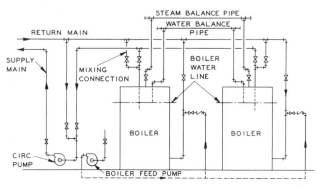

Fig. 5　Piping Connections for Two or More Boilers in HTW Systems Pressurized by Steam

• Proper safety devices for high and low water levels and excessive pressures should be incorporated in the expansion tank and interlocked with combustion safety and water flow rate controls.

The following fundamental methods, in which pressure in a given hydraulic system can be kept at a desired level, amplify the discussion in Chapter 12 (Blossom and Ziel 1959, National Academy of Sciences 1959).

1. An elevated storage tank is a simple pressurization method, but because of the great heights required for the pressure encountered, it is generally impractical.

2. Steam pressurization requires the use of an expansion vessel separate from the HTW generator. Since firing and flow rates can never be perfectly matched, some steam is always carried. Therefore, the vessel must be above the HTW generators and connected in the supply water line from the generator. This steam, supplemented by flashing of the water content in the expansion vessel, provides the steam cushion that pressurizes the system.

The expansion vessel must be equipped with steam safety valves capable of relieving the steam generated by all the generators. The generators themselves are usually designed for a substantially higher working pressure than the expansion drum, and their safety relief valves are set for the higher pressure to minimize their lift requirement.

The basic HTW pumping arrangements can be either single-pump, in which one pump handles both the generator and system loads, or two-pump, in which one pump circulates high-temperature water through the generator and a second pump circulates high-temperature water through the system (see Figures 6 and 7). The circulating pump moves the water from the expansion vessel to the system and back to the generator. The vessel must be elevated to increase the net positive suction pressure to prevent cavitation or flashing in the pump suction. This arrangement is critical. A bypass from the HTW system

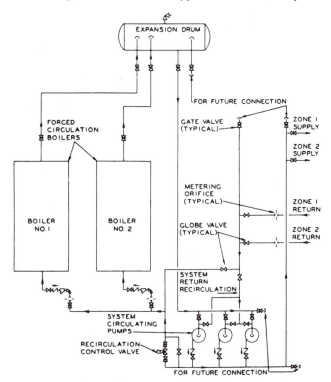

Fig. 6　HTW Piping for Combined (One-Pump) System (Steam Pressurized)

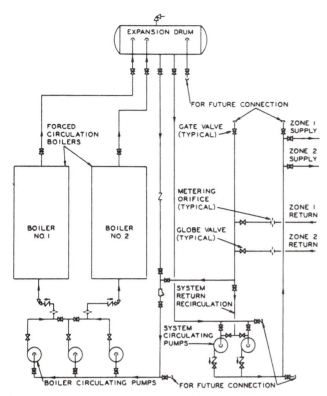

Fig. 7 HTW Piping for Separate (Two-Pump) System (Steam Pressurized)

return line to the pump suction helps prevent flashing. Cooler return water is then mixed with hotter water from the expansion vessel to give a resulting temperature below the corresponding saturation point in the vessel.

In the two-pump system, the boiler recirculation should always exceed the system circulation, since excessive cooling of the water content of the drum by the cooler return water entering the drum, in case of overcirculation, can cause pressure loss and flashing in the distribution system. Backflow into the drum can be prevented by installing a check valve in the balance line from the drum to the boiler recirculating pumps. Higher cushion pressures may be maintained by auxiliary steam from a separate generator.

Sizing. Steam-pressurized vessels should be sized for a total volume V_T, which is the sum of the volume V_1 required for the steam space, the volume V_2 required for water expansion, and the volume V_3 required for sludge and reserve. An allowance of 20% of the sum of V_2 and V_3 is reasonable for the volume V_1 required for the steam space.

The volume V_2 required for water expansion is determined from the change in water volume from the minimum to the maximum operating temperatures of the complete cycle. It is not necessary to allow for expansion of the total water volume in the system from a cold initial start. It is necessary during a start-up period to bleed off the volume of water caused by expansion from the initial starting temperature to the lowest average operating temperature.

The volume V_3 for sludge and reserve varies greatly depending on the size and design of system and generator capacity. An allowance of 40% of the volume V_2 required for water expansion is reasonable.

3. Nitrogen, the most commonly used inert gas, is used for gas pressurization. Air is not recommended because the oxygen in air contributes to corrosion in the system.

The expansion vessel is connected as close as possible to the

suction side of the HTW pump by a balance line. The inert gas used for pressurization is fed into the top of the cylinder, preferably through a manual fill connection using a reducing station connected to an inert gas cylinder. Locating the relief valve below the minimum water line is advantageous, since it is easier to keep it tightly sealed with water on the pressure side. If the valve is located above the water line, it is exposed to the inert gas of the system.

To reduce the area of contact between gas and water and the resulting absorption of gas into the water, the tank should be installed vertically. It should be located in the most suitable place in the central station. Similar to the steam-pressurized system, the pumping arrangements can be either one- or two-pump (see Figures 8 and 9).

The ratings of fittings, valves, piping, and equipment are considered in determining the maximum system pressure. A minimum pressure of about 25 to 50 psi over the maximum saturation pressure can be used. The imposed additional pressure above the vapor pressure must be large enough to prevent steaming in the HTW generators at all times, even under conditions when flow and firing rates in generators operated in parallel, or flow and heat absorption in parallel circuits within a generator, are not evenly matched. This is critical, since gas-pressurized systems do not have steam separating means and safety valves to evacuate the steam generated.

The simplest type of gas-pressurization system uses a variable gas quantity with or without gas recovery (Figure 10) (National Academy of Sciences 1959). In this system, the inert gas is relieved from the expansion vessel when the water rises and is wasted, or it is recovered in a low-pressure gas receiver from which the gas compressor pumps it into a high-pressure receiver for storage. When the water level drops in the expan-

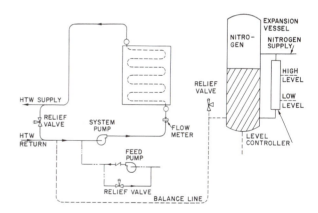

Fig. 8   Inert Gas Pressurization for One-Pump System

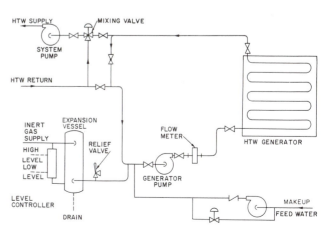

Fig. 9   Inert Gas Pressurization for Two-Pump System

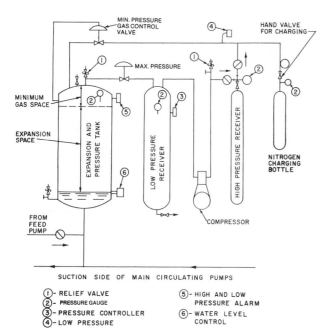

Fig. 10 Inert Gas Pressurization Using Variable Gas Quantity with Gas Recovery

sion vessel, the control cycle adds inert gas from bottles or from the high-pressure receiver to the expansion vessel to maintain the required pressure.

Gas wastage can significantly affect the operating cost. The gas recovery system should be analyzed based on the economics of each application. It is generally more applicable to larger systems.

Sizing. The vessel should be sized for a total volume V_T, which is the sum of the volume V_1 required for pressurization, the volume V_2 required for water expansion, and the volume V_3 required for sludge and reserve.

Calculations made on the basis of pressure-volume variations following Boyle's Law are reasonably accurate, assuming that the tank operates at a relatively constant temperature. The minimum gas volume can be determined from the expansion volume V_2 and from the control range between the minimum tank pressure P_1 and the maximum tank pressure P_2. The gas volume varies from the minimum V_1 to a maximum, which includes the water expansion volume V_2.

The minimum gas volume V_1 can be obtained from:

$$V_1 = P_1 V_2 / (P_2 - P_1)$$

where P_1 and P_2 are units of absolute pressure.

An allowance of 10% of the sum of V_1 and V_2 is reasonable for the sludge and reserve capacity V_3. The volume V_2 required for water expansion should be limited to the actual expansion that occurs during system operation through its minimum to maximum operating temperatures. It is necessary to bleed off water during a start-up cycle from a cold start. It is practicable on small systems, *e.g.*, under 1,000,000 Btu/h to 10,000,000 Btu/h to size the expansion vessel for the total water expansion from the initial fill temperature.

4. Pump pressurization in its simplest form consists of a feed pump and a regulator valve. The pump operates continuously, introducing water from the makeup tank into the system. The pressure regulator valve bleeds continuously back into the makeup tank. This method is usually restricted to small process heating systems. However, it can be used to temporarily pressurize a larger system to avoid shutdown during inspection of the expansion tank.

In larger central HTW systems, pump pressurization is combined with a fixed quantity gas compression tank that acts as a buffer. When the pressure rises above a preset value in the buffer tank, a control valve opens to relieve water from the balance line into the makeup storage tank. When the pressure falls below a preset second value, the feed pump is started automatically to pump water from the makeup tank back into the system. The buffer tank is designed to absorb only the limited expansion volume that is required for the pressure control system to function properly; it is usually small.

To prevent corrosion-causing elements, principally oxygen, from entering the HTW system, the makeup storage tank is usually closed and a low-pressure nitrogen cushion of 1 to 5 psig is maintained. The gas cushion is usually the variable gas quantity type with release to the atmosphere.

Direct Contact Heaters (Cascades)

High-temperature water can be obtained from direct contact heaters in which steam from turbine exhaust, extraction, or steam boilers is mixed with return water from the system. The mixing takes place in the upper part of the heater where the water cascading from horizontal baffles comes in direct contact with steam (Hansen 1966). The basic systems are shown in Figures 11 and 12.

The steam space in the upper part of the heater serves as the steam cushion for pressurizing the system. The lower part of the heater serves as the system's expansion tank. Where the water heater and the boiler operate under the same pressure, the surplus water is usually returned directly into the boiler through a pipe connecting the outlet of the high-temperature water circulating pump to the boiler.

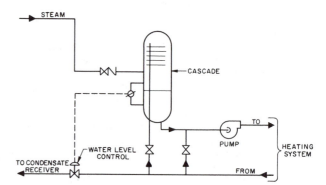

Fig. 11 Cascade HTW System

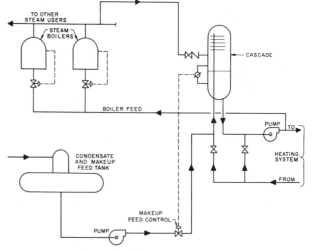

Fig. 12 Cascade HTW System Combined with Boiler Feedwater Preheating

The cascade system is also applicable where both steam and HTW services are required (Hansen and Liddy 1958). Where heat and power production are combined, the direct contact heater becomes the mixing condenser (Hansen and Perrsall 1960).

System Circulating Pumps

Forced-circulation boiler systems can be either one-pump or two-pump. These terms do not refer to the number of pumps but to the number of groups of pumps installed. In the former (see Figure 6), a single group of pumps assures both generator and distribution system circulation. In this system, both the distribution system and the generators are in series (Carter and Sturdevant 1958).

However, to ensure the minimum flow through the boiler at all times, a bypass around the distribution system must be provided. The one-pump method usually applies only to systems in which the total friction pressure is relatively low, since the energy loss of available circulating pressure from throttling in the bypass at times of reduced flow requirements in the district can substantially increase the operating cost.

In the two-pump system (see Figure 7), an additional group of recirculating pumps is installed solely to provide circulation for the generators (Carter and Sturdevant 1958). One pump is often used for each generator to draw water either from the expansion drum or the system return and to pump it through the generator into the expansion drum. The system circulating pumps draw water from the expansion drum and circulate it through the distribution system only. The supply temperature to the distribution system can be varied by mixing water from the return into the supply on the pump suction side. Where zoning is required, several groups of pumps can be used with a different pressure and different temperature in each zone. The flow rate can also be varied without affecting the generator circulation and without using a system bypass.

In steam-pressurized systems, the circulating pump is installed in the supply line to maintain all parts of the distributing system at pressures exceeding boiler pressure. This minimizes the danger of flashing into steam.

It is common practice to install a mixing connection from the return to the pump suction that bypasses the HTW generator. This connection is used for start-up and for modulating the supply temperature; it should not be relied on for increasing the pressure at the pump inlet. Where it is impossible to provide the required submergence by proper design, a separate small bore premixing line should be provided.

Hansen (1966) describes push-pull pumping, which divides the circulating pressure equally between two pumps in series. One is placed in the supply and is sized to overcome frictional resistance in the supply line of the heat distribution system. The second pump is in the return and is sized to overcome frictional resistance in the return. The expansion tank pressure is impressed on the system between the pumps. The HTW generator is either between the pumps or in the supply line from the pumps to the distribution system (see Figure 13).

In the push-pull system, the pressures in the supply and the return mains are symmetrical in relation to a line representing the pressure imposed on the system by the pressurizing source (expansion tank). This pressure becomes the system pressure when the pumps are stopped. The heat supply to using equipment or secondary circuits is controlled by two equal regulating valves, one on the inlet and the other on the outlet side, instead of the customary single valve on the leaving side. Both valves are operated in unison from a common controller; there are equal frictional resistances on both sides. Therefore, the pressure in the user circuits or equipment is maintained at all times halfway between the pressures in the supply and return mains.

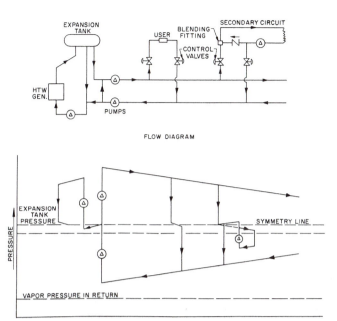

Fig. 13 Typical MTW System with Push-Pull Pumping

Since the halfway point is located on the symmetry line, the pressure in the user equipment or circuits is always equal to that of the pressurizing source (expansion tank) plus or minus static pressure caused by elevation differences. In other words, no system distribution pressure is reflected against the user circuit or equipment.

While the pressure in the supply system is higher than that of the expansion tank, the pressure in the return system, being symmetrical to the former, is lower. Therefore, the push-pull method is applicable only in systems where the temperature in the return is always significantly lower than that in the supply. Otherwise, flashing could occur. This is critical and requires careful investigation of the temperature-pressure relationship at all points. The push-pull method is not applicable in reverse-return systems.

Push-pull pumping permits use of standard 125 psi fittings and equipment in many MTW systems. Such systems, combined with secondary pumping, can be connected directly to low-temperature terminal equipment in building radiation. Temperature drops normally obtainable only in HTW systems can be achieved with MTW.

For example, 330°F water can be generated at 90 psig and distributed at less than 125 psig. Its temperature can be reduced to 200°F by secondary pumping. The pressure in the terminal equipment then is 90 psig and the MTW is returned to the primary system at 180°F. The temperature difference between supply and return in the primary MTW system is 150°F, which is comparable to that of an HTW system. In addition, conventional heat exchangers, expansion tanks, and water makeup equipment are eliminated from the secondary systems.

DISTRIBUTION PIPING DESIGN

Data for pipe friction are presented in Chapter 33 of the 1989 ASHRAE *Handbook—Fundamentals.* These pipe friction and fitting loss tables are for a 60°F water temperature. When applied to HTW systems, the values obtained are excessively high. The data should be used for preliminary pipe sizing only. Final pressure drop calculations should be made using the fundamental Darcy-Weisbach equation (on page 33.1 of the 1989 ASHRAE *Handbook—Fundamentals*) in conjunction with friction factors, pipe roughness, and fitting loss coefficients presented on page 2.10 of the 1989 ASHRAE *Handbook—Fundamentals*.

The conventional conduit or tunnel distribution systems are used with similar techniques for installation (see Chapter 11). A small valved bypass connection between the supply and the return pipe should be installed at the end of long runs to maintain a slight circulation in the mains during periods of minimum or no demand.

All pipe, valves, and fittings used in HTW systems should comply with the requirements of the ANSI/ASME *Standard B-31.1-89*, Power Piping, and the *National Fuel Gas Code* (NFPA 54-88 or AGA Z223.1-88). These codes state that hot-water systems should be designed for the highest pressure and temperature actually existing in the piping under normal operation. This pressure equals cushion pressure plus pump pressure plus static pressure. Schedule 40 steel pipe is applicable to most HTW systems with welded steel fittings and steel valves. A minimum number of joints should be used. In many installations, all valves in the piping system are welded or brazed. Flange connections used at major equipment can be serrated, raised flange facing, or ring joint. It is desirable to have backseating valves with special packing suitable for this service.

The ratings of valves, pipe, and fittings must be checked to determine the specific rating point for the given application. The pressure rating for a standard 300 psi steel valve operating at 400°F is 665 psi. Therefore, it is generally not necessary to use steel valves and fittings over 300 psi ratings in HTW systems.

Since high-temperature water is more penetrating than low-temperature water, leakage caused partly by capillary action should not be ignored because even a small amount of leakage vaporizes immediately. This slight leakage becomes noticeable only on the outside of the gland and stem of the valve where thin deposits of salt are left after evaporation. Avoid screwed joints and fittings in HTW systems. Pipe unions should not be used in place of flange connections, even for small bore piping and equipment.

Individual heating equipment units should be installed with separate valves for shutoff. These should be readily accessible. If the unit is to be isolated for service, valves will be needed in both the supply and return piping to the unit. Valve trim should be stainless steel or a similar alloy. Do not use brass and bronze.

High points in piping should have air vents for collecting and removing air, and low points should have provision for drainage. Loop-type expansion joints, in which the expansion is absorbed by deflection of the pipe loop, are preferable to the mechanical type. Mechanical expansion joints must be properly guided and anchored.

HEAT EXCHANGERS

Heat exchangers or converters commonly use steel shells with stainless steel, admiralty metal, or cupronickel tubes. Copper should not be used in HTW systems above 250°F. Material must be chosen carefully, considering the pressure-temperature characteristics of the particular system. All connections should be flanged or welded. On larger exchangers, water box-type construction is desirable to remove the tube bundle without breaking piping connections. Normally, HTW is circulated through the tubes, and because the heated water contains dissolved air, the baffles in the shell should be constructed of the same material as the tubes to control corrosion.

AIR HEATING COILS

In HTW systems over 400°F, coils should be cupronickel or all-steel construction. Below this point, other materials (*e.g.*, red brass) can be used after determining their suitability for the temperatures involved. Coils in outdoor air connections need freeze protection by damper closure or fan shutdown controlled by a thermostat. It is also possible to set the control valve on the preheat coil to a minimum position. This protects against freezing, as long as there is no unbalance in the tube circuits where parallel paths of HTW flow exist. A better method is to provide constant flow through the coil and to control heat output with face and bypass dampers or by modulating the water temperature with a mixing pump.

SPACE HEATING EQUIPMENT

In industrial areas, space heating equipment can be operated with the available high-temperature water. Convectors and radiators may require water temperatures in the low- and medium temperature range 120 to 180°F or 200 to 250°F, depending on their design pressure and proximity to the occupants. The water velocity through the heating equipment affects its capacity. This must be considered in selecting the equipment because, if a large water temperature drop is used, the circulation rate is reduced and consequently the flow velocity may be reduced enough to appreciably lower the heat transfer rate.

Convectors, specially designed to provide low surface temperatures, are now available to operate with water temperatures from 300 to 400°F.

These high temperatures are suitable for direct use in radiant panel surfaces. Since radiant output is a fourth-power function of the surface temperature, the surface area requirements are reduced over low-temperature water systems. The surfaces can be flat panels consisting of a steel tube, usually 0.38 or 0.5 in., welded to sheet steel turned up at the edges for stiffening. Several variations are available. Steel pipe can also be used with an aluminum or similar reflector to reflect the heat downward and to prevent smudging the surfaces above the pipe.

Instrumentation and Controls

Pressure gages should be installed in the pump discharge and suction and at locations where pressure readings will assist operation and maintenance. Thermometers (preferably dial-type) or thermometer wells should be installed in the flow and return pipes, the pump discharge, and at any other points of major temperature change or where temperatures are important in operating the system. It is desirable to have thermometers and gages in the piping at the entrance to each building converter.

On steam-pressurized cycles, the temperature of the water leaving the generator should control the firing rate to the generator. A master pressure control operating from the steam pressure in the expansion vessel should be incorporated as a high limit override. Inert gas-pressurized systems should be controlled from the generator discharge temperature. Combustion controls are discussed in detail in Chapter 27.

In the water-tube generators most commonly used for HTW applications, the flow of water passes through the generator in seconds. The temperature controller must have a rapid response to maintain a reasonably uniform leaving water temperature. In steam-pressurized units, the temperature variation must not exceed the antiflash pressure margin. At 300°F, a 5°F temperature variation corresponds to a 5 psi variation in the vapor pressure. At 350°F, the same temperature variation results in a vapor pressure variation of 10 psi. At 400°F, the variation increases to 15 psi and at 450°F to 22 psi. The permissible temperature swing must be reduced as the HTW temperature increases, or the pressure margin must be increased to avoid flashing.

Keep the controls simple. The rapid response through the generator makes it necessary to modulate the combustion rate on all systems with a capacity of over a few million Btu/h. In the smaller size range, this can be done by high-low firing. In large systems, particularly those used for central heating applications, full modulation of the combustion rate is desirable through at least 20% of full capacity. On-off burner control is generally not used in steam-pressurized cycles because the system loses pressurization during the off cycle, which can cause flashing and cavitation at the HTW pumps.

All generators should have separate safety controls to shut down the combustion apparatus when the system pressure or water temperature is high. HTW generators require a minimum water flow at all times to prevent tube failure. Means should be provided to measure the flow and to stop combustion if the flow falls below the minimum value recommended by the generator manufacturer. For inert gas-pressurized cycles, a low-pressure safety control should be included to shut down the combustion system if pressurization is lost. Figure 14 shows the basic schematic control diagram for an HTW generator.

Valve selection and sizing are important because of relatively high temperature drops and smaller flows in HTW systems. The valve must be sized so that it is effective over its full range of travel. The valve and equipment must be sized to absorb, in the control valve at full flow, not less than half the available pressure difference between supply and return mains where the equipment is served. A valve with equal percentage flow characteristics is needed. Sometimes two small valves provide better control than one large valve. Stainless steel trim is recommended, and all valve body materials and packing should be suitable for high temperatures and pressures. The valve should have a close-off rating at least equal to the maximum pressure produced by the circulating pump. Generally, two-way valves are more desirable than three-way valves because of equal percentage flow characteristics and the smaller capacities available in two-way valves. Single-seated valves are preferable to double-seated valves since the latter do not close tightly.

Control valves are commonly located in the return lines from heat transfer units to reduce the valve operating temperature and to prevent plug erosion caused by high-temperature water flashing to steam at lower discharge pressure. A typical application is to control the temperature of water being heated in a heat exchanger where the heating medium is high-temperature water. The temperature measuring element of the controller is installed on the secondary side and should be located where it can best detect changes to prevent overheating of outlet water. When the measuring element is located in the outlet pipe, there must be a continuous flow through the exchanger and past the element. The controller regulates the HTW supply to the primary side by means of the control valve in the HTW return. If the water leaving the exchanger is used for space heating, the set point of the thermostat in this water can be readjusted according to outdoor temperature.

Another typical application is to control a low- or medium pressure steam generator, usually less than 50 psig, using high-temperature water as the heat source. In this application, a proportional pressure controller measures the steam pressure on the secondary side and positions the HTW control valve on the primary side to maintain the desired steam pressure. For general information on automatic controls, refer to Chapter 41 of the 1991 ASHRAE *Handbook—HVAC Applications.*

Where submergence is sufficient to prevent flashing in the *vena contracta,* control valves can be in the HTW supply instead of in the return to water heaters and steam generators (see Figure 15). When used in conjunction with a check valve in the return, this arrangement shuts off the high-temperature water supply to the heat exchangers if a tube bundle leaks or ruptures.

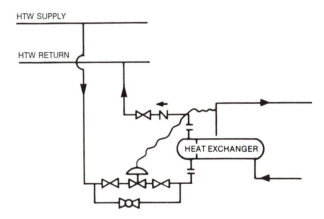

Fig. 15 Heat Exchanger Connections

WATER TREATMENT

Water treatment for HTW systems should be referred to a specialist. Oxygen introduced in makeup water immediately oxidizes steel at these temperatures, and over a period of time the corrosion can be substantial. Other impurities can also harm boiler tubes. Solids in impure water left by invisible vapor escaping at packings increase maintenance requirements. The condition of the water and the steel surfaces should be checked periodically in systems operating at these temperatures.

HEAT STORAGE

The high heat storage capacity in water produces a flywheel effect in most HTW systems that evens out load fluctuations. Systems with normal peaks can obtain as much as 15% added capacity through such heat storage. Excessive peak and low loads of a cyclic nature can be eliminated by an HTW accumulator, based on the principle of stratification. Heat storage in an extensive system can sometimes be increased by bypass-

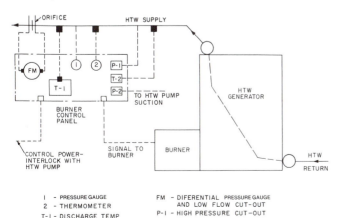

I - PRESSURE GAUGE
2 - THERMOMETER
T-I - DISCHARGE TEMP
CONTROLLER
(PROPORTIONAL)

FM - DIFERENTIAL PRESSURE GAUGE
AND LOW FLOW CUT-OUT
P-I - HIGH PRESSURE CUT-OUT
T-2 - HIGH TEMPERATURE CUT-OUT
P-2 - LOW PRESSURE CUT-OUT

Fig. 14 Control Diagram for HTW Generator

ing water from the supply into the return at the end of the mains, or by raising the temperature of the returns during periods of low load.

SAFETY CONSIDERATIONS

A properly engineered and operated HTW system is safe and dependable. Careful selection and arrangement of components and materials are important. Piping must be designed and installed to prevent undue stress. When HTW is released to atmospheric pressure, flashing takes place, which absorbs a large portion of the energy. Turbulent mixing of the liquid and vapor with room air reduces the temperature well below 212°F. With low mass flow rates, the temperature of the escaping mixture can fall to 125 to 140°F within a short distance, compared with the temperature of the discharge of a low-temperature water system, which remains essentially the same as the temperature of the working fluid (Hansen 1959, Armstrong and Harris 1966).

When large mass flow rates of HTW are released to atmospheric pressure in a confined space, *e.g.,* rupture of a large pipe or vessel, a hazardous condition could exist, similar to that occurring with the rupture of a large steam main. Failures of this nature are rare if good engineering practice is followed in system design.

REFERENCES

Applegate, G. 1958. British and European design and construction methods. ASHRAE Journal Section, *Heating, Piping & Air Conditioning* (March):169.

Armstrong, C.P. and W.S. Harris. 1966. Temperature distributions in steam and hot water jets simulating leaks. ASHRAE *Transactions* 72(1):147.

Blossom, J.S. and P.H. Ziel. 1959. Pressurizing high-temperature water systems. ASHRAE *Journal* 1 (November):47.

Carter, C.A. and B.L. Sturdevant. 1958. Design of high temperature water systems for military installations. ASHRAE Journal Section, *Heating, Piping & Air Conditioning* (February):109.

Hansen, E.G. 1959. Safety of high temperature hot water. *Actual Specifying Engineer* (July).

Hansen, E.G. 1966. Push-pull pumping permits use of MTW in building radiation. *Heating, Piping & Air Conditioning* (May):97.

Hansen, E.G. and W. Liddy. 1958. A flexible high pressure hot water and steam boiler plant. *Power* (May):109.

Hansen, E.G. and N.E. Perrsall. 1960. Turbo-generators supply steam for high-temperature water heating. *Air Conditioning, Heating and Ventilating* (June):90.

National Academy of Sciences. 1959. *High temperature water for heating and light process loads.* Federal Construction Council Technical Report No. 37. National Research Council Publication No. 753.

CHAPTER 15

INFRARED RADIANT HEATING

RADIANT principles discussed in this chapter apply to equipment with radiant source temperatures ranging from below room temperature to 5000 °F. Radiant source temperatures are categorized into four groups as follows:

- Low temperature
- Low intensity
- Medium intensity
- High intensity

Low-temperature or panel heating and cooling systems have source temperatures up to 300 °F. Typical low-temperature sources are the ceiling and/or floor of the conditioned space. The source of energy for this application can be electrical resistance wire or film element, hot water, or warm air. Low-temperature radiant heating is used in residential applications and in office, commercial, or industrial buildings. These systems are often applied in conjunction with variable air volume (VAV) systems. Chapter 6 has further information on low-temperature (panel heating and cooling) systems.

Low-intensity source temperatures range from 300 to 1200 °F. A typical low-intensity heater is mounted on the ceiling. It may be constructed of a 4 in. steel tube 20 to 30 ft long. A gas burner inserted into the end of the tube raises the tube temperature, and, because most units are equipped with a reflector, the radiant energy emitted is directed down to the conditioned space.

Medium-intensity source temperatures range from 1200 to 1800 °F. Typical sources include porous matrix, gas-fired infrared or metal sheathed, electric units.

High-intensity radiant source temperatures range from 1800 to 5000 °F. A typical high-intensity unit is an electrical reflector lamp with resistor temperatures of 4050 °F.

Low-, medium-, and high-intensity infrared heaters are frequently applied in aircraft hangars, factories, warehouses, foundries, greenhouses, and gymnasiums. They are applied to such open areas as loading docks, racetrack stands, under marquees, outdoor restaurants, and around swimming pools. Infrared heaters are also used for snow control, condensation control, and industrial process heating. Reflectors are frequently used to control the distribution of radiation in specific patterns.

When infrared is used, the environment is characterized by:

1. A high-temperature directional radiant field created by the infrared heaters
2. A low-temperature radiant field consisting of the walls and/or enclosing surfaces
3. Ambient air temperatures often lower than those found with conventional convective heaters

Convection heat loss from the radiantly heated floor, sealed objects, and the radiant heat source increases the ambient temperature. Ultimately, the combined action of these factors de-

termines occupant comfort and thermal acceptability of the environment.

ENERGY CONSERVATION

Infrared heating units are effective for spot heating. However, because of efficient performance, they are also used for total heating of large areas and entire buildings (Buckley 1989). Radiant heaters transfer energy directly to solid objects. Little energy is lost during transmission because air is a poor absorber of infrared energy. Since an intermediate transfer medium (such as air or water) is not needed, fans or pumps are not required.

As infrared energy warms floors and objects, they, in turn, release heat to the air by convection. Reradiation to surrounding objects also contributes to the comfort in the area. An energy saving advantage is that radiant heat can be turned off when it is not needed; when it is turned on again, it is effective in minutes.

Human comfort is determined by the average of mean radiant and dry-bulb temperatures. With radiant heating, the dry-bulb temperature may be kept lower for a given comfort level than with other forms of heating (ASHRAE 1981). As a result, the heat lost to ventilating air and via conduction through the shell of the structure is proportionally smaller, as is energy consumption. Infiltration, which is a function of temperature, is also reduced.

In some situations, radiant heating saves energy by reducing the temperature stratification from the equipment to the floor.

Buckley and Seel (1987) compared energy savings of infrared heating with other types of heating systems. A New York State report (1973) identified annual fuel savings as high as 50%. Recognizing the reduced fuel requirement for these applications, Buckley (1988) notes that it is common for manufacturers of radiant equipment to recommend installation of equipment with a rated output that is 80 to 85% of the heat loss calculated by methods described in Chapter 25 of the 1989 ASHRAE *Handbook—Fundamentals*.

INFRARED ENERGY GENERATORS

Gas Infrared

Modern gas-fired infrared heaters burn gas to heat a specific radiating surface. The surface is heated by direct flame contact or with combustion gases. Studies by the Gas Research Board of London (1944), Plyler (1948), and Haslam *et al.* (1925) reveal that only 10 to 20% of the energy produced by open combustion of a gaseous fuel is infrared radiant energy, whereas wavelength span can be controlled by design. The specific radiating surface of a properly designed unit increases radiant release efficiency and directs radiation toward the load. Heaters are available in the following types (see Table 1 for characteristics):

Indirect infrared radiation units (Figures 1a, 1b, and 1c) are internally fired and have the radiating surface between the hot gases and load. Combustion takes place within the radiating elements, which operate with surface temperatures up to 1200 °F. The

---

The preparation of this chapter is assigned to TC 6.5, Radiant Space Heating and Cooling.

elements may be tubes or panels, or they may have metal or ceramic components. Indirect infrared radiation units are usually vented and may require eductors, as is shown in Figure 1b.

Porous matrix infrared radiation units (Figure 1d) have a refractory material, which may be porous ceramic, drilled port ceramic, stainless steel, or a metallic screen. The units are enclosed, except for the major surface facing the load. A combustible gas-air mixture enters the enclosure, flows through the refractory material to the exposed face, and is distributed evenly by the porous character of the refractory. Combustion occurs evenly on the exposed surface. The flame recedes into the matrix, which adds radiant energy to the flame. If the refractory porosity is suitable, an atmospheric burner can be used, resulting in a surface temperature approaching 1650°F. Power burner operation may be required if refractory density is high. However, the resulting surface temperature may also be higher (1800°F).

Catalytic oxidation infrared radiant units (Figure 1e) are similar to the porous matrix units in construction, appearance, and operation, but the refractory material is usually glass wool, and the radiating surface is a catalyst that causes oxidation to proceed without visible flames.

Electric Infrared

Electric infrared heaters use heat produced by current flowing in a high-resistance wire, graphite ribbon, or film element. The following types are most commonly used (see Table 2 for characteristics):

Metal sheath infrared radiation elements (Figure 2a) are composed of a nickel-chromium heating wire embedded in an electrical insulating refractory, which is encased by a metal tube. These elements have excellent resistance to thermal shock, vibration, and impact, and they can be mounted in any position. At full voltage, the elements attain a sheath surface temperature of 1200 to 1800°F. Higher temperatures are obtained by such configurations as a hairpin shape. These units generally contain a reflector, which directs radiation to the load. Higher efficiency is obtained if these elements are shielded from wind.

Reflector lamp infrared radiation units (Figure 2b) have a coiled tungsten filament, which approximates a point source radiator. The filament is enclosed in a heat-resistant, clear, frosted, or red glass envelope, which is partially silvered inside to form an efficient reflector. Common units may be screwed into a 120-V light socket.

Quartz-tube infrared radiant units (Figure 2c) have a coiled nickel-chromium wire lying unsupported within an unevacuated fused quartz tube, which is capped (not sealed) by porcelain or metal terminal blocks. These units are easily damaged by impact and vibration but stand up well to thermal shock and splashing.

They must be mounted in a horizontal position to minimize coil sag, and they are usually mounted in a fixture that contains a reflector. Normal operating temperatures range from 1300 to 1800°F for the coil and about 1200°F for the tube.

Tubular quartz lamp units (Figure 2d) consist of a 0.38-in. diameter fused quartz tube containing an inert gas and a coiled tungsten filament held in a straight line and away from the tube by tantalum spacers. Filament ends are embedded in sealing material at the ends of the envelope. Lamps must be mounted horizontally, or nearly so, to minimize filament sag and overheating of the sealed ends. At normal design voltages, quartz lamp filaments operate at about 4050°F, while the envelope operates at about 1100°F.

Low-temperature radiant heating panels (Figure 2e) consist of a 1-in. thick galvanized steel or aluminum framed panel with a graphite or nichrome wire heating element. Panels come in various dimensions ranging in widths from 6 in. to 4 ft and lengths from 1 ft to 8 ft. Maximum watt density is 50 to 125 W/ft². Normal operating temperatures of radiating surface are 180 to 300°F. Units can be laid in a T-bar grid system or in a recessed frame, or they can be surface mounted with a separate frame or fastened directly to the ceiling.

Oil Infrared

Oil-fired infrared radiation heaters are similar to gas-fired indirect infrared radiation units (Figures 1a, 1b, and 1c). Oil-fired units are vented.

SYSTEM EFFICIENCY

Because many factors contribute to the specific performance of an infrared system, a single criterion should not be used to evaluate comparable systems. Therefore, use at least two of the following indicators when evaluating system performance:

Radiation generating ratio equals infrared energy generated divided by total energy input.

Fixture efficiency is an index of a fixture's ability to emit the radiant energy developed by the infrared source; it is usually based on total energy input. The housing, reflector, and other parts of a fixture absorb some infrared energy and convert it to heat, which is lost through convection. A fixture that controls direction and distribution of energy effectively may have a lower fixture efficiency.

Pattern efficiency is an index of a fixture's effectiveness in directing the infrared energy into a specific pattern. This effectiveness, plus effective application of the pattern to the load, influences the total effectiveness of the system (Boyd 1963). Typical radiation-generating ratios of gas infrared generators are indicated in Table 1. Limited test data indicate that the amount of radiant energy

Table 1 Characteristics of Typical Gas-Fired Infrared Heaters

Characteristics	Type 1	Type 2	Type 3
Operating temperature	To 1200°F	1600 to 1800°F	650 to 700°F
Relative heat intensity,[a] Btu/h·ft²	Low to 7500	Medium 17,000 to 32,000	Low 800 to 3000
Response time (heat-up)	180 s	60 s	300 s
Radiation generating ratio[b]	0.35 to 0.55	0.35 to 0.60	No data
Thermal shock resistance	Excellent	Excellent	Excellent
Vibration resistance	Excellent	Excellent	Excellent
Color blindness[c]	Excellent	Very good	Excellent
Luminosity (visible light)	To dull red	Yellow red	None
Mounting height	9 to 50 ft	12 to 50 ft	To 10 ft
Wind or draft resistance	Good	Fair	Very good
Venting	Optional	Nonvented	Nonvented
Flexibility	Good	Excellent—wide range of heat intensities and mounting possibilities available	Limited to low heat intensity applications

[a]Heat intensity emitted at burner surface.
[b]Ratio of radiant output to input.

[c]Color blindness refers to absorptivity by various loads of energy emitted by the different sources.

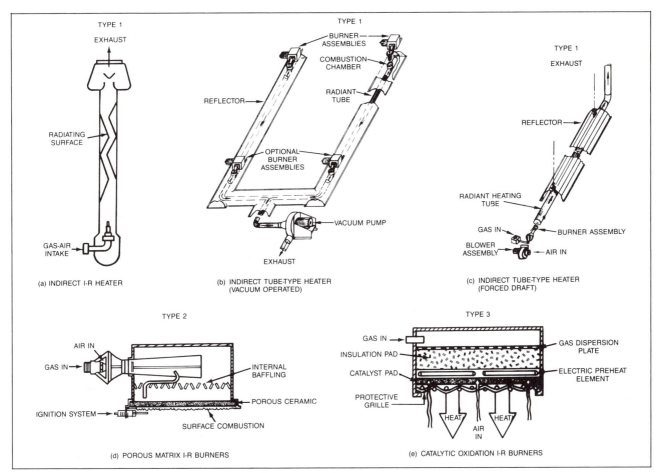

Fig. 1   Types of Gas-Fired Heaters

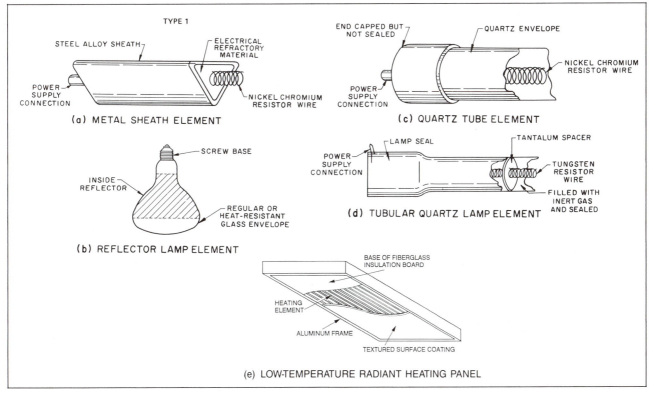

Fig. 2   Common Electric Infrared Heaters

emitted from gas infrared units, compared to the amount of convective energy, ranges from 35 to 60%. The Stefan-Boltzmann law can be used to estimate the infrared output capability, if reasonably accurate values of true surface temperature, emitting area, and surface emittance are available (DeWirth 1960). DeWirth (1962) also addresses the spectral distribution of energy curves for several gas sources.

Table 2 lists typical radiation-generating ratios of electric infrared generators. Fixture efficiencies are typically 80 to 95% of the radiation-generating ratios.

Infrared heaters should be operated at rated input. A small reduction in input causes a larger decrease in radiant output because of the fourth power dependence of radiant output on radiator temperature. Because a variety of infrared units with a variety of reflectors and shields are available, the manufacturers' information should be consulted.

REFLECTORS

Radiation from most infrared heating devices is directed by the emitting surface and can be concentrated by reflectors. Mounting height and the choice between spot or total heating usually determines what type of reflector will achieve the desired heat-flux pattern at floor level. Four types of reflectors can be used: (1) parabolic, which produces essentially parallel beams of energy; (2) elliptical, which directs all energy that is received or generated at the first focal point through a second focal point; (3) spherical, a special class of elliptical reflectors with coincident foci; and (4) flat, which redirects the emitted energy without concentrating or collimating the rays.

Energy data furnished by the manufacturer should be consulted to apply a heater properly.

CONTROLS

Normally, all controls (except the thermostat) are built into gas-fired infrared heaters, whereas electric infrared fixtures usually do not have built-in controls. Because of the effects of direct radiation, higher mean radiant temperature (MRT), and decreased ambient temperature compared to warm air systems, infrared heating requires careful selection and location of the thermostat sensor. Some installers recommend placing the thermostat or sensor in the radiation pattern. The nature of the system, the type of infrared heating units used, and the nature of the thermostat or sensor dictate the appropriate approach. Furthermore, no single location appears to be equally effective during the periods after a cold start and after substantial operation. The most desirable cycling rate for thermostats controlling infrared heaters has not been fully defined (Walker 1962).

An infrared heating system controlled by low-limit thermostats can be used for freeze protection. A thermostat usually controls an automatic valve on gas-fired infrared units to provide *on-off* control of gas flow to all burners. If a unit has a pilot flame, a sensing element prevents the flow of gas to burners only (or to both burners and pilot) when the pilot is extinguished. Electrical ignition may be used with provision for manual or automatic reignition of the pilot if it goes out. Electric spark ignition may also be used.

Gas and electric infrared systems for full building heating may have a zone thermostatic control system in which a thermostat

Table 2   Characteristics of Five Electric Infrared Elements

Characteristic	Type 1: Metal Sheath	Type 2: Reflector Lamp	Type 3: Quartz Tube	Type 4: Quartz Lamp	Type 5: Low-Temperature
Resistor material	Nickel-chromium alloy	Tungsten wire	Nickel-chromium alloy	Tungsten wire	Graphite or nichrome-wire
Relative heat intensity	Medium, 60 W/in., 0.5 in. dia	High, 125 to 375 W/spot	Medium to high, 75 W/in, 0.5 in. dia	High, 100 W/in., 3/8 in. dia	Low, 50 to 125 W/ft^2
Resistor temperature	1750°F	4050°F	1700°F	4050°F	180 to 350°F
Envelope temperature (in use)	1550°F	525 to 575°F	1200°F	1100°F	160 to 300°F
Radiation generating ratio[a]	0.58	0.86	0.81	0.86	0.7 to 0.8
Response time (heat-up)	180 s	A few seconds	60 s	A few seconds	240 to 600 s
Luminosity (visible light)	Very low (dull red)	High 8 lm/W	Low (orange)	High 7.5 lm/W	None
Thermal shock resistance	Excellent	Poor to excellent (heat-resistant glass)	Excellent	Excellent	Excellent
Vibration resistance	Excellent	Medium	Medium	Medium	Excellent
Impact resistance	Excellent	Medium	Poor	Poor	Excellent
Resistance to drafts or wind[b]	Medium	Excellent	Medium	Excellent	Poor
Mounting position	Any	Any	Horizontal[c]	Horizontal	Any
Envelope material	Steel alloy	Regular or heat-resistant glass	Translucent quartz	Clear, translucent, or frost quartz and integral red filter glass	Steel alloy or aluminum
Color blindness	Very good	Fair	Very good	Fair	Very good
Flexibility	Good—wide range of watt density, length, and voltage practical	Limited to 125-250 and 375 watt at 120 V	Excellent—wide range of watt density, diameter, length, and voltage practical	Limited. 1 to 3 wattages for each voltage; 1 length for each capacity	Good—wide range of watt density, length, and voltage practical
Life expectancy	Over 5000 h	5000 h	5000 h	5000 h	Over 10,000 h

[a]Ratio or radiant output to watt input (elements only).
[b]May be shielded from wind effects by louvers, deep-drawn fixtures, or both.

[c]May be provided with special internal supports for other than horizontal use.

representative of one outside exposure operates heaters along that outside wall. Two or more zone thermostats may be required for extremely long wall exposures. Heaters for an internal zone may be grouped around a thermostat representative of that zone. Manual switches or thermostats are usually used for spot or area heating, but input controllers may also be used.

Input controllers control electric infrared heating units effectively with metal sheath or quartz tube elements. An input controller is a motor-driven cycling device whose *on* time per cycle can be set. A 30-s cycle is normal. When a circuit's capacity exceeds an input controller's rating, the controller can be used to cycle a pilot circuit of contactors adequate for the load.

Input controllers work well with metal sheath heaters because the sheath mass smooths the pulses into even radiation. The control method decreases the efficiency of infrared generation slightly. Quartz tube elements, which have a warm-up time of several seconds, have perceptible, but not normally disturbing, pulses of infrared, with only moderate reduction in generation efficiency when controlled with these devices.

Input controllers should not be used with quartz lamps because the cycling luminosity would be distracting. Instead, output from a quartz lamp unit can be controlled by changing the voltage to the lamp element. This control can be done by modulating transformers or by switching the power supply from hot-to-hot to hot-to-ground potential.

Power drawn by the tungsten filament of the quartz lamp varies approximately as the 1.5 power of the voltage, while that of the metal sheath or quartz tube elements (using nickel-chromium wire) varies as the square of the voltage. Multiple circuits for electric infrared systems can be manually or automatically switched to provide multiple stages of heat.

Three circuits or control stages are usually adequate. For areas with fairly uniform radiation, one circuit should be controlled with input control or voltage variation control on electric units, while the other two are on full *on* or *off* control. This arrangement gives flexible, staged control with maximum efficiency of infrared generation. The variable circuit alone provides zero to one-third capacity. Adding another circuit at full *on* provides one-third to two-thirds capacity, and adding the third circuit provides two-thirds to full capacity.

PRECAUTIONS

Precautions for the application of infrared heaters include the following:

- All infrared heaters covered in this chapter have high surface temperatures when they are operating and should, therefore, not be used when the atmosphere contains ignitable dust, gases, or vapors in hazardous concentrations.
- Manufacturers' recommendations for clearance between a fixture and combustible material should be followed. If combustible material is being stored, warning notices defining proper clearances should be posted near the fixture.
- Manufacturers' recommendations for clearance between a fixture and personnel areas should be followed to prevent personnel stress from local overheating.
- Infrared fixtures should not be used if the atmosphere contains gases, vapors, or dust that decompose to hazardous or toxic materials in the presence of high temperature and air. For example, infrared units should not be used in an area with a degreasing operation that uses trichlorethylene, unless the area has a suitable exhaust system that isolates the contaminant. Trichlorethylene, when heated, forms phosgene (a toxic compound) and hydrogen chloride (a corrosive compound).
- Humidity must be controlled in areas with unvented gas-fired infrared units, because water formed by combustion increases humidity. Sufficient ventilation (AGA 1988), direct venting, or insulation on cold surfaces helps control moisture problems.
- Adequate makeup air must be provided to replace the air used by combustion-type heaters, regardless of whether units are direct-vented or not.
- If unvented combustion-type infrared heaters are used, the area must have adequate ventilation to ensure that products of combustion in the air are held to an acceptable level (Prince 1962).
- To be kept comfortable with infrared heating equipment, personnel should be protected from substantial wind or drafts. Suitable wind shields seem to be more effective than increased radiation density (Boyd 1960).

MAINTENANCE

Electric infrared systems require little care beyond the cleaning of reflectors.

Quartz and glass elements must be handled carefully because they are fragile, and fingerprints must be removed (preferably with alcohol) to prevent etching at operating temperature, which causes early failure.

Gas- and oil-fired infrared heaters require periodic cleaning to remove dust, dirt, and soot. Reflecting surfaces must be kept clean to remain efficient. An annual cleaning of heat exchangers, radiating surfaces, burners, and reflectors with compressed air is usually sufficient. Chemical cleaners must not leave a film on reflector surfaces.

Both main and pilot air ports of gas-fired units should be kept free of lint and dust. The nozzle, draft tube, and nose cone of oil-fired unit burners are designed to operate in a particular combustion chamber, so they must be replaced carefully when they are removed.

DESIGN CONSIDERATIONS FOR BEAM RADIANT HEATERS

Chapter 48 of the 1991 ASHRAE *Handbook—HVAC Applications* introduces the principles of design for spot beam radiant heating. The effective radiant flux (ERF) represents the radiant energy absorbed by an occupant from all temperature sources different from the ambient. ERF is defined as:

$$\text{ERF} = h_r \, (\bar{t}_r - t_a) \tag{1}$$

where

ERF = effective radiant flux, Btu/h·ft^2
h_r = linear radiation transfer coefficient, Btu/h·ft^2·°F
$\bar{t}_r$ = mean radiant temperature affecting occupant, °F
t_a = ambient air temperature near occupant, °F

ERF may be measured as the heat absorbed at the clothing and skin surface:

$$\text{ERF} = \alpha_k \, I_k \, (A_p/d^2)/A_D \tag{2}$$

where

α_k = absorption of skin-clothing surface at emitter temperature (Figure 3), °F
I_k = irradiance from beam heater, Btu/h·sr
A_p/d^2 = solid angle subtended by projected area of occupant from beam A_p heater, which is treated as a point source
A_D = body surface area of person, ft^2

The value of A_D has been defined by Du Bois as follows:

$$A_D = 0.0621 \, W^{0.425} \, H^{0.725}$$

where

W = mass of person, lb
H = height of person, ft

Two radiation area factors are defined as:

$$f_{eff} = A_{eff}/A_D \qquad (3)$$

$$f_p = A_p/A_{eff} \qquad (4)$$

where A_{eff} is the effective radiating area of the total body surface. Equation (2) becomes:

$$ERF = \alpha_K \times f_{eff} \times f_p \times I_K/d_2 \qquad (5)$$

Fanger (1973) developed precision optical methods to evaluate the angle factors (f_{eff} and f_p) for both sitting and standing positions and for males and females. An average value for f_{eff} of 0.71 for both sitting and standing is accurate within $\pm 2\%$. The variations in angle factor f_p over various azimuths and elevations while seated or standing are illustrated in Figures 4 and 5 and, according to Fanger, they apply equally to males and females.

Infrared heating equipment manufacturers usually supply performance specifications for their equipment (Gagge *et al.* 1967). The relation between color temperature of heaters and the applied voltage or wattage is also available. Gas-fired radiators usually operate at constant emitting temperatures of 1340 to 1700 °F. Figure 3 relates the absorptance α_K to the radiating temperature of the radiant source. Manufacturers also supply the radiant flux distribution of beam heaters with and without reflectors. Figures 6 through 9 illustrate the radiant flux distribution for four typical electric and gas-fired radiant heaters. In general, electrical beam heaters produce as much as 70 to 80% of their total energy output as radiant heat in contrast to 40% for gas-fired types. In practice, the designer should choose a beam heater that will illuminate the subject with acceptable uniformity. Even with complete illumination by a beam 0.109 sr (21.6°) wide, Figure 6 shows that only 8% (100% × 1225 × 0.109/1620) of the initial input wattage to the heater is usable for specifying the necessary I_K in Equation (5). The corresponding percentages for Figures 7, 8, and

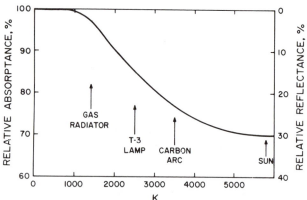

Fig. 3 **Relative Absorptance and Reflectance of Skin and Typical Clothing Surfaces**

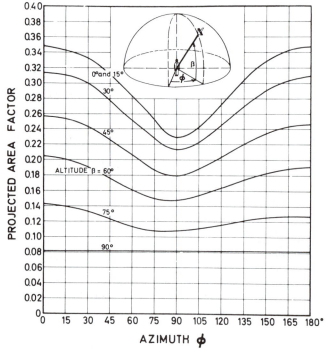

Fig. 5 **Projected Area Factor for Standing Persons, Nude and Clothed**
(Fanger 1973)

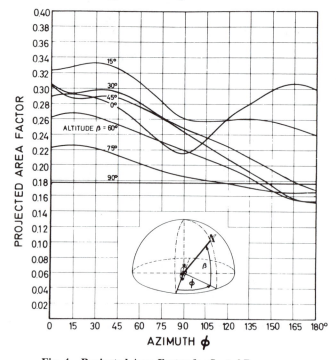

Fig. 4 **Projected Area Factor for Seated Persons, Nude and Clothed**
(Fanger 1973)

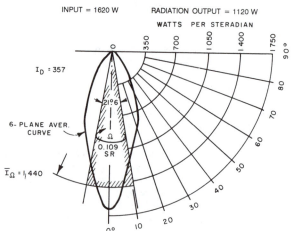

Fig. 6 **Radiant Flux Distribution Curve of Typical Narrow-Beam High Intensity Electric Infrared Heaters**

9 are 5%, 4%, and 2%, respectively. The last two are for gas-fired beams. Finally, the balance of the input energy to the beam heater not used for irradiating the occupant directly will ultimately increase the ambient air temperature and mean radiant temperature of the room. This increase will reduce the original ERF required for comfort and acceptability. The continuing reradiation and convective heating of surrounding walls and the presence of air movement ultimately make precise calculations of radiant heat exchange difficult.

The basic principles of beam heating are illustrated by the following examples.

Example 1. Determine the beam radiation intensity required for comfort from a quartz lamp (Figure 6) when the worker is sedentary, lightly clothed (0.5 clo), and seated. The ambient t_a is 15 °C, with air movement 0.15 m/s. The lamp is mounted on the 2.4 m-high ceiling and is directed at the back of the seated person so that the elevation angle β is 45° and the azimuth ϕ is 180°. Assume the ambient and mean radiant temperatures of the unheated room are equal.

Solution: The ERF for comfort can be calculated as 73 W/m² by procedures outlined in Chapter 48 of the 1991 ASHRAE *Handbook—HVAC Applications*. At 240 V operation, $\alpha_K = 0.85$ at 2200 K (from Figure 3); $f_{eff} = 0.71$; $f_p = 0.17$ (Figure 5); $d = (2.4 - 0.6)$ or 1.8 m; where 0.6 is sitting height of occupant.

By Equation (5), the irradiation I_K from the beam heater necessary for comfort is:

$$I_K = ERF\ [d^2/(\alpha_K \times f_{eff} \times f_p)]$$
$$= 73(1.8)^2/(0.85 \times 0.71 \times 0.17) = 2305\ \text{W/sr}$$

Example 2. For the same occupant in Example 1, when two beams located on the ceiling are directed downward at the subject at 45° and at azimuth angle 90° on each side, what would be the I_K required from each heater?

The ERF for comfort from each beam is 73/2 or 36.5 W/m². The value of f_p is 0.25 (from Figure 4). Hence, the required irradiation from each beam is:

$$I_K = 36.5(1.8)^2/(0.9 \times 0.71 \times 0.25)$$
$$= 740\ \text{W/sr (at half power } V = 155V, K \cong 2000, \alpha_K \cong 0.9.)$$

This estimate indicates that two beams similar to Figure 7, each operating at half of rated voltage, can produce the necessary ERF for comfort. A comparison between the I_K requirements in Examples 1 and 2 shows that irradiating the back while sitting is much less efficient than irradiating from the side.

Example 3. A broad beam gas-fired radiator is mounted 5 m above the floor. The radiator is directed 45° downward toward a standing subject 4 m away (see Heater #1 in Figure 10).

Question (1): What is the resulting ERF from the beam acting on the subject?

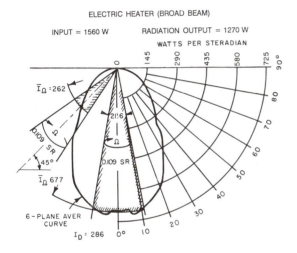

Fig. 7 **Radiant Flux Distribution Curve of Typical Broad-Beam High Intensity Electric Infrared Heaters**

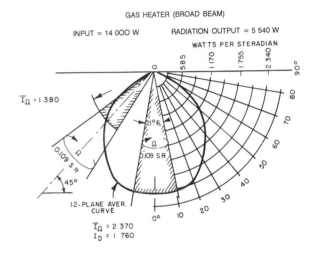

Fig. 9 **Radiant Flux Distribution Curve of Typical Broad-Beam High Intensity Atmospheric Gas-Fired Infrared Heaters**

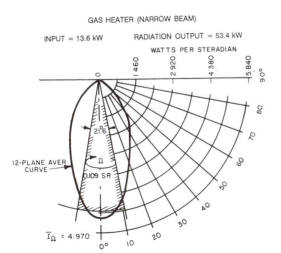

Fig. 8 **Radiant Flux Distribution Curve of Typical Narrow-Beam High Intensity Atmospheric Gas-Fired Infrared Heaters**

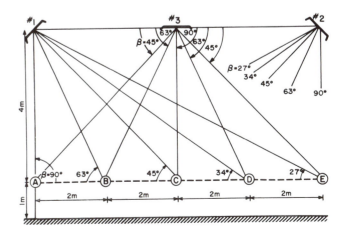

Fig. 10 **Calculation of Total ERF From Three Gas-Fired Heaters on Worker Standing at Positions A through E**

Solution: By Equation (5):

$$ERF = \alpha_K f_{eff} f_p I_K/d^2 = 13.1 \text{ W/m}^2$$

where

$\alpha_K = 0.97$ (Figure 3)
$f_{eff} = 0.71$
$f_p = 0.26$ (Figure 5 at $\beta = 40°$ and $\phi = 0°$)
$d^2 = 4^2 + 4^2 = 32 \text{ ft}^2$ (The center of the standing man is 1 m off the floor)
$I_K = 2340$ W/sr (Figure 9)

If the heater was 3 m from the standing subject (4 m above floor), the ERF would be 23.3 W/m^2.

Question (2): How does the ERF vary along the 0° azimuth, every 2 m beginning at a point directly under the heater (#1 in Figure 10) and for elevations $\beta = 90°$ at A, 63.4° at B, 45° at C, 33.6° at D, and 26.6° at E. The values for f_p for the five positions are (A) 0.08, (B) 0.19, (C) 0.26, (D) 0.30, and (E) 0.33.

Solution: Since the beam is directed 45° downward, the deviation from the beam center for a person standing at the five positions (A-E) are respectively 45°, 18.4°, 0°, 11.4°, and 18.4°; the corresponding I_K values from Figure 9 are 1450, 2340, 2340, 2340, and 2340 W/m^2. The respective d^2 are 16, 16 + 4, 16 + 16, 16 + 36, and 16 + 64. The ERF for a person standing in the five positions are (A) 5; (B) 14; (C) 13; (D) 9.5; and (E) 6.5 W/m^2 (rounding to nearest 0.5 W/m^2).

Question (3): How will the total ERF at each of the five locations A-E vary if two additional heaters (#2 and #3 in Figure 10) are added 5 m above the floor over positions C and E? The center heater is directed downward, the outer one directed as above, 45° towards the center of the room.

Solution: At each of five room locations (A, B, C, D, E), add the ERF from each of the three radiators to determine the total ERF affecting the person standing.

A	5 + 8 + 6.5	or	19.5
B	14 + 13.5 + 9.5	or	37
C	13 + 8 + 13	or	34
D	9.5 + 13.5 + 14	or	37
E	6.5 + 8 + 5	or	19.5

REFERENCES

AGA. 1988. *National fuel gas code.* ANSI Z223.1-1988. American Gas Association, Cleveland, OH.

ASHRAE. 1981. Thermal environmental conditions for human occupancy. ANSI/ASHRAE *Standard* 55-1981.

Boyd, R.L. 1960. What do we know about infrared comfort heating? *Heating, Piping and Air Conditioning* (November):133.

Boyd, R.L. 1963. Control of electric infrared energy distribution. *Electrical Engineering* (February):103.

Buckley, N.A. 1989. Applications of radiant heating saves energy. ASHRAE *Journal* 31(9):17.

Buckley, N.A. and T. Seel. 1987. Engineering principles support an adjustment factor when sizing gas-fired low-intensity infrared equipment. ASHRAE *Transactions* 93(1).

Buckley, N.A. and T. Seel. 1988. Case studies support adjusting heat loss calculations when sizing gas-fired, low-intensity, infrared equipment. ASHRAE *Transactions* 94(1B).

Buckley, N.A. 1989. Applications of radiant heating saves energy. ASHRAE *Journal* 31(9):17.

DeWerth, D.W. 1960. Literature review of infra-red energy produced with gas burners. *Research Bulletin* No. 83. American Gas Association, Cleveland, OH.

DeWerth, D.W. 1962. A study of infra-red energy generated by radiant gas burners. *Research Bulletin* No. 92. American Gas Association, Cleveland, OH.

Fanger, P.O. 1973. *Thermal comfort.* McGraw Hill Book Co., New York.

Gagge, A.P., G.M. Rapp, and J.D. Hardy. 1967. The effective radiant field and operative temperature necessary for comfort with radiant heating. ASHRAE *Transactions* 73(1) and ASHRAE *Journal* 9:63-66.

Gas Research Board of London. 1944. The use of infra-red radiation in industry. Information Circular No. 1.

Haslam, W.G., *et al.* 1925. Radiation from non-luminous flames. *Industrial and Engineering Chemistry* (March).

New York State. 1973. *Energy efficiency in large buildings.* Interdepartmental Fuel and Energy Committee ad hoc report.

Plyler, E.K. 1948. Infra-red radiation from bunsen flames. *Journal of Research*, National Bureau of Standards 40(February):113.

Prince, F.J. 1962. Selection and application of overhead gas-fired infrared heating devices. ASHRAE *Journal* (October):62.

Walker, C.A. 1962. Control of high intensity infrared heating. ASHRAE *Journal* (October):66.

DUCT CONSTRUCTION

THIS chapter covers the construction of heating, ventilating, air-conditioning, and exhaust duct systems for residential, commercial, and industrial applications. Technological advances in duct construction should be judged relative to the construction requirements herein and to appropriate codes and standards. While the construction details shown in this chapter may coincide, in part, with industry standards, they do not constitute an ASHRAE standard.

BUILDING CODE REQUIREMENTS

In the private sector, each new construction or renovation project is normally governed by state laws or local ordinances that require compliance with specific health, safety, property protection, and energy conservation regulations. Figure 1 illustrates relationships between laws, ordinances, codes, and standards that can affect the design and construction of HVAC duct systems; however, Figure 1 may not list all applicable regulations and standards for a specific locality. Specifications for federal government construction are promulgated by such agencies as the Federal Construction Council, the General Services Administration, the Department of the Navy, and the Veterans Administration.

Model code changes require long cycles for approval by the consensus process. Since the development of safety codes, energy codes and standards proceed independently; the most recent edition of a code or standard may not have been adopted by a local jurisdiction. HVAC designers must know which code compliance obligations affect their designs. If a provision conflicts with the design intent, the designer should resolve the issue with local building officials. New or different construction methods can be accommodated by the provisions for equivalency that are incorporated into codes. Staff engineers from the model code agencies are available to assist in the resolution of conflicts, ambiguities, and equivalencies.

Smoke control is covered in Chapter 47 of the 1991 ASHRAE *Handbook—HVAC Applications.* The designer should consider flame spread, smoke development, and toxic gas production from duct and duct insulation materials. Code documents for ducts in certain locations within buildings rely on a criterion of *limited combustibility* (see definitions, NFPA 90A), which is independent of the generally accepted criteria of 25 flame spread and 50 smoke development; however, certain duct construction protected by extinguishing systems may be accepted with higher levels of combustibility by code officials.

Combustibility and toxicity ratings are normally based on tests of new materials; little research is reported on ratings of duct materials that are aged or of systems that are poorly maintained for cleanliness. Fibrous and other porous materials exposed to airflow in ducts may accumulate more dirt than nonporous materials.

CLASSIFICATION OF DUCTS

Duct construction is classified in terms of application and pressure. HVAC systems in public assembly, business, educational, general factory, and mercantile buildings are usually designed as commercial systems. Air pollution control systems, industrial exhaust systems, and systems outside the pressure range of commercial system standards are classified as industrial systems.

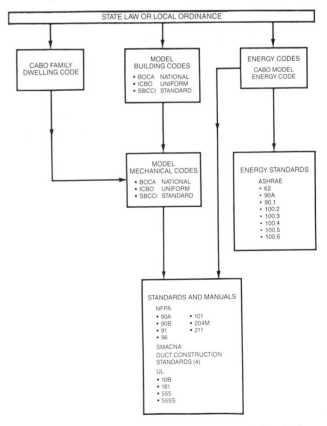

Fig. 1 Typical Hierarchy of Building Codes and Standards

The preparation of this chapter is assigned to TC 5.2, Duct Design.

Residences: ±0.5 in. of water
                                 ±1 in. of water
Commercial Systems   ±0.5 in. of water
                                 ±1 in. of water
                                 ±2 in. of water
                                 ±3 in. of water
                                 +4 in. of water
                                 +6 in. of water
                                 +10 in. of water
Industrial systems: any pressure

Air conveyed by a duct imposes both air pressure and velocity pressure loads on the duct's structure. The load resulting from mean static-pressure differential across the duct wall normally dominates and is generally used for duct classification. Turbulent airflow introduces relatively low but rapidly pulsating loading on the duct wall.

Static pressure at specific points in an air distribution system is not necessarily the static pressure rating of the fan; the actual static pressure in each duct section must be obtained by computation. Therefore, the designer should specify the pressure classification of the various duct sections in the system. All modes of operation must be taken into account, especially systems used for smoke control and those with fire dampers that must close with the system running.

DUCT SYSTEM LEAKAGE

Predicted leakage rates for unsealed and sealed ducts are reviewed in Chapter 32 of the 1989 ASHRAE *Handbook—Fundamentals*. Project specifications should define allowable duct leakage, the need for leakage testing, and require the ductwork installer to perform a leakage test after installing an initial portion of the duct system. Procedures in the *HVAC Air Duct Leakage Test Manual* (SMACNA 1985) should be followed for leakage testing. If a test indicates excess leakage, corrective measures should be taken to ensure quality control.

Responsibility for proper assembly and sealing belongs to the installing contractor. The most cost-effective way to control leakage is to follow proper installation procedures. Because access for repairs is usually limited, poorly installed duct systems that must later be resealed can cost more than a proper installation.

RESIDENTIAL DUCT CONSTRUCTION

NFPA *Standard* 90B, CABO's *One- and Two-Family Dwelling Code*, or a local code is used for duct systems in single-family dwellings. Generally, local authorities use NFPA *Standard* 90A for multifamily homes.

Supply ducts may be steel, aluminum, or a material with a UL *Standard* 181 rating. Sheet metal ducts should be constructed of minimum thickness as shown in Table 1, and installed in accordance with *HVAC Duct Construction Standards—Metal and Flexible* (SMACNA 1985). Fibrous glass ducts should be installed in accordance with the *Fibrous Glass Duct Construction Standards* (SMACNA 1979). For return duct systems, the use of alternate materials, and other exceptions, consult NFPA *Standard* 90B.

Table 1   Residential Metal Duct Construction

Shape of Duct and Exposure	Galvanized Steel Minimum Thickness, in.	Aluminum (3003) Nominal Thickness, in.
Enclosed rectangular ducts[a]		
14 in. or less	0.0127	0.016
Over 14 in.	0.0157	0.020
Rectangular and round ducts	Consult *HVAC Duct Construction Standards—Metal and Flexible* (SMACNA 1985)	

[a]Data based on nominal thickness, NFPA 90B-89.

COMMERCIAL DUCT CONSTRUCTION

Materials

NFPA *Standard* 90A is frequently used as a guide standard by many building code agencies. NFPA *Standard* 90A invokes UL *Standard* 181, which classifies ducts as follows:

Class 0—zero flame spread, zero smoke developed
Class 1—25 flame spread, 50 smoke developed

NFPA *Standard* 90A states, in summary, that ducts must be iron, steel, aluminum, concrete, masonry, or clay tile. However, ducts may be UL *Standard* 181 Class 1 materials when they are not used as vertical risers serving more than two stories or in systems with air temperature higher than 250°F. Many manufactured flexible and fibrous glass ducts are UL approved and listed as Class 1. For galvanized ducts, a G90 coating is recommended (see ASTM A525). The minimum thickness and weight of sheet metal sheets are given in Table 2.

Duct-reinforcing members are formed from sheet metal or made from hot-rolled or extruded structural shapes. The size and weights of commonly used members are given in Table 3.

Rectangular and Round Ducts

Rectangular metal ducts. Table 4 lists construction requirements for rectangular steel ducts and includes combinations of duct thicknesses, reinforcement, and maximum distance between reinforcements. *HVAC Duct Construction Standards—Metal and Flexible* (SMACNA 1985) gives the functional criteria on which Table 4 is based. Transverse joints (*e.g.*, standing drive slips, pocket locks, and companion angles) and, when necessary, intermediate structural members are designed to reinforce the duct system. Ducts larger than 96 in. require internal tie rods to maintain their structural integrity. Table 4 also shows alternative tie rod construction for positive pressure ducts over 48 in. Tie rods allow the use of smaller reinforcements than would otherwise be required. *Rectangular Industrial Duct Construction Standards* (SMACNA 1980) gives construction details for ducts up to 168 in. in width and pressure up to ±30 in. of water.

Acceptable transverse joints and intermediate reinforcement members are listed in Tables 5 and 6 for various duct pressures. Other joint systems and intermediate reinforcement must meet the rigidity requirement given in Table 6 for each class of reinforcement.

Fittings must be reinforced similarly to sections of straight duct. On size change fittings, the greater fitting dimension determines material thickness. Where fitting curvature or internal member attachments provide equivalent rigidity, such features may be credited as reinforcement.

Round metal ducts. Round ducts are inherently strong and rigid, and are generally the most efficient and economical ducts for air systems. The dominant factor in round duct construction is the ability of the material to withstand the physical abuse of installation and negative pressure requirements. Construction requirements as a function of static pressure, type of seam (spiral or longitudinal), and diameter are listed in Table 7.

Nonferrous ducts. *HVAC Duct Construction Standards—Metal and Flexible* (SMACNA 1985) lists construction requirements for rectangular (±3 in. of water) and round (±2 in. of water) aluminum ducts. *Round Industrial Duct Construction Standards* (SMACNA 1977) gives construction requirements for round aluminum duct systems with pressures greater than ±2 in. of water.

Construction details. *HVAC Duct Construction Standards—Metal and Flexible* (SMACNA 1985) gives construction details for rectangular and round ducts operating with static pressures from −3 to +10 in. of water.

Table 2a Galvanized Sheet Thickness

Galvanized Sheet Gage	Thickness, in.		Nominal Weight, lb/ft²
	Nominal	Minimum[a]	
30	0.0157	0.0127	0.656
28	0.0187	0.0157	0.781
26	0.0217	0.0187	0.906
24	0.0276	0.0236	1.156
22	0.0336	0.0296	1.406
20	0.0396	0.0356	1.656
18	0.0516	0.0466	2.156
16	0.0635	0.0575	2.656
14	0.0785	0.0705	3.281
13	0.0934	0.0854	3.906
12	0.1084	0.0994	4.531
11	0.1233	0.1143	5.156
10	0.1382	0.1292	5.781

[a]Minimum thickness is based on thickness tolerances of hot-dip galvanized sheets in cut lengths and coils (per ASTM *Standard* A525). Tolerance is valid for 48-in. and 60-in. wide sheets.

Table 2b Uncoated Steel Sheet Thickness

Manufacturers' Standard Gage	Thickness, in.			Nominal Weight lb/ft²
		Minimum[a]		
	Nominal	Hot-Rolled	Cold-Rolled	
28	0.0149		0.0129	0.625
26	0.0179		0.0159	0.750
24	0.0239		0.0209	1.000
22	0.0299		0.0269	1.250
20	0.0359		0.0329	1.500
18	0.0478	0.0428	0.0438	2.000
16	0.0598	0.0538	0.0548	2.500
14	0.0747	0.0677	0.0697	3.125
13	0.0897	0.0827	0.0847	3.750
12	0.1046	0.0966	0.0986	4.375
11	0.1196	0.1116	0.1136	5.000
10	0.1345	0.1265	0.1285	5.625

Note: Table is based on 48-in. width coil and sheet stock. 60-in. coil has same tolerance, except that 16-gage is ±0.007 in. in hot-rolled coils and sheets.
[a]Minimum thickness is based on thickness tolerances of hot-rolled and cold-rolled sheets in cut lengths and coils (per ASTM *Standards* A366, A568, and A569).

Table 2c Stainless Steel Sheet Thickness

Gage[a]	Thickness, in.		Nominal Weight, lb/ft²	
			Stainless Steel	
	Nominal	Minimum[b]	300 Series	400 Series
28	0.0151	0.0131	0.634	0.622
26	0.0178	0.0148	0.748	0.733
24	0.0235	0.0205	0.987	0.968
22	0.0293	0.0253	1.231	1.207
20	0.0355	0.0315	1.491	1.463
18	0.0480	0.0430	2.016	1.978
16	0.0595	0.0535	2.499	2.451
14	0.0751	0.0681	3.154	3.094
13	0.0900	0.0820	3.780	3.708
12	0.1054	0.0964	4.427	4.342
11	0.1200	0.1100	5.040	4.944
10	0.1350	0.1230	5.670	5.562

[a]Stainless sheet gage has no standard nominal thickness among manufacturers.
[b]Minimum thickness is based on thickness tolerances for hot-rolled sheets in cut lengths and cold-rolled sheets in cut lengths and coils (per ASTM *Standard* A480).

Table 3 Steel Angle Weights per Unit Length (Approximate)

Angle Size, in.	Weight, lb/ft
3/4 × 3/4 × 1/8	0.59
1 × 1 × 0.0466 (minimum)	0.36
1 × 1 × 0.0575 (minimum)	0.44
1 × 1 × 1/8	0.80
1 1/4 × 1 1/4 × 0.0466 (minimum)	0.45
1 1/4 × 1 1/4 × 0.0575 (minimum)	0.55
1 1/4 × 1 1/4 × 0.0854 (minimum)	0.65
1 1/4 × 1 1/4 × 1/8	1.01
1 1/2 × 1 1/2 × 0.0575 (minimum)	0.66
1 1/2 × 1 1/2 × 1/8	1.23
1 1/2 × 1 1/2 × 3/16	1.80
1 1/2 × 1 1/2 × 1/4	2.34
2 × 2 × 0.0575 (minimum)	0.89
2 × 2 × 1/8	1.65
2 × 2 × 3/16	2.44
2 × 2 × 1/4	3.19
2 1/2 × 2 1/2 × 3/16	3.07
2 1/2 × 2 1/2 × 1/4	4.10

Flat-Oval Ducts

Table 8 gives flat-oval duct construction requirements. Seams and transverse joints are the same as those permitted for round ducts. Reinforcement is determined from Tables 4 and 6 using dimension "F" defined in Table 8. Reinforcement ends on opposite sides are tied at pressures of 4 to 10 in. of water.

HVAC Duct Construction Standards—Metal and Flexible (SMACNA 1985) has additional construction details. Hanger designs and installation details for rectangular ducts generally apply to flat-oval ducts.

Fibrous Glass Ducts

Fibrous glass ducts are a composite of rigid fiberglass and a factory-applied facing (typically aluminum or reinforced aluminum), which serves as a finish and vapor barrier. This material is available in molded round sections or in board form for fabrication into rectangular or polygonal shapes. Duct systems of round and rectangular fibrous glass are generally limited to 2400 fpm and ±2 in. of water. Molded round ducts are available in higher pressure ratings. *Fibrous Glass Duct Construction Standards* (SMACNA 1979, TIMA 1989) and manufacturers' installation instructions give details on fibrous glass duct construction. The SMACNA standard also covers duct and fitting fabrication, closure, and installation, including installation of duct-mounted HVAC appurtenances (*e.g.*, volume dampers, turning vanes, register and grille connections, diffuser connections, access doors, fire damper connections, and electric heaters).

Flexible Ducts

Flexible ducts connect mixing boxes, light troffers, diffusers, and other terminals to the air distribution system. *HVAC Duct Construction Standards—Metal and Flexible* (SMACNA 1985) lists installation details. Because unnecessary length, offsetting, and compression of these ducts significantly increases airflow resistance, they should be kept as short as possible and fully extended (see Chapter 32 of the 1989 ASHRAE *Handbook—Fundamentals*).

UL *Standard* 181 covers testing of materials used to fabricate flexible ducts that are categorized as air ducts and connectors. NFPA *Standard* 90A defines the acceptable use of these products. The flexible duct connector has less resistance to flame penetration, lower puncture and impact resistance, and is subject to many restrictions listed in NFPA *Standard* 90A. Only flexible ducts that are air duct rated should be specified. Tested products are listed in the UL *Gas and Oil Directory* published annually.

Table 4a   Rectangular Ferrous Metal Duct Construction for Commercial Systems[*][a],[b]

Duct Dimen., in.	0.0575							0.0466							0.0356				
	±0.5	±1	±2	±3	+4	+6	+10	±0.5	±1	±2	±3	+4	+6	+10	±0.5	±1	±2	±3	+4
≤10																			
12	Reinforcement not required[d]							Reinforcement not required[d]						A-8	Reinforcement not required[d]				
14							B-8							B-8					A-10
16							B-8						B-10	B-5				A-8	A-10
18						C-10	C-8					B-10	C-10	C-5				A-8	B-8
20					C-10	C-10	D-8				B-10	C-10	C-8	C-5			B-10	B-8	C-8
22						C-10	C-5			B-10	C-10	C-10	C-8	C-5		A-10	B-10	B-8	B-5
24					D-10	D-8	D-5			C-10	C-10	D-10	D-8	D-5			B-10	C-10	B-5
26			C-10	D-10	D-10	D-8	D-5			C-10	D-10	D-10	D-5	D-5		B-10	C-10	C-5	C-5
28			C-10	D-10	E-10	E-8	E-5		C-10	C-10	D-10	E-8	D-5	E-5	B-10	C-10	C-8	C-5	D-5
30				D-10	D-10	E-10	E-5		C-10	D-10	D-8	E-8	D-5	E-4	B-10	C-10	D-8	C-5	D-5
36		D-10	E-10	E-8	E-5	F-5	F-5	C-10	D-10	E-8	E-5	E-5	F-5	F-4	C-10	D-10	D-5	E-5	E-5
42	D-10	E-10	E-8	E-5	F-5	G-5	H-4	D-10	E-10	E-5	E-5	F-5	G-4	G-3	D-10	D-8	E-5	E-5	F-4
48	E-10	F-10	G-8	G-5	G-5	H-4	H-3	E-10	E-8	F-5	G-5	G-5	H-4	H-3	E-8	E-5	F-5	F-4	F-3
54	E-10	G-10[e]	G-5[e]	H-5[e]	H-5[e]	H-4[e]	I-3[e]	E-10	F-8	G-5[e]	H-5[e]	H-4[e]	H-3[e]	H-2.5[e]	E-8	E-5	F-4	G-3[e]	G-3[e]
60	F-10	G-8[e]	H-5[e]	H-5[e]	I-5[e]	H-3[e]	J-3[e]	F-10	G-8[e]	H-5[e]	H-4[e]	H-3[e]	H-3[e]	I-2.5[e]	F-8	F-5	G-4[e]	G-3[e]	H-3[e]
72	H-10[e]	H-5[e]	I-5[e]	I-4[e]	I-3[e]	J-3[e]	K-2.5[e]	G-8[e]	H-5[e]	H-4[e]	H-3[e]	I-3[e]	J-2.5[e]	K-2[e]	F-5	G-4[e]	H-3[e]	H-3[e]	I-2.5[e]
84	H-8[e]	I-5[e]	J-4[e]	J-3[e]	K-3[e]	L-2.5[e]	H-2 plus rods	H-5[e]	I-5[e]	J-4[e]	J-3[e]	J-2.5[e]	K-2[e]	—	H-5[e]	H-4[e]	I-3[e]	J-2.5[e]	J-2[e]
96	I-8[e]	J-5[e]	K-4[e]	L-3[e]	L-2.5[e]	L-2[e]	H-2 plus rods	H-5	I-4	K-3[e]	K-2.5[e]	K-2[e]	L-2[e]	—	H-5[e]	I-3[e]	J-2.5[e]	J-2[e]	K-2[e]
>96	H-5	H-2.5 plus rods[g]	H-2.5 plus rods[g]	H-2.5 plus rods[f]	H-2.5 plus rods[f]	H-2.5 plus rods[f]	H-2 plus rods[g]	H-5	H-2.5 plus rods[g]	H-2.5 plus rods[g]	H-2.5 plus rods[g]	H-2.5 plus rods[g]		H-2 plus rods[g]	—	—	—	—	—

[*]See Table 4b for notes.

Table 4b   Rectangular Ferrous Metal Duct Construction for Commercial Systems[a],[b]

Duct Dimen., in.	0.0356		0.0296							0.0236							0.0187		
	+6	+10	±0.5	±1	±2	±3	+4	+6	+10	±0.5	±1	±2	±3	+4	+6	+10	±0.5	±1	±2
≤8			Reinforcement not required[d]													A-5	Reinforcement not required[d]		
10	A-10	A-5						A-5	A-5						A-5	A-4			
12	A-10	A-5					A-10	A-5	A-5					A-8	A-5	A-5			A-8
14	A-10	A-5				A-8	A-5	A-5	A-4					A-5	A-5	A-3		A-10	A-5
16	A-5	B-5			A-10	A-8	A-5	A-5	B-4			A-10	A-8	A-5	A-5	B-3		A-8	A-5
18	B-5	C-5			A-10	A-8	A-5	B-5	B-4			A-10	A-8	A-5	B-4	B-3		A-8	A-5
20	B-5	C-4		A-10	B-8	A-5	B-5	B-5	B-3		A-10	A-5	A-5	B-5	B-4	B-3	A-10	A-8	A-5
22	C-5	C-4		A-10	B-8	B-5	B-5	C-5	C-3	A-10	A-10	A-5	B-5	B-4	C-4	C-3	A-10	A-5	A-5
24	C-5	D-4		B-10	C-8	B-5	C-5	C-5	C-3	A-10	B-10	B-5	B-5	C-4	C-3	C-3	A-10	A-5	B-5
26	D-5	D-4	A-10	B-10	C-8	C-5	C-5	C-4	D-3	A-10	B-8	B-5	C-5	C-4	C-3	C-2.5	A-10	A-5	B-5
28	D-5	D-4	B-10	C-10	C-5	C-5	D-5	D-4	D-3	B-10	C-8	C-5	C-4	D-4	C-3	D-2.5	B-8	B-5	B-4
30	D-5	D-3	B-10	C-10	C-5	C-5	D-5	D-3	D-3	B-10	C-8	C-5	C-4	D-4	D-3	D-2.5	B-8	B-5	C-4
36	E-4	F-3	C-10	D-8	D-5	D-4	E-4	E-3	E-2.5	C-8	C-5	D-4	D-4	E-2.5	E-2	—	C-5	C-5	—
42	F-3	G-2.5	D-8	D-5	E-5	E-4	E-3	E-2.5	F-2	D-8	D-5	E-4	E-3	E-2.5	—	—	D-5	D-4	—
48	G-2.5	G-2	D-8	E-5	E-4	E-3	F-3	G-2.5	G-2	D-5	E-5	E-3	E-2.5	E-2	—	—	D-5	D-4	—
54	H-2.5[e]	H-2[e]	D-5	E-5	F-3	G-3[e]	G-2.5[e]	G-2[e]	—	D-5	E-4	F-3	E-2.5	F-2	—	—	D-5	—	—
60	H-2.5[e]	I-2[e]	E-5	F-5	G-3[e]	G-2.5[e]	H-2.5[e]	H-2[e]	—	E-5	F-4	G-2.5[e]	G-2.5[e]	G-2[e]	—	—	E-4	—	—
72	I-2[e]	—	F-5	G-4[e]	H-3[e]	H-2.5[e]	H-2[e]	—	—	F-4	H-2[e]	H-2[e]	H-2[e]	—	—	—	—	—	—
84	—	—	H-5[e]	—	—	—	—	—	—	G-4[e]	—	—	—	—	—	—	—	—	—
96	—	—	H-4[e]	—	—	—	—	—	—	—	—	—	dashes indicate not allowed		—	—	—	—	—
>96	—	—	—	—	—	—	—	—	—	—	—	—	—	—	—	—	—	—	—

[a] Table 4 is based on Tables 1-3 through 1-9 in *HVAC Duct Construction Standards—Metal and Flexible* (SMACNA 1985). For tie rod details, refer to this standard.
[b] For a given duct thickness, numbers indicate maximum spacing (feet) between duct reinforcement; letters indicate type (rigidity class) of duct reinforcement (see Tables 5 and 6).

Transverse joint spacing is unrestricted on unreinforced ducts. To qualify joints on reinforced ducts, select transverse joints from Table 5. Select intermediate reinforcement from Table 6. Tables are based on steel construction. Designers should specify galvanized, uncoated, or painted steel joint and intermediate reinforcement.

Use the same metal duct thickness on all duct sides. Evaluate duct reinforcement on each duct side separately. When required on four sides for +4, +6, and +10 in. of water pressure systems, corners must be tied. When required on two sides, corners must be tied with rods or angles at the ends for +4, +6, and +10 in. of water pressure systems.

Duct sides over 18-in. width with less than 0.0356-in. thickness, which have more than 10 ft² of unbraced panel area, must be cross-broken or beaded, unless they are lined or insulated externally. Lined or externally insulated ducts are not required to have cross-breaking or beading.
[c] The reinforcement tables are based on galvanized steel of the indicated thickness. They apply to galvanized, painted, uncoated, and stainless steel whenever the base metal thickness is not less than 0.0015 in. below that indicated for galvanized steel.
[d] Blank spaces indicate that no reinforcement is required.
[e] See SMACNA's publication for alternative reinforcements using tie rods or tie straps for positive pressure.
[f] Sheet metal 0.0466 in. thick is acceptable.
[g] Tie rods with a minimum diameter of 0.375 in. (or 0.25 in. if the maximum length is 36 in.) must be used on these constructions. The rods for positive pressure ducts are spaced a maximum of 60 in. apart along joints and reinforcements.

Plenums and Apparatus Casings

HVAC Duct Construction Standards—Metal and Flexible (SMACNA 1985) shows details on field-fabricated plenum and apparatus casings. Sheet metal thickness and reinforcement for plenum and casing pressures outside the range of −3 to +10 in. of water can be based on *Rectangular Industrial Duct Construction Standards* (SMACNA 1980).

Carefully analyze plenums and apparatus casings on the discharge side of a fan for maximum operating pressure in relation to the construction detail being specified. On the suction side of a fan, plenums and apparatus casings are normally constructed to withstand negative air pressures at least equal to the total upstream static pressure losses. The accidental stoppage of intake airflow can apply a negative pressure as great as the fan shutoff pressure. Conditions such as malfunctioning dampers or clogged louvers, filters, or coils can collapse a normally adequate casing. To protect large casing walls or roofs from damage, it is more economical to provide fan safety interlocks, such as damper end switches or pressure limit switches, instead of heavier sheet metal construction.

Apparatus casings can perform two acoustical functions. If the fan is completely enclosed within the casing, the transmission of fan noise through the fan room to adjacent areas is reduced substantially. An acoustically lined casing also reduces airborne noise levels in connecting ductwork. Acoustical treatment may consist of a single metal wall with a field-applied acoustical liner or thermal insulation, or a double-wall panel with an acoustical liner and a perforated metal inner liner. Double-wall casings are marketed by many manufacturers who publish data on structural, acoustical, and thermal performance and also prepare designs on a custom basis.

Acoustical Treatment

Metal ducts are frequently lined with acoustically absorbent materials to reduce aerodynamic noise. Although many materials are acoustically absorbent, duct liners must also be resistant to erosion and fire and have properties compatible with the ductwork fabrication and erection process. For high-velocity duct systems, double-wall construction using a perforated metal inner liner is frequently specified. Chapter 42 of the 1991 ASHRAE *Handbook—HVAC Applications* addresses design considerations. ASTM *Standard* C423 covers laboratory testing of duct liner materials to determine their sound absorption coefficients. Designers should review all of the tests incorporated in ASTM *Standard* C1071. A wide range of performance attributes, including erosion resistance, vapor adsorption, temperature resistance, and fungi resistance, are covered in the standard. Health and safety precautions are addressed, and manufacturer's certifications of compliance are also covered.

Table 5a   Transverse Joint Reinforcement*[a]

Minimum Rigidity Class[c]	STANDING DRIVE SLIP	STANDING S		STANDING S	STANDING S	STANDING S (BAR REINFORCED) / STANDING S (ANGLE REINFORCED)
	$H_s \times T$ (min), in.	W, in.	$H_s \times T$ (min), in.	$H_s \times T$ (min), in.	$H_s \times T$ (min), in.	$H_s \times T$ (min) plus Reinforcement ($H \times T$), in.[d]
A	Use Class B		Use Class C	$1/2 \times 0.0187$	Use Class D	Use Class F
B	$1\ 1/8 \times 0.0187$			$1/2 \times 0.0296$		
C	$1\ 1/8 \times 0.0296$	—	1×0.0187	1×0.0187		
D	—	—	1×0.0236	1×0.0236	$1\ 1/8 \times 0.0187$	
E	—	3/16	$1\ 1/8 \times 0.0356$	—	$1\ 1/8 \times 0.0466$	
F	—	3/16	$1\ 5/8 \times 0.0296$	—	$1\ 1/2 \times 0.0236$	$1\ 1/2 \times 0.0236$ plus $1\ 1/2 \times 1/8$ bar
G	—	3/16	$1\ 5/8 \times 0.0466$	—	$1\ 1/2 \times 0.0466$	$1\ 1/2 \times 0.0296$ plus $1\ 1/2 \times 1/8$ bar
H	—	—	—	—	—	$1\ 1/2 \times 0.0356$ plus $1\ 1/2 \times 1\ 1/2 \times 3/16$ angle
I	—	—	—	—	—	2×0.0356 plus $2 \times 2 \times 1/8$ angle
J	—	—	—	—	—	2×0.0356 plus $2 \times 2 \times 3/16$ angle

The table header spans "4 in. of Water Maximum[b]".

*See Table 5c for notes.

Rectangular duct liners should be secured by mechanical fasteners and installed in accordance with *HVAC Duct Construction Standards—Metal and Flexible* (SMACNA 1985). Adhesives should be Type I, in conformance to ASTM *Standard* C916 and should be applied to the duct, with at least 90% coverage of mating surfaces. Quality workmanship prevents delamination of the liner and possible blockage of coils, dampers, flow sensors, or terminal devices. Avoid uneven edge alignment at butted joints to minimize unnecessary resistance to airflow (Swim 1978).

Rectangular metal ducts are susceptible to rumble from flexure in the duct walls during startup and shutdown. If a designer wants systems to go on and off frequently (as an energy conservation measure) during the times in which buildings are occupied, duct construction that reduces objectionable noise should be specified.

Hangers

Tables 9, 10, and 11 and *HVAC Duct Construction—Metal and Flexible* (SMACNA 1985) describe commercial HVAC system hangers for rectangular, round, and flat-oval ducts. When special analysis is required for larger ducts or loads than are given in the tables, or for other hanger configurations, AISC (1989) and AISI (1989) design manuals should be consulted. To hang or support fibrous glass ducts, the methods detailed in *Fibrous Glass Duct Construction Standards* (SMACNA 1979) are recommended. UL *Standard* 181 involves maximum support intervals for UL listed ducts.

INDUSTRIAL DUCT CONSTRUCTION

NFPA *Standard* 91 is widely used for duct systems conveying particulates and removing flammable vapors (including paint spraying residue), and corrosive fumes. Particulate-conveying duct systems are classified as follows:

- Class 1 includes nonparticulate applications, including makeup air, general ventilation, and gaseous emission control.
- Class 2 is imposed if moderately abrasive particulate in light concentration, such as that produced by buffing and polishing, wood-working, and grain-handling operations.
- Class 3 consists of highly abrasive material in low concentration, such as that produced from abrasive cleaning operations, driers and kilns, boiler breaching, and sand handling.

Table 5b   Transverse Joint Reinforcement[*][a]

Minimum Rigidity Class[c]	Standing Seam $H_s \times T$ (min), in.[f]	0.0187 through 0.0296 Duct, in. H_s, in.	0.0187 through 0.0296 Duct, in. $H \times T$, in.[d]	0.0356 through 0.0575 Duct, in. H_s, in.	0.0356 through 0.0575 Duct, in. $H \times T$, in.[d]	Welded Flange $H_s \times T$(min), in.[f]	3 in. of Water Maximum Lock Type	3 in. of Water Maximum H_s, in.	Thickness, in.[e] Lock	Thickness, in.[e] Reinforcement $(H \times T)$
A	$1/2 \times 0.0236$					$1/2 \times 0.0296$				
B	$3/4 \times 0.0236$	Use Class D				$1/2 \times 0.0575$ $3/4 \times 0.0296$	Use Class D			
C	1×0.0236					$3/4 \times 0.0466$ 1×0.0296				
D	$3/4 \times 0.0575$ 1×0.0356	1	1×0.0575	Use Class E		1×0.0466 $1\,1/4 \times 0.0296$	A	1	0.0296	None
E	1×0.0575 $1\,1/2 \times 0.0236$	1	$1 \times 1/8$	1	1×0.0575	$1\,1/4 \times 0.0466$ $1\,1/2 \times 0.0296$	B	1	0.0296	$1 \times 1/8$ Bar
F	$1\,1/2 \times 0.0356$	1 1/2	$1\,1/2 \times 0.0575$	1 1/4	$1\,1/4 \times 0.0575$	$1\,1/4 \times 0.0575$ $1\,1/2 \times 0.0356$	A	1 1/2	0.0296	None
G	$1\,1/2 \times 0.0466$	1 1/2 1 1/2	$1\,1/2 \times 1/8$ 2×0.0575	1 1/2	$1\,1/2 \times 1/8$	$1\,1/2 \times 0.0575$	B	1 1/2	0.0296	$1\,1/2 \times 1/8$ Bar
H	—	1 1/2	$2 \times 1/8$	1 1/2 1 1/2	$1\,1/2 \times 3/16$ 2×0.0575	—	C	1 1/2	0.0356	$1\,1/2 \times 3/16$ Angle
I	—	Use Class J		1 1/2	$2 \times 1/8$	—	C	1 1/2	0.0356	$2 \times 1/8$ Angle
J	—	1 1/2	$2 \times 3/16$	1 1/2	$2 \times 3/16$	—	C	1 1/2	0.0356	$2 \times 3/16$ Angle
K	—	1 1/2	$2\,1/2 \times 3/16$	Use Class L		—	C	1 1/2	0.0356	$2\,1/2 \times 3/16$ Angle
L	—	1 1/2	$2\,1/2 \times 1/4$	1 1/2	$2\,1/2 \times 3/16$	—	—	—	—	—

*See Table 5c for notes.

- Class 4 is composed of highly abrasive particulates in high concentrations, including materials conveying high concentrations of particulates listed under Class 3.

For contaminant abrasiveness ratings, see *Round Industrial Duct Construction Standards* (SMACNA 1977).

Materials

Galvanized steel, uncoated carbon steel, or aluminum are most frequently used for industrial air-handling systems. Aluminum ductwork is not used for systems conveying abrasive materials; when temperatures exceed 400°F, galvanized steel is not recommended. Ductwork materials for systems handling corrosive gases, vapors, or mists must be selected carefully. For the application of metals and use of protective coatings in corrosive environments, consult *Accepted Industry Practice for Industrial Duct Construction* (SMACNA 1975), the *Pollution Engineering Practice Handbook* (Cheremisinoff and Young 1975), and the publications of the National Association of Corrosive Engineers (NACE).

Round Ducts

Round Industrial Duct Construction Standards (SMACNA 1977) gives information for the selection of material thickness and reinforcement members for nonspiral industrial systems. The tables in this manual are presented as follows:

Class. Steel—Classes 1, 2, 3, and 4; Aluminum—Class 1 only.
Pressure classes for steel and aluminum. ±2 to ±30 in. of water, in increments of 2 in. of water.
Duct diameters for steel and aluminum. 4 to 60 in., in increments of 2 in. Equations are available for calculating the construction requirements for diameters greater than 60 in.

For spiral duct applications, consult manufacturer's construction schedules, such as those listed in the *Industrial Duct Engineering Data and Recommended Design Standards* (United McGill Corporation 1985).

Rectangular Ducts

Rectangular Industrial Duct Construction Standards (SMACNA 1980) is available for selecting material thickness and reinforcement members for industrial systems. The data in this manual give the duct construction for any system pressure class and panel width. Each side of a rectangular duct is considered a panel. Usually, the four sides of a rectangular duct are built of material with the same thickness. Ducts are sometimes built with the bottom plate thicker than the other three sides (usually in ducts with heavy particulate accumulation) to save material.

The designer selects a combination of panel thickness, reinforcement, and reinforcement member spacing to limit the deflection of the duct panel to a design maximum. Any shape transverse joint or intermediate reinforcement member that meets the minimum requirement of both section modulus and moment of inertia may be selected. The SMACNA data, which may also be used for designing apparatus casings, limit the combined stress in either the panel or structural member to 24 kpsi and the maximum allowable deflection of the reinforcement members to 1/360 of the duct width.

Construction Details

Recommended manuals for other construction details are *Industrial Ventilation* (ACGIH 1988), NFPA 91 (1990), and *Accepted Industry Practice for Industrial Ventilation* (SMACNA

Table 5c   Transverse Joint Reinforcement[a]

Minimum Rigidity Class[c]	$H_s \times T$ (min), in.	Min Cap Thickness, in.	$H \times T$ (nominal), in.	$H_s \times T$ (min), in.	Consult SMACNA or manufacturers to establish ratings.
A	Use Class B				
B	3/4 × 0.0187	0.0236	Use Class E	Use Class C	
C	1 × 0.0236	0.0236		1 × 0.0236	
D	1 × 0.0296	0.0296		1 × 0.0296	
E	1 1/2 × 0.0236	0.0296	1 × 1/8	1 × 0.0575 1 1/2 × 0.0236	
F	1 1/2 × 0.0356	0.0356	Use Class G	1 1/2 × 0.0296	
G	Use Class H		1 1/4 × 1/8	1 1/2 × 0.0466	
H	2 × 0.0575	0.0356	1 1/2 × 1/8	2 × 0.0466	
I	—	—	1 1/2 × 3/16	2 × 0.0575	
J	—	—	1 1/2 × 1/4	—	
K	—	—	2 × 3/16	—	

[a]Table 5 is based on Tables 1-11 through 1-13 of *HVAC Duct Construction Standards—Metal and Flexible* (SMACNA 1985). For assembly details and other construction alternatives using tie rods, refer to this standard.

[b]Standing S slip length limits: 30 in. at 4 in. of water, 36 in. at 3 in. of water, none at 2 in. of water or less.

[c]EI index for rigidity class is in Table 6.

[d]T-dimension for formed angles is minimum thickness; nominal for structural angles.

[e]T-dimension for locks and ducts is minimum thickness; nominal for bars and angles.

[f]Duct reinforcement (rigidity class) may have to be increased from that obtained from Table 4 to match duct thickness.

1975). The transverse reinforcing of ducts subject to negative pressures below −3 in. of water should be welded to the duct wall rather than relying on mechanical fasteners to transfer the static loading to the reinforcing material.

Hangers

The *Manual of Steel Construction* (AISC 1989) and the *Cold-Formed Steel Design Manual* (AISI 1989) give design information for industrial duct hangers and supports. The SMACNA standards for rectangular and round industrial ducts (SMACNA 1980, 1977) and manufacturers' schedules include duct design information for supporting ducts at intervals of up to 35 ft.

DUCT CONSTRUCTION FOR GREASE- AND MOISTURE-LADEN VAPORS

The installation and construction of ducts used for the removal of smoke or grease-laden vapors from cooking equipment should be in accordance with NFPA *Standard* 96 and SMACNA's rectangular and round industrial duct construction standards (SMACNA 1977, 1980). Kitchen exhaust ducts that conform to NFPA 96 must (1) be constructed from carbon steel with a minimum thickness of 0.054 in. or stainless steel sheet with a minimum thickness of 0.043 in.; (2) have all longitudinal seams and transverse joints con-

tinuously welded; and (3) be installed without dips or traps that may collect residues, except where traps with continuous or automatic removal of residues are provided. Since fires may occur in these systems (producing temperatures in excess of 2000 °F), provisions are necessary for expansion in accordance with the following table. Ducts that must have a fire resistance rating are usually encased in materials with appropriate thermal and durability ratings.

Kitchen Exhaust Duct Material	Duct Expansion 2000 °F, in/ft
Carbon steel	0.19
Type 304 stainless steel	0.23
Type 430 stainless steel	0.13

Ducts that convey moisture-laden air must have construction specifications that properly account for corrosion resistance, drainage, and waterproofing of joints and seams. No nationally recognized standards exist for applications in areas such as kitchens, swimming pools, shower rooms, and steam cleaning or washdown chambers. Galvanized steel, stainless steel, aluminum, and plastic materials have been used. Wet and dry cycles increase corrosion of metals. Chemical concentrations affect corrosion rate significantly. Chapter 43 of the 1991 ASHRAE *Handbook—HVAC Applications* addresses material selection for corrosive

Table 6   Intermediate Reinforcement[a]

Class	Minimum Rigidity EI × 10^5, lb-in.^{2}	ANGLE H × T, in.[b]	ZEE OR CHANNEL H × B × T, in.[b]	HAT SECTION H × B × D × T, in.[b]	CHANNEL H × B × T, in.[b]
A	0.5	Use Class C	Use Class B	Use Class F	Use Class C
B	1.0	Use Class C	3/4 × 1/2 × 0.0356	Use Class F	Use Class C
C	2.5	1 × 0.0466 / 3/4 × 1/8	3/4 × 1/2 × 0.0466	Use Class F	3/4 × 3 × 0.0466
D	5	1 1/4 × 0.0466 / 1 × 1/8	1 × 3/4 × 0.0356	Use Class F	1 1/8 × 3 1/4 × 0.0466
E	10	1 1/4 × 0.0854 / 1 1/2 × 0.0575	1 × 3/4 × 0.0854 / 1 1/2 × 3/4 × 0.0356	Use Class F	1 × 2 × 1/8
F	15	1 1/2 × 1/8	1 × 3/4 × 1/8 / 1 1/2 × 3/4 × 0.0466	1 1/2 × 3/4 × 5/8 × 0.0356 / 1 1/2 × 1 1/2 × 3/4 × 0.0356	1 1/2 × 3 × 0.0575
G	25	1 1/2 × 3/16	1 1/2 × 3/4 × 1/8 / 2 × 1 1/8 × 0.0356	1 1/2 × 3/4 × 5/8 × 0.0575 / 1 1/2 × 1 1/2 × 3/4 × 0.0466	1 1/4 × 3 × 1/8
H	50	2 × 1/8	2 × 1 1/8 × 0.0575	1 1/2 × 1 1/2 × 3/4 × 0.0854 / 2 × 1 × 3/4 × 0.0466	1.4 × 3 × 0.22
I	75	2 × 3/16	2 × 1 1/8 × 0.0854	2 × 1 × 3/4 × 0.0854 / 2 1/2 × 2 × 3/4 × 0.0575	2 × 2 × 1/8
J	100	2 × 1/4 / 2 1/2 × 1/8	2 × 1 1/8 × 1/8 / 3 × 1 1/8 × 0.0575	2 × 1 × 3/4 × 1/8 / 2 1/2 × 2 × 3/4 × 0.0854	1.6 × 4 × 0.28
K	150	2 1/2 × 3/16	3 × 1 1/8 × 0.0854	2 1/2 × 2 × 3/4 × 1/8 / 3 × 1 1/2 × 3/4 × 0.0575	—
L	200	2 1/2 × 1/4	3 × 1 1/8 × 1/8	3 × 1 1/2 × 3/4 × 0.0854	—

[a] Table 6 is based on Table 1-10 of *HVAC Duct Construction Standards—Metal and Flexible* (SMACNA 1985).

[b] T-dimension for formed shapes is minimum thickness; nominal for structural shapes.

environments. Conventional duct construction standards are frequently modified to require welded or soldered joints, which are generally more reliable and durable than sealant-filled, mechanically locked joints. The number of transverse joints should be minimized, and longitudinal seams should not be located on the bottom of the duct. Risers should drain and horizontal ducts should pitch in the direction most favorable for moisture control. *Industrial Ventilation* (ACGIH 1988) covers hood design.

PLASTIC DUCTS

The *Thermoplastic Duct Construction Manual* (SMACNA 1974) covers thermoplastic (polyvinyl chloride, polyethylene, polypropylene, acrylonitrile butadiene styrene) and thermosetting (glass-fiber-reinforced polyester) plastic ducts used in commercial and industrial installations. SMACNA's manual provides comprehensive construction detail for positive or negative 2, 6, and 10 in. of water polyvinyl chloride duct systems. NFPA *Standard* 91 provides construction details and application limitations for plastic ducts, and references the Department of Commerce *Standard* PS 15-69 (NBS 1969) for reinforced thermosetting plastic duct fabrication details. For a listing of ASTM and commercial product standards concerning thermoplastic materials, see NFPA *Standard* 91.

UNDERGROUND DUCTS

No comprehensive standards exist for the construction of underground air ducts. Coated steel, asbestos cement, plastic, tile, concrete, reinforced fiberglass, and other materials have been used. Underground duct and fittings should always be round and have a minimum thickness as listed in Table 7. Thickness above the minimum may be needed for individual applications. Specifications for the construction and installation of underground ducts should account for the following: water tables, ground surface flooding, the need for drainage piping beneath ductwork, temporary or permanent anchorage to resist flotation, frost heave, backfill loading, vehicular traffic load, corrosion, cathodic protection, heat loss or gain, building entry, bacterial organisms, degree of water and air tightness, inspection or testing prior to backfill, and code compliance. Corrosion considerations for soil environments are addressed in Chapter 43 of the 1991 ASHRAE *Handbook—HVAC Applications. Criteria and Test Procedures for Combustible Materials Used for Warm Air Ducts Encased in Concrete Slab Floors* (NRC) provides criteria and test procedures for fire resistance, crushing strength, bending strength, deterioration, and odor. *Installation Techniques for Perimeter Heating and Cooling* (ACCA 1990) covers residential systems and gives five classifications of duct material related to particular performance characteristics. Residential installations may also be subject to the requirements in NFPA *Standard* 90B. Commercial systems also normally require compliance with NFPA *Standard* 90A.

DUCTS OUTSIDE BUILDINGS

The location and construction of ducts exposed to outdoor atmospheric conditions are generally regulated by building codes. Exposed ducts and their sealant/joining systems must be evaluated for the following: (1) waterproofing; (2) resistance to external loads (wind, snow, and ice); (3) degradation from corrosion,

Table 7a Round Ferrous Metal Duct Construction for Commercial Systems*[a]

| Duct Diameter, in. | Minimum Galvanized Steel Thickness[b], in. | | | | | | Type of Joint[d] |
| | − 2 in. of Water Maximum | | | + 2 in. of Water Maximum | | | |
	Spiral Seam Duct	Longitudinal Seam Duct[c]	Fittings	Spiral Seam Duct	Longitudinal Seam Duct[c]	Fittings	
⩽ 8	0.0157	0.0236	0.0236	0.0157	0.0157	0.0187	Beaded slip
14	0.0187	0.0236	0.0236	0.0157	0.0187	0.0187	Beaded slip
26	0.0236	0.0296	0.0296	0.0187	0.0236	0.0236	Beaded slip
36	0.0296	0.0356	0.0356	0.0236	0.0296	0.0296	Beaded slip
50	0.0356	0.0466	0.0466	0.0296	0.0356	0.0356	Flange
60	0.0466	0.0575	0.0575	0.0356	0.0466	0.0466	Flange
84	0.0575	0.0705	0.0705	0.0466	0.0575	0.0575	Flange

*See Table 7b for notes.

Table 7b Round Ferrous Metal Duct Construction for Commercial Systems[a]

| Duct Diameter, in. | Minimum Galvanized Steel Thickness[b], in. | | | | | | Type of Joint[d] |
| | − 3 in. of Water Maximum | | | + 10 in. of Water Maximum | | | |
	Spiral Seam Duct	Longitudinal Seam Duct[e]	Fittings	Spiral Seam Duct	Longitudinal Seam Duct[e]	Fittings	
⩽ 8	0.0187	0.0236	0.0236	0.0187	0.0236	0.0236	Beaded slip
14	0.0187	0.0236	0.0236	0.0187	0.0236	0.0236	Beaded slip
26	0.0236	0.0296	0.0296	0.0236	0.0296	0.0296	Beaded slip
36	0.0296	0.0356	0.0356	0.0296	0.0356	0.0356	Beaded slip
50	0.0356	0.0356	0.0356	0.0356	0.0356	0.0356	Flange
60	—	0.0466	0.0466	0.0466	0.0466	0.0466	Flange
84	—	0.0575	0.0575	0.0466	0.0575	0.0575	Flange

[a]Table 7 is based on *HVAC Duct Construction Standards—Metal and Flexible* Table 3-2, (SMACNA 1985).
[b]Table 7 may be used for galvanized, painted, uncoated, and stainless steel whenever the base metal thickness is not less than 0.0015 in. below those in the table.
[c]For seam details and limitations, refer to *HVAC Duct Construction Standards— Metal and Flexible*, Figure 3-1.
[d]Recommended joints are listed. For alternate joints, details, and limitations, refer to *HVAC Duct Construction Standards—Metal and Flexible*, Figure 3-2.
[e]Only butt weld or flat lock (grooved seam, pipe lock) longitudinal seams are permitted.

ultraviolet radiation, or thermal cycles; (4) heat transfer; (5) susceptibility to physical damage; (6) hazards at air inlets and discharges; and (7) maintenance needs. In addition, support systems must be custom designed for rooftop, wall-mounted, and bridge or ground-based applications. Specific requirements must also be met for insulated and uninsulated ducts.

SEISMIC QUALIFICATION

Seismic analysis of duct systems may be required by building codes or federal regulations. Provisions for seismic analysis are provided by the Federal Emergency Management Agency (FEMA

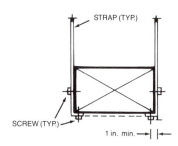

Fig. 2   Strap Hanger—Rectangular Ducts

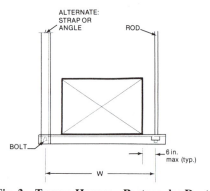

Fig. 3   Trapeze Hanger—Rectangular Ducts

1985), the National Institute of Standards (NBS 1981), and the Department of Defense (NAVFAC 1982). Ducts, duct hangers, fans, fan supports, and other duct-mounted equipment are generally evaluated independently. Chapter 49 of the 1991 ASHRAE *Handbook—HVAC Applications* gives design details. The Sheet Metal Industry Fund of Los Angeles has also published guidelines for seismic restraints of mechanical systems (SMIFLA 1982).

SHEET METAL WELDING

Specification for Welding of Sheet-Metal (ASW 1990) covers sheet metal arc welding and braze welding procedures. This specification also addresses the qualification of welders and welding operators, workmanship, and the inspection of production welds.

THERMAL INSULATION

Insulation materials for ducts, plenums, and apparatus casings are covered in Chapter 21 of the 1989 ASHRAE *Handbook—*

Table 8   Flat-Oval Duct Construction for Positive Pressure Commercial Systems[a]

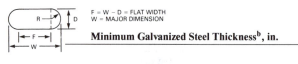

Minimum Galvanized Steel Thickness[b], in.

Major Axis (W), in.	Spiral Seam Duct	Longitudinal Seam Duct	Fittings	Type of Joint
⩽ 24	0.0236	0.0356	0.0356	Beaded slip
36	0.0296	0.0356	0.0356	Beaded slip
48	0.0296	0.0466	0.0466	Flange
60	0.0356	0.0466	0.0466	Flange
70	0.0356	0.0575	0.0575	Flange
> 70	0.0466	0.0575	0.0575	Flange

[a]Table 8 is based on *HVAC Duct Construction Standards—Metal and Flexible*, Table 3-4 (SMACNA 1985).
[b]Use Table 8 for galvanized, painted, uncoated, and stainless steel whenever the base metal thickness is not less than 0.0015 in. below those in the table.
W = Major dimension   F = W − D = Flat width

Table 9   Rectangular Duct Hangers—Commercial Systems[a,b]

Half of Duct Perimeter (P/2)[d], in.	Galvanized Straps (see Figures 2 and 3) Width × Thickness (min), in.				Rods (see Figure 3) Diameter, in.			
	Maximum Hanger Spacing, ft							
	10	8	5	4	10	8	5	4
⩽ 30	1 × 0.0296	1 × 0.0296	1 × 0.0296	1 × 0.0296	0.135	0.135	0.106	0.106
72	1 × 0.0466	1 × 0.0356	1 × 0.0296	1 × 0.0296	3/8	1/4	1/4	1/4
96	1 × 0.0575	1 × 0.0466	1 × 0.0356	1 × 0.0296	3/8	3/8	3/8	1/4
120	1 1/2 × 0.0575	1 × 0.0575	1 × 0.0466	1 × 0.0356	1/2	3/8	3/8	1/4
168	1 1/2 × 0.0575	1 × 0.0575	1 × 0.0575	1 × 0.0466	1/2	1/2	3/8	3/8
192	—	1 1/2 × 0.0575	1 × 0.0575	1 × 0.0575	1/2	1/2	3/8	3/8
> 192	—	—	—	—	◄————Special Analysis Required[e]————►			

[a]Table 9 is based on *HVAC Duct Construction Standards—Metal and Flexible* Table 4-1 (SMACNA 1985).
[b]Table based on (1) 0.0575-in. thick ducts; (2) reinforcements and trapeze weights; (3) 1 lb/ft² for insulation; and (4) no external loads.
[c]For trapeze angle size, see Table 10. For hanger attachments to structures, typical upper attachments, lower hanger attachments, and strap splicing details, consult the *HVAC Duct Construction Standards—Metal and Flexible*, Table 4-1 and Figures 4-1 through 4-4.
[d]When dimension (W) exceeds 60 in., P/2 maximum is 1.25 W.
[e]Consult AISC Manual (1989) for hanger design criteria.

Maximum allowable strap, wire, and rod loads are:

Strap W × T(min), in.	Load, lb	Rod or Wire (dia), in.	Load, lb
1 × 0.0296	260	0.106	80
1 × 0.0356	320	0.135	120
1 × 0.0466	420	0.162	160
1 × 0.0575	700	1/4	270
1 1/2 × 0.0575	1100	3/8	680
		1/2	1250
		5/8	2000
		3/4	3000

Table 10 Allowable Trapeze Angle Loads[a]

Maximum Hanger Load[b], lb

Angle[c], in.

Width of Hanger (W)[d], in.	1 × 1 × 0.0575[e]	1 × 1 × 1/8	1 1/2 × 1 1/2 × 0.0575[e]	1 1/2 × 1 1/2 × 1/8	1 1/2 × 1 1/2 × 3/16	1 1/2 × 1 1/2 × 1/4 or 2 × 2 × 1/8	2 × 2 × 3/16	2 × 2 × 1/4	2 1/2 × 2 1/2 × 3/16	2 1/2 × 2 1/2 × 1/4
≤ 18	80	150	180	350	510	650	940	1230	1500	1960
24	75	150	180	350	510	650	940	1230	1500	1960
30	70	150	180	350	510	650	940	1230	1500	1960
36	60	130	160	340	500	620	920	1200	1480	1940
42	40	110	140	320	480	610	900	1190	1470	1930
48	—	80	110	290	450	580	870	1160	1440	1900
54	—	40	70	250	400	540	840	1120	1400	1860
60	—	—	—	190	350	490	780	1060	1340	1800
66	—	—	—	100	270	400	700	980	1260	1720
72	—	—	—	—	190	320	620	900	1180	1640
78	—	—	—	—	80	210	500	790	1070	1530
84	—	—	—	—	—	80	380	660	940	1400
96	—	—	—	—	—	—	—	320	600	1060
108	—	—	—	—	—	—	—	—	150	610

[a]Table 10 is based on *HVAC Duct Construction Standards—Metal and Flexible*, Table 4-3 (SMACNA 1985).
[b]Loads in this table assume rods or other structural members are 6-in. maximum from duct sides (see Figure 3).
[c]Steel yield stress is based on 25,000 psi minimum.
[d]For W-dimension, see Figure 3.
[e]Galvanized formed angles; thickness is minimum.

Table 11 Round Duct Hangers—Commercial Systems[a,b]

Duct Diameter, in.	Max. Hanger Spacing[c], ft.	Hanger Configuration[d] (see Figure 4)	Galvanized Ring Size (min), in.	Minimum Hanger Size, in.		
				Wire	Strap	Rod
≤ 10	12	Type 1 or 3	1 × 0.0296	0.106	1 × 0.0296	1/4
18	12	Type 1 or 3	1 × 0.0296	0.162	1 × 0.0296	1/4
24	12	Type 1 or 3	1 × 0.0296	—	1 × 0.0296	1/4
		Type 2 or 4	1 × 0.0296	0.135	1 × 0.0296	1/4
36	12	Type 1 or 3	1 × 0.0356	—	1 × 0.0356	3/8
		Type 2 or 4	1 × 0.0356	0.162	1 × 0.0296	1/4
50	12	Type 4	1 × 0.0356	—	1 × 0.0356	3/8
60	12	Type 4	1 × 0.0466	—	1 × 0.0466	3/8
84	12	Type 4	1 × 0.0575	—	1 × 0.0575	3/8

[a]Table 11 is based on SMACNA's publication *HVAC Duct Construction Standards—Metal and Flexible*, Table 4-2 (SMACNA 1985).
[b]Table 11 is for duct construction per Table 7 plus one lb/ft² insulation. For heavier ductwork, size hanger members so allowable load (see Table 9) is not exceeded.
[c]The 12-ft hanger spacing may be increased for risers and other conditions with sub-stantial horizontal support.
[d]For hanger attachments to structures, typical upper attachments, lower hanger attachments, and strap splicing details, consult the *HVAC Duct Construction Standards—Metal and Flexible*, Table 4-1 and Figures 4-1 through 4-4.

Fundamentals. Codes generally limit factory-insulated ducts to UL 181, Class 0 or Class 1. *Commercial and Industrial Insulation Standards* (MICA 1988) gives insulation details.

MASTER SPECIFICATIONS

Master specifications for duct construction and most other elements in building construction are produced and regularly updated by several organizations. Two documents are *Masterspec* by the American Institute of Architects (AIA) and *Spectext* by the Construction Specifications Institute (CSI). These documents are model project specifications that require little editing to customize each application for a project.

REFERENCES

ACCA. 1990. *Installation techniques for perimeter heating and cooling.* Manual No. 4. Air Conditioning Contractors of America, Washington, D.C.

ACGIH. 1988. *Industrial ventilation—A manual of recommended practice.* American Conference of Governmental Industrial Hygienists, Lansing, MI.

AISC. 1989. *Manual of steel construction.* American Institute of Steel Construction, Chicago, IL.

AISI. 1989. *Cold-formed steel design manual.* American Iron and Steel Institute, Washington, D.C.

ANSI. 1984. Preferred metric sizes for flat metal products. *Standard* B32.3M-84. American National Standards Institute, New York.

ASHRAE. 1975. Energy conservation in new building design. ASHRAE/IES *Standard* 90B-1975.

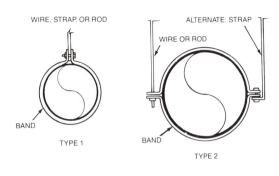

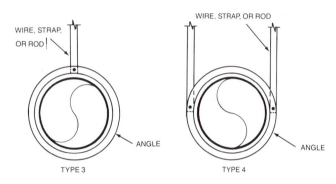

Fig. 4   Round Duct Hangers

ASHRAE. 1977. Energy conservation in new building design. ASHRAE *Standard* 90C-1977.

ASHRAE. 1980. Energy conservation in new building design. ANSI/ASHRAE/IES *Standard* 90A-1980.

ASHRAE. 1981. Energy conservation in existing buildings—High rise residential. ANSI/ASHRAE/IES *Standard* 100.2-1981.

ASHRAE. 1981. Energy conservation in existing buildings—Institutional. ANSI/ASHRAE/IES *Standard* 100.5-1981.

ASHRAE. 1981. Energy conservation in existing buildings—Public assembly. ANSI/ASHRAE/IES *Standard* 100.6-1981.

ASHRAE. 1984. Energy conservation in existing facilities—Industrial. ANSI/ASHRAE/IES *Standard* 100.4-1984.

ASHRAE. 1985. Energy conservation in existing buildings—Commercial. ANSI/ASHRAE/IES *Standard* 100.3-1985.

ASHRAE. 1989. Energy efficient design of new buildings except low-rise residential buildings. ASHRAE/IES *Standard* 90.1-1989.

AHSRAE. 1989. Ventilation for acceptable indoor air quality. ASHRAE *Standard* 62-1989.

ASTM. 1986. Standard specification for thermal and acoustical insulation (mineral fiber, duct lining material). ASTM *Standard* C1071-86.

ASTM. 1990. Specifications for adhesives for duct thermal insulation. ASTM *Standard* C916-85 (reaffirmed 1990).

ASTM. 1990. Test method for sound absorption and sound absorption coefficients by the reverberation room method. ASTM *Standard* C423-90a. American Society for Testing and Materials, Philadelphia, PA.

ASTM. 1991. General requirements for flat-rolled stainless and heat-resisting steel plate, sheet, and strip. ASTM *Standard* A480/A480M-91a.

ASTM. 1991. General requirements for steel sheet, zinc-coated (galvanized) by the hot-dip process (metric). ASTM *Standard* A525M-91.

ASTM. 1991. Specification for steel, carbon (0.15 maximum, percent), hot-rolled sheet and strip, commercial quality. ASTM *Standard* A569/A569M-91a.

ASTM. 1991. Specification for steel, sheet, carbon, cold-rolled commercial quality. ASTM *Standard* A366/A366M-91.

ASTM. 1991. Standard specification for general requirements for steel, sheet, carbon and high-strength, low-alloy, hot-rolled and cold-rolled. ASTM *Standard* A568/A568M-91.

AWS. 1990. Specification for welding of sheet metal. AWS D9.1-90. American Welding Society, Miami, FL.

BOCA. 1990. *National building code.* Building Officials and Code Administrators International, Country Club Hills, IL.

BOCA. 1990. *National mechanical code.*

CABO. 1989. *Model energy code.* Council of American Building Officials, Falls Church, VA.

CABO. 1989. *One- and two-family dwelling code.* Amended 1990 and 1991.

Cheremisinoff, P.N. and R.A. Young. 1975. *Pollution engineering practice handbook.* Ann Arbor Science Publishers, Inc., Ann Arbor, MI.

FEMA. 1985. NEHRP (National Earthquake Hazards Reduction Program) Recommended provisions for the development of seismic regulations for new buildings. Building Seismic Safety Council for the Federal Emergency Management Agency, Washington, D.C.

ICBO. 1991. *Uniform building code.* International Conference of Building Officials, Whittier, CA.

ICBO. 1991. *Uniform mechanical code.*

MICA. 1988. Commercial and industrial insulation standards. Midwest Insulation Contractors Association, Omaha, NE.

NAVFAC. 1982. *Technical manual—Seismic design for buildings.* Army TM 5-809-10, Navy NAVFAC P-355, Air Force AFM 88-3 (Chapter 13). U.S. Government Printing Office, Washington, D.C.

NBS. 1969. Voluntary product standard for custom contact-molded reinforced-polyester chemical-resistant process equipment. NBS *Standard* PS 15-69. National Bureau of Standards, Department of Commerce, Washington, D.C. (Available from the National Institute of Standards and Technology.)

NBS. 1981. Draft seismic standard for federal buildings. Publication NBSIR 81-2195.

NFPA. 1988. Chimneys, fireplaces, vents and solid fuel burning appliances. ANSI/NFPA *Standard* 211. National Fire Protection Association, Quincy, MA.

NFPA. 1989. Installation of air conditioning and ventilating systems. NFPA *Standard* 90A.

NFPA. 1989. Warm air heating and air conditioning systems. ANSI/NFPA *Standard* 90B.

NFPA. 1990. Blower and exhaust systems. ANSI/NFPA *Standard* 91.

NFPA. 1991. Life safety code. ANSI/NFPA *Standard* 101.

NFPA. 1991. Smoke and heat venting. ANSI/NFPA *Standard* 204M.

NFPA. 1991. Vapor removal from cooking equipment. ANSI/NFPA *Standard* 96.

NRC. Criteria and test procedures for combustible materials used for warm air ducts encased in concrete slab floors. National Research Council, National Academy of Sciences, Washington, D.C.

SBCCI. 1991. *Standard building code.* Southern Building Code Congress International, Inc., Birmingham, AL.

SBCCI. 1991. *Standard mechanical code.*

SMACNA. 1974. *Thermoplastic duct (PVC) construction manual.* Sheet Metal and Air Conditioning Contractors' National Association, Inc., Chantilly, VA.

SMACNA. 1975. *Accepted industry practice for industrial duct construction.*

SMACNA. 1977. *Round industrial duct construction standards.*

SMACNA. 1979. *Fibrous glass duct construction standards.*

SMACNA. 1980. *Rectangular industrial duct construction standards.*

SMACNA. 1985. *HVAC Air duct leakage test manual.*

SMACNA. 1985. *HVAC Duct construction standards—Metal and flexible.*

SMIFLA. 1985. *Guidelines for seismic restraints of mechanical systems and plumbing piping systems,* 1st ed. (1982). Sheet Metal Industry Fund of Los Angeles. The Plumbing and Piping Industry Council, Los Angeles, CA. Addendum published in 1985.

Swim, W.B. 1978. Flow losses in rectangular ducts lined with fiberglass. ASHRAE *Transactions* 84(2).

TIMA. 1989. *Fibrous glass duct construction standards.* Thermal Insulation Manufacturers Association, Alexandria, VA.

UL. 1983. Leakage rated dampers for use in smoke control systems. UL *Standard* 555S-83. Underwriters Laboratories, Inc., Northbrook, IL.

UL. 1986. Fire tests of door assemblies. ANSI/UL *Standard* 10B-86.

UL. 1990. Factory-made air ducts and connectors. UL *Standard* 181-90.

UL. 1990. Fire dampers and ceiling dampers. UL *Standard* 555-90.

UL. Gas and oil directory. (Published annually.)

United McGill Corporation. 1985. Industrial duct engineering data and recommended design standards, Form No. SMP-IDP (February): 24-28. Westerville, OH.

CHAPTER 17

AIR-DIFFUSING EQUIPMENT

SUPPLY air outlets and diffusing equipment introduce air into a conditioned space to obtain a desired indoor atmospheric environment. Return and exhaust air is removed from a space through return and exhaust inlets. Various types of diffusing equipment are available as standard manufactured products. This chapter describes this equipment and details its proper use. Chapter 31 of the 1989 ASHRAE *Handbook—Fundamentals* also covers the subject.

SUPPLY AIR OUTLETS

Supply outlets, properly sized and located, control the air pattern within a given space to obtain proper air motion and temperature equalization. The following basic supply outlets are commonly available: (1) grille outlets, (2) slot diffuser outlets, (3) ceiling diffuser outlets, and (4) perforated ceiling panels. These types differ in their construction features, physical configurations, and in the manner in which they diffuse or disperse supply air and induce or entrain room air into a primary airstream.

Accessories used with these outlets regulate the volume of supply air and control its flow pattern. For example, an outlet cannot discharge air properly and uniformly unless the air is conveyed to it in a uniform flow. Accessories may also be necessary for proper air distribution in a space, so they must be selected and used in accordance with manufacturers' recommendations. Outlets should be sized to project air so that its velocity and temperature reach acceptable levels before entering the occupied zone.

Outlets with high induction rates move (throw) air short distances but have rapid temperature equalization. Ceiling diffusers with radial patterns have shorter throws and obtain more rapid temperature equalization than slot diffusers. Grilles, which have long throws, have the lowest diffusion and induction rates. Therefore, round or square ceiling diffusers deliver more air to a given space than grilles and slot diffuser outlets that require room velocities of 25 to 35 fpm. In some spaces, higher room velocities can be tolerated, or the ceilings may be high enough to permit a throw long enough to result in the recommended room velocities.

Outlets with high induction characteristics can also be used advantageously in air-conditioning systems with low supply air temperatures and consequent high-temperature differentials between room air temperature and supply air temperatures. Therefore, ceiling diffusers may be used in systems with cooling temperature differentials up to 30 to 35 °F and still provide satisfactory temperature equalizations within the spaces. Slot diffusers may be used in systems with cooling temperature differentials as high as 25 °F. Grilles may generally be used in well-designed systems with cooling temperature differentials up to 20 °F.

Surface Effect

The induction or entrainment characteristics of a moving airstream cause a surface effect. An airstream moving adjacent to, or in contact with, a wall or ceiling surface creates a low-pressure area immediately adjacent to that surface, causing the air to remain in contact with the surface substantially throughout the length of throw. The surface effect counteracts the drop of horizontally projected cool airstreams.

Ceiling diffusers exhibit surface effect to a high degree because a circular air pattern blankets the entire ceiling area surrounding each outlet. Slot diffusers, which discharge the airstream across the ceiling, exhibit surface effect only if they are long enough to blanket the ceiling area. Grilles exhibit varying degrees of surface effect, depending on the spread of the particular air pattern.

In many installations, the outlets must be mounted on an exposed duct and discharge the airstream into free space. In this type of installation, the airstream entrains air on both its upper and lower surfaces; as a result, a higher rate of entrainment is obtained and the throw is shortened by about 33%. Airflow per unit area for these types of outlets can, therefore, be increased. Because there is no surface effect from ceiling diffusers installed on exposed ducts, the air drops rapidly to the floor. Thus, temperature differentials in air-conditioning systems must be restricted to a range of 15 to 20 °F. Airstreams from slot diffusers and grilles show a marked tendency to drop because of the lack of surface effect.

Smudging

Smudging is a problem with ceiling and slot diffusers. Dirt particles held in suspension in the room air are subjected to turbulence at the outlet face. This turbulence is primarily responsible for smudging. Smudging can be expected in areas of high pedestrian traffic, *e.g.*, lobbies and stores. When ceiling diffusers are installed on smooth ceilings (such as plaster, mineral tile, and metal pan), smudging is usually in the form of a narrow band of discoloration around the diffuser. Antismudge rings effectively reduce this type of smudging. On highly textured ceiling surfaces (such as rough plaster and sprayed-on composition), smudging occurs over a more extensive area.

PROCEDURE FOR OUTLET SELECTION

The following procedure is generally used in selecting outlet locations and types:

1. Determine the amount of air to be supplied to each room. (Refer to Chapters 25 and 26 of the 1989 ASHRAE *Handbook—Fundamentals* to determine air quantities for heating and cooling.)
2. Select the type and quantity of outlets for each room, considering such factors as air quantity required, distance available

---

The preparation of this chapter is assigned to TC 5.3, Room Air Distribution.

Table 1   Guide to Use of Various Outlets

Type of Outlet	Air Loading of Floor Space, cfm/ft^2	Approx. Max. Air Changes per Hour for 10-ft Ceiling
Grille	0.6 to 1.2	7
Slot	0.8 to 2.0	12
Perforated panel	0.9 to 3.0	18
Ceiling diffuser	0.9 to 5.0	30
Perforated ceiling	1.0 to 10.0	60

for throw or radius of diffusion, structural characteristics, and architectural concepts. Table 1 is based on experience and typical ratings of various outlets. It may be used as a guide to the outlets applicable for use with various room air loadings. Special conditions, such as ceiling heights greater than the normal 8 to 12 ft and exposed duct mounting, as well as product modifications and unusual conditions of room occupancy, can modify this table. Manufacturers' rating data should be consulted for final determination of the suitability of the outlets used.

3. Locate outlets in the room to distribute the air as uniformly as possible. Outlets may be sized and located to distribute air in proportion to the heat gain or loss in various portions of the room.

4. Select proper outlet size from manufacturers' ratings according to air quantities, discharge velocities, distribution patterns, and sound levels. Note manufacturers' recommendations with regard to use. In an open space configuration, the interaction of airstreams from multiple diffuser sources may alter single-diffuser throw data or single-diffuser air temperature/air velocity data, and it may not be sufficient to predict particular levels of air motion in a space. Also, obstructions to the primary air distribution pattern require special consideration.

Chapter 31 of the 1989 ASHRAE *Handbook—Fundamentals* gives the complete procedure for selecting the size and locations of the diffusers.

Sound Level

The sound level of an outlet is a function of the discharge velocity and the transmission of systemic noise, which is a function of the size of the outlet. Higher frequency sounds can be the result of excessive outlet velocity but may also be generated in the duct by the moving airstream. Lower pitched sounds are generally the result of mechanical equipment noise transmitted through the duct system and outlet. The cause of higher frequency sounds can be pinpointed as outlet or systemic sounds by removing the outlet during operation. A reduction in sound level indicates that the outlet is causing noise. If the sound level remains essentially unchanged, the system is at fault. Chapter 42 of the 1991 ASHRAE *Handbook—HVAC Applications* has more information on design criteria, acoustic treatment, and selection procedures.

GRILLE OUTLETS

Adjustable bar grille. This is the most common type of grille used as a supply outlet. The single-deflection grille consists of a frame enclosing a set of vertical or horizontal vanes. Vertical vanes deflect the airstream in the horizontal plane; horizontal vanes deflect the airstream in the vertical plane. The double-deflection grille has a second set of vanes installed behind and at right angles to the face vanes. This grille controls the airstream in both the horizontal and vertical planes.

Fixed bar grille. This type of grille is similar to the single-deflection grille, except that the vanes are not adjustable—they may be straight or set at an angle. The angle at which the air is discharged from this grille depends on the type of deflection vanes.

Stamped grille. This grille is stamped from a single sheet of metal to form a fretwork through which air can pass. Many designs are used, varying from square or rectangular holes to intricate ornamental designs.

Variable area grille. This type of grille is similar to the adjustable double-deflection grille, but can vary the discharge area to achieve an air volume change (variable volume outlet) at constant pressure, so that the variation in throw is minimized for a given change in supply air volume.

Applications

Properly selected grilles operate satisfactorily from high side and perimeter locations in the sill, curb, or floor. Ceiling-mounted grilles, which discharge the airstream down, are generally unacceptable in comfort air-conditioning installations in interior zones and may cause draft in perimeter applications.

High side wall. The use of a double-deflection grille usually provides the most satisfactory solution. The vertical face louvers of a well-designed grille deflect the air approximately 50° to either side and amply cover the conditioned space. The rear louvers deflect the air at least 15° in the vertical plane, which is ample to control the elevation of the discharge pattern.

Perimeter installations. The grille selected must fit the specific job. When small grilles are used, adjustable vane grilles improve the coverage of perimeter surfaces. Where the perimeter surface can be covered with long grilles, the fixed vane grille is satisfactory. Where grilles are located more than 8 in. from the perimeter surface, it is usually desirable to deflect the airstream toward the perimeter wall. This can be done with adjustable or fixed deflecting vane grilles.

Ceiling installation. Such installation is generally limited to grilles having curved vanes, which, because of their design, provide a horizontal pattern. Curved vane grilles may also be used satisfactorily in high side wall or perimeter installations.

Accessories

Various accessories, designed to improve the performance of grille outlets, are available as standard equipment.

• Opposed-blade dampers can be attached to the backs of grilles (the combination of a grille and a damper is called a register) or installed as separate units in the duct (Figure 1A). Adjacent blades of this damper rotate in opposite directions and may receive air from any direction, discharging it in a series of jets without adversely deflecting the airstream to one side of the duct.

• Multishutter dampers have a series of gang-operated blades that rotate in the same direction (Figure 1B). This uniform rotation deflects the airstream when the damper is partially open. Most dampers are operated by removable keys or fixed or removable levers.

• Gang-operated turning vanes are installed in collar connections to branch ducts. The device shown in Figure 1C has vanes that pivot and remain parallel to the duct airflow, regardless of the setting. The second device has a set of fixed vanes (Figure 1D). Both devices restrict the area of the duct in which they are installed. They should be used only when the duct is wide enough to allow the device to open to its maximum position without causing undue restriction of airflow in the duct, which might limit downstream airflow.

• Individually adjusted turning vanes are used in the device shown in Figure 1E. Two sets of vanes are used. The downstream set equalizes flow across the collar, while the upstream set turns the air. They can also be adjusted, at various angles, to act as a

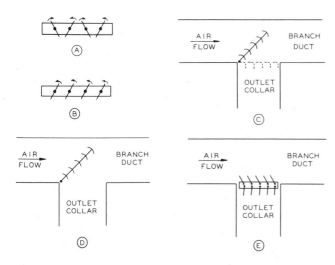

Fig. 1   Accessory Controls for Grille and Register Outlets

damper, but it is not practicable to use this device as a damper because its balancing requires removal of the grille to gain access and reinstallation to measure airflow.

- Other miscellaneous accessories available as standard equipment include remote-control devices to operate the dampers from locations other than the grille face and maximum-minimum stops to limit damper travel.

SLOT DIFFUSER OUTLETS

A slot diffuser is an elongated outlet consisting of a single or multiple number of slots. It is usually installed in long continuous lengths. Outlets with dimensional aspect ratios of 25 to 1 or greater and a maximum height of approximately 3 in. generally meet the performance criteria for slot diffusers.

Applications

High side wall installation. The perpendicular-flow slot diffuser is best suited to high side wall installations and perimeter installations in sills, curbs, and floors. The air discharged from a perpendicular slot diffuser will not drop if the diffuser is located within 6 to 12 in. from the ceiling and is long enough to establish surface effect. Under these conditions, air travels along the ceiling to the end of the throw. If the slot diffuser is mounted 1 to 2 ft below the ceiling, an outlet that deflects the air up to the ceiling must be used to achieve the same result. If the slot is located more than 2 ft below the ceiling, premature drop of cold air into the occupied zone will probably result.

Ceiling installation. The parallel-flow slot diffuser is ideal for ceiling installation because it discharges across the ceiling. The perpendicular-flow slot diffuser may be mounted in the ceiling; however, the downward discharge pattern may cause localized areas of high air motion. This device performs satisfactorily when installed adjacent to a wall or over an unoccupied or transiently occupied area. Care should be exercised in using a perpendicular-flow slot diffuser in a downward discharge pattern because variations of supply air temperature cause large variations in throw.

Sill installation. The perpendicular-flow slot diffuser is well suited to sill installation, but it may also be installed in the curb and floor. When the diffuser is located within 8 in. of the perimeter wall, the discharged air may be either directed straight toward the ceiling or deflected slightly toward the wall. When the diffuser distance from the wall is greater than 8 in., the air should generally be deflected toward the wall at an angle of approximately 15°;

deflections as great as 30° may be desirable in some cases. The air should not be deflected away from the wall into the occupied zone.

To perform satisfactorily, outlets of this type must be used only in installations with carefully designed duct and plenum systems (refer to the section, Discharge from a Long Slot, Chapter 31 of the 1989 ASHRAE *Handbook—Fundamentals*).

Slot diffusers are generally equipped with accessory devices for uniform supply air discharge along the entire length of the slot. While accessory devices help correct the airflow pattern, proper approach conditions for the airstream are also important for satisfactory performance. When the plenum supplying a slot diffuser is being designed, the transverse velocity in the plenum should be less than the discharge velocity of the jet, as recommended by the manufacturer and also as shown by experience.

If tapered ducts are used for introducing supply air into the diffuser, they should be sized to maintain a velocity of approximately 500 fpm and tapered to maintain constant static pressure.

Air-light fixtures. Slot diffusers having a single-slot discharge and nominal 2, 3, and 4-ft lengths are available for use in conjunction with recessed fluorescent light troffers. A diffuser mates with a light fixture and is entirely concealed from the room. It discharges air through suitable openings in the fixture and is available with fixed or adjustable air discharge patterns, air distribution plenum, inlet dampers for balancing, and inlet collars suitable for flexible duct connections. Light fixtures adapted for slot diffusers are available in styles to fit common ceiling constructions. Various slot diffuser and light fixture manufacturers can furnish products compatible with one another's equipment.

Accessories

Dampers. Dampers are normally available as integral equipment with slot diffusers and are used for minor flow adjustments. Installation of balancing dampers in the distribution plenum is not recommended. For accessibility, balancing dampers should be located as far from the outlet as is practical.

Flow equalizing vanes. Each vane consists of one set of individually adjustable blades installed behind the slot openings at right angles to the slot. If designed and installed correctly, the blades improve the discharge pattern of the diffuser. These blades may also be used to increase spread of the airstream for greater coverage. Dampers and flow-equalizing vanes can usually be adjusted without removing the diffuser.

CEILING DIFFUSER OUTLETS

Multipassage ceiling diffusers. These diffusers consist of a series of flaring rings or louvers, which form a series of concentric air passages. They may be round, square, or rectangular. For easy installation, these diffusers are usually made in two parts: an outer shell with duct collar and an easily removable inner assembly.

Flush and stepped-down diffusers. In the flush unit, all rings or louvers project to a plane surface, whereas in the stepped-down unit, they project beyond the surface of the outer shell.

Common variations of this diffuser type are the adjustable-pattern diffuser and the multipattern diffuser. In the adjustable-pattern diffuser, the air-discharge pattern may be changed from a horizontal to a vertical or downblow pattern. Special construction of the diffuser or separate deflection devices allow adjustment. Multipattern diffusers are square or rectangular and have special louvers to discharge the air in one or more directions.

Other outlets available as standard equipment are half-round diffusers, supply and return diffusers, and light fixture-air diffuser combinations.

Perforated-face ceiling diffusers. These meet architectural demands for air outlets that blend into perforated ceilings. Each has a perforated metal face with an open area of 10 to 50%, which determines its capacity. Units are usually equipped with deflection devices to obtain multipattern horizontal air discharge.

Variable area ceiling diffusers. Such diffusers may be round, square, or linear and have parallel or concentric passages or a perforated face. In addition, they feature a means of effectively varying the discharge area to achieve an air volume change (variable volume outlet) at a constant pressure so that the variation in throw is minimized for a given change in supply air volume.

Applications

Ceiling installations. Ceiling diffusers should be mounted in the center of the space that they serve when they discharge the supply air in all directions. Multipattern diffusers can be used in the center of the space or adjacent to partitions, depending on the discharge pattern. By using different inner assemblies, the air pattern can be changed to suit particular requirements.

Side wall installations. Half-round diffusers and multipattern diffusers, when installed high in side walls, should generally discharge the air toward the ceiling.

Exposed duct installation. Some ceiling diffuser types, particularly stepped-down units, perform satisfactorily on exposed ducts. Consult manufacturers' catalogs for specific types.

Adjustable-pattern diffusers. Surface effect is important in the performance of adjustable-pattern diffusers. In fact, this effect is so pronounced that usually only two discharge patterns are possible with adjustable-pattern diffusers mounted directly on the ceiling. When a diffuser is changed from a horizontal pattern position toward the downblow pattern position, the surface effect maintains the horizontal discharge pattern until the discharge airstream is effectively deflected at the diffuser face, resulting in a vertical pattern. However, when adjustable-pattern ceiling diffusers are mounted on exposed duct, and there is no surface effect, the air may assume any pattern between horizontal and vertical discharge. Directional or segmented horizontal air patterns can usually be obtained by adjusting internal baffles or deflectors.

Accessories

Dampering and air-straightening accessories of various types are available as standard equipment.

Multilouver dampers. Consisting of a series of parallel blades mounted inside a round or square frame, multilouver dampers are installed in the diffuser collar or the takeoff. The blades are usually arranged in two groups rotating in opposite directions and are key-operated from the face of the diffuser (Figure 2A). This arrangement equalizes airflow in the diffuser collar.

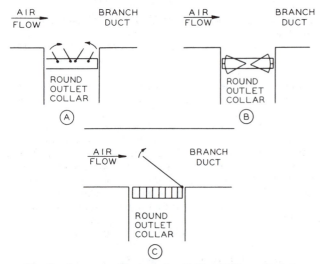

Fig. 2   Accessory Controls for Ceiling Diffuser Outlets

Opposed-blade dampers. Used for round ceiling diffusers, these dampers usually consist of a series of pie-shaped vanes mounted inside a round frame installed in the diffuser collar or the takeoff duct. The vanes pivot about a horizontal axis and are arranged in two groups, with adjacent vanes rotating counter to each other (Figure 2B). The vanes are key-operated from the diffuser face. Another opposed-blade design is similar in construction to the damper shown in Figure 1A and has either a round or square frame.

Splitter damper. A splitter damper is a single-blade sheet metal plate hinged at one edge and usually located at the branch connection of a duct or outlet (Figure 2C). It is easy to operate, but often causes irregular airflow in the duct. When used in conjunction with and near an outlet, additional directional control is required.

Equalizing devices. Such devices provide uniform airflow in the diffuser collar and consist of individually adjustable parallel blades mounted within a frame. They are installed in the diffuser collar or the takeoff duct. If used in pairs at right angles to each other, they serve as both equalizing and directional control devices.

Blankoff baffles. These baffles are used for minor adjustments of the airflow from a diffuser. They blank off a sector of the diffuser and prevent the supply air from striking an obstruction such as a column, partition, or the wall of the conditioned space by reducing flow in a given direction. Blankoff baffles generally reduce the area and increase supply air velocity, which must be considered when selecting diffuser size. Pattern control in diffusers having removable directional cores may be accomplished by rearranging the cores, generally without a change in area or increase in velocity.

Antismudge rings. These are round or square metal frames attached to, and extending 4 to 12 in. beyond, the outer edge of the diffuser. Their purpose is to minimize ceiling smudging.

AIR DISTRIBUTING CEILINGS

The air distributing ceiling uses the confined space above the ceiling as a supply plenum that receives air from stub ducts. The plenum should be designed to achieve uniform plenum pressure, resulting in uniform delivery of air to the conditioned space below.

Air is delivered through round holes or slots in the ceiling material or suspension system. These holes and slots vary in shape and size among manufacturers. Various manufacturers have developed a number of products based on the principle of a supply plenum, with sizes ranging mainly from 1 ft by 1 ft tile for a concealed grid to 2 ft by 4 ft lay-in panels for an exposed grid. Sometimes, the slots are equipped with adjustable dampers to facilitate changing the open area after installation.

The upper limit of plenum pressure must be that recommended by the ceiling manufacturer. It generally ranges from 0.10 to 0.15 in. of water, dictated by resistance to sag, to a lower pressure limit of about 0.01 in. of water, where uniformity of plenum pressure becomes more important. The range of air rates extends from about 15 cfm down to about 1 cfm per ft^2 of floor area. High flow rates are recommended only for low-temperature differentials (Nevins and Ward 1968; Miller and Nevins 1969).

Active portions of the air distributing ceilings should be located with respect to room load distribution, with higher airflow rates at the exterior exposures. This method of air distribution is particularly suited to large zones of uniform room temperature. Where different room temperatures are desired, a separate ceiling plenum is required for each zone. Construction of the ceiling plenum requires care with regard to air tightness, obstructions causing unequal plenum pressure and temperature, heat storage effect of the structure, and the influence of a roof or the areas surrounding the plenum (Hemphill 1966).

OUTLETS IN
VARIABLE AIR VOLUME SYSTEMS

The performance of a particular outlet or diffuser is generally independent of the terminal box that is upstream. For a given supply air volume and temperature differential (to meet a particular load), a standard outlet does not recognize whether the terminal box is of a constant volume, variable volume, or induction type. However, any diffuser, or system of diffusers, gives optimum air diffusion at some particular load condition and air volume. For a VAV system, as the load changes, so does the level of air movement in the conditioned space. At minimum load condition, when the volume of air is lowest, the level of air movement in the conditioned space may be low.

RETURN- AND EXHAUST-AIR INLETS

Return-air inlets may either be connected to a duct or be simple vents that transfer air from one area to another. Exhaust-air inlets remove air directly from a building and, therefore, are always connected to a duct. Whatever the arrangement, inlet size and configuration determine velocity and pressure requirements for the required airflow (see Return and Exhaust Inlets, Chapter 31 of the 1989 ASHRAE *Handbook—Fundamentals* for further discussion of these factors and the effect of inlet location on the system).

In general, the same type of equipment, grilles, slot diffusers, and ceiling diffusers used for supplying air may also be used for air return and exhaust. Inlets do not require the deflection, flow equalizing, and turning devices necessary for supply outlets. Dampers, however, are necessary when it is desirable to balance the airflow in the return-duct system.

Types of Inlets

Adjustable bar grilles. The same grilles used for air supply are used to match the deflection setting of the bars with that of the supply outlets.

Fixed bar grilles. The same grilles described in the section, Grille Outlets, are used. This grille is the most common return-air inlet. Vanes are straight or set at a certain angle, the latter being preferred when appearance is important.

V-bar grille. Made with bars in the shape of inverted *v*'s stacked within the grille frame, this grille has the advantage of being sight-proof; it can be viewed from any angle without detracting from appearance. Door grilles are usually v-bar grilles. The capacity of the grille decreases with increased sight tightness.

Lightproof grille. This grille is used to transfer air to or from darkrooms. The bars of this type of grille form a labyrinth and are painted black. The bars may take the form of several sets of v-bars or be of some special interlocking louver design to provide the required labyrinth.

Stamped grilles. Stamped grilles are also frequently used as return and exhaust inlets, particularly in rest rooms and utility areas.

Ceiling and slot diffusers. These diffusers may also be used as return and exhaust inlets.

Applications

Return and exhaust inlets may be mounted in practically any location, *e.g.*, ceilings, high or low side walls, floors, and doors.

The dampers shown in Figure 1A are used in conjunction with grille return and exhaust inlets. The type of damper does not affect the performance of the inlet. Usually, no other accessory devices are required.

EQUIPMENT FOR
AIR DISTRIBUTION SYSTEMS

In high-velocity pressure and low-velocity pressure air distribution systems, the duct velocities and static pressures are such that special control and acoustical equipment may be required for proper introduction of conditioned air into the space to be served. This section deals with control equipment for single-duct and dual-duct air-conditioning systems. Chapter 32 of the 1989 ASHRAE *Handbook—Fundamentals* has information on the design of high-velocity ducts, and Chapter 16 of this volume has duct construction details. Chapter 17 has further information on control and functions of terminal boxes. Chapter 42 of the 1991 ASHRAE *Handbook—HVAC Applications* includes information on sound control in air-conditioning systems and sound-rating methods for air outlets.

Control equipment for high-velocity pressure and low-velocity pressure systems may be classified as pressure-reducing valves or as terminal boxes. Terminal boxes may be further classified as single- or dual-duct boxes, reheat boxes, variable volume boxes, or ceiling induction boxes. They provide all or some of the following functions: pressure reduction, volume control, temperature modulation, air mixing, and sound attenuation.

Pressure-Reducing Valves

Pressure-reducing or air valves consist of a series of gang-operated vane sections mounted within a rigid casing and gasketed to reduce as much air leakage as possible between valve and duct. They are usually installed in rectangular ducts between a high-pressure trunk duct and a low-pressure branch duct for manual, remote manual, pneumatic, or electrical control.

Pressure is reduced by partially closing the valve and results in high-pressure drop through the valve. This action generates noise, which must be attenuated in the low-pressure discharge duct. The length and type of duct lining depend on the amount and frequency of noise to be attenuated.

Volume control is obtained by adjustment of the valve manually, mechanically, or automatically. Automatic adjustment is achieved by a pneumatic or electric control motor actuated by a pressure regulator or a thermostat.

Pressure-reducing valves are generally equal in size to the low-pressure branch duct connected to the valve discharge. This arrangement provides minimum pressure drop with valves opened fully.

Terminal Boxes

The terminal box is a factory-made assembly for air distribution. A terminal box, without altering the composition of the treated air from the distribution system, manually or automatically fulfills one or more of the following functions: (1) controls the velocity, pressure, or temperature of the air; (2) controls the rate of airflow; (3) mixes airstreams of different temperatures or humidities; and (4) mixes, within the assembly, air at high velocity and/or high pressure with air from the treated space. To achieve these functions, terminal box assemblies are made from an appropriate selection of the following component parts: casing, mixing section, manual damper, heat exchanger, induction section, and flow rate controller.

A terminal box commonly integrates a sound chamber to reduce noise generated within it by the manual damper or flow rate controller reducing the high-velocity, high-pressure inlet air to low-velocity, low-pressure air. The sound attenuation chamber is typically lined with thermal and sound-insulating material and is equipped with baffles. Special sound attenuation in the air discharge ducts is not usually required in smaller boxes. In larger capacity boxes, additional sound absorption materials may be required in the low-pressure distribution ducts connected to the discharge of the box. Manufacturers' catalogs should be consulted for specific performance information.

Terminal boxes are typically categorized according to the function of their flow rate controllers, which are generally categorized as constant flow rate or variable flow rate devices. They are further categorized as being pressure dependent, where the airflow rate through the assembly varies in response to changes in system pressure, or as pressure independent (or pressure compensating), where the airflow rate through the device does not vary in response to changes in system pressure. Constant flow rate controllers may be of the mechanical volume control type, which are actuated by means of the static pressure in the primary duct system and require no outside source of power. Constant flow rate controllers may also be of the pneumatic or electric volume-regulator type, which are actuated by an outside motive power and typically require internal differential pressure sensing, selector devices, and pneumatic or electric motors for operation.

Variable flow rate controllers may be of the mechanical pneumatic or electric volume-control type, which incorporates a means to reset the constant volume regulation automatically to a different control point within the range of the control device in response to an outside signal, such as from a thermostat. Boxes with this feature are pressure independent and may be used with reheat components. Variable flow rate may also be obtained by using a modulating damper ahead of a constant volume regulator. This arrangement typically allows for variations in flow between high and low limits or between a high limit and shutoff. These boxes are pressure dependent and volume limiting in function. Pneumatic variable volume may be either pressure independent, volume limiting, or pressure dependent, according to the equipment selected.

Terminal boxes can be further categorized as being system powered, wherein the assembly derives all of the energy necessary for its operation from the supply air within the distribution system, or as externally powered, wherein the assembly derives part or all of the energy necessary for its operation from a pneumatic or electric outside source. In addition, assemblies are self-contained (when they are furnished with all necessary controls for their operation, including actuators, regulators, motors, and thermostats), as opposed to non-self-contained assemblies (where part or all of the necessary controls for operation are furnished by someone other than the assembly manufacturer). In this latter case, the controls may be mounted on the assembly by the assembly manufacturer or may be mounted by others after delivery of the equipment.

The manual damper or flow rate controller within the box can be adjusted manually, automatically, or by a pneumatic or electric motor actuated by a thermostat or pressure regulator, all depending on the desired function of the box.

Air from the box may be discharged through a single rectangular opening suitable for low-pressure rigid branch duct connection or a supply outlet connected directly to the discharge end or bottom of the box. In addition, air may be discharged through several round outlets suitable for connection to flexible ducts.

Reheat Boxes

Boxes of this type add sensible heat to the supply air. Water or steam coils or electric resistance heaters are located within or attached directly to the air discharge end of the box. These boxes typically are single duct; operation can be either constant or variable volume. However, if they are variable volume, they must maintain some minimum airflow to accomplish the reheat function. This type of equipment can accomplish localized individual reheat without a central equipment station or zone change.

Dual-Duct Boxes

Dual-duct boxes are typically under control of a room thermostat. They receive warm and cold air from separate air supply ducts in accordance with room requirements to obtain room control without zoning. Pneumatic and electric volume-regulated boxes often have individual modulating dampers and operators to regulate the amount of warm and cool air. When a single modulating damper operator regulates the amount of both warm and cold air, a separate pressure-reducing damper or volume controller (either pneumatic or mechanical) is needed in the box to reduce pressure and limit airflow. Specially designed baffles may be required within the unit or at the box discharge to mix varying amounts of warm and cold air and/or to provide uniform flow downstream. Dual-duct boxes can be equipped with constant flow rate or variable flow rate devices to be either pressure independent or pressure dependent to provide a number of volume and temperature control functions.

Ceiling Induction Boxes—Air to Air

The air-to-air ceiling induction box provides either primary air or a mixture of primary air and relatively warm air to the conditioned space. It accomplishes this function by permitting the primary air to induce air from the ceiling plenum or via ducted return air from individual rooms. A single duct supplies primary air at a temperature cool enough to satisfy all zone cooling loads. The ceiling plenum air induced into the primary air is at a higher temperature than the room because heat from recessed lighting fixtures enters the plenum directly.

The induction box contains damper assemblies controlled by an actuator in response to a thermostat to control the amount of cool primary air and warm induced air. As reduction in cooling is required, the primary airflow rate is gradually reduced and the induced air rate is generally increased.

Reheat coils can be used in the primary air supply and/or in the induction opening to meet occasional interior and perimeter load requirements. Either hot water or electric reheat coils may be used.

Ceiling Induction Boxes—Fan Assisted

Fan-assisted ceiling induction boxes differ from air-to-air induction boxes in that they are equipped with a blower. This blower, generally driven by a small motor, draws air from the space or ceiling plenum (secondary air) to be mixed with the cool primary air. The advantage of fan induction boxes over straight VAV boxes is that for a small energy expenditure to the terminal fan, constant air circulation can be maintained in the space. Fan induction boxes operate at lower primary air static pressure than air-to-air induction boxes, and perimeter zones can be heated without operating the primary fan during unoccupied periods. The warm air in the plenum can be used for low- to medium-heating loads (depending on construction of the building envelope). As the load increases, heating coils in the perimeter boxes can be activated to heat the recirculated plenum air to the necessary level. Fan-assisted induction boxes can be divided into two categories: constant volume and bypass-type units.

Constant volume, fan-assisted induction boxes are used when constant air circulation is desired in the space. The unit has two inlets—one for cool primary air from the central fan system and one for the secondary or plenum air. All air delivered to the space passes through the blower. The blower operates continuously whenever the primary air fan is on and can be cycled to deliver heat, as required, when the primary fan is off.

As the cooling load decreases, a damper throttles the amount of primary air delivered to the blower. The blower makes up for this reduction of primary air by drawing air in from the space or ceiling plenum through the return or secondary air opening.

Bypass-type, fan-assisted induction boxes are called bypass boxes because the cool primary air bypasses the blower portion of the unit and is delivered directly to the space. The blower section draws in plenum air only and is mounted in parallel with the primary air damper. A back-draft damper prevents primary air from flowing into the blower section when the blower is not energized. The blower in these units is generally energized after the damper in the primary air is partially or completely throttled. Some electronically controlled units gradually increase the fan speed as the primary air damper is throttled to maintain constant airflow, while permitting the fan to shut off when it is in the full-cooling mode.

System Static-Pressure Control

To prevent static-pressure imbalance in single-duct and dual-duct high-pressure systems, some control of static pressure is necessary in systems operating with variable airflow.

In single-duct systems, variations in static pressure can be limited by the following: (1) static-pressure controllers, operating dampers in the air distribution system; (2) static-pressure controllers, operating inlet vane dampers on the fan; (3) zoning and changing air supply temperature in response to static-pressure changes; (4) constant volume regulators within the terminal device; and (5) mechanical and electrical variable speed fan controls.

In dual-duct systems, the methods of controlling static pressure are as follows: (1) dampers operated by static-pressure regulators located at critical points in the air distribution system; (2) static-pressure controllers, regulating warm and cold air temperatures to limit the variations in the airflow in individual ducts; and (3) constant-volume regulators in each individual air-mixing valve or acoustical terminal device.

AIR CURTAINS

In its simplest application, an air curtain is a continuous broad stream of air circulated across a doorway of a conditioned space.

It inhibits the penetration of unconditioned air into a conditioned space by forcing an air layer of predetermined thickness and velocity over the entire entrance. The layer moves at such a velocity and angle that the air that tries to penetrate the curtain is entrained or opposed. The air layer or jet can be redirected to compensate for pressure changes across the opening. If air is forced inward because of a difference in pressure, the jet can be redirected outward to equalize the pressure differential.

Both vertical flow (usually downward) and horizontal flow air curtains are available. The vertical flow type may have either a ducted or nonducted return.

The energy requirements of an air curtain should be considered carefully. The effectiveness of an air curtain in preventing infiltration through an entrance generally ranges from 60 to 80%. The effectiveness is the comparison of infiltration rate or heat flux through an opening when using an air curtain as opposed to the transmission that would take place through a simple opening with no restriction.

Two important factors influence the pressure differential against which an air curtain must work: the height of the structure and the structure's orientation. In high-rise structures, the practicality of using an air curtain depends mainly on the magnitude of the pressure differential caused by height or the stack effect. The orientation of the particular building and the location of adjacent buildings should also be studied and considered.

REFERENCES

Hemphill, J.M. 1966. Ventilating ceiling application. *Heating, Piping and Air Conditioning* (May):139.

Miller, P.L. and R.G. Nevins. 1969. Room air distribution with an air distributing ceiling—Part II. ASHRAE *Transactions* 75(1):118.

Nevins, R.G. and E.D. Ward. 1968. Room air distribution with an air distributing ceiling. ASHRAE *Transactions* 74(1):VI.2.1-VI.2.14.

FANS

THE fan is an air pump that creates a pressure difference and causes airflow. The impeller does work on the air, imparting to it both static and kinetic energy, varying in proportion depending on the fan type.

Fan efficiency ratings are based on ideal conditions; some fans are now rated at more then 90% total efficiency. However, actual connections often make it impossible to achieve ideal efficiencies in the field.

SYMBOLS AND DEFINITIONS

A = fan outlet area, ft^2

D = fan size or impeller diameter

N = rotational speed, revolutions per minute (revolutions per second)

Q = volume flow rate moved by fan at fan inlet conditions, cfm

p_{tf} = fan total pressure rise: fan total pressure at outlet minus fan total pressure at inlet, in. of water

p_{vf} = fan velocity pressure: pressure corresponding to average velocity determined from the volume flow rate and fan outlet area, in. of water

p_{sf} = fan static pressure rise: fan total pressure rise diminished by fan velocity pressure, in. of water. The fan inlet velocity head is assumed equal to zero for fan rating purposes.

p_{sx} = static pressure at given point, in. of water

p_{vx} = velocity pressure at given point, in. of water

p_{tx} = total pressure at given point, in. of water

V = fan inlet or outlet velocity, ft/min

W_o = power output of fan: based on fan volume flow rate and fan total pressure, horsepower

W_i = power input to fan: measured by power delivered to fan shaft, horsepower

η_t = mechanical efficiency of fan (or fan total efficiency): ratio of power output to power input ($\eta_t = W_o/W_i$)

η_s = static efficiency of fan: mechanical efficiency multiplied by ratio of static pressure to fan total pressure, $\eta_s = (p_s/p_t)\eta_t$

ρ = gas density, lb/ft^3

TYPES OF FANS

Fans are generally classified as centrifugal fans or axial flow fans according to the direction of airflow through the impeller. Figure 1 shows the general configuration of a centrifugal fan. A similar description of an axial flow fan is shown in Figure 2. Table 1 compares typical characteristics of some of the most common fan types.

Two modified versions of the centrifugal fan are being used but are not listed in Table 1 as separate fan types. Unhoused centri-

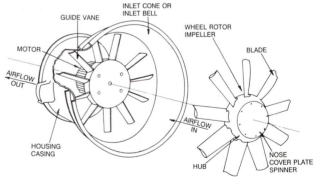

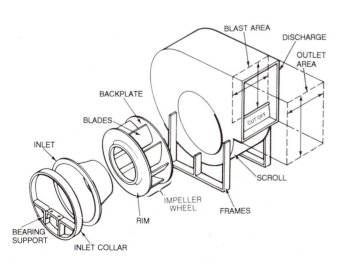

Fig. 1   Centrifugal Fan Components

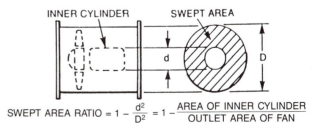

SWEPT AREA RATIO = $1 - \dfrac{d^2}{D^2} = 1 - \dfrac{\text{AREA OF INNER CYLINDER}}{\text{OUTLET AREA OF FAN}}$

Note: The swept area ratio in axial fans is equivalent to the blast area ratio in centrifugal fans.

Fig. 2   Axial Fan Components

The preparation of this chapter is assigned to TC 5.1, Fans.

Table 1   Types of Fans

	TYPE		IMPELLER DESIGN	HOUSING DESIGN
CENTRIFUGAL FANS	AIRFOIL		• Highest efficiency of all centrifugal fan designs. • Ten to 16 blades of airfoil contour curved away from direction of rotation. Deep blades allow for efficient expansion within blade passages. • Air leaves impeller at velocity less than tip speed. • For given duty, has highest speed of centrifugal fan designs.	• Scroll-type design for efficient conversion of velocity pressure to static pressure. • Maximum efficiency requires close clearance and alignment between wheel and inlet.
	BACKWARD-INCLINED BACKWARD-CURVED		• Efficiency only slightly less than airfoil fan. • Ten to 16 single-thickness blades curved or inclined away from direction of rotation. • Efficient for same reasons as airfoil fan.	• Uses same housing configuration as airfoil design.
	RADIAL		• Higher pressure characteristics than airfoil, backward-curved, and backward-inclined fans. • Curve may have a break to left of peak pressure and fan should not be operated in this area. • Power rises continually to free delivery.	• Scroll. Usually narrowest of all centrifugal designs. • Because wheel design is less efficient, housing dimensions are not as critical as for airfoil and backward-inclined fans.
	FORWARD-CURVED		• Flatter pressure curve and lower efficiency than the airfoil, backward-curved and backward-inclined. • Do not rate fan in the pressure curve dip to the left of peak pressure. • Power rises continually toward free delivery. Motor selection must take this into account.	• Scroll similar to and often identical to other centrifugal fan designs. • Fit between wheel and inlet not as critical as for airfoil and backward-inclined fans.
AXIAL FANS	PROPELLER		• Low efficiency. • Limited to low-pressure applications. • Usually low cost impellers have two or more blades of single thickness attached to relatively small hub. • Primary energy transfer by velocity pressure.	• Simple circular ring, orifice plate, or venturi. • Optimum design is close to blade tips and forms smooth airfoil into wheel.
	TUBEAXIAL		• Somewhat more efficient and capable of developing more useful static pressure than propeller fan. • Usually has 4 to 8 blades with airfoil or single thickness cross section. • Hub usually less than transfer by velocity pressure.	• Cylindrical tube with close clearance to blade tips.
	VANEAXIAL		• Good blade design gives medium- to high-pressure capability at good efficiency. • Most efficient of these fans have airfoil blades. • Blades may have fixed, adjustable, or controllable pitch. • Hub is usually greater than half fan tip diameter.	• Cylindrical tube with close clearance to blade tips. • Guide vanes upstream or downstream from impeller increase pressure capability and efficiency.
SPECIAL DESIGNS	TUBULAR CENTRIFUGAL		• Performance similar to backward-curved fan except capacity and pressure are lower. • Lower efficiency than backward-curved fan. • Performance curve may have a dip to the left of peak pressure.	• Cylindrical tube similar to vaneaxial fan, except clearance to wheel is not as close. • Air discharges radially from wheel and turns 90° to flow through guide vanes.
	POWER ROOF VENTILATORS	CENTRIFUGAL	• Low-pressure exhaust systems such as general factory, kitchen, warehouse, and some commercial installations. • Provides positive exhaust ventilation which is an advantage over gravity-type exhaust units. • Centrifugal units are slightly quieter than the axial unit described below.	• Normal housing not used, since air discharges from impeller in full circle. • Usually does not include configuration to recover velocity pressure component.
		AXIAL	• Low-pressure exhaust systems such as general factory, kitchen, warehouse, and some commercial installations. • Provides positive exhaust ventilation which is an advantage over gravity-type exhaust units.	• Essentially a propeller fan mounted in a supporting structure. • Hood protects fan from weather and acts as safety guard. • Air discharges from annular space at bottom of weather hood.

Table 1   Types of Fans (*Concluded*)

PERFORMANCE CURVES	PERFORMANCE CHARACTERISTICS*	APPLICATIONS*
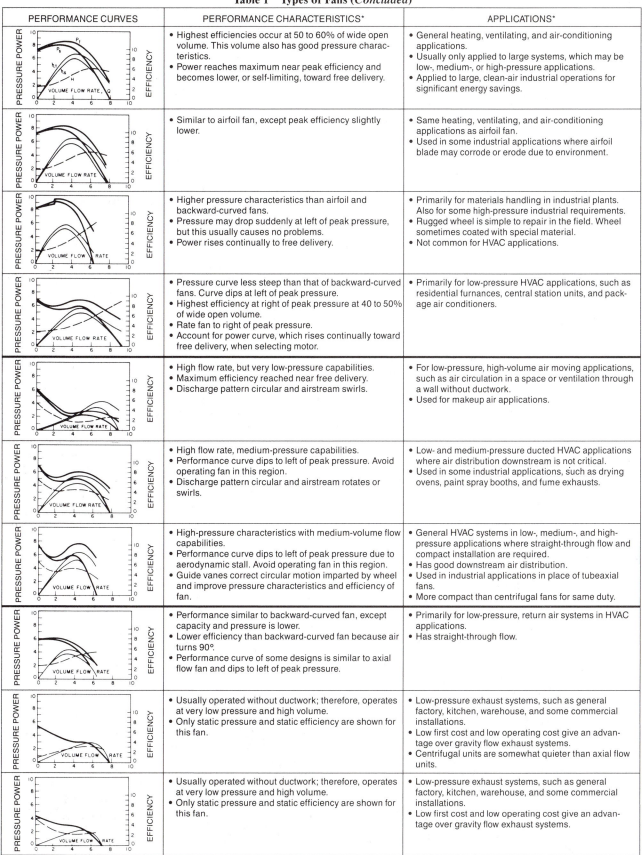	• Highest efficiencies occur at 50 to 60% of wide open volume. This volume also has good pressure charac- teristics. • Power reaches maximum near peak efficiency and becomes lower, or self-limiting, toward free delivery.	• General heating, ventilating, and air-conditioning applications. • Usually only applied to large systems, which may be low-, medium-, or high-pressure applications. • Applied to large, clean-air industrial operations for significant energy savings.
	• Similar to airfoil fan, except peak efficiency slightly lower.	• Same heating, ventilating, and air-conditioning applications as airfoil fan. • Used in some industrial applications where airfoil blade may corrode or erode due to environment.
	• Higher pressure characteristics than airfoil and backward-curved fans. • Pressure may drop suddenly at left of peak pressure, but this usually causes no problems. • Power rises continually to free delivery.	• Primarily for materials handling in industrial plants. Also for some high-pressure industrial requirements. • Rugged wheel is simple to repair in the field. Wheel sometimes coated with special material. • Not common for HVAC applications.
	• Pressure curve less steep than that of backward-curved fans. Curve dips at left of peak pressure. • Highest efficiency at right of peak pressure at 40 to 50% of wide open volume. • Rate fan to right of peak pressure. • Account for power curve, which rises continually toward free delivery, when selecting motor.	• Primarily for low-pressure HVAC applications, such as residential furnances, central station units, and pack- age air conditioners.
	• High flow rate, but very low-pressure capabilities. • Maximum efficiency reached near free delivery. • Discharge pattern circular and airstream swirls.	• For low-pressure, high-volume air moving applications, such as air circulation in a space or ventilation through a wall without ductwork. • Used for makeup air applications.
	• High flow rate, medium-pressure capabilities. • Performance curve dips to left of peak pressure. Avoid operating fan in this region. • Discharge pattern circular and airstream rotates or swirls.	• Low- and medium-pressure ducted HVAC applications where air distribution downstream is not critical. • Used in some industrial applications, such as drying ovens, paint spray booths, and fume exhausts.
	• High-pressure characteristics with medium-volume flow capabilities. • Performance curve dips to left of peak pressure due to aerodynamic stall. Avoid operating fan in this region. • Guide vanes correct circular motion imparted by wheel and improve pressure characteristics and efficiency of fan.	• General HVAC systems in low-, medium-, and high- pressure applications where straight-through flow and compact installation are required. • Has good downstream air distribution. • Used in industrial applications in place of tubeaxial fans. • More compact than centrifugal fans for same duty.
	• Performance similar to backward-curved fan, except capacity and pressure is lower. • Lower efficiency than backward-curved fan because air turns 90°. • Performance curve of some designs is similar to axial flow fan and dips to left of peak pressure.	• Primarily for low-pressure, return air systems in HVAC applications. • Has straight-through flow.
	• Usually operated without ductwork; therefore, operates at very low pressure and high volume. • Only static pressure and static efficiency are shown for this fan.	• Low-pressure exhaust systems, such as general factory, kitchen, warehouse, and some commercial installations. • Low first cost and low operating cost give an advan- tage over gravity flow exhaust systems. • Centrifugal units are somewhat quieter than axial flow units.
	• Usually operated without ductwork; therefore, operates at very low pressure and high volume. • Only static pressure and static efficiency are shown for this fan.	• Low-pressure exhaust systems, such as general factory, kitchen, warehouse, and some commercial installations. • Low first cost and low operating cost give an advan- tage over gravity flow exhaust systems.

*These performance curves reflect general characteristics of various fans as commonly applied. They are not intended to provide complete selection criteria, since other parameters, such as diameter and speed, are not defined.

fugal fan impellers are used as circulators in some industrial applications such as heat treating ovens and are identified as plug fans. In this case, there is no duct connection to the fan since it simply circulates the air within the oven. In some HVAC installations, the unhoused fan impeller is located in a plenum chamber with the fan inlet connected to an inlet duct from the system. Outlet ducts are connected to the plenum chamber. This fan arrangement is identified as a plenum fan.

PRINCIPLES OF OPERATION

All fans produce pressure by altering the velocity vector of the flow. The fans produce pressure and/or flow because the rotating blades of the impeller impart kinetic energy to the air by changing its velocity. Velocity change is the result of tangential and radial velocity components in the case of centrifugal fans, and of axial and tangential velocity components in the case of axial flow fans.

Centrifugal fan impellers produce pressure from (1) the centrifugal force created by rotating the air column enclosed between the blades and (2) the kinetic energy imparted to the air by virtue of its velocity leaving the impeller. This velocity in turn is a combination of rotative velocity of the impeller and airspeed relative to the impeller. When the blades are inclined forward, these two velocities are cumulative; when backward, oppositional. Backward-curved blade fans are generally more efficient than forward-curved blade fans.

Axial flow fans produce pressure from the change in velocity passing through the impeller, with none being produced by centrifugal force. These fans are divided into three types: propeller, tubeaxial, and vaneaxial. Propeller fans, customarily used at or near free air delivery, usually have a small hub-to-tip ratio impeller mounted in an orifice plate or inlet ring. Tubeaxial fans usually have reduced tip clearance and operate at higher tip speeds, which give them a higher total pressure capability than the propeller fan. Vaneaxial fans are essentially tubeaxial fans with guide vanes and reduced running blade tip clearance, which give improved pressure, efficiency, and noise characteristics.

Table 1 includes typical performance curves for various types of fans. These performance curves show the general characteristics of the various fans as they are normally used; they do not reflect the characteristics of these fans reduced to such common denominators as constant speed or constant propeller diameter, since fans are not selected on the basis of these constants. The efficiencies and power characteristics shown are general indications for each type of fan. A specific fan (size, speed) must be selected by evaluating actual characteristics.

FAN TESTING AND RATING

Fans are tested in accordance with the strict requirements of ANSI/ASHRAE *Standard* 51 and ANSI/AMCA *Standard* 210. The ASHRAE standard specifies the procedures and test setups to be used in testing the various types of fans and other air-moving devices.

Figure 3 diagrams one of the most common procedures for developing the characteristics of a fan in which it is tested from *shutoff* conditions to nearly *free delivery* conditions. At shutoff, the duct is completely blanked off; at free delivery, the outlet resistance is reduced to zero. Between these two conditions, various flow restrictions are placed on the end of the duct to simulate various conditions on the fan. Sufficient points are obtained to define the curve between shutoff point and free delivery conditions. Pitot tube traverses of the test duct are performed with the fan operating at constant speed. The point of rating may be any point on the fan performance curve. For each case, the specific point on the curve must be defined by referring to the flow rate and the corresponding total pressure.

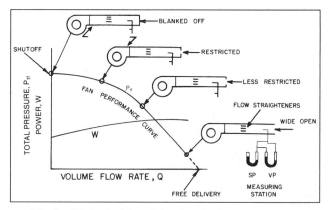

Fig. 3   Method of Obtaining Fan Performance Curves

Other test setups, also described in AMCA *Standard* 210 and ASHRAE *Standard* 51, should produce the same performance curve.

Fans designed for use with duct systems are tested with a length of duct between the fan and the measuring station. This length of duct smooths the flow of the fan and provides stable, uniform flow conditions at the plane of measurement. The measured pressures are corrected back to fan outlet conditions. Fans designed for use without ducts, including almost all propeller fans and power roof ventilators, are tested without ductwork.

Not all sizes are tested for rating. Test information may be used to calculate the performance of larger fans that are geometrically similar, but such information should not be extrapolated to smaller fans. For the performance of one fan to be determined from the known performance of another, the two fans must be dynamically similar. Strict dynamic similarity requires that the important nondimensional parameters vary in only insignificant ways. These nondimensional parameters include those that affect aerodynamic characteristics such as Mach number, Reynolds number, surface roughness, and gap size. (For more specific information, the manufacturer's application manual or engineering data should be consulted.)

FAN LAWS

The fan laws (see Table 2) relate the performance variables for any dynamically similar series of fans. The variables are fan size D; rotational speed N; gas density ρ; volume flow rate Q; pressure p_t or p_s; power W; either air W_e or shaft W_i; and mechanical efficiency η_t. *Fan Law* 1 shows the effect of changing size,

Table 2   Fan Laws

No.	Dependent Variables		Independent Variables
1a	$Q_1 = Q_2$	×	$(D_1/D_2)^3 \, (N_1/N_2)$
1b	$p_1 = p_2$	×	$(D_1/D_2)^2 \, (N_1/N_2)^2 \, \rho_1/\rho_2$
1c	$W_1 = W_2$	×	$(D_1/D_2)^5 \, (N_1/N_2)^3 \, \rho_1/\rho_2$
2a	$Q_1 = Q_2$	×	$(D_1/D_2)^2 \, (p_1/p_2)^{1/2} \, (\rho_2/\rho_1)^{1/2}$
2b	$N_1 = N_2$	×	$(D_2/D_1) \, (p_1/p_2)^{1/2} \, (\rho_2/\rho_1)^{1/2}$
2c	$W_1 = W_2$	×	$(D_1/D_2)^2 \, (p_1/p_2)^{3/2} \, (\rho_2/\rho_1)^{1/2}$
3a	$N_1 = N_2$	×	$(D_2/D_1)^3 \, (Q_1/Q_2)$
3b	$p_1 = p_2$	×	$(D_2/D_1)^4 \, (Q_1/Q_2)^2 \, \rho_1/\rho_2$
3c	$W_1 = W_2$	×	$(D_2/D_1)^4 \, (Q_1/Q_2)^3 \, \rho_1/\rho_2$

Notes:
1. Subscript 1 denotes the variable for the fan under consideration. Subscript 2 denotes the variable for the tested fan.
2. For all fans laws $(\eta_t)_1 = (\eta_t)_2$ and (Point of rating)$_1$ = (Point of rating)$_2$.
3. p equals either p_{tf} or p_{sf}.

speed, or density on volume flow, pressure, and power level. *Fan Law* 2 shows the effect of changing size, pressure, or density on volume flow rate, speed, and power. *Fan Law* 3 shows the effect of changing size, volume flow, or density on speed, pressure, and power.

The fan laws apply only to a series of aerodynamically similar fans at the same point of rating on the performance curve. They can be used to predict the performance of any fan when test data are available for any fan of the same series. Fan laws may also be used with a particular fan to determine the effect of speed change. However, caution should be exercised in these cases, since they apply only when all flow conditions are similar. Changing the speed of a given fan changes parameters that may invalidate the fan laws.

Unless otherwise identified, fan performance data are based on dry air at standard conditions—14.696 psi and 70°F (0.075 lb/ft³). In actual applications, the fan may be required to handle air or gas at some other density. The change in density may be because of temperature, composition of the gas, or altitude. As indicated by the fan laws, fan performance is affected by gas density. With constant size and speed, the power and pressure vary in accordance with the ratio of gas density to the standard air density.

Figure 4 illustrates the application of the fan laws for a change in fan speed N for a specific size fan. The computed p_t curve is derived from the base curve. For example, point E ($N_1 = 650$) is computed from point D ($N_2 = 600$) as follows:

At D,
$$Q_2 = 6000 \text{ cfm and } p_{tf_2} = 1.13 \text{ in. of water}$$

Using *Fan Law* 1a at point E
$$Q_1 = 6000 \times 650/600 = 6500 \text{ cfm}$$

Using *Fan Law* 1b
$$p_{tf_1} = 1.13 \, (650/600)^6 = 1.33 \text{ in. of water}$$

The completed total pressure curve p_{tf_1} at $N = 650$ curve thus may be generated by computing additional points from data on the base curve, such as point G from point F.

If equivalent points of rating are joined, as shown by the dotted lines in Figure 4, these points will form parabolas which are defined by the relationship expressed in Equation (1).

Each point on the base p_{tf} curve determines only one point on the computed curve. For example, point H cannot be calculated from either point D or point F. Point H is, however, related to some point between these two points on the base curve, and only that point can be used to locate point H. Furthermore, point D cannot be used to calculate point F on the base curve. The entire base curve must be defined by test.

FAN AND SYSTEM PRESSURE RELATIONSHIPS

As previously stated, a fan impeller imparts static and kinetic energy to the air. This energy is represented in the increase in total pressure and can be converted to static or velocity pressure. These two quantities are interdependent; fan performance cannot be evaluated by considering one or the other alone. The conversion of energy, indicated by changes in velocity pressure to static pressure and vice versa, depends on the efficiency of conversion. Energy conversion occurs in the discharge duct connected to a fan being tested in accordance with AMCA *Standard* 210 and ASHRAE *Standard* 51, and the efficiency is reflected in the rating.

Fan total pressure p_{tf} is a true indication of the energy imparted to the airstream by the fan. System pressure loss (ΔP) is the sum of all individual total pressure losses imposed by the air distribution system duct elements on both the inlet and outlet sides of the fan. An energy loss in a duct system can be defined only as a total pressure loss. The measured static pressure loss in a duct element equals the total pressure loss only in the special case where air velocities are the same at both the entrance and exit of the duct element. By using total pressure for both fan selection and air distribution system design, the design engineer is assured of proper design. These fundamental principles apply to both high- and low-velocity systems. (Chapter 32 of the 1989 ASHRAE *Handbook—Fundamentals* has further information.)

Fan static pressure rise p_{sf} is often used in low-velocity ventilating systems where the fan outlet area essentially equals the fan outlet duct area, and little energy conversion occurs. When fan performance data are given in terms of p_{sf}, the value of p_{tf} may be calculated from the catalog data.

To specify the pressure performance of a fan, the relationship of p_t, p_s, and p_v must be understood, expecially when negative pressures are involved. Most importantly, p_s is a defined term in AMCA *Standard* 210 and ASHRAE *Standard* 51 as $p_s = p_t - p_v$. Except in special cases, p_s is not necessarily the measured difference between static pressure on the inlet side and static pressure on the outlet side.

Figures 5 through 8 illustrate the relationships among these various pressures. Note that, as defined, $p_{tf} = p_{t2} - p_{t1}$. Figure 5 illustrates a fan with an outlet system but no connected inlet system. In this particular case, the fan static pressure p_{sf} equals the static pressure rise across the fan. Figure 6 shows a fan with an inlet system but no outlet system. Figure 7 shows a fan with both an inlet system and an outlet system. In both cases, the measured difference in static pressure across the fan ($p_{s2} - p_{s1}$) is not equal to the fan static pressure.

All of the systems illustrated in Figures 5 to 7 have inlet or outlet ducts that match the fan connections in size. Usually the duct size is not identical to the fan outlet or the fan inlet, so that a further complication is introduced. To illustrate the pressure relationships in this case, Figure 8 shows a diverging outlet cone, which is a commonly used type of fan connection. In this case, the pressure relationships at the fan outlet do not match the pressure rela-

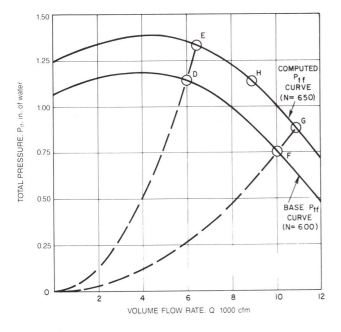

Fig. 4   Example Application of Fan Laws

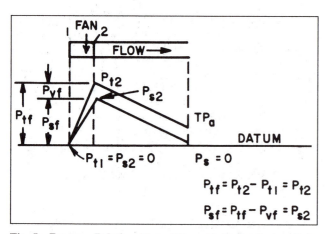

Fig. 5   Pressure Relationships of Fan with Outlet System Only

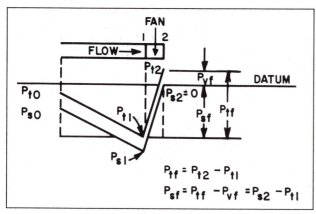

Fig. 6   Pressure Relationships of Fan with Inlet System Only

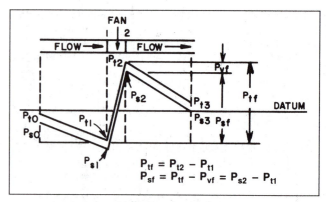

Fig. 7   Pressure Relationships of Fan with Equal-Sized Inlet and Outlet Systems

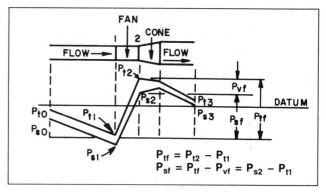

Fig. 8   Pressure Relationships of Fan with Diverging Cone Outlet

tionships in the flow section. Furthermore, the static pressure in the cone actually increases in the direction of flow. The static pressure changes throughout the system, depending on velocity. The total pressure, which, as noted in the sketch, decreases in the direction of flow, more truly represents the loss introduced by the cone or by flow in the duct. Only the fan changes this trend (*i.e.*, the decrease of total pressure in the direction of flow). Total pressure, therefore, is a better indication of fan and duct system performance. In this rather normal fan situation, the static pressure across the fan ($p_{s2} - p_{s1}$) does not equal the fan static pressure (p_{sf}).

DUCT SYSTEM CHARACTERISTICS

Figure 9 shows a simplified duct system with three 90° elbows. These elbows represent the resistance offered by the ductwork, heat exchangers, cabinets, dampers, grilles, and other system components. A given rate of airflow through a system requires a definite total pressure in the system. If the rate of flow is changed, the resulting total pressure required will vary, as shown in Equation (1), which is true for turbulent airflow systems. Heating, ventilating, and air-conditioning systems generally follow this law very closely.

$$(\Delta p_2 / \Delta p_1) = (Q_2 / Q_1)^2 \qquad (1)$$

This chapter only covers turbulent flow—the flow regime in which most fans operate. In some systems, particularly constant or variable-volume air conditioning, the air-handling devices and associated controls may produce effective system resistance curves that deviate widely from Equation (1), even though each element of the system may be described by this equation.

Equation (1) permits plotting a turbulent flow system's pressure loss (Δp) curve from one known operating condition (see Figure 4). The fixed system must operate at some point on this system curve as the volume flow rate changes. As an example, at point A, curve A, Figure 10, when the flow rate through a duct system such as shown in Figure 9 is 10,000 cfm, the total pressure drop is 3 in. of water. If these values are substituted in Equation (1) for Δp_1 and Q_1, other points of the system's Δp curve (see Figure 10) can be determined.

For 6000 cfm (Point D on Figure 10):

$$\Delta p_2 = 3(6000/10,000)^2 = 1.08 \text{ in. of water}$$

If a change is made within the system so that the total pressure at design flow rate is increased, the system will no longer operate on the previous Δp curve, and a new curve will be defined.

For example, in Figure 11, an additional elbow has been added to the duct system shown in Figure 9, which increases the total pressure of the system. If the total pressure at 10,000 cfm is increased by 1.00 in. of water, the system total pressure drop at this point will now be 4.00 in. of water, as shown by point B in Figure 10.

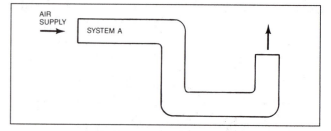

Fig. 9   Simple Duct System with Resistance to Flow Represented by Three 90° Elbows

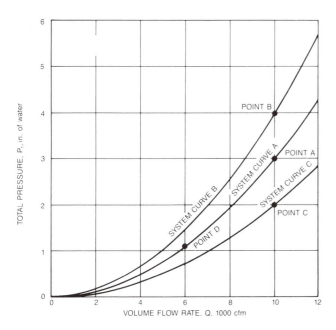

Fig. 10   Example System Total Pressure Loss (Δp) Curves

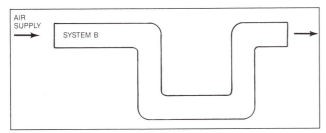

Fig. 11   Resistance Added to Duct System of Figure 9

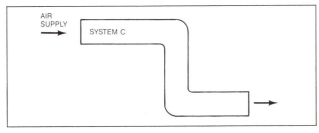

Fig. 12   Resistance Subtracted from Duct System of Figure 9

If the system in Figure 9 is changed by removing one of the schematic elbows (see Figure 12), the resulting system total pressure will drop below the total pressure resistance, and the new Δp curve will be shown in curve C of Figure 10. For curve C, a total pressure reduction of 1.00 in. of water has been assumed when 10,000 cfm flows through the system; thus, the point of operation will be at 2.00 in. of water, as shown by point C.

These three Δp curves all follow the relationship expressed in Equation (1). These curves result from changes within the system itself and do not change the fan performance. During the design phase, such system total pressure changes may occur because of studies of alternate duct routing, studies of differences in duct sizes, allowance for future duct extensions, or the effect of the design safety factor being applied to the system.

In an actual operating system, these three Δp curves can represent three system characteristic lines caused by three different positions of a throttling control damper. Curve C is the most open position, and curve B is the most closed position of the

three positions illustrated. A control damper forms a continuous series of these Δp curves as it moves from a wide open position to a completely closed position and covers a much wider range of operation than is illustrated here. Such curves can also represent the clogging of turbulent flow filters in a system.

SYSTEM EFFECTS

Normally, a fan is tested with open inlets and a section of straight duct is attached to the outlet. This setup results in uniform flow into the fan and efficient static pressure recovery on the fan outlet. If good inlet and outlet conditions are not provided in the actual installation, the performance of the fan will suffer. To select and apply the fan properly, these effects must be considered and the pressure requirements of the fan, as calculated by standard duct design procedures, must be increased.

These calculated system effect factors are only an approximation, however. Fans of different types and even fans of the same type, but supplied by different manufacturers, do not necessarily react to a system in the same way. Therefore, judgment based on experience must be applied to any design. Chapter 32 of the 1989 ASHRAE *Handbook—Fundamentals* gives information on calculating the system effect factors and lists loss coefficients for a variety of fittings. Clarke *et al.* (1978) and AMCA *Publication* 201 provide further information.

FAN SELECTION

After the system pressure loss curve of the air distribution system has been defined, a fan can be selected to meet the requirements of the system (Graham 1966, 1977). Fan manufacturers present performance data in either the graphic (curve) form (see Figure 13) or the tabular form (multirating tables). Multirating tables usually provide only performance data within the recommended operating range. The optimum selection area or peak efficiency point is identified in various ways by different manufacturers.

Performance data as tabulated in the usual fan tables are based on arbitrary increments of flow rate and pressure. In these tables, adjacent data, either horizontally or vertically, represent differ-

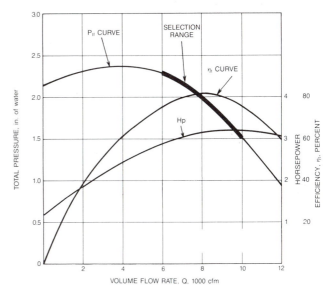

(Curve shows performance of a fixed fan size running at a fixed speed)

Fig. 13   Conventional Fan Performance Curve Used by Most Manufacturers

ent points of operation (*i.e.*, different points of rating) on the fan performance curve. These points of rating depend solely on the fan's characteristics; they cannot be obtained one from the other by the fan laws. However, these points of operation listed in the multirating tables are usually close together, so immediate points may be interpolated arithmetically with adequate accuracy for fan selection.

The selection of a fan for a particular air distribution system requires that the fan pressure characteristics fit the system pressure characteristics. Thus, the total system must be evaluated and the flow requirements, resistances, and system effect factors existing at the fan inlet and outlet must be known (see Chapter 32 of the 1989 ASHRAE *Handbook—Fundamentals*). Fan speed and power requirements are then calculated, using one of the many methods available from fan manufacturers. These may consist of the multirating tables or of single or multispeed performance curves or graphs.

In using curves, it is necessary that the point of operation selected (see Figure 14) represent a desirable point on the fan curve, so maximum efficiency and resistance to stall and pulsation can be attained. On systems where more than one point of operation is encountered during operation, it is necessary to look at the range of performance and evaluate how the selected fan reacts within this complete range. This analysis is particularly necessary for variable-volume systems, where not only the fan undergoes a change in performance, but the entire system deviates from the relationships defined in Equation (1). In these cases, it is necessary to look at actual losses in the system at performance extremes.

PARALLEL FAN OPERATION

The combined performance curve for two fans operating in parallel may be plotted by using the appropriate pressure for the ordinates and the sum of the volumes for the abscissas. When two fans having a pressure reduction to the left of the peak pressure point are operated in parallel, a fluctuating load condition may result if one of the fans operates to the left of the peak static point on its performance curve.

The p_t curves of a single fan and two identical fans operating in parallel are shown in Figure 15. Curve A-A shows the pressure

characteristics of a single fan. Curve C-C is the combined performance of the two fans. The unique figure-8 shape is a plot of all possible combinations of volume flow at each pressure value for the individual fans. All points to the right of CD are the result of each fan operating at the right of its peak point of rating. Stable performance results for all systems with less obstruction to airflow than is shown on the Δp curve D-D. At points of operation to the left of CD, it is possible to satisfy system requirements with one fan operating at one point of rating while the other fan is at a different point of rating. For example, consider Δp E-E, which requires a pressure of 1.00 in. of water and a volume of 5000 cfm. The requirements of this system can be satisfied with each fan delivering 2500 cfm at 1.00 in. of water pressure, Point CE. The system can also be satisfied at Point CE′ by one fan operating at 1400 cfm at 0.9 in. of water, while the second fan delivers 3400 cfm at the same 0.90 in. of water.

Note that system curve E-E passes through the combined performance curve at two points. Under such conditions, unstable operation can result. Under conditions of CE′, one fan is underloaded and operating at poor efficiency. The other fan delivers most of the requirements of the system and uses substantially more power than the underloaded fan. This imbalance may reverse and shift the load from one fan to the other.

FAN NOISE

Fan noise is a function of the fan design, volume flow rate Q, total pressure p_t, and efficiency η_t. After a decision has been made regarding the proper type of fan for a given application (keeping in mind the system effects), the best size selection of that fan must be based on efficiency, since the most efficient operating range for a specific line of fans is normally the quietest. Low outlet velocity does not necessarily ensure quiet operation, so selections made on this basis alone are not appropriate. Also, noise comparisons of different types of fans, or fans offered by different manufacturers, made on the basis of rotational or tip speed are not valid. The only valid basis for comparison are the actual sound power levels generated by the different types of fans when they are all producing the required volume flow rate and total pressure.

The data are reported by fan manufacturers as *sound power levels* in eight octave bands. These levels are determined by using a reverberant room for the test facility and comparing the noise generated by the fan to the noise generated by a noise source of known sound power. The measuring technique is described in AMCA *Standard* 300, Reverberant Room Method for Sound Testing of Fans. ASHRAE *Standard* 68/AMCA *Standard* 330, Laboratory Method of Testing In-Duct Sound Power Measurement Procedure for Fans, describes an alternate test method to

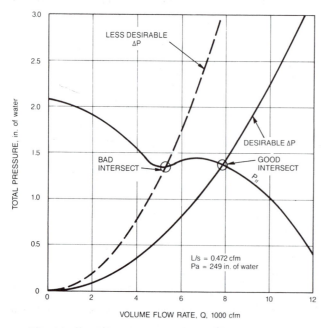

Fig. 14　Desirable Combination of p_{tf} and Δp Curves

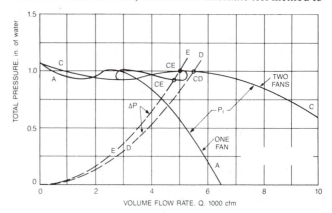

Fig. 15　Two Forward Curve Centrifugal Fans in Parallel Operation

determine the sound power a duct fan radiates into a supply and/or return duct terminated by an anechoic chamber. These standards do not fully evaluate the pure tones generated by some fans; these tones can be quite objectionable when they are radiated into occupied spaces. On critical installations, special allowance should be made by providing extra sound attenuation in the octave band containing the tone.

Typical sound power level values for various types of fans may be found in Chapter 42 of the 1991 ASHRAE *Handbook—HVAC Applications*; these are typical values and significant variations can be expected as a result of differences in impeller and housing designs. Therefore, the values should be used only for general calculations and should not be used where critical noise requirements must be met, such as in auditoriums, conference rooms, and lecture halls. Where possible, sound power level data should be obtained from the fan manufacturer for the specific fan being considered.

FAN ARRANGEMENT AND INSTALLATION

Direction of rotation is determined from the drive side of the fan. On single-inlet centrifugal fans, the drive side is usually considered the side opposite the fan inlet. (AMCA has published standard nomenclature to define positions.)

Fan Isolation

In air-conditioning systems, ducts should be connected to fan outlets and inlets with unpainted canvas or other flexible material. Access should be provided in the connections for periodic removal of any accumulations tending to unbalance the rotor. When operating against high resistance or when noise level requirements are low, it is preferable to locate the fan in a room removed from occupied areas or in one acoustically treated to prevent sound transmission. The lighter building constructions, which are common today, make it desirable to mount fans and driving motors on resilient bases designed to prevent transmission of vibrations through floors to the building structure. Conduits, pipes, and other rigid members should not be attached to fans. Noises that result from obstructions, abrupt turns, grilles, and other items not connected with the fan may be present. Treatments for such problems, as well as the design of sound and vibration absorbers, are discussed in Chapter 42 of the 1991 ASHRAE *Handbook— HVAC Applications*.

FAN CONTROL

In many heating and ventilating systems, the volume of air handled by the fan varies. The choice of the proper method for varying flow for any particular case is influenced by two basic considerations: (1) the frequency with which changes must be made and (2) the balancing of reduced power consumption against increases in first cost.

To control flow, the characteristic of either the system or the fan must be changed. The system characteristic curve may be altered by installing dampers or orifice plates. This technique reduces flow by increasing the system pressure required and, therefore, increases power consumption. Figure 10 shows three different system curves, A, B, and C, such as would be obtained by changing the damper setting or orifice diameter. Dampers are usually the lowest first cost method of achieving flow control; they can be used even in cases where essentially continuous control is needed.

Changing the fan characteristic (p_t curve) for control can reduce power consumption. From the standpoint of power consumption, the most desirable method of control is to vary the fan speed to produce the desired performance. If the change is infrequent, belt-driven units may be adjusted by changing the pulley

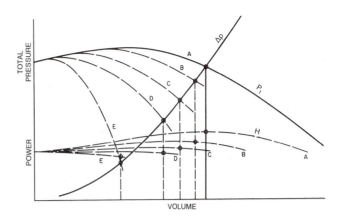

Fig. 16   Effect of Inlet Vane Control on Backward Curve Centrifugal Fan Performance

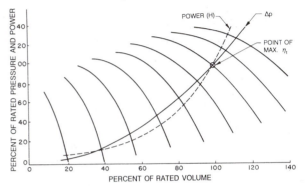

Fig. 17   Effect of Blade Pitch on Controllable Pitch Vaneaxial Fan Performance

on the drive motor of the fan. Variable-speed motors or variable-speed drives, whether electrical or hydraulic, may be used when frequent or essentially continuous variations are desired. When speed control is used, the revised p_t curve can be calculated by the fan laws.

Inlet vane control is frequently used. Figure 16 illustrates the change in fan performance with inlet vane control. Curves A, B, C, D, and E are the pressure and power curves for various vane settings between wide open (A) and nearly closed (E).

Tubeaxial and vaneaxial fans are made with adjustable pitch blades to permit balancing of the fan against the system or to make infrequent adjustments. Vaneaxial fans are also produced with controllable pitch blades (*i.e.*, pitch that can be varied while the fan is in operation) for frequent or continuous adjustment. Varying pitch angle retains high efficiencies over a wide range of conditions. Figure 17 performance is from a typical fan with variable pitch blades. From the standpoint of noise, variable speed is somewhat better than variable blade pitch; however, both of these control methods give high operating efficiency control and generate appreciably less noise than inlet vane or damper control.

FAN MOTOR SELECTION

Whenever an electric motor is selected as the prime mover for a centrifugal fan, both the fan's maximum power and starting torque should be considered. In some cases, where larger centrifugal fans are operated with relatively small motors, the motor may not accelerate the fan/impeller within some allowable starting time. Excessive starting time raises the temperature of the motor windings to a point where circuit breakers may trip out or the motor may be damaged. The user must consider this problem when selecting fan and motor combinations.

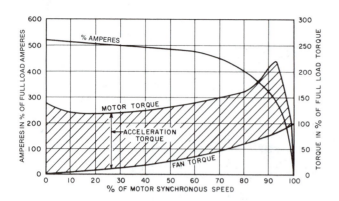

Fig. 18   Typical Drip-proof Motor Performance Test Curve

The two main factors that control starting acceleration are the fan/impeller inertia (WR^2 in the fan industry) and the starting torque characteristic of the electric motor. Curves for the motor starting torque as a percentage of full load torque at any speed are available from the motor supplier.

In most normal fan applications, the fan is not run at the motor speed but through some drive combination, and the WR^2 must be corrected to represent the apparent WR^2 as seen by the motor. Along with the WR^2 of the fan/impeller, the additional effect of the WR^2 of the shaft and fan sheave and the inertia of the motor rotor must be considered.

Because the available motor torque must be divided between the fan load and the energy needed for acceleration, the acceleration torque at any point can be represented by $T_A = T_m - T_F$, where T_m is motor torque and T_F is fan torque. This acceleration torque is graphically represented in Figure 18 as the vertical distance between the motor torque and the fan torque curves at any speed.

Normally, a maximum allowable value for acceleration time is available from the motor manufacturer; a typical time for induction motors is around 10 s. Other factors, such as ambient conditions and the number of starts per day, also affect this allowable acceleration time.

REFERENCES

AMCA. 1966. Motor positions for belt or chain drive centrifugal fans. *Standards Handbook* 99-2407-66. Air Movement and Control Association, Arlington Heights, IL.

AMCA. 1973. Fans and systems, *Standard* 201-73. Air Movement and Control Association, Arlington Heights, IL.

AMCA. 1978. Drive arrangements for centrifugal fans, *Standards Handbook* 99-2404-78. Air Movement and Control Association, Arlington Heights, IL.

AMCA. 1982. Drive arrangements for tubular centrifugal fans, *Standards Handbook* 99-2410-82. Air Movement and Control Association, Arlington Heights, IL.

AMCA. 1983. Designation for rotation & discharge of centrifugal fans, *Standards Handbook* 99-2406-83. Air Movement and Control Association, Arlington Heights, IL.

AMCA. 1985. Laboratory methods of testing fans for rating. ANSI/AMCA *Standard* 210-85, ANSI/ASHRAE *Standard* 51-1985. Air Movement and Control Association, Arlington, Heights, IL.

AMCA. 1985. Reverberant room method for sound testing of fans. *Standard* 300-85. Air Movement and Control Association, Arlington Heights, IL.

Buffalo Forge Co. 1983. *Fan engineering*, 8th ed. R. Jorgensen, ed. Buffalo, NY.

Clark, M.S., J.T. Barnhart, F.J. Bubsey, and E. Neitzel. 1978. The effects of system connections on fan performance. ASHRAE *Transactions* 84(2):227.

Graham, J.B. 1971. Methods of selecting and rating fans. ASHRAE *Symposium Bulletin*, Fan application, testing and selection.

Graham, J.B. 1966. Fan selection by computer techniques. *Heating, Piping and Air Conditioning* (April):168.

CHAPTER 19

EVAPORATIVE AIR COOLING

THIS chapter addresses direct and indirect evaporative air-cooling equipment and air washers used for air cooling, humidification, dehumidification, and air cleaning and their associated systems. Residential and industrial humidification equipment are covered in Chapter 20.

Packaged evaporative coolers, air washers, indirect air coolers, evaporative condensers, vacuum cooling apparatus, and cooling towers exchange sensible heat for latent heat. This equipment falls into two general categories: (1) apparatus for air cooling and (2) apparatus for heat rejection. Air-cooling equipment is addressed in this chapter.

Evaporative air cooling evaporates water into an airstream. Figure 1 illustrates thermodynamic changes that occur between the air and water that are in direct contact in a moving airstream. The continuously recirculated water reaches an equilibrium temperature equal to the wet-bulb temperature of the entering air. The heat and mass transfer between the air and water lowers the air dry-bulb temperature and increases the humidity ratio at a constant wet-bulb temperature.

The extent to which the leaving air temperature approaches the thermodynamic wet-bulb temperature of the entering air or the extent to which complete saturation is approached is expressed as a percentage evaporative cooling or saturation effectiveness and is defined in Equation (1) as:

$$e_c = 100 \; \frac{(t_1 - t_2)}{(t_1 - t')} \tag{1}$$

where

e_c = evaporative cooling or saturation effectiveness, percent
t_1 = dry-bulb temperature of the entering air
t_2 = dry-bulb temperature of the leaving air
t' = thermodynamic wet-bulb temperature of entering air

The term *effectiveness* is used rather than *efficiency*, since efficiency implies energy, power, or work. However, the term efficiency (evaporative cooling, saturation, or humidifying) may be used.

If warm or cold unrecirculated water is used, the air can be heated and humidified or cooled and dehumidified.

Evaporative air-cooling equipment can be classified as either direct or indirect. Direct evaporative equipment cools air by direct contact with the water, either by an extended wetted-surface material (as in packaged air coolers) or with a series of sprays (as in an air washer). Indirect systems cool air in a heat exchanger, which transfers heat to either a secondary airstream that has been evaporatively cooled (air-to-air) or to water that has been evaporatively cooled (by a cooling tower).

Combination systems can involve both direct and indirect principles as well as heat exchangers and cooling coils. In such systems, air may exit below the initial wet-bulb temperature. While such systems can be complex, the high cost of energy may justify their use in certain geographic areas.

The preparation of this chapter is assigned to TC 5.7, Evaporative Cooling.

DIRECT EVAPORATIVE AIR COOLERS

Wetted-Media Air Coolers

These coolers contain evaporative pads, usually of aspen wood fibers (Figure 2). A water-circulating pump lifts the sump water to a distributing system, and it flows down through the pads back to the sump.

A fan within the cooler pulls the air through the evaporative pads and delivers it to the space to be cooled. The fan discharges either through the side of the cooler cabinet or through the sump bottom. Wetted-pad packaged air coolers are made in sizes ranging from 2000 to 20,000 cfm.

Wetted-pad packaged air coolers are made as small tabletop coolers (50 to 200 cfm), window units (100 to 4500 cfm), and standard duct-connected coolers (5,000 to 18,000 cfm). The industry assigns an "industry standard capacity" to coolers; this is typically about 50% larger than its free-air capacity.

When clean and well maintained, commercial wetted-media air coolers operate at an effectiveness of approximately 80% and remove particulate matter 10 μm and larger. In some units, supple-

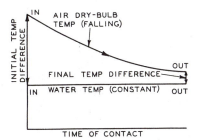

Fig. 1 Interaction of Water and Air in Evaporative Air Cooler

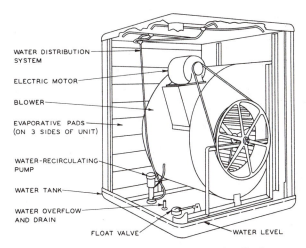

Fig. 2 Typical Wetted-Pad Evaporative Cooler

mentary filters ahead of or following the evaporative pads remove additional particulates. These filters prevent particulates from entering the cooling system when it is operated without water to circulate fresh air. The evaporative pads may be chemically treated to increase wettability. An additive may be included in the fibers to help them resist attack by bacteria, fungi, and other microorganisms.

The coolers are usually designed for an evaporative pad face velocity of 100 to 250 fpm, with a pressure drop of 0.1 in. of water. The aspen fibers are packed to approximately 0.3 to 0.4 lb/ft^2 of face area based on a 2-in. pad. Pads are mounted in removable louvered frames, which are usually made of galvanized steel or molded plastic. Troughs distribute water to the pads. A centrifugal pump with a submerged inlet pumps the water through tubes that provide an equal flow of water to each trough. The sump or water tank has a water makeup connection, float valve, overflow pipe, and drain. Provisions for the bleed-off of water to prevent the buildup of minerals and scale are usually incorporated in the design. Alternately, if the water supply is unlimited, fresh water may be used with no recirculation. This practice reduces mineral and scale buildup but wastes water.

Water usage depends on the airflow, the effectiveness of the evaporator pad, and the wet-bulb depression of the incoming air. The humidity ratio (mass of water vapor per mass of dry air) for the entering and leaving air can be determined from a psychrometric chart. Along with the airflow rate, the actual water consumption can be calculated from the difference. An approximation of water usage is 1.0 gph of water per 1000 cfm for each 10°F reduction in dry-bulb temperature.

The blower is usually a forward-curved, centrifugal fan, complete with motor and drive. The V-belt drive may include an adjustable pitch motor sheave to facilitate balancing air delivery against the static pressure requirements of the supply air duct system and to permit operation of the motor below its rated ampere draw (FLA). The motor enclosure may be drip-proof, totally enclosed, or a semi-open type specifically designed for evaporative coolers.

Rigid-Media Air Coolers

Sheets of rigid, corrugated material makeup the wetted surface of another type of evaporative cooler (Figure 3). Materials include cellulose and fiberglass that have been treated chemically with anti-rot and rigidifying resins. The sheets are laid with specific angle corrugations running in alternating directions so that the air and water flow in opposite directions. In the direction of airflow, the depth of fill is commonly 12 in., but it may be between 4 and 24 in. The media have the desirable characteristics of low resistance to airflow, high saturation effectiveness, and self-cleaning by flushing the front face of the pad.

Air washers or evaporative coolers using this material can be built in sizes to handle as much as 600,000 cfm with or without fans. Saturation effectiveness varies from 70 to over 95%, depending on media depth and air velocity. Air flows horizontally while the recirculating water flows vertically over the media surfaces by gravity feed from a flooding header and water distribution chamber. A pump recirculates the water from a lower reservoir, which is constructed of heavy gage material protected by a corrosion-resistant coating or of plastic. The reservoir is also fitted with overflow and drain connections. The upper media enclosure is fabricated of reinforced galvanized steel, of other corrosion-resistant sheet metal, or of plastic.

Flanges at the entering and leaving faces allow the unit to be connected to ductwork. A float valve maintains proper water level in the reservoir, makes up water evaporated, and supplies fresh water for dilution to prevent overconcentration of solids and minerals. Because the water recirculation rate is low and because spray nozzles are not needed to saturate the media, pumping power is low when compared to spray-filled air washers with equivalent evaporative cooling effectiveness performance.

Slinger Packaged Air Coolers

Slinger air coolers consist of an evaporative cooling section and a fan section (Figure 4). The fan is usually a forward-curved, double-inlet, double-width centrifugal fan that is V-belt driven by an electric motor. In the cooling section, outdoor air is drawn through a water spray, an evaporative filter pad, and a pad that collects entrained moisture. The spray is created by a motor-driven, vertical, clog-proof disc that is partially immersed in the water sump. The rated evaporative cooling effectiveness may be up to 80%. Air capacities range up to 30,000 cfm. Higher capacities may be obtained by using multiple cooling sections with one or more fans discharging air into a distribution system. In addition to the spray apparatus and pads, the cooling section includes a housing, a bleed-off valve, a fresh-water makeup float valve, an overflow pipe, and a drain plug. An air inlet filter, inlet louver, and rain hood are optional. The evaporator pad and moisture eliminator pad may be of latex-coated fiber, glass fiber, or coated nonferrous metal construction; from 0.75 to 2 in. deep; and washable. The face air velocity across the pad surface may be from 300 to 600 fpm, depending on the cooling effectiveness desired. Chemical treatment may be used to increase wettability and resistance to fungi, bacteria, and fire.

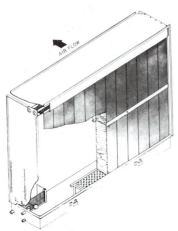

Fig. 3 Typical Rigid-Media Air Cooler

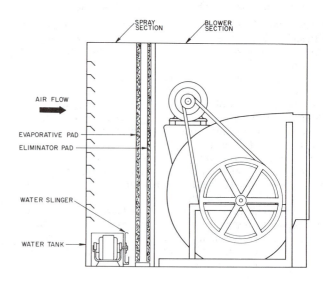

Fig. 4 Typical Slinger-Type Evaporative Cooler

The air-washing capability of the spray and filter arrangement as well as the nonclogging characteristic of the water spray disc minimize maintenance requirements.

Packaged Rotary Air Coolers

Rotary coolers (Figure 5) wet and wash the evaporative pad by rotating it through a water bath. The evaporative pad and other parts in contact with the water are made of corrosion-resistant materials. Like the slinger cooler, the evaporative pad consists of a cooling section and a blower compartment. The air-handling part of the cooler is similar to that of the wetted-pad and slinger coolers. Packaged rotary air coolers are built in capacities ranging from 2000 to 12,000 cfm.

There are two types of packaged rotary air coolers. One uses a rotating drum, which is partially submerged in the water reservoir. Air passes through the thickness dimension of the drum. The second rotary cooler has a continuously looped pad. The lower end of the pad is submerged in the water reservoir; the upper end is looped around a horizontal drive shaft at the top of the cooler. Several pad mediums are used, with varying thicknesses (depending on the medium material). Pad face velocities of 100 to 600 fpm are typical, with pressure drops in the range of 0.5 in. of water.

Both types of rotary cooler may be furnished with an automatic flush valve and timer to flush the reservoir water periodically, thus minimizing buildup of minerals, salts, and solids in the water. The timer setting is usually adjustable.

Remote Pad Evaporative Cooling Equipment

Greenhouses, animal shelters, automotive paint booths, and similar applications use a system of exhaust fans in the wall or roof of a structure, with wetted pads placed so that outdoor air is drawn through the enclosed space. The pads are wetted from above by a perforated trough or similar device, with the excess water draining to a gutter from which it may be wasted or collected for recirculation. The pad should be sized for an air velocity of approximately 150 fpm for standard fiber pads, 250 fpm for 4-in. rigid pads, and 425 fpm for 6-in. rigid pads.

INDIRECT EVAPORATIVE AIR COOLERS

Indirect Packaged Air Coolers

In indirect evaporative air coolers, outdoor air or exhaust air from the conditioned space passes through one side of a heat exchanger. This air (the secondary airstream) is cooled by evaporation by one of several methods: (1) direct wetting of the heat exchanger surface; (2) passing through wetted pad media; (3)

atomizing spray; (4) disc evaporator, etc. The surfaces of the heat exchanger are cooled by the secondary airstream. On the other side of the heat exchanger surface, the primary airstream (conditioned air to be supplied to the space) is sensibly cooled by the heat exchanger surfaces.

Although the primary air is cooled by the evaporatively cooled secondary air, no moisture is added to the primary air. Hence, the process is known as *indirect* evaporative air cooling. The supply (primary) air may be recirculated room air or outside air, or a mixture. The enthalpy of the primary airstream decreases because no moisture is added to it. This process contrasts with direct evaporative cooling, which is essentially adiabatic (constant enthalpy). The usefulness of indirect evaporative cooling is related to the wet-bulb depression of the secondary air below the dry-bulb temperature of the entering primary air.

Since neither the evaporative cooling nor the heat transfer effectiveness can reach 100%, the leaving dry-bulb temperature of the primary air must always be above the entering wet-bulb temperature of the secondary airstream. Dehumidifying in the primary airstream can occur only when the dew point of the primary airstream is several degrees higher than the wet-bulb temperature of the secondary airstream.

A packaged indirect evaporative air cooler includes a heat exchanger, wetting apparatus, secondary air fan assembly, secondary air inlet louver, and enclosure. The heat exchanger may be constructed with folded metal or plastic sheets, with or without a corrosion-resistant or moisture-retaining coating; or it may be constructed with tubes, so that one airstream flows inside the tubes and the other flows over the exterior tube surfaces. Air filters should be placed upstream from both the primary and secondary heat exchangers to minimize fouling by dust, insects, or other airborne contaminants.

Since minerals in the water being evaporated would increase the concentration of minerals in the water being recirculated, continuous waste and dilution by adding of fresh water is necessary. Water treatment may be necessary to control corrosion of heat exchanger surfaces and other metal parts.

The packaged indirect evaporative air cooler may either be self-contained, with its own primary air supply fan assembly, or part of a built-up or more complete packaged air-handling system. The cooling system could use a single stage of indirect evaporative cooling, or it could include indirect evaporative cooling as the first stage, with additional direct evaporative cooling and/or refrigerated (chilled water or direct expansion) cooling stages. When the indirect evaporative cooler is placed in series (upstream) with a conventional refrigerated coil, it reduces the sensible load on the coil and refrigeration system (Figure 6). Energy required for the indirect precooling stage includes the pump and secondary air fan motor, as well as some additional fan energy to overcome resistance added in the primary air. The energy consumed by the indirect evaporative cooling stage is less than the energy saved by

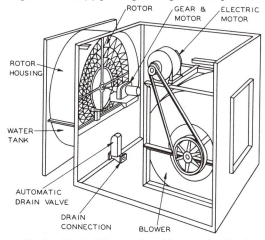

Fig. 5   Typical Rotary-Type Evaporative Cooler

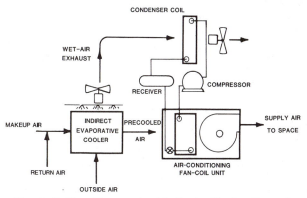

Fig. 6   Indirect Evaporative Air Cooler Used as Precooler

reduction in load on the refrigeration apparatus. As a result, the overall system efficiency may increase dramatically because energy costs and demands are reduced. Another saving could result from the reduction in size of the refrigeration equipment required. Indirect evaporative cooling may also reduce the yearly hours during which the refrigeration equipment must be operated.

Evaporatively cooled air can be discharged across air-cooled refrigeration condenser coils to improve the efficiency of the condenser and refrigeration system. Chapter 46 of the 1991 ASHRAE *Handbook—HVAC Applications* includes sample calculations covering energy-saving evaporative cooling systems. Manufacturers' data should be consulted for equipment selection, cooling performance, pressure drops, and space requirements.

Cooling performance of indirect evaporative cooling equipment is expressed as a *performance factor* (PF), i.e., the reduction in dry-bulb temperature in the primary airstream divided by the difference between the initial dry-bulb temperature of the primary airstream minus the entering wet-bulb temperature of the secondary air. The term *indirect evaporative cooling effectiveness* (as a percentage) is also used.

Manufacturers' ratings require careful interpretation. The basis of ratings should be specified, since, for the same apparatus, performance is affected by changes in primary and secondary air velocities and mass flow rates, wet-bulb temperature, altitude, etc.

Typically, the air resistance on both primary and secondary sides ranges between 50 to 500 Pa. The ratio of secondary air to conditioned primary air may range from less than 0.6 to greater than 1.0 and has an effect on performance. Based on manufacturers' ratings, available equipment may be selected for indirect evaporative cooling effectiveness ranging from 40 to 80%.

Cooling Tower/Coil Systems

The combination of a cooling tower or other evaporative water cooler with a water-to-air heat exchanger coil and water circulating pump is another type of indirect evaporative cooling system. Water is pumped from the reservoir of the cooling tower to the coil and returns to the upper distribution header of the tower. Both open-water and closed-loop systems are used. Coils should be cleanable in open systems.

The recirculated water is evaporatively cooled to within a few degrees of the wet-bulb temperature as it flows over the wetted surfaces of the cooling tower. As the cooled water flows through the tubes of the coil in the conditioned airstream, it picks up heat from the conditioned air. The temperature of the water increases, and the primary air is cooled without the addition of moisture to the primary air. The water is again cooled as it recirculates through the cooling tower. A float valve controls the fresh-water makeup, which replaces the evaporated water and prevents excessive concentration of minerals in the recirculated water. Air filtration on the air inlet of the cooling tower is also recommended.

One advantage of this system, especially for retrofit applications, is that the cooling tower may be located remote from the cooling coil. Also, the system is accessible for maintenance. Overall indirect evaporative cooling effectiveness (PF) may range between 0.55 and 0.75% or higher.

If return air is sent to the water cooler of an indirect cooling system (before being discharged outside), a water cooler should be specifically designed for this purpose. These coolers have a wetted surface media, which has a high ratio of wetted surface area per unit of fill volume. Performance depends on depth of fill, air velocity over the fill surface, water flow to airflow ratio, wet-bulb temperature, and water-cooling range. Because of the closer approach of the water temperature to the wet-bulb temperature, the overall system PF may be higher as compared to a conventional cooling tower.

Other Indirect Evaporative Cooling Apparatus

Other combinations of evaporative coolers and heat exchange apparatus can accomplish indirect evaporative cooling. Heat exchangers finding application in these systems include heat pipes and rotary heat wheels, in addition to the plate, pleated media, and shell-and-tube types. If the conditioned (primary) air and the exhaust or outside (secondary) airstream are side by side, a heat pipe apparatus or heat wheel can transfer heat from the warmer air to the cooler air. Countercurrent flow of the primary and secondary air is best for indirect evaporative cooling effectiveness. Evaporative cooling of the secondary airstream by spraying water directly on the surfaces of the heat exchanger or by an evaporative cooler upstream of the heat exchanger may cool the primary air indirectly by transferring heat from it to the secondary air.

INDIRECT/DIRECT COMBINATIONS

In a two-stage combination, indirect/direct evaporative cooling system, a first-stage indirect evaporative cooler lowers both the dry- and the wet-bulb temperature of the incoming supply air. After leaving the indirect stage, the supply air passes through a second-stage direct evaporative cooler. Figure 7 shows the process on the psychrometric chart. First-stage cooling follows a line of constant humidity ratio, since no moisture is added to the primary airstream. The second stage follows the wet-bulb line at the condition of the air leaving the first stage.

The indirect evaporative cooling stage may be any of the types described in the previous paragraphs. Figure 8 shows a system using a rotary heat wheel. The secondary air may be exhaust air from the conditioned space or outdoor air. When the secondary air passes through the evaporative cooler, the dry-bulb temperature is lowered by evaporative cooling. As this air passes through the heat wheel, the mass of the media is cooled to a temperature approaching the wet-bulb temperature of the secondary air. The heat wheel rotates so that its cooled mass enters the primary air and, in turn, sensibly cools the primary (supply) air. Following the heat wheel, a direct evaporative cooler further reduces the dry-bulb temperature of the primary air. This approach may achieve supply air dry-bulb temperatures of 3 °C or more below the secondary air wet-bulb temperature.

In areas where the 1% mean coincident wet-bulb design temperature is 19 °C or lower, average annual cooling power consumption of indirect/direct systems may be as low as 0.06 kW/kW. When the 1% mean coincident wet-bulb temperature is as high as 23 °C, an indirect/direct cooling system can have an average annual cooling power consumption as low as 0.23 kW/kW. By

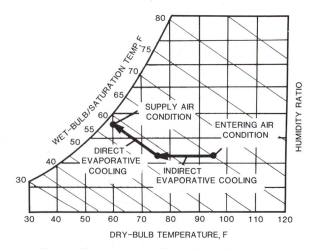

Fig. 7   Combination Indirect/Direct Evaporative Cooling Process

son, the typical refrigeration system with an air-cooled condenser may have an average annual power consumption of greater than 1.0 kW/ton.

In dry environments, indirect/direct evaporative cooling systems are usually designed to supply 100% outdoor air to the conditioned spaces of a building. In these once-through applications, space latent loads and return air sensible loads are exhausted from the building rather than returned to the conditioning equipment. Consequently, the cooling capacity required from these systems may be less than that required from a conventional refrigerated cooling system.

In areas of higher wet-bulb design temperatures or when system design requires a supply air temperature lower than that attainable using indirect/direct evaporative cooling systems, a third cooling stage may be required. This stage may be a refrigerated direct expansion or chilled water coil located either upstream or downstream from the direct evaporative cooling stage, but always downstream from the indirect evaporative stage. Refrigerated cooling is energized only when evaporative stages cannot achieve the required supply air temperature.

Figure 9 shows a schematic of a three-stage configuration (indirect/direct, with optional third-stage refrigerated cooling). The third-stage refrigerated cooling coil is located downstream from the direct evaporative cooler. In this arrangement, the conventional air-conditioning equipment operates less efficiently because a lower coil surface temperature is required to condense the water vapor introduced by the direct evaporative cooler. Locating the third-stage refrigerated coil upstream of the direct evaporative cooler improves the operating efficiency of the air conditioner but requires the coil to be larger because the coil surface will be dry.

The designer should consider options using exhaust and/or outside air as secondary air for the indirect evaporative cooling stage. If the latent load in the space is significant, the wet-bulb temperature of the exhaust air in the cooling mode may be higher than that of the outside air. In this case, outside air may be used more effectively as secondary air to the indirect evaporative cooling stage.

Custom configurations are available for indirect/direct and three-stage systems to permit many choices for locating the return, exhaust, and outside air, and mixing of airstreams, bypass of components, or variable volume control.

In direct, indirect, indirect/direct, and other staged evaporative cooling systems, the elements of the system that may be controlled include modulating outside air and return air mixing dampers; secondary air fans and recirculating pumps of an indirect evaporative stage; recirculating pumps of a direct evaporative cooling stage; face and bypass dampers for the direct stage; chilled water, temperature, or refrigerant flow for a refrigerated stage; and system or individual terminal volume by the use of variable volume terminals, fan variable inlet vanes, adjustable pitch, or variable fan speed.

For sequential control in indirect/direct evaporative cooling systems, the indirect evaporative cooler is energized for first-stage cooling, the direct evaporative cooler for second-stage cooling, and the refrigeration coil for third-stage cooling. In some applications, reversing the sequence of the direct evaporative cooler and indirect evaporative cooler may reduce the first-stage energy requirement.

AIR WASHERS

Spray-Type Air Washers

Spray-type air washers consist of a chamber or casing containing a spray nozzle system, a tank for collecting spray water as it falls, and an eliminator section for removing entrained drops of water from the air. A pump recirculates water at a rate higher than the evaporation rate. Intimate contact between the spray water and the airflow causes heat and mass transfer between the air and the water (Figure 10).

Washers are commonly available from 2000 to 250,000 cfm capacity, but specially constructed washers can be made in any size. No standardization exists; each manufacturer publishes tables giving physical data and ratings for specific products. Therefore, air velocity, water-spray density, spray pressure, and other design factors must be considered for each application.

The simplest design has a single bank of spray nozzles with a casing that is usually 4 to 7 ft long. This type of washer is applied primarily as an evaporative cooler or humidifer. It is sometimes used as an air cleaner when the dust is wettable, although the air-cleaning efficiency is relatively low. Cleaning efficiency may be increased by the addition of flooding nozzles to wash the elimina-

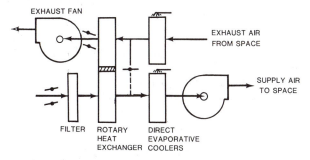

Fig. 8   Indirect/Direct Evaporative Cooling System Using Rotary Heat Wheel

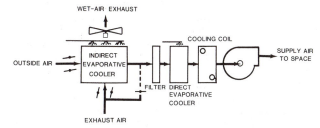

Fig. 9   Three-Stage Indirect/Direct Evaporative Cooling System

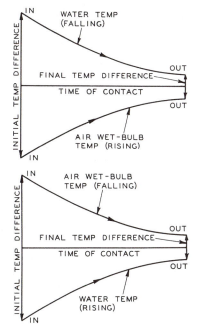

Fig. 10   Interaction of Air and Water in Air Washer Heat Exchanger

tor plates. Two or more spray banks are generally used when a very high degree of saturation is necessary and for cooling and dehumidification applications that require chilled water. Two-stage washers are used for dehumidification when the quantity of chilled water is limited or when the water temperature is above that required for the single-stage design. Arranging the two stages for counterflow of the water permits a small quantity of water with a greater water temperature rise.

The lengths of washers vary considerably. Spray banks are approximately spaced from 2.5 to 4.5 ft; the first and last banks of sprays are located about 1 to 1.5 ft from the entering or leaving end of the washer. In addition, air washers may be furnished with heating or cooling coils within the washer chamber, which may affect the overall length of the washer.

Figure 11 shows the construction features of conventional spray-type air washers. Essential requirements in air washer operations are (1) uniform distribution of the air across the spray chamber; (2) an adequate amount of spray water broken up into fine droplets; (3) good spray distribution across the airstream; (4) sufficient length of travel through the spray and wetted surfaces; and (5) elimination of free moisture from the outlet air. The cross-sectional area is determined by the design velocity of the air through the spray chamber. The units usually have an air velocity of 300 to 600 fpm; however, with special eliminators, velocities as high as 1500 fpm may be used.

Spray water requirements for spray-type air washers used for washing or evaporative cooling vary from 4 gpm per 1000 cfm with a single bank to 10 gpm per 1000 cfm for double banks. Pumping pressures usually range from 55 to 100 ft of water, depending on nozzle pressure, height of apparatus, pressure losses in pipe and strainers, etc. Approximately 10% of the water handled by the pump is bled off to reduce the chemical buildup from the evaporated water, thereby reducing the incidence of nozzle clogging or mineral buildup on wetted surfaces. A higher percentage of bleed-off may be required with heavily mineralized water.

Spray nozzles produce a finely atomized spray and are spaced to give uniform coverage of the chamber through which the air passes. Nozzle pressures normally vary from 20 to 40 psig, depending on the duty. Small orifices at pressures of up to about 40 psig produce a fine spray, necessary for high saturation effectiveness, while larger orifices with pressures of about 25 psig are common for dehumidification. When the water contains large amounts of chemicals that can clog the nozzles, larger orifice nozzles should be installed, even if larger pumping capacity is required. Self-cleaning nozzles are available. Spray-nozzle capacities vary from about 1 to 3.75 gpm per nozzle, and spray densities are usually 1 to 5 gpm per square foot of cross-sectional area per bank. Spacing varies from about 0.75 to 2.5 nozzles per square foot per bank. For lower spray densities, smaller orifices should be used to avoid bypassing air because of poor spray coverage. Flooding nozzles that are used for washing the eliminators discharge about 1.0 gpm per nozzle in a flat stream at 3 to 5 psig pressure.

Strainers, usually made of fine mesh copper or brass screen, extend across the width of the tank. They are sometimes placed in a cylindrical shape over the pump suction connection within the tank. Strainers should be readily removable for inspection and cleaning. The openings in the screen should be smaller than the spray orifice to minimize nozzle clogging. Belt-type, automatic, and other specially designed strainers are available for use in textile mills and other industries where lint or heavy concentrations of dust are present. Accessories that should be included are (1) a float valve to maintain the minimum water level automatically, (2) a quick-fill connection, (3) a trapped overflow, (4) a tank drain opening, (5) a suction connection, and (6) a marine light on the spray chamber.

The resistance to airflow through an air washer varies with the type and number of baffles, eliminators, and wetted surfaces; the number of spray banks and their direction and air velocity; the size and type of other components, such as cooling and heating coils; and other factors, such as air density. Pressure drop may be as low as 0.25 in. of water or as high as 1 in. of water. The manufacturer should be consulted regarding the resistance of any particular washer design combination.

The casing and the tank may be constructed of various materials. One or more doors are commonly provided for inspection and access. The tank is normally at least 16 in. high with a 14-in. water level; it may extend beyond the casing on the inlet end to make the suction strainer more accessible. The tank may be partitioned by the installation of a weir, usually in the entering end, to permit recirculation of spray water for control purposes in dehumidification work. The excess then returns over the weir to the central water chilling machine.

Eliminators consist of a series of vertical plates that are spaced about 0.75 to 2 in. on centers at the exit of the washer. The plates are formed with a number of bends to deflect the air and obtain impingement on the wetted surfaces. Hooks on the edge of the plates improve moisture elimination. Perforated plates may be installed on the inlet end of the washer to obtain more uniform air distribution through the spray chamber. Louvers, which prevent the backlash of spray water, may also be installed for this purpose.

High-Velocity Spray-Type Air Washers

High-velocity air washers generally operate at air velocities in the range of 1200 to 1800 fpm. Some have been applied as high as 2400 fpm, but 1200 to 1600 fpm is the most accepted range for optimum application. The reduced cross-sectional area allows air washers to be used in smaller apparatus than those of lower velocities. This space reduction makes air washers the primary heat transfer means for cooling and dehumidifying (as well as humidifying) industrial air-conditioning systems. High capacities per unit of space available from high-velocity spray devices permit practical packaging of prefabricated central station units in either completely assembled and transportable form or, for large capacity units, easily handled modules. Manufacturers supply units with capacities of up to 150,000 cfm shipped in one piece, including spray system, eliminators, pump, fan, dampers, filters, and other functional com-

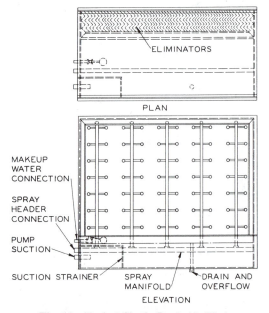

PLAN

MAKEUP
WATER
CONNECTION

SPRAY
HEADER
CONNECTION

PUMP
SUCTION

SUCTION STRAINER   SPRAY   DRAIN AND
                    MANIFOLD   OVERFLOW

ELEVATION

Fig. 11   Typical Single-Bank Air Washer

ponents. Such units are self-housed, prewired, prepiped, and ready for hoisting into place (Figure 12).

The number and arrangement of nozzles vary with different capacities and manufacturers. Adequate values of saturation effectiveness and heat transfer effectiveness are achieved by using higher spray densities.

Eliminator blades come in varying shapes, but most are a series of aerodynamically clean, sinusoidal shapes. Collected moisture flows down grooves or hooks designed into their profiles, then drains into the storage tank. Washers may be built with shallow drain pans and connected to a central storage tank. High-velocity washers are rectangular in cross section and, except for the eliminators, are similar in appearance and construction to conventional lower velocity types. Pressure losses are in the 0.5 to 1.5 in. of water range. These washers are available either as freestanding separate devices for incorporation into field-built central stations or in complete preassembled central station packages from the factory.

Cell-Type Air Washers

These washers obtain intimate air-water contact by passing the air through cells packed with glass, metal, or fiber screens (Figure 13). Water passes over cells arranged in tiers. Behind the cells

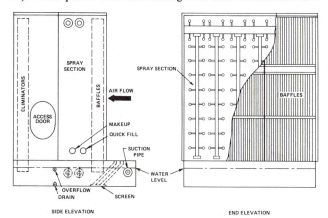

Fig. 12   High-Velocity Spray Washer

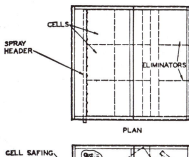

PLAN

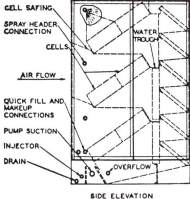

SIDE ELEVATION

Fig. 13   Typical Cell Air Washer

are blade-type or glass-mat eliminators. Most cell washers are arranged for concurrent air and water flow. They are also constructed for special duty with countercurrent flow characteristics or in a combination of both arrangements. Cell washers come in many sizes of insulated or uninsulated construction. Standard washers are available up to 10 cells high by 12 cells wide with a capacity of up to 210,000 cfm. They are also available up to 30,000 cfm, complete with fan, motor, pump, and external spray piping.

Atomization of the spray water is not required in cell washers, but good water distribution over the face of the cell is essential. A saturation effectiveness of 90 to 97% is possible with units that have 6-in. deep cells using fine fibers. Units with vertically positioned cells (usually 2 in. deep) and coarser fibers have 70 to 80% saturation effectiveness. Water requirements vary from 0.75 to 1.5 gpm per 1000 cfm of airflow, with resistance of airflow ranging from 0.15 to 0.65 in. of water, depending on the amount of water circulated and the air velocity through the cells. Although classified as air washers, they should not be used as heat exchangers because of the low volume of water flowing over the cells.

Each washer consists of a number of cells, normally 20 in. square, arranged in tiers. A typical cell is a metal frame packed with glass fiber strands. The glass occupies only 3 to 6% of the volume of the cell but, when sprayed, presents a total wetted area of approximately 100 to 120 ft^2. Wire mesh screens at both faces of the frame hold the glass pack. Each tier is independent of the others and has its own spray header, drain sheet, and, except for the lowest tier, conduit to the tank below.

Eliminators downstream from the cells remove entrained moisture from the airstream. They may be a metal-blade type of eliminator for deflection of the airstream, with hooks for trapping impinged droplets or a glass mat arranged in tiers similar to the cell. The glass mat is 2 in. deep and is randomly packed with glass fiber in a metal frame.

Connections are provided for internal spray headers, drain, overflow, a steam injector for cleaning the cells in place, pump suction, water quick fill, and makeup water controlled by a float valve.

The dynel/polyester mat cell washer operates at a basic spray rate of 0.75 gpm of water per 1000 cfm and airflow resistance of 0.7 in. of water. Where water is not recirculated, the sprays operate on a once-through basis, carrying away solids, which eliminates the need for a pump.

Saturation effectiveness may be selected from 82 down to 70% by controlling air velocity through the cell. In this shallow cell system, because moisture carryover is prevented by the fine fiber media and mat design, moisture eliminators are not needed. Air-cleaning results are typical of comparable impingement filters, with some gain from wetting action and the self-flushing of particulate matter.

Humidification with Air Washers

Air can be humidified with an air washer in three ways: (1) using recirculated spray water without prior treatment of the air, (2) preheating the air and washing it with recirculated spray water, and (3) using heated spray water. In any air washing installation, the air should not enter the washer with a wet-bulb temperature of less than 39°F, otherwise the spray water may freeze.

Recirculated spray water. Except for the small amount of outside energy added by the recirculating pump in the form of shaft work and the small amount of heat leakage into the apparatus from outside (including the pump and its connecting piping), the process is strictly adiabatic. Evaporation from the liquid is recirculated. Its temperature should adjust to the thermodynamic wet-bulb temperature of the entering air.

The whole airstream is not brought to complete saturation, but its state point should move along a line of constant thermodynamic wet-bulb temperature. As defined in Equation (1), the extent to which the leaving air temperature approaches the thermodynamic wet-bulb temperature of the entering air is expressed by the saturation effectiveness ratio. In humidifiers, this ratio is often referred to as the humidifying effectiveness.

The following is representative of the saturation or humidifying effectiveness of a spray air washer for these spray arrangements.

Bank Arrangement	Length, ft	Effectiveness, %
1 downstream	4	50 to 60
1 downstream	6	60 to 75
1 upstream	6	65 to 80
2 downstream	8 to 10	80 to 90
2 opposing	8 to 10	85 to 95
2 upstream	8 to 10	90 to 98

The degree of saturation depends on the extent of contact between air and water. Other conditions being equal, a low-velocity airflow is conducive to higher humidifying effectiveness.

Humidification with Rigid Media

Rigid media may be used for finite humidity control by arranging the media in one or more banks in the depth or height. Each bank is activated independently of the others to achieve the desired humidity.

Preheating the air. Preheating the air increases both the dry- and wet-bulb temperatures and lowers the relative humidity, but it does not alter the humidity ratio (mass ratio, water vapor to dry air). At a higher wet-bulb temperature, but with the same humidity ratio, more water can be absorbed per unit mass of dry air in passing through the washer (if the humidifying effectiveness of the washer is not adversely affected by operation at the higher wet-bulb temperature). The analysis of the process that occurs in the washer is the same as that for recirculated spray water. The final preferred conditions are achieved by adjusting the amount of preheating to give the required wet-bulb temperature at the entrance to the washer.

Heated spray water. Even if heat is added to the spray water, the mixing in the washer may still be regarded as adiabatic. The state point of the mixture should move toward the specific enthalpy of the heated spray. It is possible (by elevating the water temperature) to raise the air temperature, both dry and wet bulb, above the dry-bulb temperature of the entering air.

The relative humidity of the leaving air may be controlled by (1) bypassing some of the air around the washer and remixing the two airstreams downstream or (2) automatically reducing the number of operating spray nozzles by operating valves in the different spray header branches.

Dehumidification and Cooling with Air Washers

Air washers are also used to cool and dehumidify air. Heat and moisture removed from the air raise the water temperature. If the entering water temperature is below the entering wet-bulb temperature, both the dry- and wet-bulb temperatures are lowered. Dehumidification results if the leaving water temperature is below the entering dew-point temperature. Moreover, the final water temperature is determined by the sensible and latent heat pickup and the quantity of water circulated. However, this final temperature must not exceed the final required dew point, with one or two degrees below the dew point being common practice.

The air leaving a spray-type dehumidifier is substantially saturated. Usually, the spread between dry- and wet-bulb temperatures is less than 1°F. The spread between leaving air and leaving water depends on the difference between entering dry- and wet-bulb temperatures and on certain design features, such as the length and

height of spray chamber, air velocity, quantity of water, and character of the spray pattern. The rise in water temperature is usually between 6 and 12°F, although higher rises have been used successfully. The lower rises are ordinarily selected when the water is chilled by mechanical refrigeration because of possible higher refrigerant temperatures. It is often desirable to make an economic analysis of the effect of higher refrigerant temperature compared to the benefits of a greater rise in water temperature. For systems receiving water from a well or other source at an acceptable temperature, it may be desirable to design on the basis of a high temperature rise and minimum water flow.

The most common air washer arrangement for cooling and dehumidifying air has two spray banks and is 8 to 9 ft long. If the air washer can cool and dehumidify the entering air to a wet-bulb temperature equal to that of the leaving water temperature, it is convenient to assign such a washer a performance factor of 1.0. The actual performance factor of any washer F_p is the actual enthalpy change divided by the enthalpy change in a washer of 1.0 performance:

$$F_p = \frac{(h_1 - h_2)}{(h_1 - h_3)} \tag{2}$$

where

h_1 = enthalpy at wet bulb of entering air
h_2 = enthalpy at wet bulb of leaving air at actual condition
h_3 = enthalpy at wet-bulb temperature leaving a washer with $F_p = 1.0$

By knowing the performance factor of a particular air washer, the actual conditions of operation can be graphically determined (Figure 14). Points 1 and 2 are plotted on the saturation curve, which represent total heat at the entering and leaving air wet-bulb temperature t_1' and t_2'. Point 5 represents the condition at which leaving air wet bulb and leaving water temperature would be the same. This point is determined by solving Equation (2) for h_3:

$$h_3 = h_1 - \frac{(h_1 - h_2)}{F_p} \tag{3}$$

A diagonal line is drawn through Point 5 with a negative slope equal to the water-to-air weight ratio. Points 3 and 4, at which the diagonal line intersects horizontal lines through Points 1 and 2, show the required entering and leaving water temperatures, t_{w1} and t_{w2}. A check of the solution can be made from the fundamental heat balance expression — heat absorbed by the water equals heat removed from the air.

The graphical method can be used to solve air washer cooling and dehumidifying problems when there are a number of

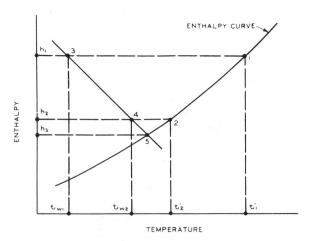

Fig. 14 Graphical Solution of Conditions Produced by Dehumidifying Air Washer

unknown factors, including quantity of water to be used and entering and leaving water temperatures. The actual performance factor of a particular washer, however, must be obtained from the manufacturer's data.

Another method expresses the performance of the dehumidifier as a relationship of the leaving to entering spread between air and water temperatures:

$$F_{p1} = 1 - \frac{(t_2' - t_{w2})}{(t_1' - t_{w1})} \qquad (4)$$

where

t_1' = thermodynamic wet-bulb temperature of entering air
t_2' = thermodynamic wet-bulb temperature of leaving air
t_{w1} = entering water temperature
t_{w2} = leaving water temperature

Performance factors expressed in these terms normally vary from about 70% for a 6-ft long single-bank washer to nearly 100% for an 8-ft long unit with two opposed spray banks.

A third method used to express the performance of a spray dehumidifier is given by the following formula, which, although similar to Equation (2), is numerically different because of a change in determining the value of h_3 as:

$$F_p = \frac{(h_1 - h_2)}{(h_1 - h_3)} \qquad (5)$$

where

h_1 = enthalpy at wet-bulb temperature of entering air
h_2 = enthalpy at wet-bulb temperature of leaving air
h_3 = enthalpy of air at the leaving water temperature

Note that the performance factors determined by the second and third methods are not equal; the difference can be seen in Figure 15.

The representative curves in Figure 15 are based on entering wet-bulb temperatures of 65 and 70°F, with leaving difference between air and water of 1.5°F and a water temperature rise of 10°F. The calculation of F_p is independent of rise in water temperature. However, the equation for F_{p1} does involve the difference between entering and leaving water temperature. For example, the performance factor for 65°F entering wet-bulb temperature and 1.5°F leaving difference falls from 0.925 for 10°F rise in water to 0.912 for 7°F rise, assuming that the leaving difference is still 1.5°F.

The performance concept being applied should be stated when factors are being considered. As an added precaution, the washer requirement should also be given in terms of leaving temperature difference between air and water for a given set of conditions.

In this section, F_p describes dehumidification performance factor as with spray-filled air washers circulating chilled water. This is a historic usage of the term. As indirect evaporative cooling has gained extensive application, the term Performance Factor (PF) has been used to describe indirect evaporative cooling effectiveness. As standards for testing and rating evaporative cooling equipment develop, standard terminology should eliminate the confusion caused by these similar terms.

Air Cleaning with Air Washers

The dust removal efficiency of air washers depends largely on the size, density, wettability, and solubility of the dust particle. Larger, more wettable particles are the easiest to remove. Separation is largely a result of the impingement of particles on the wetted surface of the eliminator plates. Since the force of impact increases with the size of the solid, the impact (together with the adhesive quality of the wetted surface) determines the washer's usefulness as a dust remover. The spray is relatively ineffective in removing most atmospheric dusts.

Air washers are of little use in removing soot particles because of the absence of an adhesive effect from the greasy surface. They are also ineffective in removing smoke, because the inertia of the small particles (less than 1 μm) does not allow them to impinge and be held on the wet plates. Instead, the particles follow the air path between the plates because they are unable to pierce the water film covering the plates.

However, cell washers are efficient air cleaners. In practice, the air-cleaning results are typical of comparable impingement filters. When cell washers remain in the airstream without being wetted or are used where large amounts of fibrous materials are present in the air, they become plugged with airborne dust in a short time unless highly efficient filters are placed upstream from the cells. The cells should be replaced if they are filled with dirt in a dry condition.

If water is not required on the cells, it may be possible to operate the water flow for a few minutes out of each hour to wash the dirt from the cell before the airflow is blocked. With some kinds of dirt, the wetted surface may increase the blockage, unless there is constant water flow.

MAINTENANCE AND WATER TREATMENT

Regular inspection and maintenance of evaporative coolers, air washers, and ancillary equipment ensures proper service and efficiency of the system. Users should follow current recommendations for maintenance and operational procedures to eliminate the possible distribution of airborne contaminants. Water lines, water distribution troughs or pans, pumps, and pump filters must be clean and free of dirt, scale, and debris. Inadequate water flow causes dry areas on the evaporative media and a reduction in the cooling effectiveness. Water and air filters should be cleaned or replaced, as required. Proper sump water level or spray pressure must be maintained. Bleed-off is the most practical means to minimize scale accumulation. The bleed-off rate should be 50 to 100% of the evaporation rate, depending on water hardness. A flush-out cycle which runs fresh water through the pad every 24 h when the fan is off, may be used. This water should run for 3 min for every foot of media height. Regular inspections should be made to ensure that the bleed-off rate is adequate and is maintained.

The use of very pure water from reverse osmosis or deionization processing should not be used in media-based coolers. This water will not wet-out well on a pad and can actually deteriorate a fibrous pad due to the water's aggressive nature. The same

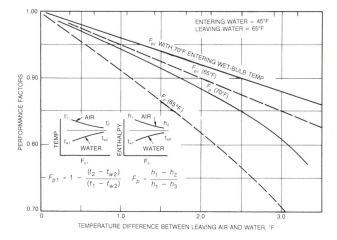

Fig. 15 Comparison of Performance Factors

problem can occur in once-through water distribution systems if the tap water is very pure.

Pretreatment of a water supply with quaternary salts or other chemicals intended to hold dissolved material in suspension is best prescribed by a local water treatment specialist. Using water treated by a zeolite ion exchange softener is not recommended because the zeolite exchange of calcium for sodium results in a soft, voluminous scale that is actively corrosive to galvanized steel. Any chemical agents used should not harm the cabinet, pads, or heat exchanger materials.

Periodic checks should be made for algae, slime, and bacterial growth. If required, an EPA-registered biocide should be added. Algae can be minimized by reducing the media and sump exposure to light sources, by keeping the bottom of the media out of water, and by allowing the media to completely dry out every 24 h.

Motors and bearings should be lubricated and fan drives checked, as required.

Units that have heat exchangers with a totally wetted surface on the outside of the tubes and materials that are not harmed by chemicals can be descaled periodically with a commercial descaling agent and then flushed out. Mineral scale deposits on a wetted indirect unit heat exchanger are usually soft and allow wetting through to the tube and evaporation at the surface of the tube. Excess scale thickness causes a loss in heat transfer and should be removed.

The air washer spray system requires the most attention. Partially clogged nozzles are indicated by a rise in spray pressure, while a fall in pressure is symptomatic of eroded orifices. Strainers can minimize this problem. Continuous operation requires either a bypass around pipe line strainers or duplex strainers. Air washer tanks should be drained and dirt deposits removed regularly. Eliminators and baffles should be periodically inspected and repainted to prevent corrosion damage.

With cell washers, it is also necessary to clean the glass media. A differential draft gage can be used to measure the air resistance across the cells to determine the frequency of cleaning needed. Mat eliminators in these washers should be removed and cleaned with a detergent solution. Any glass mats that have been eroded by the sprays should be replaced.

In colder climates, evaporative coolers must be protected from freezing. This is usually done seasonally by simply draining the cooler and the water supply line to it with solenoid valves. It is important that drain solenoid valves be of the zero differential design. If a coil is used in the system, the coil tubes must be horizontal so that they may be drained at the lowest part of their manifold.

BIBLIOGRAPHY

Anderson, W.M. 1986. Three-stage evaporative air conditioning versus conventional mechanical refrigeration. ASHRAE *Transactions* 92(1B):358.

ASHRAE. 1989. *Position statement on Legionnaires' disease.*

Eskra, V. 1980. Indirect/Direct evaporative cooling systems. ASHRAE *Journal* (May): 21.

Hendrickson, H.M. 1954. How to calculate air washer performance. *Heating, Piping and Air Conditioning* (September): 116.

Schofield, M. and N. Deschamps. 1980. EBTR Compliance and comfort too. ASHRAE *Journal* 22(6):61

Supple, R.G. 1982. Evaporative cooling for comfort. ASHRAE *Journal* (August): 36.

Watt, J.R. 1986. *Evaporative air conditioning handbook.* Chapman and Hall, London.

HUMIDIFIERS

THE selection and application of humidification equipment considers (1) the environmental conditions of the occupancy or process and (2) the characteristics of the building enclosure. Since these may not always be compatible, a compromise solution is necessary, particularly in the case of existing buildings.

ENVIRONMENTAL CONDITIONS

Environmental conditions for a particular occupancy or process dictate a specific relative humidity, a required range of relative humidity, or certain limiting maximum or minimum values. The following classifications explain the effects of relative humidity and provide guidance as to requirements for most applications.

Human Comfort

The effect of relative humidity on all aspects of human comfort has not yet been completely established. Generally, higher temperatures are considered necessary for thermal comfort to offset decreased relative humidity (see ASHRAE *Standard* 55-1981).

Low relative humidity may be undesirable for reasons other than those based on thermal comfort. Low levels increase evaporation from the membranes of the nose and throat and cause drying of the skin and hair. Some medical opinions attribute the increased incidence of respiratory complaints to the drying of mucous membranes by low indoor humidities in winter. However, no well-documented evidence indicates a serious hazard to health resulting from exposure to low humidity.

Extremes of humidity are, however, undesirable and affect human comfort, productivity, and health. Indoor relative humidity should not exceed 60% nor be less than 30% (Sterling *et al.* 1984).

Electronic data-processing equipment requires controlled relative humidity. High relative humidity may cause condensation in the equipment, while low relative humidity may promote static electricity. Also, rapid changes in relative humidity should be avoided due to their effects on punched cards, magnetic tapes, discs, and data-processing equipment. Generally, computer systems have a recommended design and operating range of 35 to 55% rh. However, the manufacturer's recommendations should be adhered to for specific equipment operation parameters.

Process Control and Materials Storage

The relative humidity required by a process is usually specific and related to the control of moisture content or regain, the rate of chemical or biochemical reactions, the rate of crystallization, product accuracy or uniformity, corrosion, and static electricity. Typical conditions of temperature and relative humidity for the storage of certain commodities and the manufacturing and processing of others may be found in Chapter 12 of the 1991 ASHRAE *Handbook—HVAC Applications.*

Low humidities in winter may cause drying and shrinking of furniture, wood floors, and interior trim. Winter humidification may be desirable to maintain a relative humidity closer to that experienced during manufacture or installation.

For storing hygroscopic materials, a constant humidity is as important as humidity level. Temperature control is also important because of the danger of condensation on products through a transient lowering of temperature.

Static Electricity

Electrostatic charges are generated when materials of high electrical resistance move one against the other. The accumulation of such charges may result in (1) unpleasant sparks to people walking over carpets; (2) difficulties in handling sheets of paper, fibers, and fabric; (3) the objectionable clinging of dust to oppositely charged objects; (4) destruction of data stored on magnetic discs and tapes in electronic data-processing centers; and (5) dangerous situations when explosive gases are present. Increasing the relative humidity of the environment prevents the accumulation of such charges, but the optimum level of humidity depends, to some extent, on the materials involved.

Relative humidities of 45% or more usually reduce or eliminate electrostatic effects in many materials, but wool and some synthetic materials may require still higher humidities (Sereda and Feldman 1964).

Hospital operating rooms, where explosive mixtures of anesthetics are used, constitute a special and critical case regarding electrostatic charges. A relative humidity of 50% or more is usually required with special grounding arrangements and restrictions on the types of clothing worn by occupants. Conditions of 72°F and 55% rh are usually recommended for comfort and safety.

Prevention and Treatment of Disease

Relative humidity has a significant effect on the control of airborne infection. At 50% rh, the mortality rate of certain organisms is highest, and the influenza virus loses much of its virulence. The mortality rate decreases both above and below this value. High humidities can support the growth of pathogenic or allergenic organisms. Relative humidity in habitable spaces should preferably be maintained between 30 and 60%.

The preparation of this chapter is assigned to TC 8.7, Humidifying Equipment.

Miscellaneous

The air absorption of sound waves is at a maximum at 15 to 20% rh and increases with frequency (Harris 1963). A marked reduction in absorption is obtained at 40% rh; above 50%, the effect of air absorption is negligible. Air absorption does not affect speech significantly, but may merit consideration in large halls or auditoriums where optimum acoustics are required for musical performances.

Certain microorganisms are occasionally present in poorly maintained household humidifiers. To deter the propagation and spread of these detrimental microorganisms, periodic cleaning of the humidifier and draining of the reservoir (particularly at the end of the heating season) are required. Cold water atomizing types have been banned as room humidifiers in some hospitals because of germ propagation.

Laboratories and test chambers, in which precise control of a wide range of relative humidity is desired, require special attention. Because of the interrelation between temperature and relative humidity, precise humidity control requires equally precise temperature control. The cooling load should be reduced and the method of heat removal should be considered if high humidities are to be maintained (Solvason and Hutcheon 1965).

ENCLOSURE CHARACTERISTICS

The extent to which a building may be humidified in winter depends on the ability of its walls, roof, and other elements to prevent or tolerate condensation. Condensed moisture or frost on surfaces exposed to the building interior (visible condensation) can deteriorate the surface finish, cause mold growth and subsequent indirect moisture damage and nuisance, and reduce visibility through windows. If the walls and roof have not been specifically designed to prevent the entry of moist air or vapor from inside, concealed condensation within these constructions is likely to occur, even at fairly low interior humidities, and cause serious deterioration.

Chapter 21 of the 1989 ASHRAE *Handbook—Fundamentals* addresses the design of enclosures to prevent condensation.

Visible Condensation

Condensation forms on any interior surface when the dew-point temperature of the air in contact with it exceeds the temperature of the surface. The maximum permissible relative humidity that may be maintained without condensation is thus influenced by the thermal properties of the enclosure and the interior and exterior environment.

Average surface temperatures may be calculated by the methods outlined in Chapter 22 of the 1989 ASHRAE *Handbook—Fundamentals* for most insulated constructions. However, localized cold spots result from high conductivity paths such as through-the-wall framing, projected floor slabs, and metal window frames that have no thermal breaks. The vertical temperature gradient from airspace and surface convection in windows and similar sections results in lower air and surface temperatures at the sill or floor. Drapes and blinds closed over windows lower surface temperature further, while heating units under windows raise the temperature significantly.

Windows present the lowest surface temperature in most buildings and provide the best guide to permissible humidity levels for no condensation. While calculations based on overall thermal coefficients provide reasonably accurate temperature predictions at mid-height, actual minimum surface temperatures are best determined by test. Wilson and Brown (1964) related the characteristics of windows with a temperature index, which is defined as

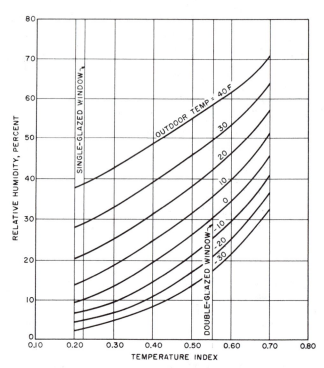

Fig. 1 Limiting Relative Humidity for No Window Condensation

$(t - t_o)/(t_i - t_o)$, where t is the inside window surface temperature, t_i is the indoor air temperature, and t_o is the outdoor air temperature.

The results of limited tests on actual windows indicate that the temperature index at the bottom of a double, residential-type window with a full thermal break is between 0.55 and 0.57, with natural convection on the warm side. Sealed, double-glazed units exhibit an index from 0.33 to 0.48 at the junction of glass and sash, depending on sash design. The index is likely to rise to values of 0.53 or greater only 1 in. above the junction.

With continuous under-window heating, the minimum index for a double window with full break may be as high as 0.60 to 0.70. Under similar conditions, windows with poor thermal breaks may be increased similarly.

Figure 1 shows the relationship between temperature index and the relative humidity and temperature conditions at which condensation occurs. The limiting relative humidities for various outdoor temperatures intersect a vertical line representing the particular temperature index. A temperature index of 0.55 has been selected to represent an average for double-glazed, residential windows and indicates the limiting relative humidities shown in Table 1. The values for single-glazing in the table are based on a calculated temperature index of 0.22.

Table 1 Maximum Relative Humidity within a Space for No Condensation on Windows

Natural Convection, Indoor Air at 74°F		
Outdoor Temperature, °F	**Single Glazing**	**Double Glazing**
40	39	59
30	29	50
20	21	43
10	15	36
0	10	30
−10	7	26
−20	5	21
−30	3	17

Concealed Condensation

The humidity level a given building is able to tolerate without serious concealed condensation may be much lower than that indicated by visible condensation criteria. The migration of water vapor through the inner envelope by diffusion or air leakage brings the vapor into contact with surfaces at temperatures approaching the outside temperature. Unless the building has been designed to eliminate or reduce this possibility effectively, the permissible humidity may be determined by the ability of the building enclosure to handle internal moisture rather than prevent its occurrence. Increased humidification for commercial buildings in cold climates should be approached with caution.

ENERGY CONSIDERATIONS

When calculating energy requirements, the effect of the dry air environment should be considered for any material supplying it with moisture. This conversion of moisture (in its liquid state) from containment in a hygroscopic material to the vapor state is an evaporative process and, therefore, requires energy. The source of energy is the heat energy contained in the air. Thus, heat lost from the air to evaporate moisture equals the heat necessary to produce an equal amount of moisture vapor with an efficient humidifier. Therefore, in some cases, no new energy is required to maintain desired humidity levels.

This apparent tradeoff has its costs in the destructive effects of moisture migration from hygroscopic materials if proper humidity levels are not maintained.

The true energy required for a humidification system must be calculated from what the actual humidity level will be in the building, not from the theoretical level. Without a humidification system, methods of calculating actual humidity have not been established.

A study of residential heating and cooling systems showed a correlation between infiltration and inside relative humidity, indicating a significant energy saving from increasing the inside relative humidity, which reduced infiltration of outside air by up to 50% during the heating season (Luck and Nelson 1977). This reduction is apparently due to the sealing of window cracks by the formation of frost.

LOAD CALCULATIONS

The humidification load depends primarily on the rate of natural ventilation of the space to be humidified or the amount of outside air introduced by mechanical means. Other sources of moisture gain or loss should also be considered, however. The humidification load can be calculated as follows:

For ventilation systems (natural infiltration)

$$H = \rho V R (W_i - W_o) - S + L \qquad (1)$$

For fixed quantity of outside air systems (mechanical ventilation)

$$H = 60\rho Q_o (W_i - W_o) - S + L \qquad (2)$$

For variable quantity of outside air systems

$$H = 60\rho Q_t (W_i - W_o) \left(\frac{t_i - t_m}{t_i - t_o} \right) - S + L \qquad (3)$$

where

H = humidification load, lb of water/h
V = volume of the space to be humidified, ft^3
R = ventilation rate, air changes/h
Q_o = quantity of outside air, cfm
Q_t = total quantity of air (outside air plus return air), cfm
t_i = temperature of air indoor design, °F
t_m = temperature of mixed air design, °F
t_o = temperature of outside air design, °F
W_i = humidity ratio at indoor design conditions, lb of water/lb of dry air
W_o = humidity ratio at outdoor design conditions, lb of water/lb of dry air
S = contribution of internal moisture sources, lb of water/h
L = other moisture losses, lb of water/h
ρ = density of air, 0.074 lb/ft^3

Design Conditions

Interior design conditions are dictated by the occupancy or process, as discussed in the sections Environmental Conditions and Enclosure Characteristics. Outdoor design data may be obtained from Chapter 24 of the 1989 ASHRAE *Handbook—Fundamentals*. Outdoor humidity can be assumed at 70 to 80% below 32°F or 50% above 32°F in winter for most areas. Corresponding absolute humidity values can subsequently be obtained either from Chapter 6 of the 1989 ASHRAE *Handbook—Fundamentals* or from an ASHRAE Psychrometric Chart.

For systems handling quantities of outside air fixed by natural infiltration rates or mechanical ventilation, load calculations are based on outdoor design conditions. Equation (1) should be used for natural infiltration; Equation (2) for mechanical ventilation.

For economizer systems that achieve a fixed mixed air temperature by varying outside air, special considerations are needed to determine the maximum humidification load. This load occurs at an outside air temperature other than the lowest design temperature, since it is a function of the amount of outside air introduced and the existing moisture content of the air. Equation (3) should be used, repeated at various outside air temperatures, to determine the maximum humidification load.

In residential load calculations, the actual outdoor design conditions of the locale are usually taken as 20°F and 70% rh, while indoor conditions are taken as 70°F and 35% rh. These values yield an absolute humidity difference ($W_i - W_o$) of 0.0040 lb/lb of dry air for use in Equation (1). With the required lowering of interior relative humidity with outdoor temperature, indicated in Table 1, these values represent the point of maximum load.

Ventilation Rate

Ventilation of the humidified space may be the result of natural infiltration, either alone or in combination with intentional mechanical ventilation. Natural infiltration varies according to the indoor-outdoor temperature difference, wind velocity, and tightness of construction, as discussed in Chapter 23 of the 1989 ASHRAE *Handbook—Fundamentals*. The rate of mechanical ventilation may be determined from building design specifications or estimated from fan performance data (see ASHRAE *Standard* 62-1989).

In load calculations, the water vapor removed from the air during cooling by air-conditioning or refrigeration equipment must be considered. This moisture may have to be replaced by humidification equipment to maintain the desired level of relative humidity in certain industrial projects where the load may be greater than that required for ventilation and heating.

Estimates of infiltration rate are made in calculating heating and cooling loads for buildings; these values also apply to humidification load calculations. For residences where such data are not available, it may be assumed that a tight house will have an air change rate of 0.5 per hour; an average house, 1 per hour; and a loose house may have as many as 1.5 air changes per hour. A tight house is assumed to be well-insulated and to have vapor barriers, tight storm doors, windows with weather stripping, and a dampered fireplace. An average house is insulated and has vapor barriers, loose storm doors and windows, and a dampered fireplace. A loose house is generally one constructed before 1930 with little or no insulation, no storm doors or windows, no weather stripping, no vapor barriers, and often, a fireplace without an effective damper. For building construction, refer to local codes and building specifications.

Additional Moisture Losses

Hygroscopic materials, which have a lower moisture content than materials in the humidified space, absorb moisture and place an additional load on the humidification system. An estimate of this load depends on the absorption rate of the particular material selected. Chapter 12 of the 1991 ASHRAE *Handbook—HVAC Applications* lists the equilibrium moisture content of hygroscopic materials at various relative humidities.

In cases where humidity levels need to be maintained regardless of condensation on exterior windows and walls, the dehumidifying effect of these surfaces constitutes a load that may need to be considered, if only on a transient basis. The loss of water vapor by diffusion through enclosing walls to the outside or to areas at a lower vapor pressure may also be involved in some applications. The properties of materials and flow equations given in Chapter 20 of the 1989 ASHRAE *Handbook—Fundamentals* can be applied in such cases. Normally, this constitutes a small load, unless openings exist between the humidified space and adjacent rooms at lower humidities.

Internal Moisture Gains

The introduction of a hygroscopic material can cause moisture gains to the space if its moisture content is above that of the space

Table 2 Moisture Production from Various Residential Operations

(Hite and Dry 1948)

Operation			Moisture, lb
Floor mopping—80 ft² kitchen			2.40
Clothes drying (not vented)*			26.40
Clothes washing*			4.33
Cooking (not vented)*	*From food*	*From gas*	
Breakfast	0.34	0.56	0.90
Lunch	0.51	0.66	1.17
Dinner	1.17	1.52	2.69
Bathing—shower			0.50
Bathing—tub			0.12
Dishwashing*			
Breakfast			0.20
Lunch			0.15
Dinner			0.65
Human Contribution—Adults, per hour			
When resting			0.20
Working hard			0.60
Average			0.40
Houseplants			0.04

*Based on family of four.

conditions. Similarly, moisture may diffuse through walls separating the space from areas of higher vapor pressure or move by convection through openings in these walls (Brown *et al.* 1963).

Moisture contributed by human occupancy is affected by the number of occupants and their degree of physical activity. Estimates of the rate of moisture gain from activities in residences may be made from the data presented in Table 2 (Hite and Dry 1948). As a guide to residential applications, the average rate of moisture production for a family of four may be taken as 0.7 lb/h. Unvented heating devices produce about 1 lb vapor for each pound of fuel burned.

Industrial processes constitute additional moisture sources. Single-color offset printing presses, for example, give off 0.45 lb/h of water. Information on process contributions can best be obtained from the manufacturer of specific equipment.

EQUIPMENT

Generally, humidification equipment can be classified as either residential or industrial, although residential humidifiers can be used for small industrial applications, and small industrial units can be used in large homes. Equipment designed for use in central air systems also differs from that for space humidification, although some units are adaptable to both.

Air washers and evaporative coolers may be used as humidifiers, but they are usually selected for some additional function such as air cooling or air cleaning, as discussed in Chapter 19.

Residential and industrial humidifiers are divided into two broad classifications according to their principle of operations. Energy is added, and evaporation occurs in the humidifier. The other type takes energy from and evaporation occurs in the space.

The output rates for residential humidifiers are generally based on 24-h operation; output rates for industrial and commercial humidifiers are based on 1-h operation. During normal usage, the time of operation can be considerably less, a fact that should be considered in design. Published evaporation rates are established by equipment manufacturers based on certain test criteria that are not consistent among manufacturers. Rates and test methods should be evaluated when selecting equipment. The Air-Conditioning and Refrigeration Institute (ARI) has developed *Standard* 610-89 for residential central system humidifiers and *Standard* 640-90 for commercial and industrial humidifiers. ANSI/AHAM *Standard* HU-1-1987 addresses self-contained units.

Residential Humidifiers for Central Air Systems

Residential humidifier units depend on airflow in the heating system for evaporation and distribution. General principles of operation and description of equipment are as follows:

Pan-type. Output varies with temperature, humidity, and airflow in the system.

- *Basic pan.* A shallow pan is normally installed within the furnace plenum. A flow control device connected to the household water supply maintains a constant water level in the pan. Humidification rate is low.
- *Electrically heated pan type.* An electric heater increases the water temperature in the pan, thus increasing the rate of evaporation. Humidification rate is positive.
- *Pans with wicking plates.* This type is similar to the basic pan, except that a number of vertical, water-absorbent plates are added to the pan to increase the wetted surface area (Figure 2). Humidification rate is low.

Wetted elements. Output varies with temperature, humidity, and airflow in the system.

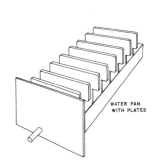

A. Pan Humidifier

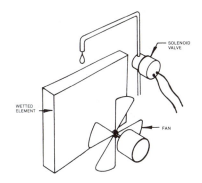

B. Power Wetted Element Humidifier

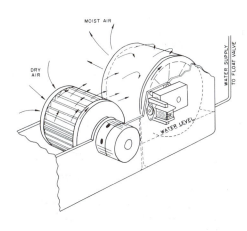

C. Wetted Drum Humidifier

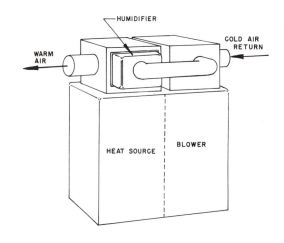

D. Bypass Wetted Element Humidifier

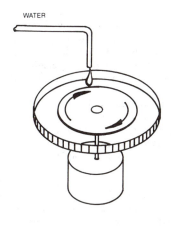

E. Atomizing Humidifier

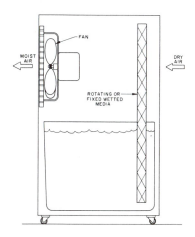

F. Appliance Portable Humidifier

Fig. 2   Residential Humidifiers

Air circulates over or through an open-textured, wetted media. The evaporating surface may be a fixed pad, wetted by either sprays or water flowing by gravity, or the pad may be a paddle wheel, drum, or belt rotating through a water reservoir.

Air flows through such units in one of the following ways:

- *Fan-type.* A small fan or blower draws air from the furnace plenum, through the wetted pad, and back to the plenum. A fixed pad may be used as is illustrated in Figure 2B, or a rotating drum-type pad as in Figure 2C.
- *Bypass type.* These units do not have their own fan but rather are mounted on the supply or return plenum of the furnace with an air connection to the return plenum (Figure 2D). The difference in static pressure created by the furnace blower circulates air through the unit.
- *Duct-mounted.* These units are designed for installation within the furnace plenum or ductwork with a drum element rotated by either the air movement within the duct or a small electric motor.

Atomizing types. Output of atomizing units does not depend on the conditions of the air. The ability of the air to absorb moisture depends on temperature, airflow, and moisture content of the air moving through the system. In this type of humidifier, small particles of water are introduced directly into the airstream by the following:

- A spinning disc or cone, which breaks the water into a fine mist (Figure 2E)
- Spray nozzles that rely on water pressure to create fine droplets
- A rotating disc, which slings water droplets into the airstream from a water reservoir
- Spray nozzles that use compressed air to create a fine mist
- Ultrasonic vibrations used as the atomizing force

Residential Humidifiers for Nonducted Applications

Many portable or room humidifiers are used in residences heated by such nonducted systems as hydronic or electric or in residences where the occupant is prevented from making a permanent installation. Most of these humidifiers are equipped with humidity controllers.

Portable units evaporate water by any of the previously described means, such as fixed or moving wetted element, atomizing spinning disc, or heated pan. They may be tabletop size or a larger, furniture-styled appliance (Figure 2F). A multispeed motor on the fan or blower may be used to adjust output. Usually, the portable humidifiers require periodic filling from a bucket or filling hose.

Some portable units are offered with an auxiliary package for semipermanent water supply. This package includes a manual shutoff valve, a float valve, copper or other tubing with fittings, etc. The lack of provision for water overflow to drain may result in water damage.

Some units, using the same principles, may be recessed in the wall between studs, mounted on wall surfaces, or installed below floor level. These units are permanently installed in the structure and use forced-air circulation. They may have an electric element for reheat, when desired. Other types for use with hydronic systems involve a simple pan or pan plate, either installed within a hot-water convector or using the steam from a steam radiator.

Industrial/Commercial Humidifiers for Central Air Systems

Humidifiers must be installed where the air can absorb the vapor and will not be cooled below the dew point downstream.

Heated pan. These units offer a broad range of capacity and may be heated by an electrical element, steam, or hot-water coil (see Figure 3A). Others are installed remote from the ductwork, and their steam output is carried via steam hose to the duct airstream. Electric pan humidifiers are usually provided with a low water level cut-off switch as a protection device for the heating elements.

Steam coils are commonly used in pan humidifiers. At steam pressures above 15 psig, moisture carryover occurs because of splashing caused by nucleate boiling. To prevent boiling over, baffle splash eliminators should be used according to manufacturers' instructions. Eliminators are essential where steam pressure is greater than 15 psig.

Steam. Direct steam injection humidifiers cover a wide range of designs and capacities. Since water vapor is steam at low pressure and temperature, the whole process can be simplified by introducing steam directly into the air to be humidified. This method is essentially an isothermal process because the temperature of the air remains constant as the moisture is added in vapor form. The steam control valve may be modulating or two-position in response to a humidity controller. The steam may be either used from an external source with enclosed grid, cup, or jacketed dry steam humidifiers or produced within the humidifier, as in the self-contained type. When the steam is supplied from a separate source at a constant supply pressure, it responds quickly to system demand. Steam units must be installed where the air can absorb the vapor, otherwise condensation can occur in the duct. For proper psychrometric calculations, refer to Chapter 6 of the 1989 ASHRAE *Handbook—Fundamentals.*

- *Enclosed steam grid humidifiers* (see Figure 3B) should be used on low steam pressures—under 12 psig—to prevent splashing of condensate in the duct. The drain should be located on the side opposite the control valve. A drip leg—dimension H, a minimum of 12 in. (Figure 3B)—should provide the pressure to flow the condensate through the trap.
- *A cup or pot-type steam humidifier* (Figure 3C) is usually attached under a system duct. Steam is attached tangentially to the inner periphery of the cup by one or more steam inlets, depending on the capability of the unit. The steam supply line should have a suitable steam trap. There may be a tendency toward supersaturation due to stratification along the bottom of the duct. Multiple units may be required to produce satisfactory distribution. Under certain conditions, droplets of condensate may be injected into the airstream.
- *A jacketed steam humidifier* uses an integral steam valve with a steam-jacketed duct-traversing dispersing tube and condensate separator to prevent condensate from being introduced into the airstream (Figure 3D). An inverted bucket-type steam trap is required to drain the separating chamber. This humidifier may be used without the jacketed tube in nonducted installations.

The aforementioned humidifiers inject steam directly from the boiler into the space or duct system. Some boiler treatment chemicals can be discharged, which can affect indoor air quality. Care should be taken to avoid contamination from boiler water or steam supply additions (ASHRAE 1989).

- *A self-contained steam humidifier* converts tap water to steam by electrical energy using either the electrode boiler principle or resistance heating. This steam is injected into the duct system through a dispersion manifold (Figure 3E), or the humidifier may be freestanding for nonducted applications.

Atomizing humidifiers with optional filter eliminator (Figure 3F). Centrifugal atomizers use a high-speed disc, which slings water through a fine comb to create a fine mist that is introduced

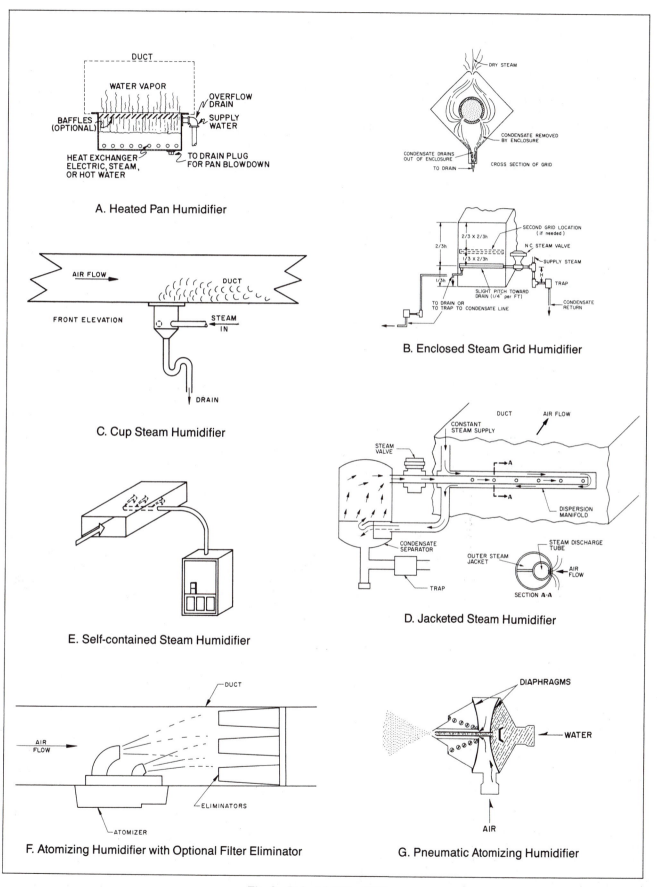

A. Heated Pan Humidifier

B. Enclosed Steam Grid Humidifier

C. Cup Steam Humidifier

D. Jacketed Steam Humidifier

E. Self-contained Steam Humidifier

F. Atomizing Humidifier with Optional Filter Eliminator

G. Pneumatic Atomizing Humidifier

Fig. 3   Industrial Humidifiers

directly into the air where it is evaporated. The ability of the air to absorb the moisture depends on temperature, air velocity, and moisture content.

Where mineral fallout from hard water is a problem, optional filter eliminators may be added to remove mineral dust from humidified air, or water demineralizers may be installed.

Additional atomizing methods use nozzles; one uses water pressure and the other uses both air and water, as shown in Figure 3G. Mixing air and water streams at combined pressures atomizes water into a fine mist, which is evaporated in the room or air duct.

Ultrasonic nozzles place air and water under pressure to atomize the water into a fine mist. Accurate psychrometric calculations must be made to ensure that the water droplets are absorbed in the duct airstream.

Wetted element humidifier. Wetted element humidifiers have a wetted media, sometimes in modular configurations, through or over which air is circulated to evaporate water. This unit depends on airflow for evaporation; the rate varies with temperature, humidity, and velocity of the air.

Industrial Humidifiers for Nonducted Applications

All the aforementioned methods of humidifying are available as unitary equipment. To achieve quicker absorption, these units may include an integral air-moving device for better movement of the humidified air.

WATER SUPPLY

In areas with water of high mineral content, precipitated solids may be a problem. The precipitated solids clog nozzles, tubes, evaporative elements, and controls. In addition, solids allowed to enter the airstream leave a fine layer of white dust over furniture, floors, and carpets. If the water is softened by ion exchange, the residue increases because the magnesium or calcium salts are exchanged for those of sodium. Some wetted-element humidifiers bleed off and replace all or some of the water passing through the element to reduce the concentration of salts in the water reservoir.

Dust, scaling, biological organisms, and corrosion are all problems associated with water in humidifiers. Stagnant water provides a fertile breeding ground for algae and bacteria, which have been linked to odor and respiratory ailments. Bacterial slime reacts with sulfates in the water to produce hydrogen sulfide and its characteristic bad odor. Periodic maintenance and regular use of odorless microbiocides and other additives eliminate organisms from the water (Sriramamurty 1977).

Scaling

Industrial pan humidifiers, when supplied with naturally low hardness water, require little maintenance provided a surface skimmer bleedoff is used.

Water softening is an effective means of eliminating mineral precipitation in a pan-type humidifier. However, the concentration of sodium left in the pan as a result of water evaporation must be held below the point of precipitation by flushing and diluting the tank with new softened water. The frequency and duration of dilution depends on the water hardness and the rate of evaporation. Dilution is usually accomplished automatically by a timer-operated drain valve and a water makeup valve.

Demineralized or reverse osmosis (RO) water may also be used. The construction materials of the humidifier and the piping must withstand the corrosive effects of this water. Commercial demineralizers or RO equipment remove hardness and other total dissolved solids completely from the makeup water to the humidifier. They are more expensive than water softeners, but no humidifier purging is required. The sizing is based on the maximum required

water flow to the humidifier and the amount of total dissolved solids in the makeup water.

HUMIDITY CONTROLS

Many humidity-sensitive materials are available. Most are organic, such as nylon, human hair, wood, and animal membranes. Sensors that change electrical resistance with humidity are also available.

Mechanical Controls

Mechanical sensors depend on a change in the length or size of the sensor as a function of relative humidity. The most commonly used sensors are synthetic polymers or human hair. They can be attached to a mechanical linkage to control the mechanical, electrical, or pneumatic switching element of a valve or motor. This design is suitable for most human comfort applications, but it may lack the necessary accuracy needed in industrial applications.

A humidity controller is normally designed to control at a set point selected by the user. Some controllers have a set-back feature that lowers the relative humidity set point as outdoor temperatures lower to reduce condensation within the structure.

Electronic Controllers

As the humidity changes, electrical sensors change electrical resistance. They typically consist of two conductive materials separated by a humidity-sensitive hygroscopic insulating material (polyvinyl acetate, polyvinyl alcohol, or a solution of certain salts). Small changes are detected as air passes over the sensing surface.

The use of electronic humidity control is common in laboratory or process applications where precise control is required. It is also used to control fan speed on portable humidifiers to control humidity in the space more closely and to reduce noise and draft to a minimum.

CONTROL LOCATION

In centrally humidified structures, the humidity controller is most commonly mounted in a space under control. Another method is to mount the humidity controller in the return air duct of an air-handling system to sense average relative humidity. Reference should be made to manufacturers' instructions regarding the controller's use on counterflow furnaces, because reverse airflow after the fan turns off can substantially shift the humidity control point in a home.

In central systems, a relatively low-temperature supply air might be required to compensate for the sensible heat gain of the room. This required temperature may be at a dew point below that necessary to maintain the desired relative humidity. This means the cooling air is saturated and cannot absorb more moisture. Operating the humidifier under these conditions causes condensation in the ducts and fogging in the room. High-capacity humidifiers installed in central air-conditioning systems should have a high limit humidistat downstream of the humidifier to prevent condensation.

REFERENCES

ASHRAE. 1981. Thermal environmental conditions for human occupancy. *Standard* 55-1981.

ASHRAE. 1989. Ventilation for acceptable indoor air quality. *Standard* 62-1989.

Brown, W.G., K.R. Solvason, and A.G. Wilson. 1963. Heat and moisture flow through openings by convection. ASHRAE *Journal* 5(9):49.

Harris, C.M. 1963. Absorption of sound in air in the audio-frequency range. *Journal of the Acoustical Society of America* 35(January).

Hite, S.C. and J.L. Dry. 1948. Research in home humidity control. *Experiment Station Research Series*, No. 106, Purdue University, West Lafayette, IN.

Luck, J.R. and L.W. Nelson. 1977. The variation of infiltration rate with relative humidity in a frame building. ASHRAE *Transactions* 83(1).

Sereda, P.J. and R.F. Feldman. 1964. Electrostatic charging on fabrics at various humidities. *Journal of the Textile Institute* 55(5):T288.

Solvason, K.R. and N.B. Hutcheon. 1965. Principles in the design of cabinets for controlled environments. *Humidity and Moisture, Measurement and Control in Science and Industry* 2:241. Reinhold Publishing Corporation, New York.

Sriramamurty, D.V. 1977. Application and control of commercial and industrial humidifiers. ASHRAE *Transactions* 83(1).

Sterling, E.M., A. Arundel, and T.D. Sterling. 1985. Criteria for human exposure to humidity in occupied buildings. ASHRAE *Transactions* 91(1).

Wilson, A.G. and W.P. Brown. 1964. Thermal characteristics of double windows. *Canadian Building Digest* No. 58, Division of Building Research, National Research Council, Ottawa, Ontario.

BIBLIOGRAPHY

Bender, D.O. 1973. Humidification in pressrooms. ASHRAE *Symposium Bulletin* NO-72-12.

Heiman, R.I. and O.E. Ulrich. 1973. Residential-commercial-industrial humidifiers. ASHRAE *Symposium Bulletin* NO-72-12.

Howery, J.F. and R.M. Pasch. 1977. Applying residential humidifiers to central systems. ASHRAE *Transactions* 83(1).

O'Dell, L.R. 1977. Considerations for sizing and installing commercial and industrial steam humidifiers. ASHRAE *Transactions* 83(1).

Smith, G.D. 1973. Considerations in building humidification. ASHRAE *Symposium Bulletin* NO-72-12.

Spethmann, D.H. 1973. Humidity control in textile mills. ASHRAE *Symposium Bulletin* NO-72-12.

Wilkes, J.F. 1973. Commercial-industrial water and steam treatment—effect on humidification. ASHRAE *Symposium Bulletin* NO-72-12.

AIR-COOLING AND DEHUMIDIFYING COILS

MOST of the equipment now used for cooling and dehumidifying an airstream under forced convection incorporates a coil section which, at minimum, contains one or more cooling coils assembled in a coil bank arrangement. Such coil selections are used extensively as components in room terminal units, on up through the larger, factory-assembled, self-contained air conditioners, central station air handlers, and field built-up systems. The applications of each type of coil are limited to the field within which the coil is rated. Other limitations are imposed by code requirements, proper choice of materials for the fluids used, the configuration of the air handler, and economic analysis of the possible alternatives for each installation.

USES FOR COILS

Coils are used for air cooling with or without accompanying dehumidification. Examples of cooling applications without dehumidification are precooling coils that use well water or other relatively high-temperature water to reduce the load on the refrigerating equipment, and chilled water coils that remove sensible heat from chemical moisture-absorption apparatus. The heat-pipe coil is also used as a supplementary heat exchanger for air-side sensible cooling preconditioning (see Chapter 44).

Most coil sections provide air sensible cooling and dehumidification simultaneously. The unit assembly usually includes the means for cleaning air to protect the coil from accumulation of dirt and to keep dust and foreign matter out of the conditioned space. Although cooling and dehumidification are their principal functions, cooling coils can also be wetted with water or a hygroscopic liquid to aid in air cleaning, odor absorption, or frost prevention. Coils are also evaporatively cooled with a water spray to improve efficiency or capacity. For general comfort conditioning, cooling, and dehumidifying, the extended surface (finned) cooling coil design is the most popular and practical.

COIL CONSTRUCTION AND ARRANGEMENT

In finned coils, the external surface of the tubes is primary and the fin surface is secondary. The primary surface generally consists of rows of round tubes or pipes that may be staggered or placed in line with respect to the airflow. Flattened tubes, or those with other nonround internal passageways, are sometimes used. The inside surface of the tubes is usually smooth and plain, but some refrigerant coil designs have various forms of internal fins

or turbulence promoters (either fabricated or extruded) to enhance performance. The individual tube passes in a coil are usually interconnected by return bends (or with hairpin bend tubes) to form the serpentine arrangement of multipass tube circuits. A variety of circuit arrangements, and combinations thereof, for varying the number of parallel water flow passes within the tube core are usually available (Figure 1).

Cooling coils for water, steam, aqueous glycol, or halocarbon refrigerants usually have aluminum fins and copper tubes, although copper fins on copper or copper nickel tubes and aluminum fins on aluminum tubes (excluding water) are also used. Typically, fin tube joints are made by mechanical expansion or soldering. Header connections and return bend joints are normally sealed by brazing or welding. Adhesives are sometimes used to bond header connections, return bends, and fin-tube joints, particularly for aluminum-to-aluminum joints. Many makes of cooling coils of the lightweight extended-surface type for both heating and cooling are available. Common core tube outside diameters are 5/16, 3/8, 1/2, 5/8, 3/4, and 1 in., with fins spaced 4 to 18 per inch. Tube spacing ranges from 0.6 to 3.0 in. on equilateral (staggered) or rectangular (in-line) centers, depending on the width of individual fins and other performance considerations. Fins should be spaced according to the job to be performed with special attention given to air friction, possibility of lint accumulation, and frost accumulation, especially at lower temperatures.

Tube wall thickness and the required use of alloys other than copper are determined mainly by the coil's working pressure and safety factor for hydrostatic burst (pressure). Fin-type and header construction also play a large part in this determination. Local job site codes and applicable national safety standards should be consulted in design and application of these coils.

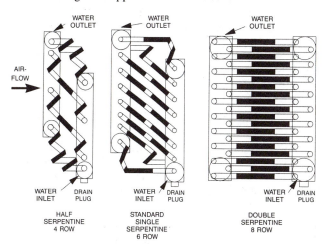

Fig. 1 Typical Water Circuit Arrangement

The preparation of this chapter is assigned to TC 8.4, Air-to-Refrigerant Heat Transfer Equipment.

Water and Aqueous Glycol Coils

Good performance of water-type coils requires the elimination of all air and water traps within the water circuit, and the proper distribution of water. Unless properly vented, air may accumulate in the coil tube circuits, reducing thermal performance and possibly causing noise or vibration in the piping system. Air vent and drain connections are usually provided on the coil's water headers. However, this does not eliminate the need to install and maintain the coil tube core in a level position. Individual coil vents and drain plugs are often incorporated on the headers (Figure 1). Water traps within the tubing of a properly leveled coil are usually a result of (1) improper nondraining circuit design and/or (2) center-of-coil downward sag. Such a situation may cause tube damage, *e.g.*, freeze-up in cold climates or tube erosion due to untreated mineralized water.

Depending on performance requirements, the water velocity inside the tubes usually ranges from approximately 1 to 8 ft/s, and the design water pressure drop across the coils varies from about 5 to 50 ft of water head. For nuclear HVAC applications, the ANSI/ASME AG-1 Code on Nuclear Air and Gas Treatment requires a minimum tube velocity of 2 ft/s. The water may contain considerable sand and other foreign matter—as in precooling coils using well water, or in applications where minerals in the cooling water deposit on and foul the (internal) tube surface. It is best to filter out such sediment. Removable water header plates or a removable plug for each tube allows the tube to be cleaned, which ensures a continuation of rated performance after the cooling units have been in service. Where buildup of scale deposits or fouling of the water-side surface is expected, a scale factor is sometimes included when calculating thermal performance of the coils. Cupronickel, red brass, bronze, and other tube alloys help protect against corrosion and erosion deterioration primarily caused by internal fluid flow abrasive sediment. The core tubes of properly designed and installed coils should feature circuits that are (individually) of: (1) equally developed line length; (2) the self-draining type, by means of gravity during the (water pump) off cycle; (3) minimum pressure drop to aid in water distribution from the supply header, but not requiring an excessive pumping head; and (4) equal feed and return by the supply and return header. Design for the proper in-tube water velocity will determine the circuitry style required. Multirow coils usually are of the cross-counterflow arrangement, as well as top-outlet/bottom-feed connection orientation.

Direct-Expansion Coils

Coils for halocarbon refrigerants present more complex cooling fluid distribution problems than do water or brine coils. The coil should cool effectively and uniformly throughout, with even refrigerant distribution. Halocarbon coils are used on flooded and direct-expansion refrigerated systems.

A flooded system is used mainly when a small temperature difference between the air and refrigerant is desired. Chapter 3 of the 1990 ASHRAE *Handbook—Refrigeration* describes flooded systems in more detail.

For direct-expansion systems, two of the most commonly used refrigerant liquid metering arrangements are the capillary tube assembly (or restrictor orifice) and the thermostatic expansion valve device. The capillary tube is applied in factory-assembled, self-contained air conditioners up to approximately 10-tons capacity but is most widely used on smaller capacity models, such as window- or room-type units. In this system, the bore and length of a capillary tube are sized so that at full load, under design conditions, just enough liquid refrigerant is metered from the condenser to the evaporator coil to be evaporated completely. While this type of metering arrangement does not operate over a wide

range of conditions as efficiently as a thermostatic expansion valve system, its performance is targeted for design conditions.

A thermostatic expansion valve system is commonly used for all direct-expansion coil applications described in this chapter, particularly field-assembled coil sections, as well as those used in central air-handling units and the larger, factory-assembled hermetic air conditioners. This system depends on the thermostatic expansion valve (TXV) to regulate automatically the rate of refrigerant liquid flow to the coil in direct proportion to the evaporation rate of refrigerant liquid in the coil, thereby maintaining optimum performance over a wide range of conditions. The superheat at the coil suction outlet is continually maintained within the usual predetermined limits of 6 to 10 °F. Because the thermostatic expansion valve responds to the superheat at the coil outlet, this superheat should be produced with the least possible sacrifice of active evaporating surface.

The length of each coil's refrigerant circuits, from the TXV's distributor feed tubes through the suction header, should be equal. The length of each circuit should be optimized to provide good heat transfer and oil return, and a complementary pressure drop across the circuit. In addition to installing the coil level, coil circuitry needs to self-drain by gravity toward the suction header connection.

To ensure reasonably uniform refrigerant distribution in multicircuit coils, a distributor is placed between the thermostatic expansion valve and coil inlets to divide the refrigerant equally among the coil circuits. The refrigerant distributor must be effective in distributing both liquid and vapor, because the refrigerant entering the coil is usually a mixture of the two (although it is mainly liquid by weight). Distributors can be placed in either the vertical or the horizontal position; however, the vertical down position usually distributes refrigerant between coil circuits better than the horizontal position.

The individual coil circuit connections from the refrigerant distributor to the coil inlet are made of small diameter tubing; they are all the same length and diameter so that the same flow occurs between each refrigerant distributor tube and each coil circuit. To approximate uniform refrigerant distribution, the refrigerant should flow to each refrigerant distributor circuit in proportion to the load on that circuit. The heat load must be distributed equally to each refrigerant circuit to obtain optimum coil performance. If the coil load cannot be distributed uniformly, the coil should be recircuited and connected with more than one thermostatic expansion valve to feed the circuits (separate coil sections may also help). In this way, the refrigerant distribution matches the unequal circuit loading. Unequal circuit loading may be caused by such variables as uneven air velocity across the face of the coil, uneven entering air temperature, improper coil circuiting, oversized orifice size in distributor, or the TXV not being directly connected (close coupled) to the distributor.

Control of Coils

Cooling capacity of water coils is controlled either by varying the rate of water flow or airflow. Water flow can be controlled by a three-way mixing, modulating, and/or throttling valve. For airflow control, face and bypass dampers are used. When cooling demand decreases, the coil face damper starts to close and the bypass damper opens. In some cases, airflow is varied by controlling the fan capacity with speed controls, inlet vanes, or discharge dampers.

Chapter 41 of the 1991 ASHRAE *Handbook—HVAC Applications* addresses controlling air-cooling coils to meet system or space requirements and factors to consider when sizing automatic valves for water coils. The selection and application of refrigerant flow-control devices—thermostatic expansion valves, capillary tube types, constant pressure expansion valves, evaporator pressure regulators, suction pressure regulators, and solenoid valves—as used with direct-expansion coils are discussed in Chapter 19 of the 1988 ASHRAE *Handbook—Equipment*.

For factory-assembled, self-contained packaged systems or field-assembled systems employing direct-expansion coils equipped with thermostatic expansion valves, a single valve is sometimes used for each coil; in other cases, two or more valves are used. The thermostatic expansion valve controls the refrigerant flow rate through the coil circuits so the refrigerant vapor at the coil outlet is superheated properly. Superheat is obtained with a suitable design of the coil and proper valve selection. Unlike water flow control valves, standard pressure/temperature-type thermostatic expansion valves alone do not control the refrigeration system's capacity or the temperature of the leaving air, nor do they maintain ambient conditions in specific spaces. However, some of the newer electronically controlled TXVs have these attributes.

To match the refrigeration load requirements for the conditioned space with the cooling capacity of the coil(s), a thermostat, located in the conditioned space(s) or in the return air, temporarily interrupts refrigerant flow to the direct-expansion cooling coils by stopping the compressor(s) and/or closing the solenoid liquid-line valve(s). Other solenoids unload compressors by means of suction control. For jobs with only a single zone of conditioned space, the compressor's on-off control is frequently used to modulate coil capacity. The selection and application of evaporator pressure regulators and similar regulators that are temperature-operated and respond to the temperature of the conditioned air are covered in Chapter 19 of the 1988 ASHRAE *Handbook—Equipment*.

Applications with multiple zones of conditioned space often use liquid solenoid valves to vary coil capacity. These valves should be used where thermostatic expansion valves feed certain types (or sections) of evaporator coils, which may (according to load variations) require a temporary, but positive, interruption of refrigerant flow. This applies particularly to multiples of evaporator coils in a unit where one or more must be shut off temporarily to regulate its zone capacity. In such cases, the solenoid valve should be installed directly upstream of the thermostatic expansion valve(s). As more than one expansion valve may feed a particular zone coil, they may be controlled by a single solenoid valve.

For coils with multiple refrigerant expansion valves, there are three arrangements: (1) face control, in which the coil is divided across its face; (2) row control; and (3) interlaced circuit (Figure 2).

Face control, which is most widely used because of its simplicity, equally loads all refrigerant circuits within the coil. Face control has the disadvantage of permitting reevaporation of condensate on the coil portion not operating and bypassing air into the conditioned space during partial load conditions when some of the thermostatic expansion valves are on an off cycle. However, if properly arranged, some single-zone humidity control advantages are realized with air bypasses through the inactive top portion of the coil, while the bottom coil portion is cooling.

Row control, seldom available as standard equipment, eliminates air bypassing during partial load operation and minimizes conden-

sate reevaporation. Close attention is required for accurate calculation of row-depth capacity, circuitry design, and TXV sizing.

Interlaced circuit control uses whole face area and depth of coil when some of the expansion valves are shut off. Without a corresponding drop in airflow, modulating the refrigerant flow to an interlaced coil produces an increased coil surface temperature, thereby necessitating compressor protection, *e.g.*, suction pressure regulators or compressor multiplexing.

Flow Arrangement

In the air-conditioning process, the relation of the fluid flow arrangement (within the coil tubes) to the coil depth greatly influences the heat transfer surface performance. Generally, air-cooling and dehumidifying coils are multirow and circuited for counterflow arrangement. The inlet air is applied at right angles to the coil's tube face (coil height), which is also at the coil's outlet header location. The air exits at the opposite face (side) of the coil where the corresponding inlet header is located. Counterflow can produce the highest possible heat exchange within the shortest possible (coil row) depth because it has the fewest temperature relationships between tube fluid and air at each (air) side of the coil; temperature of the entering air more closely approaches the leaving fluid, as in the leaving air to the entry fluid. The potential of realizing the highest possible mean temperature difference is thus arranged for optimum performance.

Most direct expansion coils also follow this general scheme of thermal counterflow, but the circuiting requirements for proper superheat control may result in requiring a hybrid combination of parallel flow and counterflow. (Air flows in the same direction as the refrigerant in parallel flow operation.)

Quite often, optimum design for large coils is parallel flow arrangement in the coil's initial (entry) boiling region followed by counterflow in the superheat (exit) region. Parallel flow evaporator circuitry is commonly used for process applications that require a low temperature difference.

Coil hand refers to either the right hand (RH) or left hand (LH) for counterflow arrangement of a multirow counterflow coil. There is no convention of what constitutes LH or RH, so manufacturers usually establish a convention for their own coils. Most manufacturers designate the location of the inlet water header or refrigerant distributor as the coil hand reference point. Figure 3 illustrates the more widely accepted coil hand designation for multirow water or refrigerant coils.

Applications

A typical arrangement of coils in a field built-up central station system is shown in Figure 4. All air should be filtered to prevent dirt, insects, and foreign matter from accumulating on the coils. The cooling coil (and humidifier when used) should include a drain pan under each coil to catch the condensate formed during

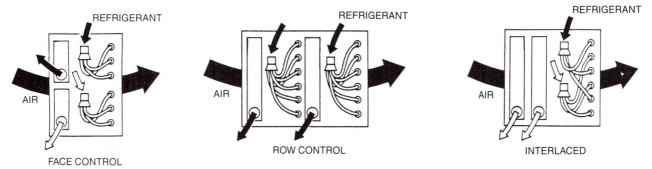

Fig. 2 Arrangements for Coils with Multiple Thermostatic Expansion Valves

the cooling cycle (and the excess water from the humidifier). The pan's drain connection should be on the downstream side of the coils, made of ample size, have accessible cleanouts, and discharge to an indirect waste or storm sewer. The drain also requires a deep-seal trap so sewer gas cannot enter the system. Precautions must be taken if the drain might freeze. The drain pan, unit casing, and water piping should be insulated to prevent sweating.

Factory-assembled central station air-handling units incorporate the design features outlined above. Generally, these packaged units will accommodate various sizes, types, and row depths of cooling and heating coils to meet most job requirements. This usually eliminates the need for field built-up central systems, except on very large jobs.

The design features of the coil (fin spacing, tube spacing, face height, type of fins), together with the amount of moisture on the coil and the degree of surface cleanliness, determine the air velocity at which condensed moisture is blown off the coil. Generally, condensate water begins to be blown off a plate fin coil face at air velocities above 600 fpm. Water blown off the coils into air ductwork external to the air-conditioning unit should be avoided. However, water blowoff from the coils is not usually a problem if coil fin heights are limited to 45 in. and arrangements are included within the unit to catch and dispose of the condensate. When a number of coils are stacked one above another, the condensate is carried into the airstream as it endeavors to drip from one coil to the next. To prevent this, a downstream eliminator section could be used, but it is preferable to use an intermediate drain pan and/or condensate trough to collect the condensate and conduct it directly to the main drain pan (Figure 5). Extending downstream of the coil, each drain pan length should be least equal to one-half the coil height; drain pans should be somewhat longer when coil airflow face velocities and/or humidity levels are higher.

When water is likely to carry over from the air-conditioning unit into external air ductwork, and no other means of prevention is provided, eliminator plates should be installed on the downstream side of the coils. Usually, eliminator plates are not included in packaged units because other means of preventing carryover are included in the design, such as space made available within the unit design for longer drain pan(s). On sprayed coil units, eliminators are usually included in the design. Such cooling and dehumidifying coils are sometimes sprayed with water to increase the rate of heat transfer, provide outlet air approaching saturation, and continually wash the surface of the coil. Coil sprays require a collecting tank, eliminators, and a recirculating pump (Figure 6). Also shown in Figure 6 is an air bypass, which helps a thermostat control hold the humidity ratio by bypassing a portion of the return air.

In field-assembled systems or factory-assembled central station air-handling units, the fans are usually positioned downstream from the coil(s) in a draw-through arrangement, which provides uniform airflow over the entire coil frontal area. For some multizone systems, particularly factory-assembled, self-contained unit designs, the fans are located upstream from the coil in a blow-through arrangement that may require air baffles or diffuser plates between the fan and the cooling coil to obtain uniform air distribution over the coil face.

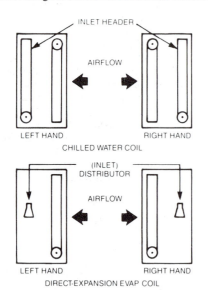

Fig. 3   Typical Coil Hand Designation

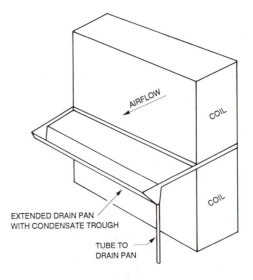

Fig. 5   Coil Bank Arrangement with Intermediate Condensate Pan

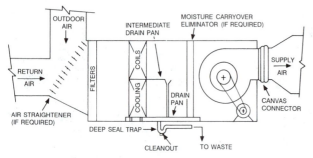

Fig. 4   Typical Arrangement of Cooling Coil Assembly in Built-Up or Packaged CSAC Unit

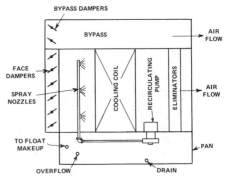

Fig. 6   Sprayed Coil System with Air Bypass

Where suction line risers are used for air-cooling coils in direct-expansion refrigeration systems, the suction line must be sized properly to ensure oil return from coil to compressor at minimum load conditions. The oil return provision is normally intrinsic with factory-assembled, self-contained air conditioners, but must be considered for factory-assembled central station units or field-installed cooling coil banks where suction line risers are required and are assembled at the job site. Sizing, design, and arrangement of suction lines and their risers are described in Chapter 3 of the 1990 ASHRAE *Handbook—Refrigeration*.

Air cooling and dehumidifying coil frames, as well as all drain pans and troughs, should be of an acceptable corrosive-resistant material suitable for the job and its expected useful service life. The air handler's coil section enclosure should be corrosion-resistant, properly double-wall insulated, and have adequate access doors for the changing of air filters, the cleaning of coils, the adjusting of flow control valves, and the maintenance of motors.

COIL SELECTION

When selecting a coil, the following factors should be considered:

1. The job requirements, *i.e.*, cooling, dehumidifying, and the capacity required to properly balance with other system components, such as compressor equipment in the case of direct-expansion coils.
2. The temperature conditions of the entering air.
3. Available cooling media and operating temperatures.
4. Space and dimensional limitations.
5. Air quantity and limitations.
6. Allowable frictional resistances in air circuit (including coils).
7. Allowable frictional resistances in cooling media piping system (including coils).
8. Characteristics of individual coil designs and circuitry possibilities.
9. Individual installation requirements, *e.g.*, type of automatic control to be used, presence of corrosive atmosphere, design pressures, as well as durability of tube, and fins and frame material.

Air quantity is affected by such factors as design parameters, codes, space, and equipment. The resistance through the air circuit influences the fan power and speed. This resistance may be limited to allow the use of a given size fan motor, to keep the operating expense low, or because of sound-level requirements. The air friction loss across the cooling coil—in summation with other series air pressure drops for such elements as air filters, water sprays, heating coils, air grilles, and ductwork—determines the static pressure requirement for the complete airway system. The static pressure requirement is used in selecting the fans and drives to obtain the design air quantity under operating conditions. See Chapter 18 for a description of fan selection.

The conditioned air face velocity is determined by economic evaluation of initial and operating costs for the complete installation as influenced by (1) heat transfer performance of the specific coil surface type for various combinations of face areas and row depths as a function of the air velocity; (2) air-side frictional resistance for the complete air circuit (including coils), which affects fan size, power, and sound-level requirements; and (3) condensate water carryover considerations. The allowable friction through the water or brine coil circuitry may be dictated by the head available from a given size pump and pump motor, as well as the same economic factors governing the air-side made applicable to the water side. Additionally, the adverse effects of high cooling water velocities on erosion-corrosion of tube walls is a major factor in sizing and circuiting to keep tube velocity below the recommended maximums for the tube material and water quality. On larger coils, water pressure drop limits of 15 to 20 ft usually keep such velocities within acceptable limits.

Coil ratings are based on uniform entry air conditions. Interference with uniform airflow through the coil affects performance. Airflow interference may be caused by the entrance of air at odd angles or by inadvertent blocking of a portion of the coil face. To obtain rated performance, the volumetric airflow quantity must be adjusted on the job to correspond to the application ratings of the coil selection and must be kept at that value. At the time of start-up for air balance, the most common causes of incorrect airflow are the lack of altitude correction to standard air (where applicable) and ductwork problems. The most common causes of an air quantity reduction are fouling of the filters and collection of dirt or frost on the coils. These difficulties can be avoided by proper design, start-up checkout, and regular servicing.

The required total heat capacity of the cooling coil should be in balance with the capacity of other refrigerant system components, such as the compressor, water chiller, condenser, and refrigerant liquid metering device. Methods of estimating balanced system capacity under various operating conditions when using direct expansion coils for both factory and field-assembled systems are described in Chapter 18 of the 1988 ASHRAE *Handbook—Equipment*.

In the case of dehumidifying coils, it is important that the proper amount of surface area be installed to obtain the ratio of air-side sensible-to-total heat required for maintaining the air dry-bulb and wet-bulb temperatures in the conditioned space. This is an important consideration when preconditioning is done by heat pipe arrangement. The method for calculating the sensible and total heat loads and the leaving air conditions at the coil to satisfy the sensible-to-total heat ratio required for the conditioned space is given in Chapter 26 of the 1989 ASHRAE *Handbook—Fundamentals*.

The same room air conditions can be maintained with different air quantities (including outside and return air) through a coil. However, for a given total air quantity with fixed percentages of outside and return air, there is only one set of air conditions leaving the coil which will precisely maintain the room's design air conditions. Once the air quantity and leaving air conditions at the coil have been selected, there is usually only one combination of face area, row depth, and air face velocity for a given coil surface which will precisely maintain the required room air conditions. Therefore, in making final coil selections, it is necessary to recheck the initial selection to ensure that the leaving air conditions, as calculated by a coil selection computer program or other procedure, will match those as determined from the cooling load estimate.

Coil ratings and selections can be obtained from manufacturers' catalogs. Most catalogs contain extensive tables giving the performance of coils at various air and water velocities and entering humidity and temperatures. Most manufacturers provide computerized coil selection to potential customers. The final choice can then be made based on system performance and economic requirements.

Performance and Ratings

The desired long-term performance of an extended surface air-cooling and dehumidifying coil depends on its correct design to specified conditions and material specifications, as well as its proper matching to other system components, proper installation, and maintenance as required.

In accordance with ARI *Standard* 410-87, Forced-Circulation Air-Cooling and Air-Heating Coils, dry surface (sensible cooling) coils and dehumidifying coils (which accomplish both cooling and dehumidification), particularly those used for field-assembled coil banks or factory-assembled packaged units using different combinations of coils, are usually rated within the following parameters:

Entering air dry-bulb temperature: 65 to 100 °F
Entering air wet-bulb temperature: 60 to 85 °F (Because air is not dehumidified for this application, select coils based on sensible heat transfer.)
Air face velocity: 200 to 800 fpm
Evaporator refrigerant saturation temperature: 30 to 55 °F at coil suction outlet (refrigerant vapor superheat at coil suction outlet is 6 °F or higher)
Entering chilled water temperature: 35 to 65 °F
Water velocity: 1 to 8 ft/s
For cold ethylene glycol solution: 1 to 6 ft/s, 0 to 90 °F entering dry-bulb temperature, 60 to 80 °F entering wet-bulb temperature, 10 to 60% aqueous glycol concentration by weight.

The ratio of air-side sensible-to-total heat removed by dehumidifying coils varies in practice from about 0.6 to 1.0, *i.e.*, sensible heat is from 60 to 100% of the total, depending on the application (see Chapter 26 of the 1989 ASHRAE *Handbook—Fundamentals*). For a given coil surface design and arrangement, the required sensible heat ratio may be satisfied by wide variations in and combinations of air face velocity, in-tube temperature and flow rate, entering air temperature, coil depth, etc., although the variations may be self-limiting. The maximum coil air face velocity should be limited to a value that prevents water carryover into the air ductwork. Dehumidifying coils for comfort application are frequently selected in the range of 400 to 500 fpm air face velocity.

The operating ratings of dehumidifying coils for factory-assembled, self-contained air conditioners are generally determined in conjunction with laboratory testing for the system capacity of the complete unit assembly. As an example, a standard rating point has been 33.4 cfm/1000 Btu/h (or 400 cfm/ton of refrigeration effect). Refrigerant duty, *e.g.*, R-22, would be for an appropriate balance at 45 °F saturated suction, 6 to 10 °F superheat. For water coils, circuitry operating at 4 ft/s, 42 °F inlet water, 12 °F rise (or 2 gal per ton of refrigeration effect). The standard ratings at 80 °F dry bulb and 67 °F wet bulb are representative of the entering air conditions encountered in many comfort operations. While the indoor conditions are usually lower than 67 °F wet bulb, the introduction of outdoor air usually brings the mixture of air to the cooling coil up to about 80 °F dry bulb/67 °F wet bulb entering air design conditions.

Dehumidifying coils for field-assembled projects and central station air-handling units are usually selected according to coil rating tables. Selecting coils from the load division indicated by the load calculation works satisfactorily for the usual human comfort applications. Additional design precautions and refinements are necessary for more exacting industrial applications and for all types of air conditioning in humid areas. One such refinement uses a separate cooling coil to cool and dehumidify the ventilation air before mixing it with recirculated air. This dual-path air process removes what is usually the main source of moisture, *i.e.*, makeup outside air. Reheat is required for some industrial applications as well as finding greater use in commercial and comfort applications.

Airflow ratings are based on standard air of 0.075 lb/ft³ at 70 °F and a barometric pressure of 29.92 in. Hg. In some developed, mountainous areas with a sufficiently large market, coils are rated for their particular altitude.

When checking the operation of dehumidifying coils in various air-conditioning installations, climatic conditions must be considered. Most problems are encountered at light load conditions, when the cooling requirement is considerably less than at design conditions. In hot, dry climates, where the outdoor dew points are consistently low, dehumidifying is not generally a problem, and the light load design point condition does not pose

any special problems. In hot, humid climates, the light load condition has a higher proportion of moisture and a correspondingly lower proportion of sensible heat. The result is higher dew points in the conditioned spaces during light load conditions, unless special means for controlling the inside dew points are used, *e.g.*, reheat or dual path.

Fin surface freezing at light loads should be avoided. Freezing occurs when a dehumidification coil's surface temperature falls below 32 °F. Freezing does not occur with standard coils for comfort installations, unless the refrigerant evaporating temperature at the coil outlet is below 25 to 28 °F saturated; the exact value depends on the design of the coil, its operating dew point, and the amount of loading. Although it is not customary to choose coil and condensing units to balance at low temperatures at peak loads, freezing may occur when the load suddenly decreases. The danger of this type of surface freezing is increased if a bypass is used, causing less air to be passed through the coil at light loads.

AIRFLOW RESISTANCE

A cooling coil's airflow resistance (air friction) depends on the tube pattern and fin geometry (tube size and spacing, fin configuration, and number of in-line or staggered rows), the coil face velocity, and the amount of moisture on the coil. The coil air friction may also be affected by the degree of aerodynamic cleanliness of the coil core; burrs on fin edges may increase coil friction and increase the tendency to pocket dirt or lint on the faces. A completely dry coil, removing only sensible heat, offers less resistance to airflow than a dehumidifying coil removing both sensible and latent heat.

For a given surface and airflow, an increase in the number of rows, or number of fins (a decrease in fin spacing) increases the airflow resistance. Therefore, the final selection involves the economic balancing of the initial cost of the coil against the operating costs of the coil geometry combinations available to adequately meet the performance requirements.

HEAT TRANSFER

The heat transmission rate of air passing over a clean tube (with or without extended surface) to a fluid flowing within it, is impeded principally by three thermal resistances. The first is from the air to the surface of the exterior fin and tube assembly known as the surface air-side film thermal resistance. The second is the combination of the fin and tube for the metal thermal resistance to the conductance of heat through the exterior fin and tube assembly. The third is the intube fluid-side film thermal resistance that impedes the flow of heat between the internal surface of the metal and the fluid flowing within the tube. For some applications, an additional thermal resistance is factored in to account for external and/or internal surface fouling. Usually, the combination of the metal and tube-side film resistance is low compared to the air-side surface resistance. Where an application allows an economical use of materials, the mass and size of the coil can be reduced. In addition, the exterior surface resistance is reduced to nearly the same as the fluid-side resistance. External as well as internal tube fins (or internal turbulators) can economically decrease overall heat transfer surface resistances. Water sprays applied to a particular coil surface may increase the overall heat transfer slightly, although they may better serve other purposes such as air and coil cleaning.

The transfer of heat between the cooling medium and the airstream is influenced by the following variables:

- Temperature difference between fluids
- Design and surface arrangement of the coil
- Velocity and character of the airstream
- Velocity and character of the coolant in the tubes

With water coils, only the water temperature rises. With coils for volatile refrigerants, an appreciable pressure drop and a corresponding change in evaporating temperature through the refrigerant circuit often occurs. The rating of direct-expansion coils is further complicated because refrigerant evaporates in part of the circuit and superheats the remainder. Thus, for halocarbon refrigerants, a cooling coil is tested and rated with a specific distributing and liquid-metering device, and the capacities are stated for a given superheat condition of the leaving vapor.

At the same air mass velocity, varying performance can be obtained, depending on the turbulence of airflow into the coil and the uniformity of air distribution over the coil face. The latter is necessary to obtain reliable test ratings and realize rated performance in actual installations. The air resistance through the coils assists in distributing the air properly, but the effect is frequently inadequate where the inlet duct connections are brought in at sharp angles to the coil face. Reverse air currents may pass through a portion of the coils. These currents reduce the capacity but can be avoided with proper inlet air vanes or baffles. Air blades may also be required. Remember that coil performance ratings represent optimum conditions resulting from adequate and reliable laboratory tests, i.e., ARI *Standard* 410-87. For cases when available data must be extended, or for the design of a single, unique installation or understanding the calculation progression, the following material and illustrative examples for calculating cooling coil performance are useful guides.

PERFORMANCE OF SENSIBLE COOLING COILS

The performance of sensible cooling coils depends on the following factors:

1. The overall coefficient U_o of sensible heat transfer between airstream and coolant fluid
2. The mean temperature difference Δt_m between airstream and coolant fluid
3. The physical dimensions of and data for the coil (such as A_a and A_o) with characteristics of the heat transfer surface

The sensible heat cooling capacity of a given coil is expressed by the following equation:

$$q_{td} = U_o F_s A_a N_r \Delta t_m \qquad (1a)$$

where

$$F_s = A_o/(A_a)N_r \qquad (1b)$$

Assuming no extraneous heat losses, the same amount of sensible heat is lost from the airstream:

$$q_{td} = w_a c_p (t_{a1} - t_{a2}) \qquad (2)$$

where

$$w_a = 60\rho_a A_a V_a$$

The same amount of sensible heat is absorbed by the coolant, and for a nonvolatile type is:

$$q_{td} = w_r c_r (t_{r2} - t_{r1}) \qquad (3)$$

The mean temperature difference in Equation (1) for a nonvolatile coolant in thermal counterflow with the air, based on the logarithmic value, is expressed as:

$$\Delta t_m = \frac{(t_{a1} - t_{r2}) - (t_{a2} - t_{r1})}{\ln[(t_{a1} - t_{r2})/(t_{a2} - t_{r1})]} \qquad (4)$$

The proper temperature differences for various crossflow situations are given in many texts, including that by Mueller (1973). These calculations are based on various assumptions, among which is that U is constant throughout. While this assumption generally is not valid for multirow coils, the use of crossflow temperature differences from Mueller (1973) or other texts should be preferable to Equation (4), which applies only to the counterflow. However, the use of the log mean temperature difference is widespread.

The overall heat transfer coefficient U_o for a given coil design, whether bare-pipe or finned-type, with clean, nonfouled surfaces consists of the combined effect of three individual heat transfer coefficients:

1. The film coefficient of sensible heat transfer f_a between air and the external surface of the coil
2. The unit conductance $(1/R_{md})$ of the coil material, i.e., tube wall, fins, etc.
3. The film coefficient of heat transfer (f_r) between the internal coil surface and the coolant fluid within the coil

These three individual coefficients acting in series form an overall coefficient of heat transfer in accordance with the material given in Chapters 3 and 20 of the 1989 ASHRAE *Handbook—Fundamentals*. For a bare-pipe coil, the overall coefficient of heat transfer for sensible cooling (without dehumidification) can be expressed by a simplified basic equation:

$$U_o = \frac{1}{(1/f_a) + (D_o - D_i)/24k + (B/f_r)} \qquad (5a)$$

When pipe or tube walls are thin and made of material with high conductivity (as in typical heating and cooling coils), the term $(D_o - D_i)/24k$ in Equation (5a) frequently becomes negligible and generally is disregarded. (This effect in typical bare-pipe cooling coils seldom exceeds 1 to 2% of the overall coefficient.) Thus, in its simplest form, for bare pipe:

$$U_o = \frac{1}{(1/f_a) + (B/f_r)} \qquad (5b)$$

For finned coils, the equation for the overall coefficient of heat transfer can be written:

$$U_o = \frac{1}{(1/\eta f_a) + (B/f_r)} \qquad (5c)$$

where η, called the *fin effectiveness*, allows for the resistance to heat flow encountered in the fins. It is defined as:

$$\eta = (EA_s + A_p) / A_o \qquad (6)$$

For typical cooling surface designs, the surface ratio B ranges from about 1.03 to 1.15 for bare-pipe coils and 10 to 30 for finned coils. Chapter 3 in the 1989 ASHRAE *Handbook—Fundamentals* describes how to estimate fin efficiency E.

The tube-side heat transfer coefficient f_r for nonvolatile fluids can be calculated with the usual equations for flow inside tubes, as shown in Chapter 3 of the 1989 ASHRAE *Handbook—Fundamentals*. Estimation of the air-side heat transfer coefficient f_a is more difficult because well-verified general predictive techniques are not available. Hence, direct use of experimental data is usually necessary. For plate fin coils, some correlations that satisfy several data sets are available (McQuiston 1981, Kusuda 1969). Webb (1980) has reviewed the air-side heat transfer and pressure drop correlations for various geometries. Mueller (1973) and Chapter 3 of the 1989 ASHRAE *Handbook—Fundamentals* provide guidance on this subject.

For analyzing a given heat exchanger, the concept of *effectiveness* is useful. Expressions for effectiveness have been derived for various flow configurations and can be found in Mueller (1973) and Kusuda (1969). The cooling coils covered in this chapter actually involve various forms of crossflow. However, the case of counterflow is addressed here to illustrate the value of this concept. The effectiveness for counterflow heat exchangers is given by the following equations:

$$q_{td} = w_a c_p (t_{a1} - t_{r1}) E_a \qquad (7a)$$

where

$$E_a = \frac{t_{a1} - t_{a2}}{t_{a1} - t_{r1}} \qquad (7b)$$

or

$$E_a = \frac{(1 - e^{-c_o(1 - M)})}{1 - M e^{-c_o(1 - M)}} \qquad (7c)$$

$$c_o = \frac{A_o U_o}{w_a c_p} = \frac{F_s N_r U_o}{60 \rho_a V_a c_p} \qquad (7d)$$

and

$$M = \frac{w_a c_p}{w_r c_r} = \frac{60 \rho_a A_a V_a c_p}{w_r c_r} \qquad (7e)$$

Note the two following special conditions:

If $M = 0$, then $E_a = 1 - e^{-c_o}$
If $M \geq 1$, then

$$E_a = \frac{1}{(1/c_o) + 1}$$

With a given design and arrangement of heat transfer surface used as cooling coil core material for which basic physical and heat transfer data are available to determine U_o from Equations (5a), (5b), and (5c), the selection, sizing, and performance calculation of sensible cooling coils for a particular application generally fall within either of two categories:

1. The heat transfer surface area A_o or the coil row depth N_r for a specific coil size is required and initially unknown. The sensible cooling capacity q_{td}, flow rates for both air and coolant, entrance and exit temperatures of both fluids, and mean temperature difference between fluids are initially known or can be assumed or determined from Equations (2) to (4). The required surface area A_o or coil row depth N_r can then be calculated directly from Equation (1a).
2. The sensible cooling capacity q_{td} is initially unknown and required for a specific coil. The face area and heat transfer surface area are known or can be readily determined. The flow rates for air and coolant and the entering temperatures of each are also known. The mean temperature difference Δt_m is unknown, but its determination is unnecessary to calculate the sensible cooling capacity q_{td}, which can be found directly by solving Equation (7a). Equation (7a) also provides a basic means to determine q_{td} for a given coil or related family of coils, over the complete rating ranges of both air and coolant flow rates and operating temperatures.

The two categories of application problems are illustrated in Examples 1 and 2, respectively:

Example 1. Standard air flowing at a mass rate equivalent to 9000 cfm is to be cooled from 85 to 75 °F, using 330 lb/min chilled water supplied at 50 °F in thermal counterflow arrangement. Assuming an air face velocity of $V_a = 600$ fpm and no air dehumidification, calculate the coil face area A_a, sensible cooling capacity q_{td}, required heat transfer surface area A_o, coil row depth N_r, and coil air-side pressure drop Δp_{st}, using a clean, nonfouled, thin-walled bare copper tube surface design for which the following physical and performance data have been predetermined:

$$\begin{aligned}
B &= 1.07 \text{ surface ratio} \\
\rho_a &= 0.075 \text{ lb/ft}^3 \\
c_p &= 0.24 \text{ Btu/lb} \cdot {}^\circ\text{F} \\
c_r &= 1.0 \text{ Btu/lb} \cdot {}^\circ\text{F} \\
F_s &= 1.34 \text{ (external surface area)/(face area)(rows deep)} \\
f_a &= 15 \text{ Btu/h} \cdot \text{ft}^2 \cdot {}^\circ\text{F} \\
f_r &= 800 \text{ Btu/h} \cdot \text{ft}^2 \cdot {}^\circ\text{F} \\
\Delta p_{st}/N_r &= 0.027 \text{ in. of water/rows deep}
\end{aligned}$$

Solution: Calculate the coil face area required.

$$A_a = 9000/600 = 15 \text{ ft}^2$$

Neglecting the effect of tube wall, from Equation (5b):

$$U_o = \frac{1}{(1/15) + (1.07/800)} = 14.7 \text{ Btu/h} \cdot \text{ft}^2 \cdot {}^\circ\text{F}$$

From Equation (2), the sensible cooling capacity is:

$$q_{td} = 60 \times 0.075 \times 15 \times 600 \times 0.24(85 - 75) = 97,200 \text{ Btu/h}$$

From Equation (3):

$$t_{r2} = 50 + 97,200/(330 \times 60 \times 1) = 54.9 {}^\circ\text{F}$$

From Equation (4):

$$\Delta t_m = \frac{(85 - 54.9) - (75 - 50)}{\ln[(85 - 54.9)/(75 - 50)]} = 27.5 {}^\circ\text{F}$$

From Equations (1a) and (1b), the surface area required is:

$$A_o = 97,200/(15 \times 27.5) = 236 \text{ ft}^2 \text{ external surface}$$

From Equation (1b), the required row depth is:

$$N_r = 236/(1.34 \times 15) = 11.7 \text{ rows deep}$$

The installed 15 ft^2 coil face, 12 rows deep, slightly exceeds the required capacity. The air-side pressure drop for the installed row depth is then:

$$\Delta p_{st} = (\Delta p_{st}/N_r)N_r = 0.027 \times 12 = 0.32 \text{ in. of water at } 70 {}^\circ\text{F}$$

In Example 1, for some applications where such items as V_a, w_r, t_{r1}, f_r, etc., may be arbitrarily varied with a fixed design and arrangement of heat transfer surface, a trade-off between coil face area A_a and coil row depth N_r is sometimes made to obtain alternate coil selections that would produce the same sensible cooling capacity q_{td}. In the above solution, e.g., an eight-row coil could be selected, but would require a larger face area A_a with lower air face velocity V_a and a lower air-side pressure drop Δp_{st}.

Example 2. An air-cooling coil using a finned tube-type heat transfer surface has physical data as follows:

$$\begin{aligned}
A_a &= 10 \text{ ft}^2 \\
A_o &= 800 \text{ ft}^2 \text{ external} \\
B &= 20 \text{ surface ratio} \\
F_s &= 27 \text{ (external surface area)/(face area)(rows deep)} \\
N_r &= 3 \text{ rows deep}
\end{aligned}$$

Air at a face velocity of $V_a = 800$ ft/min and 95 °F entering air temperature is to be cooled by 15 gal/min of well water supplied at 55 °F (8.34 lb/gal). Calculate the sensible cooling capacity q_{td}, leaving air temperature t_{a2}, leaving water temperature t_{r2}, and air-side pressure drop Δp_{st}. Assume clean and nonfouled surfaces, thermal counterflow between air and water, no air dehumidification, standard barometric air pressure, and that the following data are available or can be predetermined:

c_p = 0.24 Btu/lb·°F
ρ_a = 0.075 lb/ft³
ρ_w = 62.4 lb/ft³
f_a = 17 Btu/h·ft²·°F
f_r = 500 Btu/h·ft²·°F
η = 0.9 fin effectiveness
$\Delta p_{st}/N_r$ = 0.22 in. of water/rows deep

Solution: From Equation (5c),

$$U_o = \frac{1}{1/(0.9 \times 17) + (20/500)} = 9.49 \text{ Btu/h·ft}^2 \cdot °F$$

From Equation (7d):

$$c_o = \frac{800 \times 9.49}{60 \times 0.075 \times 10 \times 800 \times 0.24} = 0.879$$

From Equation (7e):

$$M = \frac{60 \times 0.075 \times 10 \times 800 \times 0.24}{60 \times 15 \times 8.34 \times 1} = 1.15$$

From Equation (7c):

$$-0.879(1 - 1.15) = 0.133$$

and

$$E_a = \frac{1 - e^{0.133}}{1 - 1.15e^{0.133}} = 0.454$$

From Equation (7a), the sensible cooling capacity is:

$$q_{td} = 60 \times 0.075 \times 10 \times 800 \times 0.24(95 - 55) \times 0.454 = 157{,}000 \text{ Btu/h}$$

From Equation (2), the leaving air temperature is:

$$t_{a2} = 95 - \frac{157{,}000}{60 \times 0.075 \times 10 \times 800 \times 0.24} = 76.8 °F$$

From Equation (3), the leaving water temperature is:

$$t_{r2} = 55 + \frac{157{,}000}{15 \times 60 \times 8.34 \times 1} = 75.9 °F$$

The air-side pressure drop is:

$$\Delta p = 0.22 \times 3 = 0.66 \text{ in. of water}$$

The preceding equations and two illustrative examples demonstrate the method for calculating thermal performance of sensible cooling coils that operate with a dry surface. However, when cooling coils operate wet or act as dehumidifying coils, the performance cannot be predicted unless the effect of air-side moisture transport (or latent heat removal) is included.

PERFORMANCE OF DEHUMIDIFYING COILS

A dehumidifying coil normally removes both moisture and sensible heat from entering air. In most air-conditioning processes, the air to be cooled is a mixture of water vapor and dry gases. Both lose sensible heat during contact with the first part of the cooling coil, which functions as a dry cooling coil. Moisture is removed only in the part of the coil that is below the dew point of the entering air. Figure 1 in Chapter 2 shows the assumed psychrometric conditions of this process. As the leaving dry-bulb temperature is lowered below the entering dew-point temperature, the difference between the leaving dry-bulb temperature and the leaving dew point for a given coil, airflow, and entering air condition is lessened.

When the coil starts to remove moisture, the cooling surfaces carry both the sensible and latent heat load. As the air approaches saturation, each degree of sensible cooling is nearly matched by a corresponding degree of dew-point decrease. In contrast, the latent heat removal per degree of dew-point change varies considerably. The following table compares the amount of moisture removed from saturated air that is cooled from 60 to 59°F and air that is cooled from 50 to 49°F.

Dew Point	W_s, lb/lb	Dew Point	W_s, lb/lb $\times 10^3$
60°F	0.0011087	50°F	7.661
59°F	0.0010692	49°F	7.378
Difference	0.000395	Difference	0.283

These numerical values conform to Table 1 in Chapter 6 of the 1989 ASHRAE *Handbook—Fundamentals*.

The combination of the coil with its refrigerant distributor equipment must be tested at the higher and lower capacities of its rated range. Testing at the lower capacities checks whether the refrigerant distributor provides equal distribution and if the control is able to modulate without hunting. Testing at the higher capacities checks the maximum feeding capacity of the flow control device at the greater pressure drop that occurs in the coil system.

Most manufacturers develop and produce their own performance rating tables from data obtained from suitable tests. ASHRAE *Standard* 33 specifies the acceptable method of testing coils. ARI *Standard* 410-87 gives the acceptable method for rating thermal performance or dehumidifying coils by extending test data to other operating conditions, coil sizes, and row depths. To account for simultaneous transfer of both sensible and latent heat from the airstream to the surface, ARI *Standard* 410-87 uses essentially the same method for arriving at cooling and dehumidifying coil thermal performance, as determined by McElgin and Wiley (1940).

The potential or driving force for transferring total heat from airstream to tube-side coolant is composed of two components in series heat flow, *i.e.*, (1) an air-to-surface air enthalpy difference and (2) a surface-to-coolant temperature difference.

Figure 7 is a typical thermal diagram for a coil in which the air and a nonvolatile coolant are arranged in counterflow. The top and bottom lines in the diagram indicate, respectively, changes across the coil in the airstream enthalpy h_a and the coolant temperature t_r. To illustrate continuity, the single middle line in Figure 7 represents both surface temperature t_s and the corresponding saturated air enthalpy h_s, although the temperature and air enthalpy scales do not actually coincide as shown. The differential surface area dA_w represents any specific location within the coil thermal diagram where operating conditions are such that the air-surface interface temperature t_s is lower than the local air dew-point temperature. Under these conditions, both sensible and latent heat are removed from the airstream, and the cooler surface actively condenses water vapor. The driving forces for the total heat q_t transferred from air-to-coolant are (1) an air enthalpy difference $(h_a - h_s)$ between airstream and surface interface and (2) a temperature difference $(t_s - t_r)$ across the surface metal and into the coolant.

Neglecting the enthalpy of the condensed water vapor leaving the surface and any radiation and convection losses, the total heat lost from the airstream in flowing over dA_w is:

$$dq_t = -w_a(dh_a) \tag{8}$$

The total heat transferred from the airstream to the surface interface, according to McElgin and Wiley (1940), is:

$$dq_t = \frac{(h_a - h_s)dA_w}{c_p R_{aw}} \tag{9}$$

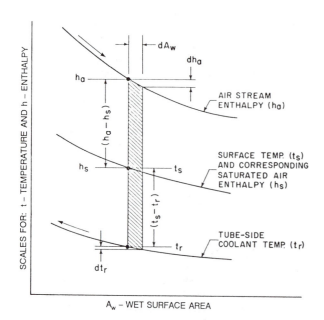

**Fig. 7   Two-Component Driving Force between
Dehumidifying Air Coolant**

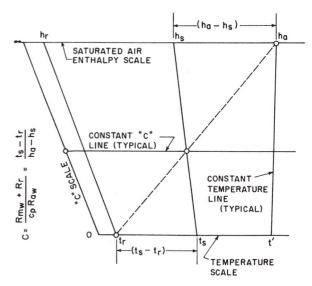

Fig. 8   Surface Temperature Chart

The total heat transferred from the air-surface interface across the surface elements and into the coolant is the same as given in Equations (8) and (9):

$$dq_t = \frac{(t_s - t_r)dA_w}{R_{mw} + R_r} \qquad (10)$$

The same quantity of total heat is also gained by the nonvolatile coolant in passing across dA_w:

$$dq_t = -w_r c_r (dt_r) \qquad (11)$$

If Equations (9) and (10) are equated and the terms rearranged, the expression for the coil characteristic C is obtained:

$$C = \frac{R_{mw} + R_r}{c_p R_{aw}} = \frac{t_s - t_r}{h_a - h_s} \qquad (12)$$

Equation (12) shows the basic relationship of the two components of the driving force between air and coolant in terms of the three principal thermal resistances. For a given coil, these three resistances of air, metal, and intube fluid (R_{aw}, R_{mw}, and R_r) are usually known or can be determined for the particular application, which gives a fixed value for C. Equation (12) can then be used to determine point conditions for the interrelated values of airstream enthalpy h_a; coolant temperature t_r; surface temperature t_s; and the enthalpy h_s of saturated air corresponding to the surface temperature. When both t_s and h_s are unknown, a trial-and-error solution is necessary; however, this can be solved graphically by a surface temperature chart, such as that shown in Figure 8.

Figure 9 shows a typical thermal diagram for a portion of the coil surface when it is operating dry. The illustration is for counterflow, using a halocarbon refrigerant. The diagram at the top of the figure illustrates a typical coil installation in an air duct with tube passes circuited countercurrent to airflow. Locations of the entering and leaving conditions for both air and coolant are shown.

The thermal diagram in Figure 9 is of the same type as that in Figure 7, showing three lines to illustrate local conditions for the air, surface, and coolant throughout a coil. The dry-wet boundary

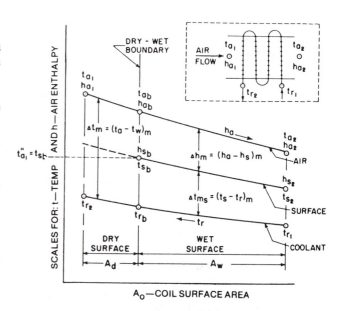

**Fig. 9   Thermal Diagram for General Case when Coil
Surface Operates Partially Dry**

conditions are located where the coil surface temperature t_{sb} equals the entering air dew-point temperature t''_{a1}. Thus, the surface area A_d to the left of this boundary is dry, with the remainder of the coil surface area A_w operating wet.

When using halocarbon refrigerants in a thermal counterflow arrangement, as illustrated in Figure 9, the dry-wet boundary conditions can be determined from the following relationships:

$$y = \frac{t_{r1} - t_{r2}}{h_{a1} - h_{a2}} = \frac{w_a}{w_r c_r} \qquad (13)$$

$$h_{ab} = \frac{t''_{a1} - t_{r2} + y h_{a1} + C h''_{a1}}{C + y} \qquad (14)$$

The value of h_{ab} from Equation (14) serves as an index of whether the coil surface is operating fully wetted, partially dry, or completely dry, according to the following three limits:

1. If $h_{ab} \geqslant h_{a1}$, all surface is fully wetted.
2. If $h_{a1} > h_{ab} > h_{a2}$, the surface is partially dry.
3. If $h_{ab} \leqslant h_{a2}$, all surface is completely dry.

Other dry-wet boundary properties are then determined as:

$$t_{sb} = t_{a1} \tag{15}$$

$$t_{ab} = t_{a1} - (h_{a1} - h_{ab}/c_p) \tag{16}$$

$$t_{rb} = t_{r2} - yc_p(t_{a1} - t_{ab}) \tag{17}$$

The dry surface area A_d requirements and capacity q_{td} are calculated by conventional sensible heat transfer relationships, as follows:

The overall thermal resistance (R_o) is comprised of three basic elements:

$$R_o = R_{ad} + R_{md} + R_r \tag{18}$$

where

$$R_r = B/f_r \tag{19}$$

The mean temperature difference between air dry bulb and coolant, using symbols from Figure 9, is:

$$\Delta t_m = \frac{(t_{a1} - t_{r2}) - (t_{ab} - t_{rb})}{\ln[(t_{a1} - t_r)/(t_{ab} - t_{rb})]} \tag{20}$$

The dry surface area required is:

$$A_d = \frac{q_{td}R_o}{\Delta t_m} \tag{21}$$

The air-side total heat capacity is:

$$q_{td} = w_a c_p(t_{a1} - t_{ab}) \tag{22a}$$

or from the coolant side:

$$q_{td} = w_a c_r(t_{r2} - t_{rb}) \tag{22b}$$

The wet surface area A_w and capacity q_{tw} are determined by the following relationships, using terminology in Figure 9:

For a given coil size, design, and arrangement, the fixed value of the coil characteristic C can be determined from the ratio of the three prime thermal resistances for the job conditions:

$$C = \frac{R_{mw} + R_r}{c_p R_{aw}} \tag{23}$$

Knowing the coil characteristic C for point conditions, the interrelations between the airstream enthalpy h_a, the coolant temperature t_r, the surface temperature t_s, and its corresponding enthalpy of saturated air h_s can be determined by use of a surface temperature chart (Figure 8) or by a trial-and-error procedure using Equation (24):

$$C = \frac{t_{sb} - t_{rb}}{h_{ab} - h_{sb}} = \frac{t_{s2} - t_{r1}}{h_{a2} - h_{s2}} \tag{24}$$

The mean effective difference in air enthalpy between airstream and surface from Figure 9 is:

$$\Delta h_m = \frac{(h_{ab} - h_{sb}) - (h_{a2} - h_{s2})}{\ln[(h_{ab} - h_{sb})/(h_{a2} - h_{s2})]} \tag{25}$$

Similarly, the mean temperature difference between surface and coolant is:

$$\Delta t_{ms} = \frac{(t_{sb} - t_{rb}) - (t_{s2} - t_{r1})}{\ln[(t_{sb} - t_{rb})/(t_{s2} - t_{r1})]} \tag{26}$$

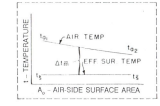
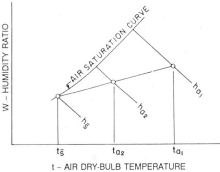

Fig. 10 Leaving Air Dry-Bulb Temperature Determination for Air-Cooling and Dehumidifying Coils

The wet surface area required, calculated from air-side enthalpy difference, is:

$$A_w = \frac{q_{tw}R_{aw}c_p}{\Delta h_m} \tag{27a}$$

or from the coolant side temperature difference:

$$A_w = \frac{q_{tw}(R_{mw} + R_r)}{\Delta t_{ms}} \tag{27b}$$

The air-side total heat capacity is:

$$q_{tw} = w_a[h_{ab} - (h_{a2} + h_{fw})] \tag{28a}$$

where

$$h_{fw} = (W_1 - W_2)c_{pw}(t_{a2}' - 32) \tag{28b}$$

and

$$c_{pw} = \text{specific heat of water} = 1.0 \text{ Btu/lb} \cdot {}^\circ\text{F}$$

Note that h_{fw} for normal air-conditioning applications is about 1% of the airstream enthalpy difference $(h_{a1} - h_{a2})$ and is neglected in many cases.

The coolant side heat capacity is:

$$q_{tw} = w_r c_r(t_{rb} - t_{r1}) \tag{28c}$$

The total surface area requirement of the coil is:

$$A_o = A_d + A_w \tag{29}$$

The total heat capacity for the coil is:

$$q_t = q_{td} + q_{tw} \tag{30}$$

The leaving air dry-bulb temperature is found by the method illustrated in Figure 10, which represents part of a psychrometric chart showing the air saturation curve and lines of constant air enthalpy closely corresponding to constant wet-bulb temperature lines.

For a given coil and air quantity, a straight line projected through the entering and leaving air conditions intersects the air saturation curve at a point denoted as the effective coil surface temperature $t_{\bar{s}}$. Thus, for fixed entering air conditions t_{a1}, h_{a1} and a given effective surface temperature $t_{\bar{s}}$, the leaving air dry bulb t_{a2} increases but is still located on this straight line if the air quantity is increased or if the coil depth is reduced. Conversely, a decrease in air quantity, or an increase in coil depth, produces a lower t_{a2} that is still located on the same straight line segment.

An index of the air-side effectiveness is the heat transfer exponent c, which is defined as:

$$c = \frac{A_o}{w_a c_p R_{ad}} \tag{31}$$

This exponent c, sometimes called the number of air-side transfer units NTU_a, is also defined as:

$$c = (t_{a1} - t_{a2})/\Delta t_{\tilde{m}} \tag{32}$$

The temperature drop $(t_{a1} - t_{a2})$ of the airstream and the mean temperature difference $\Delta t_{\tilde{m}}$ between air and effective surface in Equation (32) are illustrated in the small diagram at the top of Figure 10.

Knowing the exponent c and the entering h_{a1} and leaving h_{a2} enthalpies for the airstream, the enthalpy of saturated air $h_{\tilde{s}}$ corresponding to the effective surface temperature $t_{\tilde{s}}$ is calculated as follows:

$$h_{\tilde{s}} = h_{a1} - \frac{(h_{a1} - h_{a2})}{1 - e^{-c}} \tag{33}$$

After finding the value of $t_{\tilde{s}}$, which corresponds to $h_{\tilde{s}}$ from the saturated air enthalpy tables, the leaving air dry-bulb temperature can then be determined as:

$$t_{a2} = t_{\tilde{s}} + e^{-c}(t_{a1} - t_{\tilde{s}}) \tag{34}$$

The air-side sensible heat ratio can then be calculated as:

$$\mathrm{SHR} = \frac{c_p(t_{a1} - t_{a2})}{(h_{a1} - h_{a2})} \tag{35}$$

For the thermal performance of a coil to be determined from the foregoing relationships, values of the three principal resistances to heat flow between air and coolant (listed as follows) must be known:

1. The total metal thermal resistance across the fin R_f and tube R_t assembly for both dry R_{md} and wet R_{mw} surface operation
2. The air-film thermal resistance for dry R_{ad} and wet R_{aw} surface
3. The tube-side coolant film thermal resistance R_r

In ARI *Standard* 410-87, the metal thermal resistance R_m is calculated, knowing the physical data, material, and arrangement of the fin and tube elements, together with the fin efficiency E for the specific fin configuration. The metal resistance R_m is variable as a weak function of the effective air-side heat transfer coefficient f_a for a specific coil geometry, as illustrated in Figure 11.

For wetted surface application, Brown (1954), with certain simplifying assumptions, showed that the effective air-side coefficient f_a is directly proportional to the rate-of-change m'' of enthalpy for saturated air h_s with the corresponding surface temperature t_s. This slope m'' of the air enthalpy saturation curve is illustrated in the small inset graph at the top of Figure 11.

The abscissa for f_a in the main graph of Figure 11 is an effective value, which, for dry surface, is the simple thermal resistance reciprocal $1/R_{ad}$. For a wet surface, f_a is the product of the thermal resistance reciprocal $1/R_{aw}$ and the multiplying factor m''/c_p. ARI *Standard* 410-87 outlines a method for obtaining a mean value of (m''/c_p) for a given coil and job condition. The total metal resistance R_m in Figure 11 includes the resistance R_t across the tube wall. For most coil designs, R_t is quite small compared to the resistance through the fin metal R_f.

The air-side thermal resistances for dry R_{ad} and wet R_{aw} surfaces, together with their respective air-side pressure drops, $\Delta p_{st}/N_r$ and $\Delta p_{sw}/N_r$, are determined from tests on a represented

coil model over the full range in the rated airflow. Typical plots of the experimental data for these four performance variables versus coil air face velocity V_a at 70 °F are illustrated in Figure 12.

The heat transfer coefficient f_r of water used as the tube-side coolant is calculated from data in the literature [Equation (8), ARI *Standard* 410-87]. For evaporating refrigerants, many predictive techniques for calculating f_r are listed in Chapter 3 of the 1989 ASHRAE *Handbook—Fundamentals*. The most verified predictive technique is the Shah correlation (Shah 1976, 1982). A series of tests is specified in ARI *Standard* 410-87 to obtain heat transfer data for direct-expansion refrigerants inside tubes of a given diameter.

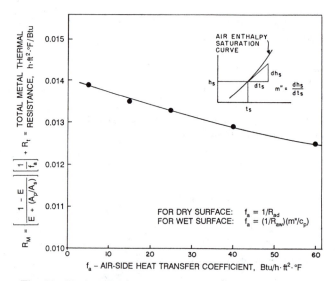

Fig. 11   Typical Total Metal Thermal Resistance of Fin and Tube Assembly

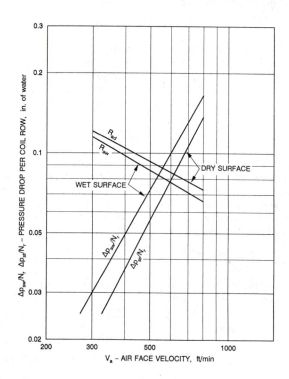

Fig. 12   Typical Air-Side Application Rating Data Experimentally Determined for Cooling and Dehumidifying Water Coils

ASHRAE *Standard* 33 specifies the laboratory apparatus and instrumentation and the procedure for conducting tests on representative coil prototypes to obtain basic performance data. Procedures are available in ARI *Standard* 410-87 for reducing this test data to the performance parameters necessary to rate a line or lines of various air coils.

The following example illustrates a method for selecting a coil size, row depth, and performance data to satisfy initially specified job requirements. The application is for a typical cooling and dehumidifying coil selection under conditions in which a part of the coil surface on the entering air side operates dry, with the remaining surface wetted with condensing moisture. Figure 9 shows the thermal diagram, dry-wet boundary conditions, and terminology used in the problem solution.

Example 3. Air at a mass rate equivalent to 6700 cfm (based on 70 °F standard conditions) enters a coil at 80 °F dry-bulb temperature t_{a1}' and 67 °F wet-bulb temperature t_{a1}' using 40 gpm (20,000 lb/h) of chilled water, supplied to the coil at an entering temperature t_{r1} of 44 °F, in thermal counterflow arrangement. Assume a standard coil air face velocity of $V_a = 558$ fpm and a clean, nonfouled, finned tube heat transfer surface in the coil core, for which the following physical and performance data (such as illustrated in Figures 11 and 12) can then be predetermined, thus:

$$B = 25.9 \text{ surface ratio}$$
$$c_p = 0.243 \text{ Btu/lb} \cdot °F$$
$$F_s = 32.4 \frac{\text{(external surface area)}}{\text{(face area)(row deep)}}$$
$$f_r = 750 \text{ Btu/h} \cdot \text{ft}^2 \cdot °F$$
$$\Delta p_{st}/N_r = 0.165 \text{ in. of water/row deep—dry surface}$$
$$\Delta p_{sw}/N_r = 0.27 \text{ in. of water/row deep—wet surface}$$
$$R_{ad} = 0.073 °F \cdot \text{ft}^2 \cdot \text{h/Btu}$$
$$R_{aw} = 0.066 °F \cdot \text{ft}^2 \cdot \text{h/Btu}$$
$$R_{md} = 0.021 °F \cdot \text{ft}^2 \cdot \text{h/Btu}$$
$$R_{mw} = 0.0195 °F \cdot \text{ft}^2 \cdot \text{h/Btu}$$

Referring to the symbols and typical diagram for applications in which only a part of the coil surface operates wet in Figure 9, determine: coil face area A_a; total refrigeration load q_t; leaving coolant temperature t_{r2}; dry-wet boundary conditions; heat transfer surface area required for dry A_d and wet A_w sections of the coil core; leaving air dry-bulb temperature t_{a2}; total number of installed coil rows deep N_{ri}; and dry Δp_{st} and wet Δp_{sw} coil air friction.

Solution: The psychrometric properties and enthalpies for dry and moist air are based on the ASHRAE Psychrometric Chart 1 and Tables 1 and 2 of Chapter 6 of the 1989 ASHRAE *Handbook—Fundamentals* as follows—

$h_{a1} = 31.52 \text{ Btu/lb}_a$	$h_{a1}'' = 26.67 \text{ Btu/lb}_a$
$W_1 = 0.0112 \text{ lb}_w/\text{lb}_a$	$^*h_{a2} = 23.84 \text{ Btu/lb}_a$
$t_{a1}'' = 60.3 °F$	$^*W_2 = 0.0095 \text{ lb}_w/\text{lb}_a$

*As an approximation, assume leaving air is saturated (*i.e.*, $t_{a2} = t_{a2}'$).

Calculate coil face area required:

$$A_a = 6700/558 = 12 \text{ ft}^2$$

From Equation (28b), find condensate heat rejection:

$$h_{fw} = (0.0112 - 0.0095)(56 - 32) = 0.04 \text{ Btu/lb}_a$$

Compute the total refrigeration load from the following Equation:

$$q_t = (60 \text{ min/h})\rho_a(\text{cfm})[h_{a1} - (h_{a2} + h_{fw})]$$
$$= 60 \times 0.075 \times 6700[31.52 - (23.84 + 0.04)] = 230,000 \text{ Btu/h}$$

From Equation (3), calculate coolant temperature leaving coil:

$$t_{r2} = t_{r1} + q_t/(20,000 \text{ lb/h})$$
$$= 44 + 230,000/(20,000) = 55.5 °F$$

From Equation (19), determine coolant film thermal resistance:

$$R_r = 25.9/750 = 0.0346 °F \cdot \text{ft}^2 \cdot \text{h/Btu}$$

Calculate the wet coil characteristic from Equation (23):

$$C = \frac{0.0195 + 0.0346}{0.243 \times 0.066} = 3.37 \text{ lb}_a \cdot °F/\text{Btu}$$

Calculate from Equation (13):

$$y = \frac{4.5 \times 6700}{20,000 \times 1} = 1.51 \text{ lb}_a \cdot °F/\text{Btu}$$

The dry-wet boundary conditions are determined as follows: The boundary airstream enthalpy from Equation (14) is:

$$h_{ab} = \frac{60.3 - 55.5 + 1.51 \times 31.52 + 3.37 \times 26.67}{3.37 + 1.51}$$
$$= 29.15 \text{ Btu/lb}_a$$

According to the limit (2) under Equation (14), a part of the coil surface on the entering air side will be operating dry, as shown in Figure 8, since $h_{a1} > 29.15 > h_{a2}$.

The boundary airstream dry-bulb temperature from Equation (16) is:

$$t_{ab} = 80 - (31.52 - 29.15)/0.243 = 70.25 °F$$

The boundary surface conditions are:

$$t_{sb} = t_a'' = 60.3 °F \quad \text{and} \quad h_{sb} = h_{a1}'' = 26.67 \text{ Btu/lb}_a$$

The boundary coolant temperature from Equation (17) is:

$$t_{rb} = 55.5 - 1.51 \times 0.243(80 - 70.25) = 51.92 °F$$

The refrigeration load for the dry surface part of the coil is now calculated from Equation (22b):

$$q_{td} = 20,000(55.5 - 51.92) = 71,600 \text{ Btu/h}$$

The overall thermal resistance for the dry surface section from Equation (18) is:

$$R_o = 0.073 + 0.021 + 0.0346 = 0.1286 °F \cdot \text{ft}^2 \cdot \text{h/Btu}$$

The mean temperature difference between air dry bulb and coolant for the dry surface section, from Equation (20), is:

$$\Delta t_m = \frac{(80 - 55.5) - (70.25 - 51.92)}{\ln[(80 - 55.5)/(70.25 - 51.92)]} = 21.25 °F$$

The dry surface area required is calculated from Equation (21) as:

$$A_d = 71,600 \times 0.1286/21.25 = 433 \text{ ft}^2$$

The refrigeration load for the wet surface section of the coil, calculated from Equation (30), is:

$$q_{tw} = 230,000 - 71,600 = 158,400 \text{ Btu/h}$$

Knowing C, h_{a2}, and t_{r1}, the surface condition at the leaving air side of the coil is calculated from Equation (24):

$$3.37 = (t_{s2} - 44)/(23.84 - h_{s2})$$

The numerical values for t_{s2} and h_{s2} are then determined directly by use of a surface temperature chart, as shown in Figure 8, or by trial-and-error procedure, using saturated air enthalpies from Table 1 of Chapter 6 of the 1989 ASHRAE *Handbook—Fundamentals*, thus:

$$t_{s2} = 52.03 °F \text{ and } h_{s2} = 21.46 \text{ Btu/lb}_a$$

The mean effective difference in air enthalpy between airstream and surface from Equation (25) is:

$$\Delta h_m = \frac{(29.15 - 26.67) - (23.84 - 21.46)}{\ln[(29.15 - 26.67)/(23.84 - 21.46)]} = 2.43 \text{ Btu/lb}_a$$

The wet surface area required, calculated from the air-side enthalpy difference from Equation (27a), is:

$$A_w = 158,400 \times 0.066 \times 0.243/2.43 = 1045 \text{ ft}^2$$

The net total surface area requirements for the coil, from Equation (29), is then:

$$A_o = 433 + 1045 = 1478 \text{ ft}^2 \text{ external}$$

The net air-side heat transfer exponent from Equation (31) is:

$$c = 1478/(4.5 \times 6700 \times 0.243 \times 0.073) = 2.76$$

The enthalpy of saturated air corresponding to the effective surface temperature from Equation (33) is:

$$h_{\bar{s}} = 31.52 - (31.52 - 23.84)/(1 - e^{-2.76}) = 23.32 \text{ Btu/lb}_a$$

The effective surface temperature, which corresponds to $h_{\bar{s}}$, is then obtained from the saturated air enthalpy tables as $t_{\bar{s}} = 55.15\,°F$. The leaving air dry-bulb temperature is calculated from Equation (34) as:

$$t_{a2} = 55.15 + e^{-2.76}(80 - 55.15) = 56.7\,°F$$

The air-side sensible heat ratio is then determined from Equation (35) as:

$$\text{SHR} = \frac{0.243(80 - 56.7)}{31.52 - 23.84} = 0.737$$

The calculated coil row depth to match job requirements is:

$$N_{rc} = A_o/A_aF_s = 1478/(12 \times 32.4) = 3.80 \text{ rows deep}$$

In most coil selection problems of the type illustrated, the initial calculated row depth to satisfy job requirements is usually a noninteger value. In many cases, there is sufficient flexibility in fluid flow rates and operating temperature levels to recalculate the required row depth of a given coil size to match an available integer row depth more closely. For example, if the initial calculated row depth were $N_{rc} = 3.5$ and three or four row deep coils were commercially available, operating conditions and/or fluid flow rates could possibly be changed to recalculate a coil depth close to either three or four rows. Also, alternate coil selections for the same job are often desirable by trading off coil face size and row depth.

Most manufacturer's have computer programs to run the iteration needed to predict operating values for the specific row depth chosen.

For Example 3, assume the initial coil selection, actually requiring 3.8 rows deep, is sufficiently refined to require no recalculation. Then select the next highest integral row depth to provide some safety factor (about 5% in this case) in surface area selection. Thus, the installed row depth N_{ri} is:

$$N_{ri} = 4 \text{ rows deep}$$

The completely wetted and completely dry air-side frictions are, respectively:

$$\Delta p_{sw} = (\Delta p_{sw}/N_{ri})N_{ri} = 0.27 \times 4 = 1.08 \text{ in. of water}$$

and

$$\Delta p_{st} = (\Delta p_{st}/N_{ri})N_{ri} = 0.165 \times 4 = 0.66 \text{ in. of water}$$

The amount of heat transfer surface area installed is:

$$A_{oi} = A_aF_sN_{ri} = 12 \times 32.4 \times 4 = 1555 \text{ ft}^2 \text{ external}$$

In summary, the final coil size selected and calculated performance data for Example 3 follows:

$A_a = 12 \text{ ft}^2$ coil face area
$N_{ri} = 4$ rows installed coil depth
$A_{oi} = 1555 \text{ ft}^2$ installed heat transfer surface area
$A_o = 1478 \text{ ft}^2$ required heat transfer surface area
$q_t = 230,000 \text{ Btu/h}$ total refrigeration load
$t_{r2} = 55.5\,°F$ leaving coolant temperature
$t_{a2} = 56.7\,°F$ leaving air dry-bulb temperature
$\text{SHR} = 0.737$ air sensible heat ratio
$\Delta p_{sw} = 1.08$ in. of water wet coil surface air friction
$\Delta p_{st} = 0.66$ in. of water dry coil surface air friction

DETERMINING REFRIGERATION LOAD

The following calculation of the refrigeration load shows a division of the true sensible and latent heat loss of the air, which is accurate within the limitations of the data. This division will not correspond to load determination obtained from approximate factors or constants.

The total refrigeration load q_t of a cooling and dehumidifying coil (or air washer) per unit mass of dry air is indicated on Figure 13 and consists of the following components:

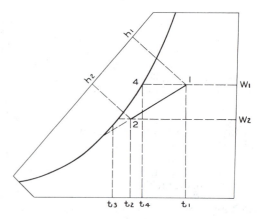

Fig. 13 Psychrometric Performance of Cooling and Dehumidifying Coil

1. The sensible heat q_s removed from the dry air and moisture in cooling from entering temperature t_1 to leaving temperature t_2
2. The latent heat q_e removed to condense the moisture at the dew-point temperature t_4 of the entering air
3. The heat required q_w to cool the condensate additionally from its dew point t_4 to its leaving condensate temperature t_3

Items 1, 2, and 3 are related as:

$$q_t = q_s + q_e + q_w \tag{36}$$

If only the total heat value is desired, it may be computed by:

$$q_t = (h_1 - h_2) - (W_1 - W_2)h_{f3} \tag{37}$$

where

h_1 and h_2 = enthalpy of air at points 1 and 2, respectively
W_1 and W_2 = humidity ratio at points 1 and 2, respectively
h_{f3} = enthalpy of saturated liquid at final temperature t_3

If a breakdown into latent and sensible heat components is desired, the following relations may be used.

The latent heat may be found from:

$$q_e = (W_1 - W_2)h_{fg4} \tag{38}$$

where

h_{fg4} = enthalpy representing the latent heat of water vapor at the condensing temperature t_4

The sensible heat may be shown to be:

$$q_s + q_w = (h_1 - h_2) - (W_1 - W_2)h_{g4} + (W_1 - W_2)(h_{f4} + h_{f3}) \tag{39a}$$

or

$$q_s + q_w = (h_1 - h_2) - (W_1 - W_2)(h_{fg4} + h_{f3}) \tag{39b}$$

where

$h_{g4} = h_{fg4} + h_{f4}$ = enthalpy of saturated water vapor at the condensing temperature t_4

h_{f4} = enthalpy of saturated liquid at the condensing temperature t_4

The last term in Equation (39a) is the heat of subcooling the condensate from the condensing temperature t_4 to its final temperature t_3. Then:

$$q_w = (w_1 - w_2)(h_{f4} - h_{f3}) \tag{40}$$

The final condensate temperature t_3 leaving the system is subject to substantial variations, depending on the method of coil installation as affected by coil face orientation, airflow direction, and air duct insulation. In practice, t_3 is frequently the same as the leaving wet-bulb temperature. Within the normal air-conditioning range, precise values of t_3 are not necessary as heat of the condensate $(W_1 - W_2)h_{f3}$, removed from the air, usually represents about 0.5 to 1.5% of the total refrigeration load.

MAINTENANCE

If the coil is to deliver its full cooling capacity, both its internal and external surfaces must be clean. The tubes generally stay clean in pressurized water or brine systems. Should large amounts of scale form when untreated water is used as coolant, chemical or mechanical cleaning of internal surfaces at frequent intervals is necessary. When coils use evaporating refrigerants, oil accumulation is possible and occasional checking and oil drainage is desirable. While outer tube surfaces can be cleaned in a number of ways, they are often washed with low-pressure water and mild detergent. The surfaces can also be brushed and cleaned with a vacuum cleaner. In cases of marked neglect—especially in restaurants, where grease and dirt have accumulated—it is sometimes necessary to remove the coils and wash off the accumulation with steam, compressed air and water, or hot water. The best practice, however, is to inspect and service the filters frequently.

LETTER SYMBOLS

A_a = coil face or frontal area, ft^2
A_d = dry external surface area, ft^2
A_o = total external surface area, ft^2
A_p = exposed external prime surface area, ft^2
A_s = external secondary surface area, ft^2
A_w = wet external surface area, ft^2
B = ratio of external to internal surface area, dimensionless
C = coil characteristic as defined in Equations (12) and (23), $\text{lb} \cdot °\text{F/Btu}$
c = heat transfer exponent, or NTU_a, as defined in Equations (31) and (32), dimensionless
c_o = heat transfer exponent, as defined in Equation (7d), dimensionless
c_p = air humid specific heat, $\text{Btu/lb}_a \cdot °\text{F}$
c_r = specific heat of nonvolatile coolant, $\text{Btu/lb}_a \cdot °\text{F}$
D_i = tube inside diameter, in.
D_o = tube outside diameter, in.
E = fin efficiency, dimensionless
E_a = air-side effectiveness defined in Equation (7b), dimensionless
$\ln$ = Naperian logarithm base, dimensionless
F_s = coil core surface area parameter = (external surface area/face area)(No. of rows deep)
f = convection heat transfer coefficient, $\text{Btu/h} \cdot \text{ft}^2 \cdot °\text{F}$
h = air enthalpy (actual in airstream or saturation value at surface temperature), Btu/lb of air
Δh_m = mean effective difference of air enthalpy, as defined in Equation (25), Btu/lb of air
k = thermal conductivity of tube material, $\text{Btu/h} \cdot \text{ft} \cdot °\text{F}$
M = ratio of nonvolatile coolant-to-air temperature changes for sensible heat cooling coils, as defined in Equation (7e), dimensionless
n'' = rate-of-change of air enthalpy at saturation with air temperature, $\text{Btu/lb} \cdot °\text{F}$
N_r = number of coil rows deep in airflow direction, dimensionless
η = fin effectiveness, as defined in Equation (6), dimensionless
Δp_{st} = isothermal dry surface air-side friction at standard 70°F, in. of water
Δp_{sw} = wet surface air-side friction at standard 70°F, in. of water
q = heat transfer capacity, Btu/h
ρ_a = air density = 0.075 lb/ft^3 at 70°F

R = thermal resistance, referred to external area A_o, $\text{h} \cdot °\text{F} \cdot \text{ft}^2/\text{Btu}$
SHR = ratio of air sensible heat to air total heat, dimensionless
t = temperature, °F
Δt_m = mean effective temperature difference, air dry-bulb-to-coolant, °F
Δt_{ms} = mean effective temperature difference, surface-to-coolant, °F
$\Delta t_{\tilde{m}}$ = mean effective temperature difference, air dry-bulb to effective surface temperature, $t_{\tilde{s}}$, °F
U_o = overall sensible heat transfer coefficient, $\text{Btu/h} \cdot \text{ft}^2 \cdot °\text{F}$
V_a = coil air face velocity at 70°F, fpm
W = air humidity ratio, pounds of water per pound of air
w = mass flow rate, lb/h
y = ratio of nonvolatile coolant temperature rise to airstream enthalpy drop, as defined in Equation (13), $\text{lb}_a \cdot °\text{F/Btu}$

Superscripts:

$'$ = wet bulb
$''$ = dew point

Subscripts:

$_1$ = condition entering coil
$_2$ = condition leaving coil
$_a$ = airstream
$_b$ = dry-wet surface boundary
$_f$ = fin (with R)
$_i$ = installed, selected (with A_o, N_r)
$_m$ = metal (with R) and mean (with other symbols)
$_o$ = overall (except for A)
$_r$ = coolant
$_s$ = surface
$_{\tilde{s}}$ = effective surface
$_t$ = tube (with R) and total (with q)
$_{ab}$ = air, dry-wet boundary
$_{ad}$ = dry air
$_{aw}$ = wet air
$_{md}$ = dry metal
$_{mw}$ = wet metal
$_{rb}$ = coolant dry-wet boundary
$_{sb}$ = surface dry-wet boundary
$_{td}$ = total heat capacity, dry surface
$_{tw}$ = total heat capacity, wet surface

REFERENCES

ARI. 1987. Forced-circulation air-cooling and air-heating coils. ARI *Standard* 410-81, Air-Conditioning and Refrigeration Institute, Arlington, VA.

ASHRAE. 1991. Method of testing for rating forced-circulation air-cooling and heating coils. ASHRAE *Standard* 33.

ASME. 1988. *Code on nuclear air and gas treatment*. ANSI/ASME AG-1-1988. American Society of Mechanical Engineers, New York.

McElgin, J. and D.C. Wiley. 1940. Calculation of coil surface areas for air cooling and dehumidification. *Heating, Piping and Air Conditioning* (March):195.

Brown, G. 1954. Theory of moist air heat exchangers. Royal Institute of Technology *Transactions* No. 77, Stockholm, Sweden, 12.

Mueller, A.C. 1973 (Revised 1986). Heat exchangers. *Handbook of Heat Transfer*, W.M. Rohsenow and J.P. Hartnett, eds., McGraw Hill Book Company, New York.

Kusuda, T. 1969. Effectiveness method for predicting the performance of finned tube coils. *Heat and Mass Transfer to Extended Surfaces*, ASHRAE.

Shah, M.M. 1976. A new correlation for heat transfer during boiling flow through pipes. ASHRAE *Transactions* 82(2).

Shah, M.M. 1982. CHART Correlation for saturated boiling heat transfer, equations and further study. ASHRAE *Transactions* 88(1).

Shah, M.M. 1978. Heat transfer, pressure drop, and visual observation test data for ammonia evaporating inside tubes. ASHRAE *Transactions* 84(2).

Webb, R.L. 1980. Air-side heat transfer in finned tube heat exchangers. *Heat Transfer Engineering* 1(3):33.

CHAPTER 22

DESICCANT DEHUMIDIFICATION AND PRESSURE DRYING EQUIPMENT

DEHUMIDIFICATION is the removal of water vapor from air, gases, or other fluids. There is no limitation of pressure in this definition, and sorption dehumidification equipment has been designed and operated successfully for system pressures ranging from subatmospheric to as high as 6000 psi. The term dehumidification is normally limited to equipment that operates at essentially atmospheric pressures and is built to standards similar to other types of air-handling equipment. For drying gases under pressure, or liquids, the term *drier* or *dehydrator* is normally used.

This chapter is primarily concerned with equipment and systems that dehumidify air rather than those that dry other gases or liquids.

Both liquid and solid desiccant materials are used in dehumidification equipment. They either adsorb the water on the surface of the desiccant (adsorption) or chemically combine with water (absorption). In regenerative equipment, the mechanism for water removal is reversible.

Nonregenerative equipment uses hygroscopic salts such as calcium chloride, urea, or sodium chloride. Regenerative systems usually use a form of silica or alumina gel, activated alumina, molecular sieves, lithium chloride salt, lithium chloride solution, or glycol solution. The choice of desiccant depends on the requirements of the installation, the design of the equipment, and chemical compatibility with the gas to be treated or impurities in the gas. Additional information on desiccant materials and how they operate may be found in Chapter 19 of the 1989 ASHRAE *Handbook—Fundamentals*.

Some of the more important commercial applications of dehumidification include the following:

- Lowering the relative humidity to facilitate manufacturing and handling of hygroscopic materials
- Lowering the dew point to prevent condensation on products manufactured in low-temperature processes
- Providing protective atmospheres for the heat treatment of metals
- Maintaining controlled humidity conditions in warehouses and caves used for storage
- Preserving ships, aircraft, and other industrial equipment that would otherwise deteriorate
- Maintaining a dry atmosphere in a closed space or container for numerous static applications, such as the cargo hold of a ship
- Eliminating condensation and subsequent corrosion
- Drying air to speed drying of heat-sensitive products, such as candy, seeds, and photographic film
- Drying natural gas

- Drying gases that are to be liquefied
- Drying instrument air and plant air
- Drying process and industrial gases
- Dehydration of liquids

This chapter covers (1) the types of dehumidification equipment for liquid and solid desiccants, including high-pressure equipment; (2) performance curves; (3) variables of operation; and (4) some typical applications. The use of desiccants for the drying of refrigerants is addressed in Chapter 7 of the 1990 ASHRAE *Handbook—Refrigeration*.

METHODS OF DEHUMIDIFICATION

Air may be dehumidified by (1) cooling or increasing its pressure, which reduces its capacity to hold moisture, or by (2) removing moisture by attracting the water vapor with a liquid or solid desiccant. Frequently, systems employ a combination of these methods to maximize operating efficiency and minimize installed cost.

Figure 1 illustrates three methods by which dehumidification with desiccant materials or desiccant equipment may be accomplished. Air in the condition at Point A is dehumidified and cooled to Point B. This is done in a liquid desiccant system with intercooling directly. In a solid desiccant unit, this process can be completed by precooling and dehumidifying from Point A to Point C, then desiccating from Point C to Point E and, finally, cooling to Point B. It could also be accomplished with solid desiccant equipment by dehumidifying from Point A to Point D and then by cooling from Point D to Point B.

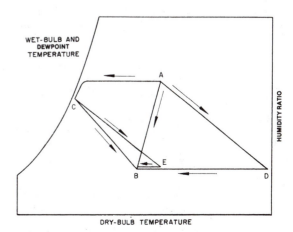

Fig. 1 Methods of Dehumidification

The preparation of this chapter is assigned to TC 3.5, Sorption.

Compressing air reduces its capacity to hold moisture. The resulting condensation reduces the moisture content of the air in absolute terms, but produces a saturated condition—100% relative humidity at elevated pressure. In applications at atmospheric pressure, this method is too expensive. However, in pressure systems such as instrument air, dehumidification by compression is worthwhile. Other dehumidification equipment, such as coolers or desiccant dehumidifiers, often follows the compressor to avoid problems associated with high relative humidity in compressed air lines.

Refrigeration of the air below its dew point is the most common method of dehumidification. This method is advantageous when the gas is comparatively warm, has a high moisture content, and when the outlet dew point desired is above 40 °F. Frequently, refrigeration is used in combination with desiccant dehumidifiers to obtain an extremely low dew point at minimum cost.

Liquid desiccant dehumidification systems pass the air through sprays of a liquid desiccant such as lithium chloride or a glycol solution. The desiccant in active state has a vapor pressure below that of the gas to be dehumidified and absorbs moisture from the gas stream. The desiccant solution, during the process of absorption, becomes diluted with moisture. During regeneration, this moisture is given up to an outdoor airstream in which the solution is heated. In a closed circuit, a partial bleedoff of the solution between the spraying or contactor unit and the regenerator unit continuously reconcentrates the desiccant.

Solid sorption passes the air through a bed of granular desiccant or through a structured packing impregnated with desiccant. Humid air passes through the desiccant, which in its active state has a vapor pressure below that of the humid air. This vapor pressure differential drives the water vapor from the air onto the desiccant. After becoming saturated with moisture, the desiccant is reactivated (dried out) by heating, which raises the vapor pressure of the material above that of the surrounding air. With the vapor pressure differential reversed, water vapor moves from the desiccant to a small purge airstream called the reactivation air, which carries the moisture away from the equipment.

DESICCANT DEHUMIDIFICATION EQUIPMENT

As defined in the preceding section, liquid desiccants or solid desiccants with granular or fixed structure may be used in the dehumidification equipment. Both types of desiccants, liquid and solid, may be used in equipment designed for the drying of air and gases at atmospheric or elevated pressures. Regardless of pressure levels, basic principles remain the same, and only the desiccant towers or chambers require special design consideration.

Desiccant capacity and actual dew-point performance depend on the specific equipment used, the characteristics of the various desiccants, initial temperature and moisture content of the gas to be dried, reactivation methods, etc. Factory-assembled units are available up to a capacity of about 60,000 cfm. Greater capacities can be obtained with field-erected units.

Tests reveal that both liquid and solid dehumidifiers effectively remove bacteria from air passing through the system with varying degrees of efficiency, depending on equipment type. Significant lowering of bacterial levels, combined with simultaneous dehumidification, is especially important for hospitals, laboratories, pharmaceutical manufacturing, etc.

LIQUID DESICCANT EQUIPMENT

Figure 2 shows a flow diagram for a typical liquid desiccant system with extended-surface contactor coils. For dehumidifying, the

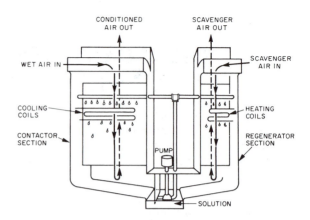

Fig. 2   Flow Diagram for Liquid-Absorbent Dehumidifier

strong absorbent solution is pumped from the sump of the unit and sprayed over the contactor coils. Air to be conditioned passes over the contactor coils and comes in intimate contact with the hygroscopic solution. Airflow can be parallel with, or counter to, the sprayed solution flow, depending on the space and application requirements.

The degree of dehumidification depends on the concentration, temperature, and characteristics of the hygroscopic solution. Moisture is absorbed from the air by the solution because of the vapor-pressure difference between the air and the liquid desiccant. The moisture content of the outlet air can be precisely maintained by varying the coolant flow in the coil to control the desiccant solution contact temperature. Continuous regeneration maintains the desiccant solution at the proper concentration.

The heat generated in absorbing moisture from the air consists of the latent heat of condensation of water vapor and the heat of solution, or the heat of mixing, of the water and the desiccant. The heat of mixing varies with the liquid desiccant used and with the concentration and temperature of the desiccant. The solution is maintained at the required temperature by cooling with city, well, refrigerated, or cooling tower water, or by refrigerant flowing inside the tubes of the contactor coil.

The total heat that must be removed by the conditioner coil consists of the heat of absorption, sensible heat removed from the air, and the residual heat load added by the regeneration process. This residual heat load can be reduced substantially by using a two-sump economizer system or a liquid-to-liquid heat exchanger. In the two-sump economizer system, a small amount of the cool-dilute absorbent solution is metered to the regeneration system and replaced by a small amount of warm, highly concentrated solution. This system reduces the heat load on the conditioner cooling coil, thus reducing the amount of coolant required.

Since the contactor coil is the heat transfer surface, proper selection of a specific desiccant solution concentration and temperature can create the desired space temperature and humidity conditions.

The dry-bulb temperature of the air leaving the liquid absorbent contactor at constant flow rate is a function of the temperature of the liquid desiccant and the amount of contact surface between the air and the solution.

The liquid desiccant is maintained at the proper concentration for moisture removal by automatically removing the water from it. A small quantity of the solution, usually 10 to 20% of the flow to the contactor coils, is passed over the regenerator coil, where the liquid is heated with steam or another heating medium. The liquid desiccants commonly used may be regenerated with low-pressure steam or hot water. The vapor pressure of the liquid desiccant at elevated temperatures is considerably higher than that of the outdoor air. The hot solution, at relatively high vapor pressure, contacts outdoor air in the regenerator. The vapor-pressure difference between the outdoor air and the hot solution causes the air to absorb the water from the solution. The hot moist air from the regenerator is discharged to the outdoors, and the concentrated solution flows to the sump, where it is mixed with the dilute solution. The solution is then ready for another cycle.

The steam flow to the regenerator coil is regulated by a control that responds to the concentration of the solution circulated over the contactor coils, such as a level control, specific gravity control, or boiling-point control.

For humidifying operations, the liquid desiccant is maintained at the required temperature by adding heat in proportion to the water absorbed by the air from the solution. Water is added to the solution automatically to maintain the proper concentration. Figure 3 shows drying performance for a liquid-desiccant dehumidifier with a lithium chloride solution.

Figure 4 indicates existing applications in which liquid desiccants apply. Medium temperature range is defined as leaving dry-bulb temperature between 80 and 100 °F, with moisture content

of 60 gr/lb of dry air or lower. Applications in this range include gelatin and glue driers, cake-icing driers or trolley rooms of bakeries, drying basic vitamins, pharmaceuticals, the application of candy coatings on gums, and the handling of many other products during manufacturing if these products are spoiled or have a shortened shelf life (because the temperature rises during production above a certain critical level).

The air-conditioning range is defined as leaving dry-bulb temperature between 45 and 80 °F, with moisture content below 60 gr/lb. This is the basic range for comfort and process air conditioning. In these installations, the liquid desiccants are generally used in combination with well water or refrigeration to handle the sensible heat load requirements.

In the refrigeration range, a leaving dry-bulb temperature of 0 to 45 °F with moisture content below 45 gr/lb is required of the equipment. Liquid desiccants, in combination with cold brines and direct-expansion or flooded refrigeration, are used. Typical applications include beer cellars, beer fermentation rooms, meat storage, penicillin processing, sugar cooling and conveying, and the manufacture of chewing gum.

Below the ranges shown in Figure 4, there are additional applications from −94 to 32 °F, in which the dew point of the leaving air will be at or near the dry-bulb temperature. Examples are wind tunnels, environmental laboratories, tunnel cooling of candy bars, and blast freezers.

SOLID SORPTION

Solid desiccants, such as silica gel, zeolites (molecular sieves), activated alumina, or hygroscopic salts are generally used to dehumidify large volumes of moist air, and in such cases, the desiccant is continuously reactivated. There are also two other methods of dehumidifying air using dry desiccants—with nonreactivated, disposable packages and with periodically reactivated desiccant cartridges.

Disposable packages of solid desiccant are often sealed into packaging for consumer electronics, pharmaceutical tablets, and military supplies. Disposable desiccant packages rely entirely on vapor diffusion to dehumidify, as air is not forced through the desiccant. This method is used only in applications where there is no anticipated moisture load at all (such as hermetically-sealed packages) because the moisture absorption capacity of any nonreactivated desiccant is rapidly exceeded if a continuous moisture load enters the dehumidified space. Disposable packages generally serve as a form of insurance against unexpected, short-term leaks in small, sealed packages.

Periodically reactivated cartridges of solid desiccant are used where the expected moisture load is continuous, but very small. A common example is the *breather*, a tank of desiccant through which air can pass, compensating for changes in liquid volume in petroleum storage tanks or drums of hygroscopic chemicals. The air is dried as it passes through the desiccant, so moisture will not contaminate the stored product. When the desiccant becomes saturated, the cartridge is removed and heated in an oven to restore its moisture sorption capacity. Desiccant cartridges are used where there is no requirement for a constant humidity control level, and where the moisture load is likely to exceed the capacity of a small, disposable package of desiccant.

Desiccant dehumidifiers, which are used for drying liquids and gases other than air, often use a variation of this reactivation technique. Two or more pressurized containers of dry desiccant are arranged in parallel, and air is forced through one container for drying, while desiccant in the other container is reactivated. These units are often called *dual-tower* or *twin-tower* dehumidifiers.

Continuous reactivation dehumidifiers are the most common type used in high moisture load applications such as humidity

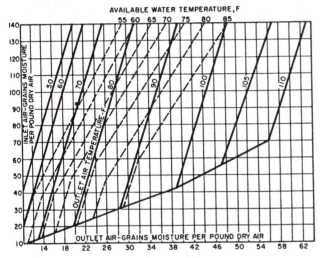

Fig. 3 Performance Data for Liquid-Absorbent Dehumidifier Using Lithium Chloride

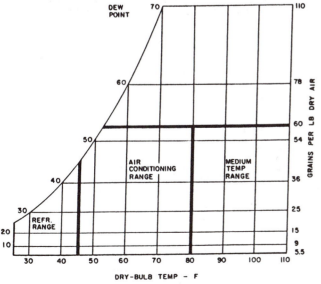

Fig. 4 Range of Conditions for Applications Using Liquid Desiccants

control systems for buildings and industrial process. In these units, humid air, generally referred to as the *process air*, is dehumidified in one part of the desiccant bed while a different part of the bed is dried for reuse by a second airstream known as the *reactivation air*. The desiccant generally rotates slowly between these two airstreams, so that dry, high capacity desiccant leaving the reactivation air is always available to remove moisture from the moist process air. This type of solid desiccant equipment is generally called a rotary desiccant dehumidifier. As it is most commonly used in building air-handling systems, the following sections describe its function in greater detail.

ROTARY SOLID DESICCANT DEHUMIDIFIERS

Figure 5 illustrates the principle of operation and the arrangement of major components of a typical rotary solid desiccant dehumidifier. The desiccant can be either beads of granular material packed into a bed, or it can be finely divided and impregnated throughout a structured media. Such media resembles corrugated cardboard rolled into a drum, so that air can pass freely through flutes aligned lengthwise through the drum.

In both granular and structured media units, the desiccant itself can be either a single material, such as silica gel, or desiccants used in combination, such as dry lithium chloride mixed with zeolites. Manufacturers provide options in the choice of desiccants, because the wide range of applications for dehumidification systems requires this flexibility to minimize operating and installed costs.

In rotary desiccant dehumidifiers, there are more than 22 variables that affect performance. In general, equipment manufacturers fix most of these in order to provide predictable performance in the more common applications for desiccant systems. Primary variables left to the system designer to define include:

Process air
• Inlet air temperature
• Moisture content
• Velocity at the face of the desiccant bed

Reactivation air
• Inlet air temperature
• Moisture content
• Velocity at the face of the desiccant bed

In any system, these variables change because of weather, variations in moisture load, and fluctuations in reactivation energy levels. It is useful for the system designer to understand the effect of these normal variations on the performance of the dehumidifier.

To illustrate these effects, Figures 6 through 10 show changes in the process air temperature and moisture leaving a "generic" rotary desiccant dehumidifier as modeled by a finite difference analysis program (Worek and Zheng 1991). The performance of commercial units differs from this model, since such units are generally optimized for very deep drying. However, for illustration purposes, the model accurately reflects the relationships between the key variables.

The desiccant used for the model is silica gel, the bed is a structured, fluted media, the bed depth is 400 mm in the direction of airflow, and the ratio of process air to reactivation air is approximately 3:1. The process air enters the machine at normal comfort conditions of 70°F, 50% rh (56 grains per pound of dry air).

Process air velocity through the desiccant bed strongly affects the leaving moisture condition. As shown in Figure 6, if the entering moisture is 56 gr/lb and all other variables are held constant, the outlet moisture varies from 22 gr/lb at 250 fpm to 40 gr/lb at 700 fpm. In other words, if air passes through the bed slowly, it is dried more deeply.

For the designer, this relationship means that if air must be dried very deeply, a large unit (slower air velocities) must be used.

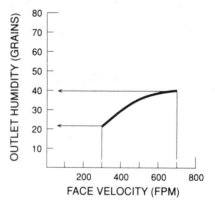

Fig. 6   Effect of Changes in Process Air Velocity on Dehumidifier Outlet Moisture Condition

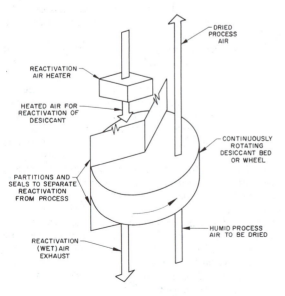

Fig. 5   Typical Rotary Dehumidification Unit

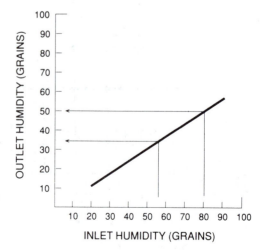

Fig. 7   Effect of Changes in Process Air Inlet Moisture on Dehumidifier Outlet Moisture Condition

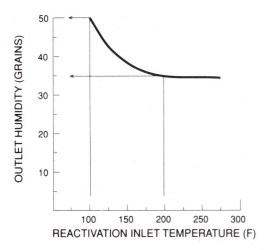

Fig. 8   Effect of Changes in Reactivation Air Inlet Temperature on Dehumidifier Outlet Moisture Condition

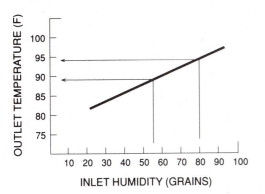

Fig. 9   Effect of Changes in Process Air Inlet Moisture on Dehumidifier Outlet Temperature Condition

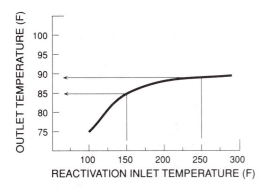

Fig. 10   Effect of Changes in Reactivation Air Inlet Temperature on Dehumidifier Outlet Temperature Condition

Process air inlet moisture content affects the outlet moisture condition. Thus, if the air is more humid entering the dehumidifier, it will be more humid leaving the unit. For example, Figure 7 indicates that for an inlet condition of 56 gr/lb, the outlet condition will be 35 gr/lb. If the inlet moisture content rises to 80 gr/lb, the outlet condition rises to 50 gr/lb.

For the system designer, one consequence is that if a constant outlet condition is necessary, the dehumidifier will need to have capacity control except in the rare circumstance where the process inlet airstream does not vary in temperature or moisture content throughout the year.

Reactivation air inlet temperature will change the outlet moisture content of the process air. In the range of temperatures from 100 to 250°F, as more heat is added to the reactivation air, the desiccant dries more completely, which means that it can attract more moisture from the process air (Figure 8). If the reactivation air is only heated to 100°F, the process outlet moisture is 50 gr/lb, or only 6 gr/lb lower than the entering condition. In contrast, if the reactivation air is heated to 200°F, the outlet moisture is 35 gr/lb, so that almost 40% of the moisture it contained originally is now removed.

Two important consequences follow from this relationship. If the designer needs dry air, it is generally more economical to use high reactivation air temperatures. Conversely, if the leaving condition from the dehumidifier need not be especially low, inexpensive, low-grade heat sources, such as waste heat, cogeneration heat, or rejected heat from refrigeration condensers, can be used to reactivate the desiccant at a low cost.

Process air outlet temperature is higher than the inlet air temperature primarily because the heat of sorption of the moisture removed from the air is converted to sensible heat. The heat of sorption is composed of the latent heat of condensation of the removed moisture, plus additional chemical heat which varies depending on the type of desiccant and the process air outlet humidity. Additionally, some heat is carried over to the process air from the reactivation sector since the desiccant is warm as it enters the relatively cooler process air. Generally, between 80 to 90% of the temperature rise of the process air is due to the heat of sorption, and the balance is due to heat carried over from reactivation.

Process outlet temperature versus inlet humidity is illustrated in Figure 9. Note that as more moisture is removed (higher inlet humidity), the outlet temperature rises. Air entering at room comfort conditions of 70°F, 56 gr/lb will leave the dehumidifier at a temperature of 89°F. If the dehumidifier removes more moisture, such as when the inlet humidity is 80 gr/lb, the outlet temperature will rise to 94°F. The increase in temperature rise is roughly proportional to the increase in moisture removal.

Process outlet temperature versus reactivation temperature is shown in Figure 10, which shows the effect of increasing reactivation temperature if the moisture content of the process inlet air stays constant. If the reactivation sector is heated to elevated temperatures, more moisture will be removed on the process side, so the temperature rise due to latent-to-sensible heat conversion will be slightly greater. In this constant-moisture inlet situation, if the reactivation sector is very hot, more heat will be carried from reactivation to process as the desiccant mass rotates from reactivation to process. Figure 10 shows that if reactivation is heated to 150°F, the process air leaves the dehumidifier at 85°F. If reactivation is heated to 250°F, the process air outlet temperature rises to 89°F. The 4°F increase in process air temperature is due primarily to the increase in heat carried over from reactivation.

One consequence of this relationship is that desiccant equipment manufacturers constantly seek to minimize the "waste mass" in a desiccant dehumidifier, to avoid heating and cooling extra, nonfunctional material such as heavy desiccant support structures or extra desiccant that air cannot reach. Theoretically, the most efficient desiccant dehumidifier has an infinitely large effective desiccant surface combined with an infinitely low mass.

Use of Cooling

In process drying applications, desiccant dehumidifiers are sometimes used without additional cooling, because the increase in temperature from the dehumidification process is helpful to the drying process. In semiprocess applications such as controlling frost formation in supermarkets, excess sensible cooling capacity may be present in the system as a whole, so that warm air from a desiccant unit is not a major consideration. However, in most other applications for desiccant dehumidifiers, provision must be made to remove excess sensible heat from the process air following dehumidification.

In a liquid desiccant system, heat is removed by cooling the liquid desiccant itself, so the process air emerges from the desiccant media at the appropriate temperature. In a solid desiccant system, cooling is accomplished downstream of the desiccant bed with cooling coils. The source of this cooling can affect the operating economics of the system.

In some systems, postcooling is accomplished in two stages, with cooling tower water as the primary source and compression or absorption cooling following the first stage. Alternately, various combinations of indirect and direct evaporative cooling equipment are used to cool the dry air leaving the desiccant unit.

In other systems where there is a time difference between the peaks of the latent and sensible load peaks, the sensible cooling capacity of the basic air-conditioning system is sufficient to handle the process air temperature rise without additional equipment. Systems in moderate climates with high ventilation requirements often combine high latent loads in the morning, evening, and night with high sensible loads at midday, so that desiccant subsystems to handle latent loads are especially economical.

Use of Units in Series

Dry desiccant dehumidifiers are often used to provide air at low dew points. Applications requiring large volumes of air at moisture contents of 5 gr/lb (0 °F dew point) are quite common and can be easily achieved by rotary desiccant units in a single pass beginning with inlet moisture contents as high as 45 gr/lb (45 °F dew point). Certain types of dry desiccant units commonly deliver air at 2 gr/lb (−18 °F dew point) without special design considerations. Where extremely low dew points must be achieved, or where air leakage inside the unit may be a concern, two desiccant dehumidifiers can be placed in series, with the dry air from a first unit feeding both the process and reactivation air to a second unit. The second unit delivers very dry air, since there is reduced risk of moisture being carried over from reactivation to process air when dry air is used to reactivate the second unit.

Industrial Rotary Desiccant Dehumidifier Performance

Figures 6 through 10 are based on the generalized model of a desiccant dehumidifier as described by the UIC IMPLICIT com-

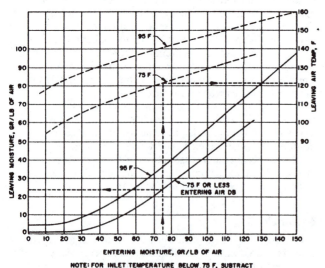

NOTE: FOR INLET TEMPERATURE BELOW 75 F, SUBTRACT TEMPERATURE DIFFERENCE FROM OUTLET TEMPERATURE

Fig. 11   Typical Performance Data for Rotary Solid Desiccant Dehumidifier

puter program (Worek and Zheng 1991). The model, however, differs somewhat from commercial products. Figure 11 is included here to show the typical performance of industrial desiccant dehumidifiers.

APPLICATIONS FOR ATMOSPHERIC PRESSURE DEHUMIDIFICATION

Preservation of Materials in Storage

Special moisture-sensitive materials are sometimes kept in dehumidified warehouses for long-term storage. Tests by the Bureau of Supplies and Accounts of the U.S. Navy concluded that 40% rh is a safe level to control deterioration of materials. Others have indicated that 60% rh is low enough to control microbiological attack. With storage at 40% rh, no undesirable effects on metals or rubber-type compounds have been noted. Some organic materials such as sisal, hemp, and paper may lose flexibility and strength, but they recover these characteristics when moisture content is regained.

Commercial storage relies on similar equipment for applications that include beer fermentation rooms, meat storage, and penicillin processing, as well as storage of machine tools, candy, food products, furs, furniture, seeds, paper stock, and chemicals. For recommended conditions of temperature and humidity, refer to the food refrigeration section of the 1990 ASHRAE *Handbook—Refrigeration.*

Process Dehumidification

The requirements for dehumidification in industrial processes are many and varied. Some of these processes are as follows:

- Metallurgical processes, in conjunction with the controlled atmosphere annealing of metals
- Conveying of hygroscopic materials
- Film drying
- Candy, chocolate, and chewing gum manufacture
- Manufacture of drugs and chemicals
- Manufacture of plastic materials
- Manufacture of laminated glass
- Packaging moisture-sensitive products
- Assembly of motors and transformers
- Solid propellent mixing
- Manufacture of electronic components, such as transistors and microwave components

For information concerning the effect of low dew point air on drying processes, refer to Chapters 18, 20, 22, and 28 of the 1991 ASHRAE *Handbook—HVAC Applications.*

Condensation Prevention

Many applications require moisture control to prevent condensation. Moisture in the air will condense on cold cargo in a ship's hold when it reaches moist climates. Moisture will condense on the ship when the moist atmosphere in a cargo hold is cooled by the hull and deck plates, while the ship goes from a warm to a cold climate.

A similar problem occurs with military and commercial aircraft when the cold air frame descends from high altitudes into a high dew point ground level environment. Desiccant dehumidifiers are used to prevent condensation inside the air frame and avionics which leads to structural corrosion and failure of electronic components.

In pumping stations and sewage lift stations, moisture condenses on piping, especially in the spring when the weather warms and water in the pipes is still cold. Dehumidification is also used to prevent moisture in the air from dripping into oil and gasoline tanks and into open fermentation tanks.

Electronic equipment is often cooled by refrigeration, and dehumidifiers are required to prevent internal condensation of moisture. Electronic and instrument compartments in missiles are purged with low dew point air prior to launching to prevent malfunctioning due to condensation.

Wave guides and radomes are also usually dehumidified, as are telephone exchanges and relay stations. Missile and radar sites, for the proper operation of their components, depend (to a large extent) on the prevention of condensate on interior surfaces.

Dry Air-Conditioning Systems

Cooling based air-conditioning systems remove moisture from air by condensing it onto cooling coils, producing saturated air at a lower absolute moisture content. In many circumstances, however, there is a benefit to using a desiccant dehumidifier to remove the latent load from the system, avoiding problems caused by condensation, frost, and high relative humidity in air distribution systems.

For example, low-temperature product display cases in supermarkets operate less efficiently when the humidity in the store is high, because they freeze condensate on their cooling coils, increasing operating costs. Desiccant dehumidifiers remove moisture from the air, using rejected heat from refrigeration condensers to reduce the cost of desiccant reactivation. The combined effect of using desiccants and conventional cooling can have a beneficial effect on installation and operating costs (Calton 1985). For guidance concerning the effect of humidity on refrigerated display cases, refer to Chapter 2 of the 1991 ASHRAE *Handbook—HVAC Applications*.

The air-conditioning systems in hospitals, nursing homes, and other medical facilities are particularly sensitive to biological contamination in condensate drain pans, filters, and porous insulation inside ductwork. These systems often benefit by using a desiccant dehumidifier to dry the ventilation air before final cooling. Condensate does not form on cooling coils or drain pans, and filters and duct lining stay dry so that mold and mildew cannot grow inside the system. Refer to ASHRAE *Standard* 62-1989 for guidance concerning maximum relative humidity in air distribution systems.

Hotels and large condominium buildings historically suffer from severe mold and mildew problems caused by excessive moisture in the building structure. Desiccant dehumidifiers are sometimes used to dry ventilation air so it can act as a sponge to remove moisture from walls, ceilings, and furnishings (AH&MA 1991).

Like supermarkets, ice rinks have large exposed cold surfaces that condense and freeze moisture in the air, particularly during spring and summer operations. Desiccant dehumidifiers remove excess humidity from air above the rink surface, preventing fog and improving both the ice surface and operating economics of the refrigeration plant. For guidance concerning the effect of relative humidity on heat transfer to the ice surface, refer to Chapter 34 of the 1990 ASHRAE *Handbook—Refrigeration*.

Indoor Air Quality Contaminant Control

Desiccant sorption is not restricted to water vapor. Both liquid and solid desiccants collect both water and large organic molecules at the same time (Hines *et al.* 1991). As a result, desiccant systems can be used to remove VOC emissions (volatile organic compounds) from building air systems.

In addition to preventing the growth of mold, mildew, and bacteria through keeping buildings dry, desiccant systems are used to supplement filters to remove bacteria from the air itself. This is particularly useful for hospitals, medical facilities, and related biomedical manufacturing facilities where airborne microorganisms can cause costly problems. The utility of certain liquid and solid desiccants in such systems stems from their ability to either kill microorganisms, or to avoid sustaining their growth (Batelle Project No. N-0914-5200-1971 and SUNY Buffalo School of Medicine).

Testing

Many test procedures require dehumidification with sorption equipment. Frequently, other means of dehumidification may be used in conjunction with sorbent units, but the low moisture content requirements can be obtained only by liquid or solid sorbents. Some of the typical testing applications are as follows:

- Wind tunnels
- Spectroscopy rooms
- Paper and textile testing
- Bacteriological and plant growth rooms
- Dry boxes
- Environmental rooms and chambers

DESICCANT DRIERS FOR ELEVATED PRESSURES

The same sorption principles that pertain to atmospheric dehumidification apply to drying of high-pressure air and process or other gases. The sorbents described previously can be used with equal effectiveness.

Absorption

Solid absorption systems use a calcium chloride desiccant that dissolves calcium chloride, generally in a single-tower unit that requires periodic replacement of the desiccant as it dissolves with the absorbed moisture. Normally, the inlet air or gas temperature does not exceed 90 to 100°F saturated. The rate of replacement of the desiccant is proportional to the moisture in the inlet process flow. A dew-point depression of 20 to 40°F at pressure can be obtained when the system is operated in the range of 60 to 100°F saturated entering temperature and 100 psig operating pressure. At lower pressures, the ability to remove moisture reduces proportionally as a ratio of absolute pressure. Such units do not require a power source for operation because the desiccant is not regenerated. However, additional desiccant must be added to the system periodically.

Adsorption

Drying with an adsorptive desiccant such as silica gel, activated alumina, or a molecular sieve usually incorporates regeneration equipment, so the desiccant can be reactivated and reused. These desiccants can be readily reactivated by heat, by purging with dry gas, or by a combination of both. Depending on the desiccant selected, the dew point performance expected would be in the range of −40 to −100°F measured at the operating pressure with inlet conditions of 90 to 100°F saturated and 100 psig. Figure 12 shows typical performance using activated alumina or silica gel desiccant.

Equipment design may vary considerably in detail, but most basic adsorption units use twin-tower construction for continuous operation, with an internal or external heat source, with air or process gas as the reactivation purge for liberating moisture adsorbed previously. A single adsorbent bed may be used for intermittent drying requirements. Adsorption units are generally constructed in the same manner as atmospheric pressure units, except that the vessels are suitable for the operating pressure. Units have been operated successfully at pressures as high as 6000 psig.

Prior compression or cooling by water, brine, or refrigeration to below the dew point of the gas to be dried reduces the total moisture load to be handled by the sorbent, thus permitting the use of smaller drying units. An economic balance must be made to determine if removal of some water before adsorption is desir-

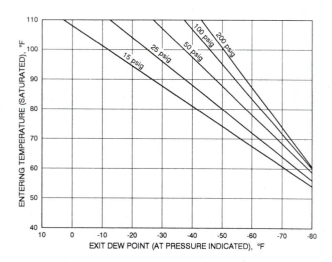

Fig. 12   Typical Performance Data for Solid Desiccant Driers at Elevated Pressures

able. The cost of compression, cooling, or both must be balanced against the cost of a larger adsorption unit.

The many different drier designs can be grouped into the following basic types:

Heat-reactivated, purge-type driers. Normally operating on 4-h (or longer) adsorption periods, these driers are generally designed with heaters embedded in the desiccant. They use a small portion of the dried process gas as a purge to remove the moisture liberated during reactivation heating.

Heatless-type driers. These driers operate on a short adsorption period (usually 60 to 300 s). Depressurization of the gas in the desiccant tower lowers the vapor pressure, so the adsorbed moisture is liberated from the desiccant and removed by a high purge rate of the dried process gas. The use of an ejector reduces the purge gas requirements.

Convection-type driers. These driers usually operate on 4-h (or longer) adsorption periods and are designed with an external heater and cooler as the reactivation system. Some designs circulate the reactivation process gas through the system by a blower, while other designs divert a portion or all of the process gas flow through the reactivation system prior to adsorption. Both heating and cooling are by convection.

Radiation-type driers. Also operating on 4-h (or longer) adsorption periods, radiation-type driers are designed with an external heater and blower to force heated atmospheric air through the desiccant tower for reactivation. Cooling of the desiccant tower is by radiation to atmosphere.

Figure 13 illustrates a typical adsorption-type desiccant drier of the heat-reactivated, purge type.

APPLICATIONS FOR DRYING AT ELEVATED PRESSURE

Preservation of Materials

Generally, materials in storage are preserved at atmospheric pressure, but a few materials are stored at elevated pressures, especially when the dried medium is an inert gas. These materials deteriorate when they are subjected to high relative humidity or oxygen content in the surrounding media. The drying of high-

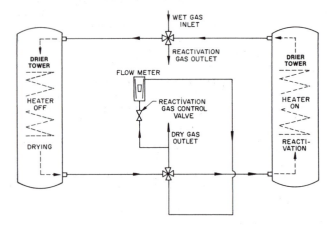

Fig. 13   Typical Adsorption Drier for Elevated Pressures

pressure air, which is subsequently reduced to 3.5 to 10 psig, has been used most effectively in pressurizing coaxial cables to eliminate electrical shorts caused by moisture infiltration. This same principle, at somewhat lower pressures, is also used in wave guides and radomes to prevent moisture film on the envelope.

Process Drying of Air and Other Gases

Drying of instrument air to a dew point of −40 °F, particularly in areas where the air lines are outdoors or are exposed to temperatures below the dew point of the air leaving the aftercooler, prevents condensation or freezeup in instrument control lines.

To prevent condensation and freezing, it is necessary to dry the plant air used for pneumatically operated valves, tools, and other equipment in areas where the piping is exposed to low ambient temperatures. Additionally, dry air prevents rusting of the air lines, which produces abrasive impurities, causing excessive wear on tools.

Drying of industrial gases or fuels such as natural gas has been accepted. For example, fuels (including natural gas) are cleaned and dried before storage underground to ensure that valves and transmission lines do not freeze from condensed moisture during extraordinarily cold weather, when the gas is most needed. Propane must also be clean and dry to prevent ice accumulation. Other gases, such as bottled oxygen, nitrogen, hydrogen, and acetylene, must have a high degree of dryness. In the manufacture of liquid oxygen and ozone, the weather air supplied to the particular process must be clean and dry.

Drying of air or inert gas for conveying hygroscopic materials in a liquid or solid state ensures continuous, trouble-free plant operation. Normally, gases for this purpose are dried to a −40 °F dew point. Purging and blanketing operations in the petrochemical industry depend on the use of dry inert gas for the reduction of explosive hazards, the reaction of chemicals with moisture or oxygen, and other such problems.

Testing of Equipment

Dry, high-pressure air is used extensively for the testing of refrigeration condensing units to ensure tightness of components and to prevent moisture infiltration. Similarly, dry inert gas is used in the testing of copper tubing and coils to prevent corrosion or oxidation. The manufacture and assembly of solid-state circuits and other electronic components require exclusion of all moisture, and final testing in dry boxes must be carried out in moisture-free atmospheres. The simulation of dry high-altitude atmospheres for testing of aircraft and missile components in wind tunnels requires extremely low dew-point conditions.

REFERENCES

AH&MA. 1991. *Mold and mildew in hotels and motels*. Executive Engineers Committee Report. Hotel and Motel Association, Washington, D.C.

Batelle Memorial Institute. 1971. Project number N-0914-5200-1971. Batelle Memorial Institute, Columbus, OH.

Calton, D.S. 1985. Application of a desiccant cooling system to supermarkets. ASHRAE *Transactions* 91(1).

Hines, A.J., T.K. Ghosh, S.K. Loyalka, and R.C. Warder, Jr. 1991. Investigation of co-sorption of gases and vapors as a means to enhance indoor air quality. ASHRAE *Research Project* 475 RP and Gas Research Institute *Project* GRI-90/0194. Gas Research Institute. ASHRAE *Transactions* 97(2). Chicago, IL.

SUNY Buffalo School of Medicine. Effects of glycol solutions on microbiological growth. Niagara Blower Report No. 03188.

Worek, W. and W. Zheng. 1991. UIC IMPLICIT Rotary desiccant dehumidifier finite difference program. The University of Illinois at Chicago, Department of Mechanical Engineering, Chicago, IL.

BIBLIOGRAPHY

ASHRAE. 1975. Symposium on sorption dehumidification. ASHRAE *Transactions* 81(1):606-38.

ASHRAE. 1980. Symposium on energy conservation in air systems through sorption dehumidifier techniques. ASHRAE *Transactions* 86(1):1007-36.

ASHRAE. 1985. Symposium on changes in supermarket heating, ventilating and air-conditioning systems. ASHRAE *Transactions* 91(1B):423-68.

ASHRAE. 1991. Desiccant cooling and dehumidification. Special publication.

Harriman, L.G., III. 1990. *The dehumidification handbook*. Munters Cargocaire, Amesbury, MA.

Jones, B.W., B.T. Beck, and J.P. Steele. 1983. Latent loads in low humidity rooms due to moisture. ASHRAE *Transactions* 89(1A):35-55.

Mei, V.C. and F.C. Chen. 1991. An assessment of desiccant cooling and dehumidification technology. Report No ORNL/CON-309. Prepared for the Office of Building and Community Services, U.S. Department of Energy, by Energy Division, Oak Ridge National Laboratory, Oak Ridge, TN.

Pesaran, A.A., T.R. Penney, and A.W. Czanderna. 1991. *Desiccant cooling state of the art assessment*. Prepared for the Office of Building and Community Services, U.S. Department of Energy, by Thermal, Fluid and Optical Science Branch, Solar Energy Research Institute, Golden, CO.

FORCED-CIRCULATION AIR COOLERS

FORCED-CIRCULATION unit coolers and product coolers are designed to operate continuously within refrigerated enclosures. A cooling coil and motor-driven fan make up the basic components of these coolers. These components provide the cooling or freezing temperatures and proper airflow to the room. Coil defrost equipment is added for low-temperature operations.

Any unit, such as a blower coil, unit cooler, product cooler, cold diffuser unit, or air-conditioning air handler is considered a forced-air cooler when operated under refrigeration conditions. Many design and construction choices are available, including: (1) finned or bare tube coils; (2) electric, gas, air, or water defrosting; (3) discharge air velocity; (4) centrifugal or propeller fans, either belt- or direct-driven; (5) ducted or nonducted, and/or (6) freestanding or ceiling suspended.

TYPES OF FORCED-CIRCULATION AIR COOLERS

Sloped Front Unit Coolers

These units, which range from 5 to 10 in. high, are commonly used in back-bar and under-the-counter fixtures, as well as in vertical, self-serve, glass door reach-in enclosures. These are small unit coolers, with airflows usually not exceeding 150 cfm per fan. The sloped fronts are designed for horizontal top mounting as a single unit, or installation as a group of parallel connected units. Direct-drive fans are sloped to fit within the restricted return air sweep rising across the enclosure access doors.

Low Air Velocity Unit Coolers

These units can have a half-round appearance, although long, narrow, dual-coil units are used in meat-cutting rooms and in meat and floral walk-in coolers. The unit has an amply finned coiled surface to maintain high humidities in the room. Discharge air velocities at the coil face range from 85 to 200 fpm.

Standard Air Velocity Unit Coolers

Low silhouette units are 12 to 15 in. high. Medium or midheight units are 18 to 48 in. high or greater. Some units can be classified as high silhouette unit coolers. The air velocity at the coil face can be as high as 600 fpm, but generally is 300 to 400 fpm.

High Air Velocity Product Coolers

These units are used in blast tunnel freezing and special cooling of products that are not adversely affected by moderate dehydration during rapid cooling. They generally draw air through the cooler at discharge velocities of about 2000 fpm.

The preparation of this chapter is assigned to TC 8.4, Air-to-Refrigerant Heat-Transfer Equipment.

Forced-circulation coolers direct air over a refrigerated coil in the enclosure. The coil lowers the air temperature below its dew-point, which causes condensate or frost to form on the coil surface. However, the normal refrigeration load is a sensible heat load, i.e., dry-bulb referenced, and the coil surface is considered dry. Rapid and frequent defrosting on a timed cycle can maintain this dry-surface condition.

Sprayed Coil Product Coolers

Spray coils feature a saturated coil surface that can cool the processed air closer to the coil surface temperature than can a dry coil. In addition, the spray continuously defrosts the low-temperature coil. Unlike unit coolers, spray coolers are usually floor mounted and discharge air vertically. The unit sections include a drain pan/sump, coil with spray section, moisture eliminators, and fan with drive. The eliminators remove airborne droplets of spray to prevent their discharge into the refrigerated area. Typically, belt-driven centrifugal fans draw air through the coil at 600 fpm or less.

Water can be used as the spray medium for coil surfaces with temperatures above freezing. For coil surfaces with temperatures below freezing, a suitable material such as listed below must be added to the water to lower the freezing point to 12 °F or lower than the coil surface temperature. Some recirculating solutions include:

- Sodium chloride. This solution is limited to a room temperature of 10 °F or higher. Its minimum freezing point is −6 °F.
- Calcium chloride. This can be used for enclosure temperatures down to about −10 °F, but its use may be prohibited in enclosures containing food products.
- Aqueous glycol solutions. These solutions are commonly used in water and/or sprayed coil coolers operating below freezing. Food grade propylene glycol solutions are commonly used because of their low oral toxicity, but they generally become too viscous to pump at temperatures below −13 °F. Ethylene glycol solutions may be pumped at temperatures down to −40 °F. Because of its toxicity, sprayed ethylene glycol in other than sealed tunnels or freezers (no human access allowed during process) is usually prohibited by most jurisdictions. When a glycol mix is sprayed in food storage rooms, any carryover of the spray must be maintained within the limits prescribed by the USDA, the Bureau of Animal Industries, and all applicable local codes.

All brines are hygroscopic; they absorb condensate and become progressively weaker. This dilution can be corrected by continually adding salt to the solution to maintain sufficient below-freezing temperature. Salt is extremely corrosive, so it must be contained in the sprayed coil unit with suitable corrosive-resistant materials, which must receive periodic inspection and maintenance.

Sprayed coil units are usually installed in refrigerated enclosures requiring high humidity, e.g., chill coolers. Paradoxically, the same sprayed coil units can be used in special applications requiring low relative humidity. For such dehydration applications, both a high brine concentrate (near its eutectic point) and a wide difference between the process air and the refrigerant temperature are maintained. Process air reheat downstream of the sprayed coil corrects the dry-bulb temperature.

COMPONENTS

Draw-through and Blow-through Airflow

Unit fans may draw air through the cooling coil and discharge it through the fan outlet into the enclosure; or the fans may blow air through the cooling coil and discharge it from the coil face into the enclosure. Blow-through units have a slightly higher thermal efficiency because heat from the fan is removed from the forced airstream by the coil. Draw-through fan energy adds to the heat load of the refrigerated enclosure, but neither load from the fractional horsepower fan motors is significant. Selection depends more on a manufacturer's design features for the unit size required, as well as the air throw required within the particular enclosure.

Blow-through design has a lower discharge air velocity stream because the entire coil face area is usually the discharge opening (grilles and diffusers not withstanding). An air throw of 33 ft or less is common for the average standard air velocity from a blow-through unit. Greater throw, in excess of 100 ft, is normal for draw-through centrifugal fan units. The propeller fan in the high silhouette draw-through unit cooler is popular for intermediate ranges of air throw.

Fan Assemblies

Direct-drive propeller fans (motor plus blade) are popular because they are simple, economical, and can be installed in multiple assemblies in a unit cooler housing. Additionally, they require less motor power for a given airflow rate capacity.

The centrifugal fan assembly usually includes belts, bearings, sheaves, and coupler drives along with each of their inherent problems. Yet, this design is necessary for applications having high air distribution static pressure losses. These applications include enclosures with ductwork runs, tunnel conveyors, and high density stacking of products. Centrifugal fan-equipped units are also used in produce ripening rooms, where a large air blast is required during the ripening process.

Casing

Casing materials are selected for compatibility with the enclosure environment in which they are installed. Construction usually features coated aluminum or galvanized steel. Stainless steel is also used in food storage or preparation enclosures where sanitation must be maintained. On larger cooler units, internal framing is fabricated of sufficiently substantial material, such as galvanized steel, and casings are usually made with similar material. Some plastic casings are used in small unit coolers, while some large, ceiling-suspended units may feature all aluminum construction.

Coil Construction

Coil construction varies from uncoated (all) aluminum tube and fin to hot-dipped galvanized (all) steel tube and fin, depending on the type of refrigerant used and the environmental exposure to the coil surface. The most popular unit coolers have copper tube/aluminum fin construction. Ammonia refrigerant equipment is not constructed from copper tube coil. Also, sprayed coils are not constructed from aluminum fin and tube materials.

Fin spacings vary from 6 to 8 fins per inch for coils with surfaces above 32 °F and between 0 and 32 °F when latent loads are insignificant. Otherwise, 3 to 6 fins per inch is the accepted spacing for coil surfaces below 32 °F, with a spacing of 4 fins per inch when latent loads exceed 15% of the total load.

Defrosting

Coils must be defrosted when frost accumulates on their surfaces. The frost (or ice) is usually greatest at the air entry side of the coil; therefore, the required defrost cycle is determined by the inlet surface condition. In contrast, a reduced secondary surface-to-primary surface ratio produces greater frost accumulations at the coil outlet face. In theory, the accumulation of greater frost at the coil entry air surface improves the heat transfer capacity of the coil. However, accumulated coil frost usually has two negative effects: (1) it impedes heat transfer because of its insulating effect and (2) it reduces airflow because it restricts the free air area within the coil.

Depending on the defrost method, as much as 80% of the defrost heat load could be transferred into the enclosure. This heat load is not normally included as part of the enclosure heat gain calculation, but it is usually accounted for by the factor that estimates the hours per day of refrigeration running time.

A longer time between defrost cycles can be achieved by using more coil tube rows and wider fin spacing. Ice accumulation should be avoided to reduce defrost time. For example, in low-temperature applications having high latent loads, unit coolers should not be located above freezer doors.

Controls

Electromechanical controls cycle to maintain the desired enclosure temperature. Modulating control valves such as evaporator pressure regulators are also used. In its simplest form, the temperature control is a thermostat mounted in the enclosure that either cycles the compressor on and off or a liquid line feed solenoid that opens and closes. A suction pressure switch can substitute for the wall-mounted thermostat.

An electronically based energy management system (EMS) maintains optimum energy control by equating suction pressure transducer readings with the signal from a temperature diode sensor in the enclosure. EMS controls are commonly used in large warehouses and supermarkets.

Defrost options operate either on demand or at fixed time intervals. Fixed time defrost, which is widely used, is initiated by a 24-h clock at predetermined defrost intervals and durations. The defrost cycle terminates when the coil temperature rises sufficiently to melt the frost. A rise in evaporator pressure or an elapsed time clock are also used to terminate the defrost cycle. On many applications, a mix of these various defrost control methods is common and adapts well to the system control.

AIR MOVEMENT AND DISTRIBUTION

Air distribution and velocity are important concerns in selecting and locating a unit cooler. The direction of the air and air throw should be such that air moves where there is a heat gain; this principle applies to the enclosure walls and ceiling, as well as to the product. Unit coolers should be placed (1) so they do not discharge air at any doors or openings; (2) away from doors that do not incorporate an entrance vestibule or pass to another refrigerated enclosure to keep from inducing additional infiltration into the enclosure; and (3) away from the airstream of another unit to avoid defrost difficulties.

The velocity and relative humidity of air passing over an exposed product affect the amount of surface drying and weight loss. Air velocities of 500 fpm over the product are typical for most freezer applications. Higher velocities require additional fan power and, in some cases, only slightly decrease the cooling time. For example, air velocities in excess of 500 fpm for freezing plastic-wrapped bread reduce freezing time very little. However, increasing the air velocity from 500 to 1000 fpm over unwrapped pizza reduces the freezing time and product exposure almost in half.

This variation shows the importance of product testing to design the special enclosures intended for blast freezing accurately and/or automated food processing. Sample tests should yield the following information: ideal air temperature, air velocity, product weight loss, and dwell time. With this information, the proper unit or product coolers, as well as the supporting refrigeration equipment and controls, can be selected.

RATING UNITS

Currently, no industry standard exists for rating unit and product coolers. Part of the difficulty in developing a workable standard is that many variables are encountered. Cooler coil performance and capacities should be based on a fixed set of conditions and greatly depend on (1) air velocity, (2) refrigerant velocity, (3) temperature difference, (4) frost condition, and (5) superheating adjustment. The most significant items are refrigerant velocity, as related to refrigerant feed through the coil, and frost condition and defrosting in low-temperature applications. The following sections address some of the problems involved in arriving at a common rating.

Refrigerant Velocity

Depending on the commercial refrigerant feed method used, both the capacity ratings of the cooler and the refrigerant velocity vary. The following feed methods are used.

Dry expansion. In this system, a thermostatically controlled, direct-expansion valve allows just enough liquid refrigerant into the cooling coil to ensure that it vaporizes at the outlet. In addition, 5 to 15% of the coil surface is used to superheat the vapor. The direct expansion (DX) coil ratings are usually the lowest of the various feed methods.

Recirculated refrigerant. This system is similar to a dry expansion feed because it has a hand expansion valve or metering device to control the flow of the entering liquid refrigerant. The coil is intentionally overfed to eliminate superheating of the refrigerant vapor. The amount of liquid refrigerant pumped through the coil may be two to six times that of a dry expansion coil. As a result, this coil's capacity is higher than that for a dry expansion feed (see Chapter 2 of the 1990 ASHRAE *Handbook—Refrigeration*).

Flooded. This system has a liquid reservoir (surge drum) located adjacent to each unit or set of units. The surge drum is filled with a subcooled refrigerant and connected to the cooler coil. To ensure gravity flow of this refrigerant and a completely wet internal coil surface, the liquid level in the surge drum must be higher than the top of the coil. The capacity of gravity-recirculated feed is usually the highest attainable.

Brine. This term encompasses any liquid or solution that absorbs heat within the coil without a change in state. Ethylene glycol and water, propylene glycol and water, and R-11 are used, in addition to solutions of calcium chloride or sodium chloride and water. The capacity ratings for brine coils depend on many variables such as flow, viscosity, specific heat, and density. As a result, this rating is obtained by special request from a coil manufacturer and generally is 10 to 40% less than its flooded rating application.

Frost Condition

Frost condition and defrosting are perhaps the most indeterminate variables that affect the capacity rating of forced-air cooler coils. Any unit operating below 32°F coil temperature accumulates frost or ice. Although a light frost accumulation slightly improves heat transfer of the coil, continuous accumulation negatively affects coil performance. Ultimately, defrosting is the only solution.

Enclosure air temperature above 35°F. Whenever the enclosure air is 35°F or slightly warmer, it can be used to defrost the coil. However, some of the moisture on the coil surface evaporates into the air, which is undesirable for low-humidity application. The following methods of control are commonly used.

1. If the refrigeration cycle is interrupted by a defrost timer, the continually circulating air melts the coil frost and ice. The timer can operate either the compressor or a refrigerant solenoid valve.
2. An oversized unit cooler controlled by a wall thermostat defrosts during its normal *off* and *on* cycling. The thermostat can control a refrigeration solenoid in a multiple-coil system or the compressor in a unitary installation.
3. Pressure control can be used for slightly oversized unitary systems. A low-pressure switch connected to the compressor suction line is set at a cut-out point such that the design suction pressure corresponds to the saturated temperature required to handle the maximum enclosure load. The suction pressure at the compressor drops and causes the compressor motor to stop as the enclosure load fluctuates or as the oversized compressor overcomes the maximum loading.

   A thermostatic expansion valve on the unit cooler controls the liquid refrigerant flow into the coil, which varies with the load. The cut-in point, which starts the compressor motor, should be set at the suction pressure, which corresponds to the equivalent saturated temperature of the desired refrigerated enclosure air. The pressure differential between the cut-in and cut-out points corresponds to the temperature difference between the enclosure air and the coil temperatures. The pressure settings should allow for the pressure drop in the suction line.

Enclosure air temperature below 35°F. Whenever the enclosure air is below 35°F, supplementary heat must be introduced into the enclosure to defrost the coil surface and drain pan. Unfortunately, some of this defrost heat remains in the enclosure until the unit starts operation after the defrost cycle. No two defrost methods have the same timing input.

1. Hot-gas defrost can be the fastest and most efficient method if an adequate supply of hot gas is available. The hot refrigerant discharge gas internally cleans the tubes and aids in returning the lubricants back to the compressor. It can be used for small commercial units and large industrial systems, and for low-temperature applications. Hot-gas defrost can also improve capacity because it can remove some of the load from the condenser when used to defrost multiple evaporators alternately on a large, continuously operating compressor system. This method of defrost puts the least amount of heat into the enclosure air, especially when latent gas defrosting is used.
2. Electric defrost effectiveness depends on the location of the electric heating elements. The electric heating elements can be placed either in contact with the finned coil surface or inserted into the interior of the coil element. In this latter method, either special fin holes or dummy tubes are used. Electric defrost can be efficient and rapid; it is simple to operate and maintain, but it does dissipate the most heat into the enclosure.
3. Heated air may be used as a defrost method. Some unit coolers are constructed to isolate the frosted coil from the cold enclosure air. Once isolated, the air around the coil is heated by hot gas or electric heating elements and is circulated to hasten the defrost. This heated air also must heat a drain pan, which is needed in all enclosures at temperatures of 34°F or less. Some units have specially constructed housings and ducting to run warm air from adjoining areas.
4. Water defrost is the quickest method of defrosting a unit. It is efficient and effective for rapid cleaning of the complete coil surface. Water defrost can be performed manually or on an

automatic timed cycle. This method becomes less desirable as the enclosure temperature decreases much below freezing, but it has been successfully used in cases as low as −40°F. Water defrost is used more for industrial product cooler units than for small ceiling-suspended units.

5. Hot brine can be used as a defroster, as brine-cooled coils can have a heater to heat brine for the defrost cycle. This system heats from within the coil and is as rapid as hot-gas defrost. The heat source can be steam, electric resistance elements, or condensing water.

Defrost control. Each defrosting method is done with the fan turned off. Frequently, two methods are run simultaneously to shorten the defrost cycle. Both inadequate defrost time and over-defrosting can degrade overall performance; thus, a defrost cycle is best ended by monitoring temperature. A thermostat may be mounted within the cooler coil to sense a rise in the temperature of the finned or tubed surface. A temperature of at least 40°F indicates the removal of frost and automatically returns the unit to the cooling cycle.

Fan operation is delayed, usually by the same thermostat, until the coil surface temperature approaches its normal operating level. This practice prevents unnecessary heating of the enclosure after defrost and also prevents drops of defrost water from being blown off the coil surface. In some applications, fan delay after defrost is essential to prevent a rapid buildup of air pressure, which could structurally damage the enclosure.

Initiation of defrost can be automated by time clocks, running time monitors, air pressure differential controls, or by monitoring the air temperature difference through the coil (which increases as the airflow is reduced by frost accumulation). Adequate supplementary heat for the drain pan and drain lines should be considered. Also, drain lines should be pitched properly and trapped outside the cold area to keep them from freezing.

Basic Cooling Capacity

Most rating tables state gross capacity and assume that the fan assembly or defrost heat is included as part of the enclosure load calculation. Some manufacturers' cooler coil ratings may appear as sensible, while some may be listed as total capacity, which includes the sensible and latent capacity. Some ratings involve reduction factors to account for frost accumulation in low-temperature applications or for some unusual condition. Some include multiplication factors for various refrigerant types.

The rating, known as the basic cooling capacity, is based on the temperature difference between the inlet air and the refrigerant within the coil; Btu/h per degree TD (temperature difference) is the dimension used. The coil inlet air temperature is considered the same as the enclosure air temperature. This practice is common for smaller enclosure applications. For larger installations, as well as heavy-use process work, manufacturer's published ratings should be applied on the average of the coil inlet-to-outlet temperatures. This is regarded as the average enclosure temperature. The refrigerant temperature is usually the temperature equivalent to the saturated pressure at the coil outlet.

The TD necessary to obtain the unit cooler capacity varies with each application. It may be as low as 8°F for wet storage coolers and as high as 25°F for workrooms. The TD can be related to the desired humidity requirements. The smaller the TD, the smaller the amount of dehumidification caused by the coil. The following information gives general guidance for selecting a proper TD.

Medium-temperature applications above 25°F saturated suction:

- For a very high relative humidity (about 90%), a temperature difference of 8 to 10°F is common.
- For a high relative humidity (approximately 80%), a temperature difference of 10 to 12°F is recommended.
- For a medium relative humidity (approximately 75%), a temperature difference of 12 to 16°F is recommended.
- Temperature differences beyond these limits usually result in low enclosure humidities, which dry the product. However, for packaged products and workrooms, a TD of 25 to 30°F is not unusual. Paper storage or similar products also require a low humidity level. Here, a TD of 20 to 30°F may be necessary.

Low-temperature applications below 25°F saturated suction:

- The temperature difference is generally kept below 15°F because of system economics and frequency of defrosting, rather than for humidity control.

INSTALLATION AND OPERATION

Whenever possible, refrigerating air-cooling units should be located away from enclosure entrance doors and passageways. This practice helps reduce coil frost accumulation, as well as fan blade icing. The cooler manufacturer's installation, start-up, and operation instructions generally give the best information. Upon installation, the unit nameplate data (model, refrigerant type, electrical data, warning notices, certification emblems, and so forth) should be recorded. This information should be compared to the job specifications, as well as the manufacturer's instructions for correctness.

GENERAL INFORMATION

Additional information on the selection, ratings, installation, and maintenance of cooler coil units is available from the various unit manufacturers. Also, the 1990 ASHRAE *Handbook—Refrigeration* section on Food Refrigeration has specific product cooling information.

REFERENCES

ARI. 1989. *Standard* 420, Unit coolers for refrigeration. Air Conditioning and Refrigeration Institute, Arlington, VA.

ASHRAE. 1990. *Standard* 25, Method of testing forced and natural convection air coolers for refrigeration.

CHAPTER 24

AIR-HEATING COILS

AIR-HEATING coils are used to heat air under forced convection. The total coil surface may consist of a single coil section or several coil sections assembled into a bank. The coils described in this chapter only apply to comfort heating and air conditioning using steam or hot water.

COIL CONSTRUCTION AND DESIGN

Extended-surface coils consist of a primary and a secondary heat transfer surface. The primary surface is the surface of the round tubes or pipes that are arranged in a repetitive pattern with respect to the airflow. The secondary surface (fins) consists of thin metal plates or a spiral ribbon uniformly spaced or wound along the length of the primary surface. It is in intimate contact with the primary surface for good heat transfer. This bond must be maintained permanently to ensure continuation of rated performance.

The heat transfer bond between the fin and the tube may be achieved in numerous ways. Bonding is generally accomplished by expanding the tubes into the tube holes in the fins to obtain a permanent mechanical bond. The tube holes frequently have a formed fin collar, which provides the area of thermal contact and may serve to space the fins uniformly along the tubes.

The fins of spiral or ribbon-type fin coils are tension-wound onto the tubes. In addition, an alloy with a low melting point, such as solder, may be used to provide a metallic bond between fins and tube. Some types of spiral fins are knurled into a shallow groove on the exterior of the tube. Sometimes, the fins are formed out of the material of the tube.

The fin designs most frequently used for heating coils are flat plate fins, plate fins of special shape, and spiral or ribbon fins.

Copper and aluminum are the materials most commonly used in the fabrication of extended-surface coils. Tubing made of steel or various copper alloys is used in applications where corrosive forces might attack the coils from either inside or outside. The most common combination for low-pressure applications is aluminum fins on copper tubes. Low-pressure steam coils are usually designed to operate up to 150 to 200 psig. Above that point, tube materials such as red brass, admiralty, or cupronickel are selected.

Steam Coils

For proper performance of steam-heating coils, condensate and air or other noncondensables must be eliminated rapidly and the steam must be distributed uniformly to the individual tubes. Noncondensable gases (such as carbon dioxide) remaining in a coil cause chemical corrosion and result in early coil failure.

Uniform steam distribution is accomplished by methods such as the following:
- Individual orifices at the entrance to each tube
- Distribution plates in the steam headers

The preparation of this chapter is assigned to TC 8.4, Air-to-Refrigerant Heat-Transfer Equipment.

- Special perforated, small diameter, steam distribution tubes that extend into the larger diameter tubes of the primary surface

Freeze-resistant coils of the perforated inner-tube type are constructed with arrangements such as the following:
- Supply and return on one end, with incoming steam used to heat the leaving condensate
- Supply and return on opposite ends
- For coils with long fins, supply and return on one end and a supply on the opposite end

Particular care in the design and installation of piping, controls, and installation is necessary to protect the coils from freeze-up because of incomplete draining of condensate. When the entering air temperature is at freezing or below, the steam supply to the coil should not be modulated. Coils located in series in the airstream, with each coil sized to be on or completely off in a specific sequence (depending on the entering air temperature), are not likely to freeze. The use of face bypass dampers is also common. During part-load conditions, air is bypassed around the steam coil with full steam flow to the coil. In this system, high-velocity jets of low-temperature air must not impinge on the coil when the face dampers are in a partially closed position. See Chapter 41 of the 1991 ASHRAE *Handbook—HVAC Applications* for more details.

Water Coils

The performance of water coils for heating depends on the elimination of air or other noncondensables from the system and the proper distribution of the water to the individual tube circuits. Air elimination in system piping is described in Chapter 12.

To produce the desired capacity without excessive water-pressure drop through the coil, various circuit arrangements are used. A single-tube serpentine circuit can be used on small booster heaters requiring small water quantities up to a maximum of approximately 4 to 5 gpm. With this arrangement, a single tube, carrying the entire water quantity, makes a number of passes across the airstream in a plane normal to the airflow.

The most common circuiting arrangement is often called single-row serpentine or standard circuiting. With this arrangement, all tubes in each coil row are supplied with an equal amount of water through a manifold, commonly called the coil header. When the water volume is small, so that water flow in the tubes is laminar, turbulators are sometimes installed in each tube circuit to produce turbulent water flow.

Special circuiting (serpentine) arrangements accomplish the same effect as mechanical turbulators when the water volume supplied to the coil is small. Serpentines reduce the number of circuits, thus increasing the water velocity in each circuit and creating turbulent flow.

Figure 1 illustrates commonly selected two-row water-circuit arrangements. When hot water coils are used with entering air temperatures below freezing (an antifreeze solution would be preferable to water), piping the coil for parallel flow rather than

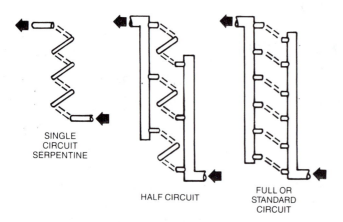

Fig. 1   Water Circuit Arrangements—Two-Row Heating Coils

counterflow, as shown, should be considered. This arrangement places the highest water temperature on the entering air side. Coils piped for counterflow have the water enter the tube row on the exit air side of the coil. Coils piped for parallel flow have the water entering the tube row on the enter air side of the coil.

Water coils are usually designed to be self-venting by supplying water to the coil so that the water flows upward in the coil, forcing air out through the return water connection. Such a design ensures that the coil is always completely filled with water, regardless of the water volume supplied, and that the coil may be drained completely.

Methods for controlling water coils to produce uniform leaving air temperatures are discussed in Chapters 12, 14, and 41 of the 1991 ASHRAE *Handbook—HVAC Applications*.

Flow Arrangement

The relative directions of fluid flows influence the performance of heat transfer surfaces. In air-heating coils with only one row of tubes, the air flows at right angles relative to the heating medium. In coils with more than one row of tubes in the direction of the airflow, the heating medium in the tubes may be circuited in various ways.

Crossflow is common in steam-heating coils. The steam temperature within the tubes remains substantially uniform, and the mean temperature difference is basically the same, whatever the direction of flow relative to the air.

Parallel flow and counterflow arrangements are common in water coils. Counterflow is the preferred arrangement to obtain the highest possible mean temperature difference. The mean temperature potential determines the heat transfer driving force of the coil. The greater the mean temperature difference, the greater is the heat transfer capacity of the coil.

Electric Heating Coils

An electric heating coil consists of a length of resistance wire (commonly nickel/chromium) to which a voltage is applied. The resistance wire may be bare or sheathed. The sheathed coil is a resistance wire encased by an electrically insulating layer such as magnesium oxide, which is encased in a finned steel tube. The sheathed coils are more expensive, have a higher air-side pressure drop, and require more space (41,200 Btu/h·ft² of face area compared to 101,000 Btu/h·ft² for bare coils). The outer surface temperature of sheathed coils is lower, the coils are mechanically stronger, and contact with body or housing is not dangerous. Sheathed coils are generally preferred for dust-laden or potentially explosive atmospheres or where there is a high probability of direct contact of personnel with coils.

Applications

Airflow through heating coils is usually vertical or horizontal; the latter arrangement is most common.

For steam heating, the coils may be installed with the tubes in a vertical or horizontal position. Horizontal tube coils should be pitched or inclined toward the return connection to drain the condensate. Water-heating coils generally have horizontal tubes to avoid air and water pockets. Where water coils may be exposed to below-freezing temperatures, drainability must be considered, or an antifreeze fluid such as glycol must be put in the coil. If a coil is to be drained and then exposed to below-freezing temperatures, it should first be flushed with an antifreeze solution.

When the leaving air temperature is controlled by modulation of the steam supply to the coil, steam distribution tube-type coils provide the most uniform exit air temperatures (see section on Steam Coils).

Correctly designed steam distributing tube coils limit the exit air temperature stratification to a maximum of 6°F over the entire length of the coil, even when the steam supply is modulated to a small fraction of the full-load capacity.

To minimize the danger of freezing in both steam- and water-heating coils, the outdoor-air inlet dampers usually close automatically when the fan is stopped (system shutdown). In steam systems with very low temperature outdoor air conditions (*e.g.*, −20°F or below), it is desirable to have the steam valve go to full open position when the system is shut down. If outside air is used for proportioning building makeup air, its damper should be an opposing blade design.

Heating coils are designed to allow for expansion and contraction resulting from the temperature ranges within which they operate. Care must be taken to prevent imposing strains from the piping on the coil connections. This is particularly important on high-temperature hot water applications. Expansion loops, expansion or swing joints, or flexible connections usually provide this protection (see Chapter 10).

It is good practice to support banked coils individually in an angle-iron frame or a similar supporting structure. With this arrangement, the lowest coil is not required to support the weight of the coils stacked above. This design also facilitates the removal of individual coils in a multiple-coil bank for repair or replacement.

Low-pressure steam systems, coils controlled by modulating the steam supply, or both, should have a vacuum breaker or be drained through a vacuum-return system to ensure proper condensate drainage. It is good practice to install a closed vacuum breaker (where one is required) connected to the condensate return line through a check valve. This unit breaks the vacuum by equalizing the pressure, yet minimizes the possibility of air bleeding into the system. Steam traps should be located at least 12 in. below the condensate outlet to enable the coil to drain properly. Also, coils supplied with low-pressure steam or controlled by modulating the steam supply should not be trapped directly to overhead return lines. Condensate can be lifted to overhead returns only when sufficient pressure is available to overcome the condensate head and any return-line pressure. If overhead returns are necessary, the condensate must be pumped to the higher elevation.

COIL SELECTION

The following factors should be considered in coil selection:

1. Required duty or capacity, considering other system components
2. Temperature of air entering the coil
3. Available heating media
4. Space and dimensional limitations
5. Air quantity and coil face velocity

6. Permissible resistances for both the air and heating media
7. Characteristics of individual coil designs
8. Individual installation requirements, such as the type of control to be used

The duties required of the coil may be determined from information in Chapters 23, 25, and 26 of the 1989 ASHRAE *Handbook—Fundamentals*. There may be a choice of heating media, as well as operating temperatures, depending on whether the installation is new or is being modified. The air handled may be limited by the use of ventilating ducts for air distribution, or it may be determined by requirements for satisfactory air distribution or ventilation.

The resistance through the air circuit influences fan power and speed. This resistance may be limited to allow the use of a given size of fan motor or to keep the operating expense low. It may also be limited because of sound-level requirements.

The permissible water resistance of a hot water coil may be dictated by the available pressure from a given size of pump and motor. This is usually controlled within limits by careful selection of the coil header size and the number of tube circuits. The performance of a heating coil depends on the correct choice of the original equipment and on proper application and maintenance. For steam coils, performance and selection of the correct type and size of steam trap is of the utmost importance.

Coil ratings are based on uniform face velocity. Nonuniform airflow through the coil will affect performance. Nonuniform airflow may be caused by air entrance at odd angles or by inadvertent blocking of a portion of the coil face. To obtain rated performance, air quantity in the field must correspond with design quantity and must always be maintained.

Complete mixing of return and outdoor air is essential to the proper operation of a coil. Mixing damper design is critical to the operation of a system. Systems in which the air passes through a fan before flowing through a coil do not ensure proper air mixing.

Coils are commonly cleaned by washing them with water. They can sometimes be brushed and cleaned with a vacuum cleaner. In extreme cases of neglect, especially in restaurants where grease and dirt have accumulated, it is sometimes necessary to remove the coils and wash off the accumulation with steam, compressed air and water, or hot water and a suitable detergent. Often, outside makeup air coils have no upstream air filters. Overall, it is best to inspect and service on a regular schedule.

Coil Ratings

Steam and hot water coils are usually rated within the following limits, which may be exceeded for special applications:

Air face velocity. Between 200 and 1500 fpm, based on air density of 0.075 lb/ft³.
Entering air temperature. Minus 20 to 100°F for steam coils; 0 to 100°F for hot water coils.
Steam pressures. From 2 to 250 psi (gage) at the coil steam supply connection (pressure drop through the steam control valve must be considered).
Hot water temperatures. Between 120 and 250°F.
Water velocities. From 0.5 to 8 fps.

Individual installations vary widely, but the following values can be used as a guide.

The most common air face velocities used are between 500 and 1000 fpm. Delivered air temperatures vary from about 72°F for ventilation only to about 150°F for complete heating. Steam pressures vary from 2 to 15 psi (gage), with 5 psi (gage) being the most common. A minimum steam pressure of 5 psi (gage) is recommended for systems with entering air temperatures below freezing. Water temperatures for comfort heating are commonly between 180 and 200°F, with water velocities between 4 and 6 fps.

Water quantity is usually based on about 20°F temperature drop through the coil. Air resistance is usually limited to 0.4 to 0.6 in. of water for commercial buildings and to about 1 in. for industrial buildings. High-temperature water systems have water temperatures commonly between 300 and 400°F, with up to 100°F drops through the coil.

The selection of heating coils is relatively simple because it involves dry-bulb temperatures and sensible heat only, without the complication of simultaneous latent heat loads, as in cooling coils. Heating coils are usually selected from charts or tables giving final air temperatures at various air velocities, entering air temperatures, and steam or water temperatures. Oversizing of modulating steam valves and steam coils in this system must be avoided because such oversizing makes control more difficult.

The selection of hot water heating coils is more complicated because of the added water velocity variable and the fact that the water temperature decreases as it flows through the coil circuits. Therefore, in selecting hot water heating coils, the water velocity and the mean temperature difference (between the hot water flowing in the tubes and the air passing over the fins) must be considered.

Most coil manufacturers have their own methods of producing performance rating tables from a suitable number of coil performance tests. A method of testing air heating coils is given in ASHRAE *Standard* 33-78, Method of Testing for Rating Forced-Circulation Air Cooling and Heating Coils. A basic method of rating to provide a fundamental means for establishing thermal performance of air-heating coils by extension of test data, as determined from laboratory tests on prototypes, to other operating conditions, coil sizes, and row depths of a particular surface design and arrangement is given in ARI *Standard* 410-87, Forced-Circulation Air-Cooling and Air-Heating Coils.

HEAT TRANSFER AND PRESSURE DROP

For air-side heat transfer and pressure drop, the information given in Chapter 21 for sensible cooling coils is applicable. For hot water coils, the information given in Chapter 21 for water-side heat transfer and pressure drop is also applicable here. For steam coils, the heat transfer coefficient of condensing steam has to be calculated. Chapter 3 of the 1989 ASHRAE *Handbook—Fundamentals* lists several equations for this purpose. Shah (1981) reviewed the available information on heat transfer during condensation in horizontal, vertical, and inclined tubes, and has made design recommendations. For estimation of pressure drop of condensing steam, see Chapter 5 of the 1989 ASHRAE *Handbook—Fundamentals*.

Parametric Effects

The heat transfer performance of a given coil can be changed by varying the flow rates of the air and/or the temperature of the heating medium. Understanding the interaction of these parameters is necessary for designing satisfactory coil-capacity and control. A review of manufacturers' catalogs, many of which are listed in the Air-Conditioning and Refrigeration Institute (ARI) Directory CHC: *Certified Air-Cooling and Air-Heating Coils*, shows the effects of varying these parameters.

REFERENCE

Shah, M.M. 1981. Heat transfer during film condensation in tubes and annuli: A review of the literature. ASHRAE *Transactions*, 87 (1): 1086-1105.

CHAPTER 25

AIR CLEANERS FOR PARTICULATE CONTAMINANTS

THIS chapter discusses the cleaning of both ventilation air and recirculated air for the conditioning of building interiors. Complete air cleaning may require the removal of airborne particles, microorganisms, and gaseous pollutants. This chapter addresses only the removal of airborne particles. The total suspended particulate concentration in this case seldom exceeds 2 mg/m³ and is usually less than 0.2 mg/m³ of air (National Air Pollution Control Administration 1969). This contrasts to exhaust gases from processes, flue gases, etc., where dust concentration typically is from 200 to 40,000 mg/m³. Chapter 40 of the 1991 ASHRAE *Handbook—HVAC Applications* covers the removal of gaseous contaminants; Chapter 26 of this volume covers exhaust-gas pollution control.

With certain exceptions, the air cleaners addressed in this chapter do not apply to exhaust gas streams, principally because of extreme differences in dust concentration and temperatures. However, the principles of air cleaning covered here apply to exhaust streams, and air cleaners discussed in the chapter are used extensively in supplying gases of low particulate concentration to industrial processes.

ATMOSPHERIC DUST

Atmospheric dust is a complex mixture of smokes, mists, fumes, dry granular particles, and natural and synthetic fibers. (When suspended in a gas, this mixture is called an *aerosol*.) A sample of atmospheric dust usually contains soot and smoke, silica, clay, decayed animal and vegetable matter, organic materials in the form of lint and plant fibers, and metallic fragments (Moore *et al.* 1954). It may also contain living organisms, such as mold spores, bacteria, and plant pollens, which may cause diseases or allergic responses. (Chapter 11 of the 1989 ASHRAE *Handbook—Fundamentals* contains further information on atmospheric contaminants.) A sample of atmospheric dust gathered at any point generally contains materials common to that locality, together with other components that originated at a distance but were transported by air currents or diffusion. These components and their concentrations vary with the geography of the locality (urban or rural), the season of the year, weather, the direction and strength of the wind, and proximity of dust sources.

Aerosol sizes range from 0.01 μm and smaller for freshly formed combustion particulates and radon progeny, to 0.1 μm for aged cooking and cigarette smokes, to 0.1 to 10 μm for airborne dusts, and up to 100 μm and larger for airborne soil, pollens, and allergens.

Concentrations of atmospheric aerosols are generally at a maximum at submicrometre sizes and decrease rapidly as particulate size increases above 1 μm. For a given size, the concentration can

vary by several orders of magnitude over time and space, particularly in the vicinity of an aerosol source. Such sources include human activities, equipment, furnishings, and pets (McCrone *et al.* 1967). This wide range of particulate size and concentration makes it impossible to design one cleaner to serve all applications.

VENTILATION AIR CLEANING

Different fields of application require different degrees of air cleaning effectiveness. In industrial ventilation, removing only the larger dust particles from the airstream may be necessary for cleanliness of the structure, protection of mechanical equipment, and employee health. In other applications, surface discoloration must be prevented. Unfortunately, the smaller components of atmospheric dust are the worst offenders in smudging and discoloring building interiors. Electronic air cleaners or medium to high efficiency dry filters are required to remove smaller particles, especially the respirable fraction, which often must be controlled for health reasons. In cleanroom applications or when radioactive or other dangerous particles are present, high or ultra-high efficiency filters should be selected.

The characteristics of aerosols that most affect the performance of an air filter include particle size and shape, mass, concentration, and electrical properties. The most important of these is particle size. Figure 1 of Chapter 11 of the 1989 ASHRAE *Handbook—Fundamentals* gives data on the sizes and characteristics of airborne particulate matter and the wide range of particle size that may be encountered.

Particle size may be defined in numerous ways. Particles less than 2.5 μm in diameter are generally referred to as the *fine* mode, with those greater than 2.5 μm being considered as the *coarse* mode. The fine and coarse mode particles typically originate by separate mechanisms, are transformed separately, have different chemical compositions, and require different control strategies. Fine mode particles generally originate from condensation processes or are directly emitted as combustion products. These particles are less likely to be removed by gravitational settling, and are just as likely to deposit on vertical surfaces as on horizontal surfaces. Coarse mode particles are typically produced by mechanical actions such as erosion and frictional wear. Coarse particles are more easily removed by gravitational settling, and thus have a shorter lifetime in the airborne state.

From an industrial hygiene perspective, particles 10 μm or greater are considered the *nonrespirable* fraction of dust. Those particles less than 10 μm are considered the *respirable* fraction. Note that particle size in this discussion refers to aerodynamic particle size. Therefore, larger particles with lower densities could be found in the lungs. Also note that fibers are a different case than particles in that fiber shape, diameter, and density all affect where a fiber will settle in the body (NIOSH 1973).

The preparation of this chapter is assigned to TC 2.4, Particulate Air Contaminants and Particulate Contaminant Removal Equipment.

Cleaning efficiency with certain media is affected to a negligible extent by the velocity of the airstream. Major factors influencing filter design and selection include: (1) degree of air cleanliness required, (2) specific particle size range or aerosols that require filtration, and (3) aerosol concentration.

RATING AIR CLEANERS

In addition to criteria affecting the degree of air cleanliness, factors such as cost (initial investment and maintenance), space requirements, and airflow resistance have encouraged the development of a wide variety of air cleaners. Accurate comparisons of different air cleaners can be made only from data obtained by standardized test methods.

The three operating characteristics that distinguish the various types of air cleaners are *efficiency, airflow resistance*, and *dust-holding capacity*. Efficiency measures the ability of the air cleaner to remove particulate matter from an airstream. Average efficiency during the life of the filter is the most meaningful for most filters and applications. However, because the efficiency of many dry-type filters increases with dust load, in applications with low dust concentrations the initial (clean filter) efficiency should be considered for design. Airflow resistance (or simply resistance) is the static pressure drop across the filter at a given airflow rate. The term pressure drop is used interchangeably with resistance. Dust-holding capacity defines the amount of a particular type of dust that an air cleaner can hold when it is operated at a specified airflow rate to some maximum resistance value or before its efficiency drops seriously as a result of the collected dust.

Complete evaluation of air cleaners therefore requires data on efficiency, resistance, dust-holding capacity, and the effect of dust loading on efficiency and resistance. When applied to automatic renewable media devices (roll filters), the rating system must evaluate the rate at which the media is supplied to maintain constant resistance when standard dust is fed at a specified rate. When applied to electronic air cleaners, the effect of dust buildup on efficiency must be evaluated.

Air filter testing is complex and no individual test adequately describes all filters. Ideally, performance testing of equipment should simulate the operation of the device under operating conditions and furnish performance ratings of the characteristics important to the equipment user. In the case of air cleaners, the wide variations in the amount and type of particulate matter in the air being cleaned make rating difficult. Another complication is the difficulty of closely relating measurable performance to the specific requirements of users. Recirculated air tends to have a larger proportion of lint than does outside air. However, these difficulties should not obscure the principle that tests should simulate actual use as closely as possible.

In general, four types of tests, together with certain variations, determine air cleaner efficiency:

Arrestance. A standardized synthetic dust consisting of various particle sizes is fed into the air cleaner, and the weight fraction of the dust removed is determined. In the ASHRAE test standard summarized later in this chapter, this type of efficiency measurement is named *synthetic dust weight arrestance* to distinguish it from other efficiency values.

Dust spot efficiency. Atmospheric dust is passed into the air cleaner, and the discoloration effect of the cleaned air on filter paper targets is compared with that of the incoming air. This type of measurement is called *atmospheric dust spot efficiency.*

Fractional efficiency or **penetration.** Uniform-sized particles are fed into the air cleaner and the percentage removed by the cleaner is determined, typically by a photometer or condensation nuclei counter.

Particle size efficiency. Atmospheric dust is fed to the air cleaner, and air samples taken upstream and downstream are drawn through a particle counter to obtain efficiency versus particle size.

The indicated weight arrestance of air filters, as in the arrestance test, depends greatly on the particle size distribution of test dust, which, in turn, must consider its state of agglomeration. Therefore, this filter test requires a high degree of standardization of the test dust, the dust dispersion apparatus, and other elements of test equipment and procedures. This test is particularly suited to low and medium efficiency air filters that are most commonly used on recirculating systems. It does not distinguish between filters of higher efficiency.

The dust spot test measures the ability of a filter to reduce the soiling of fabrics and building interior surfaces. Since these effects depend mostly on fine particles, this test is most useful for high efficiency filters. A disadvantage is the variety and variability of atmospheric dust (McCrone *et al.* 1967; Whitby *et al.* 1958; and Horvath 1967), which may cause the same filter to test at different efficiencies at different locations (or even at the same location at different times). This variation is most apparent in low efficiency filters.

ASHRAE *Standard* 52.1 specifies both a weight test and a dust spot test and requires that values for both be reported. These results allow a comparison between air filter devices.

In fractional efficiency tests, use of uniform particle size aerosols has proven to be an accurate measure of the particle size versus efficiency characteristic of filters over a wide atmospheric size spectrum. The method is time-consuming and has been used primarily in research. However, the DOP test for HEPA filters is widely used for production testing at a narrow particle size range.

Many indoor and outdoor air pollution control strategies have focused on the need for efficiencies versus size data in the submicron range. To meet this need, several manufacturers publish efficiencies for a size range of atmospheric dust (Figure 3). The data presented may be for a clean filter or for an average over the life of the filters. No test standard currently exists for particle size efficiency testing.

The exact measurement of true *dust-holding capacity* is complicated by the variability of atmospheric dust; therefore, standardized synthetic dust is normally used. Such a dust also shortens the dust-loading cycle to hours instead of weeks.

Since synthetic dusts are not the same as atmospheric dusts, dust-holding capacity as measured by these accelerated tests may be different from that achieved by tests using atmospheric dust. The exact life of a filter in field use is impossible to determine by laboratory testing. However, tests of filters under standard conditions do provide a rough guide to the relative effect of dust on performance of various units.

Reputable laboratories perform accurate and repeatable filter tests. Differences in reported values generally lie within the variability of the test aerosols and dusts. Since most media are made of random air- or water-laid fibrous materials, the inherent media variations affect filter performance. Awareness of these variations prevents misunderstanding and specification of impossibly close performance tolerances. Caution must be exercised in interpreting published efficiency data, since the performance of two cleaners tested by different procedures generally cannot be compared. Values of air cleaner efficiency can be related only approximately to the rate of soiling of a space or of mechanical equipment or to the intensity of other objectionable effects.

AIR CLEANER TEST METHODS

Air cleaner test methods have been developed in several areas: the heating and air-conditioning industry, the automotive industry, the atomic energy industry, and government and military agencies. Several tests have become standard in general ventilation applications in the United States. In 1968, the test techniques developed by the U.S. National Bureau of Standards (now the National Institute of Standards and Technology) and the Air Filter Institute (AFI) were unified (with minor changes) into a single test procedure, ASHRAE *Standard* 52.1. Dill (1938), Whitby *et al.* (1956), and Nutting and Logsdon (1953) give details of the original codes.

In general, the ASHRAE *Weight Arrestance Test* parallels the AFI, making use of a similar test dust. The ASHRAE *Atmospheric Dust Spot Efficiency Test* parallels the AFI and NBS *Atmospheric Dust Spot Efficiency Tests* and specifies a dust-loading technique.

Arrestance Test

ASHRAE *Standard* 52.1 specifies an ASHRAE synthetic test dust as follows:

> This dust (commonly referred to as "ASHRAE test dust") is composed by weight of 72% Standardized Air Cleaner Test Dust, Fine, 23% of powdered carbon; and 5% of cotton linters.

A known amount of the prepared test dust at a known and controlled rate is fed into the test unit. The concentration of dust in the air leaving the filter is determined by passing the entire airflow through a high efficiency after-filter and measuring the gain in filter weight. Arrestance (in percent) = 100 [1 − (weight gain of after-filter/weight of dust fed)].

Atmospheric dust particles range in size from a small fraction of a micrometre up to particles tens of micrometres in diameter. The artificially generated dust cloud used in the ASHRAE weight arrestance method is considerably coarser than typical atmospheric dusts. It tests the ability of a filter to remove the largest atmospheric dust particles and gives little indication of the filter performance in removing the smallest particles. But, where the mass of dust in the air is the primary concern, this is a valid test because most of the mass is contained in the larger particles. Where extremely small particles are objectionable, the weight arrestance method of rating does not differentiate between filters.

Atmospheric Dust Spot Efficiency Test

One objectionable characteristic of the finer airborne dust particles is the capacity to soil walls and other interior surfaces. The discoloring rate of white, filter-paper targets (microfine glass fiber HEPA filter media) filtering samples of air constitutes an accelerated simulation of this effect. By measuring the change in light transmitted by these targets, the efficiency of the filter in reducing the soiling of surfaces may be computed.

ASHRAE *Standard* 52.1 specifies two equivalent atmospheric dust spot test procedures, each taking a different approach to correct for the nonlinearity of the relation between the discoloration of target papers and their dust load. In the first procedure (called the intermittent flow method), samples of untreated atmospheric air are drawn upstream and downstream of the tested filter. These samples are drawn at equal flow rates through identical targets of glass fiber filter paper. The downstream sample is drawn continuously; the upstream sample is interrupted in a timed cycle so that the average rate of discoloration of the upstream and downstream targets is approximately equal. The percentage of off-time approximates the efficiency of the filter. The filter efficiency in percent is:

$$E = 100 \left(1 - \frac{Q_1}{Q_2} \frac{(T_{20} - T_{21})}{(T_{10} - T_{11})} \frac{(T_{10})}{(T_{20})}\right)$$

where

Q_1 = total quantity of air drawn through upstream target
Q_2 = total quantity of air drawn through downstream target

T_{10} = initial light transmission of upstream target
T_{11} = final light transmission of upstream target
T_{20} = initial light transmission of downstream target
T_{21} = final light transmission of downstream target

In the alternate procedure (called the constant-flow method), untreated atmospheric air samples are also drawn at equal flow rates through equal-area glass fiber filter paper targets upstream and downstream, but without interrupting either sample. Discoloration of the upstream target is therefore greater than for the downstream target. Sampling is halted when the upstream target light transmission has dropped by at least 10% but no more than 40%. The *opacity* (Y), or percent change in light transmission, is then calculated for both targets:

$$Y_1 = 100 (T_{10} - T_{11})/T_{10} \quad \text{(upstream)}$$
$$Y_2 = 100 (T_{20} - T_{21})/T_{20} \quad \text{(downstream)}$$

These opacities are next converted into *opacity indices*, Z_1 and Z_2, which corrects for nonlinearity. The relation between Z and Y is essentially independent of the type of atmospheric dust sampled, and is given by the expression:

$$Z = -73.27 \ln (1 - Y/66.07)$$

Dust spot efficiency is then:

$$E = 100 (1 - Z_2/Z_1)$$

The advantage of the constant-flow method is that it takes the same length of time to run regardless of the efficiency of the filter, whereas the intermittent-flow method takes longer for higher efficiency filters. For example, an efficiency test run on a 90% efficient filter using the intermittent-flow method requires ten times as long as a test run using the constant-flow method.

The standard allows dust spot efficiencies to be taken at intervals during a synthetic dust-loading procedure. This characterizes the change of dust spot efficiency as dust builds up on the filter in service.

Dust-Holding Capacity Test

In ASHRAE *Standard* 52.1, the same synthetic test dust mentioned in the section above is fed to the filter. The pressure drop across the filter (its resistance) rises as dust is fed. The test is normally terminated when the resistance reaches the maximum operating resistance set by the manufacturer. However, not all filters of the same type retain collected dust equally well. The test, therefore, requires that arrestance be measured at least four times during the dust-loading process and that the test be terminated when two consecutive arrestance values of less than 85%, or one value equal to or less than 75% arrestance of the maximum arrestance, have been measured. The ASHRAE *Dust-Holding Capacity* is, then, the integrated amount of dust held by the filter up to the time the dust-loading test was terminated. A typical set of curves for an ASHRAE air filter test report on a fixed cartridge-type filter is shown in Figure 1. Both synthetic dust weight arrestance and atmospheric dust spot efficiencies are shown. The standard also specifies how self-renewable devices are to be loaded with dust to establish their performance under standard conditions. Figure 2 shows the results of such a test on an automatic roll media filter.

DOP Penetration Test

For high efficiency filters of the type used in cleanrooms and nuclear applications (HEPA filters), the normal test in the United States is the Thermal DOP method, as outlined in U.S. Military Standard, MIL-STD-282 (1956), and U.S. Army document 136-300-175A (1965). In this method, a smoke cloud of DOP droplets condenses from DOP vapor (DOP = di-octyl phthalate, or bis-[2-ethylhexyl] phthalate, an oily, high boiling-point liquid).

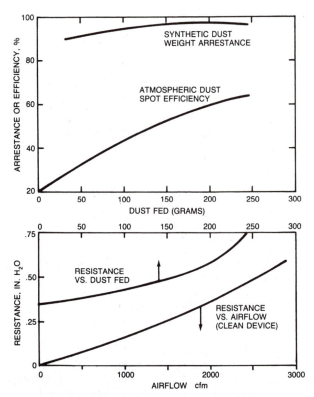

Fig. 1 Typical Performance Curves for a Fixed Cartridge-Type Filter by ASHRAE *Standard* 52.1

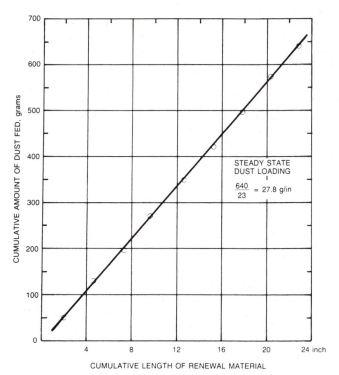

Note: Calculate loading per unit by multiplying steady-state dust loading by 144 and dividing by the media width, which is normally 24 in. For example: 27.8 × 144/24 =167 g/ft².

Fig. 2 Typical Dust-Loading Graph for Self-Renewable Air Filter

The count median diameter for DOP aerosols is about 0.18 μm, while the mass median diameter is about 0.27 μm with a cloud concentration of approximately 80 mg/m³ under properly controlled conditions. The procedure is sensitive to the mass median diameter, and DOP test results are commonly referred to as efficiency on 0.3 micrometre particles.

This smoke cloud is fed to the filter, which is held in a special test chuck. Any smoke that penetrates the body of the filter or leaks through gasket cracks passes into the region downstream from the filter where it is thoroughly mixed. The air leaving the chuck thus contains the average concentration of penetrating smoke. This concentration, as well as the upstream concentration, is measured by a light-scattering photometer. The filter penetration P is:

$$P = 100 \left(\frac{\text{Downstream Concentration}}{\text{Upstream Concentration}}\right) \%$$

Penetration, not efficiency, is usually specified in the test procedure because HEPA filters have efficiencies so near 100%. The two terms are related by the equation $E = 100 (1 - P) \%$.

U.S. Government specifications frequently call for the testing of HEPA filters at both rated flow and 20% of rated flow. This procedure helps reveal the presence of gasket leaks and pinholes in media that would otherwise escape notice. Such defects, however, are not located by this test.

The Institute of Environmental Sciences has published a *Recommended Practice for HEPA Filters*, covering five levels of performance and two grades of construction (IES-1986).

Leakage (Scan) Tests

In the case of HEPA filters, leakage tests are sometimes desirable to show that no small pinhole leaks exist or to locate and patch any that may exist. Essentially the same technique as the DOP Penetration Test is performed, except that the downstream con-

centration is measured by scanning the face of the filter and its gasketed perimeter with a moving probe. The exact point of smoke penetration can then be located and repaired. This same test (often called the *Cold DOP Test*) can be performed after the filter is installed; in this case, a portable aspirator-type DOP generator is used instead of the bulky thermal generator. The smoke produced by a portable generator is not uniform in size, but its average diameter can be approximated as 0.6 μm. Particle diameter is less critical for leak location than for penetration measurement.

Particle Size Efficiency Tests

No standard currently exists for determining the efficiency of air cleaners as a function of particle size. Such measurements depend heavily on the type of aerosol used as a filter challenge and, to a lesser extent, on the type of particle spectrometer being used. Figure 3 presents particle size efficiency curves for typical air filters compiled and averaged from manufacturers' data. These data are for general guidance only and must not be used to specify air filters.

One arrangement for obtaining such data is run concurrently with (or in place of) the dust spot test of ASHRAE *Standard* 52.1. For these measurements, sampling probes upstream and downstream of the test filter are located near the upstream and downstream dust spot target holders. With atmospheric or synthetic dust as the challenge, particles are counted both upstream and downstream to obtain an average count efficiency. The natural temporal variations in the test dust make it essential to choose up and down count cycles for statistical validity. Both laser and white light optical particle counters are used. The major advantage of the laser counter is its ability to sense particles in the size range of 0.1 to 0.3 μm, the latter being a typical lower limit for white light counters.

The Anderson-type cascade sampler is an accepted standard for measuring particle mass concentration as a function of particle size. This sampler requires the selection of proper filters, the

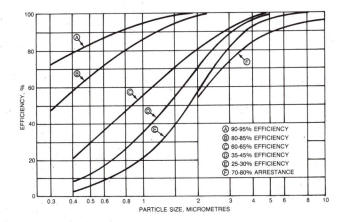

Curves are approximations compiled from manufacturers' recent data and are for general guidance only. Efficiency and arrestance per ASHRAE *Standard* 52.1 test methods. See also Bauer *et al.* (1973)

Caution: Do not apply curves' values to specify air filters, since no generally recognized test standard currently applies.

Fig. 3   Approximate Efficiency versus Particle Size for Typical Air Filters

careful measuring of filter mass change, and careful control of the sampled air volume. A real time, 10-stage cascade impactor with quartz crystal microbalance mass monitors in each stage and a microprocessor is also available. The nominal aerodynamic diameters for particles are, in micrometres: 0.05, 0.1, 0.2, 0.4, 0.8, 1.6, 3.2, 6.4, 12.5, and 25.0. Near real-time devices are available to record particulate mass concentrations continuously over a range of sizes using condensation nuclei-counting or light-scattering principles.

Specialized Performance Test

AHAM *Standard* AC-1-1986 method measures the ability of portable household air cleaners to reduce generated particulate matter suspended in the air in a room-size test chamber. The procedure measures the natural decay of three contaminants: dust, smoke, and pollen and then compares it with the air cleaner in operation.

Miscellaneous Performance Tests

Several other countries have developed standardized tests for air cleaners. Examples include several British tests (BSI 1969, 1971, 1985), a French test (AFNOR 1972), and a European Community Eurovent 4/5 test (CEN 1979). In addition, special test standards have been developed for respirator air filters (NIOSH/MSHA 1977).

Environmental Tests

Air cleaners may be subjected to fire, high humidity, a wide range of temperatures, mechanical shock, vibration, and other environmental stress. Several standardized tests exist for evaluating these environmental effects on air cleaners. MIL-STD-282 includes shock tests (shipment rough handling) and filter media water resistance tests. Several U.S. Atomic Energy Commission agencies (now part of the U.S. Department of Energy) specify humidity and temperature-resistance tests (Peters 1962, 1965).

Underwriters Laboratories has two major standards for air cleaner flammability. The first, for commercial applications,

determines flammability and smoke production. UL (UL *Standard* 900) Class 1 filters are those that, when clean, do not contribute fuel when attacked by flame and emit only negligible amounts of smoke. UL *Standard* 900 Class 2 filters are those that, when clean, burn moderately when attacked by flame or emit moderate amounts of smoke, or both. In addition, UL *Standard* 586 for flammability of HEPA filters has been established. The UL tests do not evaluate the effect of collected dust on filter flammability; depending on the dust, this effect may be severe. UL *Standard* 867 applies to electronic air cleaners.

ARI Standards

The Air-Conditioning and Refrigeration Institute has published ARI *Standard* 680-86 and ARI *Standard* 850-84 for air filter equipment. Although these standards are not widely used, they establish (1) definitions and classification; (2) requirements for testing and rating (performance test methods are per ASHRAE *Standard* 52.1); (3) specification of standard equipment; (4) performance and safety requirements; (5) proper marking; (6) conformance conditions; and (7) literature and advertising requirements.

MECHANISMS OF PARTICLE COLLECTION

In the collection of particulate matter, air cleaners rely on the following five main principles or mechanisms:

Straining. The coarsest kind of filtration strains particles through a membrane opening smaller than the particulate being removed. It is most often observed as the collection of large particles and lint on the filter surface. The mechanism is not adequate to explain the filtration of submicron aerosols through fibrous matrices, which occurs through other theoretical mechanisms, as follows.

Direct interception. The particles follow a fluid streamline close enough to a fiber that the particle contacts the fiber and remains there. The process is nearly independent of velocity.

Inertial deposition. Particles in the airstream are large enough or of large enough density that they cannot follow the fluid streamlines around a fiber; thus, they cross over streamlines, contact the fiber, and remain there. At high velocities (where these inertia effects are most pronounced), the particle may not adhere to the fiber because drag and bounce forces are so high. In this case, a viscous coating applied to the fiber obtains the full benefit and is the predominant mechanism in an adhesive-coated, wire screen impingement filter.

Diffusion. Very small particles have random motion about their basic streamlines (Brownian motion), which contributes to deposition on the fiber. This deposition creates a concentration gradient in the region of the fiber, further enhancing filtration by diffusion. The effects increase with decreasing particle size and velocity.

Electrostatic effects. Particle or media charging can produce changes in the collection of dust.

Some progress has been made in calculating theoretical filter media efficiency from the physical constants of the media by considering the effects of the collection mechanisms as described previously (Lee and Liu 1982a, 1982b; Liu and Rubow 1986).

TYPES OF AIR CLEANERS

Common air cleaners are broadly grouped as follows:

Fibrous media unit filters, in which the accumulating dust load causes pressure drop to increase up to some maximum recommended value. During this period, efficiency normally increases.

However, at high dust loads, dust may adhere poorly to filter fibers and efficiency will drop due to offloading. Filters in such condition should be replaced or reconditioned, as should filters that have reached their final (maximum recommended) pressure drop. This category includes viscous impingement and dry-type air filters, available in low efficiency to ultra-high efficiency construction.

Renewable media filters, in which fresh media is introduced into the airstream as needed to maintain essentially constant resistance and, consequently, constant efficiency.

Electronic air cleaners, which, if maintained properly by regular cleaning, have relatively constant pressure drop and efficiency.

Combination air cleaners of the above types are used. For example, an electronic air cleaner may be used as an agglomerator with a fibrous media downstream to catch the agglomerated particles blown off the plates. Electrode assemblies have been installed in air-handling systems, making the filtration system more effective (Frey 1985, 1986). Also, a renewable media filter may be used upstream of a high efficiency unit filter to extend its life. Charged media filters are also available that increase particle deposition on media fibers by an induced electrostatic field. In this case, pressure loss increases like it does on a fibrous media filter. The benefits of combining different air cleaning processes vary, but ASHRAE *Standard* 52.1 test methods may be used to compare performance.

FILTER TYPES AND THEIR PERFORMANCE

Panel Filters

Viscous impingement filters. These are panel filters made up of coarse fibers with a high porosity. The filter media are coated with a viscous substance, such as oil (also known as adhesive), which causes particles that impinge on the fibers to stick to them. Design air velocity through the media is usually in the range of 200 to 800 fpm. These filters are characterized by low pressure drop, low cost, and good efficiency on lint but low efficiency on normal atmospheric dust. They are commonly made 1/2 to 4 in. thick. Unit panels are available in standard and special sizes up to about 24 in. by 24 in. This filter is commonly used in residential furnaces and air conditioning and is often used as a prefilter for higher efficiency filters.

A number of different materials are used as the filtering medium, including coarse (15 to 60 μm diameter) glass fibers, coated animal hair, vegetable fibers, synthetic fibers, metallic wools, expanded metals and foils, crimped screens, random-matted wire, and synthetic open-cell foams. The arrangement of the medium in this type of filter involves three basic configurations:

1. *Sinuous media.* The filtering medium (consisting of corrugated metal or screen strips) is held more or less parallel to the airflow. The direction of airflow is forced to change rapidly in passing through the filter, thus giving inertial impingement of dust on the metal elements.
2. *Formed-Screen media.* Here, the filter media (screens or expanded metal) are crimped to produce high porosity media that avoid collapsing. Air flows through the media elements and dust impinges on the wires. The relatively open structure allows the filter to store substantial quantities of dust and lint without plugging.
3. *Random fiber media.* Fibers, either with or without bonding material, are formed into mats of high porosity. Media of this type are often designed with fibers packed more densely on the leaving air side than on the entering air side. This arrangement permits both the accumulation of larger particles and lint near

the air-entering face of the filter and the filtration of finer particles on the more closely packed air-leaving face. Fiber diameters may also be graded from coarse at the air-entry face to fine at the exit face.

Although viscous-impingement filters usually operate in the range of 300 to 600 fpm they may be operated at higher velocities. The limiting factor, other than increased flow resistance, is the danger of blowing off agglomerates of collected dust and the viscous coating on the filter.

The loading rate of a filter depends on the type and concentration of the dirt in the air being handled and the operating cycle of the system. Manometers, static pressure gages, or pressure transducers are often installed to measure the pressure drop across the filter bank and thereby indicate when the filter requires servicing. The final allowable pressure drop may vary from one installation to another; but, in general, unit filters are serviced when their operating resistance reaches 0.5 in. of water. The decline in filter efficiency (which is caused by the absorption of the viscous coating by dust, rather than by the increased resistance because of dust load) may be the limiting factor in operating life.

The manner of servicing unit filters depends on their construction and use. Disposable viscous-impingement, panel-type filters are constructed of inexpensive materials and are discarded after one period of use. The cell sides of this design are usually a combination of cardboard and metal stiffeners. Permanent unit filters are generally constructed of metal to withstand repeated handling. Various cleaning methods have been recommended for permanent filters; the most widely used involves washing the filter with steam or water (frequently with detergent) and then recoating it with its recommended adhesive by dipping or spraying. Unit viscous filters are also sometimes arranged for in-place washing and recoating.

The adhesive used on a viscous-impingement filter requires careful engineering. Filter efficiency and dust-holding capacity depend on the specific type and quantity of adhesive used; this information is an essential part of test data and filter specifications. Desirable adhesive characteristics, in addition to efficiency and dust-holding capacity, are (1) a low percentage of volatiles to prevent excessive evaporation; (2) a viscosity that varies only slightly within service temperature range; (3) the ability to inhibit growth of bacteria and mold spores; (4) a high capillarity or the ability to wet and retain the dust particles; (5) a high flash point and fire point; and (6) freedom from odorants or irritants.

Typical performance of viscous-impingement unit filters operating within typical resistance limits is shown as Group I in Figure 4.

Dry-type extended-surface filters. The media used in dry-type air filters is random fiber mats or blankets of varying thicknesses, fiber sizes, and densities. Media of bonded glass fiber, cellulose fibers, wool felt, synthetics, and other materials have been used commercially. The media in filters of this class is frequently supported by a wire frame in the form of pockets, or V-shaped or radial pleats. In other designs, the media may be self-supporting because of inherent rigidity or because airflow inflates it into extended form such as with bag filters. Pleating of the media provides a high ratio of media area to face area, thus allowing reasonable pressure drop.

In some designs, the filter media is replaceable and is held in position in permanent wire baskets. In most designs, the entire cell is discarded after it has accumulated its maximum dust load.

The efficiency of dry-type air filters is usually higher than that of panel filters, and the variety of media available makes it possible to furnish almost any degree of cleaning efficiency desired. Modern dry-type filter media and filter configurations also give dust-holding capacities generally higher than panel filters.

Coarse prefilters placed ahead of extended surface filters sometimes may be economically justified by the longer life of the

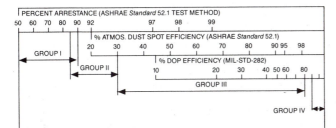

Fig. 4   Comparative Performance of Viscous Impingement and Dry Media Filters

main filters. Economic considerations should include the prefilter material cost, changeout labor, and increased fan system power use. Generally, prefilters should be considered only where they will substantially reduce the part of the dust that may plug the protected filter, which is usually one with an atmospheric dust spot efficiency of 70% or more. Temporary prefilters are worthwhile during building construction to capture heavy loads of coarse dust. HEPA-type filters of 95% DOP efficiency and greater should always be protected by prefilters of 80% or greater ASHRAE atmospheric dust spot efficiency. A single filter gage may be installed when a panel prefilter is placed adjacent to a final filter. Since the prefilter is normally changed on a schedule, the final filter pressure drop can be read without the prefilter in place evey time the prefilter is changed.

Typical performance of some types of filters in this group, when they are operated within typical rated resistance limits and over the life of the filters, is shown as Groups II and III in Figure 4.

The initial resistance of an extended surface filter varies with the choice of media and the filter geometry. Commercial designs typically have an initial resistance from 0.1 to 1.0 in. of water. It is customary to replace the media when the final resistance of 0.5 in. of water is reached for the low resistance units and 2.0 in. of water for the highest resistance units. Dry media providing higher orders of cleaning efficiency have a higher resistance to airflow. The operating resistance of the fully dust-loaded filter must be considered in the system design, since that is the maximum resistance against which the fan will operate. Variable air volume systems and systems with constant air volume controls prevent abnormally high airflows or possible fan motor overloading from occurring when filters are clean.

Flat panel filters with media velocity equal to duct velocity are possible only in the lowest efficiency units of the dry type (open cell foams and textile denier nonwoven media). Initial resistance of this group, at rated airflow, is mainly between 0.05 and 0.25 in. of water. They are usually operated to a final resistance of 0.50 to 0.70 in. of water.

In extended-surface filters of the intermediate efficiency ranges, the filter media area is much greater than the face area of the filter; hence, velocity through the filter media is substantially lower than the velocity approaching the filter face. Media velocities range from 6 to 90 fpm, although the approach velocities run to 750 fpm. Depth in direction of airflow varies from 2 to 36 in.

Filter media used in the intermediate efficiency range include those of (1) fine glass fibers, 0.7 to 10 μm in diameter, in mat form up to 1/2-in. thick; (2) thin nonwoven mats of fine glass fibers, cellulose, or cotton wadding; and (3) nonwoven mats of comparatively large diameter fibers (more than 30 μm) in greater thicknesses (up to 2 in.).

Electret filters are composed of electrostatically charged fibers. The charges on the fibers augment the collection of smaller particles by Brownian diffusion with Coulombic forces caused by the charges on the fibers. There are three types of these filters: resin wool, electret, and an electrostatically sprayed polymer. The charge on the resin wool fibers is produced by friction during the carding process. During production of the electret, a corona discharge injects positive charges on one side of a thin polypropylene film and negative charges on the other side. These thin sheets are then shredded into fibers of rectangular cross-section. The third process spins a liquid polymer into fibers in the presence of a strong electric field, which produces the charge separation. The efficiency of the charged-fiber filters is due to both the normal collection mechanisms of a media filter and the strong local electrostatic effects. The effects induce efficient preliminary loading of the filter to enhance the caking process. However, dust collected on the media can reduce the efficiency of electret filters.

Very high efficiency dry filters, HEPA (high efficiency particulate air), and ULPA (ultra low penetration air) filters, are made in an extended surface configuration of deep space folds of submicron glass fiber paper. Such filters operate at duct velocities near 250 fpm, with resistance rising from 0.5 to more than 2.0 in. of water over their service life. These filters are the standard for cleanroom, nuclear, and toxic-particulate applications.

Membrane filters are used predominantly for air sampling and specialized small-scale applications where their particular characteristics compensate for their fragility, high resistance, and high cost. They are available in many pore diameters and resistances and in flat sheet and pleated forms.

Renewable Media Filters

Moving-curtain viscous impingement filters. Automatic moving-curtain viscous filters are available in two main types. In one type, random-fiber media is furnished in roll form. Fresh media is fed manually or automatically across the face of the filter, while the dirty media is rewound onto a roll at the bottom. When the roll is exhausted, the tail of the media is wound onto the takeup roll, and the entire roll is thrown away. A new roll is then installed and the cycle is repeated.

Moving-curtain filters may have the media automatically advanced by motor drives on command from a pressure switch, timer, or media light-transmission control. A pressure switch control measures the pressure drop across the media and switches *on* and *off* at chosen upper and lower set points. This control saves media, but only if the static pressure probes are located properly and unaffected by modulating outside air and return air dampers. Most pressure drop control systems do not work well in practice. Timers and media light-transmission controls help to avoid these problems; their duty cycles can usually be adjusted to provide satisfactory operation with acceptable media consumption.

Filters of this replaceable roll design are generally constructed to be fail-safe by having a signal indicating when the roll of media is nearly exhausted. At the same time, the drive motor is deenergized so that the filter cannot run out of media. The normal service requirements involve insertion of a clean roll of media at the top of the filter and disposing of the loaded dirty roll. Automatic filters of this design are not, however, limited in application to the vertical position. Horizontal arrangements are available for use with makeup air units and air-conditioning units. Adhesives must have qualities similar to those for panel-type viscous impingement filters, plus the ability to withstand media compression and endure long storage.

The second type of automatic viscous-impingement filter consists of linked metal mesh media panels installed on a traveling curtain that intermittently passes through an adhesive reservoir. In there, the panels give up their dust load and, at the same time, take on a new coating of adhesive. The panels thus form a continuous curtain that moves up one face and down the other face. The media curtain, continually cleaned and renewed with fresh adhesive, lasts the life of the filter mechanism. The precipitated dirt must be removed periodically from the adhesive reservoir.

The resistance of both types of viscous impingement automatically renewable filters remains approximately constant as long as proper operation is maintained. A resistance of 0.40 to 0.50 in. of water at a face velocity of 500 fpm is typical of this class.

Moving-curtain, dry-media filters. Random-fiber (nonwoven) dry media of relatively high porosity are also used in moving-curtain (roll) filters for general ventilation service. Operating duct velocities in the area of 200 fpm are generally lower than for viscous-impingement filters.

Special automatic dry filters are also available, which are designed for the removal of lint in textile mills and dry-cleaning establishments and the collection of lint and ink mist in press rooms. The medium used is extremely thin and serves only as a base for the buildup of lint, which then acts as a filter medium. The dirt-laden media is discarded when the supply roll is used up.

Another form of filter designed specifically for dry lint removal consists of a moving curtain of wire screen, which is vacuum cleaned automatically at a position out of the airstream. Recovery of the collected lint is sometimes possible with such a device.

Performance of renewable media filters. ASHRAE arrestance, efficiency, and dust-holding capacities for typical viscous-impingement and dry renewable media filters are listed in Table 1.

Electronic Air Cleaners

Electronic air cleaners can be highly efficient filters using electrostatic precipitation to remove and collect particulate contaminants such as dust, smoke, and pollen. The designation electronic air cleaner denotes a precipitator for HVAC air filtration. The filter consists of an ionization section and a collecting plate section.

In the ionization section, small diameter wires with a positive direct current potential of between 6 and 25 kV DC are suspended equidistant between grounded plates. The high voltage on the wires creates an ionizing field for charging particles. The positive ions created in the field flow across the airstream and strike and adhere to (charge) the particles, which then pass into the collecting plate section.

The collecting plate section consists of a series of parallel plates equally spaced with a positive direct current voltage of 4 to 10 kV DC applied to alternate plates. Plates that are not charged are at ground potential. As the particles pass into this section, they are forced to the plates by the electric field on the charges they carry, and thus they are removed from the airstream and collected by the plates. Particulate retention is a combination of electrical and intermolecular adhesion forces and may be augmented by special oils or adhesives on the plates. Figure 5 shows a typical electronic air cleaner cell.

In lieu of positive direct current, a negative potential also functions on the same principle, but more ozone is generated.

With voltages of 4 to 25 kV DC, safety measures are required. A typical arrangement makes the air cleaner inoperative when the doors are removed for cleaning the cells or servicing the power pack.

Electronic air cleaners typically operate from a 120- or 240-V AC single-phase electrical service. The high voltage supplied to the air cleaner cells is normally created with solid-state power supplies. The electrical power consumption ranges from 20 to 40 watts per 1000 cfm of air cleaner capacity.

This type of air filter can remove and collect airborne contaminants with average efficiencies of up to 98% at low airflow velocities (150 to 350 fpm) when tested per the ASHRAE *Standard* 52.1. Efficiency decreases (1) as the collecting plates become loaded with particulates, (2) with higher velocities, or (3) with nonuniform velocity.

As with most air filtration devices, the duct approaches to and from the air cleaner housing should be arranged so that the airflow is distributed uniformly over the face area. Panel prefilters should also be used to help distribute the airflow and to trap large particles that might short out or cause excessive arcing within the high-voltage section of the air cleaner cell. Electronic air cleaner design parameters of air velocity, ionizer field strength, cell plate

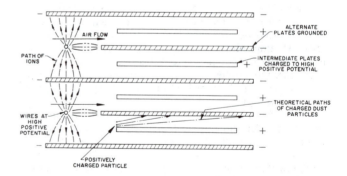

Fig. 5   Cross Section of Ionizing Electronic Air Cleaner

Table 1   Performance of Renewable Media Filters (Steady-State Values)

Description	Type of Media	ASHRAE Weight Arrestance, %	ASHRAE Atmospheric Dust Spot Efficiency, %	ASHRAE Dust-Holding Capacity, g/ft²	Approach Velocity, fpm
20 to 40 μm glass and synthetic fibers, 2 to 2 1/2 in. thick	Viscous impingement	70 to 82	<20	60 to 180	500
Permanent metal media cells or overlapping elements	Viscous impingement	70 to 80	<20	NA (permanent media)	500
Coarse textile denier nonwoven mat, 1/2 to 1 in. thick	Dry	60 to 80	<20	15 to 70	500
Fine textile denier nonwoven mat, 1/2 to 1 in. thick	Dry	80 to 90	<20	10 to 50	200

spacing, depth, and plate voltage must match the application requirements. These include contaminant type, particle size, volume of air, and required efficiency. Many units are designed for installation into central heating and cooling systems for total air filtration. Other self-contained units are furnished complete with air movers for source control of contaminants in specific applications that need an independent system.

Electronic air cleaner cells must be cleaned periodically with detergent and hot water. Some designs either incorporate automatic wash systems that clean the cell in place or the cells must be removed for cleaning. The frequency of cleaning (washing) the cell depends on the contaminant and the concentration. Industrial applications may require cleaning every 8 h, while a residential unit may only require cleaning at one to three month intervals. The timing of the cleaning schedule is important to keep the unit performing at peak efficiency. For some contaminants, special attention must be given to cleaning the ionizing wires.

Optional features are often available for electronic air cleaners. After filters such as roll filters collect particulates that agglomerate and blow off the cell plates. These are used mainly where heavy contaminant loading occurs and extension of the cleaning cycle is desired. Cell collector plates may be coated with special oils, adhesives, or detergents to improve both particle retention and removal during cleaning. High efficiency, dry-type, extended media area filters are also used as afterfilters in special designs. The electronic air cleaner used in this system improves the service life of the dry filter and collects small particles such as smoke.

Another device, a negative ionizer, uses the principle of particle charging but does not use a collecting section. Particulates enter the ionizer of the unit and receive an electrical charge. They then migrate to a grounded surface closest to the travel path. If use is continued in an area of heavy particulate concentration, a space charge can be built up, which tends to drive the charged particulate to surfaces throughout the room.

Space charge. Particulates that pass through an ionizer and are charged, but not removed, carry the electrical charge into the space. If continued on a large scale, a space charge will be built up, which tends to drive these charged particles to walls and interior surfaces. Thus, a low-efficiency electronic air cleaner used in areas of high ambient dirt concentrations or a malfunctioning unit can blacken walls faster than if no cleaning device were used (Penney and Hewitt 1949, Sutton *et al.* 1964).

Ozone. All high-voltage devices are capable of producing ozone, which is toxic and damaging to paper, rubber, and other materials. When properly designed and maintained, electronic air cleaners produce ozone concentrations that only reach a fraction of the levels acceptable for continuous human exposure and are less than those prevalent in many American cities (EPA). Continuous arcing and brush discharge in an electronic air cleaner may yield ozone levels that are annoying or mildly toxic; this will be indicated by a strong ozone odor. Although the nose is sensitive to ozone, only actual measurement of the concentration can determine that a hazardous condition exists. ASHRAE *Standard* 62-1989, Ventilation for Acceptable Indoor Air Quality, defines acceptable concentrations of oxidants, of which ozone is a major contributor. The United States Environmental Protection Agency specifies a 1-h average maximum allowable exposure to ozone to be 0.12 ppm for outside ambient air. The United States Department of Health and Human Services specifies the maximum allowable continuous exposure to ozone to be 0.05 ppm for contaminants of indoor origin. Sutton *et al.* (1976) showed that indoor ozone levels are only 30% of the outdoor level with ionizing air cleaners operating, although Weschli *et al.* (1989) found this level would increase when outdoor airflow is increased during an outdoor event that creates ozone.

SELECTION AND MAINTENANCE

To evaluate filters and air cleaners properly for a particular application the following factors should be considered:

- Degree of air cleanliness required
- Disposal of dust after it is removed from the air
- Amount and type of dust in the air to filtered
- Operating resistance to airflow (pressure drop)
- Space available for filtration equipment
- Cost of maintaining or replacing filters
- Initial cost of the system

Savings—from reduction in housekeeping expenses, protection of valuable property and equipment, ability to carry on dust-free manufacturing processes, improved working conditions, and even health benefits—should be credited against the cost of installing and operating an adequate system. The capacity and physical size of the required unit may emphasize the need for low maintenance cost. Operating costs, predicted life, and efficiency are as important as initial cost because air cleaning is a continuing process.

Panel filters do not have efficiencies as high as can be expected from extended-surface filters, but their initial cost and upkeep are generally low. They require more careful attention than the moving-curtain type if the resistance is to be maintained within reasonable limits.

If higher efficiencies are required, extended-surface filters or electronic air cleaners should be considered. The use of very fine glass fiber mats or other materials in extended-surface filters has made these available in the highest efficiency ranges.

Initial costs of extended-surface filters are lower than for electronic types, but higher than for panel types. Operating and maintenance costs of some extended surface filters may be higher than for panel types and electronic air cleaners, but the efficiencies are always higher than for panel types, and the cost/benefit ratio must be considered. Pressure drop of media type filters is greater than electronic type and slowly increases during their useful life. The advantages include the fact that no mechanical or electrical services are required. Choice should be based on both initial and operating costs, as well as on the degree of cleaning efficiency and maintenance requirements.

While electronic air cleaners have a higher initial cost, they exhibit high initial efficiencies in cleaning atmospheric air—largely because of their ability to remove fine particulate contaminants. System resistance remains unchanged as particles are collected, and the resulting residue is disposed of directly to prepare the equipment for further duty. The manufacturer must supply information on maintenance or cleaning.

Table 2 lists some applications of filters classified according to their efficiencies and type.

FILTER INSTALLATION

Many air cleaners are available in units of convenient size for manual installation, cleaning, and replacement. A typical unit filter may be 20 to 24 in. square, from 1 to 40 in. thick, and of either the dry or viscous-impingement types. In large systems, the frames in which these units are installed are bolted or riveted together to form a filter bank. Automatic filters are constructed in sections offering several choices of width up to 70 ft and generally range in height from 40 to 200 in. in 4 to 6 in. increments. Several sections may be bolted together to form a filter bank.

Several manufacturers provide side-loading filter sections for various types of filters. Filters are changed from outside the duct, making service areas in the duct unnecessary and thus saving cost and space.

Table 2   Typical Filter Applications Classified by Filter Efficiency and Type[a]

Application	System Designator[b]	Prefilter		Prefilter 2/Filter		Final Filter	Application Notes
Warehouse, storage, shop and process areas, mechanical equipment rooms, electrical control rooms, protection for heating and cooling coils	A1	None	None	50 to 85% arrestance	Panel-type or automatic roll	None	Reduce larger particle settling. Protect coils from dirt and lint.
	A2	None	None	25 to 30% dust spot	Pleated panel or extended surface	None	
Special process areas, electrical shops, paint shops, average general offices and laboratories	B1	None	None	75 to 90% arrestance 35 to 60% dust spot	Extended surface, cartridge, bag-type, or electronic (manually cleaned or replaceable media)	None	Average housecleaning. Reduces lint in air stream. Reduces ragweed pollen >85% at 35%. Removes all pollens at 60%, somewhat effective on particles causing smudge and stain.
Analytical laboratories, electronics shops, drafting areas, conference rooms, above-average general offices	C1	75 to 85% arrestance 25 to 40% dust spot	Extended surface, cartridge, or bag-type	>98% arrestance 80 to 85% dust spot	Bag-type, cartridge, or electronic (semi-automatic cleaning)	None	Above average housecleaning. No settling particles of dust. Cartridge and bag types very effective on particles causing smudge and stain, partially effective on tobacco smoke. Electronic types quite effective on smoke.
	C2	None	None	>98% arrestance 80 to 85% dust spot	Electronic (agglomerator) with bag or cartridge section	None	
Hospitals, pharmaceutical R&D and manufacturing (nonaseptic areas only), some clean ("gray") rooms	D1	75 to 85% arrestance 25 to 40% dust spot	Extended surface, cartridge, or bag-type	>98% arrestance 80 to 85% dust spot	Bag-type, cartridge, electronic (semi-automatic cleaning)	95% DOP disposable cell	Excellent housecleaning. Very effective on particles causing smudge and stain, smoke and fumes. Highly effective on bacteria.
	D2	None	None	>98% arrestance 80 to 95% dust spot	Electronic (agglomerator) with bag or cartridge section	None	
Aseptic areas in hospital and pharmaceutical R&D and manufacturing. Cleanrooms in film and electronics manufacturing, radioactive areas, etc.[c]	E1	75 to 85% arrestance 25 to 40% dust spot	Extended surface, cartridge, bag-type	>98% arrestance 80 to 85% dust spot	Bag-type, cartridge, electronic (semi-automatic cleaning)	≥99.97% DOP disposable cell	Protects against bacteria, radioactive dusts, toxic dusts, smoke, and fumes.

[a]Adapted from a similar table courtesy of E.I. du Pont de Nemours & Company.
[b]System designators have no significance other than their use in this table.
[c]Electronic agglomerators and air cleaners are not usually recommended for clean room applications.

The in-service efficiency of an air filter is, of course, sharply reduced if air leaks through either bypass dampers or poorly designed frames. The higher the efficiency of a filter, the more attention must be paid to the rigidity and sealing effectiveness of the frame. In addition, high efficiency filters must be handled and installed with care. Gilbert and Palmer (1965) suggest some of the precautions needed for filters of the HEPA type.

Air cleaners may be installed in the outdoor-air intake ducts of buildings and residences and in the recirculation and bypass air ducts, but they are always placed ahead of heating or cooling coils and other air-conditioning equipment in the system to protect them from dust. The dust captured in an outdoor air intake duct is likely to be mostly particulate matter of a greasy nature, while lint may predominate in dust from within the building.

Where high efficiency filters protect critical areas such as cleanrooms, it is important that the filters be installed as close to the room as possible to prevent the pickup of particles between the filters and the outlet. The ultimate is the so-called *laminar flow room*, in which the entire ceiling or one entire wall becomes the final filter bank.

The published performance data for all air filters are based on straight-through unrestricted airflow. Filters should be installed so that the face area is at right angles to the airflow whenever possible. Eddy currents and dead air spaces should be avoided; air should be distributed uniformly over the entire filter surface using baffles, diffusers, or air blenders, if necessary. Filters are sometimes damaged if higher-than-normal air velocities impinge directly on the face of the filter.

Failure of air filter installations to give satisfactory results can, in most cases, be traced to faulty installation or improper maintenance or both. The most important requirements of a satisfactory and efficiently operating air filter installation are as follows:

- The filter must be of ample capacity for the amount of air and dust load it is expected to handle. An overload of 10 to 15% is regarded as the maximum allowable. When air volume is subject to future increase, a larger filter bank should be installed initially.
- The filter must be suited to the operating conditions, such as degree of air cleanliness required, amount of dust in the entering air, type of duty, allowable pressure drop, operating temperatures, and maintenance facilities.
- The filter type should be the most economical for the specific application. The initial cost of the installation should be balanced against efficiency and depreciation, as well as expense and convenience of maintenance.

The following recommendations apply to filters installed with central fan systems:

- Duct connections to and from the filter should change size or shape gradually to ensure even air distribution over the entire filter area.
- Sufficient space should be provided in front of or behind the filter, or both, depending on its type, to make it accessible for inspection and service. A distance of 20 to 40 in. is required, depending on the filter chosen.
- Access doors of convenient size should be provided to the filter service areas.
- All doors on the clean-air side should be gasketed to prevent infiltration of unclean air. All connections and seams of the sheet-metal ducts on the clean-air side should be as airtight as possible. The filter bank must be caulked to prevent bypass of unfiltered air, especially when high efficiency filters are used.
- Electric lights should be installed in the plenum in front of and behind the air filter.
- Filters installed close to an air inlet should be protected from the weather by suitable louvers or inlet hood. A large mesh wire bird screen should be placed in front of the louvers or in the hood.

- Filters, other than electronic air cleaners, should have permanent indicators to give a warning when the filter resistance reaches too high a value or is exhausted, as with automatic roll media filters.
- Electronic air cleaners should have an indicator or alarm system to indicate when high voltage is off or shorted out.

SAFETY REQUIREMENTS

Safety ordinances should be investigated when the installation of an air cleaner is contemplated. Combustible filtering media may not be permitted in accordance with some existing local regulations. Combustion of dust and lint on filtering media is possible, although the media itself may not burn. This may cause a substantial increase in filter combustibility. Smoke detectors and fire sprinkler systems may be considered for filter bank locations.

In some cases, depending on the contaminant, hazardous material procedures must be followed during removal and disposal of the spent filter.

REFERENCES

AFNOR. 1972. Method of measurement of the efficiency of filters by means of an aerosol of uranine (fluorescein). AFNOR X44-011, Paris.

AHAM. 1986. Method of measuring performance of portable household electric cord-connected room air cleaners. AHAM *Standard* AC-1-1986. Association of Home Appliance Manufacturers, Chicago, IL.

ARI. 1984. Commercial and industrial air filter equipment. ARI *Standard* 850-84. Air-Conditioning and Refrigeration Institute, Arlington, VA.

ARI. 1986. Residential air filter equipment. ARI *Standard* 680-86.

ASHRAE. 1992. Methods of testing air cleaning devices used in general ventilation for removing particulate matter. ASHRAE *Standard* 52-1.

Bauer, E.J., B.T. Reagor, and C.A. Russell. 1973. Use of particle counts for filter evaluation. ASHRAE *Journal* (October):53.

BSI. 1969. Methods of test for low penetration air filters. B.S. 3928. British Standards Institute, London.

BSI. 1971. Methods of test for air filters used in air conditioning and general ventilation. B.S. 2831. British Standards Institute, London.

BSI. 1985. Air filters used in air conditioning and general ventilation. Part 1: Methods of test for atmospheric dust spot efficiency and synthetic dust weight arrestance. B.S. 6540. British Standards Institute, London.

CEN. 1979. Method of testing air filters used in general ventilation. EUROVENT 4/5. European Committee of Standards, Antwerp, Belgium.

Dill, R.S. 1938. A test method for air filters. ASHRAE *Transactions* 44:379.

EPA. Air quality criteria for photochemical oxidants. Environmental Protection Agency, Office of Air Noise and Radiation, Office of Air Quality Planning and Standards, Durham, NC.

Frey, A.H. 1985. Modification of aerosol size distribution by complex electric fields. *Bulletin of Environmental Contamination and Toxicology* 34:850-57.

Frey, A.H. 1986. The influence of electrostatics on aerosol deposition. ASHRAE *Transactions* 92(1B):55-64.

Gilbert, H. and J. Palmer. High efficiency particulate air filter units. USAEC, TID-7023. Available from NTIS, Springfield, VA.

Horvath, H. 1967. A comparison of natural and urban aerosol distribution measured with the aerosol spectrometer. *Environmental Science and Technology* (August):651.

IES. 1986. Recommended practice for HEPA filters. Institute of Environmental Sciences Publication IES-RP-CC-001-86, Mount Prospect, IL.

Lee, K.W. and B.T. Liu. 1982a. Experimental study of aerosol filtration by fibrous filters. *Aerosol Science and Technology* (1):35-46.

Lee. K.W. and B.T. Liu. 1982b. Theoretical study of aerosol filtration by fibrous filters. *Aerosol Science and Technology* (1):147-61.

Liu, B.Y. and K.L. Rubow. 1986. Air filtration by fibrous filter media. In *Fluid-Filtration: Gas*. ASTM STF973, Philadelphia, PA.

McCrone, W.C., R.G. Draftz, and J.G. Delley. 1967. *The particle atlas*. Ann Arbor Science Publishers, Ann Arbor, MI.

Moore, C.E., R. McCarthy, and R.F. Logsdon. 1954. A partial chemical analysis of atmospheric dirt collected for study of soiling properties. *Heating, Piping and Air Conditioning* (October):145.

National Air Pollution Control Administration. 1969. Air quality criteria for particulate matter. Obtainable from NTIS, Springfield, VA.

NIOSH. 1973. *The industrial environment—Its evaluation and control.* S/N 017-001-00396-4, U.S. Government Printing Office, Washington, D.C.

NIOSH/MSHA. 1977. U.S. Federal mine safety and health act of 1977. National Institute for Occupational Safety and Health, Columbus, OH and Mine Safety and Health Administration Department of Labor. Title 30 CFR Parts 11, 70, Washington, D.C.

Nutting, A. and R.F. Logsdon. 1953. New air filter code. *Heating, Piping and Air Conditioning* (June):77.

Penney, G.W. and G.W. Hewitt. 1949. Electrically charged dust in rooms. AIEE *Transactions* (68):276-82.

Peters, A.H. 1962. Application of moisture separators and particulate filters in reactor containment. USAEC-DP812, Washington, D.C.

Peters, A.H. 1965. Minimal specification for the fire-resistant high efficiency filter unit. USAEC *Health and Safety Information*, Issue No. 212, Washington, D.C.

Sutton, D.J. *et al.* 1964. Performance and application of electronic air cleaners in occupied spaces. ASHRAE *Journal* (June):55.

Sutton, D.J., K.M. Nodolf, and H.K. Makino. 1976. Predicting ozone concentrations in residential structures. ASHRAE *Journal* (September): 21.

UL. 1980. Electrostatic air cleaners. UL *Standard* 867-80. Underwriters Laboratories, Inc., Northbrook, IL.

UL. 1985. High efficiency, particulate, air filter units. UL *Standard* 586-85.

UL. 1977. Test performance of air filter units. UL *Standard* 900-77.

Weschler, C.J., H.C. Shields, and D.V. Noik. 1989. Indoor ozone exposures. *Journal of the Air Pollution Control Association* 39(12):1562.

Whitby, K.T., A.B. Algren, and R.C. Jordan. 1958. Size distribution and concentration of airborne dust. ASHRAE *Transactions* 64:129.

Whitby, K.T., A.B. Algren, and R.C. Jordan. 1956. The dust spot method of evaluating air cleaners. *Heating, Piping and Air Conditioning* (December):151.

BIBLIOGRAPHY

ASHRAE. 1965. Symposium on applied air cleaner evaluation.

Davies, C.N. 1973. *Aerosol filtration.* Academic Press, London.

Dennis, R. 1976. *Handbook on aerosols.* Technical Information Center, Energy Research and Development Administration. Oak Ridge, TN.

Dorman, R.G. 1974. *Dust control and air cleaning.* Pergamon Press, New York.

Drinker, P. and T. Hatch. *Industrial dust.* McGraw-Hill Book Co., New York.

Duncan, S.F. 1963. Effect of filter media microstructure on dust collection. ASHRAE *Journal* 6(April):37.

Engle, P.M. Jr. and C.J. Bauder. 1964. Characteristics and applications of high performance dry filters. ASHRAE *Journal* 6(May):72.

Gieske, J.A., E.R. Blosser, and R.B. Rief. 1975. A study of techniques for evaluation of airborne particulate matter. Battelle Institute Report to ASHRAE, Project No. TRP-97.

Hinds, W.C. 1982. *Aerosol technology: Properties, behavior, and measurement of airborne particles.* John Wiley and Sons, New York.

Hunt, C.M. 1972. An analysis of roll filter operation based on panel filter measurements. ASHRAE *Transactions* 78(2):227.

Jorgensen, R., ed. 1961. *Fan engineering,* 6th ed. Buffalo Forge Company, Buffalo, NY.

Licht, W. 1980. *Air pollution control engineering: Basic calculations for particulate collection.* Marcel Dekker, Inc., New York.

Lioy, P.J. and M.J.Y. Lioy, eds. 1983. *Air sampling instruments for evaluation of atmospheric contaminants,* 6th ed. American Conference of Governmental Industrial Hygienists, Cincinnati, OH.

Liu, B.Y.H. 1976. *Fine particles.* Academic Press, New York.

Lundgren, D.A., F.S. Harris, Jr., W.H. Marlow, M. Lippmann, W.E. Clark, and M.D. Durham, eds. 1979. *Aerosol measurement.* University Presses of Florida, Gainesville, FL.

Matthews, R.A. 1963. Selection of glass fiber filter media. *Air engineering* 5(October):30.

McNall, P.W. Jr. Indoor air quality—A status report. ASHRAE *Journal* 28(June):39.

National Research Council. 1981. *Indoor pollutants.* National Academy Press, Washington, D.C.

Nazaroff, W.W. and G.R. Cass. 1989. Mathematical modeling of indoor aerosol dynamics. *Environmental Science and Technology* 23(2):157-66.

Ogawa, A. 1984. *Separation of particles from air and gases,* Volumes I and II. CRC Press, Boca Raton, FL.

Penney, G.W. and N.G. Ziesse. 1968. Soiling of surfaces by fine particles. ASHRAE *Transactions* 74(1).

Rivers, R.D. 1988. Interpretation and use of air filter particle-size-efficiency data for general-ventilation applications. ASHRAE *Transactions* 88(1).

Rose, H.E. and A.J. Wood. 1966. *An introduction to electrostatic precipitation.* Dover Publishing Company, New York.

Stern, A.C., ed. 1977. *Air pollution,* 3rd ed. Vol. 4, Engineering control of air pollution. Academic Press, New York.

Swanton, J.R., Jr. 1971. Field study of air quality in air-conditioned spaces. ASHRAE *Transactions* 77(1):124.

U.S. Army. 1965. Instruction manual for the installation, operation, and maintenance of penetrometer, filter testing, DOP, Q107. Document 136-300-175A. Edgewood Arsenal, MD.

U.S. Government. 1988. Clean room and work station requirements, controlled environments. Federal *Standard* 209D, amended 1991. Available from GSA, Washington, D.C.

USAEC/ERDA/DOE. Air cleaning conference proceedings. Available from NTIS, Springfield, VA.

Walsh, P.J., C.S. Dudney, and E.D. Copenhaver, eds. 1984. *Indoor air quality.* CRC Press, Boca Raton, FL.

Whitby, K.T. 1965. Calculation of the clean fractional efficiency of low media density filters. ASHRAE *Journal* 7(September): 56.

White, P.A.F. and S.E. Smith, eds. 1964. *High efficiency air filtration.* Butterworth, London, England.

Yocom, J.E. and W.A. Cote. 1971. Indoor/Outdoor air pollutant relationships for air-conditioned buildings. ASHRAE *Transactions* 77(1):61.

Ziesse, N.G. and G.W. Penney. 1968. The effects of cigarette smoke on space charge soiling of walls when air is cleaned by a charging type electrostatic precipitator. ASHRAE *Transactions* 74(2).

CHAPTER 26

INDUSTRIAL GAS CLEANING AND AIR POLLUTION CONTROL

A N industrial gas-cleaning installation performs one or more of the following functions:

- Maintains compliance of an industrial process with the laws or regulations for air pollution.
- Reduces nuisance or physical damage from contaminants to individuals, equipment, products, or adjacent properties.
- Prepares cleaned gases for processes.
- Reclaims usable materials, heat, or energy.
- Reduces fire, explosion, or other hazards.

Equipment that removes particulate matter from a gas stream may also remove or create some gaseous contaminants; on the other hand, equipment that is primarily intended for removal of gaseous pollutants might also remove or create objectionable particulate matter to some degree. In all cases, gas-cleaning equipment changes the process stream, and it is therefore essential that the engineer evaluate the consequences of those changes on the plant's overall operation.

Equipment Selection

In selecting industrial gas-cleaning equipment, plant operations and the use or disposal of materials captured by that equipment must be considered. As the cost of gas-cleaning equipment affects manufacturing costs, alternative processes should be evaluated early to minimize the impact the equipment may have on the total cost of a product. An alternate manufacturing process may reduce the cost of or eliminate the need for gas-cleaning equipment. However, even when gas-cleaning equipment is required, process and system control should minimize the load on the collection device.

An industrial process may be changed from dirty to clean by substituting some process material (*e.g.,* switching to a cleaner burning fuel or pretreating the existing fuel source). Equipment redesign, such as enclosing pneumatic conveyors or recycling noncondensable gases, may also clean the process. Occasionally, additives, such as chemical dust suppressants used in quarrying or liquid animal fat applied to dehydrated alfalfa prior to grinding, reduce the potential for air pollution or concentrate the pollutants so that a smaller, more concentrated process stream might be treated.

Gas streams containing pollutants should not be diluted with extraneous air, because the volume of gas to be handled is a major factor in the owning and operating costs of control equipment. Regulatory authorities generally require the levels of polluting emissions to be corrected to standard conditions, taking into account temperature, pressure, moisture content, and factors related to combustion or production rate.

In this chapter, each generic type of equipment is discussed on the basis of its primary method for gas or particulate abatement. The development of systems that incorporate several of the devices discussed here for specific industrial processes is left to the engineer.

Gas-Cleaning Regulations

In the United States, industrial gas-cleaning installations that exhaust to the outdoor environment are regulated by the U.S. Environmental Protection Agency (EPA); those that exhaust to the workplace are regulated by the Occupational Safety and Health Administration (OSHA) of the U.S. Department of Labor.

The EPA has established Standards of Performance for New Stationary Sources (New Source Performance Standards, NSPS 1991) and more restrictive State Implementation Plans (SIP 1991) and local codes as a regulatory basis to achieve air quality standards (GPO 1991). Information on the current status of the NSPSs can be obtained through the Semi-Annual Regulatory Agenda, as published in the Federal Register, and through the regional offices of the EPA.

Where air is not affected by combustion, solvent vapors, and toxic materials, it may be desirable to recirculate the air to the workplace to reduce energy costs, or to balance static pressure in a building. High-efficiency fabric or cartridge filters are typically used in general ventilation systems to reduce particle concentrations to levels acceptable for recirculated air.

The Industrial Ventilation Committee of the American Conference of Governmental Industrial Hygienists (ACGIH) and the National Institute of Occupational Safety and Health (NIOSH) have established criteria for recirculation of cleaned process air to the work area (ACGIH 1988, NIOSH 1978). Fine particle control by various dust collectors under recirculating airflows have been investigated by Kuehn and Pui (1989).

Public complaints may occur even when the effluent concentrations discharged to the atmosphere are below the permissible emission rates and visibility. Thus, the plant location, the contaminants involved, and the meteorological condition of the area, in addition to codes or regulations, must be evaluated.

In most cases, emission standards require a higher degree of gas cleaning than necessary for economical recovery of process products (if this recovery were desirable). Gas cleanliness is a priority, especially where toxic materials are involved, and cleaned gases might be recirculated to the work area.

Measuring Gas Streams and Pollutants

Stack sampling is often required to fulfill requirements of operating and installation permits for gas-cleaning devices, to establish conformance with regulations, and to commission new equipment. In addition, it can be used to establish specifications for gas-cleaning equipment and to certify that the

---

The preparation of this chapter is assigned to TC 5.4, Industrial Process Air Cleaning (Air Pollution Control).

equipment is functioning properly. The tests determine the composition and quantity of gases and particulate matter at selected locations along the process stream. The following general principles apply to a stack sampling program:

- A report from stack sampling should include a summary of the process, which should identify any deviations from normal process operations that occurred during testing. The summary should be prepared during the testing phase and certified by the process owner at completion of the tests.
- The sampling location(s) must be acceptable to all parties who will use the results.
- The sampling location(s) must meet acceptable criteria with respect to temperature, flow distribution and turbulence, and distance from disturbances to the process stream. Exceptions based on physical constraints must be identified and reported.
- Samples that are withdrawn from a duct or stack must represent typical conditions in the process stream. Proper stack traverses must be made and particulate samples withdrawn isokinetically.
- Stack samples should be performed in accordance with approved methods and established protocols whenever possible.
- Variations in the volumetric flow, temperature, and particulate or gaseous pollutant emissions along with upset conditions should be identified.
- Disposal methods for waste generated during testing must be identified before testing begins. This is especially critical where pilot plant equipment is being tested, since new forms of waste are often produced.
- The regulatory basis for the tests should be established so that the results can be presented in terms of process weight rate, consumption of raw materials, energy use, and so forth.

Analyses of the samples can provide the following types of information about the emissions:

- The distribution of particle sizes—optical, physical, aerodynamic, etc.
- The concentration of particulate matter in the gases, including average and extreme values, and a profile of concentrations in the duct or stack.
- The volumetric flow of gases, including average and extreme values, and a profile of this flow in the duct or stack. The volumetric flow is commonly expressed at actual conditions and various standard conditions of temperature, pressure, moisture, and process state.
- Chemical composition of gases and particulate matter, including recovery value, toxicity, solubility, etc.
- Particle and bulk densities of particulate matter.
- Handling characteristics of particulate matter, including erosive, corrosive, abrasive, flocculative, or adhesive/cohesive qualities.
- Flammable or explosive limits.
- Electrical resistivity of deposits of particulate matter—under stack and laboratory conditions. These data are useful for assessments of electrostatic precipitators and other electrostatically augmented technology.

EPA Reference methods. The EPA has developed methods to measure the particulate and gaseous components of emissions from many industrial processes and has incorporated these in the NSPS by reference. Appendix A of the New Source Performance Standards lists the reference methods (GPO 1991). These are updated regularly in the Federal Register. Guidance for using these reference methods can be found in the *Quality Assurance Handbook for Air Pollution Measurement Systems* (EPA 1984).

The EPA (1984) has promulgated fine particle standards for ambient air quality. Known as PM-10 Standards, these revised standards focus particulate abatement efforts towards fine particle control. Fine particles (with aerodynamic particle size smaller than 10 μm) are of concern because they penetrate deeply into the lungs. With the development of these standards, concern has arisen over the efficiency of industrial gas cleaners at various particle sizes.

ASTM Methods. The EPA has also approved other test methods, *i.e.*, ASTM methods, when cited in the NSPSs or the applicable EPA reference methods.

Other methods. Sometimes, the reference methods must be modified, with the consent of regulatory authorities, to achieve representative sampling under the less-than-ideal conditions of industrial operations. These modifications to test methods should be clearly identified and explained in test reports.

Gas Flow Distribution

The control of gas flow through industrial gas-cleaning equipment is important for good system performance. With the large gas flows commonly encountered, and often a need to retrofit equipment to existing processes, space allocations do not normally permit streamlined duct. Instead, elbow splitters, baffles, etc. are used. These components must be designed to limit dust buildup and corrosion to acceptable levels. Uniformity of gas flow is so important that vendors are frequently required to build scale models of their equipment and proposed duct to demonstrate gas flow distribution.

Monitors and Controls

Current regulatory trends anticipate or demand continuous monitoring and control of equipment to maintain optimum performance against standards. Under regulatory data requirements, operating logs provide the owner with process control information and others with a baseline for the development or service of equipment.

Larger systems include programmable controllers and computers for control, energy management, data logging, and diagnostics. Some systems even have modem connections to support monitoring and service needs from remote locations.

PARTICULATE EMISSION CONTROL

A large range of equipment for the separation of particulate matter from gaseous streams is available. Typical concentration ranges for this equipment are summarized in Table 1.

Table 1   Intended Duty of Gas-Cleaning Equipment

	Maximum Concentration
Air cleaners	<0.002 gr/scfd
Clean rooms	
Commercial/residential buildings	
Plant air recirculation	
Industrial gas cleaners	<35 gr/scfd
Product capture in pneumatic conveying	<3500 gr/scfd

Note: gr/scfd = grains per standard cubic foot on a dry basis.

HEPA and ULPA filters are often used in clean rooms to maintain nearly particulate-free environments. In commercial and residential buildings, air cleaners are used to remove nuisance dust. In other instances, air cleaners are selected to control particulate matter that constitutes a health hazard in the workplace (*e.g.*, radioactive particles, beryllium particles, or biological airborne wastes). Recirculation of air in industrial plants could use air cleaners or may require more heavy-duty equipment. Usually, secondary filtration systems (usually HEPA filter systems) are required with recirculation systems.

Air cleaners are discussed in Chapter 25. This chapter is concerned with heavy-duty equipment for the control of particulate matter from industrial processes. The particulate emissions from these processes generally have a concentration in the range of 0.01 to 35 gr/ft^3. The gas-cleaning equipment is usually installed for air pollution control.

Particulate control technology is selected to satisfy the requirements for specific processes. Available technology differs in basic design, removal efficiency, first cost, energy requirements, maintenance, land use, and operating costs. Several of the principle types of particulate control equipment include the following:

Gravity and momentum collectors	Settling chambers; louvers and baffle chambers
Centrifugal collectors	Cyclones; mechanical centrifugal collectors
Electrostatic precipitators	Tubular, plate, wet or dry electrostatic precipitators
Fabric filters	Bag houses; fabric collectors
Wet collectors	Spray scrubbers; impingement scrubbers, wet cyclones; venturi scrubbers; packed towers; absorbers; mobile bed scrubbers; electrostatically augmented scrubbers
Adsorbers	Activated carbon beds
Incinerators	Afterburner; catalytic or flare incinerator

Collector Performance

Particulate collectors may be evaluated for their ability to remove particulate matter from a gas stream and for their ability to reduce the emissions of selected particle sizes. The degree to which particulate matter is separated from a gas stream is known as the efficiency of a collector; the fraction of material escaping collection is the penetration. The reduction of particulate matter in a selected particle size range is known as the fractional efficiency for a particulate collector.

The efficiency of a collector is generally expressed as a percentage in terms of the mass flow rate of material entering and exiting the collector.

$$\eta(\%) = 100(w_i - w_o)/w_i = 100 w_c/w_i \qquad (1)$$

where:

η = efficiency of collector
w_i = mass flow rate of contaminant in gases entering collector
w_o = mass flow rate of contaminant in gases exiting collector
w_c = mass flow rate of contaminant captured by collector

The efficiency of a collector is often expressed in terms of the concentrations of particulate matter entering and exiting equipment. This approach is generally unsatisfactory because of changes that occur in the gas stream due to air leakage in the system, condensation, and temperature and pressure changes.

Penetration is usually measured, and the efficiency for a collector calculated using the following equation:

$$P(\%) = 100 \, w_o/w_i \qquad (2)$$

where:

$$\eta = 100 - P \qquad (3)$$

The fractional efficiency for a particulate collector is determined by measuring the mass rate of emissions entering and exiting the collector in selected particle size ranges. Methods for measuring the fractional efficiency of particulate collectors in industrial applications are only beginning to emerge, largely because of the need to compare the fine particle performance of collectors used under the PM-10 regulations for ambient air quality.

Other measures of performance beside efficiency should also be considered in designing industrial gas-cleaning systems. Table 2 compares some of these factors for typical equipment (IGCI 1964). Note that this table contains only nominal values and is no substitute for experience, trade studies, and an engineering assessment of requirements for specific installations.

Table 3 summarizes the types of collectors that have been used in industrial applications (Kane and Alden 1967).

Table 2   Measures of Performance for Gas-Cleaning Equipment

Type of Dust-Collecting Equipment	Particle Diameter[a], μm	Maximum Loading, gr/ft^3	Collection Efficiency, % by mass	Pressure Loss Gas, in. of Water	Pressure Loss Liquid, psi	Utilities per 1000 cfm	Comparative Energy Requirement	Superficial Velocity[b], fpm	Capacity Limits, 1000 cfm	Space Required (Relative)
Dry inertial collectors										
Settling chamber	>40	>5	50	0.1 to 0.5	—	—	1	300 to 600	None	Large
Baffle chamber	>20	>5	50	0.5 to 1.5	—	—	1.5	1000 to 2000	None	Medium
Skimming chamber	>20	>1	70	<1.0	—	—	3.0	2000 to 4000	50	Small
Louver	>10	>1	80	0.3 to 2.0	—	—	1.5 to 6.0	2000 to 4000	30	Medium
Cyclone	>15	>1	85	0.5 to 3.0	—	—	1.5 to 9.0	2000 to 4000	50	Medium
Multicyclone	>5	>1	95	2.0 to 10.0	—	—	6.0 to 20	2000 to 4000	200	Small
Impingement	>10	>1	90	1.0 to 2.0	—	—	3.0 to 6.0	2000 to 4000	None	Small
Dynamic	>10	>1	90	Provides head	—	1.0 to 2.0 hp	10 to 20	—	50	—
Electrostatic precipitators	>0.01	>0.1	99	0.2 to 1.0	—	0.1 to 0.6 kW	4.0 to 11	100 to 600	10 to 2000	Large
Fabric filters	>0.08	>0.1	99	2.0 to 6.0	—	—	6.0 to 20	1.0 to 20	200	Large
Wet scrubbers										
Gravity spray	>10	>1	70	0.1 to 1.0	20 to 100	0.5 to 2.0 gpm	5.0	100 to 200	100	Medium
Centrifugal	>5	>1	90	2.0 to 8.0	20 to 100	1 to 10 gpm	12 to 26	2000 to 4000	100	Medium
Impingement	>5	>1	95	2.0 to 8.0	20 to 100	1 to 5 gpm	9.0 to 31	3000 to 6000	100	Medium
Packed bed	>5	>0.1	90	0.5 to 10	5 to 30	5 to 15 gpm	4.0 to 34	100 to 300	50	Medium
Dynamic	>2	>1	95	Provides head	5 to 30	1 to 5 gpm, 3 to 20 hp	30 to 200	3000 to 4000	50	Small
Submerged orifice	>2	>0.1	90	2.0 to 6.0	None	No pumping	9.0 to 21	3000	50	Medium
Jet	>2	>0.1	90	Provides head	50 to 100	50 to 100 gpm	15 to 30	2000 to 20,000	100	Small
Venturi	>0.1	>0.1	99	10 to 30	5 to 30	3 to 10 gpm	30 to 300	12,000 to 42,000	100	Small

[a]The minimum particle diameter for which the device is effective.
[b]The average speed of gases flowing through the equipment's collection region.

Table 3 Collectors Used in Industry

Operation	Concentration	Particle Sizes	Cyclone	High-Efficiency Centrifugal	Wet Collectors Med. Pressure	Wet Collectors High-Energy	Fabric Filter	Electrostatic Precipitator	Notes
CERAMICS									
a. Raw product handling	Light	Fine	Rare	Seldom	Frequent	N/U	Frequent	N/U	1
b. Fettling	Light	Fine to medium	Rare	Occasional	Frequent	N/U	Frequent	N/U	2
c. Refractory sizing	Heavy	Coarse	Seldom	Occasional	Frequent	Rare	Frequent	N/U	3
d. Glaze and vitreous enamel spray	Moderate	Medium	N/U	N/U	Usual	N/U	Occasional	N/A	—
e. Glass melting	Light	Fine	N/A	N/A	Occasional	N/U	Occasional	Usual	—
f. Frit smelting	Light	Fine	N/A	N/A	N/U	Often	Often	Often	—
g. Fiberglass forming and curing	Light	Fine	N/A	N/A	Occasional	N/U	N/U	Usual	—
CHEMICALS									
a. Material handling	Light to moderate	Fine to medium	Occasional	Frequent	Frequent	Frequent	Frequent	N/U	4
b. Crushing, grinding	Moderate to heavy	Fine to coarse	Often	Frequent	Frequent	Occasional	Frequent	N/U	5
c. Pneumatic conveying	Very heavy	Fine to coarse	Usual	Occasional	Rare	Rare	Usual	N/U	6
d. Roasters, kilns, coolers	Heavy	Medium to coarse	Occasional	Usual	Usual	Frequent	Rare	Often	7
e. Incineration	Light to medium	Fine	N/U	N/U	N/U	Frequent	Rare	Frequent	8
COAL MINING AND HANDLING									
a. Material handling	Moderate	Medium	Rare	Occasional	Occasional	N/U	Usual	N/A	9
b. Bunker ventilation	Moderate	Fine	Occasional	Frequent	Occasional	N/U	Usual	N/A	10
c. Dedusting, air cleaning	Heavy	Medium to coarse	Occasional	Frequent	Occasional	N/U	Usual	N/A	11
d. Drying	Moderate	Fine	Rare	Occasional	Frequent	Occasional	N/U	N/A	12
COMBUSTION FLY ASH									
a. Coal burning:									
chain grate	Light	Fine	N/A	Rare	N/U	N/U	Frequent	N/U	13
spreader stoker	Moderate	Fine to coarse	Rare	Rare	N/U	N/U	Frequent	Rare	14
pulverized coal	Heavy	Fine	N/A	Frequent	N/U	N/U	Frequent	Usual	14
fluidized bed	Moderate	Fine	Usual	—	—	—	Frequent	Frequent	—
coal slurry	Light	—	—	—	—	—	Often	Often	—
b. Wood waste	Varies	Coarse	Usual	Usual	N/U	N/U	Occasional	Often	15
c. Municipal refuse	Light	Fine	N/U	N/U	Occasional	N/U	Usual	Frequent	—
d. Oil	Light	Fine	N/U	N/U	N/U	N/U	Usual	Often	—
e. Biomass	Moderate	Fine to coarse	N/U	N/U	Occasional	N/U	Usual	Frequent	—
FOUNDRY									
a. Shakeout	Light to moderate	Fine	Rare	Rare	Rare	Seldom	Usual	N/U	16
b. Sand handling	Moderate	Fine to medium	Rare	Rare	Usual	N/U	Rare	N/U	17
c. Tumbling mills	Moderate	Medium to coarse	N/A	N/A	Frequent	N/U	Usual	N/U	18
d. Abrasive cleaning	Moderate to heavy	Fine to medium	N/A	Occasional	Frequent	N/U	Usual	N/U	19
GRAIN ELEVATOR, FLOUR AND FEED MILLS									
a. Grain handling	Light	Medium	Usual	Occasional	Rare	N/U	Frequent	N/A	20
b. Grain drying	Light	Coarse	N/A	N/A	N/U	N/U	See Note 20	N/A	21
c. Flour dust	Moderate	Medium	Rare	Often	Occasional	N/U	Usual	N/A	22
d. Feed mill	Moderate	Medium	Often	Often	Occasional	N/U	Frequent	N/A	23
METAL MELTING									
a. Steel blast furnace	Heavy	Varied	Frequent	Rare	Frequent	Frequent	N/U	Frequent	24
b. Steel open hearth, basic oxygen furnace	Moderate	Fine to coarse	N/A	N/A	N/A	Often	Rare	Frequent	25
c. Steel electric furnace	Light	Fine	N/A	N/A	N/A	Occasional	Usual	Rare	26
d. Ferrous cupola	Moderate	Varied	N/A	N/A	Frequent	Often	Frequent	Occasional	27
e. Nonferrous reverberatory furnace	Varied	Fine	N/A	N/A	Rare	Occasional	Usual	N/U	28
f. Nonferrous crucible	Light	Fine	N/A	N/A	Rare	Rare	Occasional	N/U	29
METAL MINING AND ROCK PRODUCTS									
a. Material handling	Moderate	Fine to medium	Rare	Occasional	Usual	N/U	Considerable	N/U	30
b. Driers, kilns	Moderate	Medium to coarse	Frequent	Occasional	Frequent	Occasional	N/U	Occasional	31
c. Cement rock drier	Moderate	Fine to medium	N/A	Frequent	Occasional	Rare	N/U	Occasional	32
d. Cement kiln	Heavy	Fine to medium	N/A	Frequent	Rare	N/U	Usual	Usual	33
e. Cement grinding	Moderate	Fine	N/A	Rare	N/U	N/U	Usual	Rare	34
f. Cement clinker cooler	Moderate	Coarse	N/A	Occasional	N/U	N/U	Occasional	Occasional	35
METAL WORKING									
a. Production grinding, scratch brushing, abrasive cutoff	Light	Coarse	Occasional	Frequent	Considerable	N/U	Considerable	N/U	36
b. Portable and swing frame	Light	Medium	Rare	Frequent	Frequent	N/U	Considerable	N/U	—
c. Buffing	Light	Varied	Frequent	Rare	Frequent	N/U	Rare	N/U	37
d. Tool room	Light	Fine	Frequent	Frequent	Frequent	N/U	Frequent	N/U	38
e. Cast-iron machining	Moderate	Varied	Rare	Frequent	Considerable	N/U	Considerable	N/U	39

Table 3   Collectors Used in Industry (*Continued*)

Operation	Concentration	Particle Sizes	Cyclone	High-Efficiency Centrifugal	Wet Collectors Med. Pressure	Wet Collectors High-Energy	Fabric Filter	Electrostatic Precipitator	Notes
PHARMACEUTICAL AND FOOD PRODUCTS									
a. Mixers, grinders, weighting, blending, bagging, packaging	Light	Medium	Rare	Frequent	Frequent	N/U	Frequent	N/U	40
b. Coating pans	Varied	Fine to medium	Rare	Rare	Frequent	N/U	Frequent	N/U	41
PLASTICS									
a. Raw material processing	(See comments under CHEMICALS)								42
b. Plastic finishing	Light to moderate	Varied	Frequent	Frequent	Frequent	N/U	Frequent	N/U	43
PULP AND PAPER									
a. Recovery boilers									
direct contact	Heavy	Medium	N/U	N/U	N/U	N/U	Occasional	Usual	—
low odor	Heavy	Medium	N/U	N/U	N/U	N/U	Occasional	Usual	—
b. Lime kilns	Heavy	Coarse	N/U	N/U	N/U	N/U	Often	Often	—
c. Wood-chip driers	Varies	Fine to coarse	N/U	N/U	N/U	N/U	Occasional	Often	—
RUBBER PRODUCTS									
a. Mixers	Moderate	Fine	N/A	N/A	Frequent	N/U	Usual	N/U	44
b. Batch-out rolls	Light	Fine	N/A	N/A	Usual	N/U	Frequent	N/U	45
c. Talc dusting and dedusting	Moderate	Medium	N/A	N/A	Frequent	N/U	Usual	N/U	46
d. Grinding	Moderate	Coarse	Often	Often	Frequent	N/U	Often	N/U	47
WOOD PARTICLE BOARD AND HARD BOARD									
a. Particle driers	Moderate	Fine to coarse	Usual	Occasional	Frequent	Occasional	Rare	Occasional	48
WOODWORKING									
a. Woodworking machines	Moderate	Varied	Usual	Occasional	Rare	N/U	Frequent	N/U	49
b. Sanding	Moderate	Fine	Frequent	Occasional	Occasional	N/U	Frequent	N/U	50
c. Waste conveying, hogs	Heavy	Varied	Usual	Rare	Occasional	N/U	Occasional	N/U	51

Notes for Table 3

[1] Dust released from bin filling, conveying, weighing, mixing, pressing, forming. Refractory products, dry pan, and screening operations more severe.

[2] Operations found in vitreous enameling, wall and floor tile, pottery.

[3] Grinding wheel or abrasive cutoff operation. Dust abrasive.

[4] Operations include conveying, elevating, mixing, screening, weighing, packaging. Category covers so many different materials that recommendation will vary widely.

[5] Cyclone and high-efficiency centrifugal collectors often act as primary collectors, followed by fabric filters or wet collectors.

[6] Usual setup uses cyclone as product collector followed by fabric filter for high overall collection efficiency.

[7] Dust concentration determines need for dry centrifugal collector; plant location, product value determines need for final collectors. High temperatures are usual, and corrosive gases not unusual.

[8] Ionizing wet scrubbers are widely used.

[9] Conveying, screening, crushing, unloading.

[10] Remote from other dust-producing points. Separate collector generally used.

[11] Heavy loading suggests final high-efficiency collector for all except very remote locations.

[12] Loadings and particle sizes vary with different drying methods.

[13] Boiler blow-down discharge is regulated, generally for temperature and, in some places, for pH limits; check local environmental codes on sanitary discharge.

[14] Collection of particulate or sulfur control usually requires a scrubber (dry or wet) and a fabric filter or electrostatic precipitator.

[15] Public nuisance from settled wood char indicates collectors are needed.

[16] Hot gases and steam usually involved.

[17] Steam from hot sand, adhesive clay bond involved.

[18] Concentration very heavy at start of cycle.

[19] Heaviest load from airless blasting because of high cleaning speed. Abrasive shattering greater with sand than with grit or shot. Amounts removed greater with sand castings, less with forging scale removal, least when welding scale is removed.

[20] Operations such as car unloading, conveying, weighing, storing.

[21] Special filters are successful.

[22] In addition to grain handling, cleaning rolls, sifters, purifiers, conveyors, as well as storing, packaging operations are involved.

[23] In addition to grain handling, bins, hammer mills, mixers, feeders, conveyors, bagging operations need control.

[24] Primary dry trap and wet scrubbing usual. Electrostatic precipitators are added where maximum cleaning is required.

[25] Air pollution control is expensive for open hearth, accelerating the use of substitute melting equipment such as basic oxygen process and electric-arc furnace.

[26] Fabric filters have found extensive application for this air pollution control problem.

[27] Cupola control will vary with plant size, location, melt rate, and air pollution emission regulations.

[28] Corrosive gases can be a problem, especially in secondary aluminum.

[29] Zinc oxide plume can be troublesome in certain plant locations.

[30] Crushing, screening, conveying, storing involved. Wet ores often introduce water vapor in exhaust airstream.

[31] Dry centrifugal collectors are used as primary collectors, followed by a final cleaner.

[32] Same as No. 30.

[33] Collectors usually permit salvage of material and also reduce nuisance from settled dust in plant area.

[34] Salvage value of collected material is high. Same equipment used on raw grinding before calcining.

[35] Coarse abrasive particles readily removed in primary collector types.

[36] Roof discoloration, deposition on autos can occur with cyclones and, less frequently, with dry centrifugal. Heavy-duty air filter sometimes used as final cleaner.

[37] Linty particles and sticky buffing compounds can cause trouble in high-efficiency centrifugals and fabric filters. Fire hazard is also often present.

[38] Unit collectors extensively used, especially for isolated machine tools.

[39] Dust ranges from chips to fine floats, including graphitic carbon.

[40] Materials involved vary widely. Collector selection may depend on salvage value, toxicity, sanitation yardsticks.

[41] Controlled temperature and humidity of supply air to coating pans makes recirculation from coating pans desirable.

[42] Manufacture of plastic compounds involves operations allied to many in chemical field and varies with the basic process employed.

[43] Operations are similar to woodworking, and collector selection involves similar considerations.

[44] Concentration is heavy during feed operation. Carbon black and other fine additions make collection and dust-free disposal difficult.

[45] Often, no collection equipment is used where dispersion from exhaust stack is good and stack location is favorable.

[46] Salvage of collected material often dictates type of high-efficiency collector.

[47] Fire hazard from some operations must be considered.

[48] Granular bed filters, at times electrostatically augmented, have occasionally been used in this application.

[49] Bulky material. Storage for collected material is considerable; bridging from splinters and chips can be a problem.

[50] Production sanding produces heavy concentrations of particles too fine to be effectively captured by cyclones or dry centrifugal collectors.

[51] Primary collector invariably indicated with concentration and partial size range involved; When used, wet or fabric collectors are employed as final collectors.

Definitions

N/A—Not applicable because of inefficiency or process incompatibility.

N/U—Not widely used.

Particle size

Fine—50% in 0.5 to 7.0 μm diameter range

Medium—50% in 7.0 to 15 μm diameter range

Coarse—50% over 15 μm diameter range

Concentration of particulate matter entering collector (loading)

Light—<2.0 gr/ft³

Moderate—2.0 to 5.0 gr/ft³

Heavy—>5.0 gr/ft³

Mechanical Collectors

Settling chambers. Particulate matter will fall from suspension in a reasonable time if the particles are larger than about 40 μm. Plenums, dropout boxes, or gravitational settling chambers are thus used for the separation of coarse or abrasive particulate matter from gas streams.

Settling chambers are occasionally used in conjunction with fabric filters or electrostatic precipitators to reduce overall system cost. The settling chamber serves as a precollector to remove coarse particles from the gas stream.

Settling chambers sometimes contain baffles to distribute gas flow and to serve as surfaces for the impingement of coarse particles. Other designs use baffles to change the direction of the gas flow, thereby allowing coarse particles to be thrown from the gas stream by inertial forces.

The fractional efficiency for a settling chamber with uniform gas flow may be estimated by:

$$\eta(\%) = 100 \; u_t L / HV \tag{4}$$

where:

η = efficiency of collector for particles with settling velocity, u_t
u_t = settling velocity for selected particles, fps
L = length of chamber, ft
H = height of chamber, ft
V = superficial velocity of gases through chamber, fps

The superficial velocity of gases through the chamber is determined from measurements of the volumetric flow of gases entering and exiting the chamber and the cross-sectional area of the chamber. This average velocity must be low enough to prevent reentrainment of the deposited dust; a superficial velocity below 60 fpm is satisfactory for many materials.

Typical data on settling can be found in Table 4 (Billing and Wilder 1970). Because of air inclusions in the particle, the density of a dust particle can be substantially lower than the true density of the material from which it is made.

Inertial collectors—Louver and baffle collectors. Louvers are widely used to control particles larger than about 15 μm in diameter. The louvers cause a sudden change in direction of gas flow. By virtue of their inertia, particles move away from high-velocity gases and are collected in a hopper or trap, or are withdrawn in a concentrated sidestream. The sidestream is cleaned using a cyclone or high-efficiency collector, or it is simply discharged to the atmosphere. In general, the pressure drop across inertial collectors with louvers or baffles is greater than that for settling chambers, but this loss is balanced by higher collection efficiency and more compact equipment.

Inertial collectors are occasionally used to control mist. In some applications, the interior of the collector might be irrigated to prevent reentrainment of dry dust and to remove soluble deposits.

Typical louver and baffle collectors are shown in Figure 1.

Inertial collectors—Cyclones and multicyclones. A cyclone collector transforms a gas stream into a confined vortex, from which inertia moves suspended particles to the wall of the cyclone's body. The inertial effect of turning the gas stream, as employed in the baffle collector, is used continuously in a cyclone to gain improved collection efficiency.

A low-efficiency cyclone operates with a static pressure drop from 1 to 1.5 in. of water between its inlet and outlet and can remove 50% of the particles from 5 to 10 μm. High-efficiency cyclones operate with static pressure drops near 8 in. of water between their inlet and outlet and can remove 70% of the particulates of approximately 5 μm.

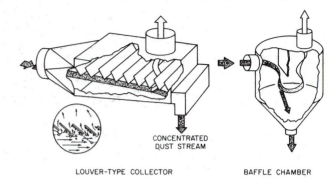

CONCENTRATED DUST STREAM

LOUVER-TYPE COLLECTOR   BAFFLE CHAMBER

Fig. 1   Typical Louver and Baffle Collectors

The efficiency of a cyclone depends on particle density, shape, and size (aerodynamic size D_p). Cyclone efficiency may be estimated from Figure 2 (Lapple 1951).

The parameter D_{pc}, known as the cut size, is defined as the diameter of particles collected with 50% efficiency. The cut size may be estimated using the following equation:

$$D_{pc} = [9 \mu b / 2 N_e V_i (\rho_p - \rho_g) \pi]^{0.5} \tag{5}$$

where:

D_{pc} = cut size, ft
μ = absolute gas viscosity, centipoise
b = cyclone inlet width, ft
N_e = effective number of turns within cyclone. The quantity N_e is approximately 5 for a high-efficiency cyclone and may be in the range from 0.5 to 10 for other cyclones.
V_i = inlet gas velocity, fpm
ρ_p = density of particle material, lb/ft³
ρ_g = density of gas, lb/ft³

At inlet gas velocity above 4800 fpm, internal turbulence limits improvements in the efficiency of a given cyclone. The pressure drop through a cyclone is proportional to the inlet

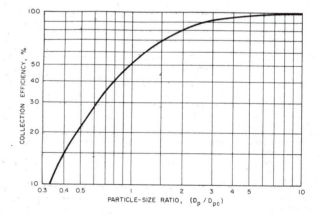

Fig. 2   Cyclone Efficiency

Table 4   Terminal Settling Velocities of Particles, m/s

Particle Density Relative to Water	Particle or Aggregate Diameter, μm									
	1	2	5	10	20	50	100	200	500	1000
0.05	1.8 E−6	7.0 E−6	4.0 E−5	1.6 E−4	7.0 E−4	4.0 E−3	1.6 E−2	5.5 E−2	2.3 E−1	5.1 E−1
0.1	3.7 E−6	1.4 E−5	8.0 E−5	3.2 E−4	1.4 E−3	8.0 E−3	3.0 E−2	1.1 E−1	3.9 E−1	8.3 E−1
0.2	7.4 E−6	2.8 E−5	1.6 E−4	6.4 E−4	2.8 E−3	1.6 E−2	5.9 E−2	1.9 E−1	6.5 E−1	1.30
0.5	1.8 E−5	7.0 E−5	4.0 E−4	1.6 E−3	7.0 E−3	3.9 E−2	1.4 E−1	4.2 E−1	1.21	2.50
1.0	3.7 E−5	1.4 E−4	8.0 E−4	3.2 E−3	1.2 E−2	7.6 E−2	2.5 E−1	7.0 E−1	1.95	3.90
2.0	7.4 E−5	2.8 E−4	1.6 E−3	6.4 E−3	2.5 E−2	1.4 E−1	4.5 E−1	1.14	3.10	6.25
5.0	1.8 E−4	7.0 E−4	4.0 E−3	1.5 E−2	6.4 E−2	3.5 E−1	9.7 E−1	2.23	5.75	11.0
10.0	3.7 E−4	1.4 E−3	8.0 E−3	3.1 E−2	1.2 E−1	6.0 E−1	1.66	3.52	8.90	17.3

Note: E−6 = 10⁻⁶ etc.

velocity pressure, and hence, the square of the volumetric flow. Koch and Licht (1977) and API (1975) developed more sophisticated theories to account for specific designs.

Dry centrifugal collector. Dry centrifugal collectors are inertial collectors used to separate particulate matter from gases by spinning. The mechanical type or skimming chamber is motor-driven and can be designed to move gases as well as clean them. Dry centrifugal collectors are made of special materials to resist corrosion and erosion; they may also be lined with refractories for high-temperature uses.

The mechanical centrifugal separator shown in Figure 3 generally is not used to control fine particles. It is a primary collector for pneumatic conveying, where the particles are substantially larger than 5 μm in diameter. It is also used in handling fly ash and dust from plastics and woodworking.

The inability of the centrifugal collector to control finer dust is apparent. However, these collectors work in many areas where dust is not easily friable and remains relatively large during the cleaning process. Centrifugal collectors are often used as precleaners to reduce the loading of more efficient pollution control devices.

Electrostatic Precipitators

Electrostatic precipitators use an electric field between oppositely charged electrodes to electrically charge particulate matter that is suspended in a gas stream, and then to separate the charged particulate matter from the gas. The particles move to the electrodes where they are adhesively/cohesively bound or agglomerated to form a deposit. Particles, or small clusters, that reentrain from the electrodes usually carry a charge that must either be reversed or carried to the electrode of opposite polarity for collection. A deposit might be removed from an electrode by mechanical vibration or by continuous or periodic washing. The deposits fall to hoppers or trays from which they may be removed from the electrostatic precipitator.

There are two general classes of electrostatic precipitators: (1) electronic air cleaners, commonly used for commercial/residential ventilation and air-conditioning applications (see Chapter 25); and (2) industrial electrostatic precipitators, used primarily for the heavier particulate emissions of industry.

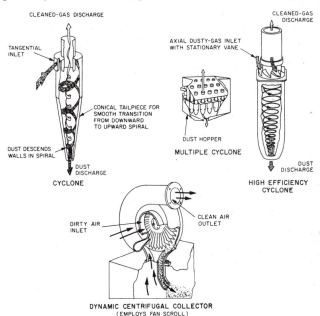

Fig. 3 Typical Dry Centrifugal Collectors

Industrial electrostatic precipitators are generally distinguished from electronic air cleaners by several attributes:

1. Industrial electrostatic precipitators use a larger separation between the high voltage and grounded electrodes to secure acceptable collection and electrode cleaning under the high dust loading and hostile conditions of industrial gas streams.
2. These precipitators employ heavy-duty construction for operation to 850°F and pressures to +30 in. of water.
3. Generally, industrial electrostatic precipitators are not required to meet ozone emission standards; therefore, they may operate with negative ionization, the polarity that gives the maximum electric field strength between the electrodes.
4. They are normally custom-engineered and assembled on location for a particular application.

Electrostatic precipitators can be designed to operate at collection efficiencies above 99.9% for closely specified conditions. Properties of the dust, such as particle size distribution and a deposit's electrical resistivity, can affect performance significantly, as can variations in gas composition and flow rate.

Figure 4 shows several types of industrial plate-type electrostatic precipitators. For each electrostatic precipitator, the charging electrodes are located between parallel collecting plates. The gas flow through these precipitators is horizontal. Full-scale precipitators based on these designs ordinarily operate with superficial velocity in the chamber between 60 and 400 fpm, and with treatment time in the range of 2 to 10 s. Operating voltage from 20 to 60 kV are applied between the charging electrode and the grounded plates. Temperatures of up to 850°F are handled routinely by precipitators consisting almost entirely of carbon steel. Special design techniques are required for temperatures above 975°F and at lower temperatures, where components of the process stream might be corrosive to carbon steel.

Principle of operation. Since the principles of electrostatic precipitation are basic to all electrostatically augmented gas-cleaning equipment, some fundamentals of electrostatic processes are provided here. Detailed information on electrostatic precipitation can be found in White (1963) and Oglesby and Nichols (1970).

In nature, positive and negative polarity objects tend to attract, leading to a neutral condition. Objects that are negative have an excess of electrons and those that are positive have an electron deficiency. The objects with electrical charge can be charged molecules (ions), particles, and surfaces (electrodes). For electrostatic precipitation to work, there must be charged particles or a means to charge the particles in an electrical field so that the particles may be attracted to a surface.

A high potential (voltage) dc-supply is used to separate electrical charge and to maintain that charge separation between the electrodes of an electrostatic precipitator. The dc-supply is usually a transformer and full-wave rectifier with one of its dc-terminals connected to ground. Direct current (dc) is normally chosen so that, once charged, particles continuously move toward one of the electrodes. Of course, there is an equal amount of charge on both the high potential and grounded electrodes.

The work done to separate the charge is simply the product of the quantity of charge that is separated (amount on either electrode) and the potential difference (voltage) between the electrodes:

Work done in joules = (Charge separated) × (Potential difference)

If the charge separation is not maintained because of some partial conduction path between the electrodes (leakage), a

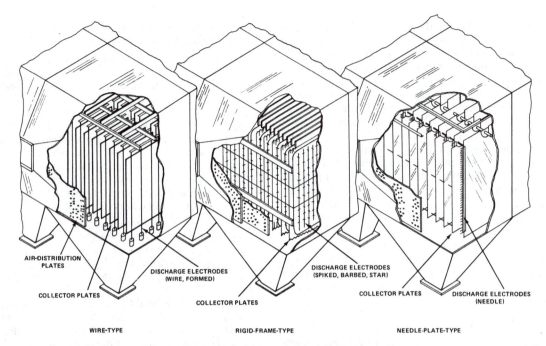

Fig. 4   Typical Industrial Electrostatic Precipitators

certain amount of power is required to maintain the potential difference between the electrodes. This power is the product of the rate of charge transfer and the potential difference between the electrodes.

Power in watts
= Work done/Unit time, J/s
= (Charge separated/Unit time) × (Potential difference)
= (Current) × (Potential difference)

The charge placed on the electrodes gives rise to forces between the charged electrodes, and between electrically charged objects in the gas stream and the electrodes. For example, if a particle has a negative charge and is placed between positively and negatively charged electrodes, the particle will be pushed by the charge on the negative electrode and pulled by the charge on the positive electrode. The particle will migrate until it strikes the electrode of polarity opposite to its own.

Space charge. The space-charge effect from charged particles and ions in the gas stream alter the electrical forces that drive the migration of particles to the electrodes. Finally, the accumulation of dust on electrodes also influences the electrical forces that determine electrostatic precipitation.

Electric field and corona current. The force density (or force per unit charge) on an object in the gas stream is called the electric field, which can be described in terms of the distribution of charge on the surface of the electrodes and the distribution of charge in the gas stream. The electric field from a single charge is given by Coulomb's Law:

$$\vec{E}_a = \frac{Q_a}{4\pi\epsilon_o} \frac{\hat{r}}{r^2} \tag{6}$$

which indicates that the strength of the electric field at a chosen point is proportional to the charge at the origin of the field, and inversely proportional to the square of the distance from the chosen point to the charge. The components from all elements of the charge distributions are combined to give the electric field at any point inside an electrostatic precipitator. The combination or superposition of electric field components is illustrated in Figure 5.

The force on a particle with charge q is determined once the electric field is known.

$$\vec{F} = q\vec{E}(\vec{r}) \tag{7}$$

The electric fields in the typical electrode systems found in electrostatic precipitators are very complex, and numerical procedures are beginning to yield insight into the collection process. The general expressions given here, however, provide a basis for understanding the specifications, design, and operation of electrostatic precipitators or other electrostatically augmented gas-cleaning equipment.

When some region of an electrode has small dimensions in comparison with the electrode of opposite polarity (*e.g.*, a wire at high potential between grounded parallel plates), the surface charge on that electrode is large. The electric field is high near this electrode, and electrically charged components of the gas stream experience strong forces; these forces promote a localized electrical breakdown or ionization of the gas stream near the electrode. The flow of ions between the electrodes is known as a corona current.

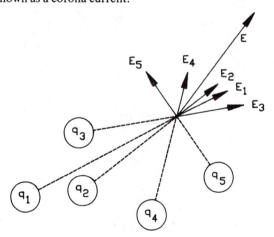

Fig. 5   Resultant Electric Field from Several Charges

Corona current is the primary charge transfer mechanism in an electrostatic precipitator. It is the dominant component of the current that is read at the meters on the electrostatic precipitator's control panel. Since the current passes between the electrodes, the transformer must perform additional work to maintain the charge separation between the electrodes.

Particulate matter is charged in an electrical corona surrounding the charging electrode. The voltage for onset of corona increases rapidly with the radius of curvature of an electrode. Therefore, small diameter wires, projection points, or needles are used as corona-producing, particle-charging electrodes. The precipitator must operate between the voltage levels at which corona is initiated and the spark-over voltage, at which point ionized, conducting channels connect the high voltage and grounded electrodes and performance deteriorates. Electrodes are designed to efficiently produce ions for particle charging, while maintaining a sufficiently strong electric field to collect the charged particles.

In common applications, only a small percentage of the ions in a corona current attach to particles. The overabundance of ions is necessary to increase the probability for the attachment of many ions to each particle. The opportunity for more efficient operation of electrostatic precipitators has led developers to test many electrode designs and develop energy management systems.

The electric field is the most important factor to consider when comparing electrode designs, since it controls the electrical charging process and the migration of particles to electrodes. Although the electric field is generally proportional to the potential difference between the electrodes (in the space charge-free case) for a given electrode design, it is not always true that the force on a particle is larger in an electrostatic precipitator that operates at a higher potential (voltage) and current.

Space charge produced by the corona will suppress the electric field near the small diameter electrode and enhance it near the large diameter electrode. It will not, however, influence the average field between the electrodes.

Electrical resistivity of deposits. Electrostatic precipitation is often adversely affected when deposits with high electrical resistivity coat the electrodes. When electrical currents must pass through these deposits, arcing within the deposit or widely varying potentials will occur. The arcing produces reverse ionization (back corona) in the gas stream and suppresses the gas flow induced by the corona; it also increases the vibrational energy in a deposit, making collection of particles difficult.

A resistive deposit is charged like a leaky capacitor. When the corona current must be conducted through a resistive deposit, a voltage drop appears across the deposit. The voltage drop can become sufficient to cause the capacitor to break down. Typical current-voltage characteristics for an electrostatic precipitator are shown in Figure 6. These illustrate the transition between conductive and insulating deposits (Noll 1984).

The electrical resistivity of many dusts depends strongly on temperature and moisture in the gas stream. Also, various electrolytic materials added to the gas stream will be adsorbed on particles to render deposits conductive. As an example, electrostatic precipitators have served utility power plant operations effectively because traces of sulfur trioxide are present in the flue gas. As little as 5 to 10 ppm by mass of sulfur trioxide at 300°F reduces the resistivity of fly ash, which makes the difference between poor performance and good performance for the equipment (White 1974).

Corona quenching. Ions, which migrate from the corona electrode, have the same polarity as the electrode around

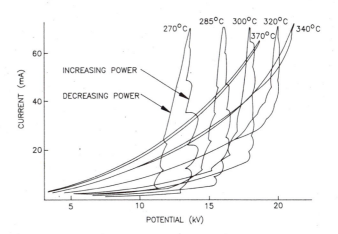

Fig. 6   Typical Current Voltage Characteristics for Electrostatic Precipitators

which they are produced. When the ions attach to particulate matter, the mobility of this charge is reduced. Since ion production continues, the space charge in the gas stream increases. This space charge produces a bucking or quenching electric field that limits further production of ions.

Under conditions of similar concentration, fine dusts have a larger surface area than coarse dusts, and, thus, can store more space charge. This is why corona quenching is usually associated with high dust concentrations or fine dusts.

Corona quenching can be observed at the control panel of electrostatic precipitators with multiple electrical sections or "fields" along the direction of gas flow. The suppression of the corona currents when all fields are energized are simply compared with the currents for singly energized fields.

Mathematical models for designing electrostatic precipitators. Particulate matter is a set of particles, each of which has its own distinguishing characteristics. The particles respond to the steady stresses of gravity, the electric field, gas flow, and (in the case of a deposit or agglomerate) the adhesive/cohesive bonds they have with other particles and an electrode. The particles also respond to random, periodic, and impulsive events which arise from electrode rapping and vibration, sound waves, sparking, particle collisions at a deposit, and turbulence in the gas stream. The collection process is complex and can be handled only by empirical models. Usually the models are exponential, which reflects their basis on random processes occurring during electrostatic precipitation. The simplest model for the efficiency of an electrostatic precipitator is the Deutsch equation:

$$\eta = e^{-\theta/\tau} \tag{8}$$

where:

η = collection efficiency, expressed as a ratio
θ = residence time for sample of gas in electrostatic precipitator
$1/\tau$ = precipitation rate parameter

The model claims that the particle collection efficiency improves when a sample of contaminated gas is moved more slowly through the collector or when a larger electrostatic precipitator is used.

The Anderson equation expands on this concept by including information on the electrode surface area and volumetric flow of gases in the chamber.

$$\eta = 1 - e^{-\omega(A/Q)} \tag{9}$$

where:

> η = collection efficiency, expressed as a ratio
> A = electrode surface area exposed to contaminated gases, ft^2
> Q = volumetric flow of contaminated gases, cfm
> ω = effective migration velocity, fpm

Typical values for the effective migration velocity (average drift velocity of particles towards the electrodes) ω are given in Table 5 (HEW 1967). The ratio A/Q is known as the specific collection area, *i.e.*, the amount of electrode area needed to clean a unit volumetric flow of gas.

Table 5   Typical Effective Migration Velocities for Electrostatic Precipitators

Application	ω, fpm
Pulverized coal (fly ash)	20 to 26
Paper mills	15
Open-hearth furnace	11
Secondary blast furnace (80% foundry iron)	25
Gypsum	31 to 38
Hot phosphorus	6
Acid mist (H_2SO_4)	11 to 15
Acid mist (TiO_2)	11 to 15
Flash roaster	15
Multiple-hearth roaster	16
Portland cement manufacturing (wet process)	20 to 22
Portland cement manufacturing (dry process)	11 to 14
Catalyst dust	15
Gray iron cupola (iron-coke ratio = 10)	6 to 7

These simple empirical models are inadequate for estimates of the performance of specific installations. They do, however, introduce common parameters for more advanced modeling. Typical models are based on theoretical concepts tempered with nonideal factors to explain empirical data (DuBard and Altman 1984, Faulkner and DuBard 1984).

Because the characteristics of emissions vary from location to location, even for similar processes, on-site pilot plant testing is often necessary to determine the proper system configuration and ESP size (Noll 1984a, Crynack and Sherow 1984).

Components of Electrostatic Precipitators

Selection of electrodes. Two generic types of electrodes are usually identified for industrial applications—single-stage and two-stage.

Single-stage electrodes combine the corona-charging region with the collection region, such as the wire-tube and wire-plate designs. Both these designs incorporate a large-diameter wire to achieve long life for the corona electrodes. The large diameter is also selected to permit operation of the equipment at the high potential needed to produce corona current and an adequate collection field. Wire-plate electrostatic precipitators are commonly used for dry dust collection. The many varieties of corona electrodes in wire-plate and wire-tube designs (aside from smooth wires) include barbed wires, rods or ribbons with sharp projections, and spiral wires.

The oldest type of single-stage electrostatic precipitator has tubular electrodes for collecting particulate matter. Wire-tube electrostatic precipitators are generally used to control mist and, depending on application, are cleaned using a continuous or intermittent flow of liquid down the inner wall of the tubes. The tubes typically have a 3 to 10 in. diameter and are 3.3 to 13 ft long. They are arranged vertically in banks with the wires, about 0.08 in. in diameter, suspended in the center of each tube with tension weights at the bottom.

Although still used for certain applications, the wire-tube electrostatic precipitator has given way to several plate types to reduce space requirements.

Two-stage electrodes separate the corona-charging region from a relatively corona-free collection region. Since the current density is low in the collection region, higher collection fields can be achieved between these electrodes than can be obtained between single-stage electrodes. Two-stage designs may include sections with chargers before each collection "field." The development of two-stage electrostatic precipitators for applications producing particulate emissions with high or variable electrical resistivity is a recent focus. Both tubular- and plate-type, two-stage electrostatic precipitators are commercially available.

No electrostatic precipitator is purely single-stage or two-stage. When corona electrodes of single-stage equipment become fouled with deposits, regions of the electrostatic precipitator begin to serve the function of collectors in two-stage equipment. Also, regions in the corona-free second stage of most two-stage equipment provide some ionization, producing a single-stage character.

These nonideal conditions have led to another classification scheme for industrial electrostatic precipitators. This classification distinguishes between weighted wire electrodes, rigid-frame electrodes, and rigid electrodes.

Electrical equipment. Electrical power supplies for electrostatic precipitators are generally designed for 55 kV(rms) service with current ratings from 100 to 1500 mA (Hall 1975). Each power supply is equipped with manual or automatic voltage controls to maintain the voltage as high as possible. Typical automatic voltage control (AVC) systems prevent mechanical damage to an electrostatic precipitator when an arc occurs. AVC systems reduce the voltage to suppress arcs, reenergize the electrostatic precipitator at a slightly lower voltage than that which had led to an arc, and then raise the voltage on each successive cycle until another spark occurs. By setting the rate of rise of voltage between arcs, the sparking rate is set to give a good balance between maintaining the voltage as high as possible and keeping the field energized. Sparking rates of about two sparks per second are typical.

Modern AVC systems include computer-controlled management of energy usage and plume visibility. These systems are commonly integrated with other controls for electrode cleaning and gas and dust handling; they can also include system diagnostics and automatic logging of performance.

Safety interlocks. Internal, high-voltage equipment must not be accessible to personnel until the power supply is disconnected and grounded. Key interlock systems, which prevent opening hatchways, for example, while the high-voltage equipment is energized, are normally provided. Also, care should be taken to ensure that sampling locations and cleanout ports are not located so as to allow probes or rodding tools to contact the high-voltage equipment. Probes should also be carefully grounded when sampling is done downstream of electrostatic precipitators.

Electrode cleaning systems. The precipitated material must be removed from the collecting electrodes to maintain performance of an electrostatic precipitator. In the case of dry dust, the collecting electrodes are vibrated or rapped, causing the material to dislodge and fall into hoppers. Since the material must fall though the gas stream to reach the hopper, reentrainment or at least transient visible emissions must be minimized. Examples of such measures include (1) designing the collecting electrodes to controlled levels of vibration; (2) minimizing the intensity and frequency of rapping; and (3) rapping only a small section of the electrostatic precipitator at

irrigation with water or other liquids. The cleaning systems of both wet and dry electrostatic precipitators are optimized in the field to yield the lowest steady-state emissions and visibility.

Stack-sampling should be done with careful recognition of the cleaning cycle of an electrostatic precipitator. Time-weighted averaging should be used to estimate steady-state performance over the entire cleaning cycle.

Fabric Filters

Fabric filters are dry dust collectors that use a stationary media to capture the suspended dust and remove it from the gas stream. This media, called fabric, can be composed of a wide variety of materials, including natural and synthetic cloth, glass fibers, and even paper.

Principle of operation. When contaminated gases pass through a fabric, particles may attach to the fibers and/or other particles and separate from the gas stream. The particles are normally captured near the inlet side of a fabric to form a deposit known as a dust cake. The dust cake is periodically removed from the fabric, thus preventing excessive resistance to the gas flow. Finer particles may penetrate more deeply into a fabric, and if not removed during cleaning, may blind it.

The primary mechanisms for particle collection are direct interception, inertial impaction, electrostatic attraction, and diffusion (Billings *et al.* 1970).

Direct interception occurs when the fluid streamline carrying the particle passes within one-half of a particle diameter of a fiber. Regardless of the particle's size, mass, or inertia, it will be collected if the streamline passes sufficiently closely.

Inertial impaction occurs when the particle would miss the fiber if it followed the streamline, but its inertia overcomes the resistance offered by the flowing gas, and the particle continues in a direct enough course to strike the fiber.

Electrostatic attraction occurs when the particle, the filter, or both possess sufficient electrical charge to cause the particles to precipitate on the fiber.

In addition to the steady stresses on the abovementioned particle, random processes drive particles from their paths near fibers. This Brownian motion or diffusion makes particles more likely to be near fibers and thus be collected. Once a particle resides on a fiber, it effectively increases the size of the fiber.

Fabric filters have several advantages over other high-efficiency dust collectors, such as electrostatic precipitators and wet scrubbers:

- They provide high efficiency with lower installed cost.
- The particulate matter is collected in the same state that it was suspended in the gas stream—a significant factor if product recovery is desired.
- Process upsets seldom result in the violation of emission standards. The mass rate of particulate matter escaping collection remains low over the life of the collector and is insensitive to large changes in the mass-loading of dust entering the collector.

Fabric filters also have some limitations:

- Moist and sticky materials and condensation blind fabrics and reduce or prevent gas flow. These actions can curtail plant operation.
- Fabric life may be shortened in the presence of acidic or alkaline components in the gas stream.
- Use of fabric filters is generally limited to temperatures below 500°F.
- Should a spark or flame accidently reach the collector, fabrics can contribute to a fire/explosion hazard.

- When a large volume of gas is to be cleaned, the large number of fabric tubes (envelopes) required and the maintenance problem of detecting, locating, and replacing a damaged element should be considered.

Pressure-volume relationships. The size of a fabric filter is based on empirical data on the amount of fabric media required to clean the desired volumetric flow of gas, with an acceptable static pressure drop across the media. The appropriate media and conditions for its use are selected by pilot test, or from experience with media in similar full-scale installations. These tests provide a recommended range of approach velocities for a specific application.

The approach velocity is stated in practical applications as a gas-to-cloth ratio that has the dimensions of a velocity. The gas-to-cloth ratio is the ratio of the volumetric flow of gases through the fabric filter to the area of fabric that participates in the filtration. It is the average approach velocity to the media.

It is difficult to estimate the correlation between pressure drop and gas-to-cloth ratio for a new installation. However, when the flow entering a media with a dust cake is laminar and uniform, the pressure drop is proportional to the approach velocity.

$$\Delta p = KV_iW = KQW/A \qquad (10)$$

where:

Δp = pressure drop, in. of water
V_i = approach velocity or gas-to-cloth ratio, fpm
W = areal density of dust cake, oz/ft^2
K = specific resistance coefficient, in. of water/fpm/oz of dust/ft^2
Q = volumetric flow of gases, cfm
A = area of cloth that intercepts gases, ft^2

Equation (10) suggests that increasing the area of the fabric during initial installation has some advantage. A larger fabric area reduces both the gas-to-cloth ratio and the thickness of the dust cake, resulting in a decreased pressure drop across the collector and reduced cleaning requirements. This results in lower operational and maintenance cost for the fabric filter system. In addition, the lower gas-to-cloth ratio allows for some expansion of the system and, more important, additional surge capacity when upset conditions, such as unusually high moisture content, occur.

The specific resistance coefficient K is usually higher for fine dusts. The use of a primary collector to remove the coarse fraction seldom causes a significant change in the pressure drop across the collector. In fact, the coarse dust fraction helps reduce operating pressure, since it results in a more porous dust cake, which provides better dust cake release.

Figure 7 illustrates the dependence of the pressure drop on time for a single-compartment fabric filter, operating through its cleaning cycle. The volumetric flow, particulate concentration, and distribution of particle sizes are assumed to remain constant. In practice, however, these assumptions are not usually valid. However, the interval between cleaning events is usually long enough that these variations are insignificant for most systems, and the dependence of pressure drop on time between cleaning events is approximately linear.

The volumetric flow of gases from a process often varies in response to changes in pressure drop across the fabric filter. The degree to which this variation is significant will depend on the operating point for the fan and the requirements for the process.

Electrostatic augmentation. Electrostatic augmentation involves establishing an electric field between the fabric and another electrode, precharging the dust particles, or both. The

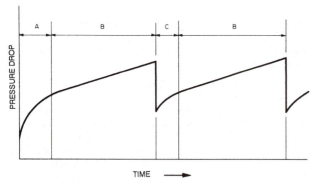

A = Initial cake formation (new fabric)
B = Deposition during homogeneous cake formation
C = Cake removal and initial cake formation
B + C = Interval between cleaning events

Fig. 7   Time Dependence of Pressure Drop across Fabric Filter

effect of electrostatic augmentation is that the interstitial openings in the fabric material function as if they were smaller, and hence smaller particles are retained. Its principle advantage has been in the more rapid buildup of the dust layer and somewhat higher efficiency for a given pressure drop. Although tested by many, this technique has not been broadly applied.

Fabrics. Commercially available fabrics, when applied appropriately, will separate 1 μm or larger particles from a gas with an efficiency of 99.9% or better. Particle size is not the major factor influencing efficiency attained from an industrial fabric filter. Most manufacturers of fabric filters will guarantee such efficiencies on applications in which they have prior experience. Lower efficiencies are generally attributed to poor maintenance (torn fabric seams, loose connections, etc.) or the inappropriate selection of lighter/higher permeability fabrics in an effort to reduce the cost of the collector.

Fabric specifications summarize information on such factors as cost, fiber diameter, type of weave, fabric weight, tensile strength, dimensional stability, chemical resistance, finish, permeability, and abrasion resistance. Usually, the comparisons are difficult, and the supplier must be relied on to select the appropriate material to meet service conditions.

Table 6 summarizes experience with the exposure of fabrics to industrial atmospheres. While higher temperatures are acceptable for short periods, reduced fabric life can be expected with continued use above the maximum temperature. Fabric-

filter media is often protected from high temperatures by thermostatically controlled air bleed-in or collector bypass dampers.

When the gases are moist or the fabric must collect hygroscopic or sticky materials, synthetic media are recommended. They are also recommended for high-temperature gases. Polypropylene has become a frequent selection. One limitation of synthetic media is their greater penetration during the cleaning cycle.

Woven fabrics are generally porous, and filtration depends on collection in a dust cake. New cloth collects poorly until particles bridge the openings in the cloth. Once the cake is formed, the initial layers become part of the fabric; they are not destroyed when the bulk of the collected material is dislodged during the cleaning cycle.

Cotton and wool fibers in woven media as well as most felted fabrics will accumulate an initial dust cake in a few minutes. Synthetic woven fabrics may require a few hours in the same application, because of the smoothness of the monofilament threads. Spun threads in the fill direction, when used, reduce the time required to build up the initial dust cake. Felted fabrics contain no straight-through openings and have a reasonably good efficiency for most particulates, even when clean. When the dust cake builds internally, as well as on the fabric's surface, it will not become porous by shaking.

Cartridges manufactured from pleated paper are also fabric filters, because they operate in the same general manner as high-efficiency fabrics. As with cloth filters, the efficiency of the filter is increased further by the formation of a dust cake on the media. Cartridge filters with pleated paper media are cleaned using a reverse pulse of compressed air.

Types of fabric filters. Fabric filters are generally classified by the method(s) used to clean the fabric. The most common cleaning methods are (1) shaking the bags, (2) reversing the flow of gas that passes through the bags, and (3) using an air pulse (pulse jet) to shock the dust cake and break it from the bags. Pulse jet fabric filters are usually cleaned on-line, whereas fabric filters using shakers or reverse air cleaners are usually cleaned off-line. Generally, large installations are compartmented and use off-line cleaning. Other cleaning methods include shake-deflate, a combination of shaker and reverse-air cleaning, and acoustically augmented cleaning.

Shaker collectors. When a single compartment is needed that can be cleaned off-line during shift-change or breaks, shaker-type fabric filters are usually the least expensive choice. The fabric media in a shaker-type fabric filter, whether formed into cylindrical tubes or envelopes, is mechanically agitated to remove the dust cake.

When the fabric filter cannot be stopped for cleaning, the collector is divided into a number of independent sections that are sequentially taken off-line for cleaning. Since it is usually difficult to maintain a good seal with dampers, relief dampers are often included. The relief dampers introduce a small volume of reverse gas to keep the gas flow at the fabric suitable for cake removal. Use of compartments with their frequent cleaning cycles does not permit a substantial increase in flow rates over that of a single-compartment unit cleaned every 4 h. The best condition for fabric reconditioning occurs with the system stopped, because even small particles will then fall into the hopper.

The gas-to-cloth ratio for a shaker-type collector is usually in the range from 2 to 4 fpm; it might be lower where the collector filters particulate matter that is predominantly less than 2 μm in size. The abatement of metallurgical fume is one example, where a shaker collector is used to control particles that are less than 1 μm in size.

Table 6   Temperature Limits and Characteristics of Fabric Filter Media

	Maximum Continuous Operating Temperature, °F	Acid Resistance	Alkali Resistance	Flex Abrasion	Cost Relative to Cotton
Cotton	180	Poor	Very Good	Very Good	1.00
Wool	200	Good	Poor	Fair to Good	2.75
Nylon* +	200	Poor	Excellent	Excellent	2.50
Nomex* +	400	Fair	Very Good	Very Good	8.00
Acrylic	260	Good	Fair	Good	3.00
Polypropylene	180	Excellent	Excellent	Very Good	1.75
Polyethylene	145	Excellent	Excellent	Very Good	2.00
Teflon +	425	Excellent	Excellent	Good	30.00
Glass fiber	500	Fair to Good	Fair to Good	Poor	5.00
Polyester* +	275	Good	Good	Very Good	2.50
Cellulose	180	Poor	Good	Good	—

*These fibers are subject to hydrolysis when they are exposed to hot, wet atmospheres.
+ DuPont trademark.

The bags are usually 4 to 8 in. in diameter and 10 to 20 ft in length, whereas envelope assemblies may be almost any size or shape. For ambient air applications, a woven cotton or polypropylene fabric is usually selected for shaker-type fabric filters. Synthetics are chosen for their resistance to elevated temperature and their chemical resistant properties.

Typical shaker-type fabric filters, employing bag and envelope media, are illustrated in Figures 8 and 9, respectively.

Figure 10 shows typical pressure diagrams for four- and six-compartment fabric filters that are continuously cleaned with mechanical shakers. Continuously cleaned units have com-

partment valves that close for the shaking cycle. This diagram is typical for a multiple compartment fabric filter, where individual compartments are cleaned off-line.

Reverse flow collectors. Reverse flow cleaning is generally chosen when the volumetric flow of gases is very large. This method of cleaning inherently requires a compartmented design, because the reverse flow needed to collapse the bags entrains dust that must be returned to on-line compartments of the fabric filter. Each compartment is equipped with one main shut-off valve and one reverse gas valve (whether the system is blown-through or drawn-through). A secondary blower and duct system is required to reverse the gas flow in the compartment to be cleaned. The reverse gas circuit increases the volumetric flow and dust loading through the collector's active compartments, when a compartment is isolated for cleaning.

The fabric media is reconditioned by reversing the direction of flow through the bags, which partially collapse. The cleaning action is illustrated in Figure 11. After reverse flow cleaning, the reintroduction of gas is delayed to allow dislodged dust to fall into the hopper.

This cleaning method reduces the number of moving parts in the fabric filter system—an advantage from a maintenance standpoint, especially when large volumetric flows are cleaned. However, the cleaning or reconditioning is less vigorous than with other methods, and the residual drag of the reconditioned fabric is higher. Reverse-flow cleaning is particularly suited for fabrics like glass cloth, where gentle cleaning is required.

Reverse-flow bags are usually 8 to 12 in. in diameter and 20 to 33 ft long and are generally operated at flow velocities in the 2 to 4 fpm range. Consequently, space requirements in plan dimensions are favorable for the reverse-flow unit.

For ambient-air applications, a woven cotton or polypropylene fabric is the usual selection for reverse-flow cleaning. For higher temperatures, woven polyester, glass fiber, or trademarked fabrics are often selected.

Pulse jet collectors. Efforts to decrease fabric filter sizes by increasing the flow rates through the fabric have concentrated on methods of frequent or continuous cleaning cycles, without taking major portions of the filter surface out of service. In the pulse jet design shown in Figure 12, a compressed air jet operating for a fraction of a second causes a rapid vibration or ripple in the fabric, which dislodges the accumulated dustcake. The pulse jet design is predominantly used because it is easier to maintain than the reverse jet mechanism.

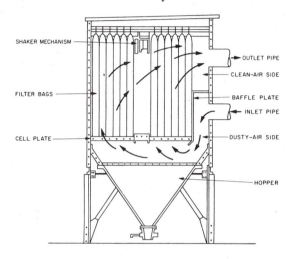

Fig. 8 Bag-Type Shaker Collector

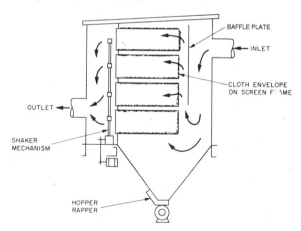

Fig. 9 Envelope-Type Shaker Collector

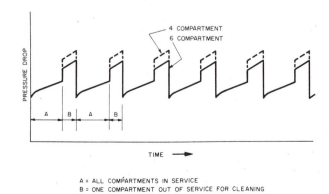

Fig. 10 Pressure Drop across Shaker Collector versus Time

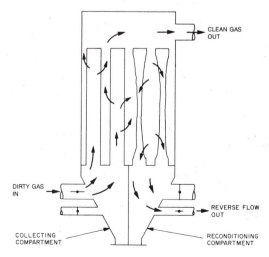

Fig. 11 Reverse Flow Cleaning of Fabric Filter

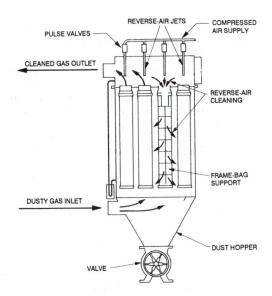

Fig. 12   Typical Pulse Jet Fabric Filter

The tubular bags are supported by wire cages during normal operation. The reverse flow pulse breaks up the dust layer on the outside of the bag and the dislodged material eventually falls to a hopper.

Flow rates for reverse jet or pulse jet designs range from 6 to 10 fpm for favorable dusts, but require greater fabric area for many materials that produce a low-permeability dust cake. Felted fabrics are generally used for these designs because the jet cleaning opens the pores of woven cloth and produces excessive leakage through the filter.

Most pulse jet designs call for 4.5 to 6 in. diameter bag, 6 to 12 ft long. The pulse jets usually require high-pressure, dry, compressed (up to 100 psi) air, leading to high operating cost.

A specialized pulse jet design uses pleated filter media formed into cartridges instead of bags (Figure 13). With the pleated construction, a large filter media surface area, in the form of either flat or circular cartridges, is included in a relatively small housing. A lower approach velocity, in the range of 1 to 3 fpm and high-efficiency media can be used because of the unique configuration of the filter media.

The filter media is a blend of cellulose, with either phenolic or resin binders. As with fabric media, the efficiency of collection is increased by the formation of a dust cake. Since the media's porosity is more accurately controlled in its manufacture, the pleated media cartridge is usually more dependable in building a surface dust cake. Fine particles penetrate less into the media, and the overall efficiency of the filter is increased.

Filter media for cartridge collectors include many felted synthetics and high cellulose blends. These are frequently selected to control nuisance particulate matter for heat-conserving recirculation systems.

Granular Bed Filters

Usually, granular bed filters employ a fixed bed of granular material that is periodically cleaned off-line. In recent years, continuously moving beds have been developed. Most of the commercial systems incorporate electrostatic augmentation to enhance fine particle control and to achieve good performance with a moving bed. Reentrainment in moving granular bed filters still significantly influences overall bed efficiency (Wade *et al.* 1978).

Principle of operation. A typical granular bed filter is shown in Figure 14. Particulate-laden gas travels horizontally through the louvers and a granular media, while the bed material flows downward. The gases typically travel with a superficial velocity near 100 to 150 fpm.

The filter media moves continuously downward by gravity to prevent a filter cake from forming on the face of the filter and to prevent a high pressure drop. To provide complete cleaning of the louver's face, the louvers are designed so that some of the media falls through each louver opening, thus preventing any bridging or buildup of particulate material.

Electrostatic augmentation gives the granular bed filter many of the characteristics of a two-stage electrostatic precipitator. The obvious disadvantage of a granular bed filter is in removal of the collected dust, which would require liquid backwash or circulation and cleaning of the filter material.

Particulate Scrubbers

Wet-type dust collectors use a liquid (usually water) to capture and separate particulate matter (dust, mist, and fumes) from a gas stream. Particle sizes, which can be controlled by a wet scrubber, range from 0.3 to 50 μm or larger. Wet collectors

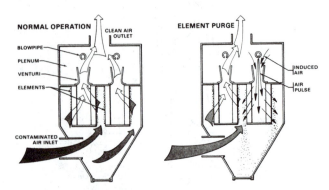

Fig. 13   Pulse Jet Design with Pleated Media Cartridges

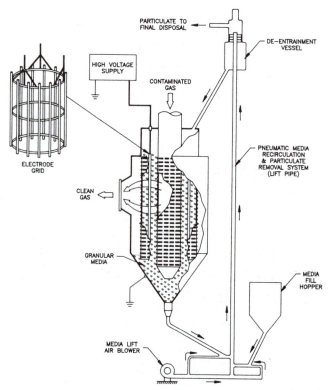

Fig. 14   Typical Granular Bed Filter

can be classified into three categories: (1) low-energy type (up to 1 W/cfm, 1 to 6 in. of water); (2) medium-energy type (1 to 3 W/cfm, 6 to 18 in. of water); and (3) high-energy type (>3 W/cfm, >18 in. of water). The performance of typical wet scrubbers is summarized in Table 2.

Wet collectors may be used for the collection of most particulates from industrial process gas streams where economics allow for collection of the material in a wet state.

Advantages of wet collectors include:

- Constant operating pressure
- No secondary dust sources
- Small spare parts requirement
- Ability to collect both gases and particulates
- Low cost
- Ability to handle high-temperature and high-humidity gas streams, as well as to reduce the possibility of fire or explosion
- Space requirements for scrubbers are reasonably small

Disadvantages include:

- High susceptibility to corrosion, and corrosion-resistant construction is expensive
- High humidity in the discharge gas stream results in visible and often objectionable steam plumes
- Large pressure drops
- High power requirement
- Disposal of waste water or clarification waste may be difficult or expensive
- Fractional efficiency for most scrubbers falls off rapidly for particles less than 1 μm in size
- Freeze protection is required in many applications in northern environments
- Under winter conditions, fog generation can cause visibility problems and neighborhood complaints

Principle of operation. The more important mechanisms involved in the capture and removal of the particulate matter are inertial impaction, Brownian diffusion, and condensation.

Inertial impaction occurs when a dust particle and a liquid droplet collide, resulting in the particle being captured. The resulting liquid/dust particle is relatively large and may be easily removed from the carrier gas stream by gravitation or impingement on separators.

Brownian diffusion occurs when the dust particles are extremely small and have motion independent of the carrier gas stream. These small particles collide with one another, making larger particles, or collide with a liquid droplet and are captured.

Condensation occurs when the gas or air is cooled below its dew point. When moisture is condensed from the gas stream,

fogging occurs, and the dust particles serve as condensation nuclei. The dust particles become larger as a result of the condensed liquid, and the probability of their removal by impaction is increased.

Wet collectors perform two individual operations. The first occurs in the contact zone, where the dirty gas comes in contact with the liquid; the second is in the separation zone, where the liquid that has captured the particulate matter is removed from the gas stream. All well-designed wet collectors use one or more of the following principles:

- High liquid-to-gas ratio
- Intimate contact between the liquid and dust particles
- Formation of larger numbers of small liquid droplets to increase the chances of impaction
- Abrupt transition from dry to wet zones to avoid particle buildup where the dry gas enters the collectors

For a given type of wet collector, the greater the power applied to the system, the higher will be the collection efficiency (Lapple and Kamack 1955). This is described as the contacting power theory. Figures 15 and 16 compare the fractional efficiencies of several wet collectors and the relationship between the pressure drop across a venturi scrubber and the abatement of particulate matter, respectively.

Spray towers and impingement scrubbers. Spray towers and impingement scrubbers are available in many different arrangements. The gas stream may be subjected to a single spray or a series of sprays, or the gas may be forced to impinge on a series of irrigated baffles. Except for packed towers, these types of scrubbers are in a low-energy category; thus, they have a relatively low degree of particulate removal.

The efficiency of a spray tower can be improved by adding high-pressure sprays. Pressures in the range from 30 to 100 psig can improve the efficiencies of spray towers from 50 to 75% to 95 to 99% for dust particles with size near 2 μm. A typical spray tower and an impingement scrubber are illustrated in Figures 17 and 18, respectively.

Centrifugal-type collectors. These collectors are characterized by a tangential entry of the gas stream into the collector. They are classed with medium-energy scrubbers. The impingement scrubber shown in Figure 18 is an example of a centrifugal-type wet collector.

Orifice-type collectors. Orifice-type collectors are also classified in the medium-energy category. Usually, the gas stream is made to impinge on the surface of the scrubbing liquid and is forced through constrictions where the gas velocity is increased and where the liquid-gaseous-particulate interaction occurs. Water usage for orifice collectors is limited to evapo-

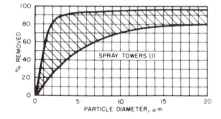
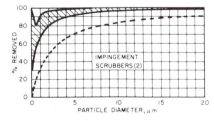
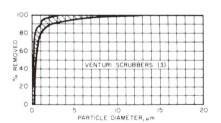

Notes:
1. Efficiency depends on liquid distribution.
2. Upper curve is for a packed tower; lower curve is for orifice-type wet collector.
   Dashed lines indicate performance of low-efficiency, irrigated baffles or rods.
3. Efficiency is directly related to fluid rate and pressure drop.

Fig. 15   Fractional Efficiency of Several Wet Collectors

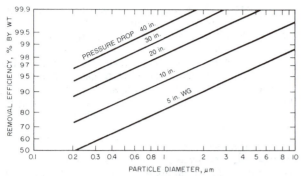

Fig. 16   Efficiency of Venturi Scrubber

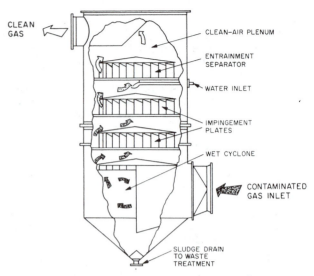

Fig. 18   Typical Impingement Scrubber

Fig. 17   Typical Spray Tower

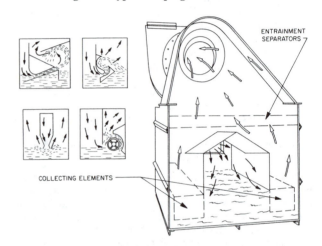

Fig. 19   Typical Orifice-Type Wet Collector

ration loss and removal of collected pollutants. A typical orifice-type wet collector is illustrated in Figure 19.

Venturi scrubber. A high-energy venturi scrubber passes the gas through a venturi-shaped orifice where the gas is accelerated to 12,000 fpm or greater. Depending on the design, the scrubbing liquid is added at, or ahead of, the throat. The rapid acceleration of the gas shears the liquid into a fine mist, increasing the chance of liquid-particle impaction. Yung has developed a mathematical model for the performance and design of venturi scrubbers (Semrau 1977). Subsequent validation experiments (Rudnick 1986) have demonstrated that this model yields the most representative prediction of venturi scrubber performance in comparison with other performance models.

In typical applications, the pressure drop for gases across a venturi is higher than for other types of scrubber. Water circulation is also higher; thus, venturi systems use water reclamation systems. One example of a venturi scrubber is illustrated in Figure 20.

Electrostatically augmented scrubbers. Several gas-cleaning devices combine electrical charging of particulate matter with wet scrubbing. Electrostatic augmentation enhances fine particle control by causing an electrical attraction between the particulate contaminants and the liquid droplets. These scrub-

bers can be constructed from corrosion-resistant materials, and can be useful for the control of both particulate matter and soluble or reactive gases such as HCl, HF, and SO_2. When compared with venturi scrubbers, electrostatically augmented scrubbers remove particles that are smaller than 1 μm in size at a much lower pressure drop.

There are three generic designs for electrostatic augmentation:

1. Unipolar charged aerosols pass through a contact chamber containing randomly oriented packing elements of dielectric material.
2. Unipolar charged aerosols pass through a low-energy venturi scrubber.
3. Unipolar charged aerosols pass into a spray chamber where they are attracted to oppositely charged liquid droplets.

A typical electrostatically augmented scrubber is shown in Figure 21.

Collection efficiencies of 50 to 90% can be achieved in a single-particle charging and collection stage, depending on the mass loading of fine particles and the superficial velocity of gases in the collector. Higher collection efficiencies can be obtained by using two or more stages. Removal efficiency of gaseous pollutants depends on the mass transfer and absorption design of the scrubber section.

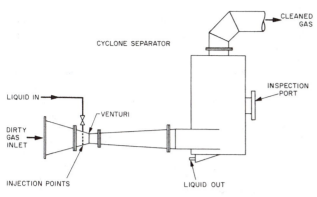

Fig. 20   Typical High-Energy Venturi Scrubber

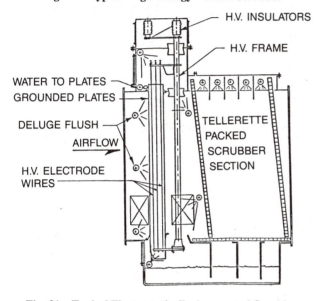

Fig. 21   Typical Electrostatically Augmented Scrubber

In most applications of electrostatically augmented scrubbers, the dirty gas stream is quenched by adiabatic cooling with liquid sprays; thus it contains a large amount of water vapor that wets the particulate contaminants. This moisture provides a dominant influence on particle adhesion and the electrical resistivity of deposits within the collector.

Electrical equipment for particle charging is similar to that for electrostatic precipitators. The scrubber section is usually equipped with a liquid recycle pump, recycle piping, and a liquid distribution system.

Electrostatically augmented scrubbers have been applied successfully to clean and reduce the opacity of gas streams from industrial incinerators that burn hazardous liquid and solid wastes, from plywood veneer driers, refractory tunnel kilns, phosphate rock defluorination reactors, coke calciners, and municipal incinerators.

GASEOUS EMISSIONS CONTROL

Many industrial processes produce large quantities of gaseous or vaporous contaminants that must be separated from gas streams. Removal of these contaminants is usually achieved through absorption into a liquid or adsorption onto a solid media. Incineration of the exhaust gas (as discussed in the next section) has also been successfully used for the removal of organic gases and vapors.

Spray Dry Scrubbing

Spray dry scrubbing is used to absorb and neutralize acidic, gaseous contaminants in hot industrial gas streams. The system uses an alkali spray to react with the acid gases to form a salt. The process heat evaporates the liquid, resulting in a dry particulate that is removed from the gas stream.

Typical industrial applications of spray dry scrubbing are the control of hydrochloric acid emissions from biological hazardous-waste incinerators, sulfuric acid and sulfur trioxide emissions from burning high-sulfur coal, and control of sulfur oxides, boric acid, and hydrogen fluoride gases from glass-melting furnaces.

Principle of operation. Spray drying involves four operations: (1) atomization, (2) gas droplet mixing, (3) drying of liquid droplets, and (4) removal and collection of a dry product. These operations are carried out in a tower or a specially designed vessel.

In any spray drier design, good mixing and efficient gas droplet contact are desirable. Drier height is largely determined by the time required to dry the largest droplets produced by the atomizer. Towers used for acid gas control typically have gas residence times of about 10 s, compared to about 3 s for towers designed for evaporative cooling. The longer residence time is needed because drying by itself is not the primary goal for the equipment. Many of the acid-alkali reactions are accelerated in the liquid state. It is, therefore, desirable to cool the gases as close as possible to the adiabatic saturation temperature (dew point), without risking condensation in downstream particulate collectors. At these low temperatures, droplets survive against evaporation far longer, thus yielding a better chance of contacting all acidic contaminants, while the alkali compounds are in their most reactive state.

Equipment. The atomizer must fulfill several important functions. First, it must disperse a liquid containing an alkali compound that will react with acidic components of the gas stream. Second, it must disperse the liquid containing an alkali compound, so that it is distributed uniformly within the drier and mixed thoroughly with the hot gases. Third, the droplets, produced by the atomizer, must be of a size that will evaporate within the chamber before striking a surface.

In typical spray driers used for acid-gas control, the droplets have diameters ranging from 50 to 200 μm. The larger droplets are of most concern, because these might survive long enough to impinge on equipment surfaces. In general, a trade must be made between the largest amount of liquid that can be sprayed and the largest droplets that can be tolerated by the equipment.

The angular distribution or "fan-out" of the spray is also important. In spray drying, the angle is often 60 to 80°, although both lower and higher angles are sometimes required. The fan-out may change with distance from the nozzle, especially at high pressures.

An important aspect of spray drier design and operation is the production and control of the gas flow patterns within the drying chamber. Because of the importance of the flow patterns, spray driers are usually classified on the basis of gas flow direction in the chamber relative to the spray. There are three basic designs: (1) *cocurrent,* in which the liquid feed is sprayed cocurrently with the hot gas; (2) *countercurrent,* in which the feed is sprayed countercurrently to the gas; and (3) *mixed flow,* in which there is a combined cocurrent and countercurrent flow.

There are several types of atomizers. High-speed rotating disks achieve atomization through centrifugal motion. Although disks are bulky and relatively expensive, they are also more flexible than nozzles in compensating for changes in particle size caused by variations in feed characteristics. Disks are also used when high-pressure feed systems are not available. They are frequently used when high volumes of liquid must be spray dried. Disks are not well suited to counterflow or horizontal flow driers.

Nozzles are also commonly used. These may be subdivided into two distinct types—*centrifugal pressure* nozzles and *two-fluid* (or pneumatic) nozzles. In the centrifugal pressure nozzle, energy for atomization is supplied solely by the pressure of the feed liquid. Most pressure nozzles are of the swirl type, in which tangential inlets or slots spin the liquid in the nozzle. The pressure nozzle satisfactorily atomizes liquids with viscosities up to several hundred centipoises. It is well suited to counterflow spray driers and to installations requiring multiple atomizers. Capacities up to 10,000 lb/h through a single nozzle are possible. Pressure nozzles have some disadvantages. For example, pressure, capacity, and orifice size are independent, resulting in a certain degree of inflexibility. Moreover, pressure nozzles (particularly those with small passages) are susceptible to erosion in applications involving abrasive materials. In such instances, tungsten carbide or a similar tough material is mandatory.

In two-fluid nozzles, air (or steam) supplies most of the energy required to atomize the liquid. Liquid, admitted under low pressure, may be mixed with the air either internally or externally. Although energy requirements for this atomizer are generally greater than for spinning disks or pressure nozzles, the two-fluid nozzle can produce very fine atomization, particularly with viscous materials.

Finally, the density and viscosity of the feed materials and how these might change at elevated temperature should be considered. Some alkali compounds do not form a solution at the concentrations needed for acid-gas control. It is not uncommon that pumps and nozzles must be chosen to handle and meter slurries.

Spray drier systems include metering valves, pumps and compressors, and controls to assure optimal chemical feed and temperature within the gas-cleaning system.

Wet-Packed Scrubbers

Packed scrubbers are used to remove gaseous contaminants and particulate matter from gas streams. Scrubbing is accomplished by impingement of particulate matter and/or by absorption of soluble gas or vapor molecules on the liquid-wetted surface of the packing. There is no limit to the amount of particulate capture, as long as the properties of the liquid film are unchanged.

Gas or vapor removal is more complex than particulate capture. The contaminant becomes a solute and has a vapor pressure above that of the scrubbing liquid. This vapor pressure typically increases with increasing concentration of the solute in the liquid, and/or with increasing liquid temperature. Scrubbing of the contaminant continues as long as the partial pressure of contaminant in the gas is above its vapor pressure with respect to the liquid. The rate of contaminant removal is a function of the difference between the partial pressure and vapor pressure, as well as the rate of diffusion of the contaminant.

In most common gas scrubbing operations, small increases in the superficial velocity of gases through the collector decrease the removal efficiency; therefore, these devices are operated at the highest possible superficial velocities consistent with acceptable contaminant control. Usually, increasing the liquid rate has little effect on efficiency; therefore, liquid flow is kept near the minimum required for satisfactory operation and removal of particulate matter. As the superficial velocity of gases in the collector increases above a certain value, there is a tendency to strip liquid from the surface of the packing and entrain the liquid from the scrubber. If the pressure drop is not the limiting factor for operation of the equipment, maximum scrubbing capacity will occur at a gas rate just below the rate that causes excessive liquid reentrainment.

Scrubber packings. Packings are designed to present a large surface area that will wet evenly with liquid. They should also have high void-fraction, so that pressure drop will be low. High-efficiency packings promote turbulent mixing of the gas and liquid. Figure 22 illustrates six types of packings that are randomly dumped into scrubbers. Packings are available in ceramic, metal, and thermoplastic materials. Plastic packings are extensively used in scrubbers because of their low mass and resistance to mechanical damage. They offer a wide range of chemical resistance to acids, alkalies, and many organic compounds; however, plastic packing can be deformed by excessive temperatures or by solvent attack.

The relative capacity of tower packings at constant pressure drop can be obtained by calculation from the packing factor. The gas-handling capacity of a packing is inversely proportional to the square root of the packing factor F.

$$G = K/\sqrt{F} \qquad (11)$$

The smaller the packing factor of a given packing, the greater will be its gas-handling capacity. Typical packing factors for scrubber packings are summarized in Table 7.

Arrangements of packed scrubbers. The four generic arrangements for wet-packed scrubbers are illustrated in Figure 23:

1. Horizontal cocurrent scrubber
2. Vertical cocurrent scrubber
3. Crossflow scrubber
4. Countercurrent scrubber

Table 7   Packing Factors F for Various Scrubber Packing Materials

Type of Packing	Material	Nominal Size, in.				
		¾	1	1.5	2	3 or 3.5
Super Intalox	Plastic		40		21	16
Super Intalox	Ceramic		60		30	
Intalox saddles	Ceramic	145	92	52	40	22
Berl saddles	Ceramic	170	110	65	45	
Raschig rings	Ceramic	255	155	95	65	37
Hy-Pak	Metal		43	26	18	15
Pall rings	Metal		48	33	20	16
Pall rings	Plastic		52	40	24	16
Tellerettes	Plastic		36		18	16
Maspac	Plastic				32	21

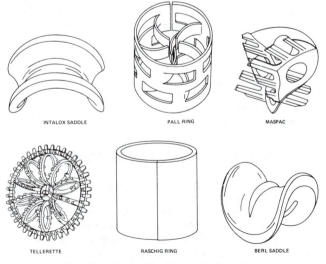

INTALOX SADDLE   PALL RING   MASPAC

TELLERETTE   RASCHIG RING   BERL SADDLE

Fig. 22   Typical Packings for Scrubbers

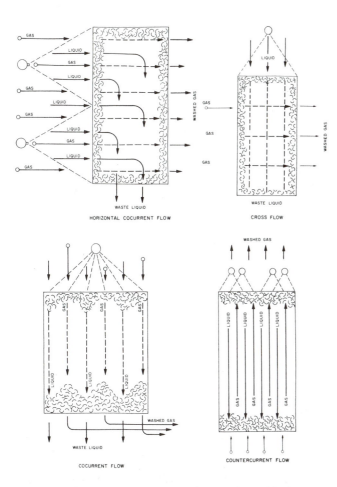

Fig. 23 Flow Arrangements through Packed Beds

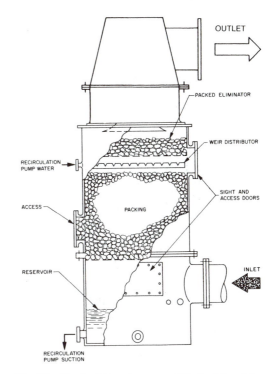

Fig. 24 Typical Countercurrent Packed Scrubber

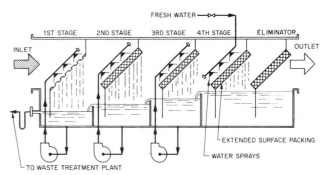

Fig. 25 Horizontal Flow Scrubber with Extended Surface

Cocurrent flow scrubbers can be operated with horizontal or vertical gas and liquid flows. A *horizontal cocurrent scrubber* depends on the gas velocity to carry the liquid into the packed bed. It operates as a wetted entrainment separator with limited gas and liquid contact time. The superficial velocity of gases in the collector is limited by liquid reentrainment to about 650 fpm. A vertical cocurrent scrubber can be operated at very high pressure drop (1 to 3 in. of water/ft of packing depth), since there is no flooding limit for the superficial velocity. The contact time in a cocurrent scrubber is a function of bed depth. The effectiveness of absorption processes is lower in cocurrent flow scrubbers than in the other arrangements, because the liquid-containing contaminate is in contact with the exit gas stream.

Cross-flow scrubbers use vertically downflowing liquid and a horizontally moving gas stream. The effectiveness of absorption processes in cross-flow scrubbers lies between those for cocurrent and countercurrent flow scrubbers.

Countercurrent scrubbers use a downward flowing liquid and upward flowing gas. The gas-handling capacity of countercurrent scrubbers is limited by pressure drop or by liquid entrainment. The contact time can be controlled by the depth of packing used. The effectiveness of absorption processes is maximized in countercurrent scrubbers, since the exiting gas is in contact with fresh scrubbing liquid.

The most broadly used arrangement is the countercurrent packed scrubber. This type of scrubber, illustrated in Figure 24, gives the best removal of gaseous contaminants, while keeping liquid consumption at a minimum. The effluent liquid has the highest contaminant concentration in countercurrent packed scrubbers.

Extended surface packings have been used successfully for the absorption of highly soluble gases such as HCl, since the required contact time is minimal. This type of packing consists of a woven mat of fine plastic fibers, formed from a plastic material that is not affected by chemical exposure. Figure 25 shows an example of a scrubber, consisting of three wetted stages of extended surface packing in series with the gas flow. A final dry mat is used as an entrainment eliminator. If solids are present in the inlet gas stream, a wetted impingement stage precedes the wetted mats to prevent plugging of the woven mats.

Figure 26 shows a scrubber with a vertical arrangement of extended surface packing. This design uses three complete stages in series with the gas flow. The horizontal mat at the bottom of each stage operates as a flooded bed scrubber. The flooded bed is used to minimize water consumption. The two inclined upper mats operate as entrainment eliminators.

Pressure drop. The pressure drop through a particular packing in countercurrent scrubbers can be calculated from the airflow and water flow per unit area. Charts, such as the one

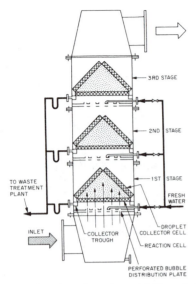

Fig. 26 Vertical Flow Scrubber with Extended Surface

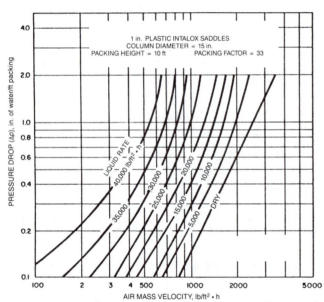

Fig. 27 Pressure Drop versus Gas Rate for Typical Packing

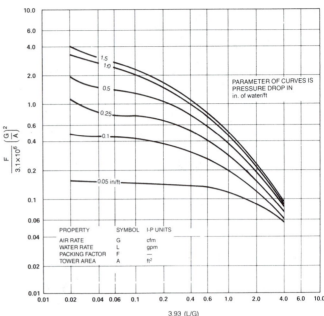

Fig. 28 Generalized Pressure Drop Curves for Packed Beds

of contaminant to be removed, distribution and amount of absorbent available for reaction, temperature, and reaction rate for absorption.

Practically all commercial packings have been tested for absorption rate (mass transfer coefficient) using standard absorber conditions—CO_2 in air, and a solution of NaOH in water. This system was selected because the interaction of the variables is well understood. Further, the mass transfer coefficients for this system are low; thus they can be determined accurately by experiment. The values of mass transfer coefficients (K_Ga) for various packings under these standard test conditions are given in Table 8.

Table 8 Mass Transfer Coefficients for Scrubber Packing Materials

		Nominal Size, in.			
Type of Packing	Material	1	1.5	2	3 or 3.5
		K_Ga, lb · mol/h · ft³ · atm			
Super Intalox	Plastic	2.19		1.44	0.887
Intalox saddles	Ceramic	1.96	1.71	1.44	0.820
Raschig rings	Ceramic	1.73	1.50	1.21	
Hy-Pak	Metal	2.20	1.87	1.69	1.09
Pall rings	Metal	2.32	1.87	1.62	0.91
Pall rings	Plastic	1.98	1.73	1.46	0.89
Tellerettes	Plastic	2.19		1.98	
Maspac	Plastic			1.44	0.89

System: CO_2 and NaOH; Liquid rate: 4 gpm/ft²; Gas rate: 110 cfm/ft².

The vast majority of wet absorbers are used to control low concentrations (less than 0.005 mol fraction) of contaminants in air. Dilute aqueous solutions of caustic soda (NaOH) are usually chosen as the scrubbing fluid. These conditions simplify the design of scrubbers somewhat. Mass transfer from the gas to the liquid is then explained by the Two Film Theory. First, the gaseous contaminant travels by diffusion from the main gas stream through the gas film, then through the liquid film, and, finally, into the main liquid stream. The relative influences of the gas and liquid films on the absorption rate depend on the solubility of the contaminant in the liquid. Sparingly soluble gases, like H_2S and CO_2, are said to be liquid film-con-

shown in Figure 27, are available from manufacturers of each type and size of packing.

The pressure drop for any packing can also be estimated by using the data on packing factors in Table 7 and the modified, generalized, pressure drop correlation shown in Figure 28. This correlation was developed specifically for a gas stream substantially of air, with water as the scrubbing liquid. It should not be used if the properties of the gas or liquid vary significantly from air or water, respectively. Countercurrent scrubbers are generally designed to operate at pressure drops between 0.25 and 0.65 in. of water/ft of packing depth. Liquid irrigation rates typically vary between 5 and 20 gpm/ft² of bed area.

Absorption efficiency. The prediction of the absorption efficiency of a packed bed scrubber is much more complex than estimating its capacity, because performance estimates involve the mechanics of absorption. Some of the factors affecting efficiency are the superficial velocity of gases in the scrubber, the liquid injection rate, packing size, type of packing, amount

trolled; highly soluble gases, such as HCl and NH$_3$, are said to be gas film-controlled. In liquid film-controlled systems, the mass transfer coefficient varies with the liquid injection rate but is only slightly affected by the superficial velocity of the gases. In gas film-controlled systems, the mass transfer coefficient is a function of both the superficial velocity of the gases and the liquid injection rate.

In the absence of leakage, the percentage of the contaminant (by volume) removed from the air is a function of the inlet and outlet concentration of contaminant in the airstream, where Y_i and Y_o are the mol fractions of contaminants entering and exiting the scrubber (expressed on a dry gas basis).

$$\% \text{ Removed} = 100 \left[1 - (Y_o/Y_i) \right] \qquad (12)$$

The driving pressure for absorption (additionally assuming negligible vapor pressure above the liquid) is controlled by the logarithmic mean of inlet and outlet concentrations of the contaminant, where p is the pressure.

$$\Delta p_{\text{ln}} = p \frac{Y_i - Y_o}{\ln (Y_i/Y_o)} \qquad (13)$$

The rate of absorption of contaminant (mass transfer coefficient) is related to the depth of packing.

$$K_G a = N/HA\Delta p_{\text{ln}} \qquad (14)$$

where:

H = depth of packing, ft
A = cross-sectional area of scrubber, ft^2
N = solute absorbed, lb · mol/h

The value of N can be determined from Equation (15), where G is the mass flow rate of gases through the scrubber in lb · mol/h.

$$N = G(Y_i - Y_o) \qquad (15)$$

The superficial velocity of gases is a function of the unit gas flow rate and the gas density as:

$$V = (TGM_v)/(C_1 A) \qquad (16)$$

where:

V = superficial gas velocity, fpm
M_v = molar volume, ft^3/lb · mol
T = exit gas temperature, °R (°F + 460)
C_1 = 460°R

By combining these equations and assuming ambient pressure, a graphical solution can be derived for both liquid film- and gas film-controlled systems. Figures 29, 30, and 31 show the height of packing required versus percent removal for various mass transfer coefficients at superficial velocities of 120, 240, and 360 fpm, respectively, with liquid film-controlled systems.

Figures 32, 33, and 34 show the height of packing versus percent removal for various mass transfer coefficients at the same superficial velocities with gas film-controlled systems. These graphs can be used to determine the height of 2.0-in. plastic Intalox saddles (Figure 22) required to give the desired percentage of contaminant removal. The height for any other type or size of packing is inversely proportional to the ratio of standard $K_G a$ taken from Table 8. Thus, if 13 ft packing depth were required for 95% removal of contaminants, the same efficiency could be obtained with a 9.5 ft depth of 1-in. plastic pall rings (Figure 22), at the same superficial velocity and liquid injection rate. However, the pressure drop would be higher for the smaller diameter packing.

Figures 29 through 34 are useful, when the value of the mass transfer coefficient is known for the particular contaminant to be removed. Table 9 contains mass transfer coefficients for 2.0-in. plastic Intalox saddles in typical liquid film-controlled scrubbers. These values can be compared with the mass trans-

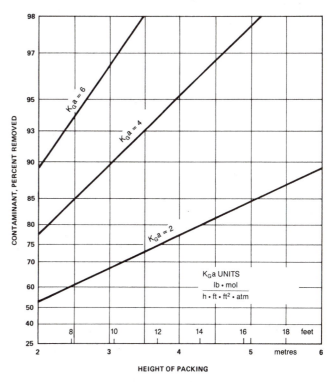

Fig. 29   Contaminant Control at Superficial Velocity = 120 fpm (Liquid Film Controlled)

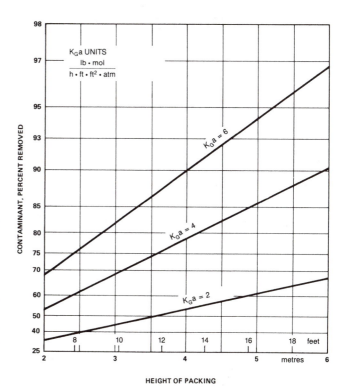

Fig. 30   Contaminant Control at Superficial Velocity = 240 fpm (Liquid Film Controlled)

fer coefficients in Table 10 for the same packing used in gas film-controlled scrubbers. When the scrubbing liquid is not water, the mass transfer coefficient in these tables can only be used if the amount of reagent in the solution exceeds by at least 33% the amount needed to completely absorb the gaseous contaminant.

When HCl is dissolved in water, there is little vapor pressure of HCl above solutions of less than 8% (by mass) concentration. On the other hand, when NH_3 is dissolved in water, there is an appreciable vapor pressure of NH_3 above solutions, even at low concentrations. The height of packing needed for NH_3 removal, obtained from Figures 32 through 34, is based on the use of dilute acid to maintain the pH of the solution below 7.

Table 9 Relative K_Ga for Various Packings

Gas Contaminant	Scrubbing Liquid	K_Ga $\dfrac{\text{lb} \cdot \text{mol}}{\text{h} \cdot \text{ft}^3 \cdot \text{atm}}$
CO_2	4% (by mass) NaOH	2.0
H_2S	4% (by mass) NaOH	5.92
SO_2	Water	2.96
HCN	Water	5.92
HCHO	Water	5.92
Cl_2	Water	4.55

Liquid film-controlled systems, 2-in. plastic Super Intalox. Temperatures: from 60 to 75°F; liquid rate: 10 gpm/ft²; gas rate: 215 cfm/ft².

Table 10 Relative K_Ga for Various Packings

Gas Contaminant	Scrubbing Liquid	K_Ga $\dfrac{\text{lb} \cdot \text{mol}}{\text{h} \cdot \text{ft}^3 \cdot \text{atm}}$
HCl	Water	18.66
HBr	Water	5.92
HF	Water	7.96
NH_3	Water	17.30
Cl_2	8 wt % NaOH	14.33
SO_2	11 wt % Na_2CO_3	11.83
Br_2	5 wt % NaOH	5.01

Gas film-controlled systems, 2-in. plastic Super Intalox. Temperatures: from 60 to 75°F; liquid rate: 10 gpm/ft²; gas rate: 215 cfm/ft².

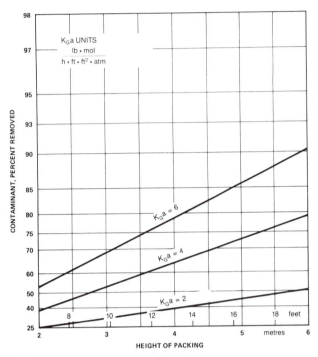

Fig. 31   Contaminant Control at Superficial Velocity = 360 fpm (Liquid Film Controlled)

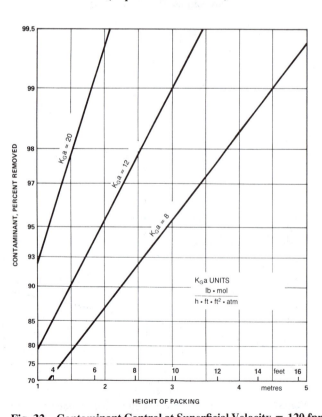

Fig. 32   Contaminant Control at Superficial Velocity = 120 fpm (Gas Film Controlled)

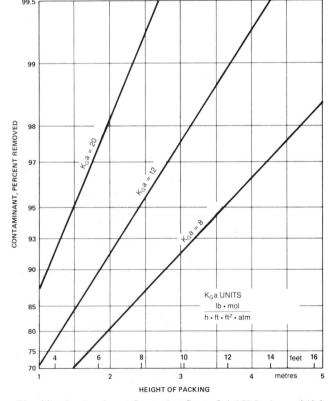

Fig. 33   Contaminant Control at Superficial Velocity = 240 fpm (Gas Film Controlled)

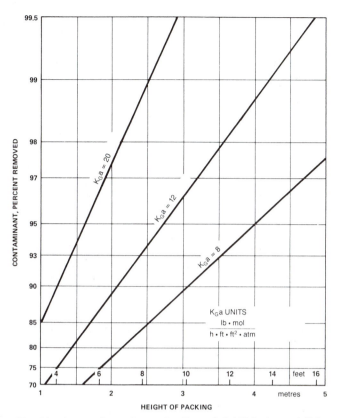

Fig. 34   Contaminant Control at Superficial Velocity = 360 fpm (Gas Film Controlled)

Typical scrubbing problem. The problem is to remove 600 ppm of HF from air at 90°F. The concentration of HF in the exhaust gas stream should not exceed 30 ppm. The concentrations are expressed on a dry gas basis. The desired percentage of HF removal is 95%, if no leakage is assumed.

Conditions for the design follow:

G = Total volumetric flow of gas = 4600 cfm
    Liquid injection rate = 3.75 gpm/ft²
    Liquid temperature = 68°F
    Packed tower diameter = 4 ft
A = Cross-sectional area of absorber = $\pi(4/2)2$ = 12.6 ft²
L = Total liquid flow rate = 3.75 × 12.6 = 47.1 gpm
F = Packing factor = 21
  abscissa (Figure 28) = 3.93 L/G = 3.93(47.1/4600) = 0.040
  ordinate (Figure 28) = (21/3.1 × 10⁶)(4600/12.6)² = 0.90

From Figure 28, the pressure drop through the packed tower is approximately 0.28 in. of water/ft of packing.

From Table 10, K_Ga for HF is 7.96. From Figure 34, the depth of packing required for 95% removal is 13 ft. Thus, the total pressure drop is 13 × 0.28 = 3.64 in. of water.

General efficiency comparisons. Figure 34 indicates, with a K_Ga equal to 8, that 90% removal of HF could be achieved with 10 ft of packing; this is 23% less packing than needed for 95% removal. Furthermore, with the same superficial velocity, both liquid film- and gas film-controlled systems would require a 43% increase in absorber depth to raise the removal efficiency from 80 to 90%.

A comparison of Figures 33 and 34 shows that increasing the superficial velocity by 50% in a gas film-controlled scrubber requires only a 12% increase in bed depth to maintain equal removal efficiencies. In the liquid film-controlled system (Figures 30 and 31), increasing the superficial velocity by 50% requires an approximately 50% increase in bed depth to maintain equal removal efficiencies.

Thus, in a gas film-controlled system, the superficial velocity can be increased significantly with only a small increase in bed depth required to maintain the efficiency. In practical terms, gas film-controlled scrubbers of fixed depth can handle an overload condition with only a minor loss of removal efficiency.

The performance of liquid film-controlled scrubbers degrades significantly under similar overload conditions. This occurs because the mass transfer coefficient is independent of superficial velocity.

Liquid effects. Some liquids tend to foam when they are contaminated with particulates or soluble salts. In these cases, the pressure drop should be kept in the lower half of the normal range—0.25 to 0.40 in. of water/ft of packing.

In the control of gaseous pollution, most systems do not destroy the pollutant but merely remove it from the air. When water is used as the scrubbing liquid, the effluent from the scrubber will contain suspended particulate or dissolved solute. Water treatment is often required to alter the pH and/or remove toxic substances before the solutions can be discharged.

Adsorption of Gaseous Contaminants

The surface of freshly broken or heated solids often contains incomplete chemical bonds that are able to physically or chemically adsorb nearby molecules in a gas or liquid. The captured molecules form a thin layer on the surface of the solid that is one to a few molecules thick. Commercial adsorbents are solids with an enormous internal surface area. This large surface area enables them to capture and hold large numbers of molecules. For example, each pound of a typical activated carbon adsorbent contains over 5,000,000 ft² of internal surface area. Adsorbents are used for removing organic vapors, water vapor, odors, and hazardous pollutants from gas streams.

The most common adsorbents used in industrial processes include activated carbons, activated alumina, silica gel, and molecular sieves. Activated carbons are derived from coal, petroleum, or coconut shells. They are primarily selected to remove organic compounds in preference to water. The other three common gas-phase adsorbents have a great affinity for water and will adsorb it to the exclusion of any organic molecules also present in a gas stream. They are used primarily as gas drying agents. Molecular sieves also find use in several specialized pollution control applications, including removal of mercury vapor, sulfur dioxide, or nitrogen oxides from gas streams.

The capacity of a particular activated carbon to adsorb any organic vapor from an exhaust gas stream is related to the concentration and molecular weight of the organic compound and the temperature of the gas stream. Higher molecular weight compounds are usually more strongly adsorbed than those with lower molecular weight. The capacity of activated carbon to adsorb any given organic compound increases with the concentration of that compound. Adsorption is also favored by reducing the temperature. Typical adsorption capacities of an activated carbon for toluene (molecular weight 92) and acetone (molecular weight 58) are illustrated in Figure 35 for various temperatures and concentrations.

Adsorption is reversible. An increase in temperature causes some or all of an adsorbed vapor to desorb. The temperature of low-pressure steam is sufficient to drive off most of a low boiling point organic compound previously adsorbed at ambient temperature. Higher boiling point organic compounds may require high-pressure steam or hot inert gas to secure good desorption. Very high molecular weight compounds can require reactivation of the carbon adsorber in a furnace at 1350°F to drive off all the adsorbed material. Regeneration of the carbon adsorbers can also be accomplished, in some in-

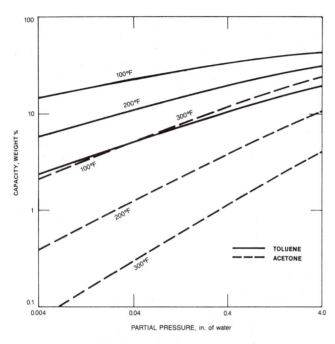

Fig. 35   Adsorption Isotherms on Activated Carbon

stances, by washing with an aqueous solution of a chemical, which will react with the adsorbed organic material, making it water soluble. An example is caustic soda washing of carbon containing adsorbed sulfur compounds.

The difference between an adsorbent's capacity under adsorbing and desorbing conditions in any application is its working capacity. Activated carbon for air pollution control is found in canisters under the hoods of most automobiles. The adsorber in these canisters captures gasoline vapors escaping from the carburetor (when the engine is stopped) and from the fuel tank's breather vent. Desorption of gasoline vapors is accomplished by pulling fresh air through the carbon canister and into the carburetor when the engine is running. Although there is no temperature difference between adsorbing and desorbing conditions in this case, the outside airflow desorbs enough gasoline vapor to give the carbon a substantial working capacity.

For applications where only traces of a pollutant must be removed from exhaust air, the life of a carbon bed is very long. In these cases, it is often more economical to replace the carbon and not invest in regeneration equipment. Larger quantities can be returned to the carbon manufacturer for high-temperature thermal reactivation. Regeneration in place by steam, hot inert gas, or washing with a solution of alkali is sometimes practiced.

Impregnated (*chemically reactive*) adsorbents are used when physical adsorption, by itself, is too weak to remove a particular gaseous contaminant from an industrial gas stream. Through impregnation, the reactive chemical is spread over the immense internal surface area of an adsorbent.

Typical applications of impregnated adsorbent include:

- Sulfur or iodine impregnated carbon removes mercury vapor from air, hydrogen, or other gases by forming mercuric sulfide or iodide.
- Metal oxide impregnated carbons remove hydrogen sulfide.
- Amine or iodine impregnated carbons and silver exchanged zeolites remove radioactive methyl iodide from nuclear power plant work areas and exhaust gases.

- Alkali impregnated carbons remove acid gases.
- Activated alumina, impregnated with potassium permanganate, removes acrolein and formaldehyde.

Equipment for adsorption. Three types of adsorbers are usually found in industrial applications: (1) fixed beds, (2) moving beds, and (3) fluidized beds.

Fixed beds are generally used in those applications where pollutant concentrations are low and where the adsorber media is not regenerated. Carbon filter elements are the most common example of fixed beds.

Moving beds use granular adsorbers placed on inclined trays or on vertical frames similar to those used in granular bed particulate collectors. Moving beds offer the user continuous contaminant control and regeneration. Often, moving bed adsorbers and regeneration equipment are integrated components in a process.

Fluidized beds contain a fine granular adsorber, which is continuously mixed with the contaminated gas by suspension in the process gas stream. The bed may be "fixed" at lower superficial velocities, highly turbulent, or conveying (circulating). Illustrations of these types of fluidized beds are shown in Figure 36.

Solvent recovery. The most common use of adsorption in stationary sources is in recovering solvent vapors from manufacturing and cleaning processes. Typical applications include solvent degreasing, rotogravure printing, dry cleaning, and the manufacture of products such as synthetic fibers, adhesive labels, tapes, coated copying paper, rubber goods, and coated fabrics.

Figure 37 illustrates the components of a typical solvent recovery system using two carbon beds. One bed is used as an adsorber while the other is regenerated with low-pressure steam. Desorbed solvent vapor and steam are recovered in a water-cooled condenser. If the solvent is immiscible with water, an automatic decanter separates the solvent for reuse. A distillation column is used for water-miscible solvents.

Adsorption time per cycle typically runs from 30 min to several hours. The adsorbing carbon bed is switched to regeneration by an automatic timer shortly before the solvent vapor breaks through from the bed, or immediately thereafter, by an organic vapor-sensing control device in the exhaust gas stream.

Low-pressure steam consumption for regeneration is generally about 3.5 lb/lb of solvent recovered (Boll 1976), but it can range from 2 to over 5 lb/lb of solvent recovered, depending on the specific solvent and its concentration in the exhaust gas stream being stripped. Steam, with only a slight superheat, is normally used so that it will condense quickly and give rapid heat transfer.

After steaming, the hot moist carbon bed is usually cooled and partially dried before being placed back on stream. Heat for drying is supplied by the cooling of the carbon and adsorber, and sometimes by an external air heater. In most cases, it is desirable to leave some moisture in the bed. When solvent vapors are adsorbed, there is an evolution of heat. For most common solvents, the heat of adsorption is 40 to 60 Btu/lb·mol. When high-concentration vapors are adsorbed in a dry carbon bed, this heat can cause a substantial temperature rise and can even ignite the bed, unless it is controlled. If the bed contains moisture, the water absorbs energy and helps to prevent an undue rise in bed temperature; also, certain applications may require heat sensors and automatic sprinklers.

As the adsorptive capacity of activated carbon depends on temperature, it is important that solvent-laden air going to a recovery unit be as cool as practicable. The exhaust gases from many solvent-emitting processes (such as drying ovens) are at elevated temperatures. Water- or air-cooled heat exchangers

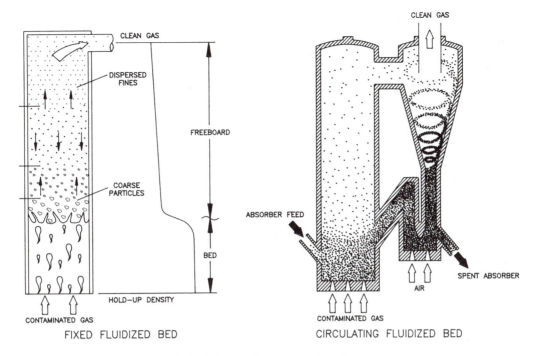

Fig. 36   Fluidized Bed Adsorption Equipment

need to be installed to reduce the temperature of the gas that enters the adsorber.

Very low solvent vapor concentrations can be recovered in an activated carbon system. The size and cost of the recovery unit, however, depend on the volume of air to be handled; it is thus important to minimize the volume of an exhaust stream and keep the solvent vapor concentration as high as possible, consistent with safety requirements. Insurance carriers specify that solvent vapor concentrations must not exceed 25% of the lower explosive limit, when intermittent monitoring is used. With continuous monitoring, concentrations as high as 50% of the low explosive limit (LEL) are permissible.

Solvent recovery systems, with gas-handling capacities up to about 11,000 cfm, are available as skid-mounted packages. Several of these packaged units can be used for larger gas flows. Custom systems can be built to handle 200,000 cfm or more of gas. Materials of construction may be painted carbon steel, stainless steel, Monel, or even titanium, depending on

the nature of the gas mixture. The activated carbon is usually placed in horizontal flat beds or vertical cylindrical beds. The latter design minimizes ground space required for the system. Other alternatives are possible; one manufacturer uses a segmented horizontal rotating cylinder of carbon in which one segment is adsorbing while others are being steamed and cooled.

Commercial-scale solvent recovery systems typically recover over 99% of the solvent contained in a gas stream. The efficiency of the collecting hoods at the source of the solvent emission is usually the determining factor in solvent recovery.

Dust filters are generally placed ahead of carbon beds to prevent blinding of the adsorber by dust. Occasionally, the carbon is removed for screening to eliminate accumulated dust and fine particles of carbon.

The working capacity of activated carbon can decrease with time, if the solvent mixture contains high boiling point components. This occurs when high boiling point organic compounds are only partially removed by low-pressure steam. In this situation, two alternatives should be considered:

1. Periodic removal of the carbon and return to its manufacturer for high-temperature furnace reactivation to near-virgin carbon activity.
2. Use of more rigorous solvent desorbing conditions in the solvent recovery system. Either high-temperature steam, hot inert gas, or a combination of electrical heating and application of a vacuum may be used. The latter method is selected, for example, to recover high boiling lithography ink solvents.

Odor control. Incineration or scrubbing are usually the most economical methods of controlling high concentrations of odorous compounds from equipment such as cookers in rendering plants. However, many odors that arise from low concentrations of vapors that are not harmful, are still offensive. The odor threshold (for 100% response) of acrolein in air, for example, is only 0.21 ppm by volume, while those for ethyl mercaptan is 0.001 ppm and for hydrogen sulfide is 0.0005 ppm

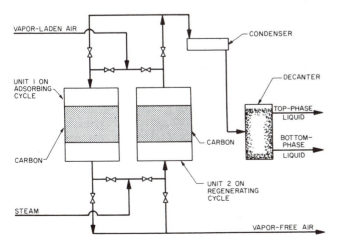

Fig. 37   Schematic of Two-Unit Fixed Bed Adsorber

(MCA 1968, AIHA 1989). Activated carbon beds effectively overcome many odor-emission problems. Activated carbon is used to control odors from chemical and pharmaceutical manufacturing operations, foundries, sewage treating plants, oil and chemical storage tanks, lacquer drying ovens, food processing plants, and rendering plants. In some of these applications, activated carbon is the sole odor-control method; in others, the carbon adsorber is applied to the exhaust from a scrubber.

Odor control systems, using activated carbon, can be as simple as a steel drum, fitted with appropriate gas inlet and outlet ducts, or as complex as a large, vertically moving bed, in which carbon is contained between louvered side panels. A typical moving bed adsorber is shown in Figure 38. In this arrangement, fresh carbon can be added from the top, while spent or dust-laden carbon is periodically removed from the bottom.

Figure 39 shows a fixed bed odor adsorber. Adsorbers of this general configuration are available as packaged systems, complete with motor and blower. Air-handling capacities range from 500 to 12,000 cfm.

The life of activated carbon in odor-control systems ranges from a few weeks to a year or more, depending on the concentration of the odorous emission.

Applications of fluidized bed adsorbers. The injection of alkali compounds into fluidized bed combustors for sulfur oxide control is one example of the use of fluidized bed adsorbers. Another example is the control of hydrogen fluoride

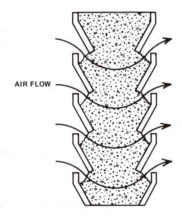

AIR FLOW

Fig. 38   Moving Bed Adsorber

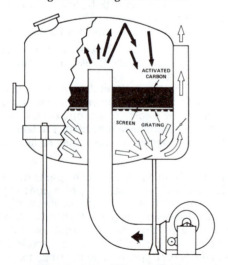

ACTIVATED CARBON

SCREEN   **GRATING**

Fig. 39   Typical Odor Adsorber

emissions from Soderberg aluminum reduction processes by a fixed or circulating fluidized bed of alumina.

INCINERATION OF GASES AND VAPORS

Incineration is the process in which volatile organic compounds (VOCs), organic aerosols, and most odorous materials in a contaminated gas stream are converted to innocuous carbon dioxide and water vapor using heat energy. Incineration is the most effective means for totally eliminating VOCs (EPA 1976). The types of incineration commonly used are thermal and catalytic.

Thermal Incinerators

Thermal incinerators, also known as afterburners or direct flame incinerators, consist of an insulated oxidation chamber in which gas and/or oil burners are typically located. The contaminated gas stream enters the chamber and comes into direct contact with the flame, which provides the heat energy necessary to promote oxidation. Proper conditions of time, temperature, and turbulence oxidize the gas stream contaminants effectively. The contaminated gas stream enters the incinerator near the burner where turbulence inducing devices are usually installed. The final contaminant conversion efficiency largely depends on good mixing within the contaminated gas stream and on the temperature of the oxidation chamber.

Supplemental fuel is used for start-up, to raise the temperature of the contaminated gas stream enough to initiate contaminant oxidation. Once oxidation begins, the temperature rises further due to the energy released by incineration of the contaminant. The supplementary fuel feed rate is then modulated to maintain the desired incinerator operating temperature. Most organic gases oxidize to approximately 90% conversion efficiency if a temperature of at least 1200°F and a residence time of 0.3 to 0.5 s is achieved within the oxidation chamber. However, incinerator temperatures are typically maintained in the range of 1400 to 1500°F to ensure conversion efficiencies of 90% or greater.

Although the efficiency of thermal incinerators can exceed 90% destruction of the contaminant, the reaction may produce undesirable products of combustion. For example, incineration of chlorinated hydrocarbons will cause the formation of hydrogen chloride, which can have an adverse effect on equipment. These new contaminants then require additional controls.

In the past, thermal incineration was not recommended due to the high operating cost associated with the supplementary fuel needed to maintain the high operating temperature. However, with the advent of widespread air pollution control requirements for VOCs, reliable heat recovery methods for use with thermal incinerators were developed. Incineration systems now incorporate primary heat recovery to preheat the incoming contaminated gas stream and, in some cases, provide secondary heat recovery for process or building heating. Primary heat recovery is almost always achieved using air-to-air heat exchangers. Use of a regenerable, ceramic media for heat recovery has increased due to superior heat recovery efficiency. Secondary heat recovery may incorporate an air-to-air heat exchanger or a waste heat boiler (DOE 1979).

Incineration systems, using conventional air-to-air heat exchangers, can achieve 80% primary efficiency. Regenerative heat exchanger units have claimed as high as 95% heat recovery and are routinely operated with 85 to 90% heat recovery efficiency. When operated at these high heat recovery efficiencies and with inlet VOC concentrations of 15 to 25% of the lower explosive limit (LEL), the incineration process approaches a self-sustaining condition, thereby requiring almost no supplementary fuel.

Catalytic Incinerators

Catalytic incinerators operate under the same principles as thermal incinerators, except that they use a catalyst to promote oxidation. The catalyst allows oxidation to occur at lower temperatures than in a thermal incinerator for the same VOC concentration. Therefore, catalytic incinerators require less supplemental fuel to preheat the contaminated gas stream and have lower overall operating temperatures.

A catalytic incinerator generally consists of a preheat chamber followed by the catalyst bed. Residence time and turbulence are not as important as with thermal incinerators, but it is essential that the contaminated gas stream be heated uniformly to the required catalytic reaction temperature. The required temperature varies, depending on the catalyst material and configuration.

The temperature of the contaminated gas stream is raised in the preheat chamber by a conventional burner. Although the contaminated gas stream contacts the burner flame, the heat input is significantly less than that for a thermal incinerator, and only a small degree of direct contaminant oxidation occurs. Natural gas is preferred to prevent catalyst contamination, which could occur with sulfur-bearing fuel oils. However, No. 2 fuel oil units have been operated successfully. The most effective catalysts contain noble metals such as platinum or palladium.

Catalysis occurs at the molecular level. Therefore, an available, active catalyst surface area is important for maintaining high conversion efficiencies. If particulate materials contact the catalyst either as discrete or partially oxidized aerosols, they can ash on the catalyst surface and blind it. This problem is usually accompanied by a secondary pollution problem, *i.e.,* odorous emissions caused by the partially oxidized organic compounds.

The greatest concern to users of catalytic incinerators is catalyst poisoning or deactivation. Poisoning is caused by specific gas stream contaminants that chemically combine or alloy with the active catalyst material. Poisons frequently cited include phosphorous, bismuth, arsenic, antimony, lead, tin, and zinc. Some organic compounds, such as polyester amides and imides, are also poisonous. The first five materials are considered fast-acting poisons and must be excluded from the contaminated gas stream. Even trace quantities of the fast-acting poisons can cause rapid catalyst deactivation. The last two materials are slow-acting poisons; catalysts are somewhat tolerant of these materials, particularly at temperatures lower than 1000°F. However, even the slow poisons should be excluded from the contaminated gas stream to ensure continuous, reliable performance. Therefore, galvanized steel, another possible source of the slow poisons, should not be used for the duct leading to the incinerator.

Sulfur and halogens are also regarded as catalyst poisons. In most cases, their chemical interaction with the active catalyst material is reversible. That is, catalyst activity can be restored by operating the catalyst without the halogen or sulfur-bearing compound in the gas stream. The potential problem of greater concern with respect to the halogen-bearing compounds is the formation of hydrogen chloride or hydrogen fluoride gas, or hydrochloric or hydrofluoric acid emissions.

Catalytic incinerators generally cost less to operate than thermal incinerators, because of their lower fuel consumption. With the exception of regenerative heat recovery techniques, primary and secondary heat recovery can also be incorporated into a catalytic incineration system to reduce operating costs

further. Maintenance costs are usually higher for catalytic units, particularly if frequent catalyst cleaning or replacement is necessary. The concern regarding catalyst life has been the major factor limiting more widespread application of catalytic incinerators.

Applications of Incinerators

Odor control. All highly odorous pollutant gases are combustible or chemically changed to less odorous pollutants when they are sufficiently heated. Often, the concentration of odorous materials in the waste gas is extremely low, and the only feasible method of control is by incineration. Odors from rendering plants, mercaptans, and organic sulfides from kraft pulping operations are examples of effluents that can be controlled by incineration. Other forms of oxidation, such as chlorination or ozonation, can achieve the same ends (see Chapter 40 of the 1991 ASHRAE *Handbook—HVAC Applications*).

Reduction in opacity of plumes. Flame afterburning is often used to destroy organic aerosols that cause visible plumes. Processes that use afterburners include coffee roasters, smokehouses, and enamel baking ovens. Such burners can also be used for heating wet stack gases, which might otherwise show a steam plume.

Reduction in emissions of reactive hydrocarbons. Some air pollution control agencies regulate the emission of organic gases and vapors because of their involvement in photochemical smog reactions. Flame afterburning is an effective way of destroying these materials.

Reduction in explosion hazard. Refineries and chemical plants are among the industries that must dispose of highly combustible or otherwise dangerous organic materials. The safest method of disposal is usually by burning in flares or in specially designed furnaces. However, special precautions and equipment design must be used in the handling of potentially explosive mixtures.

Adsorption and Incineration

Alternate cycles of adsorption and desorption in an activated carbon bed are used to concentrate solvent or odor vapors prior to incineration, thus greatly reducing the fuel required for burning organic vapor emissions. Fuel savings of 98% are possible versus direct incineration. The process is particularly useful in cases where emission levels vary from hour to hour.

Contaminated gas is passed through a carbon bed until saturation occurs. The gas stream is then switched to another carbon bed, and the exhausted bed is shut down for desorption. A hot inert gas, usually burner flue gas, is introduced to the adsorber to drive off concentrated organic vapors and to convey them to an incinerator. The volume of this desorbing gas stream is much smaller than the original contaminated gas volume, so that only a small incinerator, operating intermittently, is required (Grandjacques 1977).

AUXILIARY EQUIPMENT

Duct

Basic duct design is reviewed in Chapter 32 of the 1989 ASHRAE *Handbook—Fundamentals*. Here, only those duct components or problems that warrant special concern when handling gases that contain particulate or gaseous contaminants are covered.

Duct systems should be designed to allow thermal expansion or contraction as gases move from the process, through gas-cleaning equipment, and on to the ambient environment. Appropriately designed duct expansion joints must be located in proper relation to sliding and fixed duct supports. Besides withstanding maximum possible temperature, duct must also withstand maximum positive or negative pressure, a partially full load of accumulated dust, and reasonable amounts of corrosion. Duct supports should be designed on these overload conditions as well.

Where gases might condense and cause corrosion or sticky deposits, duct should be insulated or fabricated from materials that will survive this environment. A psychrometric analysis of the exhaust gases is useful to determine the dew point. Surface temperatures should then be held above the dew-point temperature by preheating on start-up or by insulating. Slag traps with clean-out doors, inspection doors where direction changes, and dead-end full-sized caps are required for systems having heavy particulate loading. Special attention should be given to high-temperature duct, where the duct might corrode if insulated or become encrusted when molten particulate impacts on cool, uninsulated surfaces. Water-cooled duct or refractory lining are often used where high operating temperature exceeds safe limits of low-cost materials.

Gas flow through duct should be considered as a part of overall system design. Good gas flow distribution is essential for measurements of process conditions and can lead to energy savings and increased system life. The minimum speed of gases in a duct should be sufficiently high to convey the heaviest particulate fraction with a degree of safety.

Slide gates, balance gates, equipment bypass ducts, and clean-out doors should be incorporated in the duct system to allow for maintenance of key gas-cleaning systems. In some cases, emergency bypass circuitry should be included to vent emissions and protect gas-cleaning equipment from process upsets.

Temperature controls. The temperature of gases and its control in a gas-cleaning system are often vital to a system's performance and life. Sometimes, gases are cooled to concentrate contaminants, condense gases, and recover energy. On other occasions, gas-cleaning equipment, such as fabric filters and scrubbers, can only operate at well-controlled temperatures. Cooling exhaust gases through air-to-air heat transfer has been highly successful in many applications. Controlled evaporative cooling is also employed, but it increases the dew point and the danger of acid gas condensation and/or the formation of sticky deposits. However, controlled evaporative cooling to within 50°F of dew point has been used with success. Dilution by the injection of ambient air into the duct is expensive, because it increases the volumetric flow of gas and, consequently, increases the size of collector needed to meet gas-cleaning objectives. Water-cooled duct is often used where the gas temperature exceeds safe limits of the low-cost materials.

For dilution cooling, louver-type dampers are often used to inject ambient air and provide fine temperature control. Controls can be used to provide full modulation of the damper or to provide open or closed operation. Emergency bypass damper systems and bypass duct/stacks are used where limiting excessive temperature is critical.

Fans. Since the static pressure across a gas-cleaning device varies depending on conditions, the fan should operate on the steep portion of the fan pressure-volumetric flow curve. This tends to provide less variation in the volumetric flow. An undersized fan has a steeper characteristic than an oversized fan for the same duty; however, it will be noisier.

In the preferred arrangement, the fan is located on the clean gas side of gas-cleaning equipment. Advantages arising from placing the fan at this point include:

- A fan on the clean gas side handles clean gases and has no abrasive exposure from the collected product.

- High-efficiency backward blade and air foil designs can be selected because accumulation on the fan wheel is not as great a factor.
- Escape of hazardous materials due to leaks is minimized.
- The collector can be installed inside the plant, even near the process, since any leakage in the duct or collector will be into the system and will not increase the potential for exposure. However, the fan itself should be mounted outside, so that the positive pressure duct is exterior to the work environment.

For economic reasons, a fan may be located on the contaminant-laden side of the gas-cleaning equipment if the contaminants are relatively nonabrasive and if the equipment can be preferably located outdoors. This arrangement should be avoided because of the potential for leakage of concentrated contaminants to the environment. In some instances, however, the collector housing design, duct design, and energy savings of this arrangement reduces costs.

Most scrubbers are operated on the suction side of the fan. This not only eliminates the leakage of contaminants into the work area, but also allows for servicing the unit while it is in operation. Additionally, such an arrangement minimizes corrosion of the fan. Stacks on the exhaust streams of scrubbers should be arranged to drain condensate, rather than allow it to accumulate and reenter the fan.

Fabric filters require special consideration. When new, clean fabric is installed in a collector, the resistance is low, and the fan motor may be overloaded. This overloading may be prevented during start-up by use of a temporary throttling damper in the main duct, for example, on the clean side of the filter in a pull-through system. Overloading may also be prevented by using a backward-curved blade (nonoverloading fan) on the clean side of the collector.

Dust- and Slurry-Handling Equipment

Once the particulate matter is collected, new control problems arise from the need to remove, transport, and dispose of the material from the collector. A study of all potential methods for handling the collected material, which incidentally might be a hazardous waste, must be an integral part of initial system design.

Hoppers. Dust collector hoppers are intended only to channel collected material to the hopper's outlet, where it is continuously discharged. The plates and charging electrodes of electrostatic precipitators can be shorted electrically, resulting in failure of complete electrical sections when hoppers are used to store dust. Fabric collectors, particularly those that have the gas inlet in the hopper, are usually designed on the assumption that the hopper is not used for storage and that collected waste will be continuously removed. High dust levels in the hoppers of fabric collectors often result in high dust reentrainment. This reentrainment causes increased operating pressure and the potential for fabric damage. The excessive dust levels not only expose the system to potentially corrosive conditions and fire/explosion hazards, but also place increased structural demands on the system.

Aside from misuse of hoppers for storage, common problems with dust-handling equipment include (1) plugging of hoppers, (2) blockage of dust valves with solid objects, and (3) improper or insufficient maintenance.

Hopper auxiliaries to be considered include (1) insulation, (2) dust level indicators, (3) rapper plates, (4) vibrators, (5) heaters, and (6) "poke" holes.

Hopper discharge. Dust is often removed continuously from hoppers by means of rotary valves. Alternate equipment includes the double-flap valve, or vacuum system valves. Wet electrostatic precipitators and scrubbers often use sluice valves and drains to ensure that insoluble particulate remains in suspension during discharge.

Dust conveyors. Larger dust collectors are fitted with one or more conveyors to feed dust to a central discharge location or to return it to a process. Drag, screw, and pneumatic conveyors are commonly used with dust collectors. Sequential start-up of conveyor systems is essential. Use of motion switches to monitor operation of the conveyor is useful.

Dust disposal. Several methods are available for disposal of collected dust. It can be emptied into dumpsters in its as-collected dry form, or pelletized and hauled to a landfill. It can also be converted to a slurry and pumped to a settling pond or to clarification equipment. While the advantages and disadvantages of each method are beyond the scope of this chapter, they should be evaluated for each application.

Slurry treatment. When slurry from wet collectors cannot be returned directly to the process or tailing pond, liquid clarification and treatment systems can be used for recycling the water to prevent stream pollution. Stringent stream pollution regulations make even a small discharge of bleed water a problem. Clarification equipment may include settling tanks, sludge-handling facilities, and, possibly, centrifuges or vacuum filters. Provisions must be provided for handling and disposal of dewatered sludge, so that secondary pollution problems do not develop.

OPERATION AND MAINTENANCE

A planned program for operation and maintenance of equipment is a necessity. Such programs are becoming mandatory because of the need for operators to prove continuous compliance with emission regulations. Good housekeeping and record-keeping will also help prolong the life of the equipment, support a program for positive relations with regulatory and community groups, and aid problem solving efforts, should nonroutine maintenance or service be necessary.

A typical program includes the following minimum requirements (Stern *et al.* 1984):

- A central location for filing equipment records, warranties, instruction manuals, etc.
- Lubrication and cleaning schedules
- Planning and scheduling of preventive maintenance (including inspection and major repair)
- A storeroom and inventory system for spare parts and supplies
- A listing of maintenance personnel (including supplier contacts and consultants)
- Costs and budgets for activities associated with operation and maintenance of the equipment
- Storage for special tools and equipment

Guidance on the conduct of inspections can be found in EPA (1983).

Corrosion

Because high-temperature gas-cleaning applications often involve corrosive materials, it is necessary to anticipate chemical attack on components of a system. This is especially true if the temperature in a gas-cleaning system falls below the moisture or acid dew point. Housing insulation should be such that the internal metal surface temperature is 20 to 30°F greater than the moisture and/or acid dew point at all times. In applications with fabric filters, where alkali materials are injected into the gas stream to react with acid gases, care must be taken to protect the clean side of the housing, downstream from the fabric, from corrosion.

Fires and Explosions

Industrial gas-cleaning systems often concentrate combustible materials and expose those materials to environments that are hostile and difficult to control. These environments also make fires difficult to detect and stop. Industrial gas-cleaning systems are, therefore, potentially fire or explosion hazards (Billings and Wilder 1970, Frank 1981, Edison Electric Institute 1980).

Fires and explosions in industrial process exhaust streams are not generally limited to gas-cleaning equipment. Ignition may take place in the process itself, in the duct, or in exhaust system components other than the gas-cleaning equipment. Once uncontrolled combustion begins, it may propagate throughout the system. Workers around pollution control equipment should never open access doors to gas-cleaning equipment when there is believed to be a fire in process. The fire could easily turn into an explosion.

The following devices help maintain a safe particulate control system.

Explosion doors. An explosion door or explosion relief valve permits instantaneous pressure relief for equipment, when the pressure reaches a predetermined level. Explosion doors are mandatory for certain applications to meet OSHA, insurance, or NFPA regulations.

Detectors. Temperature-actuated switches or infrared sensors can be used to detect changes in the inlet to outlet temperature difference, or a localized, elevated temperature, that might signal a fire within the gas-cleaning system or a process upset. These detectors can be used to activate bypass dampers, trigger fire alarm/control systems, and/or shut down fans.

Fire-control systems. Inert gas and water spray systems can be used to control fires in dust collectors. They are of little value in controlling explosions.

BIBLIOGRAPHY

ACGIH. 1988. *Industrial ventilation: Manual of recommended practices,* 20th ed. Committee on Industrial Ventilation, American Conference of Governmental Industrial Hygienists, Cincinnati, OH.

AIHA. 1989. Odor thresholds for chemicals with established occupational health standards. American Industrial Hygiene Association, Akron, OH.

API. 1975. API Publication 931, Chapter 11: Cyclone separators. American Petroleum Institute, Washington, D.C.

Billings, C.E. and J. Wilder. 1970. *Handbook of fabric filter technology,* Vol. 1: Fabric filter systems study. NTIS Publication PB 200 648, 2-201.

Boll, C.H. 1976. Recovering solvents by adsorption. *Plant Engineering* (January).

Crynack, R.B. and J.D. Sherow. 1984. Use of a mobile electrostatic precipitator for pilot studies. Proceedings of the Fifth Symposium on the Transfer and Utilization of Particulate Control Technology 2:3-1.

DOE. 1979. The coating industry: Energy savings with volatile organic compound emission control. Report No. TID-28706, U.S. Department of Energy, Washington, D.C.

DuBard, J.L. and R.F. Altman. 1984. Analysis of error in precipitator performance estimates. Proceedings of the Fifth Symposium on the Transfer and Utilization of Particulate Control Technology 2:2-1.

EEI. 1980. Air preheaters and electrostatic precipitators fire prevention and protection (coal fired boilers). Report of the Fire Protection Committee, Edison Electric Institute, No. 06-80-07 (September).

EPA. 1983. Coal-fired industrial boiler inspection guide. Report No. 340/1-83-025 (December).

EPA. 1984. Quality assurance handbook for air pollution measurement systems. Report No. EPA-600/4-77-27b, Environmental Protection Agency, Washington, D.C.

EPA. 1984a. Proposed fine particle standards for ambient air quality. Federal Register 48, 10408, March 30, Environmental Protection Agency, Washington, D.C.

Faulkner, M.G. and J.L. DuBard. 1984. A mathematical model of electrostatic precipitation, 3rd ed. U.S. EPA publication EPA-600/7-84-069a.

Frank, T.E. 1981. Fire & explosion control in bag filter dust collection systems. Proceedings of the Conference on the Hazards of Industrial Explosions from Dusts, New Orleans, LA (October).

Grandjacques, B. 1977. Carbon adsorption can provide air pollution control with savings. *Pollution Engineering* (August).

GPO. 1991. Code of federal regulations 40(60). U.S. Government Printing Office, Washington, D.C. This document is revised annually and published each year in July.

Hall, H.J. 1975. Design and application of high voltage power supplies in electrostatic precipitation. *Journal of the Air Pollution Control Association* 25(2).

HEW. 1967. Air pollution engineering manual. HEW Publication No. 999-AP-40, Department of Health and Human Services (formerly Department of Health, Education, and Welfare), Washington, D.C.

IGCI. 1964. Determination of particulate collection efficiency of gas scrubbers. Publication No. 1, Industrial Gas Cleaning Institute, Washington, D.C.

Kane, J.M. and J.L. Alden. 1967. *Design of industrial exhaust systems*, 4th ed. Industrial Press, New York.

Koch, W. and W. Licht. 1977. *Chemical Engineering* (November 7):80-88.

Kuehn and Pui. 1989. Dust collector recirculation for industrial operations. ASHRAE RP-531.

Lapple, C.E. 1951. Processes use many collection types. *Chemical Engineering* (May):145-51.

Lapple, C.E. and H.J. Kamack. 1955. Performance of wet scrubbers. *Chemical Engineering Progress* (March).

MCA. 1968. Odor thresholds for 53 commercial chemicals. Manufacturing Chemists Association, Washington, D.C. (October).

NIOSH. 1978. A recommended approach to recirculation of exhaust air. Publication No. 78-124, National Institute of Occupational Safety and Health, Washington, D.C.

Noll, C.G. 1984. Electrostatic precipitation of particulate emissions from the melting of borosilicate and lead glasses.

Noll, C.G. 1984a. Demonstration of a two-stage electrostatic precipitator for application to industrial processes. Proceedings of the Second International Conference on Electrostatic Precipitation, Kyoto, Japan (November):428-34.

Oglesby, S. and G.B. Nichols. 1970. A manual of electrostatic precipitator technology. NTIS PB-196-380.

Rudnick, S.N., *et al.* 1986. Particle collection efficiency in a venturi scrubber: Comparison of experiments with theory. *Environmental Science & Technology* 20(3):237-42.

Semrau, K.T. 1977. Practical process design of particulate scrubbers. *Chemical Engineering* (September): 87-91.

SIP. 1991. State implementation plans and guidance are available from the Regional Offices of the U.S. EPA, and the state environmental authorities.

Stern, A.C., R.W. Boubel, D.B. Turner, and D.L. Fox. *Fundamentals of air pollution control,* 2nd ed. Chapter 25, Control Devices and Systems.

Wade, G., J. Wigton, J. Guillory, G. Goldback, and K. Phillips. 1978. Granular bed filter development program. U.S. DOE Report No. FE-2579-19 (April).

White, H.J. 1963. *Industrial electrostatic precipitation.* Addison-Wesley Publishing Company, Inc., Reading, MA.

White, H.J. 1974. Resistivity problems in electrostatic precipitation. *Journal of the Air Pollution Control Association* 24(4).

CHAPTER 27

AUTOMATIC FUEL-BURNING EQUIPMENT

GAS-BURNING EQUIPMENT

A gas burner conveys gas (or a mixture of gas and air) to the combustion zone. Burners are of the atmospheric injection, luminous flame, or power burner types.

RESIDENTIAL EQUIPMENT

Residential gas burners are those designed for central heating plants or for unit application. Gas-designed units and conversion burners are available for the different types of central systems and for other applications where the units are installed in the heated space.

Central heating appliances include warm-air furnaces and steam or hot-water boilers.

Warm-air furnaces are available in different designs. Design choice depends on the force needed to move combustion products, the force needed to move heated supply and return air, the location within a building, and the efficiency required.

The force to move supply and return air can be supplied by the natural buoyancy of heated air in a gravity furnace (if the space to be heated is close to and/or above the furnace) or by a blower in a forced-air furnace.

The force to move the combustion products can be supplied by the natural buoyancy of hot combustion products in a natural-draft furnace, by a blower in a forced-draft or induced-draft furnace, or by the thermal expansion forces in a pulse combustion furnace.

Furnaces are also available in upflow, downflow, horizontal, and other heated air directions to the application requirements.

The efficiency required determines, to a great extent, some of the above characteristics and the need for other characteristics such as the source of combustion air, the use of vent dampers and draft hoods, the need to recover latent heat from the combustion products, and the design of the heat transfer components.

Conversion burners are not normally used in modern furnace design because the burner design is integrated with the furnace design for safety and efficiency. Older gravity furnaces usually use conversion burners more readily than can modern gravity and forced-air furnaces.

Steam or hot-water boilers are available in cast iron, steel, and nonferrous metals. In addition to supplying space heating, many boilers provide domestic hot water, using tankless integral or external heat exchangers.

Some gas furnaces and boilers are available with sealed combustion chambers. These units have no draft hood and are called *direct vent appliances*. Combustion air is piped from outdoors directly to the combustion chamber.

In some instances, the combustion air intake is in the same location as the flue gas outlet. The air and flue pipes are sometimes constructed as concentric pipes, with a terminal that exposes the air intake and flue gas outlet to a common pressure condition. No chimney or vertical vent is needed with such units. Some induced-draft combustion systems are designed to operate safely when common-vented with other appliances that have natural-draft combustion systems.

For appliances in the United States and not covered by the U.S. National Appliance Energy Conservation Act, the required minimum steady-state efficiency determined by flue loss tests as required by standards of the American National Standards Institute (ANSI) is:

1. Boilers 75%, Z21.13b-1987.
2. Gas-fired central furnaces (other than direct vent), ANSI *Standard* Z21.47a-1990

   - Forced-air 75%
   - Gravity 70%

3. Direct vent central furnaces 75%, ANSI *Standard* Z21.64-1990.

For appliances covered by the U.S. National Appliance Energy Conservation Act, the required minimum annual fuel utilization efficiency (AFUE) determined by U.S. Department of Energy tests is:

1. Boilers
   Excluding gas steam: 80% AFUE as of January 1, 1992.
   Gas steam: 75% AFUE as of January 1, 1992.

2. Furnaces
   Excluding mobile home furnaces: 78% AFUE as of January 1, 1992.
   Mobile home furnaces: 75% as of January 1, 1992.

Conversion burners are complete burner and control units designed for installation in existing boilers and furnaces. Atmospheric conversion burners may have drilled-port, slotted-port, or single-port burner heads. These burners are either upshot or inshot types. Figure 1 shows a typical upshot gas conversion burner.

Several power burners are available in residential sizes. These are of gun-burner design and are desirable for furnaces or boilers with restricted flue passages or with downdraft passages.

Conversion burners for domestic application are available in sizes ranging from 40,000 to 400,000 Btu/h input, the maximum

The preparation of this chapter is assigned to TC 3.7, Fuels and Combustion.

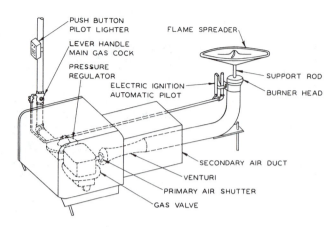

Fig. 1   Typical Single-Port Gas Conversion Burner

rate being set by ANSI/AGA *Standard* Z21.17. However, many such gas conversion burners installed in apartment buildings have input rates up to 900,000 Btu/h or more.

As the successful and safe performance of a gas conversion burner depends on numerous factors other than those incorporated in such equipment, installations of this kind must be made in strict accordance with current ANSI/AGA *Standard* Z2.8. Draft hoods conforming to current ANSI/AGA *Standard* Z21.12 should also be installed (in place of the dampers used with a solid fuel) on all boilers and furnaces converted to burn gas. Due to space limitations, a converted appliance with a breeching over 12 in. in size is often fitted with a double-swing barometric regulator instead of a draft hood.

Pulse combination systems are used in residential heating equipment. Pulse systems do not burn gas in a steady flame. Instead, the air/gas mixture is ignited at a rate of 60 to 80 times per second. Reported advantages of pulse combustion include reduced oxides of nitrogen (NO_x) emissions and improved heat transfer due to the oscillating flow. Additional discussion on pulse combustion technology can be found in Chapter 15 of the 1989 ASHRAE *Handbook—Fundamentals*.

COMMERCIAL-INDUSTRIAL EQUIPMENT

Many basic types of industrial gas burners are available, but only those commonly used for heating service are covered here. These burners may be of the atmospheric or power type. In addition to these types of burners for use in central heating systems, various gas-designed units such as unit heaters, duct furnaces, infrared heaters, and makeup heaters are available for space heating.

The installation of combustion burners larger than 400,000 Btu/h for use in large boilers is usually engineered by the burner manufacturer, the manufacturer's representative, or a local utility company. Conversion burners are available in several sizes and types. In some cases, the burner may be an assembly of multiple burner heads filling the entire firebox. For conversion burner installation in boilers requiring more than 400,000 Btu/h input, reference should be made to current ANSI/AGA *Standard* Z83.3. Conversion power burners above 400,000 Btu/h are available and should conform to UL *Standard* 795.

Atmospheric burners include an air shutter, a venturi tube, a gas orifice, and outlet ports. These burners are of inshot or upshot design. Inshot burners are placed horizontally, making them adaptable to firing Scotch-type boilers. Upshot burners are arranged vertically, making them more adaptable in firebox-type boilers.

Power-type burners use a fan to supply and control combustion air. These burners can be natural draft or forced draft. In natural-draft installations, a chimney must draw the products of combustion through the boiler or furnace; the burner fan supplies only enough power to move the air through the burner. Many natural-draft power burners have a configuration similar to that of an inshot atmospheric burner to which a fan and windbox have been added. More complex gas-air mixing patterns are possible, and combustion capabilities are thereby improved.

The fan size and speed on power burners have gradually increased, and the combustion process has been modified so that the fan not only moves air through the burner but also forces it through the boiler. Combustion occurs under pressure in controlled airflow. While a vent of only limited height may be needed to convey the combustion products to the outdoors, higher chimneys and vents are usually required to elevate the effluent further. These burners are forced-draft burners. In a forced-draft power burner, air and gas flows can be modulated by suitable burner controls provided by the manufacturer. Gas input is usually controlled by an appliance pressure regulator and a firing rate valve, both piped in series in the gas train. If the utility also provides a regulator, it is installed at the gas service entry at the gas meter location.

The gas is introduced into a controlled airstream designed to produce thorough gas-air mixing but still capable of maintaining a stable flame front. In a ring burner, the gas is introduced into the combustion airstream through a gas-filled ring just ahead of the combustion zone. In a premix burner, gas and primary air are mixed together, and the mixture is then introduced into secondary air in the combustion zone.

The power burner has superior combustion control, particularly in restricted furnaces and under forced draft. Commercial and industrial gas burners frequently operate with higher gas pressures than are intended for residential equipment. It is often necessary to determine both maximum and minimum gas pressures to be applied to the gas regulator and control trains. If the maximum gas pressure exceeds the 150 to 500 in. of water pressure (5.4 to 18.0 psig) standard for domestic equipment, selecting appropriate gas controls is necessary. All gas-control trains must be rated for the maximum expected gas pressure. Gas pressures in densely populated areas may be significantly lower than in rural areas. Gas-control trains must be sized accordingly.

Pulse combustion technology is available for commercial heating equipment. Boilers using this technology are available in sizes to 750,000 Btu/h.

Gas-fired systems for packaged fire-tube or water-tube boilers in heating plants consist of specially engineered and integrated combustion and control systems. The burners have forced- and/or induced-draft fans. The burner fan power requirements of these systems and the forced-draft equipment mentioned previously differ. The system pressure, up to and including the stack outlet, may be positive. If the boiler flue breeching has positive flue gas pressure, it must be gastight.

The integrated design of burner and boiler makes it possible, through close control of air-fuel ratios and by matching of flame patterns to boiler furnace configurations, to maintain high combustion efficiencies over a wide range of loads. Combustion space and heat transfer areas are designed for maximum heat transfer when the specific fuel for which the unit is offered is used. Most packaged units are fire-tested as complete packages prior to shipment, allowing for inspection of all components of the units to ensure that burner equipment, automatic controls, etc., function properly. These units generally bear the Underwriters Laboratories label.

Gas-fired air heaters are generally designed for use in large spaces such as airplane hangars and public garages. They are self-contained, are automatically controlled with integral means for air circulation, and are equipped with automatic electric ignition of pilots, induced or forced draft, prepurge, and fast-acting combustion safeguards. They also are used with ducts, discharge nozzles, grilles or louvers, and filters.

Unit heaters are used extensively for heating spaces such as stores, garages, and factories. These heaters consist of a burner, heat exchanger, fan for distributing the air, draft hood, automatic pilot, and controls for burners and fan. They are usually mounted in an elevated position from which the heated air is directed downward by louvers. Some unit heaters are suspended from the ceiling, while others are freestanding floor units of the heat tower type. Unit heaters are classified for use with or without ducts, depending on the applicable ANSI/AGA standard under which their design is certified. When connected to ducts, they must have sufficient blower capacity to deliver an adequate air quantity against duct resistance.

Duct furnaces are usually like unit heaters without the fan and are used for heating air in a duct system with blowers provided to move the air through the system. Duct furnaces are tested for operation at much higher static pressures than are obtained in unit heaters (ANSI/AGA *Standard* 83.9).

Infrared heaters, vented or unvented, are used extensively for heating factories, foundries, sports arenas, loading docks, garages, and other installations where convection heating is difficult to apply. Following are two general gas-fired types of infrared heaters.

1. Surface combustion radiant heaters, which have a ported refractory or metallic screen burner face through which a self-sufficient mixture of gas and air flows. The gas-air mixture burns on the surface of the burner face heating it to incandescence and releasing heat by radiation. These units operate at surface temperatures of about 1600 °F and generally are unvented. Buildings containing unvented heaters discharging the combustion gases into the space should be adequately ventilated.

2. Internally fired heaters, which consist of a heat exchanger with the exposed surface radiating heat at a surface temperature of approximately 180 °F. The exposed surface can be equipped with reflecting louvers or a single large reflector to direct the radiated energy. These units are usually vented.

These infrared heaters are usually mounted in elevated positions and radiate heat downward (see Chapter 15).

Direct-fired makeup air heaters (see Chapter 32) are used to temper the outside air supply, which replaces contaminated exhaust air. The combustion gases of the heater are mixed directly with large volumes of outside air. Such mixing is considered safe because of the high dilution ratio.

ENGINEERING CONSIDERATIONS

With gas-burning equipment, the principal engineering considerations are sizing of gas piping, adjustment of the primary air supply, and, in some instances, adjustment of secondary aeration, as with power burners and conversion burners. Consideration of input rating may be necessary to compensate for high altitude conditions. Chapter 33 of the 1989 ASHRAE *Handbook—Fundamentals* has piping and sizing details.

Combustion Process and Adjustments

Gas burner adjustment is mainly an adjustment of air supply. Most residential gas burners are of the atmospheric injection (Bunsen) type in which primary air is introduced and mixed with the gas in the throat of the mixing tube. For normal operation of most atmospheric-type burners, 40 to 60% of the theoretical air as primary air will give best operation. Slotted-port and ribbon burners may require from 50 to 80% primary air for proper operation. The amount of excess air required depends on several fac-

tors, including uniformity of air distribution and mixing, direction of gas travel from the burner, and the height and temperature of the combustion chamber. With power burners using motor-driven blowers to provide both primary and secondary air, the excess air can be closely controlled while proper combustion is secured.

Secondary air is drawn into gas appliances by natural draft. Yellow flame burners depend on secondary air for combustion.

Air shutter adjustments should be made by closing the air shutter until yellow flame tips appear and then by opening the air shutter to a final position at which the yellow tips just disappear. This type of flame obtains ready ignition from port to port and also favors quiet flame extinction.

Gas-designed equipment usually does not incorporate any means for varying the secondary air supply (and hence the CO_2). The amount of effective opening and baffling is determined by compliance with ANSI standards. Gas conversion burners, however, do incorporate means for controlling secondary air to permit adjustment over a wide range of inputs. It is desirable, through the use of suitable indicators, to determine whether or not carbon monoxide is present in flue gases. For safe operation, carbon monoxide should not exceed 0.04% (air-free basis).

Compensation for Altitude

Compensation for altitude must be made for altitudes higher than 2000 ft above sea level. The typical gas-fired central heating appliance using atmospheric burners, multiple flue ways, and an effective draft diverter must be derated at high altitudes.

All air for combustion is supplied by the chimney effect of the flue way. The geometry of these flue ways allows a given volume of air to pass through the appliance, regardless of its mass. Each burner has a gas orifice and a venturi tube designed to permit a given volume of gas to be introduced and burned during a given period. The entire system is essentially a constant volume device and must, therefore, be derated in accordance with the change in mass of these constant volumes at higher altitudes. The derating factor recommended is 4% per 1000 ft of altitude (*Gas Engineers Handbook* 1965).

Many commercial and industrial applications use a forced-draft gas-fired packaged system, including a burner with a forced-draft fan and a fuel-handling system. The burner head, the heat exchanger, and the vent act as a series of orifices downstream of the forced-draft fan. To compensate for the increased volume of air and flue products that must be forced through these orifices at higher altitudes, it is necessary to increase burner fan capacity until the required volume of air is delivered at a pressure high enough to overcome the fixed restrictions. These oversized fans and fan motors are usually offered as options.

One problem that frequently occurs in the selection of commercial gas-fired equipment, and particularly for forced-draft gas-combustion equipment, relates to the heat content of the gas as delivered at elevated locations. Natural gas usually has a heating value of just over 1000 Btu/ft^3 at standard conditions. At the lower elevations, the heating value of the gas, as delivered, is close to the heating value at the standard conditions and selection of controls is relatively simple. However, at 5000 ft elevation, 1000 Btu/ft^3 natural gas has a heat content of 850 Btu/ft^3, as delivered to the burner-control train. Some gas supplies in the mountains are enriched, and the energy content at standard conditions is higher than 1000 Btu/ft^3. Problems frequently occur in the selection of controls sized to permit the required volume of gas to be delivered for combustion. Gas specifications should indicate if the energy value and density shown are for standard conditions or for the gas as furnished at a higher elevation.

The problem of gas supply heat content is more significant for the proper application of commercial forced-draft combustion equipment than it is for smaller atmospheric burner units because of the larger gas volumes handled and the higher design pressure drops common to this type of equipment.

As an example, a gas-control train at sea level has a pressure of 14 in. of water delivered to the meter inlet. Allowing 6 in. of water pressure drop through the meter and piping to the burner-control train at full gas flow, 8 in. of water pressure remains for delivery of gas for combustion. To force the gas through the gas pressure regulator, shutoff valve, and associated manifold piping, 4 in. of water pressure is typically applied. The last 4 in. of water overcomes the resistance of the gas flow control valve, the gas ring, and the furnace pressure, which may be about 2 in. of water.

Consider this same system at 5000-ft altitude, starting from the furnace back. Because of the larger combustion gas volume handled, the furnace pressure will have increased to about 2.5 in. of water. Some changes can be made to the gas ring to allow for a larger gas flow, but there is usually some increase in the pressure drop, approximately to 2.5 in. of water, so that a pressure of 5 in. of water is now required at the gas volume control valve. Because gas pressure drop varies with the square of the volume flow rate, the manifold pressure drop will increase to about 5.5 in. of water, and the piping pressure drop will increase to 8.5 in. of water. A meter inlet pressure of 19 in. of water would now be required. Since higher gas pressures frequently are not available, it becomes necessary to increase piping and control manifold sizes.

OIL-BURNING EQUIPMENT

An oil burner is a mechanical device for preparing fuel oil to combine with air under controlled conditions for combustion. Fuel oil is prepared for combustion by being atomized at a controlled flow rate. Air for combustion is generally supplied with a forced-draft fan, although natural draft or mechanically induced draft can be used. Ignition is typically provided by an electric spark, although gas pilot flames, oil pilot flames, and hot-surface igniters may also be used. Oil burners operate from automatic temperature or pressure-sensing controls.

Oil burners may be classified by application, type of atomizer, or firing rate. They can be divided into two major groups: residential and commercial-industrial. Further breakdown is made by type of design and operation, such as pressure atomizing, air or steam atomizing, rotary, vaporizing, and mechanical atomizing. Unvented, portable kerosene heaters are not classified as residential oil burners or as oil heat appliances.

RESIDENTIAL OIL BURNERS

Residential oil burners are ordinarily used in the range of 0.5 to 3.5 gph fuel consumption rate. However, burners up to 7 gph sometimes fall in the residential classification because of basic similarities in controls and standards. (Burner capacity of 7 gph and above is classified as commercial-industrial.) No. 2 fuel oil is generally used, although burners in the residential size range can also operate on No. 1 fuel oil. In addition to applications to boilers and furnaces for space heating, burners in the 0.5 to 1 gph size range are also used for separate tank-type residential hot-water heaters, infrared heaters, space heaters, and other commercial equipment.

Over 95% of residential burner production is the high-pressure atomizing gun burner with retention-type heads (Figure 2). This type of burner supplies oil to the atomizing nozzle at pressures that range from 100 to 300 psi (the low end of this range is nearly standard). A fan supplies air for combustion, and generally, a fan inlet damper regulates the air supply at the burner. A high-voltage electric spark, which may be intermittent, ignites the fuel. This is called either *constant ignition* (on when the burner motor is on) or *interrupted ignition* (on only to start combustion). Typically, these

burners fire into a combustion chamber in which draft is maintained.

Use of retention heads and residential burner motors operating at 3500 rpm instead of 1750 rpm is widespread. The retention head is essentially a bluff-body flame stabilizer that assists combustion by providing better air-oil mixing, turbulence, and shear (Figure 3). Using 3500 rpm motors (with the retention head) has resulted in a more compact burner with equal capacity and more tolerance for varying draft conditions. In addition, most burner designs now include fans with increased static pressure (to 3 in. of water). Such burners are less sensitive to draft variations and, in most cases, can be used without a flue draft regulator. These higher static burners are particularly useful in systems that are not vented with conventional chimneys. These systems include sealed combustion and sidewall venting (induced or forced).

Other designs are still in operation but are not a significant part of the residential market. They include the following:

The *low-pressure atomizing gun burner* differs from the high-pressure type because it uses air at a low pressure to atomize the oil.

The *pressure atomizing induced-draft burner* uses the same type of oil pump, nozzle, and ignition system as the high-pressure atomizing gun burner.

Vaporizing burners are designed for use with No. 1 fuel oil. Fuel is ignited electrically or by manual pilot.

Rotary burners, usually of the vertical wall flame type.

Beyond the pressure atomizing burner, considerable developmental effort has been placed on advanced technologies, *e.g.*, air atomization, fuel prevaporization, and internal recirculation of

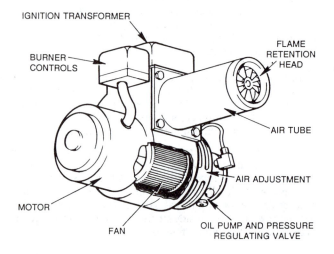

Fig. 2　High-Pressure, Atomizing-Gun, Oil Burner

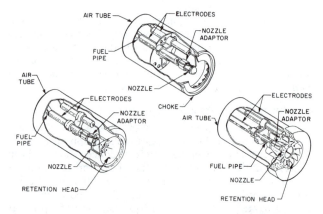

Fig. 3　Details of a High-Pressure Atomizing Oil Burner

combustion products into the flame zone. Some of these are now used commercially in Europe. Reported benefits include reduced emissions, increased efficiency, and the ability to quickly vary the firing rate. See Locklin and Hazard (1980) for a historical review of technology for residential oil burners. More recent developments are described in the *Proceedings of the Oil Heat Technology Conferences* (McDonald 1989).

Central heating appliances include warm-air furnaces and steam or hot-water boilers. Oil-burning furnaces and boilers operate essentially the same as their gas counterparts.

Steam or hot-water boilers are available in cast iron and steel. In addition to supplying space heating, many boilers are designed to provide hot water using tankless integral or external heat exchangers. Residential units designed to operate as direct vent appliances are available.

NFPA *Standard* 31 (ANSI *Standard* Z95.1), Oil Burning Equipment, prescribes correct installation practices for fuel-oil-burning appliances.

COMMERCIAL-INDUSTRIAL OIL BURNERS

Commercial and industrial oil burners are designed for use with distillate grades or residual grades of fuel oil. With slight modifications, burners designed for residual grades can use the distillate fuel oils.

The commercial-industrial burners covered here are atomizers, which inject the fuel oil into the combustion space as a fine, conical spray with the apex at the burner atomizer. The burner also forces combustion air into the oil spray, causing an intimate and turbulent mixing of air and oil. An electrical spark, spark-ignited gas, or oil igniter, applied for a predetermined time, ignites the mixture, and sustained combustion takes place.

All of these burners are capable of almost complete burning of the fuel oil, without visible smoke, when they are operated with excess air as low as 20% (approximately 12% CO_2 in the flue gases). Atomizing oil burners are generally classified according to the method used for atomizing the oil, such as pressure atomizing, return-flow pressure atomizing, air atomizing, rotary cup atomizing, steam atomizing, or mechanical atomizing burners. Descriptions of these burners are given in the following sections, together with usual capacities and applications. Table 1 lists approximate size range, fuel grade, and usual applications. All burners described are available as gas-oil (dual-fuel) burners.

Pressure Atomizing Oil Burners

This type of burner is used in most installations where No. 2 grade fuel oil is burned. The oil is pumped at pressures of 100 to 300 psi through a suitable burner nozzle orifice that breaks it into

a fine mist and swirls it into the combustion space as a cone-shaped spray. Combustion air from a fan is forced through the burner air-handling parts surrounding the oil nozzle and is directed into the oil spray.

For smaller capacity burners, ignition is usually started by an electric spark applied near the discharge of the burner nozzle. For burner capacities above 20 gph, a spark-ignited gas or oil igniter is used.

Pressure atomizing burners are designated commercially as forced-draft or natural- (or induced) draft burners. The forced-draft burner has a fan and motor with capacity to supply all the air for combustion to the combustion chamber or furnace at a pressure high enough to force the gases through the heat-exchange equipment without the assistance of an induced-draft fan or chimney draft. Mixing of the fuel and air is such that a minimum of refractory material is required in the combustion space or furnace to support combustion. The natural-draft (induced-draft) burner requires a draft in the combustion space.

Burner range, or variation in burning rate, is changed by simultaneously varying the oil pressure to the burner nozzle and regulating the airflow by a damper. This range is limited to about 1.6 to 1 for any given nozzle orifice. Burner firing mode controls for various capacity burners vary among manufacturers. Usually, larger burners are equipped with controls that provide variable heat inputs. If the burner capacity is up to 15 gph, an *on-off* control is used; if it is up to 25 gph, a modulation control is used. In the latter two cases, the low burning rate is about 60% of the full-load capacity of the burner.

For these burners, no preheating is required for burning No. 2 oil. No. 4 oil must be preheated to about 100 °F for proper burning. When properly adjusted, these burners operate well with 20% excess air (approximately 12% CO_2); no visible smoke (approximately No. 2 smoke spot number, as determined by ASTM *Standard* D2156); and only a trace of carbon monoxide in the flue gas in commercial applications.

Burners with lower firing rates used to power appliances, residential heating units, or warm-air furnaces are usually set up to operate to about 50% excess air (approximately 10% CO_2).

Good operation of these burners calls for a relatively constant draft—either in the furnace or at the breeching connection, depending on the burner selected.

Return-Flow Pressure Atomizing Oil Burner

This burner is a modification of the pressure atomizing burner; it is also called a modulating pressure atomizer. It has the advantage of wide load range for any given atomizer, about 3 to 1 turndown as against the 1.6 to 1 for the straight pressure atomizing burner (see the section Mechanical Atomizing Oil Burners).

This wide range is accomplished through a return-flow nozzle,

Table 1 Classification of Atomizing Oil Burners

Type of Oil Burner	Heat Range, 1000 Btu/h	Flow Volume, gph	Fuel Grade	Usual Application
Pressure atomizing	70 to 7000	0.5 to 50	No. 2 (less than 25 gph) No. 4 (higher)	Boilers Warm-air furnaces Appliances
Return-flow pressure atomizing or modulating pressure atomizing	3500 and above	25 and above	No. 2 and heavier	Boilers Warm-air furnaces
Air atomizing	70 to 1000	0.5 to 70	No. 2 and heavier	Boilers Warm-air furnaces
Horizontal/rotary cup	750 to 37,000	5 to 300	No. 2 for small sizes No. 4 to 6 for larger sizes	Boilers Large warm-air furnaces
Steam atomizing (register-type)	12,000 and up	80 and up	No. 2 and heavier	Boilers
Mechanical atomizing (register-type)	12,000 and up	80 and up	No. 2 to 6	Boilers Industrial furnaces
Return-flow mechanical atomizing	45,000 to 180,000	300 to 1200	No. 2 and heavier	Boilers

which has an atomizing swirl chamber just ahead of the orifice. Good atomization throughout the load range is attained by maintaining a high rate of oil flow and high pressure drop through the swirl chamber. The excess oil above the load demand is returned from the swirl chamber to the oil storage tank or to the suction of the oil pump.

Control of the burning rate is affected by varying the oil pressure in both the oil inlet and oil return lines. Except for the atomizer, load range, and method of control, the information given for the straight pressure atomizing burner applies to this burner.

Air Atomizing Oil Burners

Except for the nozzle, this burner is similar in construction to the pressure atomizing burner. Atomizing air and oil are supplied to individual parts within the nozzle. The nozzle design allows the oil to break up into small droplet form as a result of the shear forces created by the atomizing air. The atomized oil is carried from the nozzle through the outlet orifice by the airflow into the furnace.

The main combustion air from a draft fan is forced through the burner throat and mixes intimately with the oil spray inside the combustion space. The burner igniter is similar to that used on pressure atomizing burners.

This burner is well suited for heavy fuel oils, including grade No. 6, and has a wide load range, or turndown, without changing nozzles. Turndown ranges of 3 to 1 for the smaller sizes and about 6 or 8 to 1 for the larger sizes may be expected. Load range variation is accomplished by simultaneously varying the oil pressure, the atomizing air pressure, and the combustion air entering the burner. Some designs use relatively low atomizing air pressure (5 psi and lower); other designs use air pressures up to 75 psi. The burner uses from 2.2 to 7.7 ft^3 of compressed air per gallon of fuel oil (on an air-free basis).

Because of its wide load range, this burner operates well on modulating control.

No preheating is required for No. 2 fuel oil. The heavier grades of oil must be preheated to maintain proper viscosity for atomization. When properly adjusted, these burners operate well with 15 to 25% excess air (approximately 14 to 12% CO_2, respectively, at full load); no visible smoke (approximately No. 2 smoke spot number); and only a trace of carbon monoxide in the flue gas.

Horizontal Rotary Cup Oil Burners

This burner atomizes the oil by spinning it in a thin film from a horizontal rotating cup and injecting high-velocity primary air into the oil film through an annular nozzle that surrounds the rim of the atomizing cup.

The atomizing cup and frequently the primary air fan are mounted on a horizontal main shaft that is motor driven and rotates at constant speed—58 to 100 r/s—depending on the size and make of the burner. The oil is fed to the atomizing cup at controlled rates from an oil pump that is usually driven from the main shaft through a worm and gear.

A separately mounted fan forces secondary air through the burner windbox. The use of secondary air should not be introduced by natural draft. The oil is ignited by a spark-ignited gas or oil-burning igniter (pilot). The load range or turndown for this burner is about 4 to 1, making it well suited for operation with modulating control. Automatic combustion controls are electrically operated.

When properly adjusted, these burners operate well with 20 to 25% excess air (approximately 12.5 to 12% CO_2 at full load); no visible smoke (approximately No. 2 smoke spot number); and only a trace of carbon monoxide in the flue gas.

This burner is available from several manufacturers as a package comprised of burner, primary air fan, secondary air fan with separate motor, fuel oil pump, motor, motor starter, ignition system (including transformer), automatic combustion controls, flame safety equipment, and control panel.

Good operation of these burners requires relatively constant draft in the combustion space. The main assembly of the burner with motor, main shaft, primary air fan, and oil pump is arranged for mounting on the boiler front and is hinged so that the assembly can be swung away from the firing position for easy access.

Rotary burners require some refractory in the combustion space to help support combustion. This refractory may be in the form of throat cones as well as minimum combustion chambers.

Steam Atomizing Oil Burners (Register Type)

Atomization is accomplished in this burner by the impact and expansion of steam. Oil and steam flow in separate channels through the burner gun to the burner nozzle. There, they mix before discharging through an orifice, or series of orifices, into the combustion chamber.

Combustion air, supplied by a forced-draft fan, passes through the directing vanes of the burner register, through the burner throat, and into the combustion space. The vanes give the air a spinning motion, and the burner throat directs it into the cone-shaped oil spray, where intimate mixing of air and oil takes place.

Full-load oil pressure at the burner inlet is generally some 100 to 150 psi, and the steam pressure is usually kept higher than the oil pressure by about 25 psi. Load range is accomplished by varying these pressures. Some designs operate with oil pressure ranging from 150 psi at full load to 10 psi at minimum load, resulting in a range of turndown of about 8 to 1. This wide load range makes the steam atomizing burner suited to modulating control. Some manufacturers provide dual atomizers within a single register so that one can be cleaned without dropping load.

Depending on the burner design, steam atomizing burners use from 1 to 5 lbs of steam to atomize a gallon of oil. This corresponds to 0.5 to 3.0% of the steam generated by the boiler. Where no steam is available for startup, compressed air from the plant air supply may be used for atomizing. Some designs permit the use of a pressure atomizing nozzle tip for startup when neither steam nor compressed air is available.

This burner is used mainly on water-tube boilers, which generate steam at 150 psi or higher and at capacities above 12,000,000 Btu/h input.

Oils heavier than grade No. 2 must be preheated to the proper viscosity for good atomization. When properly adjusted, these burners operate well with 15% excess air (14% CO_2) at full load, without visible smoke (approximately No. 2 smoke spot number), and with only a trace of carbon monoxide in the flue gas.

Mechanical Atomizing Oil Burners (Register Type)

Mechanical atomizing, as generally used, describes a technique synonymous with pressure atomizing. Both terms designate atomization of the oil by forcing it at high pressure through a suitable stationary atomizer.

The mechanical atomizing burner has a windbox with an assembly of adjustable internal air vanes called an *air register*. Usually the forced-draft fan is mounted separately, with a duct connection between fan outlet and burner windbox.

Oil pressure of some 90 to 900 psi is used, and load range is obtained by varying the pressure between these limits. The operating range or turndown for any given atomizer can be as high as 3 to 1. Because of its limited load range, this type of burner is seldom selected for new installations.

Return-Flow Mechanical Atomizing Oil Burners

This burner is a modification of a mechanical atomizing burner; atomization is accomplished by oil pressure alone. Load ranges up to 6 or 8 to 1 are obtained on a single-burner nozzle by varying the oil pressure between 100 and 1000 psi.

The burner was developed for use in large installations such as on shipboard and in electric generating stations where wide load range is required and water loss from the system makes the use of atomizing steam undesirable. It is also used for firing large hot-water boilers. Compressed air is too expensive for atomizing oil in large burners.

This is a register burner similar to the mechanical atomizing burner. Wide range is accomplished by use of a return-flow nozzle, which has a swirl chamber just ahead of the orifice or sprayer plate. Good atomization is attained by maintaining a high rate of oil flow and high pressure drop through the swirl chamber. The excess oil above the load demand is returned from the swirl chamber to the oil storage tank or to the oil pump suction. Control of burning rate is accomplished by varying the oil pressure in both the oil inlet and the oil return lines.

Dual-Fuel Gas/Oil Burners

Dual-fuel, combination gas/oil burners are forced-draft burners that incorporate, in a single assembly, the features of the commercial-industrial grade gas and oil burners described in the preceding sections. These burners have a three-position switch that permits the manual selection of gas, oil, or a *center-off* position. This switch contains a positive center stop or delay to ensure that the burner flame relay or programmer cycles the burner through a postpurge and prepurge cycle before starting again on the other fuel. The burner manufacturers of larger boilers design the special mechanical linkages needed to deliver the correct air-fuel ratios at full-fire, low-fire, or any intermediate rate. Smaller burners may be straight *on-off* firing. Larger burners may have low-fire starts on both fuels and use a common flame scanner. Smaller dual-fuel burners usually include pressure atomization of the oil. Air atomization systems are included in large oil burners.

Automatic changeover dual-fuel burners are available for use with gas and No. 2 oil. A special temperature control mounted on an outside wall senses outdoor temperature. It is electrically interlocked with the dual-fuel burner control system. When the outdoor temperature drops to the outdoor control set point, it changes fuels automatically after putting the burner through a postpurge and a prepurge cycle. The *minimum* additional controls and wiring needed to operate automatically with an outdoor temperature control are (1) burner fuel changeover relay(s) and (2) time delay devices to ensure interruption of the burner control circuit at the moment of fuel changeover.

The fuel changeover relays replace the three-position manual fuel selection switch. A manual fuel selection switch can be retained as a manual override on the automatic feature. These control systems require special design and are generally provided by the burner manufacturer.

The dual-fuel burner is fitted with a gas train and oil piping that is connected to a two-pipe oil system following the principles of the preceding sections. A reserve of oil must be maintained at all times for automatic fuel changeover.

Boiler flue chimney connectors are equipped with a double-swing barometric draft regulator or, if required, sequential furnace draft control to operate an automatic flue damper. Dual-fuel burners and their accessories should be installed by experienced contractors to ensure satisfactory operation.

EQUIPMENT SELECTION

Economic and practical factors (such as the degree of operating supervision required by the installation) generally dictate that fuel oil must be selected by considering the maximum heat input of the oil-burning unit. For heating loads and where only one oil-burning unit is operated at any given time, the relationship is as shown in Table 2 (this table is only a guide). In many cases, a de-

Table 2 Guide for Fuel Oil Grades versus Firing Rate

Maximum Heat Input of Unit, 1000 Btu/h	Volume Flow Rate, gph	Grade of Fuel
Up to 3500	Up to 25	2
3500 to 7000	35 to 50	2, 4, 5
7000 to 15,000	50 to 100	5, 6
Over 15,000	Over 100	6

tailed analysis of operating parameters results in burning the lighter grades of fuel oils at capacities far above those indicated. Process application, especially with multiple units, may require different criteria.

Fuel Oil Storage Systems

All fuel oil storage tanks should be constructed and installed in accordance with NFPA *Standard* 31 and with local ordinances.

Storage capacity. Dependable and economical operation of oil-burning equipment requires ample and safe storage of fuel oil at the site. Design responsibility should include analysis of specific storage requirements as follows:

1. Rate of oil consumption.
2. Dependability of oil deliveries.
3. Economical delivery lots. The cost of installing larger storage capacity should be balanced against the savings indicated by accommodating larger delivery lots. Truck lots and railcar lots vary with various suppliers, but the quantities are approximated as follows:

Small truck lots in metropolitan area	500 to 2000 gal
Normal truck lots	3000 to 5000 gal
Transport truck lots	5000 to 9000 gal
Rail tanker lots	8000 to 12,000 gal

Tank size and location. Standard oil storage tanks range in size from 55 to 50,000 gal and larger. Tanks are usually built of steel; concrete construction may be used only for heavy oil. Unenclosed tanks located in the lowest story, cellar, or basement should not exceed 660 gal capacity each, and the aggregate capacity of such tanks should not exceed 1320 gal, unless each 660 gal tank is insulated in an approved fireproof room having a fire resistance rating of at least 2 h.

Storage tanks with the storage capacity at a given location exceeding about 1000 gal should be underground whenever practical and accessible for truck or rail delivery with gravity flow from the delivering carrier into storage. If the oil is to be burned in a central plant such as a boiler house, the storage tanks should be located, if possible, so that the oil burner pump (or pumps) can pump directly from storage to the burners. In case of a year-round operation, except for storage or supply capacities below 2000 gal, at least two tanks should be installed to facilitate tank inspection, cleaning, repairs, and clearing of plugged suction lines.

When the main oil storage tank is not close enough to the oil-burning units for the burner pumps to take suction from storage, a supply tank must be installed near the oil-burning units and oil must be pumped periodically from storage to the supply tank by a transport pump at the storage location. Supply tanks should be treated the same as storage tanks regarding location within buildings, tank design, etc. On large installations, it is recommended that standby pumps be installed as a protection against heat loss in case of pump failure.

Since all piping connections to underground tanks must be at the top, such tanks should not be more than 10 ft 6 in. in height from top to bottom to avoid pump suction difficulties. (This dimension may have to be less for installations at high altitudes.) The total suction head for the oil pump must not exceed 14 ft at sea level.

Connections to storage tank. All piping connections for tanks over 275-gal capacity should be through the top of the tank. Figure

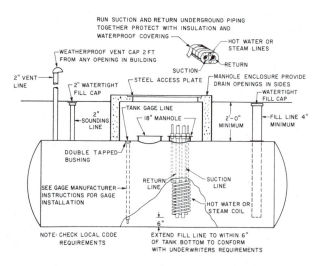

Fig. 4   Typical Oil Storage Tank (No. 6 Oil)

4 shows a typical arrangement for a cylindrical storage tank with heating coil as required for No. 5 or No. 6 fuel oils. The heating coil and oil suction lines should be located near one end of the tank.

The maximum allowable steam pressure in such a heating coil is 15 psi. The heating coil is unnecessary for oils lighter than No. 5, unless a combination of high pour point and low outdoor temperature makes heating necessary.

A watertight manhole with internal ladder provides access to the inside of the tank. If the tank is equipped with an internal heating coil, a second manhole is required and arranged to permit withdrawal of the coil.

The fill line should be vertical and should discharge near the end of the tank away from the oil suction line. The inlet of the fill line must be outside the building and accessible to the oil delivery vehicle, unless an oil transfer pump is used to fill the tank. When pos-

sible, the inlet of the fill line should be at or near grade level where filling may be accomplished by gravity. The fill line should be at least 2 in. in diameter for No. 2 oil and 6 in. in diameter for No. 4, 5, or 6 oils for gravity filling. Where filling is done by pump, the fill line for No. 4, 5, or 6 oils may be 4 in. in diameter.

An oil return line bringing recirculated oil from the burner line to the tank should discharge near the oil suction line inlet. Each storage tank should be equipped with a vent line sized and arranged in accordance with NFPA *Standard* 31.

Each storage tank must have a device for determining oil level. For tanks inside buildings, the gaging device should be designed and installed so that oil or vapor will not discharge into a building from the fuel supply system. No storage tank should be equipped with a glass gage or any gage which, when broken, would permit the escape of oil from the tank. Gaging by a measuring stick is permissible for outside tanks or for underground tanks.

Fuel-Handling Systems

The fuel-handling system consists of the pumps, valves, and fittings for moving fuel oil from the delivery truck or car into the storage tanks and from the storage tanks to the oil burners. Depending on the type and arrangement of the oil-burning equipment and the grade of fuel oil burned, fuel-handling systems vary from simple to quite complicated arrangements.

The simplest handling system would apply to a single burner and small storage tank for No. 2 fuel oil similar to a residential heating installation. The storage tank is filled through a hose from the oil delivery truck, and the fuel-handling system consists of a supply pipe between the storage tank and the burner pump. Equipment should be installed on light oil tanks to indicate visibly or audibly when the tank is full; on heavy oil tanks, a remote-reading liquid-level gage should be installed.

Figure 5 shows a complex oil supply arrangement for two burners on one oil-burning unit. For a unit with a single burner, the change in piping is obvious. For supplying oil to two or more units, the oil line downstream of the oil discharge strainer becomes a main supply header, and the branch supply line to each unit

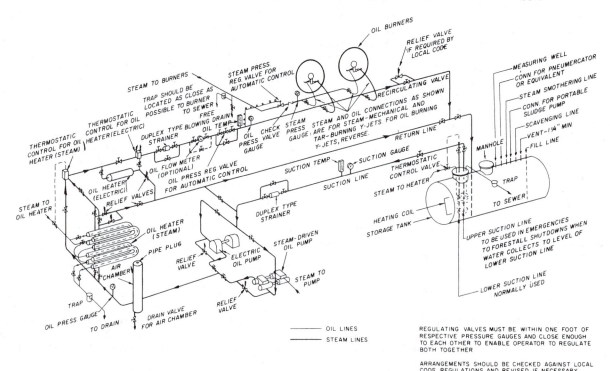

Fig. 5   Industrial Burner Auxiliary Equipment

would include oil flow meter, automatic control valve, etc. For light oils requiring no heating, all oil heating equipment shown in Figure 5 would be omitted. Both a suction and return line should be used, except for gravity flow in residential installations.

Oil pumps (steam or electrically driven) should deliver oil at the maximum rate required by the burners (this includes the maximum firing rate, the oil required for recirculating, plus a 10% margin).

The calculated suction head at the entrance of any burner pump should not exceed 10 in. Hg for installations at sea level. Higher elevations require that the suction head be reduced in direct proportion to the reduction in barometric pressure.

The oil temperature at the pump inlet should not exceed 120°F. Where oil burners with integral oil pumps (and oil heaters) are used and where the suction lift from the storage tank is within the capacity of the burner pump, each burner may take oil directly from the storage tank through an individual suction line, except where No. 6 oil is used.

Where two or more tanks are used, the piping arrangement into the top of each tank should be the same as for a single tank so that any tank may be used at any time; any tank can be inspected, cleaned, or repaired while the system is in operation.

The length of suction line between storage tank and burner pumps should not exceed 100 ft. Where the main storage tank(s) is located more than 100 ft from the pumps, in addition to the storage tanks, a supply tank should be installed near the pumps, and a transfer pump should be installed at the storage tanks for delivery of oil to the supply tank.

Central oil distribution systems comprising a central storage facility, distribution pumps or provision for gravity delivery, distribution piping, and individual fuel meters are used for residential communities—notably mobile home parks. The provisions of NFPA *Standard* 31, Installation of Oil Burning Equipment, should be followed in making a central oil distribution system installation.

Fuel-Oil Preparation System

Fuel-oil preparation systems consist of oil heater, oil temperature controls, strainers, and associated valves and piping required to maintain fuel oil at the temperatures necessary to control the oil viscosity, facilitate oil flow and burning, and remove suspended matter.

Preparation of fuel oil for handling and burning requires heating the oil if it is grade 5 or 6. This decreases its viscosity so it will flow properly through the oil system piping and can be atomized by the oil burner. Grade 4 occasionally requires heating to facilitate burning. Grade 2 requires heating only under unusual conditions.

For handling residual oil from the delivering carrier into storage tanks, the viscosity should be about 156×10^6 centistokes (cs). For satisfactory pumping, the viscosity of the oil surrounding the inlet of the suction pipe must be 444×10^6 cs or lower; in the case of high pour point oil, the temperature of the entire oil content of the tank must be above the pour point.

These storage tank heaters are usually made of pipe coils or grids using steam or hot water at not over 15 psi pressure as the heating medium. Electric heaters are sometimes used. For control of viscosity for pumping, the heated oil surrounds the oil suction line inlet. When heating high pour-point oils, the heater should extend the entire length of the tank. All heaters have suitable thermostatic controls. In some cases, storage tank heating may be accomplished satisfactorily by returning or recirculating sufficient amounts of oil to the tank after it has passed through heaters located between the oil pump and oil burner.

Heaters to regulate viscosity at the burners are installed between the oil pumps and the burners. For small packaged burners, the heaters, when required, are either assembled integrally with the

individual burners or mounted separately. The source of heat may be electricity, steam, or hot water. For larger installations, the heater is mounted separately and is often arranged in combination with central oil pumps, forming a central oil pumping and heating set.

The separate or central oil pumping and heating set is recommended for those installations burning heavy oils where the load demand is periodical, and continuous circulation of hot oil is necessary during down periods.

Another system of oil heating to maintain pumping viscosity that is occasionally used for small- or medium-sized installations consists of an electrically heated section of oil piping. Low-voltage current is passed through the pipe section, which is isolated by nonconducting flanges.

The oil heating capacity for any given installation should be approximately 10% greater than the maximum oil flow. Maximum oil flow is the maximum oil burning rate plus the rate of oil recirculation.

Controls for oil heaters must be dependable to ensure proper oil atomization and avoid overheating of oil, which results in coke deposits inside the heaters. In steam or electric heating, an interlock should be included with a solenoid valve or switch to shut off the steam or electricity. During periods when the oil pump is not operating, the oil in the heater can become overheated and deposit carbon. This also can be a problem with high-temperature hot water; in such a case, provisions must be made to avoid overheating the oil when the oil pump is not operating.

Oil heaters with low- or medium-temperature hot water are not generally subject to coke deposits. Where steam or hot water is used in oil heaters located after the oil pumps, the pressure of the steam or water in the heaters is usually lower than the oil pressure. Consequently, heater leakage between oil and steam causes oil to flow into the water or condensing steam. To avoid such oil entering the boilers, the condensed steam or the water from such heaters should be discarded from the system, or special equipment should be provided for oil removal.

Hot-water oil heaters of double-tube-and-shell construction with inert heat transfer oil and sight glass between are available. With this type of heater, oil leaks through an oil-side tube appear in the sight glass, and repairs can be made to the oil-side before a water-side tube leaks.

This discussion of oil-burning equipment applies to oil-fired boilers and furnaces. Chapters 28 and 29 address boilers and furnaces in more detail.

SOLID FUEL-BURNING EQUIPMENT

A mechanical stoker is a device that feeds a solid fuel into a combustion chamber. It supplies air for burning the fuel under automatic control and, in some cases, incorporates automatic ash and refuse removal.

CAPACITY CLASSIFICATION OF STOKERS

Stokers are classified according to their coal feeding rates. The following classification has been made by the U.S. Department of Commerce, in cooperation with the Stoker Manufacturers Association:

Class 1: capacity under 60 lb of coal per hour.
Class 2: capacity 60 to 100 lb of coal per hour.
Class 3: capacity 100 to 300 lb of coal per hour.
Class 4: capacity 300 to 1200 lb of coal per hour.
Class 5: capacity 1200 lb of coal per hour and over.

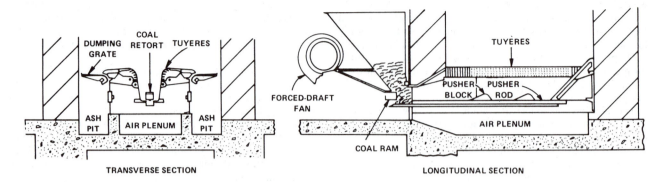

Fig. 6   Horizontal Underfeed Stoker with Single Retort

Class 1 stokers are used primarily for residential heating and are designed for quiet, automatic operation. These stokers are usually underfeed types and are similar to those shown in Figure 6, except they are usually screw feed. Although some stokers for residential application are still used, the primary thrust of stoker application is in the commercial and industrial areas. Stokers in this class feed coal to the furnace intermittently, in accordance with temperature or pressure demands. A special control is needed to ensure stoker operation in order to maintain a fire during periods when no heat is required.

Class 2 and 3 stokers are usually of the screw-feed type, without auxiliary plungers or other means of distributing the coal. They are used extensively for heating plants in apartments, hotels, and industrial plants. They are of the underfeed type and are available in both the hopper and bin-feed type. These units are also built in a plunger-feed type with an electric motor, steam, or hydraulic cylinder coal-feed drive.

Stokers in this class are available for burning all types of anthracite, bituminous, and lignite coals. The tuyere and retort design varies according to the fuel and load conditions. Stationary grates are used on bituminous models, and the clinkers formed from the ash accumulate on the grates surrounding the retort.

Anthracite stokers in this class are equipped with moving grates that discharge the ash into a pit below the grate. This ash pit may be located on one or both sides of the grate and, in some installations, is of sufficient capacity to hold the ash for several weeks of operation.

Class 4 stokers vary in details of design, and several methods of feeding coal are practiced. Underfeed stokers are widely used, although overfeed types are used in the larger sizes. Bin feed and hopper models are available in underfeed and overfeed types.

Class 5 stokers are underfeed, spreader, chain grate or traveling grate, and vibrating grate. Various subcategories reflect types of grates and methods of ash discharge.

STOKER TYPES BY FUEL-FEED METHODS

Stokers are classified according to the method of feeding fuel to the furnace, such as (1) spreader, (2) underfeed, (3) chain grate or traveling grate, and (4) vibrating grate. The type of stoker used in a given installation depends on the general system design, capacity required, and type of fuel burned. In general, the spreader stoker is the most widely used in the capacity range of 75,000 to 400,000 lb/h because it responds quickly to load changes and can burn a wide range of coals. Underfeed stokers are principally used with small industrial boilers of less than 30,000 lb/h. In the intermediate range, the large underfeed units, as well as the chain and traveling grate stokers, are being displaced by spreader and vibrating grate stokers. (Table 3 summarizes their major features.)

Spreader Stokers

Spreader stokers use a combination of suspension burning and grate burning. As illustrated in Figure 7, coal is continually projected into the furnace above an ignited fuel bed. The coal fines are partially burned in suspension. Large particles fall to the grate and are burned in a thin, fast-burning fuel bed. Because this firing method provides extreme flexibility to load fluctuations and because ignition is almost instantaneous on increased firing rate, the spreader stoker is favored over other stokers in many industrial applications.

The spreader stoker is designed to burn about 50% of the fuel in suspension. Thus, it generates much higher particulate loadings than other types of stokers and requires dust collectors to trap particulate material in the flue gas before discharge to the stack. To minimize carbon loss, fly-carbon reinjection systems are sometimes used to return this carbon into the furnace for complete burnout. Because this process increases furnace dust emissions, it can be used only with highly efficient dust collectors.

Grates for spreader stokers may be of several types. All grates are designed with high airflow resistance to avoid formation of blowholes through the thin fuel bed. Early designs were simple stationary grates from which ash was removed manually. Later designs allowed intermittent dumping of the grate either manually or by a power cylinder. Both types of dumping grates are frequently used for the small- and medium-sized boilers (see Table 3). Also, both types are sectionalized and there is a separate under-

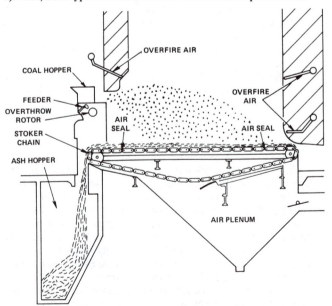

Fig. 7   Spreader Stoker, Traveling Grate Type

Table 3   Characteristics of Various Types of Stokers (Class 5)

Stoker Type and Subclass	Typical Capacity Range, lb/h	Maximum Burning Rate, Btu/h·ft²	Characteristics
Spreader			
Stationary and dumping grate	20,000 to 80,000	450,000	Capable of burning a wide range of coals; best to follow fluctuating loads; high fly ash carryover; low-load smoke.
Traveling grate	100,000 to 400,000	750,000	
Vibrating grate	20,000 to 100,000	400,000	
Underfeed			
Single- or double-retort	20,000 to 30,000	400,000	Capable of burning caking coals and a wide range of coals (including anthracite); high maintenance, low fly ash carryover; suitable for continuous-load operation.
Multiple-retort	30,000 to 500,000	600,000	
Chain grate and *traveling grate*	20,000 to 100,000	500,000	Characteristics similar to vibrating grate stokers, except that these stokers have difficulty in burning strongly caking coals.
Vibrating grate	1400 to 150,000	400,000	Low maintenance, low fly ash carryover; capable of burning wide variety of weakly caking coals; smokeless operation over entire range.

grate air chamber for each grate section and a grate section for each spreader unit. Consequently, both the air and fuel supply to one section can be temporarily discontinued for cleaning and maintenance without affecting the operation of other sections of the stoker.

For high efficiency operation, a continuous ash-discharging grate, such as the traveling grate, is necessary. The introduction of the spreader stoker with the traveling grate increased burning rates by about 70% over the stationary grate and dumping grate types. Although continuous ash-discharge grates of reciprocating and vibrating types have been developed, the traveling grate stoker is preferred because of its higher burning rates.

Fuels and fuel bed. All spreader stokers (in particular, those with traveling grates) are able to use fuels with a wide range of burning characteristics, including coals with caking tendencies, because the rapid surface heating of the coal in suspension destroys the caking tendency. High moisture, free-burning bituminous and lignite coals are commonly burned, while coke breeze can be burned in a mixture with a high volatile coal. However, anthracite, because of its low volatile content, is not a suitable fuel for spreader stoker firing. Ideally, the fuel bed of a coal-fired spreader stoker is from 2 to 4 in. thick.

Burning rates. The maximum heat release rates range from 400,000 Btu/h·ft² (a coal consumption of approximately 40 lb/h) on stationary, dumping, and vibrating grate designs) to 750,000 Btu/h·ft² on traveling grate spreader stokers. Higher heat-release rates are practical with certain waste fuels in which a greater portion of fuel can be burned in suspension than is possible with coal.

Underfeed Stokers

Underfeed stokers introduce raw coal into a retort beneath the burning fuel bed. They are classified as horizontal feed and gravity feed. In the horizontal type, coal travels within the furnace in a retort parallel with the floor; in the gravity feed type, the retort is inclined by 25°. Most horizontal feed stokers are designed with single or double retorts (rarely, with triple retorts), while gravity feed stokers are designed with multiple retorts.

In the horizontal stoker (Figure 6), coal is fed to the retort by a screw (for the smaller stokers) or a ram (for the larger units). Once the retort is filled, the coal is forced upward and spills over the retort to form and feed the fuel bed. Air is supplied through tuyeres at each side of the retort and through air ports in the side grates. Over-fire air is used to provide additional combustion air to the flame zone directly above the bed to prevent smoking, especially at low loads.

Gravity-feed units are similar in operating principle. These stokers consist of sloping multiple retorts and have rear ash discharge. Coal is fed into each retort, where it is moved slowly to the rear while simultaneously being forced upward over the retorts.

Fuels and fuel bed. Either type of underfeed stoker can burn a wide range of coal, although the horizontal type is better suited for free-burning bituminous coal. These units can burn caking coal, provided there is not an excess amount of fines. The ash-softening temperature is an important factor in selecting coals because the possibility of excessive clinkering increases at lower ash-softening temperatures. Because combustion occurs in the fuel bed, these stokers respond slowly to load change. Fuel-bed thickness in underfeed stokers is extremely nonuniform, ranging from 8 to 24 in. The fuel bed often contains large fissures separating masses of coke.

Burning rates. The single-retort or double-retort horizontal stokers are generally used to service boilers with capacities up to 30,000 lb/h. These units are designed for heat-release rates of 400,000 Btu/h·ft².

Chain Grate and Traveling Grate Stokers

Figure 8 shows a typical chain grate or traveling grate stoker. These stokers are often used interchangeably because they are fundamentally the same, except for grate construction. The essential difference is that the links of chain grate stokers are assembled so that they move with a scissors-like action at the return bend of the stoker, while in most traveling grates there is no relative movement between adjacent grate sections. Accordingly, the chain grate is

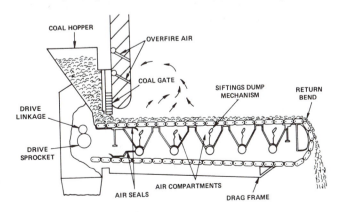

Fig. 8   Chain Grate Stoker

more suitable for handling coals with clinkering-ash characteristics than is the traveling grate unit.

The operation of each type is similar. Coal, fed from a hopper onto the moving grate, enters the furnace after passing under an adjustable gate to regulate the thickness of the fuel bed. The layer of coal on the grate entering the furnace is heated by radiation from the furnace gases or from a hot refractory arch. As volatile matter is driven off by this rapid radiative heating, ignition occurs. The fuel continues to burn as it moves along the fuel bed and the layer becomes progressively thinner. At the far end of the grate, where the combustion of the coal is completed, the ash is discharged into the pit as the grates pass downward over a return bend.

Often, furnace arches (front and/or rear) are included with these stokers to improve combustion by reflecting heat to the fuel bed. The front arch also serves as a bluff body to mix rich streams of volatile gases with air to reduce unburned hydrocarbons. A chain grate stoker with overfire air jets eliminates the need of a front arch for burning volatiles. As shown in Figure 9, the stoker was zoned, or sectionalized, and equipped with individual zone dampers to control the pressure and quantity of air delivered to the various sections.

Fuels and fuel bed. The chain grate and traveling grate stokers can burn a variety of fuels (*e.g.*, peat, lignite, sub-bituminous coal, free-burning bituminous coal, anthracite coal, and coke), provided that the fuel is sized properly. However, strongly caking bituminous coals have a tendency to mat and prevent proper air distribution to the fuel bed. Also, a bed of strongly caking coal may not be responsive to rapidly changing loads. Fuel-bed thickness varies with the type and size of coal burned. For bituminous coal, a 5 to 7 in. bed is common; for small-sized anthracite, the fuel bed is reduced to 3 to 5 in.

Burning rates. Chain grate and traveling grate stokers are offered for a maximum continuous burning rate of between 350,000 to 500,000 Btu/h·ft², depending on the type of fuel and its ash and moisture content.

Vibrating Grate Stokers

The vibrating grate stoker, as shown in Figure 9, is similar to the chain grate stoker in that both are overfeed, mass-burning, continuous ash-discharge units. However, in the vibrating stoker, the sloping grate is supported on equally spaced vertical plates that oscillate back and forth in a rectilinear direction, causing the fuel to move from the hopper through an adjustable gate into the active combustion zone. Air is supplied to the stoker through laterally exposed areas beneath the stoker formed by the individual flexing of the grate support plates. Ash is automatically discharged into a shallow or basement ash pit. The grates are water-cooled and are connected to the boiler circulating system.

The rates of coal feed and fuel-bed movement are controlled by the frequency and duration of the vibrating cycles and regulated by automatic combustion controls that proportion the air supply to optimize heat release rates. Typically, the grate is vibrated about every 90 s for durations of 2 to 3 s, but this depends on the type of coal and boiler operation. The vibrating grate stoker has found increasing acceptance because of its simplicity, inherently low fly ash carryover, low maintenance, wide turndown (10 to 1), and adaptability to multiple-fuel firing.

Fuels and fuel bed. The water-cooled vibrating grate stoker is suitable for burning a wide range of bituminous and lignite coals. Because of the gentle agitation and compaction caused by the vibratory actions, coal having a high free-swelling index can be burned, and a uniform fuel bed without blowholes and thin spots can be maintained. The uniformity of air distribution and resultant fuel-bed conditions produce good response to load swings and smokeless operation over the entire load range. Fly ash emission is probably greater than from the traveling grate because of the slight intermittent agitation of the fuel bed. The fuel bed is similar to that of a traveling grate stoker.

Burning rates of vibrating grate stokers vary with the type of fuel used. In general, however, the maximum heat release rates should not exceed 400,000 Btu/h·ft² (a coal use of approximately 40 lb/h) to minimize fly ash carryover.

CONTROLS

This section only covers the automatic controls necessary for automatic fuel-burning equipment. Chapter 41 in the 1991 ASHRAE *Handbook—HVAC Applications* addresses basic automatic control.

Automatic fuel-burning equipment requires a control system that provides a prescribed sequence of operating events and takes proper corrective action in the event of any failure in the equipment or its operation. The basic requirements for oil burners, gas burners, and coal burners (stokers) are the same. The term *burner*, unless otherwise specified, refers to all three types of fuel-burning equipment. However, the details of operation and the components used differ for the three types of burners. The controls can be classified as operating controls, limit controls, and interlocks. Operating controls include a primary sensor, secondary sensors, actuators, ignition system, firing-rate controls, draft controls, and programmers. Limit controls include flame safeguard, temperature limit, pressure limit, and water level limit controls. Several control functions are frequently included in a single component. Control systems for domestic burners and the smaller commercial burners are generally operated on low voltage (frequently 24 V). Some domestic gas burner control systems obtain their electrical power from the direct conversion of heat to low-voltage electrical energy with a thermopile. These are called self-generating systems. Line voltage and pneumatic controls are common in the larger commercial and industrial areas.

OPERATING CONTROLS

Operating controls initiate the normal starting and stopping of the burner in response to the primary sensor acting through appropriate actuators. Secondary sensors, secondary actuators, and the ignition system are all part of the operating controls systems.

A *primary sensor* is required to monitor the effect desired by the furnace. Examples of primary sensors are a room thermostat for a residential furnace; a pressure-actuated switch for a steam boiler; or a thermostat for a hot-water heater. Several sensors are sometimes connected together to average the measurement (temperature or pressure) at several points. Control on the basis of a temperature or pressure difference between two points can also be achieved. An example is the outside temperature-sensing element

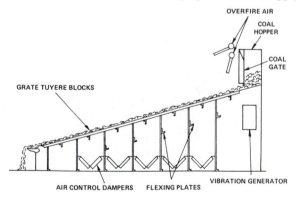

Fig. 9   Vibrating Grate Stoker

sometimes used in conjunction with the room thermostat to reset room temperature as a function of the outdoor temperature.

Secondary sensors are generally required also. An example is the fan control used on warm-air furnaces to control the fan as a function of plenum temperature. In some cases, the primary sensor operates a heat-control device such as a damper motor or a circulating pump. The burner is then operated from a secondary control such as an immersion thermostat in a hot-water system or a thermostat in the bonnet of the warm-air furnace. The primary sensor in a steam system with individual room control actuates a steam valve supplying that room or zone. The burner is then operated by a secondary control consisting of a pressure sensor located on a boiler.

An *actuator* is a device that converts the control system signal into a useful function. Actuators generally consist of valves, dampers, or relays.

Valves are required for final shutoff of the gas pilot and main gas (or oil) valves in the supply line. The type of valve used depends on the service, fuel, and characteristics of the burner. Valves are classified by types as follows:

- Solenoid gas or oil valves provide quick opening and quick closing.
- Motor-operated gas and oil valves may be of gear train, pneumatic, or hydraulically operated types. They provide relatively slow opening and quick closing. Opening time is not adjustable. They may have provision for direct or reverse acting operating levers and, in some cases, have position indicators.
- Diaphragm valves for gas are usually operated by gas pressure, although some types are steam or air operated. These valves may have provision for adjustable opening time, which may be adjusted to the desired burner operating characteristics. They may also have provision for direct- or reverse-acting lever arm.
- Manually opened safety shutoff valves for gas or oil can be opened manually only when power is available. When power is interrupted or any associated interlock is activated, they trip free for fast closure. They are normally used on semiautomatic and manually fired installations.
- Burner input control valves are required for gas or oil by some regulatory agencies in addition to safety shutoff valves. These control valves may be of the slow-opening type or may provide for high-low or modulating operation. These valves may or may not have provision for final shutoff. Requirements of local codes and approval bodies should be followed to ensure that valves selected are approved for the type of burner used and that maximum overall efficiency of the burner is obtained.

Ignition

An automatic burner igniter is necessary for safe operation and is, in most applications, an essential part of the automatic control system. The burner igniter is an electric spark that directly ignites the main fuel supply or a relatively small fuel burner that ignites the main fuel supply. Most igniters for oil burners are continuous (function continuously while burner is in operation) or intermittent (function only long enough to establish main flame). On small input gas-fired appliances, ignition of the main fuel is facilitated by a small standing gas pilot, which burns continuously.

Electric spark igniters are generally used only for ignition of distillate grades of fuel oil. The spark is generated by a transformer, which supplies up to 10,000 V to the sparking electrodes. This ignition system is used frequently on oil burners having capacities of up to 20 gph.

When pilot burners are used on larger equipment, the fuel burned may be gas, liquid petroleum (LP), or distillate fuel oil. Ignition of the pilot burner is by electric spark.

Safety Controls and Interlocks

Safety controls and interlocks are provided to ensure against furnace explosions and other hazards, such as overheating of boilers resulting from low water, depending on the type of oil-burning units. These controls shut off the fuel or prevent firing in case of the following:

- Flame failure—main flame or pilot flame
- Combustion air fan failure
- Overheating of warm-air furnace parts
- Low water level in boilers
- Low fuel-oil pressure
- Low fuel-oil temperature, when burning heavy oils
- High fuel-oil temperature, when burning heavy oils
- Low atomizing air or atomizing steam pressure
- High water temperature in water heaters and hot-water boilers
- High steam pressure
- Burner out of firing position (burner position interlock)
- Failure of rotary burner motor (rotary cup interlock)

These interlocks do not affect the normal operation of the unit. The various types of safety devices in normal use and their application and operation are described in the following sections.

Limit Controls

The safety features, which must be incorporated into the control system, function only when the system exceeds prescribed safe operating conditions. Such safety features are embodied in the limit controls. They actuate electrical switches that are normally closed and close the fuel valve in the event of an unsafe condition such as (1) excessive temperature in the combustion chamber or heat exchanger; (2) excessive pressure in a boiler or hot-water heater; (3) low water level in a boiler and in larger commercial and industrial burners; (4) high or low gas pressure; (5) low oil pressure; (6) low atomizing media pressure; and (7) low oil temperature when firing residual fuel oil. Separate limit and operating controls are always recommended. The operating control is adjusted to function within the operating range, while the limit is set at a point above the operating range to stop the burner in the event of failure of the operating control.

Flame Safeguard Controls

Loss of ignition, or flame failure while fuel is being injected into a furnace, can cause a furnace explosion. Therefore, a flame sensor is used to shut off the fuel supply in case of flame failure. The flame detector of this device senses the presence or absence of flame through temperature, flame conductance of an electric current, or flame radiation. Through suitable electrical devices, the flame detector shuts down the burner when there is loss of flame.

Modern flame sensors of the flame conducting type and flame scanner type supervise the igniter (pilot burner) and the main burner. Thus, at light-off, a safe igniter flame must be established before fuel is admitted to the main burner.

PROGRAMMING

Flame safety systems are interlocked and equipped with timing devices so that automatic lighting occurs in proper sequence with proper time intervals. In the larger units (size dependent on local authorities), these controls comprise a programmer and a flame-sensing device or flame detector.

The programmer maintains the proper safety sequence for lighting the burner, subject to supervision by the flame detector. When light-off is called for, by push button or by the combustion control, the programmer controls the sequence of operation as follows:

1. Starts the burner fan so that the furnace, or combustion space; the gas passages of the heat exchanger; and the chimney connector are purged of any unburned combustible gases. This flow of air is maintained until the entire volume of combustion space, gas passages, and chimney connector has been changed at least several times. It is commonly called *prepurge*.

2. After the prepurge, the controller initiates operation of the burner igniter and allows a short time for the flame detector to prove satisfactory operation of the igniter. If the flame detector is not satisfied, the program controller switches back to its initial start position, and an integral switch opens to prevent further operation until the electrical circuit is manually reset.

3. If the igniter operation satisfies the flame detector, the program controller starts fuel flow to the burner. The flame detector is in control from this point on until the burner is shut off. If the flame detector is not satisfied with established stable burning during the startup, it shuts down the burner and the program controller switches back to its initial position.

4. When the burner is shut off for any reason, the program controller causes the forced-draft fan to continue delivering air for a specific time to postpurge the unit of any atomized oil and combustible gases. This postpurge should amount to at least two air changes in the combustion area, gas passages, and chimney connector. (Some authorities disagree about the necessity of this function.)

Electronic programmers are available that, in addition to programming the entire boiler operating sequence, have data acquisition and storage capability to record operating conditions and flag the cause of burner shutdowns.

Other Safety Devices

Safety devices to shut off the burner in case of hazardous conditions other than flame failures consist of simple sensors such as temperature elements, water level detectors, and pressure devices. These are all designed, installed, and interlocked to shut off the fuel oil and combustion air to the burner if any of the conditions listed occurs. All safety interlocks are arranged to prevent lighting the burner until satisfactory conditions have been established.

COMBUSTION CONTROL SYSTEMS

Some modern gas- and oil-burning units are equipped with integrated devices that automatically regulate the flow of fuel and combustion air to satisfy the demand from the units. These systems, called *combustion control systems*, consist of a master controller, which senses the heat demands through room temperature or steam pressure, etc., and controlled devices, which receive impulses from the master controller. The signals change the rate of heat input by starting or stopping motors, driving oil pumps and fans, or adjusting air dampers and oil valves. The signal may be electrical or, in the case of some larger boilers, compressed air. Descriptions of various types of combustion control systems follow.

An *on-off control system* simply starts and stops the fuel burner to satisfy the heat demand and does not control a combustion system's rate of heat input. Until the controlled temperature of a room, the outlet temperature of a water heater, or the steam boiler pressure reaches a predetermined point, the burner remains on. At a set maximum condition of temperature or steam pressure, the burner shuts off.

The burner again ignites at a predetermined condition of temperature, pressure, etc. This control is normal for small units with burners using No. 2 fuel oil at capacities up to about 2,120,000 Btu/h or 15 gph. These small burners have gas controls or a single motor to drive the burner fan or the fuel oil pump. Proportioning of combustion air to fuel is accomplished by the fan design

and/or fixed position inlet air damper. For gas burners, pressure atomizing oil burners, and other types of burners burning No. 2 or No. 4 oil at capacities between 2,120,000 to 4,240,000 Btu/h at 15 to 30 gph, the on-off control system usually has a low-start feature. This device starts the burner at less than full load (perhaps 60%) and then increases to full burning rate after several seconds. This type of control is called *low-high-off*.

A *low-high-low-off* control system (often referred to as two-position firing) also provides a low-fire start. In addition, it cycles the burner back to low fire when the maximum control set point is reached.

In operation, the two-position control cycle has considerable advantage on a rapidly fluctuating load. However, on oil firing, combustion conditions are poor at the low-load rate, so this cycle is seldom recommended for cases of sustained low loads, such as heating loads. The two-position control cycle provides somewhat steadier conditions. A positioning motor, through a mechanical linkage, properly positions the oil flow control valve and the burner damper. This permits damper settings for proportioning air to fuel.

A modulating control regulates the firing rate to follow the load demands more closely than the on-off or low-high-low-off controls. The fuel oil control valve and the burner damper respond over a range of positions within the operating range capability of the burner. The simplest and least expensive modulating control systems, as frequently used for burner capacities between 4,240,000 to 70,000,000 Btu/h or 30 to 500 gph, use a positioning motor and mechanical linkage, which permit a large number of positions of the fuel control valve and the forced-draft damper. Most dampers larger than 70,000,000 Btu/h or 500 gph require more sophisticated controls that permit a continuous adjustment of valve and damper positions within the range of burner operation.

Modulating control systems that position the fuel oil control valve and the forced-draft damper in a predetermined relationship are called *parallel controls* or *positioning controls*. These valves maintain the fuel and air in proper proportion throughout the load range. Modulating controls for single-burner boilers usually include an on-off control for loads below the operating range of the burner.

A *full-metering* and *proportioning control system* is a modulating control system that maintains the proper ratio of air to fuel throughout the burner range with a fuel-air ratio controller. This system meters the airflow, usually by sensing the air pressure drop through the burner throat, and positions the air controller to suit momentary load demands. The fuel control valve immediately adjusts to match fuel flow to airflow. For some requirements, the ratio controller is reversed to match airflow to fuel flow. In unusual cases, the ratio controller is set to operate one way on load increase and the opposite way during load decrease. This control maintains optimum efficiency throughout the entire burner range, usually holding close to 15% excess air, as compared to 25% or greater with on-off controls.

Draft Controls

Combustion control systems function properly with a reasonably constant furnace draft or pressure. The main concern is the varying draft effect of high chimneys. To control this draft, various regulators are used in the flue connections at the gas outlets of the units. Draft hoods or balanced draft dampers are used for smaller units and draft damper controllers for large boilers. These regulators should be supplied as part of the combustion control equipment.

REFERENCES

ANSI. 1971. Gas utilization equipment in large boilers. ANSI *Standard* Z83.3-1971. Secretariat: American Gas Association, Cleveland, Ohio.

ANSI. 1981. Draft hoods. ANSI *Standard* Z21.12-1981, Secretariat: American Gas Association. Cleveland, Ohio.

ANSI. 1982. Gas-fired duct furnaces. ANSI *Standard* Z83.9-1982. Secretariat: American Gas Association, Cleveland, Ohio.

ANSI. 1984. Domestic gas conversion burners. ANSI *Standard* Z21.17-1984. Secretariat: American Gas Association, Cleveland, Ohio.

ANSI. 1984. Installation of domestic gas conversion burners. ANSI *Standard* Z21.8-1984. Secretariat: American Gas Association, Cleveland, Ohio.

ANSI. 1991. Direct vent central furnaces. ANSI *Standard* Z21.64-1990. Secretariat: American Gas Association, Cleveland, Ohio.

ANSI. 1991. Gas-fired central furnaces (except direct vent central furnaces). ANSI *Standard* Z21.47a-1990. Secretariat: American Gas Association, Cleveland, Ohio.

ANSI. 1991. Gas-fired low-pressure steam and hot-water boilers. ANSI *Standard* Z21.13b-1991. Secretariat: American Gas Association, Cleveland, Ohio.

Gas engineers handbook, Section 12, Chapter 2. 1965. Industrial Press, New York.

Locklin, D.W. and H.R. Hazard. 1980. Technology for the development of high-efficiency oil-fired residential heating equipment. Brookhaven National Laboratory Report BNL 51325. Upton, NY.

McDonald, R.J. (editor). 1989. Proceedings of the 1989 oil heat technology conference and workshop. Brookhaven National Laboratory Report BNL 52217, Upton, NY.

NFPA. 1987. Prevention of furnace explosions in fuel oil- and natural gas-fired single burner-furnaces. NFPA *Standard* 85A. National Fire Protection Association, Qunicy, MA.

UL. 1973. Commercial-industrial gas-heating equipment. UL *Standard* 795-73. Underwriters Laboratories, Northbrook, IL.

UL. 1980. Oil burners. UL *Standard* 296, 7th ed.

UL. 1973. Oil-fired boiler assemblies. UL *Standard* 726, 3rd ed.

UL. 1980. Oil-fired central furnaces. UL *Standard* 727, 6th ed.

BOILERS

A boiler is a pressure vessel designed to transfer heat (produced by combustion) to a fluid. The definition has been expanded to include transfer of heat from electrical resistance elements to the fluid or by direct action of electrodes on the fluid. In most boilers, the fluid is usually water in the form of liquid or steam.

If the fluid being heated is air, the heat exchange device is called a furnace, not a boiler. The firebox, or combustion space, of some boilers is also called a furnace.

For the purposes of this chapter, excluding special and unusual fluids, materials, and methods, a boiler is a cast-iron, steel, or copper pressure vessel heat exchanger, designed with and for fuel-burning devices and other equipment to (1) burn fossil fuels (or use electric current) and (2) transfer the released heat to water (in water boilers) or to water and steam (in steam boilers). Boiler heating surface is the area of fluid-backed surface exposed to the products of combustion, or the fire-side surface. Various codes and standards define allowable heat transfer rates in terms of heating surface. Boiler design provides for connections to a piping system, which delivers heated fluid to the point of use and returns the cooled fluid to the boiler.

Chapters 10 and 14 cover most applications of heating boilers.

BOILER CLASSIFICATIONS

Boilers may be grouped into classes based on working pressure and temperature, fuel used, shape and size, usage (such as heating or process), and steam or water. Most classifications are of little importance to a specifying engineer, except as they affect performance. Excluding designed-to-order boilers, significant class descriptions are given in boiler catalogs or are available from the boiler manufacturer. The following basic classifications may be helpful.

Working Temperature/Pressure

With few exceptions, all boilers are constructed to meet the ASME *Boiler and Pressure Vessel Code*, Section 4, Rules for Construction of Heating Boilers (low-pressure boilers) and the ASME Code, Section 1, Rules for Construction of Power Boilers (medium- and high-pressure boilers).

Low-pressure boilers are constructed for maximum working pressures of 15 psi steam and up to 160 psi hot water. Hot-water boilers are limited to 250 °F operating temperature. The controls and relief valves, which limit temperature and pressure, are not part of the boiler but must be installed to protect it.

Medium- and high-pressure boilers are designed to operate at above 15 psi steam or above 160 psi or 250 °F for water boilers.

Steam boilers are available in standard sizes of up to 50,000 lb steam/h (60×10^3 to 50×10^6 Btu/h), many of which are used

for space heating in both new and existing systems. On larger installations, they may also provide steam for auxiliary uses, such as hot-water heat exchangers, absorption cooling, laundry, sterilizers, etc. In addition, many steam boilers provide steam (for which there is no viable substitute) at various temperatures and pressures for a wide variety of industrial processes.

Water boilers are available in standard sizes of up to 50×10^6 Btu/h (50 MBtu/h to 50,000 MBtu/h), many of which are in the low-pressure class and are used for space heating in both new and existing systems. Some water boilers may be equipped with either internal or external heat exchangers to supply domestic (service) hot water.

Every steam or water boiler is rated at the maximum working pressure determined by the ASME Boiler Code (or other code) under which it is constructed and tested. It also must be equipped, when installed, with safety controls and pressure-relief devices mandated by such code provisions.

Fuel Used

Boilers may be designed to burn coal, wood, various grades of fuel oil, various types of fuel gas, or to operate as electric boilers. A boiler designed for one specific fuel type may not be convertible to another type of fuel. Some boiler designs can be adapted to burn coal, oil, or gas. Several designs allow firing with oil or gas by burner conversion or by using a dual-fuel burner. The variations in boiler design according to fuel used are primarily of interest to boiler manufacturers, who can furnish details to a specifying engineer. The manufacturer is responsible for performance and rating according to the code or standard for the fuel used (see the section Performance Codes and Standards).

Materials of Construction

Other than special or unusual models, most boilers are made of cast iron or steel. Some small boilers are made of copper or copper-clad steel.

Cast-iron boilers are constructed of individually cast sections, assembled into blocks (assemblies) of sections. Push or screw nipples, gaskets, or an external header join the sections pressure-tight and provide passages for the water, steam, and products of combustion. The number of sections assembled determines boiler size and energy rating. Sections may be vertical or horizontal, the vertical design being the more common (Figures 1A and 1C).

The boiler may be *dry-base* (the firebox is beneath the fluid-backed sections) (Figure 1B); *wet-base* (the firebox is surrounded by fluid-backed sections, except for necessary openings) (Figure 2A); or *wet-leg* (the firebox top and sides are enclosed by fluid-backed sections) (Figure 2B).

At the manufacturer's option, the three boiler types can be designed to be equally efficient. Testing and rating standards apply equally to all three types. The wet-base design is easiest to adapt for combustible floor installations. Applicable codes usually

---

The preparation of this chapter is assigned to TC 6.1, Hydronic and Steam Equipment and Systems.

demand a floor temperature under the boiler no higher than 90 °F, plus room temperature. A wet-base steam boiler at 215 °F or a water boiler at 240 °F cannot meet this requirement without floor insulation. Large cast-iron boilers are also made as water-tube units with external headers (Figure 2C).

Steel boilers are fabricated into one assembly of a given size and rating, usually by welding. The heat-exchange surface past the firebox usually is an assembly of vertical, horizontal, or slanted tubes. The tubes may be *fire-tube*, with flue gas inside and heated fluid outside (Figures 1D and 1E show residential units, and Figures 3A through 3D and 4A show commercial units), or *water-tube*, with fluid inside (Figures 4C and 4D). The tubes may be in one or more passes. As with cast-iron boilers, dry-base, wet-leg, or wet-base design may be used. Most small steel heating boilers are of the dry-base, vertical fire-tube type (Figure 1D).

Larger boilers usually have horizontal or slanted tubes; both fire-tube and water-tube designs are used. A popular design for medium and large steel boilers is the *Scotch*, or *Scotch Marine*, which is characterized by a central fluid-backed cylindrical firebox, surrounded by fire-tubes in one or more passes, all within the outer shell (Figures 3A through 3D).

Copper boilers are usually some variation of the water-tube type. Parallel finned copper tube coils with headers, and serpentine copper tube units are most common (Figures 1F and 1G). Some are offered as wall-hung residential boilers. The commercial bent water-tube design is shown in Figure 4B. Natural gas is the usual fuel for copper boilers.

Cast-iron boilers range in size from 35×10^3 to 10×10^6 Btu/h gross output. Steel boilers range in size from 50×10^3 Btu/h to the largest boilers made.

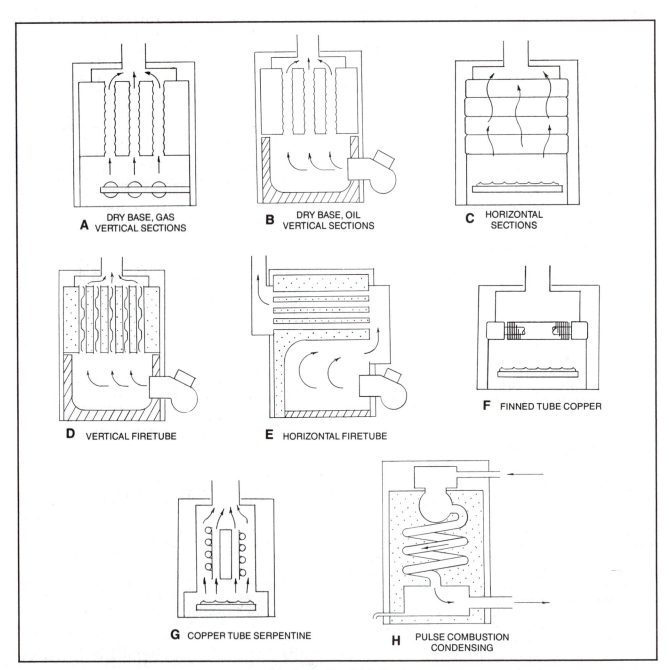

A DRY BASE, GAS VERTICAL SECTIONS

B DRY BASE, OIL VERTICAL SECTIONS

C HORIZONTAL SECTIONS

D VERTICAL FIRETUBE

E HORIZONTAL FIRETUBE

F FINNED TUBE COPPER

G COPPER TUBE SERPENTINE

H PULSE COMBUSTION CONDENSING

Fig. 1   Residential Boilers

Condensing or Noncondensing

Until recently, boilers were designed to operate without condensing flue gases in the boiler. This precaution was necessary to prevent corrosion of cast-iron or steel parts. Hot-water units were often operated at 140°F minimum return temperature to prevent rusting when natural gas was used.

Because higher boiler efficiencies can be achieved with lower water temperatures, condensing boilers purposely allow the flue gas water vapor in the boiler to condense and drain. Figure 5 shows a typical relationship of overall boiler efficiency to return water temperature. The dew point of 130°F shown varies with the percent of hydrogen in the fuel and the CO_2 or excess air in the flue gas. A pulse combustion condensing boiler is shown in Figure 1H.

Low return water temperatures and condensing boilers are particularly important because they are extremely efficient at part-load operation when high water temperatures are not required. For example, a hot-water heating system that operates under light load conditions at 80°F return water temperature has a potential overall boiler efficiency of 97% when operated with natural gas of the specifications shown in Figure 5.

Figure 6 shows how dew point varies with a change in CO_2 for natural gas. Boiler designs with combustion efficiencies and CO_2 values such that the flue gas temperatures fall between the dew point and the dew point +140°F should be avoided, unless the venting system is appropriate for condensing. With a typical CO_2 concentration in the 6 to 8% range, units with combustion efficiencies between 83 and 87% fall in this category. Chapter 31 gives further details.

The condensing portion of these boilers requires special materials to resist the corrosive effects of the condensate. Certain stainless steels are suitable, and small boilers can be fabricated of the appropriate material. Commercial boiler installations can be adapted to condensing operation by adding a condensing heat exchanger in the venting system.

The condensing medium can be (1) return heating system water, (2) service water, or (3) other water or fluid sources at temperatures in the 70 to 130°F range. The medium can also be a source of heat recovery in HVAC systems.

Electric Boilers

Electric boilers are in a separate class. Since no combustion occurs, no boiler heating surface and no flue openings are necessary. The heating surface is the surface of the electric elements or electrodes immersed in the boiler water. The design of electric boilers is largely determined by the shape and heat release rate of the electric heating elements used. Electric boiler manufacturers' literature describes available sizes, shapes, voltages, ratings, and methods of control.

Selection Parameters

Boiler selection should be based on a competent review of the following parameters:

All Boilers

- ASME (or other) Code, under which the boiler is constructed and tested
- Net boiler output capacity, Btu/h
- Total heat transfer surface, ft^2
- Water content, lb
- Auxiliary power requirements, kWh
- Internal water-flow patterns
- Cleaning provisions for all heat transfer surfaces
- Operational efficiency
- Space requirements and piping arrangement
- Water treatment requirements

Fuel-fired Boilers

- Combustion space (furnace volume), ft^3
- Internal flow patterns of combustion products
- Combustion air and venting requirements

Steam Boilers

- Steam space, ft^3
- Steam disengaging area, ft^2

The codes and standards outlined in the section Performance Codes and Standards include requirements for minimum efficiency, maximum temperature, burner operating characteristics, and safety control. For most purposes, test agency certification and labeling, published in boiler catalogs and shown on boiler rating plates, are sufficient. Some boilers not tested and rated by a recognized agency are produced and do not bear the label of any such agency. Nonrated boilers (rated and warranted only by the manufacturer) are used when codes do not demand a rating agency label. Almost without exception, both rated and nonrated boilers are of ASME Code construction and are marked accordingly.

EFFICIENCY: INPUT AND OUTPUT RATINGS

The efficiency of fuel-burning boilers is defined in two ways: combustion efficiency and overall efficiency. Overall efficiency of electric boilers is in the 92 to 96% range.

Combustion efficiency is input minus stack (chimney) loss, divided by input, and ranges from 75 to 86% for most noncondensing mechanically fired boilers. Condensing boilers operate in the range of 88 to over 95% efficiency.

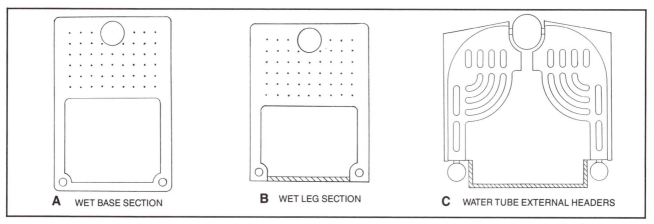

Fig. 2   Cast-Iron Commercial Boilers

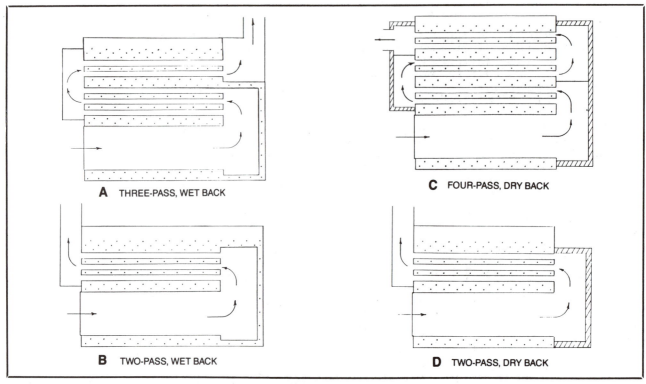

Fig. 3   Scotch-type Commercial Boilers

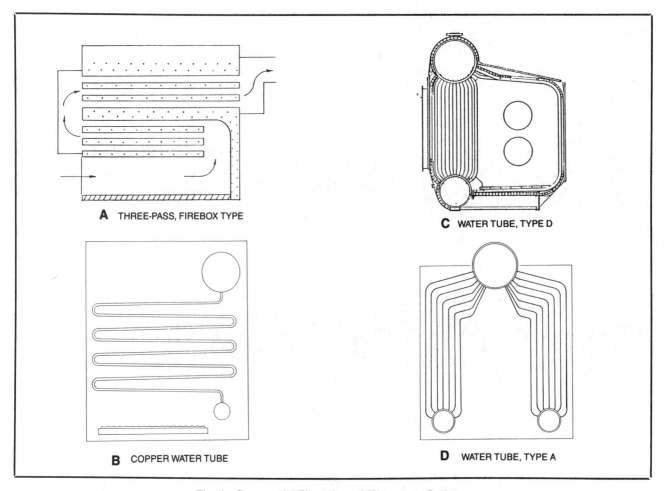

Fig. 4   Commercial Fire-tube and Water-tube Boilers

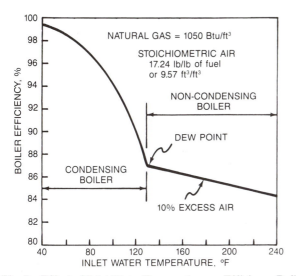

Fig. 5 Effect of Inlet Water Temperature on Efficiency Boilers

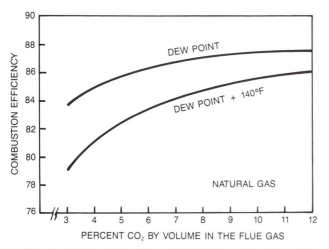

Fig. 6 Relationship of Dew Point, Carbon Dioxide, and Combustion Efficiency for Natural Gas

Overall efficiency is gross output divided by input. Gross output is measured in the steam or water leaving the boiler and depends on individual installation characteristics. Overall efficiency is lower than combustion efficiency by the percentage of heat lost from the outside surface of the boiler (this loss is usually known as radiation loss). Overall efficiency can be precisely determined only by laboratory tests under fixed test conditions. Approximate combustion efficiency can be determined under any operating condition by measuring operating flue-gas temperature and percentage CO_2 or O_2 and by consulting a chart or table for the fuel being used.

PERFORMANCE CODES AND STANDARDS

Heating boilers are usually rated according to standards developed by (1) the Hydronics Institute [formerly the Institute of Boiler and Radiator Manufacturers (I-B-R) and the Steel Boiler Institute (SBI)]; (2) the American Gas Association (AGA); and (3) the American Boiler Manufacturers Association (ABMA).

The Hydronics Institute has adopted a standard for rating cast-iron and steel heating boilers (Testing and Rating Standard for Heating Boilers) based on performance obtained under controlled test conditions. The gross output obtained by the test is limited by certain factors, such as flue gas temperature, draft, CO_2 in flue gas, and minimum overall efficiency. This standard applies primarily to oil-fired equipment but is also used for power gas ratings for dual-fueled units.

Gas boilers are design-certified by AGA based on tests conducted in accordance with the *American National Standard* Z21.13 (Gas Fired Low Pressure Steam and Hot Water Heating Boilers).

ABMA has adopted test procedures for commercial-industrial and packaged fire-tube boilers based on the ASME *Performance Test Code, Steam-Generating Equipment* (PTC 4.1). The units are tested for performance under controlled test conditions with minimum levels of efficiency required.

In 1978 the Department of Energy issued a test procedure applying to all gas- and oil-fired boilers up to 300,000 Btu/h input. This procedure supersedes industry rating methods used previously. The DOE test procedure determines both on-cycle and off-cycle losses based on a laboratory test involving cyclic conditions. The test results are applied to a computer program, which simulates an installation and results in an annual fuel utilization efficiency (AFUE). The steady-state efficiency developed during the test is similar to combustion efficiency described previously and is the basis for determining heating capacity, a term equivalent to gross output.

BOILER SIZING

Boiler sizing is the selection of boiler output capacity to meet connected load. The boiler *gross output* is the rate of heat delivered by the boiler to the system under continuous firing at rated input. *Net rating* (I-B-R rating) is gross output minus a fixed percentage to allow for an estimated average piping heat loss, plus an added load for initially heating up the water in a system (sometimes called *pickup*).

Piping loss is variable. If all piping is within the space defined as load, loss is zero. If piping runs through unheated spaces, heat loss from the piping may be much higher than accounted for by the fixed net rating factor. Pickup is also variable. When actual connected load is less than design load, the pickup factor may be unnecessary. On the design's coldest day, extra system output (boiler and radiation) is needed to pick up the load from a shutdown or low night setback. If night setback is not used, or if no extended shutdown occurs, no pickup load exists. Standby system input capacity for pickup, if needed, can be in the form of excess capacity in baseload boilers, or in a standby boiler.

If piping and pickup losses are negligible, the boiler gross output can be considered the design load. If piping loss and pickup load are large or variable, those loads should be calculated and equivalent gross boiler capacity added. Unless terminal units can deliver full boiler output under pickup conditions, at an inlet water temperature lower than boiler high limit setting, rated boiler capacity cannot be delivered. System input (boiler output) will be greater than system output, water temperature will rise, and the boiler will cycle on the high limit control, delivering an average input to the system that is much lower than the boiler gross output. Boiler capacity must be matched by terminal unit and system delivery capacity.

CONTROL OF BOILER INPUT AND OUTPUT

Boiler controls regulate the rate of fuel input in response to a control signal representing load change, so that average boiler output equals load within some accepted control tolerance. Boiler controls include safety controls that shut off fuel flow when unsafe conditions develop.

Operating Controls

Steam boilers are operated by boiler-mounted, pressure-actuated controls, which vary the input of fuel to the boiler. Common examples of controls are *on-off, high-low-off*, and *modulating*. Modulating controls infinitely vary the fuel input from 100% down to a selected minimum point. The ratio of maximum to minimum is the *turndown ratio*. The minimum input is usually between 5 and 25%, *i.e.*, 20 to 1, to 4 to 1, and depends on the size and type of fuel-burning apparatus.

Hot-water boilers are operated by temperature-actuated controls that are usually mounted on the boiler. Controls are the same as for steam boilers, *i.e.*, on-off, high-low-off, and modulating.

The boiler manufacturer should be consulted if low operating temperatures or frequent cold starts are anticipated. Thermal shock, caused by cold water returning to the boiler and impinging on the hot surfaces inside the boiler, should be avoided with steel fire-tube boilers. Many steel boiler manufacturers suggest special piping arrangements to prevent thermal shock.

Detailed descriptions of specific control systems for heating systems or for boilers are beyond the scope of this chapter. Basic control system designs and diagrams are furnished by boiler manufacturers. Control diagrams or specifications to meet special safety requirements are available from insurance agencies, governmental bodies, and ASME. Such special requirements usually apply to boilers with inputs of 400×10^3 Btu/h or greater but may also apply to smaller sizes. The boiler manufacturer may furnish installed and prewired boiler control systems, with diagrams, when the requirements are known.

When a boiler installation code applies, the details of control system requirements may be specified by the inspector after boiler installation. Special controls must then be supplied and the control design completed per the inspector's instructions. It is essential that the heating system designer or specifying engineer determine the applicable codes, establish sources of necessary controls, and provide the skills needed to complete the control system. Local organizations are often available to provide these services.

REFERENCES

ABMA. 1981. Packaged firetube ratings. American Boiler Manufacturers Association, Arlington, VA.

ANSI. 1989. Gas-fired low-pressure steam and hot water boilers. ANSI *Standard* Z21.13-87. American National Standards Institute, New York.

ASME. 1964. Steam generating units. *Performance Test Code* 4.1-64. Reaffirmed 1985. American Society of Mechanical Engineers, New York.

ASME. 1988. Controls and safety devices for automatically fired boilers. ASME *Standard* CSD-1-88.

ASME. 1989. Rules for construction of power boilers. *Boiler and Pressure Vessel Code*, Section 1-86.

ASME. 1989. Rules for construction of heating boilers. *Boiler and Pressure Vessel Code*, Section 4-86.

ASME. 1989. Recommended rules for care and operation of heating boilers. *Boiler and Pressure Vessel Code*, Section 6-86.

ASME. 1989. Recommended rules for care of power boilers. *Boiler and Pressure Vessel Code*, Section 7-86.

De Lorenzi, O. 1967. Combustion engineering: A reference book on fuel burning and steam generation. Combustion Engineering, Inc., New York.

Hydronics Institute. 1990. Testing and rating standard for heating boilers. Hydronics Institute, Berkeley Heights, NJ.

Strehlow, R.A. 1984. Combustion Fundamentals. McGraw-Hill, New York.

Woodruff, E.B., H.B. Lammers, and T.F. Lammers. 1984. Steam-plant operation, 5th ed. McGraw-Hill, New York.

CHAPTER 29

FURNACES

RESIDENTIAL FURNACES

RESIDENTIAL furnaces are available in a variety of self-enclosed appliances, which provide heated air through a system of ductwork into the space being heated. There are two types: (1) fuel-burning furnaces and (2) electric furnaces.

Fuel-burning furnaces. Combustion takes place within a combustion chamber. Circulating air passes over the outside surfaces of the heat exchanger so that it does not contact the fuel or the products of combustion, which are passed to the outside atmosphere through a vent.

Electric furnaces. A resistance-type heating element either heats the circulating air directly or through a metal sheath enclosing the resistance element.

Residential furnaces may be further categorized by (1) type of fuel, (2) mounting arrangement, (3) airflow direction, (4) combustion system, and (5) installation location.

NATURAL GAS FURNACES

Natural gas is the most common fuel supplied for residential heating, and the central system forced air furnace, such as that shown in Figure 1, is the most common way of heating with natural gas. This type of furnace is equipped with a blower to circulate air through the furnace enclosure, over the heat exchanger, and through the ductwork distribution system. A typical furnace consists of the following basic components: (1) a cabinet or casing; (2) heat exchangers; (3) combustion system including burners and controls; (4) forced draft, induced draft, or draft hood; (5) circulating air blower and motor; and (6) air filter and other accessories such as a humidifier, electronic air cleaner, air-conditioning coil, or a combination of these elements.

Casing or Cabinet

The furnace casing is most commonly formed from painted cold-rolled steel. Access panels on the front of the furnace allow access to those sections requiring service. The inside of the casing adjacent to the heat exchanger is lined with a foil-faced blanket insulation and/or a metal radiation shield to reduce heat losses through the casing and to limit the outside surface temperature of the furnace. On some furnaces, the inside of the blower compartment is lined with insulation to acoustically dampen the blower noise.

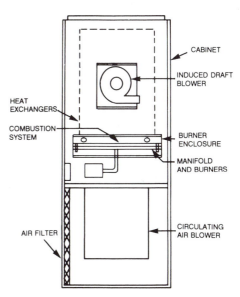

Fig. 1   Induced Draft Gas Furnace

Heat Exchangers

Heat exchangers are normally made of mirror-image formed parts and are joined together to form a *clam shell*. Heat exchangers made of finless tubes bent into a compact form are also found in some furnaces. Standard indoor furnaces are generally made of cold-rolled steel. If the furnace is exposed to clean air and the heat exchanger remains dry, this material has a long life and does not easily corrode. Some problems of heat exchanger corrosion and failure have been encountered because of exposure to halogen ions in the flue gas. These problems were caused by combustion air contaminated by substances such as laundry bleach, cleaning solvents, and halogenated hydrocarbon refrigerants.

Coated or alloy steel are used in top-of-the-line models and in furnaces for special applications. Common corrosion-resistant materials include aluminized steel, ceramic-coated cold-rolled steel, and stainless steel. Furnaces certified for use downstream of a cooling coil must have corrosion-resistant heat exchangers.

Research has been done on corrosion-resistant materials for use in condensing (secondary) heat exchangers (Stickford *et al.* 1985). The presence of chloride compounds in the condensate can cause the failure of a condensing heat exchanger, unless a corrosion-resistant material is used.

Several manufacturers produce liquid-to-air heat exchangers in which a liquid is heated and is either evaporated or pumped to a condenser section or fan coil, which heats circulating air.

The preparation of this chapter is assigned to TC 6.3, Central Forced Air Heating and Cooling Systems.

Burners and Internal Controls

Burners are most frequently made of stamped sheet metal, although cast iron is also used. Fabricated sheet metal burners may be made from cold-rolled steel coated with high-temperature paint or from a corrosion-resistant material such as stainless or aluminized steel. Burner material must meet the corrosion protection requirements of the specific application. Gas furnace burners may be either of the monoport or multiport types; the type used with a particular furnace depends on compatibility with the heat exchanger.

Furnace controls include an ignition device, gas valve, fan switch, limit switch, and other components specified by the manufacturer. These controls allow gas to flow to the burners when heat is required. The four most common ignition systems are (1) standing pilot, (2) intermittent pilot, (3) direct spark, and (4) hot surface ignition (ignites main burners directly). The Technical Data section has further details on the functions and performance of individual control components.

Venting Components

Atmospheric vent indoor furnaces are equipped with a draft hood connecting the heat exchanger flue gas exit to the vent pipe or chimney. The draft hood has a relief air opening large enough to ensure that the exit of the heat exchanger is always at atmospheric pressure. One purpose of the draft hood is to make certain that the natural draft furnace continues to operate safely without generating carbon monoxide if the chimney is blocked, if there is a downdraft, or if there is excessive updraft. Another purpose is to maintain constant pressure on the combustion system. Residential furnaces built since 1987 are equipped with a blocked vent shutoff switch to shut down the furnace in case the vent becomes blocked. Direct vent furnaces do not use a draft hood of this type; in this case, a control system shuts the furnace down if the chimney or vent becomes blocked.

Power vent furnaces use a small blower to force or induce the flue products through the furnace. These furnaces may or may not have a relief air opening, but both furnaces meet the same safety requirements.

Research into common venting of atmospheric draft appliances (water heaters) and power vent furnaces shows that nonpositive vent pressure systems may operate on a common vent. Refer to manufacturers' instructions for specific information. A venturi-aspirator approach is being developed for positive vent pressure systems (Deppish and DeWerth 1986).

ANSI *Standard* Z21.47A-1990 classifies venting systems. Central furnaces are categorized by temperatures and pressures attained in the vent as follows:

Category I—A central furnace that operates with a nonpositive vent pressure and with a vent gas temperature at least 140 °F above its dew point.

Category II—A central furnace that operates with a nonpositive vent pressure and with a vent gas temperature less than 140 °F above its dew point.

Category III—A central furnace that operates with a positive vent pressure and with a vent gas temperature at least 140 °F above its dew point.

Category IV—A central furnace that operates with a positive vent pressure and with a vent gas temperature less than 140 °F above its dew point.

Furnaces rated in accordance with ANSI *Standard* Z21.47 are marked to show that they are in one of the four venting categories listed here.

Blowers and Motors

Centrifugal blowers with forward-curved blades of the double-inlet type are used in most forced air furnaces. These blowers over-

come the resistance of the furnace air passageways, filters, and ductwork. They are usually sized to provide the additional air requirement for cooling and the static pressure required for the cooling coil. The blower may be a direct-drive type, with the blower wheel attached directly to the motor shaft, or it may be a belt-drive type, with a pulley and V-belt used to drive the blower wheel.

Electric motors used to drive furnace blowers are usually designed for that purpose. Direct-drive motors may be of the shaded pole or permanent split-capacitor type. Speed variation may be obtained by taps connected to extra windings in the motor. Belt-drive blower motors are normally split-phase or capacitor-start. The speed of belt-drive blowers is controlled by adjusting a variable-pitch drive pulley. Electronically controlled, variable-speed motors are also available.

Air Filters

An air filter in a forced air furnace removes dust from the air that could reduce the effectiveness of the blower and heat exchanger(s). Filters supplied with a forced air furnace are often disposable. Permanent filters that may be washed or vacuum cleaned and reinstalled are also used. The filter is always located in the circulating airstream ahead of the blower and heat exchanger.

Accessories

Humidifiers. These are not included as a standard part of the furnace package. However, one advantage of a forced air heating system is that it offers the opportunity to control the relative humidity of the heated space at a comfortable level. Chapter 20 addresses various types of humidifiers used with forced air furnaces.

Electronic air cleaners. These air cleaners are much more effective than the air filter provided with the furnace, and they filter out much finer particles, including smoke and pollen. Electronic air cleaners create an electric field of high-voltage direct current in which dust particles are given a charge and collected on a plate having the opposite charge. The collected material is then cleaned periodically from the collector plate by the homeowner. Electronic air cleaners are mounted in the airstream entering the furnace. Chapter 25 has detailed information on filters.

Automatic vent dampers. This device closes the vent opening on a draft hood-equipped atmospheric furnace when the furnace is not in use, thus reducing off-cycle losses. More information about the energy-saving potential of this accessory is included in the Technical Data section.

Airflow Variations

The components of a gas-fired, forced air furnace can be arranged in a variety of configurations to suit a residential heating system. The relative positions of the components in the different types of furnaces are as follows:

- **Upflow or "highboy" furnace.** As shown in Figure 2, the upflow furnace has the blower beneath the heat exchanger discharging vertically upward. Air enters through the bottom or the side of the blower compartment and leaves at the top. This furnace may be used in closets and utility rooms on the first floor or in basements, with the return air ducted down to the blower compartment entrance.

- **Downflow furnace.** In a downflow furnace (Figure 3), the blower is located above the heat exchanger discharging downward. Air enters at the top and is discharged vertically at the bottom. This furnace is normally used with a perimeter heating system in a house without a basement. It is also used in upstairs furnace closets and utility rooms supplying conditioned air to both levels of a two-story house.

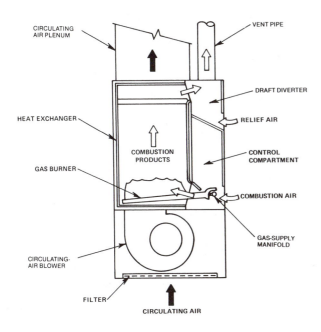

Fig. 2   Upflow Category I Furnace with Draft Hood

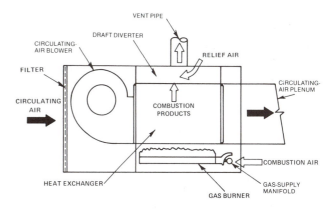

Fig. 4   Horizontal Category I Furnace with Draft Hood

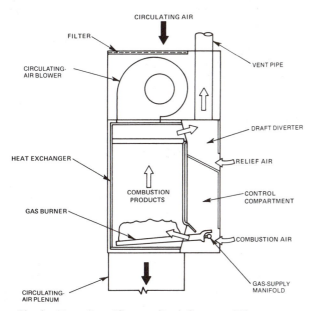

**Fig. 3   Downflow (Counterflow) Category I Furnace
with Draft Hood**

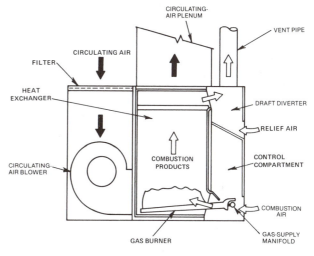

**Fig. 5   Basement (Lowboy) Category I Furnace
with Draft Hood**

• **Horizontal furnace.** The blower is located beside the heat exchanger in a horizontal furnace (Figure 4). The air enters at one end, travels horizontally through the blower, over the heat exchanger, and is discharged at the opposite end. This furnace is used for locations with limited head room such as attics and crawl spaces, or is suspended under a roof or placed above a suspended ceiling. These units are often designed so that the components may be rearranged to allow installation with airflow from left to right or from right to left.

• **Basement or "lowboy" furnace.** The basement furnace (Figure 5) is a variation of the upflow furnace and requires less head room. The blower is located beside the heat exchanger at the bottom. Air enters the top of the cabinet, is drawn down through the blower, discharged over the heat exchanger, and leaves vertically at the top.

• **Gravity furnace.** While these furnaces are available, they are not common. This furnace has larger air passages through the casing and over the heat exchanger so that buoyancy force created by the air being warmed circulates the air through the ducts. Gravity wall furnaces are discussed in Chapter 30.

Combustion System Variations

Gas-fired furnaces use an atmospheric vent or a power vent combustion system. With an atmospheric vent furnace, the buoyancy of the hot combustion products carries these products through the heat exchanger, into the draft hood, and up the chimney.

Power vent furnaces have a combustion blower, which may be located either upstream or downstream from the heat exchangers (Figure 6). If the blower is located upstream, blowing the combustion air into the heat exchangers, the system is known as a forced draft system. If the blower is downstream, the arrangement is known as an induced draft system. Power combustion systems have generally been used with outdoor furnaces; however, with the passage of the 1987 National Appliance Energy Conservation Act, power combustion is becoming more common for indoor furnaces as well. Power combustion furnaces do not require a draft hood, resulting in reduced off-cycle losses and improved efficiency.

Direct vent furnaces may be either natural draft or fan-assisted combustion systems. They do not have a draft hood and obtain combustion air from outside the structure. Mobile home furnaces must be of the direct vent type.

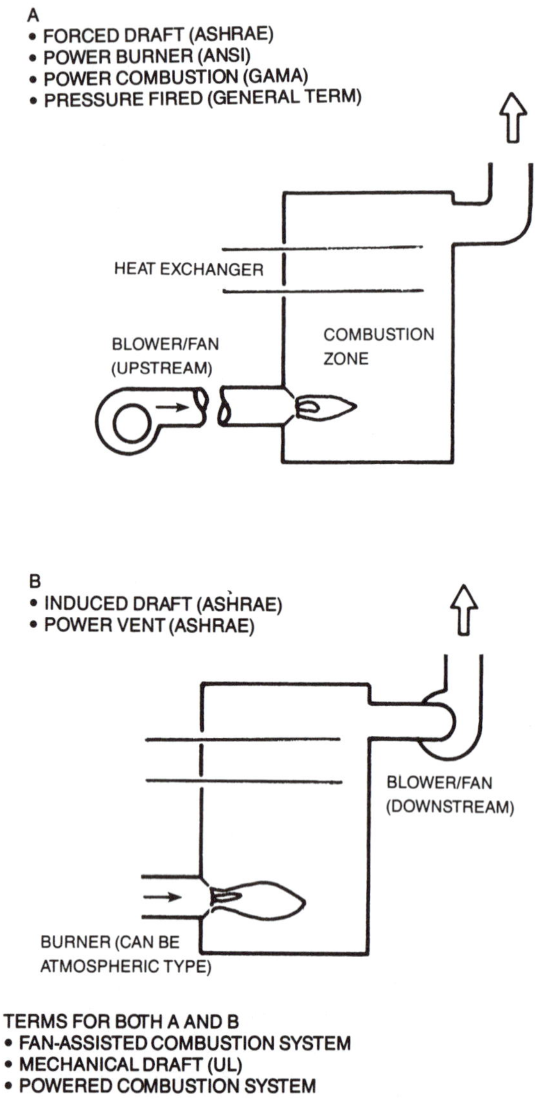

A
- FORCED DRAFT (ASHRAE)
- POWER BURNER (ANSI)
- POWER COMBUSTION (GAMA)
- PRESSURE FIRED (GENERAL TERM)

HEAT EXCHANGER

BLOWER/FAN
(UPSTREAM)

COMBUSTION
ZONE

B
- INDUCED DRAFT (ASHRAE)
- POWER VENT (ASHRAE)

BLOWER/FAN
(DOWNSTREAM)

BURNER (CAN BE
ATMOSPHERIC TYPE)

TERMS FOR BOTH A AND B
- FAN-ASSISTED COMBUSTION SYSTEM
- MECHANICAL DRAFT (UL)
- POWERED COMBUSTION SYSTEM

**Fig. 6   Terminology Used to Describe
Fan-Assisted Combustion**

Indoor-Outdoor Furnace Variations

Central system residential furnaces are designed and certified for either indoor or outdoor use. Outdoor furnaces are normally horizontal flow.

The heating-only outdoor furnace is similar to the more common indoor horizontal furnace. The primary difference is that the outdoor furnace is weatherized. The motors and controls are sealed, and the exposed components are made of corrosion-resistant materials such as galvanized or aluminized steel.

A common style of outdoor furnace is the combination package unit. This unit is a combination of an air conditioner and a gas or electric furnace built into a single casing. The design varies, but the most common combination consists of an electric air conditioner coupled with a horizontal gas or electric furnace. The advantage is that much of the interconnecting piping and wiring is included in the unit.

LPG FURNACES

Most manufacturers have their furnaces certified for both natural gas and liquefied petroleum gas (LPG). The major differ-

ence between the two furnaces is the pressure at which the gas is injected from the manifold into the burners. For natural gas, the manifold pressure is usually controlled at 3 to 4 in. of water; for LPG, the pressure is usually 10 to 11 in. of water.

Because of higher injection pressure and the greater heat content per volume of LPG, there are certain physical differences between a natural gas furnace and an LPG furnace. One difference is that the pilot and burner orifices must be smaller for LPG furnaces. The gas valve regulator spring is also different. Sometimes it is necessary to change burners, but this is not required normally. Manufacturers sell conversion kits containing both the required parts and instructions to convert furnace operation from one gas to the other.

OIL FURNACES

Indoor oil furnaces come in the same configuration as gas furnaces. They are available in upflow, downflow, horizontal, and lowboy configurations for ducted systems. Oil-fired outdoor furnaces and combination units are not common.

The major differences between oil and gas furnaces are in the combustion system, heat exchanger, and the barometric draft regulator used in lieu of a draft hood. The ducted system, oil-fired, forced air furnaces are usually forced draft and equipped with pressure-atomizing burners. The pump pressure and the orifice size of the injection nozzle regulate the firing rate of the furnace. Electric ignition lights the burners. Other furnace controls, such as the blower switch and the limit switch, are similar to those used on gas furnaces.

The heat exchangers of oil-fired furnaces are normally heavy-gage steel formed into a welded assembly. The hot flue products flow through the inside of the heat exchanger into the chimney. The conditioned air flows over the outside of the heat exchanger and into the air supply plenum.

ELECTRIC FURNACES

Electric-powered furnaces come in a variety of configurations and have some similarities to gas- and oil-fired furnaces. However, when used with an air conditioner, the cooling coil may be upstream from the blower and heaters. On gas- and oil-fired furnaces, the cooling coil is normally mounted downstream from the blower and heat exchangers.

Figure 7 shows a typical arrangement for an electric forced air furnace. Air enters the bottom of the furnace and passes through the filter, then flows up through the cooling coil section into the blower. The electric heating elements are immediately above the blower so that the high-velocity air discharging from the blower passes directly through the heating elements.

The furnace casing, air filter, and blower are similar to equivalent gas furnace components. The heating elements are made in modular form, with 5 kW capacity being typical for each module. Electric furnace controls include electric overload protection, contactor, limit switches, and a fan switch. The overload protection may be either fuses or circuit breakers. The contactor brings the electric heat modules on. The fan switch and limit switch functions are similar to those of the gas furnace, but one limit switch is usually used for each heating element.

Frequently, electric furnaces are made from modular sections; for example, the coil box, blower section, and electric heat section are made separately and then assembled in the field. Regardless of whether the furnace is made from a single-piece casing or a modular casing, it is generally a multiposition unit. Thus, the same unit may be used for either upflow, downflow, or horizontal installations.

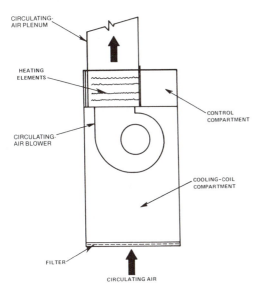

Fig. 7   Electric Forced Air Furnace

When the electric heating appliance is sold without a cooling coil, it is known as an electric furnace. The same appliance is called a fan-coil air handler when it has an air-conditioning coil already installed. When the unit is used as the indoor section of a split heat pump, it is called a heat pump fan-coil air handler. For detailed information on heat pumps, see Chapter 46.

Electric forced air furnaces are also used with packaged heat pumps and packaged air conditioners.

SYSTEM DESIGN AND EQUIPMENT SELECTION

Warm Air Furnaces

Two steps are required in selecting a warm air furnace: (1) determining the required heating capacity of the furnace and (2) selecting a specific furnace to satisfy this requirement.

Heating capacity. The heating capacity of a warm air furnace depends on several variables that operate alone or in combination. The first variable is the design heating requirement of the residence. The heat loss of the structure can be calculated using the procedures outlined in the 1989 ASHRAE *Handbook—Fundamentals*.

The additional heating required if the furnace is operating on a night setback cycle should also be considered. During the morning recovery period, additional capacity is required to bring the conditioned space temperature up to the desired level. The magnitude of this recovery capacity depends on weather conditions, the magnitude of the night setback, and the time allowed for the furnace to return room air temperature to the desired level. Another consideration similar to night setback concerns structures that only require intermittent heating, such as churches and auditoriums. Chapter 13 of the 1991 ASHRAE *Handbook—HVAC Applications* has further information.

A third variable is the influence of internal loads. Normally, the heat gain from internal loads is neglected when selecting a furnace, but if the internal loads are constant, they should be used to reduce the required capacity of the furnace, especially in nonresidential applications.

The energy required for humidification is a fourth variable. The humidification energy depends on the desired level of relative humidity and the rate at which the moisture must be supplied to maintain the specified level. Net moisture requirements must take into account internal gains as a result of people, equipment, and appliances, and losses through migration in exterior surfaces, plus air infiltration. Chapter 20 gives details on how to determine humidification requirements.

A fifth variable is the influence of off-peak storage devices. When used in conjunction with a furnace, a storage device decreases the required capacity of the furnace. The storage device can supply the additional capacity required during the morning recovery of a night setback cycle or reduce the daily peak loads to assist in load shedding. Detailed calculations can determine the contribution of storage devices.

The sixth variable is a check on the furnace's capacity to accommodate air conditioning, even if air conditioning is not planned initially. The cabinet should be large enough to accept a cooling coil that satisfies the cooling load. The blower and motor should have sufficient capacity to provide increased airflow rates typically required in air-conditioning applications. Chapter 9 includes specific design considerations.

Specific furnace selection. The second step in the selection of a warm air furnace is to choose a specific furnace that satisfies the required design capacity. The final decision depends on numerous parameters, the most significant of which is the fuel type. The second step of the furnace selection process is subdivided by fuel types.

Natural Gas Furnaces

Size selection. Historically, furnaces have been oversized because (1) the calculation procedure was not exact, especially the estimate of air infiltration; (2) weather conditions are occasionally more severe than the design conditions used to calculate the required furnace capacity; (3) the additional first cost of a slightly larger furnace was considered a good value in view of possible undersizing, which would be expensive to correct; and (4) adequate airflow for cooling was another consideration. Natural gas was relatively inexpensive so that possible inefficiencies due to oversizing were not considered detrimental. The net result led to significant oversizing.

Oversizing may increase overall energy use for new houses where vents and ducts are sized to furnace capacity. However, in retrofits (where fixed vent and duct sizes are assumed), oversizing has little effect on overall energy use. In either situation, oversizing may reduce the comfort level due to wide temperature variations in the conditioned space. In a retrofit, if a higher efficiency furnace is selected, the output capacity must match the original equipment's output. Otherwise, additional furnace oversizing results.

Chapter 25 of the 1989 ASHRAE *Handbook—Fundamentals* recommends oversizing new installations by 40%, if a 10°F night setback is prescribed, to obtain a 1-h recovery.

Performance criteria. Performance criteria or a consistent definition of efficiency must be used throughout. Some typical efficiencies encountered are (1) steady-state efficiency, (2) utilization efficiency, (3) annual fuel utilization efficiency.

These efficiencies are generally used by the furnace industry in the following manner:

- *Steady-state efficiency.* This is the efficiency of a furnace when it is operated under equilibrium conditions based on ASHRAE/ANSI *Standard* 103-88. It is calculated by measuring the energy input, subtracting the losses for exhaust gases and flue gas condensate (for condensing furnaces only), and then dividing by the fuel input (cabinet loss not included), i.e.,

$$SS\,(\%) = \frac{\text{Fuel Input} - \text{Flue Loss} - \text{Condensate Loss}}{\text{Fuel Input}} \times 100$$

For furnaces tested under the isolated combustion system (ICS) method, cabinet heat loss must also be deducted from the energy input.

$$SS (\%)(ICS) = 100 \text{ (Fuel Input } - \text{ Flue Loss}$$
$$- \text{ Condensate Loss } - \text{ Jacket Loss)/Fuel Input}$$

- *Utilization efficiency.* This is obtained from an empirical equation developed by the National Institute of Standards and Technology (NBSIR 78-1543) with 100% efficiency and deducting losses for exhausted latent and sensible heat, cyclic effects, infiltration, and pilot burner effect.
- *Annual fuel utilization efficiency* (AFUE). This value is the same as utilization efficiency, except that losses from a standing pilot during the nonheating season are deducted. This equation can also be found in NBSIR 78-1543 or ANSI/ASHRAE *Standard* 103-1988 and is used in the Federal Trade Commission's (FTC) *Energy Guide Fact Sheets.*

The AFUE is determined for residential fan-type furnaces by using ANSI/ASHRAE *Standard* 103-1988, Method of Testing for Annual Fuel Utilization Efficiency of Residential Central Furnaces and Boilers. The test procedure is also presented in the Code of Federal Regulations Title II, 10 Part 430, Appendix N, in conjunction with the amendments issued by the U.S. Department of Energy in the March 28, 1984, *Federal Register.* This version of the test method allows the rating of nonweatherized furnaces as indoor, isolated combustion systems (ICS), or both. Weatherized furnaces are rated as outdoor.

Federal law requires manufacturers of furnaces to use AFUE as determined using the isolated combustion system method, to rate efficiency. Effective January 1, 1992, all furnaces produced must meet a minimum AFUE (ICS) level of 78%. Table 1 gives efficiency values for different furnaces.

Annual energy savings may be compared with the following formula:

$$\text{Annual energy reduction (AER)} = \frac{AFUE_2 - AFUE_1}{AFUE_2}$$

where $AFUE_2$ is greater than $AFUE_1$. For example, compare items 3 and 2 of Table 1.

$$AER = \frac{78 - 69}{78} = \frac{9}{78} = 11.5\%$$

The SP43 work (see System Performance section) confirms that this is a reasonable estimate.

Table 1 Typical Values of AFUE
(Furnaces Located in Conditioned Space)

Type of Gas Furnace	AFUE Indoor, %	ICS*
1. Atmospheric with standing pilot	64.5	63.9
2. Atmospheric with intermittent ignition	69.0	68.5
3. Atmospheric with intermittent ignition and auto vent damper	78.0	68.5
4. Same basic furnace as 2, except with power combustion	80.0	78.0
5. Same as 4, except with improved heat transfer	82.0	80.0
6. Direct vent with standing pilot, preheat	66.0	64.5
7. Direct vent, power vent, and intermittent ignition	80.0	78.0
8. Power burner (induced draft)	80.0	78.0
9. Condensing	93.0	91.0

Type of Oil Furnace	Indoor, %	ICS*
1. Standard	71.0	69.0
2. Same as 1, with improved heat transfer	76.0	74.0
3. Same as 2, with auto vent damper	83.0	74.0
4. Condensing	91.0	89.0

*Isolated combustion system

Construction features and limitations. Many indoor furnaces have cold-rolled steel heat exchangers. If the furnace is exposed to clean air and the heat exchanger remains dry, this material has a long life and does not corrode easily. Many deluxe, noncondensing furnaces have a coated heat exchanger to provide extra protection against corrosion. Research by Stickford *et al.* (1985) indicates that chloride compounds in the condensate of condensing furnaces can cause the heat exchanger to fail unless it is made of specialty steel. A corrosion-resistant heat exchanger must also be used in a furnace certified for use downstream of a cooling coil.

Design life. Typically, the heat exchangers made of cold-rolled steel have a design life of approximately 15 years. Special coated or alloy heat exchangers, when used for standard applications, increase the design life to as much as 20 years. Coated or alloy heat exchangers are recommended for furnace applications in corrosive atmospheres.

Sound level. This variable must be considered in most applications. Chapter 36 of the 1991 ASHRAE *Handbook—HVAC Applications* outlines the procedures to follow in determining acceptable noise levels.

Safety. Because of open-flame combustion, the following safety items need to be considered: (1) the surrounding atmosphere should be free of dust or chemical concentrations; (2) a path for combustion air must be provided for both sealed and open combustion chambers; and (3) the gas piping and vent pipes must be installed according to the *National Fuel Gas Code* (ANSI Z223.1), local codes, and the manufacturer's instructions.

Applications. Gas furnaces are primarily applied to residential heating. The majority are used in single-family dwellings but are also applicable to apartments, condominiums, and mobile homes.

Performance versus cost. These factors must be considered in selection. Included in life cycle cost determination are initial cost, maintenance, energy consumption, design life, and the price escalation of the fuel. Procedures for establishing operating costs for use in product labeling and audits are available from the Department of Energy. For residential furnaces, fact sheets provided by manufacturers are available at the point of sale.

Other Fuels

The system design and selection criteria for LPG furnaces are identical to those for natural gas furnaces.

The design criteria for oil furnaces are also identical to those for natural gas furnaces, except that oil-fired furnaces equipped with pressure-atomizing or rotary burners should be rated in accordance with ANSI Z91.1 test methods. This requires a minimum steady-state efficiency of 75% for forced air furnaces.

Electric Furnaces

The design criteria for electric furnaces are identical to those for natural gas furnaces. The selection criteria are similar, except that an electric furnace does not have the flue losses and combustion air losses of a gas furnace. For this reason, the seasonal efficiency is usually equal to 100% when installed indoors.

The design life of electric furnaces is related to the durability of the contactors and the heating elements. The typical design life is approximately 15 years.

Safety primarily considers proper wiring techniques. Wiring should comply with the *National Electrical Code* (NEC) (ANSI/NFPA 70) and applicable local codes.

TECHNICAL DATA

Detailed technical data on furnaces are available from manufacturers, wholesalers, and dealers. The data are generally tabulated in product specification bulletins printed by the manufacturer for each furnace line. These bulletins usually include performance information, electrical data, blower and air delivery data, control system information, optional equipment information, and dimensional data.

Natural Gas Furnaces

Capacity ratings. ANSI *Standard* Z21.47 requires the heating capacity to be marked on the rating plate of commercial furnaces. The heating capacity of residential furnaces, less than 225,000 Btu/h input, is shown on the furnace fact sheet required by the FTC. Capacity is calculated by multiplying the input by the steady-state efficiency.

Residential gas furnaces are readily available with heating capacities ranging from 35,000 to 175,000 Btu/h. Some smaller furnaces are manufactured for special-purpose installations such as mobile homes. Smaller capacity furnaces are becoming common because new homes are better insulated than older homes and have a lower heat load. Larger furnaces are also available, but these are generally considered for commercial use.

Because of the overwhelming popularity of the upflow furnace, it is available in the greatest number of models and sizes. Downflow, horizontal, and combination furnaces are all readily available but are generally limited in model type and size.

Residential gas furnaces are available as heating-only and as heat-cool models. The difference is that the heat-cool model is designed to operate as the air-handling section of a split-system air conditioner. The heating-only models typically operate with enough airflow to allow a 60 to 100 °F air temperature rise through the furnace. This rise provides good comfort conditions for the heating system with a low noise level blower and low electrical energy consumption. Condensing furnaces may be designed for a lower temperature rise (as low as 35 °F).

The heat-cool furnace models have multispeed blowers with a more powerful motor capable of delivering about 400 cfm per ton of air conditioning. Models are generally available in 2, 3, 4, and 5-ton sizes, but all cooling sizes are not available for every furnace size. For example, a 60,000 Btu/h furnace would be available in models with blowers capable of handling 2 or 3 tons of air conditioning; 120,000 Btu/h models would be matched to 4 or 5 tons of air conditioning. In addition to the blower differences, controls of the heat-cool furnace models are generally designed to operate the multispeed blower motor at the most appropriate speed for either heating or cooling operation when airflow requirements vary for each mode. This feature provides optimum comfort conditions for year-round system operation.

Efficiency ratings. Currently, gas furnaces have steady-state efficiencies that vary from 75 to 95%. Atmospheric and power vent furnaces typically range from 75 to 82% efficiency, while condensing furnaces have over 90% steady-state efficiency. Koenig (1978), Gable and Koenig (1977), Hise and Holman (1977), and Bonne *et al.* (1977) found that oversizing residential gas furnaces with standing pilots reduced the seasonal efficiency of heating systems in new installations with vents and ducts sized according to furnace capacity.

The AFUE of a furnace may be improved by ways other than changing the steady-state efficiency. These improvements generally add more components to the furnace. One method replaces the standing pilot with an intermittent ignition device. Gable and Koenig (1977) and Bonne *et al.* (1976) indicate that this feature can save as much as 5.6×10^6 Btu/year per furnace. For this reason, some jurisdictions require the use of intermittent ignition devices.

Another method of improving AFUE is to take all combustion air from outside the heated space (direct vent) and preheat it. A combustion air preheater incorporated into the vent system draws combustion air through an outer pipe that surrounds the flue pipe. Such systems have been used on mobile home and outdoor furnaces. Annual energy consumption of a direct vent furnace with combustion air preheat may be as much as 9% less than a stan-

dard furnace of the same design (Bonne *et al.* 1976). Direct vent without combustion air preheat is not inherently more efficient, because the reduction in combustion induced infiltration is offset by the use of colder combustion air.

An automatic vent damper (thermal or electromechanical) is another device that saves energy on indoor furnaces. This device, which is placed after the draft hood outlet, closes the vent when the furnace is not in operation. It saves energy during the off cycle of the furnace by (1) reducing exfiltration from the house and (2) trapping residual heat from the heat exchanger within the house rather than allowing it to flow up the chimney. These savings approach 11% under ideal conditions where combustion air is taken from the heated space that is under thermostat control. However, these savings are much less (estimates vary from 0 to 4%) if combustion air is taken from outside the heated space. The isolated combustion system method of determining AFUE gives no credit to vent dampers installed on indoor furnaces since the ICS method assumes the use of outdoor combustion air with the furnace installed in an unconditioned space.

The AFUE of power draft and power vent furnaces is higher than for standard natural draft furnaces. These furnaces normally have such a high internal flow resistance that combustion airflow stops when the combustion blower is off. This characteristic results in greater energy savings than those from a vent damper. Computer studies by Gable and Koenig (1977), Bonne *et al.* (1976), and Chi (1977) have estimated annual energy savings up to 16% for power vent furnaces with electric ignition as compared to natural draft furnaces with standing pilot.

Controls. Externally, the furnace is controlled by a low-voltage room thermostat. Controls can be heating only, combination heating-cooling, multistage, or night setback. Chapter 35 of the 1991 ASHRAE *Handbook—HVAC Applications* addresses thermostats in more detail. A night setback thermostat can reduce the annual energy consumption of a gas furnace. Dual setback (setting the temperature back during the night and during unoccupied periods in the day) can save even more energy. Gable and Koenig (1977) and Nelson and MacArthur (1978) estimate that energy savings of up to 30% are possible, depending on the degree and length of setback and the geographical location. The percentage of energy saving is greater in regions with mild climates; however, the total energy saving is greatest in cold regions.

Several types of gas valves provide various operating functions within the furnace. The type of valve available relates closely to the type of ignition device used. Two-stage valves, available on some furnaces, operate at full gas input or at a reduced rate and are primarily controlled with a two-stage thermostat. They provide less heat and, therefore, less temperature variation and greater comfort during mild weather conditions when full heat output is not required. Fuel savings with two-stage firing rate systems may not be realized unless both the gas and combustion air are controlled.

The fan switch controls the circulating air blower. This switch may be temperature-sensitive and exposed to the circulating airstream within the furnace cabinet, or it may be an electrically operated relay. Blower startup is typically delayed about 1 min after the startup of the burners. This delay gives the heat exchangers time to warm up and eliminates the excessive flow of cold air when the blower comes on. Blower shutdown is also delayed several minutes after burner shutdown to remove residual heat from the heat exchangers and to improve the annual efficiency of the furnace. Constant blower operation throughout the heating season was encouraged to improve air circulation and provide even temperature distribution throughout the house. However, constant blower operation increases electrical energy consumption and overall operating cost in many instances. Electronic motors that provide continuous but variable airflow use less energy. Both strategies may be considered when air filtering performance is important.

The limit switch prevents overheating in the event of severe reduction to circulating airflow. This temperature-sensitive switch is exposed to the circulating airstream and shuts off the gas valve if the temperature of the air leaving the furnace is excessive. The fan and limit switches are often incorporated in the same housing and are sometimes operated by the same thermostatic element. The blocked vent shutoff switch and flame rollout switch have been required on residential furnaces produced since November 1989; they shut off the gas valve if the vent is blocked or when insufficient combustion air is present.

Furnaces using fan-assisted combustion feature a pressure switch to verify the flow of combustion air prior to the opening of the gas valve.

LPG Furnaces

Most residential natural gas furnaces are also available in a liquefied petroleum gas version with identical ratings. The technical data for these two furnaces are identical, except for the gas controls and the burner and pilot orifice sizes. Orifice sizes on LPG furnaces are much smaller, because LPG has a higher density and may be supplied at a higher manifold pressure. The heating value and specific gravity of typical gases are listed as follows:

Gas Type	Heating Value, Btu/ft³	Specific Gravity Air = 1.0
Natural	1030	0.60
Propane	2500	1.53
Butane	3175	2.00

Gas controls may be different because of the higher manifold pressure and the requirement on indoor furnaces for pilot gas shutoff if pilot ignition should fail. Pilot gas leakage is more critical with propane or butane gas because both are heavier than air and can accumulate to create an explosive mixture within the furnace or furnace enclosure.

Since 1978, ANSI *Standard* Z21.47 requires a gas pressure regulator as part of the LPG furnace. Prior to that, the pressure regulator was supplied only with the LP supply system.

Beside natural and LPG, a furnace may be certified for manufactured gas, mixed gas, or LPG-air mixtures; however, furnaces with these certifications are not commonly available. Mobile home furnaces are certified as convertible from natural gas to LPG.

Oil Furnaces

Oil furnaces are similar to gas furnaces in size, shape, and function, but the heat exhanger, burner, and combustion control systems are significantly different.

Input ratings are based on the oil flow rate (gal/h), and the heating capacity is calculated by the same method as that for gas furnaces. The heating value of oil is 140,000 Btu/gal. Fewer models and sizes are available for oil as are available for gas, but residential furnaces in the 64,000 to 150,000 Btu/h heating capacity are common. Air delivery ratings are similar to gas furnaces, and both heating-only and heat-cool models are available.

The efficiency of an oil furnace can drop during normal operation if the burner is not maintained and kept clean. In this case, the oil does not atomize sufficiently to allow complete combustion, and energy is lost up the chimney in the form of unburned hydrocarbons. Because most oil furnaces use power burners and electric ignition systems, the annual efficiency is relatively high.

Oil furnaces are available in upflow, downflow, and horizontal-flow models. The thermostat, fan switch, and limit switch are similar to those of a gas furnace. Oil flow is controlled by a pump and burner nozzle, which sprays the oil-air mixture into a single-chamber drum-type heat exchanger. The heat exchangers are normally heavy-gage cold-rolled steel. Humidifiers, electronic air cleaners, and night setback thermostats are available as accessories.

Electric Furnaces

Residential electric resistance furnaces are available in heating capacities of 5 to 35 kW. Air-handling capabilities are selected to provide sufficient air to meet the requirement of an air conditioner of reasonable size to match the furnace. Small furnaces supply about 800 cfm and large furnaces about 2000 cfm.

The only loss associated with an electric resistance furnace is in the cabinet—about 2% of input. Both the steady-state efficiency and the annual efficiency of an electric furnace is greater than 98%, and if the furnace is located within the heated space, the seasonal efficiency is 100%.

Although the efficiency of an electric furnace is high, electricity is a relatively expensive form of energy. The operating cost may be reduced substantially by using an electric heat pump in place of a straight electric resistance furnace. Heat pumps are discussed in Chapter 8.

Humidifiers and electronic air cleaners are available as accessories for both electric resistance furnaces and heat pumps.

Conventional setback thermostats are recommended to save energy. The electric demand used to recover from the setback, however, may be quite significant. Bullock (1977) and Schade (1978) addressed this problem by using (1) a conventional two-stage setback thermostat with staged supplemental electric heat or (2) solid-state thermostats, with programmed logic to inhibit supplemental electric heat from operating during morning recovery. Benton (1983) reported energy savings of up to 30% for these controls, although the recovery time may be extended up to several hours.

Electric furnaces are available in upflow, downflow, or horizontal models. Internal controls include overload fuses or circuit breakers, overheat limit switches, a fan switch, and a contactor to bring on the heating elements at timed intervals.

SYSTEM PERFORMANCE

Both the performance and the interaction of the furnace and the distribution system with the building determine how much fuel energy input to the furnace beneficially heats the conditioned space. System performance is defined by the space in which the performance applies. Conditioned space is defined as that space whose temperature is actively controlled by a thermostat. A building can contain other space (an attic, basement, or crawl space) that can influence the thermal performance of the conditioned space, but it is not defined as part of the conditioned space.

For houses with basements, it is important to decide whether the basement is part of the conditioned space, since it typically receives some fraction of the HVAC system output. The basement is part of the conditioned space only if it is under active thermostat control and warm air registers are provided to maintain comfort; otherwise it is not part of the conditioned space. The following system performance examples show designs for improving system efficiency, along with their effect on temperature in the unconditioned basement.

SP43 Dynamic Simulation Model

The dynamic response and interactions between components of central forced warm air systems are sufficiently complex that the effects of system options on annual fuel use are not easily evaluated. ASHRAE Special Project 43 (SP43) was initiated in 1982 to assess the effects of system component and control mode options. The resulting simulation model accounts for the dynamic and thermal interactions of equipment and loads in response to varying weather patterns. Figure 8 summarizes the energy flow paths in the SP43 model.

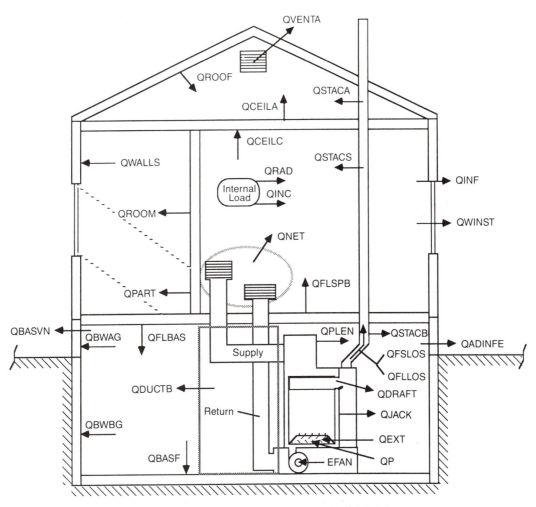

Fig. 8   Summary of Energy Flow Paths in SP43 Model

Fisher *et al.* (1984) describe the SP43 simulation model. Jakob, Fisher *et al.* (1986), Herold, Flanigan *et al.* (1987), and Jakob *et al.* (1987) describe the validation model for the heating mode through field experiments in two houses. Herold *et al.* (1986) summarize both the project and the model.

Jakob, Locklin *et al.* (1986) and Locklin *et al.* (1987) presented the model's predictions of the effect of system and control variables on the overall performance of the forced warm air heating system. These variables include furnace and venting types, furnace installation location and combustion air source, furnace sizing, night setback, thermostat cycling rate, blower operating strategy, basement insulation levels, duct sealing and insulation levels, house and foundation type, and climate.

SYSTEM PERFORMANCE FACTORS

A series of system performance factors, consisting of both efficiency factors and dimensionless energy factors, describe dynamic performance of the individual components and the overall system over any period of interest. Jakob, Locklin *et al.* (1986) and Locklin, Herold *et al.* (1987) describe the factors in detail.

Table 2 identifies the performance factors and their mathematical definitions in four main categories: (1) equipment-component efficiency factors, (2) equipment-system performance factors, (3) equipment-load interaction factors, and (4) energy cost factors. Key aspects of the factors are summarized in the following sections.

Equipment-Component Efficiency Factors

Furnace efficiency, E_F. This factor is the ratio of the energy that is delivered to the plenum during cyclic operation of the furnace to the total input energy on an annual basis. E_F includes summertime pilot losses and blower energy.

This factor is similar to the AFUE in ASHRAE and DOE standards (ASHRAE 1982, DOE 1978), which provides an estimate of annual energy, taking into account assumed system dynamics. However, E_F differs from AFUE in the following ways:

- The AFUE for a given furnace is defined by a single predetermined cyclic condition with standard dynamics; E_F is based on the integrated cyclic performance over the year.
- The AFUE does not include auxiliary electric input and it gives credit for jacket losses, except when the furnace is an outdoor unit; E_F and the other efficiency factors defined here include auxiliary electric input.
- Effects of system-induced infiltration are handled differently in the two concepts; they are covered in the factors for equipment/load interaction in the SP43 model, whereas AFUE includes an adjustment factor to account for these effects.
- The AFUE rating is a fixed value for a given furnace model and is independent of the system or dynamics over a full heating season.

Duct efficiency, E_D. This is the ratio of the energy intentionally delivered to the conditioned space through the supply registers to the energy delivered to the furnace plenum, on an annual basis.

Table 2   Definitions of System Performance Factors—Annual Basis

Equipment-Component efficiency factors

E_F = Furnace Efficiency = $100 \dfrac{\text{Furnace Output}}{\text{Total Energy Input}} = 100 \dfrac{\text{Duct Input}}{\text{Total Energy Input}}$

E_D = Duct Efficiency = Duct Output / Duct Input

Equipment-System performance factors

E_{HD} = Heat Delivery Efficiency = $\dfrac{E_F \times E_D}{100} = 100 \dfrac{\text{Duct Output}}{\text{Total Energy Input}}$

F_{MG} = Miscellaneous Gain Factor = $\dfrac{\text{Total Heat Delivered}^a}{\text{Duct Output}}$

E_S = System Efficiency = $E_{HD} \times F_{MG} = 100 \dfrac{\text{Total Heat Delivered}}{\text{Total Energy Input}}$

Equipment/Load-interaction factors

F_{IL} = Induced Load Factor[b] = $\dfrac{\text{System Induced Load}^c}{\text{Total Heat Delivered}}$

F_{LM} = Load Modification Factor = $1.0 - F_{IL} = \dfrac{\text{Total Heat Delivered} - \text{System Induced Load}}{\text{Total Heat Delivered}}$

I_S = System Index = $E_S \times F_{LM} = \dfrac{\text{Total Heat Delivered} - \text{System Induced Load}}{\text{Total Energy Input}}$

Energy cost factors

R_{AE} = Auxiliary Energy Ratio = $\dfrac{\text{Auxiliary Energy Input}}{\text{Primary Energy Input}} = \dfrac{\text{Electrical Energy Input}}{\text{Fuel Energy Input}}$

R_{CL} = Local Energy Cost Ratio = $\dfrac{\text{Electrical Cost Per Energy Unit}}{\text{Reference Fuel Cost Per Energy Unit}}$ (in common units)

F_{CR} = Cost Ratio Factor = $\dfrac{\text{Primary Energy Input} + \text{Auxiliary Energy Input}}{\text{Primary Energy Input} + R_{CL}\,(\text{Auxiliary Energy Input})} = \dfrac{1.0 + R_{AE}}{1 + R_{CL} \times R_{AE}}$

        Special Case (Electric Furnace): $F_{CR} = 1/R_{CL}$

I_{SCM} = Cost Modified System Index = $I_S \times F_{CR} = \dfrac{\text{Net Load} - \text{System Induced Load}}{\text{Primary Energy Input} + R_{CL}\,(\text{Auxiliary Energy Input})}$

Annual energy use

AEU = Annual Energy Use (fuel and electricity) predicted by SP43 model, in total energy units

Annual Fuel Used = AEU / $(1.0 + R_{AE})$

Annual Electricity Used = AEU / $(1.0 + 1/R_{AE})$

Percent savings

% Energy Savings = $100\,(I_S - (I_S)_{BC} / I_S$, where $(I_S)_{BC}$ = the I_S for the base case

% Cost Savings = $100\,[I_{SCM} - (I_{SCM})_{BC}] / I_{SCM}$, where $(I_{SCM})_{BC}$ = the cost modified I_S for both cases

Other factors for dynamic performance

AFUE = Annual Fuel Utilization Efficiency by DOE/ASHRAE efficiency rating, applicable to specific furnaces. Values in this chapter are for generic furnaces.

SSE = Steady-State Efficiency value for a given furnace by ANSI test procedure

---

Note: Energy inputs and outputs are integrated over an annual period. Efficiencies (E) are expressed as percents. Indexes (I), factors (F), and ratios (R) are expressed as fractions.

[a]The *Total Heat Delivered* is the integration over time of all the energy supplied to the conditioned space by the HVAC equipment. By definition, it is exactly equal to the space-heating load.

[b] The *Induced Load Factor* may be positive or negative, depending on the value of the load relative to the selected base case.

[c]The *System Induced Load* is the difference between the space-heating load for a particular case and the space-heating load for the base case. For the base case, the System Induced Load is, by definition, zero.

Equipment-System Performance Factors

Heat delivery efficiency, E_{HD}. This is the product of the furnace efficiency E_F and duct efficiency E_D. Heat delivery efficiency is the ratio of the energy intentionally delivered to the conditioned space to the total input energy. It is a measure of how effectively the HVAC system delivers heat directly to the conditioned space on an annual basis.

Miscellaneous gain factor, F_{MG}. The miscellaneous gain factor is the ratio of the total heat delivered to the conditioned space divided by the energy intentionally delivered to the conditioned space through the duct registers. This factor accounts for unintentional, but beneficial, heat that directly reaches the conditioned space from equipment components such as the vent or ducts. Jacket heat loss is also included as a miscellaneous gain if the furnace equipment is located within the conditioned space. For example, under certain circumstances, the conditioned space may lose heat to the vent during the off-cycle. This loss would be included in F_{MG}. The fact that the net heat delivered to the conditioned space by the heating system may be less than the heat delivered to the conditioned space through the duct registers implies that F_{MG} can be less than one.

System efficiency, E_S. This is the product of E_{HD} and F_{MG}. System efficiency indicates how much of the HVAC system's energy input is delivered to the conditioned space. It is the ratio of the total energy delivered to the conditioned space divided by the total energy input to the furnace. Thus, E_S includes intentional and unintentional energy gains.

Equipment/Load-Interaction Factors

Load modification factor, F_{LM}. The load modification factor is the ratio of the total heat delivered minus the system induced load to the total heat delivered to the conditioned space for the case of interest. The system induced load is the difference between the space-heating load for a particular case and the space-heating load for a base case. Note that the total heat delivered *equals* the space-heating load. The load modification factor F_{LM} may also be considered as the ratio of the total heat delivered for a base case to the total heat delivered for a case of interest. F_{LM} adjusts the system efficiency E_S to account for equipment operation on the heating load. It accounts for the effect of the combustion-induced infiltration and off-period infiltration due to draft hood flow, as well as the effects of changes in temperature of unconditioned spaces adjacent to the conditioned space.

F_{LM} credits the system efficiency for a heating load reduction, or debits the system efficiency if system operation increases the heating load with respect to the base case. F_{LM} is greater than 1.0 (a credit) if the specific conditioned space load is less than the base case load, as for a system with a vent damper that reduces off-cycle infiltration. F_{LM} is less than 1.0 (a debit) if the conditioned space load is greater than the base case load, such as occurs when additional infiltration is induced for combustion air during burner operation and for vent dilution during off periods.

Miscellaneous heat gains. Losses from the heating systems that go directly to the conditioned space are considered miscellaneous heat gains. For example, jacket losses become a miscellaneous gain if the equipment is located in the conditioned space.

Load-equipment interactions. These interactions can increase or decrease the load. Examples include a vent damper, which can reduce off-cycle infiltration and, therefore, the unconditioned space load, or additional infiltration that is induced for combustion air during burner operation and for ventilation during off periods. Heating system losses to unconditioned spaces like the basement, which subsequently affect that space either by positive heat flow or by reducing heat loss, are also considered load-equipment interactions.

Duct losses. The magnitude of duct air leakage and convective heat losses to (1) the conditioned space, (2) to spaces inside the structure but outside the conditioned space, such as the basement or attic, or (3) to the outside has a substantial effect on system performance.

System index, I_S. The system index is the total heat delivered minus the system induced load divided by the total energy input. It is the product of the system efficiency E_S and the load modification factor F_{LM}. I_s is an energy-based figure of merit that adjusts the system efficiency E_S for any credits or debits due to system induced loads relative to a base case load. The system index I_S is a powerful tool for comparing alternative systems. However, high values of I_S are sometimes associated with lower basement temperatures caused by designs that deliver a greater portion of the furnace output directly to the conditioned space. The ratio of the system indexes for two systems being compared is the inverse of the ratio of their annual energy use A_{EU}.

Energy Cost Factors

Table 2 also defines factors that can modify the system index I_S to account for relative costs of fuel and electrical energy and form a cost modified system index I_{SCM}. The auxiliary energy ratio A_{ER} is the ratio of electrical energy input to fuel energy input. The local energy cost ratio R_{CL} is the ratio of electrical and fuel energy costs in common units of energy. (A value of $R_{CL} = 4$ is used in the comparisons presented here.)

Key Implications

The following important implications apply to the definitions for the various system performance factors.

- The defined conditioned space is important to the comparisons of system index I_S. Because the load modification factor F_{LM} and the system index I_S are based on the same reference equipment and house configuration, the performance of various furnaces installed in basements or furnaces installed within the conditioned space (*i.e.*, closet installations) may be compared. However, performance of furnaces installed in basements or crawl spaces cannot be compared with heating systems installed only within the conditioned space.

  Since the I_S depends on a reference equipment and house configuration, it may only be used as a ranking index from which the relative benefits of different system features can be derived. That is, it can be used to compare the savings of various system features in specific applications to a base case.

- The miscellaneous gain factor F_{MG} includes only those heating system losses that go *directly* to the conditioned space.

- The equipment-system performance factors relate strictly to the subject equipment whereas the equipment/load-interaction factors draw comparisons between the subject equipment and an explicitly defined base case. This base case is a specific load and equipment configuration to which all alternative systems are compared.

Systems with the best total energy economy have the highest I_S. Systems with leaky and uninsulated ducts could have a higher system efficiency, even though fuel use would be higher, if basement duct losses that become gains to the conditioned space are included in the miscellaneous gain factor F_{MG}. The foregoing definitions prevent this possibility.

System Performance Examples

This section summarizes two phases of ASHRAE SP43, which evaluates the annual energy savings attained by certain forced warm air heating systems. The study was limited to certain house configurations and climates and to gas-fired equipment. Electric heat pump and zoned baseboard systems were not studied. For this reason, these data should not be used to compare systems or select a heating fuel. Several of the factors addressed are not references to performance; instead, they are figures of merit, which represent

Table 3   System Performance Examples

	Base Case	Alternative Case
Performance Factor	**Typical Conventional, Natural Draft Furnace with IID**	**Typical Noncondensing Power Vent Furnace**
DOE/ASHRAE AFUE (indoor) per DOE rules, %	69	81.5
Furnace efficiency E_F, %	75.5	85.5
Duct efficiency E_D, %	60.9	59.3
Heat delivery efficiency E_{HD}, %	46.0	50.7
Miscellaneous gain factor F_{MG}	1.004	0.983
System efficiency E_S, %	46.1	49.8
Load modification factor F_{LM}	1.000 (base case)	1.099
System Index I_S	0.461	0.548
Annual energy use, 10^6 Btu	73.0	61.5
Auxiliary energy ratio	0.027	0.028
Energy savings from base case, %	—	15.8
Cost savings from base case, % (with RCL = 4)	—	15.6

Notes:

1. The factors in this table should not be extrapolated to cases other than those covered in this chapter. That is, electric heat pump, radiant heating, or zoned baseboard systems were not studied in the SP43 project and should not be compared by this model.

2. The values presented here do not represent only this class of equipment; electric furnaces and heat pumps in a similar installation and under similar conditions would incur similar losses. The system index for any central air system can be improved, in comparison to the above examples, by insulating the ducts or locating the ductwork inside the conditioned space or both.

the effect of various components on a system. As such, these factors should not be applied outside the scope of these examples.

The following examples of overall system thermal performance illustrate how the furnace, vent, duct system, and building can interact. Table 3 summarizes SP43 simulation model predictions of the annual system performance for a base case (a conventional, natural-draft gas furnace with an intermittent ignition device) and an example case (noncondensing power vent furnace). Each is installed in a typical three-bedroom, ranch-style house of frame construction, located in Pittsburgh, Pennsylvania. Table 4 shows the assumptions for the thermal envelope of the house, and Table 5 shows other assumptions for these predictions. Table 6 lists energy flow descriptions and predicted energy values.

Table 7 shows the predicted annual system performance factors for the different furnaces installed in the same house in the same climate.

Table 4   Assumptions for Thermal Envelope of House in Pittsburgh

	U-Values of Thermal Envelope, $Btu/h \cdot ft^2 \cdot °F$
Per HUD-MPS 1980	
Attic	0.03
Walls	0.07
Windows	0.69
Sliding glass doors	0.69
Storm doors	None
Floors[a]	0.07
Basement conditions	
Basement ceiling insulation	None
Basement wall insulation	0.13
Design heat loss[b]	47,000 Btu/h

[a]Recommended value of floor insulation over an unheated space.

[b]Calculated by method described in ASHRAE GRP158, *Cooling and Heating Load Calculation Manual*, for a 1400 ft² ranch-style house with basement. Outdoor design temperature = 5°F.

Base case. The annual furnace efficiency E_F of the conventional natural draft furnace is predicted to be 75.5%. Air leakage and heat loss from the uninsulated duct system result in a duct efficiency E_D of 60.9%. The heat delivery efficiency E_{HD}, which is the ratio of the energy intentionally delivered to the conditioned space through the supply registers to the total input energy, is 46.0%. The miscellaneous gain factor F_{MG}, which is 1.004, accounts for the small heat gain to the conditioned space from the heated masonry chimney passing through the conditioned space. The system efficiency E_S (the ratio of the total heat delivered or the space-heating load) to the total energy input is 46.1%. Because this case is designated as the base case, the load modification factor F_{LM} is 1.0. Thus, the system index I_S is 0.461.

The duct losses and jacket losses are accounted for in the energy balance on the basement air and in the energy flow between the basement and the conditioned space. The increase in infiltration caused by the need for combustion air and vent dilution air is also accounted for in the energy balances on the living space air and basement air. In the base case, the temperature in the unconditioned basement is nearly the same (68°F) as in the first floor where the thermostat is located. This condition is caused by heat loss of the exposed ducts in the basement and by the low outdoor infiltration into the basement that was achieved by sealing normal cracks associated with construction.

Example case. The furnace efficiency E_F for the noncondensing power vent furnace being compared is 85.5%. The duct efficiency E_D is 59.3%, slightly lower than that for the duct system with the conventional natural draft furnace. Therefore, the heat delivery efficiency E_{HD} is 50.7%, which reflects the higher furnace efficiency. For the noncondensing power vent furnace, the miscellaneous gain factor F_{MG} is 0.983, reflecting the small heat loss from the conditioned space to the colder masonry chimney (due to reduced off-cycle vent flow). The system efficiency E_S, i.e., the heat delivery efficiency E_{HD} times the miscellaneous gain factor F_{MG}, is 49.8%. For this furnace system, compared to the base case system of the conventional, natural draft furnace, the

Table 5   Other Assumptions for Simulation Predictions

	Base Case
Furnace, adjustments, and controls	
Furnace size	1.4 DHL
Circulating air temperature rise	60°F
Thermostat set point	68°F
Thermostat cycling rate at 50% on-time	6 cycles/h
Night setback, 8 h	None
Blower control	
On	80 s
Off	90°F
Duct-related factors	
Insulation	None
Leakage, relative to duct flow	10%
Location	Basement
Load-related factors	
Nominal infiltration[a]	
Conditioned space	0.75 ACH
Basement	0.25 ACH
Occupancy, persons	3 during evening and night, 1 during day
Internal loads	Typical appliances, day and evening only (20 kWh/day)
Shading by adjacent trees or houses	None

[a]Model runs used variable infiltration, as driven by indoor-outdoor temperature differences, wind, and burner operation. Values shown above are nominal.

ACH = Air changes per hour

DHL = Design heat loss

Table 6   Energy Flow Descriptions and Values

Energy Flow Name	Description	Conventional, Natural Draft Furnace (Base Case)	Noncondensing Power Vent Furnace (Example Alternative Case)
		Energy Flow, 10^3 Btu	
QWALLS	Sum of convective heat flows from conditioned space to the exterior walls	8,333	8,396
QCEILC	Convective heat flow from the conditioned space to the ceiling	4,587	4,688
QFLSPB	Convective heat flow from conditioned space floor to conditioned space air	3,833	3,366
QROOM	Convective heat flow from shaded section of partition to conditioned space	1,239	1,164
QPART	Convective heat flow from unshaded section of partition to conditioned space	3,726	3,651
QWINST	Sum of solar energy transmitted through windows and absorbed by surroundings and heat conducted through window glazing	12,470	12,529
QINF	Infiltration heat flow from the conditioned space to the environment	25,196	21,356
QINC	Convective portion of internal heat gain	8,092	8,092
QSTACS	Direct heat flow from furnace vent to the conditioned space	118	−530
QNET	Net heat delivered to conditioned space through duct registers (supply-return)	33,588	31,209
QRAD	Radiative portion of internal heat gain	8,092	8,092
QBWAG	Convective heat flow from the basement air to the aboveground section of the basement wall	5,022	4,991
QBWBG	Convective heat flow from the basement air to the below-ground section of the basement wall	5,597	5,535
GBASF	Convective heat flow from basement air to basement floor	8	60
QFLBAS	Convective heat flow from conditioned space floor to basement air	8,206	8,756
QBASVN	Infiltration heat flow from basement to environment	6,561	6,512
QDUCTB	Total direct heat loss from duct system to basement (leakage + convection)	20,790	20,605
QSTACB	Direct heat flow from furnace vent to basement	2,627	1,300
QPLEN	Convective heat flow from plenum surface to basement air	781	812
QADINFE	Induced infiltration heat flow for combustion makeup air from basement to environment	498	39
QJACK	Convective heat flow from the furnace jacket to the basement air	393	410
QFSLOS	Sensible heat loss to furnace vent	11,483	3,469
QFLLOS	Latent heat loss to furnace vent	7,045	5,926
QEXT	Primary energy input to furnace	71,133	59,841
EFAN	Fan electrical consumption	1,912	1,655
QSTACA	Direct heat loss from furnace vent to attic air	807	250
QCEILA	Convective heat flow from the attic floor (conditioned space ceiling) to the attic air	2,120	2,323
QROOF	Convective heat flow from roof to attic air	122	111
QAVENT	Infiltration heat flow from attic to environment	2,796	2,676
QP	Pilot energy consumption	0	0
QDRAFT	Heat lost by spillage from draft hood into basement	0	0
FO1	Furnace output = Enthalphy change across furnace	55,182	52,605
FO2	Furnace output = Total energy input − losses FO2 = QEXT + QP + EFAN − QFSLOS − QFLLOS	54,517	52,101
FO3	Furnace output = Net heat delivered through registers + Duct loss + Plenum heat loss FO3 = QNET + QDUCTB + QPLEN	55,159	52,625
PEI	Primary energy input = QEXT + QP	71,133	59,841
AEI	Auxiliary energy input = EFAN	1,912	1,655
TEI	Total energy input = PEI + AEI	73,045	61,496
AEU	Annual energy use = TEI	73,045	61,496
DI	Duct input = Furnace output = FO1	55,182	52,605
DO	Duct output = QNET	33,588	31,209
THD	Total heat delivered = QNET + QSTACS	33,706	30,679
RNL	Reference net load = Total heat delivered for the base case	33,706	33,706
SIL	System induced load = THD − RNL	0	−3,028

Table 7   Performance Factor Values

Factor	Description	Conventional, Natural Draft Furnace (Base Case)	Noncondensing Power Vent Furnace (Example Alternative Case)
E_F	Furnace efficiency	75.5%	85.5%
E_D	Duct efficiency	60.9%	59.3%
E_{HD}	Heat delivery efficiency	46.0%	50.7%
F_{MG}	Miscellaneous gain factor	1.004	0.983
E_S	System efficiency	46.1%	49.9%
F_{LM}	Load modification factor	1.0	1.099
I_S	System index	0.461	0.548
R_{AE}	Auxiliary energy ratio	0.0269	0.0277

load modification factor F_{LM} is 1.099. Therefore, the space-heating load for the house with the noncondensing power vent furnace is 1/1.099 or 91% of the space heating load for the house with the conventional, natural draft furnace. This reduction in heating load is mainly due to the reduction in off-cycle vent flow.

For the noncondensing power vent furnace system, the system index I_S is 0.548. Note that the ratio of system indexes for these two cases (0.548/0.461 = 1.189) is the inverse of the ratio of their annual energy use AEU (61.5 × 10^6/ 73 × 10^6 = 0.842).

Table 6 lists the energy flows shown on Figure 8 for the two furnace systems. Table 3 also contains the combinations of these energy flows. Table 2 refines these energy flow combinations. Table 7 shows how the values of the performance factors listed in Table 3 are calculated from the energy flows.

Effect of Furnace Types

Table 8 summarizes the energy effects of several furnaces. Note that the system indexes I_S for both thermal and electric vent dampers are similar, although thermal vent dampers are slower reacting and less effective at blocking the vent. Also, the ratio of furnace AFUE to the base case AFUE closely matches the ratio of I_S to $(I_S)_{BC}$ for the corresponding furnaces. The exceptions are the vent damper cases, where the improvement in I_S suggests a smaller AFUE credit. In general, the study found that the furnace AFUE is a good indication of relative annual performance of furnaces in typical systems.

The results reported in Table 8 are for homes that do not include the basement in the conditioned space; that is, energy lost to the basement contributes only indirectly to the useful heating effect. If the basement is assumed to be a part of the conditioned space, the miscellaneous gain factor and subsequent calculated efficiencies and indexes should be multiplied by 1.66 to account for the beneficial effects of equipment (furnace jacket and duct system) heat losses that contribute to heating the basement. The 1.66 multiplier is the reciprocal of the electric furnace system efficiency E_S as shown in the following table, assuming that all electric heat is recovered for useful heating of the raw expanded conditioned space.

The following table shows values of system efficiency E_S and system index I_S that could be inferred by including the basement as a conditioned space.

	Conditioned Space Excludes Basement		Conditioned Space Includes Basement	
	E_S	I_S	E_S	I_S
Base case	46.0	0.461	76.4	0.765
Direct vent gas	59.2	0.622	98.3	1.033
Electric furnace	60.3	0.651	100.0	1.081

The reported values in Table 8 adhere to the definitions presented earlier. Other equipment types must be analyzed on the same common basis to draw similar conclusions. For instance, an electric baseboard heating system would have performance factors that are essentially the same as those for the electric furnace located in the conditioned space.

Effect of Climate and Night Setback

Table 9 covers the effects of climate (insulation levels change by location) and night setback on system performance for two furnaces. The improvements in system index I_S with higher percent on-time (colder climates) follow improvements in duct efficiency. Furnace efficiency E_F appears to be relatively uniform in houses representative of typical construction practices in each city and where the furnace is sized at 1.4 times the design heat loss. Also, the percentage of savings due to night setback increases in magnitude with warmer climates. The percent energy saved in the three climates varies with the magnitude of energy use (from 10 to 16% for the natural draft cases).

Effect of Furnace Sizing

Furnace sizing affects the system index I_S depending on how the vent duct system is designed. As Table 10 indicates, where the ducts and vent are sized according to the furnace size (referred to as the *new* case), performance drops about 10%, as the furnace capacity is varied between 1.0 times DHL and 2.5 times DHL for a given application. In the retrofit case, where the vent and duct system are sized at a furnace capacity of 1.4 times the design heat loss, the system index I_S changes little with increased furnace capacity, indicating little energy savings. In a new case, where the duct system is designed for cooling and the vent size does not change between furnace capacities, the SP43 study indicates that there is essentially no effect on I_S.

Table 8 Effect of Furnace Type on Annual Heating Performance

Furnace Characterization Typical Values,[a] AFUE/SS		Predicted by SP43 Model									
		Annual Performance Factors									
		E_F	E_D	F_{MG}	F_{LM}	I_S	I_{SCM} (R_{CL} = 4)	I_{SCM}[b] $(I_{SCM})_{BC}$	AEU, 10^6 Btu	Auxiliary Energy Ratio	Average Basement, °F
Conventional, natural draft											
Pilot	64.5/77	72.9	60.9	1.006	1.000	0.447	0.416	0.971	75.5	0.026	67.9
Intermittent ignition device (Base case)[c]	69/77	75.5	60.9	1.004	1.000	0.461	0.428	1.000	73.0	0.027	67.8
IID + Thermal vent damper	78/77	75.4	61.0	1.002	1.086	0.501	0.464	1.085	67.3	0.027	68.2
IID + Electric vent damper	78/77	75.4	61.2	0.988	1.105	0.504	0.467	1.091	66.9	0.027	68.3
Power vent types											
Noncondensing	81.5/82.5	85.5	59.3	0.983	1.099	0.548	0.507	1.185	61.5	0.028	67.6
Condensing[d]	92.5/93.1	95.5	62.0	1.000	1.050	0.622	0.566	1.322	54.2	0.034	66.9
Electric furnace	na	99.5	60.6	1.000	1.079	0.651	0.163	0.380	51.8	0.020[e]	67.1

Notes: The values in this table are figures of merit to be considered within the confines of the SP43 project, and they should not be applied outside the scope of these examples.

[a]AFUE = Annual Fuel Utilization Efficiency by ASHRAE/DOE standard
    SS = Steady-state efficiency by ANSI standard

[b]R_{CL} = 4.0
[c]Ranch-type house with basement in Pittsburgh, PA, climate and base conditions of 60 °F circulating air temperature rise, 6 cycles/h, no setback, 10% duct air leakage.
[d]Direct vent uses outdoor air for combustion (includes preheat).
[e]Blower energy is treated as auxiliary energy.

Table 9 Effect of Climate and Night Setback on Annual Heating Performance

Furnace Type and Location	Setback, °F	Avg. % On-Time	Avg. Bsmt., °F	Furnace Eff. E_F, %	Duct Eff. E_D, %	System Index, I_S	AEU[a] 10^6 Btu	% Energy Saved by Setback
Conventional natural draft								
Nashville	0	12.7	64.6	73.9	56.2	0.417	55.6	
	10	10.7	63.1	74.8	58.9	0.497	46.7	16.0
Pittsburgh (Base city)	0	18.0	67.8	75.5	60.9	0.461[b]	73.0	
	10	15.6	65.7	75.9	62.4	0.532	63.4	13.2
Minneapolis	0	20.9	68.0	76.8	63.2	0.487	99.1	
	10	18.7	65.7	77.0	62.4	0.546	88.4	10.7
Direct, Condensing								
Nashville	0	11.1	63.9	94.0	57.0	0.564	41.0	
	10	9.5	62.6	93.7	59.5	0.662	35.0	14.8
Pittsburgh	0	16.2	67.8	93.3	61.7	0.612	55.1	
	10	14.2	65.0	93.0	63.2	0.700	48.1	12.6
Minneapolis	0	18.2	66.8	95.1	64.5	0.654	73.7	
	10	16.4	64.7	94.8	65.6	0.729	66.2	10.2

Notes: Ranch-type house, basement, and base conditions: 60°F circulating air temperature rise, 6 cycles/h, 10% duct air leakage. Thermal envelope typical of each city; for example, no basement insulation in Nashville, TN.

The values in this table are figures of merit to be considered within the confines of the SP43 project, and they should not be applied outside the scope of these examples.

[a] AEU = Annual energy use.

[b] For a corresponding electric furnace, $I_S = 0.651$.

Table 11 shows similar results in condensing furnaces for the new case of ducts and vents sized according to the furnace capacity. In this case, the decrease in system index is smaller, about 2% over the range of 1.15 to 2.5 times DHL.

Finally, both Table 10 and 11 show that duct efficiency increases with furnace capacity, because higher capacity furnaces are on less than lower capacity furnaces.

Effects of Furnace Sizing and Night Setback

Tables 10 and 11 show the relationship between furnace sizing and night setback for the retrofit case. The energy savings due to

night setback, 8 h per day at 10°F, is nearly constant at 15% and independent of furnace size. Table 9 covers the effect of climate variations on energy savings due to night setback.

Duct treatment. Table 12 shows the effect of duct treatment on furnace performance. Duct treatment includes sealants to reduce leaks and interior or exterior insulation to reduce heat loss due to conduction. Sealing and insulation improves the system performance as indicated by the system index I_S. For cases with no duct insulation, reducing duct leakage from 10% to zero increases the system index 2.6%. Insulating the ducts also improves system performance. R5 insulation on the exterior of the ducts increases

Table 10 Effect of Sizing, Setback, and Design Parameters on Annual Heating Performance—Conventional Natural Draft Furnace

Furnace Multiplier[a]	Duct Design	Setback, °F	Annual Performance Factors					Temperature Swing, °F	Avgerage Room, °F	Recovery Time, h[b]	Average Basement, °F
			E_F	E_D	F_{MG}	F_{LM}	$I_S/(I_S)_{BC}$				
1.00	New[c]	0	76.1	59.2	1.016	1.104	1.095	3.4	67.7	n.a.[d]	67.9
1.15	New	0	75.6	59.0	1.009	1.059	1.032	4.0	67.8	n.a.	67.9
	Retrofit	0	75.5	59.0	1.002	1.032	0.998	4.0	67.8	n.a.	67.9
	Retrofit	10	76.1	60.7	0.999	1.155	1.155	4.1	65.8	2.02	65.7
1.40	New	0	75.5	60.8	1.010	1.029	1.034	4.8	68.1	n.a.	67.9
	Retrofit	0	75.5	60.9	1.004	1.000	1.000[e]	4.9	68.0	n.a.	67.8
	Retrofit	10	75.9	62.4	1.001	1.120	1.152	5.2	66.0	1.02	65.7
1.70	New	0	74.8	61.3	1.013	1.017	1.023	5.9	68.2	n.a.	68.1
	Retrofit	0	75.0	61.7	1.006	0.983	0.992	5.9	68.2	n.a.	67.9
	Retrofit	10	75.6	63.5	1.003	1.095	1.143	6.3	66.2	0.54	65.8
2.50	New	0	74.9	63.4	1.008	0.943	0.978	8.2	68.6	n.a.	68.1
	Retrofit	0	75.2	64.9	1.009	0.937	1.000	8.6	68.6	n.a.	67.8
	Retrofit	10	75.7	66.5	1.006	1.042	1.144	9.3	66.6	0.24	65.8

Notes:
1. Ranch house with basement in Pittsburgh, PA, climate with base conditions of 60°F circulating air temperature rise, 6 cycles/h, 10% duct air leakage.
2. The values in this table are figures of merit to be considered within the confines of the SP43 project, and they should not be applied outside the scope of these examples.

[a] Furnace output rating or heating capacity = (Furnace multiplier)(Design heat loss).
[b] Longest recovery time during winter (lowest outdoor temperature) = 5°F.
[c] Retrofit case was not run for furnace multiplier = 1.00.
[d] n.a. indicates not applicable.
[e] $(I_S)_{BC} = 0.4614$; for the corresponding electric furnace, $I_S/(I_S)_{BC} = 1.410$.

Table 11 Effect of Furnace Sizing on Annual Heating Performance—Condensing Furnace with Preheat

Furnace Multiplier[a]	Annual Performance Factors						Temperature Swing, °F	Average Room, °F	Average Basement, °F
	E_F	E_D	F_{MG}	F_{LM}	I_S	$I_S/(I_S)_{BC}$			
1.15	95.1	60.9	1.000	1.076	0.623	1.351[b]	3.5	67.9	66.7
1.40	95.5	62.0	1.000	1.050	0.622	1.347	4.3	68.1	66.7
2.50	95.4	64.8	1.000	0.990	0.611	1.325	7.3	68.7	66.7

Notes:
1. Ranch house with basement in Pittsburgh climate with base conditions of 60°F circulating air temperature rise, 6 cycles/h, 10% duct air leakage
2. The values in this table are figures of merit to be considered within the confines of the SP43 project, and they should not be applied outside the scope of these examples.

[a] Furnace output rating or heating capacity = (Furnace multiplier)(Design heat loss).
[b] $(I_S)_{BC} = 0.4614$; for the corresponding electric furnace, $I_S = 0.651$.

Table 12 Effect of Duct Treatment on System Performance

Case	1	2ᵃ	3	4	5	6	7
Condition							
Duct insulation	None	None	None	R5	R5	R5	R5[b]
Duct leakage, %	0	10	20	0	10	20	10
Basement insulation							
Ceiling	None	None	None	None	None	None	None
Wall	R8	R8	R8	R8	R8	R8	R8
Performance							
Burner on-time, %	17.5	18.0	18.6	16.8	17.2	17.8	16.6
Blower on-time, %	23.8	24.3	24.9	23.0	23.4	24.1	22.9
Average basement							
temperature, °F	66.8	67.8	68.8	65.2	66.3	67.4	65.1
Furnace efficiency E_F, %	75.4	75.5	75.7	75.0	75.2	75.4	75.0
Duct efficiency E_D, %	66.8	60.9	54.8	77.4	70.4	63.2	79.6
Load modification factor	0.94	1.00	1.07	0.85	0.91	0.97	0.84
$I_S/(I_S)_{BC}$[c]	1.026	1.000	0.970	1.070	1.044	1.010	1.085

[a] Case 2 is the base case.
[b] Case 7 is interior insulation (liner); Cases 1-6 are exterior insulation (wrap).
[c] $(I_S)_{BC}$ = 0.4614; for the corresponding electric furnace, I_S = 0.651.

I_S by 4.4%, and R5 insulation on the interior of the duct increases I_S by 8.5%.

Basement configuration. Table 13 covers the effect of basement configuration and duct treatment on system performance. Insulating and sealing the ducts reduces the basement temperature. More heat is then required in the conditioned space to make up for losses to the colder basement. Where ducts pass through the attic or ventilated crawl space, insulation and sealing improves duct performance, although the total system performance is poorer. On the other hand, installing the ducts within the conditioned space significantly improves the miscellaneous gain factor F_{MG} in which the duct losses are added directly to the conditioned space. In this case, the system index I_S would also improve.

COMMERCIAL FURNACES

The basic difference between residential and commercial furnaces is the size and heating capacity of the equipment. The heating capacity of a commercial furnace may range from 150,000 to over 2,000,000 Btu/h. Generally, furnaces with output capacities less than 320,000 Btu/h are classified as light commercial, and those above 320,000 Btu/h are large commercial equipment. In addition to the difference in capacity, commercial equipment is

Table 13 Effect of Duct Treatment and Basement Configuration on System Performance

Case	1	2	3	4	5
Condition					
Duct insulation	None	None	None	None	R5
Duct leakage, %	20	10	10	10	0
Basement insulation					
Ceiling	None	None	R11	R11	R11
Wall	None	None	None	R8	R8
Performance					
Burner on-time, %	23.6	22.7	21.8	18.0	16.3
Blower on-time, %	29.8	29.2	28.0	24.2	22.2
Average basement					
temperature, °F	64.4	63.3	62.4	67.9	64.3
Furnace efficiency E_F, %	75.3	75.2	75.2	75.7	75.0
Duct efficiency E_D, %	50.9	56.9	56.7	61.7	77.0
Load modification factor	0.91	0.85	0.89	0.99	0.88
$I_S/(I_S)_{BC}$[a]	0.765	0.794	0.825	1.001	1.103

Note: See Table 12 for base case.
[a] $(I_S)_{BC}$ = 0.4614

constructed from material with increased structural strength and more sophisticated control systems.

EQUIPMENT VARIATIONS

Light commercial heating equipment comes in almost as many flow arrangements and design variations as residential equipment. Some are identical to residential equipment, while others are unique to commercial applications. Some commercial units function as a part of a ducted system, and others operate as unducted space heaters.

Ducted Equipment

Upflow gas-fired commercial furnaces. These furnaces are available up to 300,000 Btu/h and supply enough airflow to handle up to 10 tons of air conditioning. They may have high static pressure, belt-driven blowers, and frequently they consist of two standard upflow furnaces tied together in a side-by-side arrangement. They are normally incorporated into a system in conjunction with a commercial split-system air-conditioning unit and are available in either LPG or natural gas. Oil-fired units may be available on a limited basis.

Horizontal gas-fired duct furnaces. Available for built-up light commercial systems, this type of furnace is not equipped with its own blower but is designed for uniform airflow across the entire furnace. Duct furnaces are normally certified for operation either upstream or downstream of an air conditioner cooling coil. If a combination blower and duct furnace is desired, a package called a blower unit heater is available. Duct furnaces and blower unit heaters are available in natural gas, LPG, oil, and electric models.

Electric duct furnaces. These furnaces are available in a large range of sizes and are suitable for operation in upflow, downflow, or horizontal positions. These units are also used to supply auxiliary heat with the indoor section of a split-heat pump.

Combination package units. The most common commercial furnace is the combination package unit, sometimes known as a combination roof-top unit. They are available as air-conditioning units with LPG and natural gas furnaces, electric resistance heaters, or heat pumps. Combination oil heat-electric cool units are not commonly available. Combination units come in a full range of sizes covering air-conditioning ratings from 5 to 50 tons with matched furnaces supplying heat-to-cool ratios of approximately 1.5 to 1.

Combination units of 15 tons and under are available as single-zone units. The entire unit must be in either the heating or cooling mode. All air delivered by the unit is at the same temperature. Frequently, the heating function is staged so that the system operates at reduced heat output when the load is small.

Large combination units in the 15 to 50-ton range are available as single-zoned units, as are small units; however, they are also available as multizoned units. A multizone unit supplies conditioned air to several different zones of a building in response to individual thermostats controlling those zones. These units are capable of supplying heating to one or more zones at the same time that cooling is supplied to other zones.

Large combination units are normally available only in a curbed configuration; that is, the units are mounted on a rooftop over a curbed opening in the roof. The supply and return air enters through the bottom of the unit. Smaller units may be available for either curbed or uncurbed mounting. In either case, the unit is usually connected to ductwork within the building to distribute the conditioned air.

Unducted Heaters

Three types of commercial heating equipment are used as unducted space heaters. One is the *unit heater*, which is available from about 25,000 to 320,000 Btu/h. They are normally mounted from ceiling hangers and blow air across the heat exchanger into

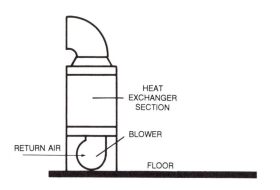

Fig. 9   Standing Floor Furnace

the heated space. Natural gas, LPG, and electric unit heaters are available. The second unducted heater used in commercial heating is the *infrared heater*. These units are mounted from ceiling hangers and transmit heat downward by radiation. Both gas and electric infrared heaters are available.

Finally, *floor (standing) furnaces* (Figure 9) are used as large area heaters and are available in capacities ranging from 200,000 to 2,000,000 Btu/h. Floor furnaces direct heated air through nozzles for task heating or use air circulators to heat large industrial spaces. Residential floor-suspended furnaces are described in Chapter 30.

SYSTEM DESIGN AND EQUIPMENT SELECTION

The procedure for design and selection of a commercial furnace is similar to that for a residential furnace. First, the design capacity of the heating system must be determined, considering structure heat loss, recovery load, internal load, humidification, off-peak storage, waste heat recovery, and backup capacity. Since most commercial buildings use setback during weekends, evenings, or other long periods of inactivity, the recovery load is important, as are internal loads and waste heat recovery.

Selection criteria differ from a residential furnace in some respects and are identical in others. Sizing criteria are essentially the same, and it is recommended that the furnace be oversized 30% above total load. Since combination units must be sized accurately for the cooling load, it is possible that the smallest gas-fired capacity available will be larger than the 30% value. This is especially true for the warmer climates of the United States.

Efficiency of commercial units is about the same as for residential units. Two-stage gas valves are frequently used with commercial furnaces, but the efficiency of a two-stage system may be lower than for a single-stage system. At a reduced firing rate, the excess combustion airflow through the burners increases, decreasing the steady-state operating efficiency of the furnace. Multistage furnaces with multistage thermostats and controls are commonly used to provide more uniform distribution of heat within the building.

The design life of commercial heating and cooling equipment is about 20 years. Most gas furnace heat exchangers are either coated steel or stainless steel. Since most commercial furnaces are made for outdoor application, the cabinets are made from corrosion-resistant coated steel, such as galvanized or aluminized. Blowers are usually belt-driven and are capable of delivering air at high static pressures.

The noise level of commercial heating equipment is important with some products and less important with others. Unit heaters, for example, are used primarily in industrial applications where noise is less important. Most other commercial equipment is used in schools, office buildings, and other commercial buildings where noise level is important. In general, the larger the furnace, the more air it handles, and the more noise it generates. However, commercial systems with longer and larger ductwork result in more sound attenuation. The net result is that quality commercial heating systems produce about the same noise level in the heated space as do residential systems.

Safety requirements are the same for light commercial systems as they are for residential systems. Above 400,000 Btu/h gas input, the ANSI Z21.47 requirements for gas controls are more stringent. A large percentage of commercial heating systems are located on rooftops or some other location outside a building. Outdoor furnaces always provide a margin of safety beyond that of an indoor furnace.

TECHNICAL DATA

Technical data for commercial furnaces are supplied by the manufacturer. Furnaces are available with heat outputs ranging from 150,000 to more than 2,000,000 Btu/h. For heating-only commercial heaters, the airflow is set to supply air with an 85°F temperature rise. Heat-cool combination units supply air equal to about 400 cfm per ton of cooling capacity. Heat-to-cool ratios are generally held at about 1.5 to 1.

The steady-state efficiency for commercial furnaces is about the same as that for residential furnaces. However, little information has been published about the annual efficiency of commercial systems. Some efficiency improvement components, such as intermittent ignition devices, are common in commercial furnaces.

GENERAL CONSIDERATIONS
INSTALLATION PRACTICES

Installation requirements call for the forced-air heating system to meet two basic criteria: (1) the system must be safe, and (2) it must provide comfort for the occupants of the conditioned space.

Indoor furnaces are sometimes installed as isolated combustion systems, *i.e.*, one in which a furnace is installed indoors and all combustion and ventilation air is admitted through grilles or is ducted from outdoors and does not interact with air in the conditioned space; for example, interior enclosures with air from attic or ducted from outdoors, exterior enclosures with air from outdoors through grilles, or enclosures in garages or carports attached to the building (NFPA 54-1984). This type of installation presents special considerations in determining efficiency.

Generally, the following three sources of installation information must be followed to ensure the safe operation of a heating system: (1) the equipment manufacturer's installation instructions, (2) local installation code requirements, and (3) national installation code requirements. While local code requirements may or may not be available, the other two are always available. Depending on the type of fuel being used, one of the following national code requirements will apply:

NFPA 54-1988 (also ANSI Z223.1-84) *National Fuel Gas Code*
NFPA 70-90 *National Electric Code*
NFPA 31-87 *Oil Burning Equipment Code*

Comparable Canadian standards bear the numbers CSA Z240.4, Z240.5, Z240.6.1, and Z240.6.2. These regulations provide complete information about construction materials, gas line sizes, flue pipe sizes, wiring sizes, and the like.

Proper design of the air distribution system is necessary for both comfort and safety. Chapter 32 of the 1989 ASHRAE *Handbook—Fundamentals*, Chapter 1 of the 1991 ASHRAE *Handbook—HVAC Applications* and Chapter 9 of this volume provide information on the design of ductwork for forced air heating systems. Forced air furnaces may provide design airflow at a static pressure as low as 0.12 in. of water for a residential unit to above 1.0 in. of water for a commercial unit. The air distribution system must handle the required volume flow rate within the pressure limits of the equipment. If the system is a combined heating-cooling installation, the air distribution system must meet the cooling requirement because more air is required for cooling than for heating. It is also important to include the pressure drop of the cooling coil. The ARI maximum allowable pressure drop for residential cooling coils is 0.3 in. of water.

AGENCY LISTINGS

The construction and performance of furnaces is controlled by several agencies.

The Gas Appliance Manufacturers Association (GAMA), in cooperation with its industry members, sponsors a certification program relating to gas and oil-fired residential furnaces and boilers. This program uses an independent laboratory to verify the furnace and boiler manufacturer's stated annual fuel utilization efficiency (AFUE) and heating capacity, as determined by testing in accordance with the Department of Energy's *Uniform Method for Measuring the Energy Consumption of Furnaces and Boilers.* Gas and oil furnaces with input ratings less than 225,000 Btu/h and gas and oil boilers with input ratings less than 300,000 Btu/h are currently included in the program.

Also included in the program is the semiannual publication of the GAMA *Consumers Directory of Certified Efficiency Ratings for Residential Heating and Water Heating Equipment,* which identifies certified products and lists the input rating, certified heating capacity, and annual fuel utilization efficiency (AFUE) for each. Participating manufacturers are entitled to use the GAMA Certification Symbol (seal). These directories are distributed to the reference departments of public libraries in the United States.

ANSI *Standard* Z21.47-90, Gas-Fired Central Furnaces, and ANSI *Standard* Z21.64-88, Direct Vent Central Furnaces (sponsored by the American Gas Association), give minimum construction, safety, and performance requirements for gas furnaces. The AGA maintains laboratories to certify furnaces and operates a factory inspection service. Furnaces tested and found to be in compliance are listed in the AGA Directory and carry the Blue Star Seal of Certification. Underwriters Laboratory and other approved laboratories can also test and certify equipment in accordance with ANSI *Standard* Z21.47/21.64.

Gas furnaces may be certified for standard, alcove, closet, or outdoor installation. Standard installation requires clearance between the furnace and combustible material of at least 6 in. Furnaces certified for alcove or closet installation can be installed with reduced clearance, as listed. Furnaces certified for either sidewall venting or outdoor installation must operate properly in a 40 mph wind. Construction materials must be able to withstand natural elements without degradation of performance and structure. Horizontal furnaces are normally certified for installation on combustible floors and for attic installation and are so marked, in which case they may be installed with point or line contact between the jacket and combustible constructions. Upflow and downflow furnaces are normally certified for alcove or closet installation. Gas furnaces may be listed to burn natural, mixed, manufactured, LP, or LP-air gases. A furnace must be ordered for the specific gas to be used, since different burners and controls, as well as orifice changes, may be required. In Canada, similar requirements have been established by the Canadian Standards Association and the Canadian Gas Association. The testing agency is the Canadian Gas Association, and certified products are marked with the CGA seal.

Sometimes oil burners and control packages are sold separately; however, they are normally sold as part of the furnace package.

Pressure-type or rotary burners should bear the Underwriters Laboratory label showing compliance with UL 296. In addition, the complete furnace should bear markings indicating compliance with UL 727. Vaporizing burner furnaces should also be listed under UL 727.

Underwriters Laboratory *Standards* UL 883, UL 1025, and UL 1096 give requirements for the listing and labeling of electric furnaces. Similarly, UL 559 gives the requirements for heat pumps.

The following list summarizes the important standards that apply to space-heating equipment and are issued by the American Gas Association, Underwriters Laboratory, and the Canadian Gas Association:

ANSI/ASHRAE 103-1988	Methods of Testing for Heating Seasonal Efficiency of Central Furnaces and Boilers
ANSI Z83.8-85	Gas Unit Heaters
ANSI Z83.9-82	Gas-Fired Duct Furnaces
ANSI Z21.47-89	Gas-Fired Central Furnaces
ANSI Z21.64-89	Direct-Vent Central Furnaces
ANSI Z21.66-85	Electrically Operated Automatic Vent-Damper Devices for Use with Gas-Fired Appliances
ANSI Z21.67-85	Mechanically Actuated Automatic Vent-Damper Devices for Use with Gas-Fired Appliances
ANSI Z21.68-85	Thermally Actuated Automatic Vent-Damper Devices for Use with Gas-Fired Appliances
ANSI Z83.4-85	Direct Gas-Fired Makeup Air Heaters
ANSI Z83.6-82	Gas-Fired Infrared Heaters
UL 296-80	Oil Burner
UL 307-78	Heating Appliances for Mobile Homes and Recreational Vehicles
UL 559-85	Heat Pumps
UL 727-80	Oil-Fired Central Furnaces
UL 883-80	Fan-Coil Units and Room Fan-Heater Units
UL 1096-81	Electric Central Air Heating Equipment
UL 1025-80	Electric Air Heaters
CGAS 2.3	Gas-Fired Gravity and Forced-Air Central Furnace with Inputs up to 400,000 Btu/h
CGAS 3.2	Gas-Fired Gravity and Forced-Air Central Furnaces with Inputs over 400,000 Btu/h
CGAS 3.7	Makeup Air Heaters
CGAS B2.5	Vented Wall Furnaces
CGAS B2.6	Unit Heaters
CGAS B2.8	Duct Furnaces
CGAS B2.16	Infrared Radiant Heaters
CGAS B2.19	Sealed Combustion System Wall Furnaces
CGAS B140.4	Oil-Fired Gravity and Forced-Air Central Furnaces

REFERENCES

Benton, R. 1983. Computer predictions and field test verification of energy savings with improved control. ASHRAE *Transactions* 89(1).

Bonne, J., J.E. Janssen, A.E. Johnson, and W.T. Wood. 1976. Residential heating equipment HFLAME Evaluation of target improvements. National Bureau of Standards, Final Report, Contract No. T62709, Center of Building Technology, Honeywell, Inc.

Bullock, E.C. 1977. Energy saving through thermostat setback with residential heat pumps. Workshop on thermostat setback. National Bureau of Standards, Gaithersburg, MD.

Chi, J. 1977. DEPAF—A computer model for design and performance analysis of furnaces. AICHE-ASME Heat Transfer Conference, Salt Lake City, UT (August).

Deppish, J.R. and D.W. DeWerth. 1986. GATC Studies common venting. *Gas Appliance and Space Conditioning Newsletter* 10 (September).

Fischer, R.D., F.E. Jakob, L.J. Flanigan, D.W. Locklin, and R.A. Cudnik. 1984. Dynamic performance of residential warm-air heating systems—Status of ASHRAE Project SP43. ASHRAE *Transactions* 90(2B):573-90.

Gable, G.K. and K. Koenig. 1977. Seasonal operating performance of gas heating systems with certain energy saving features. ASHRAE *Transactions* 83(1).

Herold, K.E., F.E. Jakob, and R.D. Fischer. 1986. The SP43 simulation model: Residential energy use. Proceedings of ASME Conference at Anaheim, 81-87.

Herold, K.E., L.J. Flanigan, R.D. Fischer, and R.A. Cudnik. 1987. Update on experimental validation of the SP43 simulation model for warm air heating systems. ASHRAE *Transactions* 93(1).

Jakob, F.E., R.D. Fischer, L.J. Flanigan, D.W. Locklin, K.E. Herold, and R.A. Cudnik. 1986. Validation of the ASHRAE SP43 dynamic simulation model for residential forced-warm-air systems. ASHRAE *Transactions* 92(2B):623-43.

Jakob, F.E., D.W. Locklin, R.D. Fischer, L.J. Flanigan, and R.A. Cudnik. 1986. SP43 Evaluation of system options for residential forced-air heating. ASHRAE *Transactions* 92(2B):644-73.

Jakob, F.E., R.D. Fischer, and L.J. Flanigan. 1987. Experimental validation of the duct submodel for the SP43 simulation model. ASHRAE *Transactions* 93(1):1499-1515.

Locklin, D.W., K.E. Herold, R.D. Fischer, F.E. Jacob, and R.A. Cudnik. 1987. Supplemental information from SP43 evaluation of system options for residential forced-air heating. ASHRAE *Transactions* 93(2):1934-58.

Nelson, L.W. and W. MacArthur. 1978. Energy saving through thermostat setback. ASHRAE *Journal* (September).

NFPA. 1984. *National fuel gas code. National Fire Code* 54-1984. National Fire Protection Association, Quincy, MD.

Schade, G.R. 1978. Saving energy by night setback of a residential heat pump system. ASHRAE *Transactions* 84(1).

Stickford, G.H., *et al.* 1985. Technology development for corrosion-resistant condensing heat exchangers. Battelle Columbus Laboratories to Gas Research Institute, GRI-85-0282.

RESIDENTIAL IN-SPACE HEATING EQUIPMENT

IN-SPACE heating equipment differs from central heating in that fuel is converted to heat in the space to be heated. Such heaters may be permanently installed or portable and may employ a combination of radiation, natural convection, and forced convection to transfer the heat produced. The energy source may be liquid, solid, gaseous, or electric.

GAS IN-SPACE HEATERS

Gas in-space heating equipment comes in a variety of styles and models.

Room Heaters

A vented, circulator room heater is a self-contained, freestanding, nonrecessed, gas-burning appliance that furnishes direct warm air to the space in which it is installed, without ducting (Figure 1). It converts the energy in the fuel gas to convected and radiant heat by transferring heat from flue gases to a heat exchanger surface without mixing flue gases and circulating heated air.

A vented radiant circulator is equipped with high-temperature glass panels and radiating surfaces to increase radiant heat transfer. Separation of flue gases from circulating air must be maintained. Vented radiant circulators range from 10,000 to 75,000 Btu/h.

Gravity-vented radiant circulators may also be equipped with an optional circulating air fan, but they perform satisfactorily with or without the fan. Fan-type vented radiant circulators are equipped with an integral circulating air fan, which is necessary for satisfactory performance.

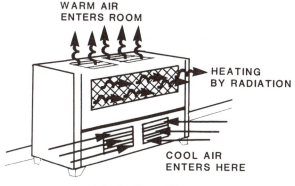

WARM AIR
ENTERS ROOM

HEATING
BY RADIATION

COOL AIR
ENTERS HERE

Fig. 1 Room Heater

The preparation of this chapter is assigned to TC 6.4, In-Space Convective Heating.

Vented room heaters are connected to a vent, chimney, or single-wall metal pipe venting system engineered and constructed to develop a positive flow to the outside atmosphere. Room heaters should not be used in a room that has limited air exchange with adjacent spaces, since combustion air is required from the space.

Unvented radiant or convection heaters range in size from 10,000 to 40,000 Btu/h and can be freestanding or wall-mounted, nonrecessed units of either the radiant or closed-front type. Unvented room heaters require an outside air intake. The size of the fresh air opening required is marked on the heater. To ensure adequate fresh air supply, unvented gas-heating equipment must, according to voluntary standards, include a device that shuts the heater off if the oxygen in the room becomes inadequate. Unvented room heaters may not be installed in hotels, motels, or rooms of institutions, such as hospitals or nursing homes.

Catalytic room heaters are fitted with fibrous material impregnated with a catalytic substance that accelerates the oxidation of a gaseous fuel to produce heat without flames. The design distributes the fuel throughout the fibrous material so that oxidation occurs on the surface area in the presence of a catalyst and room air.

Catalytic heaters transfer heat by low-temperature radiation and by convection. The surface temperature is below a red heat and is generally below 1200 °F at the maximum fuel input rate. Since they operate without a flame, catalytic heaters offer the inherent safety feature of flameless combustion, as compared with conventional flame-type gas-fueled burners. Catalytic heaters have also been used in agriculture and for industrial applications in combustible atmospheres.

Unvented household catalytic heaters are used in Europe. Most of these are portable and are typically mounted on casters in a casing that includes a cylinder of liquefied petroleum gas (LPG) so that they may be rolled from one room to another. LPG cylinders holding more than 2 lb of fuel are not permitted for indoor use in the United States. As a result, catalytic room heaters sold in the United States are generally permanently installed and fixed as wall-mounted units. Local codes and the National Fuel Gas Code (ANSI Z223.1-88/NFPA 54) should be reviewed for accepted combustion air requirements.

Wall Furnaces

A wall furnace is a self-contained vented appliance with grilles that are designed to be a permanent part of the structure of a building (Figure 2). It furnishes heated air that is circulated by natural or forced convection. A wall furnace may have boots, which cannot extend 10 in. beyond the horizontal limits of the casing through walls of normal thickness, to provide heat to adjacent rooms. Wall furnaces range from 10,000 to 90,000 Btu/h. Wall furnaces are classified as conventional or direct vent.

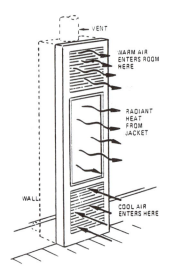

Fig. 2   Wall Furnance

Conventional vent units require approved B-1 vent pipes and are installed to comply with the National Fuel Gas Code (ANSI Z223.1-88/NFPA 54).

Some wall furnaces are available as counterflow units, which use fans to reverse the natural flow of air across the heat exchanger. Air enters at the top of the furnace and discharges at or near the floor. Counterflow systems reduce heat stratification in a room and, as with any vented unit, a minimum of inlet air for proper combustion must be supplied.

Vented-recessed wall furnaces are recessed in the wall. The decorative grillwork extending into the room allows more usable area in the room being heated. Dual-wall furnaces are two units that fit between the studs of adjacent rooms, thereby using a common vent.

Both vented-recessed and dual-wall furnaces are usually natural convection units. Cool room air enters at the bottom and is warmed as it passes over the heat exchanger, entering the room through the grillwork at the top of the heater. As long as the thermostat calls for the burners to be on, this process continues. Accessory fans assist in the movement of air across the heat exchanger and help minimize air stratification.

Direct-vent wall furnaces are constructed so that combustion air comes from outside, and all flue gases discharge into the outside atmosphere. These appliances are complete with grilles or the equivalent, and are designed to be attached to the structure permanently. Direct-vent wall heaters are normally mounted on walls with outdoor exposure.

Direct-vent wall furnaces can be used in extremely tight (well-insulated) rooms because combustion air is drawn from outside the room. There are no infiltration losses for dilution or combustion air. Most direct-vent heaters are designed for natural convection, although some may be equipped with fans. Direct-vent furnaces are available from 6000 to 65,000 Btu/h input.

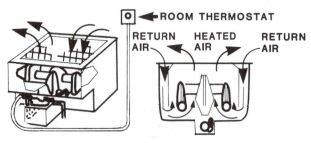

Fig. 3   Floor Furnance

Floor Furnaces

Floor furnaces are self-contained units suspended from the floor of the heated space (Figure 3). Combustion air is taken from outside, and flue gases are also vented outside. Cold air returns at the periphery of the floor register, and warm air to the room comes up through the center of the register.

United States Minimum Efficiency Requirements

The National Appliance Energy Conservation Act (NAECA) of 1987 mandates minimum efficiency requirements for gas-fired direct heating equipment. The minimums (effective as of January 1, 1990) are measured using the Department of Energy test method (March 28, 1984 Federal Register) and require that manufacturers of direct heating equipment (*i.e.*, gas-fired room heaters, wall furnaces, and floor furnaces) meet these minimums. Table 1 lists the minimums for each category.

Table 1   Gas-Fired Direct Heating Equipment Efficiency Requirements

Input, 1000 Btu/h	Minimum AFUE, %	Input, 1000 Btu/h	Minimum AFUE, %
Wall Furnance (with fan)		*Floor Furnace*	
<42	73	<37	56
>42	74	>37	57
Wall Furnace (gravity type)			
<10	59	*Room Heaters*	
>10 – <12	60	<18	57
>12 – <15	61	>18 – <20	58
>15 – <19	62	>20 – <27	63
>19 – <27	63	>27 – <46	64
>27 – <46	64	>46	65
>46	65		

CONTROLS

Valves

Gas in-space heaters are controlled by the following four types of valves:

Single-stage control valve. The full on-off, single-stage valve is controlled by a wall thermostat. Models are available that are powered by a 24-V supply or from energy supplied by the heat of the pilot light on the thermocouple (self-generating).

Two-stage control valve. The two-stage type (with hydraulic thermostat) fires at either a full input (100% of rating) or at some reduced step, which can be as low as 20% of the heating rate. The amount of time at the reduced firing rate depends on the heating load and the relative oversizing of the heater.

Step-modulating control valve. The step-modulating valve (with a hydraulic thermostat) steps on to a low fire and then either cycles off and on at the low fire, if the heating load is light, or gradually increases its heat output to meet any higher heating load that cannot be met with the low firing rate. There is an infinite number of fuel firing rates between low and high fire with this control.

Manual-control valve. The manual-control valve is controlled by the user rather than a thermostat. The user adjusts the fuel flow and, thus, the level of fire to suit heating requirements.

Thermostats

Temperature controls for gas in-space heaters are of the following two types:

Wall thermostats. These thermostats are available in 24-V and millivolt systems. The 24-V system requires an external power source and a 24-V transformer. Wall thermostats respond to temperature changes and turn the automatic valve to either full-on or full-off. The millivolt system requires no external power, since the power is generated by multiple thermocouples and may be either

250 or 750 millivolt, depending on the distance to the wall thermostat. This system also turns the automatic valve to either full-on or full-off.

Built-in hydraulic thermostats. Hydraulic thermostats are also available in two types: (1) a snap-action unit built with a liquid-filled capillary tube, which responds to changes in temperature and turns the valve to either full-on or full-off; and (2) a modulating thermostat, which is similar to the first type, except that the valve comes on and shuts off at a preset minimum input. Temperature alters the input anywhere from full-on to the minimum input. When the heating requirements are satisfied, the unit shuts off.

VENT CONNECTORS

Any vented gas-fired appliance must be installed correctly to vent combustion products. Figure 4 shows a typical vent connection method. A detailed description of proper venting techniques is found in the National Fuel Gas Code (ANSI *Standard* Z223.1-88/NFPA 54) and Chapter 31.

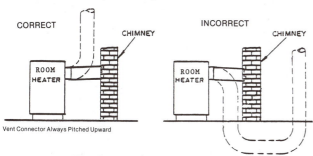

Fig. 4   Vent Connection Methods

SIZING UNITS

The size of the unit selected depends on the size of the room, the number and direction of exposures, the amount of insulation in the ceilings and walls, and the geographical location. Heat loss requirements can be calculated from procedures described in Chapter 25 of the 1989 ASHRAE *Handbook—Fundamentals.*

DeWerth and Loria (1989) studied the use of gas-fired, in-space supplemental heaters in two test houses. In this study they proposed a heater sizing guide, which is summarized in Table 2. The example energy consumption is for unvented, vented, and direct vent heaters installed in (1) a bungalow built in the 1950s with average insulation, and (2) a townhouse built in 1984 with above average insulation and tightness.

OIL AND KEROSENE IN-SPACE HEATERS

Vaporizing Oil Pot Heaters

These heaters have an oil vaporizing bowl (or other receptacle), which admits liquid fuel and air in controllable quantities; the fuel

is vaporized by the heat of combustion and mixed with the air in appropriate proportions. Combustion air may be induced by natural draft or forced into the vaporizing bowl by a fan. Indoor air is generally used for combustion and draft dilution. Window-installed units have the burner section outdoors. Both natural and forced convection heating units are available. A small blower is sold as an option on some models. The heat exchanger, usually cylindrical, is made of steel (Figure 5). These heaters are available as room units (both radiant and circulation), floor furnaces, and recessed wall heaters. They may also be installed in a window, depending on the cabinet construction. The heater is always vented to the outside. A 3 to 5 gal fuel tank may be attached to the heater, or a larger outside tank can be used.

Vaporizing burners of the pot type are equipped with a single constant-level and metering valve (see Figure 5). Fuel flows by gravity to the burner through the adjustable metering valve. Control can be manual, with an off pilot and variable settings up to maximum, or it can be thermostatically controlled, with the burner operating at a selected firing rate between "pilot" and "high."

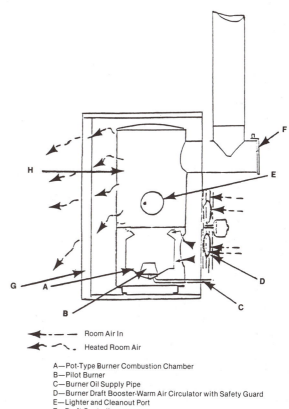

- - - - Room Air In
←～ Heated Room Air

A—Pot-Type Burner Combustion Chamber
B—Pilot Burner
C—Burner Oil Supply Pipe
D—Burner Draft Booster-Warm Air Circulator with Safety Guard
E—Lighter and Cleanout Port
F—Draft Controller
G—Perforated Metal Grill
H—Steel Drum-Type Heat Exchanger

Fig. 5   Oil-Fueled Heater with Vaporizing Pot-Type Burner

Table 2   Gas Input Required for In-Space Supplemental Heaters

Heater Type	Average AFUE, %	Steady State Eff., %	Older Bungalow[a] Outside Air Temp., °F			Energy Efficient House[b] Outside Air Temp., °F		
			5	30	50	5	30	50
Vented	54.6	73.1	6.5	3.8	1.6	2.8	1.6	0.7
Unvented	90.5	90.5	6.0	3.5	1.5	2.6	1.5	0.6
Direct vent	76.0	78.2	5.9	3.4	1.5	2.1	1.2	0.5

[a] Tested bungalow total heated volume = 6825 ft^3 and $U \approx 0.3$ to 0.5 Btu/h·°F.
[b] Tested energy efficient house total heated volume = 11,785 ft^3 and $U \approx 0.2$ to 0.3 Btu/h·°F.

Powered Atomizing Heaters

Wall furnaces, floor furnaces, and freestanding room heaters are also available with a powered gun-type burner using No. 1 or No. 2 fuel oil. For more information, refer to Chapter 27.

Portable Kerosene Heaters

Since kerosene heaters are not normally vented, precautions must be taken to provide sufficient ventilation. Kerosene heaters are of four basic types: radiant; natural convection; direct-fired, forced convection; and catalytic.

The radiant kerosene heater employs a reflector, while the natural convector heater is cylindrical in shape. Fuel vaporizes from the surface of a wick, which is immersed in an integral fuel tank of up to 2 gal capacity, similar to that of a kerosene lamp. Fuel-burning rates range from about 5,000 to 22,500 Btu/h. Radiant heaters usually have a removable fuel tank to facilitate refueling.

The direct-fired, forced convection type of portable kerosene heater has a vaporizing burner and a heat circulating fan. These heaters are available with thermostatic control and variable heat output.

Another kerosene heater is the catalytic type, which uses a metal catalyst to oxidize the fuel. It is started by lighting kerosene at the surface; however, after a few moments, the catalyst surface heats to the point where flameless oxidation of the fuel begins.

ELECTRIC IN-SPACE HEATERS

Wall, Floor, Toe Space, and Ceiling Heaters

Heaters for recessed or surface wall mounting are made with open wire or enclosed metal-sheathed elements. An inner liner or reflector is usually placed between the elements and the casing to promote circulation and minimize the rear casing temperature. Heat is distributed by both convection and radiation; the proportion of each depends on unit construction.

Ratings are usually 1000 to 5000 W at 120, 208, 240, or 277 V. Models with air circulation fans are available. Other types with similar components can be recessed into the floor. Electric convectors should be placed so that air moves freely across the elements.

Baseboard Heaters

These heaters consist of a metal cabinet containing one or more horizontal enclosed metal sheath elements. The cabinet is less than 6 in. in overall depth, the ratio of the overall length to the overall height is more than two to one, and it can be installed 18 in. above the floor.

Units are available from 2 to 12 ft in length, with ratings from 100 to 400 W/ft, and they fit together to make up any desired continuous length or rating. Electric hydronic baseboard heaters containing immersion heating elements and an antifreeze solution are made with ratings of 300 to 2000 W. The placement of any type of electric baseboard heater follows the same principles that apply to baseboard installations (see Chapter 33), since baseboard heating is primarily perimeter heating.

RADIANT HEATING SYSTEMS

Heating Panels and Heating Panel Sets

These systems have electric resistance wire or etched or graphite elements embedded between two layers of insulation. High-density thermal insulation behind the element minimizes heat loss, and the outer shell is formed steel with baked enamel finish. Heating panels provide supplementary heating by convection and radiation. They can be recessed or surface-mounted on hard surfaces or fit in standard T-bar suspended ceilings.

Units are usually rated between 250 and 1000 W in sizes varying from 24 in. by 24 in. to 24 in. by 96 in. in standard voltages of 120, 208, 240, and 277.

Embedded Cable Heat and Storage Heating Systems

Electric radiant heating systems that incorporate embedded cables, in ceilings and floors are covered in Chapter 6. Electric storage systems, including room storage heaters and floor slab systems, are covered in Chapter 39 of the 1991 ASHRAE *Handbook—HVAC Applications*.

Cord-connected Portable Heaters

Portable electric heaters are often used in areas that are not accessible to a central heating system. They are also used to maintain an occupied room at a comfortable level independent of the rest of the residence.

Portable electric heaters for connection to 120-V, 15-A outlets are available with outputs of 2050 to 5100 Btu/h (600 to 1500 W), the most common being 1320 and 1500 W. Many heaters are available with a selector switch for three wattages, *i.e.*, 1100-1250-1500 W. Heavy-duty heaters are usually connected to 240-V, 20-A outlets with outputs up to 13,700 Btu/h (4000 W), while those for connection to 240-V, 30-A outlets have outputs up to 19,100 Btu/h (5600 W). All electric heaters of the same wattage produce the same amount of heat.

Portable electric heaters transfer heat by two predominant methods: radiation and convection. Radiant heaters provide heat for people or objects. An element in front of a reflector radiates heat outward in a direct line. Conventional radiant heaters have ribbon or wire elements. Quartz radiant heaters have coil wire elements encased in quartz tubes. The temperature of a radiant wire element usually ranges between 1200 and 1600 °F.

Convection heaters warm the air in rooms or zones. Air flows directly over the hot elements and mixes with room air. Convection heaters are available with and without fans. The temperature of a convection element is usually less than 930 °F.

An adjustable, built-in bimetal thermostat usually controls the power to portable electric heaters. Fan-forced heaters usually produce better temperature control because the fan, in addition to cooling the case, forces room air past the thermostat for fast response. One built-in control uses a thermistor to signal a solid logic circuit that adjusts wattage and fan speed. Most quartz heaters use an adjustable control that operates the heater a percent of total cycle time from 0 (off) to 100% (full-on).

Controls

Low-voltage and line-voltage thermostats of on-off operation are used to control in-space electric heaters. Low-voltage thermostats, operating at 30 V or less, control relays or contactors that carry the rated voltage and current load of the heaters. Since the control current load is small (usually less than one amp), the small switch can be controlled by a highly responsive sensing element.

Line-voltage thermostats carry the full load of the heaters at rated voltage directly through their switch contacts. Most switches carry a listing by Underwriters Laboratories at 22-A (resistive), 277-V rating. While most electric in-space heating systems are controlled by remote wall-mounted thermostats, many are available with integral or built-in line-voltage thermostats.

Most low-voltage and line-voltage thermostats use small internal heaters, either fixed or adjustable in heat output, that are energized when the thermostat contacts close to provide heat anticipation. The cycling rate of the thermostat is increased by the use of anticipation heaters, resulting in more accurate control of the space temperature.

Droop is an apparent shift or lowering of the control point and is associated with line-voltage thermostats. In line-voltage

Table 3 Solid Fuel In-Space Heaters

Type[a]	Approximate Efficiency[a], %	Features	Advantages	Disadvantages
Simple fireplaces, masonry or prefabricated	−10 to +10	Open front. Radiates heat in one direction only.	Visual beauty.	Low efficiency. Heats only small areas.
High efficiency fireplaces	25 to 45	Freestanding or built-in with with glass doors, grates, ducts, and blowers.	Visual beauty. More efficient. Heats larger areas. Long service life. Maximum safety.	Medium efficiency.
Box stoves	20 to 40	Radiates heat in all directions.	Low initial cost. Heats large areas.	Fire hard to control. Short life. Wastes fuel.
Airtight stoves	40 to 55	Radiates heat in all directions. Sealed seams, effective draft control.	Good efficiency. Long burn times, high heat output. Longer service life.	Can create creosote problems.
High efficiency catalytic wood heaters	65 to 75	Radiates heat in all directions. Sealed seams, effective draft control.	Highest efficiency. Long burn times, high heat output. Long life.	Creosote problems. High purchase price.

[a]Product categories are general; product efficiencies are approximate.

thermostats, switch heating caused by large currents can add materially to the amount of droop. Most line-voltage thermostats in residential use control room heaters of 3 kW (12.5 A at 240 V) or less. With this moderate load and with properly sized anticipation heaters, the droop experienced in those applications is acceptable. Cycling rates and droop characteristics are significant to thermostat performance.

SOLID FUEL IN-SPACE HEATERS

The classification, solid fuel in-space heater (see Table 3), refers to most wood-burning and coal-burning devices, except central wood-burning furnaces and boilers. An in-space heater can be either a fireplace or a stove.

FIREPLACES

Simple Fireplaces

Simple fireplaces, especially all-masonry and noncirculating metal built-in fireplaces, produce little useful heat. They lend atmosphere and a sense of coziness to a room. Freestanding fireplaces are slightly better heat producers. Simple fireplaces have an average efficiency of about 10%. In extreme cases, more heated air is drawn up the chimney than is produced by the fire.

The addition of glass doors to the front of a fireplace has both a positive and negative effect. The glass doors restrict the free flow of indoor heated air up the chimney, but at the same time, they restrict the radiation of the heat from the fire into the room.

Factory-built Fireplaces

A factory-built fireplace is a system consisting of the fire chamber, chimney, roof assembly, and other related parts that are entirely factory made and intended for unit assembly in the field. These fireplaces have refractory lined metal, as opposed to masonry, fireboxes. Factory-built fireplaces come in both radiant and heat-circulating designs. Typical configurations are open front designs, but corner opening, three-sided units with openings either on the front or side, four-sided units, and see-through systems are available.

Radiant type. The radiant design system transmits heat energy from the firebox opening by direct radiation to the space in front of it. These fireplaces may also incorporate such features as an outside air supply and glass doors. Radiant design factory-built fireplaces are primarily used for aesthetic wood burning and typically have efficiencies similar to masonry fireplaces (0 to 10%).

Heat circulating. The heat circulating design system adds the function of heat transfer via heated air convection by circulating air around the fire chamber and releasing it to the space to be heated through grilles and louvers. Intake air is generally located below the firebox or low on the sides adjacent to the opening, and the heated air exhausts through grilles or louvers located above the firebox or high on the sides adjacent to it. Some designs employ ducts to circulate heated air to spaces other than the area of the front of the fireplace. Some circulating systems rely on natural convection, while others have electric fans or blowers to move air. The incorporation of these energy-saving features typically boosts efficiency 25 to 60%.

Freestanding Fireplaces

Freestanding fireplaces are open combustion wood-burning appliances that are not built into a wall or chase. One type of freestanding fireplace is a firepit in which the fire is open all around; smoke rises into a hood and then into a chimney. Another type is a prefabricated metal unit, which is typically shaped somewhere between a cylinder and a sphere and has an opening on one side. Because they radiate heat to all sides, freestanding fireplaces are typically more efficient than radiant fireplaces.

STOVES

Conventional Woodstoves

Woodstoves are chimney-connected, solid fuel-burning room heaters designed to be operated with the fire chamber closed. They deliver heat directly to the space in which they are located. They are not designed to accept ducts and/or pipes for heat distribution to other spaces. Woodstoves are controlled combustion appliances. Combustion air enters the firebox through the appliance's controllable air inlet; the air supply and, thus, combustion rate are controlled by the user. Conventional controlled combustion woodstoves manufactured prior to the mid-1980s typically have overall efficiencies ranging from 40 to 55%.

Most controlled combustion appliances are manufactured from steel or cast iron, or a combination of the two metals; others are constructed of soapstone or masonry. The thermal conductivity is lower for soapstone and masonry materials; however, soapstone and masonry have greater specific heats (the amount of heat that can be stored in a given mass). Additionally, other materials such as special refractories and ceramics are used in new low emission appliances. Woodstoves are classified as either radiant or convection (sometimes called circulating) heaters, depending on the way they heat interior spaces.

Radiant woodstoves are generally constructed with single exterior walls, which are heated by radiant heat from the fire contained within. A radiant appliance primarily heats by radiation of infrared energy; it heats room air only to the extent that air passes over the hot surface of the appliance.

The vertical walls of convection woodstoves use double-wall construction, with an air space between the walls. The double walls are open at the top and bottom of the appliance to permit room air to circulate through the air space. Hot air is more buoyant than cold air, and the rising hot air draws in the cooler room air at the bottom of the appliance. This air is then heated as it passes over the surface of the inner radiant wall. Some radiant heat from the inner wall is absorbed by the outer wall, but the constant introduction of room temperature air at the bottom of the appliance keeps the outer wall moderately cool. This generally enables convection woodstoves to be placed closer to combustible materials than radiant woodstoves. Some woodstoves use fans to augment the movement of heated air. Convection woodstoves generally provide more even heat distribution than radiant types.

Advanced Design Woodstoves

Strict air pollution standards prompted the development of new stove designs. These advanced technology (clean-burning) woodstoves employ either catalytic or noncatalytic technology to achieve very high combustion efficiency and to reduce creosote and the emission levels for particulate and carbon monoxide.

Catalytic combustors are currently available as an integral part of many new wood-burning appliances and are also available as add-on or retrofit units for most existing appliances. The catalyst is either platinum, palladium, or rhodium, or a combination of these elements. It is bonded to a ceramic or stainless steel substrate. A catalytic combustor's function in a wood-burning appliance is to substantially lower the ignition temperatures of unburned gases, solid and/or liquid droplets (from approximately 1000 to 500 °F). As these unburned combustibles leave the main combustion chamber and pass through the catalytic combustor, they ignite and burn rather than enter the atmosphere. For the combustor to efficiently burn the gases, the proper mixture of oxygen and a sufficient temperature to maintain ignition are required; further, the gases must have sufficient residence time in the combustor. A properly operating catalytic combustor has a temperature in the range of 1000 to 1700 °F. Catalyst-equipped woodstoves have an EPA default efficiency of 72%, although many stoves are considerably more efficient. The default efficiency is the value one standard deviation below the mean of the efficiencies from a data base of stoves, as determined in 1986 by the U.S. Environmental Protection Agency.

Another approach to increasing combustion efficiency and meeting emissions requirements is the use of technologically advanced internal appliance designs and materials. Generally, noncatalytic, low emissions wood-burning appliances have smaller fireboxes than conventional appliances and incorporate high-temperature refractory materials. The fire chamber is designed to increase temperature, turbulence, and residence time in the primary combustion zone. Secondary air is introduced to promote continued burning of the gases, solids, and liquid vapors in a secondary combustion zone. Many stoves add a third and fourth burn area within the firebox. The location and design of the air inlets is critical because air circulation patterns are the key to approaching complete combustion. Noncatalytic woodstoves have an EPA default efficiency of 63%; however, many models approach 80%.

Fireplace Inserts

Fireplace inserts are closed combustion wood-burning room heaters that are designed to be installed in an existing masonry fireplace. They combine elements of both radiant and convective woodstove designs. They have large radiant surfaces that face the room and circulating jackets on the sides to capture heat that would otherwise go up the chimney. Inserts currently being manufactured use either catalytic or noncatalytic technology to achieve clean burning.

Pellet-burning Stoves

Pellet-burning stoves burn small pellets made from wood by-products rather than logs. An electric auger feeds the pellets from a hopper into the fire chamber where air is blown through, creating very high temperatures in the firebox. The fire burns at such a high temperature that the smoke is literally burned up, resulting in a very clean burn, and no chimney is needed. Instead, the waste gases are exhausted to the outside through a vent. An air intake is operated by an electric motor; another small electric fan blows the heated air from the area around the fire chamber into the room. A microprocessor controls the operation, allowing the pellet-burning stove to be controlled by a thermostat. Pellet-burning stoves typically have the lowest emissions of all wood-burning appliances and have an EPA default efficiency of 78%. Because of the high air-to-fuel ratios used by pellet-burning stoves, these stoves are excluded from EPA woodstove emissions regulations.

GENERAL INSTALLATION PRACTICES

The criteria to ensure safe operation are normally covered by local codes and ordinances or, in rare instances, by state and federal requirements. Most codes, ordinances, or regulations refer to the following building codes and standards for in-space heating.

Building Codes	References
BOCA/National Building Code, 1990	BOCA
CABO One- and Two-Family Dwelling Code, 1988	CABO
Standard Building Code, 1988	SBCCI
Uniform Building Code, 1988	ICBO
Mechanical Codes	
BOCA/National Mechanical Code, 1990	BOCA
Uniform Mechanical Code, 1988	ICBO/IAPMO
Standard Mechanical Code, 1988	SBCCI
Electrical Codes	
National Electrical Code	ANSI/NFPA 70-1990
Canadian Electrical Code	CSA C22.1-1990
Chimneys	
Chimneys, Fireplaces, Vents and Solid Fuel-Burning Appliances	ANSI/NFPA 211-1988
Chimney, Factory-Built Residential Type and Building Heating Appliance	UL 103-89
Solid-Fuel Appliances	
Factory-Built Fireplaces	UL 127-88
Room Heaters, Solid-Fuel Type	UL 1482-88

Chapter 48, Codes and Standards, has further information, including the names and addresses of the above agencies. Safety and performance criteria are furnished by the manufacturer.

Solid-Fuel Safety

The evacuation of gases of combustion are a prime concern in the installation of solid fuel-burning equipment. NFPA 211, Chimneys, Fireplaces, Vents and Solid Fuel-Burning Appliances, lists requirements that should be followed. Because safety requirements for connector pipes (stovepipes) are not always readily available, safety requirements are summarized here.

Table 4 Chimney Connector Wall Thickness*

Size (Dia.), in.	Gage	Min. Thickness, in.
<6	26	0.019
≥6 to 10	24	0.023
≥10 to 16	22	0.029
≥16	16	0.056

*Do not use thinner connector pipe. Replace connectors as necessary. Leave at least 18 in. clearance between the connector and a wall or ceiling, unless listed for a smaller clearance or an approved clearance reduction system is used.

- Connector pipe is usually black (or blue), steel, single-wall pipe of a thickness shown in Table 4. Stainless steel is a corrosion-resistant alternative that need not meet the thicknesses listed in Table 4.
- Connectors should be installed with the crimped (male) end of the pipe toward the stove, so that creosote and water drips back into the stove.
- The pipe should be as short as is practical with a minimum of turns and horizontal runs. Horizontal runs should be pitched 1/4 in. per foot up toward the chimney.
- Chimney connectors should not pass through a ceiling, closet, alcove, or concealed space.
- When passing through a combustible interior or exterior wall, connectors must be routed through a listed wall pass-through that has been installed in accordance with the conditions of the listing, or it must use one of the home-constructed systems recognized in NFPA 211 or local building codes. Adequate clearance and protection of combustible materials is extremely important. In general, listed devices are easy to install and relatively inexpensive compared to home-constructed systems.

Creosote forms in all wood-burning systems. The rate of formation is a function of the quantity and type of fuel burned, the appliance in which it is burned, and the manner in which the appliance is operated. Thin deposits in the connector pipe and chimney do not interfere with operation, but thick deposits (greater than 1/4 in.) may ignite. Inspection and cleaning of chimneys connected to wood-burning appliances should be done on a regular basis (at least annually).

Burn only the solid fuel that is listed for the appliance. Coal should only be burned in fireplaces or stoves designed specifically for coal burning. The chimney used in coal-fired applications must also be designed and approved for coal as well as wood.

Install solid fuel appliances in strict conformance with the clearance requirements established as part of their safety listing. When clearance reduction systems are used, ensure that stoves remain at least 12 in. and connector pipe at least 6 in. from combustibles, unless smaller clearances are established as part of the listing.

Utility-furnished Energy

Those systems that rely on energy furnished by a utility are usually required to comply with local utility service rules and regulations. The utility will usually provide information on the installation and use of the equipment using their energy. Bottled gas (LPG) equipment is generally listed and tested under the same standards as natural gas. LPG equipment may be identical to natural gas equipment, but it always has a different orifice and, sometimes, a different burner and controls. The listings and examinations are usually the same for natural, mixed, manufactured, or liquid petroleum gas.

Products of Combustion

Equipment that produces products of combustion must be connected with a closed piping system from the combustion chamber outdoors. Gas-fired equipment may be vented through masonry stacks, chimneys, specifically designed venting systems, or, in some cases, venting systems incorporating forced or induced draft fans. Chapter 31 covers chimneys, gas vents, and fireplace systems in more detail.

Agency Testing

Guidelines for the construction and performance of in-space heaters are contained in the standards of several agencies. The following list summarizes the standards that apply to residential in-space heating and are coordinated or sponsored by ASHRAE, the American National Standards Institute (ANSI), Underwriters Laboratories (UL), American Gas Association (AGA), and the Canadian Gas Association (CGA). Some CGA standards have a CANI prefix.

ANSI Z21.11.1-88	Gas-Fired Room Heaters, Vented
ANSI Z21.11.2-89	Gas-Fired Room Heaters, Unvented
ANSI Z21.44-88	Gas-Fired Gravity and Fan-Type Direct-Vent Wall Furnaces
ANSI Z21.48-89	Gas-Fired Gravity and Fan-Type Floor Furnaces
ANSI Z21.49-89	Gas-Fired Gravity and Fan-Type Vented Wall Furnaces
CANI 2.1-86	Gas-Fired Vented Room Heaters
CGA 2.5-86	Gravity and Fan-Type Vented Wall Furnaces
CGA 2.19-81	Gravity and Fan-Type Sealed Combustion System Wall Furnaces
ANSI/UL 127-85	Factory-Built Fireplaces
ANSI/UL 574-85	Electric Oil Heaters
UL 647-84	Unvented Kerosene-Fired Heaters and Portable Heaters
ANSI/UL 729-87	Oil-Fired Floor Furnaces
ANSI/UL 730-86	Oil-Fired Wall Furnaces
ANSI/UL 737-88	Fireplace Stoves
UL 896-73	Oil Burning Stoves
ANSI/UL 1025-89	Electric Air Heaters
ANSI/UL 1042-86	Electric Baseboard Heating Equipment
ANSI/UL 1482-88	Heaters, Room Solid-Fuel Type
ASHRAE 106-1984	Rating Wood-Burning Fireplaces and Fireplace Stoves
ASHRAE 62-1989	Ventilation for Acceptable Indoor Air Quality

BIBLIOGRAPHY

ANSI. 1984. National fuel gas code. ANSI *Standard* Z223.1-88 (Also NFPA 54-88). American National Standards Institute, New York.

DeWerth, D.W. and R.L. Loria. 1989. In-space heater energy use for supplemental and whole house heating. ASHRAE *Transactions* 95(1).

GAMA. *Directory of gas room heaters, floor furnaces and wall furnaces.* Gas Appliance Manufacturers Association, Arlington, VA.

MacKay, S., L.D. Baker, J.W. Bartok, and J.P. Lassoie. 1985. *Burning wood and coal.* Northeast Regional Agricultural Engineering Service, Cornell University, Ithaca, NY.

NFPA. 1988. Chimneys, fireplaces, vents and solid fuel burning appliances. NFPA *Standard* 211-88. National Fire Protection Association, Quincy, MA.

U.S. Department of Energy. 1984. Uniform test method for measuring the energy consumption of vented home heating equipment. *Federal Register* 49:12, 169 (March).

Wood Heating Education and Research Foundation. 1984. *Solid fuel safety study manual for level I solid fuel safety technicians.* Washington, D.C.

CHIMNEY, GAS VENT, AND FIREPLACE SYSTEMS

A PROPERLY designed chimney controls draft and removes flue gases. This chapter describes the design of chimneys that discharge flue gases from appliance-chimney and fireplace-chimney systems.

In this chapter, *appliance* refers to any furnace, boiler, or incinerator (including the burner). The term *chimney* includes specialized vent products such as gas vents, unless the context indicates otherwise. *Draft* is negative static pressure, measured relative to atmospheric pressure; thus positive draft is negative static pressure. *Flue gas* is the mixture of gases discharged from the appliance and conveyed by the chimney or vent system.

Appliances have the following draft configurations (Stone 1971):

1. Those that require draft applied at the appliance flue gas outlet to induce air into the appliance.
2. Those that operate without draft applied at the appliance flue gas outlet, *e.g.*, a gas appliance with a draft hood in which the combustion process is isolated from chimney draft variations.
3. Those that produce positive pressure at the appliance outlet collar so that no chimney draft is needed; appliances that produce some positive outlet pressure but also need some chimney draft.

In the first two configurations, hot flue gas buoyancy, induced draft chimney fans, or a combination of both produce draft. The third configuration may not require chimney draft, but it should be considered in the design if a chimney is used. If the chimney system is undersized, draft inducers in the connector or chimney may supply appliance-chimney system draft needs. If the connector or chimney pressure requires control, draft control devices must be used.

The draft needed to overcome chimney flow resistance (Δp) is as follows:

$$\Delta p = \text{Theoretical draft} - \text{Available draft} = D_t - D_a$$

The above appliance draft configurations use this equation: in the second configuration with zero draft requirement at the appliance outlet, available draft is zero, and theoretical draft of the chimney equals the chimney flow resistance.

Available draft D_a is the draft needed at the appliance outlet. If increased chimney height and flue gas temperatures provide surplus available draft, draft control is required.

Theoretical draft D_t is the natural draft produced by the buoyancy of hot gases in the chimney relative to cooler gases in

the atmosphere. It depends on chimney height and the mean chimney flue gas temperature difference t_m, which is the temperature difference of the flue and atmospheric gases. Therefore, cooling by heat transfer through the chimney wall is a key variable in chimney design. Precise evaluation of theoretical draft is not necessary for most design calculations due to the availability of design charts, computer programs, capacity tables in the references, building codes, and manufacturers' data sheets.

Chimney temperatures and acceptable combustible material temperatures must be known to determine safe clearances between the chimney and combustible materials. Safe clearances for some chimney systems, such as Type B gas vents, are determined by standard tests and/or specified in building codes.

The following sections cover the basis of chimney design for average operating conditions. A rigorous evaluation of the flue gas temperature in the chimney can be obtained using the VENT-II, Version 4.1 computer program.

CHIMNEY FUNCTIONS

The proper chimney can be selected by evaluating such factors as draft, configuration, size, and operating conditions of the appliance; construction of surroundings; appliance usage classification; residential, low, medium, or high heat (NFPA *Standard* 211); and height of building. The chimney designer should know the applicable codes and standards to ensure acceptable construction.

In addition to chimney draft, the following factors must be considered for safe and reliable operation: air supply; draft control devices; chimney materials (corrosion and temperature resistance); flue gas temperatures, composition, and dew point; wind eddy zones; and particulate dispersion. Chimney materials must resist oxidation and condensation at both high and low fire levels.

Startup. The equations and design chart (Figure 1) may be used to determine vent or chimney size based on steady-state operating conditions. The equations and chart, however, do not consider modulation, cycling, or time to achieve equilibrium flow conditions from a cold start. While mechanical draft systems have no problem starting gas flow, gravity systems rely on the buoyancy of hot gases as the sole force to displace the cold air in the chimney. Priming follows Newton's laws of motion. The time to fill a system with hot gases, displace the cold air, and start flow is reasonably predictable and is usually a minute or less. But unfavorable thermal differentials, building-chimney interaction, mechanical equipment (exhaust fans, etc.), or wind forces that oppose the

---

The preparation of this chapter is assigned to TC 3.7, Fuels and Combustion.

normal flow of vent gases can overwhelm the buoyancy force. Then, rapid priming cannot be obtained solely from correct system design. The Vent-II computer program contains detailed analysis of gas vent and chimney priming and other cold-start considerations and allows for appliance cycling and pressure differentials that affect performance. A copy of the solution methodology, including equations (Rutz 1991), may be ordered from ASHRAE publication sales.

Air intakes. All rooms or spaces containing fuel-burning equipment must have a constant combustion air supply at adequate static pressure to ensure proper combustion. In addition, outside air is required to replace the air entering chimney systems through draft hoods and barometric draft regulators and to ventilate closely confined boiler and furnace rooms.

Because of the variable air requirements, the presence of other air-moving equipment in the building, and the variations in building construction and arrangement, no universally accepted rule specifies outdoor air openings. The minimum combustion air opening size depends on burner input with a practical minimum being the vent connector or chimney area. Any design must consider flow resistance of the combustion air supply, including register-louver resistance. Air supply openings that meet most building code requirements have a negligible flow resistance.

Vent size. Small residential and commercial natural draft gas appliances need vent diameters of 3 to 12 in. NFPA *Standard* 54 (ANSI *Standard* Z223.1), *National Fuel Gas Code*, recommends sizes or input capacities for most acceptable gas appliance venting materials. These sizes also apply to gas appliances with integral automatic vent dampers, as well as to appliances with field-installed automatic vent dampers. Field-installed automatic vent dampers should be certified for use with a specific appliance by a recognized testing agency and installed by qualified installers.

Draft control. Pressure, temperature, and other draft controls have replaced draft hoods in many residential furnaces and boilers to attain higher steady-state and seasonal efficiencies. Appliances that use pulse combustion or forced- or induced-draft fans, as well as those designed for sealed or direct venting, do not have a draft hood but may require special venting and special vent terminals. If fan-assisted burners deliver fuel and air to the combustion chamber and also overcome the appliance flow resistance, draft hoods or other control devices may be installed, depending on the design of the appliance. Category III and IV and some Category I and II appliances, as defined in ANSI *Standard* Z21.47, do not use draft hoods; in such cases, the appliance manufacturer's vent system design requirements should be followed. The section Vent and Chimney Accessories has information on draft hoods, barometric regulators, draft fans, and other draft control devices.

Frequently, a chimney must produce excess flow or draft. For example, dangerously high flue gas outlet temperatures from an incinerator may be reduced by diluting the air in the chimney with excess draft. The section Mass Flow Based on Fuel and Combustion Products briefly addresses draft control conditions.

Pollution control. Where control of pollutant emissions is impossible, the chimney should be tall enough to ensure dispersion over a wide area to prevent objectionable ground level concentrations. The chimney can also serve as a passageway to carry flue gas to pollution control equipment. This passageway must meet the building code requirements of a chimney, even at the exit of pollution control equipment, because of possible exposure to heat and corrosion. A bypass chimney should also be provided to allow continued exhaust in the event of pollution control equipment failure, repair, or maintenance.

Equipment location. Chimney materials may permit installing appliances at intermediate or all levels of a high-rise building without imposing penalties due to weight. Some gas vent systems permit individual apartment-by-apartment heating systems.

Wind effects. Wind and eddy currents affect the discharge of gases from vents and chimneys. They must expel flue gas beyond the cavity or eddy zone surrounding a building to prevent reentry through openings and fresh air intakes. A chimney and its termination can stabilize the effects of wind on appliances and their equipment rooms. In many locations, the equipment room air supply is not at neutral pressure under all wind conditions. Locating the chimney outlet well into the undisturbed wind stream and away from the cavity and wake zones around a building counteracts wind effects on the air supply pressure. It also prevents reentry through openings and contamination of fresh air intakes.

Chimney outlets below parapet or eave level, nearly flush with the wall or roof surface, or in known regions with stagnant air may be subjected to downdrafts and are undesirable. Caps for downdraft and rain protection must be installed according to their listings and the cap manufacturer's instructions, or the applicable building code.

Wind effects can be minimized by putting the chimney terminal and the combustion air inlet terminal close together in the same pressure zone.

Safety factors. Safety factors allow for uncertainties of vent and chimney operation. For example, flue gas must not spill from a draft hood or barometric regulator, even when the chimney has very low available draft. The Table 2 design condition for gas vents, *i.e.*, 300°F rise in the system at 5.3% CO_2, allows gas vents to operate with reasonable safety above or below the suggested temperature and CO_2 limits.

Safety factors may also be added to the system friction coefficient to account for a possibile extra fitting, soot accumulation, and air supply resistance. Specific gravity of flue gases can vary depending on the fuel burned. Natural gas flue gas, for example, has a density as much as 5% less than air, while coke flue gas has a density as much as 8% greater. However, these density changes are insignificant relative to other uncertainties and no compensation is needed.

STEADY-STATE CHIMNEY DESIGN EQUATIONS

Chimney design balances the forces that produce flow against those that retard flow (friction). *Theoretical draft* is the force that produces flow in gravity or natural draft chimneys. It is defined as the static pressure resulting from the difference in densities between a stagnant column of hot flue gases and an equal column of ambient air. In the design or balancing process, theoretical draft may not equal friction loss, because the appliance is frequently built to operate with some specific pressure (positive or negative) at the appliance flue gas exit. This exit pressure, or *available draft*, depends on appliance operating characteristics, fuel, and type of draft control.

Flow losses caused by friction may be estimated by several formulas for flow in pipes or ducts, such as the equivalent length method or the loss coefficient or velocity head method. Chapter 32 of the 1989 ASHRAE *Handbook—Fundamentals* covers computation of flow losses. This chapter emphasizes the loss coefficient method, because fittings usually cause the greater portion of system pressure drop in chimney systems, and conservative loss coefficients (which are almost independent of piping size) provide an adequate basis for design. Rutz *et al.* (1991) developed a computer program entitled *Vent-II (Version 4.1): An Interactive Personal Computer Program for Design and Analysis of Venting Systems for One or Two Gas Appliances*, which predicts flows, temperatures, and pressures in venting systems.

For large gravity chimneys, available draft may be calculated from Equation (1) or (2). Both equations use the equivalent length approach, as indicated by the symbol L_e in the flow-loss term (ASHVE 1941). These equations permit considering density difference between chimney gases and ambient air, as well as compensating shape factors. Mean flue gas temperature must be estimated separately.

For a cylindrical chimney:

$$D_a = 2.96HB\left(\frac{\rho_o}{T_o} - \frac{\rho_c}{T_m}\right) - \frac{0.000315w^2 T_m fL_e}{1.3 \times 10^7 d_f^5 B\rho_c} \qquad (1)$$

and, for a rectangular chimney:

$$D_a = 2.96HB\left(\frac{\rho_o}{T_o} - \frac{\rho_c}{T_m}\right) - \frac{0.000097w^2 T_m fL_e(x + y)}{1.3 \times 10^7 (xy)^3 B\rho_c} \qquad (2)$$

In these expressions, ρ_o/T_o determines theoretical draft, based on applicable gas and ambient density. The term ρ_c/T_m defines draft loss based on the factors for flow in a circular or rectangular duct system. To use these equations, the quantity of flow w for a variety of fuels and situations and the available draft needs of various types of appliances must be determined.

Equations (1) and (2) may be rearranged into a form that is more readily applied to the problems of chimney design, size, and capacity by considering the following factors:

- Mass flow of combustion products
- Chimney gas temperature and density
- Theoretical and available draft
- Allowable system pressure loss because of flow
- Chimney gas velocity
- System resistance coefficient
- Final input-volume relationships

For applications to system design, the chimney gas velocity is eliminated; however, actual velocity can be found readily, if needed.

Mass flow of combustion products (Step 1). Mass flow in a chimney or venting system may differ from that in the appliance, depending on the type of draft control or number of appliances operating in a multiple appliance system. Mass flow (rather than volume flow) is preferred because it remains constant in any continuous portion of the system, regardless of changes in temperature or pressure. For the chimney gases resulting from any combustion process, mass flow can be expressed as:

$$w = IM/1000 \qquad (3)$$

where:

- w = mass flow rate, lb/h
- I = appliance heat input, Btu/h
- M = ratio of mass flow to heat input, lb of chimney products per 1000 Btu of fuel burned. M depends on the composition of the fuel and excess air (or CO_2) percentage in the chimney.

Mean chimney gas temperature (Step 2). Chimney gas temperature, which is covered in a later section, depends on the fuel, appliance, draft control, chimney size, and configuration.

Density of gas within the chimney and theoretical draft both depend on gas temperature. While the gases flowing in a chimney system lose heat continuously, from the entrance to the exit, a single mean gas temperature must be used either in the design equation or the chart. Mean chimney gas density is essentially the same as that of air density at the same temperature. Thus, density may be found as:

$$\rho_m = 1.325B/T_m \qquad (4)$$

where:

- ρ_m = gas density, lb/ft^3
- B = local barometric pressure, in. Hg
- T_m = mean chimney gas temperature at average conditions in the system, °R

The constant in Equation (4) is a compromise value for typical humidity. The subscript m for density and temperature requires that these properties be calculated at mean gas temperature or vertical midpoint of a system (inlet conditions can be used where temperature drop is not significant).

Theoretical draft (Step 3). The theoretical draft of a gravity chimney or vent is the difference in weight between a given column of warm (light) chimney gas and an equal column of cold (heavy) ambient air. Chimney gas density or temperature, chimney height, and barometric pressure determine theoretical draft; flow is not a factor. The equation for theoretical draft assumes chimney gas density is the same as that of air at the same temperature and pressure, thus:

$$D_t = 0.2554BH\left(\frac{1}{T_o} - \frac{1}{T_m}\right) \qquad (5)$$

where:

- D_t = theoretical draft, in. of water
- H = height of chimney above datum of draft measurement, ft
- T_o = ambient temperature, °R

Theoretical draft thus increases directly with height and with the difference in density between the hot and cold columns.

System pressure loss resulting from flow (Step 4). In any chimney system, flow losses, expressed as pressure drop Δp in in. of water, absorb the energy difference between theoretical and available draft, thus:

$$\Delta p = D_t - D_a \qquad (6)$$

Available draft D_a is the static pressure defined by the appliance operating requirements as follows:

- Positive draft (negative chimney pressure) appliances: D_a, available draft is positive in Equation (6), thus $\Delta p < D_t$.
- Draft hood (neutral draft) appliances: $D_a = 0$, and $\Delta p = D_t$.
- Negative draft (positive pressure, above atmospheric, forced draft) appliances: D_a is negative in Equation (6) so that $\Delta p = D_t - (-D_a) = D_t + D_a$ and $\Delta p > D_t$.

Regardless of the sign of D_a, Δp is always positive.

In any duct system, flow losses resulting from velocity and resistance can be determined from the Bernoulli equation as:

$$\Delta p = \frac{k\rho_m V^2}{5.2(2g)} \qquad (7)$$

where:

- k = dimensionless system resistance coefficient of piping and fittings
- V = system gas velocity at mean conditions, ft/s
- g = gravitational constant 32.1740 ft/s^2

Pressure losses are thus directly proportional to the resistance factor and to the square of the velocity.

Chimney gas velocity (Step 5). Velocity in a chimney or vent varies inversely with gas density ρ_m and directly with mass flow rate w. The equation for gas velocity at mean gas temperature in the chimney is:

$$V = \frac{144 \times 4w}{3600\pi\rho_m d_i^2} \qquad (8)$$

where:

V = gas velocity, ft/s
d_i = inside diameter, in.
ρ_m = gas density, lb/ft^3

To express velocity as a function of input and chimney gas composition, w in Equation (8) is replaced by using Equation (3).

$$V = \frac{144 \times 4}{3600\pi\rho_m d_i^2} \times \frac{IM}{1000} \quad (9)$$

Thus, chimney velocity depends on the product of heat input I and the ratio of mass flow to input M.

System resistance coefficient (Step 6). The velocity head method for resistance losses assigns a fixed numerical coefficient (independent of velocity) or k factor to every fitting or turn in the flow circuit, as well as to piping.

Input, diameter, and temperature relationships (Step 7). To obtain a design equation in which all terms are readily defined, measured, or predetermined, the gas velocity and density terms must be eliminated. Thus, using Equation (4) to replace ρ and Equation (9) to replace V in Equation (7) gives:

$$\Delta p = \frac{k\rho_m V^2}{5.2(2g)} = \frac{k}{5.2(2g)}\left(\frac{T_m}{1.325B}\right)\left(\frac{144 \times 4IM}{3.6 \times 10^6}\right)^2 \quad (10)$$

Rearranging to solve for I and including the value of $2g$ gives:

$$I = 4.13 \times 10^5 \frac{d_i^2}{M}\left(\frac{\Delta p B}{k T_m}\right)^{0.5} \quad (11)$$

Solving for input using Equation (11) is a one-step process, given the diameter and configuration of the chimney. More frequently, however, input, available draft, and height are given and the diameter d_i must be found. Because system resistance is a function of the chimney diameter, a trial resistance value must be assumed to calculate a trial diameter. This method allows for a second (and usually accurate) solution for the final required diameter.

Volume of flow in chimney or system (Step 8). Volume flow may be calculated in a chimney system for which Equation (11) can be solved by solving Equation (7) for velocity at mean density (or temperature) conditions:

$$V = 18.3\sqrt{\Delta p/(k\rho_m)} \quad (12)$$

This equation can be expressed in the same terms as Equation (11) by using the density value ρ of Equation (4) in Equation (12) and then substituting Equation (12) for velocity V. Area is expressed in terms of d_i.

$$Q = 5.2d_i^2(\Delta p\, T_m/kB)^{0.5} \quad (13)$$

where:

Q = volume flow rate, cfm

The volume flow obtained from Equation (13) is at mean gas temperature T_m and at local barometric pressure B.

Equation (13) is useful in the design of forced draft and induced draft systems because draft fans are usually specified in terms of volume flow rate at some standard ambient or selected gas temperature. An induced draft fan is necessary for chimneys that are undersize, too low, or those that must be operated with draft in the manifold under all conditions.

Figure 1 is a graphic solution for Equations (11) and (13), which is accurate enough for most problems. However, to use either the equations or the chart, the details in the sections to follow should be understood so that proper choices can be made for mass flow, pressure loss, and heat transfer effects. Neither the design chart

Table 1 Mass Flow Equations for Common Fuels

	Ratio of Mass Flow to Input, M
Fuel	$M = \dfrac{\text{lb Total Products}}{\text{1000 Btu Fuel Input}}$
Natural gas	$0.705\left(0.159 + \dfrac{10.72}{\%CO_2}\right)$
LP (propane, butane, or mixture)	$0.706\left(0.144 + \dfrac{12.61}{\%CO_2}\right)$
No. 2 Oil (light)	$0.72\left(0.12 + \dfrac{14.4}{\%CO_2}\right)$
No. 6 Oil (heavy)	$0.72\left(0.12 + \dfrac{15.8}{\%CO_2}\right)$
Bituminous coal (soft)	$0.76\left(0.11 + \dfrac{18.2}{\%CO_2}\right)$
Type O waste or wood	$0.69\left(0.16 + \dfrac{19.7}{\%CO_2}\right)$

Percent CO_2 is determined in products with water condensed (dry basis).
Total products includes combustion products and excess air.

in Figure 1 nor the equations contain the same number or order of steps as the derivation; for example, a step disappears when theoretical and available draft are combined into Δp. Similarly, the examples selected vary in their sequence of solution, depending on which parameters are known and on the need for differing answers such as diameter for a given input, diameter versus height, or the amount of pressure boost from a forced draft fan.

MASS FLOW BASED ON FUEL AND COMBUSTION PRODUCTS

In chimney system design, the composition and flow rate of the flue gases must be assumed to determine the ratio of mass flow to input M. The mass flow equations in Table 1 illustrate the influence of fuel composition; however, additional guidance is needed for system design. The information provided for many heat-producing appliances is limited to whether they have been tested, certified, listed, or approved to comply with applicable standards. From this information and from the type of fuel and draft control, certain inferences can be drawn regarding the flue gases. Table 2 suggests typical values for the vent or chimney systems for gaseous and liquid fuels when specific outlet conditions for the appliance are not known. When combustion conditions are given in terms of excess air, Figure 2 can be used to estimate CO_2.

Figure 3 can be used with Figure 1 to estimate mass and volume flow. Flow conditions within the chimney connector, manifold, vent, and chimney vary with configuration and appliance design and are not necessarily the same as boiler or appliance outlet conditions. The equations and design chart (Figure 1) are based on the fuel combustion products and temperatures within the chimney system. If a gas appliance with draft hood is used, Table 2 recommends that dilution air through the draft hood reduce the CO_2 percentage to 5.3%. For equipment using draft regulators, the dilution and temperature reduction is a function of the draft regulator gate opening, which depends on excess draft. If the chimney system produces the exact draft necessary for the appliance, little dilution takes place.

For manifolded gas appliances that have draft hoods, the dilution through draft hoods of inoperative appliances must be con-

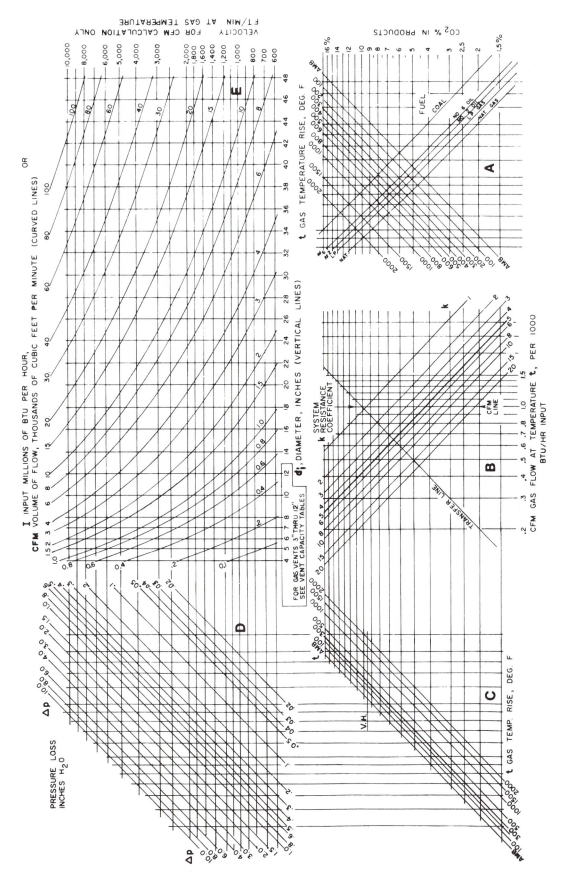

Fig. 1   Design Chart for Vents, Chimneys, and Ducts
(This solution of Equations (11) and (13) applies to combustion products and air.)

Table 2   Typical Chimney and Vent Design Conditions[a]

Fuel	Appliance	%CO_2	Temperature Rise, °F	Mass Flow/Input, M lb Total Products[b] per 1000 Btu Fuel Input	Vent Gas Density[c], lb/ft³	Flow Rate/Unit Heat Input[c], cfm per 1000 Btu/h at Gas Temperature
Natural gas	Draft hood	5.3	300	1.54	0.0483	0.531
Propane gas	Draft hood	6.0	300	1.59	0.0483	0.549
Natural gas						
Low efficiency	No draft hood	8.0	400	1.06	0.0431	0.410
High efficiency	No draft hood	7.0	240	1.19	0.0522	0.381
No. 2 oil	Residential	9.0	500	1.24	0.0389	0.532
Oil	Forced draft over 400,000 Btu/h	13.5	300	0.86	0.0483	0.293
Waste, Type O	Incinerator	9.0	1340	1.62	0.0213	1.268

[a] The values tabulated are for appliances with flue losses of 17% or more. For appliances with lower flue losses (high efficiency types), see appliance installation instructions or ask manufacturer for operating data.

[b] Total products include combustion products and excess air.

[c] At sea level and 60°F ambient temperature.

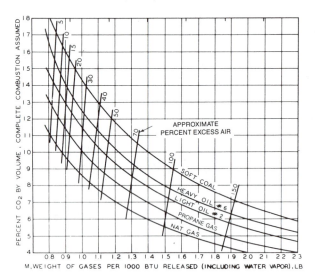

Fig. 2   Graphical Evaluation of Rate of Vent Gas Flow from Percent CO_2 and Fuel Rate

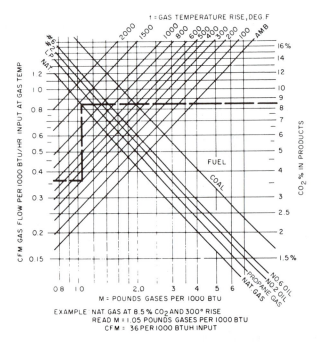

EXAMPLE: NAT. GAS AT 8.5% CO_2 AND 300° RISE
READ M = 1.05 POUNDS GASES PER 1000 BTU
CFM = 36 PER 1000 BTUH INPUT

Fig. 3   Flue Gas Weight and Volume Flow

Table 3   Mass Flow for Incinerator Chimneys

Type of Waste	Heat Value of Waste,[a] Btu/lb	Auxiliary Fuel[a] per Unit Waste, Btu/lb	Combustion Products cfm/lb Waste at 1400°F[b]	lb per lb Waste[c]	lb Products per 1000 Btu
Type 0	8500	0	10.74	13.76	1.62
Type 1	6500	0	8.40	10.80	1.66
Type 2	4300	0	5.94	7.68	1.79
Type 3	2500	1500	4.92	6.25	2.50
Type 4	1000	3000	4.14	5.33	5.33

[a] Auxiliary fuel may be used with any type of waste, depending on incinerator design.
[b] Specialized units may produce higher or lower outlet gas temperatures, which must be considered in sizing the chimney, using Equations (11) or (13) or the design chart.
[c] Multiply these values by pounds of waste burned per hour to establish mass flow.

sidered in precise system design. However, with forced draft appliances having wind box or inlet air controls, dilution through inoperative appliances may be unimportant, especially if pressure at the outlet of inoperative appliances is neutral (atmospheric level). Mass flow within incinerator chimneys must account for the probable heating value of the waste, its moisture content, and the use of additional fuel to initiate or sustain combustion. Classifications of wastes and corresponding values of M in Table 3 are based on recommendations from the Incinerator Institute of America. Combustion data given for Types 0, 1, and 2 wastes do not include any additional fuel. Where constant burner operation accompanies the combustion of waste, the additional quantity of products should be considered in the chimney design.

The system designer should obtain exact outlet conditions for the maximum rate operation of the specific appliance. This information can save chimney construction costs. For appliances with higher seasonal or steady-state efficiencies, however, special attention should be given to the manufacturer's venting recommendations because the flue products may differ in composition and temperature from conventional values.

CHIMNEY GAS TEMPERATURE AND HEAT TRANSFER

Figure 1 is based on a design ambient temperature of 60°F, and all temperatures given are in terms of rise above this ambient. Thus, the 300°F line indicates a 360°F observed vent gas temperature. Using a reasonably high ambient (such as 60°F) for design ensures improved operation of the chimney when ambient temperatures drop, because of greater temperature differentials and greater draft.

A design requires assuming an initial or inlet chimney gas temperature. In the absence of specific data, Table 4 is conservative and may be used to select a temperature. For appliances capable

**Table 4 Mean Chimney Gas Temperatures T_m
for Various Appliances**

Appliance Type	Initial Temperature T_m in Chimney, °F
Natural gas-fired heating appliance with draft hood (low efficiency)	360
LP gas-fired heating appliances with draft hood (low efficiency)	360
Gas-fired heating appliance, no draft hood	
Low efficiency	460
High efficiency	300
Oil-fired heating appliances (low efficiency)	560
Conventional incinerators	1400
Controlled air incinerators	1800 to 2400
Pathological incinerators	1800 to 2800
Turbine exhaust	900 to 1400
Diesel exhaust	900 to 1400
Ceramic kilns	1800 to 2400

Subtract 60°F ambient to obtain temperature rise for use with Figure 1.

**Table 5 Overall Heat Transfer Coefficients of
Various Chimneys and Vents**

	U, Btu/h·ft²·°F[a]		
Material	Observed	Design	Remarks
Industrial steel stacks	—	1.3	Under wet wind
Clay or iron sewer pipe	1.3 to 1.4	1.3	Used as single-wall material
Asbestos-cement gas vent	0.72 to 1.42	1.2	Tested per UL *Standard* 441
Black or painted steel stove pipe	—	1.2	Comparable to weathered galvanized steel
Single-wall galvanized steel	0.31 to 1.38	1.0	Depends on surface condition and exposure
Single-wall unpainted pure aluminum	—	1.0	No. 1100 or other bright surface aluminum alloy
Brick chimney, tile lined	0.5 to 1.0	1.0	For gas appliances in residential construction per NFPA 211
Double-wall gas vent, 1/4-in. air space	0.37 to 1.04	0.6	Galvanized steel outer pipe, pure aluminum inner pipe; tested per UL *Standard* 441
Double-wall gas vent, 1/2-in. air space	0.34 to 0.7	0.4	
Insulated prefabricated chimney	0.34 to 0.7	0.3	Solid insulation meets UL 103 when fully insulated

[a] U values based on inside area of chimney.

of operating over a range of temperatures, size should be calculated at both extremes to ensure an adequate chimney.

The drop in vent gas temperature from appliance to exit reduces capacity, particularly in sizes of 12 in. or less. In gravity Type B gas vents, which may be as small as 3 in. in diameter, and in other systems used for venting gas appliances, capacity is best determined from the *National Fuel Gas Code* (ANSI *Standard* Z223.1). In this code, the tables compensate for the particular characteristics of the chimney material involved, except for very high single-wall metal pipe. Over 12 through 18-in. diameters, the effect of heat loss diminishes greatly, because there is greater gas flow relative to system surface area. For 20-in. and greater diameters, cooling has little effect on final size or capacity.

A straight vertical vent or chimney directly off the appliance requires little compensation for cooling effects, even with smaller sizes. However, a horizontal connector running from appliance to the base of the vent or chimney has enough heat loss to diminish draft and capacity. Figure 4 is a plot of temperature correction C_u, which is a function of connector size, length, and material for either conventional single-wall metal or double-wall metal connectors.

To use Figure 4, estimate connector size and length and read the temperature multiplier. For example, 16 ft of single-wall connector, 7 in. in diameter has a multiplier of 0.61. If inlet temperature rise is 300°F above ambient, operating mean temperature rise will be 0.61 × 300 = 183°F rise. This factor adequately corrects the temperature at the midpoint of the vertical vent for heights up to 100 ft.

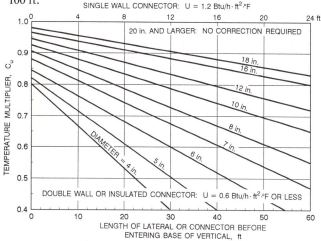

**Fig. 4 Temperature Multiplier C_u for Compensation
of Heat Losses in Connector**

The temperature multiplier must be applied to grids A and C, Figure 1 as follows:

1. In grid A, the temperature must be multiplied by 0.61. For an appliance with an outlet temperature rise above ambient of 300°F, flow in the vent is based on 300 (0.61), or 183°F rise.
2. This same 183°F rise must be used in grid C.

Determine Δp using a 183°F rise for theoretical draft to be consistent with the other two temperatures. (It is incorrect to multiply theoretical draft pressure by the temperature multiplier.)

The first trial solution for diameter, using Figure 1 or Equations (11) and (13), need not consider the cooling temperature multiplier, even for small sizes. A first approximate size can be used for the temperature multiplier for all subsequent trials because capacity is insensitive to small changes in temperature.

The correction procedure assumes the overall heat transfer coefficient of a vertical chimney is about 0.6 Btu/h·ft²·°F or less—the value for double-wall metal. This procedure does not correct for cooling in very high stacks constructed entirely of single-wall metal, especially those exposed to cold ambient temperatures. For severe exposures or excessive heat loss, a trial calculation assuming a conservative operating temperature shows whether capacity problems will be encountered.

For more precise heat loss calculations, Table 5 suggests overall heat transfer coefficients for various constructions (Segeler 1965), as installed in typical environments at usual flue gas flow velocities. For masonry, any additional thickness beyond the single course of brick plus tile liner used in residential chimneys decreases the coefficient.

Design Equations (11) and (13) do not account for the effects of heat transfer or cooling on flow, draft, or capacity. Equations (14) and (15) describe flow and heat transfer, respectively, within a venting system with $D_a = 0$.

$$q_m = 520 \sqrt{2g}\, c_p\, \frac{B}{29.92}\, \frac{A}{T_m}\, \left(\frac{H}{kT_o}\right)^{0.5}(T_m - T_o)^{1.5} \quad (14)$$

$$\frac{q}{q_m} = \frac{t_e}{t_m} = \exp\left(\pi \bar{U}\, d_f\, \frac{L_m t_m}{q_m}\right) \quad (15)$$

where:

$\exp x = e^x$
q = heat flow rate at vent inlet, Btu/s
q_m = heat flow rate at midpoint of vent, Btu/s
A = area of passage cross section, ft^2
H = height of chimney above grade or inlet, ft
$t_m = (T_m - T_o)$ = chimney gas mean temperature rise, °F
L_m = length from inlet to location (in the vertical) of mean gas temperature, ft
d_f = inside diameter, ft
$t_e = (T - T_o)$ = temperature difference entering system, °F
$\bar{U}$ = heat transfer coefficient, Btu/s·ft^2·°F
T_o = 520 °R

Assuming reasonable constancy of $\bar{U}$, the overall heat transfer coefficient of the venting system material, Equations (14) and (15) provide a solution for maximum vent gas capacity. They can also be used to develop cooling curves or calculate the length of pipe where internal moisture condenses. Kinkead (1962) details methods of solution and application to both individual and combined gas vents.

THEORETICAL DRAFT, AVAILABLE DRAFT, AND ALTITUDE CORRECTION

Equation (5) for theoretical draft is the basis for Figure 5, which can be used up to 1000 °F and 7000-ft elevation. Theoretical draft should be estimated and included in system calculations, even for appliances producing considerable positive outlet static pressure, to achieve the economy of minimum chimney size. Equation (5) may be used directly to calculate exact values for theoretical draft

Table 6 Approximate Theoretical Draft of Chimneys

Vent Gas Temperature Rise, °F	D_t per 100 ft, in. of water
100	0.2
150	0.3
200	0.4
300	0.5
400	0.6
500	0.7
600	0.8
800	0.9
1100	1.0
1600	1.1
2400	1.2

Notes: Ambient temperature = 60°F = 520°R
Chimney gas density = air density
Sea level barometric pressure = 29.92 in. Hg
Equation (5) may be used to calculate exact values for D_t at any altitude.

at any altitude. For ease of application and consistency with the design chart, Table 6 lists approximate theoretical draft for typical gas temperature rises above 60 °F ambient.

Appliances with fixed fuel beds, such as hand-fired coal stoves and furnaces, require positive available draft (negative gage pressure). Small oil heaters with pot-type burners, as well as residential furnaces with pressure atomizing oil burners, need positive available draft, which can usually be set by following the manufacturer's instructions for setting the draft regulator. Available draft requirements for larger packaged boilers or equipment assembled from components may be negative, zero (or neutral), or positive.

Compensation of theoretical draft for altitude or barometric pressure is usually necessary for appliances and chimneys functioning at elevations greater than 2000 ft. Depending on the design, the following approaches to pressure or altitude compensation are necessary for chimney sizing.

1. Design chart: Use sea level theoretical draft.
2. Equation (11): Use local theoretical draft with actual energy input or use sea level theoretical draft with energy input multiplied by ratio of sea level to local barometric pressure (Table 7 factor).
3. Equation (13): Use local theoretical draft and barometric pressure with volume flow at the local density.

Figure 1 may be corrected for altitude or reduced air density by multiplying the operating input by the factor in Table 7. Gas appliances with draft hoods, for example, are usually derated 4% per 1000-ft elevation above sea level when they are operated at 2000-ft altitude or above. The altitude correction factor derates the design input so that the vent size at altitude for derated gas equipment is effectively the same as at sea level. For other appliances where burner adjustments or internal changes might be used to adjust for reduced density at altitude, the same factors produce an adequately compensated chimney size. For example, an appliance operating at a 6000-ft elevation at 10,000,000 Btu/h input, but requiring the same draft as at sea level, should have a chimney selected on the basis of 1.25 times the operating input, or 12,500,000 Btu/h.

Theoretical draft must be estimated at sea level to calculate Δp for use with Figure 1. The altitude correction multiplier for input (Table 7) is the only method of correcting to other elevations. Reducing theoretical draft imposes an incorrect compensation on the chart.

SYSTEM FLOW LOSSES

Theoretical draft is always positive (unless chimney gases are colder than ambient air); however, available draft can be negative, zero, or positive. The pressure difference Δp, or theoretical minus available draft, overcomes the flow losses. Table 8 lists the pressure

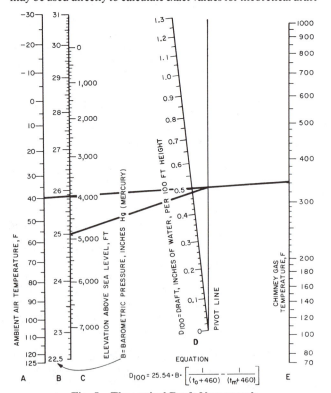

$$D_{100} = 25.54 \cdot B \cdot \left[\frac{1}{(t_o + 460)} - \frac{1}{(t_m + 460)}\right]$$

Fig. 5 Theoretical Draft Nomograph

Table 7 Altitude Correction

Altitude, ft	Barometric Pressure B, in. Hg	Factor
Sea level	29.92	1.00
2,000	27.82	1.08
4,000	25.82	1.16
6,000	23.98	1.25
8,000	22.22	1.34
10,000	20.58	1.45

Multiply operating input by the factor to obtain design input.

Table 8 Pressure Equations for Δp

Required Appliance Outlet Pressure or Available Draft D_a	Δp Equation	
	Gravity Only	Gravity Plus Inducer
1. Negative, needs positive draft	$\Delta p = D_t - D_a$	$\Delta p = D_t - D_a + D_b$
2. Zero, vent with draft hood or balanced forced draft	$\Delta p = D_t$	$\Delta p = D_t + D_b$
3. Positive, causes negative draft	$\Delta p = D_t + D_a$	$\Delta p = D_t + D_a + D_b$

Notes: Equations use absolute pressure for D_a.
D_b = static pressure boost of inducer at flue gas temperature and rated flow.

components for three draft configurations. The table applies to still air (no wind) conditions. The effect of wind on capacity or draft may be included by imposing a static pressure (either positive or negative) or by changing the vent terminal resistance loss. However, a properly designed and located vent terminal should cause little change in Δp at typical wind velocities.

Although small static draft pressures can be measured at the entrance and exit of gas appliance draft hoods, available draft at the appliance is effectively zero. Therefore, all theoretical draft energy produces chimney flow velocity and overcomes chimney flow resistance losses.

CHIMNEY GAS VELOCITY

The input capacity or diameter of a chimney may usually be found without determining flow velocity. Internal or exit velocity must occasionally be known, however, to ensure effluent dispersal or avoid flow noise. Also, the flow velocity of incinerator chimneys, turbine exhaust systems, and other appliances with high outlet pressures or velocities is needed to estimate piping loss coefficients.

Equations (7), (8), (9), and (12) can be applied to find velocity. Figure 1 may also be used to determine velocity; the right-hand scale of Group E reads directly in velocity for any combination of flow and diameter. For example, at 10,000 cfm and 34-in. diameter (6.305-ft² area), the indicated velocity is about 1600 fpm. The velocity may also be calculated by dividing volume flow rate by chimney area.

A similar calculation may be performed when the energy input is known. For example, a 34-in. chimney serving a 10 million Btu/h natural gas appliance, at 8.5% CO_2 and 300°F above ambient in the chimney, produces (from Figure 3) 0.36 cfm per 1000 Btu/h. Chimney gas flow rate is:

$$10,000,000(0.36/1000) = 3600 \text{ cfm}$$

Dividing by area to obtain velocity, $V = 3600/6.305 = 571$ fpm. The right-hand scale of Group E, Figure 1, may also be multiplied by the cfm per 1000 Btu to find velocity. For the same chimney design conditions, the scale velocity value of 1600 fpm is multiplied by 0.36 to yield a velocity 576 fpm in the chimney.

Chimney gas velocity affects the piping friction factor and also the roughness correction factor. The section on Resistance Coefficients has further information, and Example 2 illustrates how these factors are used in the velocity equations.

Chimney systems can operate over a wide range of velocities, depending on modulation characteristics of the burner equipment or the number of appliances in operation. The typical velocity in vents and chimneys ranges from 300 to 3000 fpm. A chimney design developed for maximum input and maximum velocity should be satisfactory at reduced input because theoretical draft is roughly proportional to flue gas temperature, while flow losses are proportional to the square of the velocity. So, as input is reduced, flow losses decrease more rapidly than system motive pressures.

Effluent dispersal may occasionally require a minimum upward chimney outlet velocity, such as 3000 fpm. A tapered exit cone can best meet this requirement. For example, to increase an outlet velocity from 1600 to 3000 fpm from the 34 in. chimney ($A = 6.305$ ft² area), the discharge area of the cone must be $6.305 \times 1600/3000 = 3.36$ ft², or 24.8 in. in diameter.

An exit cone avoids excessive flow losses because the entire system operates at the lower velocity, and a resistance factor is only added for the cone (Table 9). In this case, the added resistance for a gradual taper nearly equals:

$$k = (d_{i1}/d_{i2})^4 - 1 = (34/24.8)^4 - 1 = 2.53$$

Noise in chimneys may be caused by turbulent flow at high velocity or by combustion-induced oscillations or resonance.

Table 9 Resistance-Loss Coefficients

Component	Suggested Design Value Dimensionless[a]	Estimated Span and Notes
Inlet—acceleration		
Gas vent with draft hood	1.5	1.0 to 3.0
Barometric regulator	0.5	0.0 to 0.5
Direct connection	0.0	Also dependent on blocking damper position
Round elbow, 90°	0.75	0.5 to 1.5
Round elbow, 45°	0.3	—
Tee or 90° connector	1.25	1.0 to 4.0
Y connector	0.75	0.5 to 1.5
Cap, top		
Open straight	0.0	—
Low resistance (UL)	0.5	0.0 to 1.5
Other	—	1.5 to 4.5
Spark screen	0.5	—
Converging exit cone	$(d_{i1}/d_{i2})^4 - 1$	System designed using d_{i1}
Tapered reducer (d_{i1} to d_{i2})	$1 - (d_{i2}/d_{i1})^4$	System designed using d_{i2}
Increaser		See Chapter 2, 1989 ASHRAE Handbook—Fundamentals.
Piping k_L	$0.4 \dfrac{L, \text{ft}}{d_i, \text{in.}}$	Numerical coefficient from 0.2 to 0.5; see Figure 13, Chapter 2, 1989 ASHRAE Handbook—Fundamentals for size, roughness, and velocity effects.

[a] Initial assumption, when size is unknown:
  $k = 5.0$ for entire system, for first trial
  $k = 7.5$ for combined gas vents only
Note: For combined gravity gas vents serving two or more appliances (draft hoods), multiply total k (components + piping) by 1.5 to obtain gravity system design coefficient. (This rule does not apply to forced or induced draft vents or chimneys.)

Noise is seldom encountered in gas vent systems or in systems producing positive available draft, but it may be a problem with forced draft appliances. Turbulent flow noise may be avoided by designing for lower velocity, which may entail increasing the chimney size above the minimum recommended by the appliance manufacturer. Chapter 42 of the 1991 ASHRAE *Handbook—HVAC Applications* has more information on noise control.

RESISTANCE COEFFICIENTS

The resistance coefficient k which appears in Equations (11) and (13) and Figure 1 summarizes the friction loss of the entire chimney system, including piping, fittings, and configuration or interconnection factors. Capacity of the chimney varies inversely with the square root of k, while diameter varies as the fourth root of k. The insensitivity of diameter and input to small variations in k simplifies design. Analyzing such details as pressure regain, increasers and reducers, and gas cooling junction effects is unnecessary if slightly high resistance coefficients are assigned to any draft diverters, elbows, tees, terminations, and, particularly, piping.

The flow resistance of a fitting such as a tee with gases entering the side and making a 90° turn is assumed to be constant at $k = 1.25$, independent of size, velocity, orientation, inlet or outlet conditions, or whether the tee is located in an individual vent or in a manifold. Conversely, if the gases pass straight through a tee, as in a manifold, assumed resistance is zero, regardless of any area changes or flow entry from the side branch. For any chimney with fittings, the total flow resistance is a constant plus variable piping resistance—the latter being a function of centerline length divided by diameter. Table 9 suggests moderately conservative resistance coefficients for common fittings. Elbow resistance may be lowered by long radius turns; however, corrugated 90° elbows may have resistance values at the high end of the scale. The table shows resistance as a function of inlet diameter d_{i1} and outlet diameter d_{i2}.

Expressed mathematically, system resistance k may be written:

$$k = k_1 + n_2 k_2 + n_3 k_3 + k_4 + k_L + \text{etc.}$$

where:

k_1 = inlet acceleration coefficient
k_2 = elbow loss coefficient, n_2 = number of elbows
k_3 = tee loss coefficient, n_3 = number of tees
k_4 = cap, top, or exit cone loss coefficient
k_L = piping loss = FL/d (Figure 13, Chapter 2, of the 1989 ASHRAE *Handbook—Fundamentals*)

For combined gas vents using appliances with draft hoods, the summation k must be multiplied by 1.5 (see Example 5).

The resistance coefficient method adapts well to systems whose fittings cause significant losses. Even for extensive systems, an initial assumption of $k = 5.0$ gives a tolerably accurate vent or chimney diameter in the first trial solution. Using this diameter with the piping-resistance function in a second trial normally yields the final answer.

The minimum system resistance coefficient in a gas vent with a draft hood is always 1.0 because all gases must accelerate through the draft hood from almost zero velocity to vent velocity.

For a system connected directly to the outlet of a boiler or other appliance where the capacity is stated as full-rated heat input against a positive static pressure at the chimney connection, minimum system resistance is zero, and no value is added for existing velocity head in the system.

When size is unknown, the following k values may be used to run a first trial estimate:

$k = 5.0$ for the entire system
$k = 7.5$ for combined gas vents only

Note: For combined gas vents serving two or more appliances (draft hoods), multiply total k by 1.5 to obtain the gravity system design coefficient. (This multiplier does not apply to forced or induced draft vents or chimneys.)

For simplified design, the piping resistance loss function, $k_L = 0.4L/d_i$ applies for all sizes of vents or chimneys and for all velocities and temperatures. As diameter increases, this function becomes increasingly conservative, which is desirable because larger chimneys are more likely to be made of rough masonry construction or other materials with higher pressure losses. The 0.4 constant also introduces an increasing factor of safety for flow losses at greater lengths and heights.

Figure 6 is a plot of friction factor F versus velocity and diameter for commercial iron and steel pipe at a gas temperature of 300°F above ambient (Lapple 1949). The figure shows, for example, that a 48-in. diameter chimney with 80 ft/s gas velocity may have a constant as low as 0.2. In most cases, $0.3L/d_i$ gives reasonable design results for chimney sizes 18 in. and larger because systems of this size usually operate at gas velocities greater than 10 ft/s.

At 1000°F or over, the factors in Figure 6 should be multiplied by 1.2. Because Figure 6 is for commercial iron and steel pipe, an additional correction for greater or less surface roughness may be imposed. For example, the factor for a very rough 12-in. diameter pipe may be doubled at a velocity as low as 2000 fpm.

For most chimney designs, a friction factor F of 0.4 gives a conservative solution for diameter or input for all sizes, types, and operating conditions of prefabricated and metal chimneys; alternately, $F = 0.30$ is reasonable if the diameter is 18 in. or more. Because neither input nor diameter is particularly sensitive to the total friction factor, the overall value of k requires little correction.

Masonry chimneys, including those lined with clay flue tile, may have rough surfaces, tile shape variations that cause misalignment, and joints at frequent intervals with possible mortar protrusions. In addition, the inside cross-sectional area of liner shapes may be less than expected because of local manufacturing variations, as well as differences between claimed and actual size. To account for these characteristics, the estimate for the piping-loss coefficient should be on the high side, regardless of chimney size or velocity.

Computations should be made by assuming smooth surfaces and then adding a final size increase to compensate for shape factor and friction loss. It follows that performance or capacity of metal and prefabricated chimneys is generally superior to site-constructed masonry.

Configuration and Manifolding Effects

The most common configuration is the individual vent, stack, or chimney, in which one continuous system carries the products from appliance to terminus. Other configurations include the combined vent serving a pair of appliances, the manifold serving

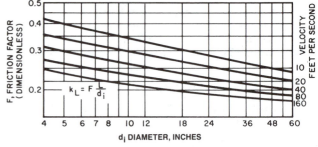

Fig. 6   Friction Factor for Commercial Iron and Steel Pipe
(Lapple 1949)

several, and branched systems with two or more lateral manifolds connected to a common vertical system. As the number of appliances served by a common vertical vent or chimney increases, the precision of design decreases because of diversity factors (variation in the number of units in operation) and the need to allow for maximum and minimum input operation (Stone 1957). For example, the vertical common vent for interconnected gas appliances must be larger than for a single appliance of the same input to allow for operating diversity and draft hood dilution effects. Connector rise, headroom, and configuration in the equipment room must be designed carefully to avoid draft hood spillage and related oxygen depletion problems.

For typical combined vents, the diversity effect must be introduced into Figure 1 and the equations by multiplying system resistance loss coefficients by 1.5 (see Table 9 note and Example 5). This multiplier compensates for junction effect and part-load operation.

Manifolds for appliances with barometric draft regulators can be designed without allowing for dilution by inoperative appliances. In this case, because draft regulators remain closed until regulation is needed, dilution under part-load is negligible. In addition, flow through any inoperative appliance is negligible because the combustion air inlet dampers are closed and the multiple-pass heat exchanger has a high internal resistance.

Multiple oil-burning appliances, for example, have a lower flow velocity and, hence, lower losses. As a result, they produce reasonable draft at part-load or with only one of several appliances in operation. Therefore, diversity of operation has little effect on chimney design. Some installers set each draft regulator at slightly different settings to avoid oscillations or hunting, possibly caused by burner or flow pulsations.

To determine the resistance coefficient of any portion of a manifold, the calculation begins with the appliance most distant from the vertical portion. All coefficients are then summed from its outlet to the vent terminus. The resistance of a series of tee joints to flow passing horizontally straight through them (not making a turn) is the same as that of an equal length of piping (as if all other appliances were off). This assumption holds, whether the manifold is tapered (to accommodate increasing input) or of constant size large enough for the accumulated input.

Coefficients are assigned only to inlet and exit conditions, to fittings causing turns, and to the piping running from the affected appliance to the chimney exit. Piping shape may initially be ignored, whether it is round, square, or rectangular and whether it is for connectors, vertical piping, or both.

Certain high-pressure, high-velocity packaged boilers require special manifold design to avoid turbulent flow noise. In such cases, manufacturers' instructions usually recommend increaser *Y* fittings, as shown in Figure 7. The loss coefficients listed in Table 9 for standard tees and elbows are higher than necessary for long radius elbows or *Y* entries. Occasionally, on equipment with high chimneys augmenting boiler outlet pressure, it may appear feasible to reduce the diameter of the vertical portion to below that recommended by the manufacturer. However, any reduction may cause turbulent noise, even though all normal design parameters have been considered.

Manufacturers' sizing recommendations, shown in Figure 7, apply to the specific appliance and piping arrangement shown. The values are conservative for long radius elbows or *Y* entries. Frequently, the boiler room layout forces the use of additional elbows. In such cases, the size must be increased to avoid excessive flow losses.

The design chart (Figure 1) can be used to calculate the size of a vertical portion smaller in area than the manifold, or a chimney connector smaller than the vertical, but with one simplifying assumption. The maximum velocity of the flue gas, which exists in the smaller of the two portions, is assumed to exist throughout the entire system. This assumption leads to a conservative design, as true losses in the larger area are lower than assumed. Further, if the size change is small, either as a contraction or enlargement, the added loss coefficient for this transition fitting (see Table 9)

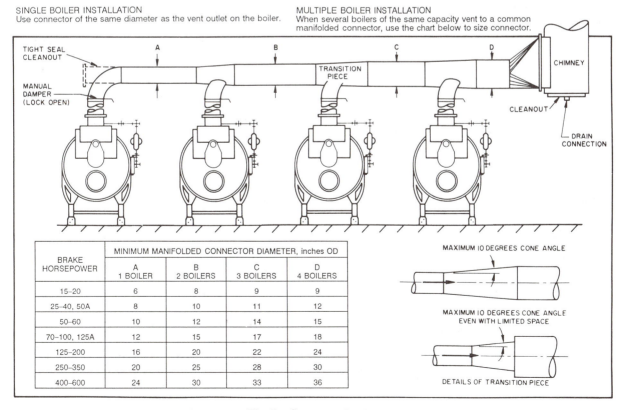

SINGLE BOILER INSTALLATION
Use connector of the same diameter as the vent outlet on the boiler.

MULTIPLE BOILER INSTALLATION
When several boilers of the same capacity vent to a common manifolded connector, use the chart below to size connector.

BRAKE HORSEPOWER	MINIMUM MANIFOLDED CONNECTOR DIAMETER, inches OD			
	A 1 BOILER	B 2 BOILERS	C 3 BOILERS	D 4 BOILERS
15–20	6	8	9	9
25–40, 50A	8	10	11	12
50–60	10	12	14	15
70–100, 125A	12	15	17	18
125–200	16	20	22	24
250–350	20	25	28	30
400–600	24	30	33	36

MAXIMUM 10 DEGREES CONE ANGLE

MAXIMUM 10 DEGREES CONE ANGLE EVEN WITH LIMITED SPACE

DETAILS OF TRANSITION PIECE

Fig. 7   Connector Design

is compensated for by reduced losses in the enlarged part of the system.

These comments on size changes apply more to individual than to combined systems because it is undesirable to reduce the vertical area of the combined type and, more frequently, it is desirable to enlarge it. If an existing vertical chimney is slightly undersized for the connected load, the complete chart method must be applied to determine whether a pressure boost is needed, as size is no longer a variable.

Sectional gas appliances with two or more draft hoods do not pose any special problems if all sections fire simultaneously. In this case, the designer can treat them as a single appliance. The appliance installation instructions either specify the size of manifold for interconnecting all draft hoods or require a combined area equal to the sum of all attached draft hood outlet areas. Once the manifold has been designed and constructed, it can be connected to a properly sized chimney connector, vent, or chimney. If the connector and chimney size is computed as less than manifold size (as may be the case with a tall chimney), the operating resistance of the manifold will be lower than the sum of the assigned component coefficients because of reduced velocity.

The general rule for conservative system design in which manifold, chimney connector, vent, or chimney are different sizes, can be stated as: *Always assign full resistance coefficient values to all portions carrying combined flow and determine system capacity from the smallest diameter carrying the combined flow.* In addition, horizontal chimney connectors or vent connectors should pitch upward toward the stack, 1/4 in. per foot minimum.

The following sample calculation, using Equation (11)—which differs from the original order of steps used with Figure 1—illustrates the direct solution for input, velocity, and volume. A calculation for input that is derived from the chart differs because of the chart arrangement.

Example 1. Find the input capacity (Btu/h) of a vertical, double-wall Type B gas vent, 24 in. in diameter, 100 ft high at sea level. This vent is used with draft hood natural gas-burning appliances.

Solution:

Step 1. Mass flow from Table 2. $M = 1.54$ lb/1000 Btu for natural gas, if no other data are given.

Step 2. Temperature from Table 4. Temperature rise $= 300°F$ and $T = 360 + 460 = 820°R$ for natural gas.

Step 3. Theoretical draft from Table 6. For 100 ft height at 300°F rise, $D = 0.5$ in. of water.
Available draft for draft hood appliances: zero.

Step 4. Flow losses from Table 8. $\Delta p = D = 0.5$; or flow losses for a gravity gas vent equal theoretical draft at mean gas temperature.

Step 5. Resistance coefficients from Table 9. For a vertical vent:

Draft hood	$k_1 = 1.5$
Vent cap	$k_4 = 1.0$
For 100 ft piping	$k_1 = 0.4 (100/24) = 1.67$
	System $k = 4.17$

Step 6. Solution for input.
Altitude: Sea level, $B = 29.92$ from Table 7. $d_i = 24$ in. These values are substituted into Equation (11) as follows:

$$I = 4.13 \times 10^5 \frac{(d_i)^2}{M} \left(\frac{\Delta p B}{k T_m} \right)^{0.5}$$

$$= 4.13 \times 10^5 \frac{(24)^2}{1.54} \left(\frac{(0.5)(29.92)}{(4.17)(820)} \right)^{0.5}$$

$$I = 10.2 \times 10^6 \text{ Btu/h input capacity.}$$

Step 7. A solution for velocity requires a prior solution for input to apply to Equation (9). First using Equation (4):

$$\rho_m = 1.325 \frac{29.92}{820} = 0.0483 \text{ lb/ft}^3$$

$$V = \frac{(10.2 \times 10^6)(1.54)}{1000} \times \frac{1}{(19.63)(0.0483)(24)^2}$$

$$= 28.8 \text{ ft/s}$$

Step 8. Volume flow can now be found, because velocity is known. The flow area of 24 in. diameter is 3.14 ft², thus:

$$Q = (60 \text{ s/min})(3.14 \text{ ft}^2)(28.8 \text{ ft/s}) = 5426 \text{ cfm}$$

No heat loss correction is needed to find the new gas temperature because the size is greater than 20 in., and this vent is vertical, with no horizontal connector.

For the same problem, Figure 1 requires a different sequence of solution. The ratio of mass flow to input for a given fuel (with parameter M) is not used directly; the chart requires selecting a CO_2 percentage in the chimney, either from Table 2 or from operating data on the appliances. Then, the temperatures are entered only as rise above ambient. The solution path is as follows:

1. Enter Group A at 5.3% CO_2, and construct line horizontally to left intersecting Nat. gas.
2. From Nat. gas intersection, construct vertical line to $t = 300$.
3. From 300 intersection in A, go horizontally left to transfer line to B.
4. Transfer line vertically to $k = 4.17$.
5. From $k = 4.17$ run horizontally left to $t = 300$ in C.
6. From 300 go up vertically to $\Delta p = 0.54$ in Group D.
7. From $\Delta p = 0.54$ in D, go horizontally right to intersection $d_i = 24$ in. in Group E.
8. Read capacity or input at 24 in. intersection in E as 10.2 million Btu/h between curved lines.

If input is known and diameter must be found, the procedure is the same as with Equation (11). A preliminary k, usually 5.0 for an individual vent or chimney, must be estimated to find a trial diameter. This diameter is used to find a corrected k, and the chart is solved again for diameter.

CHIMNEY CAPACITY CALCULATION EXAMPLES

Figures 8, 9, 10, and 11 show chimney capacity for individually vented appliances computed by the methods presented. These capacity curves may be used to estimate input or diameter for the design chart (Figure 1) or Equations (11) or (13). These capacity curves apply primarily to individually vented appliances with a lateral chimney connector; systems with two or more appliances or additional fittings require a more detailed analysis. Figures 8 through 11 assume the length of the horizontal connector is no less than 10 ft, or no longer than 50% of the height, or 50 ft, whichever is less. Chimney heights of 10 to 20 ft assume a fixed 10-ft long connector. Between 20 and 100 ft, the connector is 50% of the height. If the chimney height exceeds 100 ft, the connector is fixed at 50-ft long.

For a chimney of similar configuration but with a shorter connector, the size indicated is slightly larger than necessary. In deriving the data for Figures 8 to 11, additional conservative assumptions were used, including the temperature correction C_u for double-wall laterals (see Figure 4), and a constant friction factor for all sizes (0.4 L/d).

The loss coefficient for a low resistance cap is included in Figures 8 to 10. If no cap is installed, these figures indicate a larger size than needed.

Figure 8 applies to a gas vent with draft hood and a lateral that runs to the vertical section. Maximum static draft is developed at the base of the vertical, but friction reduces the observed value to less than the theoretical draft. Areas of positive pressure may exist

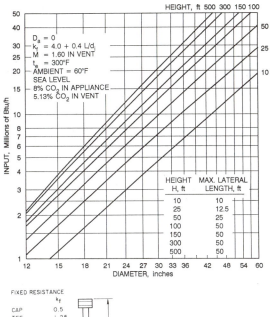

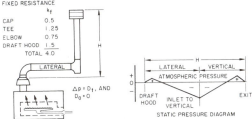

Fig. 8   Gas Vent with Lateral

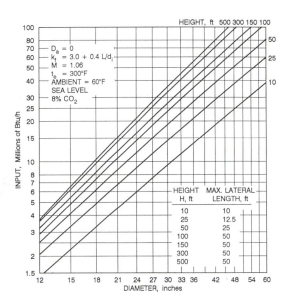

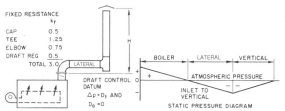

**Fig. 10   Forced Draft Appliance with Neutral
(Zero) Draft (Negative Pressure Lateral)**

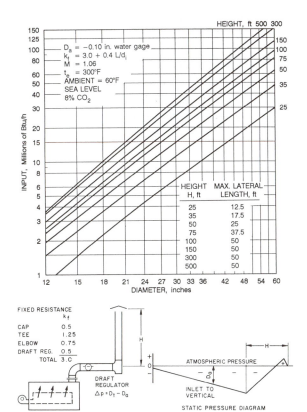

**Fig. 9   Draft Regulated Appliance with 0.10 in. (water gage)
Available Draft Required**

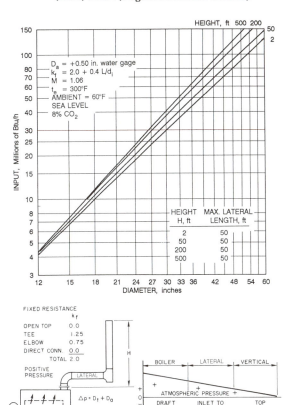

Fig. 11   Forced Draft with Positive Outlet Pressure

at the elbow above the draft hood and at the inlet to the cap. The height of the system is the vertical distance from the draft hood outlet to the vent cap.

Figure 9 applies to a typical boiler system requiring negative combustion chamber pressure, as well as negative static outlet pressure. The chimney static pressure is below atmospheric pressure, except for the minor outlet reversal caused by cap resistance. Height of this system is the difference in elevation between the point of draft measurement (or control) and the exit. (Chimney draft should not be based on the height above the boiler room floor).

Figure 10 illustrates the use of a negative static pressure connector serving a forced draft boiler. This system minimizes flue gas leakage in the equipment room. The draft is balanced or neutral, which is similar to a gas vent, with zero draft at the appliance outlet and pressure loss Δp equal to theoretical draft.

Figure 11 applies to a forced draft boiler capable of operating against a positive static outlet pressure of up to 0.50 in. The chimney system has no negative pressure, so outlet pressure may be combined with theoretical draft to get minimum chimney size. For chimney heights or system lengths less than 100 ft, the effect of adding 0.50 in. positive pressure to theoretical draft causes all curves to fall into a compressed zone, best shown as a single line. For sizes less than 24 in., height has no effect on capacity, while for sizes of 24 in. or more, heights greater than 100 ft add to capacity. The capacity line for heights from 10 to 100 ft applies also to the size of a straight horizontal discharge or exhaust system with no vertical rise and up to 100 ft long. An appliance that can produce 0.50 in. positive forced draft is adequate for venting any simple arrangement with up to 100 ft of flow path and no wind back pressure, for which additional forced draft is required.

Design Chart Examples

The following examples illustrate the use of the design chart (Figure 1) and the corresponding equations.

Example 2. Individual gas appliance with draft hood at sea level (see Figure 12).
Input = 980,000 Btu/h; natural gas; 80-ft high with 40-ft lateral; Double-wall $U = 0.6$. Find the vent diameter.

Solution: Assume $k = 5.0$. The following factors are used successively: $CO_2 = 5.3\%$ for Nat Gas; Gas temp. rise = 300°F; Transfer line: $k = 5.0$; $\Delta p = 0.537 (80/100) = 0.43$.

Preliminary solution: At $I = 0.98 \times 10^6$, read diameter, $d_i = 8.5$ in. Use next largest diameter, 9 in., to correct temperature and theoretical draft; compute new system k and Δp. From Figure 4, $C_U = 0.70$, $t = (0.70)(300) = 210°F$. This temperature determines the new value of theoretical draft, found by interpolation between 200 and 250°F in

Table 6 = 0.422, $D_t = 0.8(0.422) = 0.34$ in. of water. The four fittings of the system have a total fixed $k_f = 4.0$. Add k_L, the piping component for 120 ft of 9 in. diameter, to k_f, or $k = 4.0 + 0.4 (120/9) = 9.3$ for the system. System losses, Δp, equal theoretical draft = 0.34.

Final solution: Returning to the chart, the factors are: $CO_2 = 5.3\%$; Nat. gas; Gas temp. rise = 210°F; Transfer line: $k = 9.3$; Gas temp. rise = 210°F; $\Delta p = 0.34$. At $I = 0.98 \times 10^6$, read the diameter = 10 in., which is the correct answer. Had system resistance been found using 10 in. rather than 9 in., the final size would be less than 10 in., based on a system k less than 9.3.

Example 3. Gravity incinerator chimney (see Figure 13).
Located at 8000 ft elevation, the appliance burns 600 lb/h of Type O waste with 100% excess air at 1400°F outlet temperature. Ambient temperature is 60°F. Outlet pressure is zero at low fire, +0.10 in. of water at high fire. The chimney will be prefabricated medium heat-type with a 60-ft connector, a roughness factor of 1.2. The incinerator outlet is 18 in. in diameter, and it normally uses a 20-ft vertical chimney. Find the diameter of the chimney, the connector, and the height required to overcome flow and fitting losses.

Solution:

1. Mass flow from Table 3 is 13.76 lb gases per lb of waste or $w = 600(13.76) = 8256$ total lb/h.

2. Mean chimney gas temperature is based on 60-ft length of 18-in. diameter, double-wall chimney (see Figure 4), $C_u = 0.83$. Temperature rise = 1400 − 60 = 1340°F; thus, $t_m = 1340(0.83) = 1112°F$ rise above 60°F ambient. $T_m = 1112 + 60 + 460 = 1632°R$. Use this temperature to find gas density at 8000 ft elevation (from Table 7 B = 22.22 in. Hg):

$$\rho_m = 1.325(22.22/1632) = 0.0180 \text{ lb/ft}^3$$

3. Find the required height by finding theoretical draft per foot from Equation (5) or Figure 5, because Table 6 applies only to sea level.

$$D_t/H = (0.2554)(22.22) \left(\frac{1}{520} - \frac{1}{1632} \right)$$
$$= 0.0074 \text{ in. of water per ft of height}$$

4. Find Δp, allowable pressure loss, as if the incinerator chimney is for a positive pressure appliance (see Table 8) having an outlet pressure of +0.1 in. of water. Thus, $\Delta p = D_t + D_a$, where $D_t = 0.0074H$, $D_a = 0.10$ and $\Delta p = 0.0074H + 0.1$ in. of water.

5. Calculate flow velocity at mean temperature from Equation (8) to balance flow losses against diameter/height combinations:

$$V = \frac{8256}{(19.63)(0.018)(18)^2} = 72 \text{ ft/s}$$

This velocity exceeds the capability of a gravity chimney of moderate height and may require a draft inducer if an 18-in. chimney must be used. Verify by calculating resistance and flow losses by the following steps.

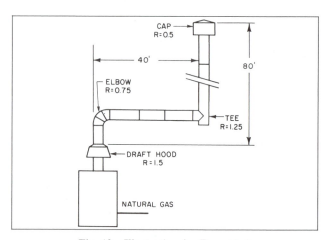

Fig. 12   Illustration for Example 2

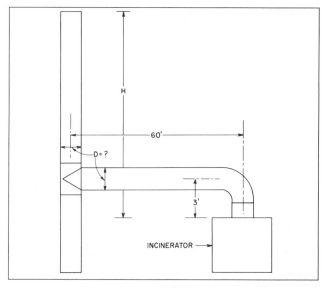

Fig. 13   Illustration for Example 3

6. Resistance coefficients for fittings from Table 9 are:

1 Tee	1.25
1 Elbow	0.75
Spark screen	0.50
Total fitting	2.50

The piping resistance, adjusted for length, diameter, and a roughness factor of 1.2 must be added to the total fitting resistance. From Figure 6, find the friction factor at 18-in. diameter and 72 ft/s as 0.22. Assuming 20 ft of height with a 60 ft lateral, piping friction loss is:

$$\frac{FL}{d_i} = \frac{(0.22)(1.2)(80)}{18} = 1.17$$

or total friction, $k = 1.17 + 2.50 = 3.67$

Use Equation (7) to find Δp, which will determine if this chimney height and diameter are suitable.

$$\Delta p = \frac{(3.67)(0.0180)(72)^2}{(5.2)(64.4)} = 1.02 \text{ in. of water flow losses}$$

For these operating conditions, available draft plus theoretical yields:

$$\Delta p = 0.0074(20) + 0.1 = 0.148 + 0.1 = 0.248 \text{ in. of water driving force}$$

Flow losses of 1.02 in. exceed the 0.248-in. driving force; thus, the selected diameter, height, or both are incorrect, and this chimney will not work. This can also be shown by comparing draft per foot with flow losses per foot for the 18-in. diameter configuration. Flow losses per foot of 18-in. chimney are:

$$1.02/80 = 0.0128 \text{ in. of water}$$

Draft per foot of height $= 0.0074$ in. of water

Regardless of how high the chimney is made, losses caused by a 72 ft/s velocity build up faster than draft.

7. A draft inducer could be selected to make up the difference between losses of 1.03 in. and the 0.25-in. driving force. Operating requirements are:

$$w/(60\rho) = 8256/(60 \times 0.0180) = 7644 \text{ cfm}$$

$$\Delta p = 1.02 - 0.248 = 0.772 \text{ in. of water at 7644 cfm and } 1112\,°F$$

If the inducer selected injects single or multiple air jets into the gas stream (see Figure 23c), it should be placed only at the chimney top or outlet. This location requires no compensation for additional air introduced by an enlargement downstream from the inducer.

Because 18 in. is too small, assume that a 24-in. diameter may work at a 20-ft height and recalculate.

1. As before, $w = 8256$ lb/h of gases.

2. At 24-in. diameter, no temperature correction is needed for the 60-ft connector. Thus, $T_m = 1400 + 460 = 1860\,°R$ (see Table 4) and density is:

$$\rho_m = (1.325)(22.22/1860) = 0.0158 \text{ lb/ft}^3$$

3. Theoretical draft per foot of chimney height:

$$D_t/H = 0.2554(22.22)\left(\frac{1}{520} - \frac{1}{1860}\right) = 0.00786 \text{ in. of water per foot}$$

4. Velocity is:

$$V = \frac{8256}{(19.63)(0.0158)(24)^2} = 46.2 \text{ ft/s}$$

and the friction factor from Figure 6 is 0.225, which when multiplied by a roughness factor of 1.2 for the piping used, is 0.27.

5. For the entire system with $k = 2.50$ for fittings and 80 ft of piping, find $k = 2.5 + 0.27(80/24) = 3.4$. Calculating:

$$\Delta p = \frac{(3.4)(0.0158)(46.2)^2}{(5.2)(64.4)} = 0.342 \text{ in. of water}$$

$D_t + D_a = (20)(0.00786) + 0.1 = 0.257$ in. of water and driving force is less than losses. The small difference indicates that while 20 ft of height is insufficient, additional height may solve the problem. The added height must make up for 0.085 additional draft. As a first approximation:

Added height = Additional draft/draft per foot
$$= 0.085/0.00786 = 10.8 \text{ ft}$$

Total height $= 20 + 10.8 = 30.8$ ft

This is less than the actual height needed because resistance changes have not been included. For an exact solution for height, the driving force can be equated to flow losses as a function of H. The complete equation is:

$$0.10 + 0.00786H = \frac{(3.175 + 0.01125H)(0.0158)(46.2)^2}{(5.2)(64.4)}$$

Solving, $H = 32.66$ ft at 24-in. diameter.

Checking by substitution, total driving force $= 0.357$ in. of water and total losses, based on a system with 92.15 ft of piping, equals 0.356 in. of water.

The value of $H = 32.66$ ft is the minimum necessary for proper system operation. Because of the great variation in fuels and firing rate with incinerators, greater height for assurance of adequate draft and combustion control should be used. An acceptable height would be from 40 to 50 ft.

Example 4. Two forced draft boilers (see Figure 14).

This example shows how multiple-appliance chimneys can be separated into subsystems. Each boiler is rated 100 boiler horsepower on No. 2 oil. Manufacturer states operation at 13.5% CO_2 and 300 °F rise, against 0.50 in. positive static pressure at outlet. Chimney has 20-ft single-wall manifold and is 50-ft high above sea level. Find size of connectors, manifold, and vertical.

Solution: First, find the capacity or size of the piping and fittings from boiler A to the tee over boiler B. Then, size the boiler B tee and all subsequent portions to carry the combined flow of A and B. Also, check the subsystem for boiler B; however, because its shorter length compensates for greater fitting resistance, its connector may be the same size as for boiler A.

Find the size for combined A and B flow either by assuming $k = 5.0$ or estimating that the size will be twice that found for boiler A operating by itself. Estimate system resistance for the combined portion by including those fittings in the B connector, as well as those in the combined portion.

Data needed for solution for boiler A using Equation (11) for No. 2 oil at 13.5% CO_2 is:

$$M = 0.72\left(0.12 + \frac{14.4}{13.5}\right) = 0.854 \text{ lb/1000 Btu} \quad \text{(Table 1)}$$

For a temperature rise of 300 °F and an ambient of 60 °F, $T_m = 300 + 60 + 460 = 820\,°R$.

Theoretical draft from Table 6 $= 0.5$ in. of water for 100 ft of height; for 50-ft height, $D_t = (0.5)50/100 = 0.25$ in. of water.

Using positive outlet pressure of 0.5 in. of water as available draft, $\Delta p = 0.5 + 0.25 = 0.75$ in. of water.

Assume $k = 5.0$, and for 80% efficiency, input is 100 times boiler horsepower or:

$$I = 100(42,000) = 4.2 \times 10^6 \text{ Btu/h}$$

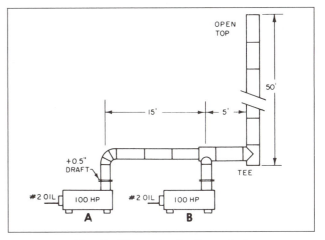

Fig. 14 Illustration for Example 4

Substitute in Equation (11):

$$4.2 \times 10^6 = 4.13 \times 10^5 \frac{(d_i)^2}{0.854} \left(\frac{(0.75)(29.92)}{(5.0)(820)} \right)^{0.5}$$

Solving, $d_i = 10.83$ in. as a first approximation. Find correct k using next largest diameter, or 12 in.

Inlet acceleration	= 0.0
90° Elbow	= 0.75
Tee	= 1.25
Piping 0.4(70/12)	= 2.33
System k	= 4.33

Note: Assume tee over boiler B has $k = 0$ in subsystem A.

Corrected temperature rise (see Figure 4 for single wall) = 0.75 (300) = 225 or $T_m = 225 + 60 + 460 = 745°R$. This temperature changes D_t to 0.425 (Table 6) per 100 ft or 0.21 for 50 ft. Δp becomes 0.21 + 0.50 = 0.71 in. of water.

$$4.2 \times 10^6 = 4.13 \times 10^5 \frac{(d_i)^2}{0.854} \left(\frac{(0.71)(29.92)}{(4.33)(745)} \right)^{0.5}$$

$d_i = 10.35$ in., or a 12-in. diameter is adequate.

For size of manifold and vertical, starting with the tee over boiler B, assume 16 in. (see also Figure 11).

System $k = 3.9$ for the 55 ft of piping from B to outlet.

Inlet acceleration	= 0.0
Two tees (boiler B subsystem)	= 2.5
Piping 0.4(55/16)	= 1.4
Total	= 3.9

Temperature and Δp will be as corrected (a conservative assumption) in the second step for boiler A. Having assumed a size, find input.

$$I = 4.13 \times 10^5 \frac{16^2}{0.854} \left(\frac{(0.71)(29.92)}{(3.9)(745)} \right)^{0.5} = 10.59 \times 10^6 \text{Btu/h}$$

A 16-in. diameter manifold and vertical is more than adequate. Solving for the diameter at the combined input, $d_i = 14.25$ in.; thus a 15- or 16-in. chimney must be used.

Note: Regardless of calculations, do not use connectors smaller than the appliance outlet size in any combined system. Applying the temperature correction for a single-wall connector has little effect on the result because positive forced draft is the predominate motive force for this system.

Example 5. Six gas boilers manifolded at 6000-ft elevation.

Each boiler is fired at 1.6×10^6 Btu/h, with draft hoods and an 80-ft long manifold connecting into a 400-ft high vertical, as per Figure 15. Each boiler is controlled individually. Find the size of the constant-diameter manifold, vertical, and connectors with a 2-ft rise. All are double-wall.

Solution: Simultaneous operation determines both the vertical and manifold sizes. Assume the same appliance operating conditions as in Example 1: $CO_2 = 5.3\%$, natural gas, temperature rise = 300°F. Initially assume $k = 5.0$ is multiplied by 1.5 for combined vent (see note at bottom of Table 9); thus, design $k = 7.5$. For gas vent at 400-ft height, $\Delta p = 4 \times 0.5 = 2.0$ in. At 6000-ft elevation, operating input must be multiplied by altitude correction (Table 7) of 1.25. Total design input is

$1.6 \times 10^6(6)(1.25) = 12 \times 10^6$Btu/h. From Table 2, $M = 1.54$ at operating conditions. $T_m = 300 + 60 + 460 = 820°R$, and $B = 29.92$ because the 1.25 input multiplier corrects back to sea level. Using Equation (11):

$$12 \times 10^6 = 4.13 \times 10^5 \frac{(d_i)^2}{1.54} \left(\frac{2 \times 29.92}{7.5 \times 820)} \right)^{0.5}$$

$$d_i = 21.3 \text{ in.}$$

Because the diameter is greater than 20 in., no temperature correction is needed.

Recompute k using 22-in. diameter:

Draft hood inlet acceleration	= 1.5
Two tees (connector and base of chimney)	= 2.5
Low resistance top	= 0.5
Piping 0.4(480/22)	= 8.8
System k	= 13.3

Combined gas vent design k = multiple vent factor 1.5(13.3) = 20.0. Substitute again in Equation (11):

$$12 \times 10^6 = 4.13 \times 10^5 \frac{(d_i)^2}{1.54} \left(\frac{2 \times 29.92}{20.0 \times 820} \right)^{0.5}$$

$$d_i = 27.2 \text{ in.; thus use 28 in.}$$

System k (fixed)	= 4.5
Piping 0.4(480/28)	= 6.9
	11.4

Design $k = 11.4(1.5) = 17.1$

Substitute in Equation (11) to obtain the third trial:

$$12 \times 10^6 = 4.13 \times 10^5 \frac{(d_i)^2}{1.54} \left(\frac{2 \times 29.92}{17.1 \times 820} \right)^{0.5}$$

$$d_i = 26.2 \text{ in.}$$

The third trial is less than the second and again shows the manifold and vertical chimney diameter to be between 26 and 28 in.

For connector size, see the *National Fuel Gas Code* for double-wall connectors of combined vents. The height limit of the table is 100 ft—do not extrapolate and read the capacity of 18 in. diameter as 1,740,000 at 2-ft rise. Use 18-in. connector or draft hood outlet, whichever is larger. No altitude correction is needed for connector size; the draft hood outlet size considers this effect.

Note: Equation (11) can also be solved at local elevation for exact operating conditions. At 6000 ft, local barometric pressure is 24 in., and assumed theoretical draft must be corrected in proportion to the reduction in pressure:

$D_t = 2 (24.0/29.92) = 1.60$ in. Operating input of 9.6×10^6 Btu/h is used to find d_i again taking final $k = 17.1$

$$9.6 \times 10^6 = 4.13 \times 10^5 \frac{(d_i)^2}{1.54} \left(\frac{1.60 \times 24.0}{17.1 \times 820} \right)^{0.5}$$

$$d_i = 26.2 \text{ (same as before)}$$

This example illustrates the equivalence of the chart method of solution with the solution by Equation (11). Equation 11 gives the correct solution using either Method 1, with the input only corrected back to sea level condition, or Method 2, using Δp at the elevation and local barometric pressure with operating input at altitude. Method 1, correcting input only, is the only choice with Figure 1 because the design chart cannot correct to local barometric pressure.

Example 6. Pressure boost for undersized chimney (not illustrated).

A natural gas boiler (no draft hood) is connected to an existing 12-in. diameter chimney. Input is 4×10^6 Btu/h operating at 10% CO_2 and 300°F rise above ambient. System resistance loss coefficient, $k = 5.0$ with 20-ft vertical height. Appliance operates with neutral outlet static draft; thus, $D_a = 0$.

a. How much draft boost is needed at operating temperature?
b. What fan rating is required at 60°F ambient temperature?
c. Where should the fan be located in the system?

Solution: Combustion data—10% CO_2 at 300°F rise indicates 0.31 cfm per 1000 Btu/h, using Figure 3.

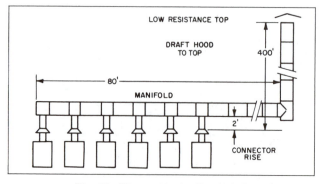

Fig. 15   Illustration for Example 5

Total flow rate = (0.31/1000)(4 × 10⁶) = 1240 cfm at chimney gas temperature. Then, using Equation (13), the only unknown is Δp.

Substituting:

$$1240 = 5.2(12)^2 \left(\frac{\Delta p (300 + 60 + 460)}{(5.0)(29.92)} \right)^{0.5}$$

$\Delta p = 0.50$ in. of water needed at 300°F rise. Pressure boost supplied by fan must equal Δp minus theoretical draft (Table 8) when available draft is zero. For 20 ft of height at 300°F rise,

$$D_t = 0.5(20/100) = 0.10 \text{ in. of water}$$

a. Draft boost = $\Delta p - D_t = 0.5 - 0.10 = 0.40$ in. of water at operating temperature.

b. The draft fans may be rated with ambient air at 60°F and 1240 cfm. Flow rate is (see Fan Laws in Chapter 18) inversely proportional to absolute gas temperature. Thus, for ambient air,

$$D_b = 0.40 \, \frac{T_m}{T_o} = 0.40 \left(\frac{300 + 60 + 460}{60 + 460} \right) = 0.63 \text{ in. of water}$$

This pressure is needed to produce 0.40 in. at operating temperature. In specifying power ratings for draft fan motors, a safe policy is to select one that operates at the required flow rate at ambient temperature and pressure (see Example 7).

c. A fan can be located anywhere from boiler outlet to chimney outlet. Regardless of location, the amount of boost is the same; however, chimney pressure relative to atmosphere will change. At boiler outlet, the fan pressurizes the entire connector and chimney. Thus, the system should be gas tight to avoid leaks. At the chimney outlet, the system is below atmospheric; any leaks will flow into the system and seldom cause problems. With an ordinary sheet metal connector attached to a tight vertical chimney, the fan may be placed close to the vertical chimney inlet. Thus, the connector leaks safely inward, while the vertical chimney is under pressure.

Example 7. Draft inducer selection (see Figure 16).

A third gas boiler must be added to a two-boiler system with an 18-in. diameter, 15-ft horizontal, with 75 ft of total height. Outlet conditions for natural gas draft hood appliances are 5.3% CO_2 at 300°F rise. Boilers are controlled individually, each with 1.6 × 10⁶ Btu/h or 4.8 × 10⁶ total input. System is now undersized for gravity full-load operation. Find capacity, pressure, size, and power rating of a draft inducer fan installed at the outlet.

Solution: Using Equation (13) requires evaluating two operating conditions: (1) full input at 300°F rise and (2) no input with ambient air. Because the boilers are controlled individually, the system may operate at nearly ambient temperature (100°F rise or less) when only one boiler oper-

ates at part load. Use the system resistance k for boiler 3 as the system value for simultaneous operation. It needs no compensating increase as with gravity multiple venting, because a fan induces flow at all temperatures. The resistance summation is:

Inlet acceleration (draft hoods)	= 1.50
Tee above boiler	= 1.25
Tee at base of vertical	= 1.25
90 ft of 18-in. dia. pipe = 0.4(90/18)	= 2.00
System total	= 6.00

At full load, $T_m = (300 + 60 + 460) = 820$°R, $B = 29.92$ in. Hg. Flow in cfm must be found for operating conditions of 1.54 lb per 1000 Btu at density ρ_m and full input (4.8 × 10⁶)/60 Btu/min.

$$\rho_m = 1.325 \, B/T_m = 1.325(29.92/820) = 0.0483 \text{ lb/ft}^3$$

$$Q = \left(\frac{1.54}{1000} \right) \left(\frac{1}{0.0483} \right) \left(\frac{4.8 \times 10^6}{60} \right) = 2551 \text{ cfm}$$

For 300°F rise, $D_t = 0.5(75/100) = 0.375$ in. of water theoretical draft in the system. Solving Equation (13) for Δp:

$$\Delta p = \frac{(6)(29.92)(2551)^2}{(5.2)^2(18)^4(820)} = 0.502 \text{ in. of water}$$

Thus, a fan is needed because Δp exceeds D_t. Required static pressure boost (Table 8) is:

$$D_b = 0.502 - 0.375 = 0.127 \text{ in. of water at } 300°F$$
$$\text{or a density of } 0.0483 \text{ lb/ft}^3$$

Fans are rated for ambient or standard air (60 to 70°F) conditions. Pressure is directly proportional to density or inversely proportional to absolute temperature. Moving 2551 cfm at 300°F against 0.127 in. of water pressure requires the ambient pressure with standard air to be:

$$D_b = 0.127(820/520) = 0.20 \text{ in. of water}$$

Thus, a fan that delivers 2551 cfm at 0.20 in. of water at 60°F is required. Figure 17 shows the operating curves of a typical fan that meets this requirement. The exact volume and pressure developed against a system k of 6.0 can be found for this fan by plotting system airflow rate versus Δp from Equation (13) on the fan curve. The solution, at Point C, occurs at 1950 cfm, where both the system Δp and fan static pressure equal 0.46 in. of water.

While some fan manufacturers' ratings are given at standard air conditions, the motors selected will be overloaded at temperatures below 300°F. Figure 17 shows that power required for two conditions with ambient air is:

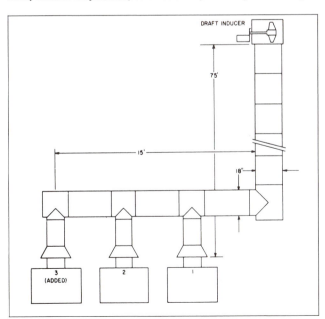

Fig. 16   Chimney System

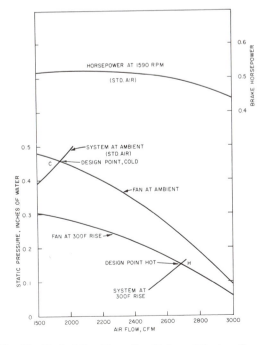

Fig. 17   Typical Fan Operating Data and System Curves

1. 1950 cfm at 0.46 in. static pressure requires 0.51 horsepower
2. 2650 cfm at 0.25 in. static pressure requires 0.50 horsepower

Thus, the minimum size motor will be at least 1/2 horsepower, running at 1590 rpm.

Manufacturers' literature must be analyzed carefully to discover if the sizing and selection method is consistent with appliance and chimney operating conditions. Final selection requires both a thorough analysis of fan and system interrelationships and consultation with the fan manufacturer to verify the capacity and power ratings.

GAS APPLIANCE VENTING

Gas-burning equipment requiring venting of combustion products must be installed and vented in accordance with ANSI *Standard* Z223.1 (NFPA *Standard* 54), the *National Fuel Gas Code*. This standard includes capacity data and definitions for commonly used gas vent systems.

Traditionally, gas vent appliances were designed with a draft hood or draft diverter and depended on natural buoyancy to vent their products of combustion. The operating characteristics of many different types of appliances were similar, allowing for generic venting guidelines to be applied to any gas appliance. These guidelines consisted of tables of maximum capacities, representing the largest appliance input rating that could safely be vented using a certain vent diameter, height, lateral, and material.

Small gas-fired appliance design changed significantly, however, with the increased furnace minimum efficiency requirements to 78% AFUE ICS imposed by the U.S. National Appliance Energy Conservation Act (NAECA). Many manufacturers developed appliances with fan-assisted combustion systems (FACS) to meet efficiency standards, which greatly changed the venting characteristics of gas appliances.

Because the FACS appliances do not entrain dilution air, the maximum capacity of a FACS appliance that can be connected to a given vent system is larger than maximum capacity of a draft hood-equipped appliance that could be connected. The dew point of the vent gases from a FACS appliance, however, remains high in the absence of dilution air, and the potential for condensation to form and remain in the vent system is much greater than with draft hood-equipped appliances. Excessive condensate dwell time (wet time) in the vent system can cause corrosion failure of the vent system, or to problems with runoff of condensate, either into the appliance or into the structure surrounding the vent.

New venting guidelines were developed to address the differences between draft hood-equipped and FACS Category I appliances. In addition to new maximum capacity values for FACS appliances, the guidelines include minimum capacity values for FACS appliances to ensure that the vent system wet times do not reach a level that would lead to corrosion or drainage problems. Because corrosion is the principal concern of condensation in the vent connector, whereas drainage is the principal concern in the vertical portion of the vent, different wet time limits were established for the vent connector and the vertical portion of the vent.

Wet time values in the vent system were determined using the VENT-II computer program to perform the transient analysis with the appliance cycling. The cycle times of 3.87 min on, 13.3 min off, 17.17 min total, were determined based on a design outdoor ambient of 42°F with an oversize factor of 1.7. The vent connector is required to dry before the end of the appliance on-cycle, while the vertical portion of the vent is required to dry out before the end of the total cycle.

When applying the new venting guidelines for FACS appliances, use of single-wall metal vent connectors is severely restricted. Also, tile-lined masonry chimneys are not recommended in most typical installations when a FACS appliance is the only appliance connected to the venting system, due to excessive condensation in the masonry chimney. FACS appliances can be used with a typical masonry chimney, however, (1) if the FACS appliance is common-vented with a draft hood-equipped appliance, or (2) if a liner listed for use with gas appliances is installed in the masonry chimney.

The 1992 *National Fuel Gas Code* lists new minimum and maximum vent capacities for FACS gas appliances and the previous maximum vent capacities for gas appliances equipped with draft hoods.

Vent Connectors

Vent connectors connect gas appliances to the gas vent, chimney, or single-wall metal pipe, except when the appliance flue outlet or draft hood outlet is connected directly to one of these. Materials for vent connectors for conversion burners or other equipment without draft hoods must have resistance to corrosion and heat not less than that of galvanized 0.0276 in. (24 gage) thick sheet steel. Where a draft hood is used, the connector must have resistance to corrosion and heat not less than that of galvanized 0.0187 in. (28 gage) thick sheet steel or Type B vent material.

Masonry Chimneys for Gas Appliances

A masonry chimney serving a gas-burning appliance should have a tile liner and should comply with applicable building codes such as NFPA *Standard* 211. An additional chimney liner may be needed to avoid slow priming and/or condensation, particularly an exposed masonry chimney with high mass and low flue gas temperature. A low temperature chimney liner may be a single-wall passage of pure aluminum or stainless steel or a double-wall Type B vent.

Type B and L Factory-Built Venting Systems

Factory prefabricated vents are listed by Underwriters' Laboratories for use with various types of fuel-burning equipment. These should be installed according to the manufacturer's instructions and their listing.

Type B gas vents are listed for vented gas-burning appliances. They should not be used for incinerators, appliances readily converted to the use of solid or liquid fuel, and combination gas-oil burning appliances. They may be used in multiple gas appliance venting systems.

Type B-W gas vents are listed for vented wall furnaces, certified as complying with the pertinent ANSI standard.

Type L venting systems are listed by Underwriters' Laboratories in 3 through 6 in. sizes and may be used for those oil- and gas-burning appliances (primarily residential) certified or listed as suitable for this type of venting. Under the terms of the listing, a single-wall connector may be used in open accessible areas between the appliance's outlet and Type L material in a manner analogous to Type B. Type L piping material is recognized in NFPA *Standards* 54 and 211 for certain connector uses between appliances such as domestic incinerators and chimneys.

Gas Equipment without Draft Hoods

Figure 1 or the equations may be used to calculate chimney size for nonresidential gas appliances with the draft configurations listed as (1) and (3) at the beginning of this chapter. Draft configurations (1) and (3) for residential gas appliances, such as boilers and furnaces, may require special vent systems. The appliance test and certification standards include evaluation of manufacturers' appliance installation instructions (including the vent system) and of operating and application conditions that affect venting. The appliance instructions must be followed strictly.

Conversion to Gas

The installation of conversion-burner equipment requires evaluating for proper chimney draft and capacity by the methods indicated in this chapter or by conforming to local regulations. The physical condition and suitability of an existing chimney must be checked before converting it from a solid or liquid fuel to gas. For masonry chimneys, local experience may indicate how well the construction withstands the lower temperature and higher moisture content of natural or liquefied petroleum gas combustion products. The Masonry Chimneys for Gas Appliances section under the Gas Appliance Venting section of this chapter has more details.

The chimney should be relined, if required, with corrosion-resistant masonry or metal to prevent deterioration. A chimney should be inspected and, if needed, cleaned. The chimney drop-leg (bottom of the chimney) must be at least 4 in. below the bottom of the connection to the chimney. A liner must extend beyond the top of the chimney. The chimney should also have a cleanout at the base.

FIREPLACE CHIMNEYS

Fireplaces with natural draft chimneys follow the same gravity fluid flow law as gas vents and thermal flow ventilation systems (Stone 1969). All thermal or buoyant energy is converted into flow, and no draft exists over the fire or at the fireplace inlet. Formulas have been developed to study a wide range of fireplace applications, but the material in this section covers general cases only.

Mass flow of hot flue gases through a vertical pipe up to some limiting value is a function of rate of heat release and the chimney area, height, and system pressure loss coefficient. A fireplace may be considered as a gravity duct inlet fitting, with a characteristic entrance-loss coefficient and an internal heat source. A fireplace functions properly (does not smoke) when adequate intake or face velocity across those critical portions of the frontal opening nullifies external drafts or internal convection effects.

The mean flow velocity into a fireplace frontal opening is nearly constant from 300 °F gas temperature rise up to any higher temperature. Local velocities vary within the opening, depending on its design, because the air enters horizontally along the hearth, flows into the fire and upward, and clings to the back wall, as shown in Figure 18. A recirculating eddy forms just inside the upper front half of the opening, induced by the high velocity or flow along the back. Restrictions or poor construction in the

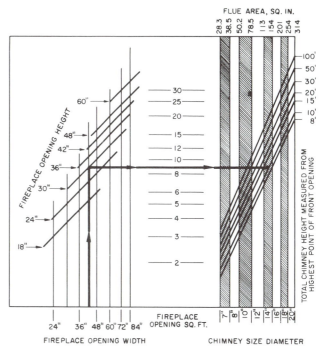

Fig. 19 Fireplace Sizing Chart for Circular Chimneys
Mean Face Velocity = 0.8 ft/s

throat area between the lintel and damper also increase the eddy. Because the eddy moves smoke from the zone of maximum velocity, its tendency to escape must be counteracted by some minimum air movement inward over the entire front of the fireplace, particularly under the lintel.

A minimum mean frontal inlet velocity of 0.8 ft/s, in conjunction with a chimney gas temperature of at least 300 to 500 °F above ambient, should control smoking in a well-constructed conventional masonry fireplace. Figure 19 shows fireplace and chimney dimensions for the specific conditions of circular flues at 0.8 ft/s frontal velocity. The chart solves readily for maximum frontal opening for a given chimney, as well as for chimney size and height with a predetermined opening. Figure 19 assumes no wind or air supply difficulties.

A damper in standard masonry construction with a free area equal to or less than the required flue area is too restrictive. Most damper information does not list damper-free area or effective throat-opening dimensions. Further, interference with lintels or other parts may cause dimensions to vary. For best results, a damper should have a free area at least twice the chimney area.

Indoor-outdoor pressure differences caused by winds, building stack effects, and operation of forced air heating systems or mechanical ventilation affect the operation of a fireplace. Thus, smoking during start-up can be caused by factors seldom related to the chimney. Frequently, in new homes (especially in high-rise multiple-family construction), fireplaces of normal design cannot cope with mechanically induced reverse flow or shortages of combustion air. In such circumstances, fireplaces should include induced draft blowers of sufficient capability to overpower other mechanized air-consuming systems. An inducer for this purpose is best located at the chimney outlet and should produce 0.8 to 1.0 ft/s fireplace face velocity of ambient air in any individual flue or a chimney velocity of 10 to 12 ft/s.

Remedies to increase frontal velocity include the following: (1) increase chimney height (using the same flue area) and extend the last tile 6 in. upward, or more; (2) decrease frontal opening by lowering the lintel or raising the hearth (glass doors may help); and (3) increase free area through the damper (ensure that it opens fully without interference).

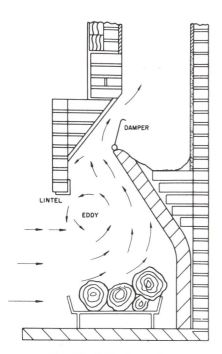

Fig. 18   Eddy Formation

AIR SUPPLY TO FUEL-BURNING EQUIPMENT

A failure to supply outdoor air may result in erratic or even dangerous operating conditions. A correctly designed gas appliance with a draft hood can function with short vents (5 ft high) using an air supply opening as small in area as the vent outlet collar. Such an orifice, when equal to vent area, has a resistance coefficient in the range of 2 to 3. If the air supply opening is as much as twice the vent area, however, the coefficient drops to 0.5 or less.

The following rules may be used as a guide:

1. Residential heating equipment installed in unconfined spaces in buildings of conventional construction does not ordinarily require ventilation other than normal air infiltration. In any residence or building that has been built or altered to conserve energy or to minimize infiltration, the heating appliance area should be considered as a confined space. The air supply should be installed in accordance with the *National Fuel Gas Code* or the recommendation given below.

2. Residential heating equipment installed in a confined space having unusually tight construction requires two permanent openings to an unconfined space or to the outdoors. An unconfined space has a volume of 50 ft^3 or more per 1000 Btu/h of the total input rating of all appliances installed in that space. Free opening areas must be greater than 1 in^2 per 4000 Btu/h input with vertical ducts or 1 in^2 per 2000 Btu/h with horizontal ducts to the outdoors. Two openings communicating directly with sufficient unconfined space must be greater than 1 in^2 per 1000 Btu/h. Upper openings should be within 12 in. of the ceiling; lower openings should be within 12 in. of the floor.

3. Complete combustion of gas or oil requires approximately 1 ft^3 of air, at standard conditions, for each 100 Btu of fuel burned, but excess air for proper burner operation is usually required.

4. The size of these air openings may be modified when special engineering ensures an adequate supply of air for combustion, dilution, and ventilation, or where local ordinances apply to boiler and machinery rooms.

5. In calculating free area of air inlets, the blocking effect of louvers, grilles, or screens protecting openings should be considered. Screens should not be smaller than 1/4-in. mesh. If the free area through a particular louver or grille is known, it should be used in calculating the size opening required to provide the free area specified. If the free area is not known, it may be assumed that wood louvers have 20 to 25% free area and metal louvers and grilles have 60 to 75% free area.

6. Mechanical ventilation systems serving the fuel-burning equipment room or adjacent spaces should not be permitted to create negative equipment room air pressure. The equipment room may require tight self-closing doors and provisions to supply air to the spaces under negative pressure so the fuel-burning equipment and venting operate properly.

7. Fireplaces may require special consideration. For example, a residential attic fan can be hazardous if it is inadvertently turned on while a fireplace is in use.

8. In buildings where large quantities of combustion and ventilation or process air are exhausted, a sufficient supply of fresh uncontaminated makeup air, warmed, if necessary, to the proper temperature should be provided. Good practice provides about 5 to 10% more makeup air than the amount exhausted.

VENT AND CHIMNEY MATERIALS

Factors to be considered when selecting chimney materials include (1) the temperature of gases; (2) their composition and propensity to condense (dew point); (3) the presence of sulfur, halogens, and other fuel and air contaminants that lead to corrosion; and (4) the operating cycle of the appliance.

Figure 1 covers materials for vents and chimneys in the 4- to 48-in. size range; these include single-wall metal, various multiwall air and mass insulated types, and precast and site-constructed masonry. While each has different characteristics, such as frequency of joints, roughness, and heat loss, the type of materials used for systems 14 in. and larger is relatively unimportant in determining draft or capacity. This does not preclude selecting a safe product or method of construction that minimizes heat loss and fire hazard in the building.

National codes and standards classify heat-producing appliances as low, medium, and high heat, with appropriate reference to chimney and vent constructions permitted with each. These classifications are primarily based on size, process use, or combustion temperature. In many cases, the appliance classification gives little information about the outlet gas temperature or venting need. The designer should, wherever possible, obtain gas outlet temperature conditions and properties that apply to the specific appliance, rather than going by code classification only.

Where building codes permit engineered chimney systems, selection based on gas outlet temperature can save space, as well as reduce structural and material costs. For example, in some jurisdictions, approved gas-burning appliances with draft hoods operating at inputs over 400,000 Btu/h may be placed in a heat-producing classification that prohibits use of Type B gas vents. An increase in input may not cause an increase in outlet temperature or in venting hazards, and most building codes recommend correct matching of appliance and vent.

Single-wall uninsulated steel stacks can be protected from condensation and corrosion internally with refractory firebrick liners or by spraying calcium aluminate cement over a suitable interior expanded metal mesh or other reinforcement. Another form of protection applies proprietary silica or other prepared refractory coatings to pins or a support mesh on the steel. The material must then be suitably cured for moisture and heat resistance.

Moisture condensation on interior surfaces of connectors, vents, stacks, and chimneys is a more serious cause of deterioration than heat. Chimney wall temperature and flue gas velocity, temperature, and dew point affect condensation. Contaminants such as sulfur, chlorides, and fluorides in the fuel and combustion air raise the flue gas dew point. Studies by Yeaw and Schnidman (1943), Pray *et al.* (1942-53), Mueller (1968), and Beaumont *et al.* (1970) indicate the variety of analysis methods, as well as difficulties in predicting the causes and probability of actual condensation.

Combustion products from any fuel containing hydrogen condenses onto cold surfaces or condenses in bulk if the main flow of flue gases is cooled sufficiently. Because flue gas loses heat through walls, condensation, which first occurs on interior wall surfaces cooled to the flue gas dew point, forms successively a dew, then a liquid film, and, with further cooling, causes liquid to flow down into zones where condensation would not normally occur.

Start-up of cold interior chimney surfaces is accompanied by transient dew formation, which evaporates on heating above the dew point. Little corrosion results from this phenomenon when very low sulfur fuels are used. Proper selection of chimney dimension and materials minimizes condensation and, thus, corrosion.

Experience shows a correlation between the sulfur content of the fuel and the deterioration of interior chimney surfaces. Figure 20 illustrates one case, which applies to any fuel gas. The figure shows that the flue gas dew point increases at 40% excess air from 127 °F with zero sulfur to 160 °F with 15 grains of sulfur per 100 ft^3 of fuel having a heating value of 550 Btu/ft^3.

Because the corrosion mechanism is not completely understood, judicious use of resistant materials, suitably insulated or jacketed to reduce heat loss, is preferable to low-cost single-wall construction. Refractory materials and mortars should be acid

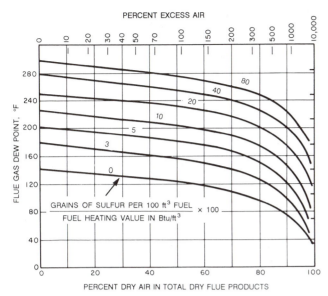

Fig. 20   Influence of Sulfur Oxides on Flue Gas Dew Point
(Stone 1969)

resistant, while steels should be resistant to sulfuric, hydrochloric, and hydrofluoric acids; pitting; and oxidation. Where low flue gas temperatures are expected, together with low ambients, an airspace jacket or mineral fiber lagging, suitably protected against water entry, helps maintain surface and flue gas temperatures above the dew point. In addition to reducing corrosion, the use of low-sulfur fuel reduces air pollution (which is required in many localities).

Type 1100 aluminum alloy or any other noncopper-bearing aluminum alloy of 99% purity or better provides satisfactory performance in prefabricated metal gas vent products. For chimney service, gas temperatures from appliances burning oil or solid fuels may exceed the melting point of aluminum; therefore, steel is required. Stainless steels such as Type 430 or Type 304 give good service in residential construction and are recognized in UL-listed prefabricated chimneys. Where more corrosive environments are anticipated, such as high sulfur fuel or chlorides from solid fuel, contaminated air, or refuse, Type 29-4C or equivalent stainless steel offers a good match of corrosion resistance and mechanical properties.

As an alternative to stainless steel, porcelain enamel offers good resistance to corrosion, if two coats of acid-resistant enamel are used on all surfaces. A single coat, which always has imperfections, will allow base metal corrosion, spalling, and early failure.

Prefabricated chimneys and venting products are available that use light corrosion-resistant materials, both in metal and masonry. The standardized, prefabricated, double-wall metal Type B gas vent has an aluminum inner pipe and a coated steel outer casing, either galvanized or aluminized. Standard airspace from 1/4 to 1/2 in. is adequate for applicable tests and a wide variety of exposures.

Air-insulated all-metal chimneys are available for low-heat use in residential construction. Thermosiphon air circulation or multiple reflective shielding with three or more walls keep these units cool. Insulated, double-wall residential chimneys are also available. The annulus between metal inner and outer walls is filled with insulation and retained by coupler end structures for rapid assembly.

Prefabricated, air-insulated, double-wall metal chimneys for multifamily residential and larger buildings, classed as building-heating appliance chimneys, are available (Figure 21). Refactory-lined prefabricated chimneys (medium heat type) are also available for this use.

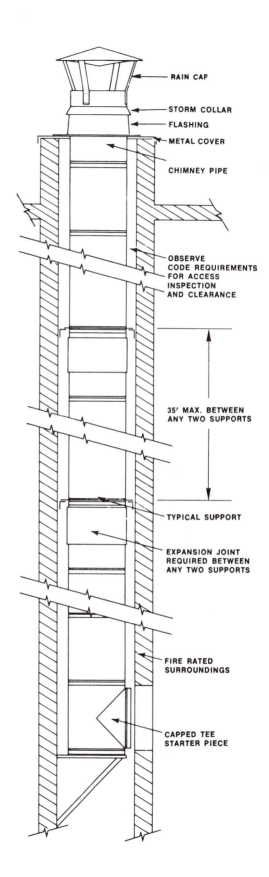

**Fig. 21   Building Heating Equipment or
Medium-Heat Chimney**

Commercial and industrial incinerators, as well as heating appliances, may be vented by prefabricated metal-jacketed cast refractory chimneys, which are listed in the medium-heat category and are suitable for intermittent gas temperatures to 2000°F. All prefabricated chimneys and vents carrying a listing by a recognized testing laboratory have been evaluated for class of service regarding temperature, strength, clearance to adjacent combustible materials, and suitability of construction in accordance with applicable national standards.

Underwriters Laboratories standards (as listed in Table 10) describe the construction and temperature testing of various classes of prefabricated vent and chimney materials. Standards for some related parts and appliances are also included in Table 10, because a listed factory-built fireplace, for example, must be used with a specified type of factory-built chimney. The temperature given for the steady-state operation of chimneys is the lowest in the test sequence. Factory-built chimneys under UL *Standard* 103 are also required to demonstrate adequate safety during a 1-h test at 1330°F rise and to withstand a 10-min. simulated soot burnout at either a 1630°F or 2030°F rise.

These product tests determine minimum clearance to combustible surfaces or enclosures, based on allowable temperature rise on combustibles. They also ensure that the supports, spacers, and parts of the product that contact combustible materials remain at safe temperatures during operation. Product markings and installation instructions of listed materials are required to be consistent with test results, refer to types of appliances that may be used, and explain structural and other limitations.

VENT AND CHIMNEY ACCESSORIES

The design of a vent or chimney system must consider the existence or need for such accessories as draft diverters, draft regulators, induced draft fans, blocking dampers, expansion joints, and vent or chimney terminals. Draft regulators include barometric draft regulators and furnace sequence draft controls, which monitor automatic flue dampers during operation. The design, materials, and flow losses of chimney and vent connectors are covered in other sections.

Draft Hoods

The draft hood isolates the appliance from venting disturbances (updrafts, downdrafts, or blocked vent) and allows combustion to start without venting action. Suggested general dimensions of

Table 10   Underwriters Laboratories Test Standards

No.	Subject	Steady-State Appliance Flue Gas Temperature Rise, °F	Fuel
103	Chimneys, factory-built, residential type (includes building heating appliances)	930	All
127	Fireplaces, factory-built	930	Solid or gas
311	Roof jacks for mobile homes	930	Oil, gas
378	Draft equipment (such as regulators and inducers)	—	All
441	Gas vents (Type B, BW)	480	Gas only
641	Low-temperature venting systems (Type L)	500	Oil, gas
959	Chimneys, factory-built medium heat	1730	All
1738	Venting systems for gas-burning appliances, Categories II, III, and IV	140 to 480	Gas only

draft hoods are given in ANSI *Standard* Z21.12, which describes certification test methods for draft hoods. In general, inlet and outlet flue pipe sizes of the draft hood should be the same as that of the appliance outlet connection. The vent connection at the draft hood outlet should have a cross-sectional area at least as large as that of the draft hood inlet.

Draft hood selection comes under the following two categories:

1. Draft hood supplied with a design-certified gas appliance— certification of a gas appliance design under pertinent ANSI standards includes its draft hood (or draft diverter). Consequently, a draft hood should not be altered, nor should it be replaced without consulting the manufacturer and local code authorities.
2. Draft hoods supplied separately for gas appliances— installation of listed draft hoods on existing vent or chimney connectors should be made by experienced installers in accordance with accepted practice standards.

Every design-certified gas appliance requiring a draft hood must be accompanied by a draft hood or provided with a draft diverter as an integral part of the appliance. The draft hood is a vent inlet fitting, as well as a safety device for the appliance, and certain assumptions can be made regarding its interaction with a vent. First, when it is operating without spillage, the heat content of gases (enthalpy relative to dilution air temperature) leaving the draft hood is almost the same as that entering. Second, safe operation is obtained with 40 to 50% dilution air. It is unnecessary to assume 100% dilution air for gas venting conditions. Third, during certification tests, the draft hood must function without spillage, using a vent with not over 5 ft of effective height and one or two elbows. Therefore, if vent heights appreciably greater than 5 ft are used, an individual vent of the same size as the draft hood outlet may be much larger than necessary.

When vent size is reduced, as with tall vents, draft hood resistance is less than design value relative to the vent, and the vent tables in the *National Fuel Gas Code* give adequate guidance for such size reductions.

Despite its importance as a vent inlet fitting, the draft hood designed for a typical gas appliance primarily represents a compromise of the many design criteria and tests solely applicable to that appliance. This permits considerable variation in resistance loss and thus, catalog data on draft hood resistance loss coefficients does not exist. The span of draft hood loss coefficients, including inlet acceleration, varies from the theoretical minimum of 1.0 for certain low-loss bell or conical shapes to 3 or 4, where the draft hood relief opening is located within a hot-air discharge (as with wall furnaces), and high resistance is needed to limit sensible heat loss into the vent.

Draft hoods must not be used on draft configurations (1) or (3) equipment operated with either power burners or forced venting, unless the equipment has fan-assisted burners that overcome some or most of the appliance flow resistance and create a pressure inversion ahead of the draft (or barometric) hood.

Gas appliances with draft hoods must have excess chimney draft capacity to draw in adequate draft hood dilution air. Failure to provide adequate combustion air can cause oxygen depletion and cause flue gases to spill from the draft hood.

Draft Regulators

Appliances requiring draft at the appliance flue gas outlet generally make use of barometric regulators for combustion stability. A balanced hinged gate in these devices bleeds air into the chimney automatically when pressure decreases. This action simultaneously increases gas flow and reduces temperature. Well-designed barometric regulators provide constant static pressure over a span of impressed draft of about 0.2 in. of water, where impressed draft is that which would exist without regulation. A

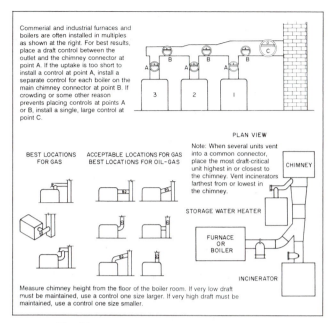

- Heating appliance is oversized.
- Chimney is too high or oversized.
- Appliance is located in heated space.
- Two or more appliances are on the same chimney (but a damper must be installed on each appliance connected to that chimney).
- Appliance is located in building zone at higher pressure than outdoors. This positive pressure can cause steady flow losses through the chimney.

Energy savings may not justify the cost of installation if one or more of the following conditions exist:

- All combustion and ventilation air is supplied from outdoors to direct vent appliances or to appliances located in an isolated, unheated room.
- Appliance is in an unheated basement that is isolated from the heated space.
- A short vent in a one-story flat-roof house is unlikely to carry away a significant amount of heated air.

For vents or chimneys serving two or more appliances, dampers should be installed on all attached appliances for maximum effectiveness. If only one damper is installed in this instance, loss of heated air through an open draft hood may negate a large portion of the potential energy savings.

Heat Exchangers or Flue Gas Heat Extractors

The sensible heat available in the flue products of properly adjusted furnaces burning oil or gas is about 10 to 15% of the rated input. Small accessory heat exchangers that fit in the connector between the appliance outlet and the chimney can recover a portion of this heat for localized use; however, they may cause some adverse effects.

All gas vent and chimney size or capacity tables assume the gas temperature or heat available to create theoretical draft is not reduced by a heat transfer device. In addition, the tables assume flow resistance for connectors, vents, and chimneys, comprising typical values for draft hoods, elbows, tees, caps, and piping with no allowance for added devices placed directly in the stream. Thus, heat exchangers or flue gas extractors should offer no flow resistance or negligible resistance coefficients when they are installed.

A heat exchanger that is reasonably efficient and offers some flow resistance may adversely affect the system by reducing both flow and gas temperature. This may cause moisture condensation in the chimney, draft hood spillage, or both. Increasing heat transfer efficiency increases the probability of both effects occurring simultaneously. An accessory heat exchanger in a solid fuel system, especially a wood stove or heater, may collect creosote or cause its formation downstream.

The retrofit of heat exchangers in gas appliance venting systems requires careful evaluation of heat recovered versus both installed cost and the potential for chimney safety and operating problems. Every heat exchanger installation should undergo the same spillage tests given a damper installation. In addition, the gas temperature should be checked to ensure it is high enough to avoid condensation between the exchanger outlet and the chimney outlet.

DRAFT FANS

The selection of draft fans, blowers, or inducers must consider (1) types and combinations of appliances, (2) types of venting material, (3) building and safety codes, (4) control circuits, (5) gas temperature, (6) permissible location, (7) noise, and (8) power cost. Besides specially designed fans and blowers, some conventional fans can be used if the wheel and housing materials are heat- and corrosion-resistant, and if blower and motor bearings are protected from adverse effects of the gas stream.

Fig. 22 Use of Barometric Draft Regulators

regulator can maintain a 0.06 in. draft for impressed drafts from 0.06 through 0.26 in. of water. If the chimney system is very high or otherwise capable of available draft in excess of the pressure span capability of a single regulator, additional or oversize regulators may be used. Figure 22 shows proper locations for regulators in a chimney manifold.

Barometric regulators are available with double-acting dampers, which also swing out to relieve momentary internal pressures or divert continuing downdrafts. In the case of downdrafts, temperature safety switches, actuated by hot gases escaping at the regulator, sense and limit malfunctions.

Vent Dampers

Electrically, mechanically, and thermally actuated automatic vent dampers can reduce energy consumption and improve the seasonal efficiency of gas- and oil-burning appliances. Vent dampers reduce the loss of heated air through gas appliance draft hoods and the loss of specific heat from the appliance after the burner has ceased firing. These dampers may be retrofit devices or integral components of some appliances.

Electrically and mechanically actuated dampers must open prior to main burner gas ignition and must not close during operation. These safety interlocks, which electrically interconnect with existing control circuitry, include an additional main control valve, if called for, or special gas pressure-actuated controls.

Vent dampers that are thermally actuated with bimetallic elements are available with spillage-sensing interlocks with burner controls for a draft hood-type gas appliance. These dampers open in response to gas temperature after burner ignition. Because thermally actuated dampers may exhibit some flow resistance, even at equilibrium operating conditions, instructions regarding allowable heat input and minimum required vent or chimney height should be observed.

Special care must be taken to ensure safety interlocks with appliance controls are installed according to instructions. Spillage-free gas venting after the damper has been installed must be verified with all types.

The energy savings of a vent damper can vary widely. Dampers reduce energy consumption under one or a combination of the following conditions:

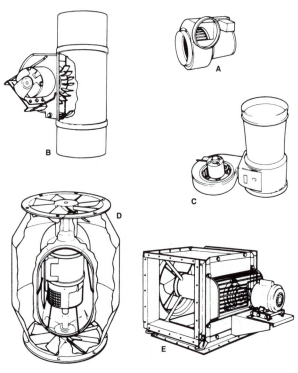

Fig. 23   Draft Inducers

Small draft inducers for residential gas appliance and unit heater use are available with direct-drive blower wheels and an integral sail switch to sense flow, as shown in Figure 23A. The control circuit for such applications must provide adequate vent flow both before and while fuel flows to the main burner. Other types of small inducers are either saddle-mounted blower wheels (Figure 23B) or venturi ejectors that induce flow by jet action (Figure 23C). An essential safety requirement for inducers serving draft hood gas appliances does not permit appliance interconnections on the discharge or outlet side of the inducer. This requirement prevents backflow through an inoperative appliance.

With prefabricated sheet metal venting products such as Type B gas vents, the vent draft inducer should be located at or downstream from the point the vent exits the building. This placement keeps the indoor system below atmospheric pressure and prevents gases from escaping through seams and joints. If the inducer cannot be placed on the roof, metal joints must be reliably sealed in all pressurized parts of the system.

Pressure capability of residential draft inducers is usually less than 1 in. of water static pressure at rated flow. Larger inducers of the fan, blower, or ejector types have greater pressure capability and may be used to reduce system size, as well as supplement available draft. Figure 23D shows one specialized axial flow fan capable of higher pressures. This unit is structurally self-supporting and can be mounted in any position in the connector or stack, because the motor is in a well, separated from the gas stream. A right-angle fan, as shown in Figure 23E, is supported by an external bracket and adapts to several inlet and exit combinations. The unit uses the developed draft and an insulated tube to cool the extended shaft and bearings.

Pressure, volume, and power curves should be obtained to match an inducer to the application. For example, in an individual chimney system (without draft hood) in which a directly connected inducer only handles combustion products, power required for continuous operation need only consider volume at operating gas temperature. An inducer serving multiple, separately controlled,

draft hood gas appliances must be powered for ambient temperature operation at full flow volume in the system. At any input, the inducer for a draft hood gas appliance must handle about 50% more standard chimney gas volume than a directly connected inducer. These demands follow the Fan Laws (Chapter 18) applicable to power venting as follows; at constant volume with a given size inducer or fan:

- Pressure difference developed is directly proportional to gas density.
- Pressure difference developed is inversely proportional to absolute gas temperature.
- Pressure developed diminishes in direct proportion to drop in absolute atmospheric pressure, as with altitude.
- Required power is directly proportional to gas density.
- Required power is inversely proportional to absolute gas temperature.

Centrifugal and propeller draft inducers in vents and chimneys are applied and installed the same as any heat-carrying duct system. Venturi ejector draft boosters involve some added consideration. An advantage of the ejector is that motor, bearings, and blower blades are outside the contaminated gas stream, thus eliminating a major source of deterioration. This advantage causes some loss of efficiency and can lead to reduced capacity because undersized systems, having considerable resistance downstream, may be unable to handle the added volume of the injected airstream without loss of performance. Ejectors are best suited to use at the exit or where there is an adequately sized chimney or stack to carry the combined discharge.

If total pressure defines outlet conditions or is used for fan selection, the relative amounts of static and velocity pressures must be factored out; otherwise, the velocity head method of calculation does not apply. To factor total pressure into its two components, either the discharge velocity in an outlet of known area or flow rate must be known. For example, if an appliance or blower produces a total pressure of 0.25 in. of water at 1500 ft/min discharge velocity, the velocity pressure component can be found from Figure 1. Enter at the horizontal line marked VH in grid C and move horizontally to operating temperature. Read up from the appropriate temperature rise line to the 1500 ft/min horizontal velocity line in the Δp block. Here the velocity pressure reads 0.14 in. (calculates to 0.143 at ambient standard conditions) so static pressure is $0.25 - 0.14 = 0.11$ in. Since this static pressure is part of the system driving force, it combines with theoretical draft to overcome losses in the system.

Draft inducer fans can be operated either continuously or on demand. In either case, a safety switch that senses flow or pressure is needed to interrupt burner controls, if adequate draft fails. Demand operation links the thermostat with the draft control motor. Once flow starts, as sensed with a flow or pressure switch, the burner can be started. A single draft inducer, operating continuously, can be installed in the common vent of a system serving several separately controlled appliances. This simplifies the circuitry because only one control is needed to sense loss of draft. However, the single fan increases boiler standby loss and heat losses via ambient air drawn through inoperative appliances.

TERMINATIONS— CAPS AND WIND EFFECTS

The vent or chimney height and method of termination is governed by a variety of considerations, including fire hazard; wind effects; entry of rain, debris, and birds; as well as by operating considerations such as draft and capacity. For example, the 3-ft height required for residential chimneys above a roof is required so that small sparks will burn out before they fall on the roof shingles.

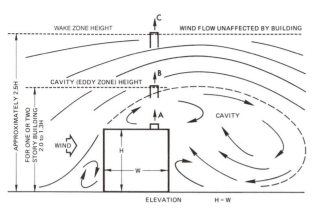

Chimney heights:

A—Discharge into cavity should be avoided because reentry will occur. Dispersion equations do not apply.

B—Discharge above cavity is good. Reentry is avoided, but dispersion may be marginal or poor from standpoint of air pollution. Dispersion equations do not apply.

C—Discharge above wake zone is best; no reentry; maximum dispersion.

Fig. 24 Wind Eddy and Wake Zones for One- or Two-Story Buildings and Their Effect on Chimney Gas Discharge

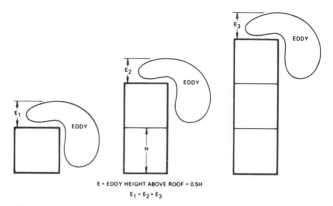

Studies found for a single cube-shaped building (length equals height) that (1) the height of the eddy above grade is 1.5 times the building height and (2) the height of unaffected air is 2.5 times height above grade. The eddy height above the roof equals 0.5H, and it does not change as building height increases in relation to building width.

Fig. 25 Height of Eddy Currents around Single High-Rise Buildings

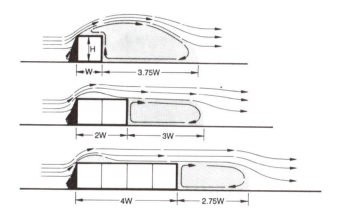

Fig. 26 Eddy and Wake Zones for Low, Wide Buildings

Many vent and chimney malfunctions are attributed to interactions of the chimney termination or its cap with winds acting on the roof or with adjoining buildings, trees, or mountains. Because winds fluctuate, no simple method of analysis or reduction to practice exists for this complex situation. Figures 24, 25, and 26 show some of the complexities of wind flow contours around simple structural shapes.

Figure 24 shows three zones with differing degrees of flue gas dispersion around a rectangular building: the cavity or eddy zone, the wake zone, and the undisturbed flow zone (Clark 1967). In addition, a fourth flow zone of intense turbulence is located downwind of the cavity. Chimney gases discharged into the wind at a point close to the roof surface in the cavity zone may be recirculated locally. Higher in the cavity zone, wind eddies can carry more dilute flue products to the lee side of the building. Gases discharged into the wind in the wake zone do not recirculate into the immediate vicinity, but may soon descend to ground level. Above the wake zone, dispersal into the undisturbed wind flow carries and dilutes the flue gases over a wider area. The boundaries of these zones vary with building configuration and wind direction and turbulence; they are influenced strongly by surroundings.

The possibility of air pollutants reentering the cavity zone due to plume spread, or air pollution intercepting downwind cavities associated with adjacent structures or downwind buildings should be considered. Thus, a design criteria of elevating the stack discharge above the cavity is not valid for all cases. A meteorologist, experienced with dispersion processes near buildings, should be consulted for complex cases and for cases involving air contaminants.

As chimney height increases through these zones, draft performance improves dispersion, while additional problems of gas cooling, condensation, and structural wind load are created. As building height increases (Figure 25), the eddy forming the cavity zone no longer descends to ground level. For a low, wide building (Figure 26), wind blowing parallel to the long roof dimension can reattach to the surface; thus, the eddy zone becomes flush with the roof surface (Evans 1957). For satisfactory dispersion with low, wide buildings, chimney height must still be determined as if $H = W$ (Figure 24).

Evans (1957) and Chien *et al.* (1951) studied pitched roofs in relation to wind flow and surface pressures. Because the typical residence has a pitched roof and probably uses natural gas or a low-sulfur fossil fuel, dispersion is not important because flue products are relatively free of pollutants. For example, a precaution found in the *National Fuel Gas Code* (NFPA 54) requires minimum distance of the gas vent termination from an air intake, but it does not require penetration above the cavity zone.

Flow of wind over a chimney termination can assist or impede draft. In regions of stagnation, on the windward side of a wall or a steep roof, winds create positive static pressures that impede established flow or cause backdrafts within vents and chimneys.

Conversely, location near the surface of a low flat roof can aid draft because the entire roof surface is under negative static pressure. Velocity is low, however, due to the cavity formed as wind sweeps up over the building. With greater chimney height, termination above the low-velocity cavity or negative pressure zone subjects the chimney exit to greater wind velocity, thereby increasing draft from two causes: (1) height and (2) wind aspiration over an open top. As the termination is moved from the center of the building to the sides, its exposure to winds and pressure also varies.

Terminations on pitched roofs may be exposed to either negative or positive static pressures, as well as to variation in wind velocity and direction. On the windward side, pitched roofs vary from complete to partial negative pressure as pitch increases from approximately flat to 30° (Chien *et al.* 1951). At a 45° pitch, the windward pitched roof surface is strongly positive, and, beyond this slope, pressures approach those observed on a vertical wall facing the wind. Wind pressure varies with its horizontal direction on a pitched roof and on the lee or sheltered side, wind velocity is very low and static pressures are usually negative. Wind velocities and pressures vary, not only with pitch, but with position between ridge and eave. Reduction of these observed external wind effects to simple rules of termination for a wide variety of chimney and venting systems requires many compromises.

In the wake zone or any higher location exposed to full wind velocity, an open top can create strong venting updrafts. The updraft effect relative to wind dynamic pressure is related to the Reynolds number. Open tops, however, are sensitive to the wind angle, as well as to rain (Clark 1967), and many proprietary tops have been designed to stabilize wind effects, and improve the performance. Because of the many compromises made in the design of a vent termination, this stability is usually achieved by sacrificing some of the updraft created by the wind. Further, the location of a vent cap in a cavity region frequently removes it from the zone where wind velocity could have a significant effect.

Studies of vent cap design undertaken to optimize performance of residential types indicate that the following performance features are important: (1) still-air resistance, (2) updraft ability with no flow, and (3) discharge resistance when vent gases are carried at low velocity in a typical wind (10 ft/s vent velocity in a 20 mph wind). Tests described in UL *Standard* 441 for proprietary gas vent caps consider these three aspects of performance to ensure adequate vent capacity.

Frequently, the air supply to an appliance room is difficult to orient to eliminate wind effects. Therefore, the vent outlet must have a certain updraft capability, which can help balance a possible adverse wind. When wind flows across an inoperative vent termination, a strong updraft develops. Appliance start-up reduces this updraft and, in typical winds, the vent cap may develop greater resistance than it would have in still air. Certain vent caps can be made with very low still-air resistance, yet exhibit excessive wind resistance, which reduces capacity. Finally, because the appliance operates whether or not there is a wind, still-air resistance must be low.

Some proprietary air ventilators have excessive still-air resistance and should be avoided on vent and chimney systems, unless a considerably oversized vent is specified. Vertical-slot ventilators, for example, have still-air resistance coefficients of about 4.5. To achieve low still-air resistance on vents and chimneys, the vertical-slot ventilator must be 50% larger than the diameter of the chimney or vent, unless it has been specifically listed for such use.

Freestanding chimneys high enough to project above the cavity zone require structurally adequate materials or guying and bracing for prefabricated products. The prefabricated metal building/heating equipment chimney places little load on the roof structure, but guying is required at 8 to 12 ft intervals to resist both overturning and oscillating wind forces. Various other expedients, such as spiral baffles on heavy gage freestanding chimneys, have been used to reduce oscillation.

The chimney height needed to carry the effluent into the undisturbed flow stream above the wake zone can be reduced by increasing the effluent discharge velocity. A 3000 ft/min discharge velocity avoids downward eddying along the chimney and expels the effluent free of the wake zone. Velocity this large can be achieved only with forced or induced draft.

The entry of rain is a problem for open, low-velocity, or inoperative systems. Good results have been obtained with drains that divert the water onto a roof or into a collection system leading to a sump. Figure 27 shows several configurations (Clark 1967 and Hama and Downing 1963). The runoff from stack drains contains acids, soot, and metallic corrosion products, which can cause roof staining. Therefore, these methods are not recommended for residential

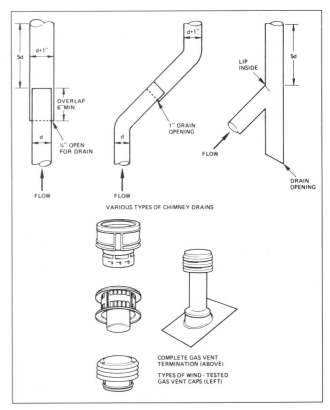

Fig. 27   Vent and Chimney Rain Protection

use. An alternate procedure is to allow all water to drain to the base of the chimney, where it is piped from a capped tee to a sump.

Rain caps prevent the vertical discharge of high-velocity gases. However, caps are preferred for residential gas-burning equipment because it is easier to exclude rain than to risk rainwater leakage at horizontal joints or to drain it. Also, caps keep out debris and bird nests, which can block the chimney. Satisfactory gas vent cap performance can be achieved in the wind by using a variety of standard configurations, including the *A* cap or a wind band ventilator, as well as proprietary designs (Figure 27). Where partial rain protection is desired without excessive flow resistance, and wind characteristics are either unimportant or the wind flow will be horizontal, a flat disc or cone cap 1.7 to 2.0 diameters across, located 0.5 diameter above the end of the pipe, has a still-air resistance loss coefficient of about 0.5.

CODES AND STANDARDS

Building and installation codes and standards prescribe the installation and safety requirements of heat-producing appliances and their vents and chimneys. Chapter 48 lists the major national building codes, one of which may be in effect in a given area. Some jurisdictions either adopt a national building code with varying degrees of revisions to suit local custom or, as in many major metropolitan areas, develop a local code, which agrees in principle but shares little common text with the national codes. Familiarity with applicable building codes is essential because of the great variation in local codes and adoption of modern chimney design practice.

The national standards listed in Table 11 give greater detail on the mechanical aspects of fuel systems and chimney or vent construction. While these standards emphasize safety aspects, especially clearances to combustibles for various venting materials, they also recognize the importance of proper flow, draft, and capacity.

Table 11 List of National Standards Relating to Installation[a]

Subject	Materials Covered	NFPA[b]	ANSI[c]
Oil-burning equipment	Type L, Listed chimneys, single wall, masonry	31[c]	Z95.1
Gas appliances and gas piping	Type B, L, Listed chimneys, single wall, masonry	54	Z223.1
Chimneys, fireplaces, and venting	All types	211	—
Fire safety for recreational vehicles	Roof jacks and vents	501C	—
Gas piping and gas equipment on industrial premises	All types	54	Z223.1
Gas conversion burners	Chimneys	None	Z21.8[d]
	Safe design		Z21.17[d]
Draft hoods	Part dimensions	None	Z21.12[d]
Automatic vent dampers for use with gas-fired appliances			
Electrically operated	Safe design		Z21.66[d]
Mechanically operated	Safe design		Z21.67[d]
Thermally operated	Safe design		Z21.68[d]

[a] These standards are subject to periodic review and revision to reflect advances in the industry, as well as for consistency with legal requirements and other codes.
[b] National Fire Protection Association, Quincy, MA.
[c] American National Standards Institute, New York.
[d] Available from American Gas Association, Arlington, VA.

CONVERSION FACTORS

The following conversion factors have been simplified for chimney design.

$$\text{Btu/h input} = 3{,}347{,}500 \frac{\text{Boiler Horsepower}}{\text{Percent Efficiency}}$$

$$= 44{,}600 \times \text{Boiler horsepower (approx. 75\% eff.)}$$
$$= 41{,}800 \times \text{Boiler horsepower (approx. 80\% eff.)}$$
$$= 37{,}200 \times \text{Boiler horsepower (approx. 90\% eff.)}$$
$$= 140{,}000 \times \text{Numbers 1 and 2 oil (gal/h)}$$
$$= 150{,}000 \times \text{Numbers 4, 5, and 6 oil (gal/h)}$$
$$= 13{,}000 \times \text{Coal (lb/h)}$$
$$= 1{,}000 \times \text{Natural gas (ft}^3\text{/h)}$$
$$= 3.412 \times \text{Watt rating}$$
$$\text{MBh} = 1{,}000 \text{ Btu/h}$$
$$\text{kW input} = \text{kW output/efficiency}$$
$$\text{kW} = 9.81 \times \text{Boiler horsepower}$$

SYMBOLS

A = area of passage cross section, ft^2
B = existing or local barometric pressure, in. Hg
C_u = temperature multiplier for heat loss, dimensionless
c_p = specific heat of gas at constant pressure, Btu/lb · °F
d_f = inside diameter, ft
d_i = inside diameter, in.
D_a = available draft, in. of water
D_t = theoretical draft, in. of water
D_b = increase in static pressure by fan, in. of water
F = friction factor for L/d_i
f = Darcy friction factor from Moody diagram
g = gravitational constant, 32.174 ft/s^2
H = height of vent or chimney system above datum of draft measurement, or system inlet, ft
I = operating heat input, Btu/h
k = system friction-loss coefficient, dimensionless or velocity head
k_f = fitting friction-loss coefficient, dimensionless
k_L = piping friction-loss coefficient, dimensionless
L = length of all piping in chimney system from draft measurement datum to exit, linear ft
L_e = total equivalent length, L plus equivalent length of elbows, etc. [for use in Equations (1) and (2) only], ft
L_m = length of system from inlet to midpoint of vertical or to location of mean gas temperature, ft
M = pounds of products per 1000 Btu of fuel burned

$\Delta p = D_t - D_a$ (see page 31.1), in. of water
q = sensible heat at a particular point in the vent, Btu/s
q_m = sensible heat at average temperature in vent, Btu/s
Q = volumetric flow rate, cfm
t = temperature difference, °F
t_e = temperature difference entering system, °F
t_m = temperature difference at average temperature location in system, °F [$t_m = T_m - T_o$]
T = absolute temperature, °R
T_m = temperature at average conditions in system, °R
T_o = ambient temperature, °R
U = overall heat transfer coefficient of vent or chimney wall material, referred to inside surface area, Btu/h · ft^2 · °F
$\bar{U}$ = heat transfer coefficient for Equation (15), Btu/s · ft^2 · °F
V = velocity of gas flow in passage, ft/s
w = mass flow of gas, lb/h
x = length of one internal side of rectangular chimney cross section, ft
y = length of other internal side of rectangular chimney cross section, ft
ρ_c = density of chimney gas at 0°F and 1 atmosphere, lb/ft^3 [0.09 lb/ft^3 in Equations (1) and (2)]
ρ_m = density of chimney gas at average temperature and local barometric pressure, lb/ft^3
ρ_o = density of air at 0°F and 1 atmosphere, lb/ft^3 (0.0863 lb/ft^3)

BIBLIOGRAPHY

ASHRAE. 1984. Method of testing for performance rating of woodburning appliances. ASHRAE *Standard* 106-1984.

ASHVE. 1941. *Heating ventilating air conditioning guide.* American Society of Heating and Ventilating Engineers. Reference available from ASHRAE.

Beaumont, M., D. Fitzgerald, and D. Sewell. 1970. Comparative observations on the performance of three steel chimneys. *Journal of the Institution of Heating and Ventilating Engineers* (July): 85.

Briner, C.F. 1986. Heat transfer and fluid flow analysis of three-walled metal factory-built chimneys. ASHRAE *Transactions* 92(1B):727-38.

Briner, C.F. 1984. *Heat transfer and fluid flow analysis of sheet metal chimney systems.* ASME Paper No. 84-WA/SOL-36, American Society of Mechanical Engineers, New York.

Chien, N., Y. Feng, H. Wong, and T. Sino. 1951. *Wind-tunnel studies of pressure on elementary building forms.* Iowa Institute of Hydraulic Research.

Clarke, J.H. 1967. Air flow around buildings. *Heating, Piping and Air Conditioning* May:145.

Evans, B.H. 1957. Natural air flow around buildings. Texas Engineering Experiment Station *Research Report* 59.

Hama, G.M. and D.A. Downing. 1963. The characteristics of weather caps. *Air Engineering* (December):34.

Hampel, T.E. Venting system priming time. *Research Bulletin* 74, Appendix E, American Gas Association Laboratories, 146.

Jakob, M. 1955. *Heat transfer*, Volume I. John Wiley and Sons, New York.

Kinkead, A. 1962. Gravity flow capacity equations for designing vent and chimney systems. Proceedings of the Pacific Coast Gas Association 53.

Lapple, C.E. 1949. Velocity head simplifies flow computation. *Chemical Engineering* (May):96.

Mueller G.R. 1968. Charts determine gas temperature drops in metal flue stacks. *Heating, Piping and Air Conditioning* (January):138.

NFPA. 1988. Standard for chimneys, fireplaces, vents, and solid-fuel burning appliances. NFPA *Standard* 211-88. National Fire Protection Association, Quincy, MA.

NFPA and ANSI. 1992. *National fuel gas code*, NFPA 54 and ANSI Z223.1. National Fire Protection Association, Quincy, MA, and American Gas Association, Arlington, VA.

Paul, D.D. 1992. Venting guidelines for category I gas appliances, GRI-89/0016 Gas Research Institute, Chicago, IL.

Pray, H.R., *et al.* 1942-1953. The corrosion of metals and materials by the products of combustion of gaseous fuels. *Battelle Memorial Institute Reports* 1,2,3, and 4 to the American Gas Association.

Ramsey, C.G. and H.R. Sleeper. 1970. *Architectural graphic standards*, 6th ed. John Wiley and Sons Inc., New York, 528.

Reynolds, H.A. 1960. Selection of induced draft fans for heating boilers. *Air Conditioning, Heating and Ventilating* (December):51.

Rutz, A.L. 1991. VENT II *Solution methodology*. Available from ASHRAE publications sales.

Rutz, A.L. and D.D. Paul. 1991. User's manual for VENT-II (Version 4.1) with diskettes: An interactive personal computer program for design and analysis of venting systems of one or two gas appliances. GRI-90/0178. Gas Research Institute, Chicago, IL. Available from National Technical Information Services, U.S. Department of Commerce, No. PB91-509950, 5285 Port Royal Road, Springfield, VA 22161.

Segeler, C.G., ed. 1965. *Gas engineers' handbook*. Section 12, Industrial Press, New York, 49.

Sepsy, C.F. and D.B. Pies. 1972. An experimental study of the pressure losses in converging flow fittings used in exhaust systems. Ohio State University College of Engineering, Columbus, OH (December).

Stone, R.L. 1957. Design of multiple gas vents. *Air Conditioning, Heating and Ventilating* (July).

Stone, R.L. 1971. A practical general chimney design method. ASHRAE *Transactions* 77:1.

Stone, R.L. 1969. Fireplace operation depends upon good chimney design. ASHRAE *Journal* (February):63.

Yeaw, J.S. and L. Schnidman. 1943. Dew point of flue gases containing sulfur. *Power Plant Engineering* 47 (I and II).

CHAPTER 32

MAKEUP AIR UNITS

A MAKEUP air unit is an assembly of elements designed to condition makeup air introduced into a space for ventilation to replace air exhausted from a building. The air exhausted may be from a process or general area exhaust, through either powered exhaust fans or gravity ventilators. If the temperature or humidity (or both) within the structure are controlled, the environmental control system must have the capacity to condition the replacement air.

Makeup air units may, therefore, be required to heat, ventilate, cool, humidify, dehumidify, and filter. Makeup air units may be required to replace the air at the space conditions or may be used for part or all of the heating, ventilating, or cooling load.

Makeup air can be at fixed airflow rates or can be variable volume outside air, under building pressurization control, to respond directly to exhaust flow rates.

Buildings under negative pressure caused by inadequate makeup air have the following symptoms:

- Gravity stacks from unit heaters and processes back-vent.
- Exhaust systems do not perform at rated volume.
- The perimeter of the building is cold in the winter due to high infiltration.
- There are severe indrafts at exterior doors.
- Exterior doors are hard to open.
- Heating systems cannot maintain uniform comfort conditions throughout the building since the center core area becomes overheated.

APPLICATIONS

Makeup air systems can be considered simply as a means of supplying heated replacement air to compensate for air being exhausted from the building. In this context, the arrangement and distribution of the units can be kept very simple. For straight makeup air service, the units are generally installed with minimal distribution ductwork, often using discharge nozzles located in the building truss space (Figure 1). For other applications, such as for heating and ventilating service, where comfort, velocity, or other critical criteria exist, extensive ductwork distribution may be necessary. Similarly, where pollutants may accumulate to the underside of the building roof, distribution ductwork should supply air below the level of contaminants. The volume of makeup air should generally exceed the total building exhaust, so that the building is maintained under slight positive pressure. This reduces cold air infiltration in winter, reducing drafts at perimeter walls and doors. In some applications, automatic pressurization control can be used to vary the volume of outside air introduced to meet a specific set point, often in the area of 0.01 to 0.03 in. of water.

While the above discussion addresses only replacement of exhaust air, makeup air units can also be used for environmental control, as door heaters, for process system ventilation, to reduce

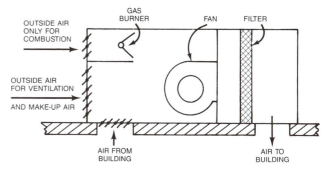

Fig. 1   Direct-Fired Makeup Air Unit

contaminants, or to prevent infiltration. The configuration of the units, associated controls, and air distribution method should be tailored to meet specific design objectives.

When used as part of a building's heating, ventilating, or air-conditioning system, the unit arrangement and distribution should comply with guidelines for industrial and special air conditioning and ventilation (see Chapters 12 and 25 of the 1991 ASHRAE *Handbook—HVAC Applications*).

Makeup air heating is typically designed with a low temperature differential between the room or space ambient temperature and the heater discharge temperature. Based on the heat loss of the building and its ventilation requirement, the burner or coil and fan should be sized so that the average discharge difference would be between only 5 and 10 °F. Maximum allowable discharge temperature should be below 100 °F. The low-temperature discharge prevents the normally expected stratification of air caused by standard heating systems that discharge air at 120 °F or more. Even in a tall building, a properly designed makeup air system produces stratification of only 2 to 4 °F.

Improved summer cooling can be obtained by adding an air washer to the makeup air unit. Chapter 4 of this volume and Chapter 46 of the 1991 ASHRAE *Handbook—HVAC Applications* have further details.

Makeup air for environmental control may be provided by unit ventilators with the air-handling capacity to meet summer requirements.

Makeup air units are also used as door heaters. The units operate continuously, delivering outside air to the building at, or slightly above, room temperature. When the door opens, a door switch signals the unit to discharge the air at a higher temperature. Units used for this purpose must have the capacity to provide final temperatures of about 120 °F at the local design temperature. The supply duct must produce an air curtain across the face of the door opening. The purpose of the air curtain is not to prevent the entry of outside air but to mix hot air with air entering the doorway.

Spot cooling in high heat areas can be done with makeup air. Velocity is an important component of the design, and special attention should be given to the effective temperature of the air

The preparation of this chapter is assigned to TC 5.8, Industrial Ventilation.

supply. In the winter, sensible temperatures are elevated to prevent discomfort.

Process air requirements for closed systems, such as spray booths and ovens, are provided by the same type of equipment as for the general area. In enclosed spray booths, the air-change rate is generally so great that the entire environment of the booth is that of the air supply. Again, it is necessary to consider the effective temperature for personnel comfort.

SELECTION

Functions to consider in the selection of makeup air units are type of unit, heating and cooling, filters, automatic controls, sound levels, and capacities.

Types

Makeup air units are available in blow-through or draw-through configurations for either indoor or outdoor installation. The components used with other types of air-handling units are also used with makeup air units.

Fans are centrifugal (single or double width, forward- or backward-inclined, flat-plate or airfoil) or axial-flow (propeller, ducted-propeller, vaneaxial, tube axial, or axial-centrifugal).

Indoor units. Indoor units may be truss-mounted, floor-mounted, or wall-mounted.

Outdoor (weatherproof) units. Rooftop locations for makeup air units, which remove them from truss spaces, eliminate many problems of crowding and conflict with piping, conveyors, or lighting. But even with extensive catwalks, access platforms, and ladders, maintenance of overhead equipment is often difficult and, sometimes, neglected.

Moving the equipment to the outside also eliminates many of the restrictions on size and enables the manufacturer to design greater flexibility into the equipment (Figure 2).

Packaged makeup air equipment. Packaged makeup air units are manufactured in capacities from 2,000 to 100,000 cfm. Equipment ranges from basic units of fans, heating elements, outside air/return air mixing dampers, and controls for truss mounting to packaged penthouse units, which include filters, air washers, or cooling coils as complete assemblies. Walk-in service compartments, which are heated, ventilated, and lighted, can also be included. Makeup air units are available in packages, using any heat source with optional components and configurations, which permit custom arrangement. Equipment manufacturers can provide a packaged, engineered product, including power and control wiring, piping, and temperature controls.

Heating and Cooling Media

Heat sources include gas (direct- or indirect-fired), oil (indirect-fired), steam, hot water, electric, heat transfer fluids, and exhaust air (air-to-air heat exchangers).

Cooling sources include refrigerants and chilled water coils, evaporative cooling, sprayed coils, air washers, and air-to-air heat exchangers.

Gas direct-fired units. The rapid response and ability to modulate over the complete range (turn down from 25 or 30 to 1) makes the direct-fired burner particularly suited for heating makeup air. The burner releases all the heat from combustion into the airstream. However, a portion of the available energy is used for latent heat of vaporization of the water, which is a product of combustion. While this would increase the wet-bulb temperature and, as a result, the enthalpy, the sensible heat available for a dry-bulb temperature rise is approximately 92%. For 100% outside air units, air can be drawn through the unit. For mixed air units, only

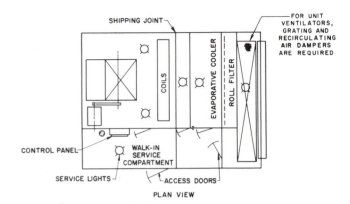

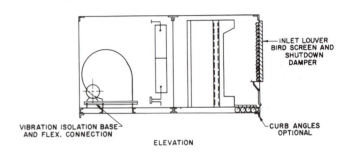

Fig. 2   Typical Packaged Penthouse Makeup Air Heater

the outside air portion should pass through the burner (as shown in Figure 1) but not the return air from the building. This practice eliminates health hazards caused by chemicals released in the building from being converted in the burner to carbon monoxide or other toxic gases.

Gas and oil indirect-fired units. Gas and oil indirect-fired units (with air-to-air heat exchangers) are used for makeup air, but their limited turndown of approximately 5 to 1 and the condensation danger of flue gases in tubes have limited their use as 100% makeup air heaters. However, their use as recirculating heaters, with a percentage of outside air, has made them an important part of the makeup air field.

Steam units. To permit cleaning, coil-fin spacing should be no less than 0.13 in. Steam or hot water coil makeup air units should always be specified with filters to protect the coils from dirt and capacity loss.

Steam makeup air heaters require careful design to prevent freeze-up of coils. Design considerations include (1) large tubes of the steam-distributing type, preferably mounted vertically; (2) adequately sized steam traps located for easy servicing; (3) sufficient static pressure on traps; and (4) an adequate condensate return system.

An effective temperature control is the use of preheat and reheat coils, with full steam pressure on the preheat coil and modulating control on the reheat coil. In areas where design temperatures reach $-30\,°F$, it may be necessary to use two preheat coils, each controlled separately.

Effective freeze protection may be achieved by face and bypass control systems with full steam pressure in the coils at all times. Face and bypass dampers should be designed to prevent stratification of air to the fans and to eliminate spin at the fan inlets. Bypass dampers must be sized to handle the full air capacity of the unit. It is important to limit radiation and air velocity over the coils on light loads. This can be done by installing face dampers on the downstream side of the coils.

Hot water units. The design of hot water heating coils is described in Chapter 24. Hot water coils are not as prone to freezing as steam coils, because the cooler water can be removed from the hot water coil more readily. Nonetheless, freeze-ups do occur, and precautions must be taken.

When freezing is a possibility, parallel flow circuiting should be specified. Counterflow circuiting, particularly with long coils, is dangerous because the coldest air is in contact with the coldest water.

Proper control is important in freeze protection. Overly hot water and oversized valves and pumps may cause hunting of the control. Freezing can occur when the flow control valve is shut.

Hot water flow must be maintained through the unit during freezing weather, even if the unit is not in operation or the coil should be drained. If outside air dampers do not close tightly, some leakage of cold air across the coil can be expected.

Air-to-air heat exchangers. Because of increases in fuel costs, these types of energy recovery makeup air units are important. For a description, see Chapter 44.

Filters

Filters may be automatic or manual roll types, throwaway or cleanable panels, bag-type, electrostatic, or a combination of these types (see Chapter 25).

Automatic Controls

Makeup air units require modulating temperature controls to maintain the discharge temperature near room temperature with a large change in intake temperatures. Design conditions may vary from −40 to 35 °F. Outdoor air imposes an instantaneous load similar to solar heat loads on windows; therefore, capacity must be based on maximum loads.

The basic temperature control system for a makeup air heater is discharge control. Because of the wide load variation, reset is necessary to prevent offset or droop. In direct-fired gas heaters, the discharge control sensor must act quickly because of the rapid response of the burner. Turndown ratios of 25 or 30 to 1 are needed to prevent overheating as outdoor temperatures rise close to the ambient.

While control of discharge temperature is basic, other control systems may include a room thermostat control, where the control point of the discharge controller is changed to satisfy room conditions or a room pressure control to maintain a slight building pressure.

Sound Level

Noise in the work environment is limited by law.

Capacities and Ratings

The capacity of a makeup air unit is primarily rated by air volume. Unlike heating systems where loads change with the seasons of the year, exhaust systems usually operate at a constant volume. Standard air should be used in capacity ratings, and all supply and exhaust volumes should be reduced to standard conditions. Fan data, as shown on fan curves, are for bare fans and must be corrected to account for location in an enclosure. Performance testing of the assembly of the components in a complete unit is necessary to have true ratings. The sum of the individual resistances of components as tested under ideal conditions may be less than the system resistance when the components are placed together.

Ratings of a makeup air unit include the following: (1) air volume; (2) total static pressure; (3) external static pressure; (4) fan static efficiency at point of rating; (5) sound power level at point of rating; and (6) heating and cooling capacities, including inlet and outlet dry- and wet-bulb temperatures.

CODES AND STANDARDS

Government health departments are becoming more aware of the health problems associated with buildings under negative air pressure. Back-venting of combustion equipment has become a serious health hazard, and several cities have passed ordinances requiring the installation of tempered makeup air equipment.

Design engineers must be familiar with codes, particularly those regarding the use of direct gas-firing, which is subject to special regulation in most places. Some codes limit direct gas-firing to areas not containing sleeping quarters; prohibit recirculation of air from the space through the burner; require interlocked exhaust fans; and/or stipulate adequate exhaust, proper filtration, and specific clearances from combustible materials.

MAINTENANCE

Maintenance requirements are similar to those of unit heaters.

REFERENCES

ACGIH. 1988. *Industrial ventilation guide.* American Conference of Government Industrial Hygenists, Cincinnati, OH.

ANSI. 1987. *Standard* Z83.18-87. Direct gas-fired industrial heaters. American National Standards Institute, New York.

ASHRAE. 1989. Ventilation for acceptable air quality. ASHRAE *Standard* 62-1989.

ICBO. 1985. *Uniform mechanical code.* International Conference of Building Officials, Whittier, CA. Also International Association of Plumbing and Mechanical Officials, Los Angeles.

CHAPTER 33

HYDRONIC RADIATORS

RADIATORS, convectors, and baseboard and finned-tube units are heat distributing devices used in steam and low-temperature water heating systems. They supply heat through a combination of radiation and convection and maintain the desired air temperature in a space without the use of fans. Figures 1 and 2 show sections of typical heat distributing units.

DESCRIPTION

Radiators

The term radiator, while generally confined to sectional cast-iron column, large-tube, or small-tube units, also includes flat panel types and fabricated steel sectional types. Small-tube radiators, with a length of only 1.75 in. per section, occupy less space than column and large-tube units and are particularly suited to installation in recesses (see Table 1). Column, wall-type, and large-tube radiators are no longer manufactured, although many of these units are still in use. Refer to Tables 2, 3, and 4 in Chapter 28 of the 1988 ASHRAE *Handbook—Equipment* or Byrley (1978) for principal dimensions and average ratings of these units.

A variety of radiators is used. The most common types are sectional radiators, panel radiators, tubular steel radiators, and specialty radiators.

Sectional radiators are fabricated from welded sheet metal sections (generally 2, 3, or 4-tube wide), and resemble freestanding cast-iron radiators.

Panel radiators consist of fabricated flat panels (generally 1, 2, or 3 deep), with or without an exposed extended fin surface attached to the rear for increased output. These radiators are most common in Europe.

Tubular steel radiators consist of supply and return headers with interconnecting parallel steel tubes in a wide variety of lengths and heights. Some are used in bathroom towel-heating applications, whereas others are used in shapes specially adapted to coincide with the building structure.

Specialty radiators are fabricated of welded steel or extruded aluminum for installation in ceiling grids or floor-mounting arrangements. An array of shapes different from conventional radiators is available.

Pipe Coils

Pipe coils have largely been replaced by finned-tube radiation. See Table 5, Chapter 28 of the 1988 ASHRAE *Handbook—Equipment* for the heat emission of such pipe coils.

Convectors

A convector is a heat distributing unit that operates with gravity-circulated (natural convection) air. It has a heating element with a large amount of secondary surface and contains two or more

---

The preparation of this chapter is assigned to TC 6.1, Hydronic and Steam Equipment and Systems.

tubes headered at both ends. The heating element is surrounded by an enclosure with an air inlet opening below and an air outlet opening above the heating element.

Convectors are made in a variety of depths, sizes, lengths, and in enclosure or cabinet types. The heating elements are available in fabricated ferrous and nonferrous metals. The air enters the enclosure below the heating element, is heated in passing through the element, and leaves the enclosure through the outlet grille located above the heating element. Factory-assembled units comprised of a heating element and enclosure have been widely used. These may be freestanding, wall-hung, or recessed and may have outlet grilles and arched inlets or inlet grilles, as desired.

Baseboard Units

Baseboard (or baseboard radiation) units are designed for installation along the bottom of walls, in place of the conventional baseboard. They may be made of cast iron, with a substantial portion of the front face directly exposed to the room, or with a finned-tube element in a sheet metal enclosure. They operate with gravity-circulated room air.

Baseboard heat distributing units are divided into three types: radiant, radiant-convector, and finned tube. The radiant unit, which is made of aluminum, has no openings for air to pass over the wall side of the unit. Most of this unit's heat output is by radiation.

The radiant-convector baseboard is made of cast iron or steel. The units have air openings at the top and bottom to permit circulation of room air over the wall side of the unit, which has extended surface to provide increased heat output. A large portion of the heat emitted is transferred by convection.

The finned-tube baseboard has a finned-tube heating element concealed by a long, low sheet metal enclosure or cover. A major portion of the heat is transferred to the room by convection. The output varies over a wide range, depending on the physical dimensions and the materials used. A unit with too high an output per unit length should be avoided. Optimum comfort for room occupants is obtained when units are installed along as much of the exposed wall as possible.

Finned-Tube Units

Finned-tube (or fin-tube) units are fabricated from metallic tubing, with metallic fins bonded to the tube. They operate with gravity-circulated room air. Finned-tube elements are available in several tube sizes, in either steel or copper—1 to 2 in. NPS or 3/4 to 1 1/4 in. nominal copper—with various fin sizes, spacings, and materials. The resistance to the flow of steam or water is the same as that through standard distribution piping of equal size and type.

Finned-tube elements installed in occupied spaces generally have covers or enclosures in a variety of designs. When human contact is unlikely, they are sometimes installed bare or are provided with an expanded metal grille for minimum protection.

33.1

A portion of the front skirt of a cover is made of solid material. It can be mounted with clearance between the wall and the cover, and without completely enclosing the rear of the finned-tube element. A cover may have a top, front, or inclined outlet. An enclosure is a shield of solid material that completely encloses both the front and rear of the finned-tube element. An enclosure may have an integral back or may be installed tightly against the wall so that the wall forms the back, and it may have a top, front, or inclined outlet.

Heat Emission

These heat distributing units emit heat by a combination of radiation to the space and convection to the air within the space.

Chapter 3 of the 1989 ASHRAE *Handbook—Fundamentals* covers the heat transfer processes and the factors that influence them. Those units with a large portion of their heated surface exposed to the space (*i.e.*, radiator and cast-iron baseboard) emit more heat by radiation than do units with completely or partially concealed heating surfaces (*i.e.*, convector, finned-tube, and finned-tube type baseboard). Also, finned-tube elements constructed of steel emit a larger portion of heat by radiation than do finned-tube elements constructed of nonferrous materials.

The output ratings of these heat distributing units are expressed in Btu/h, MBh (1000 Btu/h), or in square feet (ft^2) equivalent direct radiation (EDR). By definition, 240 Btu/h = 1 ft^2 EDR with 1 psig steam.

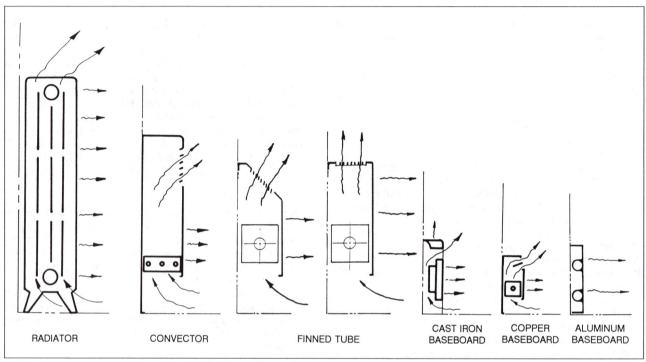

Fig. 1 Terminal Units

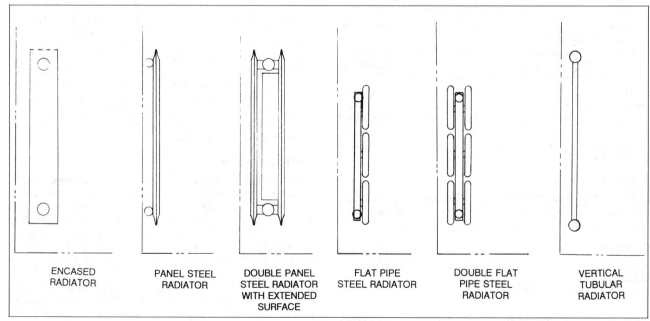

Fig. 2 Typical Radiators

Table 1 Small-Tube Cast-Iron Radiators

| Number of Tubes per Section | Catalog Rating per Section[a], ft² | Btu/h | Section Dimensions | | | | |
			A Height, in.[b]	B Width, in. Min.	B Width, in. Max.	C Spacing, in.[c]	D Leg Height, in.[b]
3	1.6	384	25	3.25	3.50	1.75	2.50
4	1.6	384	19	4.44	4.81	1.75	2.50
	1.8	432	22	4.44	4.81	1.75	2.50
	2.0	480	25	4.44	4.81	1.75	2.50
5	2.1	504	22	5.63	6.31	1.75	2.50
	2.4	576	25	5.63	6.31	1.75	2.50
6	2.3	552	19	6.81	8	1.75	2.50
	3.0	720	25	6.81	8	1.75	2.50
	3.7	888	32	6.81	8	1.75	2.50

[a]These ratings are based on steam at 215 °F and air at 70 °F. They apply only to installed radiators exposed in a normal manner, not to radiators installed behind enclosures, grilles, or under shelves. For Btu/h ratings at other temperatures, divide table values by factors found in Table 3.

[b]Overall height and leg height, as produced by some manufacturers, are 1 in. (greater than shown in Columns A and D. Radiators may be furnished without legs. Where greater than standard leg heights are required, this dimension shall be 4.5 in.

[c]Length equals number of sections multiplied by 1.75 in.

RATINGS OF HEAT DISTRIBUTING UNITS

Radiators

Current methods for rating radiators were established by the U.S. National Bureau of Standards publication, *Simplified Practice Recommendation R174-65, Cast-Iron Radiators*, which has been withdrawn (see Table 1).

Convectors

The generally accepted method of testing and rating ferrous and nonferrous convectors was given in *Commercial Standard CS 140-47, Testing and Rating Convectors* (U.S. Dept. of Commerce 1947), but it has been withdrawn. This standard contained details covering construction and instrumentation of the test booth or room, and procedures for determining steam and water ratings.

Under the provisions of *Commercial Standard CS 140-47*, the rating of a top outlet convector was established at a value not in excess of the test capacity (which is the heat extracted from the steam or water in the convector under standard test conditions). For convectors with other types of enclosures or cabinets, a percentage that varies up to a maximum of 15% (depending on the height and type of enclosure or cabinet) was added for heating effect (Brabbee 1927, Willard *et al.* 1929). The additions made for heating effect must be shown in the manufacturer's literature.

The testing and rating procedure set forth by *Commercial Standard CS 140-47* does not apply to finned-tube or baseboard radiation.

Baseboard

The generally accepted method of testing and rating baseboards is covered in the I = B = R *Testing and Rating Standard for Baseboard Radiation* (Hydronics Institute 1990). This standard contains details covering construction and instrumentation of the test booth or room, procedures for determining steam and hot water ratings, and licensing provisions for obtaining approval of these ratings.

Baseboard ratings include an allowance for heating effect of 15%, added to the test capacity. The addition made for heating effect must be shown in the manufacturer's literature.

Finned-Tube Units

The generally accepted method of testing and rating finned-tube units is covered in the I = B = R *Testing and Rating Standard for Finned-Tube (Commercial) Radiation* (Hydronics Institute 1990). This standard contains details covering construction and instrumentation of the test booth or room, procedures for determining steam and water ratings, and licensing provisions for obtaining approval of these ratings. Table 2 gives factors for converting steam ratings to hot-water ratings at various average water temperatures. These factors apply only when the water velocity in the element is 3.0 ft/s (see Figure 3).

The rating of a finned-tube unit in an enclosure that has a top outlet is established at a value not in excess of the test capacity (which is the heat extracted from the steam or water in the unit under standard test conditions). For finned-tube units with other types of enclosures or covers, a percentage is added for heating effect, which varies up to a maximum of 15%, depending on the height and type of enclosure or cover. The additions made for heating effect must be shown in the manufacturer's literature (Pierce 1963).

Other Heat Distributing Units

Unique radiators and radiators from other countries generally are tested and rated for heat emission in accordance with prevailing standards. These other testing and rating methods have been basically the same procedures as the I = B = R standard, which is used in the United States.

Corrections for Nonstandard Conditions

The heat output of a radiator, convector, baseboard, or finned-tube heat distributing unit is an exponential function of the tem-

Table 2 Factors to Convert Finned-Tube Steam Ratings to Hot-Water Ratings

Avg. Water Temperature, °F	Factor	Avg. Water Temperature, °F	Factor
100	0.15	185	0.73
110	0.20	190	0.78
120	0.26	195	0.82
130	0.33	200	0.86
140	0.40	205	0.91
150	0.45	210	0.95
155	0.49	215	1.00
160	0.53	220	1.05
165	0.57	225	1.09
170	0.61	230	1.14
175	0.65	235	1.20
180	0.69	240	1.25

perature difference between the air in the room and the heating medium in the room-heating unit, shown as:

$$q = c (t_s - t_a)^n \qquad (1)$$

where

q = heat output, Btu/h
c = constant determined by test
t_s = average temperature of heating medium, °F. For hot water, the arithmetic average of the entering and leaving water temperatures is used.
t_a = room air temperature, °F. Air temperature 60 in. above the floor is generally used for radiators, while the entering air temperature is used for all other types of heating units.
n = exponent that equals 1.3 for cast-iron radiators, 1.4 for baseboard radiation, and 1.5 for convectors. For finned-tube units, n varies with air and heating medium temperatures. Correction factors to convert outputs at standard rating conditions to outputs at other conditions are given in Table 3.

DESIGN CONSIDERATIONS

Effect of Water Velocity

Designing for high temperature drops through the system—drops of as much as 60° to 80°F in low-temperature systems and as high as 200°F in high-temperature systems, can result in low water velocities in the finned-tube or baseboard element. Application of very short runs designed for conventional temperature drops, i.e., 20°F, can also result in low velocities.

Figure 3 shows the effect of water velocity on the heat output of typical sizes of finned-tube elements. The figure is based on work done by Harris and Pierce and tests at the Hydronics Institute. The velocity correction factor F_v, based on Equation (2), is:

$$F_v = (V/3.0)^{0.04} \qquad (2)$$

where

V = water velocity, ft/s

Heat output varies little over the range from 0.5 to 3 ft/s where factors range from 0.93 to 1.00. The factor drops rapidly below 0.5 ft/s because the flow changes from turbulent to streamline at around 0.1 ft/s. Such low velocities should be avoided since the output is difficult to predict accurately when designing a system. In addition, the curve is so steep in this region that small changes in actual flow have a significant effect on output. Not only does

the heat transfer rate change, but the temperature drop and, therefore, the average water temperature changes (assuming a constant inlet temperature).

The designer should check water velocity throughout the system and select finned-tube or baseboard elements on the basis of velocity as well as average temperature. Manufacturers of finned-tube and baseboard elements offer a variety of tube sizes—ranging from 0.5-in. copper tubes for small baseboard elements to 2 in. for large finned-tube units—to aid in maintenance of turbulent flow conditions over a wide range of flows.

Effect of Altitude

The effect of altitude on output varies depending on the material used and the portion of the unit's output that is radiant rather than convective. The reduced air density affects the convective portion. Figure 4 shows the reduction in output with air density (Sward and Decker 1965). The approximate correction factor F_A for determining the reduced output of typical units is:

$$F_A = (p/p_o)^n \qquad (3)$$

where

p = local station pressure
p_o = standard atmospheric pressure
n = 0.9 for copper baseboard or finned tube
n = 0.5 for steel finned-tube or cast-iron baseboard
n = 0.2 for radiant baseboard

The value of p/p_o at various altitudes is as follows:

Altitude, ft	p/p_o
2000	0.93
4000	0.86
5000	0.83
6000	0.80

Sward and Harris (1970) has shown that some components of heat loss are affected in the same manner.

Effect of Mass

Mass of the terminal unit (typically cast-iron versus copper-aluminum finned element) affects the heat-up and cool-down rates of the equipment. It is important that high- and low-mass radiation not be mixed in the same zone. High-mass systems have historically been favored for best comfort and economy, but

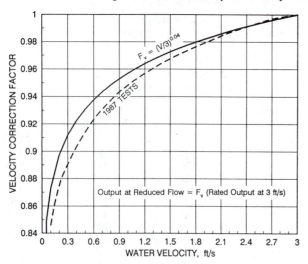

Fig. 3 Water Velocity Correction Factor for Baseboard and Finned-Tube Radiators

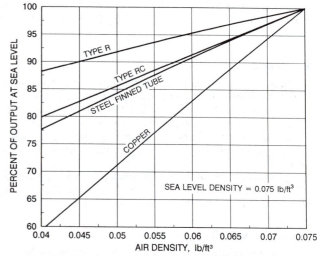

Fig. 4 Effect of Air Density on Radiator Output

tests by Harris (1970) have shown no measurable difference. The thermostat, or control system, can compensate by changing the cycle rate of the burner. The effect of mass is further reduced by constantly circulating modulated temperature water to the unit. The only time that mass can have a significant effect is in response to a massive shift in load. In that situation, the low-mass unit will respond faster.

Performance at Low Water Temperatures

Tables 2 and 3 summarize the performance of baseboard and finned-tube units with an average water temperature down to 100 °F. Solar-heated water or heat pump system cooling water are typical applications in this range.

Effect of Enclosure and Paint

An enclosure placed around a direct radiator restricts the airflow and diminishes the proportion of output resulting from radiation. However, enclosures of proper design may improve the heat distribution within the room as compared to the heat distribution obtained with an unenclosed radiator (Willard *et al.* 1929, Allcut 1933).

For a radiator or cast-iron baseboard, the finish coat of paint affects the heat output. Oil paints of any color give about the same results as unpainted black or rusty surfaces, but an aluminum or a bronze paint reduces the heat emitted by radiation. The net effect may reduce the total heat output of the radiator by 10% or more (Rupert 1937, Severns 1927, Allen 1920).

APPLICATIONS

Radiators

Radiators can be used with steam or hot-water systems. They are installed in areas of greatest heat loss—under windows, along cold walls, or at doorways. They can be installed freestanding, semirecessed, or with decorative enclosures or shields, although the enclosures or shields affect the output (Willard *et al.* 1929).

Unique and imported radiators are generally not suitable for steam applications although they have been used extensively in low-temperature water systems with valves and connecting piping left exposed. Various combinations of supply and return locations are possible, which may alter the heat output. Although long lengths may be ordered for linear applications, lengths may not be reduced or increased by field modification. The small cross-sectional areas often inherent in unique radiators require careful evaluation of flow requirements, water temperature drop, and pressure drops.

Convectors

Convectors can be used with steam or hot-water systems. Like radiators, they should be installed in areas of greatest heat loss. They are particularly applicable where wall space is limited, such as entryways and kitchens.

Baseboard Radiation

Baseboard units are used almost exclusively with hot-water systems. When used with one-pipe steam systems, tube sizes of 1.25 in. NPS must be used to allow drainage of condensate counterflow to the steam flow.

The basic advantage of the baseboard unit is that its normal placement is along the cold walls and under areas where the greatest heat loss occurs. Other advantages are (1) it is inconspicuous, (2) it offers minimal interference with furniture placement, and (3) it distributes the heat near the floor. This last characteristic reduces the floor-to-ceiling temperature gradient to about 2 to 4 °F and tends to produce uniform temperatures throughout the room. It also makes baseboard heat distributing units adaptable to homes without basements, where cold floors are common (Kratz and Harris 1945).

Heat-loss calculations for baseboard heating systems are the same as those used for other types of heat distributing units. The procedure for designing baseboard heating systems is given in I = B = R *Installation Guide* No. 200 (1989).

Finned-Tube Radiation

The finned-tube unit can be used with steam or hot-water systems. It is advantageous for heat distribution along the entire outside wall, thereby preventing downdrafts along the walls in buildings such as schools, churches, hospitals, offices, airports, and factories. It may be the principal source of heat in a building or a supplementary heater to combat downdrafts along the exposed walls in conjunction with a central conditioned air system.

Normal placement of a finned tube is along the walls where the heat loss is greatest. If necessary, the units can be installed in two or three tiers along the wall. Hot-water installations requiring two or three tiers should run a serpentine water flow, because a header connection with parallel flow may (1) permit the water to short circuit along the path of least resistance, (2) suffer reduced capac-

Table 3 Correction Factors for Various Types of Heating Units

Steam Pressure (approx.) Gage in. Hg.	Absolute psi	Steam or Water Temp., °F	Cast-Iron Radiators Room Temp., °F					Convectors Inlet Air Temp., °F					Finned-Tube Inlet Air Temp., °F					Baseboard Inlet Air Temp., °F				
			80	75	70	65	60	75	70	65	60	55	75	70	65	60	55	75	70	65	60	55
		100											0.10	0.12	0.15	0.17	0.20	0.08	0.10	0.13	0.15	0.18
		110											0.15	0.17	0.20	0.23	0.26	0.13	0.15	0.18	0.21	0.25
		120											0.20	0.23	0.26	0.29	0.33	0.18	0.21	0.25	0.28	0.31
		130											0.26	0.29	0.33	0.36	0.40	0.25	0.28	0.31	0.34	0.38
Vacuum		140											0.33	0.36	0.40	0.42	0.45	0.31	0.34	0.38	0.42	0.45
22.4	3.7	150	0.39	0.42	0.46	0.50	0.54	0.35	0.39	0.43	0.46	0.50	0.40	0.42	0.45	0.49	0.53	0.38	0.42	0.45	0.49	0.53
20.3	4.7	160	0.46	0.50	0.54	0.58	0.62	0.43	0.47	0.51	0.54	0.58	0.45	0.49	0.53	0.57	0.61	0.45	0.49	0.53	0.57	0.61
17.7	6.0	170	0.54	0.58	0.62	0.66	0.69	0.51	0.54	0.58	0.63	0.67	0.53	0.57	0.61	0.65	0.69	0.53	0.57	0.61	0.65	0.69
14.6	7.5	180	0.62	0.66	0.69	0.74	0.78	0.58	0.63	0.67	0.71	0.76	0.61	0.65	0.69	0.73	0.78	0.61	0.65	0.69	0.72	0.78
10.9	9.3	190	0.69	0.74	0.78	0.83	0.87	0.67	0.71	0.76	0.81	0.85	0.69	0.73	0.78	0.81	0.86	0.69	0.73	0.78	0.82	0.86
6.5	11.5	200	0.78	0.83	0.87	0.91	0.95	0.76	0.81	0.85	0.90	0.95	0.77	0.81	0.86	0.90	0.95	0.81	0.86	0.92	0.95	1.00
psi																						
1	15.6	215	0.91	0.95	1.00	1.04	1.09	0.90	0.95	1.00	1.05	1.10	0.91	0.94	1.00	1.06	1.11	0.91	0.95	1.00	1.05	1.09
6	21	230	1.04	1.09	1.14	1.18	1.23	1.05	1.10	1.15	1.20	1.26	1.03	1.08	1.14	1.19	1.24	1.04	1.09	1.14	1.19	1.25
15	30	250	1.23	1.28	1.32	1.37	1.43	1.27	1.32	1.37	1.43	1.47	1.20	1.26	1.31	1.37	1.43	1.22	1.27	1.32	1.37	1.43
27	42	270	1.43	1.47	1.52	1.56	1.61	1.47	1.54	1.59	1.67	1.72	1.38	1.44	1.50	1.56	1.62	1.43	1.47	1.52	1.59	1.64
52	67	300	1.72	1.75	1.82	1.89	1.92	1.85	1.89	1.96	2.04	2.08	1.67	1.73	1.79	1.86	1.92	1.75	1.82	1.89	1.92	1.96

Use these correction factors to determine output ratings for radiators, convectors, and finned-tube and baseboard units at operating conditions other than standard. Standard conditions for a radiator are 215 °F heating medium temperature and 70 °F room temperature (at the center of the space and at the 5-ft level). Standard conditions for convectors and finned-tube and baseboard units are 215 °F heating medium temperature and 65 °F inlet air temperature. Inlet air at

65 °F for convectors and finned-tube or baseboard units represent the same room comfort conditions as 70 °F room air temperature for a radiator.

To determine the output of a heating unit under conditions other than standard, mutiply the standard heating capacity by the appropriate factor for the actual operating heating medium and room or inlet air temperatures.

ity because of low water velocity in each tier, or (3) become air-bound in one or more of the tiers.

Many enclosures have been developed to meet building design requirements. The wide variety of finned-tube elements (tube size and material, fin size, spacing, fin material, and multiple tier installation), along with the various heights and designs of enclosures, give great flexibility of selection for finned-tube units that meet the needs of load, space, and appearance.

In areas where zone control rather than individual room control can be applied, all finned-tube units in the zone should be in series. In such a series loop installation, however, temperature drop must be considered in selecting the element for each separate room in the loop.

REFERENCES

Allcut, E.A. 1933. Heat output of concealed radiators. *School of Engineering Research Bulletin* 140, University of Toronto, Toronto, Canada.

Allen, J.R. 1920. Heat loss from direct radiation. ASHVE *Transactions* 26:11.

Brabbee, C. 1927. The heating effect of radiators. ASHVE *Transactions* 26:11.

Byrley, R.R. 1978. *Hydronic rating handbook*. Color Art Inc., St. Louis, MO.

Department of Commerce. 1947. *Commercial standard for testing and rating convectors*. CS 140-47. Washington, D.C.

Harris, W.H. 1957. Factor affecting baseboard rating test results. Engineering Experiment Station *Bulletin* 444. University of Illinois, Champaign-Urbana, IL.

Harris, W.H. 1970. Operating characteristics of ferrous and non-ferrous baseboard. I=B=R #8. University of Illinois, Champaign-Urbana, IL.

Hydronics Institute. 1989. I=B=R Installation guide for residential hydronic heating systems, No. 200, 1st ed. Hydronics Institute, Berkeley Heights, NJ.

Hydronics Institute. 1990. I=B=R Testing and rating standard for baseboard radiation, 11th ed. Hydronics Institute, Berkeley Heights, NJ.

Hydronics Institute. 1990. I=B=R Testing and rating standard for finned tube (commercial) radiation, 5th ed. Hydronics Institute, Berkeley Heights, NJ.

Kratz, A.P. 1931. Humidification for residences. Engineering Experiment Station *Bulletin* 230:20, University of Illinois, Champaign-Urbana, IL.

Kratz, A.P. and W.S. Harris. 1945. A study of radiant baseboard heating in the I=B=R research home. Engineering Experiment Station *Bulletin* 358. University of Illinois, Champaign-Urbana, IL.

Laschober, R.R. and G.R. Sward. 1967. Correlation of heat output of unenclosed single- and multiple-tier finned tube units. ASHRAE *Transactions* 73:V.3.1-15.

Pierce, J.S. 1963. Application of fin tube radiation to modern hot water systems. ASHRAE *Journal* (February):72.

Rubert, E.A. 1937. Heat emission from radiators. Engineering Experiment Station *Bulletin* 24. Cornell University, Ithaca, NY.

Severns, W.H. 1924. Comparative tests of radiator finishes. ASHVE *Transactions* 33:41.

Sward, G.R. and A.S. Decker. 1965. Symposium on high altitude effects on performance of equipment. ASHRAE.

Sward, G.R. and W.S. Harris. 1970. Effect of air density on the heat transmission coefficients of air films and building materials. ASHRAE *Transactions* 76:227-39.

Willard, A.C., A.P. Kratz, M.K. Fahnestock, and S. Konzo. 1929. Investigation of heating rooms with direct steam radiators equipped with enclosures and shields. ASHVE *Transactions* 35:77 or Engineering Experiment Station *Bulletin* 192. University of Illinois, Champaign-Urbana, IL.

Willard, A.C, A.P. Kratz, M.K. Fahnestock, and S. Konzo. 1931. Investigation of various factors affecting the heating of rooms with direct steam radiators. Engineering Experiment Station *Bulletin* 223. University of Illinois, Champaign-Urbana, IL.

CHAPTER 34

SOLAR ENERGY EQUIPMENT

COMMERCIAL and industrial solar energy systems are generally classified according to the heat transfer fluid used in the collector loop—namely, air or liquid. While both share the basic fundamentals associated with conversion of solar radiant energy, the equipment used in each is entirely different. Air systems are primarily limited to forced-air space heating and industrial and agricultural drying processes. Liquid systems are suitable for a broader range of applications, such as hydronic space heating, service water heating, industrial process water heating, energizing absorption air conditioning, pool heating, and as a heat source for series-coupled heat pumps. Because of this wide range in capability, liquid systems are more commonly found in commercial industrial applications.

AIR HEATING SYSTEMS

Air heating systems circulate air through ducting to and from an air heating collector (Figure 1). Air systems are effective for space heating applications because a heat exchanger is not required and the collector inlet temperature is low throughout the day (approximately room temperature). Air systems do not need protection from freezing, overheat, or corrosion. Furthermore, air costs nothing and does not cause disposal problems. However, air ducts and air-handling equipment require more space than pipes and pumps, ductwork is hard to seal, and leaks are difficult to detect. Fans consume more power than the pumps of a liquid system, but if the unit is installed in a facility that uses air distribution, only a slight power cost is chargeable against the solar space heating system.

Most air space heating systems also preheat domestic hot water through an air-to-liquid heat exchanger. In this case, tightly fitting dampers are required to prevent reverse thermosiphoning at night, which could freeze water in the heat exchanger coil. When this system only heats water in the summer, the parasitic power consumption must be charged against the solar energy system, because no space heating is involved and there are no comparable energy costs associated with conventional water heating. In some situations, solar water-heating air systems could be more expensive than conventional water heaters, particularly if electrical energy costs are high. To reduce the parasitic power consumption, some systems use the low speed of a two-speed fan.

LIQUID HEATING SYSTEMS

Freezing is the principal cause of liquid system failure. For this reason, freeze tolerance dominates the design and selection of equipment. A solar collector radiates heat to the cold sky and

The preparation of this chapter is assigned to TC 6.7, Solar Energy Utilization.

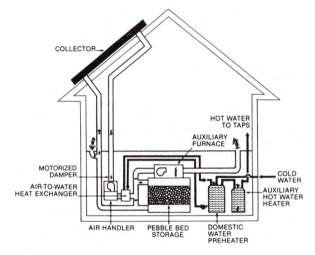

COLLECTOR

HOT WATER
TO TAPS

AUXILIARY
FURNACE

COLD
WATER

AUXILIARY
HOT WATER
HEATER

MOTORIZED
DAMPER

AIR-TO-WATER
HEAT EXCHANGER

AIR HANDLER   PEBBLE BED
STORAGE

DOMESTIC
WATER
PREHEATER

Fig. 1   Air Heating Space and Domestic Water Heater System

freezes at air temperatures well above 32 °F. Where freezing conditions are rare, small solar heating systems may be equipped with low cost, low reliability protection devices that depend on electrical and/or mechanical components such as electronic controllers and automatic valves. With increasing consumer/government demands for extended warranties, solar designers and installers must consider reliable designs. Because of the large investment associated with most commercial and industrial installations, reliable freeze protection is essential, even in the warmest climates.

Open and Closed Systems

In an open liquid solar energy system, the collector loop is open to the city water supply, and fresh water circulates through the collector. In a closed system, the collector loop is isolated from the high-pressure city water supply by a heat exchanger. In areas of poor water quality, isolation protects the collectors from fouling due to minerals in the water. Closed-loop systems also offer greater freeze protection, so they are used almost exclusively in commercial and industrial applications.

Closed-loop systems use two methods of freeze protection: (1) nonfreezing fluids and (2) drainback.

Nonfreezing fluid freeze protection. The most popular solar energy system for commercial application is the closed-loop system. It contains a nonfreezing heat transfer fluid to transmit heat from the solar collectors to storage (Figure 2). The most common heat transfer fluids are water/ethylene glycol and water/propylene glycol, although other heat transfer fluids such as silicone oils, hydrocarbon oils, or refrigerants can be used. Because the collector loop is closed and sealed, the only contribution to pump-

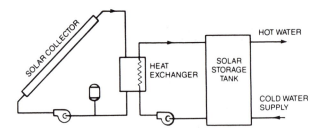

Fig. 2   Simplified Schematic of Nonfreezing System

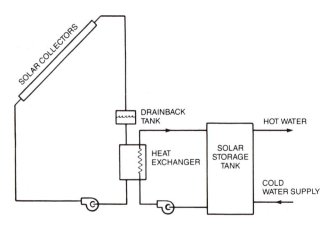

**Fig. 3   Simplified Schematic of Drainback Freeze
Protection Systems**

pressure is friction loss; therefore, the location of the solar collectors relative to the heat exchanger and storage tank is not critical. Traditional hydronic sizing methods can be used for selecting pumps, expansion tanks, heat exchangers, and air removal devices, as long as the thermal properties of the heat transfer liquid are considered.

When the control system senses an increase in solar panel temperature, the pump circulates the heat transfer liquid and energy is collected. The same controller also activates a pump on the domestic water side that circulates water through the heat exchanger where it is heated by the heat transfer fluid. This mode continues until the temperature differentials between the collector and the tank are too slight for meaningful energy to be collected. At this point, the control system shuts the pumps off. During low-temperature conditions, the nonfreezing fluid protects the solar collectors and related piping from bursting. Because the heat transfer fluid can affect system performance, reliability, and maintenance requirements, the fluid selected should be carefully considered.

Because the collector loop of the nonfreezing system remains filled with fluid, it gives flexibility for routing pipes and locating components. However, a double-separation heat exchanger is generally required (by local building codes) to prevent contamination of the domestic water in the event of a leak. The double-wall heat exchanger also protects the collectors from freeze damage if water leaks into the collector loop. However, the double-wall heat exchanger reduces efficiency by forcing the collector to operate at higher temperatures. The heat exchanger can be placed inside the tank, or an external heat exchanger can be used, as shown in Figure 2. The collector loop is closed and, therefore, requires an expansion tank and pressure relief valve. Air purge is also necessary to expel air during filling and to remove air that has been absorbed into the heat transfer fluid.

Over-temperature protection is necessary to prevent the collector fluid from corroding or decomposing. For maximum reliability, the glycol should be replaced every few years. In other cases, systems have failed because the collector fluid in the loop thermosiphoned and froze the water in the heat exchanger. Such a situation is disastrous and must be avoided by design if the water side is exposed to the city water system, because the collector loop eventually fills with water and all freeze protection is lost.

Drainback Freeze Protection

A drainback solar water heating system uses ordinary water as the heat transport medium between the collectors and thermal energy storage. Reverse draining (or back siphoning) the water into a drainback tank protects the system from freezing whenever the controls turn off the circulator pump (Figure 3) or a power outage occurs.

The drainback tank can be a sump with a volume slightly greater than the collector loop, or it can be the thermal energy storage tank. The collector loop can be vented to the atmosphere or not.

Many designers prefer the nonvented approach because makeup water is not required and the corrosive effects of the air that would otherwise be ingested into the collector loop are eliminated. However, the pressure is still low enough to avoid the use of ASME code tanks.

Service hot water may be warmed through a single separation (single-wall) heat exchanger, because the system contains only potable fluids. The drainback system is virtually fail-safe because it automatically reverts to a safe condition whenever the circulator pump stops. Furthermore, a 20 to 30% glycol solution can be added to drainback loops for added freeze protection in case of controller or sensor failure. Because the glycol is not exposed to stagnation temperatures, it does not decompose.

A drainback system requires space for the pitching of collectors and pipes necessary for proper drainage. Also, a nearby heated area must have a room for the pumps and the drainback tank. Plumbing exposed to freezing conditions drains to the drainback tank, making the drainback design unsuitable for sites where the collector cannot be elevated above the storage tank.

Both dynamic and static pressure losses comprise the pumping power required of a drainback system. The dynamic pressure loss is due to the friction of fluid motion in the pipes and is a function of fluid viscosity, pipe diameter, and fluid velocity. In drainback systems, an additional static pressure loss is associated with the distance the water must drop. It is this differential pressure in the downcoming pipe that creates the potential energy between the collector supply and return leg and causes the fluid in the collector loop to siphon back through the collector pump when it is turned off. In many systems, this drop is relatively short—the distance between the top of the drainback tank and the surface of the fluid in the tank. Upon start-up, an open drop will occur in the downcomer until steady flow is established, thus requiring a greater pumping pressure for a few minutes until the collector loop is charged. To ensure that the collector loop will drain, some systems have an oversized downcomer, which results in a relatively long drop. These open drop systems consume a greater amount of pump energy because of the continued presence of this high hydrostatic pressure loss.

Drainback performs better than other systems in areas that are marginal for solar applications. Drainback has an advantage because time and energy are not lost in reheating a fluid mass left in the collector and associated piping (as in the case of antifreeze systems). Also, water has a higher heat transfer capacity and is less viscous than other heat transfer fluids, resulting in smaller parasitic energy use and higher overall system efficiency. Closed return designs also consume less parasitic energy for pumping because water is the heat transfer fluid. Drainback systems can be worked on safely under stagnation conditions, but they should be restarted at night to avoid unnecessary thermal stress on the collector.

SOLAR ENERGY COLLECTORS

Solar energy system design requires careful attention to detail. Solar radiation is a low intensity form of energy, and the equipment to collect and use it is expensive. Imperfections in design and installation can lead to poor cost-effectiveness or to complete system failure. Chapter 30 in the 1991 ASHRAE *Handbook—HVAC Applications* covers solar energy use. Books on design, installation, operation, and maintenance are also available (ASHRAE 1988, 1990, 1991).

Solar energy and HVAC systems often use the same components and equipment. This chapter only covers the following elements that are exclusive to solar energy applications:

- Collectors and collector arrays
- Thermal energy storage
- Heat exchangers
- Controls

Collectors

Solar collectors can be divided into liquid heating and air heating. The most common type for commercial, residential, and low-temperature industrial applications (<200°F) is the flat-plate collector. Figure 4 shows cross sections of flat-plate air and liquid collectors. A flat-plate collector contains an absorber plate covered with a black surface coating and one or more transparent covers. The covers are transparent to incoming solar radiation and relatively opaque to outgoing (long-wave) radiation, but their principal purpose is to reduce heat losses caused by convection. The collector box is insulated to prevent conduction heat loss from the back and edge of the absorber plate. This type of collector can supply hot water or air at temperatures up to 200°F, although relative efficiency diminishes rapidly at temperatures above 160°F. The advantages of flat-plate collectors are simple construction, low relative cost, no moving parts, relative ease of repair, and durability. They also absorb diffuse radiation, which is a distinct advantage in cloudy climates.

Collectors also take the form of concentric glass cylinders, with the space between cylinders evacuated (Figure 5). This vacuum envelope reduces convection and conduction losses, so the cylinders can operate at higher temperatures than flat-plate collectors. As with flat-plate collectors, they collect both direct and diffuse radiation. However, their efficiency is higher at low-incidence angles than at the normal position of the sun. This effect tends to give an evacuated tube collector an advantage in day-long performance over the flat-plate collector. Because of its high-temperature capability, the evacuated tube collector is favored for energizing absorption air-conditioning equipment.

Flat-plate and evacuated-tube collectors are usually mounted in a fixed position. Concentrating collectors are available that must be arranged to track the movement of the sun. These are mainly used for high-temperature industrial applications above 240°F.

Air heating collectors, similar to their liquid heating counterparts, are contained in a box, covered with one or more glazings, and insulated on the sides and back. The primary differences are in the design of the absorber plate and flow passages. Because the working fluid (air) has poor heat transfer characteristics, it flows over the entire absorber plate, and sometimes on both the front and back of the plate in order to provide greater heat transfer surface. In spite of the larger surface area, air collectors generally have poorer overall heat transfer than liquid collectors. However, they are usually operated at a lower temperature for space heating applications because they require no intervening heat exchangers.

As with liquid collectors, there is a trend toward the use of spectrally selective surface treatments in combination with a single glazing of low-iron glass. Also, air collectors are being designed for flow rates in the range of 4 scfm/ft^2, whereas early models had suggested flows of 2 scfm/ft^2. While these modifications increase air collector efficiency and lower manufacturing costs, they also increase fan power consumption and lower output air temperature. In some applications, natural-convection air collectors are cost-effective. These collectors are self-regulating, can be designed not to reverse at night, and use no fan power.

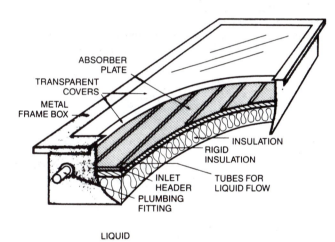

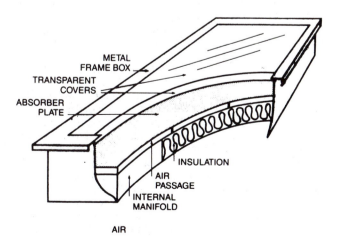

Fig. 4   Solar Flat-Plate Collectors

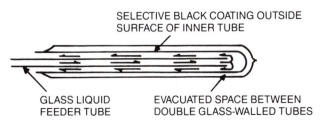

Fig. 5   Evacuated Tube Collector

Collector Performance

Under steady conditions, the useful heat delivered by a solar collector is equal to the energy absorbed in the heat transfer fluid minus the heat losses from the surface directly and indirectly to the surroundings. This principle can be stated in the relationship:

$$Q_\mu = A_c [I_t \tau\alpha - U_L (\bar{t}_p - t_a)] \qquad (1)$$

where

Q_u = useful energy delivered by collector, Btu/h
A_c = total collector area, ft^2
I_t = total (direct plus diffuse) solar energy incident on upper surface of sloping collector structure, Btu/h • ft^2
τ = fraction of incoming solar radiation that reaches absorbing surface, transmissivity (dimensionless)
α = fraction of solar energy reaching surface that is absorbed, absorptivity (dimensionless)
U_L = overall heat loss coefficient, Btu/h • ft^2 • °F
$\bar{t}_p$ = average temperature of the absorbing surface of the absorber plate, °F
t_a = atmospheric temperature, °F

With the exception of average plate temperature, these terms can be readily determined. For convenience, Equation (1) can be modified by substituting inlet fluid temperature for the average plate temperature, if a suitable correction factor is applied. The resulting equation is:

$$Q_\mu = F_R A_C [I_t \tau\alpha - U_L (t_i - t_a)] \qquad (2)$$

where

t_i = temperature of fluid entering collector, °F
F_R = correction factor, or heat-removal efficiency factor, having a value less than 1.0

The heat removal factor F_R can be considered the ratio of the heat actually delivered to that delivered if the collector plate were at uniform temperature equal to that of the entering fluid. A F_R of 1.0 is theoretically possible if (1) the fluid is circulated at such a high rate that its temperature rises a negligible amount and (2) the heat transfer coefficient and fin efficiency are so high that the temperature difference between the absorber surface and the fluid is negligible.

In Equation (2), the temperature of the inlet fluid depends on the characteristics of the complete solar heating system and the heat demand of the building. However, F_R is affected only by the solar collector characteristics, the fluid type, and its flow rate through the collector.

Solar air heaters remove substantially less heat than liquid collectors. However, their lower collector inlet temperature makes their system efficiency comparable to liquid systems for space heating applications.

Equation (2) may be rewritten in terms of the efficiency of total solar radiation collection by dividing both sides of the equation by $I_t A_c$. The result is:

$$\eta = F_R \tau\alpha - F_R U_L \frac{(t_i - t_a)}{I_t} \qquad (3)$$

Equation (3) plots as a straight line on a graph of efficiency versus the heat loss parameter $(t_i - t_a)/I_t$. The intercept (intersection of the line with the vertical efficiency axis) equals $F_R \tau\alpha$ and the slope of the line. That is, any efficiency difference divided by the corresponding horizontal scale difference equals $-F_R U_L$. If experimental data on collector heat delivery at various temperatures and solar conditions are plotted on a graph, with efficiency as the vertical axis and $(t_i - t_a)/I_t$ as the horizontal axis, the best straight line through the data points correlates collector performance with solar and temperature conditions. The intersection of the line with the vertical axis is the ambient temperature, where collector efficiency is at its maximum. At the intersection of the line with the horizontal axis, collection efficiency is zero. This condition corresponds to such a low radiation level, or to such a high temperature of the fluid into the collector, that heat losses equal solar absorption, and the collector delivers no useful heat. This condition is normally called *stagnation* and usually occurs when no coolant flows to a collector.

Equation (3) includes all important design and operational factors affecting steady-state performance, except collector flow rate and solar incidence angle. Flow rate indirectly affects performance through the average plate temperature. If the heat removal rate is reduced, the average plate temperature increases, and more heat is lost. If the flow is increased, collector plate temperature and heat loss decrease.

These relationships assume that the sun is perpendicular to the plane of the collector, which rarely occurs. For glass cover plates, specular reflection of radiation occurs, thereby reducing the t_a product. The incident angle modifier $K_{\tau\alpha}$, defined as the ratio of t_a at some incidence angle θ to $\tau\alpha$ at normal radiation $(\tau\alpha)_n$, is described by the following expression for specular reflection:

$$K_{\tau\alpha} = \frac{(\tau\alpha)}{(\tau\alpha)_n} = 1 + b_o \left[\frac{1}{\cos\theta} - 1 \right] \qquad (4)$$

For a single glass cover, b_o is approximately -0.10. Many flat-plate collectors, particularly evacuated tubes, have some limited focusing capability. The incident angle modifiers for these collectors are not modeled well by the simple expression for specular reflection, which is a linear function of $(1/\cos\theta) - 1$.

Equation (3) is not convenient for air collectors when it is desirable to present data based on collector outlet temperature rather than inlet temperature, which is common for liquid systems. The relationship between the heat removal factors for these two cases follows:

$$F_R (\tau\alpha) = \frac{(F_R \tau\alpha)'}{1 + (F_R U_L)'/\dot{m}c_p} \qquad (5a)$$

$$F_R U_L = \frac{(F_R U_L)'}{1 + (F_R U_L)'/\dot{m}c_p} \qquad (5b)$$

where $\dot{m}c_p$ is mass times specific heat of air, $F_R U_L$ and $F_R(\tau\alpha)$ applies to $t_i - t_a$ in Equation (3), and $(F_R U_L)'$ and $(F_R \tau\alpha)'$ applies to $t_{out} - t_a$ in Equation (3).

Testing Methods

ASHRAE *Standard* 93-1986 gives information on testing solar energy collectors using single-phase fluids and no significant internal storage. The data can be used to predict performance in any location and under any conditions where load, weather, and insolation are known.

The standard presents efficiency in a modified form of Equation (3). It specifies that the efficiency be reported in terms of gross collector area A_g, rather than aperature collector area A_c. The reported efficiency is lower than the efficiency based on net area, but the total energy collected does not change by this simplification. Therefore, gross collector area must be used when analyzing performance of collectors based on experiments that determine $F_R \tau\alpha$ and $F_R U_L$, according to ASHRAE *Standard* 93-1986.

Also, the standard suggests testing be done at 0.03 gpm per square foot of gross collector area for liquid systems and that the test fluid be water. While it is acceptable to use lower flow rates or a heat transfer fluid other than water, the designer must adjust the F_R for a different heat removal rate based on mass and specific heat mc_p. The following approximate approach may used to estimate small changes in mc_p.

$$\frac{(F_R U_L)_2}{(F_R U_L)_1} = \frac{1 - exp\ [- A_c (F_R U_L)'/(\dot{m} c_p)_2]}{1 - exp\ [- A_c (F_R U_L)'/(\dot{m} c_p)_1]} \qquad (6)$$

Air collectors are tested at a flow rate of 2 scfm/ft² and the same relationship applies for adjusting to other flow rates.

Annual compilations of collector test data that meet the criteria of ASHRAE *Standard* 93-1986 may be obtained from the following organization:

Solar Rating and Certification Corporation
777 N. Capitol St.
Washington, D.C. 20002

Another source of this information is the collector manufacturer. However, a manufacturer sometimes publishes efficiency data at a much higher flow rate than the recommended design value, so collector data should be obtained from an independent laboratory qualified to conduct the testing prescribed by ASHRAE *Standard* 93-1986.

Operational Results

Logee and Kendall (1984) analyzed data from actual sites with regard to collector and collector control performance. The systems were classified with respect to load (service hot water, building space heating, air conditioning); collector design (glazing, absorber surface, evacuated tube, concentrating); and heat transfer fluid (air, water, oil, or glycerol). Collector efficiency curves generated from the field data represented one month's operation during each of the four seasons. Most of the collectors (18) in the 33 systems studied performed below ASHRAE test results. Eleven systems had similar collector efficiencies as the test panel results, and four were slightly more efficient.

Collector Construction

Absorber plates. The key component of a flat-plate collector is the absorber plate. It contains the heat transfer fluid and serves as a heat exchanger by converting radiant energy into thermal energy. It must maintain structural integrity at temperatures ranging from below freezing to well above 300 °F. Chapter 30 in the 1991 ASHRAE *Handbook—HVAC Applications* illustrates typical liquid collectors and shows the wide variety of absorber plate designs in use.

Materials for absorber plates and tubes are usually highly conductive metals such as copper, aluminum, and steel, although low-temperature collectors for swimming pools are usually made from extruded elastomeric material such as EPDM and PVC. Flow passages and fins are usually copper, but aluminum fins are sometimes inductively welded to copper tubing. Occasionally, fins are mechanically attached, but the potential for corrosion exists with this design. A few manufacturers produce all-aluminum collectors, but they must be checked carefully to see if they have corrosion protection in the collector loop.

Figure 6 shows a plan view of typical absorber plates. The serpentine design is used less frequently because it is difficult to drain and imposes a high pressure drop. Most manufacturers use absorber plates similar to the example shown in Figures 6d and 6e.

In liquid collectors, manifold selection is important because the design can restrict the array piping configuration. The manifold must be drainable and free floating, with generous allowance for thermal expansion. Some manufacturers provide a choice of manifold connections to give designers flexibility in designing arrays. One such product can be obtained with side, back, or end connectors or a combination of these.

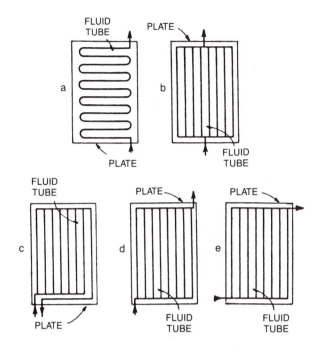

LIQUID COLLECTORS

Fig. 6 Plan View of Liquid Collector Absorber Plates

In air collectors, most manufacturers increase the heat transfer area via fins, matrices, or corrugated surfaces. Many of these designs increase air turbulence, which improves the collector efficiency (at the expense of increased fan power).

Figure 7 shows cross sections of typical air collectors. Fins on the back of the absorber (Figure 7a) increase the convection heat transfer surface. Air flowing across a corrugated absorber plate (Figure 7b) creates turbulence along the plate, which increases the convective heat transfer coefficient. A box frame (Figure 7c) creates airflow passages between the vanes. The vanes conduct from the absorbing surface plate to the back plate. Heat is transferred to the air by all of the surfaces of each boxed airflow channel. A matrix absorber plate (Figure 7d) is formed by stacking several sheets of metal mesh such as expanded metal plastering lath. Placing the mesh diagonally in the collector forces the air through the matrix so it does not contact the glazing after being heated.

A well-designed collector manifold with a series/parallel connection minimizes leaks and reduces the operating costs of the fans. Manufactured collectors are made in modular sizes, typically 3 ft by 7 ft, and these are connected to form an array. Because most commercial-scale systems involve upwards of 1000 ft² of collector, there can be numerous ducting connections, depending on design.

Figure 8 shows the simplest collector manifolding. Each collector has its own inlet and outlet, and each requires two branch connections per collector (and possibly a balancing damper). Some models are specifically designed for either series or parallel connections, and some collectors have built-in manifold connections to simplify their connection as an array. Figure 9 shows an example of an internally manifolded collector in a combination series-parallel arrangement. The number of ducts connecting the collectors to the trunk ducts are reduced to minimize leakage and reduce ducting and installation costs.

Absorber plates may be coated with spectrally selective or spectrally nonselective materials. Selective coatings are more efficient, but they also cost more than nonselective, or flat black, coatings. The Argonne National Laboratory has published a detailed

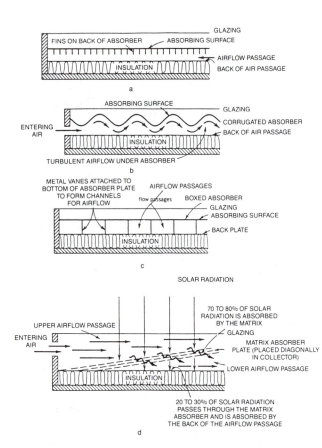

Fig. 7   Cross Section of Air Collectors

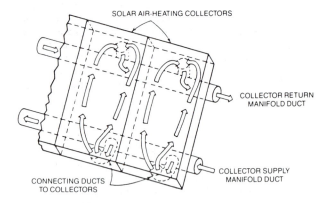

Fig. 8   Air Collectors with External Manifolds

discussion of the various coating materials used in solar applications (ANL 1979).

Housing. The collector housing is the container that provides structural integrity for the collector assembly. The housing must be structurally sound, weathertight, fire-resistant, and capable of being connected mechanically to a substructure to form an array. Collector housing materials include the following:

• Galvanized or painted steel
• Aluminum folded sheet stock or extruded wall materials
• Various plastics, either molded or extruded
• Composite wood products
• Standard elements of the building

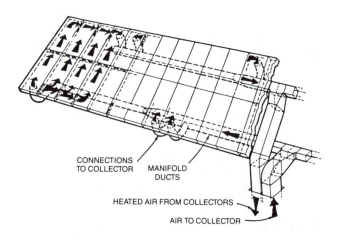

Fig. 9   Air Collectors with Internal Manifolds

Extruded anodized aluminum offers durability and ease of fabrication. Grooves are sometimes included in the channels to accommodate proprietary mounting fixtures. The high temperatures of a solar collector deteriorate wood housings; consequently, wood is often forbidden by fire codes.

Glazing. Solar collectors for domestic hot water are usually single-glazed to reduce absorber plate convective and radiative losses. Some collectors have double glazing to further reduce these losses; however, these should be restricted to applications at the higher value of $(t_i - t_a)/I$, e.g., for space heating or activating absorption refrigeration. Glazing materials are either plastic, plastic film, or glass. Glass can absorb the long-wave thermal radiation emitted by the absorber coating, but it is not affected by ultraviolet radiation. Because of their impact tolerance, only tempered, low-iron glass covers should be considered. These covers have a solar transmission rating of 86%. With acid etching, this transmission rating can be increased to 90%.

If the probability of vandalism is high, polycarbonate, which has high impact resistance, should be considered. Unfortunately, its transmittance is not as high as low-iron glass and it is susceptible to long-term ultraviolet (UV) degradation.

Insulation. Collector enclosures must be well insulated to minimize heat losses. The insulation must withstand temperatures up to 400°F and, most important, must not produce volatiles within this range. Many insulation materials designed for construction applications are not suitable for solar collectors because the binders outgas volatiles at normal collector operating temperatures.

Solar collector insulation is typically made of mineral fiber, ceramic fiber, glass and plastic foam, and fiberglass. Fiberglass is the least expensive insulation and is widely used in solar collectors. For high-temperature applications, rigid fiberglass board with a minimum of binder is recommended. Also, a layer of polyisocyanurate foam in collectors is often used because of its superior R factor. Because it can outgas at high temperatures, the foam must not be allowed to contact the collector plate.

Figure 10 illustrates the preferred method of combining fiberglass and foam insulations to combine high efficiency with durability. Note that the absorber plate should be free-floating to avoid thermal stresses. Regardless of attempts to make collectors watertight, moisture is always present in the interior. This moisture can physically degrade mineral wool and reduce the R value of fiberglass, so drainage and venting are crucial.

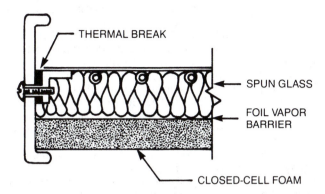

**Fig. 10   Cross Section of Suggested Insulation to Reduce
Heat Loss from Back Surface of Absorber**

Collector Test Results and Initial Screening Methods

Final selection of a collector should be made only after energy analyses of the complete system, including realistic weather conditions and loads, have been conducted for one year. Also, a preliminary screening of collectors with various performance parameters should be conducted to identify those that best match the load. The best way to accomplish this is to identify the expected range of $(t_i - t_a)/I$ for the load and climate on a plot of the efficiency as a function of heat loss parameter, as indicated in Figure 11.

Ambient temperature during the swimming season may vary from 18 °F below pool temperature to 18 °F above. The corresponding parameter values range from 0.15 on cool overcast days (low I) to as low as −0.15 on hot overcast days. For most swimming pool heating, the unglazed collector offers the highest performance and is the least expensive collector available.

The heat loss parameter for service water heating can range from 0.05 to 0.35, depending on the climate at the site and desired hot water delivery temperature. Space heating requires an even greater collector inlet temperature than hot water, and the primary load coincides with lower ambient temperature. In many areas of the United States, space heating operation coincides with low radiation values, which further increases the heat loss parameter. However, many space heating systems are accompanied by water heating at a lower value of $(t_i - t_a)/I_t$.

Air conditioning with solar activated absorption equipment is only economical when the cooling season is long and solar radiation levels are high. These devices require at least 180 °F water. Thus, on an 80 °F day with radiation at 300 Btu/h · ft², the heat loss parameter is 0.3. Higher operating temperatures are desirable to prevent excessive derating of the air conditioner. In this application, only the most efficient (low $F_R U_L$) collector is suitable.

Collector efficiency curves may be used as an initial screening device. However, efficiency curves only illustrate the instantaneous performance of a collector. They do not include incidence angle effects (which vary throughout the year), heat exchanger effects, probabilities of occurance of t_i, t_a, and I_t, system heat loss, or control strategies. The final selection requires determining the long-term energy output of a collector combined with cost-effectiveness studies. Estimating the annual performance of a particular collector and system requires the aid of appropriate analysis tools such as FCHART (Beckman *et al.* 1977) or TRNSYS (SEL 1983).

Generic Test Results

While details of construction may vary, it is possible to group collectors into generic classifications. Using the performance characteristics of each classification, the designer can select the category best suited to a particular application.

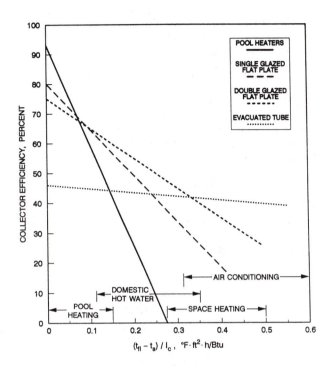

Fig. 11   Collector Efficiencies of Various Liquid Collectors

Huggins and Block (1983) identified eight such classifications for liquid collectors from a sample of 270 collectors tested by various laboratories. Following is a list of the classifications:

Number of cover plates. The majority of collectors used in the South and far West of the United States have one cover plate. Double glazed, *i.e.*, two-cover plates, are more common in colder climates. The trend in cold climates has been toward single glazing combined with a selective surface. However, with the advent of reflectivity coatings on plastic films, triple and quadruple glazings may become common.

Cover plate material. The following is a list of common cover plate materials:

Glass is the most widely used glazing material. It has a high transmittance and long-term durability.

Fiber reinforced plastic (FRP) is the second most widely used material.

Thin film plastics serve as a glazing on some collectors.

Absorber plate coating. The most common coatings are listed below:

Selective surface coatings such as black chrome, black nickel, and copper oxide have high absorptance and low emittance properties. The infrared emissivity of these surfaces is below 0.2.

Moderately selective surface coatings are special paints that have moderately selective surface properties. The emissivity of the surfaces ranges from 0.2 to 0.7.

Flat black paints are nonselective and have high heat resistance. The emissivity of these surfaces ranges from 0.7 to 0.98.

Absorber materials. Copper, aluminum, and stainless steel may be used in combinations of tubes and fins or integral tubes in plates.

Absorber configuration. The absorber may be configured with parallel pipes, series or serpentine pipes, a parallel and series combination, or plate flow.

Enclosure. The frame holding the collector components may be either metallic or nonmetallic.

Insulation materials. Fiberglass and foam insulation, or a combination of both, may be used to keep heat from escaping from the back and sides of the collector.

Table 1 lists the eight generic solar collectors in which five or more collectors were tested. The first three columns specify the generic type; the fourth column shows the number of collectors in each category; and the final columns lists the mean, the standard deviation, and the maximum and minimum values of the intercept and slope. Logee and Kendall (1984) showed that collectors in actual applications can perform differently than they do on test stands. This difference emphasizes the need for good system designs and installation practices.

MODULE DESIGN

Piping Configuration

Most commercial and industrial systems require a large number of collectors. Connecting the collectors with one set of manifolds makes it difficult to ensure drainability, balanced flow,

and low-pressure drop. An array usually includes many individual groups of collectors, called modules, to provide the necessary flow characteristics. Modules can be grouped into (1) parallel flow or (2) combined parallel/series flow. Parallel flow is the most frequently used because it is inherently balanced, has low pressure drop, and is drainable. Figure 12 illustrates various collector header designs and the reverse return method of forming a parallel flow module (the flow is parallel but the collectors are connected in series). When arrays must be greater than one panel high, a combination of series/parallel flow may be used. Figure 13 illustrates one method of connecting collectors in a two-high module.

Generally, flat-plate collectors are made to connect to the main piping in one of two methods shown in Figure 12. The *external manifold* collector has a small-diameter connection meant to carry only the flow for one collector. It must be connected individually to the manifold piping, which is not part of the collector panel, as depicted in Figure 12a and Figure 13.

The *internal manifold* collector incorporates the manifold piping integral with each collector (Figures 12b and 12c). Headers at either end of the collector distribute flow to the risers. Several collectors with large headers can be placed side by side to form a

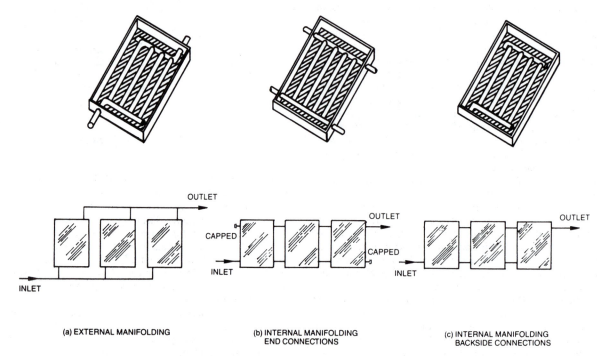

Fig. 12   Collector Manifolding Arrangements for Parallel Flow Module

Table 1   Collector Intercept and Slope by Generic Type
(Huggins and Block 1983)

Glazing and Cover Material	Absorber Material and Type	Absorber Coating	Number of Collectors	Intercept, $F_R\tau\alpha$		Slope $F_R U_L$, Btu/h·ft²·°F	
				Mean	Standard Deviation	Mean	Standard Deviation
Single glass	Copper tubes and fins	Flat black paint	47	67.2	5.0	−115	14
Single glass	Copper tubes and fins	Moderately selective	9	73.0	3.6	−112	11
Single glass	Copper tubes and fins	Selective surface	58	71.7	3.3	−83	11
Single glass	Copper tubes and aluminum fins	Flat black paint	22	69.1	6.0	−116	12
Single glass	Copper sheet integral tubes	Selective surface	6	70.5	5.1	−89	17
Single FRP	Copper tubes and fins	Flat black paint	26	61.9	5.5	−117	15
Single FRP	Copper tubes and aluminum fins	Flat black paint	11	57.1	6.2	−114	10
Double glass	copper tubes and fins	Flat black paint	9	59.7	6.7	−84	9

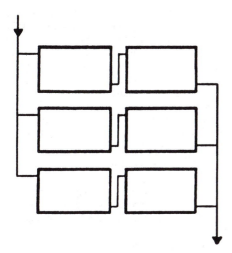

**Fig. 13   Example of Collector Manifolding Arrangements
for Combined Series/Parallel Flow Modules**

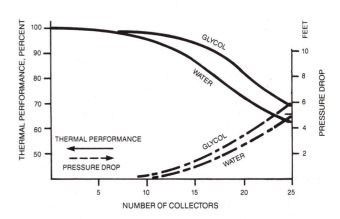

**Fig. 14   Pressure Drop/Thermal Performance of Collectors with
Internal Manifolds**

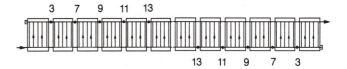

Numbers indicate restrictor hole diameter in sixteenths of an inch.

Fig. 15   Flow Pattern in Long Collector Row with Restrictions

continuous supply and return manifold. With 1-in. headers, four to six 40-ft² collectors can be placed side by side. Collectors with 1-in. headers can be mounted in a module without producing unbalanced flow. Most collectors have four plumbing connections, some of which may be capped if the collector is located on the end of the array. Internally manifolded collectors have the following advantages:

- Piping costs are lower because fewer fittings and less insulation is required.
- Heat loss is less because less piping is exposed.
- Installation is more attractive.

Some of the disadvantages of the internally manifolded collectors are as follows:

- The entire module must be pitched for drainback systems, thus complicating mounting.
- Flow may be imbalanced if too many collectors are connected in parallel.
- Removing the collector for servicing may be difficult.
- Stringent thermal expansion requirements must be met if too many collectors are combined in a module.

Velocity Limitations

Fluid velocity limits the number of internally manifolded collectors that can be contained in a module. For 1-in. headers, up to eight 20-ft² collectors can usually be connected for satisfactory performance. If too many are connected in parallel, the middle collectors will not receive enough flow, and performance will decrease. Also, connecting too many collectors increases pressure drop. Figure 14 illustrates the effect of collector number on performance and pressure drop for one particular design. Newton (1983) describes a general method to determine the number of internally manifolded collectors that can be connected.

Flow restrictors can be used to accommodate a large number of collectors in a row. The flow distribution in the twelve collectors of Figure 15 would not be satisfactory without the flow restrictors shown at the interconnections. The flow restrictors are barriers with a hole drilled the size of the diameter indicated. Some manufacturers calculate the required hole diameters and provide the predrilled restrictors.

Chapter 33 of the 1989 ASHRAE *Handbook—Fundamentals* gives information on sizing piping for self-balancing flow in exter-

nal manifolded collectors. Knowles (1980) provides the following expression for the minimum acceptable header diameter:

$$D = 0.24(Q/p)^{0.45} N^{0.64} \qquad (7)$$

where

D = header diameter, in.
N = number of collectors in module
Q = recommended flow rate for collector, gpm
p = pressure drop across collector at recommended flow rate, psi

Because piping is available in a limited number of diameters, selection of the next larger size ensures balanced flow. Usually, the size of supply and return are graduated to maintain the same pressure drop while minimizing piping cost. Complicated configurations may require a hydraulic static regain calculation.

Thermal Expansion

Thermal expansion will affect the module shown in Figure 16. Thermal expansion (or contraction) of a module of collectors in parallel may be estimated by the following equation.

$$\Delta = 0.000335n(t_c - t_i) \qquad (8)$$

where

Δ = expansion or contraction of collector array, in.
n = number of collectors in an array
t_c = collector temperature, °F
t_i = installation temperature of the collector array, °F

Because absorbers are rigidly connected, the absorber must have sufficient clearance from the side frame to allow the expansion indicated in Equation (8).

ARRAY DESIGN

Piping Configuration

To maintain balanced flow, an array or field of collectors should be built from identical modules configured as described in previous sections. Whenever possible, modules must be connected in reverse return fashion (Figure 16). The reverse return ensures the array is self-balanced. With proper care, an array can drain, which is an essential requirement for drainback freeze protection.

Piping to and from the collectors must be sloped properly in a draindown system. Typically, piping and collectors must slope to drain at 1/4 in. per lineal foot. Elevations throughout the array should be noted on the drawings, especially the highest and lowest point of the piping.

The external manifold collector has different mounting and plumbing considerations than the internal manifold collector (Figure 17). A module of external manifolded collectors can be mounted horizontally, as shown in Figure 17. The lower header *must* be pitched as shown. The pitch of the upper header can be horizontal or pitched toward the collectors, so it can drain back through the collectors.

Arrays with internal manifolds pose a greater challenge designing and installing the collector mounting system. For these collectors to drain, the entire bank must be tilted, as shown in Figure 17.

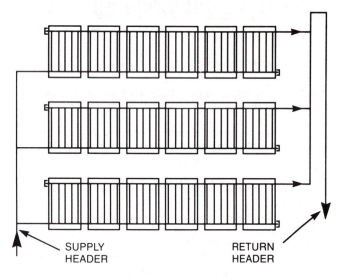

SUPPLY HEADER   **RETURN HEADER**

Fig. 16 Reverse Return Array Piping

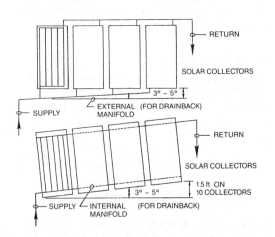

Fig. 17 Mounting for Drainback Collector Modules

Reverse return always implies an extra pipe run. Sometimes, it is more convenient to use a direct return configuration (Figure 18). In this case, balancing valves are needed to ensure uniform flow through the modules. The balancing valves *must* be connected at the module outlet to provide the flow resistance necessary to ensure filling of all modules on pump start-up.

It is often impossible to configure parallel arrays because of the presence of rooftop equipment, roof penetrations, or other building-imposed constraints. Though the following list is not complete, the following requirements should be considered when developing the array configuration:

- Strive for a self-balancing configuration.
- For drainback systems, design the modules, subarrays, and arrays to be individually and collectively drainable.
- Always locate collectors or modules with high flow resistance at the outlet to improve flow balance and ensure filling of draindown system.
- Minimize flow and heat transfer losses.

In general, it is easier to configure complex array designs for nonfreezing fluid systems. Newton (1983) provides some typical examples. However, with careful attention to the criteria mentioned above, it is also possible to design successful large drainback arrays.

Air distribution is the most critical feature of an air system because pressure drop has a critical effect on fan power. Each collector and the other system components must have proper air distribution for effective operation. Balancing dampers, automatic dampers, back-draft dampers, and fire dampers are usually needed. Air leaks (both into and out of the ducting and from component to component) must be kept to a minimum.

For example, some air collector systems contain a water coil for preheating water. Despite the inclusion of automatic dampers and back-draft dampers, leakage within the system can freeze the coils. One possible solution is to position the coil near the warm end of the storage bin. Another is to circulate an antifreeze solution in the coil. Whatever the solution, the designer must remember that even the lowest leakage dampers will leak if installed or adjusted improperly.

As with liquid systems, the main supply and return ducts should be connected in a reverse return fashion, with balancing dampers on each supply branch to the collector modules. If reverse return is not feasible, the main ducts should include balancing dampers at strategic locations. Here too, fewer branch ducts reduce balancing needs and costs.

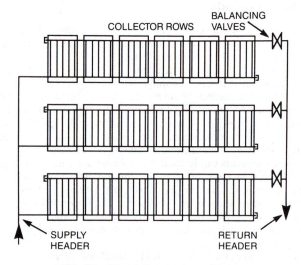

SUPPLY HEADER   **RETURN HEADER**

Fig. 18 Direct Return Array Piping

Unlike liquid systems, air collectors can be built on site. While material and cost savings can be substantial with site-built collectors, extreme care must be taken to ensure long life, low leakage, and proper air distribution. Quality control in the field can be a problem, so well-trained designers and installers are critical to the success of these systems.

The impact of the array and air distribution system designs must be considered in the overall performance of the system. Beckman *et al.* (1977) give standard procedures for estimating the impact of series connection and duct thermal losses. The impact on fan operation and fan power is more difficult to determine. For unique system designs and more detailed performance estimates, including fan power, an hourly simulation like TRNSYS (SEL 1983) can be used.

Shading

When large collector arrays are mounted on flat roofs or level ground, multiple rows of collectors are usually installed in a sawtooth fashion. These multiple rows should be spaced so they do not shade each other at low sun angles. However, it is usually not desirable to avoid mutual shading altogether. It is sometimes possible to add additional rows to a roof or other constrained area; this increases the solar fraction but sacrifices efficiency. Kutscher (1982) presents a method of estimating the energy delivered annually when there is some row-to-row shading within the array. Figure 19 provides a factor F_{shad}, which corrects the annual performance (on an unshaded basis) of the field of shaded collectors. Since the first row is unshaded, the following equation is used to compute the average shading factor for the entire field.

$$F_{shad, field} = \frac{1 + (n - 1) F_{shad}}{n} \qquad (9)$$

where: n = number of rows in the field.

Thermal Energy Storage

Design and selection of the thermal storage equipment is one of the most neglected elements of solar energy systems. In fact, the energy storage system has an enormous impact on overall system cost, performance, and reliability. Furthermore, the storage

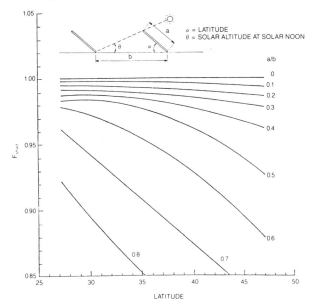

Fig. 19   Annual Row-to-Row Shading Loss Modifiers for Flat Plate Collectors Tilted at Latitude
(Kutscher 1982)

system design is highly interactive with other system elements such as the collector loop and the thermal distribution system. As such, it should be considered within the context of the total system.

Energy can be stored in liquids, solids, or phase-change materials (PCM). Water is the most frequently used liquid storage medium for liquid systems, although the collector loop may contain water, oils, or aqueous/glycol as a collection fluid. For service-water heating applications and most building space heating, water is normally contained in some type of tank. Air systems typically store water in rocks or pebbles, but sometimes the structural mass of the building is used. Chapter 30 of the 1991 ASHRAE *Handbook—HVAC Applications* and ASHRAE (1991) cover this topic in more detail.

Air Storage

The most common storage media for air collectors are rocks or a regenerator matrix made from concrete masonry units (CMUs). Other possible media include phase-change materials (PCMs), water, and the inherent building mass. Gravel is widely used as a storage medium because it is plentiful and relatively inexpensive.

In places where large interior temperature swings are tolerable, the inherent mass of the building may be sufficient for thermal storage. Storage may also be eliminated where the array output seldom exceeds the concurrent demand. These types of applications are usually the most cost-effective applications of air collectors, and heated air from the collectors can be distributed directly to the space.

The three main requirements for a gravel storage system (or any storage system for air collectors) are good insulation, low air leakage, and low pressure drop. Many different designs can fulfill these needs. The container is usually constructed from masonry or frame, or a combination of the two. Airflow can be vertical or horizontal, whichever is most convenient.

Because of the restricted airflow paths in a pebble bed, there is little natural air movement, and temperature can stratify without the effect of gravity. A vertical flow bed that has solar-heated air enter at the bottom and exit from the top can work as effectively as a horizontal flow bed. However, it is important to heat the bed with airflow in one direction and to retrieve the heat with flow in the opposite direction. In this manner, pebble beds perform as effective counterflow heat exchangers. Conversely, properly designed and applied wash-through (one-way flow) beds can be effective, as can underflow beds that charge through airflow and discharge by flow radiation. These are less expensive because they do not require elaborate ductwork or complicated controls.

Rocks for pebble beds range from 0.75 in. nominal to over 4 in. in size, depending on airflow rates, bed geometry, and desired pressure drops. The volume of rock needed depends on the fraction of the collector output that must be stored. For residential systems, storage volume is typically in the area of 0.5 to 1 ft³/ft² of collector area. For commercial systems, these values can be used as guidelines, but a more detailed analysis should be performed. Pebble bed size can be quite large for large arrays, and location and their large mass can create problems.

Though other storage options exist for air systems, applied knowledge of these techniques is limited. Hollow-core concrete decking appears attractive because of the potential to reduce storage and distribution costs. PCMs are also functionally attractive because of their high volumetric heat storage capabilities, which typically require a tenth of the volume of a pebble bed.

Water can also be used as a storage medium for air collectors through the use of a conventional heating coil to transfer heat from the air to the water in the storage tank. Advantages of water storage include compatibility with hydronic heating systems and relatively compact storage (roughly one-third the volume requirement of pebble beds).

Liquid System Thermal Storage

For units large enough for commercial liquid systems, the following factors should be considered.

- Pressurized versus unpressurized storage
- External heat exchanger versus internal heat exchanger
- Single versus multiple tanks
- Steel tank versus nonmetallic tanks
- Type of service, *e.g.,* service hot water (SHW), building space heating (BSH), or a combination of the two
- Location, space, accessibility constraints imposed by architectural limitations
- Interconnect constraints imposed by the existing mechanical systems
- Limitations imposed by equipment availability

In the following sections, examples of the more common configurations will be presented.

Pressurized storage. Defined here as storage that is open to the city water supply, pressurized storage is preferred for small service-water heating systems because it is convenient and provides an economical way of meeting ASME Pressure Vessel Code requirements with off-the-shelf equipment. Typical storage size is about 1.0 to 2 gal/ft² of collector area. The largest size off-the-shelf water heater is 120 gal; however, no more than three of these should be connected in parallel. Hence, the largest size system that can be considered with off-the-shelf water heater tanks is about 360 gal. For larger solar hot water and combined systems, the following concerns are important when selecting storage:

- The higher cost per unit volume of ASME rated tanks in sizes greater than 120 gal
- Handling difficulties due to large mass
- Accessibility to locations suitable for storage
- Interfacing with existing SHW and BSH systems
- Corrosion protection of steel tanks

The choice of pressurized storage for intermediate size systems is based on the availability of suitable, low-cost tanks near the site. Identification of a suitable supplier of low-cost tanks can extend the advantages of pressurized storage to larger SHW installations.

Storage pressurized at city water supply pressure is not practical for building space heating (BSH), except for small applications such as residences, apartments, and small commercial buildings.

With pressurized storage, the heat exchanger is always located on the collector side of the tank. Either the internal or the external heat exchanger configuration can be used. Figure 20 illustrates the three principal types of internal heat exchanger concepts, an immersed coil, a wraparound jacket, and a tube bundle. Small tanks (less than 120 gal) are available with either of the first two heat exchangers already installed. For larger tanks, a large assortment of tube bundle heat exchangers are available that can be incorporated into the tank design by the manufacturer.

Sometimes, more than one tank is needed to meet design requirements. Additional tank(s) result in the following benefits:

- Added storage volume
- Increased heat exchanger surface
- Reduced pressure drop in the collection loop

Figure 21 illustrates the multiple tank configuration for pressurized storage. The exchangers are connected in reverse return fashion to minimize flow imbalance. A third tank may be added. Additional tanks have the following disadvantages compared to a single tank of the same volume:

- Installation costs are higher
- Greater room is required
- Heat losses are higher so performance is reduced

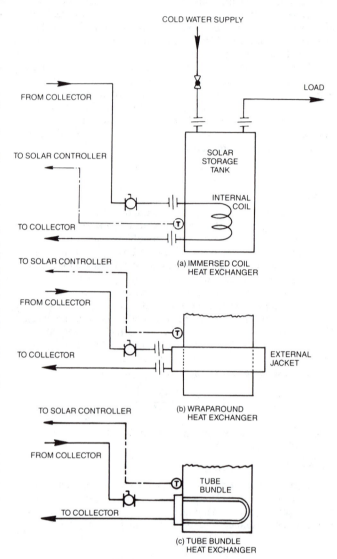

Fig. 20 Pressurized Storage with Internal Heat Exchanger

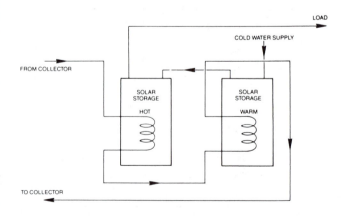

Fig. 21 Multiple Storage Tank Arrangement with Internal Heat Exchangers

An external heat exchanger provides greater flexibility, because the tank and the heat exchanger can be selected independently of each other (Figure 22). Flexibility is not achieved without cost, however, because an additional pump and its parasitic energy requirements are required.

When selecting an external heat exchanger for a system protected by a nonfreezing liquid, the following factors related to start-up after at least one night in extremely cold conditions should be considered:

- Freeze-up of the water side of the heat exchanger
- Performance loss due to extraction of heat from storage

For small systems, an internal heat exchanger/tank arrangement prevents the water side of the heat exchanger from freezing. However, the energy required to maintain the water above freezing must be extracted from storage, thereby decreasing overall performance. With the external heat exchanger/tank combination, a bypass can be arranged to bypass cold fluid around the heat exchanger until it has been heated to an acceptable level, such as 80 °F. When the heat transfer fluid has warmed to this level, it can enter the heat exchanger without causing freezing or extraction of heat from storage. If necessary, this arrangement can also be used with internal heat exchangers to improve performance.

Unpressurized storage. For systems sized greater than about 1000 ft² (1500 gal storage volume minimum), unpressurized storage is usually more cost-effective. As used in this chapter, the term *unpressurized* means tanks at or below the pressure expected in an unvented drainback loop.

Unpressurized storage for water and space heating implies a heat exchanger on the load side of the tank to isolate the high-pressure (potable water) loop from the low-pressure collector loop. Figure 23 illustrates unpressurized storage with an external heat exchanger. In this configuration, heat is extracted from the top of the solar storage tank, and the cooled water is returned to the bottom of the tank. On the load side of the heat exchanger, the water to be heated flows from the bottom of the backup storage tank, and heated water returns to the top. The heat exchanger may have a double wall to protect a potable water supply. A differential temperature controller controls the two pumps on either side of the heat exchanger. When small pumps are used, both may be controlled by the same controller without overloading it.

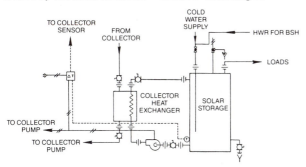

Fig. 22   External Heat Exchanger for Pressurized Storage

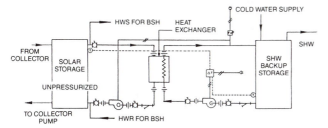

Fig. 23   Unpressurized Storage System with External Heat Exchanger

The external heat exchanger shown in Figure 23 provides good system flexibility and freedom in component selection. Occasionally, system cost and parasitic power consumption may be reduced by an internal heat exchanger. At times, heat exchangers fabricated in the field use coiled soft copper tube. For larger systems, where custom fabrication is more feasible, tanks can be supplied with a specified heat exchanger installed at the top.

Storage Tank Construction

For most liquid solar energy systems, steel is the preferred material for thermal energy storage construction. Steel tanks are (1) relatively easy to fabricate to ASME Pressure Vessel Code requirements, (2) readily available, and (3) easily attached by pipes and fittings.

Steel tanks used for pressures of 30 psi and above must be ASME rated. Because water main pressure is usually above this level, open systems must use ASME code tanks. These pressure-rated storage tanks are more expensive than nonpressurized types. Significant cost reduction can usually be realized if a nonpressurized tank can be used.

Steel tanks are subject to corrosion, however. Because corrosion rates accelerate at increased temperature, the designer must be particularly aware of corrosion protection methods for solar energy applications. A steel tank must be protected against (1) electrochemical corrosion, (2) oxidation (rusting), and (3) galvanic corrosion (ANL 1979).

Electrochemical corrosion. The pH of the liquid and the electric potential of the metal primarily govern electromechanical corrosion. A sacrificial anode fabricated from a metal more reactive than steel can protect against this type. Magnesium is recommended for solar applications. Since protection ends when the anode has dissolved, the anode must be inspected annually.

Oxidation. Oxygen can enter the tank through the air dissolved in the water entering the tank or through an air vent. In pressurized storage, oxygen is continually replenished by the incoming water. Besides causing rust, the oxygen catalyzes other types of corrosion. Unpressurized storage systems are not as susceptible to corrosion caused by oxygen because they can be designed as unvented systems and, corrosion is limited to the small amount of oxygen contained in the initial fill water.

Coatings applied to the inside of the tank protect it from oxidation. The most common coatings used include:

- *Phenolic epoxy* should be applied in four coats.
- *Baked-on epoxy* is preferred over painted-on epoxy.
- *Glass lining* offers more protection than the epoxies and can be used under severe water conditions.
- *Hydraulic stone* provides the best protection against corrosion and increases the heat-retention capabilities of the tank. Its mass may cause handling problems in some installations.

Regardless of the lining used, it should be flexible enough to withstand extreme thermal cycling, or it should have the same coefficient of expansion as the steel tank. All linings used for potable water tanks should be FDA approved for the maximum temperature expected in the tank.

Galvanic corrosion. Dissimilar materials in an electrolyte (water) are in electrical contact with each other and corrode by galvanic action. Copper fittings screwed into a steel tank corrode the steel, for example. Galvanic corrosion can be minimized by using dielectric bushings to connect pipes to tanks.

Fiberglass reinforced plastic (FRP) tanks offer the advantages of low mass, high corrosion resistance, and low cost. Premium quality resins permit operating temperatures as high as 210 °F, well above the temperatures imposed by flat-plate solar collectors. Before accepting delivery of an FRP tank, it should be inspected for damage that may have occurred during shipment. Also, the gel coat on the inside must be intact, and no glass fibers should be exposed.

Concrete vessels, lined with a waterproofing membrane, may also be used to contain a liquid thermal storage medium in vented systems only. Concrete storage tanks are inexpensive, and they can be shaped to fit almost any retrofit application. Also, they possess excellent resistance to the loading that occurs when they are placed in a below-grade location. Concrete tanks must have smooth corners and edges.

Concrete storage vessels have some disadvantages. Seepage often occurs unless a proper waterproofing surface is applied. Waterproofing paints are generally unsatisfactory because the concrete often cracks after settling. Other problems may occur due to poor workmanship, poor location, or poor design. Usually, the tanks should stand alone and not be integrated with a building or other structure. Also, careful attention should be given to the expansion joints and seams, since they are particularly difficult to seal. Sealing at pipe taps can also be a problem. Penetrations should be above the liquid level, if possible.

Finally, concrete tanks are heavy, and they may be more difficult to support in a proper location. Their weight may also make insulation of the tank bottom more expensive and difficult.

Storage Tank Insulation

Heat loss from storage tanks and appurtenances is one of the major causes of poor system performance. The average R value of storage tanks in solar applications is about half the insulation design value because of poorly insulated supports. Different standards recommend various design criteria for tank insulation. The Sheet Metal and Air Conditioning Contractors National Association (SMACNA) standard of a 2% loss in 12 h is generally accepted because it maintains a more stringent requirement than other standards. The following equation can be used to calculate the insulation R factor for this requirement:

$$\frac{1}{R} = \frac{fQ}{At}\frac{1}{(t_{avg} - t_a)} \qquad (10)$$

where

t_{avg} = average temperature of tank
t_a = air temperature surrounding tank

The insulation factor fQ/At is found from Table 2 for various tank shapes (ANL 1979).

Most solar water heating systems use large steel pressure vessels, which are usually shipped uninsulated. Materials suitable for field insulation include fiberglass, rigid foam, and flexible foam blankets.

Fiberglass is easy to transport and make fire retardant, but it requires significant labor to apply and seal. Another widely used insulation system consists of rigid sheets of polyisocyanurate foam cut and taped around the tank. Material that is 3 to 4 in. thick can provide R-20 to R-30 insulation value.

Rigid foam insulation is sprayed directly onto the tank from a foaming truck or in a shop. It bonds well to the tank surface (no air space between the tank and the insulation), and insulates better than an equivalent thickness of fiberglass. When most foams are exposed to flames, they ignite and/or produce toxic gases. When located in, or adjacent to, a living space, they often must be protected by a fire barrier and/or a sprinkler system. Some tank manufacturers and suppliers offer custom tank jackets or flexible foam with zipper-like connections that provide quick installation and neat appearance. Some of the same fire considerations apply to these foam blankets.

The supports of a tank are a major source of heat loss. To provide suitable load-bearing capability, the supports must be in direct contact with the tank wall or attached to it. Thermal breaks must be provided between the supports and the tank (Figure 24). If insulating the tank from the support is impractical, the external surface of the supports must be insulated, and the supports

Table 2   Insulation Factor fQ/At for Cylindrical Water Tanks

Horizontal Tank Insulation Factor, Btu/h · ft²

Size, gal	D	2D	4D	6D
250	3.63	3.46	3.05	2.77
500	4.57	4.36	4.84	3.49
750	5.24	4.99	4.40	3.99
1000	5.76	5.49	4.84	4.39
1500	6.60	6.28	5.54	5.03
2000	7.26	6.92	6.10	5.53
3000	8.31	7.92	6.98	6.33
4000	9.15	8.71	7.68	6.97
5000	9.86	9.39	8.28	7.51

Vertical Tank Insulation Factor, Btu/h · ft²

Size, gal	D to 3D	D/2	D/3	D/4
80	2.10	1.88	2.15	1.97
120	2.39	2.15	2.46	2.26
250	3.07	2.74	2.46	2.88
500	3.87	3.46	3.96	3.63
750	4.43	3.96	4.53	4.16
1000	4.87	4.36	4.99	4.57
1500	5.58	4.99	5.71	5.24
2000	6.13	5.49	6.28	5.76
3000	7.03	6.28	7.19	6.60
4000	7.73	6.92	7.92	7.26
5000	8.33	7.45	8.53	7.82

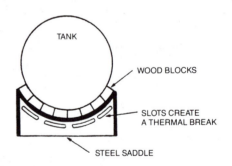

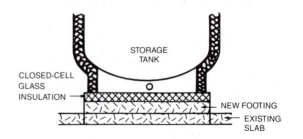

Fig. 24   Typical Tank Support Detail

must be placed on insulative material capable of supporting the compressive load. Wood, foamed glass, and closed-cell foam can be used, depending on the compressive load.

Stratification and Short Circuiting

Because hot water rises and cold water sinks in a vessel, the pipe to the collector inlet should always be connected to the bottom of the tank. The collector operates at its best efficiency because it is always at the lowest possible temperature. The return from the collector should always run near the top of the tank, the location of the hottest fluid. Similarly, the load should be extracted from the top of the tank, and cold-water makeup should be introduced at the bottom. Because of the increased static pressure, vertical storage tanks enhance stratification better than horizontal tanks.

If a system has a rapid tank turnover rate or if the buffer tank of the system is closely matched to the load, thermal stratification does not offer any advantages. However, in most solar energy systems, thermal stratification increases performance because solar energy collecting systems operate over relatively small temperature differences. Thus, any enhancement of temperature differentials improves heat transfer efficiency.

Thermal stratification can be enhanced by the following actions:

- Using a tall vertical tank
- Situating the inlet and outlet piping of a horizontal tank to minimize vertical fluid mixing
- Sizing the inlets and outlets such that exhaust flow velocity is less than 2 fps
- Using flow diffusers (Figure 25)
- Plumbing multiple tanks in series

Multiple tanks generally yield the greatest temperature difference between cold water inlet and hot water outlet. With the piping/tank configuration shown in Figure 21, the difference can approach 30°F. The additional costs of having two tanks must be considered. Extra footings, insulation, and sensors are required in addition to the cost of the second tank. However, multiple tanks may save labor at the site due to their small diameter and shorter length. The smaller tanks require less demolition to install in retrofit (*e.g.,* access through doorways, hallways, and windows).

Sizing

The following is a list of specific site factors related to system performance or cost.

Solar fraction. Most solar energy systems are designed to produce about 50% of the energy required to satisfy the load. If a system is sized to produce less than this amount, there is greater coincidence between load and available solar energy. Therefore, smaller, lower cost, storage can be used with low solar fraction systems without impairing system performance.

Load matching. The load profile of a commercial application may be such that less storage can be used without incurring performance penalties. For example, a company cafeteria or a restaurant serving large luncheon crowds has its greatest hot water usage during, and just after, midday. Because the load coincides with the availability of solar radiation, less storage is required for carryover. For loads that peak during the morning, collectors can be rotated toward the southeast, sometimes improving output for a given size array.

Storage cost. If a solar storage tank is insulated adequately, performance generally improves with increases in storage volume, but the savings must justify the investment. The cost of storage can be kept at a minimum by using multiple, low-cost water heaters and unpressurized storage tanks made from such materials as fiberglass or concrete.

The performance and cost relationships between solar availability collector design, storage design, and load are highly interactive. Furthermore, the designer should maximize the benefits from the investment for at least one year, rather than for one or two months. For this reason, solar energy system performance models, combined with economic analyses, should be used to optimize storage size for a specific site. Models that can be used to determine system performance include FCHART, SOLCOST, and TRNSYS. FCHART (Beckman *et al.* 1977) is readily available in either the work-sheet form for hand calculations or as a reasonably priced computer program; however, it contains a built-in daily hot water load profile specifically for residential applications. This limitation is not very serious; studies have shown the daily hot water profile has little effect on the system performance.

For daily profiles with extreme deviations from the residential profile, such as the company cafeteria example used previously, FCHART may not provide satisfactory results. SOLCOST can accommodate user-selected daily profiles, but it is not widely available. Neither program can accommodate special days, *e.g.,* holidays and weekends, when no service hot water loads are present. TRNSYS (SEL 1983), which has been modified for operation on personal computers, is readily available and well-supported. It is tedious to set up and run for the first time, but after it is set up, various design parameters, such as collector area and storage size, can be easily evaluated for a given system configuration.

The previous sections have centered on service hot water (SHW) and conventional hydronic space heating loads. Many hydronic solar energy systems are used with a series-coupled heat pump or an absorption air-conditioning system. In the case of a heat pump, lower storage temperatures are desirable, thereby creating higher collector efficiencies. In this case, the collector size may be greater than recommended for conventional SHW or hydronic building space heating (BSH) systems. In contrast, absorption air conditioners require much higher temperatures than BSH or hydronic BSH to activate the generator (>175°F). Therefore, the storage collector ratio can be much less than recommended for BSH. For absorption air conditioning, thermal energy storage may act as a buffer between the collector and generator, which prevents cycling due to frequent changes in insolation level.

HEAT EXCHANGERS

Requirements

The heat exchanger transfers heat from one fluid to another. In closed solar energy systems, it also isolates circuits operating at different pressure levels and separates fluids that must not be mixed. Heat exchangers are used in solar applications to separate the heat transfer fluid in the collector loop from the domestic water supply in the storage tank (pressurized storage) or the domestic water supply from the storage (unpressurized storage).

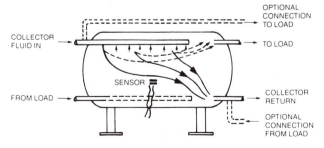

Fig. 25   Tank Plumbing Arrangements to Minimize Short Circuiting and Mixing
(Krieder 1982)

The selection of a heat exchanger involves the following considerations:

Performance. Heat exchangers always degrade the performance of the solar collector; therefore, the selection of an adequate size is important. When in doubt, an oversized heat exchanger should be selected.

Guaranteed separation of fluids. Many code authorities require a vented, double-wall heat exchanger to ensure fluid isolation. System protection requirements may also dictate the need for guaranteed fluid separation.

Thermal expansion. The temperature in a heat exchanger may vary from below freezing to the boiling temperature of water. The design must withstand these thermal cycles without failing.

Materials. Galvanic corrosion is always a concern in liquid solar energy systems. Consequently, the piping, collectors, and other hydronic component materials must be compatible.

Space constraints. Often, limited space is available for mounting and servicing the heat exchanger. Physical size and configuration must be considered when selection is made.

Serviceability. The water side of a heat exchanger is exposed to the scaling effects of dissolved minerals, so the design must provide access for cleaning and scale removal.

Pressure loss. Energy consumed in pumping fluids reduces system performance. The pressure drop through the heat exchanger should be limited to 1 to 2 psi to minimize energy consumption.

Pressure capability. Because the heat exchanger is exposed to cold water supply pressure, it should be rated for pressures above 75 psig.

Internal Heat Exchanger

The two basic heat exchangers for solar applications are placed either inside or outside the storage or drainback tank. The internal heat exchanger can be a coil inside the tank or a jacket wrapped around the pressure vessel (Figure 20). Several manufacturers supply tanks with either type of internal heat exchanger. However, the maximum size of pressurized tanks with internal heat exchangers is usually limited to about 120 gal. Heat exchangers may be installed inside nonpressurized tanks that open from the top with relative ease. Figures 26 and 27 illustrate methods of achieving double-wall protection with either type of internal heat exchanger.

For installations requiring larger tanks or heat exchangers, a tube bundle is required. However, it is not always possible to find a heat exchanger of the desired area that will fit within the constraint imposed by the tank diameter. Consequently, a horizontal tank with the tube bundle inserted from the tank end can be used. A second option is to place a shroud around the tube bundle and to pump fluid around it (Figure 28). Such an approach combines the performance of an external heat exchanger with the compactness of an internal heat exchanger. Unfortunately, tank mixing and loss of stratification result from this approach.

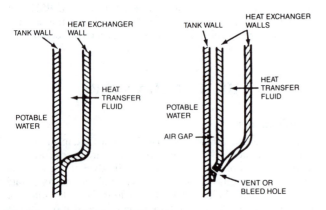

Fig. 26   Cross Section of Wraparound-Shell Heat Exchangers

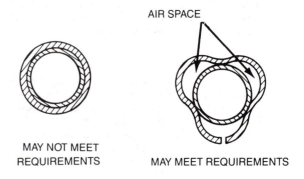

Fig. 27   Double-Walled Tubing

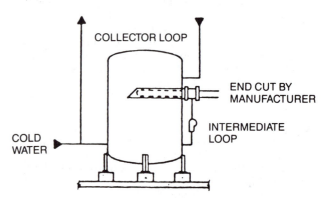

Fig. 28   Tube Bundle Heat Exchanger with Intermediate Loop

External Heat Exchanger

The external heat exchanger offers a greater degree of design flexibility than the internal heat exchanger because it is detached from the tank. For this reason, it is preferred for most commercial applications. Shell-in-tube, tube-in-tube, and plate-and-frame heat exchangers are used in solar applications. Shell-in-tube heat exchangers are found in many solar designs because they are economical, easy to obtain, and are constructed with suitable material. One limitation of the shell-in-tube is that it normally does not have double-wall protection, which is often required for potable water-heating applications. A number of manufacturers produce tube-in-tube heat exchangers that offer high performance and the double-wall safety required by many code authorities, but they are usually limited in size. The plate-and-frame heat exchanger is more cost-effective for large potable water applications where positive separation of heat transfer fluids is required. These heat exchangers are compact and offer excellent heat transfer performance.

Plate and frame. These heat exchangers are suitable for pressures up to 300 psig and temperatures to 400 °F. They are economically attractive when a quality and heavy-duty construction material is required or if a double-wall, shell-and-tube heat exchanger is needed for leak protection. Typical applications include food industries or domestic water heating when any possibility of product contamination must be eliminated. This exchanger also gives added protection to the collector loop.

Contamination is not possible when an intermediate loop is used (Figure 29) and the integrity of the plates is maintained. A colored fluid in the intermediate loop gives a visual means of detecting a leak through changes in color. The sealing mechanism of the plate-and-frame heat exchanger prevents cross-contamination of heat transfer fluids. The plate-and-frame heat exchanger cost is comparable to, or cheaper than, equivalent shell-and-tube heat exchangers constructed of stainless steel.

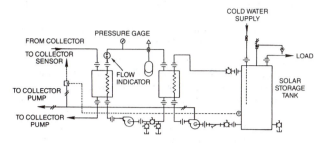

**Fig. 29   Double-Wall Protection Using
Two Heat Exchangers in Series**

Shell-and-tube. Shell-and-tube heat exchangers accommodate large heat exchanger areas in a compact volume. The number of shell-side passes and tube-side passes (*i.e.,* the number of times the fluid changes directions from one end of the heat exchanger to the other) is a major variable in shell-and-tube heat exchanger selection. Because the exchanger must compensate for thermal expansion, flow in and out of the tube side are generally at the same end of the exchanger. Therefore, the number of tube-side passes is even. By appropriate baffling, two, four, or more tube passes may be created. However, as the number of passes increases, the path length grows, resulting in greater pressure drop of the tube-side fluid. Unfortunately, shell-and-tube heat exchangers are hard to find in the double-wall configuration.

Tube-and-tube. Fluids in the tube-and-tube heat exchanger run counterflow, which gives closer approach temperatures. The exchanger is also compact and only limited by system size. Several may be piped in parallel for higher flow, or in series to provide approach temperatures as close as 15°F. Many manufacturers offer the tube-in-tube with double-wall protection.

The counterflow configuration operates at high efficiency. For two heat exchangers in series, the effectiveness may reach 0.80. For single heat exchangers or multiple heat exchangers in parallel, the effectiveness may reach 0.67. The cost of tube-and-tube heat exchangers is low, making them cost-effective for smaller residential systems.

Heat Exchanger Performance

Because of the high value of the heat being transferred, heat exchangers used in solar energy systems must be overdesigned when compared with usual standards. This is especially true with collectors of medium to low performance.

Solar collectors perform less efficiently at high fluid inlet temperatures. Heat exchangers require a temperature difference between the two streams to transfer heat. The smaller the heat transfer surface area, the greater the temperature difference required to transfer the same amount of heat. Consequently, the collector inlet temperature must be higher for a given tank temperature. As the solar collector is forced to operate at the progressively higher temperatures associated with smaller heat exchangers, its efficiency is reduced.

In addition to size and surface area, heat exchanger configuration is important for achieving maximum performance. Heat exchanger performance is characterized by its effectiveness (a type of efficiency), which is defined as:

$$E = \frac{q}{(\dot{m}c_p)_{min}(t_{hi} - t_{ci})} \tag{11}$$

where

q = heat transfer rate, Btu/h
$(\dot{m}c_p)_{min}$ = minimum capacitance rate in heat exchanger (mass flow rate times fluid specific heat)
t_{hi} = hot (collector loop) stream inlet temperature
t_{ci} = cold (storage) stream inlet temperature

For heat exchangers located in the collector loop, the minimum flow usually occurs on the collector side rather than the tank side.

The effectiveness is the ratio between the heat actually transferred and the maximum heat that could be transferred for given flow and fluid inlet temperature conditions. The effectiveness is relatively insensitive to temperature, but it is a strong function of heat exchanger design.

A designer must decide what heat exchanger effectiveness is required for the specific application. A method that incorporates heat exchanger performance into a collector efficiency equation [Equation (3)] uses storage tank temperature t_s as the collector inlet temperature with an adjusted heat removal factor F_R'. Equation (12) relates F_R' and heat exchanger effectiveness:

$$\frac{F_R'}{F_R} = \frac{1}{1 + (F_R U_L / \dot{m}c_p)(A\dot{m}c_p / Ec_{min} - 1)} \tag{12}$$

where c_{min} is the smaller of the two fluid capacities ($\dot{m}c_p$) in the heat exchanger.

The heat exchanger effectiveness must be converted into heat transfer surface area. SERI (1981) provides details of shell-and-tube heat exchangers and heat transfer fluids.

CONTROLS

Applications

The heart of an active solar energy system is the automatic temperature control. Numerous studies and reports of operational systems show faulty controls are usually the cause of poor solar system performance. Reliable solar system controllers are available, and with proper understanding of each system function, proper control systems can be designed. In general, control systems should be simple; additional controls are not a good solution to a problem that can be solved by better mechanical design. The following key considerations pertain to control system design:

- Collector sensor location/selection
- Storage sensor location
- Over-temperature sensor location
- Proper on-off controller characteristics
- Selection of reliable solid-state devices, sensors, controllers, etc.
- Control panel location in heated space
- Connection of controller according to manufacturer's instructions
- Design of control system for all possible system operating modes, including heat collection, heat rejection, power outage, freeze protection, auxiliary heating, etc.
- Selection of alarm indicators for critical applications including pump failure, low temperatures, high temperatures, loss of pressure, controller failure, and nighttime operation

The following five categories need to be considered when designing automatic controls for solar energy systems:

1. Collection to storage
2. Storage to load
3. Auxiliary energy to load
4. Alarms
5. Miscellaneous (*e.g.*, for heat rejection, freeze protection, draining, and over-temperature protection)

Differential Temperature Controller

Most controls used in solar energy systems are similar to those for HVAC systems. The major exception is the differential temperature controller (DTC), which is the basis of solar energy system control. The DTC is a comparing controller with at least two temperature sensors that controls one or several devices. Typically, the sensors are located at the solar collectors and storage tank

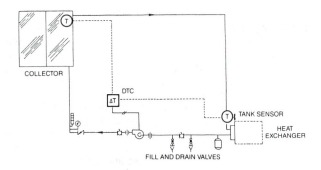

**Fig. 30 Basic Nonfreezing Collector Loop for
Building Service Hot Water Heating—
Nonglycol Heat Transfer Fluid**

(Figure 30). On unpressurized systems, other DTCs may control the extraction of heat from the storage tank.

The DTC compares the temperature difference and when the temperature of the panel exceeds that of the storage by the predetermined amount (generally by 8 to 20 °F), the DTC switches to turn on the actuating devices. When the temperature of the panel drops to 3 to 10 °F above the storage temperature, the DTC, either directly or indirectly, stops the pump. Indirect control through a control relay may operate one or more pumps and possibly perform other control functions, such as the actuation of control valves.

The manufacturer's predetermined set point of the DTC may be adjustable or fixed at a specific temperature differential. If the controller has a fixed differential, the controller selection should correspond to the requirements of the system. An adjustable differential set point makes the controller more flexible and allows it to be adjusted to the specific system. The optimum differential *off* temperature should be the minimum possible; the minimum depends on whether the system circuitry has a heat exchanger between the collectors and storage.

If the system requires a heat exchanger, the energy transferred between two fluids raises the differential temperature set point. The minimum, or *off* temperature differential, is the point at which pumping the energy costs as much as the value of the energy being pumped. For systems with heat exchangers, the *off* set point is generally between 5 and 10 °F. If the system does not have a heat exchanger, a range of 3 to 6 °F is acceptable for the *off* set point. The heat lost in the piping and the power required to operate the pump should also be considered.

The optimum differential *on* set point is difficult to calculate because of the changing variables and conditions. Typically, the *on* set point is 10 to 15 °F above the *off* set point. The optimum *on* set point trades between optimum energy collection and avoiding short cycling of the pump. ASHRAE's *Active Solar Thermal Design Manual* (Newton 1983) describes techniques for minimizing short cycling.

Over-Temperature Protection

Overheating may occur during periods of high insolation and low load; thus, all portions of the solar system require protection against overheating. Liquid expansion or excessive pressure may burst piping or storage tanks, and steam or other gases within a system may restrict the liquid flow, making the system inoperable. Glycols break down and become corrosive if subjected to temperatures greater than 240 °F. The system can be protected from overheating by (1) stopping circulation in the collection loop until the storage temperature decreases, (2) discharging the overheated water from the system and replacing it with cold make-up

water, or (3) using a heat exchanger coil as a means of heat rejection to ambient air.

The following questions should be answered to determine if over-temperature protection is necessary.

1. Is the load ever expected to be off, such that the solar input will be much higher than the load? The designer must determine possibilities based on the owner's needs and a computer analysis of system performance.
2. Do individual components, pumps, valves, circulating fluids, piping, tanks, and liners need protection? The designer must examine all components and base the over-temperature protection set point on the component that has the lowest maximum operating temperature specification. This may be a valve or pump with a 180 to 300 °F maximum operating temperature. Sometimes, this factor may be overcome by selecting components with higher operating temperature capabilities.
3. Is the formation of steam or discharging boiling water at the tap possible? If the system has no mixing valve that mixes cold water with the solar-heated water before it enters the tap, the water must be maintained below boiling temperature. Otherwise, the water will flash to steam as it exits the tap and, most likely, scald the user. Some city codes require a mixing valve to be placed in the system for safety.

Differential temperature controllers are available that sense over-temperatures. Depending on the controller used, the sensor may be mounted at the bottom or top of the storage tank. If it is mounted at the bottom of the tank, the collector-to-storage differential temperature sensor can be used to sense over-temperatures. Input to a DTC mounted at the top of the tank is independent of the bottom-mounted sensor, and the sensor monitors the true high temperature.

The normal action taken when the DTC senses an over-temperature is to turn off the pump to stop heat collection. After the panels in a drainback system are drained, they will attain stagnation temperatures. While drainback is not desirable, the panels used for these systems should be designed and tested to withstand these conditions. In addition, draindown panels should withstand the thermal shock of start-up when relatively cool water enters the panels while they are at stagnation temperatures. The temperature difference can range from 75 to 300 °F. Such a difference could warp panels made with two or more materials of different thermal expansion coefficients. If the solar panels cannot withstand the thermal shock, an interlock should be incorporated into the control logic to prevent this situation. One method uses a high-temperature sensor mounted on the collector absorber that prevents the pump from operating until the collector temperature drops below the sensor set point.

If circulation stops in a closed-loop antifreeze system that has a heat exchanger, high stagnation temperatures will occur. These temperatures could break down the glycol heat transfer fluid. To prevent damage or injury due to excessive pressure, a pressure relief valve must be installed in the loop, and a means of rejecting heat from the collector loop must be provided. The next section describes a common way to relieve pressure. When water-based absorber fluids are used, pressure builds up from boiling. However, pressure increases due to the thermal expansion of any fluid.

The pressure-relief valve should be set to relieve at or below the maximum operating pressure of any component in the closed-loop system. Typical settings are around 50 psig, corresponding to a temperature of approximately 300 °F. However, these settings should be checked. When the pressure-relief valve does open, it discharges expensive antifreeze solution. Glycol antifreeze solutions damage many types of roof membranes. The discharge can be piped to large containers to save the antifreeze, but this design can create dangerous conditions because of the high pressures and temperatures involved.

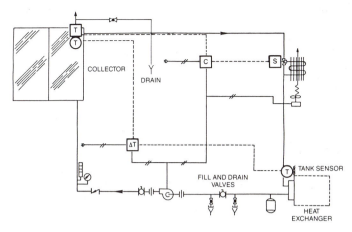

Fig. 31 Heat Rejection from Nonfreezing System Using Liquid-to-Air Heat Exchanger

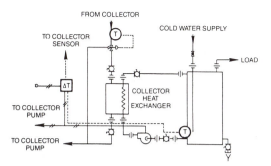

Fig. 32 Nonfreezing System with H-X Bypass

If a collector loop containing glycol stagnates, chemical decomposition raises the fusion point of the liquid, and freezing becomes possible. An alternate method continues fluid circulation but diverts the flow from storage to a heat exchanger that dumps heat to the ambient air or other source (Figure 31). This wastes energy, but it protects the system. A sensor on the solar collector absorber plate that turns on the heat rejection equipment can provide control. The temperature sensor set point is usually 200 to 250 °F and depends on the system components. When the sensor reaches the high-temperature set point, it turns on pumps, fans, alarms, or whatever is necessary to reject the heat and warn of the overtemperature. The dump continues to operate until the overtemperature control in the collector loop DTC senses an acceptable drop in tank temperature and is reset to its normal state.

Hot Water Dump

If water temperatures above 200 °F are allowed, the standard temperature-pressure (210 °F, 125 psig) safety relief valve may operate occasionally. If these temperatures are reached, the valve opens, and some of the hot water vents out. However, these valves are designed for safety purposes, and after a few openings, they leak hot water. Thus, they should not be relied on as the only control device. An aquastat that controls a solenoid, pneumatic, or electrically actuated valve should be used instead.

Heat Exchanger Freeze Protection

The following factors should be considered when selecting an external heat exchanger for a system protected by a nonfreezing fluid that is started after an overnight, or longer, exposure to extremely cold conditions.

- Freeze-up of the water side of the heat exchanger
- Performance loss due to extraction of heat from storage.

An internal heat exchanger/tank has been placed on the water side of the heat exchanger of small systems to prevent freezing. However, the energy required to maintain the water above freezing must be extracted from storage, which decreases overall performance. With the external heat exchanger/tank combination, a bypass can be installed, as illustrated in Figure 32. The controller positions the valve to bypass the heat exchanger until the fluid in the collector loop attains a reasonable level, for example 80 °F. When the heat transfer fluid has warmed to this level, it can enter the heat exchanger without freezing or extracting heat from storage. The arrangement illustrated in Figure 32 can also be used with an internal heat exchanger, if necessary, to improve performance.

REFERENCES

ANL. 1979. *Design and installation manual for thermal energy storage.* ANL-79-15, Argonne National Laboratories, Argonne, IL.

ANL. 1979. *Reliability and materials design guidelines for solar domestic hot water systems.* ANL/SDP-9 Argonne National Laboratories.

ASHRAE. 1988. *Active solar heating systems design manual.*

ASHRAE. 1990. *Guide for preparing active solar heating systems operation and maintenance manuals.*

ASHRAE. 1991. *Active solar heating systems installation manual.*

Beckman, W. A., S. Klein, and J.A. Duffie. 1977. *Solar heating design by the F-Chart method.* Wiley-Interscience, New York.

Huggins, J.C. and D.L. Block. 1983. Thermal performance of flat plate solar collectors by generic classification. Proceedings of the ASME Solar Energy Division Fifth Annual Conference, Orlando, Florida.

ICBO. 1985. *Uniform building code.* International Conference of Building Officials, Whittier, CA.

Knowles, A.S. 1980. *A simple balancing technique for liquid cooled flat plate solar collector arrays.* International Solar Energy Society, Phoenix, AZ.

Krieder, J. 1982. *The solar heating design process—Active and passive systems.* McGraw-Hill, New York.

Kutscher, C.F. 1982. *Design approaches for industrial process heat systems.* SERI/TR-253-1356, Solar Energy Research Institute, Golden, CO.

Logee, T.L. and P.W. Kendall. 1984. *Component report performance of solar collector arrays and collector controllers in the national solar data network.* SOLAR/0015-84/32, Vitro Corporation, Silver Spring, MD.

SEL. 1983. TRNSYS *A transient simulation program. Engineering Experiment Station Report 38-12, Solar Energy Laboratory,* University of Wisconsin-Madison, Madison, WI.

SERI. 1981. *Solar design workbook.* Solar Energy Research Institute, Golden, CO.

COMPRESSORS

THE compressor is one of the six essential parts of the compression refrigeration system; the others include the condenser, evaporator, and expansion device. The compressor circulates refrigerant through the system in a continuous cycle.

There are two basic types of compressors: positive displacement and dynamic. Positive displacement compressors increase the pressure of refrigerant vapor by reducing the volume of the compressor chamber through work applied to the compressor's mechanism: reciprocating, rotary (rolling piston, rotary vane, single screw, and twin screw), scroll, and trochoidal.

Dynamic compressors increase the pressure of refrigerant vapor by a continuous transfer of angular momentum from the rotating member to the vapor followed by the conversion of this momentum into a pressure rise. Centrifugal compressors function based on these principles.

This chapter describes the design features of several categories of commercially available refrigerant compressors.

POSITIVE DISPLACEMENT COMPRESSORS

PERFORMANCE

Compressor performance is the result of design compromises involving physical limitations of the refrigerant, compressor, and motor, while attempting to provide the following:

- Greatest trouble-free life expectancy
- Most refrigeration effect for the least power input
- Lowest applied cost
- Wide range of operating conditions
- Acceptable vibration and sound level

Two useful measures of compressor performance are capacity (which is related to compressor volume displacement) and efficiency.

System capacity is the rate of heat removal by the refrigerant pumped by the compressor in a refrigerating system. Capacity equals the product of the mass flow of refrigerant produced by the compressor and the difference in specific enthalpies of the refrigerant vapor at its thermodynamic state when it leaves the evaporator and the refrigerant liquid vapor mixture entering the evaporator.

The coefficient of performance (COP) for a hermetic compressor includes the combined operating efficiencies of the motor and the compressor.

$$COP \text{ (hermetic)} = \frac{\text{Capacity, Btu/h}}{\text{Input power to motor, Btu/h}}$$

The COP for an open compressor does not include motor efficiency.

$$COP \text{ (open)} = \frac{\text{Capacity, Btu/h}}{\text{Input power to shaft, Btu/h}}$$

Ideal Compressor

The capacity of a compressor at a given operating condition is a function of the mass of gas compressed per unit time. Ideally, the mass flow rate is equal to the product of the compressor displacement per unit time and the gas density, as shown in Equation (1):

$$\omega = \rho V_d \qquad (1)$$

where:

ω = ideal mass rate of gas compressed, lb/h
ρ = density of gas entering compressor, lb/ft^3
V_d = geometric displacement of compressor, ft^3

The ideal refrigeration cycle is addressed in Chapter 1 of the 1989 ASHRAE *Handbook—Fundamentals*; the following quantities can be determined from the pressure-enthalpy diagram in Figure 6 of that chapter:

$$Q_{refrigeration \, effect} = (h_1 - h_4)$$
$$Q_{work \, of \, compression} = (h_2 - h_1)$$

Using ω, the mass flow rate of gas as determined by Equation (1):

$$\text{Ideal capacity} = \omega \, Q_{refrigeration \, effect}$$
$$\text{Ideal power input} = \omega \, Q_{work}$$

Actual Compressor Performance

Deviations from ideal performance of positive displacement compressors occur because of various losses, with a resulting decrease in capacity and an increase in power input. Depending on the type of compressor, some or all of the following factors have a major effect on compressor performance.

1. *Pressure drops within the compressor*:
   - Through shutoff valves (suction, discharge, or both)
   - Across suction strainer
   - Across motor (hermetic compressor)
   - In manifolds (suction and discharge)
   - Through valves and valve ports (suction and discharge)
   - In internal muffler

The preparation of this chapter is assigned to TC 8.1, Positive Displacement Compressors, and TC 8.2, Centrifugal Machines.

2. *Heat gain to refrigerant from*:
   - Hermetic motor
   - Friction
   - Heat of compression; heat exchange within compressor
3. *Valve inefficiencies due to imperfect mechanical action*
4. *Internal gas leakage*
5. *Oil circulation*
6. *Reexpansion.* The volume of gas remaining in the compression chamber after discharge, which reexpands into the compression chamber during the suction cycle and limits the mass of fresh gas that can be brought into the compression chamber.
7. *Deviation from isentropic compression.* When considering the ideal compressor, an isentropic compression cycle is assumed. In the actual compressor, the compression cycle deviates from isentropic compression primarily because of fluid and mechanical friction and heat transfer within the compression chamber. The actual compression cycle and the work of compression must be determined from measurements.
8. *Over- and undercompression.* In fixed volume ratio rotary, screw, and orbital compressors, overcompression occurs when pressure in the compression chamber reaches discharge pressure before reaching the discharge port, and undercompression occurs when the compression chamber reaches the discharge port prior to achieving discharge pressure.

All of these losses are difficult to consider individually. They can, however, be grouped together and considered by category. Their effect on ideal compressor performance is measured by the following efficiencies:

Compression efficiency considers only what occurs within the compression volume and is a measure of the deviation of actual compression from isentropic compression. It is defined as the ratio of the work required for isentropic compression of the gas to the work delivered to the gas within the compression volume (as obtained by measurement).

Mechanical efficiency is the ratio of the work delivered to the gas (as obtained by measurement) to the work input to the compressor shaft.

Volumetric efficiency is the ratio of actual volume of gas entering the compressor to the theoretical displacement of the compressor.

Isentropic (Adiabatic) efficiency is the ratio of the work required for isentropic compression of the gas to work input to the compressor shaft.

Actual capacity is a function of the ideal capacity and the overall volumetric efficiency e_o of the actual compressor.

$$\text{Actual capacity} = e_o \, \omega \, Q_{refrigeration \, effect} \qquad (2)$$

Actual shaft power is a function of the power input to the ideal compressor and the compression, mechanical, and volumetric efficiencies of the compressor, as shown in Equation (3).

$$P_{bp} = \frac{P_{tp} \, e_0}{e_2 \, e_3} = \frac{P_{tp} \, e_0}{e_4} \qquad (3)$$

where:

P_{bp} = actual shaft power, Btu/h
P_{tp} = ideal power input = ωQ_{work}, Btu/h
e_0 = volumetric efficiency
e_2 = compression efficiency
e_3 = mechanical efficiency
e_4 = isentropic efficiency

Liquid subcooling is not accomplished by the compressor. However, the effect of liquid subcooling is included in the ratings for compressors by most manufacturers.

Heat rejection is the sum of the refrigeration effect and the heat equivalent of the power input to the compressor. The quantity of heat rejection must be known in order to size condensers and/or coolers, if they are used.

PROTECTIVE DEVICES

Compressors are provided with one or more of the following devices for protection against abnormal conditions and to comply with various codes.

1. High-pressure protection as required by Underwriters Laboratories and per ARI standards and ANSI/ASHRAE *Standard* 15-1989, Safety Code for Mechanical Refrigeration. This may include the following:
   a. A high-pressure cutout.
   b. A high- to low-side internal relief valve or rupture member. To comply with the ANSI/ASHRAE *Standard* 15-1989, all compressors 50 cfm displacement and larger must be equipped with either an internal or external high- to low-side relief valve to prevent rupture of the crankcase or housing. The differential pressure setting depends on the refrigerant used and the operating conditions. Care must be taken to ensure that the relief valve will not accidentally blow on a fast pulldown. Many welded hermetic compressors have an internal high- to low-pressure relief valve to limit maximum system pressure in units not equipped with other high-pressure control devices.
   c. A relief valve or rupture member on the compressor casing similar to those used on other pressure vessels.
2. High-temperature control devices to protect against overheating and oil breakdown.
   a. Motor overtemperature protective devices, which are addressed in the section Hermetic Motor Protection in Chapter 40.
   b. To protect against lubricant and refrigerant breakdown, a temperature sensor is sometimes used to stop the compressor when discharge temperatures exceed safe values. The switch may be placed internally (near the compression chamber) or externally (on the discharge line).
   c. On larger compressors, lubricant temperatures are controlled by cooling with either a heat exchanger or direct liquid injection, or the compressor may shut down on high lubricant temperature.
   d. Where lubricant sump heaters are used to maintain a minimum lubricant sump temperature, a thermostat is usually used to limit the maximum lubricant temperature.
3. Low-pressure protection may be provided for:
   a. *Suction gas.* Many compressors or systems are limited to a minimum suction pressure by a protective switch. Motor cooling, freeze-up, or pressure ratio usually determine the pressure setting.
   b. *Compressor.* Lubricant pressure protectors are used with forced feed lubrication systems to prevent the compressor from operating with insufficient pressure.
4. Time delay, or lockouts with manual resets, prevent damage to both compressor motor and contactors from repetitive rapid-starting cycles.
5. Low voltage and phase loss or reversal protection is used on some systems. Phase reversal protection is used with multiphase devices to ensure the proper direction of rotation.
6. Suction line strainer. Most compressors are provided with a strainer at the suction inlet to remove any dirt that might exist in the suction line piping. Factory-assembled units with all system parts cleaned at the time of assembly may not require the suction line strainer. A suction line strainer is normally required in all field-assembled systems.

Liquid Hazard

A gas compressor is not designed to handle large quantities of liquid. The damage that may occur depends on the quantity of liquid, the frequency with which it occurs together with the type of compressor. Slugging, floodback, and flooded starts are three ways in which liquid can damage a compressor. Generally, reciprocating compressors, because they are equipped with discharge valves, are more susceptible to damage from slugging than various rotary and orbital compressors.

Slugging is the short-term pumping of a large quantity of liquid refrigerant or oil. It can occur just after start-up if refrigerant accumulated in the evaporator during shutdown returns to the compressor. It can also occur when system operating conditions change radically, such as during a defrost cycle. Slugging can also occur with quick changes on compressor loading.

Floodback is the continuous return of liquid refrigerant mixed with the suction gas. It is a hazard to compressors with a dependence on the maintenance of a certain amount of oil in the compression chamber.

Flooded start occurs during system shutdown when refrigerant is allowed to migrate to the compressor.

Compressors that are easily susceptible to damage by liquid can be protected by using a suction accumulator.

Sound Level

An acceptable sound level is a basic requirement of good design and application. Chapter 42 in the 1991 ASHRAE *Handbook—HVAC Applications* covers design criteria in more detail.

Vibration

Compressor vibration results from gas-pressure pulses and inertia forces associated with the moving parts. The problems of vibration can be handled in the following ways:

Isolation. With this common method, the compressor is resiliently mounted in the unit by springs, synthetic rubber mounts, etc. In hermetic compressors, the internal compressor assembly is usually spring-mounted within the welded shell, and the entire unit is externally isolated.

Reduction of amplitude. The amount of movement can be reduced by adding mass to the compressor. Mass is added either by rigidly attaching the compressor to a base condenser, or chiller, or by providing a solid foundation. When structural transmission is a problem, particularly with large machines, the entire assembly is then resiliently mounted.

Chapter 42 in the 1991 ASHRAE *Handbook—HVAC Applications* has further information.

Shock

In designing for shock, three types of dynamic loads are recognized:

- Suddenly applied loads of short duration
- Suddenly applied loads of long duration
- Sustained periodic varying loads

Since the forces are primarily inertial, the basic approach is to maintain low equipment mass and make the strength of the carrying structure as great as possible. The degree to which this practice is followed is a function of the amount of shock loading.

Commercial units. The major shock loading to these units occurs during shipment or when they operate on commercial carriers. Train service provides a severe test because of low forcing frequencies and high shock load. Shock loads as high as 10 g have been recorded; 5 g can be expected.

Trucking service results in higher forcing frequencies, but shock loads can be equal to, or greater than, those for rail transportation. Aircraft service forcing frequencies generally fall in the range of 20 to 60 Hz with shocks to 3 g.

Military units. The requirements are given in detail in specifications that exceed anything expected of commercial units. In severe applications, deformation of the supporting members and

shock isolators may be tolerated, provided that the unit performs its function.

Basically, the compressor must be made of components rigid enough to avoid misalignment or deformation during shock loading. Therefore, structures with low natural frequencies should be avoided.

Testing

Testing for ratings must be in accordance with ASHRAE *Standard* 23-78, Methods of Testing for Rating Positive Displacement Refrigerant Compressors. Compressor tests are of two types: the first determines capacity, efficiency, sound level, motor temperatures, etc.; the second determines the probable life of the machine.

Standard rating conditions, which are usually specified by compressor manufacturers, include the maximum possible compression ratio, the maximum operating pressure differential, and the maximum permissible discharge pressure.

Lubrication requirements, which are prescribed by the compressor manufacturers, include the type, viscosity, and other characteristics of the oils to be used with the many different operating levels and the type of refrigerant being used.

Power requirements for compressor starting, pulldown, and operation vary because unloading means differ in the many styles of compressors available. Manufacturers supply full information covering the various methods employed.

HERMETIC MOTORS

Motors for positive-displacement compressors range from fractional horsepower to several thousand horsepower. When selecting a motor for driving a compressor, the following factors should be considered:

- Horsepower and rotational speed
- Starting and pull-up torques
- Ambient and maximum rise temperatures
- Cost and availability
- Insulation system
- Efficiency and performance
- Starting currents
- Voltage and phase
- Type of protection required
- Multispeed or variable speed

Large, industrial open compressors can be run from 900 to 3600 rpm and voltages ranging from 230 to 4160 volts. Due to local codes, requirements of local utilities, or end-user requirements, motors can be started across-the-line, part winding, Wye-Delta, autotransformer, or solid state to limit the starting currents. Caution must be taken that the starting method will supply enough torque to accelerate the motor and overcome the torque required for compression.

Hermetic motors can be more highly loaded than comparable open motors because of the refrigerant cooling used.

To use a hermetic motor effectively, the maximum design load should be as close as possible to the breakdown torque at the lowest voltage used. This approach yields a motor design that operates better at lighter loads and higher voltage conditions. The limiting factor at high loads is normally the motor temperature, while at light loads the limiting factor is the discharge gas temperature. Overdesign of the motor increases discharge gas temperature at light loads.

The single-phase motor presents more design problems than the polyphase, because the relationship between main and auxiliary windings becomes critical together with the necessary starting equipment.

The locked-rotor rate of temperature rise must be kept low enough to prevent excessive motor temperatures with the motor protection available. The maximum temperature under these onditions should be held to 300°F. With better protection, a higher rate of rise can be tolerated, and a less expensive motor can be used.

The materials selected for these motors must have high dielectric strength, resist fluid and mechanical abrasion, and be compatible with an atmosphere of refrigerant and oil.

The types of hermetic motors selected for various applications are as follows:

Refrigeration compressors—Single phase
   Low to medium torque—Split phase or PSC (Permanent Split Capacitor)
   High torque—CSCR (Capacitor Start-Capacitor Run) and CSIR (Capacitor Start-Induction Run)
Room air-conditioner compressors—Single phase
   PSC or CSCR
   2 speed, pole switching variable speed
Central air-conditioning and commercial refrigeration
   1 phase, PSC and CSCR to 6 hp
   3 phase, 2 hp and above, across-the-line start
   10 hp and above; part winding; Wye-Delta and across-the-line start
   2 speed, pole switching variable speed

For further information on motors and motor protection, see Chapter 40.

RECIPROCATING COMPRESSORS

Most reciprocating compressors are single-acting, using pistons that are driven directly through a pin and connecting rod from the crankshaft. Double-acting compressors that use piston rods, crossheads, stuffing boxes, and oil injection are not used extensively and, therefore, are not covered here.

Single-stage compressors can achieve saturated suction temperatures to −50°F at 95°F saturated condensing temperature by using R-502. Chapters 3 and 4 of the 1990 ASHRAE *Handbook—Refrigeration* have additional information on other halocarbon and ammonia systems.

Booster compressors are typically used for low-temperature applications with R-22 or ammonia. Minus 85°F saturated suction can be achieved by using R-22, and −65°F saturated suction is possible using ammonia.

The booster raises the refrigerant pressure to a level where further compression can be achieved with a high-stage compressor, without exceeding the compression-ratio limitations of the respective machines.

Since superheat is generated as a result of compression in the booster, intercooling is normally required to reduce the refrigerant stream temperature to the practical level required at the inlet to the high-stage unit. Intercooling methods include controlled liquid injection into the intermediate stream, gas bubbling through a liquid reservoir, and use of a liquid-to-gas heat exchanger where no fluid mixing occurs.

Integral two-stage compressors achieve low temperature (−80°F, using R-22 or ammonia) within the frame of a single compressor. The cylinders within the compressor are divided into respective groups so that the combination of volumetric flow and pressure ratios are balanced to achieve booster and high-stage performance effectively. Refrigerant connections between the high-pressure suction and low-pressure discharge stages allow an interstage gas cooling system to be connected to remove superheat between stages. This interconnection is similar to the methods used for individual high-stage and booster compressors.

Capacity reduction is typically achieved by cylinder unloading, as in the case of single-stage compressors. Special consideration must be given to maintaining the correct relationship between high- and low-pressure stages.

The most widely used compressor is the halocarbon compressor, which is manufactured in three types of design: (1) open, (2) semihermetic or bolted hermetic, and (3) welded-shell hermetic.

Ammonia compressors are manufactured only in the open design because of the incompatibility of the refrigerant and hermetic motor materials.

Open-type compressors are those in which the shaft extends through a seal in the crankcase for an external drive.

Hermetic compressors are those in which the motor and compressor are contained within the same pressure vessel, with the motor shaft integral with the compressor crankshaft and the motor in contact with the refrigerant.

A **semihermetic compressor** (bolted, accessible, or serviceable) is a hermetic compressor of bolted construction amenable to field repair.

In **welded-shell hermetic compressors** (sealed) the motor-compressor is mounted inside a steel shell, which, in turn is sealed by welding. (Combinations of design features used are shown in Table 1. Typical performance values for halocarbon compressors are given in Table 2.)

Performance Data

Figure 1 presents a typical set of capacity and power curves for a four-cylinder semihermetic compressor, 2.38-in. bore, 1.75-in. stroke, 1720 rpm, operating with Refrigerant 22. Figure 2 shows the heat rejection curves for the same compressor. Compressor curves should be labeled with the following information:

- Compressor identification
- Degrees subcooling or statement that data have been corrected to zero degrees subcooling

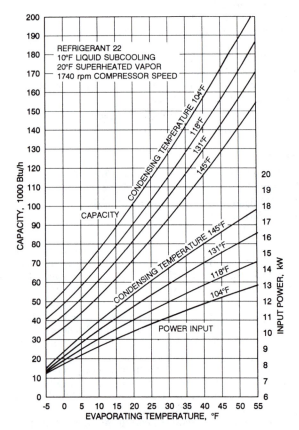

Fig. 1   Capacity and Power-Input Curves for Typical Hermetic Reciprocating Compressor

Table 1 Typical Design Features of Reciprocating Compressors

Item	Open	Semi-hermetic	Welded Hermetic	Open
	Halocarbon Compressor			**Ammonia Compressor**
1. Number of Cylinders—one to:	16	12	6	16
2. Power range	0.17 hp and up	0.5 to 150 hp	0.17 to 25 hp	10 hp and up
3. Cylinder arrangement				
a. Vertical, V or W, radial	X	X		
b. Radial, horizontal opposed			X	
c. Horizontal, vertical V or W				X
4. Drive				
a. Hermetic Compressors, induction electric motor)		X	X	
b. Open compressors—direct drive, V belt chain, gear, by electric motor or engine)	X			X
5. Lubrication—splash or force feed, floor)	X	X	X	X
6. Suction and discharge valves—ring plate or ring or reed flexing)	X	X	X	X
7. Suction and discharge valve arrangement				
a. Suction and discharge valves in head	X	X	X	X
b. Uniflow—suction valves in top of piston, suction gas entering through cylinder walls. Discharge valves in head	X			X
8. Cylinder cooling				
a. Suction gas cooled	X	X	X	X
b. Water jacket cylinder wall, head, or cylinder wall and head	X			X
c. Air cooled	X	X	X	X
d. Refrigerant cooled heads	X			X
9. Cylinder head				
a. Spring loaded	X	X	X	X
b. Bolted head	X	X	X	X

Item	Open	Semi-hermetic	Welded Hermetic	Open
	Halocarbon Compressor			**Ammonia Compressor**
10. Bearings				
a. Sleeve, antifriction	X	X	X	X
b. Tapered roller	X			X
11. Capacity control, if provided—manual or automatic				
a. Suction valve lifting	X	X	X	X
b. Bypass-cylinder heads to suction	X	X	X	X
c. Closing inlet	X	X		X
d. Adjustable clearance	X	X		X
e. Variable speed	X	X	X	X
12. Materials				
Motor insulations and rubber materials must be compatible with refrigerant and oil mixtures. Otherwise, no restrictions	X	X	X	
No copper or brass				X
13. Oil return				
a. Crankcase separated from suction manifolds, oil return check valves, equalizers, spinners, foam breakers	X	X		X
b. Crankcase common with suction manifold			X	
14. Synchronous speeds (50 to 60 Hz)	250 to 3600	1500 to 3600	1500 to 3600	250 to 3600
15. Pistons				
a. Aluminum or cast iron	X	X	X	X
b. Ringless	X	X	X	X
c. Compression and oil control rings	X	X	X	X
16. Connecting rod				
Split rod with removable cap or solid eccentric strap	X	X	X	X
17. Mounting				
Internal spring mount		X	X	
External spring mount		X	X	
Rigidly mounted on base	X	X		X

Table 2 Typical Performance Values

Compressor Size and Type	R-12, R-500, R-502 Evap. Temp. = −40 °F Cond. Temp. = 105 °F Suction Gas = 65 °F Subcooling = 0 °F	R-12, R-500, R-502 Evap. Temp. = 0 °F Cond. Temp. = 110 °F Suction Gas = 65 °F Subcooling = 0 °F	R-12, R-22, R-500, R-502 Evap. Temp. = 40 °F Cond. Temp. = 105 °F Suction Gas = 55 °F Subcooling = 0 °F	R-12, R-22, R-500, R-502 Evap. Temp. = 45 °F Cond. Temp. = 130 °F Suction Gas = 65 °F Subcooling = 0 °F
	Operating Conditions and Refrigerants			
Large, over 25 hp				
Open	0.21 tons/hp	0.40 tons/hp	0.91 tons/hp	0.74 tons/hp
Hermetic	3.15 Btu/h per W	6.00 Btu/h per W	13.12 Btu/h per W	9.90 Btu/h per W
Medium, 5 to 25 hp				
Open	0.19 tons/hp	0.37 tons/hp	0.83 ton/hp	0.65 tons/hp
Hermetic	2.89 Btu/h per W	5.60 Btu/h per W	12.04 Btu/h per W	9.15 Btu/h per W
Small, under 5 hp				
Open	—	—	—	—
Hermetic	—	3.80 Btu/h per W	10.14 Btu/h per W	7.76 Btu/h per W

- Compressor speed
- Type refrigerant
- Suction gas superheat
- Compressor ambient
- External cooling requirements (if any)
- Maximum power or maximum operating conditions
- Minimum operating conditions at fully loaded and fully unloaded operation

Motor Performance

The motor efficiency is usually the result of a compromise between cost and size. Generally, the larger a motor is for a given rating, the more efficient it can be made. Accepted efficiencies range from approximately 80% for a 3 hp motor to 92% for a 100 hp motor. Uneven loading has a marked effect on motor efficiency. It is important that cylinders be spaced evenly. Also, the more cylinders there are, the smaller the impulses become. Greater moments of inertia of moving parts and higher speeds reduce the impulse effect. Small and evenly spaced impulses also help reduce noise and vibration.

Since many compressors start against load, it is desirable to estimate starting torque requirements. The following equation is for a single cylinder compressor. It neglects friction, the additional torque required to force discharge gas out of the cylinder, and the fact that the tangential force at the crankpin is not always equal to the normal force at the piston. This equation also assumes considerable gas leakage at the discharge valves but little or no leakage past the piston rings or suction valves. It yields a conservative estimate.

$$T_s = \frac{(p_2 - p_1)As}{(2N_2/N_1)} \qquad (4)$$

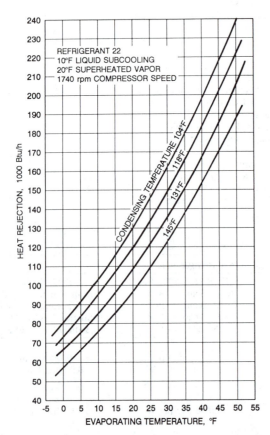

240
230
220 REFRIGERANT 22
210 10°F LIQUID SUBCOOLING
     20°F SUPERHEATED VAPOR
     1740 rpm COMPRESSOR SPEED

Fig. 2 Heat Rejection Curves for Typical Hermetic Reciprocating Compressor

where:

T_s = starting torque, lb·in.
p_2 = discharge pressure, psi
p_1 = suction pressure, psi
A = area of cylinder, in²
s = stroke of compressor, in.
N_2 = motor speed, rpm
N_1 = compressor speed, rpm

Equation (4) shows that when pressures in the system are balanced or almost equal ($p_2 = p_1$), torque requirements are considerably reduced. Thus, a pressure-balancing device on an expansion valve or a capillary tube that equalizes pressures at shutdown allows the compressor to be started without excessive effort. For multicylinder compressors, an analysis must be made of both the number of cylinders that might be on a compression stroke and the position of the rods at start. Since the force needed to push the piston to the top dead center is a function of how far the rod is away from the cylinder centerline, the worst possible angles these might assume need to be graphically determined by torque-effort diagrams. The torques for some arrangements are shown below:

No. Cylinders	Arrangement of Cranks	Angle between Cylinders	Approx. Torque from Equation (4)
1	Single		T_s
2	Single	90°	$1.025T_s$
2	180° apart	0° or 180°	T_s
3	Single	60°	$1.225T_s$
3	120° apart	120°	T_s
4	180°, 2 rods/crank	90°	$1.025T_s$
6	180°, 3 rods/crank	60°	$1.23T_s$

Pull-up torque is as important as starting torque. It is required to accelerate the compressor from rest, overcoming both inertia and gas forces to bring itself to operating speed. The greatest pull-up torque requirement comes when starting a compressor at a pressure ratio of about 2:1.

Features

Crankcases. The crankcase, or in a welded hermetic compressor, the cylinder block, is usually of cast iron. Aluminum is also used, particularly in small open and welded hermetic compressors. Open and semihermetic crankcases enclose the running gear, oil sump and, in the latter case, the hermetic motor. Access openings with removable covers are provided for assembly and service purposes. Welded hermetic cylinder blocks are often just skeletons, consisting of the cylinders, the main bearings, and either a barrel into which the hermetic motor stator is inserted or a surface to which the stator can be bolted.

The cylinders can be integral with the crankcase or cylinder block, in which case a material that provides a good sealing surface and resists wear must be provided. In aluminum crankcases, cast-in liners of iron or steel are usual. In large compressors, premachined cylinder sleeves inserted in the crankcase are common. With halocarbon refrigerants, excessive cylinder wear or scoring is not much of a problem and the choice of integral cylinders or inserted sleeves is often based on manufacturing considerations.

Crankshafts. Crankshafts are made of either forged steel with hardened bearing surfaces finished to 8 microinches or iron castings. Grade 25 to 40 (25,000 to 40,000 psi) tensile gray iron can be used where the lower modulus of elasticity can be tolerated. Nodular iron shafts approach the stiffness, strength, and ductility of steel and should be polished in both directions of rotation to 16 microinches maximum for best results. Crankshafts often include counterweights and should be dynamically and/or statically balanced.

While a safe maximum stress is important in shaft design, it is equally important to prevent excessive deflection that can edge-load bearings to failure. In hermetics, deflection can permit motor air gap to become eccentric, which affects starting, reduces efficiency, produces noise, and further increases bearing edge-loading.

Generally, the harder the bearing material used, the harder the shaft. With bronze bearings, a journal hardness of 350 Brinell is usual, while unhardened shafts at 200 Brinell in babbitt bearings are typical. Other combinations of materials and hardnesses have been used successfully.

Main bearings. Both the crank and drive means may be overhung with the bearings between; however, usual practice places the cylinders between the main bearings and, in a hermetic, overhangs the motor. Main bearings are made of steel-backed babbitt, steel-backed or solid bronze, or aluminum. In an aluminum crankcase, the bearings are usually integral. By automotive standards, unit loadings are low; however, the oil-refrigerant mixture frequently provides only marginal lubrication and 8000 h/year operation in commercial refrigeration service is quite possible. For conventional shaft diameters and speeds, 600 psi, main bearing loading based on projected area is not unusual. Running clearances average 0.001 in./in. of diameter with steel-backed babbitt bearings and a steel or iron shaft. Bearing oil grooves placed in the unloaded area are usual. Feeding oil to the bearing is only one requirement; another is the venting of evolved refrigerant gas and oil escape from the bearing to carry away heat.

In most compressors, crankshaft thrust surfaces (with or without thrust washers) must be provided in addition to main bearings. Thrust washers may be steel-backed babbitt, bronze, aluminum, hardened steel, or polymer and are usually stationary. Oil grooves are often included in the thrust face.

Connecting rods and eccentric straps. Connecting rods have the large end split and a bolted cap for assembly. Unsplit eccentric straps require the crankshaft to be passed through the big bore at assembly. Rods or straps are of steel, aluminum, bronze, nodular iron, or gray iron. Steel or iron rods often require inserts of such bearing material as steel-backed babbitt or bronze, while aluminum and bronze rods can bear directly on the crankpin and piston pin. Refrigerant compressor service limits unit loadings to 3000 psi based on projected area with a bronze bushing in the rod small bore and a hardened steel piston pin. Aluminum rod loadings at the piston pin of 2000 psi have been used. Large end unit loadings are usually under 1000 psi.

The Scotch yoke-type of piston-rod assembly has also been used. In small compressors, it has been fabricated by hydrogen brazing steel components. Machined aluminum components have been used in large hermetic designs.

Piston, piston ring, and piston pin. Pistons are usually made of cast iron or aluminum. Cast-iron pistons with a running clearance of 0.0004 in./in. of diameter in the cylinder will seal adequately without piston rings. With aluminum pistons, rings are required because a running clearance in the cylinder of 0.002 in./in. or more of diameter may be necessary, as determined by tests at extreme conditions. A second or third compression ring may add to power consumption with little increase in capacity; however, it may help oil control, particularly if drained. Oil scraping rings with vented grooves may also be used. Cylinder finishes are usually obtained by honing, and a 12 to 40 μin. range will give good ring seating. An effective oil scraper can often be obtained with a sharp corner on the piston skirt.

The minimum piston length is determined by the side thrust and is also a function of running clearance. Where clearance is large, pistons should be longer to prevent slap. An aluminum piston (with ring) having a length equal to 0.75 times the diameter, with a running clearance of 0.002 in./in. of diameter, and a rod length to crank arm ratio of 4.5, has been used successfully.

Piston pins are steel, case-hardened to Rockwell C 50 to 60 and ground to a 8 μin. finish or better. Pins can be restrained against rotation in either the piston bosses or the rod small end, be free in both, or be full-floating, which is usually the case with aluminum pistons and rods. Retaining rings prevent the pin from moving endwise and abrading the cylinder wall.

There is no well-defined limit to piston speed; average velocities of 1200 fpm, determined by multiplying twice the stroke in feet by the rpm, have been used successfully.

Suction and discharge valves. The most important components in the reciprocating compressor are the suction and discharge valves. Successful designs provide long life and low pressure losses. The life of a properly made and correctly applied valve is determined by the motion and stress it undergoes in performing its function. Excessive pressure losses across the valve result from high gas velocities, poor mechanical action, or both.

For design purposes, gas velocity is defined as being equal to the bore area multiplied by the average piston speed and divided by the valve area. Permissible gas velocities through the restricted areas of the valve are left to the discretion of the designer and depend on the level of volumetric efficiency and performance desired. In general, designs with velocities up to 12,000 fpm with ammonia and up to 9000 fpm with Refrigerants 12 and 22 have been successful.

An ideal valve system would meet the following requirements:

- Large flow areas with shortest possible path
- Straight gas flow path, no directional changes
- Low valve mass combined with low lift for quick action
- Symmetry of design with minimum pressure imbalance
- No increase in clearance volume
- Durability
- Low cost
- Tight sealing at ports
- Minimum valve flutter

Most valves in use today fall in one of the following groups:

1. **Free-floating reed valve,** with backing to limit movement, seats against a flat surface with circular or elongated ports. It is simple, and stresses can be readily determined, but it is limited to relatively small ports; therefore, multiples are often used. Totally backed with a curved stop, it is a valve that can stand considerable abuse.
2. **Reed, clamped at one end,** with full backstop support or a stop at the tip to limit movement, has a more complex motion than a free-floating reed; the resulting stresses are far greater than those calculated from the curvature of the stop. Considerable care must be taken in the design to ensure adequate life.
3. **Ring valve** usually has a spring return. A free-floating ring is seldom used because of its high-leakage losses. Improved performance is obtained by using spring return, in the form of coil springs or flexing backup springs, with each valve. Ring-type valves are particularly adaptable to compressors using cylinder sleeves.
4. **Valve formed as a ring** has part of the valve structure clamped. Generally, full rings are used with one or more sets of slots arranged in circles. By clamping the center, alignment is ensured and a force is obtained that closes the valve. To limit stresses, the valve proportions, valve stops, and supports are designed to control and limit valve motion.

Lubrication. Lubrication systems range from a simple splash system to the elaborate forced-feed systems with filters, vents, and equalizers. The type of lubrication required depends largely on bearing loads and application.

For low to medium bearing loads and factory-assembled systems where cleanliness can be controlled, the **splash system** gives excellent service. Bearing clearances must be larger, however;

otherwise, oil does not enter the bearing readily. Thus, the splashing effect of the dippers in the oil and the freer bearings cause the compressor to operate somewhat noisily. Furthermore, the splash at high speed encourages frothing and oil pumping; this is not a problem in package-type equipment, but may be in remote systems where gas lines are long.

A **flooded system** includes disks, screws, grooves, oil-ring gears, or other devices that lift the oil to the shaft or bearing level. These devices flood the bearing and are not much better than splash systems, except that the oil is not agitated as violently, so that quieter operation results. Since little or no pressure is developed by this method, it is not considered forced feed.

In **forced-feed lubrication**, a pump gear, vane, or plunger develops pressure, which forces oil into the bearing. Smaller bearing clearances can be used because adequate pressure feeds oil in sufficient quantity for proper bearing cooling. As a result, the compressor may be quieter in operation.

Gear pumps are used to a large extent. Spur gears are simple but tend to promote flashing of the refrigerant dissolved in the oil because of the sudden opening of the tooth volume as two teeth disengage. This disadvantage is not apparent in internal-type eccentric gear or vane pumps where a gradual opening of the suction volume takes place. The eccentric gear pump, the vane pump, or the piston pump therefore give better performance than simple gear pumps when the pump is not submerged in the oil.

Oil pumps must be made with minimum clearances to pump a mixture of gas and oil. The discharge of the pump should have provision to bleed a small quantity of oil into the crankcase. A bleed vents the pumps, prevents excess pressure, and ensures faster priming.

A strainer should be inserted in the suction line to keep foreign substances from the pump and bearings. If large quantities of very fine particles are present and bearing loadings are high, it may be necessary to add an oil filter to the discharge side of the pump.

Oil must return from the suction gas into the compressor crankcase. A flow of gas from piston leakage opposes this oil flow, so the velocity of the leakage gas must be low to permit oil to separate from the gas. A separating chamber may be built as part of the compressor to help separate oil from the gas.

In many designs, a check valve is inserted at the bottom of the oil return port to prevent a surge of crankcase oil from entering the suction. This check valve must have a bypass, which is always open, to permit the check valve to open wide after the oil surge has passed. When a separating chamber is used, the oil surge is trapped before it can enter the suction port, thus making a check valve less essential.

Seals. Stationary and rotary seals have been used extensively on open-type reciprocating compressors. Older stationary seals usually used metallic bellows and a hardened shaft for a wearing surface. Their use has diminished because of high cost.

The rotary seal costs less and is more reliable. A synthetic seal tightly fitted to the shaft prevents leakage and seals against the back face of a carbon nose. The front face of this carbon nose seals against a stationary cover plate. This design has been used on shafts up to 4 in. in diameter. The rotary seal should be designed so that the carbon nose is never subjected to the full thrust of the shaft; the spring should be designed for minimum cocking force; and materials should be such that a minimum of swelling and shrinking is encountered.

Special Devices

Capacity control. An ideal capacity control system would have the following operating characteristics (not all of these benefits can occur simultaneously):
- Continuous adjustment to load
- Full-load efficiency unaffected by the mechanism

- No loss in efficiency at part load
- Reduction of starting torque requirement
- No reduction in compressor reliability
- No reduction of compressor operating range
- No increase in compressor vibration and sound level at part load

Capacity control may be obtained by (1) controlling suction pressure by throttling; (2) controlling discharge pressure; (3) returning discharge gas to suction; (4) adding reexpansion volume; (5) changing the stroke; (6) opening a cylinder discharge port to suction while closing the port to discharge manifold; (7) changing compressor speed; (8) closing off cylinder inlet, and (9) varying suction valve opening.

The most commonly used methods are the opening of the suction valves by some external force, gas bypassing within the compressor, and gas bypassing outside the compressor.

When capacity control compressors are used, system design becomes more important and the following must be considered:

- Possible increase in compressor vibration and sound level at unloaded conditions
- Minimum operating conditions as limited by discharge or motor temperatures (or both) at part-load conditions
- Good oil return at minimum operating conditions when fully unloaded
- Rapid cycling of unloaders
- Refrigerant feed device capable of controlling at minimum capacity

Crankcase heaters. During shutdown, refrigerants tend to migrate to the coldest part of the refrigeration system. In cold weather, the compressor oil sump is usually the coldest area. When the system refrigerant charge is large enough to dilute the oil excessively and cause flooded starts, a crankcase heater should be used. The heater should maintain the oil at a minimum of 20 °F above the rest of the system at shutdown and well below the breakdown temperature of the oil at any time.

Internal centrifugal separators. Some compressors are equipped with antislug devices in the gas path to the cylinders. This device centrifugally separates oil and liquid refrigerant from the flow of foam during a flooded start and thus protects the cylinders. It does not eliminate the other hazards caused by liquid refrigerant in gas compressors.

Application

To operate through the entire range of conditions for which the compressor was designed and to obtain the desired service life, it is important that the mating components in the system be correctly designed and selected. Suction superheat must be controlled, oil must return to the compressor, and adequate protection must be provided against abnormal conditions.

Chapters 21 through 25 of the 1990 ASHRAE *Handbook— Refrigeration* cover system design. Chapter 7 in that volume gives details of system cleanup in the event of a hermetic motor burnout.

Suction superheat. No liquid refrigerant should be present in the suction gas entering the compressor because it causes oil dilution and gas formation in the lubrication system. If the liquid carryover is severe enough to reach the cylinders, excessive wear of valves, stops, pistons, and rings can occur; liquid slugging may actually break valves.

Suction gas without superheat is not harmful to the compressor as long as no liquid refrigerant is entrained; some systems are even designed to operate this way, although suction superheat is intended in the design of most systems. Measuring suction superheat can be difficult, and the indication of a small amount does not necessarily mean that liquid is not present. An effective suction separator may be necessary to remove all liquid with either type of system.

High superheat may result in dangerously high discharge temperatures and, in hermetics, high motor temperatures.

Normal values of superheat are usually 10 to 20°F in the air-conditioning range. For R-12 and R-22, an increase of about 0.5% in compressor capacity and a decrease of 0.5% in power can be expected for every 10°F increase in superheat obtained in the evaporator.

Automatic oil separators. Oil separators are used most often to reduce the amount of oil discharged into the system by the compressor and to return oil to the crankcase. They are recommended when the oil is not miscible with the refrigerant (ammonia and R-13, in particular). Because oil readily mixes with halocarbon refrigerants, oil separators are usually unnecessary, except in special cases such as low-temperature or flooded systems.

Parallel operation. Where multiple compressors are used, the trend is toward completely independent refrigerant circuits. This has an obvious advantage in the case of a hermetic motor burnout.

Parallel operation of compressors in a single system has some operational advantage at part load. Careful attention must be given to apportioning returned oil to the multiple compressors so that each always has an adequate quantity. Figure 3 shows the method most widely used. Line A connects the tops of the crankcases and equalizes the pressure above the oil, while line B permits oil equalization at the normal level. Lines of generous size must be used. In addition to lines A and B, the high-pressure sides of the compressor oil pumps can be cross-connected so that if one pump ceases to function the other will supply lubricant to both machines. Generally, line A is a large diameter, while line B is a small diameter, which limits the possible blowing of oil from one crankcase to the other.

A central reservoir for returned oil may also be used with means (such as crankcase float valves) for maintaining the proper levels in the various compressors. With staged systems, the low-stage compressor oil pump can sometimes deliver a measured amount of oil to the high-stage crankcase. The high-stage oil return is then sized and located to return a slightly greater quantity of oil to the low-stage crankcase. Where compressors are at different elevations and/or staged, the use of pumps in each oil line is necessary to maintain adequate crankcase oil levels. In both cases, proper gas equalization must be provided.

ROTARY COMPRESSORS

ROLLING PISTON COMPRESSORS

Rolling piston, or fixed vane, rotary compressors are used in household refrigerators and air-conditioning units in sizes up to about 3 hp (Figure 4). This type of compressor uses a roller mounted on the eccentric of a shaft with a single vane or blade suitably positioned in the nonrotating cylindrical housing, generally called the cylinder block. The blade reciprocates in a slot machined in the cylinder block. This reciprocating motion is caused by the eccentrically moving roller.

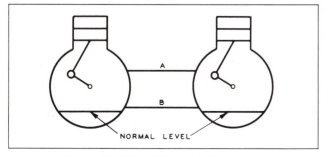

Fig. 3   Modified Oil-Equalizing System

Displacement for this compressor can be calculated from:

$$V_d = \pi H(A^2 - B^2)/4 \qquad (5)$$

where:

V_d = displacement
H = cylinder block height
A = cylinder diameter
B = roller diameter

The drive motor stator and compressor are solidly mounted in the compressor housing. This design feature is possible due to low vibrations associated with the rotary compression process, as opposed to reciprocating designs which employ spring isolation between the compressor parts and the housing.

Suction gas is directly piped into the suction port of the compressor, and the compressed gas is discharged into the compressor housing shell. This high-side shell design is used because of the simplicity of its lubrication system and the absence of oiling and compressor cooling problems. Compressor performance is also improved because this arrangement minimizes heat transfer to the suction gas and reduces gas leakage areas.

Internal leakage is controlled through hydrodynamic sealing and selection of mating parts for optimum clearances. Hydrodynamic sealing depends on clearances, surface speed, surface finishes, and oil viscosity. Close tolerance and low surface finish machining is necessary to support hydrodynamic sealing and to reduce gas leakage losses.

Performance

Rotary compressors have a high volumetric efficiency because of the small clearance volume and correspondingly low reexpansion losses inherent in its design. Figure 5 and Table 3 show performance of a typical rolling piston rotary compressor, commercially available for air-conditioning and heat pump applications.

An acceptable sound level is important in the design of any small compressor. Figure 6 illustrates a convenient method of analyzing and evaluating sound output of a rotary compressor designed for the home refrigerator. Since gas flow is continuous and no suction valve is required, rotary compressors can be relatively quiet. The sound level for a rotary compressor of the same design is directly related to its power input.

Features

Shafts. Shaft deflection, under load, is caused by compression gas loading of the roller and the torsional and side pull loading of the motor rotor. Design criteria must allow minimum oil film under the maximum run and starting loads. The motor rotor

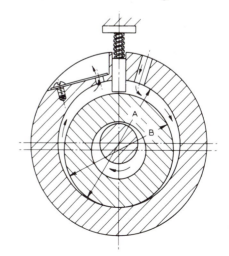

Fig. 4   Fixed-Vane, Rolling-Piston Rotary Compressor

should have minimal deflection to eliminate starting problems under extreme conditions of torque.

The shafts are generally made from steel forging and nodular cast iron. Depending on the materials chosen, a relative hardness should be maintained between the bearing and the journal. Journals are ground round to high precision and polished to a finish of 10 AA or better.

Journals and bearings. The bearing must support the rotating member under all conditions. Powdered metal has been extensively used for these components, due to its porous properties, which help in lubrication. This material can also be formed into complex bearing shapes with little machining required.

Vanes. Vanes are designed for reliability by the choice of materials and lubrication. The vanes are hardened, ground, and polished to the best finish obtainable. Steel, powdered metal, and aluminum alloys have been used. Powder metal vanes have a particular hardness typically in the 60 to 80 Rockwell C (R_c) range, and apparent hardness in the 35 to 55 R_c range.

To obtain good sealing and proper lubrication, the shape of the vane tip must conform to the surface of generation along the roller wall in accordance with hydrodynamic theory.

Table 3 Typical Rolling Piston Compressor Performance

Compressor speed	3450 rpm
Refrigerant	HCFC-22
Condensing temperature	130 °F
Liquid temperature	115 °F
Evaporator temperature	45 °F
Suction pressure	90.7 psia
Suction gas temperature	65 °F
Evaporator capacity	12,000 Btu/h
Energy efficiency ratio	11.0 Btu/W·h
Coefficient of performance	3.22
Input power	1090 W

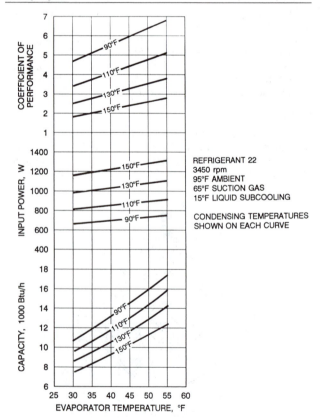

Fig. 5 Performance Curves for Typical Rolling-Piston Compressor

Vane springs. Vane springs force the vane to stay in contact with the roller during start-up. The spring rate depends on the inertia of the vane.

Valves. Only discharge valves are required by rolling piston rotary compressors. They are usually simple reed valves made of high-grade steel.

Lubrication. A good lubrication system circulates an ample supply of clean oil to all working surfaces, bearings, blades, blade slots, and seal faces. High-side pressure in the housing shell ensures a sufficient pressure differential across the bearings; passageways distribute oil to the bearing surfaces.

Mechanical efficiency. High mechanical efficiency depends on minimizing friction losses. Friction losses occur in the bearings and between the vane and slot wall, vane tip, and roller wall, and roller and bearing faces. The amount and distribution of these losses vary based on the geometry of the compressor.

Motor selection. Breakdown torque requirements depend on the displacement of the compressor, the refrigerant, and the operating range. Domestic refrigerator compressors, using R-12 refrigerant, require a breakdown torque of about 36 to 38 oz·ft per cubic inch of compressor displacement per revolution. Similarly, larger compressors using R-22 for window air conditioners use about 67 to 69 oz·ft per cubic inch of breakdown torque.

Rotary machines do not usually require complete unloading for successful starting. The starting torque of standard split-phase motors is ample for small compressors. Permanent split capacitor motors for air conditioners of various sizes provide sufficient starting and improve the power factor to the required range.

ROTARY VANE COMPRESSORS

Rotary vane compressors have a low weight-to-displacement ratio, which, in combination with compact size, makes them suitable for transport application. Small compressors in the 3 to 50 hp range are single staged, for a saturated suction temperature range of −40 to 45 °F at saturated condensing temperatures of up to 140 °F. By employing a second stage, low-temperature applications down to −60 °F are possible. Currently, R-22, R-502, and R-717 refrigerants are used.

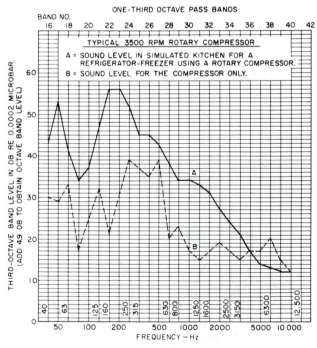

Fig. 6 Sound Level of Combination Refrigerator-Freezer with Typical Rotary Compressor

Figure 7 is a cross-sectional view of an eight-bladed compressor. The eight discrete volumes are referred to as cells. A single shaft rotation produces eight distinct compression strokes. While conventional valves are not required for this compressor, suction and discharge check valves are recommended to prevent reverse rotation and oil logging during shutdown.

Design of the compressor results in a fixed built-in compression ratio. This ratio depends on the compressor volume ratio and the refrigerant. Compressor volume ratio is determined by the relationship between the volume of the cell as it is closed off to the suction port to its volume before it opens to the discharge port. Compression ratio can be calculated from the following relationship:

$$CR = (V_i)k \qquad (6)$$

where:

CR = compression ratio
V_i = compression volume ratio
k = specific heat constant for refrigerant

The compressors currently available are of an oil-flooded, open drive design, which require use of an oil separator. Single-stage separators are used in close-coupled, high-temperature, direct expansion systems, where oil return is not a problem. Two-stage separators with a coalescing-type second stage are used in low-temperature systems, ammonia systems, and in flooded evaporators likely to trap oil.

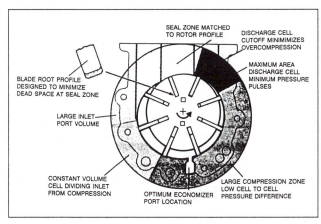

Fig. 7   Rotary Vane Compressor

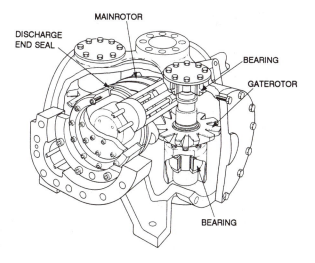

Fig. 8   Section of Single-Screw Refrigeration Compressor

SINGLE-SCREW COMPRESSORS

Screw compressors belong to the broad class of rotary positive-displacement compressors. Screw compressors currently in production for refrigeration and air-conditioning applications comprise two distinct types—single screw and twin screw. Both are conventionally used in the fluid injection mode where sufficient fluid cools and seals the compressor. Single-screw compressors have the capability to operate at pressure ratios above 20:1 single stage. The capacity range currently available is from 20 to 1300 tons.

Description

The single-screw compressor consists of a single cylindrical main rotor that works with a pair of gaterotors. Both the main rotor and gaterotors can vary widely in terms of form and mutual geometry. Figure 8 shows the design normally encountered in refrigeration.

The main rotor has six helical grooves, with a cylindrical periphery and a globoid (or hourglass shape) root profile. The two identical gaterotors each have 11 teeth and are located on opposite sides of the main rotor. The casing enclosing the main rotor has two slots, which allow the teeth of the gaterotors to pass through them. Two diametrically opposed discharge ports use a common discharge manifold located in the casing.

The compressor is driven through the main rotor shaft, and the gaterotors follow by direct meshing action at 6:11 ratio of the main rotor speed. The geometry of the single-screw compressor is such that 100% of the gas compression power is transferred directly from the main rotor to the gas. No power (other than small frictional losses) is transferred across the meshing points to the gaterotors.

Compression Process

The operation of the single-screw compressor can be divided into three distinct phases: suction, compression, and discharge. With reference to Figure 9, the process is as follows:

Suction. During rotation of the main rotor, a typical groove in open communication with the suction chamber gradually fills with suction gas. The tooth of the gaterotor in mesh with the groove acts as an aspirating piston.

Compression. As the main rotor turns, the groove engages a tooth on the gaterotor and is covered simultaneously by the cylindrical main rotor casing. The gas is trapped in the space formed by the three sides of the groove, the casing, and the gaterotor tooth. As rotation continues, the groove volume decreases and compression occurs.

Discharge. At the geometrically fixed point where the leading edge of the groove and the edge of the discharge port coincide, compression ceases, and the gas discharges into the delivery line until the groove volume has been reduced to zero.

Mechanical Features

Rotors. The screw rotor is normally made of cast iron, and the mating gaterotors are made from an engineered plastic. The inherent lubricating quality of the plastic, as well as its compliant nature, allow the single-screw compressor to achieve close clearances with conventional manufacturing practices.

The gaterotors are mounted on a metal support designed to carry the differential pressure between discharge pressure and suction pressure. The gaterotor function is equivalent to a piston in that it sweeps the groove and causes compression to occur. Furthermore, the gaterotor is in direct contact with the screw groove flanks and thus also acts as a seal. Each gaterotor is attached to its support by a simple spring and dashpot mechanism allowing the gaterotor, with a low moment of inertia, to have an angular degree of freedom from the larger mass of the support. This method of

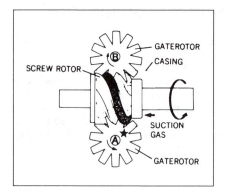
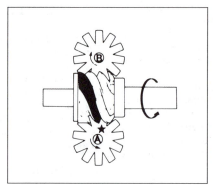
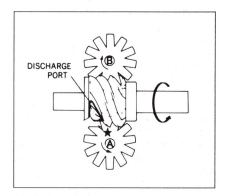

Suction. During rotation of the main rotor, a typical groove in open communication with the suction chamber gradually fills with suction gas. The tooth of the gaterotor in mesh with the groove acts as an aspirating piston.

Compression. As the main rotor turns, the groove engages a tooth on the gaterotor and is covered simultaneously by the cylindrical main rotor casing. The gas is trapped in the space formed by the three sides of the groove, the casing, and the gaterotor tooth. As rotation continues, the groove volume decreases and compression occurs.

Discharge. At the geometrically fixed point where the leading edge of the groove and the edge of the discharge port coincide, compression ceases, and the gas discharges into the delivery line until the groove volume has been reduced to zero.

Fig. 9   Sequence of Compression Process in Single-Screw Compressor

attachment allows the gaterotor assemblies to be true idlers by allowing them to dampen out transients without damage or wear.

Bearings. In a typical open or semihermetic single-screw compressor, the main rotor shaft contains one pair of angular contact ball bearings (an additional angular contact or roller bearing is used for some heat pump semihermetics). On the opposite side of the screw, one roller is used.

Note that the compression process takes place simultaneously on each side of the main rotor of the single-screw compressor. This balanced gas pressure results in zero loads on the rotor bearing during full load and while symmetrically unloaded as shown in Figure 10. Should the compressor be unloaded asymmetrically (see economizer operation below 50% capacity), the designer is not restricted by the rotor geometry and can easily add bearings with a long design life to handle the load. Axial loads are also low because the grooves terminate on the outer cylindrical surface of the rotor and suction pressure is vented to both ends of the rotor (Figure 10).

The gaterotor bearing must overcome a small moment force due to the gas acting on the compression surface of the gaterotor. Each gaterotor shaft has at least one bearing for axial positioning (usually a single angular contact ball bearing can perform the axial positioning and carry the small radial load at one end), and one roller or needle bearing at the other end of the support shaft also carries the radial load. Since the single-screw compressor's physical geometry places no constraints on bearing size B10, lives of 200,000 h are typical.

Cooling, sealing, and bearing lubrication. A major function of injecting a fluid into the compression area is the removal of the heat of compression. Also, since the single-screw compressor (like all screw compressors) has fixed leakage areas, the fluid helps seal potential leakage paths. Fluid is normally injected into a closed groove through ports in the casing or in the moving capacity control slide. Most single-screw compressors allow the use of many different injection fluids to suit the nature of the gas being compressed.

Oil-injected compressors. Oil is used in single-screw compressors to seal, cool, lubricate, and actuate capacity control. It gives a flat efficiency curve over a wide compression ratio and speed range, thus decreasing discharge temperatures and reducing noise levels. In addition, the compressor can handle some liquid floodback because it tolerates oil.

Oil-injected single-screw compressors operate at high head pressures using common high-pressure refrigerants such as R-12, R-502, R-114, R-134a, and R-717. They also operate effectively at high pressure ratios because the injected oil cools the compression process. Currently, compressors with capacities in the 20 to 1500 hp range are manufactured.

Oil injection requires an oil separator to remove the oil from the high pressure refrigerant (see Figure 11). For those applications with exacting demands for low oil carryover, separation equipment is available to leave less than 5 ppm oil in the circulated refrigerant.

With most compressors, oil can be injected automatically without a pump because of the pressure difference between the oil reservoir (discharge pressure) and the reduced pressure in a flute or bearing assembly during compression. A continuously running oil pump is used in some compressors to generate oil pressure

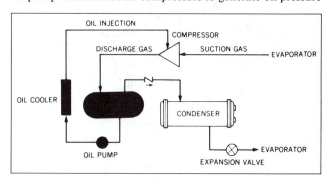

Fig. 10   Radial and Axially Balanced Main Rotor   **Fig. 11   Oil and Refrigerant Schematic of Oil Injection System**

30 to 45 psi over compressor discharge pressure. This pump requires 0.3 to 1.0% of the compressor's motor power.

Several methods of oil cooling include:

- Direct injection of liquid refrigerant into the compression process. Injection is controlled directly from the compressor discharge temperature, and loss of compressor capacity is minimized as injection takes place in a closed flute just before discharge occurs. This method requires very little power (typically less than 5% of compressor power).
- A small refrigerant pump draws liquid from the receiver and injects it directly into the compressor discharge line. The injection rate is controlled by sensing discharge temperature and modulating the pump motor speed. The power penalty in this method is the pump power (about 1 hp for compressors up to 1000 hp), which can result in energy savings over refrigerant injected into the compression chamber.
- External oil cooling between the oil reservoir and the point of injection is possible. Various heat exchangers are available to cool the oil: (1) separate water supply, (2) chiller water on a package unit, (3) condenser water on a package unit, (4) water from an evaporative condenser sump, (5) forced air-cooled oil cooler, and (6) high-pressure liquid recirculation (thermosiphon).

The heat added to the oil during compression is the amount usually removed in the oil cooler.

Oil injection-free compressors. While the single-screw compressor operates well with oil injection, it also operates with equal or better efficiencies in an oil injection-free (OIF) mode with many common halocarbon refrigerants. This means that the fluid injected into the compression chamber is the condensate of the fluid being compressed. For air conditioning and refrigeration, where pressure ratios are in the range of 2 to 8, the oil normally injected into the casing is replaced by liquid refrigerant. No lubrication is required because the only power transmitted from the screw to the gaterotors is that needed to overcome small frictional losses. Thus, the refrigerant need only cool and seal the compressor. The liquid refrigerant may still contain a small amount of oil to lubricate the bearings (0.1 to 1%, depending on compressor design). A typical OIF circuit is depicted in Figure 12.

Oil injection-free operation has the following advantages:

- Single-screw compressors require no discharge oil separators.
- Semihermetic compressors require no oil or or refrigerant pumps.
- External coolers are not required.
- The compressor accepts liquid refrigerant entering the suction, increasing the reliability and significantly reducing the size requirements of direct expansion evaporators.

Economiziers. Screw compressors are available with a secondary suction port between the primary compressor suction and discharge port. This port, when used with an economizer, improves compressor-useful refrigeration and compressor efficiency (Figure 13).

In operation, gas is drawn into the rotor grooves in the normal way from the suction line. The grooves are then sealed off in sequence, and compression begins. An additional charge is added to the closed flute through a suitably placed port in the casing by an intermediate gas source at a slightly higher pressure than that reached in the compression process at that time. The original and the additional charge are then compressed together to discharge conditions. The pumping capacity of the compressor at suction conditions is not affected by this additional flow through the economizer port.

When the port is used with an economizer, most of the flash gas generated by the expansion of the high-pressure liquid to low-pressure liquid enters a closed flute. Since this does not affect the suction capacity of the compressor, the effective refrigerating capacity of the economized compressor is increased over the noneconomized compressor by the amount of the flash gas (volume) plus the increased heat absorption capability H of the liquid entering the evaporator. Furthermore, the only additional mass flow the compressor must handle is the flash gas entering a closed flute, which is above suction pressure. Thus, under most conditions, the capacity improvement is accompanied by an efficiency improvement (Figure 13). Economizers become effective when the system pressure ratios are equal to 3.5 and above.

Figure 14 shows a pressure-enthalpy diagram for a flash tank economizer system. In it, high-pressure liquid passes through an expansion device and enters a tank at an intermediate pressure between suction and discharge. This pressure is maintained by the pressure in the compressor's closed flute (closed from suction).

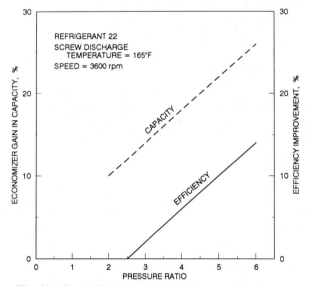

Fig. 13   Typical Improvement in Efficiency and Capacity with Economizer

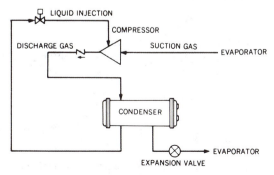

Fig. 12   Schematic of Oil Injection Free Circuit

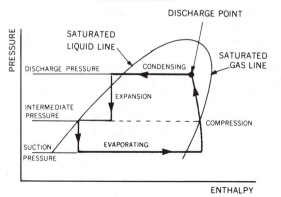

Fig. 14   Theoretical Economizer Cycle

The gas generated from the expansion enters the compressor through the economizer port. When passed to the evaporator, the liquid (which is now saturated at the intermediate pressure) gives a larger refrigeration capacity per pound. In addition, the compressor has more useful volume capacity at suction because the flash is eliminated at the suction, and the increase in power input is a lower percentage than the capacity increase (due to the flash gas entering the closed flute and being compressed to discharge).

As screw compressors are unloaded, the economizer pressure falls toward suction pressure. As a result, the additional capacity and improved efficiency of the economizer system falls to zero at 70 to 80% of full-load capacity.

The single-screw compressor has two compression chambers, each with its own slide valve. Each slide valve can be operated independently, which allows economizer gas to be introduced into one side of the compressor. By operating the slide independently, the chamber without the economizer gas can be unloaded to 0% capacity (50% capacity of the compressor). The other chamber remains at full capacity and retains the full economizer effect, making the economizer effective below 50% compressor capacity.

The secondary suction port may also be used for (1) a system side load or (2) a second system evaporator that operates at a temperature level above the system evaporator.

Centrifugal economizer. Some single-screw compressor designs incorporate a patented centrifugal economizer system. The centrifugal economizer replaces the force of gravity in a flash-type economizer system with centrifugal force to separate the flash gas generated at an intermediate pressure from the liquid refrigerant prior to liquid entering the evaporator. The centrifugal economizer system thereby uses a much smaller pressure vessel and, in some designs, the economizer fits within the envelope of a standard motor housing without having to increase its size.

The separation is achieved by a centrifugal impeller mounted on the compressor shaft (see Figure 22); a special valve maintains a uniformly thick liquid ring around the circumference of the impeller, assuring that no gas leaves with the liquid going to the evaporator. The flash gas is then ducted to a closed groove in the compression cycle. Some designs use the flash gas with a similar liquid refrigerant to cool the motor prior to introduction into the closed compression groove.

Volume ratio. The degree of compression within the rotor grooves is predetermined for a particular port configuration on screw compressors having fixed suction and discharge ports. A characteristic of the compressor is the volume ratio V_i, which is defined as the ratio of the volume of the groove at the start of compression to the volume of the same groove when it first begins to open to the discharge port. Hence, the volume ratio is determined by the size and shape of the discharge port.

For maximum efficiency, the pressure generated within the grooves during compression should exactly equal the pressure in the discharge line at the moment when the groove opens to it. If this is not the case, either over- or undercompression occurs, both resulting in internal losses. Although such losses cause no harm to the compressor, they increase power consumption and reduce efficiency.

Screw compressors equipped with economizers and properly selected volume ratios can reduce and, in some cases (such as air-conditioning and heat pump applications with pressure ratio ranges from 2 to 6), eliminate these power increases due to an increase in pressure ratio. This is because as the pressure ratio increases, the flash gas supercharging the closed groove also increases, thus reducing any undercompression.

Volume ratios can be specified on most compressors. Selection should be made according to operating conditions. The built-in pressure ratio of a screw compressor is a function of the volume ratio as follows:

$$p_i = V_i^{n} \qquad (7)$$

where n is the isentropic exponent for the refrigerant being used.

Compressors equipped with slide valves (for capacity modulation) usually locate the discharge port at the discharge end of the slide valve. Alternative port configurations yielding the required volume ratios are then designed into the capacity control components, thus providing ease of interchangeability, both during construction and after installation (although partial disassembly is required).

Recently, single-screw compressors in refrigeration and process applications have been equipped with the capability to vary the volume ratio of the compressor while the compressor is running. This is accomplished by designing a simple slide valve that advances or delays the discharge port opening. Note that a separate slide has been designed to modulate the capacity independently of the volume ratio slide (see Figures 18 and 19). Having the independent modulation of volume ratio (through discharge port control) and capacity modulation (through a completely independent slide that only varies the position where compression begins) allows the single-screw compressor to achieve efficient volume ratio control when capacity is less than full load.

Capacity control. As with all positive displacement compressors, both speed modulation and suction throttling can be used. Ideal capacity modulation for any compressor includes (1) continuous modulation from 100 to less than 10%, (2) good part-load efficiencies, (3) unloaded starting, and (4) unchanged reliability.

Variable compressor displacement, the most common means for meeting these criteria, usually takes the form of two movable slide valves in the compressor casing (the single-screw compressor has two gaterotors forming two compression areas). At part load, each slide valve produces a slot that delays the point at which compression begins. This causes a reduction in groove volume, and hence, in compressor throughput. As the suction volume is displaced before compression takes place, little or no thermodynamic losses occur. However, if no other steps were taken, this mechanism would result in an undesirable drop in the effective volume ratio in undercompression and inefficient part-load operation.

This problem is avoided either by arranging that the capacity modulation valve reduces the discharge port area at the same time as the bypass slot is created (Figure 15) or having one valve control capacity only and a second valve independently modulate volume ratio (Figures 16, 17, and 18). A full modulating mechanism is provided in most large single-screw compressors, while two-position slide valves are used where requirements allow. The

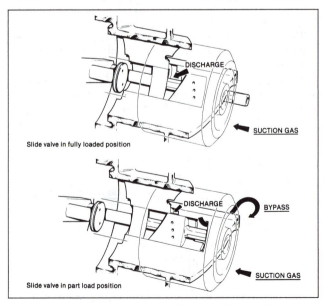

Slide valve in fully loaded position

DISCHARGE

SUCTION GAS

DISCHARGE   BYPASS

SUCTION GAS

Slide valve in part load position

Fig. 15   Section Depicting Capacity Control Slide Valve Operation

specific part-load performance will be affected by a compressor's built-in volume ratio V_i, evaporator temperature, and condenser temperature, and whether the slide valves are symmetrically or asymmetrically unloaded.

Detailed design of the valve mechanism differs between makes of compressors but usually consists of an axial sliding valve along each side of the rotor casing (Figure 15). This mechanism is usually operated by a hydraulic or gas piston and cylinder assembly located within the compressor itself or by a positioning motor. The piston is actuated either by oil, discharge gas, or high-pressure liquid refrigerant at system discharge pressure driven in either direction according to the operation of a four-way solenoid valve.

Figure 17 shows a capacity slide valve (top) and a variable volume ratio slide (bottom). The capacity slide is in the full-load position, and the volume ratio slide is at a moderate volume ratio. Figure 18 depicts the same system as shown in Figure 17, except that the capacity slide is in a partially loaded position, and the volume ratio slide has moved to a position to match the new system conditions.

The single-screw compressor's two compression chambers, each having its own capacity slide valve that can be operated independently, permits one slide valve to be unloaded to 0% capacity (50%

compressor capacity) while the other slide valve remains at full capacity. Operation in this manner (asymmetrical) realizes an improvement in part-load efficiency below the 50% capacity point and further part-load efficiency gains are realized when the economizer gas is only entered into a closed groove on the side that is unloaded second (see explanation in the section on economizers). Figure 19 demonstrates the effect of asymmetrical unloading of a single-screw compressor.

Performance. Figures 20 and 21 show typical efficiencies of all single-screw compressor designs. High isentropic and volumetric efficiencies are the result of internal compression, the absence of suction or discharge valves and their losses, and extremely small clearance volumes. The curves show the importance of selecting the correct volume ratio in fixed volume ratio compressors.

Manufacturer's data for operating conditions or speed should not be extrapolated. Screw compressor performance at reduced speed is usually significantly different from that specified at the normally rated point. Performance data normally include information about the degree of liquid subcooling and suction superheating assumed in the data.

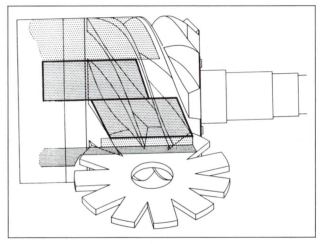

Fig. 18   Capacity Slide in Part-Load Position and Volume Ratio Slide Positioned to Maintain System Volume Ratio

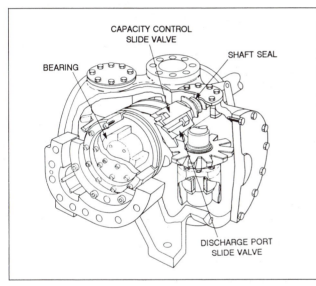

Fig. 16   Section Showing Refrigeration Compressor Equipped with Variable Capacity Slide Valve and Variable Volume Ratio Slide Valve

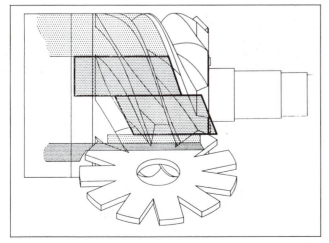

Fig. 17   Capacity Slide in Full-Load Position and Volume Ratio Slide in Intermediate Position

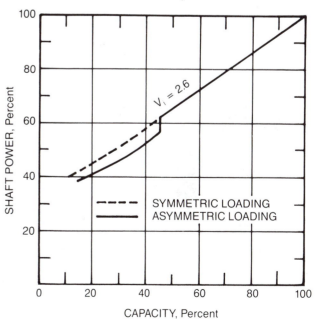

Fig. 19   Part-Load Effect of Symmetrical and Asymmetrical Capacity Control

Applications. Single-screw compressors have been widely applied as refrigeration compressors, using halocarbon refrigerants, ammonia, and hydrocarbon refrigerants. A single gaterotor semihermetic version is increasingly being applied in large supermarkets.

Oil injected and oil injection-free (OIF) semihermetic compressors are widely used for air-conditioning and heat pump service, with compressor sizes ranging from 40 to 500 tons.

Semihermetic design. Figure 22 shows a semihermetic single-screw compressor. Figure 23 exhibits a semihermetic single-screw compressor using only one gaterotor. This design has found application in large supermarket rack systems. The single gaterotor compressor exhibits high efficiency and has been designed for long bearing life, which compensates for the unbalanced load on the screw rotor shaft with increasing bearing size.

Noise and Vibration

The inherently low noise and vibration characteristics of single-screw compressors are due to small torque fluctuation and no valving required in the compression chamber. In particular, the advent of OIF technology eliminates the need for oil separators that have traditionally created noise.

TWIN SCREW COMPRESSORS

Twin screw is a common designation for double helical rotary screw compressors. A twin screw compressor consists of two mating helically grooved rotors—male (lobes) and female (flutes or gullies) in a stationary housing with inlet and outlet gas ports (Figure 24). The flow of gas in the rotors is mainly in an axial direction. Frequently used lobe combinations are 4+6, 5+6, and 5+7 (male + female). For instance, with a four-lobe male rotor, the driver rotates at 3600 rpm; the six-lobe female rotor follows at 2400 rpm. The female rotor can be driven through synchronized timing gears or directly driven by the male rotor on a light oil film. In some applications, it is practical to drive the female rotor, which results in a 50% speed and displacement increase over the male-driven compressor, assuming a 4+6 lobe combination. Geared speed increasers are also used on some applications to increase the capacity delivered by a particular compressor size.

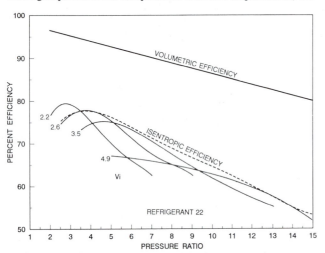

Fig. 20   Typical Compressor Performance on R-22

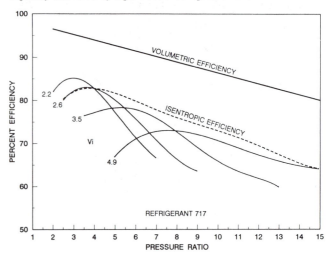

Fig. 21   Typical Compressor Performance on R-717

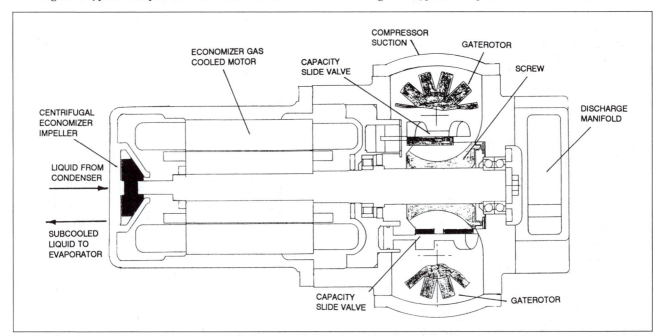

Fig. 22   Typical Semihermetic Single-Screw Compressor

Double helical screws find application in many air-conditioning, refrigeration, and heat pump applications, typically in the industrial and commercial market. Machines can be designed to operate at high or low pressure levels and are often applied below 2:1 and above 20:1 compression ratios single stage. Commercially available compressors are suitable for application on all normally used high-pressure refrigerants.

Compression Process

Compression is obtained by direct volume reduction with pure rotary motion. For clarity, the following description of the three basic compression phases is limited to one male rotor lobe and one female rotor interlobe space (Figure 25).

Suction. As the rotors begin to unmesh, a void is created on both the male side (male thread) and the female side (female thread), and gas is drawn in through the inlet port. As the rotors continue to turn, the interlobe space increases in size, and gas flows continuously into the compressor. Just prior to the point at which the interlobe space leaves the inlet port, the entire length of the interlobe space is completely filled with gas.

Compression. Further rotation starts the meshing of another male lobe with another female interlobe space on the suction end and progressively compresses the gas in the direction of the discharge port. Thus, the occupied volume of the trapped gas within the interlobe space is decreased and the gas pressure consequently increased.

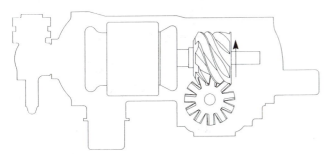

Fig. 23   Single Gaterotor Semihermetic Single-Screw Compressor

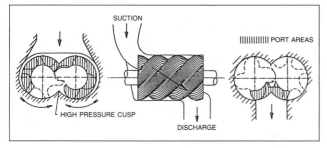

Fig. 24   Twin Screw Compressor

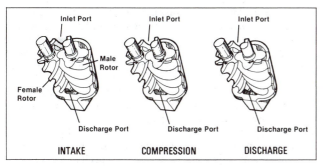

Fig. 25   Compression Process

Discharge. At a point determined by the designed built-in volume ratio, the discharge port is uncovered and the compressed gas is discharged by further meshing of the lobe and interlobe space.

During the remeshing period of compression and discharge, a fresh charge is drawn through the inlet on the opposite side of the meshing point. With four male lobes rotating at 3600 rpm, four interlobe volumes are filled and give 14,400 discharges per minute. Since the intake and discharge cycles overlap effectively, a smooth continuous flow of gas results.

Mechanical Features

Rotor profiles. Helical rotor design started with an asymmetric point-generated rotor profile. This profile was only used in compressors with timing gears (dry compression). The symmetric, circular rotor profile was introduced because it was easier to manufacture than the preceding profile, and it could be used without timing gears for wet or oil-flooded compression.

Current rotor profiles are normally asymmetric and line-generated profiles, giving higher performance due to better rotor dynamics and decreased leakage areas. This design allowed the possibility for female rotor drive, as well as the conventional male drive. Rotor profile, blowhole, length of sealing line, quality of sealing line, torque transmission between rotors, rotor-housing clearances, interlobe clearances, and lobe combinations are optimized for specific pressures, temperatures, speeds, and wet or dry operation. Optimal rotor tip speed is 3000 to 8000 ft/min for wet operation (oil-flooded) and 12,000 to 24,000 ft/min for dry operation.

Rotor contact and loading. Contact between the male and female rotors is mainly rolling, primarily at a contact band on each rotor's pitch circle. Rolling at this contact band means that virtually no rotor wear occurs.

Gas forces. On the driven rotor, the internal gas forces always create a torque in a direction opposite to the direction of rotation. This is known as positive or braking torque. On the undriven rotor, the design can be such that the torque is positive, negative, or zero, except on female drive designs, where zero or negative torque does not occur. Negative torque occurs when internal gas forces tend to drive the rotor. If the average torque on the undriven rotor is near zero, this rotor is subjected to torque reversals as it goes through its phase angles. Under certain conditions, this can cause instability problems. Torque transmitted between the rotors does not create problems because the rotors are mainly in rolling contact.

Male drive. The transmitted torque from male rotor to female rotor is normally 5 to 25% of input torque.

Female drive. The transmitted torque from female rotor to male rotor is normally 50 to 60% of input torque.

Rotor loads. The rotors in an operating compressor are subjected to radial, axial, and tilting loads. Tilting loads are radial loads caused by axial loads outside of the rotor center line. The axial load is normally balanced with a balancing piston for larger high-pressure machines (rotor diameter above 6 in. and discharge pressures above 160 psi). Balancing pistons are typically close tolerance, labyrinth-type devices with high-pressure oil or gas on one side and low pressure on the other. They are used to produce a thrust load to offset some of the primary gas loading on the rotors, thus reducing the amount of thrust load the bearings support.

Bearings. Twin-screw compressors normally have either four or six bearings, depending on whether one or two bearings are used for the radial and axial loads. Some designs incorporate multiple rows of smaller bearings per shaft to share the loads. Sleeve bearings have been used historically to support radial loads in machines with male rotor diameters larger than 6 in., while antifriction bearings are generally applied to smaller machines. However, improvements in antifriction designs and materials have led to compressors with up to 14 in. rotor diameter with full antifriction bearing designs. Cylindrical and tapered roller bearings and various types of ball bearings are used in screw compressors for carrying radial loads. The most common thrust or axial load-carrying bearings are angular contact ball bearings, although tapered rollers or tilting pad bearings are used in some machines.

General design. Screw compressors are often designed for particular pressure ranges. Low-pressure compressors have long, high displacement rotors and adequate space to accommodate bearings to handle the relatively light loads. They are frequently designed without thrust balance pistons, since the bearings alone can handle the low thrust loads and still maintain good life.

High-pressure compressors have short and strong rotors (shallow grooves) and, therefore, have space for large bearings. They are normally designed with balancing pistons for high thrust bearing life.

Rotor materials. Rotors are normally made of steel, but aluminum, cast iron, and nodular iron are used in some applications.

Capacity Control

As with all positive displacement compressors, both speed modulation and suction throttling can reduce the volume of gas drawn into a screw compressor. Ideal capacity modulation for any compressor would be (1) continuous modulation from 100% to less than 10%, (2) good part-load efficiency, (3) unloaded starting, and (4) unchanged reliability. However, not all applications need ideal capacity modulation. Variable compressor displacement and variable speed are the best means for meeting these criteria. Variable compressor displacement is the most common capacity control method used. Various mechanisms achieve variable displacement, depending on the requirements of a particular application.

Capacity slide valve. A slide valve for capacity control is a valve with sliding action parallel to the rotor bores. They are placed within or close to the high-pressure cusp region, face one or both rotor bores, and bypass a variable portion of the trapped gas charge back to suction, depending on their position. Within this definition, there are two types of capacity slide valves.

1. *Capacity slide valve regulating discharge port.* This type of slide valve is located within the high-pressure cusp region. It controls capacity as well as the location of the radial discharge port at part load. The axial discharge port is designed for a volume ratio giving good part-load performance without losing full-load performance. Figure 26 shows a schematic view of the most common arrangement.
2. *Capacity slide valve not regulating discharge port.* A slide valve outside the high-pressure cusp region controls only capacity and not the radial discharge port.

The first type is the most common arrangement. It is generally the most efficient of the available capacity reduction methods, due to its indirect correction of built-in volume ratio at part load and its ability to give large volume reductions without large movement of the slide valve.

Capacity slot valve. A capacity slot valve consists of a number of slots that follow the rotor helix and face one or both rotor bores. The slots are gradually opened or closed with a plunger or turn valve (Figure 26). These recesses in the casing wall increase the volume of the compression space and also create leakage paths over the lobe tips. The result is somewhat lower full-load performance when compared to a design without slots.

Capacity lift valve. Capacity lift valves or plug valves are movable plugs in one or both rotor bores (with radial or axial lifting action) that regulate the actual start of compression. These valves control capacity in a finite number of steps, rather than by the infinite control of a conventional slide valve (Figure 27).

Neither slot valves nor lift valves offer quite as good efficiency at part load as a slide valve, because they do not relocate the radial discharge port. Thus, undercompression losses at part load can

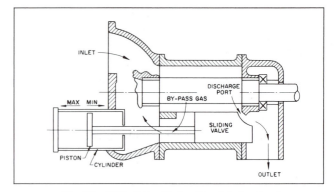

Fig. 26   Slide Valve Unloading Mechanism

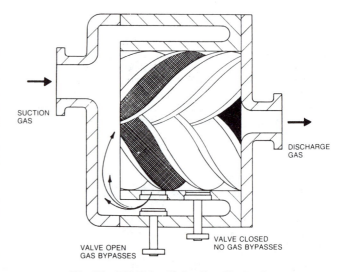

Fig. 27   Lift-Valve Unloading Mechanism

be expected if the machines have the correct volume ratio for full-load operation and the compression ratio at part load does not reduce.

Volume Ratio

In all positive displacement rotary compressors with fixed port location, the degree of compression within the rotor thread is determined by the location of the suction and discharge ports. The built-in volume ratio V_i of screw compressors is defined as the ratio of volume of the thread at the start of the compression process to the volume of the same thread when it first begins to open to the discharge port. The suction port must be located to trap the maximum suction charge; hence, the volume ratio is determined by the location of the discharge port.

Only the suction pressure and volume ratio of the compressor determine the internal pressure achieved before opening to discharge. However, the condensing and evaporating temperatures determine the system discharge pressure and the compression ratio in the piping that leads to the compressor. Any mismatch between the internal and system discharge pressures results in under- or overcompression losses and in lower efficiency.

If the operating conditions of the system seldom change, it is possible to specify a fixed volume ratio compressor that will give good efficiency. Compressor manufacturers normally make compressors with three or four possible discharge port locations that correspond to system conditions encountered frequently. Generally, the system designer is responsible for specifying a compressor that most closely matches expected pressure conditions.

The required volume ratio for a particular application can be determined as follows:

First, determine the compression ratio of a given refrigerating system:

$$CR = p_d/p_s \qquad (8)$$

where:

CR = system compression ratio
p_s = expected suction pressure in absolute units
p_d = expected discharge pressure in absolute units

Then, determine the internal pressure ratio of the available compressors by approximating compression as an isentropic process as follows:

$$p_i = V_i{}^k \qquad (9)$$

where:

p_i = internal pressure ratio
V_i = compressor volume ratio
k = ratio of specific heats for refrigerant used

And, finally, the compressor should be selected, as nearly as possible, to match the internal pressure ratio of the compressor to the system compression ratio:

$$p_i = CR \qquad (10)$$

Usually, in slide valve-equipped compressors, the radial discharge port is located in the discharge end of the slide valve. A short slide valve gives a low volume ratio, and a long slide valve gives a higher volume ratio. The difference in length basically locates the discharge port earlier or later in the compression process. Different length slide valves allow changing the volume ratio of a given compressor, although disassembly is required.

Variable volume ratio. While operating, some twin screw compressors adjust the volume ratio of the compressor to the most efficient ratio for whatever system pressures are encountered.

In fixed volume ratio compressors, the motion of the slide valve toward the inlet end of the machine is stopped when it comes in contact with the rotor housing in that area. In the most common

of the variable volume ratio machines, this portion of the rotor housing has been replaced with a second slide, the movable slide stop, which can be actuated to different locations in the slide valve bore (Figure 28).

By moving the slides back and forth, the radial discharge port can be relocated during operation to match the compressor volume ratio to the optimum. This added flexibility allows operation at different suction and discharge pressure levels while still maintaining maximum efficiency. The comparative efficiencies of fixed and variable volume ratio screw compressors are shown in Figure 29 for full-load operation on ammonia and R-22 refrigerants. The figure shows that a variable volume ratio compressor efficiency curve encompasses the peak efficiencies of compressors with fixed volume ratios over a wide range of system pressure ratios. Following are other secondary effects of a variable volume ratio:

- Less oil foam in oil separator (no overcompression)
- Less oil carried over into the refrigeration system (because of less oil foam in oil separator)
- Extended bearing life; minimized load on bearings
- Extended efficient operating range with economizer discharge port corrected for flash gas from economizer, as well as gas from suction
- Lower noise levels
- Lower discharge temperatures and oil cooler heat rejection

The greater the change in either suction or condensing pressure a given system experiences, the more benefits are possible with a variable volume ratio. Efficiency improvements as high as 30% are possible, depending on the application, refrigerant, and system operating range.

Oil Injection

Two primary types of compressor lubrication systems are employed in twin screw compressors—dry and oil flooded.

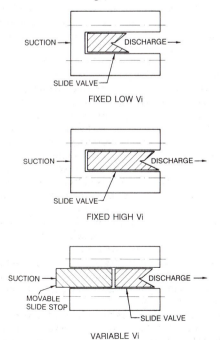

Fig. 28   View of Fixed and Variable-Volume Ratio (V_i) Slide Valves from Above

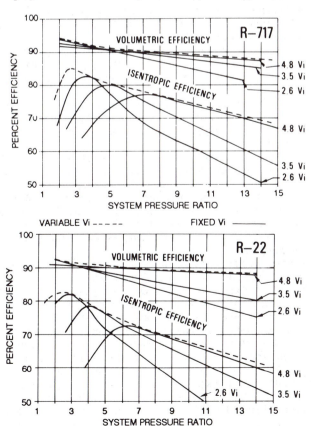

Fig. 29   Twin Screw Compressor Efficiency Curves

Dry operation (no rotor contact). Since the two rotors in twin screw compressors are parallel, timing gears are a practical means of synchronizing the rotors so that they do not touch each other. Eliminating rotor contact eliminates the need for lubrication in the compression area. Initial screw compressor designs were based on this approach, and dry screws still find application in the gas process industry.

Synchronized twin screw compressors once required high rotor tip speed and were, therefore, noisy. However, with current profile technology, the synchronized compressor can run at lower tip speeds and higher pressure ratios, giving quieter operation. The added cost of timing gears and internal seals generally make the dry screw more expensive than an oil-flooded screw for normal refrigeration or air conditioning.

Oil-flooded operation. The oil-flooded twin screw compressor is the most common type of screw used in refrigeration and air conditioning. Compressor capacities range from 6 to 6000 cfm. Oil-flooded compressors typically have oil supplied to the compression area at a volume rate of about 0.5% of the displacement volume. Part of this oil is used for lubrication of the bearings and shaft seal prior to injection. Typically, paraffinic or napthenic-based mineral oils are used, although synthetics are used on some applications. The oil is normally injected into a closed thread through ports in the moving slide valve or through stationary ports in the casing.

The oil fulfills three primary purposes—sealing, cooling, and lubrication. It also tends to fill any leakage paths between and around the rotors. This keeps volumetric efficiency (VE) high, even at high compression ratios. Normal compressor VE exceeds 85% even at 25:1 single stage (ammonia, 7.6 in. rotor diameter). It also gives flat efficiency curves with decreasing speeds, where desired, for quiet operation. Oil transfers much of the heat of compression from the gas to the oil, keeping the typical discharge temperature below 190°F, which allows high compression ratios without the danger of breaking down the refrigerant or the oil. The lubrication function of the oil protects bearings, seals, and the rotor contact areas.

The ability of a screw compressor to tolerate oil also permits the compressor to handle a certain amount of liquid floodback, as long as the liquid quantity is not large enough to lock the rotors hydraulically.

Oil separation and cooling. Oil injection requires an oil separator to remove oil from the high-pressure refrigerant. Separation equipment routinely gives less than 5 ppm oil in the circulated refrigerant.

Oil injection is normally achieved by one of two methods: (1) with a continuously running oil pump capable of generating an oil pressure of 30 to 45 psi over compressor discharge pressure, representing 0.3 to 1.0% of compressor motor power; or (2) with some compressors, oil can be injected automatically, without a pump because of the pressure difference between the oil reservoir (discharge pressure) and the reduced pressure in a thread during the compression process.

Since the oil absorbs a significant amount of the heat of compression in an oil-flooded operation, it must be cooled to maintain low discharge temperatures. One cooling method is by direct injection of liquid refrigerant into the compression process. The amount of injected liquid refrigerant corresponds to 0.02% of displacement volume. The amount of liquid injected is normally controlled by sensing the discharge temperature and injecting enough liquid to maintain a constant temperature. Some of the injected liquid mixes with the oil and leaks to lower pressure threads, where it tends to raise pressures and reduce the amount of gas the compressor can draw in. Also, any of the liquid that has time to absorb heat and expand to vapor must be recompressed, which tends to raise absorbed power levels. Compressors are designed with the liquid injection ports as late as possible in the compression to minimize capacity and power penalties. Typical penalties for liquid injection are in the 1 to 10% range, depending on the compression ratio.

Another method of oil cooling draws liquid from the receiver with a small refrigerant pump and injects it directly into the compressor discharge line. The power penalty in this method is the pump power (about 1 hp for compressors up to 1000 hp).

In the third method, the oil can be cooled outside the compressor between the oil reservoir and the point of injection. Various configurations of heat exchangers are available for this purpose, and the oil cooler heat rejection can be accomplished by (1) separate water supply, (2) chiller water on a package unit, (3) condenser water on a package unit, (4) water from an evaporative condenser sump, (5) forced air-cooled oil cooler, (6) liquid refrigerant, and (7) high-pressure liquid recirculation (thermosyphon).

External oil coolers using water or other means from a source independent of the condenser allow for the condenser to be reduced in size by an amount corresponding to the oil cooler capacity. Where oil cooling is carried out from within the refrigerant system by means such as (1) direct injection of liquid refrigerant into the compression process or the discharge line, (2) direct expansion of liquid in an external heat exchanger, (3) using chiller water on a package unit, (4) recirculation of high-pressure liquid from the condenser, or (5) water from an evaporative condenser sump, the condenser must be sized for the total heat rejection, *i.e.*, evaporator load plus shaft power.

With an external oil cooler, the mass flow rate of oil injected into the compressor is usually determined by the desired discharge temperature rather than by the compressor sealing requirements, since the oil acts predominantly as a heat transfer medium. Conversely, with direct liquid injection cooling, the oil requirement is dictated by the compressor lubrication and sealing needs.

Economizers

Twin screw compressors are available with a secondary suction port between the primary compressor suction and discharge ports. This port can accept a second suction load at a pressure above the primary evaporator, or flash gas from a liquid subcooler vessel, known as an economizer.

In operation, gas is drawn into the rotor thread from the suction line. The thread is then sealed in sequence and compression begins. An additional charge may be added to the closed thread through a suitably placed port in the casing. The port is connected to an intermediate gas source at a pressure slightly higher than that reached in the compression process at that time. Both original and additional charges are then compressed to discharge conditions.

When the port is used as an economizer, a portion of the high-pressure liquid is vaporized at the side port pressure and subcools the remaining high-pressure liquid nearly to the saturation temperature at the operating side port pressure. Since this has little effect on the suction capacity of the compressor, the effective refrigerating capacity of the compressor is increased by the amount of the flash gas (volume) plus the increased heat absorption capacity *H* of the liquid entering the evaporator. Furthermore, the only additional mass flow the compressor must handle is the flash gas entering a closed thread, which is above suction pressure. Thus, under most conditions, the capacity improvement is accompanied by an efficiency improvement.

Economizers become effective when system pressure ratios are equal to about 2 and above (depending on volume ratio). The subcooling can be made with a heat exchanger, flash tank, or other separation means.

As twin screw compressors are unloaded, one problem may occur. As the economizer pressure falls toward suction pressure, the additional capacity and improved efficiency of the economizer system is no longer available below a certain percentage of capacity, depending on design.

Hermetic Compressors

Hermetic screw compressors are commercially available through 400 tons of refrigeration effect using R-22. The hermetic motors can operate under discharge, suction, or intermediate pressure. Motor cooling can be with gas, oil, and/or liquid refrigerant. Oil separation for these types of compressors may be accomplished either with an integrated oil separator or with a separately mounted oil separator in the system. Figures 30 and 31 show two types of hermetic twin screw compressors.

Performance Characteristics

Figure 29 gives full-load efficiencies of a modern twin screw compressor. Both fixed and variable volume ratio compressors

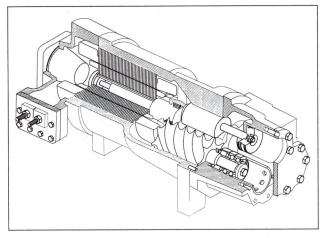

**Fig. 30   Semihermetic Twin Screw with Suction
Gas-Cooled Motor**

without economizers are shown. High isentropic and volumetric efficiencies are the result of internal compression, the absence of suction or discharge valves, and small clearance volumes. The curves show that while volumetric efficiency depends little on the choice of volume ratio, isentropic efficiency depends strongly on it.

Performance data usually note the degree of liquid subcooling and suction superheating assumed. If an economizer is used, the liquid temperature approach and pressure drop to the economizer should be specified.

ORBITAL COMPRESSORS

SCROLL COMPRESSORS

Description

Scroll compressors are rotary motion, positive displacement machines that compress with two interfitting, spiral-shaped scroll members (Figure 32). They are currently used in residential and commercial air-conditioning and heat pump applications as well as in automotive air-conditioning systems. Capacities range from 10,000 to 170,000 Btu/h. To function effectively, the scroll compressor requires close tolerance machining of the scroll members, which is possible due to the recent advances in manufacturing technology. This positive displacement, rotary motion compressor includes performance features, such as high efficiency and low noise.

Scroll members are typically a geometrically matched pair, assembled 180° out of phase. Each scroll member is open on one end of the vane and bound by a base plate on the other. The two scrolls are fitted to form pockets between their respective base plate and various lines of contact between their vane walls. One scroll is held fixed, while the other moves in an orbital path with respect

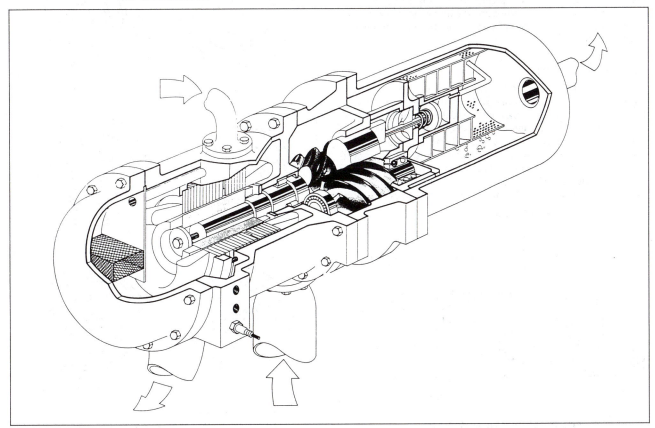

Fig. 31   Semihermetic Twin Screw with Motor Housing Used as Economizer; Built-in Oil Separator

to the first. The flanks of the scrolls remain in contact, although the contact locations move progressively inward. Relative rotation between the pair is prevented by an interconnecting coupling. An alternate approach creates relative orbital motion via two scrolls synchronously rotating about noncoincident axes. As in the former case, an interconnecting coupling maintains a relative angle between the pair of scrolls (Morishita *et al.* 1988).

Compression is accomplished by sealing suction gas in pockets of a given volume at the outer periphery of the scrolls and progressively reducing the size of those pockets as the scroll relative motion moves them inwards towards the discharge port. Figure 33 shows the sequence of suction, compression, and discharge phases. As the outermost pockets are sealed off (Figure 33a), the trapped gas is at suction pressure and has just entered the compression process. At stages (b) through (f), orbiting motion moves the gas toward the center of the scroll pair, and pressure rises as pocket volumes are reduced. At stage (g), the gas reaches the central discharge port and begins to exit the scrolls. Stages (a) through (h) in Figure 33 show that two distinct compression paths operate simultaneously in a scroll set. The discharge process is nearly continuous, since new pockets reach the discharge stage very shortly after the previous discharge pockets have been evacuated.

Scroll compression embodies a fixed, built-in volume ratio that is defined by the geometry of the scrolls and by discharge port location. This feature provides the scroll compressor with different performance characteristics than those of reciprocating or conventional rotary compressors.

Both high-side and low-side shells are available. In the former, the entire compressor is at discharge pressure, except for the outer areas of the scroll set. Suction gas is introduced into the suction port of the scrolls through piping, which keeps it discrete from the rest of the compressor. Discharge gas is directed into the compressor shell, which acts as a plenum. In the low-side type, most of the shell is at suction pressure, and the discharge gas exiting the scrolls is routed outside the shell, sometimes through a discrete or integral plenum.

Mechanical Features

Scroll members. Gas sealing is critical to the performance advantage of scroll compressors. Sealing within the scroll set must be accomplished at the flank contact locations and between the vane tips and bases of the intermeshed scroll pair. Tip/base sealing is generally considered more critical than flank sealing. The method used to seal the scroll members tends to separate scroll compressors into compliant and noncompliant designs.

Noncompliant designs. In designs lacking compliance, the orbiting scroll takes a fixed orbital path. In the radial direction,

sealing small irregularities between the vane flanks (due to flank machining variation) can be accomplished with oil flooding. In the axial direction, the position of both scrolls remains fixed, and flexible seals fitted into machined grooves on the tips of both scrolls accomplish tip sealing. The seals are pressure loaded to enhance uniform contact (McCullough *et al.* 1976, Sauls 1983).

Radial compliance. This feature enhances flank sealing and allows the orbiting scroll to follow a flexible path defined by its own contact with the fixed scroll. In one type of radial compliance, a sliding "unloader" bushing is fitted onto the crankshaft eccentric pin in such a way that it directs the radial motion of the orbiting scroll. The orbiting scroll is mounted over this bushing through a drive bearing, and the scroll may now move radially in and out to accommodate variations in orbit radius caused by machining and assembly discrepancies. This feature tends to keep the flanks constantly in contact, and reduces impact on the flanks that can result from intermittent contact. Sufficient clearance in the

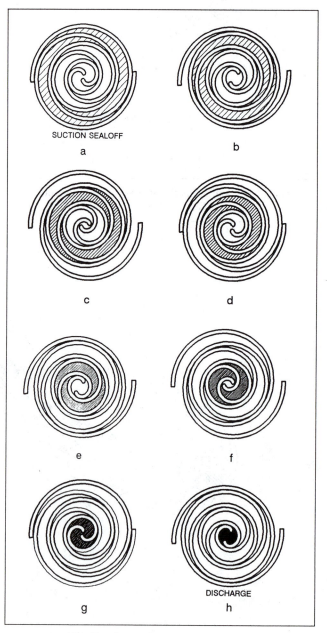

Fig. 33 Scroll Compression Process
(Purvis 1987)

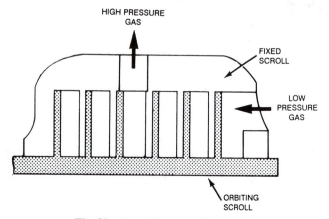

Fig. 32 Interfitted Scroll Members
(Purvis 1987)

pin/unloader assembly allows the scroll flanks to separate fully when desired.

In some designs, the mass of the orbiting scroll is selected so that centrifugal force overcomes radial gas compression forces that would otherwise keep the flanks separated.

In some other designs, the drive is designed so that the influence of centrifugal force is reduced, and the drive force overcomes the radial gas compression force (McCullough 1975). Radial compliance has the added benefit of increasing resistance to slugging and contaminants, since the orbiting scroll can "unload" to some extent as it encounters obstacles or nonuniform hydraulic pressures (Bush *et al.* 1988).

Axial compliance. With this feature, an adjustable axial pressure force maintains sealing contact between the scroll tips and bases while running. This force is released when the unit is shut down, allowing the compressor to start unloaded and to approach full operational speed before a significant load is encountered. This scheme obviates the use of tip seals, eliminating them as a potential source of wear and leakage. With the scroll tips bearing directly on the opposite base plates and with suitable lubrication, sealing tends to improve over time. Axial compliance can either be implemented on the orbiting scroll or the fixed scroll (Tojo *et al.* 1982, Caillat *et al.* 1988). The use of axial compliance requires auxiliary sealing of the discharge side with respect to the suction side of the compressor.

Antirotation coupling. To ensure relative orbital motion, the orbiting scroll must not rotate in response to gas loading. This rotation is most commonly accomplished by an Oldham coupling mechanism, which physically connects the scrolls and permits all planar motion, except relative rotation, between them.

Bearing system. The bearing system consists of a drive bearing mounted in the orbiting scroll and generally one of two main bearing systems. The main bearings are either of the cantilevered type (main bearings on same side of the motor as the scrolls) or consists of a main bearing on either side of the motor (Figure 34). All bearing load vectors rotate through a full 360° due to the nature of the drive load.

The orbiting scroll is supported axially by a thrust bearing, on a housing which is part of the internal frame or is mounted directly to the compressor shell.

Capacity Control

Two different capacity control mechanisms are currently being used by the scroll compressor industry:

Variable speed scroll compressor. The conventional air-conditioning system uses constant speed motor to drive the compressor. The variable speed scroll compressor uses an inverter drive to convert a fixed frequency alternating current into one with adjustable voltage and frequency, which allows the variation of the rotating speed of the compressor motor. The compressor

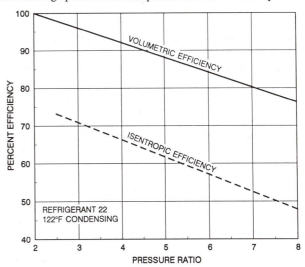

**Fig. 35   Volumetric and Isentropic Efficiency versus
Pressure Ratio for Scroll Compressors**
(Elson *et al.* 1990)

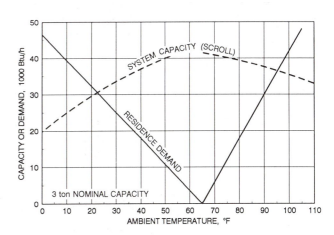

**Fig. 36   Scroll Capacity versus Residence Demand
(3-ton Nominal Capacity)**
(Purvis 1987)

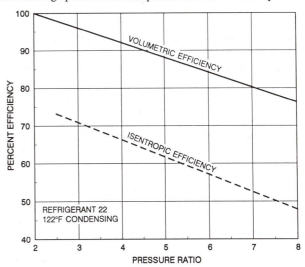

Fig. 34   Bearings and Other Components of Scroll Compressor
(Elson *et al.* 1990)

uses either an induction or a permanent magnet motor. Typical operating frequencies vary between 15 and 150 Hz. The capacity provided by the machine is nearly directly proportional to its running frequency. Thus, virtually infinite capacity steps are possible for the system with a variable speed compressor. The variable speed scroll compressor is now widely used in Japan.

Variable displacement scroll compressor. This capacity control mechanism incorporates porting holes in the fixed scroll member. The control mechanism disconnects or connects compression chambers to the suction side by respectively closing or opening the porting holes. When all porting holes are closed, the compressor runs at full capacity; opening of all porting holes to the suction side yields the smallest capacity. Thus, by opening or closing a different number of porting holes, variable cooling or heating capability is provided to the system. The number of different capacities and the extent of the capacity reduction available is governed by the locations of the ports in reference to full capacity suction seal-off.

Performance

Scroll technology offers an advantage in performance for a number of reasons. Large suction and discharge ports reduce pressure losses incurred in the suction and discharge processes. Also, physical separation of these processes reduces heat transfer to the suction gas. The absence of valves and reexpansion volumes and the continuous flow process results in high volumetric efficiency over a wide range of operating conditions. Figure 35 illustrates this effect. The built-in volume ratio can be designed for lowest over-or undercompression at typical demand conditions (2.5 to 3.5 pressure ratio for air conditioning). Isentropic efficiency in the range of 70% is possible at such pressure ratios, and it remains quite close to the efficiency of other compressor types at high pressure ratios (Figure 35). Scroll compressors offer a flatter capacity versus outdoor ambient curve than reciprocating products, which means that they can more closely approach indoor requirements at high demand conditions. As a result, the heat pump mode requires less supplemental heating; the cooling mode is more comfortable, because cycling is less as demand decreases (Figure 36).

Scroll compressors available for the North American market are typically specified as producing ARI operating efficiencies (COP) in the range of 3.10 to 3.34.

Noise and Vibration

The scroll compressor inherently possesses a potential for low sound and vibration characteristics. It includes a minimal number of moving parts compared to other compressor technologies. Since scroll compression requires no valves, impact noise and vibration are completely eliminated. The presence of a continuous suction-compression-discharge process and low gas pressure pulsation help to keep vibration levels low. A virtually perfect dynamic balancing of the orbiting scroll inertia with counterweights eliminates possible vibrations due to the rotating parts.

Also, smooth surface finish and accurate machining of the vane profiles and base plates of both scroll members (requirement for small leakage) aids in minimal impact of the vanes. A typical sound spectrum of the scroll compressor is shown in Figure 37.

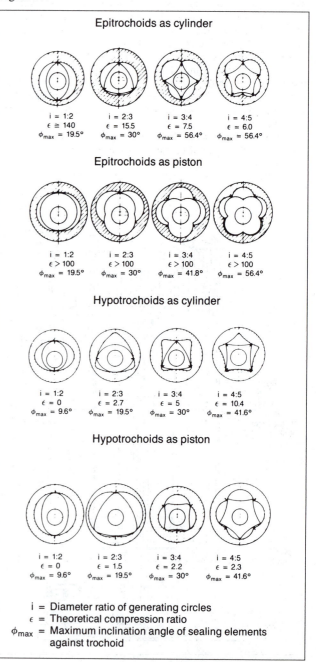

i = Diameter ratio of generating circles
ϵ = Theoretical compression ratio
ϕ_{max} = Maximum inclination angle of sealing elements against trochoid

Fig. 38   Possible Versions of Epitrochoidal and Hypotrochoidal Machines

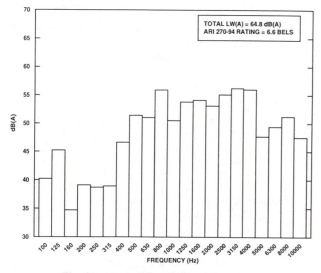

Fig. 37   Typical Scroll Sound Spectrum

Operation and Maintenance

Most scroll compressors used today are of the hermetic type, which require virtually no maintenance. However, the compressor manufacturer's operation and application manual should be followed.

TROCHOIDAL COMPRESSORS

The trochoidal compressor is a small, rotary, positive displacement compressor which can run at high speeds of up to 9000 rpm. They are manufactured in various configurations. Trochoidal curvatures can be produced by the rolling motion of one circle outside or inside the circumference of a basic circle, producing either epitrochoids or hypotrochoids, respectively. Both types of trochoids can be used either as a cylinder or piston form, so that four types of trochoidal machines can be designed (Figure 38).

In each case, the counterpart of the trochoid member always has one apex more than the trochoid itself. In the case of a trochoidal cylinder, the apexes of the piston show a slipping motion along the inner cylinder surface; for trochoidal piston design, the piston shows a gear-like motion. As seen in Figure 39, a built-in theoretical pressure ratio disqualifies many configurations as valid concepts for refrigeration compressor design. Because of additional valve ports, clearances, etc., and the resulting decrease in the built-in maximum theoretical pressure ratio, only the first two types with epitrochoidal cylinders, and all candidates with epitrochoi-

dal pistons, can be used for compressor technology. The latter, however, require sealing elements on the cylinder as well as on the side plates, which does not allow the design of a closed sealing borderline.

In the past (Cooley 1901, Planche 1920, Millard 1943), trochoidal machines were designed much like those of today. However, like other positive displacement rotary concepts that could not tolerate oil injection, the early trochoidal equipment failed because of sealing problems. The invention of a closed sealing border by Wankel changed this (Figure 39). Today, the Wankel trochoidal compressor with a three-sided epitrochoidal piston (motor) and two-envelope cylinder (casing) is built in capacities of up to 2 tons.

Description and Performance

Compared to other compressors of similar capacity, trochoidal compressors have many advantages typical of reciprocating compressors. Because of the closed sealing border of the compression space, these compressors do not require extremely small and expensive manufacturing tolerances; neither do they need oil for sealing, keeping them at low pressure side with the advantage of low solubility and high viscosity of the oil-refrigerant mixture. Valves are usually used on a high-pressure side while suction is ported. A valveless version of the trochoidal compressor can also be built. Figure 40 shows the operation of the Wankel rotary compressor (2:3 epitrochoid) for a system with discharge reed valves.

The Wankel compressor performance compares favorably with the reciprocating piston compressors at a higher speed and moderate pressure ratio range. A smaller number of moving parts, less friction, and the resulting higher mechanical efficiency improves the overall isentropic efficiency. This can be observed at higher speeds when sealing is better, and in the moderate pressure ratio range when the influence of the clearance volume is limited.

CENTRIFUGAL COMPRESSORS

Centrifugal compressors, sometimes called turbocompressors, belong to a family of turbomachines that includes fans, propellers, and turbines. These machines continuously exchange angular momentum between a rotating mechanical element and a steadily flowing fluid. Because their flows are continuous, turbomachines have greater volumetric capacities, size for size, than do positive displacement devices. For effective momentum exchange, their rotative speeds must be higher, but little vibration or wear results because of the steadiness of the motion and the absence of contacting parts.

Centrifugal compressors are well suited for air-conditioning and refrigeration applications because of their ability to produce high pressure ratios (as compared to other turbocompressors). The suction flow enters the rotating element, or impeller, in the axial direction and is discharged radially at a higher velocity. The change in

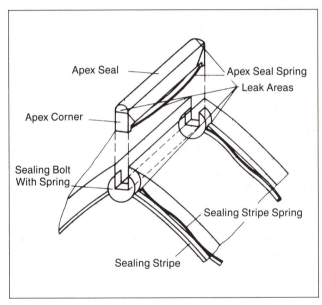

Fig. 39 Wankel Sealing System for Trochoidal Compressors

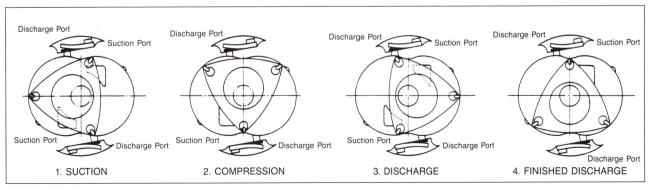

Fig. 40 Sequence of Operation of Wankel Rotary Compressor

diameter through the impeller increases the velocity of the gas flow. This dynamic head is then converted to static head, or pressure, through a diffusion process, which generally begins within the impeller and ends in a radial diffuser and scroll outboard of the impeller.

Centrifugal compressors are used in a variety of refrigeration and air-conditioning installations. Suction flow rates range between 60 and 30,000 cfm, with rotational speeds between 1800 and 90,000 rpm. However, the high angular velocity associated with a low volumetric flow establishes a minimum practical capacity for most centrifugal applications. The upper capacity limit is determined by physical size, a 30,000 cfm compressor being about 6 or 7 ft in diameter.

Suction temperatures are usually between 50 and −150°F, with suction pressures between 2 and 100 psia and discharge pressures up to 300 psia. Pressure ratios range between 2 and 30. Almost any refrigerant can be used.

Refrigeration Cycle

Figure 41 illustrates a simple vapor compression cycle in which a centrifugal compressor operates between states 1 and 2. Typical applications might involve a single-, two-, or three-stage halocarbon compressor or a seven-stage ammonia compressor.

Figure 42 shows a more complex cycle, with two stages of compression and interstage liquid flash cooling. This cycle has a higher coefficient of performance than the simple cycle and is frequently used with two- through four-stage halocarbon and hydrocarbon compressors.

More than one stage of flash cooling can be applied to compressors with more than two impellers. Liquid subcooling and interstage desuperheating can also be advantageously used. For information on refrigeration cycles, see Chapter 1 of the 1989 ASHRAE *Handbook—Fundamentals*.

Angular Momentum

The momentum exchange, or energy transfer, between a centrifugal impeller and a flowing refrigerant is expressed by:

$$W_i = u_i c_u / g \qquad (11)$$

where:

W_i = impeller work input per unit mass of refrigerant
u_i = impeller blade tip speed
c_u = tangential component of refrigerant velocity leaving impeller blades
g = gravitational constant

These velocities are shown in Figure 43, where refrigerant flows out from between the impeller blades with relative velocity *b* and absolute velocity *c*. The relative velocity angle β is a few degrees less than the blade angle because of a phenomenon known as slip.

Equation (11) assumes that the refrigerant enters the impeller without any tangential velocity component or swirl. This is generally the case at design flow conditions. If the incoming refrigerant was already swirling in the direction of rotation, the impeller's ability to impart angular momentum to the flow would be reduced. A subtractive term would then be required in the equation.

Some of the work done by the impeller increases the refrigerant pressure, while the remainder only increases its kinetic energy. The ratio of pressure-producing work to total work is known as the impeller reaction. Since this varies from about 0.4 to about 0.7, an appreciable amount of kinetic energy leaves the impeller with magnitude $c^2/2g$.

To convert this kinetic energy into additional pressure, a diffuser is located after the impeller. Radial vaneless diffusers are most common, but vaned, scroll, and conical diffusers are also used.

In a multistage compressor, the flow leaving the first diffuser is guided to the inlet of the second impeller and so on through the machine, as can be seen in Figure 44. The total compression work input per unit mass of refrigerant is the sum of the individual stage inputs

$$W = \Sigma W_i \qquad (12)$$

provided that the mass flow rate is constant throughout the compressor.

Description

A centrifugal compressor can be single stage, having only one impeller, or it can be multistage, having two or more impellers mounted in the same casing as shown in Figure 44. For process refrigeration applications, a compressor can have as many as ten stages.

The suction gas generally passes through a set of adjustable inlet guide vanes or an external suction damper before entering the impeller. The vanes (or suction damper) are used for capacity control as will be described later.

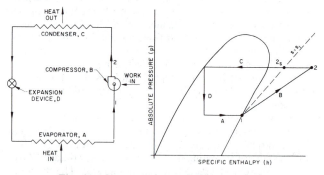

Fig. 41   Simple Vapor Compression Cycle

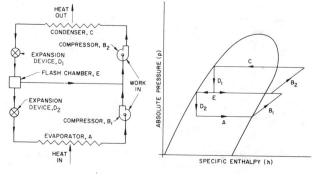

Fig. 42   Compression Cycle with Flash Cooling

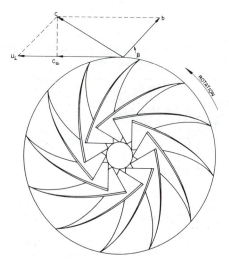

Fig. 43   Impeller Exit Velocity Diagram

The high-velocity gas discharging from the impeller enters the radial diffuser which can be vaned or vaneless. Vaned diffusers are typically used in compressors designed to produce high heads. These vanes are generally fixed, but they can be adjustable. Adjustable diffuser vanes can be used for capacity modulation either in lieu of or in conjunction with the inlet guide vanes.

For multistage compressors, the gas discharged from the first stage is directed to the inlet of the second stage through a return channel. The return channel contains a set of fixed flow straightening vanes or an additional set of adjustable inlet guide vanes. Once the gas reaches the last stage, it is discharged in a volute or collector chamber. From there, the high-pressure gas passes through the compressor discharge connection.

When multistage compressors are used, side loads can be introduced between stages so that one compressor performs several functions at several temperature levels. Multiple casings can be connected in tandem to a single driver. These can be operated in series, in parallel, or even with different refrigerants.

ISENTROPIC ANALYSIS

The static pressure resulting from a compressor's work input or, conversely, the amount of work required to produce a given pressure rise, depends on the efficiency of the compressor and the thermodynamic properties of the refrigerant. For an adiabatic process, the work input required is minimal if the compression is isentropic. Therefore, actual compression is often compared to an isentropic process, and the performance thus evaluated is based on an isentropic analysis.

The reversible work required by an isentropic compression between states 1 and 2s in Figure 41 is known as the adiabatic work, as measured by the enthalpy difference between the two points:

$$W_s = h_{2s} - h_1 \qquad (13)$$

The irreversible work done by the actual compressor is

$$W = h_2 - h_1 \qquad (14)$$

assuming negligible cooling occurs. Flash-cooled compressors cannot be analyzed by this procedure unless they are subdivided into uncooled segments with the cooling effects evaluated by other means. Compressors with side flows must also be subdivided. In Figure 42, the two compression processes must be analyzed individually.

Equation (14) also assumes a negligible difference in the kinetic energies of the refrigerant at states 1 and 2. If this is not the case, a kinetic energy term must be added to the equation. All the thermodynamic properties throughout the section Centrifugal

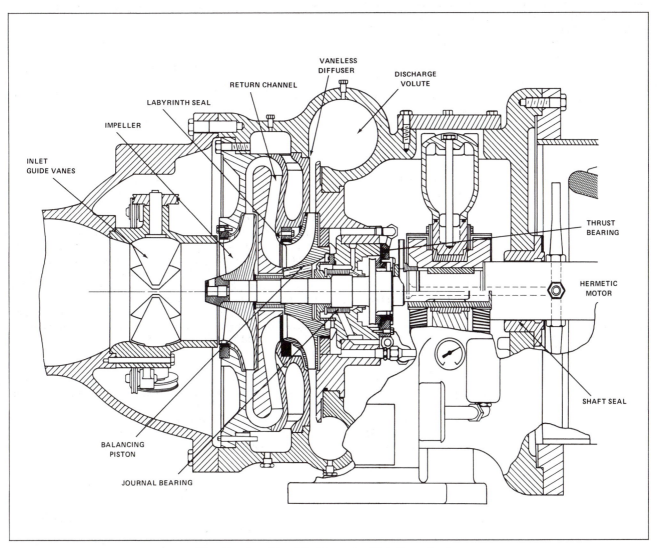

Fig. 44 Centrifugal Refrigeration Unit Cross Section

Compressors are static properties as opposed to stagnation properties; the latter includes kinetic energy.

The ratio of isentropic work to actual work is the adiabatic efficiency:

$$\eta_s = \frac{W_s}{h_2 - h_1} \qquad (15)$$

This varies from about 0.62 to about 0.83, depending on the application. Because of the thermodynamic properties of gases, a compressor's overall adiabatic efficiency does not completely indicate its individual stage performance. The same compressor produces different adiabatic results with different refrigerants and also with the same refrigerant at different suction conditions.

In spite of its shortcomings, isentropic analysis has a definite advantage in that adiabatic work can be read directly from thermodynamic tables and charts similar to those presented in Chapter 17 of the 1989 ASHRAE *Handbook—Fundamentals*. Where these are unavailable for the particular gas or gas mixture, they can be accurately calculated and plotted using thermodynamic relationships and a computer.

POLYTROPIC ANALYSIS

Polytropic pressures and efficiencies are more consistent from one application to another, because a reversible polytropic process duplicates the actual compression between states 1 and 2 in Figure 41. Therefore, values calculated by the polytropic analysis have greater versatility than those of the isentropic analysis.

The path equation for this reversible process is:

$$\eta = \nu(dp/dh) \qquad (16)$$

where η is the *polytropic efficiency* and ν is the specific volume of the refrigerant. The reversible work done along the polytropic path is known as the polytropic work and is given by:

$$W_p = \int_{p_1}^{p_2} \nu \, dp \qquad (17)$$

It follows from Equations (14), (16), and (17) that the polytropic efficiency is the ratio of reversible work to actual work:

$$\eta = \frac{W_p}{h_2 - h_1} \qquad (18)$$

Thermodynamic relations for an ideal gas

$$p\nu^n = p_1 \nu_1{}^n = p_2 \nu_2{}^n \qquad (19)$$

can be used to permit integration so that Equation (17) can be written:

$$W_p = \frac{n}{(n-1)} p_1 \nu_1 \left[\left(\frac{p_2}{p_1} \right)^{(n-1)/n} - 1 \right] \qquad (20)$$

Further manipulation eliminates the exponent

$$W_p = \left[\frac{p_2 \nu_2 - p_1 \nu_1}{\ln(p_2 \nu_2 / p_1 \nu_1)} \right] \ln(p_2/p_1) \qquad (21)$$

For greater accuracy in handling gases with properties known to deviate substantially from those of a perfect gas, a more complicated procedure is required.

Equation (16) can be approximated by using Equation (19) and:

$$\frac{p_m}{T} = \frac{p_1{}^m}{T_1} = \frac{p_2{}^m}{T_2} \qquad (22)$$

where

$$m = \frac{ZR}{c_p} \left(\frac{1}{\eta} + X \right) = \frac{(k - 1/k)(1/\eta + X)Y}{(1 + X)^2} \qquad (22a)$$

$$n = \frac{1}{Y - (ZR/c_p)(1/\eta + X)(1 + X)}$$
$$= \frac{1 + X}{Y[(1/k)(1/\eta + X) - (1/\eta - 1)]} \qquad (22b)$$

and:

$$X = \frac{T}{\nu} \left(\frac{\partial \nu}{\partial p} \right)_p - 1 \qquad (23a)$$

$$Y = -\frac{p}{\nu} \left(\frac{\partial \nu}{\partial p} \right)_T \qquad (23b)$$

$$Z = \frac{p\nu}{RT} \qquad (23c)$$

Also, R is the gas constant and k is the ratio of specific heats; all properties are at temperature T.

The accuracy with which Equations (19) and (22) represent Equation (16) depends on the constancy of m and n along the polytropic path. Because these exponents usually vary, mean values between states 1 and 2 should be used.

Compressibility functions X and Y have been generalized for gases in corresponding states by Schultz (1962) and their equivalents are listed by Edminster (1961). For usual conditions of refrigeration interest, *i.e.*, for $p < 0.9 \, p_c$, $T < 1.5 \, T_c$, and $0.6 < Z$, these functions can be approximated by:

$$X = 0.1846 \, (8.36)^{1/Z} - 1.539 \qquad (24a)$$

$$Y = 0.074 \, (6.65)^{1/Z} + 0.509 \qquad (24b)$$

The compressibility factor Z has been generalized by Edminster (1961) and Hougen *et al.* (1959), among others. Generalized corrections for the specific heat at constant pressure c_p can also be found in these works.

Equations (19) and (22) make possible the integration of Equation (17):

$$W_p = f \left(\frac{n}{n-1} \right) p_1 \nu_1 \left[\left(\frac{p_2}{p_1} \right)^{(n-1)/n} - 1 \right] \qquad (25)$$

In Equation (25), the polytropic work factor f corrects for whatever error may result from the approximate nature of Equations (19) and (22). Since the value of f is between 1.00 and 1.02 in most refrigeration applications, it is generally neglected.

Once the polytropic work has been found, the efficiency follows from Equation (18). Polytropic efficiencies range from about 0.70 to about 0.84, a typical value being 0.76.

The highest efficiencies are obtained with the largest compressors and the densest refrigerants because of a Reynolds number effect discussed by Davis *et al.* (1951). A small number of stages is also advantageous because of parasitic losses associated with each stage.

Overall, polytropic work and efficiencies are more consistent from one application to another because they represent an average stage aerodynamic performance.

Instead of using Equations (16) through (25), it is easier and often more desirable to determine the adiabatic work by an isentropic analysis and then to convert to polytropic work by

$$\frac{W_p}{W_s} \cong \eta \left[\frac{(p_2/p_1)^{(k-1)/k\eta} - 1}{(p_2/p_1)^{(k-1)/k} - 1} \right] \qquad (26)$$

Equation (26) is strictly correct only for an ideal gas, but because it is a ratio involving comparable errors in both numerator and denominator, it is of more general utility. Equation (26) is plotted in Figure 45 for $n = 0.76$. To obtain maximum accuracy,

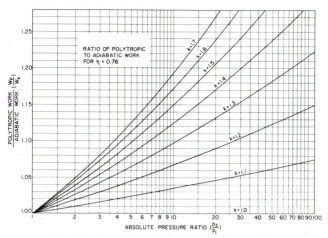

Fig. 45 Ratio of Polytropic to Adiabatic Work

the ratio of specific heats k must be a mean value for states 1, 2, and 2_s. If c_p is known, k can be determined by:

$$k = \frac{1}{1 - (ZR/c_p)(1 + X)^2/Y} \quad (27)$$

The gas compression power is:

$$P = wW \quad (28)$$

where w is the mass flow rate. To obtain total shaft power, add the mechanical friction losses. Friction losses vary from less than 1% of the gas power to more than 10%. A typical estimate is 3%.

Nondimensional Coefficients

Several nondimensional performance parameters used to describe centrifugal compressor performance are flow coefficient, polytropic work coefficient, Mach number, and specific speed.

Flow coefficient. Desirable impeller diameters and rotational speeds are determined from blade tip velocity by a dimensionless flow coefficient (Q/ND^3) in which Q is the volumetric flow rate. Practical values for this coefficient range from 0.02 to 0.35, with good performance falling between 0.11 and 0.21. Optimum results occur between 0.15 and 0.18. Impeller diameter D_i and rotational speed N follow from:

$$(Q/ND^3) = \pi (Q_i/u_iD_i^2) = \pi^3 (Q_iN^2/u_i^3) \quad (29)$$

where u_i is the tip speed.

The maximum flow coefficient in multistage compressors is found in the first stage and the minimum in the last stage (unless large side loads are involved). For high-pressure ratios, special measures may be necessary to increase the last stage (Q/ND^3) to a practical level. Side loads are beneficial in this respect, but interstage flash cooling is not.

Polytropic work coefficient. Polytropic work and polytropic head cannot be used interchangeably. Since this chapter is concerned with polytropic work, that term will be used exclusively. Polytropic head is related to it by $H_p = W_p/g$.

Besides the power requirement, polytropic work also determines impeller blade tip speed and number of stages. For an individual stage, the stage work is related to speed by:

$$W_{pi} = \mu_i u_i^2 \quad (30)$$

where μ_i is the stage work coefficient.

The overall polytropic work is the sum of the stage works:

$$W_p = \Sigma W_{pi} \quad (31)$$

and the overall work coefficient is

$$\mu = gW_p/\Sigma u_i^2 \quad (32)$$

Values for μ (and μ_i) range from about 0.42 to about 0.74, with 0.55 representative for estimating purposes. Compressors designed for modest work coefficients have backward-curved impeller blades. These tend to have greater part-load ranges and higher efficiencies than do radial-bladed designs.

Maximum tip speeds are limited by strength considerations to about 84,600 fpm. For cost and reliability, 59,000 fpm is a more common limitation. On this basis, the maximum polytropic work capability of a typical stage is about 15,000 ft·lb$_f$/lb.

A greater restriction on stage work capability is often imposed by the impeller Mach number M_i. For adequate performance, M_i must be limited to about 1.8 for stages with impellers overhung from the ends of shafts and to about 1.5 for impellers with shafts passing through their inlets. For good performance, these values must be even lower. Such considerations limit maximum stage work to about 1.5 a_i^2, where a_i is the acoustic velocity at the stage inlet.

Specific speed. This nondimensional index of optimum performance characteristic of geometrically similar stages is defined by:

$$N_s = N\sqrt{Q_i}/W_{pi}^{0.75} = (1/\pi^3\mu_i^{0.75})\sqrt{Q_i/ND_i^3} \quad (33)$$

The highest efficiencies are generally attained in stages with specific speeds between 600 and 850.

Mach Number

Two different Mach numbers are used: The flow Mach number M is the ratio of flow velocity c, to acoustical velocity a at a particular point in the fluid stream:

$$M = c/a \quad (34a)$$

where

$$a = v\sqrt{-(\partial p/\partial v)_s} \sqrt{n_s pv} \quad (34b)$$

Values of acoustical velocity for a number of saturated vapors at various temperatures are presented in Table 4.

The flow Mach numbers in a typical compressor vary from about 0.3 at the stage inlets and outlets to about 1.0 at the impeller exits. With increasing flow Mach numbers, the losses increase because of separation, secondary flow, and shock waves.

Table 4 Acoustical Velocities of Saturated Vapors, ft/s*

Refrigerant	Evaporator Temperature, °F								
	−300	−250	−200	−150	−100	−50	−0	50	100
11					386	410	430	446	456
12				388	413	433	445	447	437
13			390	417	433	434	419	384	
13B1			324	349	369	381	382	370	345
22				471	500	523	535	535	522
23			490	526	550	558	547	516	
113						343	362	377	388
114				315	338	357	372	381	380
123					361	383	402	417	425
124				358	382	403	417	423	417
125			353	381	405	419	421	407	373
134a				418	446	469	482	484	470
142b				421	450	475	494	503	501
152a				530	565	594	614	621	612
500				429	458	479	493	494	483

*Data from REFPROP, Office of Standard Reference Data, National Institute of Standards and Technology, Gaithersburg, MD.

The impeller Mach number, which is a pseudo Mach number, is the ratio of impeller tip speed to acoustical velocity at the stage inlet:

$$M_i = u_i/a_i \qquad (35)$$

Performance

From an applications standpoint, more useful parameters than μ and (Q/ND^3) are Ω and Θ (Sheets 1952):

$$\Omega = gW_p/a_i^2 = m\,(\Sigma u_i^2/a_i^2) \qquad (36a)$$

$$\Theta = Q_1/a_1 D_1^2 = (M_1/\pi)(Q_1/ND_1^3) \qquad (36b)$$

They are as general as the customary test coefficients and produce performance maps like the one in Figure 46, with speed expressed in terms of first-stage impeller Mach number.

A compressor user with a particular installation in mind may prefer more explicit curves, such as pressure ratio and power versus volumetric flow rate at constant rotational speeds. Plots of this sort may require fixed suction conditions to be entirely accurate, especially if discharge pressure and power are plotted against mass flow rate or refrigeration effect.

A typical compressor performance map is shown in Figure 46 where percent of rated work is plotted with efficiency contours against percent of rated volumetric flow at various speeds. Point A is the design point at which the compressor operates with maximum efficiency. Point B is the selection or rating point at which the compressor is being applied to a particular system. From the application or user's point of view, Ω and Θ have their 100% values at Point B.

To reduce first cost, refrigeration compressors are selected for heads and capacities beyond their peak efficiency, as shown in Figure 46. The opposite selection would require larger impellers and additional stages. Refrigerant acoustical velocity and the ability to operate at a high enough Mach number are also of concern. If the compressor shown in Figure 46 were of a multistage design, M_1 would be about 1.2; for a single-stage compressor, it would be about 1.5.

Another acoustical effect is seen on the right of the performance map, where increasing speeds do not produce corresponding increases in capacity. The maximum flow rates at M_1 and $1.1\,M_1$ approach a limit determined by the relative velocity of the refrigerant entering the first impeller. As this velocity approaches a sonic value, the flow becomes choked and further increases become impossible. Another commonly used term for this phenomenon is *stonewalling*; it represents the maximum capacity of an impeller.

Testing

When a centrifugal compressor is tested, overall μ and η versus Q_1/ND_1^3 at constant M_1 are plotted. They are useful because test results with one gas are sometimes converted to field performance with another. When side flows and cooling are involved, the overall work coefficient is found from Equations (31) and (32) by evaluating the mixing and cooling effects between stages separately. The *overall efficiency* in such cases is:

$$\eta = \frac{\Sigma w_i W_{pi}}{\Sigma w_i W_i} \qquad (37)$$

Testing with a fluid other than the design refrigerant is a common practice known as *equivalent performance testing*. Its need arises from the impracticability of providing test facilities for the complete range of refrigerants and input power for which centrifugal compressors are designed. Equivalent testing is possible because a given compressor produces the same μ and η at the same (Q/ND^3) and M_i with any fluids whose volume ratios (v_1/v_2) and Reynolds numbers are the same.

The thermodynamic performance of a compressor can be evaluated according to either the stagnation or the static properties of the refrigerant, and it is important to distinguish between these concepts. The *stagnation efficiency*, for example, may be higher than the *static efficiency*. The safest procedure is to use static properties and evaluate kinetic energy changes separately.

Surging

Part-load range is limited on the other side of the performance map by a *surge envelope*. Satisfactory compressor operation to the left of this line is prevented by unstable *surging* or *hunting*, in which the refrigerant alternately flows backward and forward through the compressor, accompanied by increased noise, vibration, and heat. Prolonged operation under these conditions can damage the compressor.

The flow reverses during surging about once every 2 seconds. Small systems surge at higher frequencies and large systems at lower. Surging can be distinguished from other kinds of noise and vibration by the fact that its flow reversals alternately unload and load the driver. Motor current varies markedly during surging, and turbines alternately speed up and slow down.

Another kind of instability, *incipient surge* or *stall*, may occur slightly to the right of the true surge envelope. This phenomenon involves the formation of rotating stall pockets or cells in the diffuser. It produces a roaring noise at a frequency determined by the number of cells formed and the impeller running speed. The driver load is steady during incipient surge, which is harmless to the compressor, but it may, however, vibrate system components excessively.

System Balance

If a refrigeration system characteristic is superimposed on a compressor performance map, it shows the speeds and efficiencies at which the compressor operates in that particular application. A typical brine cooling system curve is plotted in Figure 46, passing through Points B, C, D, E, F, G, and H. With increased speed, the compressor at Point H will produce more than its rated capacity; with decreased speeds at Points C and D, it will produce less. Because of surging, the compressor cannot be operated satisfactorily at Points E, F, or G.

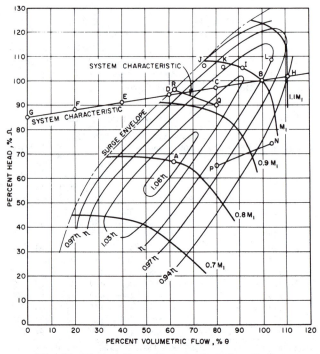

Fig. 46 Typical Compressor Performance Curves

The system can be operated at these capacities, however, by using a hot gas bypass. The volume flow at the compressor suction must be at least that for Point D in Figure 46; this volume flow is reached by adding hot gas from the compressor discharge to the evaporator, or compressor suction piping. When hot gas bypass is used, no further power reduction is realized as the load decreases. The compressor is being artificially loaded to stay out of the surge envelope. The increased volume due to hot gas recirculation performs no useful refrigeration effect.

Capacity Control

When the driver speed is constant, a common method of altering capacity is to swirl the refrigerant entering one or more impellers. Adjustable inlet guide vanes, or *prerotation vanes* (Figure 44), produce the swirl. Setting these vanes to swirl the flow in the direction of rotation produces a new compressor performance curve without any change in speed. Controlled positioning of the vanes can be accomplished by pneumatic, electrical, or hydraulic means.

Typical curves for five different vane positions are shown in Figure 47 for the compressor in Figure 46 at the constant speed M_1. With the prerotation vanes wide open, the performance curve is identical to the M_1 curve in Figure 46. The other curves are different, as are the efficiency contours and the surge envelope.

The same system characteristic has been superimposed on this performance map, as in Figure 46, to provide a comparison of these two modes of operation. In Figure 47, Point E can be reached with prerotation vanes, Point H cannot. Theoretically, turning the vanes against rotation would produce a performance line passing through Point H, but sonic relative inlet velocities prevent this, except at low Mach numbers. Hot gas bypass is still necessary at Points F and G with prerotation vane control, but to a lesser extent than with variable speed.

The gas compression powers for both control methods are listed in Table 5. For the compressor and system assumed in this example, Table 5 shows that speed control requires less gas compression power down to about 55% of rated capacity. Prerotation vane control requires less power below 55%. For a complete analysis, friction losses and driver efficiencies must also be considered.

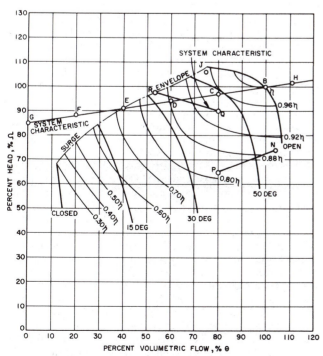

Fig. 47 Typical Compressor Performance with Various Prerotation Vane Settings

Table 5 Typical Part-Load Gas Compression Power Input for Speed and Vane Controls

System Volumetric Flow, %	Power Input, %	
	Speed Control	Vane Control
111	120	—
100	100	100
80	76	81
60	59	64
40	55	50
20	51	46
0	47	43

Variable speed control systems require decreasing polytropic work (head) with decreasing flow (capacity) to perform more efficiently.

Decreasing polytropic work is generally accomplished by reducing condensing pressure. Impeller tip speed must remain constant if the lift requirements do not change.

Refrigeration capacity is directly related to compressor speed, but the compressor's ability to produce head is a function of the square of a change in compressor speed. For example, a 50% reduction in compressor speed will result in a 50% reduction in refrigeration capacity; however, the available pressure for the compressor will be 25% of its value at 100% speed. If this is not consistent with actual operating conditions, hot gas bypass will be required to false load the compressor and prevent it from entering into a surge condition.

In Figure 47, for example, the compressor could operate at 15% of its rated capacity with 28% of its rated power if the system polytropic work requirement could be reduced by 29%.

Since fixed-speed motors are the most common drivers of centrifugal compressors, prerotation vane control is more prevalent than speed variation. This is generally the best control mode for applications where pressure requirements do not vary significantly at part load. Less common control methods are (1) suction throttling, (2) adjustable diffuser vanes, (3) movable diffuser walls, (4) impeller throttling sleeve, and (5) combinations of these with prerotation vanes and variable speed. Each method has advantages and disadvantages in terms of performance, complexity, and cost.

APPLICATION

Critical Speeds

Centrifugal compressors are designed so that the first lateral critical speed is either well above or well below the operating speed. Operation at speeds between 0.8 and 1.1 times the first lateral speed is generally unacceptable from a reliability standpoint. The second lateral critical speed should be at least 25% above the operating speed of the machine.

The operating speeds of hermetic compressors are fixed, and each manufacturer has full responsibility for making sure the critical speeds are not too close to the operating speeds. For open-drive compressors, however, operating speed depends on the required flow rate of the application. Thus, the designer must make sure that the critical speeds are sufficiently far away from the operating speeds.

In applying open-drive machines, it is also necessary to consider torsional critical speeds, which are a function of the designs of the compressor, the drive turbine or motor, and the coupling(s). In geared systems, the gearbox design is also involved. Manufacturers of centrifugal compressors use computer programs to calculate the torsional natural frequencies of the entire system, including the driver, the coupling(s), and the gears, if any. Responsibility for performing this calculation and ascertaining that the torsional natural frequencies are sufficiently far away from torsional exciting frequencies should be shared between the compressor manufacturer and the system designer.

For engine drives, it may be desirable to use a fluid coupling to isolate the compressor (and gear set) from engine torque pulsations. Depending on compressor bearing design, there may be other speed ranges that should be avoided to prevent the nonsynchronous shaft vibration commonly called oil whip or oil whirl.

Vibration

Excessive vibration of a centrifugal compressor is an indication of malfunction, which may lead to failure. Periodic checks of the vibration spectrum at suitable locations or continuous monitoring of vibration at such locations are, therefore, useful in ascertaining the operational health of the machine. The relationship between internal displacements and stresses and external vibration is different for each compressor design. In a given design, this relationship also differs for the various causes of internal displacements and stresses, such as imbalance of the rotating parts (either inherent or caused by deposits, erosion, corrosion, looseness, or thermal distortion), bearing instabilities, misalignments, distortion because of piping loads, broken motor rotor bars, or cracked impeller blades. It is, therefore, impossible to establish universal rules for the level of vibration considered excessive.

To establish meaningful criteria for a given machine or design, it is necessary to have baseline data indicative of proper operation. Significant increases of any of the frequency components of the vibration spectrum above the baseline will then indicate a deterioration in the machine's operation; the frequency component for which this increase occurs is a good indication of the part of the machine deteriorating. Increases in the component at the fundamental running frequency, for instance, are usually because of deterioration of balance. Increases at approximately one-half the fundamental running frequency are due to bearing instabilities, and increases at twice the running frequency are usually the result of deterioration of alignment, particularly coupling alignment.

As a general guide to establishing satisfactory vibration levels, a constant velocity criterion is sometimes used. In many cases, a velocity amplitude of 0.2 in/s constitutes a reasonable criterion for vibration measured on the bearing housing.

Although measurement of the vibration amplitude on the bearing housing is convenient, the value of such measurements is limited because the stiffness of the bearing housing in typical centrifugal compressors is generally considerably larger than that of the oil film. Thus, vibration monitoring systems often use noncontacting sensors, which measure the displacement of the shaft relative to the bearing housing, either instead of, or in addition to, monitoring the vibration of the bearing housing (Mitchell 1977). Such sensors are also useful for monitoring the axial displacement of the shaft relative to the thrust bearing.

In some applications, compressor vibration levels, which are perfectly acceptable from a reliability standpoint, can cause noise problems if the machine is not isolated properly from the building. Obtaining factory-run tested vibration levels will give a base level comparison for future reference of field-tested vibration levels.

Noise

The satisfactory application of centrifugal compressors requires careful consideration of noise control, especially if compressors are to be located near a noise-sensitive area of a building. The noise of centrifugal compressors is primarily of aerodynamic origin, constituted principally of gas pulsations associated with the impeller frequency and gas flow noise. Most of the predominant noise sources are of a sufficiently high frequency (above 1000 Hz) so that significant noise reductions can be achieved by carefully designed acoustical and structural isolation of the machine. While the noise originates within the compressor proper, most is usually radiated from the discharge line and the condenser shell. Reductions of equipment room noise by up to 10 dB can be obtained by lagging the discharge line and the condenser shell. In geared compressors, gear-mesh noise may also contribute to the high-frequency noise; however, these frequencies are often above the audible range. This noise can be reduced by the application of sound insulation material to the gear housing.

There are two aspects of importance in noise considerations for applications of centrifugal compressors. In the equipment room, OSHA regulations specify employer responsibilities with regard to exposure to high sound levels. Increasing liability concerns in this regard are making designers more aware of compressor sound level considerations. Another important consideration is noise travel beyond the immediate equipment room.

Noise problems with centrifugal refrigeration equipment can occur in noise-sensitive parts of the building, such as a nearby office or conference room. The cost of controlling the transmission of compressor noise to such areas should be considered in the building layout and weighed against cost factors for alternative locations of the equipment in the building.

If the equipment room is to be located close to noise-sensitive building areas, it is usually cost-effective to have the noise and vibration isolation designed by an experienced acoustical consultant, since small errors in design or execution can make the results unsatisfactory (Hoover 1960).

Blazier (1972) covers general information on typical noise levels near centrifugal refrigeration machines. Data on the noise output of a specific machine should be obtained from the manufacturer; the request should specify that the measurements are to be in accordance with the current edition of ARI *Standard* 575-79, Method of Measuring Machinery Sound within Equipment Rooms.

Drivers

Centrifugal compressors are driven by almost any prime mover—a motor, turbine, or engine. Power requirements range from 33 to 12,000 hp. Sometimes the driver is coupled directly to the compressor; often, however, there is a gear set between them, usually because of low driver speed. Flexible couplings are required to accommodate the angular, axial, and lateral misalignments that may arise within a drive train. Additional information on prime movers may be found in Chapters 40 and 41.

Many specialized applications are made of centrifugal refrigeration compressors, an outstanding example being their use in hermetic water chilling systems of 80 to 2000-ton capacity. These units use standardized single-, two-, and three-stage compressors driven by integral motors operating in refrigerant atmospheres. Liquid or gaseous refrigerants cool the motors, making them quieter and less costly than conventional open designs and eliminating the need for any mechanical shaft seal.

Hermetic compressors operating at rotative speeds higher than two-pole motor synchronous speed (3600 rpm) are driven by internal speed-increasing gears. These also operate in refrigerant atmospheres. A discussion of centrifugal water chilling systems is contained in Chapter 17 of the 1988 ASHRAE *Handbook—Equipment*. Standardized single- and two-stage compressors with nonhermetic drivers are also used in water chilling systems of 80 to 10,000 tons. Internal gears, which are quieter, less costly, and more compact than external gearboxes, are available in compressors for 80 to 1400 tons.

A hermetic compressor must absorb the heat of the motor since the motor is cooled by the refrigerant. This keeps the motor running at a cooler temperature and in a controlled, clean environment. The design of the refrigeration system must take the rejection of this heat into account. Conversely, an open motor is cooled by the air in the equipment room so that the refrigeration system need not provide cooling for the motor heat. Provisions must be made, however, for removing the heat from the equipment room environment, typically by mechanical ventilation and/or cooling. In the unusual occurrence of a motor burnout, the hermetic system may require thorough cleaning, which would not be necessary with an open motor. However, when replaced or serviced, the open motor must be carefully and properly aligned in order to assure reliable performance.

Starting torque must be considered in selecting a driver, particularly a motor or single-shaft gas turbine. Compressor torque is roughly proportional to both speed squared and to the refrigerant density. The latter is often much higher at start-up than at rated operating conditions. If prerotation vanes or suction throttling cannot provide sufficient torque reduction for starting, the system standby pressure must be lowered by some auxiliary means.

In certain applications, a centrifugal compressor drives its prime mover backward at shutdown. The compressor is driven backward by refrigerant equalizing through the machine. The extent to which reverse rotation occurs depends on the kinetic energy of the drive train relative to the expansive energy in the system. Large installations with dense refrigerants are most susceptible to running backward, a modest amount of which is harmless if suitable provisions have been made. Reverse rotation can be minimized or eliminated by closing discharge valves, side-load valves, and prerotation vanes at shutdown and opening hot gas bypass valves and liquid refrigerant drains.

Paralleling

The problems associated with paralleling turbine-driven centrifugal compressors at reduced load are illustrated by Points I and J in Figure 46. These represent two identical compressors connected to common suction and discharge headers and driven by identical turbines. A single controller sends a common signal to both turbine governors so that both compressors should be operating at part-load Point K (full load is at Point L). The I machine runs 1% faster than its twin because of their respective governor adjustments, while the J compressor works against 1% more pressure difference because of the piping arrangement. The result is a 20% discrepancy between the two compressor loadings.

One remedy is to readjust the turbine governors so that the J compressor runs 0.5% faster than the other unit. A more permanent solution, however, is to eliminate one of the common headers and to provide either separate evaporators or separate condensers. This increases the compression ratio of whichever machine has the greater capacity, decreases the compression ratio of the other, and shifts both toward Point K.

The best solution of all is to install a flow meter in the discharge line of each compressor and to use a master-slave control system in which the original controller signals only one turbine, the master, while a second controller causes the slave unit to match the master's discharge flow.

The problem of imbalance, associated with turbine-driven centrifugal compressors, is minimal in fixed-speed compressors with vane controls. A loading discrepancy comparable to the abovementioned example would require a 25% difference in vane positions.

The paralleling of centrifugal compressors offers advantages in redundancy and improved part-load operation. This arrangement provides the capability of efficiently unloading to a lower percentage of total system load. When the unit requirement reduces to 50%, one compressor can carry the complete load and will be operating at a higher percent volumetric flow and efficiency than a single large compressor.

Means must be provided to prevent refrigeration flow through the idle compressor to prevent an inadvertent flow of hot gas bypass through the compressor. In addition, isolation valves should be provided on each compressor to allow removal or repair of either compressor.

Other Special Applications

Other specialized applications of centrifugal compressors are found in petroleum refineries and in the chemical industry, as discussed in Chapter 36 of the 1990 ASHRAE *Handbook—Refrigeration*. Marine requirements are detailed in ASHRAE *Standard* 26-1985, Recommended Practice for Mechanical Refrigeration Installations on Shipboard.

MECHANICAL DESIGN

Impellers

Impellers without covers, such as the one shown in Figure 43, are known as open or unshrouded designs. Those with covered blades (Figure 44), are known as shrouded impellers. Open models must operate in close proximity to contoured stationary surfaces to avoid excessive leakage around their vanes. Shrouded designs must be fitted with labyrinth seals around their inlets for a similar purpose. Labyrinth seals behind each stage are required in multistage compressors.

Impellers must be shrunk, clamped, keyed, or bolted to their shafts to prevent loosening from thermal and centrifugal expansions. Generally, they are made of cast or brazed aluminum or of cast, brazed, riveted, or welded steel. Aluminum has a higher strength-weight ratio than steel, up to about 300°F, which permits higher rotating speeds with lighter rotors. Steel impellers retain their strength at higher temperatures and are more resistant to erosion. Lead-coated and stainless steels can be selected in corrosive applications.

Casings

Centrifugal compressor casings are about twice as large as their largest impellers, with suction and discharge connections sized for flow Mach numbers between 0.1 and 0.3. They are designed for the pressure requirements of ASHRAE *Standard* 15-1989, Safety Code for Mechanical Refrigeration. A hydrostatic test pressure 50% greater than the maximum design working pressure is customary.

Cast iron is the most common casing material, having been used for temperatures as low as −150°F and pressures as high as 300 psia. Nodular iron and cast or fabricated steel are also used for low temperatures, high pressures, high shock, and hazardous applications. Multistage casings are usually split horizontally, although unsplit barrel designs can also be used.

Lubrication

Like motors and gears, the bearings and lubrication systems of centrifugal compressors can be internal or external, depending on whether or not they operate in refrigerant atmospheres. For reasons of simplicity, size, and cost, most air-conditioning and refrigeration compressors have internal bearings, as shown in Figure 44. In addition, they often have internal oil pumps, driven either by an internal motor or the compressor shaft; the latter arrangement is typically used with an auxiliary oil pump for starting and/or backup service.

Most refrigerants are soluble in lubricating oils, the extent increasing with refrigerant pressure and decreasing with oil temperature. A compressor's oil may typically contain 20% refrigerant (by weight) during idle periods of high system pressure and 5% during normal operation. Thus, refrigerant will come out of solution and foam the oil when such a compressor is started.

To prevent excessive foaming from cavitating the oil pump and starving the bearings, oil heaters minimize refrigerant solubility during idle periods. Standby oil temperatures between 130 and 150°F are required, depending on system pressures. Once a compressor has started, its oil should be cooled to increase oil viscosity and maximize refrigerant retention during the pulldown period.

A sharp reduction in system pressure before starting tends to supersaturate the oil. This produces more foaming at start-up than would the same pressure reduction after the compressor has started. Machines designed for pressure ratios of 20 or more may reduce system pressures so rapidly that excessive oil foaming cannot be avoided, except by maintaining a low standby pressure. Additional information on the solubility of refrigerants in oil can be found in Chapters 2 and 8 of the 1990 ASHRAE *Handbook—Refrigeration*.

External bearings avoid the complications of refrigerant-oil solubility at the expense of some oil-recovery problems. Any non-hermetic compressor must have at least one shaft seal. Mechanical seals are commonly used in refrigeration machines because they are leak-tight during idle periods. These seals require some lubricating oil leakage when operating, however. The amounts range up to 20 drops per minute, depending on seal design, face, velocities, and refrigerant pressures. Shaft seals leak oil out of compressors with internal bearings and into compressors with external bearings. Means for recovering seal oil leakage with a minimal loss of refrigerant must be provided in external lubrication systems.

Bearings

Centrifugal compressor bearings are generally of a hydrodynamic design, with sleeve bearings (one-piece or split) being the most common for radial loads; tilting pad, tapered land, and pocket bearings are customary for thrust. The usual materials are aluminum, babbit-lined, and bronze.

Thrust bearings tend to be the most important in turbomachines, and centrifugal compressors are no exception. Thrust comes from the pressure behind an impeller exceeding the pressure at its inlet. In multistage designs, each impeller adds to the total, unless some are mounted backward to achieve the opposite effect. In the absence of this opposing balance, it is customary to provide a balancing piston behind the last impeller, with pressures on the piston thrusting opposite to the stages. To avoid axial rotor vibration, some net thrust must be retained in either balancing arrangement.

Accessories

The minimum accessories required by a centrifugal compressor are an oil filter, an oil cooler, and three safety controls. Oil filters are usually rated for 15 to 20 μm or less. They may be built into the compressor but are more often externally mounted. Dual filters can be provided so that one can be serviced while the other is operating.

Single or dual oil coolers usually use condenser water, chilled water, refrigerant, or air as their cooling medium. Water- and refrigerant-cooled models may be built into the compressor, and refrigerant-cooled oil coolers may be built into a system heat exchanger. Many oil coolers are mounted externally for maximum serviceability.

Safety controls, with or without anticipatory alarms, must include a low oil pressure cutout, a high oil temperature switch, and high discharge and low suction pressure (or temperature) cutouts. A high motor temperature device is necessary in a hermetic compressor. Other common safety controls and alarms sense discharge temperature, bearing temperature, oil filter pressure differential, oil level, low oil temperature, shaft seal pressure, balancing piston pressure, surging, vibration, and thrust bearing wear.

Pressure gages and thermometers are useful indicators of the critical items monitored by the controls. Suction, discharge, and oil pressure gages are the most important, followed by suction, discharge, and oil thermometers. Suction and discharge instruments are often attached to system components rather than to the compressor itself, but they should be provided. Interstage pressures and temperatures can also be helpful, either on the compressor or on the system.

OPERATION AND MAINTENANCE

Reference should be made to the compressor manufacturer's operating and maintenance instructions for recommended procedures. A planned maintenance program, as described in Chapter 35 of the 1991 ASHRAE *Handbook—HVAC Applications*, should be established. As part of this program, an operating log should be kept, tabulating pertinent unit temperatures, pressures, flows, fluid levels, electrical data, and refrigerant added. These can be compared periodically with values recorded for the new unit. Gradual changes in data can be used to signify the need for routine maintenance; abrupt changes should indicate system or component difficulty. A successful maintenance program requires the operating engineer to be able to recognize and identify the reason for these data trends. In addition, by having a knowledge of the system component parts and their operational interaction, the designer will be able to use these symptoms to prescribe the proper maintenance procedures.

The following items deserve attention in establishing a planned compressor maintenance program:

1. A tight system is important. Leaks on compressors operating at subatmospheric pressures allow noncondensables and moisture to enter the system, adversely affecting system operation and component life. Leakage in higher pressure systems allows oil and refrigerant loss. ASHRAE *Guideline* 3-1990, Reducing Emission of Fully Halongenated CFC Refrigerants in Refrigeration and Air-Conditioning Equipment and Applications, can be used as a guide to ensure system tightness. The existence of vacuum leaks can be detected by a change in operational pressures not supported by corresponding refrigerant temperature data or the frequency of purge unit operation. Pressure leaks are characterized by symptoms related to refrigerant charge loss such as low suction pressures and high suction superheat. Such leaks should be located and fixed to prevent component deterioration.

2. Compliance with the manufacturer's recommended oil filter inspection and replacement schedule allows visual indication of the condition of the compressor lubrication system. Repetitive clogging of filters can mean system contamination. Periodic oil sample analysis can monitor acid, moisture, and particulate levels to assist in problem detection.

3. Operating and safety controls should be checked periodically and calibrated to ensure system reliability.

4. The electrical resistance of hermetic motor windings between phases and to ground should be checked (megged) regularly, following the manufacturer's outlined procedure. This will help detect any internal electrical insulation deterioration or the formation of electrical leakage paths before a failure occurs.

5. Water-cooled oil coolers should be systematically cleaned on the water side (depending on water conditions), and the operation of any automatic water control valves should be checked.

6. For some compressors, periodic maintenance, such as manual lubrication of couplings and other external components, and shaft seal replacement, is required. Prime movers and their associated auxiliaries all require routine maintenance. Such items should be made part of the planned compressor maintenance schedule.

7. Vibration analysis, when performed periodically, can locate and identify trouble (*e.g.*, unbalance, misalignment, bent shaft, worn or defective bearings, bad gears, mechanical looseness, and electrical unbalance). Without disassembly of the machine, such trouble can be found in its early stages before machinery failure or damage can occur. Dynamic balancing can restore rotating equipment to its original, efficient and quiet operating mode. The benefits of such testing help avoid costly emergency repairs, pinpoints irregularities before becoming major problems, and can increase the useful life of system components.

8. The necessary steps for preparing the unit for prolonged shutdown, *i.e.*, winter, and specified instruction for starting after this standby period, should both be part of the program. With compressors that have internal lubrication systems, provisions should be made to have their oil heaters energized continuously throughout this period or to have their oil charges replaced prior to putting them back into operation.

REFERENCES

API. 1979. Centrifugal compressors for general refinery services. API *Standard* 617-79. American Petroleum Institute, Washington, D.C.

ARI. 1983. Centrifugal water-chilling packages. ARI *Standard* 550-83.

ASHRAE. 1968. The operation and maintenance of centrifugal units. ASHRAE Symposium LP-68-2.

ASHRAE. 1969. Centrifugal heat pump systems. ASHRAE Symposium DV-69-4.

ASHRAE. 1973. Centrifugal chiller noise in buildings. ASHRAE Symposium LO-73-9.

ASME. 1965. Compressors and exhausters. ASME Performance Test Code 10-65. American Society of Mechanical Engineers, New York.

ASME. 1983. Pressure vessels division 1. SI Sec 8-D-1-83. American Society of Mechanical Engineers, New York.

Blazer, W.E., Jr. 1972. Chiller noise: Its impact on building design. ASHRAE *Transactions* 78(1):268.

Bush, J. and J. Elson. 1988. Scroll compressor design criteria for residential air conditioning and heat pump applications. Proceedings of the 1988 International Compressor Engineering Conference (1):83-97 (July). Office of Publications, Purdue University, West Lafayette, IN.

Caillat, J., R. Weatherston, and J. Bush. Scroll-type machine with axially compliant mounting. U.S. Patent 4,767,292, 1988.

Creux, L. Rotary engine. U.S. Patent 801,182, 1905.

Davis, H., H. Kottas, and A.M.G. Moody. 1951. The influence of Reynolds number on the performance of turbomachinery. ASME *Transactions* (July):499.

Edmister, W.C. 1961. *Applied hydrocarbon thermodynamics.* Gulf Publishing Company, Houston, TX, 22 and 52.

Elson, J., G. Hundy, and K. Monnier. 1990. Scroll compressor design and application characteristics for air conditioning, heat pump, and refrigeration applications. Proceedings of the Institute of Refrigeration (London), 2.1-2.10 (November).

Hoover, R.M. 1960. Noise levels due to a centrifugal compressor installed in an office building penthouse. *Noise Control* (6):136.

Hougen, O.A., K.M. Watson, and R.A. Ragatz. 1959. *Chemical process principles*, Part II—Thermodynamics. John Wiley and Sons, Inc., New York, 579 and 611.

Kerschbaumer, H.G. 1975. An investigation into the parameters determining energy consumption of centrifugal compressors operating in heat recovery systems. International Congress of Refrigeration, Moscow, Paper B.2.61.

McCullough, J. Positive fluid displacement apparatus. U.S. Patent 3,924,977, 1975.

McCullough, J. and R. Shaffer. Axial compliance means with radial sealing for scroll type apparatus. U.S. Patent 3,994,636, 1976.

Military Specification, Refrigerating Unit, Centrifugal for Air Conditioning. 1968. MIL-R-24085A, Ships (March).

Mitchell, J.S. 1977. Monitoring machinery health. Power Vol. 121, Part I (March):46; Part II (May):87; Part III (July):38.

Morishita, E., Y. Kitora, T. Suganami, S. Yamamoto, and M. Nishida. 1988. Rotating scroll vacuum Pump. Proceedings of the 1988 International Compressor Engineering Conference (July). Office of Publications, Purdue University, West Lafayette, IN.

Purvis, E. Scroll compressor technology. 1987. Heat Pump Conference, New Orleans.

Sato, T. and K. Terauchi. Scroll type fluid displacement apparatus with centrifugal force balanceweight. U.S. Patent 4,597,724, 1986.

Sauls, J. Involute and laminated tip seal of labyrinth type for use in a scroll machine. U.S. Patent 4,411,605, 1983.

Schultz, J.M. 1962. The polytropic analysis of centrifugal compressors. ASME *Transactions* (January and April):69 and 222.

Sessler, S.M. 1973. Acoustical and mechanical considerations for the evaluation of chiller noise. ASHRAE *Journal* 15(10):39.

Sheets, H.E. 1952. Nondimensional compressor performance for a range of Mach numbers and molecular weights. ASME *Transactions* (January):93.

Tojo, K., T. Hosoda, M. Ikegawa, and M. Shiibayashi. Scroll compressor provided with means for pressing an orbiting scroll member against a stationary scroll member and self-cooling means. U.S. Patent 4,365,941, 1982.

CONDENSERS

THE condenser in a refrigeration system is a heat exchanger that usually rejects all the heat from the system. This heat consists of heat absorbed by the evaporator plus the heat equivalent of the energy input to the compressor. The compressor discharges hot, high-pressure refrigerant gas into the condenser, which rejects heat from the gas to some cooler medium. Thus, the cool refrigerant condenses back to the liquid state and drains from the condenser to continue in the refrigeration cycle.

The common forms of condensers may be classified on the basis of the cooling medium as (1) water-cooled, (2) air-cooled, and (3) evaporative (air- and water-cooled).

WATER-COOLED CONDENSERS

HEAT REMOVAL

The heat rejection rate in a condenser for each unit of heat removal produced in the evaporator may be estimated from the graph in Figure 1. The theoretical values shown are based on Refrigerant 22 with 10°F suction superheat, 10°F liquid subcooling, and 80% compressor efficiency. Actually, the heat removed is slightly higher or lower than these values, depending on compressor efficiency. Usually, the heat rejection requirement q_o can be accurately determined from known values of evaporator load q_i and the heat equivalent of the actual power required q_w for compression (obtained from the compressor manufacturer's catalog):

$$q_o = q_i + q_w \tag{1}$$

Note: q_w is reduced by any independent heat-rejection processes (oil cooling, motor cooling, etc.).

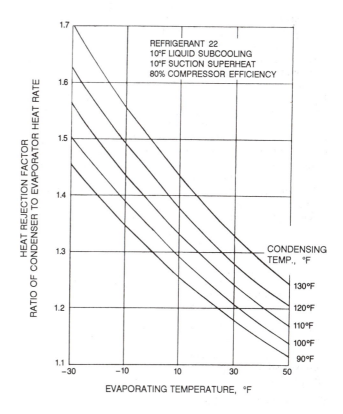

REFRIGERANT 22
10°F LIQUID SUBCOOLING
10°F SUCTION SUPERHEAT
80% COMPRESSOR EFFICIENCY

CONDENSING TEMP., °F

HEAT REJECTION FACTOR
RATIO OF CONDENSER TO EVAPORATOR HEAT RATE

EVAPORATING TEMPERATURE, °F

Fig. 1   Heat Removed in Condenser

The preparation of this chapter is assigned to TC 8.4, Air-to-Refrigerant Heat-Transfer Equipment; TC 8.5, Liquid-to-Refrigerant Heat Exchangers; and TC 8.6, Cooling Towers and Evaporative Condensers.

The volumetric flow rate of condensing water required may be found from Equation (2):

$$Q = q_o/[\rho c_p(t_2 - t_1)] \tag{2}$$

where:

Q = volumetric flow rate of water, ft³/h
(multiply ft³/h by 0.125 to obtain gpm)
q_o = heat-rejection rate, Btu/h
ρ = density of water, lb/ft³
t_1 = temperature of water entering condenser, °F
t_2 = temperature of water leaving condenser, °F
c_p = specific heat of water, Btu/(lb·°F)

Example 1. Estimate the volumetric flow rate of condensing water required for the condenser of a Refrigerant 22 water-cooled unit operating at a condensing temperature of 105°F, an evaporating temperature of 40°F, 10°F liquid subcooling, and 10°F suction superheat. Water enters the condenser at 86°F and leaves at 95°F. The refrigeration load is 100 tons.

Solution: From Figure 1, the heat rejection factor for these conditions is about 1.19.

q_o = 100(1.19) = 119 tons
ρ = 62.1 lb/ft³ at 90.5°F
c_p = 1.0 Btu/(lb·°F)

From Equation (2):

Q = 1496(119)/62.1(95 − 86)1.0 = 319 gpm

Note: The value 1496 is a unit's conversion factor.

HEAT TRANSFER

A water-cooled condenser transfers heat in three stages: sensible cooling in the gas desuperheating and condensate subcooling stages, and transfer of latent heat in the condensing stage. Condensing is by far the dominant process in normal refrigeration system application, accounting for 83% of the heat rejection in the previous example. Because the tube wall temperature is normally lower than the condensing temperature at all locations in the condenser, condensation takes place throughout the condenser.

The effect of changes in the entering gas superheat are typically insignificant due to an inverse proportional relationship between temperature difference and heat transfer coefficient. As a result, an average overall heat transfer coefficient and the mean temperature difference (calculated from the condensing temperature corresponding to the saturated condensing pressure and the entering and leaving water temperatures) give reasonably accurate predictions of performance.

Subcooling has an effect on the average overall heat transfer coefficient when tubes are submerged in liquid.

The heat rejection rate is then determined as:

$$q = UA\Delta t_m \tag{3}$$

where:

q = total heat transfer rate, Btu/h
U = overall heat transfer coefficient, Btu/(h·ft²·°F)
A = heat transfer surface area associated with U, ft²
Δt_m = mean temperature difference, °F

Chapter 3 of the 1989 ASHRAE *Handbook—Fundamentals* shows how to calculate Δt_m. When a detailed analysis of integral desuperheating and/or subcooling processes in condenser design is desired, computational schemes have been developed (Bell 1972).

Overall Heat Transfer U_o

The overall heat transfer in a water-cooled condenser with water inside the tubes may be computed from calculated or test-derived heat transfer coefficients of the water and refrigerant sides, from physical measurements of the condenser tubes, and from a fouling factor on the water side, by using Equation (4).

$$U_o = \frac{1}{(A_o/A_i)/h_w + (A_o/A_i)r_{fw} + (t/k)(A_o/A_m) + 1/(h_r\phi_s)} \tag{4}$$

where:

U_o = overall heat transfer coefficient, based on the external surface and the mean temperature difference, between the external and internal fluids, Btu/(h·ft²·°F)
A_o/A_i = ratio of external to internal surface area
h_w = internal or water-side film coefficient, Btu/(h·ft²·°F)
r_{fw} = fouling resistance on water side, (ft²·h·°F)/Btu
t = thickness of tube wall, ft
k = thermal conductivity of tube material, Btu/(h·ft·°F)
A_o/A_m = ratio of external to mean heat transfer surface areas of metal wall
h_r = external, or refrigerant side, coefficient, Btu/(h·ft²·°F)
ϕ_s = surface fin efficiency (100% for bare tubes)

For tube-in-tube condensers or other condensers where the refrigerant flows inside the tubes, the equation for U_o, in terms of water-side surface, becomes:

$$U_o = \frac{1}{(A_o/A_i)/h_r + r_{fw} + (t/k) + 1/h_w} \tag{5}$$

where:

h_r = internal or refrigerant-side coefficient, Btu/(h·ft²·°F)
h_w = external or water-side coefficient, Btu/(h·ft²·°F)

Water-Side Film Coefficient h_w

Values of the water-side coefficient may be calculated from equations in Chapter 3 of the 1989 ASHRAE *Handbook—Fundamentals*. For turbulent flow, at Reynolds numbers exceeding 10,000 in horizontal tubes and using average water temperatures, the general equation (McAdams 1954) is:

$$h_w D/k = 0.023(DG/\mu)^{0.8}(c_p\mu/k)^{0.4} \tag{6}$$

where:

D = inside tube diameter, ft
k = thermal conductivity of water, Btu/(h·ft·°F)
G = mass velocity of water, lb/(h·ft²)
μ = viscosity of water, lb/(ft·h)
c_p = specific heat of water at constant pressure, Btu/(lb·°F)

The constant (0.023) in Equation (6) reflects plain ID tubes. Bergles (1973, 1976) discusses numerous water-side enhancement methods, which increase the value of the constant in Equation (6).

Because of its strong influence on the value of h_w, water velocity should generally be maintained as high as possible without initiating erosion or excessive pressure drop. Typical maximum velocities from 6 to 10 ft/s are common with clean water. Experiments by Sturley (1975) at velocities up to approximately 26 ft/s show no damage to copper tubes after long operation. Regarding erosion potential, water quality is the key factor (Ayub and Jones 1987). A minimum velocity of 3 ft/s is good practice when the water quality is such that noticeable fouling or corrosion could result. With clean water, the velocity may be lower if it must be conserved or if it has a low temperature. In some cases, the minimum flow may be determined by a lower Reynolds number limit.

Refrigerant-Side Film Coefficient h_r

Factors influencing the value of h_r are listed below:

- Type of refrigerant being condensed
- Geometry of condensing surface (plain tube OD and finned tube fin spacing, height, and cross-section profile)

- Condensing temperature
- Condensing rate in terms of mass velocity or rate of heat transferred
- Arrangement of tubes in bundle and inlet and outlet connection locations
- Vapor distribution and rate of flow
- Condensate drainage
- Liquid subcooling

Values of the refrigerant-side coefficients may be estimated from correlations shown in Chapter 4 of the 1989 ASHRAE *Handbook—Fundamentals*. Information on the effects of the type of refrigerant, condensing temperature, and loading (temperature drop across the condensate film) on the condensing film coefficient can be found in the section Condensing in the same chapter. Katz *et al.* (1947); Katz and Robinson (1947); and Kratz *et al.* (1930) give further information.

Actual values of h_r for a given physical condenser design can be determined from test data by use of a *Wilson Plot* (McAdams 1954, Briggs and Young 1969).

The type of condensing surface has a considerable effect on the condensing coefficient. Most halocarbon refrigerant condensers use finned tubes where the fins are integral with the tube. Water velocities normally used are large enough for the resulting high water-side film coefficient to justify using an extended external surface to balance the heat transfer resistances of the two surfaces. Pearson and Withers (1969) compared the refrigerant condensing performance of some integral finned tubes with different fin spacings. Some other refrigerant-side enhancements are described by Bergles (1976) and Webb (1984a). Condensing coefficient data for R-11 are given by Webb and Murawski (1990) for several tube geometries. The effect of fin shape on the condensing coefficient is addressed by Kedzierski and Webb (1990).

Physical aspects of a given condenser design—such as tube spacing and orientation, shell-side baffle arrangement, orientation of multiple water-pass arrangements, refrigerant connection locations, and the number of tubes high in the bundle—affect the refrigerant-side coefficient by influencing vapor distribution and flow rate through the tube bundle and condensate drainage from the bundle. Butterworth (1977) reviewed correlations accounting for these variables in predicting the heat transfer coefficient for shell-side condensation. These effects are also surveyed by Webb (1984b). Kistler *et al.* (1976) developed analytical procedures for design use within these parameters.

As refrigerant condenses on the tubes, it falls on the tubes in lower rows. Due to the added resistance of this liquid film, the effective film coefficient for lower rows should be less than that for upper rows. Therefore, the average overall refrigerant film coefficient should decrease as the number of tube rows increase. Webb and Murawski (1990) present row effect data for five tube geometries. However, the additional compensating effects of added film turbulence and direct contact condensation on the subcooled liquid film make actual row effect uncertain.

Liquid refrigerant may be subcooled by raising the condensate level to submerge a desired number of tubes. The refrigerant film coefficient associated with the submerged tubes is less than the condensing coefficient. If the refrigerant film coefficient in Equation (4) is an average based on all tubes in the condenser, its value decreases as a greater portion of the tubes is submerged.

Tube-Wall Resistance (t/k)

Most refrigeration condensers, with the exception of ammonia, use relatively thin-walled copper tubes. Where these are used, the temperature drop or gradient across the tube wall is not significant. If the tube metal has a high thermal resistance, as does 70/30 cupronickel, considerable temperature drop will occur or, conversely, an increase in the mean temperature difference Δt_m or in surface area is required to transfer the same amount of heat, compared to copper. Although the term (t/k) in Equations (4) and (5) is an approximation, as long as the wall thickness is not more than 14% of the tube diameter, the error will be less than 1%. To improve the accuracy of the wall resistance calculation for cases of heavy tube wall or low-conductivity material, see Chapter 3 of the 1989 ASHRAE *Handbook—Fundamentals*.

Surface Efficiency ϕ_s

For a finned tube, a temperature gradient exists from the root of a fin to its tip because of the thermal resistance of the fin material. The surface efficiency can be calculated from the fin efficiency, which accounts for this effect (see Chapter 3 of the 1989 ASHRAE *Handbook—Fundamentals*.) For tubes with low conductivity material, high fins, or high values of fin pitch, the fin efficiency becomes increasingly significant. Young and Ward (1957), along with the UOP *Engineering Data Book* II (1984), describe methods of evaluating these effects.

Fouling Factor r_{fw}

Manufacturers' ratings are based on commercially clean equipment with an allowance for the possibility of water-side fouling. This allowance often takes the form of a fouling factor, which is a thermal resistance referenced to the water-side area of the heat transfer surface. Thus, the temperature penalty imposed on the condenser is equal to the heat flux at the water-side area, multiplied by the fouling factor. Increased fouling increases overall heat transfer resistance, due to the parameter $(A_o/A_i)r_{fw}$ in Equation (4). Fouling increases the Δt_m required to obtain the same capacity—with a corresponding increase in condenser pressure and system power—or lowers system capacity.

Allowance for a given fouling factor has a greater influence on equipment selection than simply increasing the overall resistance (Starner 1976). Required increases in surface area result in lower water velocities. Consequently, the increase in heat transfer surface required for the same performance is due to both the fouling resistance and the additional resistance which results from lower water velocities.

For a given tube surface, load, and water temperature range, the tube length can be optimized to give a desired condensing temperature and water-side pressure drop. The solid curves in Figure 2 show the effect on water velocity, tube length, and overall surface required due to increased fouling. A fouling factor of 0.00072 ft²·h·°F/Btu doubles the required surface area compared to that with no fouling allowance.

A worse case occurs when an oversized condenser must be selected to meet increased fouling requirements but without the flexibility of increasing tube length. As shown by the dashed lines in Figure 2, the water velocities decrease more rapidly as the total surface increases to meet the required performance. Here, the required surface area doubles with a fouling factor of only 0.00049 ft²·h·°F/Btu. If the application can afford more pumping power, the water flow rate may be increased to obtain a higher velocity, which increases the water film heat transfer coefficient. This factor, plus the lower leaving water temperature, reduces the condensing temperature.

Fouling is a major unresolved problem in heat exchanger design (Taborek *et al.* 1972). The major uncertainty is which fouling factor to choose for a given application or water condition to obtain expected performance from the condenser; use of too low a fouling factor wastes compressor power, while too high a factor wastes heat exchanger material.

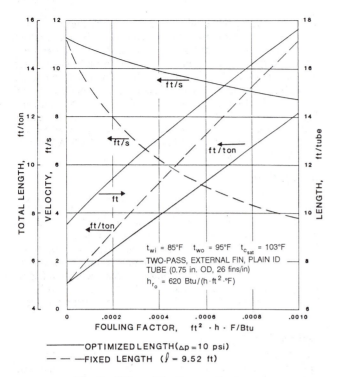

t_{wi} = 85°F t_{wo} = 95°F $t_{c_{sat}}$ = 103°F
TWO-PASS, EXTERNAL FIN, PLAIN ID TUBE (0.75 in. OD, 26 fins/in)
h_{r_o} = 620 Btu/(h·ft²·°F)

Fig. 2   Effect of Fouling on Condenser

Fouling may result from sediment, biological growths, and corrosion products. Scale results from the deposition of chemicals from the cooling water onto the warmer surface of the condenser tube. Chapter 43 of the 1991 ASHRAE *Handbook—HVAC Applications* discusses water chemistry and water treatment factors that are important in controlling corrosion and scale in condenser cooling water.

Tables of fouling factors are available; however, in many cases, the values are greater than necessary (TEMA 1988). Extensive research has been conducted into the causes, rates, and values of water-side fouling. Numerous models have been proposed. These studies generally found that fouling resistance reaches an asymptotic value with time (Suitor *et al.* 1976). Many fouling research data are based on surface temperatures that are considerably higher than those found in air-conditioning and refrigeration condensers. Lee and Knudsen (1979) and Coates and Knudsen (1980) found that, in the absence of suspended solids or biological fouling, long-term fouling of condenser tubes does not exceed 0.0002 ft²·h·°F/Btu, and short-term fouling does not exceed 0.0001 ft²·h·°F/Btu (ASHRAE 1982). Periodic cleaning of condenser tubes (mechanically or chemically) usually maintains satisfactory performance, except in severe environments. The appropriate edition of ARI *Standard* 450 for water-cooled condensers should be referred to when reviewing manufacturers' ratings. This standard gives methods for correcting ratings for different values of fouling.

WATER PRESSURE DROP

The water (or other fluid) pressure drop is important for designing or selecting condensers. Where a cooling tower cools the condensing water, the water pressure drop through the condenser is generally limited to about 10 psi. If the condenser water comes from another source, the pressure drop through the condenser should be lower than the available pressure to allow for pressure fluctuations and additional flow resistance caused by fouling.

Pressure drop through horizontal condensers includes the loss through the tubes, tube entrance and exit losses, and losses through the heads or return bends (or both). The effect of the coiling of the tubes must be considered in shell-and-coil condensers.

Expected pressure drop through tubes can be calculated from a modified Darcy-Weisbach equation:

$$\Delta p = N_p[K_H + f(L/D)](\rho V^2/2g_c) \qquad (7)$$

where:
Δp = pressure drop, psi
N_p = number of tube passes
K_H = entrance and exit flow resistance and flow reversal coefficient, number of velocity heads ($V^2/2g$)
f = friction factor
L = length of tube, ft
D = inside tube diameter, ft
ρ = fluid density, lb/ft³
V = fluid velocity, ft/s
g_c = gravitational constant, 32.17 $lb_m \cdot ft/(lb_f \cdot s^2)$

For tubes with smooth inside diameters, the friction factor may be determined from a Moody chart or various relations, depending on the flow regime and wall roughness (see Chapter 2 of the 1989 ASHRAE *Handbook—Fundamentals*). For tubes with internal enhancement, the friction factor should be obtained from the tube manufacturer.

The value of K_H depends on tube entry and exit conditions and the flow path between passes. As a minimum, a value of 1.5 is recommended. This factor is more critical with short tubes.

Predicting pressure drop for shell-and-coil condensers is more difficult than it is for shell-and-tube condensers because of design curvature of the coil and tube flattening or kinking during coil bending. Seban and McLaughlin (1963) discuss the effect of curvature or bending of pipe and tubes on the pressure drop.

LIQUID SUBCOOLING

The amount of condensate subcooling provided by the condensing surface in a shell-and-tube condenser is small, generally less than 2°F. When a specific amount of subcooling is required, it may be obtained by submerging tubes in the condensate. Tubes in the lower portion of the bundle are used for this purpose. If the condenser is multipass, then the subcooling tubes should be included in the first pass to gain exposure to the coolest water.

Subcooling benefits are achieved at the expense of both condensing surface and refrigerant inventory. An optimum design should be based on these opposing factors.

When means are provided to elevate the condensate level for the desired submergence of the subcooler tubes, heat is transferred principally by natural convection.

Subcooling performance can be improved by enclosing tubes in a separate compartment within the condenser to obtain the benefits of forced convection of the liquid condensate.

Segmental baffles may be provided to produce flow across the tube bundle. Kern and Kraus (1972) describe how heat transfer performance can be estimated analytically by use of longitudinal or crossflow correlations, but it is more practically determined by test because of the large number of variables. It is important to keep the refrigerant pressure drop incurred along the flow path from exceeding the pressure difference permitted by the saturation pressure of the subcooled liquid.

CIRCUITING

Varying the number of water-side passes in a condenser can affect the saturated condensing temperature significantly, which affects system performance. Figure 3 shows the change in condensing temperature for one, two, or three passes in a particular condenser. As an example, at a loading of 22,000 Btu/(h · tube), a two-pass condenser with a 10°F range would have a condensing temperature of 102.3°F. At the same loading with a one-pass, 5°F range, this unit would have a condensing temperature of 99.3°F. The one-pass option does, however, require twice the water flow rate with an associated increase in pumping power. A three-pass design may be favorable when the costs associated with water flow outweigh the system gains from lower condensing temperature. Hence, different numbers of passes (if an option) and ranges should be weighed against other parameters (water source, pumping power, cooling tower design, etc.) to optimize overall system performance and cost.

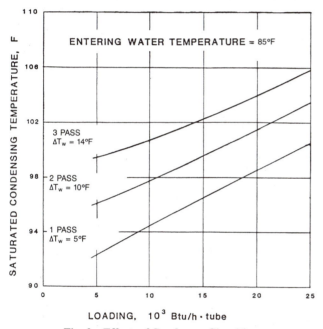

LOADING, 10^3 Btu/h · tube

Fig. 3   Effects of Condenser Circuiting

CONDENSER TYPES

The most common types of water-cooled refrigerant condensers are (1) shell-and-tube; (2) shell-and-coil; (3) tube-in-tube; and (4) brazed plate. The type selected depends on the size of the cooling load, the refrigerant used, the quality and temperature of the available cooling water, the amount of water that can be circulated, the location and space allotment, the required operating pressures (water and refrigerant sides), cost, and maintenance considerations.

Shell-and-Tube Condensers

Built in sizes from 1 to 10,000 tons, the refrigerant in these condensers condenses outside the tubes and the cooling water circulates through the tubes in a single or multipass circuit. Fixed tube sheet, straight tube construction is usually used, although U-tubes that terminate in a single tube sheet are sometimes used. Typically, shell-and-tube condenser tubes run horizontally. Where floor installation area is limited, the condenser tubes may be oriented vertically. However, vertical

tubes have poor condensate draining, which reduces the refrigerant film coefficient. Vertical condensers with open water systems have been used with ammonia.

Gas inlet and liquid outlet nozzles should be located carefully. The proximity of these nozzles may adversely affect condenser performance by requiring excessive amounts of liquid refrigerant to seal the outlet nozzle from inlet gas flow. This effect can be diminished by the addition of baffles at the inlet and/or outlet connection.

Halocarbon refrigerant condensers have been made with many types of material specifications, including all prime surface or finned, ferrous, or nonferrous tubes. Common tubes are nominal 0.75 and 1.0 in. OD copper tubes with integral fins on the outside. These tubes are often available with fin heights from 0.035 to 0.061 in. and fin spacings of 19, 26, and 40 fins/in. For ammonia condensers, prime surface steel tubes, 1.25 in. OD, with 0.095 in. average wall thickness, are commonly used.

An increased number of tubes designed for enhanced heat transfer are available (Bergles 1976). On the inside of the tube, common enhancemments include longitudinal or spiral grooves and ridges, internal fins, and other devices to promote turbulence and augment heat transfer. On the refrigerant side, condensate surface tension and drainage are important in design of the tube outer surface. Tubes are available with the outsides machined or formed specifically to enhance the condensation and promote drainage. Heat transfer design equations should be obtained from the manufacturer.

Because the water and refrigerant film resistances with enhanced tubes are reduced, the effect of fouling becomes relatively great. Where high levels of fouling occur, the fouling resistance may easily account for over 50% of the total. In such cases, the advantages of enhancement may diminish. On the other hand, water-side augmentation, which creates turbulence, may reduce fouling. The actual value of fouling resistances depends on the particular type of enhancement and the service conditions (Starner 1976, Watkinson et al. 1974).

Similarly, refrigerant-side enhancements may not show as much benefit in very large tube bundles as in smaller bundles. This is due to the row effect addressed previously in the section on refrigerant film coefficient.

The tubes are either brazed into thin copper, copper alloy, or steel tube sheets, or they are rolled into heavier nonferrous or steel tube sheets. Straight tubes with a maximum OD less than the tube hole diameter and rolled into tube sheets are removable. This construction facilitates field repair in the event of tube failure.

The required heat transfer area for a shell-and-tube condenser can be found by solving Equations (2), (3), and (4). The mean temperature difference is the logarithmic mean temperature difference, with the entering and leaving refrigerant temperatures taken as the saturated condensing temperature. Depending on the parameters fixed, an iterative solution may be required.

Shell-and-U-tube condenser design principles are the same as those outlined for horizontal shell-and-tube units, with one exception: the water pressure drop through the U-bend portion of the U-tube is generally less than that through the compartments in the water head where the direction of water flow is reversed. The pressure loss is a function of the inside tube diameter and the ratio of the inside tube diameter to bending centers. Pressure loss should be determined by test.

Shell-and-Coil Condensers

Shell-and-coil condensers have the cooling water circulated through one or more continuous or assembled coils contained within the shell. The refrigerant condenses outside the tubes. Capacities range from 0.5 to 15 tons. Due to the type of construction, the tubes are neither replaceable nor mechanically cleanable.

Again, Equations (2), (3), and (4) may be used for performance calculations, with the saturated condensing temperature used for the entering and leaving refrigerant temperatures in the logarithmic mean temperature difference. The value of h_w (the water-side film coefficient) and, especially, the pressure loss on the water side require close attention; laminar flow can exist at considerably higher Reynolds numbers in coils than in straight tubes. Because the film coefficient for turbulent flow is greater than that for laminar flow, values of h_w, as calculated from Equation (6), will be too high if the flow is not turbulent. Once the flow has become turbulent, the film coefficient will be greater than that for a straight tube (Eckert 1963). Pressure drop through helical coils can be much greater than that through smooth straight tubes for the same length of travel. The section Water Pressure Drop outlines the variables that make an accurate determination of the pressure loss difficult. The pressure loss and heat transfer rates should be determined by test because of the large number of variables inherent in this condenser.

Tube-in-Tube Condensers

These condensers consist of one or more assemblies of two tubes, one within the other, in which the refrigerant vapor is condensed in either the annular space or the inner tube. These units are built in sizes from 0.3 to 50 tons.

Equations (2) and (3) would be used to size a tube-in-tube condenser. Because the refrigerant may have a significant pressure loss through its flow path, the refrigerant temperatures used in calculating the mean temperature difference should be selected carefully. The refrigerant temperatures should be consistent with the model used for the refrigerant film coefficient. Logarithmic mean temperature difference for either counterflow or parallel flow should be used, depending on the piping connections. Equation (4) can be used to find the overall heat transfer coefficient when the water flows in the tubes, and Equation (5) may be used when the water flows in the annulus.

Tube-in-tube condenser design differs from those outlined previously, depending on whether the water flows through the inner tube or through the annulus. Condensing coefficients are more difficult to predict when condensation occurs within a tube or annulus, because the mechanism differs considerably from condensation on the outside of a horizontal tube. Where the water flows through the annulus, disagreement exists regarding the appropriate method used to calculate the water-side film coefficient and the water pressure drop. The problem is further complicated if the tubes are also formed in a spiral.

The water side is mechanically cleanable only when the water flows inside straight tubes and cleanout access is provided. Tubes are not replaceable.

Brazed Plate Condensers

Brazed plate condensers are constructed of plates brazed together to make up an assembly of separate channels. Capacities range from 0.5 to 100 tons.

The plates, typically stainless steel, are usually configured with a wave-style pattern resulting in high turbulence and low susceptibility to fouling. The design has some ability to withstand freezing and, because of the compact design, allows low refrigerant charge. The construction does not allow mechanical cleaning, and internal leaks cannot typically be repaired.

Performance calculations are similar to other types of condensers; however, very few correlations are available for the heat transfer coefficients.

NONCONDENSABLE GASES

When first assembled, most refrigeration systems contain gases, usually air and water vapor. As addressed later in this chapter, these gases are detrimental to condenser performance, and it is thus important to evacuate the entire refrigeration system before operation.

For low-pressure refrigerants, where the operating pressure of the evaporator is less than ambient pressure, even slight leaks can be a continuing source of noncondensables. In such cases, a purge system, which automatically expels noncondensable gases, is recommended. Figure 4 shows the refrigerant loss associated with the use of purging devices at various operating conditions.

When present, noncondensable gases collect on the high-pressure side of the system and raise the condensing pressure above that corresponding to the temperature at which the refrigerant is actually condensing. The increased condensing pressure increases power consumption and reduces capacity. Also, if oxygen is present at a point of high discharge temperature, the oil may oxidize.

The excess pressure is caused by the partial pressure of the noncondensable gas. These gases form a resistance film over some of the condensing surface, thus lowering the heat transfer coefficient. Webb *et al.* (1980) showed how a small percentage of noncondensables can cause major decreases in the refrigerant film coefficient in shell-and-tube condensers. See also Henderson and Marchello (1969) and Chapter 4 of the 1989 ASHRAE *Handbook—Fundamentals*. The noncondensable situation of a given condenser is difficult to characterize because such gases tend to accumulate in the coldest and least agitated part of the condenser or in the receiver. Thus, a fairly high percentage of noncondensables can be tolerated if the gases are confined to areas far from the heat transfer surface. One way to account for noncondensables is to treat them as a

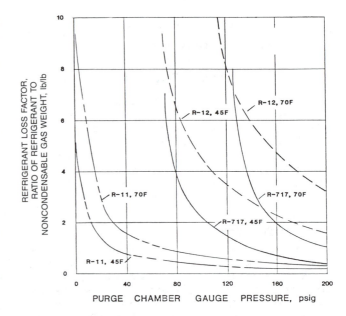

Fig. 4   Loss of Refrigerant during Purging at Various Gas Temperatures and Pressures

refrigerant or gas-side fouling resistance in Equation (4). Some predictions are presented by Wanniarachchi and Webb (1982).

As an example of the effect on system performance, experiments performed on a 250-ton R-11 chiller condenser reveal that 2% noncondensables by volume caused a 15% reduction in condensing coefficient. Also, 3 and 8% noncondensables by volume caused power increases of 2.6 and 5%, respectively.

The presence of noncondensable gases can be tested by shutting down the refrigeration system, while allowing the condenser water to flow long enough for the refrigerant to reach the same temperature as the water. If the condenser pressure is higher than the pressure corresponding to the refrigerant temperature, noncondensable gases are present. This test may not be sensitive enough to detect the presence of small amounts of noncondensables, which can, nevertheless, decrease shell-side condensing coefficients.

CONSTRUCTION AND TEST CODES

Pressure vessels must be constructed and tested under the rules of national, state, and local codes. The introduction of the current ASME Boiler and Pressure Vessel Code, Section VIII, gives guidance on rules and exemptions.

The more common applicable codes and standards are as follows:

ARI *Standard* 450-87, Standard for Water-Cooled Refrigerant Condensers, Remote Type, covers industry criteria for standard equipment, standard safety provisions, marking, fouling factors, and recommended rating points for water-cooled condensers.

ASHRAE *Standard* 22-92, Methods of Testing for Rating Water-Cooled Refrigerant Condensers, covers recommended testing methods.

ANSI/ASHRAE *Standard* 15-89, Safety Code for Mechanical Refrigeration, specifies design criteria, use of materials, and testing. It refers to the ASME Boiler and Pressure Vessel Code, Section VIII, for refrigerant-containing sides of pressure vessels, where applicable. Factory test pressures are specified, and minimum design working pressures are given by this code. This code requires pressure-limiting and pressure-relief devices on refrigerant-containing systems, as applicable, and defines the setting and capacity requirements for these devices.

ASME Boiler and Pressure Vessel Code, Unfired Pressure Vessels, Section VIII, covers the safety aspects of design and constructions. Most states require condensers to meet the requirements of the ASME if they fall within scope of the ASME code. Some of the exceptions from meeting the ASME requirements listed in the ASME code are as follows:

- Condenser shell ID is 6 in. or less.
- 15 psig or less.
- The fluid (water) portion of the condenser need not be built to the requirements of the ASME code if the fluid is water, the design pressure does not exceed 300 psig, and the design temperature does not exceed 210°F.

Condensers meeting the requirements of the ASME code will have an ASME stamp. The ASME stamp is a *U* or *UM* inside a three-leaf clover. The *U* can be used for all condensers; the *UM* can be used (considering local codes) for those with net refrigerant-side volume less than 1.5 ft³ if less than 600 psig, or less than 5 ft³ if less than 250 psig.

UL *Standard* 207-86, Refrigerant-Containing Components and Accessories, covers specific design criteria, use of materials, testing, and initial approval by Underwriters Laboratories. A condenser with the ASME *U* stamp does not require UL approval.

Design Pressure

Refrigerant side, as a minimum, should be saturated pressure for the refrigerant used at 105°F, or 15 psig, whichever is greater. Standby temperatures and temperatures encountered during shipping of units with a refrigerant charge should also be considered.

Required fluid (water) side pressure varies, depending largely on the following conditions: static head, pump head, transients due to pump start-up, and valve closing. A common water-side design pressure is 150 psig, although with taller building construction, requirements for 300 psig are not uncommon.

OPERATION AND MAINTENANCE

When a water-cooled condenser is selected, anticipated operating conditions, including water and refrigerant temperatures, have usually been determined. Standard practice allows for a fouling factor in the selection procedure. A new condenser, therefore, operates at a condensing temperature lower than the design point because it has not yet fouled. Once a condenser starts to foul or scale, economic considerations determine how frequently the condenser should be cleaned. As the scale builds up in a condenser, the condensing temperature and subsequent power consumption increase, while the unit capacity decreases. This effect can be seen in Figure 5 for a condenser with a design fouling factor of 0.00025 ft²·h·°F/Btu. At some point, the increased cost of power can be offset by the labor cost of cleaning.

Local water conditions, as well as the effectiveness of chemical water treatment, if used, make the use of any specific maintenance schedule difficult.

Cleaning can be done either mechanically with a brush or chemically with an acid solution. In applications where water-side fouling may be severe, on-line cleaning can be accomplished by brushes installed in cages in the water heads. By using valves, flow is reversed at set intervals, propelling the

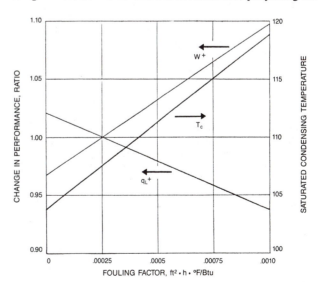

q_L^+ = chiller actual capacity/chiller design capacity
W^+ = compressor actual kW/compressor design kW
T_c = saturated condensing temperature
Design condenser fouling factor = 0.00025 ft²·h·°F/Btu
Cooler leaving water temperature = 44°F
Condenser entering water temperature = 85°F

Fig. 5   Effect of Fouling on Chiller Performance

brushes through the tubes (Kragh 1975). The most effective method depends on the type of scale formed. Competent advice in selecting the particular method of tube cleaning is advisable.

Occasionally, one or more tubes may develop leaks because of corrosive impurities in the water or through improper cleaning procedures. These leaks must be found and repaired as soon as possible; this can normally be done by replacing the leaky tubes, a procedure requiring tools and skills that are best found through the original condenser manufacturer. In large condensers, where the contribution of a single tube is relatively insignificant, a simpler approach may be to seal the ends of the leaking tube.

If the condenser is located where the water can freeze during the winter, special precautions should be taken when it is idle. Opening all vents and drains may be sufficient, but water heads should be removed and tubes blown free of water.

If refrigerant vapor is to be released from the condenser and there is water in the tubes, the pumps should be on and the water flowing. Otherwise, freezing can easily occur.

Finally, the condenser manufacturer's installation recommendations on orientation, piping connections, space requirements for tube cleaning or removal, and other important factors should be followed.

AIR-COOLED CONDENSERS

In an air-cooled condenser, heat is transferred in three main phases: (1) desuperheating, (2) condensing, and (3) subcooling. Figure 6 shows the changes of state of Refrigerant 12 passing through the condenser coil and the corresponding temperature change of the cooling air as it passes through the coil. Desuperheating, condensing, and subcooling zones vary from 5 to 10%, depending on the entering gas temperature and the leaving liquid temperature, but Figure 6 is typical for most common refrigerants.

Condensing takes place in approximately 85% of the condenser area at a substantially constant temperature. The indicated drop in condensing temperature results from the friction loss through the condenser coil.

An air-cooled condenser may be located either adjacent or remote from the compressor. Additional variations will be either vertical (top) or horizontal (side) air discharge orientating as well as designed for indoor or outdoor application.

COIL CONSTRUCTION

A condenser coil with optimum circuiting requires the smallest heat transfer surface and has the fewest operational problems, such as trapping oil and refrigerant, high discharge pressure, etc. Coils are commonly constructed of copper, aluminum, or steel tubes, ranging from 0.25 to 0.75 in. in diameter. Copper is easy to use in manufacturing and requires no protection against corrosion. Aluminum requires exact manufacturing methods and special protection in the case of aluminum-to-copper joints. Steel tubing requires weather protection.

Tube diameter is chosen as a compromise between factors, such as manufacturing facilities, cost, header difficulties, air resistance, and refrigerant flow resistance. Where a choice exists, the smaller diameter gives more flexibility in coil circuit design and results in lower system refrigerant charge.

Occasionally, a base tube condenser is used in applications where high airborne dirt loading is expected to be excessive and more so when coil cleaning accessibility is severely limited. When conditions are normal, including high saline atmosphere situations, a finned surface coil is used. The most common of these is the plate fin with extruded tube collars, each of which is fastened onto the coil core tubes by mechanical expansion. This manufacturing process of tube expansion results in a rigid coil assembly that has maximum tube-to-fin thermal conductivity. Spiral-fin type surface coils as well as the newer spline fin, each of which are tightly wound onto the individual tubes that make up the coil tube core, are also used. Common fin spacings for each type range from 8 to 20 fins/in.

During the manufacturing process, tube sheets or end plates are fastened onto the finned tube core at each end to complete the coil assembly. These provide a means of mounting in the condenser enclosure or, in themselves, can form a part of the condenser unit enclosure. Center supports of much the same design also provide condenser cabinetry fastening areas as well as fan section compartmentalizing possibilities within the unit.

Condenser units may be constructed with an additional smaller coil section in the air entry side that is dedicated to refrigerant subcooling. In this arrangement, the refrigerant usually exits the main condenser coil and flows to the receiver. From the receiver, the refrigerant liquid travels back to the condenser for its flow through the subcooling section coil and then to the expansion valve(s). Another means of achieving the same subcooling benefit and not losing any of the main coil's liquid subcooling in a receiver is to integrate a subcooling section within the condenser's main coil assembly. There the coil's circuitry at the liquid area of the coil is constricted to provide more passes through fewer tube circuits. This design is known as tri-pot circuitry, where two parallel circuits are joined to form one single circuit. Condenser coils with integrated subcooling passes have several converging circuits.

The amount of subcooling depends on the heat transfer characteristics of the condenser and the temperature difference between the condenser and the entering air. Overall system capacity increases about 0.5% per 1°F subcooling at the same suction and discharge pressure. In systems with or without receivers, increasing condensing pressure increases the amount of subcooling, but the resultant liquid temperature is warmer, which usually has a negative effect on net capacity and system energy efficiency ratio (EER). Proper circuitry design can increase subcooling with minimal condensing pressure drop. This, in effect, increases capacity by added subcooling that is negated by only a slight increase in pressure. Increasing subcooling within limits keeps the liquid refrigerant temperature at the expansion valve below saturation temperature. This eliminates refrigerant flashing at the expansion valve.

FANS AND AIR REQUIREMENTS

Condenser coils can be cooled by natural convection, wind, or fans of various designs such as propeller, centrifugal, and vaneaxial. Because efficiency of the material used is sharply

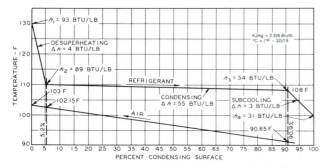

**Fig. 6   Temperature and Enthalpy Changes in
Air-Cooled Condenser**

increased by increased air velocity across the coil, forced convection, powered by fans, is predominantly used.

For a given condensing temperature, the lower limit of air quantity occurs when the leaving air temperature approaches the entering saturated refrigerant condensing temperature. The upper limit is usually determined by air resistance in the coil and casing, noise, fan power, or fan size. As with the coil, a system optimum must be reached by balancing operating cost and first cost with size and sound requirements. Commonly used values are 600 to 1200 cfm/ton. Temperature differences (TD) at the normal cooling condition between condensing refrigerant and entering air range from 10 to 30°F. Fan power requirements generally range from 0.1 to 0.2 hp/ton.

The type of fan depends primarily on static pressure and unit shape requirements. Propeller fans are well-suited to units with low internal pressure drop and free air discharge. Speeds are selected in the range of 850 to 1750 rpm, depending on size and sound requirements. The airflow pattern has a marked effect on the sound level. A partially obstructed fan inlet or outlet can drastically increase noise level. For example, support brackets close to the fan blade have the same effect. Propeller fans are often mounted directly on the motor shaft for simplicity, but care must be taken that the proper motor bearings and mountings are used. The belt drive offers more flexibility but usually requires more maintenance. Propeller fans at free blow are more efficient than centrifugal fans.

Centrifugal fans perform better at higher static pressures. They are used almost exclusively where any but the simplest duct is required. Vaneaxial fans are often more efficient than centrifugal fans, but they are generally not easily adaptable to condenser unit mounting. Because of the static capabilities, belt drives are used more on centrifugal fans than on propeller fans. Additionally, propeller fans are used more as direct-drive motor-to-blade assemblies (and in multiple assemblies) in the larger size condenser units. Such assemblies have propeller fan blades up to 36 in. in diameter and motors down to 8-pole synchronous speeds.

HEAT TRANSFER AND PRESSURE DROP

The overall heat transfer coefficient of air-cooled condensers can be expressed by Equation (5c) in Chapter 21. For estimation of air-side heat transfer coefficients and fin effectiveness, the information in Chapter 21 (regarding sensible cooling coils) also applies to condensers. Heat transfer on the refrigerant side is complex. Refrigerant enters the condenser as superheated vapor and cools by single-phase convection to saturation temperature, after which condensation starts. If the tube-wall temperature is lower than the saturation temperature, condensation occurs close to the entry gas area of the condenser coil. Eventually, in the last part of the condenser, all the vapor condenses to liquid. Along with this change of state are the substantial reduction of refrigerant flow velocity and volume. In general usage, design provisions are made for adequate heat transfer surface area and complementary pressure drop to realize subcooling of the refrigerant prior to its leaving the condenser. Chapter 3 in the 1989 ASHRAE *Handbook—Fundamentals* lists equations for heat transfer coefficients in the single-phase flow sections. For condensation from superheated vapors, there is no verified general predictive technique. For condensation from saturated vapors, one of the best methods is the Shah correlation (Shah 1979, 1981). For an estimation of two-phase pressure drop, see Chapter 4 of the 1989 ASHRAE *Handbook—Fundamentals*.

CONDENSERS REMOTE FROM COMPRESSOR

Remote air-cooled condensers are used for refrigeration systems from 0.5 ton to over 500 tons. Forced air condensers are the most common. The main components consist of (1) a finned condensing coil, (2) one or more fans and motors, and (3) an enclosure or frame.

The coil-fan arrangement can take almost any form, ranging from the horizontal coil with upflow air (top discharge) to the vertical coil with horizontal airflow (front discharge). The choice of design depends primarily on the intended application.

CONDENSERS AS PART OF CONDENSING UNIT

When the condenser coil is included with the compressor as a packaged assembly, it is referred to as a condensing unit. The condenser coil is then further categorized as an indoor or outdoor unit, depending on its rain-protective cabinetry, electrical controls, and code approvals. Factory-assembled condensing units consist of one or more compressors, a condenser coil, electrical controls in a panel, a receiver, shut-off valves, and switches. On smaller sized units, precharged interconnecting refrigerant lines may be offered as a kit. Open, semihermetic or full-hermetic compressors are used for single or parallel (multiplexing) piping connections, in which case an oil separator and its crankcase oil return arrangement is likely to be included in the package. Indoor units afford ease of service access to all components, while outdoor unit componentry access is obtained by removing service access cabinet panels.

Sound is a major consideration of both the design and installation of condensing units. This nuisance item should be dealt with carefully in the following manner:

1. Avoid a straight-line path from the compressor to the listener.
2. Acoustical material may be used inside the cabinet and air passages should be streamlined as much as possible; a top-discharge unit is usually quieter.
3. Mount the compressor carefully on the base.
4. A light gage base is often superior to a heavier, stiffer base, which readily transmits vibration to other panels.
5. Natural frequencies of panels and refrigerant lines must be different from basic compressor and fan frequencies.
6. Refrigerant lines with many bends produce numerous pulsation forces and natural frequencies.
7. Sometimes, a condenser with a very low refrigerant pressure drop has one or two passes resonant with the compressor discharge pulsations.
8. Fan and motor supports can be made to isolate noise and vibration.
9. Fan selection is a major factor; steep propeller blades and unstable centrifugal fans can be serious noise producers.
10. Reduced speed fan operation is often used at night when noise is more objectionable.

Comparisons between water- and air-cooled equipment are often made. Where small units (less than 3 hp) are used and abundant low-cost water is available, both first cost and operating costs may be lower for water-cooled equipment. A 20% larger compressor and/or longer running time is generally required for air-cooled operation; thus operating cost is higher. With water shortages common in many areas, however, the factors change. When cooling towers are used to produce condenser cooling water, the initial cost of a water-cooled system may be higher. Also, the lower operating cost of the water-cooled system may be offset by cooling tower pump and fan power and maintenance costs. When making a comparison, the complete system operation and maintenance factors should be considered.

APPLICATION AND RATING OF CONDENSERS

Condensers are rated in terms of total heat rejection (THR), *i.e.*, the total heat removed in desuperheating, condensing, and subcooling the refrigerant. This value is the product of the weight rate of refrigerant flow and the difference in enthalpy of the refrigerant vapor entering the condenser coil and the enthalpy of the leaving refrigerant liquid.

A condenser may also be rated in terms of net refrigeration effect (NRE), or as net heat rejection (NHR), *i.e.*, the total heat rejection less the heat of compression added to the refrigerant in the compressor. This is the practical expression of the capacity of a refrigeration system.

For open compressors, the THR is the sum of the actual power input to the compressor and the NRE. For hermetic compressors, the THR is obtained by adding the NRE to the total motor power input and subtracting the heat losses from the surface of the compressor and discharge line. The surface heat losses are generally 0 to 10% of the power consumed by the motor. All quantities must be expressed in consistent units. Table 1 recommends factors for converting condenser THR ratings to NRE for both open and hermetic reciprocating compressors.

Ratings of air-cooled condensers are based on the temperature difference (TD) between the dry-bulb temperature of the air entering the coil and the saturated condensing temperature corresponding to the pressure at the inlet. Typical TD values are 10 to 15°F for low-temperature systems at a −20 to −40°F evaporator temperature, 15 to 20°F for medium temperature systems at a 20°F evaporator temperature, and 25 to 30°F for air-conditioning systems at a 45°F evaporator temperature. The THR capacity of the condenser is considered proportional to the TD. The capacity at 30°F TD is about 50% greater than the same condenser selected for 15°F TD.

In determining air temperature entering the condenser coil, refer to the weather data presented in Chapter 24 of the 1989 ASHRAE *Handbook—Fundamentals*. The 2.5% value is suggested for summer operation.

The specifying engineer must choose the specific design dry-bulb temperature carefully, especially for refrigeration serving process cooling. Entering air temperature that is higher than expected quickly causes higher than design compressor pressure and power. Both these factors may cause unexpected system shutdown, usually when it can least be tolerated. Congested or unusual locations may create entering air temperatures that are higher than general ambient conditions. Recirculated condenser air is usually a problem in inadequate locations.

The capacity of an air-cooled condenser equipped with an integral subcooling circuit varies depending on the refrigerant charge. The charge is greater when the subcooling circuit is full of liquid, which increases subcooling. When the subcooling circuit is used for condensing, the refrigerant charge is lower, the condensing capacity is greater, and the liquid subcooling is reduced. Publication of ratings should be made as a standard rating and application range in accordance with ARI *Standard* 460-80 and ASHRAE *Standard* 20-70, Methods of Testing for Rating Remote Mechanical Draft Air-Cooled Refrigerant Condensers.

Condenser ratings are based on THR capacity for a given temperature difference and the specified refrigerant. When application ratings are given in terms of NRE, they must always be defined with respect to the suction temperature, tempera-

Table 1   Net Refrigeration Effect Factors for Air-Cooled and Evaporative Condensers

Open Compressors[a]											
Saturated Suction Temperature, °F	**Condensing Temperature, °F**										
	85	90	95	100	105	110	115	120	125	130	135
−40	0.71	0.70	0.69	0.68	0.67	0.65	0.64	0.63	0.62	0.60	—
−20	0.77	0.76	0.74	0.73	0.72	0.71	0.70	0.69	0.67	0.66	—
0	0.82	0.80	0.79	0.78	0.77	0.76	0.75	0.74	0.73	0.71	—
+20	0.86	0.85	0.84	0.83	0.82	0.81	0.79	0.78	0.77	0.76	0.75
+40	0.91	0.90	0.89	0.87	0.86	0.85	0.84	0.83	0.82	0.81	0.80

[a]Factors based on 15°F of superheat entering the compressor for Refrigerant 22; actual suction gas temperature of 65°F for saturated suction temperature down to −10°F, 55°F at −20°F, 35°F at −40°F, for Refrigerants 12, 500, and 502; 10°F for ammonia (R-717).

Sealed Compressors[b]											
Saturated Suction Temperature, °F	**Condensing Temperature °F**										
	85	90	95	100	105	110	115	120	125	130	135
−40	0.55	0.54	0.53	0.52	0.51	0.50	0.49	0.47	0.46	0.44	—
−20	0.65	0.64	0.62	0.61	0.60	0.59	0.58	0.55	0.53	0.51	—
0	0.72	0.71	0.70	0.69	0.67	0.66	0.64	0.62	0.60	0.58	—
+20	0.77	0.76	0.75	0.74	0.72	0.71	0.69	0.68	0.66	0.64	0.62
+40	0.81	0.80	0.79	0.78	0.77	0.75	0.74	0.72	0.71	0.70	0.68
+50	0.83	0.82	0.81	0.80	0.79	0.78	0.76	0.75	0.74	0.73	0.72

[b]Factors based on 15°F of superheat entering the compressor for Refrigerant 22; actual suction gas temperature of 65°F for saturated suction temperature down to −10°F, 55°F at −20°F, 35°F at −40°F, for Refrigerants 12, 500, and 502.

Notes:
1. These factors should be used only for air-cooled and evaporative condensers connected to reciprocating compressors.
2. Condensing temperature is that temperature corresponding to the saturation pressure as measured at the discharge of the compressor.
3. Net Refrigeration Effect Factors are an aproximation only. Represent the net refrigeration capacity as *Approximate Only* in published ratings.
4. For more accurate condenser selection, and for condensers connected to centrifugal compressors, use the total heat rejection effect of the condenser and the compressor manufacturer's total rating, which includes the heat of compression.

ture difference, type of compressor (either open or hermetic), and the refrigerant. Condensers should be selected by matching the THR ratings of a particular compressor to the THR effect of the condenser.

The capacity of available compressors and condensers are seldom the same at a given set of tabulated conditions. The capacity balance point of these two components can be determined either graphically by plotting condenser THR against compressor THR, as shown in Figure 7, or by computer simulation.

When graphing, the compressor THR Q_d is plotted for a given saturated suction temperature at various saturated discharge temperatures. The saturated discharge temperature at the compressor must be corrected for pressure drop in the hot gas line as well as the condenser's internal pressure drop to estimate a more accurate working saturated condensing temperature. For an R-22 system, the pressure drip through a properly designed discharge line less than 100 ft long equals a saturated temperature drop of about 10°F, and the condenser internal pressure drop equals about 2°F. The corrected curve Q_c, compressor NRE, and power input are also plotted.

Then the condenser THR is plotted for various saturated condensing temperatures for a given entering air temperature. The balanced compressor and condenser THR and the saturated condensing temperature may be read at the intersection of the corrected compressor curve Q_c and the condenser THR curve. The compressor saturated discharge is read from the compressor curve Q_d for the same THR. The net refrigeration capacity and power input are read at the saturated discharge temperature.

Figure 7 is a plot of a compressor operating at a 90°F ambient with two sizes of an air-cooled condenser. The compressor will operate at a saturated discharge temperature of 129°F, with a saturated condensing temperature of 126°F when using the smaller condenser. The THR is 320,000 Btu/h. The NRE is 250,000 Btu/h and the power consumption is 28 kW. With the larger condenser, the compressor operates at a saturated discharge temperature of 120°F and a saturated condensing temperature of 118°F. The THR is 340,000 Btu/h, and the power consumption is 26.5 kW.

The condenser selected must satisfy the anticipated cooling requirements at design ambient conditions at the lowest possible total compressor-condenser system operating cost, including an evaluation of the cost of control equipment.

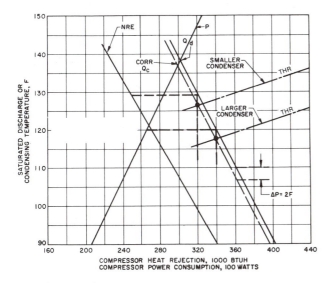

Fig. 7   Performance of Condenser-Compressor Combination

CONTROL OF AIR-COOLED CONDENSERS

For a refrigeration system to function properly, the condensing pressure and temperature must be maintained within certain limits. An increase in condensing temperature causes a loss in capacity, requires extra power, and may overload the compressor motor. Low condensing pressures hinder flow through conventional liquid feed devices; this hindrance starves the evaporator and causes loss of capacity, possible ice stripping of a low-temperature freezer coil's fin face surface, and trip out on low pressure.

Some systems have low-pressure drop thermostatic expansion valves. These low head (usually surge-receiving) systems require additional means of control to assure liquid line subcooled refrigerant entering the expansion valve. Supplemental electronic controls are usually employed. These are often found on year-round operating units performing medium and low-temperature refrigeration for food preservation.

To prevent excessively low head pressure during winter operation, two basic control methods are used: (1) refrigerant-side control, and (2) air-side control. Many methods accomplish head pressure control in each of these two categories, the most common of which are:

1. (a) Refrigerant-side control is accomplished by modulating the amount of active condensing surface available for condensing by flooding the coil with liquid refrigerant. This method requires a receiver and a larger charge of refrigerant. Several valving arrangements give the required amount of flooding to meet the variable needs. Both temperature and pressure actuation are used.

   (b) Refrigerant-side control by going to one-half condenser operation. The condenser is initially designed with two equal parallel sections, each accommodating 50% of the load during normal summer operation.

   During the winter, solenoid or 3-way valves block off one of the condenser sections as well as its pump-down-to-suction. This saves the flooding overcharge and also allows the shutdown of the fans on the inactive condenser side (Figure 8).

2. The air side may be controlled by one of three methods, or a combination of two of them: (1) cycling of fans, (2) modulating dampers, and (3) fan speed control. Any control of airflow must be oriented so that the prevailing wind does not cause adverse operating conditions.

Fan cycling in response to outdoor ambient temperature eliminates rapid cycling but is limited to use with multiple fan units or is supplemental to other control methods. A common method for control of a two-fan unit is to cycle only one fan. A three-fan unit may cycle two fans. Further reduction in condenser airflow is possible by modulating the airflow through the uncontrolled fan section with either speed control or dampers on the air intake or discharge side (Figure 9). In multiple, direct drive motor-propeller arrangements, idle (off-cycle) fans should not be allowed to rotate backward; otherwise, air will short circuit through them. Also, motor-starting torque may not be sufficient to overcome this reverse rotation.

Air dampers controlled in response to either receiver pressure or ambient temperature are also used to control compressor head pressure. These devices throttle the airflow through the condenser coil from 100% to zero. Propeller fans for such units should have flat power characteristics so that the fan motor does not overload when the damper is nearly closed.

Automatic control of fan speed can be used to control compressor pressure. Because fan power increases in proportion to the cube of fan speed, energy consumption can be reduced substantially by slowing the fan at low ambient tem-

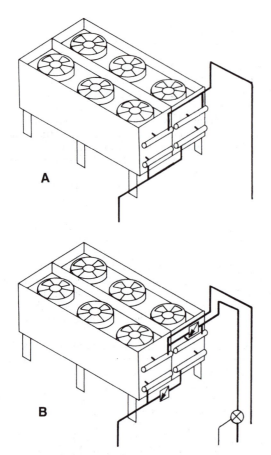

Fig. 8   Equal-Sized Condenser Sections Connected in Parallel and for Half-Condenser Operation during Winter

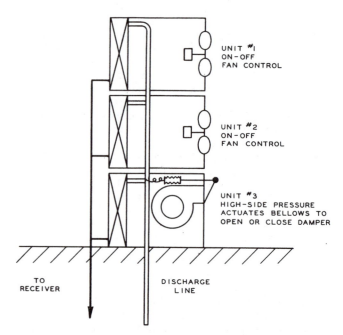

Fig. 9   Unit Condensers Installed in Parallel with Combined Fan Cycling and Damper Control

When the liquid temperature coming from the outdoor condensed is lower than the evaporator operating temperature, a noncondensable charged thermostatic expansion valve bulb should be considered for use. This type of bulb prevents erratic operation of the expansion valve by eliminating the possibility of condensation of the thermal bulb charge on the head of the expansion valve.

INSTALLATION AND MAINTENANCE

The installation and maintenance of remote condensers require little labor because of their relatively simple design. Remote condensers are located as close as possible to the compressor, either indoors or outdoors. They may be located above or below the level of the compressor, but always above the level of the receiver. Various considerations relating to installation and maintenance are:

Indoor condensers. When condensers are located indoors, conduct the warm discharge air to the outdoors. An outdoor air intake opening near the condenser is provided and may be equipped with shutters. Indoor condensers can be used for space heating during the winter and for ventilation during the summer (Figure 10).

Outdoor installations. In outdoor installations of vertical face condensers, ensure that prevailing winds blow toward the

peratures and during part-load operation. Solid-state controls can modulate frequency to vary the speed of an ac motor. Two-speed fan motors can give similar energy savings.

Parallel operation of condensers, especially with capacity control devices, requires careful design. Connecting only identical condensers in parallel reduces operational problems. The Multiple Condenser Installation section has further information. (Figures 14 and 15 also apply to air-cooled condensers.)

The control methods described maintain sufficient compressor pressure for proper expansion valve operation. During off cycles, when the outdoor temperature is lower than the temperature of the indoor space to be conditioned or refrigerated, the refrigerant migrates from the evaporator and receiver and condenses in the cold condenser. The system pressure drops and may correspond to the outdoor temperature. On start-up, insufficient feeding of liquid refrigerant to the evaporator because of low head pressure causes cycling of the compressor, until the suction pressure reaches an operating level above the setting of the low pressurestat. At extremely low outdoor temperatures, the pressure in the system may be below the cut-in point of the low pressurestat and the compressor will not start. This difficulty is solved by either of two methods—(1) bypassing the low-pressure switch on start-up, or (2) using the condenser isolation method of control.

1. On system start-up, a time-delay relay may bypass the low-pressure switch for approximately 180 s to permit the head pressure to build up and allow compressor operation.
2. The condenser isolation method uses a valve arrangement to isolate the condenser from the rest of the system to prevent the refrigerant from migrating to the condenser coil during the off cycle.

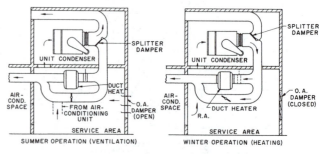

Fig. 10   Air-Cooled Unit Condenser for Winter Heating and Summer Ventilation

air intake, or discharge shields should be installed to deflect opposing winds.

Piping. Piping practice with condensers is identical to that established by experience with other remote condensers. Size discharge piping for a total pressure drop equivalent to a maximum of 2°F saturated temperature drop. Follow standard piping procedures.

Receiver. A condenser may have its receiver installed close by, or it may be separate and thus remotely installed. Most water-cooled systems function without a receiver, since it is an integral part of the unit's water-cooled condenser. If a receiver is used and located in a comparatively warm ambient temperature, size the liquid drain line from the condenser to the receiver for a liquid velocity of 100 fpm and design for gravity drainage as well as for venting to the condenser. The receiver should be equipped with a safety vent that meets applicable codes, sight glass(es), and positive action, in and out shut-off valves. On a system using an air-cooled condenser with integral subcooling circuits, do not install a receiver as part of the operating system located between the subcooling coil outlet and the expansion valve(s).

Maintenance. Schedule periodic lubrication of fan motor and fan bearings. Do not over-oil motors. Follow the instruction either on the motor nameplate and/or the user's operation manual for this equipment. Adjust belt tension as necessary after installation and at yearly intervals.

Condensers also require periodic removal of lint, leaves, dirt, and other airborne materials from their inlet coil surface. This is accomplished by a long bristle brush used on the air entry side of the coil. Washing with a low-pressure water hose or surface blow-off with compressed air is more effectively done from the opposite side of the coil's air entry face. A mild cleaning solution is required to remove restaurant-type grease from the fin surface. Indoor condensers in dusty locations, such as processing plants and supermarkets, should have adequate access for unrestricted coil cleaning and fin surface inspection. In all cases, extraneous material on fins reduces

equipment performance, shortens operating life, increases running time, and uses greater energy.

EVAPORATIVE CONDENSERS

As with water-cooled and air-cooled condensers, evaporative condensers reject heat from a condensing vapor into the environment. In an evaporative condenser, hot, high-pressure vapor from the compressor discharge circulates through a condensing coil that is continually wetted on the outside by a recirculating water system. As seen in Figure 11, air is simultaneously directed over the coil, causing a small portion of the recirculated water to evaporate. This evaporation removes heat from the coil, thus cooling and condensing the vapor.

Evaporative condensers reduce the water pumping and chemical treatment requirements associated with cooling tower/refrigerant condenser systems. In comparison with an air-cooled condenser, an evaporative condenser requires less coil surface and airflow to reject the same heat, or alternately, greater operating efficiencies can be achieved by operating at a lower condensing temperature.

The evaporative condenser can operate at a lower condensing temperature than an air-cooled condenser because the air-cooled condenser is limited by the ambient dry-bulb temperature. In the evaporative condenser, heat rejection is limited by the ambient wet-bulb temperature, which is normally 14 to 25°F lower than the ambient dry bulb. The evaporative condenser also provides lower condensing temperatures than the cooling tower/water-cooled condenser because the heat transfer/mass transfer steps are reduced from two (between the refrigerant and the cooling water and between the water and ambient air) to one step (refrigerant directly to ambient air).

While both the water-cooled condenser/cooling tower combination and the evaporative condenser use evaporative heat rejection, the former has added a second step of nonevaporative heat transfer from the condensing refrigerant to the circulating water, requiring more surface area. Evaporative condensers are, therefore, the most compact for a given capacity.

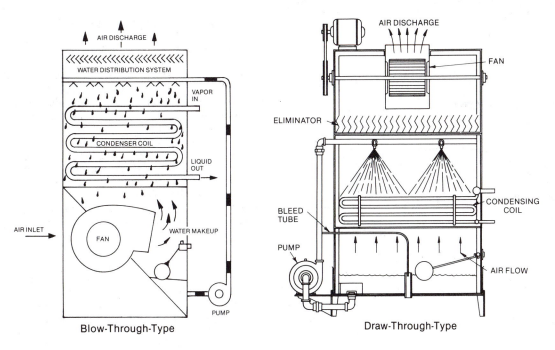

Blow-Through-Type   Draw-Through-Type

Fig. 11   Functional View of Evaporative Condenser

HEAT TRANSFER

In an evaporative condenser, heat flows from the condensing refrigerant vapor inside the tubes, through the tube wall, to the water film outside the tubes, and then from the water film to the air. Figure 12 shows temperature trends in a counterflow evaporative condenser. The driving potential in the first step of heat transfer is the temperature difference between the condensing refrigerant and the surface of the water film, whereas the driving potential in the second step is a combination of temperature and water vapor enthalpy difference between the water surface and the air. Sensible heat transfer between the water stream and the airstream at the water-air interface occurs because of the temperature gradient, while mass transfer (evaporation) of water vapor from the water-air interface to the airstream occurs because of the enthalpy gradient. Commonly, a single enthalpy driving force between the air saturated at the temperature of the water-film surface and the enthalpy of air in contact with that surface is applied to simplify an analytical approach. The exact formulation of the heat- and mass-transfer process requires consideration of the two driving forces simultaneously.

Because the performance of an evaporative condenser can not be represented solely by a temperature difference or an enthalpy difference, simplified predictive methods can only be used for interpolation of data between test points or between tests of different size units, provided that the air velocity, water flow rate, refrigerant velocity, and tube bundle configuration are comparable.

The rate of heat flow from the refrigerant through the tube wall and to the water film can be expressed as:

$$q = U_s A(t_c - t_s) \qquad (6)$$

where:
  q = rate of heat flow, Btu/h
  U_s = heat-transfer coefficient, Btu/(h · ft² · °F)
  t_c = saturation temperature at the pressure of refrigerant entering condenser, °F
  t_s = temperature of water film surface, °F
  A = outside surface area of condenser tubes, ft²

The rate of heat flow from the water-air interface to the airstream can be expressed as:

$$q = U_c A(h_s - h_e) \qquad (7)$$

where:
  q = heat input to condenser, Btu/h
  U_c = transfer coefficient from the water-air interface to the airstream, Btu/(h · ft² · enthalpy difference in Btu/lb)
  h_s = enthalpy of air saturated at t_c, Btu/lb
  h_e = enthalpy of air entering condenser, Btu/lb

Equations (6) and (7) have three unknowns: U_s, U_c, and t_s (h_s is a function of t_s). Consequently, the solution requires an iterative procedure that estimates one of the three unknowns and then solves for the remaining.

Leidenfrost and Korenic (1979, 1982), Korenic (1980), and Leidenfrost *et al.* (1980) have evaluated heat transfer performance of evaporative condensers by analysis of internal conditions within a coil. They use a one-dimensional steady-state analysis and finite element technique to calculate the change in the state of the air and water along their respective passages of the coil. The analytical model is based on an exact graphical presentation of the heat- and mass-transfer processes between the air and the water at their interface (Bosnjakovic 1965).

Principal components of an evaporative condenser include the condensing coil, the fan(s), the spray water pump, the water distribution system, cold water sump, drift eliminators, and water makeup assembly.

COILS

Evaporative condensers generally use bare pipe or tubing without fins. The high rate of energy transfer from the wetted external surface to the air eliminates the need for extended surface. Furthermore, bare coils sustain performance better because they are less susceptible to fouling and are easier to clean. The high rate of energy transfer from the wetted external surface to the air makes finned coils uneconomical when used exclusively for wet operation. However, finned coils provide advantages in the areas of plume elimination, minimizing water consumption, and winter operation alternatives.

Coils are usually fabricated from steel tubing, copper tubing, iron pipe, or stainless steel tubing. Ferrous materials are generally hot-dip galvanized for exterior protection.

METHOD OF COIL WETTING

The pump moves water from the cold water sump to the distribution system located above the coil. The water descends through the air circulated by the fan(s), over the coil surface, and eventually returns to the pan sump. Water distribution systems are designed for complete and continuous wetting of the full coil surface. This ensures the high rate of heat transfer achieved with wet tubes and prevents excessive scaling, which is more likely to occur on intermittently or partially wetted surfaces. Such scaling is undesirable because it decreases the heat transfer efficiency (which tends to raise the condensing temperature) of the unit. Water lost through evaporation and blowdown from the cold water sump is replaced through an assembly which typically consists of a mechanical float valve or solenoid valve and float switch combination.

AIRFLOW

Most evaporative condensers have fan(s) to either blow or draw air through the unit (Figure 11). Typically, the fans are either the centrifugal or propeller type, depending on the external pressure needs, permissible sound levels, and energy requirements.

Drift eliminators recover entrained moisture from the airstream. While these eliminators strip most of the water from

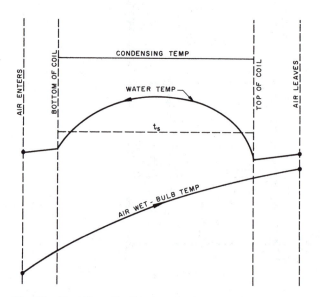

Fig. 12   Heat Transfer Diagram for Evaporative Condenser

the discharge airstream, a certain amount is discharged as drift. The rate of drift loss from an evaporative condenser is a function of the unit configuration, eliminator design, airflow through the evaporative condenser, and the water flow rate. Generally, an efficient eliminator design can reduce drift loss to a range of 0.001 to 0.2% of the water circulation rate.

If the air inlet is near the sump, louvers or deflectors may be installed to prevent water from splashing out of the unit.

CONDENSER LOCATION

Most evaporative condensers are located outdoors, frequently on the roofs of machine rooms. They may also be located indoors and ducted to the outdoors. Generally, centrifugal fan models must be used for indoor applications to overcome the external static resistance of the duct system.

Evaporative condensers installed outdoors can be protected from freezing in cold weather by a remote sump arrangement in which the water and pump are located in a heated space that is remote to the condensers (Figure 13). Piping is arranged so that whenever the pump stops, all of the water drains back into the sump to prevent freezing. Where remote sumps are not practical, reasonable protection can be provided by sump heaters, such as electric immersion heaters, steam coils, or hot water coils. Water pumps and lines must also be protected, for example with electric heat tracing tape and insulation.

Where the evaporative condenser is ducted to the outdoors, moisture from the warm saturated air can condense in condenser discharge ducts, especially if the ducts pass through a cool space. Some condensation may be unavoidable even with short, insulated ducts. In such cases, the condensate must be drained. Also, in these ducted applications, the drift eliminators must be highly effective.

MULTIPLE CONDENSER INSTALLATIONS

Large refrigeration plants may have several evaporative condensers connected in parallel or evaporative condensers in parallel with shell-and-tube condensers. In such systems, unless all condensers have the same refrigerant-side pressure

loss, refrigerant liquid will load those condensers with the highest pressure loss. Also, in periods of light load when some condenser fans are off, liquid will load the active condensers.

Trapped drop legs, as illustrated in Figures 14 and 15, provide proper control of such multiple installations. Effective height of drop legs H must equal the pressure loss through the condenser at maximum loading. Particularly in cold weather, when condensing pressure is controlled by shutting down some condenser fans, active condensers may be loaded considerably above nominal rating, with higher pressure losses. The drop leg should be high enough to anticipate this condition.

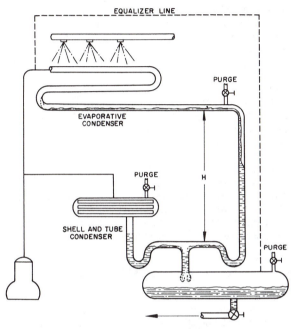

The pressure created by fluid height (H) above the trap must be greater than the internal resistance of the condenser. Note purge connections at both condensers and receiver.

Fig. 14 Parallel Operation of Evaporative and Shell-and-Tube Condensers

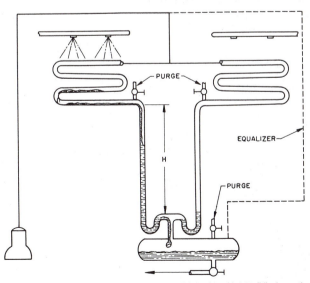

Either condenser can operate independent of the other. Height (H) above the trap, for either condenser, must be not less than the internal resistance of the condenser. Note purge connections at both condensers and the receiver.

Fig. 15 Parallel Operation of Two Evaporative Condensers

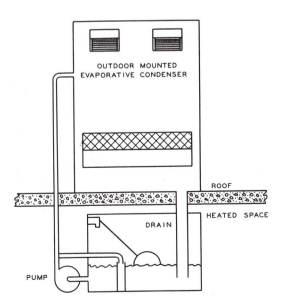

Fig. 13 Evaporative Condenser Arranged for Year-Round Operation

RATINGS

Heat rejected from an evaporative condenser is generally expressed as a function of the saturated condensing temperature and entering air wet bulb. The type of refrigerant has considerable effect on ratings; this effect is handled by separate tables (or curves) for each refrigerant or by a correction factor when the difference is small. Superheat of refrigerant entering the condenser may also affect the rating; test standards such as ASHRAE *Standard* 64-1989, Methods of Testing Remote Mechanical-Draft Evaporative Refrigerant Condensers, establish superheat at a typical value.

It is common to select evaporative condensers in terms of system refrigeration capacity. Manufacturers' ratings include evaporative temperatures and generally assume a nominal heat of compression. Suction gas-cooled hermetic and semihermetic compressors add heat. The total power input to such compressors should be added to the refrigeration load to obtain the total heat rejected by the condenser.

Rotary screw compressors use oil to lubricate the moving parts and provide a seal between the rotors and the compressor housing. This oil is subsequently cooled either in a separate heat exchanger or by refrigerant injection. In the former case, the amount of heat removed from the oil in the heat exchanger is subtracted from the sum of the refrigeration load and the compressor brake horsepower to obtain the total heat rejected by the evaporative condenser. When liquid injection is used, the total heat rejection is the sum of the refrigeration load and the brake power. Heat rejection rating data together with any ratings based on refrigeration capacity should be included.

DESUPERHEATING COILS

A desuperheater is an air-cooled finned coil usually installed in the discharge airstream of an evaporative condenser (Figure 16). The primary function of the desuperheater is to increase the condenser capacity by removing some of the superheat from the discharge vapor before the vapor enters the wetted condensing coil. The amount of superheat removed is a function of the desuperheater surface, condenser airflow, and the temperature difference between the refrigerant and the air. In practice, a desuperheater is limited to reciprocating compressor ammonia installations where discharge temperatures are relatively high (250 to 300°F).

REFRIGERANT LIQUID SUBCOOLERS

The refrigerant pressure at the expansion device feeding the evaporator(s) can be lower than the receiver pressure because of liquid line pressure losses. If the evaporator is located above the receiver, the static head difference further reduces the pressure at the expansion device. To avoid liquid line flashing where these conditions exist, it is necessary to subcool the liquid refrigerant after it leaves the receiver. The minimum amount of subcooling required is the temperature difference between the condensing temperature and the saturation temperature corresponding to the saturation pressure at the expansion device. Subcoolers are often used with halocarbon systems but are seldom used with ammonia systems for the following reasons:

- Because ammonia has a relatively low liquid density, liquid line static head losses are small.
- Ammonia has a very high latent heat; hence, the amount of flash gas resulting from typical pressure losses in the liquid line is extremely small.
- Ammonia is seldom used in a direct-expansion feed system where subcooling is critical to proper expansion valve performance.

One method commonly used to supply subcooled liquid for halocarbon systems places a subcooling coil section in the evaporative condenser below the condensing coil (Figure 17). Depending on the design wet-bulb temperature, condensing temperature, and subcooling coil surface area, the subcooling coil section normally furnishes 10 to 15°F of liquid subcooling. As shown in Figure 17, a receiver must be installed between the condensing coil and subcooling coil to provide a liquid seal for the subcooling circuit.

MULTICIRCUIT CONDENSERS AND COOLERS

Evaporative condensers and evaporative fluid coolers, which are essentially the same, may be multicircuited to condense different refrigerants or cool different fluids simultaneously, providing the condensing temperatures or leaving temperatures are the same. Typical multicircuits (1) condense different refrigerants found in food market units where R-12 and R-502 are condensed in separate circuits, (2) condense and cool different fluids in separate circuits such as condensing the refrigerant from a screw compressor in one circuit and cooling the water or glycol for its oil cooler in another circuit, and (3) cool different fluids in separate circuits such as in many industrial applications where it is necessary to keep the fluids from different heat exchangers in separate circuits.

A multicircuit unit is usually controlled by sensing sump water temperature and modulating an air volume damper or by cycling a fan motor. Two-speed motors or separate high-

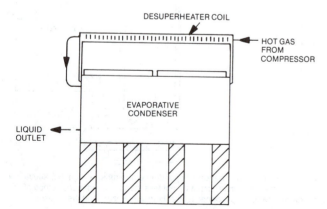

Fig. 16 Evaporative Condenser with Desuperheater Coil

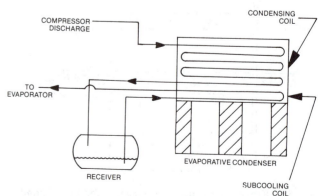

Fig. 17 Evaporative Condenser with Liquid Subcooling Coil

and low-speed motors are also used. Using sump water temperature has an averaging effect on control. The total sump water volume is recirculated once every minute or two, making it a good indicator of changes in load.

WATER TREATMENT

As the recirculated water evaporates in an evaporative condenser, the dissolved solids originally present in the make-up water remain in the system and are continually increased as more water is added. Continued concentration of these dissolved solids can lead to scaling and/or corrosion problems. In addition, airborne impurities and biological contaminants are often introduced into the recirculated water. If these impurities are not controlled, they can cause sludge or biological fouling. Accordingly, a water treatment program is needed to control all potential contaminants. While simple bleedoff may be adequate for control of scale and corrosion, it will not control biological contamination, which must be addressed in any treatment program. (Chapter 43 of the 1991 ASHRAE *Handbook—HVAC Applications* has additional information.) Specific recommendations on water treatment can be obtained from any competent water treatment supplier.

WATER CONSUMPTION

For the purpose of sizing the makeup water piping, all of the heat rejected by an evaporative condenser is assumed to be latent heat (approximately 1050 Btu/lb of water evaporated). The heat rejected depends on operating conditions, but it can range from 14,000 Btu/h per ton of air conditioning to 17,000 Btu/h per ton of freezer storage. The evaporated water ranges from about 1.6 to 2 gph/ton of refrigeration. In addition, a small amount of water can be lost in the form of drift through the eliminators. With good, quality makeup water, the bleed rates may be as low as one-half the evaporation rate, and the total water consumption would range from 2.4 gph/ton for air conditioning to 3 gph/ton for refrigeration.

CAPACITY MODULATION

To ensure operation of expansion valves and other refrigeration system components, extremely low condensing pressures must be avoided. Capacity reduction, controlled by a pressure switch, is obtained in several ways: (1) intermittent operation of the fan(s), (2) a modulating damper in the airstream to reduce the airflow (centrifugal fan models only), and (3) multispeed fan motors. Multispeed fan motors are available as two-speed motors or variable frequency motors. Two-speed fan motors usually operate at 100 and 50% fan speed, which provides 100% and approximately 60% condenser capacity, respectively. With the fans off and the water pump operating, condenser capacity is approximately 10%.

Often, two-speed fan motors provide sufficient capacity control, because it is seldom necessary to hold condensing pressure to a very tight tolerance other than to maintain a certain minimum condensing pressure to ensure refrigerant liquid feed pressure for the system low side and/or sufficient pressure for hot gas defrost requirements. Variable-speed motors are more expensive but offer a full range of speed control for those applications requiring close control on condensing pressure. Careful evaluation of possible critical speed problems should be reviewed with the manufacturer on all variable speed motor applications.

Modulating air dampers also offer closer control on condensing pressure, but they do not offer as much fan power reduction at part load as fan speed control. Water pump cycling for capacity control is not recommended because the periodic drying of the tube surface promotes a buildup of scale.

PURGING

Refrigeration systems that are operating below atmospheric pressure and systems that are opened for service may require purging to remove air that causes high condensing pressures. With the system operating, purging should be done from the top of the outlet connection. On multiple-coil condensers or multiple-condenser installations, one coil at a time should be purged. Purging two or more coils at one time equalizes coil outlet pressures and can cause refrigerant liquid to back up in one or more of the coils, thus reducing the system operating efficiency. Purging may also be done from the high point of the evaporative condenser refrigerant feed but is only effective when the system is not operating. During normal operation, noncondensables are dispersed throughout the high-velocity vapor, and excessive refrigerant would be lost if purging were done from this location.

MAINTENANCE

Evaporative condensers are often installed in remote locations and may not receive the routine attention of operating and maintenance personnel. Therefore, programmed maintenance is essential; where manufacturers' recommendations are not available, the following list may be used as a guide.

Maintenance Item	Frequency
1. Check fan and motor bearings and lubricate, if necessary. Check tightness and adjustment of thrust collars on sleeve-bearing units and locking collars on ball-bearing units.	Q
2. Check belt tension.	M
3. Clean strainer. If air is extremely dirty, strainer may need frequent cleaning.	W
4. Check, clean and flush sump, as required.	M
5. Check operating water level in sump, and adjust makeup valve, if required.	W
6. Check water distribution, and clean as necessary.	W
7. Check bleed water line to ensure it is operative and adequate as recommended by manufacturer.	W
8. Check fans and air inlet screens and remove any dirt or debris.	D
9. Inspect unit carefully for general preservation and cleanliness, and make any needed repairs immediately.	R
10. Check operation of controls such as modulating capacity control dampers.	M
11. Check operation of freeze control items such as pan heaters and their controls.	Y
12. Check the water treatment system for proper operation.	W
13. Inspect entire evaporative condenser for spot corrosion. Treat and refinish any corroded spot.	Y

D = Daily; W = Weekly; M = Monthly; Q = Quarterly; S = Semiannually; Y = Yearly; R = As required.

CODES AND STANDARDS

If state or municipal codes do not take precedence, design pressures, materials, welding, tests, and relief devices should be in accordance with the ASME *Boiler and Pressure Vessel Code,* Section VIII, Div. 1, and ANSI/ASHRAE *Standard* 15-1978, Safety Code for Mechanical Refrigeration. Evaporative condensers are exempt, however, from the ASME Code on the basis of Item (c) of the Scope of the Code, which states if the ID of the condenser shell is 6 in. or less, it is not governed by the Code.

REFERENCES

ARI. 1987. Standard for water-cooled refrigerant condensers, remote type. ARI *Standard* 450-87. Air-Conditioning and Refrigeration Institute, Arlington, VA.

Ayub, Z.H. and S.A. Jones. 1987. Tubeside erosion/corrosion in heat exchangers. *Heating/Piping/Air Conditioning* (December):81.

ASHRAE. 1982. Waterside fouling resistance inside condenser tubes. Research Note 31 (RP 106). ASHRAE *Journal* 24(6):61.

Bell, K.J. 1972. Temperature profiles in pure component condensers with desuperheating and/or subcooling. AICHE *Symposium Paper* 14a (February). American Institute of Chemical Engineers, New York.

Bergles, A.E. 1973. Recent developments in convective heat transfer augmentation. *Applied Mechanics Reviews* 26(6):675.

Bergles, A.E. 1976. Survey of augmentation of two-phase heat transfer. ASHRAE *Transactions* 82(1):881.

Bosnjakovic, F. 1965. *Technical thermodynamics.* Holt, Rinehart, and Winston, New York.

Briggs, D.E. and E.H. Young. 1969. Modified Wilson plot techniques for obtaining heat transfer correlations for shell and tube heat exchangers. *Chemical Engineering Progress Symposium Series* 65(92):35.

Butterworth, D. 1977. Developments in the design of shell-and-tube condensers. ASME Paper No. 77-WA/HT-24. American Society of Mechanical Engineers, New York.

Coates, K.E. and J.G. Knudsen. 1980. Calcium carbonate scaling characteristics of cooling tower water. ASHRAE *Transactions* 86(2).

Eckert, E.R.G. 1963. *Heat and mass transfer.* McGraw-Hill, New York, 143.

Henderson, C.L. and J.M. Marchello. 1969. Film condensation in the presence of a noncondensable gas. ASME *Journal of Heat Transfer,* Transactions (August):447.

Katz, D.L., R.E. Hope, S.C. Datsko, and D.B. Robinson. 1947. Condensation of Freon-12 with finned tubes. *Refrigerating Engineering* (March):211 and (April):315.

Katz, D.L. and D.B. Robinson. 1947. Condensation of refrigerants and finned tubes. *Heating and Ventilating,* Reference Section (November).

Kedzierski, M.A. and R.L. Webb. 1990. Practical fin shapes for surface drained condensation. *Journal of Heat Transfer* 112:479-85.

Kern, D.Q. and A.D. Kraus. 1972. *Extended surface heat transfer,* Chapter 10. McGraw-Hill, New York.

Kistler, R.S., A.E. Kassem, and J.M. Chenoweth. 1976. Rating shell-and-tube condensers by stepwise calculations. ASME Paper No. 76-WA/HT-5. American Society of Mechanical Engineers, New York.

Korenic, B. 1980. Augmentation of heat transfer by evaporative coolings to reduce condensing temperatures. PhD Thesis, Purdue University, West Lafayette, IN.

Kragh, R.W. 1975. Brush cleaning of condenser tubes saves power costs. *Heating/Piping/Air Conditioning* (September).

Kratz, A.P., H.J. MacIntire, and R.E. Gould. 1930. Heat transfer in ammonia condensers. Bulletin No. 209, Part III (June). University of Illinois, Engineering Experiment Station.

Lee, S.H. and J.G. Knudsen. 1979. Scaling characteristics of cooling tower water. ASHRAE *Transactions* 85(1).

Leidenfrost, W. and B. Korenic. 1979. Analysis of evaporative cooling and enhancement of condenser efficiency and of coefficient of performance. *Warme und Stoffubertragung* 12.

Leidenfrost, W. and B. Korenic. 1982. Experimental verification of a calculation method for the performance of evaporatively cooled condensers. *Brennstoff-Warme-Kraft* 34(1):9. VDI Association of German Engineers, Dusseldorf.

Leidenfrost, W., K.H. Lee, and B. Korenic. 1980. Conservation of energy estimated by second law analysis of a power-consuming process. *Energy*:47.

McAdams, W.H. 1954. *Heat transmission,* 3rd ed. McGraw-Hill, New York, 219 and 343.

Pearson, J.F. and J.G. Withers. 1969. New finned tube configuration improves refrigerant condensing. ASHRAE *Journal* (June):77.

Seban, R.A. and E.F. McLaughlin. 1963. Heat transfer in tube coils with laminar and turbulent flow. *International Journal of Heat and Mass Transfer* 6:387.

Shah, M.M. 1979. A general correlation for heat transfer during film condensation in tubes. *International Journal of Heat and Mass Transfer* 22(4):547.

Shah, M.M. 1981. Heat transfer during film condensation in tubes and annuli: A review of the literature. ASHRAE *Transactions* 87(1).

Starner, K.E. 1976. Effect of fouling factors on heat exchanger design. ASHRAE *Journal* (May):39.

Sturley, R.A. 1975. Increasing the design velocity of water and its effect in copper tube heat exchangers. Paper No. 58, The International Corrosion Forum, Toronto, Ontario (April).

Suitor, J.W., W.J. Marner, and R.B. Ritter. 1976. The history and status of research in fouling of heat exchangers in cooling water service. Paper No. 76-CSME/CS Ch E-19, National Heat Transfer Conference, St. Louis, MO (August).

Taborek, J., F. Voki, R. Ritter, J. Pallen, and J. Knudsen. 1972. Fouling—The major unresolved problem in heat transfer. Chemical Engineering Progress Symposium Series 68, Parts I and II, Nos. 2 and 7.

TEMA. 1988. *Standards of the Tubular Exchanger Manufacturers' Association,* 7th ed. Tubular Exchanger Manufacturers' Association, New York.

UOP, Wolverine Division. 1984. *Engineering data book* II, Section 1, 30.

Wanniarachchi, A.S. and R.L. Webb. 1982. Noncondensible gases in shell-side refrigerant condensers. ASHRAE *Transactions* 88(2):170-84.

Watkinson, A.P., L. Louis, and R. Brent. 1974. Scaling of enhanced heat exchanger tubes. *The Canadian Journal of Chemical Engineering* 52:558.

Webb, R.L., A.S. Wanniarachchi, and T.M. Rudy. 1980. The effect of noncondensible gases on the performance of an R-11 centrifugal water chiller condenser. ASHRAE *Transactions* 86(2):57.

Webb, R.L. 1984a. Shell-side condensation in refrigerant condensers. ASHRAE *Transactions* 90(1B):5-25.

Webb, R.L. 1984b. The effects of vapor velocity and tube bundle geometry on condensation in shell-side refrigeration condensers. ASHRAE *Transactions* 90(1B):39-59.

Webb, R.L. and C.G. Murawski. 1990. Row effect for R-11 condensation on enhanced tubes. *Journal of Heat Transfer* 112(3).

Young, E.H. and D.J. Ward. 1957. Fundamentals of finned tube heat transfer. *Refining Engineer* I (November).

COOLING TOWERS

MOST air-conditioning systems and industrial processes generate heat that must be removed and dissipated. Water is commonly used as a heat transfer medium to remove heat from refrigerant condensers or industrial process heat exchangers.

In the past, this was accomplished by drawing a continuous stream of water from a natural body of water or a utility water supply, heating it as it passed through the process, and then discharging the water directly to a sewer or returning it to the body of water. Water purchased from utilities for this purpose has now become prohibitively expensive because of increased water supply and disposal costs. Similarly, cooling water drawn from natural sources is relatively unavailable because the increased temperature of the discharge water disturbs the ecology of the water source.

Air-cooled heat exchangers may be used to cool the water by rejecting heat directly to the atmosphere, but the first cost and fan energy consumption of these devices are high. They are economically capable of cooling the water to within approximately 20°F of the ambient dry-bulb temperature. Such temperature levels are often too high for the cooling water requirements of most refrigeration systems and many industrial processes.

Cooling towers overcome most of these problems and, as such, are commonly used to dissipate heat from water-cooled refrigeration, air-conditioning, and industrial process systems. The water consumption rate of a cooling tower system is only about 5% of that of a once-through system, making it the least expensive system to operate with purchased water supplies. Additionally, the amount of heated water discharged (blowdown) is very small, so the ecological effect is reduced greatly. Lastly, cooling towers can cool water to within 5 to 10°F of the ambient wet-bulb temperature or about 35°F lower than air-cooled systems of reasonable size.

PRINCIPLE OF OPERATION

A cooling tower cools water by a combination of heat and mass transfer. The water to be cooled is distributed in the tower by spray nozzles, splash bars, or filming-type fill, which exposes a very large water surface area to atmospheric air. Atmospheric air is circulated by (1) fans, (2) convective currents, (3) natural wind currents, or (4) induction effect from sprays. A portion of the water absorbs heat to change from a liquid to a vapor at constant pressure. This heat of vaporization at atmospheric pressure is transferred from the water remaining in the liquid state into the airstream.

Figure 1 shows the temperature relationship between water and air as they pass through a counterflow cooling tower. The curves indicate the drop in water temperature (Point A to Point B) and the rise in the air wet-bulb temperature (Point C to Point D) in their respective passages through the tower. The temperature difference between the water entering and leaving the cooling tower (A minus

B) is the *range*. For a system operating in a steady state, the range is the same as the water temperature rise through the load heat exchanger, provided the flow rate through the cooling tower and heat exchanger are the same. Accordingly, the range is determined by the heat load and water flow rate, not by the size or capability of the cooling tower.

The difference between the leaving water temperature and the entering air wet-bulb temperature (B minus C) in Figure 1 is the *approach to the wet bulb* or simply the *approach* of the cooling tower. The approach is a function of cooling tower capability, and a larger cooling tower produces a closer approach (colder leaving water) for a given heat load, flow rate, and entering air condition. Thus, the amount of heat transferred to the atmosphere by the cooling tower is always equal to the heat load imposed on the tower, while the temperature level at which the heat is transferred is determined by the thermal capability of the cooling tower and the entering wet-bulb temperature.

The thermal performance of a cooling tower depends principally on the entering air wet-bulb temperature. The entering air dry-bulb temperature and relative humidity, taken independently, have an insignificant effect on thermal performance of mechanical-draft cooling towers, but they do affect the rate of water evaporation within the cooling tower. A psychrometric analysis of the air passing through a cooling tower illustrates this effect (Figure 2). Air enters at the ambient condition Point A, absorbs heat and mass (moisture) from the water, and exits at Point B in a saturated condition (at very light loads, the discharge air may not be saturated). The amount of heat transferred from the water to the air is proportional to the difference in enthalpy of the air between the

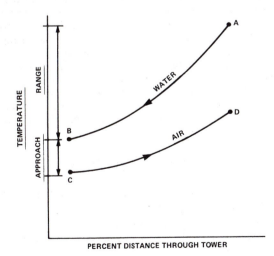

Fig. 1 Temperature Relationship between Water and Air in Counterflow Cooling Tower

---

The preparation of this chapter is assigned to TC 8.6, Cooling Towers and Evaporative Condensers.

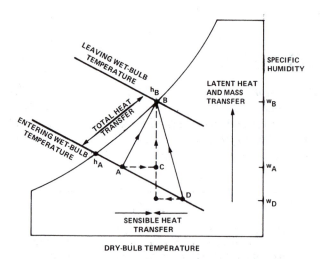

Fig. 2   Psychrometric Analysis of Air Passing through Cooling Tower

entering and leaving conditions ($h_B - h_A$). Because lines of constant enthalpy correspond almost exactly to lines of constant wet-bulb temperature, the change in enthalpy of the air may be determined solely by the change in wet-bulb temperature of the air.

Vector AB in Figure 2 may be separated into component AC, which represents sensible air heating (sensible water cooling), and component CB, which represents latent air heating (latent water cooling). If the entering air condition is changed to Point D at the same wet-bulb temperature but at a higher dry-bulb temperature, the total heat transfer remains the same, but the sensible and latent components change. AB represents sensible cooling of the water by evaporation and sensible and latent heating of the air. Case BD still represents sensible cooling of the water by evaporation, but also represents sensible cooling and latent heating of the air. Thus, for the same water-cooling load, the amount of evaporation depends on the amount of sensible heating or cooling of the air.

The ratio of latent to sensible heat is important in analyzing the water usage of a cooling tower. Mass transfer (evaporation) occurs only in the latent portion of the heat transfer process and is proportional to the change in specific humidity. Because the entering air dry-bulb temperature or relative humidity affects the latent to sensible heat transfer ratio, it also affects the rate of evaporation. In Figure 2, the rate of evaporation is less in Case AB ($W_B - W_A$) than in Case DB ($W_B - W_D$), because the latent heat transfer (mass transfer) represents a smaller portion of the total.

The evaporation rate at typical design conditions is approximately 1% of the water flow rate for each 12.6 °F of water temperature range. The actual annual evaporation rate is less than the design rate because the sensible component of total heat transfer increases as the entering air temperature decreases.

In addition to water loss from evaporation, losses also occur because of liquid carryover into the discharge airstream and blowdown to maintain acceptable water quality. Both of these factors are addressed later in this chapter.

DESIGN CONDITIONS

The thermal capability of any cooling tower may be defined by the following parameters:

1. Entering and leaving water temperatures
2. Entering air wet-bulb or entering air wet-bulb and dry-bulb temperatures
3. Water flow rate

The entering air dry-bulb temperature affects the amount of water evaporated from the water cooled in any evaporative-type cooling tower. It also affects airflow through hyperbolic towers and directly establishes thermal capability within any indirect-contact cooling tower component operating in a dry mode. Variations in tower performance associated with changes in the remaining parameters are covered in the Performance Curves section of this chapter.

The thermal capability of cooling towers for air conditioning is identified in nominal tonnage, based on heat dissipation of 15,000 Btu/h per condenser ton and a water circulation rate of 3 gpm per ton cooled from 95 to 85 °F at 78 °F wet-bulb temperature. For specific applications, however, nominal tonnage ratings are not used, and the thermal performance capability is usually stated in terms of flow rate at specified operating conditions (entering and leaving water temperatures and entering air wet-bulb and/or dry-bulb temperatures).

TYPES OF COOLING TOWERS

Two basic types of evaporative cooling devices are used. The first type involves direct contact between heated water and atmosphere (see Figure 3). The direct-contact device (cooling tower) exposes water directly to the cooling atmosphere, thereby transferring the source heat load directly to the air. The second type involves indirect contact between heated fluid and atmosphere (see Figure 4).

Indirect-contact towers (closed-circuit fluid coolers) contain two separate fluid circuits: (1) the external circuit in which water is exposed to the atmosphere as it cascades over the tubes of a coil bundle, and (2) an internal circuit in which the fluid to be cooled circulates inside the tubes of the coil bundle. In operation, heat flows from the internal fluid circuit, through the tube walls of the coil, to the external water circuit, which is cooled evaporatively. As the internal fluid circuit never contacts the atmosphere, this

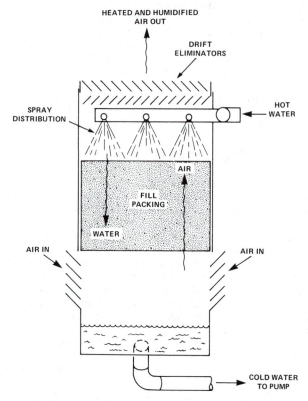

Fig. 3   Direct-Contact Evaporative Cooling Tower

unit can be used to cool fluids other than water and/or to prevent contamination of the primary cooling circuit with airborne dirt and impurities.

Spray-filled towers expose water to air without using a heat transfer medium, which is the most rudimentary method of exposing water to air in direct-contact devices. The amount of water surface exposed to the air depends on the efficiency of the sprays, and the time of contact depends on the elevation and pressure of the water distribution system.

To increase contact surfaces, as well as time of exposure, a heat transfer medium, or *fill*, is installed below the water distribution system, in the path of the air. The two types of fill in use are splash-type and film-type (Figure 5). Splash-type fill maximizes con-

tact area and time by forcing the water to cascade through successive elevations of splash bars arranged in staggered rows. Film-type fill achieves the same effect by causing the water to flow in a thin layer over closely-spaced sheets (principally PVC) that are arranged vertically.

Either type of fill is applicable to both counterflow and crossflow towers. For thermal performance levels typically encountered in air conditioning and refrigeration, the tower with film-type fill is usually more compact. However, splash-type fill is less sensitive to initial air and water distribution and is usually the fill of choice for water qualities conducive to plugging.

Types of Direct-Contact Cooling Towers

Nonmechanical-draft towers. Aspirated by sprays or density differential, these towers do not contain fill and do not use a mechanical device (fan). The aspirating effect of the water spray, either vertically or horizontally (Figures 6 and 7), induces airflow through the tower in a parallel-flow pattern.

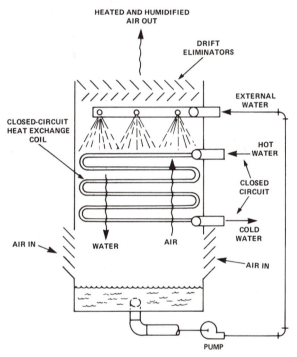

Fig. 4   Indirect-Contact Evaporative Cooling Tower

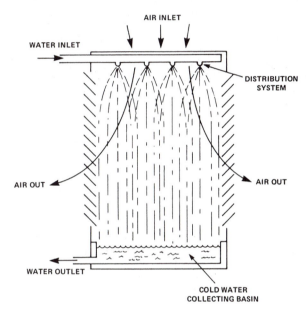

Fig. 6   Vertical Spray Tower

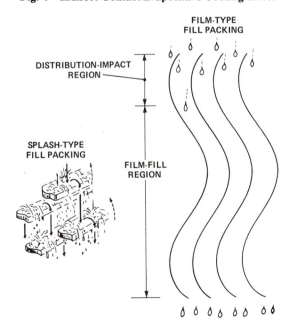

Fig. 5   Types of Fill

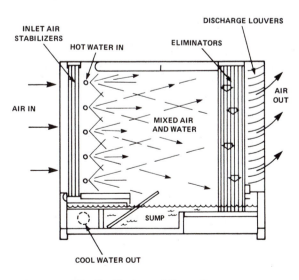

Fig. 7   Horizontal Spray Tower

Because air velocities (both entering and leaving) are relatively low, such towers are susceptible to adverse wind effects and, therefore, are normally used to satisfy a low cost requirement when operating temperatures are not critical to the system. Some horizontal spray towers use high-pressure sprays to induce large air quantities and improve air/water contact. Multispeed or staged pumping systems are normally recommended to reduce energy use in periods of reduced load and ambient conditions.

Chimney towers (hyperbolic) have been used primarily for larger power installations, but may be of generic interest (Figure 8). The heat transfer mode may be either counterflow, crossflow, or parallel flow. Air is induced through the tower by the air density differentials, which exist between the lighter heat-humidified chimney air and the outside atmosphere. Fills are typically the splash- or film-type.

Primary justification of these high first-cost products comes through reduction in auxiliary power requirements (elimination of fan energy), reduced property acreage, and elimination of recirculation, and/or vapor plume interference. Materials used in chimney construction have been primarily steel-reinforced concrete, while early-day timber structures had limitations of size.

Mechanical-draft towers. The fans on mechanical-draft towers (conventional towers) may be on the inlet air side (forced-draft) or exit air side (induced-draft) (Figure 9). The centrifugal or propeller fan is chosen, depending on external pressure needs, permissible sound levels, and energy usage requirements. Water is downflow, while the air may be upflow (counterflow heat transfer) or horizontal-flow (crossflow heat transfer). Air may be single-entry (in one side of tower) or double-entry (in two sides of tower). All four combinations have been produced in various sizes, for example, (1) forced-draft counterflow, (2) induced-draft counterflow, (3) forced-draft crossflow, and (4) induced-draft crossflow.

Towers are typically classified as either factory-assembled (Figure 10), when the entire tower or a few large components are factory-assembled and shipped to the site for installation, or field-erected (Figure 11), where the tower is completely constructed on-site.

Most factory-assembled towers are of metal construction, usually galvanized steel. Other constructions include treated wood, stainless steel, and plastic towers or components. Field-erected towers are predominantly framed of preservative-treated redwood or treated fir with fiberglass-reinforced plastic used for

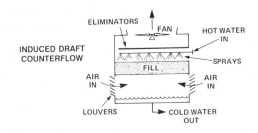

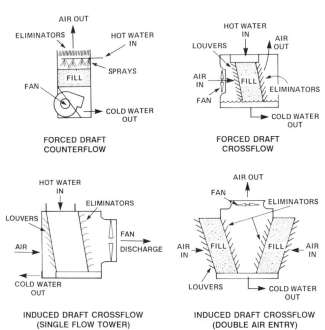

Fig. 9   Conventional Mechanical-Draft Cooling Towers

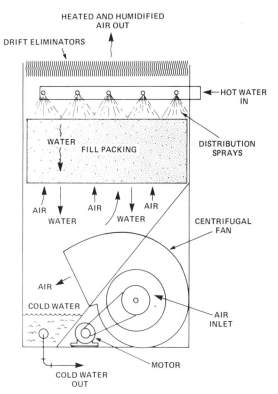

Fig. 10   Counterflow Forced-Draft Tower

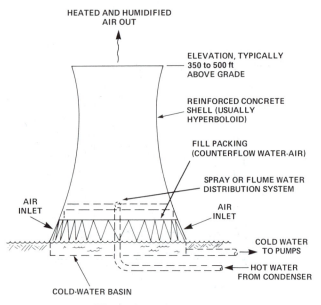

Fig. 8   Hyperbolic Tower

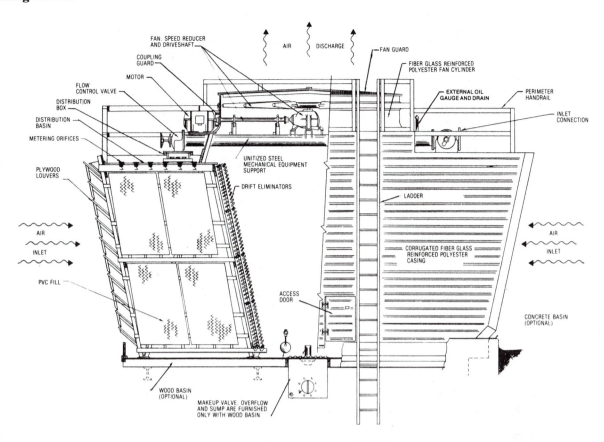

Fig. 11 Crossflow Mechanical-Draft Tower

special components and the casing. Field-erected towers may also be constructed of galvanized steel, stainless steel, or fiberglass. Coated metals, primarily steel, are used for complete towers or components. Concrete or ceramic materials are usually restricted to the largest sizes (see the Materials of Construction section of this chapter).

Special purpose towers containing a conventional mechanical-draft unit in combination with an air-cooled (finned-tube) heat exchanger are *wet/dry towers* (Figure 12). They are used either for vapor plume reduction or water conservation. The hot, moist plumes discharged from cooling towers are especially dense in cooler weather. On some installations, limited abatement of these plumes is required to avoid restricted visibility on roadways, bridges, and buildings.

A vapor plume abatement tower usually has a relatively small air-cooled component, which tempers the leaving airstream to reduce the relative humidity and thereby minimize the fog-generating potential of the tower. Conversely, a water conservation tower usually requires a large, air-cooled component to provide significant savings in water consumption. Some designs can handle heat loads entirely by the nonevaporative air-cooled heat exchangers during reduced ambient conditions.

A variant of the wet/dry tower is an evaporatively precooled/air-cooled heat exchanger. It uses an adiabatic saturator (air precooler/humidifier) to enhance the summer performance of an air-cooled exchanger, thus conserving water in comparison to conventional cooling towers (annualized) (Figure 13). Evaporative fill sections usually operate only during specified summer periods, while full dry operation is expected below 50 to 70 °F ambient conditions. Integral water pumps return the lower basin water to upper distribution systems of the adiabatic saturators in a manner similar to the closed-circuit fluid cooler and evaporative condenser products.

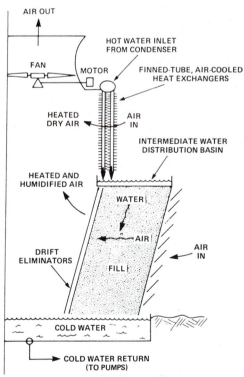

Fig. 12 Combination Wet-Dry Tower

Ponds, spray ponds, spray module ponds, and channels. Heat dissipates from the surface of a body of water by evaporation, radiation, and convection. Captive lakes or ponds (man-made

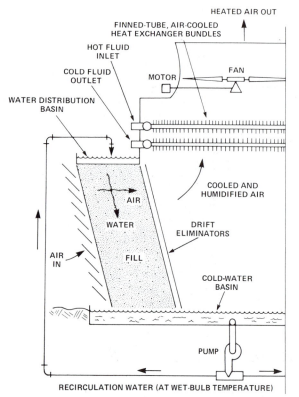

Fig. 13   Adiabatically Saturated Air-Cooled Heat Exchanger

or natural) are sometimes used to dissipate heat by natural air currents and wind. This system is usually used in large plants where real estate is not limited.

A pump-spray system above the pond surface improves heat transfer by spraying the water in small droplets, thereby extending the water surface and bringing it into intimate contact with the air. The heat transfer is largely the result of evaporative cooling (see the Cooling Tower Theory section of this chapter). The system is a piping arrangement using branch arms and nozzles to spray the circulated water into the air. The pond acts largely as a collecting basin. Control of temperatures, real estate demands, limited approach to the wet-bulb temperature, and winter operational difficulties have ruled out the spray pond in favor of more compact and more controllable mechanical- or natural-draft towers.

Equation (1) can be used to estimate cooling pond area. Because of variations in wind velocity and solar radiation, a substantial margin of safety should be added to the results.

$$w_p = \frac{A(95 + 0.425v)}{h_{fg}}[p_w - p_a] \qquad (1)$$

where

w_p = evaporation of water, lb/h
A = area of pool surface, ft^2
v = air velocity over water surface, fpm
h_{fg} = latent heat required to change water to vapor at surface water temperature, Btu/lb
p_a = saturation pressure at room air dew point, in. Hg
p_w = saturation vapor pressure taken at the surface water temperature, in. Hg

Types of Indirect-Contact Towers

Closed-circuit fluid coolers (mechanical draft). Both counterflow- and crossflow arrangements are used in forced and induced fan arrangements. The tubular heat exchangers are typically serpentine bundles, usually arranged for free-gravity internal

drainage. Pumps are integrated in the product to transport water from the lower collection basin to the upper distribution basins or sprays. The internal coils can be fabricated from any of several materials, but galvanized steel or copper predominate. Closed-circuit fluid coolers, which are similar to evaporative condensers, are used increasingly on heat-pump systems and screw compressor oil pump systems.

The indirect-contact towers require a closed-circuit heat exchanger (usually tubular serpentine coil bundles) that is exposed to air/water cascades similar to the fill of a cooling tower (Figure 4). Some types include supplemental film or splash fill sections to augment the external heat exchange surface area.

Coil shed towers (mechanical draft). Coil shed towers usually consist of isolated coil sections (nonventilated), which are located beneath a conventional cooling tower (Figure 14). Counterflow- and crossflow-types are available with either forced- or induced-fan arrangements. Redistribution water pans, located at the tower's base, feed cooled water by gravity-flow to the tubular heat exchange bundles (coils). These units are similar in function to closed-circuit fluid coolers, except that supplemental fill is always required, and the airstream is directed only through the fill regions of the tower. Typically, these units are arranged as field-erected, multifan cell towers and are used primarily in the process cooling industry.

MATERIALS OF CONSTRUCTION

Materials found in cooling tower construction are usually selected to resist the corrosive water and atmospheric conditions.

Wood. Wood has been used extensively for all static components except hardware. Redwood and fir predominate, usually with postfabrication pressure treatment of waterborne preservative chemicals, typically chromated-copper-arsenate (CCA) or acid-copper-chromate (ACC). These microbiocide chemicals prevent the attack of wood-destructive organisms, such as termites or fungi.

Metals. Steel with galvanized zinc is used for small- and medium-size installations. Hot-dip galvanizing after fabrication is used for larger weldments. Hot-dip galvanizing and cadmium and zinc plating are used for hardware. Brasses and bronzes are selected for special hardware, fittings, and tubing material. Stainless steels (principally 302, 304, and 316) are often used for sheet

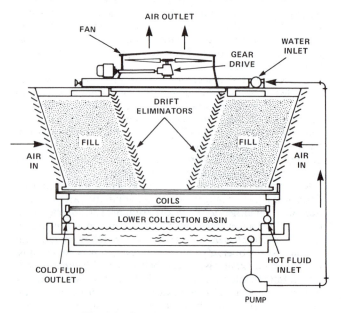

Fig. 14   Coil Shed Cooling Tower

metal, drive shafts, and hardware in exceptionally corrosive atmospheres. Cast iron is a common choice for base castings, fan hubs, motor or gear reduction housings, and piping-valve components. Metals coated with polyurethane and polyvinyl-chloride are used selectively for special components. Epoxy-coal tar compounds and epoxy-powdered coatings are also used for key components or entire cooling towers.

Plastics. Fiberglass-reinforced polyester materials are used for components such as piping, fan cylinders, fan blades, casing, louvers, and structural connecting components. Polypropylene and ABS are specified for injection-molded components, such as fill bars and flow orifices. PVC is increasingly used as fill, eliminator, and louver materials. Reinforced plastic mortar is used in larger piping systems, coupled by neoprene O-ring-gasketed ball and socket joints.

Concrete, masonry, and tile. Concrete is typically specified for cold-water basins of field-erected cooling towers and is used in piping, casing, and structural systems of the largest towers, primarily in the power industry. Special tiles and masonry are used when aesthetic considerations are important.

SELECTION CONSIDERATIONS

Selecting the proper water-cooling equipment for a specific application requires consideration of (1) cooling duty, (2) economics, (3) required services, and (4) environmental conditions. Many of these factors are interrelated, but should be evaluated individually.

Because a wide variety of water-cooling equipment may meet the required cooling duty, such items as size, height, length, width, plan area, volume of airflow, fan and pump energy consumption, materials of construction, water quality, and availability influence the final equipment selection.

The optimum choice is generally made after an economic evaluation. Chapter 33 of the 1991 ASHRAE *Handbook—HVAC Applications* describes two common methods of economic evaluation—life-cycle costing and payback analysis. Each of these procedures compares equipment on the basis of total owning, operating, and maintenance costs.

Initial cost comparisons consider the following factors:

1. Erected cost of equipment
2. Costs of interface with other subsystems, which include items such as:
   a. Basin grillage and value of the space occupied
   b. Cost of pumps and prime movers
   c. Electrical wiring to pump and fan motors
   d. Electrical controls and switchgear
   e. Cost of piping to and from the tower (Some designs require more inlet and discharge connections than others, thus affecting the cost of piping.)
   f. The tower basin, sump screens, overflow piping, and make-up lines, when they are not furnished by the manufacturer
   g. Shutoff and control valves, when they are not furnished by the manufacturer
   h. Walkways, ladders, etc., providing access to the tower

In evaluating owning and maintenance costs, major items to consider are as follows:

1. System energy costs (fans, pumps, etc.) on the basis of operating hours per year
2. Demand charges
3. Expected equipment life
4. Maintenance and repair costs
5. Money costs

Other factors are (1) safety features and safety codes; (2) conformity to building codes; (3) general design and rigidity of structures; (4) relative effects of corrosion, scale, or deterioration on service life; (5) availability of spare parts; (6) experience and reliability of manufacturers; and (7) operating flexibility for economical operation at varying loads or during seasonal changes. In addition, equipment vibration, sound levels, acoustical attenuation, and compatibility with the architectural design are important. The following section details many of these more important considerations.

APPLICATION

This chapter describes some of the major design considerations, but the manufacturer of the cooling tower should be consulted for more detailed recommendations.

Siting

When a cooling tower can be located in an open space with free air motion and unimpeded air supply, siting is normally not a problem in obtaining a satisfactory installation. However, towers are often situated indoors, against walls, or in enclosures. In such cases, the following factors must be considered:

1. Sufficient free and unobstructed space should be provided around the unit to ensure an adequate air supply to the fans and to allow proper servicing.
2. The tower discharge air should not be deflected in any way that might promote recirculation [a portion of the warm, moist discharge air reentering the tower (Figure 15)]. Recirculation raises the entering wet-bulb temperature, causing increased hot- and cold-water temperatures and, during cold weather operation, can promote the icing of air intake areas. The possibility of air recirculation should be particularly considered on multiple-tower installations.

Additionally, cooling towers should be located to prevent the introduction of the warm discharge air and any associated drift, which may contain chemical and/or biological contaminants, into the fresh air intake of the building that the tower is serving or those of adjacent buildings.

Location of the cooling tower is usually a result of one or more of the following: (1) structural support requirements, (2) rigging limitations, (3) local codes and ordinances, (4) cost of bringing auxiliary services to the cooling tower, and (5) architectural compatibility. Sound, fog, and drift considerations are also best handled by proper site selection during the planning stage.

Piping

Piping should be adequately sized according to standard commercial practice. All piping should be designed to allow expansion and contraction. If the tower has more than one inlet connection, balancing valves should be installed to balance the flow to each cell properly. Positive shutoff valves should be used, if necessary, to isolate individual cells for servicing.

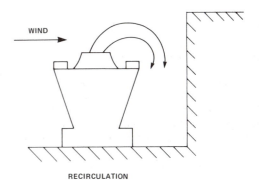

RECIRCULATION

Fig. 15 Discharge Air Reentering Tower

When two or more towers are operated in parallel, an equalizer line between the tower sumps handles imbalances in the piping to and from the units and changing flow rates that arise from such obstructions as clogged orifices and strainers. All heat exchangers, and as much tower piping as possible, should be installed below the operating water level in the cooling tower to prevent overflowing of the cooling tower at shutdown and to ensure satisfactory pump operation during start-up. Tower basins need to carry the proper amount of water during operation to prevent air entrainment into the water suction line. Tower basins should also have enough reserve volume between the operating and overflow levels to fill riser and water distribution lines on start-up and to fulfill the water-in-suspension requirement of the tower.

Capacity Control

Most cooling towers encounter substantial changes in ambient wet-bulb temperature and load during the normal operating season. Accordingly, some form of capacity control may be required to maintain prescribed condensing temperatures or process conditions.

Fan cycling is the simplest method of capacity control on cooling towers and is often used on multiple-unit or multiple-cell installations. In nonfreezing climates, where close control of the exit water temperature is not essential, fan cycling is an adequate and inexpensive method of capacity control. However, motor burnout from too-frequent cycling is a concern.

Two-speed fan motors, in conjunction with fan cycling, can double the number of steps of capacity control, when compared to fan cycling alone. This is particularly useful on single-fan motor units, which would have only one step of capacity control by fan cycling. Two-speed fan motors are commonly used on cooling towers as the primary method of capacity control, and they provide the added advantage of reduced energy consumption at reduced load conditions.

Modulating dampers in the discharge of centrifugal blower fans are used for cooling tower capacity control, as well as for energy management. In many cases, modulating dampers are used in conjunction with two-speed motors. The frequency-modulating controls currently being developed permit a multiplicity of fan speeds and give promise of finite capacity control and energy management, as do the newer automatic variable pitch propeller fans. Careful evaluation of possible critical speed problems should be reviewed with the manufacturer on all variable speed motor applications.

Cooling towers that inject water to induce the airflow through the cooling tower have various pumping arrangements for capacity control. Multiple pumps in series or two-speed pumping provide capacity control and also reduce energy consumption.

Modulating water bypasses for capacity control should be used only after consultation with the cooling tower manufacturer. This is particularly important at low ambient conditions in which the reduced water flows can promote freezing within the tower.

Free Cooling

With an appropriately equipped and piped system, reduced load and/or reduced ambient conditions can significantly reduce energy consumption by using the tower for free cooling. Because the tower's cold water temperature drops as the load and ambient temperature drop, the water temperature will eventually be low enough to serve the load directly, allowing the energy-intensive chiller to be shut off. Figures 16, 17, and 18 outline three methods of free cooling, but do not show all of the piping, valving, and controls that may be necessary for the functioning of a specific system.

Indirect free cooling. This type of cooling keeps the condenser water and chilled-water circuits separate and may be accomplished in the following ways:

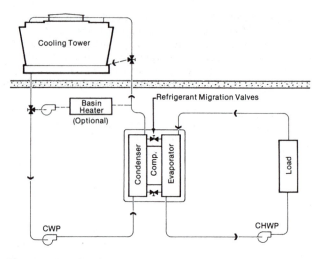

Fig. 16 Free Cooling by Use of Refrigerant Vapor Migration

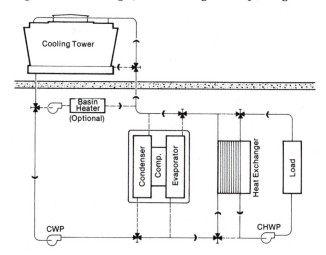

Fig. 17 Free Cooling by Use of Auxiliary Heat Exchanger

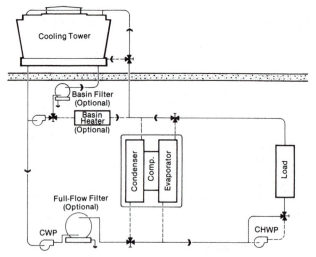

Fig. 18 Free Cooling by Interconnection of Water Circuits

1. In the vapor migration system (Figure 16), bypasses between the evaporator and the condenser permit the migratory flow of refrigerant vapor to the condenser; they also permit gravity flow of liquid refrigerant back to the evaporator, without operation of the compressor. Not all chiller systems are adaptable to this

arrangement, and those that are may offer limited load capability under this mode of operation. In some cases, auxiliary pumps enhance refrigerant flow and, therefore, load capability.

2. A separate heat exchanger (Figure 17) in the system (usually of the plate-and-frame type) allows heat to transfer from the chilled-water circuit to the condenser water circuit by total bypass of the chiller system.

3. An indirect-contact evaporative cooling tower (Figure 4) also permits indirect free cooling. Its use is covered under Direct Free Cooling.

Direct free cooling. This type of cooling involves interconnecting the condenser water and chilled-water circuits so the cooling tower water serves the load directly (Figure 18). In this case, the chilled-water pump is normally bypassed so design water flow can be maintained to the cooling tower. The primary disadvantage of the direct free-cooling system is that it allows the relatively dirty condenser water to contaminate the clean chilled-water system. Although filtration systems (either side-stream or full-flow) minimize this contamination, many specifiers consider it to be an overriding concern. As aforementioned, the use of a closed-circuit cooling tower eliminates this concern for contamination. During summer, the water from the tower is circulated in a closed loop through the condenser. During winter, the water from the tower is circulated in a closed loop directly through the chilled-water circuit.

Maximum use of the free-cooling mode of system operation occurs when a reduction in the ambient temperature reduces the need for dehumidification. Therefore, higher temperatures in the chilled-water circuit can normally be tolerated during the free-cooling season and are beneficial to the system's heating-cooling balance. In many cases, typical 45 °F chilled-water temperatures are allowed to rise to 55 °F or higher in the free-cooling mode of operation. This maximizes tower usage and minimizes system energy consumption.

Winter Operation

When a cooling tower is to be used in subfreezing climates, the following design and operating considerations are necessary:

1. The open circulating water in a cooling tower
2. The closed circulating water in a closed-circuit evaporative fluid cooler
3. Sump water in both a cooling tower or closed-circuit evaporative cooler

Open circulating water. Cooling towers that operate in freezing climates can be winterized by a suitable method of capacity control. This capacity control maintains the temperature of the water leaving the tower well above freezing. In addition, during cold weather, regular visual inspections of the cooling tower should be made to ensure all controls are operating properly.

On induced-draft propeller fan towers, fans may be periodically operated in reverse to deice the air intake areas. Forced-draft centrifugal fan towers should be equipped with capacity control dampers to minimize the possibility of icing.

Closed circulating water. In addition to the previously mentioned protection for the open circulating water, precautions must be taken to protect the fluid inside the heat exchanger of a closed-circuit fluid cooler. When system design permits, the best protection is to use an antifreeze solution. When this is not possible, the system must be designed to provide supplemental heat to the heat exchanger, and the manufacturer should be consulted concerning the amount of heat input required.

All exposed piping to and from the cooler should be insulated and heat traced. In case of a power failure during subfreezing weather, the heat exchanger should include an emergency draining system.

Sump water. Freeze protection for the sump water in an idle tower or closed-circuit fluid cooler can be obtained by various means. A good method for protecting the sump water is to use an auxiliary sump tank located within a heated space. When a remote sump is impractical, auxiliary heat must be supplied to the tower sump to prevent freezing. Common sources are electric immersion heaters and steam and hot-water coils. The tower manufacturer should be consulted for the exact heat requirements to prevent freezing at design winter temperatures.

All exposed water lines susceptible to freezing should be protected by electric tape or cable and insulation. This precaution applies to all lines or portions of lines that have water in them when the tower is shut down.

Sound

Sound has become an important consideration in the selection and siting of outdoor equipment, such as cooling towers and other evaporative cooling devices. Many communities have enacted legislation that limits allowable sound levels of outdoor equipment. Even if legislation does not exist, people who live and work near a tower installation may object if the sound intrudes on their environment. Because the cost of correcting a sound problem may exceed the original cost of the cooling tower, sound should be considered in the early stages of system design.

To determine the acceptability of tower sound in a given environment, the first step is to establish a noise criterion for the area of concern. This may be an existing or pending code or an estimate of sound levels that will be acceptable to those living or working in the area. The second step is to estimate the sound levels generated by the tower at the critical area, taking into account the effects of geometry of the tower installation and the distance from the tower to the critical area. Often, the tower manufacturer can supply sound rating data on a specific unit, which serves as the basis for this estimate. Lastly, the noise criterion is compared to the estimated tower sound levels to determine the acceptability of the installation.

In cases where the installation may present a sound problem, several potential solutions are available. It is good practice to situate the tower as far as possible from any sound-sensitive areas. Two-speed fan motors should be considered to reduce tower sound levels (by a nominal 12 dB) during light load periods, such as at night, if these correspond to critical sound-sensitive periods. However, fan motor cycling should be held to a minimum, since a fluctuating sound usually is more objectionable than a constant sound level.

In critical situations, effective solutions may include barrier walls between the tower and the sound-sensitive area or acoustical treatment of the tower. Attenuators specifically designed for the tower are available from most manufacturers. It may be practical to install a tower larger than would normally be required and lower the sound levels by operating the unit at reduced fan speed. For additional information on sound control, see Chapter 42 of the 1991 ASHRAE *Handbook—HVAC Applications*.

Drift

Water droplets become entrained in the airstream as it passes through the tower. While eliminators strip most of this water from the discharge airstream, a certain amount discharges from the tower as drift. The rate of drift loss from a tower is a function of tower configuration, eliminator design, airflow rate through the tower, and water loading. Generally, an efficient eliminator design reduces drift loss to a range of 0.002 to 0.2% of the water circulation rate.

Because drift contains the minerals of the makeup water (which may be concentrated three to five times) and, often, water treatment chemicals, cooling towers should not be placed near parking areas, large windowed areas, or architectural surfaces sensitive to staining or scale deposits.

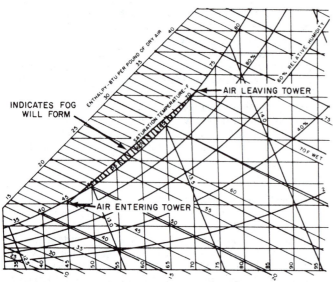

Fig. 19 Fog Prediction Using Psychrometric Chart

Fogging (Cooling Tower Plume)

The warm air discharged from a cooling tower is essentially saturated. Under certain operating conditions, the ambient air surrounding the tower cannot absorb all of the moisture in the tower discharge airstream, and the excess condenses as fog.

Fogging may be predicted by projecting a straight line on a psychrometric chart from the tower entering air conditions to a point representing the discharge conditions (Figure 19). A line crossing the saturation curve indicates fog generation; the greater the area of intersection to the left of the saturation curve, the more intense the plume. Fog persistence depends on the original intensity and the degree of mechanical and convective mixing with ambient air that dissipates the fog.

Methods of reducing or preventing fogging have taken many forms, including heating the tower exhaust with natural gas burners or steam coils, installing precipitators, or spraying chemicals at the tower exhaust. However, such solutions generally are costly to operate and are not always effective.

On larger, field-erected installations, combination wet-dry cooling towers, which combine the normal evaporative portion of a tower with a finned-tube, dry surface heat exchanger section, in series or in parallel, afford a more practical means of plume control. In such units, the saturated discharge air leaving the evaporative section is mixed within the tower with the warm, relatively dry air off the finned-coil section to produce a subsaturated air mixture leaving the tower.

Often, however, the most practical solution to tower fogging is to locate the tower where visible plumes, should they form, will not be objectionable. Accordingly, when selecting cooling tower sites, the potential for fogging and its effect on tower surroundings, such as large windowed areas or traffic arteries, should be considered.

Maintenance

Usually, the tower manufacturer furnishes operating and maintenance manuals that include parts lists for a specific unit; manufacturer's instructions should be followed. When these are unavailable, the schedule of services in Table 1 can guide the operator to establish a reasonable program of inspection and maintenance.

When a cooling tower or other evaporative cooling device is operated in subfreezing ambient temperatures, the unit should be

Table 1 Typical Inspection and Maintenance Schedule[a]

	Fan	Motor	Gear Reducer	V-Belt Drives	Fan Shaft Bearings	Drift Eliminators	Fill	Cold-Water Basin	Distribution System	Structural Members	Casing	Float Valve	Bleed Rate	Drive Shaft	Flow Control Valves	Suction Screen
1. Inspect for clogging						W	W		W							W
2. Check for unusual noise or vibration	D	D	D							Y				D		
3. Inspect keys and set screws		S	S	S	S									S		
4. Lubricate		Q		Q											S	
5. Check oil seals			S													
6. Check oil level			W													
7. Check oil for water and dirt			M													
8. Change oil (at least)			S													
9. Adjust tension				M												
10. Check water level								W	W							
11. Check flow rate													M			
12. Check for leakage								S	S			S				
13. Inspect general condition				M		Y	Y	Y			S	Y	Y		S	S
14. Tighten loose bolts	S	S	S		S						S	Y			S	S
15. Clean	R	R	R			R	R	R	R	R	R		R	R	R	W
16. Repaint	R	R	R				R	R		R				R		
17. Completely open and close															S	
18. Make sure vents are open		M														

D—daily; W—weekly; M—monthly; Q—quarterly; S—semiannually; Y—yearly; R—as required.
[a]More frequent inspections and maintenance may be desirable.

winterized (see Winter Operation). During cold weather operation, frequent visual inspections and routine maintenance services ensure all controls are operating properly and can discover any icing conditions before they become serious.

The efficient performance of a cooling tower depends not only on mechanical maintenance, but also on cleanliness. Accordingly, cooling tower owners should incorporate the following as a basic part of their maintenance program.

1. Periodic inspection of mechanical equipment, fill, and both hot- and cold-water basins to ensure they are maintained in a good state of repair.
2. Periodic draining and cleaning of wetted surfaces and areas of alternate wetting and drying to prevent the accumulation of dirt, scale, or biological organisms such as algae and slime, in which bacteria may develop.
3. Proper treatment of the circulating water for biological control and corrosion, in accordance with accepted industry practice.
4. Systematic documentation of operating and maintenance functions. This is extremely important because without it, no policing can be done to determine whether an individual has actually adhered to a maintenance policy.

Water Treatment

The quality of water circulating through an evaporative cooling system has a significant effect on the overall system efficiency, the degree of maintenance required, and the useful life of system components. Because the water is cooled primarily by evaporation of a portion of the circulating water, the concentration of dissolved solids and other impurities in the water can increase rapidly. Also, appreciable quantities of airborne impurities, such as dust and gases, may enter during operation. Depending on the nature of the impurities, they can cause scaling, corrosion, and/or silt deposits.

To limit the concentrations of impurities, a small percentage of the circulating water is wasted (called blowdown or bleedoff). The number of concentrations thus obtained in the circulating water can be calculated from the equation:

$$\text{No. of Concentrations} = \frac{\text{Evaporation} + \text{Drift} + \text{Blowdown}}{\text{Drift} + \text{Blowdown}}$$

The entries in this equation may be expressed as quantities or as percentages of the circulating rate. The evaporation loss averages approximately 1% for each 12.5°F of cooling range, while drift loss on a mechanical-draft tower is usually less than 0.2% of the circulating rate. Accordingly, for a tower operating with a 12.5°F range and using a figure of 0.1% for drift, a blowdown rate of 0.9% of the circulating rate would be required to maintain a level of two concentrations.

In addition to blowdown, evaporative cooling systems are often treated chemically to control scale, inhibit corrosion, restrict biological growth, and control the collection of silt. Scale formation occurs whenever the dissolved solids and gases in the circulating water reach their limit of solubility and precipitate out onto piping, heat transfer surfaces, and other parts of the system. Simple blowdown can often control scale formation, but where this is inadequate, or it is desirable to reduce the rate of blowdown, chemical scale inhibitors can be added to increase the level of concentrations at which precipitation occurs. Typical scale inhibitors include acids, inorganic phosphates, and similar compounds.

Corrosion and scale are controlled by blends of phosphates, phosphonate, molybdate, zinc, silicate, and various polymers. Tolyltriazole or benzotriazole are added to these blends to protect copper and copper alloys from corrosion. Chromates were previously used for corrosion control, but as of January 1990, the EPA has banned their use in cooling tower treatment in comfort cooling systems. When selecting a treatment product, consult local, state, and federal environmental guidelines and requirements. Further information on water treatment is given in Chapter 43 of the 1991 ASHRAE *Handbook—HVAC Applications*. The phosphate compounds, while nontoxic, tend to promote algae growth; their use, therefore, may be limited in the future.

Algae, slimes, fungi, and other microorganisms grow readily in evaporative cooling systems and can (1) form an insulating coating on heat transfer surfaces, (2) restrict fluid flow, (3) promote corrosion, and/or (4) attack organic components within the system (such as wood). The common method of control is to shock treat the system on a periodic basis with a toxic agent such as chlorine or other biocide. Normally, two different biocides are added on an alternating basis to ensure microorganisms do not develop a resistance to any one compound.

During normal cooling tower operation, large quantities of airborne dirt can enter and subsequently settle out as silt deposits. Such deposits can promote corrosion, harbor microorganisms detrimental to the system, and obstruct fluid flow. Silt is normally controlled by adding polymers, which keep the silt in suspension while it flows through the system. Eventually, it settles in the tower basin where deposits can be more easily removed during periodic maintenance.

In most cases where chemical water treatment is practiced, a competent water treatment specialist can be invaluable because system requirements can vary widely.

PERFORMANCE CURVES

The combination of flow rate and heat load dictates the range a cooling tower must accommodate. The entering wet-bulb temperature and required system temperature level combine with cooling tower size to balance the heat rejected at a specified approach. The performance curves in this section are typical and may vary from project to project.

Cooling towers can accommodate a wide diversity of temperature levels, ranging as high as 150 to 160°F hot-water temperature in the hydrocarbon processing industry. In the air-conditioning and refrigeration industry, towers are generally applied in the range of 90 to 115°F hot-water temperature. A typical standard design condition for such cooling towers is 95°F hot-water to 85°F cold-water temperature, and 78°F wet-bulb temperature.

A means of evaluating the relative performance of a cooling tower used for a typical air-conditioning system is shown in Figures 20 through 23. The example tower was selected for a flow rate of 3 gpm per nominal ton when cooling water from 95 to 85°F at 78°F entering wet-bulb temperature (Figure 20).

When operating at other wet bulbs or ranges, the curves may be interpolated to find the resulting temperature level (hot- and cold-water) of the system. When operating at other flow rates, this same tower performs at the levels described by the following performance curve titles: 2, 4, and 5 gpm per nominal ton. Intermediate flow rates may be interpolated between charts to find resulting operating temperature levels.

The format of these curves is similar to the predicted performance curves supplied by manufacturers of cooling towers; the exception is predicted performance curves usually are families of only three specific ranges—80%, 100%, and 120% of design range—and only three charts are provided, covering 90%, 100%, and 110% of design flow. Such performance curves, therefore, bracket the acceptable tolerance range of test conditions and may be interpolated for any specific test condition within the scope of the curve families and chart flow rates.

The charts may also be used to identify the feasibility of varying the parameters to meet specific applications. For example, the subject tower can handle a greater heat load (flow rate) when operating in a lower ambient wet-bulb region. This may be seen by comparing the intersection of the 10°F range curve with 73°F wet bulb at 85°F cold water to show the tower is capable of rejecting 33% more heat load at this lower ambient temperature (Figure 22).

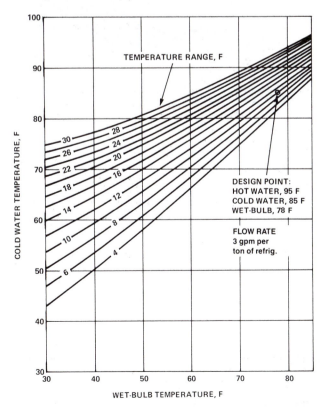

Fig. 20 Cooling Tower Performance—100% Design Flow

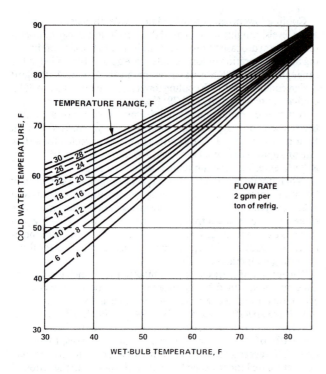

Fig. 21   Cooling Tower Performance—67% Design Flow

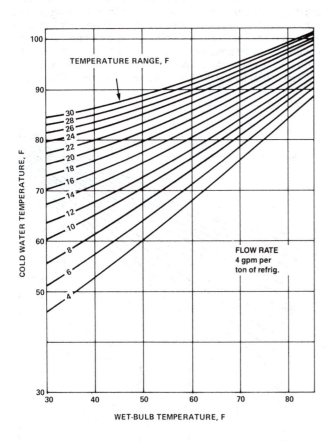

Fig. 22   Cooling Tower Performance—133% Design Flow

Similar comparisons and crossplots identify relative tower capacity or degree of difficulty for a wide range of variables. The curves produce accurate comparisons within the scope of the

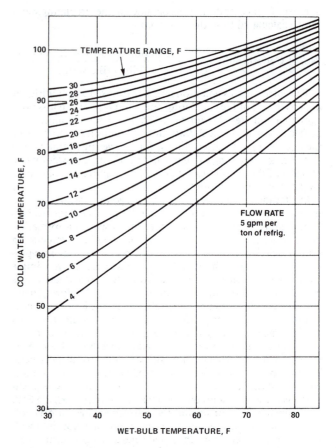

Fig. 23   Cooling Tower Performance—167% Design Flow

information presented but should not be extrapolated outside the field of data given. Also, the curves are based on a typical mechanical-draft, film-filled, crossflow, medium size, air-conditioning cooling tower. Other types and sizes of towers produce somewhat different balance points of temperature level. However, the curves may be used to evaluate a tower for year-round or seasonal use, if they are restricted to the general operating characteristics described. (See specific manufacturer's data for maximum accuracy when planning for test or critical temperature needs.)

As stated, the water cooling tower, when selected for a specified design condition, operates at other temperature levels when the ambient temperature is off-design or when heat load or flow rate varies from the design condition. When flow rate is held constant, range falls as heat load falls, causing temperature levels to fall to a closer approach. Hot- and cold-water temperature levels fall when the ambient wet bulb falls at constant heat load, range, and flow rate. As water loading to a particular tower falls, while holding the ambient wet bulb and range constant, the tower cools the water to a lower temperature level or closer approach to the wet bulb.

COOLING TOWER THERMAL PERFORMANCE

Chapter 48, Codes and Standards, lists those codes and standards that pertain to field test and certification of cooling towers. ASME *Standard* PTC 23 and Cooling Tower Institute *Bulletin* ATC-105 are commonly followed to field test cooling towers. ASME *Standard* PTC 23, Atmospheric Water Cooling Equipment, considers both entering and ambient wet-bulb temperatures but emphasizes the entering wet-bulb temperature, which can exceed the ambient wet-bulb temperature if a portion of the warm, moist discharge air reenters the tower (see section on Siting). Hot-water temperatures are normally measured in the distribution basin or inlet piping instead of near the heat exchanger (heat source), which generally has large temperature gradients. Test data is evaluated by interpolating the manufacturer's performance curves.

Cooling Tower Institute *Bulletin* ATC-105 (1990) dictates inlet wet-bulb temperature measurement be used with the wet-bulb stations located within 4 ft of the air intakes. A prior agreement between test participants is required to define the acceptable number and location of measurement stations to ensure the test average temperature is an accurate representation of the true weighted average inlet wet-bulb temperature.

Data is evaluated by either the characteristic curve or performance curve method. Tower capability is established by comparing the actual flow rate of water measured during the test (adjusted for fan power) to the flow rate predicted from the performance curves at the test conditions.

Several factors must be considered if valid tower performance data are to be gathered. The conditions required to run a valid performance test include a steady heat load combined with steady circulating water flow, both as near design as possible. Weather conditions should be reasonably stable, with prevailing winds of 10 mph or less. The tower should be clean and adjusted for proper water distribution, with all fans operating at design speed. Both the CTI and ASME codes indicate exact maximum allowable deviations from design for temperature, flow, and heat load for an acceptable test.

Accurate data with calibrated instrumentation at representative locations must be obtained. Water temperatures to the tower are normally measured in the distribution basin or inlet piping. To assure complete mixing, water temperatures from the tower are usually measured at the discharge of the circulating pumps. Wet-bulb temperatures are determined by a mechanically aspirated psychrometer. Finally, water flow to the tower should be measured by calibrated pitot-tube traverse of the tower piping or by calibrated in-line flow meters.

To reduce the complexity and expense of the test, the participants may agree to (1) use sling psychrometers or static wet- and dry-bulb thermometers, which are positioned in the entering airstream in lieu of mechanically aspirated psychrometers; and (2) measure circulating water rate by comparing head above gravity flow orifices in hot-water basins to calibrated head/flow data curves or by comparing the hot-water manifold spray pressure to calibrated spray distribution nozzle data curves.

In lieu of field testing:

1. Cooling tower thermal performance can be certified under the provisions of CTI *Standard* 201. This standard:

   - Applies to mechanical-draft water cooling towers for cooling fresh water
   - Is based on entering wet-bulb temperature
   - Applies to applications where external factors do not impose limitations on airflow

2. Alternatively, the purchaser may obtain a performance bond from the manufacturer, which provides for seeking redress from a surety in the event the tower has been shown to fail to meet the manufacturers rated thermal performance level.

COOLING TOWER THEORY

Baker and Shryock (1961) developed the following theory. Consider a cooling tower having one square foot of plan area; cooling volume V, containing a ft^2 of extended water surface per

cubic foot; water rate L and air rate G, both in lb/h. Figure 24 schematically shows the processes of mass and energy transfer. The bulk water at temperature t is surrounded by the bulk air at dry-bulb temperature t_a, having enthalpy h_a and humidity ratio W_a. The interface is assumed to be a film of saturated air with an intermediate temperature t'', enthalpy h'', and humidity ratio W''. Assuming a constant value of 1 Btu/lb·°F for the specific heat of water c_p, the total energy transfer from the water to the interface is:

$$dq_w = Lc_p dt = K_L a \, (t - t'') \, dV \qquad (2)$$

where

q_w = rate of heat transfer, bulk water to interface, Btu/h
K_L = unit conductance, heat transfer, bulk water to interface, Btu/h·ft^2·°F
V = cooling volume, ft^2
a = area of interface, ft^2/ft^3

The heat transfer from interface to air is:

$$dq_s = K_G a \, (t'' - t_a) \, dV \qquad (3)$$

where

q_s = rate of sensible heat transfer, interface to airstream, Btu/h
K_G = overall unit conductance, sensible heat transfer between interface and main airstream, Btu/h·ft^2·°F

The diffusion of water vapor from film to air is:

$$dm = K'a \, (W' - W_a) \, dV \qquad (4)$$

where

m = mass transfer rate, interface to airstream, lb/h
K' = unit conductance, mass transfer, interface to main airstream, lb/h·ft^2 (lb/lb)

Considering the latent heat of evaporation a constant r, the heat rate is:

$$rdm = dq_L = rK'a \, (W' - W_a) \, dV \qquad (5)$$

The process will reach equilibrium when $t_a = t$ and the air becomes saturated with moisture at that temperature. Under adiabatic conditions, equilibrium is reached at the temperature of adiabatic saturation or at the thermodynamic wet-bulb temperature of the air. This is the lowest attainable temperature in a cooling tower. The circulating water rapidly approaches this temperature when a tower operates without heat load. The process is the same when a heat load is applied, but the air enthalpy increases as it

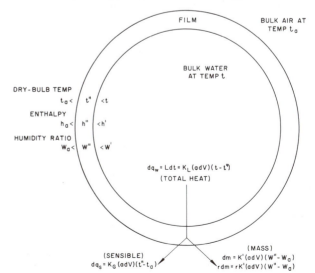

Fig. 24   Heat and Mass Transfer Relationships between Water, Interfacial Film, and Air

(Baker and Shryock 1961)

moves through the tower so the equilibrium temperature increases progressively. The approach of the cooled water to the entering wet-bulb temperature is a function of the capability of the tower.

Merkel (1925) assumed the Lewis relationship to be equal to one (1) in combining the transfer of mass and sensible heat into an overall coefficient based on enthalpy difference as the driving force:

$$K_G/(K' c_{pm}) = 1 \qquad (6)$$

where

c_{pm} = humid specific heat of moist air, Btu/lb (dry air)·°F

The relationship also explains why the wet-bulb thermometer closely approximates the temperature of adiabatic saturation in an air-water vapor mixture. Simplification yields:

$$Lc_p dt = Gdh = K'a (h'' - h_a) dV \qquad (7)$$

The equation considers the transfer from the interface to the airstream, but the interfacial conditions are indeterminate. If the film resistance is neglected and an overall coefficient K is postulated, based on the driving force of enthalpy h' at the bulk water temperature t, the equation becomes:

$$Lc_p dt = Gdh = Ka (h' - h_a) dV \qquad (8)$$

or

$$KaV/L = \int_{t_1}^{t_2} c_p dt/(h' - h_a) \qquad (9)$$

and

$$KaV/G = \int_{h_1}^{h_2} dh/(h' - h_a) \qquad (10)$$

In cooling tower practice, the integrated value of Equation (9) is commonly referred to as the number of transfer units or NTU. This value gives the number of times the average enthalpy potential $(h' - h_a)$ goes into the temperature change of the water (Δt) and is a measure of the difficulty of the task. Thus, one transfer unit has the definition of $c_p \Delta t/(h' - h_a)_{avg.} = 1$.

The equations are not self-sufficient and are not subject to direct mathematical solution. They reflect mass and energy balance at any point in a tower and are independent of relative motion of the two fluid streams. Mechanical integration is required to apply the equations, and the procedure must account for relative motion. The integration of Equation (9) gives the NTU for a given set of conditions.

Counterflow Integration

The counterflow cooling diagram is based on the saturation curve for air-water vapor (Figure 25). As water is cooled from t_{w1} to t_{w2}, the air film enthalpy follows the saturation curve from A to B. Air entering at wet-bulb temperature t_{aw} has an enthalpy h_a corresponding to C'. The initial driving force is the vertical distance BC. Heat removed from the water is added to the air so the enthalpy increase is proportional to water temperature. The slope of the air operating line CD equals L/G.

Counterflow calculations start at the bottom of a tower, the only point where the air and water conditions are known. The NTU is calculated for a series of incremental steps, and the summation is the integral of the process.

Example 1. Air enters the base of a counterflow cooling tower at 75°F wet-bulb temperature, water leaves at 85°F, and L/G (water-to-air) ratio is 1.2, so $dh = 1.2 \times 1 \times dt$, where 1 is the specific heat c_p of water. Calculate the NTU for various cooling ranges.

Solution: The calculation is shown in Table 2. Water temperatures are shown in Column 1 for 1°F increments from 85 to 90°F and 2°F increments from 90 to 100°F. The corresponding film enthalpies are obtained from psychrometric tables and are shown in Column 2.

The upward air path is in Column 3. The initial air enthalpy is 38.6 Btu/lb, corresponding to a 75°F wet bulb and increases by the relationship $\Delta h = 1.2 \times 1 \times \Delta t$.

The driving force, $h' - h_a$, at each increment is listed in Column 4. The reciprocals $1/(h' - h_a)$ are calculated (Column 5), and the average for each increment is multiplied by $c_p \Delta t$ to obtain the NTU for each increment (Column 6). The summation of the incremental values (Column 7) represents the NTU for the summation of the incremental temperature changes, which is the cooling range given in Column 8.

Because of the slope and position of CD relative to the saturation curve, the potential difference increases progressively from the bottom to the top of the tower in this example. The degree of difficulty decreases as this driving force increases, which is reflected as a reduction in the incremental NTU proportional to a variation in incremental height. This procedure determines the temperature gradient with respect to tower height.

Table 2　Counterflow Integration Calculations

1 Water Temperature t, °F	2 Enthalpy of Film h', Btu/lb	3 Enthalpy of Air h_a, Btu/lb	4 Enthalpy Difference $h' - h_a$, Btu/lb	5 $1/(h' - h_a)$	6 NTU = $c_P\Delta t/(h' - h_a)$ (Average)	7 ΣNTU	8 Cooling Range °F
85	49.4	38.6	10.8	0.0926			
					0.0921	0.0921	1
86	50.7	39.8	10.9	0.0917			
					0.0917	0.1838	2
87	51.9	41.0	10.9	0.0917			
					0.0913	0.2751	3
88	53.2	42.2	11.0	0.0909			
					0.0901	0.3652	4
89	54.6	43.4	11.2	0.0893			
					0.0889	0.4541	5
90	55.9	44.6	11.3	0.0885			
					0.1732	0.6273	7
92	58.8	47.0	11.8	0.0847			
					0.1653	0.7925	9
94	61.8	49.9	12.4	0.0806			
					0.1569	0.9493	11
96	64.9	51.8	13.1	0.0763			
					0.1477	1.097	13
98	68.2	54.2	14.0	0.0714			
					0.1376	1.2346	15
100	71.7	56.6	15.1	0.0662			

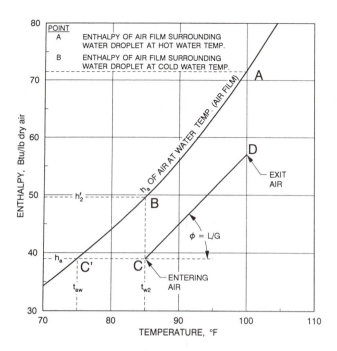

Fig. 25   Counterflow Cooling Diagram

The procedure of Example 1 considers increments of temperature change and calculates the coincident values of NTU, which correspond to increments of height. Baker and Mart (1952) developed a unit-volume procedure that considers increments of NTU (representing increments of height) with corresponding

temperature changes calculated by iteration. The unit-volume procedure is more cumbersome but is necessary in crossflow integration because it accounts for temperature and enthalpy change, both horizontally and vertically.

Crossflow Integration

In a crossflow tower (Figure 26) water enters at the top, and the solid lines of constant water temperature show its temperature distribution. Air enters from the left and the dotted lines show constant enthalpies. The cross section is divided into unit-volumes in which dV becomes $dxdy$ and Equation (8) becomes:

$$c_p L dt dx = G dh dy = Ka (h' - h_a) dxdy \qquad (11)$$

The overall L/G ratio applies to each unit-volume by considering $dx/dy = w/z$. The cross-sectional shape is automatically considered when an equal number of horizontal and vertical increments are used. Calculations start at the top of the air inlet and proceed down and across. Typical calculations are shown in Figure 27 for water entering at 100 °F, air entering at 75 °F wet-bulb temperature, and $L/G = 1.0$. Each unit-volume represents 0.1 NTU. Temperature change vertically in each unit is determined by iteration from:

$$\Delta t = 0.1 (h' - h_a)_{av} \qquad (12)$$

$c_p(L/G)dt = dh$ determines the horizontal change in air enthalpy. With each step representing 0.1 NTU, two steps down and across equal 0.2 NTU, etc., for conditions corresponding to the average leaving water temperature.

Figure 26 shows that air flowing across any horizontal plane is moving toward progressively hotter water, with entering hot-water temperature as a limit. Water falling through any vertical section is moving toward progressively colder air that has the entering wet-bulb temperature as a limit. This is shown in Figure 28, which is a plot of the data in Figure 27. Air enthalpy follows the family of curves radiating from Point A. Air moving across the top of the tower tends to coincide with OA. Air flowing across the bottom of a tower of infinite height follows a curve that coincides with the saturation curve AB.

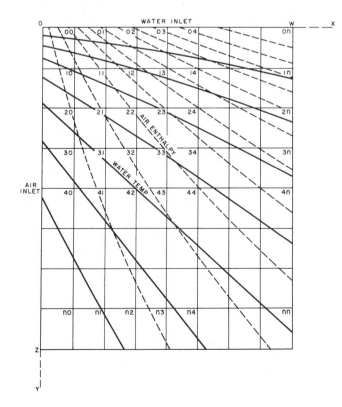

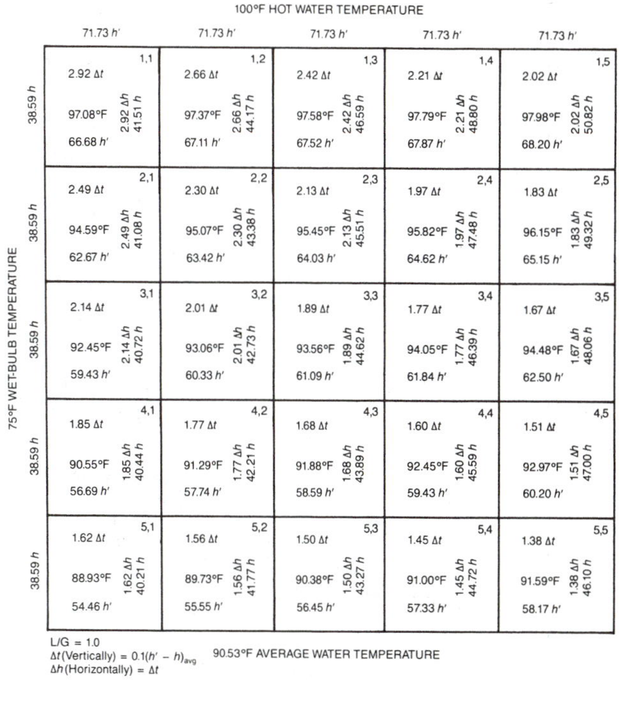

Fig. 26   Water Temperature and Air Enthalpy Variation through Crossflow Cooling Tower
(Baker and Shryock 1961)

Fig. 27   Crossflow Calculations
(Baker and Shryock 1961)

Water temperatures follow the family of curves radiating from Point B, between the limits of BO at the air inlet and BA at the outlet of a tower of infinite width. The single operating line CD of the counterflow diagram in Figure 25 is replaced in the crossflow diagram by a zone represented by the area intersected by the two families of curves.

TOWER COEFFICIENTS

Calculations can reduce a set of conditions to a numerical value representing degree of difficulty. The NTU corresponding to a set of hypothetical conditions is called the required coefficient and evaluates degree of difficulty. When test results are being considered, the NTU represents the available coefficient and becomes an evaluation of the equipment tested.

The calculations consider temperatures and the L/G ratio. The minimum required coefficient for a given set of temperatures occurs at $L/G = 0$, corresponding to an infinite air rate. No increase in air enthalpy occurs, so the driving force is maximum and the degree of difficulty is minimum. Decreased air rate (increase in L/G) decreases the driving force, and the greater degree of difficulty shows as an increase in NTU. This situation is shown for counterflow in Figure 29. Maximum L/G (minimum air rate) occurs when CD intersects the saturation curve. Driving force becomes zero, and NTU is infinite. The point of zero driving force may occur at the air outlet or at an intermediate point because of the curvature of the saturation curve.

Similar variations occur in crossflow cooling. Variations in L/G vary the shape of the operating area. At $L/G = 0$, the operating area becomes a horizontal line, which is identical to the counterflow diagram, and both coefficients are the same. An increase in L/G causes an increase in the height of the operating area and a decrease in the width. This continues as the areas extend to Point A as a limit. This maximum L/G always occurs when the wet-bulb temperature of the air equals the hot-water temperature and not at an intermediate point, as may occur in counterflow.

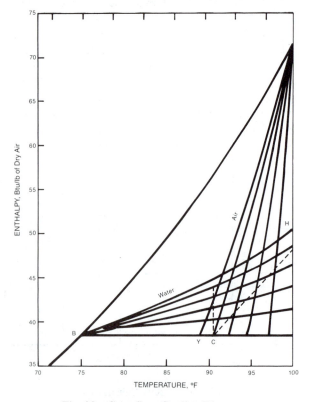

Fig. 28   Crossflow Cooling Diagram

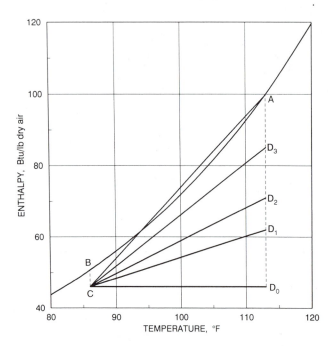

Fig. 29   Counterflow Cooling Diagram for Constant Conditions, Variable L/G Ratios

Both types of flow have the same minimum coefficient at $L/G = 0$, and both increase to infinity at a maximum L/G. The maximums are the same if the counterflow potential reaches zero at the air outlet, but the counterflow tower will have a lower maximum L/G when the potential reaches zero at an intermediate point, as in Figure 29. A cooling tower can be designed to operate at any point within the two limits, but most applications limit the design to much narrower limits determined by air velocity.

A low air rate requires a large tower, while a high air rate in a smaller tower requires greater fan power. Typical limits in air velocity are about 300 to 700 fpm in counterflow and 350 to 800+ fpm in crossflow.

Available Coefficients

A cooling tower can operate over a wide range of water rates, air rates, and heat loads, with variation in the approach of the cold water to the wet-bulb temperature. Analysis of a series of test points shows the available coefficient is not a constant, but varies with the operating conditions, as shown in Figure 30.

Figure 30 is a typical correlation of a tower characteristic showing the variation of available KaV/L with L/G for parameters of constant air velocity. Recent fill developments and more accurate test methods have shown that some of the characteristic lines are curves, rather than a series of straight, parallel lines on logarithmic coordinates.

Ignoring the minor effect of air velocity, a single average curve may be considered, which corresponds to:

$$KaV/L \sim (L/G)^n \qquad (13)$$

The exponent n varies within a range of about -0.35 to -1.1 but averages between -0.55 and -0.65. Within the range of testing, -0.6 has been considered sufficiently accurate.

The family of curves corresponds to the relation:

$$KaV/L \sim (L)^n (G)^m \qquad (14)$$

where n is as above and m varies numerically slightly from n and is a positive exponent.

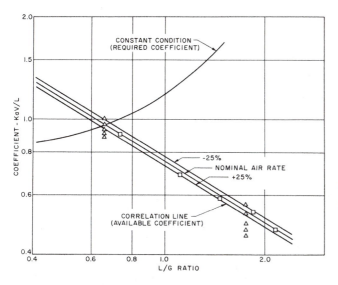

Fig. 30 Tower Characteristic, *KaV/L* versus *L/G*
(Baker and Shryock 1961)

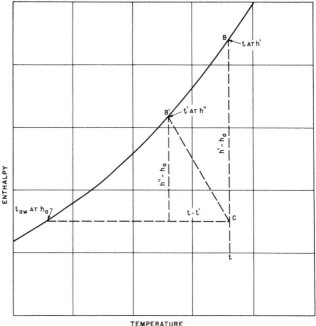

Fig. 31 True versus Apparent Potential Difference
(Baker and Shryock 1961)

The triangular points in Figure 30 show the effect of varying temperature at nominal air rate. The deviations result from simplifying assumptions and may be overcome by modifying the integration procedure. Usual practice, as shown in Equation (10), ignores evaporation and assumes that:

$$Gdh = c_p L dt \tag{15}$$

The exact enthalpy rise is greater than this because a portion of the heat in the water stream leaves as vapor in the airstream. The correct heat balance is (Baker and Shryock 1961):

$$Gdh = c_p L dt + L_E (t_{w2} - 32) \tag{16}$$

This reduces the driving force and increases the NTU.

Evaporation causes the water rate to decrease from L at the inlet to $L - L_E$ at the outlet. The water-to-air ratio varies from L/G at the water inlet to $(L - L_E)/G$ at the outlet. This results in an increase in NTU.

Basic theory considers the transfer from the interface to the airstream. As the film conditions are indeterminate, the film resistance is neglected as assumed in Equation (8). The resulting coefficients show deviations closely associated with hot-water temperature and may be modified by an empirical hot-water correction factor (Baker and Mart 1952).

The effect of film resistance (Mickley 1949) is shown in Figure 31. Water at temperature t is assumed to be surrounded by a film of saturated air at the same temperature, Point B, at enthalpy h', on the saturation curve. The film is actually at a lower temperature, Point B', at enthalpy h''. The surrounding air at enthalpy h_a corresponds to Point C. The apparent potential difference is commonly considered $(h' - h_a)$, but the true potential difference is $(h'' - h_a)$ (Mickley 1949). From Equations (2) and (7):

$$h'' - h_a/c_p(t' - t) = K_L/K' \tag{17}$$

The slope of CB' is the ratio of the two coefficients. No means to evaluate the coefficients has been proposed, but a slope of -11.1 for crossflow towers has been reported (Baker and Shryock 1961).

Establishing Tower Characteristics

Maximum performance in a given volume of fill is obtained with uniform water distribution and constant air velocity through-

out. Water distribution changes have a significant effect on the characteristic. Air is channeled by the cell structure, more severely in a counterflow tower, because of restricted air inlet and change in airflow direction.

Some cooling occurs in the spray chamber above the filling and also in the open space below the filling, but it is all credited to the filled volume. The indicated available coefficient reflects the performance of the entire assembly. The characteristic of a fill pattern varies widely, particularly in counterflow towers. A true tower characteristic must be developed from full-scale manufacturer's tests of the actual assembly.

REFERENCES

ASME. 1958. Atmospheric water cooling equipment, ASME PTC 23-58:11. American Society of Mechanical Engineers, New York.

Baker, D.R. 1962. Use charts to evaluate cooling towers. *Petroleum Refiner* (November).

Baker, D.R. and L.T. Mart. 1952. Analyzing cooling tower performance by the unit-volume coefficient. *Chemical Engineering* (December):196.

Baker, D.R. and H.A. Shryock. 1961. A comprehensive approach to the analysis of cooling tower performance. ASME *Transactions, Journal of Heat Transfer* (August):339.

CTI. 1990. Acceptance test code for water cooling towers. CTI *Bulletin* ATC-105, Cooling Tower Institute, Houston, TX.

Fluor Products Company. 1958. Evaluated weather data for cooling equipment design.

Kohloss, F.H. 1970. Cooling tower application. ASHRAE *Journal* (August).

Landon, R.D. and J.R. Houx, Jr. 1973. Plume abatement and water conservation with the wet-dry cooling tower, The Marley Company.

Merkel, F. 1925. Verduftungskuhlung. *Forschungarbeiten*, No. 275.

Mickley, H.S. 1949. Design of forced-draft air conditioning equipment. *Chemical Engineering Progress* 45:739.

LIQUID COOLERS

A liquid cooler (hereafter called a cooler) is a component of a refrigeration system in which the refrigerant is evaporated, thereby producing a cooling effect on a fluid (usually water or brine). This chapter addresses the performance, design, and application of coolers. It briefly describes various types of liquid coolers and the refrigerants commonly used as listed in Table 1.

TYPES OF LIQUID COOLERS

Direct-Expansion

Refrigerant evaporates inside tubes of a direct-expansion cooler. These coolers are usually used with positive-displacement compressors, such as reciprocating, rotary, or rotary screw compressors, to cool water or brine. Shell-and-tube is the most common arrangement, although tube-in-tube and brazed plate coolers are also available.

Figure 1 shows a typical shell-and-tube cooler. A series of baffles channels the fluid throughout the shell side. The baffles increase the velocity of the fluid, thereby increasing its heat transfer coefficient. The velocity of the fluid flowing perpendicular to the tubes should be at least 2 ft/s to clean the tubes and less than 10 ft/s to prevent erosion.

Distribution is critical in direct-expansion coolers. If some tubes are fed more refrigerant than others, they tend to bleed liquid refrigerant into the suction line. Since most direct-expansion coolers are controlled to a given suction superheat, the remaining tubes must produce a higher superheat to evaporate the liquid bleeding through. This unbalance causes poor overall heat transfer. Uniform distribution is usually achieved by a spray distributor or by keeping the volume of the refrigerant inlet head to a minimum. Both methods create sufficient turbulence to mix the liquid and vapor refrigerant entering the cooler so that each tube gets the same mixture of liquid and vapor.

The number of refrigerant passes is another important item in the performance of a direct-expansion cooler. A single-pass cooler must evaporate all the refrigerant before it reaches the end of the tubes; this requires long tubes or enhanced inside tube surfaces. A multiple-pass cooler can have less or no surface enhancement, but after the first pass, good distribution is difficult to obtain.

A tube-in-tube cooler is similar to a shell-and-tube design, except that it consists of one or more pairs of coaxial tubes. The fluid usually flows inside the inner tube while the refrigerant flows in the annular space between the tubes. In this way, the fluid side can be mechanically cleaned if access to the header is provided.

Brazed (or welded) plate coolers are constructed of plates brazed together to make up an assembly of separate channels. Space requirements are minimal due to their compactness;

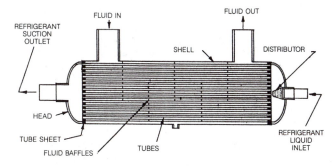

Fig. 1 Direct-Expansion Shell-and-Tube Cooler

Table 1 Types of Coolers

Type of Cooler	Usual Refrigerant Feed Device	Usual Range of Capacity, Tons	Commonly Used with Refrigerant Numbers
Flooded shell-and-tube	Low-pressure float High-pressure float Fixed orifice(s), Weir	25 to 2000	11, 12, 22, 113, 114, 123, 134a, 500, 502, 717
Spray-type shell-and-tube[a]	Low-pressure float High-pressure float	50 to 10,000	11, 12, 13B1, 22, 113, 114, 123, 134a
Direct-expansion shell-and-tube	Thermal expansion valve	2 to 350	12, 22, 134a, 500, 502, 717
Baudelot (flooded)	Low-pressure float	10 to 100	717
Baudelot (direct-expansion)	Thermal expansion valve	5 to 25	12, 22, 134a, 717
Tube-in-tube[b]	Thermal expansion valve	5 to 25	12, 22, 134a, 717
Shell-and-coil	Thermal expansion valve	2 to 10	12, 22, 134a, 717
Brazed or welded plate[b]	Thermal expansion valve	0.5 to 100	12, 22, 134a, 500, 502, 717

[a]See Flooded Shell-and-Tube section.   [b]See Direct-Expansion Cooler section.

The preparation of this chapter is assigned to TC 8.5, Liquid-to-Refrigerant Heat Exchangers.

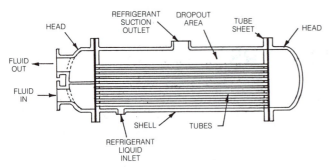

Fig. 2   Flooded Shell-and-Tube Cooler

however, the construction does not allow mechanical cleaning, and internal leaks typically cannot be repaired.

Most direct-expansion coolers are designed for horizontal mounting. If mounted vertically, performance may vary considerably from that predicted.

Flooded Shell-and-Tube

In a flooded cooler, the refrigerant vaporizes on the outside of tubes, which are submerged in liquid refrigerant within a closed shell. The fluid flows through the tubes as shown in Figure 2. Flooded coolers are usually used with rotary screw or centrifugal compressors to cool water or brine.

Refrigerant liquid/vapor mixture usually feeds into the bottom of the shell through a distributor that distributes the refrigerant vapor equally under the tubes. The relatively warm fluid in the tubes heats the refrigerant liquid surrounding the tubes, causing it to boil. As bubbles rise up through the space between tubes, the liquid surrounding the tubes becomes increasingly bubbly (or foamy, if much oil is present).

The refrigerant vapor must be separated from the mist generated by the boiling refrigerant. The simplest separation method is provided by a dropout area between the top row of tubes and the suction connectors. If this dropout area is insufficient, a coalescing filter may be required between the tubes and connectors. Perry and Green (1984) give additional information on mist elimination.

The size of tubes, number of tubes, and number of passes should be determined to maintain the fluid velocity typically between 3 and 10 ft/s. Velocities beyond these limits may be used if the fluid is free of suspended abrasives and fouling substances (Sturley 1975, Ayub and Jones 1987). In some cases the minimum flow may be determined by a lower Reynolds number limit.

One variation of this cooler is the spray-type shell-and-tube cooler. In large diameter coolers with a refrigerant that has a heat transfer coefficient that is adversely affected by the head of the refrigerant, a spray can be used to cover the tubes with liquid rather than flooding them. A mechanical pump circulates liquid from the bottom of the cooler to the spray heads.

Flooded shell-and-tube coolers are generally unsuitable for other than horizontal orientation.

Baudelot

Baudelot coolers (Figure 3) are used to cool a fluid to near its freezing point in industrial, food, and dairy applications. In this type of cooler, the fluid is circulated over the outside of a number of columns of horizontal tubes or vertical plates, making them easy to clean. The inside surface of the tubes or plates is cooled by evaporating the refrigerant. The fluid to be cooled is distributed uniformly along the top of the heat exchanger and then flows by gravity to a collection pan below. The cooler may be enclosed by insulated walls to avoid unnecessary loss of refrigeration effect.

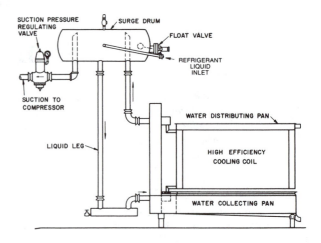

Fig. 3   Baudelot Cooler

Refrigerant 717 (ammonia) is commonly used with the Baudelot cooler and arranged for flooded operation, using a conventional gravity feed system with a surge drum. A low-pressure float valve maintains a suitable refrigerant liquid level in the surge drum. Baudelot coolers using other common refrigerants are generally of the direct-expansion type, with thermostatic expansion valves.

Shell-and-Coil

A shell-and-coil cooler is a tank containing the fluid to be cooled with a simple coiled tube used to cool the fluid. This type of cooler has the advantage of cold water storage to offset peak loads. In some models, the tank can be opened for cleaning. Most applications are at low capacities for bakeries, photographic laboratories, and to cool drinking water.

The coiled tube containing the refrigerant can be either inside the tank (Figure 4) or attached to the outside of the tank in such a manner as to permit heat transfer.

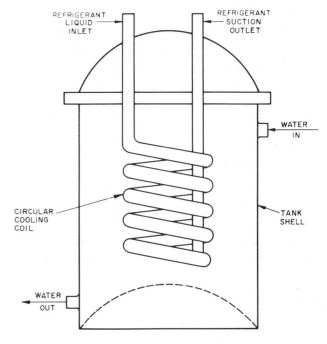

Fig. 4   Shell-and-Coil Cooler

HEAT TRANSFER

Heat transfer for liquid coolers can be expressed by the following steady-state heat transfer equation:

$$q = U A \Delta t_m \qquad (1)$$

where

q = total heat transfer rate, Btu/h
Δt_m = mean temperature difference, °F
A = heat transfer surface area associated with U, ft^2
U = overall heat transfer coefficient, Btu/(h·ft^2·°F)

The area A can be calculated if the geometry of the cooler is known. Chapter 3 of the 1989 ASHRAE *Handbook—Fundamentals* describes the calculation of the mean temperature difference.

This chapter only discusses the components of U, but not in depth. U may be calculated by one of the following equations.

Based on inside surface area

$$U = \frac{1}{1/h_i + [A_i/(A_o h_o)] + (t/k)(A_i/A_m) + r_{fi}} \qquad (2)$$

Based on outside surface area

$$U = \frac{1}{[A_o/(A_i h_i)] + 1/h_o + (t/k)(A_o/A_m) + r_{fo}} \qquad (3)$$

where

h_i = inside heat transfer coefficient based on inside surface area, Btu/(h·ft^2·°F)
h_o = outside heat transfer coefficient based on outside surface area, Btu/(h·ft^2·°F)
A_o = outside heat transfer surface area, ft^2
A_i = inside heat transfer surface area, ft^2
A_m = mean heat transfer area of metal wall, ft^2
k = thermal conductivity of heat transfer material, Btu/(h·ft·°F)
t = thickness of heat transfer surface, ft
r_{fi} = fouling factor of fluid side based on inside surface area, ft^2·h·°F/Btu
r_{fo} = fouling factor of fluid side based on outside surface area, ft^2·h·°F/Btu

Note: If fluid is on inside, multiply r_{fi} by A_o/A_i to find r_{fo}.
If fluid is on outside, multiply r_{fo} by A_i/A_o to find r_{fi}.

Heat Transfer Coefficients

The refrigerant side coefficient usually increases under the following conditions: (1) increase in cooler load, (2) decrease in suction superheat, (3) decrease in oil concentration, and (4) increase in saturated suction temperature. The amount of increase or decrease varies, depending on the type of cooler. Based on ASHRAE-sponsored research, RP-469, Schalger *et al.* (1989) discuss the effects of oil in direct expansion coolers. Flooded coolers have a relatively small change in heat transfer coefficient as a result of a change in load, whereas a direct expansion cooler shows a significant increase in heat transfer coefficient with an increase in load. A Wilson Plot of test data (McAdams 1954, Briggs and Young 1969) can show actual values for the refrigerant side coefficient of a given cooler design. Young and Ward (1957) give additional information on refrigerant side heat transfer coefficients.

The fluid side coefficient is determined by cooler geometry, fluid flow rate, and fluid properties (viscosity, specific heat, thermal conductivity, and density) (Palen and Taborek 1969, Wolverine 1984). For a given fluid, the fluid side coefficient increases with an increase in fluid flow rate due to increased turbulence, and an increase in fluid temperature due to improvement of fluid properties as temperature increases.

The heat transfer coefficient in direct-expansion and flooded coolers increases significantly with fluid flow. The effect of flow is smaller for Baudelot and shell-and-coil coolers. Many of the listed references give additional information on fluid side heat transfer coefficients.

To increase the heat transfer coefficients of coolers, an enhanced heat transfer surface can help in the following ways:

- It increases heat transfer area, thereby increasing overall heat transfer rate, even if refrigerant side heat transfer coefficient is unchanged.
- Where the flow of fluid or refrigerant is low, it improves heat transfer coefficients by increasing turbulence at the surface and mixing the fluid at the surface with fluid away from the surface.
- In flooded coolers, an enhanced refrigerant side surface may provide more and better nucleation points to promote boiling of refrigerant.

Webb and Pias (1991a) describe many enhanced surfaces used in flooded coolers. The enhanced surface geometries provide substantially higher boiling coefficients than do integral fin tubes. Nucleate pool boiling data are provided by Webb and Pias (1991b). The boiling process that occurs in the tube bundle of a flooded cooler may be enhanced by forced convection effects. This is basically an additive effect, in which the local boiling coefficient is the sum of the nucleate boiling coefficient and the forced convection effect. As part of ASHRAE-sponsored research, RP-392, Webb *et al.* (1989) describe a theoretical model to predict the performance of flooded coolers recommending row-by-row calculations.

Based on the model, Webb and Apparao (1990) present the results of calculations using a computer program. These results show some performance differences of various internal and external surface geometries. As an example, Figure 5 shows the contribution of nucleate pool boiling to the overall refrigerant heat transfer coefficient for an integral fin tube and an enhanced tube as a function of the tube row. The forced convection effect predominates with the integral fin tube.

Fouling Factors

Most fluids over time foul the fluid side heat transfer surface, thus reducing the overall heat transfer coefficient of the cooler. If fouling is expected to be a problem, a mechanically cleanable

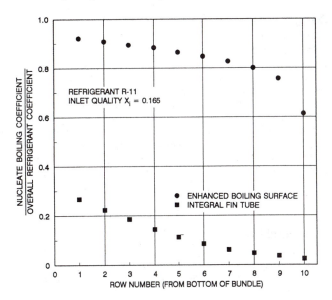

Fig. 5 Nucleate Boiling Contribution to Total Refrigerant Heat Transfer

cooler should be used, such as a flooded, Baudelot, or cleanable direct-expansion tube-in-tube cooler. Direct-expansion shell-and-tube, shell-and-coil, and brazed plate coolers can be cleaned chemically. Flooded coolers and direct-expansion tube-in-tube coolers with enhanced fluid side heat transfer surfaces have a tendency to be self-cleaning due to high fluid turbulence, and a smaller fouling factor can probably be used for these coolers. Water quality in closed chilled water loops has been studied as part of ASHRAE-sponsored research (RP-560). Haider *et al.* (1991) found little potential for fouling in such systems. ARI *Standard* 480 discusses fouling calculations.

The refrigerant side of the cooler is not subject to fouling, and a fouling factor need not be included for the that side.

Tube Wall Resistance

Typically the t/k term in Equations (2) and (3) is negligible. However, with low thermal conductivity materials or thick wall tubing, it may become significant. Refer to Chapter 3 of the 1989 ASHRAE *Handbook—Fundamentals* and to Chapter 36 for further details.

PRESSURE DROP

Fluid Side

Pressure drop is usually minimal in Baudelot and shell-and-coil coolers, but must be considered in direct-expansion and flooded coolers. Both the direct-expansion and flooded coolers rely on turbulent fluid flow to improve heat transfer. This turbulence is obtained at the expense of pressure drop. Increasing the pressure drop increases heat transfer.

For air-conditioning service, the pressure drop is commonly limited to 10 psi to keep pump size and energy costs reasonable. For flooded coolers, see Chapter 36 for a discussion of pressure drop for flow in tubes. Pressure drop for fluid flow in shell-and-tube direct expansion coolers depends greatly on tube and baffle geometry. The following equation projects the change in pressure drop due to a change in flow.

$$\text{New Pressure Drop} = \text{Original Pressure Drop} \left[\frac{\text{New Flow}}{\text{Original Flow}} \right]^{1.8} \quad (4)$$

Refrigerant Side

The refrigerant side pressure drop must be considered for the following coolers: direct-expansion, shell-and-coil, brazed plate, and, sometimes, Baudelot. When there is a pressure drop on the refrigerant side, the refrigerant inlet and outlet pressures and corresponding saturated temperature are different. This difference causes a change in the mean temperature difference, which affects the total heat transfer rate. If the pressure drop is high, operation of the expansion valve may be affected due to reduced pressure drop across the valve. This pressure drop varies, depending on the refrigerant used, operating temperature, and type of tubing (Wallis 1969, Martinelli and Nelson 1948).

VESSEL DESIGN

Mechanical Requirements

Pressure vessels must be constructed and tested under the rules of national, state, and local codes. The introduction of the current ASME Boiler and Pressure Vessel Code, Section VIII, gives guidance on rules and exemptions.

The more common applicable codes and standards are as follows:

1. ARI *Standard* 480-87, Remote Type Refrigerant Cooled Liquid Coolers, covers industry criteria for standard equipment, standard safety provisions, marking, and recommended rating requirements.
2. ASHRAE *Standard* 24-89, Methods of Testing for Rating Liquid Coolers, covers the recommended testing methods for measuring the capacity of liquid coolers.
3. ANSI/ASHRAE *Standard* 15-89, Safety Code for Mechanical Refrigeration, involves specific design criteria, use of materials, and testing. It refers to the ASME Boiler Code, Section VIII, for refrigerant-containing sides of pressure vessels, where applicable. Factory test pressures are specified, and minimum design working pressures are given. This code requires pressure limiting and pressure relief devices on refrigerant-containing systems, as applicable, and defines the setting and capacity requirements for these devices.
4. ASME Boiler and Pressure Vessel Code, Unfired Pressure Vessels, Section VIII, covers the safety aspects of design and construction. Most states require coolers to meet the requirements of the ASME if they fall within scope of the ASME code. Some of the exceptions from meeting the ASME requirements listed in the ASME code are as follows:

   - Cooler shell ID of 6 in. or less.
   - 15 psi gage or less.
   - The fluid (water) portion of the cooler need not be built to the requirements of the ASME code if the fluid is water, the design pressure does not exceed 300 psig, and the design temperature does not exceed 210 °F.

   Coolers meeting the requirements of the ASME code will have an ASME stamp, which is a *U* or *UM* inside a three-leaf clover. The *U* can be used for all coolers and the *UM* can be used for small coolers.
5. Underwriters Laboratories *Standard* UL-207-89, Refrigerant-Containing Components and Accessories, involves specific design criteria, use of materials, testing, and initial approval by Underwriters Laboratories. A cooler with the ASME *U* stamp does not require UL approval.

Design pressure. On the refrigerant side, design pressure as a minimum should be the saturated pressure for the refrigerant used at 80 °F as per ANSI/ASHRAE *Standard* 15-1989. Standby temperatures and temperatures encountered during shipping of chillers with a refrigerant charge should also be considered.

Required fluid (water) side pressure varies depending largely on the following conditions: (1) static head, (2) pump head, (3) transients due to pump start-up, and (4) valve closing.

Chemical Requirements

The following chemical requirements are given by Perry and Green (1984) and NACE (1974):

Refrigerant 717 (Ammonia). Carbon steel and cast iron are the most widely used materials for ammonia systems. Stainless steel alloys are satisfactory but more costly. Copper and high copper alloys are avoided because they are attacked by ammonia when moisture is present. Aluminum and aluminum alloys may be used with caution with ammonia.

Halocarbon refrigerants. Almost all the common metals and alloys are used satisfactorily with these refrigerants. Exceptions include magnesium and aluminum alloys containing more than 2% magnesium, where water may be present. Zinc is not recommended for use with Refrigerant 113; it is more chemically reactive than other common construction metals and, therefore, is usually avoided when other halogenated hydrocarbons are used as a refrigerant. Under some conditions with moisture present, halocarbon refrigerants form acids that attack steel and even nonferrous metals. This problem does not commonly occur in properly cleaned and dehydrated systems. ANSI/ASHRAE *Standard* 15-89, paragraph 7.1.2, states that aluminum, zinc, or magnesium shall not be used in contact with methyl chloride nor magnesium alloys with any halogenated refrigerant.

Water. Relatively pure water is satisfactory with both ferrous and nonferrous metals. Brackish or sea water, and some river waters, are quite corrosive to iron and steel and also with copper, aluminum, and many alloys of these metals. A reputable water consultant, who knows the local water condition, should be contacted. Chemical treatment by pH control, inhibitor applications, or both, may be required. Where this is not feasible, more noble construction materials or special coatings must be used.

Brines. Ferrous metals and a few nonferrous alloys are almost universally used with sodium chloride and calcium chloride brines. Copper base alloys can be used, if adequate quantities of sodium dichromate are added and caustic soda is used to neutralize the solution. Even with ferrous metals, these brines should be treated periodically to hold the pH value near the neutral point.

Ethylene glycol and propylene glycol are stable compounds that are less corrosive than chloride brines.

Electrical Requirements

When fluid being cooled is electrically conductive, the system must be grounded to prevent electrical chemical corrosion.

APPLICATION CONSIDERATIONS

Refrigerant Flow Control

Direct-expansion coolers. The constant superheat thermal expansion valve is the most common control used. It is located directly upstream of the cooler. A thermal bulb strapped to the suction line leaving the cooler senses refrigerant temperature. The valve can be adjusted to produce a constant suction superheat during steady operation.

The thermal expansion valve adjustment is commonly set at a suction superheat of 10 °F, which is sufficient to ensure that liquid is not carried into the compressor. Direct expansion cooler performance is affected greatly by superheat setting. Reduced superheat improves cooler performance; thus, suction superheat should be set as low as possible while avoiding liquid carryover to the compressor.

Flooded coolers. As the name implies, flooded coolers must have good liquid refrigerant coverage of the tubes to achieve good performance. Liquid level control in a flooded cooler becomes the principal issue in flow control. Some systems are designed *critically charged*, so that when all the liquid refrigerant is delivered to the cooler, it is just enough for good tube coverage. In these systems, an orifice is often used as the throttling device between condenser and cooler.

A float valve is another method of flooded cooler control. A high-side float valve can be used, with the float-sensing condenser liquid level, to drain the condenser. For more exact control of liquid level in the cooler, a low side float valve is used, where the valve senses cooler liquid level and controls the flow of entering refrigerant.

Freeze Prevention

Freeze prevention must be considered for coolers operating near the freezing point of the fluid. Freezing of the fluid in some coolers causes extensive damage. Two methods can be used for freeze protection: (1) hold the saturated suction pressure above the fluid freezing point or (2) shut the system off if the temperature of the fluid approaches its freezing point.

A suction pressure regulator can hold the saturated suction pressure above the freezing point of the fluid. A low-pressure cutout can shut the system off before the saturated suction pressure drops

to below the freezing point of the fluid. The leaving fluid temperature can be monitored to cut the system off before a danger of freezing, usually about 10 °F above the fluid freezing temperature. It is recommended that both methods be used.

Baudelot, shell-and-coil, brazed plate, and direct-expansion shell-and-tube coolers are all somewhat resistant to damage caused by freezing, and ideal for applications where freezing may be a problem.

If a cooler is installed in an unconditioned area, possible freezing due to low ambient temperature must be considered. If the cooler is used only when the ambient temperature is above freezing, the fluid should be drained from the cooler for cold weather. As an alternate to draining, if the cooler is in use year-round, the following methods can be used:

- A heat tape or other heating device can be used to keep the cooler above freezing.
- For water, adding an appropriate amount of ethylene glycol will prevent freezing.
- Continuous pump operation may also prevent freezing.

Oil Return

Most compressors discharge a small percentage of oil in the discharge gas. This oil mixes with the condensed refrigerant in the condenser and flows to the cooler. Since the oil is nonvolatile, it does not evaporate and may collect in the cooler.

In direct-expansion coolers, the gas velocity in the tubes and the suction gas header is usually sufficient to carry the oil from the cooler into the suction line. From there, with proper piping design, it can be carried back to the compressor. At light loads and low temperature, oil may gather in the superheat section of the cooler, detracting from performance. For this reason, operation of refrigerant circuits at light load for long periods should be avoided, especially at low-temperature conditions.

In flooded coolers, vapor velocity above the tube bundle is usually insufficient to return oil up the suction line and oil tends to accumulate in the cooler. With time, depending on compressor oil loss rate, the oil concentration in the cooler may become large. When concentration exceeds about 5%, heat transfer performance may be adversely affected.

It is common in flooded coolers to take some oil-rich liquid and return it to the compressor on a continuing basis, in order to establish a rate of return equal to compressor oil loss rate.

Maintenance

Maintenance of coolers centers around two areas: (1) safety and (2) cleaning of the fluid side. The cooler should be inspected periodically for any weakening of its pressure boundaries. The inspection should include visual inspection for corrosion, erosion, and any deformities. Any pressure relief device should also be inspected. The insurer of the cooler may require regular inspection of the cooler. If the fluid side is subjected to fouling, it may require periodic cleaning. Cleaning may be by either mechanical or chemical means. The manufacturer or service organization experienced in cooler maintenance should have details for cleaning.

Insulation

A cooler operating at a saturated suction temperature lower than the dew point of the surrounding air should be insulated to prevent condensation. Chapter 20 of the 1989 ASHRAE *Handbook—Fundamentals* describes insulation in more detail.

REFERENCES

Ayub, Z.H. and S.A. Jones. 1987. Tubeside erosion/corrosion in heat exchangers. *Heating/Piping/Air Conditioning* (December):81.

Briggs, D.E. and E.H. Young. 1969. Modified Wilson Plot techniques for obtaining heat transfer correlations for shell and tube heat exchangers. *Chemical Engineering Progress Symposium Series* 65(92):25.

Haider, S.I., R.L. Webb, and A.K. Meitz. 1991. A survey of water quality and its effect on fouling in flooded water chiller evaporators. ASHRAE *Transactions* 97(1).

Martinelli, R.C. and D.B. Nelson. 1948. Prediction of pressure drop during forced circulation boiling of water. ASME *Transactions* (August):695.

McAdams, W.H. 1954. *Heat transmission*, 3rd ed. McGraw-Hill Book Company, New York.

NACE. 1974. *Corrosion data survey*, 5th ed. Compiled by N.E. Hamner for the National Association of Corrosion Engineers, Houston, TX.

Palen, J.W. and J. Taborek. 1969. Solution of shell side pressure drop and heat transfer by stream analysis method. *Chemical Engineering Progress Symposium Series* 65(92).

Perry, J.H. and R.H. Green. 1984. *Chemical engineers handbook*, 6th ed. McGraw Hill Book Company, New York.

Schlager, L.M., M.B. Pate, and A.E. Bergles. 1989. A comparison of 150 and 300 SUS oil effects on refrigerant evaporation and condensation in a smooth tube and a micro-fin tube. ASHRAE *Transactions* 95(1).

Sturley, R.A. 1975. Increasing the design velocity of water and its effect on copper tube heat exchangers. Paper No. 58, The International Corrosion Forum, Toronto, Canada.

Wallis, G.B. 1969. *One dimensional two phase flow*. McGraw Hill Book Company, New York.

Webb, R.L. and T. Apparao. 1989. A theoretical model to predict the heat duty and pressure drop in flooded refrigerant evaporators. ASHRAE *Transactions* 95(1):326-38.

Webb, R.L. and T. Apparao. 1990. Performance of flooded refrigerant evaporators with enhanced tubes. *Heat Transfer Engineering* 11(2):29-43.

Webb, R.L. and C. Pias. 1991a. Literature survey of pool boiling on enhanced surfaces. ASHRAE *Transactions* 97(1).

Webb, R.L. and C. Pias. 1991b. Nucleate boiling data for five refrigerants on three tube geometries. ASHRAE *Transactions* 97(1).

Wolverine Division of UOP, Inc. 1984. *Engineering data book II*.

Young, E.H. and D.J. Ward. 1957. Fundamentals of finned tube heat transfer, Part I. *Refining Engineer* (November).

CENTRIFUGAL PUMPS

CENTRIFUGAL pumps recirculate hot water in heating systems and chilled water cooling systems to establish a predetermined rate of flow between the boiler/chiller (for thermal storage tanks) and the space conditioning terminal units. The influence of pump performance on installation, system controllability, and seasonal operating costs is covered in Chapter 12.

Other pump applications on hydronic systems include (1) condenser water circuits to cooling towers and water source heat pumps, (2) boiler feed, and (3) condensate return. Pumps are required with boiler feed and condensate return only if a steam boiler is included in the system. In such cases, the boiler manufacturer defines the specific pumping requirements. When a cooling tower rejects heat for a chilled water plant, the condenser water pumps are selected on the basis of the flow rate specified by the refrigeration equipment manufacturer and the location of the tower relative to the condenser.

In centrifugal pumps, a driver converts part of the output torque into pressure energy by centrifugal force, which is a function of the impeller vane peripheral velocity. Impeller rotation adds energy to a liquid after it enters the eye of the impeller. The casing collects the liquid as it leaves the impeller and guides it out of the discharge nozzle. The pressure energy added by the pump (1) overcomes the friction caused by the flow through heating and air-conditioning equipment, i.e., piping, valves, coils, chillers, or boilers; and (2) raises the water to higher elevations such as to the top of a cooling tower.

PUMP TYPES

Most centrifugal pumps used in hydronic systems are single stage with a single- or double-entry impeller. Double suction pumps are generally used for high flow applications, but either form is available with similar performance characteristics and efficiencies. Selection can be based on installed cost and personal preferences.

These pumps have either volute or diffuser types of casings. The volute types include all pumps that collect the water from the impeller and discharge it perpendicularly to the pump shaft. Diffuser-type casings collect the water from the impeller and discharge it parallel to the pump shaft. All pumps described here are the volute type, except the vertical turbine pump, which is a diffuser type.

Pumps can be classified by method of connection to the electric motor and can be close-coupled or flexible-coupled. The close-coupled pump has the impeller mounted directly on a motor shaft extension, while the flexible-coupled pump has an impeller shaft supported by a frame or bracket that is connected to the electric motor through a flexible coupling.

Pumps are also classified by their mechanical features and installation arrangement. *Circulator* is a generic term for pipe-mounted, low-pressure, low flow units, and may be either wet rotor or conventional flexible-coupled open-type motor driven. In addition to their application in residential and small commercial buildings, circulators recirculate flow of terminal unit coils to enhance heat transfer efficiencies and improve the management of large systems.

One-horsepower and larger pumps are available as close-coupled or base mounted. The close-coupled pumps are *end suction* for horizontal mounting or *vertical inline* for direct installation in the piping. The base-mounted pumps are *end-suction-frame mounted* or *double suction* horizontally split case units. Double-suction pumps can also be arranged in a vertical position on a support frame with the motor vertically mounted on a bracket above the pump unit.

Pumps are labeled by their mounting position, either horizontal or vertical. Significant types of pumps used in heating and air-conditioning or hydronic systems are (1) circulator; (2) close-coupled, end suction; (3) frame-mounted or flexible-coupled, end suction; (4) double suction, horizontal split case, single-stage; (5) horizontal split case, multistage; (6) vertical in-line; and (7) vertical turbine. Table 1 lists these pump types and summarizes their design features. Figure 1 shows the general configuration of these pumps and lists their typical applications. Many variations of these pumps are offered by pump manufacturers for particular applications.

Table 1 Mechanical Features of Centrifugal Pumps

Pump Type	Impeller Type	Number of Impellers	Casing	Motor Connection	Motor Mounting Position
Circulator	Single suction	1	Volute	Flexible coupled	Horizontal
Close-coupled, end suction	Single suction	1 or 2	Volute	Close-coupled	Horizontal
Frame mounted, end suction	Single suction	1 or 2	Volute	Frame coupled	Horizontal
Double suction, split casing	Double suction	1	Volute	Flexible coupled	Horizontal or vertical
Vertical in-line	Single suction	1	Volute	Flexible or close-coupled	Vertical
Vertical turbine	Single suction	1 to 20	Diffuser	Flexible coupled	Vertical

The preparation of this chapter is assigned to TC 8.10, Pumps and Hydronic Piping.

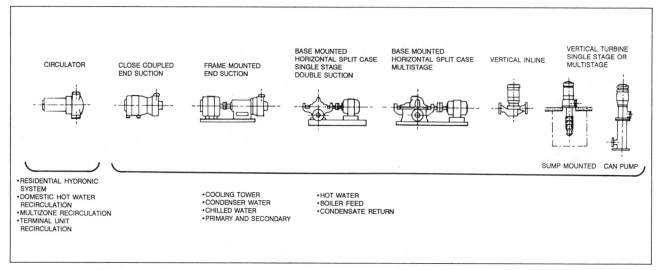

Fig. 1   Applications for Centrifugal Pumps Used in Hydronic Systems

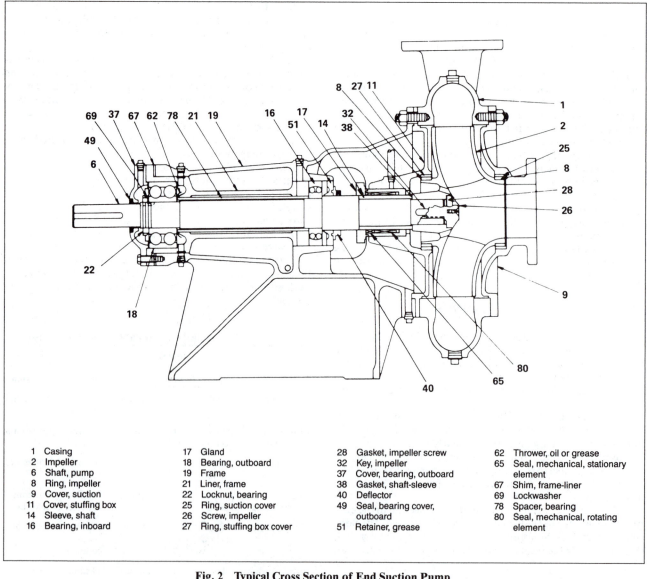

1	Casing	17	Gland	28	Gasket, impeller screw	62	Thrower, oil or grease

1	Casing	17	Gland	28	Gasket, impeller screw	62	Thrower, oil or grease
2	Impeller	18	Bearing, outboard	32	Key, impeller	65	Seal, mechanical, stationary element
6	Shaft, pump	19	Frame	37	Cover, bearing, outboard		
8	Ring, impeller	21	Liner, frame	38	Gasket, shaft-sleeve	67	Shim, frame-liner
9	Cover, suction	22	Locknut, bearing	40	Deflector	69	Lockwasher
11	Cover, stuffing box	25	Ring, suction cover	49	Seal, bearing cover, outboard	78	Spacer, bearing
14	Sleeve, shaft	26	Screw, impeller			80	Seal, mechanical, rotating element
16	Bearing, inboard	27	Ring, stuffing box cover	51	Retainer, grease		

Fig. 2   Typical Cross Section of End Suction Pump
(Courtesy Hydraulic Institute)

GENERAL CONSTRUCTION FEATURES

The following construction features of centrifugal pumps are shown in Figure 2.

Materials. Centrifugal pumps are generally offered in bronze-fitted, all bronze, or iron-fitted construction. In bronze-fitted construction, the impeller, shaft sleeve (if used), and wear rings are bronze, and the casing is cast iron. These construction materials refer to the liquid end of the pump (those parts of the pump that contact the liquid being pumped).

The *stuffing box* is that portion of the pump where the rotating shaft enters the pump casing. To seal leaks at this point, a mechanical seal or packing is used in the stuffing box.

Mechanical seals are used predominately in hydronic applications. There are unbalanced and balanced (for higher pressures) seals. Inside seals operate inside the stuffing box, while outside seals have their rotating element outside the box. Pressure and temperature limitations vary depending on the liquid being pumped and the style of seal.

Packing is used also, particularly where abrasive substances included in the water are not detrimental to system operation. Some leakage at the packing gland is needed to lubricate and cool the area between the packing material and shaft.

Shaft sleeves protect the motor or pump shaft.

Wearing rings are for the impeller and/or casing. They are replaceable and prevent wear to the impeller or casing.

Ball bearings are most frequently used, except in circulators, where motor and pump bearings are of the sleeve type.

The *balance ring* is placed on the back side of a single-inlet, enclosed impeller to reduce the axial load. Double-inlet impellers are inherently axially balanced.

Nominal operating speeds of motors may be selected in the range between 600 and 3600 rpm. (Consult pump manufacturers for optimum pump speed for each specific pumping requirement, with due consideration for efficiency, cost, noise, and maintenance.)

PUMP TERMS, EQUATIONS, AND LAWS

Table 2 lists terms and equations for pumping and Table 3 lists the affinity laws for pumps. These laws describe the relationships among the changes of pump impeller diameter, speed, and specific gravity. Without knowledge of the system curve, the laws should not be used to predict the pump performance of a particular hydronic system. Figure 3 describes pump performance at 1750 and 1150 rpm, in accordance with the affinity laws for constant impeller diameter and viscosity.

If the hydronic system has a system curve as shown in Figure 3, curve A, the pump at 1150 rpm will operate at point 1, not at point 2, as the affinity laws predict. If the system curve is the same as curve B in this figure, at 1150 rpm, the pump will run at shutoff head and will not deliver water, thus demonstrating that the affinity laws should be used to develop new pump curves, but not to predict performance unless the system curve is known. (System curves are covered in Hydronic System Characteristics.)

PUMP PERFORMANCE CURVES

Performance of a pump is most commonly shown by graphs, as in Figure 4, which relate the flow, the head produced, the power required, the efficiency, the shaft speed, and the net positive suction head (absolute) required for pumps with various impeller diameters. (See the section Pump Suction Characteristics for a description of net positive suction head.) For many small pumps, some of this information is omitted. Pump curves present the average results obtained from testing several pumps of the same design under standard test conditions. Consult manufacturers for pump applications that differ considerably from ordinary practice.

Table 2 Common Pump Terms, Symbols, and Formulas

Term	Symbol	Units	Formula
Velocity	v	ft/s	
Volume	V	ft^3	
Flow rate	Q_v	gpm	
Pressure	p	psi	
Density	ρ	lb/ft^3	
Acceleration of gravity	g	32.17 ft/s^2	
Specific gravity	SG	—	$\dfrac{\text{Mass of liquid}}{\text{Mass of water at 39°F}}$
Speed	n	rpm	
Head	H	ft	2.31 p/SG
Net positive suction head (NPSH)	H	ft	
Efficiency (percent)			
Pump	η_p		
Electric motor	η_m		
Variable speed drive	η_v		
Equipment (constant speed pumps)	η_e		$\eta_e = \eta_p \eta_m / 100$
Equipment (variable speed pumps)	η_e		$\eta_e = 10^{-4}\, \eta_p \eta_m \eta_v$
Utilization Q_D = design flow Q_A = actual flow H_D = design head H_A = actual head	η_u		$\eta_u = 100\, \dfrac{Q_D H_D}{Q_A H_A}$
System Efficiency Index (decimal)			$\text{SEI} = 10^{-4}\, \eta_e \eta_u$
Output power (pump)	P_o	hp	$Q_v H \text{SG}/3960$
Shaft power	P_s	hp	$100\, P_o / \eta_p$
Input power	P_i	kW	$74.6\, P_s / \eta_m$

Table 3 Affinity Laws for Pumps

Impeller Diameter	Speed	Specific Gravity (SG)	To Correct for	Multiply by
Constant	Variable	Constant	Flow	$\left(\dfrac{\text{New Speed}}{\text{Old Speed}}\right)$
			Head	$\left(\dfrac{\text{New Speed}}{\text{Old Speed}}\right)^2$
			Power	$\left(\dfrac{\text{New Speed}}{\text{Old Speed}}\right)^3$
Variable	Constant	Constant	Flow	$\left(\dfrac{\text{New Diameter}}{\text{Old Speed}}\right)$
			Head	$\left(\dfrac{\text{New Diameter}}{\text{Old Speed}}\right)^2$
			Power	$\left(\dfrac{\text{New Diameter}}{\text{Old Speed}}\right)^3$
Constant	Constant	Variable	Power	$\left(\dfrac{\text{New SG}}{\text{Old SG}}\right)$

The head-capacity curve for a centrifugal pump describes the head produced by the pump, from maximum flow to the shutoff or no-flow condition. The pump curve is considered flat if the shutoff head is about 1.10 to 1.20 times the head at the best efficiency point. If the head at shutoff exceeds 1.20 times the head at the best efficiency point, it is called a steep-curved pump.

The need for energy conservation and the use of two- and variable-speed pumps require particular attention to the efficiency curves of a pump. As shown in Figure 4, the best efficiency point

is 400 gpm and 44 ft of head. The efficiency curves form an eye, with the maximum efficiency being the inner circle. The affinity laws can be used to predict the best efficiency point at other pump speeds; this best efficiency point follows a parabolic curve to zero as the pump speed is decreased. The best efficiency curve for the pump of Figure 4 is described in Figure 5, which shows the movement of the best efficient point from high to low speed.

PUMP SUCTION CHARACTERISTICS (NPSH)

Particular attention must be given to the condition of the liquid as it enters a pump in condenser, condensate, and boiler feed systems. If the absolute pressure on the liquid at the suction nozzle approaches the vapor pressure of the liquid, cavitation occurs, and vapor pockets form in the impeller passages. The collapse of the vapor pockets could progressively damage the impeller.

The amount of pressure in excess of the vapor pressure required to prevent the formation of vapor pockets is the net positive suction head required (NPSHR). NPSHR is a characteristic of a given pump, and it varies considerably with pump speed and flow. NPSHR is determined by testing individual pumps; it increases rapidly at high flows, as shown in Figure 4.

Particular attention must be given to NPSHR when a pump is operating with hot liquids. The vapor pressure increases with water temperature and reduces the net positive suction head available (NPSHA). While each pump has its own particular NPSHR, each installation has its own particular NPSHA, which is the total useful energy above the vapor pressure of the liquid available to the pump at the suction connection. To calculate the NPSHA, use one of the following:

$$\text{NPSHA in proposed installation} = h_p + h_z - h_{vpa} - h_f$$

$$\text{NPSHA in existing installation} = p_a + p_s + v_2/2g - h_{vpa} - h_f$$

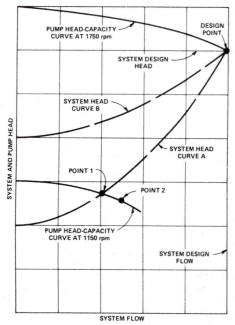

Fig. 3   Application of Affinity Laws

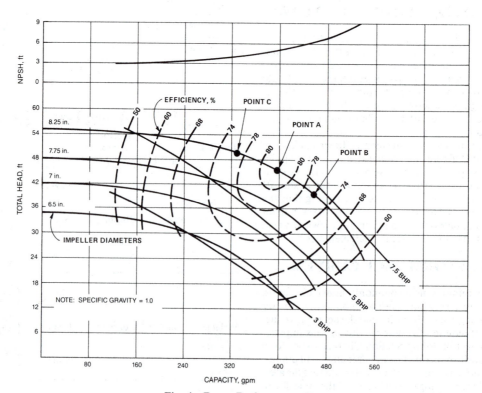

Fig. 4   Pump Performance Curves

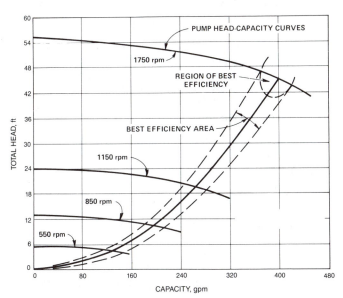

Fig. 5   Pump Best Efficiency Curves

where

$NPSHA$ = net positive suction head available at pump suction nozzle

h_p = absolute head on surface of liquid where pump takes suction, ft of liquid

h_z = static elevation of liquid above centerline of pump, ft of liquid; h_z is negative if liquid level is below pump centerline

h_f = friction and entrance head losses in suction piping, ft of liquid

h_{vpa} = absolute vapor pressure at pumping temperature, ft of liquid

p_a = atmospheric head for elevation of installation, ft of liquid

p_s = gage head at suction flange of pump corrected to centerline of pump, ft of liquid; p_s is negative if below atmospheric pressure

$v_2/2g$ = velocity head at point of measurement of p_s, ft of liquid

If NPSHA is less than NPSHR, cavitation, noise, inadequate pumping, and mechanical problems occur. NPSH is normally not a factor with hot water and chilled water pumps because sufficient system fill pressure is usually exerted on the pump suction.

HYDRONIC SYSTEM CHARACTERISTICS

Hydronic systems in HVAC applications are all of the loop type (open or closed), which circulate water through the system and return it to the pumps. No appreciable amount of water is lost from the system, except in the cooling tower, where evaporative cooling occurs.

Hydronic systems are either full flow or throttling flow. Full-flow systems are usually found on residential or small commercial systems where pump motors are small and the energy waste caused by constant flow is not appreciable. Larger systems use flow control valves that control flow in the system in accordance with the heating or cooling load imposed. Hydronic systems can also operate continuously or intermittently. Most hot water or chilled water systems operate continuously as long as a heating or cooling load exists on the system. Condensate and boiler feed pumps are often of the intermittent type, starting and stopping as the water level changes in condensate tanks or boilers.

SYSTEM CURVES

The friction loss of a piping system depends on the flow rate through it. If one set of friction pressure/flow rate data is available, a characteristic curve may be developed for the system by using the principle that friction pressure varies directly with the square of the flow.

After the appropriate temperature difference is determined, the design flow rate is established. The form of the piping circuits and their piping sizes are developed next, and the design friction loss is calculated. When these two values are known, friction losses corresponding to other flow rates are calculated using the formula $H_1/H_2 = (Q_1/Q_2)^2$. Plotting the known and calculated points generates the system pressure curve as shown in Figure 6. This curve is typical for a constant flow recirculated system (friction pressure loss only), and no static or minimum pressure is maintained.

The second example of the system curve is a system that has both static pressure and friction loss. Such a system might be the piping circuit between a refrigeration plant condenser and its cooling tower, with the elevation difference between the water level in the tower pan and spray pressureer pipe creating the static pressure. With this pumping system, a friction pressure does not develop until the flow begins, and flow does not begin until the pressure developed by the pump exceeds the static pressure. The static pressure is not part of the system curve computation and is, therefore, an independent pressure as shown by Figure 7.

In some hydronic systems, flow rates are changed by variable-speed pumping or by the sequencing of constant-speed multiple pumps. Such systems must be accurately balanced, so all circuits and all terminal units receive equivalent, proportional flow. If individual control valves are not used on the terminal units, the static pressure curve is similar to the one shown in Figure 6. This system is seldom used because control at part-load operation is difficult. Diversity is not provided, so the system is not economical for a large multiuse or multibuilding system.

Most variable-volume systems have individual two-way valves on each terminal unit that permit full diversity or random loading from zero to full load. Regardless of the load required, the designer establishes a minimum pressure difference to ensure that any terminal and its control valve receive design flow at full output demand. When graphing a system curve for a nonsymmetrically loaded variable-volume system, the Δp (minimum maintained pressure) is treated like a static pressure condition and becomes the starting datum for the system curve (Figure 7).

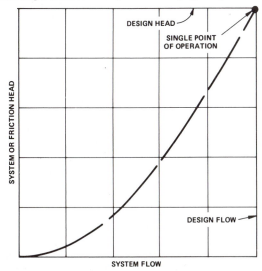

Fig. 6   Constant Flow System Curve

A variable-volume system, such as the one described, can randomly load or unload its terminals from zero to full load in any sequence. A system that picks up terminal loads in the order of the distance of the terminals from the main equipment room operates with a low slope curve (Figure 8). A system that starts loading progressively with terminals remote from the equipment room establishes a steep slope curve.

The net vertical difference between the low and high curves in Figure 9 is the difference in friction loss developed by the distribution mains for the two extremes of possible load patterns. The low and high curves represent the boundaries for the operation of the system. The area in which the system operates depends on the diverse loading or unloading imposed on the terminal units.

Friction losses for pipe and fittings are available in Chapter 42 and in Chapter 33 of the 1989 ASHRAE *Handbook—Fundamentals* as well as in the *Pipe Friction Manual* of the Hydraulic Institute.

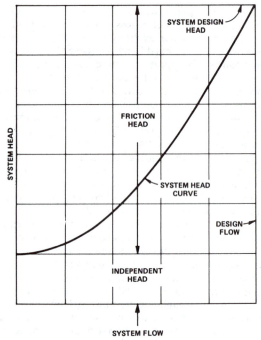

Fig. 7   Typical System Curve

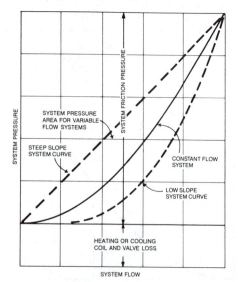

**Fig. 8   System Head for Heating or Cooling System
with Control Valves**

SELECTION AND ARRANGEMENT OF PUMPS

The selection of pumps for a particular hydronic system requires a substantial amount of data to ensure that an adequate, reliable, and efficient pump is selected. The following is some of the information required. A specific application may have conditions that require further information and an economic evaluation.

- Maximum and minimum flow in system
- System pressure at maximum and minimum flows
- Continuous or intermittent operation
- System operating pressures and temperatures
- Pump environmental conditions, including ambient temperature
- Number of pumps desired and percent of standby required for emergency operation
- Electrical current characteristics
- Electrical service starting limitations
- Special electrical controls
- Water chemistry that may affect materials selection

The selection should consider changes of flow in the system. Figure 4 describes a typical head capacity curve and efficiency pattern for a centrifugal pump. For a pump running with little change in volume, the desirable selection point is Point A in the region of best efficiency. If the flow is normally less than design, the pump should operate at Point B, so that the pump performance passes through the region of best efficiency as flow is throttled in the system. Other design criteria such as NPSH must be considered when a pump is selected to operate at Point B.

In some cases, operation at flows greater than design may occur; for such applications, the pump should be selected at Point C, so that the performance passes through the region of best efficiency as flow increases in the system. In addition, the designer should evaluate the effects of flow or pressure variation on pump efficiency.

The point of operation of a pump on its head capacity curve must also be considered. Figure 9 describes pump-operating points that may result when a pump is selected for a specific hydronic system. Point 1 indicates the selection point for the pump; at this point, the capacity of the pump is equal to the design flow and head of the system. Design contingencies may include

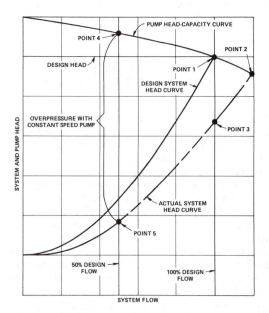

Fig. 9   Pump Operating Points

friction factors that do not exist when the system is put into operation. This fact is displayed in Figure 9 by the two system curves, one for design and the other for the actual system.

In Figure 9, if the system is free flowing without control valves and, with the actual system curve shown, the pump will operate at Point 2, not Point 1; the pump produces a higher flow rate than design flow rate. If the system is controlled with two-way valves on all heating or cooling coils, the pump will operate at Point 1 at design flow and will create an overpressure on the coils and control valves equal to the pressure difference between Points 1 and 3. If the system flow is reduced to 50% of design on such a system, the overpressure will increase to the amount between Points 4 and 5. Pump operation can be summarized as follows:

1. On a system without control valves, the pump always operates at the point of intersection of the pump head capacity curve and the system curve.
2. On controlled flow systems, the pump follows its head capacity curve. The difference between pump pressure and system pressure is converted into overpressure, which is consumed by the control valves.

Because overpressure can occur in controlled flow systems when coils are equipped with two-way control valves, selection and application of pumps for such systems must keep overpressure to a minimum by the following methods:

• Multiple pumps operating in parallel
• Multiple pumps operating in series
• Multispeed pumps
• Variable-speed pumps

The actual method used on a specific hydronic system depends on economics. The results for a particular system can be determined by developing the system curve and plotting pump head capacity curves on the same graph with the system curve.

Overpressure is most commonly eliminated by using multiple pumps in parallel. Figure 10 shows two pumps piped in parallel and includes the head capacity curves for single- and two-pump operation. The figure shows that one pump at 50% system flow will reduce the overpressure caused by two-pump operation or one pump designed to handle maximum design flow and pressure.

Figure 11 illustrates two pumps piped in series with bypasses for single-pump operation. The figure shows series pumping on a hydronic system with a large amount of system friction. For such systems, series pumping can greatly reduce the overpressure on a controlled flow system. Series pumping should not be used on hydronic systems with flat system curves similar to the one shown

in Figure 11. For such a system, one-pump operation with series connection results in the pump running at shutoff pressure and producing no flow in the system.

Standard two-speed motors are available in 1750/1150 rpm, 1750/850 rpm, 1150/850 rpm, and 3500/1750 rpm speeds. These motors can reduce overpressuring at reduced system flow. Figure 5 describes the effect of two-speed motors on pump curves, particularly for a 1750/1150 rpm motor.

Variable-speed drives have a similar effect on head capacity curves as two-speed motors. These drives normally have an infinitely variable speed range, so that the pump, with proper controls, can follow the system curve without any overpressure. Figure 5 describes the operation of a variable-speed pump at 1750, 1150, 850, and 550 rpm.

MOTIVE POWER

Electric motors drive most centrifugal pumps for hydronic systems. Internal combustion engines or turbines power some pumps, especially in central power houses for large installations; they are discussed in Chapter 41. Following is a discussion of electric motor drives.

Electric motors for centrifugal pumps can be any of the horizontal or vertical electric motors described in Chapter 40. As discussed earlier, many of the centrifugal pumps for hydronic systems are close-coupled with the pump impeller mounted on a motor shaft extension; others are flexibly coupled through a pump mounting bracket or frame to the electric motor.

The sizing of electric motors is critical because of the increased costs of electrical power. In the past, recommended practice specified the use of nonoverload motors; *i.e.*, the motor nameplate rating did not exceed the pump brake power (kW) at any point on the pump head capacity curve for that particular speed and impeller. A centrifugal pump on a hydronic system with variable flow has a broad range of power requirements, which results in low motor loading at low flow conditions. Figure 12 is a typical

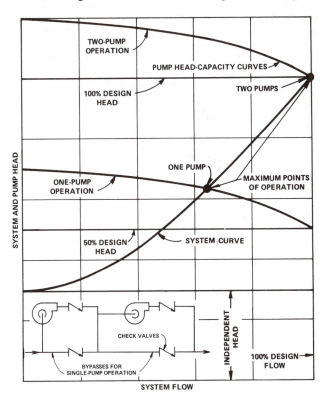

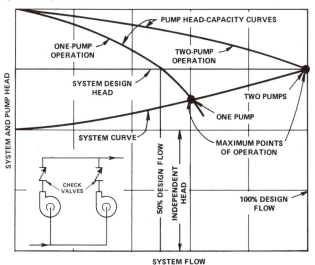

Fig. 10   Pump and System Curves for Parallel Pumping

Fig. 11   Pump and System Curves for Series Pumping

efficiency curve for an induction motor; it shows that operation at low loads can result in low motor efficiencies and high operating costs. To lessen the effect of low efficiency motors operating at low loads, pump motors should operate within the service factor of the motor, providing the service factor is acceptable to the pump manufacturer.

A number of variable-speed drive devices are available for operating centrifugal pumps on hydronic systems. These include fluid coupling, SCR variable frequency, direct current, wound rotor, and eddy current drives. Each drive has specific design features, which should be evaluated for use with hydronic pumps; in each case, the efficiency range should be investigated from minimum to maximum speed. Figure 13 describes the overall range of efficiencies for variable-speed drives.

HYDRONIC APPLICATIONS

Four major types of hydronic systems that use centrifugal pumps are (1) chilled water, (2) hot water, (3) condenser water, and (4) condensate or boiler feedwater systems. The following is a brief description of system head curves affecting pump selection; more detailed information is included in Chapter 12.

Chilled water and hot water pumps are used in closed recirculating systems. Hot water pumps, particularly high-temperature hot water units, require special pump features, such as high-temperature seals and thermal expansion means. The hydraulic application is similar for chilled and hot water pumps.

Figure 8 illustrates a typical system curve for a variable-volume, chilled or hot water system. Flow is regulated through the heating or cooling coils by two-way valves. The independent pressure, as shown, is the pressure drop of the heating and cooling coils and their control valves. The system friction pressure is much greater than the coil and valve loss, resulting in a steep system curve. Such a system lends itself to multiple pumps, two-speed motors, or variable-speed drives on the pumps to avoid overpressuring the system.

Three-way valves on the heating and cooling coils also eliminate overpressuring, but they waste energy because the pumps must deliver system design flow at all times, even though a small heating or cooling load exists on the system. Pumps and controls should be selected and installed so pump flow and head match, as closely as is economically feasible, the required pressure and flow of the system from minimum to maximum load.

In cooling towers, the independent head is the static rise to the top of the cooling tower, as shown in Figure 14. The suction conditions of condenser water pumps must be considered to ensure that the NPSHA from the system is always greater than the NPSHR by the pumps. To avoid the negative effects of suction, turbulence, and friction, the condenser protection strainer should be placed on the discharge side of the pump. Also, condenser water pipes may become rusted, eroded, or coated with material; thus pipe friction must be evaluated on a new and old pipe basis.

Condensate and boiler feed systems, unlike chilled, hot, or condenser water systems, have flat system curves as shown in Figure 15. The boiler pressure is the independent head and, in most cases, is much greater than the system friction head. These systems lend themselves to parallel pumping and seldom require two-speed or variable-speed pumps. Because these units pump hot condensate from open tanks, care should be taken to ensure that the NPSHA from the system is always greater than the NPSHR by the pumps.

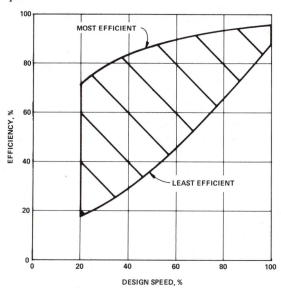

Fig. 13   Efficiency Range of Variable-Speed Drives

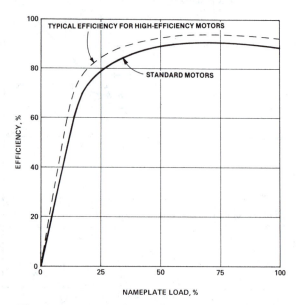

Fig. 12   Typical Efficiency Curve for Induction Motor

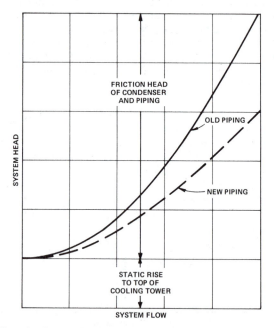

Fig. 14   System Head Curve for Condenser Water System

ENERGY CONSERVATION IN PUMPING

The continuous operation of many pumps in heating and air conditioning requires appreciable energy consumption, even with relatively small pumps. Therefore, pump efficiency must be studied when pumps are selected for a particular application.

Economical use of energy depends on the efficiency of pumping equipment and drivers, as well as the use of pumping energy in the water system. To clarify these two phases of energy use, two different efficiencies are used in the pump industry. One is the overall efficiency of the pumping equipment, called the equipment efficiency or wire-to-water efficiency. For electric motor-driven, constant-speed pumps, the equipment efficiency is the product of the pump efficiency η_p and the motor efficiency η_m. The equation for equipment efficiency, in percent, is

$$\eta_e = \eta_p \eta_m / 100$$

If the pumps are variable speed, the variable-speed drive efficiency η_v must be included in the equipment efficiency equation, which becomes

$$\eta_e = 10^{-4} \eta_p \eta_m \eta_v$$

The equipment efficiency shows how much of the energy applied results in useful energy in the water. A more difficult to define efficiency is that of energy use delivered to the water by the pump. To understand the efficiency of utilization η_u for a hydronic system, the perfect pump and control system must be defined. Such a system would deliver the needed amount of water at the right pressure to all parts of the hydronic system at all times. The system curve would describe the flow-pressure relationship for a hydronic system at all flow rates in the system.

If control valves are used on all heating or cooling coils so that the water flow in the system is the same as the design water flow, and if the pump delivers only the pressure required by the hydronic system at all flow conditions, then the system curve would represent the perfect hydronic system, and the efficiency of use would be 100%. Unfortunately, calculated system curves using (1) friction tables for piping and fittings and (2) manufacturers' data for friction losses of control valves and equipment often contain contingency factors. Thus, the calculated friction head is greater than the actual system friction head. The true system curve, therefore, can be determined only by testing the system after it is installed, which makes exact energy calculations difficult. For typical designs, the designer of hydronic systems can accept calculated friction losses as those for a perfect hydronic system and use the following equation to determine the efficiency of utilization, in percent:

$$\eta_u = 100 Q_s H_s / (Q_a H_a)$$

where Q_s and H_s are the flow and pressure required by the system at a specific heating or cooling load, while Q_a and H_a are the actual flow and head of the pump at that heating or cooling load. This equation demonstrates that three-way or bypass valves that waste energy by increasing the flow through the system do reduce the efficiency of utilization. Pumps reduce this efficiency if they provide more head than is needed by the system. Overflow and overpressure must be avoided to increase the efficiency of utilization.

The hydronic System Efficiency Index (SEI) is the useful energy or system load divided by the input power. It is often described as the *wire-to-system* efficiency, which indicates the ratio of the useful energy to the power input to an electric motor. Therefore, the following equation applies for pumping systems:

$$\text{SEI} = \eta_q \eta_u / 10^4, \text{ with SEI expressed as a decimal.}$$

This equation can be a guideline for achieving energy conservation in pumping and control systems. For proper evaluation of a hydronic system, the SEI must be calculated from minimum to maximum flow, not just at design condition. Figure 16 describes one SEI curve for a hydronic system from minimum to maximum flow. The SEI curve may be as smooth as that shown in Figure 16, or it may be jagged, representing changes in pump operation.

For pumping systems in heating and air conditioning, the SEI can be as low as 0.10 with maximums of about 0.75 to 0.78. This variation demonstrates the need for energy evaluation of pumping systems to ensure the maximum SEI is achieved at all possible loads on the hydronic system, not just at design load.

INSTALLATION AND OPERATION

The installation of a centrifugal pump should consider ambient, hydraulic, electrical, chemical, metallurgical, and acoustical

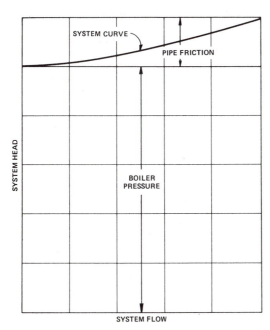

Fig. 15   System Head Curve for Condensate or Boiler Feed System

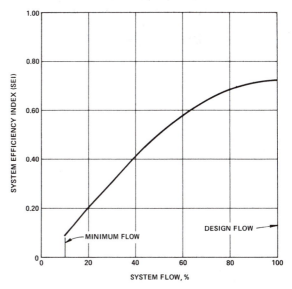

Fig. 16   Curve for System Efficiency Index (SEI)

Table 4   Pumping System Trouble Analysis Guide

Complaint	Possible Cause	Recommended Action	Complaint	Possible Cause	Recommended Action
Pump or system noise	Shaft misalignment	• Check and realign.	Inadequate or no circulation	Pump running backward (3-phase)	• Reverse any two-motor leads.
	Worn coupling	• Replace and realign.		Broken pump coupling	• Replace and realign.
	Worn pump/motor bearings	• Replace, check manufacturer's lubrication recommendations. • Check and realign shafts.		Improper motor speed	• Check motor nameplate wiring and voltage.
	Improper foundation or installation	• Check foundation bolting or proper grouting. • Check possible shifting because of piping expansion/contraction. • Realign shafts.		Pump (or impeller diameter) too small	• Check pump selection (impeller diameter) against specified system requirements.
	Pipe vibration and/or strain caused by pipe expansion/contraction	• Inspect, alter, or add hangers and expansion provision to eliminate strain on pump(s).		Clogged strainer(s)	• Inspect and clean screen.
				Clogged impeller	• Inspect and clean.
	Water velocity	• Check actual pump performance against specified, and reduce impeller diameter as required. • Check for excessive throttling by balance valves or control valves.		System not completely filled	• Check setting of PRV fill valve. • Vent terminal units and piping high points.
				Balance valves or isolating valves improperly set	• Check settings and adjust as required.
	Pump operating close to or beyond end point of performance curve	• Check actual pump performance against specified, and reduce impeller diameter as required.		Air-bound system	• Vent piping and terminal units. • Check location of expansion tank connection line relative to pump suction. • Review provision for air elimination.
	Entrained air or low suction pressure	• Check expansion tank connection to system relative to pump suction. • If pumping from cooling tower sump or reservoir, check line size. • Check actual ability of pump against installation requirements. • Check for vortex entraining air into suction line.		Air entrainment	• Check pump suction inlet conditions to determine if air is being entrained from suction tanks or sumps.
				Insufficient NPSHR	• Check NPSHR of pump. • Inspect strainers and check pipe sizing and water temperature.

conditions that exist for a specific installation. The instructions for installation, operation, and maintenance of centrifugal pumps in the *Hydraulic Institute Standards,* provide detailed information to help the designer of hydronic systems. Table 4 describes common pumping problems and presents solutions.

BIBLIOGRAPHY

ASHRAE. 1983. *Water system design and retrofit for energy/cost effectiveness.* Professional Development Seminar Text.

Hicks and Edwards. 1970. *Pump application engineering.* McGraw-Hill Book Company, Inc., New York.

Hydraulic Institute. 1983. *Hydraulic Institute standards,* 14th ed. Cleveland, OH.

Karassik, I.J., W.C. Krutzch, J.P. Messina, and W.H. Fraser. 1986. *Pump handbook,* 2nd ed. McGraw Hill Book Company, Inc., New York.

Stepanoff, A.J. 1957. *Centrifugal and axial flow pumps: Theory, design and application,* 2nd ed. John Wiley and Sons, Inc., New York.

CHAPTER 40

MOTORS AND MOTOR CONTROLS

THE alternating-current (ac) motor is available in many different types. The direct-current (dc) motor is also used, but to a limited degree. Complete technical information is available on all types of ac and dc motors in NEMA *Standard* MG 1-78.

ALTERNATING-CURRENT POWER SUPPLY

Important characteristics of an ac power supply include: (1) voltage, (2) number of phases, (3) frequency, (4) voltage regulation, and (5) continuity of power.

Various voltage systems, such as 115-V, 230-V, etc., are used. Standard voltage ratings for 60-Hz electric power systems and equipment have been established by the American National Standards Institute (ANSI C84.1-77).

The *nominal voltage* (or *service voltage*) of a circuit or system is the value assigned to the circuit or system to designate its voltage class. It is the voltage at the connection between the electric systems of the supplier and the user. The *utilization voltage* is the voltage at the line terminals of the equipment.

Single-phase and three-phase motor and control voltage ratings shown in Table 1 are adapted to the nominal system voltages indicated. Motors with these ratings are considered suitable for ordinary use on their corresponding systems; for example, a 230-V motor should generally be used on a nominal 240-V system. Operation of 230-V motors on a nominal 208-V system is not recommended because the utilization voltage is commonly below the tolerance on the voltage rating for which the motor is designed. Such operation generally results in excessive overheating and serious reduction in torque.

Motors are usually guaranteed to operate satisfactorily and to deliver their full power at the rated frequency and at a voltage 10% above or below rating, or at the rated voltage and plus or minus 5% frequency variation. Table 2 shows the effect of voltage and frequency variation on induction motor characteristics.

The phase voltages of three-phase motors should be balanced. If not, a small voltage imbalance can produce a greater current imbalance and a much greater temperature rise, which can result in nuisance overload trips or motor failures. Motors should not be operated where the voltage imbalance is greater than 1% without checking with the manufacturer. Voltage imbalance is defined in NEMA *Standard* MG 1-78, para. 14.34 as:

$$\% \text{ Voltage Imbalance} = 100 \left[\frac{\text{Maximum Voltage Deviation from Average Voltage}}{\text{Average Voltage}} \right]$$

In addition to voltage imbalance, current imbalance can be present in a system where Y-Y transformers without tertiary windings are used, even if the voltage is in balance. As stated previously, this current imbalance is not desirable. If this current imbalance exceeds either 10% or the maximum imbalance recommended by the manufacturer, corrective action should be taken (see NFPA *Standard* 70-84).

$$\% \text{ Current Imbalance} = 100 \left[\frac{\text{Maximum Current Deviation from Average Current}}{\text{Average Current}} \right]$$

Another cause of current imbalance is normal winding impedance imbalance, which adds or subtracts from the current imbalance caused by voltage imbalance.

Table 1 Motor and Motor-Equipment Voltages

Applicable to All Nominal System Voltages Containing this Voltage	All Motor and Motor-Control Equipment Nameplate Voltage Ratings Containing this Voltage			
	Integral hp		Fractional hp	
	Three-Phase	Single-Phase	Three-Phase	Single-Phase
120	—	115	—	115
208	200	—	200	—
240	230	230	230	230
277	—	265	—	265
480	460	—	460	—
600[a]	575	—	575	—
2400	2300	—	—	—
4160	4000	—	—	—
4800	4600	—	—	—
6900	6600	—	—	—
13,800	13,200	—	—	—

[a]Certain control and protective equipment have a maximum voltage limit of 600 V; the manufacturer or power supplier or both should be consulted to ensure proper application.

The preparation of this chapter is assigned to TC 8.11, Electric Motors—Open and Hermetic.

CODES AND STANDARDS

Codes are requirements for public health or safety; *standards* are uniform methods of rating, sizing, and measuring the performance of electrical equipment to provide a guideline for use and comparison. Codes and standards are sponsored at a national level and are usually voluntary. Those for electrical equipment used in the air-conditioning and refrigerating industries are listed in Chapter 48. The *National Electrical Code* (NEC) (NFPA 70-90) and the *Canadian Electrical Code,* Part I (CSA *Standard* C22.1-89) are important in the United States and Canada.

The *National Electrical Code* contains the minimum recommendations considered necessary to ensure safety of electrical installations and equipment. It is also referred to in Sub-Part *S* (Electrical) of the Occupational Safety and Health Acts (OSHA) of 1970 and, therefore, becomes part of the OSHA requirements where they apply. In addition, practically all communities in the United States have adopted it as a minimum electrical code.

Underwriters Laboratories, Inc. (UL) promulgates standards for various types of equipment. Underwriters' standards for electrical equipment cover construction and performance for the safety of such equipment and interpret requirements to ensure compliance with the intent of the NEC. A complete list of available standards may be obtained from UL, which also publishes lists of equipment that comply with their standards. These listed products bear the UL label and are recognized by local authorities.

The *Canadian Electrical Code,* Part I, is a standard of the Canadian Standards Association. It also is a voluntary code with minimum requirements for electrical installations in buildings of every kind.

The *Canadian Electrical Code,* Part II, contains specifications for the construction and performance of electrical equipment, in compliance with Part I. Underwriters Laboratories' standards and standards of the Canadian Standards Association for electrical equipment are similar, so equipment designed to meet the requirements of one code may also meet the requirements of the other. However, there is not complete agreement between the codes, so individual standards must be checked when designing equipment for use in both countries. The Canadian Standards Association examines and tests material and equipment for compliance with the *Canadian Electrical Code.*

Copies of electrical standards may be obtained directly from the sponsoring organizations (see Chapter 48).

MOTOR EFFICIENCY

The many factors affecting motor efficiency include (1) sizing of the motor to the load, (2) type of motor specified, (3) motor design speed, and (4) type of bearing specified.

Oversizing of motor to load may result in inefficiency. As shown in the performance characteristic curves for single-phase motors in Figures 1, 2, and 3, the efficiency falls off rapidly at loads lighter

Table 2 Effect of Voltage and Frequency Variation on Induction Motor Characteristics[a]

Voltage and Frequency Variation		Starting and Maximum Running Torque	Synchronous Speed	% Slip	Full-Load Speed	Efficiency		
						Full Load	0.75 Load	0.5 Load
Voltage variation	120% Voltage	Increase 44%	No change	Decrease 30%	Increase 1.5%	Small increase	Decrease 0.5 to 2%	Decrease 7 to 20%
	110% Voltage	Increase 21%	No change	Decrease 17%	Increase 1%	Increase 0.5 to 1%	Practically no change	Decrease 1 to 2%
	Function of voltage	$(Voltage)^2$	Constant	$\dfrac{1}{(Voltage)^2}$	(Synchronous speed slip)	—	—	—
	90% Voltage	Decrease 19%	No change	Increase 23%	Decrease 1.5%	Decrease 2%	Practically no change	Increase 1 to 2%
Frequency variation	105% Frequency	Decrease 10%	Increase 5%	Practically no change	Increase 5%	Slight increase	Slight increase	Slight increase
	Function of frequency	$\dfrac{1}{(Frequency)^2}$	Frequency	—	(Synchronous speed slip)	—	—	—
	95% Frequency	Increase 11%	Decrease 5%	Practically no change	Decrease 5%	Slight decrease	Slight decrease	Slight decrease

Voltage and Frequency Variation		Power Factor			Full-Load Current	Starting Current	Temperature Rise, Full Load	Maximum Overload Capacity	Magnetic Noises, No Load in Particular
		Full Load	0.75 Load	0.5 Load					
Voltage variation	120% Voltage	Decrease 5 to 15%	Decrease 10 to 30%	Decrease 15 to 40%	Decrease 11%	Increase 25%	Decrease 9 to 11 °F	Increase 44%	Noticeable increase
	110% Voltage	Decrease 3%	Decrease 4%	Decrease 5 to 6%	Decrease 7%	Increase 10 to 12%	Decrease 5.4 to 7.2 °F	Increase 21%	Increase slightly
	Function of voltage	—	—	—	—	Voltage	—	$(Voltage)^2$	—
	90% Voltage	Increase 3%	Increase 2 to 3%	Increase 4 to 5%	Increase 11%	Decrease 10 to 12%	Increase 11 to 13 °F	Decrease 19%	Decrease slightly
Frequency variation	105% Frequency	Slight increase	Slight increase	Slight increase	Decrease slightly	Decrease 5 to 6%	Decrease slightly	Decrease slightly	Decrease slightly
	Function of frequency	—	—	—	—	$\dfrac{1}{Frequency}$	—	—	—
	95% Frequency	Slight decrease	Slight decrease	Slight decrease	Increase slightly	Increase 5 to 6%	Increase slightly	Increase slightly	Increase slightly

[a]These variations are general and will differ for specific ratings.

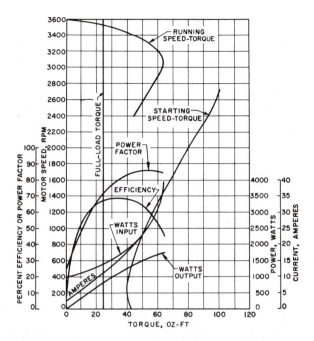

Fig. 1 Typical Performance Characteristics of Capacitor Start/Induction-Run Two-Pole General Purpose Motor, 1 hp

than the rated full load. Polyphase motors usually reach peak efficiency at loads slightly lighter than full load (Figure 4). Motor performance curves are available from the motor manufacturer and may help in applying motors optimally for an application. Larger output motors are more efficient at rated load than are smaller motors. Also, higher speed induction motors are more efficient.

The type of motor specified is significant because a permanent split-capacitor motor is more efficient than a shaded-pole fan motor. A capacitor-start/capacitor-run motor is more efficient than either a capacitor-start or a split-phase motor. In polyphase motors, the lower the locked rotor torque specified, the higher the efficiency obtained in the design.

The motor industry now offers high-efficiency design motors. These generally incorporate the more efficient type of motor for an application and more material than a standard efficiency design of the same type. More information on motor efficiency can be found in NEMA *Standards* MG 10-83 and MG 11-77.

NONHERMETIC MOTORS

Types

The electrical industry classifies motors as *small kilowatt* (*fractional horsepower*) or *integral kilowatt* (*integral horsepower*). Note that in this context, *kilowatt* refers to power output of the motor. Small kilowatt motors have ratings of less than 1 hp at 1700 to 1800 rpm for four-pole and 3500 to 3600 rpm for two-pole machines. These motors are available in a wide variety of types and sizes, depending on their application.

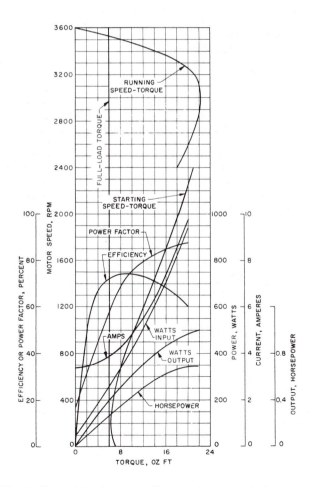

Fig. 2 Typical Performance Characteristics of a Resistance-Start Split-Phase Two-Pole Hermetic Motor, 0.25 hp

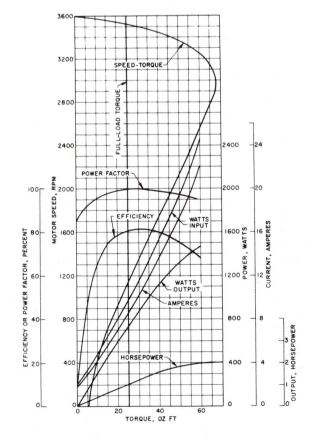

Fig. 3 Typical Performance Characteristics of a Permanent Split-Capacitor Two-Pole Motor, 1 hp

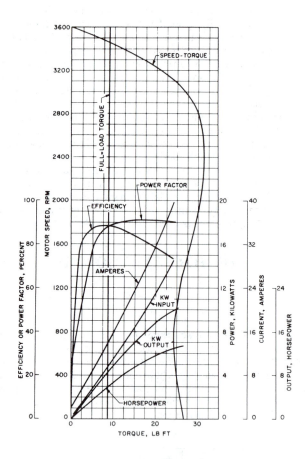

Fig. 4   Typical Performance Characteristics of a Polyphase Two-Pole Motor, 5 hp

Single-phase motors are available through 5 hp and are most common through 0.75 hp. Motors larger than 0.75 hp are usually polyphase.

Table 3 lists motors by types indicating the normal kilowatt range and the type of power supply used. All motors listed are suitable for either direct or belt drive, except shaded-pole motors (limited by low starting torque). Application of permanent split-capacitor motors to belt drives must be done carefully because of lower starting torque.

Application

When applying an electric motor, the following characteristics are important: (1) mechanical arrangement, including position of the motor and shaft, type of bearing, portability desired, drive connection, mounting, and space limitations; (2) speed range desired; (3) power requirement; (4) torque; (5) inertia; (6) frequency of starting; and (7) ventilation requirements. Motor characteristics that are frequently applied are generally presented in curves (see Figures 1 through 4).

Torque. The torque required to operate the driven machine at all times between initial breakaway and final shutdown is important in determining the type of motor. The torque available at zero speed or standstill (the *starting torque*) may be less than 100% or as high as 400% of full-load torque, depending on motor design. The *starting current,* or *locked-rotor current,* is usually 400 to 600% of the current at rated full load.

Full-load torque. This is the torque developed to produce the rated hp at the rated speed. **Full-load speed** also depends on the design of the motor. For induction motors, a speed of 1725 rpm

Table 3   Motor Types

Type	Range, hp	Type of Power Supply
Fractional sizes		
Split-phase	0.05 to 0.5	single-phase
Capacitor-start	0.05 to 1.5	single-phase
Repulsion-start	0.13 to 1.5	single-phase
Permanent split-capacitor	0.05 to 1.5	single-phase
Shaded-pole	0.01 to 0.25	single-phase
Squirrel cage induction	0.17 to 1.5	polyphase
Direct current	0.5 to 1.5	dc
Integral sizes		
Capacitor-start/capacitor-run	1 to 5	single-phase
Capacitor-start	1 to 5	single-phase
Squirrel cage induction (normal torque)	1 and up	polyphase
Slip-ring	1 and up	polyphase
Direct current	1 and up	dc
Permanent split-capacitor	1 to 5	single-phase

is typical for four-pole motors and a speed of 3450 rpm is typical for two-pole motors at 60 Hz.

Maximum or **breakdown torque.** Motors have a maximum or breakdown torque, which cannot be exceeded. The relation between breakdown torque and full-load torque varies widely, depending on motor design.

Power. The power delivered by a motor is a product of its torque and speed. Since a given motor delivers increasing power up to maximum torque, a basis for power rating is needed. The National Electrical Manufacturers Association (NEMA) bases power rating (hp) on breakdown torque limits for single-phase motors, 10 hp and less. All others are rated at their power capacity within voltage and temperature limits as listed by NEMA.

Full-load rating is based on maximum winding temperature. If the nameplate marking includes the maximum ambient temperature for which the motor is designed and the insulation system designation, the maximum temperature rise of the winding may be determined from the appropriate section of NEMA *Standard* MG-1.

Service factor. This factor is the maximum overload that can be applied to general-purpose motors and certain definite purpose motors without exceeding the temperature limitation of the insulation system. When the voltage and frequency are maintained at the values specified on the nameplate and the ambient temperature does not exceed 104°F, the motor may be loaded up to the power obtained by multiplying the rated hp by the service factor shown on the nameplate.

The power rating is normally established on the basis of tests run in still air. However, most direct-drive, air-moving applications are checked with air over the motor. If the motor nameplate marking does not specify a service factor, refer to the appropriate section of NEMA *Standard* MG-1. Characteristics of ac motors are given in Table 4.

HERMETIC MOTORS

An hermetic motor consists of a stator and a rotor, without shaft, end shields, or bearings, for installation in hermetically sealed refrigeration compressor units. With the motor and compressor sealed in a common chamber, the insulation system must be impervious to the action of the refrigerant and lubricating oil. Hermetic motors are used in both welded and accessible hermetic (semihermetic) compressors.

Application

Domestic refrigeration. Hermetic motors up to 0.33 hp are used. They are split-phase, permanent split-capacitor, or capacitor-start motors for medium or low starting torque compressors and capacitor-start and special split-phase motors for high starting torque compressors.

Room air conditioners. Motors from 0.33 to 3 hp are in use. They are permanent split-capacitor or capacitor-start/capacitor-run types. These designs have high power factor and efficiency and meet the need for low current draw, particularly on 115-V circuits.

Central air-conditioning systems (including **heat pumps**). Both single-phase (6 hp and below) and polyphase (1.5 hp and above) motors are used. The single-phase motors are permanent split-capacitor or capacitor-start/capacitor-run types.

Small commercial refrigeration. Practically all are below 5 hp, with single-phase being the most common. Capacitor-start/induction-run motors are normally used up to 0.75 hp because of starting torque requirements. Capacitor-start/capacitor-run motors are used for larger sizes because they provide high starting torque and high full-load efficiency and power factor.

Large commercial refrigeration. Most motors are three-phase and larger than 5 hp.

Horsepower ratings of motors for hermetic compressors do not necessarily have a direct relationship to the thermodynamic output of a system. Designs are tailored to match the compressor characteristics and specific applications. For additional information pertaining to hermetic motor application, see Chapter 35.

INTEGRAL THERMAL PROTECTION

The *National Electrical Code* and UL standards cover motor protection requirements. Separate, external protection devices include the following:

Thermal protectors. These are protective devices for assembly as an integral part of a motor or hermetic refrigerant motor compressor. They protect the motor against overheating caused by overload, failure to start, or excessive operating current. Thermal protectors are required to protect polyphase motors from overheating because of an open phase in the primary circuit of the supply transformer. Thermal protection is accomplished by either line-break devices or control circuit thermal sensing systems.

The protection of hermetic motor-compressors has some unique aspects compared to nonhermetic motor protection. The refrigerant cools the motor and compressor, so the thermal protector may be required to prevent overheating from loss of refrigerant charge, low suction pressure and high superheat at the compressor, obstructed suction line, or malfunction of the condensing means.

The *National Electrical Code, Article* 440, limits maximum continuous currents on motor-compressors to 156% of rated load current if an integral thermal protector is used. NEC *Article* 430 limits maximum continuous current on nonhermetic motors to different percentages of full-load current as a function of size. If separate overload relays or fuses are used for protection, *Article* 430 limits maximum continuous current to 140% and 125%, respectively, of rated load.

UL *Standard* 984 specifies that the compressor enclosure must not exceed 300 °F under any conditions. The motor winding temperature limits are set by the compressor manufacturer based on individual compressor design considerations. UL *Standard* 547 sets the limits for motor winding temperature for open motors as a function of the class of the motor insulation used.

Line-break thermal protectors. Integral with a motor or motor-compressor, sensing both current and temperature, line-break thermal protectors are connected electrically in series with the motor; their contacts interrupt the total motor line-current. These protectors are used in fractional and small integral single-phase and polyphase motors up through 15 hp.

Table 4   Characteristics of AC and DC Motors (Nonhermetic)

	Split-Phase	Permanent Split-Capacitor	Capacitor-Start Induction-Run	Capacitor-Start Capacitor-Run	Shaded-Pole	Polyphase, 60-Hz
Connection diagram						
Speed torque curves						
Starting method	Centrifugal Switch	None	Centrifugal Switch	Centrifugal Switch	None	Motor Controller
Ratings, hp	0.05 to 0.5	0.05 to 5	0.05 to 5	0.05 to 5	0.01 to 0.25	0.5 and up
Full-load speeds at 60-Hz (two-pole, four-pole)	3450 to 1725	3450 to 1725	3450 to 1725	3500 to 1750	3100 to 1550	3500 to 1750
Torque[a] Locked rotor Breakdown	125 to 150% 250 to 300%	25% 250 to 300%	250 to 350% 250 to 300%	250% 250%	25% 125%	150 to 350% 250 to 350%
Speed classification	Constant	Constant	Constant	Constant	Constant or adjustable	Constant
Full-load power factor	60%	95%	65%	95%	60%	80%
Efficiency	Medium	High	Medium	High	Low	High-Medium

[a]Expressed as percent of rated horsepower torque.

Protectors installed inside a motor-compressor are hermetically sealed, since exposed arcing cannot be tolerated in the presence of refrigerant. They provide better protection than the external type for loss of charge, obstructed suction line, or low voltage on the stalled rotor. This is due to low currents associated with these fault conditions, hence the need to sense the motor temperature increase by thermal contact. Protectors used inside the compressor housing must withstand pressure requirements established by Underwriters Laboratories.

Protectors mounted externally on motor-compressor shells, sensing only shell temperature and line current, are typically used on smaller compressors, such as those used in household refrigerators and small room air conditioners. One benefit occurs during high head pressure starting conditions, which can occur if voltage is lost momentarily or if the user inadvertently turns off the compressor with the temperature control and then turns it back on immediately. Usually, these units will not start under these conditions. When this happens, the protector takes the unit off the line and resets automatically when the compressor cools and pressures have equalized to a level that allows the compressor to start.

Protectors installed in nonhermetic motors may be attached to the winding or may be mounted off the windings but within the motor housing. Those protectors placed on the winding are generally installed prior to varnish dip and bake, and their construction must prevent varnish from entering the contact chamber.

Since the protector carries full motor line current, its size is based on adequate contact capability to interrupt the stalled current of the motor on continuous cycling for periods specified in UL *Standards* 984 and 547.

The compressor or motor manufacturer applies and selects appropriate motor protection in cooperation with the protector manufacturer. Any change in protector rating, by other than the specifying manufacturer after the proper application has been made, may result in either overprotection and frequent nuisance tripouts or underprotection and burnout of the motor windings. Connections to protector terminals, including lead wire sizes, should not be changed, and no additional connections should be made to the terminals. Any change in connection changes the terminal conditions and affects protector performance.

Control circuit thermal protection systems approved for use with a motor or motor compressor, either sensing both current and temperature or sensing temperature only, are used with integral hp single-phase and three-phase motors.

The current and temperature system uses bimetallic temperature sensors installed in the motor winding in conjunction with thermal overload relays. The sensors are connected in series with the control circuit of a magnetic contactor that interrupts the motor current. Thermostat sensors of this type, depending on their size and mass, are capable of tracking motor winding temperature for running overloads. On locked rotor, in cases where the rate of change in winding temperature is rapid, the temperature lag is usually too great for such sensors to provide protection when they are used alone. However, when they are used in conjunction with separate thermal overload or magnetic time-delay relays that sense motor current, the combination provides protection; on locked rotor, the thermal or magnetic relays protect for the initial heating cycle, and the combined functioning of relay and thermostat protects for subsequent cycles.

Thermistor sensors may also be used with electronic current-sensing devices. The temperature-only system uses sensors that undergo a change in resistance with temperature. The resistance change provides a switching signal to the electronic circuit, whose output is in series with the control circuit of a magnetic contactor used to interrupt the motor current. The output of the electronic protection circuitry (module) may be an electromechanical relay or a power triac. The sensors may be installed directly on the stator winding end turns or buried inside the windings. Their small size and good thermal transfer allow them to track the temperature of the winding for locked rotor, as well as running overload.

The sensors may be of three types. One type uses a ceramic material with positive temperature coefficient of resistance; the material exhibits a large abrupt change in resistance at a particular design temperature. This change occurs at what is known as the *anomaly point* and is inherent in the sensor. The anomaly point remains constant once the sensor is manufactured; sensors are produced with anomaly points at different temperatures to meet different application requirements. However, a single module calibration can be supplied for all anomaly temperatures of a given sensor type.

Another sensor type uses a metal wire, which has a linear increase in resistance with temperature. The sensor assumes a specific value of resistance corresponding to each desired value of response or operating temperature. It is used with an electronic protection module calibrated to a specific resistance. Modules supplied with different calibrations are used to achieve various values of operating temperature.

A third type is a negative temperature coefficient of resistance sensor, which is integrated with electronic circuitry similar to that used with the metal wire sensor.

More than one sensor may be connected to a single electronic module in parallel or series, depending on design. However, the sensors and modules must be of the same system design and intended for use with the particular number of sensors installed and the wiring method employed. Electronic protection modules must be paired only with sensors specified by the manufacturer, unless specific equivalency is established and identified by the motor or compressor manufacturer.

MOTOR CONTROL

In general, motor control equipment may (1) disconnect the motor and controller from the power supply, (2) start and stop the motor, (3) protect against short circuits, (4) protect from overheating, (5) protect the operator, (6) control motor speed, and (7) protect motor branch circuit conductors and control apparatus.

Separate Motor Protection

Most air-conditioning and refrigeration motors or motor compressors, whether open or hermetic, are equipped with integral motor protection by the equipment manufacturer. If this is not the case, separate motor-protection devices, sensing current only, must be used. These consist of thermal or magnetic relays, similar to those used in industrial control, that provide running overload and stalled-rotor protection. Because hermetic motor windings heat rapidly due to the loss of the cooling effect of refrigerant gas flow when the rotor is stalled, *quick-trip* devices must be used.

Thermostats or thermal devices are sometimes used to supplement current sensing devices. Such supplements are necessary (1) when automatic restarting is required after trip or (2) to protect from abnormal running conditions that do not increase motor current. These devices are covered in the section Integral Thermal Protection.

Protection of Control Apparatus and Branch Circuit Conductors

In addition to protection for the motor itself, Articles 430 and 440 of the *National Electric Code* require the control apparatus and branch circuit conductors to be protected from overcurrent resulting from motor overload or failure to start. This protection can be given by some thermal protective systems that do not permit a continuous current in excess of required limits. In other cases, a current-sensing device, such as an overload relay, a fuse, or a circuit breaker, is used.

Circuit breakers. These devices are used for disconnecting, as well as circuit protection and are available in ratings for use with small household refrigerators as well as in large commercial installations. Manual switches for disconnecting and fuses for short-circuit protection are also used. For single-phase motors up to 3 hp, 230-V, an attachment plug is an acceptable disconnecting device.

Controllers. The motor control used is determined by the size and type of motor, the power supply, and the degree of automation. Control may be manual, semiautomatic, or fully automatic.

Central system air conditioners are generally located at a distance from the controlled space environment control, such as room thermostats and other control devices. Therefore, *magnetic controllers* must be used in these installations. Also, all dc and all large ac installations must be equipped with in-rush *current-limiting controllers*, which are discussed later. *Synchronous motors* are sometimes used to improve the power factor. *Multispeed motors* provide flexibility for many applications.

Manual control. For an ac or dc motor, manual control is usually located near the motor. When so located, an operator must be present to start and stop or change the speed of the motor by adjusting the control mechanism.

Manual control is the simplest and least expensive control method for small ac motors, both single-phase and polyphase, but it is seldom used with hermetic motors. The manual controller usually consists of a set of main line contacts, which are provided with thermal overload relays for motor protection.

Manual speed controllers can be used for large air-conditioning systems using *slip-ring motors;* they may also provide reduced-current starting. Different speed points are used to vary the amount of cooling provided by the compressor.

Across-the-line magnetic controllers. These controllers are widely used in central air-conditioning systems. They are applicable to motors of all sizes, provided power supply and motor are suitable to this type of control. Across-the-line magnetic starters may be used with automatic control devices for starting and stopping. Where push buttons are used, they may be wired for either low-voltage release or low-voltage protection.

Full-voltage starting. For motors, full-voltage starting is preferable because of its lower initial cost and simplicity of control. Except for dc machines, most motors are mechanically and electrically designed for full-voltage starting. The starting in-rush current, however, is limited in many cases by power company regulations made because of voltage fluctuations, which may be caused by heavy current surges. It is, therefore, often necessary to reduce the starting current below that obtained by across-the-line starting in order to meet the limitations of power supply. One of the simplest means of accomplishing this reduction is by using resistors in the primary circuit. As the motor accelerates, the resistance is cut out by the use of timing or current relays.

Another method of reducing the starting current for an ac motor uses an *autotransformer* motor controller. Starting voltage is reduced, and, when the motor accelerates, it is disconnected from the transformer and connected across-the-line by means of timing or current relays. Primary resistor starters are generally smaller and less expensive than autotransformer starters for motors of moderate size. However, primary resistor starters require more line current for a given starting torque than do autotransformer starters.

Star-delta motor controllers. These controllers limit current efficiently, but they require motors designed for this type of starting. They are particularly suited for centrifugal, rotary screw, and reciprocating compressor drives starting without load.

Part-winding motor controllers. Part-winding motor controllers (or **incremental start controllers**) are used to limit line disturbances by connecting only part of the motor winding to the line and connecting the second motor winding to the line after a time interval of 1 to 3 s. If the motor is not heavily loaded, it accelerates when the first part of the winding is connected to the line; if it is too heavily loaded, the motor may not start until the second winding is connected to the line. In either case, the voltage *dip* will be less than the dip that would result if a standard squirrel cage motor with a cross-the-line starter were used. Part-winding motors may be controlled either manually or magnetically. The magnetic controller consists of two contactors and a timing device for the second contactor.

Multispeed motor controllers. Multispeed motors provide flexibility in many types of drives in which variation in capacity is needed. Two types of multispeed motors are used: (1) motors with one reconnectable winding and (2) motors with two separate windings. Motors with separate windings need a contactor for each winding and only one contactor can be closed at any time. Motors with a reconnectable winding are similar to motors with two windings, but the contactors and motor circuits are different.

Slip-ring motor controllers. Slip-ring ac motors provide variable speed. The *wound rotor* of these motors functions in the same manner as in the squirrel cage motor, except that the rotor windings are connected through slip rings and brushes to external circuits with resistance to vary the motor speed. Increasing the resistance in the rotor circuit reduces motor speed, and decreasing the resistance increases motor speed. When the resistance is shorted out, the motor operates with maximum speed, efficiency, and power factor. On some large installations, manual drum controllers are used as speed-setting devices. Complete automatic control can be provided with special control devices for selecting motor speeds.

Controllers for direct-current motors. These motors have favorable speed-torque characteristics, and their speed is easily controlled. Controllers for dc motors are more expensive than those for ac motors, except for very small motors. Large dc motors are started with resistance in the armature circuit, which is reduced step-by-step until the motor reaches its base speed. Higher speeds are provided by weakening the motor field.

Single-Phase Motor-Starting Methods

Motor-starting switches and relays for single-phase motors must provide a means for disconnecting the *starting winding* of split-phase or capacitor-start/induction-run motors or the start capacitor of capacitor-start/capacitor-run motors. Open machines usually have a centrifugal switch mounted on the motor shaft, which disconnects the starting winding at about 70% of full-load speed.

The starting methods by use of relays are as follows:

Thermally-operated relay. When the motor is started, a contact that is normally closed applies power to the starting winding. A thermal element that controls these contacts is in series with the motor and carries line current. Because of the current flow through this element, it is heated until, after a definite period of time, it is warmed sufficiently to open the contacts and remove power from the starting winding. The running current then heats the element enough to keep the contacts open. The setting of the time for the starting contacts to open is determined by tests on the system components, *i.e.,* the relay, the motor, and the compressor, and is based on a prediction of the time delay required to bring the motor up to speed.

An alternate form of a thermally operated relay is a PTC (Positive Temperature Coefficient of resistance) starting device. This uses a ceramic element of low resistance, at room temperature, whose value increases about 1000 times the room temperature resistance when it is heated to a predetermined temperature. It is placed in series with the start winding of split-phase motors and allows current flow when power is applied, permitting the motor to start. After a definite period of time, the self-heating of the PTC resistive element causes it to reach its high-resistance state, reducing current flow in the start winding to milliamp level during running. The residual current maintains the PTC element in the high-resistance state while the motor is running. A PTC starting device may also be connected in parallel with a run capacitor, and the combination may be connected in series with the starting winding. It allows the motor to start like a split-phase motor and, when the PTC element reaches the high-resistance state, it allows the motor to operate as a capacitor-run motor. The PTC element remains hot while power is applied to the motor. When power is removed, the PTC element must be allowed to cool to its low resistance state before restarting the motor.

Current-operated relay. In this type of connection, a relay coil carries the line current going to the motor. When the motor is started, the inrush current to the running winding passes through the relay coil, causes the normally open contacts to close, and applies power to the starting winding. As the motor comes up to speed, the current decreases until, at a definite calibrated value of current corresponding to a preselected speed, the magnetic force of the coil diminishes to a point that allows the contacts to open to remove power from the starting winding. This relay takes advantage of the *main winding current* versus *speed* characteristics of the motor. The current-speed curve varies with line voltage, so that the starting relay must be selected for the voltage range likely to be encountered in service. Ratings established by the manufacturer should not be changed because this may result in undesirable starting characteristics. They are selected to disconnect the starting winding or start capacitor at approximately 70 to 90% of synchronous speed for four-pole motors.

Voltage-operated relay. Capacitor-start and capacitor-start/capacitor-run hermetically sealed motors above 0.5 hp are usually started with a normally closed contact voltage relay. In this method of starting, the relay coil is connected in parallel with the starting winding. When power is applied to the line, the relay does not operate because it is calibrated to operate at a higher voltage. As the motor comes up to speed, the voltage across the starting winding and relay coil increases in proportion to the motor speed. At a definite voltage corresponding to a preselected speed, the relay operates and opens its contacts, thereby opening the starting-winding circuit or disconnecting the starting capacitor. The relay then keeps these contacts open because there is sufficient voltage induced in the starting winding, when the motor is running, to hold the relay in the open contact position.

REFERENCES

ANSI. 1982. *Electrical power systems and equipment—Voltage ratings (60 Hz).* ANSI C84.1-82. American National Standards Institute, New York.

CSA. 1982. *Canadian Electrical Code.* Canadian Standards Association *Standard* C22.1-82, p. 8. Rexdale, Ontario, Canada.

IEEE. 1986. *Procedures for testing single-phase and polyphase induction motors for use in hermetic compressors. Standard* 839-1986, IEEE. New York.

NEMA. 1977. *Energy management guide for selection and use of single-phase motors.* MG 11-77. National Electrical Manufacturers Association, Washington, D.C.

NEMA. 1987. *Motors and generators.* MG 1-87. National Electrical Manufacturers Association, Washington, DC.

NEMA. 1983. *Energy management guide for selection and use of polyphase motors. Standard* MG 10-83, NEMA. Washington, D.C.

NFPA. 1987. *National Electrical Code.* NFPA-70-87. National Fire Protection Association, Quincy, MA.

CHAPTER 41

ENGINE AND TURBINE DRIVES

THIS chapter addresses drives (other than electric motors) for reciprocating and centrifugal compressors, including gas engines and steam and gas turbines. Engines are also used to drive generators in standby power and continuous-duty power applications, either with or without the useful application of recoverable engine heat (cogeneration). These systems are covered in Chapter 7. Electric motors are discussed in Chapter 40.

ENGINES

ENGINE APPLICATIONS

Engine-Driven Reciprocating Compressors

Engine-driven, reciprocating compressor, water-chiller units are usually field-assembled from commercially available equipment for comfort service, low-temperature refrigeration, and heat pump applications. Both direct-expansion coils and liquid chillers are used. Some models achieve a low operating cost and a high degree of flexibility by combining speed variation with cylinder unloading. These units achieve capacity control by reducing engine speed to about 30 to 50% of rated speed; further capacity modulation may be achieved by unloading the compressor in increments. Engine speed should not be reduced below the minimum specified by the manufacturer for adequate lubrication or good fuel economy.

Most engine-driven reciprocating compressors are equipped with a cylinder loading mechanism for idle (unloaded) starting. This arrangement may be required because the starter may not have sufficient torque to crank both the engine and the loaded compressor. With some compressors, not all the cylinders unload (e.g., 4 out of 12) and, in this case, a bypass valve has to be installed for a fully unloaded start. The engine first speeds to one-half or two-thirds of full speed. Then, a gradual cylinder load is added, and the engine speed increases over a period of 120 to 180 s. In some applications, such as an engine-driven heat pump, low-speed starting may cause oil accumulation and sludge. As a result, a high-speed start is required.

These systems operate at specific fuel consumptions (SFC) of approximately 8 to 13 ft^3 of pipeline quality natural gas (heat value $\approx$ 1000 Btu/ft^3) per horsepower-hour (hp·h) in sizes down to 25 tons. Comparable heat rates for diesel engines run from 7000 to 9000 Btu/hp·h. Smaller units are not commercially available. Coolant pumps can also be driven by the engine. These direct connected pumps never circulate tower water through the engine jacket. Figure 1 illustrates the fuel economy effected by varying prime

mover speed with load (reciprocating compressor) until the machine is operating at about half its capacity. Below this level, the load is reduced at essentially constant engine speed by unloading the compressor cylinders.

Frequent operation of the system at low engine idling speed may require an auxiliary oil pump for the compressor. To reduce wear and assist in starts, a tank-type lube oil heater or a crankcase heater and a motor-driven auxiliary lubricating oil pump should be installed to lubricate the engine with warm oil when it is not running. Engine-driven refrigerant piping practices are the same as those for motor-driven units.

Engine-Driven Centrifugal Compressors

Packaged engine-driven centrifugal chillers up to 500 tons are beginning to appear on the market and do not require field assembly. Automotive derivative engines modified for use on natural gas are typical on this type of package because of their compact size and mass. Larger, complete packaged systems are not commercially available and require field assembly. These units may be equipped with either manual or automatic start-stop systems and engine speed controls.

Externally driven centrifugal chillers are usually field assembled and normally include a compressor mounted on an individual base and coupled by means of flanged pipes to an evaporator and a condenser.

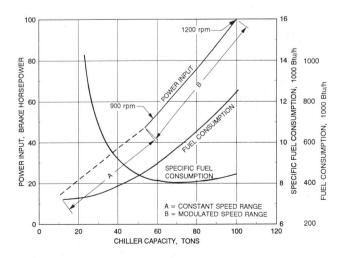

Fig. 1 Performance Curve for Typical 100-Ton, Gas Engine-Driven Reciprocating Chiller

---

The preparation of this chapter is assigned to TC 9.5, Cogeneration Systems and TC 8.2, Centrifugal Machines.

The centrifugal compressor is driven through a speed increaser. Many of these compressors operate at about six times the speed of the engines; compressor speeds of up to 14,000 rpm have been used. To effect the best compromise between the initial cost of the equipment (engine, couplings, and transmission) and the maintenance cost, engine speeds between 900 and 1200 rpm are generally used. Engine output can be modulated by reducing engine speed. If the operation at 100% of rated speed produces 100% of the rated output, approximately 60% of rated output is available at 75% of rated speed. Capacity control of the centrifugal compressor can be achieved by either variable inlet guide vane control with constant compressor speed or a combination of variable speed control and inlet guide vane control, the latter providing the greatest operating economies.

Engine-Driven Heat Pumps

An additional economic gain can result from operating an engine-driven air-conditioning unit as a heat pump to provide heat during the winter. Using the same equipment for both heating and cooling reduces capital investment. A gas engine drive for heat pump operation also makes it possible to recover heat from the engine exhaust gas and jacket water (see Chapter 7 for guidelines on heat recovery).

Figure 2 shows the total energy available from a typical engine-driven heat pump operating at 1 bhp/ton. Table 1 lists the coefficients of performance (COP) for various configurations.

Engine-Driven Screw Compressors

Engine driven screw-type chiller packages are commercially available for refrigeration applications. Manufacturers offer water chillers that use screw-type compressors driven directly by an engine fueled by natural gas. Capacity control is achieved by varying the engine speed and adjusting the slide valve on the compressor. Units have cooling COPs near 1.4 at rated cooling load.

ENGINE FUELS

Fuel Selection

Engines may be fueled with gasoline, natural gas, propane, sludge gas, or diesel and heavier oils. Multifueled engines, using diesel oil as one of the fuels, are available from several manufacturers in engine sizes larger than 220 hp. When gas is used as the fuel in a diesel engine, a small amount of diesel oil is used as pilot oil or as the ignitor.

Gasoline engines are not generally used because of fuel storage hazards, fuel cost, and maintenance problems caused by combustion product deposits on internal parts, such as pistons and valves.

Methane-rich sludge gas obtained from sewage treatment processes at treatment plants can be used as a fuel for both engines and other heating services. The fuel must be dried and cleaned prior to its injection into the engine. In addition, because of the lower heat content of the fuel gas (approximately 600 to 700 Btu/ft^3). The engine must be fitted with a larger carburetor. Dual carburetors are often installed with sewage gas as the primary fuel and natural gas as backup. The large amount of hydrogen sulfide in the fuel requires the use of special materials, such as aluminum in the bearings and bushings and low friction plastic in the O-rings and gaskets. The final choice of fuels should be based on fuel availability, cost, storage requirements, and fuel rate, because, except for gasoline engines, maintenance costs tend to be comparable.

FUEL HEATING VALUE

Fuel consumption data may be reported in terms of either higher heating value (HHV) or lower heating value (LHV). HHV is preferred by the gas utility industry and is the basis for most gaseous fuel considerations. Most natural gases have an LHV/HHV factor of 0.9 to 9.5. For other gaseous fuels, however, this factor may range from 0.87 for hydrogen to 1.0 for carbon monoxide. For fuel oils, this factor ranges from 0.96 (heavy oils) to 0.93 (light oils); the HHV is customarily used for oil (including pilot oil in dual-fuel engines).

The engine power ratings of many manufacturers are based on an LHV of approximately 900 Btu/ft^3. The LHV is calculated for each fuel to deduct from the HHV the energy necessary to keep the water produced during combustion in the vapor state. This latent heat does no work on the piston. Manufacturers sometimes suggest derating engine output about 2% per 10% decrease in fuel heating value below the base specified by the manufacturer. The engineer should evaluate engine performance based on LHV and net power. Additional power required to drive auxiliary equipment that does not drive compressors, pumps, or generators should not be counted.

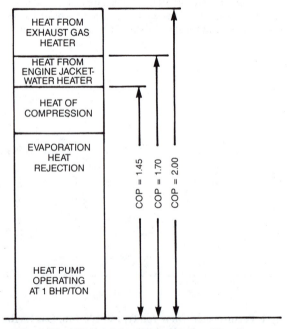

Fig. 2 Heat Balance for Engine-Driven Heat Pump

Table 1   Coefficient of Performance for Engine-Driven Heat Pump

Item	Heat Source		
	Refrigerant Condenser Only, Btu/ton·h	Refrigerant Condenser and Jacket Water, Btu/ton·h	Refrigerant Condenser, Jacket Water, Exhaust Gas, Btu/ton·h
Total heat input to engine	10,000	10,000	10,000
Cooler heat rejection to condenser (from building load)	12,000	12,000	12,000
Heat of compression	2,545	2,545	2,545
Heat from engine jacket water heater	—	2,500	2,500
Heat from exhaust gas heater	—	—	3,000
Total heat to heating circuit	14,545	17,045	20,045
Coefficient of performance	1.45	1.70	2.00

(Fig. 2 labels, top to bottom:) HEAT FROM EXHAUST GAS HEATER; HEAT FROM ENGINE JACKET-WATER HEATER; HEAT OF COMPRESSION; EVAPORATION HEAT REJECTION; HEAT PUMP OPERATING AT 1 BHP/TON; COP = 1.45; COP = 1.70; COP = 2.00

ENGINE SIZING

In sizing an engine, proper evaluation of the various heat gains and losses and their time of occurrence avoids overestimating the actual load. Chapter 26 of the 1989 ASHRAE *Handbook—Fundamentals* presents methods for calculating cooling load. Chapter 7 covers cogeneration.

Proper evaluation of operating cost requires knowledge of consumption rates. Specific data can be obtained from the manufacturer; however, a range of consumption rates is given in Table 2. Atmospheric corrections included in Equation (1) may be used to match prime movers to loads at various ambient conditions.

$$P_u = P_{max} \text{ (Rating or derating factor)} \qquad (1)$$

where

P_u = usable kilowatts of power
P_{max} = kilowatt rating under ideal conditions, based on individual manufacturers' laboratory performance data (usually a dynamometer test), corrected in accordance with manufacturers' specifications; for example:

    60°F and 29.92 in. Hg or
    80°F and 1000 ft elevation (NEMA) or
    85°F and 500 ft elevation (SAE) or
    90°F and 1500 ft elevation (DEMA)

Rating factor = [100 − (Percent altitude correction
    − Percent temperature correction
    − Percent heating value correction
    − Percent reserve)]/100

where

altitude correction = 3% per 1000 ft above a specified level for naturally aspirated engines; 2% per 1000 ft for turbocharged engines.

temperature correction = 1% per 10°F rise above a specified base temperature for air intake.

heating value correction = see section on fuel heating value.

reserve = a percentage allowance (safety factor) to permit design output under unforeseen operating conditions that would reduce output, such as a dusty environment, poor maintenance, higher ambient temperature, or lowered cooling efficiency. Recommended values can be found in Table 3.

DESIGN CONSIDERATIONS

The following factors should be considered in designing any engine-driven system:

- Exhaust lines must be vented outdoors.
- Noise levels must be acceptable by OSHA standards (see Chapter 42 of the 1991 ASHRAE *Handbook—HVAC Applications.*)

Table 2 Fuel Consumption Rates

Fuel	Heating Value, Btu/gal	Range of Consumption, Btu/hp·h
Fuel oil	137,000 to 156,000	7000 to 9000
Gasoline	130,000	10,000 to 14,000

Typical consumption for different types of gas engines	Compression Ratio	Gas Consumption, Btu/hp·h
Turbocharged	10.5:1	8,100
Naturally aspirated	10.5:1	9,200
Naturally aspirated	7.5:1	10,250

- Cooling towers should be sized to handle the engine water-jacket heat rejection. Exact figures for engine heat rejection should be obtained from the manufacturer (see Chapter 37).
- Inlet air for gas turbines and engines should be taken from outside where it is not likely to be diluted with higher temperature exhaust air.
- Engine rooms should have adequate ventilation to prevent excessive room temperatures caused by radiated engine heat. If radiated heat is 8 to 10% of the fuel input, an air cooler may be required.

INSTALLATION

Noise Control

Engine-driven machines installed indoors, even where the background noise level is high, usually require noise attenuation and isolation from adjoining areas. Air-cooled radiators, noise radiated from surroundings, and exhaust heat recovery boilers may also require silencing. Boilers that operate dry do not require separate silencers. Installations in more sensitive areas may be isolated, receive sound treatment, or both.

Basic attenuation includes (1) turning air intake and exhaust openings away (usually up) from the potential listener; (2) limiting blade-tip speed (if forced-draft air cooling is used) to 12,000 fpm for industrial applications, 10,000 fpm for commercial applications, and 8000 fpm for critical locations; (3) acoustically treating the fan shroud and plenum between blades and coils; (4) isolating (or covering) moving parts, including the unit, from its shelter (where used); (5) properly selecting the gas meter and regulator(s) to prevent singing; and (6) adding sound traps or silencers on ventilation air intake, exhaust, or both.

Further attenuation means include (1) lining the intake and exhaust manifolds with sound-absorbing materials; (2) mounting the unit, particularly a smaller engine, on vibration isolators, thereby reducing foundation vibration; (3) installing a barrier between the prime mover and the listener (often, a concrete block enclosure); (4) enclosing the unit with a cover of absorbing material; and (5) locating the unit in a building constructed of massive materials, paying particular attention to the acoustics of the ventilating system and the doors.

Exhaust Systems

Engine exhaust must be safely conveyed from the engine through piping and any auxiliary equipment to the atmosphere at an allowable pressure drop and noise level. Allowable back pressures, which vary with engine design, run from 2 to 25 in. of water. For low-speed engines, this limit is typically 6 in. of water; for high-speed engines, it is typically 12 in. of water. Adverse effects of excessive pressure drops include power loss, poor fuel economy, and excessive valve temperatures, all of which result in shortened service life and jacket water overheating.

Table 3 Percent Minimum Engine Reserves for Air Conditioning and Refrigeration

Altitude, ft	Naturally Aspirated		Turbocharged Aftercooled	
	Air Conditioning	Refrigeration	Air Conditioning	Refrigeration
Sea level	15	20	20	30
1000	12	17	18	28
2000	10	14	16	26
3000	10	11	14	24
4000	10	10	12	22
5000	10	10	10	20
10,000	10	10	10	10

General installation recommendations include the following:

- Installing a high-temperature, flexible connection between engine and exhaust piping. Exhaust gas temperature does not normally exceed 1200 °F; however, 1400 °F may be reached for short periods. An appropriate stainless-steel connector may be used.
- Adequately supporting the exhaust system weight downstream of the connector. At maximum operating temperature, no weight should be exerted on the engine or the exhaust outlet.
- Minimizing the distance between silencer and engine.
- Using a 30 to 45 ° tail-pipe angle to reduce turbulence.
- Specifying tail-pipe length (in the absence of other criteria) in odd multiples of $12.5\ T_e^{0.5}/P$, where T_e is the temperature of the exhaust gas (Rankine), and P is exhaust frequency, pulses per second.

The value of P is calculated as follows:

P = rpm/120 [or (r/s)/2] for four-stroke engines
P = rpm/60 [or r/s] for two-stroke engines

where rpm (r/s) is equal to the engine speed for V-engines with two exhaust manifolds. A second but less desirable exhaust arrangement is a Y-connection with branches entering the single pipe at about 60 °; a T-connection should never be used, as the pulses of one branch would interfere with the pulses from the other.

- Using an engine-to-silencer pipe length that is 25% of the tail pipe length.
- Using a separate exhaust for each engine to reduce the possibility of condensation in the off-engine.
- Using individual silencers to reduce condensation that results from an off-engine.
- Limiting heat radiation from exhaust piping by means of a ventilated sleeve surrounding the pipe or high-temperature insulation.
- Using fittings large enough to maintain adequate pressure.
- Providing for thermal expansion in exhaust piping, about 0.09 in/ft of length.
- Specifying muffler pressure drops to be within engine back-pressure limits.
- Not connecting engine exhaust pipe to chimney that serves natural-draft gas appliances.
- Sloping the exhaust away from the engine to prevent condensate backflow. Drain plugs in silencers and drip legs in long, vertical exhaust runs may also be required. The use of raincaps may prevent the entrance of moisture, but might add back pressure.

The exhaust pipe may be routed between an interior engine installation and a roof-mounted muffler through (1) an existing unused flue or one serving power-vented gas appliances only (a flue should not be used when exhaust gases may be returned from the interior); (2) an exterior fireproof wall, with provision for condensate drip to the vertical run; or (3) the roof, provided that a galvanized thimble with flanges having an annular clearance of 4 to 5 in. is used. A clearance of 1 to 2 in. between the flue terminal and the rain cap on the pipe permits the venting of the flue. A clearance of 30 in. between the muffler and roof is common. Vent passages and chimneys should be checked for resonance.

When interior mufflers must be used, minimize the distance between the muffler and the engine and insulate inside the muffler portion of the flue. Flue runs exceeding 25 ft may require power venting.

Flexible connections. The following design and installation features are recommended:

Material—Convoluted steel (Grade 321 stainless steel is recommended). Stainless steel is favored for interior installations.

Location—Principal imposed motion (vibration) should be at right angles to the connector axis.

Assembly—The connector (not an expansion joint) should not be stretched or compressed; it should be secured without bends, offsets, or twisting (the use of float flanges is recommended).

Anchor—The exhaust pipe should be rigidly secured immediately downstream of the connector in line with the downstream pipe.

Exhaust piping. Some alloys and standard steel alloy or steel pipe may be joined by fittings of malleable cast iron. Table 4 shows exhaust pipe size. The exhaust pipe should be at least as large as the engine exhaust connection. Stainless steel, double-wall liners may be used.

Ventilation Air

Sufficient ventilation offsets heat losses from engine-driven equipment, muffler, and exhaust piping, and protects against minor fuel supply leaks except rupture of the supply line. Equation (2) may be used to calculate ventilation air requirements.

$$V = UA_T(t_1 - t_2)/1.08\,(t_2 - t_3) \qquad (2)$$

where

V = ventilation air required, cfm
U = 2 Btu/h·ft² ·°F
A_T = total engine surface, $2[h\,(l + w) + lw]$, plus insulated exhaust piping area, ft²
t_1 = weighted mean engine and exposed exhaust surface temperature, °F
t_2 = engine room temperature, °F
t_3 = maximum outdoor air temperature, generally between 90 and 105 °F
h = engine height, ft
l = engine length, ft
w = engine width, ft

Also note that:

$(t_2 = t_3) = 20$ °F, when $t_3 = 90$ °F; 10 °F when $t_3 = 120$ °F (generally).

Table 5 is sometimes used for minimum ventilation air requirements. Ventilation may be provided by a fan that induces the draft through a sleeve surrounding the exhaust pipe. A slight positive pressure is maintained in the engine room.

Combustion Air

The following factors apply to combustion air requirements:

1. Supply an airflow of 2 to 5 cfm/bhp, depending on type, design, and size. Two-cycle units consume about 40% more air than four-cycle units.
2. Avoid heated air because power output varies by $(T_r/T)^{0.5}$, where T_r is the temperature at which the engine is rated, and T is the engine air intake temperature, both in °R.
3. Locate the intake away from contaminated air sources.
4. Install properly sized air cleaners that can be readily inspected (indicators are available) and maintained.

Table 4 Minimum Exhaust Pipe Diameter to Limit Engine Exhaust Back Pressure to 8 in. of Water

Output Power, hp	Equivalent Length of Exhaust Pipe			
	25 ft	50 ft	75 ft	100 ft
	Minimum Pipe Diameter, in.			
25	2	2	3	3
50	2.5	3	3	3.5
75	3	3.5	3.5	4
100	3.5	4	4	5
200	4	5	5	6
400	6	6	8	8
600	6	8	8	8
800	8	8	10	10

Table 5 Ventilation Air for Engine Equipment Room

Engine Room Air Temp Rise[a]	Muffler and Exhaust Pipe[b]	Muffler and Exhaust Pipe[c]	Engine, Air-Cooled or Radiator-Cooled[d]
°F	cfm/hp	cfm/hp	cfm/hp
10	140	280	550
20	70	140	280
30	50	90	180

[a]Exhaust minus inlet   [c]Not insulated
[b]Insulated or enclosed in ventilated duct   [d]Heat discharged in engine room

Engine room air-handling systems may include supply and exhaust fans, louvers, shutters, bird screens, and air filters. The total static pressure opposing the fan should be 0.35 in. of water maximum.

When shutters and louvers are used, they may be either manually or motor operated to control the quantity of intake or exhaust air. The vent area should be increased 25 to 50% to allow for free area. Thermostatically controlled shutters regulate airflow to maintain the desired temperature range. In cold climates, louvers should be closed when the engine is shut down to help maintain engine ambient temperature at a safe level. A crankcase heater can be installed on backup systems located in unheated spaces.

Air cleaners minimize cylinder wear and piston ring fouling. About 90% of valve, piston ring, and cylinder wall wear is the result of dust. Both dry and wet cleaners are used. Wet cleaners, if oversized or operated below their capacity, are generally inefficient; if they are too small, the resultant oil pullover reduces filter life. Filters may also serve as flame arresters.

Alignment

Proper alignment of the driving force to the compressor is important in preventing undue stresses to the compressor's mechanical shaft seal. Installation instructions usually suggest that the alignment of the assembly be performed and measured at maximum load condition on the compressor and maximum heat input to the turbine or engine.

Foundations

Modern multicylinder, medium-speed engines may not require massive concrete foundations, although concrete offers advantages in cost and in maintaining alignment for some driven equipment (see Chapter 42 of the 1991 ASHRAE *Handbook—Applications*).

Fabricated steel bases are satisfactory for direct-coupled, self-contained units, such as electric sets. Steel bases mounted on vibration isolators, steel-spring type or equal, are adequate and need no special foundation other than a floor designed to accommodate the weight.

Concrete bases are also satisfactory for such units, provided the bases are equally well isolated from the supporting floor or subfloor. Glass fiber blocks are effective as isolation material for concrete bases, which should be thick enough to prevent deflection. Excessively thick bases only serve to increase subfloor or soil loading and should always be supported by a concrete subfloor; in addition, some acceptable isolation material should be placed between the base and the floor. To avoid the transmission of vibration, an engine base or foundation should never rest directly on natural rock formations.

Engine Cooling

An engine converts fuel to shaft power and heat. Methods of dissipating this heat include (1) a jacket water system; (2) exhaust gas, which includes latent heat; (3) lubrication and piston cooling oil; (4) turbocharger and air intercooler; and (5) radiation from engine surfaces.

The manner and amount of heat rejection varies with the type, size, make of engine, and the extent of engine loading. Installation includes the following:
- An outside air entrance at least as large as the radiator face and 25 to 50% larger if protective louvers impede airflow.
- Auxiliary means, *e.g.*, a hydraulic, pneumatic, or electric actuator, to open the louvers blocking the heated air exit, rather than a gravity-operated actuator.
- Control of jacket water temperature by louvers in lieu of a bypass arrangement.
- Positioning the engine so that the face of the radiator is in a direct line with an air exit leeward to the prevailing wind.
- An easily removable shroud so that exhaust air cannot reenter the radiator.
- Separating the units in a multiple installation to avoid air recirculation among them.
- Low-temperature protection against snow and ice formation.

Propeller fans cannot be attached to long ducts because they can only achieve low static pressure. Radiator cooling air directed over the engine promotes good circulation around the engine; thus it runs cooler than for airflows in the opposite direction.

Water cooling. In most installations, heat in the engine coolant is removed by heat exchange with a separate water system.

Recirculated water is cooled in cooling towers where water is added to make up for evaporation. Closed systems should be treated with a rust inhibitor and/or antifreeze to protect the engine jacket from corrosion.

Since engine coolant is best kept in a closed loop, it is usually circulated on the shell side of an exchanger. A minimum fouling factor of 0.002 should be assigned to the tube side.

Operating temperatures. Water jackets outlet and inlet temperature ranges of 175 to 190 °F and 165 to 175 °F, respectively, are generally recommended, except when the engines are used with a heat recovery system. These temperatures are maintained by one or more thermostats that act to bypass water, as required. A 10 to 15 °F temperature rise is usually accompanied by a circulating water rate of about 0.5 to 0.7 gpm per engine horsepower.

Installation involves (1) sizing water piping according to the engine manufacturer's recommendations, (2) avoiding restrictions in the water pump inlet line, (3) never connecting piping rigidly to the engine, and (4) providing shutoffs to facilitate maintenance.

MAINTENANCE

Maintenance of engine-driven refrigeration compressors helps ensure continuous and economical system operation. Maintenance for the compressor section is similar to the maintenance of electric-drive units, with the exception of periodic gearbox oil changing and the operation and care of external lubricating pumps.

Engines require periodic servicing and replacement of parts, depending on usage and the type of engine. Records should be kept of all servicing; checklists should be used for this purpose.

Table 6 shows ranges of typical maintenance routines for both diesel and gas-fired engines, based on the number of hours run. The actual intervals vary according to the cleanliness of the combustion air, cleanliness of the engine room, the manufacture of the engine, the number of engine starts and stops, and lubricating conditions indicated by the oil analysis. With some engines and some operating conditions, the intervals between procedures, as listed in Table 6, may be extended.

A preventive maintenance program should also include inspections for (1) leaks (a visual inspection facilitated by a clean engine); (2) abnormal sounds and odors; (3) unaccountable speed changes; and (4) condition of fuel and lubricating oil filters. Daily logs should be kept on all pertinent operating parameters, such as water

Table 6 Recommended Engine Maintenance

Procedure	Hours between Procedure	
	Diesel Engine	**Gas Engine**
1. Take lubricating oil sample	1 per month plus 1 at each oil change	1 per month plus 1 at each oil change
2. Change lubricating oil filters	350 to 750	500 to 1000
3. Clean air cleaners, fuel	350 to 750	350 to 750
4. Clean fuel filters	500 to 750	N.A.
5. Change lubricating oil	500 to 1000	1000 to 2000
6. Clean crankcase breather	350 to 700	350 to 750
7. Adjust valves	1000 to 2000	1000 to 2000
8. Lubricate tachometer, fuel priming pump, and auxiliary drive bearings	1000 to 2000	1000 to 2000 (fuel pump N.A.)
9. Service ignition system, adjust breaker gap, timing, spark plug gap, and magneto	N.A.	1000 to 2000
10. Check transistorized magneto	N.A.	6000 to 8000
11. Flush lubrication oil piping system	3000 to 5000	3000 to 5000
12. Change air cleaner	2000 to 3000	2000 to 3000
13. Replace turbocharger seals and bearings	4000 to 8000	4000 to 8000
14. Replace piston rings, cylinder liners (if applicable), connecting rod bearings, and cylinder heads; recondition or replace turbochargers; replace gaskets and seals	8000 to 12,000	8000 to 12,000
15. Same as item 14, plus recondition or replace crankshaft, replace all bearings	24,000 to 36,000	24,000 to 36,000

and lubricating oil temperatures, individual cylinder compression pressures, which are useful in indicating blowby, and changes in valve tappet clearance, which indicate the extent of wear in the valve system.

Lubricating oil analysis is a low-cost method of determining the physical condition of the engine and a guide to maintenance procedures. Commercial laboratories providing this service are widely available. The analysis should measure the concentration of various elements found in the lubricating oil, such as bearing metals, silicates, and calcium. It should also measure the dilution of the oil, suspended and nonsuspended solids, water, and oil viscosity. The laboratory can often assist in interpreting the readings and alert the user to impending problems.

Lubrication

Manufacturers' recommendations should be followed. Both the crankcase oil and oil filter elements should be changed at least once every six months.

CODES AND STANDARDS

In addition to applicable local codes, the following National Fire Protection Association (NFPA) standards should be consulted:

National Electrical Code
Article 440, Air Conditioning and Refrigeration Equipment
Article 445, Generators
Article 700-12(b), Emergency System Set
Articles 701 and 702, Stand-by Power

National Fire Protection Association Standards
30 Flammable and Combustible Liquids Code
31 Installation of Oil Burning Equipment
37 Installation and Use of Stationary Combustion Engines and Gas Turbines
54 National Fuel Gas Code
58 Storage and Handling of Liquefied Petroleum Gases
59 Storage and Handling of Liquefied Petroleum Gases at Utility Gas Plants
59A Production, Storage and Handling of Liquefied Natural Gas (LNG)
90A Installation of Air Conditioning and Ventilation Systems
211 Chimneys, Fireplaces, Vents and Solid Fuel Burning Appliances

Gas codes cover service at all pressures. ASHRAE *Standard* 15-1989, Safety Code for Mechanical Refrigeration, covers the refrigeration section of the system and should be followed. The standards of the National Engine Manufacturers Association (NEMA) should also be consulted.

CONTROLS AND ACCESSORIES

Line-Type Gas-Pressure Regulator

Turbocharged (and aftercooled) engines, as well as many naturally aspirated units, are equipped with line regulators designed to control the gas pressure to the engine regulator, as shown in Table 7. The same regulators (both line and engine) used on naturally aspirated gas engines may be used on turbocharged equipment.

Line-type gas-pressure regulators are commonly called service regulators (and field regulators). They are usually located just upstream of the engine regulator to ensure the required pressure range exists at the inlet to the latter control. A remote location is sometimes specified; authorities having jurisdiction should be consulted. Although this intermediate regulation does not constitute a safety device, it does permit initial regulation (by the gas utility at meter inlet) at a higher outlet pressure, thus allowing an extra cushion of gas between the line regulator and the meter for both full gas flow at engine start-ups and delivery to any future branches from the same supply line. The engine manufacturer specifies the size, type, orifice size, and other regulator characteristics based on the anticipated gas-pressure range.

Engine-Type Gas-Pressure Regulator

This engine-mounted pressure regulator, also called a carburetor regulator (and sometimes a secondary regulator or a B regulator), controls the fuel pressure to the carburetor. Regulator construction may vary with the fuel used. The unit is similar to a zero governor.

Air-Fuel Control

The quantity flow of air-fuel mixtures in definite ratios must be controlled under all the load and speed conditions required of engines.

Table 7 Line Regulator Pressures

Line Regulator	Turbocharged Engine, psig	Naturally Aspirated Engines psig
Inlet[a]	14 to 20	2 to 50
Outlet[b]	12 to 15[c]	7 to 10 in. of water

[a]Overall ranges, not the variation for individual installations.
[b]Also inlet to engine regulator.
[c]Turbocharger boost plus 7 to 10 in. of water.

Air-fuel ratios. High-rated, naturally aspirated, spark-ignited engines require closely controlled air-fuel ratios. Excessively lean mixtures cause excessive oil consumption and engine overheating. Engines using pilot oil ignition can run at rates above 0.18 ppm/hp without misfiring. Air rates may vary with changes in compression ratio, valve timing, and ambient conditions.

Carburetors. In these venturi-type devices, the airflow mixture is controlled by a governor-actuated butterfly valve. This air-fuel control has no moving parts other than the butterfly valve.

The motivating force in naturally aspirated engines is the vacuum created by the intake strokes of the pistons. Turbocharged engines, on the other hand, supply the additional energy of pressurized air and pressurized fuel.

Ignition

An electrical system or pilot oil ignition may be used. Electrical systems are either low-tension (make-and-break) or high-tension (jump spark). Systems with breakerless ignition distribution are also in use.

Governors

A governor senses speed (and sometimes load), either directly or indirectly, and acts by means of linkages to control the flow of gas and air through engine carburetors or other fuel-metering devices to maintain a desired speed. Speed control gained from electronic, hydraulic, or pneumatic governors extends engine life by minimizing forces on engine parts, permits automatic throttle response without operator attention, and prevents destructive overspeeding.

Use of a separate overspeed device (*e.g.*, a maximum speed type of governor, sometimes called an overspeed stop) prevents runaway in the event of a failure that renders the governor inoperative.

Both constant and variable engine speed controls are available. For constant speed, the governor is set at a fixed position. These positions can be reset manually.

Safety Controls and Considerations

Both a low-lubrication pressure switch and a high jacket water temperature cutout are standard for most gas engines. Other safety controls used include (1) an engine speed governor; (2) ignition current failure shutdown (battery-type ignition only); and (3) the safety devices associated with a driven machine. These devices shut down the engine to protect it against mechanical damage. They do not necessarily shut off the gas fuel supply unless they are specifically set to do so. Shutting off fuel to the problem engine may be desirable. For example, a control may stop one engine and simultaneously energize the cranking circuit of its standby.

GAS LEAKAGE PREVENTION

The first method of avoiding gas leakage that results from engine regulator failure is a solenoid shutdown valve with a positive cutoff, installed either upstream or downstream of the engine regulator. The second method is a sealed combustion system that carries any leakage gas directly to the outdoors (*i.e.*, all combustion air is ducted to the engine directly from the outdoors).

HEAT RECOVERY

Exhaust heat may be used to make steam or used directly for drying or other processes. The steam provides space heating, hot water, and absorption refrigeration, which may supply air conditioning and process refrigeration. Heat recovery systems generally involve equipment specifically tailored for the job, although conventional fire-tube boilers are sometimes used. Exhaust heat

recovery from reciprocating engines may be accomplished by ebullient cooling of the jacket water and the use of a muffler-type exhaust heat recovery unit. Table 8 gives the temperature levels normally required for various heat recovery applications.

Ebullient systems. These pressurized reciprocating engine cooling heat recovery systems are also referred to as high-temperature systems, because they operate with a few degrees of temperature differential in the range of 212 to 250 °F.

Engine builders have approved various conventionally cooled models for ebullient cooling application. Azeotropic antifreezes should be used to ensure constant coolant composition in boiling. However, the engine power outputs are generally derated by the manufacturer when the jackets are cooled ebulliently.

Ebullient cooling or hot water cooling equipment usually replaces the radiator, the belt-driven fan, and the gear-driven water pump. Many heat recovery systems use the jacket water in water-to-liquid exchangers when the thermal loads are at temperature levels below 180 °F, particularly when their load profiles can absorb most or all of the jacket heat. Such systems are easier to design and have historically developed fewer problems than those with ebullient cooling.

Heat from exhaust gases. In some engines, exhaust heat rejection exceeds jacket water rejection. Generally, gas engine exhaust temperatures run from 700 to 1200 °F, as shown in Table 9. Cooling exhaust gases to about 300 °F prevents condensation in the exhaust line. Thus, about 50 to 75% of the sensible heat in the exhaust may be considered recoverable.

The economics of exhaust heat boiler design often limits the temperature differential between exhaust gas and generational steam to a minimum of 100 °F. Therefore, in low-pressure steam boilers, the gas temperature can be reduced to 300 to 350 °F; the corresponding final exhaust temperature range in high-pressure steam boilers is 400 to 500 °F. Table 8 gives application data.

Gas engine exhaust may be used directly in a waste heat absorption chiller. Units are available in sizes ranging from 100 to 1500 tons. Using oil/diesel exhaust for this purpose has not been successful due to fouling and corrosion problems.

Table 8 Temperature Levels Normally Required for Various Heating Applications

Application	Temperature, °F
Absorption refrigeration machines	190 to 245
Space heating	120 to 250
Water heating (domestic)	120 to 200
Process heating	150 to 250
Evaporation (water)	190 to 250
Residual fuel heating	212 to 330
Auxiliary power producers, with steam turbines or binary expanders	190 to 350

Table 9 Approximate Full-Load Exhaust Mass Flows and Temperatures for Various Types of Engines

Types of Engines	Mass Flow, lb/bhp·h	Temperature, °F
Two-Cycle		
Blower charged gas	16	700
Turbocharged gas	14	800
Blower charged diesel	18	600
Turbocharged diesel	16	650
Four-Cycle		
Naturally aspirated gas	9	1200
Turbocharged gas	10	1200
Naturally aspirated diesel	12	750
Turbocharged diesel	13	850
Gas turbine, nonregenerative	60	1050

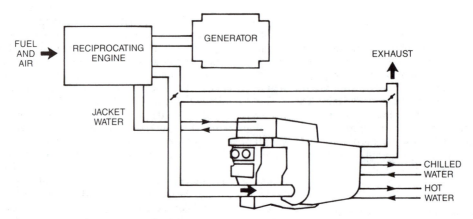

Fig. 3 Exhaust Gas Chiller-Heater with Jacket Heat Reclaim

It should be noted that two-cycle engines have higher mass flow rates and lower temperatures than four-cycle engines, because of higher airflows with cylinder purging, sometimes making them less desirable for heat recovery. Gas turbines have even larger mass flow, but at adequately high temperatures to generate high-temperature recovery streams, making them more advantageous when thermal loads can be efficiently used.

HEAT RECOVERY ABSORPTION CHILLERS

Two-stage absorption chillers have been developed to take advantage of the dual temperatures available from engine exhaust and jacket water. The use of the engine exhaust heat to provide cooling is described in the previous section, Heat Recovery. A class of absorption machines has been designed specifically for heat recovery in cogeneration applications. These units make direct use of both the jacket water and the exhaust gas. Ebullient cooling of the engine is not required (Figure 3). These chillers range in size from approximately 50 to 200 tons.

CONTINUOUS DUTY APPLICATIONS

An engine that drives a refrigeration compressor can be switched over automatically to drive an electrical generator in the event of a power failure (Figures 4 and 5). This plan assumes that loss of the compressor service can be tolerated in an emergency.

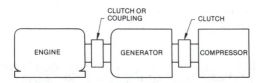

Generator rotor requires little energy while generator is under no load. In essence, the rotor is a flywheel during this period.

Fig. 4 Dual-Service Application with Generator Rotor Attached to Engine Crankshaft

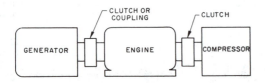

Fig. 5 Dual-Service Application Using Double-Ended Engine

If the engine capacity of such a dual-service system equals 150 or 200% of the compressor load, the power available from the generator can be delivered to the utility grid during normal operation. While induction generators may be used for this application, a synchronous generator is required for emergency operation if there is a grid outage.

Dual-service arrangements have the following advantages:
- Lower capital investment and reduced space and maintenance are required when compared to two engines in single service, even after allowing for the additional controls needed with dual-service installations.
- Dual-service engines are more reliable than emergency (reserved) single-service engines because of continuous or regular use.
- Engines that are in service and running can be switched over to emergency power generation with a minimum loss of continuity.

EXPANSION RECIPROCATING ENGINES

These engines are used mainly for cryogenic applications to about $-320°F$, *e.g.*, oxygen for steel mills, low-temperature chemical processes, and the space program.

Relatively high-pressure air or gas expanded in an engine drives a piston and is cooled in the process. At the shaft, about 42 Btu/hp are removed. Available units, developing as much as 600 hp, handle flows ranging from 100 to 10,000 scfm. Throughput at a given pressure is controlled by varying the cutoff point, the engine speed, or both. The conversion efficiency of heat energy to shaft work ranges from 65 to 85%. A 5 to 1 pressure ratio and an inlet pressure of 3000 psig are recommended. Outlet temperatures of $-450°F$ have been handled satisfactorily.

STEAM AND GAS TURBINES

STEAM TURBINES

Application

Steam turbines are principally used in the air-conditioning and refrigeration field to drive centrifugal compressors. Such compressors are usually part of a water or brine-chilling system using one of the halogenated hydrocarbon refrigerants. In addition, many industrial processes employ turbine-driven centrifugal compressors with a variety of other refrigerants such as ammonia, propane, and butane.

Associated applications of steam turbines include driving chilled water and condenser water circulating pumps and serving as prime movers for electrical generators in total energy systems.

In industrial applications, the steam turbine may be advantageous, serving either as a work-producing steam pressure reducer or as a scavenger, employing otherwise wasted low-pressure steam. Many steam turbines are used in urban areas where commercial buildings are served with steam from a central public utility or municipal source. Finally, the institutional field, where large central plants serve a multitude of buildings with heating and cooling, uses steam turbine-driven equipment. Chapter 7 covers the application of steam turbines in total energy and cogeneration systems.

Types

Mechanical-drive steam turbines are generally condensing and noncondensing, denoting the condition of the exhaust steam. Figure 6 shows basic types of turbines. NEMA Publication SM23 (1985) defines these and further subdivisions of their basic families.

Noncondensing turbine. A steam turbine designed to operate with an exhaust steam pressure at any level that may be required for a downstream process, with all condensing taking place there.

Condensing turbine. A steam turbine with an exhaust steam pressure below atmospheric pressure, that is directly condensed.

Automatic-extraction turbine. A steam turbine that has both an opening(s) in the turbine casing for the extraction of steam and means for directly regulating the extraction steam pressure.

Nonautomatic-extraction turbine. A steam turbine that has an opening(s) in the turbine casing for the extraction of steam but does not have means for controlling the pressure of the extracted steam.

Induction (mixed-pressure) turbine. A steam turbine with separate inlets for steam at two pressures, has an automatic device for controlling the pressure of the secondary steam inducted into the turbine, and means for directly regulating the flow of steam to the turbine stages below the induction opening.

Induction-extraction turbine. A steam turbine with the capability of either exhausting or admitting a supplemental flow of steam through an intermediate port in the casing, thereby maintaining a process heat balance.

Turbines of the extraction and induction-extraction type may have several casing openings, each passing steam at a different pressure.

Most steam turbines driving centrifugal compressors for air conditioning are of the multistage condensing type (Figure 7). Such a turbine gives good steam economy at reasonable initial cost. Usually, steam is available at a gage pressure of 50 psig or higher, and there is no demand for exhaust steam. However, turbines may work equally well where an abundance of low-pressure steam is available. The wide range of application of this turbine is shown by at least one industrial firm that drives a sizable capacity of water-chilling centrifugals with an initial steam gage pressure of less than 4 psig, thus balancing summer cooling against winter heating with steam from generator-turbine exhausts.

Aside from wide industrial use, the noncondensing (or back-pressure) turbine is most often used in water-chilling plants for driving a centrifugal compressor, which shares the cooling load with one or more absorption units (Figure 8). The exhaust steam from the turbine, commonly at about 15 psig, serves as the heat source for the absorption unit's generator (concentrator). This dual use of the heat energy in the steam generally results in a lower energy input per unit of refrigeration effect output than is attainable operating alone. An important aspect in the design of such combined systems is the need to balance the turbine exhaust steam flow with the absorption input steam requirements over the full range of load.

Extraction and mixed pressure turbines are used mainly in industry or in large central plants. Extracted steam is often used for boiler feedwater heating or other processes where steam with lower heat content is needed.

Steam energy in a turbine produces shaft power by the following methods: the necessary rotative force may be imposed on the turbine through the velocity, or pressure energy of the steam, or both. If velocity energy is employed, the movable wheels are usually fitted with crescent-shaped blades. A row of fixed nozzles,

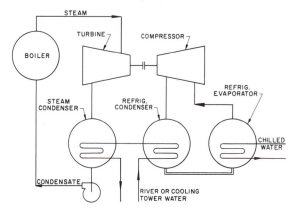

Fig. 7 Condensing Turbine-Driven Centrifugal Refrigeration Machine

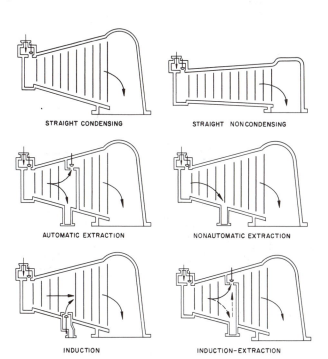

Fig. 6 Basic Turbine Types

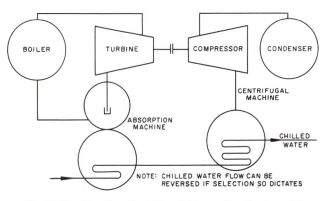

Fig. 8 Combination Centrifugal-Absorption System with Back-Pressure Turbine

which increases the steam velocity, precedes the blades and causes the wheels to rotate with little or no steam pressure drop across their blades. Such combinations of nozzles and velocity-powered wheels are characteristic of an impulse turbine. A reaction turbine uses alternate rows of fixed and moving blades, generally of an airfoil shape, with steam velocity increasing in the fixed portion and dropping in the movable portion and steam pressure dropping through both the fixed and movable blades.

The power capabilities of a reaction turbine are at maximum when the moving blades travel at about the velocity of the steam passing through them; in the impulse turbine, maximum power is produced with a blade velocity of about 50% of steam velocity. Steam velocity is related directly to pressure drop. To achieve the desired relationship between steam velocity and blade velocity without resorting to large wheel diameters or high rotative speeds, most turbines include a series of impulse or reaction stages, or both, thus dividing the total steam pressure drop into manageable increments. A typical commercial turbine may have two initial rows of rotating impulse blading with an intervening stationary row (called a Curtis stage), followed by several alternating rows of fixed and movable blading of either the impulse or reaction type. Most multistage turbines use some degree of reaction.

Performance

The energy input to a steam turbine is denoted by steam rate (or water rate). Actual steam or design rate is the theoretical steam rate of the turbine, divided by its efficiency. Typical efficiencies for mechanical steam turbines range from 55 to 80%. As defined by NEMA, the theoretical steam rate is the quantity of steam per unit of power required by an ideal Rankine cycle heat engine. Therefore:

$$w = 2546/(h_1 - h_2)$$

where

w = theoretical steam flow rate, lb/hp·h
h_1 = enthalpy of steam at its initial temperature and pressure, Btu/lb
h_2 = enthalpy of steam at exhaust steam pressure and initial entropy, Btu/lb

Turbine performance tests should be conducted in accordance with appropriate ASME Performance Test Code 6 (1985), 6S (1970), and 6A (1982).

The steam rate of a turbine is reduced with higher turbine speeds, a greater number of stages, larger turbine size, and a higher difference in heat content between entering and leaving steam conditions. Often, one or more of these factors can be improved with only a nominal increase in initial capital cost. Centrifugal water-chilling compressor applications range from approximately 100 to 10,000 hp and 3000 to 10,000 rpm, with the higher speeds generally associated with lower power outputs, and lower speeds with higher power outputs.

The higher ranges of centrifugal compressor speeds, 1800 to 90,000 rpm, cannot always be handled by direct connected steam turbines without a severe cost penalty. Turbines with their wheels shrunk or keyed to the shaft (or both) are suited to the lower and intermediate compressor speed requirements. Most turbine designs, however, change markedly above a high-speed threshold of about 7500 to 8500 rpm. The resultant stresses and critical speed characteristics demand solid rotor construction, with the wheels acting as an integral part of the shaft material. To avoid this cost penalty, an intermediate gear speed changer is often used. Such gears may be either a separate component, connected by flexible couplings to the compressor and turbine, or built integrally with the steam turbine. In the latter case, the manufacturer rates power output at the gearshaft, thus taking into account the 2 to 3% power lost by the gear. (Some typical characteristics of turbines driving centrifugal water chillers are shown in Figures 9 to 12.)

While initial steam pressures commonly fall in the 100 to 250 psig range, wide variations are possible. Back pressures associated with noncondensing turbines generally range from 50 psig to atmospheric, depending on the use for the exhaust steam. Raising the initial steam temperature by superheating improves steam rates.

NEMA standards govern allowable deviations from design steam pressures and temperatures. Because of possible unpredictable variations in steam conditions and load requirements, turbines are selected for a power capability of 105 to 110% of design shaft output and a speed capability of 105% of design rpm.

Governors

One important aspect of turbine application and specification is the type and quality of speed-governing apparatus. The wide variety of available governing systems permits the selection of a governor ideally matched to the characteristics of the driven machine and the load profiles.

The principal and most common function of a steam turbine governing system is to maintain constant turbine speed despite load fluctuations or minor variations in supply steam pressure. This arrangement assumes that close control of the output of the driven component, such as a centrifugal compressor in a water-chilling system, is primary to plant operation, and that the compressor can adjust its capacity to varying loads.

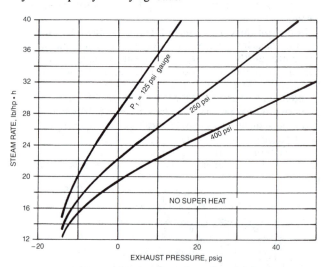

Fig. 9 Typical Effect of Exhaust Pressure on Noncondensing Turbine

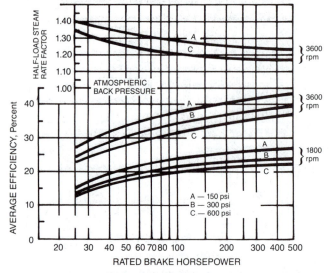

Fig. 10 Efficiency of Typical Single-Stage Noncondensing Turbine

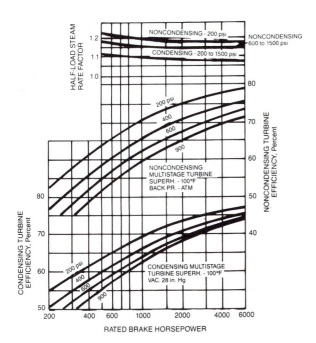

Fig. 11 Efficiency of Typical Multistage Turbines

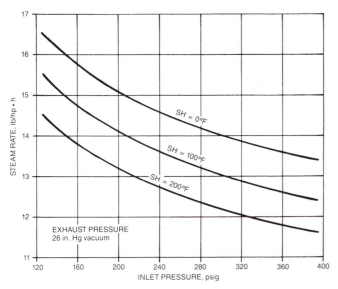

Fig. 12 Effect of Inlet Pressure and Superheat on Condensing Turbine

Often it is desirable to vary the turbine speed in response to an external signal. In centrifugal water chilling systems, for example, reduced speed generally reduces steam rate at partial load. This control is usually an electric or pneumatic device responsive to the temperature of the fluid leaving the water-chilling heat exchanger (evaporator). To avoid compressor surge and to optimize the steam rate, the speed is controlled initially down to some part load, then controlled in conjunction with the compressor's built-in capacity control (*i.e.*, inlet vanes). Process applications frequently require placing an external signal on the turbine-governing system, overriding its inherent constant-speed capability.

Such external signals may be necessary to maintain a fixed compressor discharge pressure, regardless of load or condenser water temperature variations. Plants relying on a closely maintained heat balance may control turbine speed to maintain an optimum pressure level of steam entering, being extracted from, or exhausting from the turbine. One example is the combination turbine absorption plant, where control of pressure of the steam exhausting from the turbine (and feeding the absorption unit) is an integral part of the plant-control system.

The steam turbine governing system consists of (1) a speed governor (mechanical, hydraulic, electrical); (2) a speed control mechanism (relays, servomotors, pressure or power amplifying devices, levers, and linkages); (3) governor-controlled valve(s); (4) a speed changer; and (5) external control devices, as required.

The speed governor responds directly to turbine speed and initiates action of the other parts of the governing system. The simplest speed governor is the direct-acting flyball type, which depends on changes in centrifugal force for proper action. Capable of adjusting speeds through an approximate 20% range, it is widely used on single-stage mechanical-drive steam turbines with speeds of up to 5000 rpm and steam pressures of up to 600 psig.

The speed governor used most frequently on centrifugal water-chilling system turbines is the oil-pump type. In its direct-acting form, oil pressure, produced by a pump either directly mounted on the turbine shaft or in some form responsive to turbine speed, actuates the inlet steam valve.

The oil relay hydraulic governor, as shown in Figure 13, has greater sensitivity and effective force. Here, the speed-induced oil-pressure changes are amplified in a servomotor or pilot-valve relay to produce the motive effort required to reposition the steam inlet valve or valves.

The least expensive turbine has a single governor-controlled steam admission valve, perhaps augmented by one or more small auxiliary valves (usually manually operated), which close off nozzles supplying the turbine steam chest for better part-load efficiency. Figure 14 shows the effect of auxiliary valves on part-load turbine performance.

For more precise speed-governing and maximum efficiency without manual valve adjustment, multiple automatic nozzle control is used (Figure 15). Its principal application is in larger turbines where a single governor-controlled steam admission valve would be too large to permit sensitive control. The greater power required to actuate the multiple-valve mechanism dictates the use of hydraulic servomotors.

Speed changers adjust the setting of the governing system while the turbine is in operation. Usually, they comprise either a means of changing spring tension or of regulating oil flow by a needle valve. The upper limit of a speed changer's capability should not

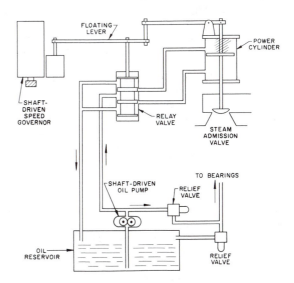

Fig. 13 Oil Relay Governor

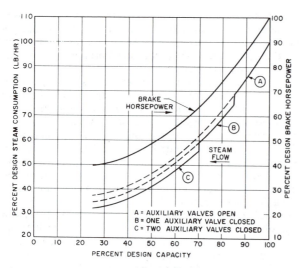

Fig. 14 Typical Part-Load Turbine Performance Showing Effect of Auxiliary Valves

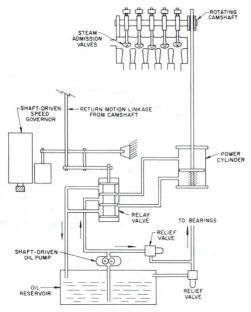

Fig. 15 Multivalve Oil Relay Governor

exceed the rated turbine speed. Such speed changers, while usually mounted on the turbine, may sometimes be remotely located at a central control point.

As stated previously, external control devices are often used when some function other than turbine speed is controlled. In such cases, an electric, hydraulic, or pneumatic signal overrides the turbine speed governor's action, and the latter assumes a speed-limiting function. The external signal controls steam admission either by direct inlet-valve positioning or by adjustment of the speed-governor setting. The valve-positioning method either exerts mechanical force on the valve-positioning mechanism or, if power has to be amplified, regulates the pilot valve in a hydraulic servo-motor system. Where more precise control is required, the speed governor adjusting method is preferable. Although the external signal continually resets the governor, as required, the speed governor always provides ideal turbine speed control. Thus, it maintains the particular set speed, regardless of load or steam-pressure variations.

NEMA classifies steam turbine governors as shown in Table 10. Range of speed changer adjustment, expressed as a percentage of rated speed, is the rang ethrough which the turbine speed may be

Table 10 NEMA Classification of Speed Governors

Class of Governor	Adjustable Speed Range, %	Maximum Steady-State Speed Regulation, %	Maximum Speed Variation, % Plus or Minus	Maximum Speed Rise, %	Trip Speed, % (Above Rated Speed)
A	10 to 65	10	0.75	13	15
B	10 to 80	6	0.50	7	10
C	10 to 80	4	0.25	7	10
D	10 to 90	0.50	0.25	7	10

adjusted downward from rated speed by the speed changer, with the turbine operating under the control of the speed governor and passing a steam flow equal to the flow at rated power output and speed.

The range of the speed changer adjustment, expressed as a percentage of rated speed, is derived from the following equation:

$$\text{Range (\%)} = \frac{\left(\begin{array}{c}\text{Rated} \\ \text{speed}\end{array}\right) - \left(\begin{array}{c}\text{Minimum speed} \\ \text{setting}\end{array}\right)}{\text{Rated speed}} 100$$

Steady-state speed regulation, expressed as a percentage of rated speed, is the change in sustained speed when the power output of the turbine is gradually changed from rated power output to zero power output under the following conditions:

- The steam conditions (initial pressure, initial temperature, and exhaust pressure) are set at rated values and held constant.
- The speed changer is adjusted to give rated speed with rated power output.
- Any external control device is rendered inoperative and blocked in the open position to allow the free flow of steam to the governor-controlled valve(s).

The steady-state speed regulation, expressed as a percentage of rated speed, is derived from the following equation:

$$\text{Regulation (\%)} = \frac{\left(\begin{array}{c}\text{Speed at zero} \\ \text{power output}\end{array}\right) - \left(\begin{array}{c}\text{Speed at rated} \\ \text{power output}\end{array}\right)}{\text{Speed at rated power output}} 100$$

Steady-state speed regulation of automatic extraction or mixed pressure-type turbines is derived with zero extraction or induction flow and with the pressure-regulating system(s) inoperative and blocked in the position corresponding to rated extraction or induction pressure(s) at rated power output.

Speed variation, expressed as a percentage of rated speed, is the total magnitude of speed change or fluctuations from the speed setting. It is defined as the difference in speed variation between the governing system in operation and the governing system blocked to be inoperative, with all other conditions constant. This characteristic includes dead band and sustained oscillations. Expressed as a percentage of rated speed, the speed variation is derived from the following equation:

$$\begin{array}{c}\text{Speed} \\ \text{Variation} \\ \text{(\%)}\end{array} = \frac{\left(\begin{array}{c}\text{Change in} \\ \text{speed above} \\ \text{set speed}\end{array}\right) - \left(\begin{array}{c}\text{Change in} \\ \text{speed below} \\ \text{set speed}\end{array}\right)}{\text{Rated speed}} 100$$

Dead band is a characteristic of the speed-governing system (referred to as wander). It is the insensitivity of the speed-governing system and the total speed change during which the governing valve(s) does not change position to compensate for the speed change.

Stability is a measure of the ability of the speed-governing system to position the governor-controlled valve(s); thus sustained oscillations of speed are not produced during a sustained load demand or following a change to a new load demand.

Speed oscillations are characteristics of the speed governing system (referred to as hunt). The ability of a governing system to keep sustained oscillations to a minimum is measured by its stability.

Maximum speed rise, expressed as a percentage of rated speed, is the maximum momentary increase in speed obtained when the turbine is developing rated power output at rated speed and the load is suddenly and completely reduced to zero. The maximum speed rise, expressed as a percentage of rated speed, is derived from the following equation:

$$\text{Speed Rise (\%)} = \frac{\left(\begin{array}{c}\text{Maximum speed at}\\\text{zero power output}\end{array}\right) - \left(\begin{array}{c}\text{Rated}\\\text{speed}\end{array}\right)}{\text{Rated speed}} \, 100$$

Trip speed is the speed at which the overspeed protective device operates.

Other Controls

In addition to speed-governing controls, certain safety devices are required on steam turbines. These include an overspeed mechanism, which acts through a quick-tripping valve, independent of the main governor valve, to shut off the steam supply to the turbine, and a pressure-relief valve in the turbine casing. Overspeed trip devices may act directly through linkages to close the steam valve or hydraulically by relieving oil pressure, allowing the valve to close. Also, the turbine must shut down should other safety devices, such as oil pressure failure or any of the refrigeration protective controls, so dictate. These devices usually act through an electrical interconnection to close the turbine trip valve mechanically or hydraulically.

To shorten the coast-down time of a tripped condensing-type turbine, a vacuum breaker in the turbine exhaust opens to admit air on receiving the trip signal.

Lubrication

Small turbines may often have only simple oil rings to handle bearing lubrication, but most turbines in refrigeration service have a complete pressure-lubrication system. Basic components include a shaft-driven oil pump, an oil filter, an oil cooler, a means of regulating oil pressure, a reservoir, and interconnecting piping. Turbines having a hydraulic governor may use oil from the lubrication circuit or, with certain types of governors, have a self-contained governor oil system. To ensure an adequate supply of oil to bearings during acceleration and deceleration periods, many turbines include an auxiliary motor- or turbine-driven oil pump. Oil pressure-sensing devices act in two ways: (1) to stop the auxiliary pump once the shaft-driven pump has attained proper flow and pressure; or (2) to start the auxiliary pump if the shaft-driven pump fails or loses pressure when decelerating.

In some industrial applications, the lubrication systems of the turbine and the driven compressor are integrated. Proper oil pressure, temperature, and compatibility of the lubricant qualities must be maintained.

Construction

Turbine manufacturers' standards prescribe casing materials for various limits of steam pressure and temperature. The use of built-up or solid rotors is a function of turbine speed or inlet steam temperature. Water must drain from pockets within the turbine casing to prevent damage caused by condensate accumulation. Carbon rings or closely fitted labyrinths prevent leakage of steam between pressure stages of the turbine, outward steam leakage, and inward air leakage at the turbine glands. Erosive and corrosive effect of moisture entering with the supply steam must be considered. Heat loss is prevented (often at the manufacturer's plant) by thermal insulation and protective metal jacketing on the hotter portions of the turbine casing.

GAS TURBINES

The gas turbine has achieved an increasingly important position as a prime mover for centrifugal refrigeration compressor drives in applications ranging from 800 to 20,000 hp.

It has the following advantages over other internal combustion engine drivers:

- Small size
- High power-to-weight ratio
- Ability to burn a variety of fuels, though more limited than the reciprocating engine
- Ability to meet stringent pollution standards
- High reliability
- Available in self-contained packages
- Instant power; no warmup required
- No cooling water required
- Vibration-free operation
- Easy maintenance
- Low installation cost
- Clean, dry exhaust
- Lubrication oil not contaminated by combustion oil

Application

Figure 16 shows a typical gas turbine refrigeration cycle. A gas turbine must be brought up to speed by an auxiliary starter, which may be a reciprocating engine. With a single-shaft turbine, the air compressor turbine, speed reducer gear, and refrigeration compressor must all be started and accelerated by this starter. The refrigeration compressor must be unloaded to ease this starting requirement. Sometimes, this may be done by making sure the capacity control vanes close tightly. At other times, it may be necessary to depressurize the refrigeration system to get started.

With a split-shaft design, only the air compressor and the gas-producer turbine must be started and accelerated. The rest of the unit starts rotating whenever enough energy is supplied to the blades of the power turbine. At this time, the gas-producer turbine is up to speed, and the fuel supply is ignited.

Electric starters are usually available as standard equipment. Reciprocating engines, steam turbines, and hydraulic or pneumatic motors may also be used.

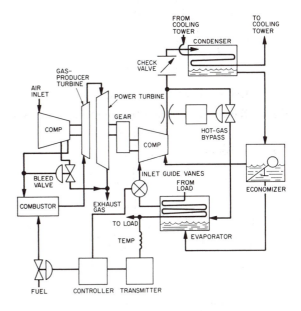

Fig. 16 Typical Gas Turbine Refrigeration Cycle

Theory

The basic gas turbine cycle (Figure 17) is the Brayton cycle (open cycle), which consists of adiabatic compression, constant pressure heating, and adiabatic expansion. Figure 17 shows that the thermal efficiency of a gas turbine falls below the ideal value because of inefficiencies in the compressor and turbine, and because of duct losses. Increases in entropy occur during the compression and expansion processes, and a reduction in area enclosed by points 1, 2, 3, and 4 exists. This loss of area is a direct measure of the loss in efficiency of the cycle.

Nearly all turbine manufacturers present gas turbine engine performance in power terms and specific fuel consumption in terms of Btu/hp·h. A comparison of fuel consumption in specific terms is the quickest way to compare gas turbine overall thermal efficiencies.

Components

Figure 18 presents the major components of the gas turbine unit—the air compressor, the combustor, and the turbine. Atmospheric air is compressed to about 175 to 600 psi by the air compressor. Fuel is then injected into the airstream and ignited in the combustor, with the leaving gases attaining temperatures between 1400 and 1800°F. These high-pressure hot gases are then expanded through a turbine, which provides not only the power required by the air compressor, but also power to drive the refrigeration machine.

Gas turbines are available in two major classifications—single-shaft (Figure 18) and split-shaft (Figures 19 and 20). The single-shaft turbine has the air compressor, the gas-producer turbine, and the power turbine on the same shaft. If the turbine is split, with the section required for the air compression on one shaft and the section producing output power on a separate shaft, it is known as a split- or dual-shaft design. For a split-shaft turbine, the portion that includes the compressor, combustion chamber, and first turbine section is the hot gas generator or gas producer. The second turbine section is the power turbine.

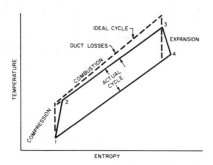

^aIdeal cycle = dashed lines; solid lines = actual cycle.

Fig. 17 Temperature-Entropy Diagram for Brayton Cycle

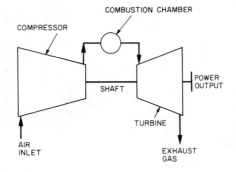

THERMAL EFFICIENCY RANGE 18-36%

Fig. 18 Single-Shaft Turbine

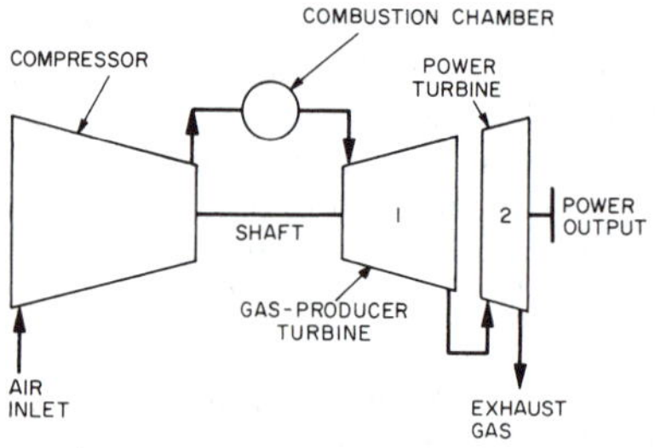

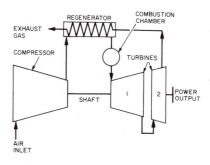

THERMAL EFFICIENCY RANGE 18-36%

Fig. 19 Split-Shaft Turbine

THERMAL EFFICIENCY RANGE 28-38%

Fig. 20 Split-Shaft Turbine with Regenerator

The turbine used depends on job requirements. Single-shaft engines are usually selected when constant-speed drive is required, as in generator drives, and when starting torque requirements are low. A single-shaft engine can be used to drive centrifugal compressors, but the starting system and the compressor match point must be considered.

Split-shaft engines allow for variable speed at full load. An additional advantage of the split-shaft engine is that it can easily be started with a high torque load connected to the power output shaft.

Performance

Most centrifugal refrigeration compressors are driven by constant-speed drivers, such as electric motors, and refrigeration capacity is controlled by variable inlet guide vanes on the compressor. Governors can maintain steam turbines at a constant speed to control the steam supply. Also, a fuel-supply regulator can maintain single-shaft gas turbines at a constant speed.

With the split-shaft design, the output shaft of the turbine unit runs at the speed required by the refrigeration compressor. The temperature of the chilled water or brine leaving the cooler of the refrigeration machine controls the fuel.

The rating of a gas turbine is affected by both inlet pressure to the air compressor and exhaust pressure from the turbine. In most applications, filters and silencers have to be installed in the air inlet. Silencers, waste heat boilers, or both are used on the exhaust. The pressure drop of these accessories and piping losses must be considered when determining the power output of the unit. The power output is also affected by altitude and atmospheric temperature. Gas turbine ratings are usually given at ISO conditions (59°F and sea level pressure) at the inlet flange of the air compressor and the exhaust flange of the turbine. Corrections for other conditions must be obtained from the manufacturer, as they vary with each model. Inlet air cooling has been used as a mechanism to increase capacity. Some approximations follow:

- Inlet temperature. Each 18°F rise in inlet temperature decreases the power output by 9%, and vice versa.

- An increase of 1000 ft in altitude decreases the power output by approximately 3.5%.
- Inlet pressure loss in filter, silencer, and ducting decreases power output by approximately 0.5% for each inch of water pressure loss.
- Discharge pressure loss in exhaust heat boiler, silencer, and ducting decreases power output approximately 0.3% for each inch of water pressure loss.

Figure 21 shows a typical performance curve for a 10,000 hp turbine engine. For a given air inlet temperature, 86 °F, for example, the engine develops its maximum power at about 82% of maximum speed.

EXHAUST HEAT USE

The overall thermal efficiency of the prime mover is 18 to 36%, with exhaust gases from the turbine ranging from 806 to 986 °F. If the exhaust heat can be used, overall thermal efficiencies can be increased.

Figure 20 shows a regenerator that uses the heat of the exhaust gases to heat the air from the compressor prior to combustion. Overall thermal efficiency can be increased to between 28 and 38% by using a regenerator.

If process heat is required, the exhaust can satisfy a portion of that heat, and the combined system is a cogeneration system. The exhaust can be used (1) directly as a source of hot air, (2) in a large boiler or furnace as a source of preheated combustion air (as the exhaust contains about 16% oxygen), or (3) to heat a process or working fluid such as the steam system shown in Figure 22. Thermal efficiencies of these systems vary from 50 to 90% and over. The exhaust of a gas turbine has about 4000 to 8000 Btu/hp of available heat.

Additionally, because of the high oxygen content, the exhaust stream can support the combustion of an additional 30,000 Btu/hp of fuel. This additional heat can then be used in general manufacturing processes.

Operation

Gas turbines operate with a wide range of fuels. For refrigeration service, a natural gas system is usually provided with an option of a standby No. 1 or 2 grade fuel oil system.

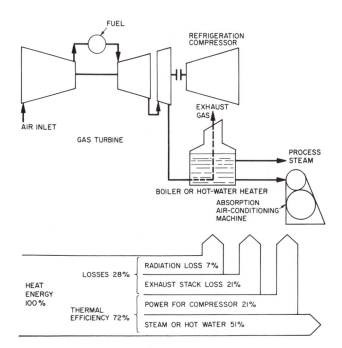

Fig. 22 Gas Turbine Refrigeration System Using Exhaust Heat

When selecting gas turbines, the output shaft must rotate in the direction required of the refrigeration compressor (in many cases, the manufacturer of split-shaft engines can provide the power turbines section with either direction of rotation).

At low loads, both the gas-turbine unit and the centrifugal refrigeration machine are affected by surge, a characteristic of all centrifugal and axial-type compressors. At a certain pressure ratio or head, a minimum flow through the compressor is necessary to maintain stable operation. In the unstable area, a momentary backward flow of gas occurs through the compressor. Stable operation can be maintained, however, by the use of a hot-gas bypass valve.

The turbine manufacturer normally includes automatic surge protection, either as a bleed valve that bypasses a portion of the air directly from the axial compressor into the exhaust duct or by changing the position of the axial compressor stator vanes. Both methods are employed in some cases.

The assembly should be prevented from rotating backward, which may occur if the unit is suddenly stopped by one of the safety controls. The difference in pressure between the refrigeration condenser and cooler can make the compressor suddenly become a turbine and cause it to rotate in the opposite direction. This rotation can force hot turbine gases back through the air compressor, causing considerable damage. Reverse flow through the refrigeration compressor may be prevented in a variety of ways, depending on the system's components. When there is no refrigerant receiver, quick-closing inlet guide vanes are usually satisfactory because there is very little high-pressure refrigerant to cause reverse rotation. A refrigeration system with a receiver has a substantial amount of energy available to cause reverse rotation. This can be reduced by opening the hot-gas bypass valve on shutdown and by employing a discharge check valve on the compressor.

Safety controls usually supplied with a gas turbine include the following:

- Overspeed
- Compressor surge
- Over-temperature during operation under load
- Low oil pressure
- Failure to light off during start cycle
- Underspeed during operation under load

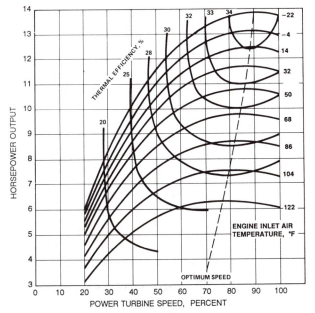

Fig. 21 Typical Turbine Engine Performance Characteristics

Noise. On air-conditioning installations with gas turbine drivers, the control of noise cannot be overlooked. Gas turbine manufacturers have developed sound-attenuated enclosures that cover the turbine and gear package. Turbine drivers, when properly installed with a sound-attenuated enclosure, an inlet silencer, and an exhaust silencer, will meet the strictest noise standards. However, the turbine manufacturer should be consulted for detailed noise level data and recommendations on the least expensive method of attenuation for a particular installation.

Pollution. Gas turbine power plants have relatively low emission levels compared to other internal combustion engines; however, for each application, the gas turbine manufacturer should be consulted to ensure that local and state codes are met.

Maintenance

Industrial gas turbines are designed to operate for 12,000 to 30,000 h between overhauls, with normal maintenance. Normal maintenance includes checking filter and oil level, inspecting for leaks, etc., all of which can be done by the operator with ordinary mechanics' tools. However, the engine requires factory-trained service personnel to inspect engine components such as combustors and nozzles. These inspections, depending on the manufacturer's recommendations, are required as frequently as every 4000 h of operation.

Most gas turbines are maintained by condition monitoring and inspection rather than by specific overhaul intervals. Gas turbines specifically designed for industrial applications may have an indefinite life for the major housings, shafts, and low-temperature components. Hot section repair intervals for combustor and turbine components can vary from 10,000 to 100,000 h.

The total cost of maintaining a gas turbine includes (1) cost of operator time, (2) normal parts replacement, (3) lubricating oil, (4) filter changes, (5) overhauls, and (6) factory service time (to conduct engine inspections). The cost of all these items can be estimated by the manufacturer and must be taken into account to determine the total operating cost.

REFERENCES

ASME. 1985. Guidance for evaluation of measurement uncertainty in performance tests of steam turbines. ASME PTC 6 Report 85, American Society of Mechanical Engineers, New York.

ASME. 1982. Appendix A to Test Code for Steam Turbines. PTC 6A-82.

ASME. 1988. Procedures for routine performance tests of steam turbines. PTC 6S-88 Report.

NEMA. 1985. Steam turbines for mechanical drive service. NEMA *Standard* SM 23. National Electrical Manufacturers Association, Washington, D.C.

Transamerica Delaval, Inc. 1983. *Transamerica Delaval engineering handbook*, 4th ed. McGraw-Hill, New York.

CHAPTER 42

PIPES, TUBES, AND FITTINGS

THIS chapter covers the selection, application, and installation of pipe, tubes, fittings, and valves commonly used for heating, air-conditioning, and refrigerating systems. Pipe hangers and pipe expansion are also addressed. When selecting and applying these components, applicable local codes, state or provincial codes, and voluntary industry standards (some of which have been adopted by code jurisdictions) must be followed.

The following U.S. organizations issue codes and standards for piping systems and components:

ASME — American Society of Mechanical Engineers
ASTM — American Society for Testing Materials
NFPA — National Fire Protection Association
BOCA — Building Officials and Code Administrators, International, Inc.
MSS — Manufacturers Standardization Society of the Valve and Fitting Industry, Inc.
ANSI — American National Standards Institute
AWWA — American Water Works Association

Parallel federal specifications also have been developed by government agencies and are used for many public works projects. Chapter IV of ASME *Standard* B31.9, Building Services Piping, lists applicable U.S. codes and standards for HVAC piping. In addition, it gives the requirements for safe design and construction of piping systems for building heating and air conditioning.

PIPE

Steel Pipe

Steel pipe is manufactured by several processes. Seamless pipe, made by piercing or extruding, has no longitudinal seam. Other manufacturing methods roll skelp into a cylinder and weld a longitudinal seam. A continuous-weld (CW) furnace-butt-welding process forces and joins the edges together at high temperature. An electric current welds the seam of electric resistance welded (ERW) pipe. ASTM *Standards* A106 and A53 specify steel pipe. Both specify an A and B grade. The A grade has a lower tensile strength and is not widely used.

The ASME pressure piping codes require that a longitudinal joint efficiency factor E be applied to each type of seam when calculating the allowable stress, as listed in Table 1. ASME *Standard* B36.10 specifies the dimensional standard for steel pipe. Through 12 in. diameter, nominal pipe sizes (NPS) are used, which do not match the internal or external diameters. For 14 in. and larger pipe, the size corresponds to the outside diameter.

Steel pipe is manufactured with wall thicknesses identified by schedule and weight. Although schedule numbers and weight designations are related, they are not constant for all pipe sizes. Standard weight (STD) and Schedule 40 pipe have the same wall thickness through 10 in. NPS. For 12 in. and larger standard weight pipe, the wall thickness remains constant at 0.375 in., while Schedule 40 wall thicknesses increase with each size. A similar equality exists between Extra Strong (XS) and Schedule 80 pipe up through 8 in.; afterward, XS pipe has a 0.500 in. wall, while Schedule 80 increases in wall thickness. Table 2 lists properties of representative steel pipe.

Joints in steel pipe are made by welding or using threaded, flanged, grooved, or welded outlet fittings. Unreinforced welded in branch connections weaken a main pipeline, and added reinforcement is necessary, unless the excess wall thickness of both mains and branches are sufficient to sustain the pressure.

ASME *Standard* B31.1, Power Piping, gives formulas for determining when reinforcement is required. Such calculations are seldom needed in HVAC applications because (1) standard weight pipe through 20 in. NPS at 300 psig requires no reinforcement; full-size branch connections are not recommended; and (2) fittings such as tees and reinforced outlet fittings provide inherent reinforcement.

Copper Tube

Because of their inherent resistance to corrosion and ease of installation, copper and copper alloys are often used in heating, air-conditioning, refrigeration, and water supply installations. There are two principal classes of copper tube. ASTM *Standard* B88 includes Types K, L, M, and DWV for water and drain service. ASTM *Standard* B280 specifies ACR (air-conditioning and refrigeration) tube for refrigeration service.

Types K, L, M, and DWV designate descending wall thicknesses for copper tube. All types have the same outside diameter for corresponding sizes. Table 3 lists properties of ASTM B88 copper tube. In the plumbing industry, tube of nominal size approximates the inside diameter. The heating and refrigeration trades specify copper tube by the outside diameter (OD). ACR tubing has a different set of wall thicknesses. Types K, L, and M tube may be hard drawn or annealed (soft) temper.

Copper tubing is joined with soldered or brazed wrought or cast copper capillary socket-end fittings. Table 4 lists the pressure-temperature ratings of soldered and brazed joints. Small copper tube is also joined by flare or compression fittings.

Hard-drawn tubing has a higher allowable stress value than annealed, but if hard tubing is joined by soldering or brazing, the annealed allowable stress value should be used.

Brass pipe and copper pipe are also made in steel pipe thicknesses for threading. High cost has eliminated these materials from the market, except for special applications.

The preparation of this chapter is assigned to TC 8.10, Pumps and Hydronic Piping.

Table 1 Allowable Stresses for Pipe and Tube

Specification	Grade	Type	Manufacturing Process	Available Sizes, in.	Minimum Tensile Strength, ksi	Basic Allowable Stress, S ksi	Joint Efficiency Factor, E	Allowable Stress, $S_E{}^a$ ksi	Allowable Stress Range, $S_A{}^b$ ksi
ASTM A53 Steel	—	F	Cont. Weld	1/2 to 4	45.0	11.25	0.6	6.8	16.9
ASTM A53 Steel	B	S	Seamless	1/2 to 26	60.0	15.0	1.0	15.0	22.5
ASTM A53 Steel	B	E	ERW	2 to 20	60.0	15.0	0.85	12.8	22.5
ASTM A106 Steel	B	S	Seamless	1/2 to 26	60.0	15.9	1.0	15.0	22.5
ASTM B-88 Copper	—	—	Hard Drawn	1/4 to 12	36.0	9.0	1.0	9.0	13.5

Note: Listed stresses are for temperatures to 650°F for steel pipe and 250°F for copper tubing.
[a] To be used for internal pressure stress calculations in Equations (1) and (2).
[b] To be used only for piping flexibility calculations; Equations (3) and (4).

The heating and air-conditioning industry generally uses Types L and M tubing, which have a higher internal working pressure rating than the solder joints used at fittings. Type K may be used with brazed joints for higher pressure-temperature requirements or for direct burial. Type M should be used with care where exposed to potential external damage.

Copper and brass should not be used in ammonia refrigerating systems. The Special Systems section covers other limitations on refrigerant piping.

Ductile Iron and Cast Iron

Cast-iron soil pipe comes in XH or service weight. It is not used under pressure because the pipe is not suitable and the joints are not restrained. Cast-iron pipe and fittings typically have bell and spigot ends for lead and oakum joints or elastomer push-on joints. Cast-iron pipe and fittings are also furnished with *no-hub* ends for joining with *no-hub* clamps. Local plumbing codes specify permitted materials and joints.

Ductile iron has now replaced cast iron for pressure pipe. Ductile iron is stronger, less brittle, and similar to cast iron in corrosion resistance. It is commonly used for buried pressure water mains or in other locations where internal or external corrosion is a problem. Joints are made with flanged fittings, mechanical joint (MJ) fittings, or elastomer gaskets for bell and spigot ends. Bell and spigot and MJ joints are not self-restrained. Restrained MJ systems are available. Ductile iron pipe is made in seven thickness classes for different service conditions. AWWA *Standard* C150/A2l.50, Thickness Design of Ductile Iron Pipe, covers the proper selection of pipe classes.

FITTINGS

The following standards give dimensions and pressure ratings for fittings, flanges, and flanged fittings. This data is also available from manufacturers' catalogs.

Applicable Standards for Fittings

Steel[a] .... **ASME Std.**

Flanges and Flanged Fittings (Classes 150 and 300) B16.5
Factory-made Wrought Steel Butt-Welding Fittings B16.9
Forged Steel Fittings, Socket Welded and Threaded B16.11
Wrought Steel Butt-Welding Short Radius Elbows B16.28

Cast Iron, Malleable Iron, Ductile Iron[b]

Cast-Iron Pipe Flanges and Flanged Fittings B16.1
Malleable Iron Threaded Fittings, Classes 150 and 300 ..... B16.3
Cast-Iron Threaded Fittings, Classes 125 and 250 .... B16.4
Cast-Iron Screwed Drainage Fittings B16.12
Ductile Iron Pipe Flanges and Flanged Fittings,
  Classes 150 and 300 .... B16.42

Copper and Bronze[c] .... **ASTM Std.**

Cast Bronze Threaded Fittings, Classes 125 and 250 .... B16.15
Cast Copper Alloy Solder Joint Pressure Fittings B16.18
Wrought Copper and Copper Alloy Solder
  Joint Pressure Fittings B16.22
Cast Copper Alloy Solder Joint Drainage Fittings, DWV B16.23
Bronze Pipe Flanges and Flanged Fittings,
  Classes 150 and 300 .... B16.24
Cast Copper Alloy Fittings for Flared Copper Tubes B16.26
Wrought Copper Solder Joint Drainage Fittings B16.29

Nonmetallic[d]

Threaded PVC Plastic Pipe Fittings, Schedule 80 .... D2464
Threaded PVC Plastic Pipe Fittings, Schedule 40 .... D2466
Socket-Type PVC Plastic Pipe Fittings, Schedule 80 .... D2467
Reinforced Epoxy Resin Piping Gas Pressure Fittings D2517
Threaded CPVC Plastic Pipe Fittings, Schedule 80 .... F437
Socket-Type CPVC Plastic Pipe Fittings, Schedule 40 ...... F438
Socket-Type CPVC Plastic Pipe Fittings, Schedule 80 ...... F439
Socket Heat Fusion Polybutylene Fittings D3309
Insert Fittings for Polybutylene Tubing F845/D3309
PVC Solvent Cements D2564
CPVC Solvent Cements F493

[a] Wrought steel butt-welding fittings are made to match steel pipe wall thicknesses and are rated at the same working pressure as seamless pipe. Flanges and flanged fittings are rated by working steam pressure classes. Forged steel fittings are rated from 2000 to 6000 psi in classes and are used for high temperature and pressure service for small pipe sizes.
[b] The class numbers refer to the maximum working saturated steam gage pressure (in psi). For liquids at lower temperatures, higher pressures are allowed. Groove-end fittings of these materials are made by various manufacturers who publish their own ratings.
[c] The classes refer to maximum working steam pressure (in psi). At ambient temperatures, higher liquid pressures are allowed. Solder joint fittings are limited by the strength of the soldered or brazed joint (Table 4).
[d] Ratings of plastic fittings match the pipe of corresponding schedule number.

JOINING METHODS

Threading is the most commonly used method for joining small diameter steel or brass pipe, as shown in ASME *Standard* B1.20.1. Pipe with a wall thickness less than standard weight should not be threaded. ASME *Standard* B31.5, Refrigeration Piping, limits the threading for various refrigerants and pipe sizes.

Soldering and Brazing

Copper tube is usually joined by soldering or brazing socket end fittings. Brazing materials melt at temperatures over 1000°F and produce a stronger joint than solder. Table 4 lists soldered and brazed joint strengths. Health concerns have caused many jurisdictions to ban solders containing lead or antimony for joining pipe in potable water systems. Lead-based solders, in particular, must not be used for potable water systems.

Table 2 Steel Pipe Data

Nominal Size and Pipe OD, in.	Schedule Number or Weight[a]	Wall Thickness, in. t	Inside Diameter, in. d	Surface Area Outside, ft²/ft	Surface Area Inside, ft²/ft	Metal Area, in²	Flow Area, in²	Weight of Pipe, lb/ft	Weight of Water, lb/ft	Mfr. Process	Joint Type	Working Pressure[b] ASTM A53 B to 400°F psig
1/4	40 ST	0.088	0.364	0.141	0.095	0.125	0.104	0.424	0.045	CW	Thread	188
D = 0.540	80 XS	0.119	0.302	0.141	0.079	0.157	0.072	0.535	0.031	CW	Thread	871
3/8	40 ST	0.091	0.493	0.177	0.129	0.167	0.191	0.567	0.083	CW	Thread	203
D = 0.675	80 XS	0.126	0.423	0.177	0.111	0.217	0.141	0.738	0.061	CW	Thread	820
1/2	40 ST	0.109	0.622	0.220	0.163	0.250	0.304	0.850	0.131	CW	Thread	214
D = 0.840	80 XS	0.147	0.546	0.220	0.143	0.320	0.234	1.087	0.101	CW	Thread	753
3/4	40 ST	0.113	0.824	0.275	0.216	0.333	0.533	1.13	0.231	CW	Thread	217
D = 1.050	80 XS	0.154	0.742	0.275	0.194	0.433	0.432	1.47	0.187	CW	Thread	681
1	40 ST	0.133	1.049	0.344	0.275	0.494	0.864	1.68	0.374	CW	Thread	226
D = 1.315	80 XS	0.179	0.957	0.344	0.251	0.639	0.719	2.17	0.311	CW	Thread	642
1 1/4	40 ST	0.140	1.380	0.435	0.361	0.669	1.50	2.27	0.647	CW	Thread	229
D = 1.660	80 XS	0.191	1.278	0.435	0.335	0.881	1.28	2.99	0.555	CW	Thread	594
1 1/2	40 ST	0.145	1.610	0.497	0.421	0.799	2.04	2.72	0.881	CW	Thread	231
D = 1.900	80 XS	0.200	1.500	0.497	0.393	1.068	1.77	3.63	0.765	CW	Thread	576
2	40 ST	0.154	2.067	0.622	0.541	1.07	3.36	3.65	1.45	CW	Thread	230
D = 2.375	80 XS	0.218	1.939	0.622	0.508	1.48	2.95	5.02	1.28	CW	Thread	551
2 1/2	40 ST	0.203	2.469	0.753	0.646	1.70	4.79	5.79	2.07	CW	Weld	533
D = 2.875	80 XS	0.276	2.323	0.753	0.608	2.25	4.24	7.66	1.83	CW	Weld	835
3	40 ST	0.216	3.068	0.916	0.803	2.23	7.39	7.57	3.20	CW	Weld	482
D = 3.500	80 XS	0.300	2.900	0.916	0.759	3.02	6.60	10.25	2.86	CW	Weld	767
4	40 ST	0.237	4.026	1.178	1.054	3.17	12.73	10.78	5.51	CW	Weld	430
D = 4.500	80 XS	0.337	3.826	1.178	1.002	4.41	11.50	14.97	4.98	CW	Weld	695
6	40 ST	0.280	6.065	1.734	1.588	5.58	28.89	18.96	12.50	ERW	Weld	696
D = 6.625	80 XS	0.432	5.761	1.734	1.508	8.40	26.07	28.55	11.28	ERW	Weld	1209
8	30	0.277	8.071	2.258	2.113	7.26	51.16	24.68	22.14	ERW	Weld	526
D = 8.625	40 ST	0.322	7.981	2.258	2.089	8.40	50.03	28.53	21.65	ERW	Weld	643
	80 XS	0.500	7.625	2.258	1.996	12.76	45.66	43.35	19.76	ERW	Weld	1106
10	30	0.307	10.136	2.814	2.654	10.07	80.69	34.21	34.92	ERW	Weld	485
	40 ST	0.365	10.020	2.814	2.623	11.91	78.85	40.45	34.12	ERW	Weld	606
D = 10.75	XS	0.500	9.750	2.814	2.552	16.10	74.66	54.69	32.31	ERW	Weld	887
	80	0.593	9.564	2.814	2.504	18.92	71.84	64.28	31.09	ERW	Weld	1081
12	30	0.330	12.090	3.338	3.165	12.88	114.8	43.74	49.68	ERW	Weld	449
	ST	0.375	12.000	3.338	3.141	14.58	113.1	49.52	48.94	ERW	Weld	528
D = 12.75	40	0.406	11.938	3.338	3.125	15.74	111.9	53.48	48.44	ERW	Weld	583
	XS	0.500	11.750	3.338	3.076	19.24	108.4	65.37	46.92	ERW	Weld	748
	80	0.687	11.376	3.338	2.978	26.03	101.6	88.44	43.98	ERW	Weld	1076
14	30 ST	0.375	13.250	3.665	3.469	16.05	137.9	54.53	59.67	ERW	Weld	481
	40	0.437	13.126	3.665	3.436	18.62	135.3	63.25	58.56	ERW	Weld	580
D = 14.00	XS	0.500	13.000	3.665	3.403	21.21	132.7	72.04	57.44	ERW	Thread	681
	80	0.750	12.500	3.665	3.272	31.22	122.7	106.05	53.11	ERW	Weld	1081
16	30 ST	0.375	15.250	4.189	3.992	18.41	182.6	62.53	79.04	ERW	Weld	421
D = 16.00	40 XS	0.500	15.000	4.189	3.927	24.35	176.7	82.71	76.47	ERW	Weld	596
	ST	0.375	17.250	4.712	4.516	20.76	233.7	70.54	101.13	ERW	Weld	374
18	30	0.437	17.126	4.712	4.483	24.11	230.3	81.91	99.68	ERW	Weld	451
D = 18.00	XS	0.500	17.000	4.712	4.450	27.49	227.0	93.38	98.22	ERW	Weld	530
	40	0.562	16.876	4.712	4.418	30.79	223.7	104.59	96.80	ERW	Weld	607
20	ST	0.375	19.250	5.236	5.039	23.12	291.0	78.54	125.94	ERW	Weld	337
D = 20.00	30 XS	0.500	19.000	5.236	4.974	30.63	283.5	104.05	122.69	ERW	Weld	477
	40	0.593	18.814	5.236	4.925	36.15	278.0	122.82	120.30	ERW	Weld	581

[a] Numbers are schedule numbers per ASTM B36.10; ST = Standard Weight; XS = Extra Strong.
[b] Working pressures have been calculated per ASME/ANSI B31.9 using furnace butt-weld (continuous weld, CW) pipe through 4 in. and electric resistance weld (ERW) thereafter. The allowance A has been taken as:
  (a) 12.5% of t for mill tolerance on pipe wall thickness, plus

(b) An arbitrary corrosion allowance of 0.025 in. for pipe sizes through NPS 2 and 0.065 in. from NPS 2½ through 20, plus
(c) A thread cutting allowance for sizes through NPS 2.
Because the pipe wall thickness of threaded standard pipe is so small after deducting the allowance A the mechanical strength of the pipe is impaired. It is good practice to limit standard weight threaded pipe pressures to 90 psig for steam and 125 psig for water.

Table 3 Copper Tube Data

Nominal Diameter	Type	Wall Thickness t, in.	Diameter Outside D, in.	Diameter Inside d, in.	Surface Area Outside, ft²/ft	Surface Area Inside, ft²/ft	Cross-Sectional Metal Area, in²	Cross-Sectional Flow Area, in²	Weight of Tube, lb/ft	Weight of Water, lb/ft	Working Pressure[a,b,c] ASTM B88 to 250°F Annealed, psig	Working Pressure ASTM B88 to 250°F Drawn, psig
1/4	K	0.035	0.375	0.305	0.098	0.080	0.037	0.073	0.145	0.032	851	1596
	L	0.030	0.375	0.315	0.098	0.082	0.033	0.078	0.126	0.034	730	1368
3/8	K	0.049	0.500	0.402	0.131	0.105	0.069	0.127	0.269	0.055	894	1676
	L	0.035	0.500	0.430	0.131	0.113	0.051	0.145	0.198	0.063	638	1197
	M	0.025	0.500	0.450	0.131	0.008	0.037	0.159	0.145	0.069	456	855
1/2	K	0.049	0.625	0.527	0.164	0.138	0.089	0.218	0.344	0.094	715	1341
	L	0.040	0.625	0.545	0.164	0.143	0.074	0.233	0.285	0.101	584	1094
	M	0.028	0.625	0.569	0.164	0.149	0.053	0.254	0.203	0.110	409	766
5/8	K	0.049	0.750	0.652	0.196	0.171	0.108	0.334	0.418	0.144	596	1117
	L	0.042	0.750	0.666	0.196	0.174	0.093	0.348	0.362	0.151	511	958
3/4	K	0.065	0.875	0.745	0.229	0.195	0.165	0.436	0.641	0.189	677	1270
	L	0.045	0.875	0.785	0.229	0.206	0.117	0.484	0.455	0.209	469	879
	M	0.032	0.875	0.811	0.229	0.212	0.085	0.517	0.328	0.224	334	625
1	K	0.065	1.125	0.995	0.295	0.260	0.216	0.778	0.839	0.336	527	988
	L	0.050	1.125	1.025	0.295	0.268	0.169	0.825	0.654	0.357	405	760
	M	0.035	1.125	1.055	0.295	0.276	0.120	0.874	0.464	0.378	284	532
1 1/4	K	0.065	1.375	1.245	0.360	0.326	0.268	1.217	1.037	0.527	431	808
	L	0.055	1.375	1.265	0.360	0.331	0.228	1.257	0.884	0.544	365	684
	M	0.042	1.375	1.291	0.360	0.338	0.176	1.309	0.682	0.566	279	522
	DWV	0.040	1.375	1.295	0.360	0.339	0.168	1.317	0.650	0.570	265	497
1 1/2	K	0.072	1.625	1.481	0.425	0.388	0.351	1.723	1.361	0.745	404	758
	L	0.060	1.625	1.505	0.425	0.394	0.295	1.779	1.143	0.770	337	631
	M	0.049	1.625	1.527	0.425	0.400	0.243	1.831	0.940	0.792	275	516
	DWV	0.042	1.625	1.541	0.425	0.403	0.209	1.865	0.809	0.807	236	442
2	K	0.083	2.125	1.959	0.556	0.513	0.532	3.014	2.063	1.304	356	668
	L	0.070	2.125	1.985	0.556	0.520	0.452	3.095	1.751	1.339	300	573
	M	0.058	2.125	2.009	0.556	0.526	0.377	3.170	1.459	1.372	249	467
	DWV	0.042	2.125	2.041	0.556	0.534	0.275	3.272	1.065	1.416	180	338
2 1/2	K	0.095	2.625	2.435	0.687	0.637	0.755	4.657	2.926	2.015	330	619
	L	0.080	2.625	2.465	0.687	0.645	0.640	4.772	2.479	2.065	278	521
	M	0.065	2.625	2.495	0.687	0.653	0.523	4.889	2.026	2.116	226	423
3	K	0.109	3.125	2.907	0.818	0.761	1.033	6.637	4.002	2.872	318	596
	L	0.090	3.125	2.945	0.818	0.771	0.858	6.812	3.325	2.947	263	492
	M	0.072	3.125	2.981	0.818	0.780	0.691	6.979	2.676	3.020	210	394
	DWV	0.045	3.125	3.035	0.818	0.795	0.435	7.234	1.687	3.130	131	246
3 1/2	K	0.120	3.625	3.385	0.949	0.886	1.321	8.999	5.120	3.894	302	566
	L	0.100	3.625	3.425	0.949	0.897	1.107	9.213	4.291	3.987	252	472
	M	0.083	3.625	3.459	0.949	0.906	0.924	9.397	3.579	4.066	209	392
4	K	0.134	4.125	3.857	1.080	1.010	1.680	11.684	6.510	5.056	296	555
	L	0.110	4.125	3.905	1.080	1.022	1.387	11.977	5.377	5.182	243	456
	M	0.095	4.125	3.935	1.080	1.030	1.203	12.161	4.661	5.262	210	394
	DWV	0.058	4.125	4.009	1.080	1.050	0.741	12.623	2.872	5.462	128	240
5	K	0.160	5.125	4.805	1.342	1.258	2.496	18.133	9.671	7.846	285	534
	L	0.125	5.125	4.875	1.342	1.276	1.963	18.665	7.609	8.077	222	417
	M	0.109	5.125	4.907	1.342	1.285	1.718	18.911	6.656	8.183	194	364
	DWV	0.072	5.125	4.981	1.342	1.304	1.143	19.486	4.429	8.432	128	240
6	K	0.192	6.125	5.741	1.603	1.503	3.579	25.886	13.867	11.201	286	536
	L	0.140	6.125	5.845	1.603	1.530	2.632	26.832	10.200	11.610	208	391
	M	0.122	6.125	5.881	1.603	1.540	2.301	27.164	8.916	11.754	182	341
	DWV	0.083	6.125	5.959	1.603	1.560	1.575	27.889	6.105	12.068	124	232
8	K	0.271	8.125	7.583	2.127	1.985	6.687	45.162	25.911	19.542	304	570
	L	0.200	8.125	7.725	2.127	2.022	4.979	46.869	19.295	20.280	224	421
	M	0.170	8.125	7.785	2.127	2.038	4.249	47.600	16.463	20.597	191	358
	DWV	0.109	8.125	7.907	2.127	2.070	2.745	49.104	10.637	21.247	122	229
10	K	0.338	10.125	9.449	2.651	2.474	10.392	70.123	40.271	30.342	304	571
	L	0.250	10.125	9.625	2.651	2.520	7.756	72.760	30.054	31.483	225	422
	M	0.212	10.125	9.701	2.651	2.540	6.602	73.913	25.584	31.982	191	358
12	K	0.405	12.125	11.315	3.174	2.962	14.912	100.554	57.784	43.510	305	571
	L	0.280	12.125	11.565	3.174	3.028	10.419	105.046	40.375	45.454	211	395
	M	0.254	12.125	11.617	3.174	3.041	9.473	105.993	36.706	45.863	191	358

[a] When using soldered or brazed fittings, the joint determines the limiting pressure.
[b] Working pressures calculated using ASME B31.9 allowable stresses. A 5% mill tolerance has been used on the wall thickness. Higher tube ratings can be calculated using the allowable stress for lower temperatures.
[c] If soldered or brazed fittings are used on hard drawn tubing, use the annealed ratings. Full-tube allowable pressures can be used with suitably rated flare or compression-type fittings.

Table 4 Internal Working Pressure for Copper Tube Joints

(Based on ANSI/ASME *Standard* B31.9, Building Services Piping)

Alloy Used for Joints	Service Temperature, °F	Water and Noncorrosive Liquids and Gases[a] Internal Working Pressure, psi					Sat. Steam and Condensate
		Nominal Tube Size, Types K, L, M, in.					
		1/4 to 1	1-1/4 to 2	2-1/2 to 4	5 to 8[a]	10 to 12[a]	1/4 to 8
50-50	100	200	175	150	130	100	—
Tin-lead[b]	150	150	125	100	90	70	—
Solder	200	100	90	75	70	50	—
ASTM B32 Gr 50A	250	85	75	50	45	40	15
95-5	100	500	400	300	270	150	—
Tin-antimony[c]	150	400	350	275	250	150	—
Solder	200	300	250	200	180	140	—
ASTM B32 Gr 50TA	250	200	175	150	135	110	15
Brazing alloys	100 to 200	d	d	d	d	d	—
melting at or	250	300	210	170	150	150	—
above 1000°F	350	270	190	150	150	150	120

[a] Solder joints are not to be used for:
    (a) Flammable or toxic gases or liquids.
    (b) Gas, vapor, or compressed air in tubing over 4 in., unless maximum pressure is limited to 20 psig.

[b] Lead solders must not be used in potable water systems.
[c] Tin-antimony solder is allowed for potable water supplies in some jurisdictions.
[d] Rated pressure for temperatures up to 200°F is that of the tube being joined.

Flared and Compression Joints

Flared and compression fittings can be used to join copper, steel, stainless steel, and aluminum tubing. Properly rated fittings can keep the joints as strong as the tube.

Flanges

Flanges can be used for large pipe and all piping materials. They are commonly used to connect to equipment, valves, and wherever it may be necesssary to open the joint to permit service or replacement of components. For steel pipe, flanges are available in pressure ratings to 2500 psig. For welded pipe, weld neck, slip-on, or socket weld connections are available. Thread-on flanges are available for threaded pipe.

Flanges are generally flat faced or raised face. Flat-faced flanges with full-faced gaskets are most often used with cast iron and materials that cannot take high bending loads. Raised-face flanges with ring gaskets are preferred with steel pipe because they facilitate increasing the sealing pressure on the gasket to help prevent leaks. Other facings, such as O-ring and ring joint, are available for special applications.

All flat-faced, raised-face, and lap-joint flanges require a gasket between the mating flange surfaces. Gaskets are made from rubber, synthetic elastomers, cork, fiber, plastic, teflon, metal, and a combination of these materials. The gasket must be compatible with the flowing media and the temperatures at which the system is operating.

Welding

Welding steel pipe joints over 2 in. in diameter offer the following advantages:

1. They do not age, dry out, or deteriorate as do gasketed joints.
2. Welded joints can accommodate greater vibration and water hammer and higher temperatures and pressures than other joints.
3. For critical service, pipe joints can be tested by any of several nondestructive examination (NDE) methods, such as by radiography or ultrasound.
4. Welded joints provide maximum long-term reliability.

The applicable section of ASME *Standard* B31, Code for Pressure Piping, and the ASME Boiler and Pressure Vessel Code guide rules for welding. The Code for Pressure Piping series requires that all welders and welding procedure specifications (WPS) be quali-

fied. Separate WPS are needed for different welding methods and materials. The qualifying tests and the variables requiring separate procedure specifications are set forth in the ASME Boiler and Pressure Vessel Code, Section IX. The manufacturer, fabricator, or contractor is responsible for the welding procedure and welders. Section B31.9, Building Services Piping, of the Code for Pressure Piping requires visual examination of welds and outlines limits of acceptability.

The following welding processes are often used in the HVAC industry:

SMAW—Shielded Metal Arc Welding (stick welding). The molten weld metal is shielded by the vaporization of the electrode coating.

GMAW—Gas Metal Arc Welding, also called MIG. The electrode is a continuously fed wire, which is shielded by argon or carbon dioxide gas from the welding gun nozzle.

GTAW—Gas Tungsten Arc Welding, also called TIG or Heliarc. This process uses a nonconsumable tungsten electrode surrounded by a shielding gas. The weld material may be provided from a separate noncoated rod.

Reinforced Outlet Fittings

Reinforced outlet fittings are used to make branch and take-off connections and are designed to permit welding directly to pipe without supplemental reinforcing. Fittings are available with threaded, socket, or butt-weld outlets.

Other Joints

Grooved joint systems require that a shallow groove be cut or rolled into the pipe end. These joints can be used with steel, cast iron, ductile iron, and plastic pipes. A segmented clamp engages the grooves, and the seal is provided by a special gasket designed so that internal pressure tightens the seal. Some clamps are designed with clearance between tongue and groove to accommodate misalignment and thermal movements, while others are designed to limit movement and provide a rigid system. Manufacturers' data gives temperature and pressure limitations.

Another form of mechanical joint consists of a sleeve slightly larger than the outside diameter of the pipe. The pipe ends are inserted into the sleeve, and gaskets are packed into the annular space between the pipe and coupling and held in place by retainer rings. This type of joint can accept some axial misalignment, but it must be anchored or otherwise restrained to prevent axial pull-out or lateral movement. Manufacturers provide pressure-temperature data.

Ductile iron pipe is furnished with a spigot end adapted for a gasket and retainer ring. This joint is also not restrained.

Unions

Unions allow disassembly of threaded pipe systems. Unions are three-part fittings with a mating machined seat on the two parts that thread onto the pipe ends. A threaded locking ring holds the two ends tightly together. A union also allows threaded pipe to be turned at the last joint connecting two pieces of equipment. Companion flanges (a pair) for small pipe serve the same purpose.

VALVES

The types of valves used in heating, plumbing, and air-conditioning systems are described in Chapter 43. Manufacturers' catalogs list the pressure-temperature ratings.

SPECIAL SYSTEMS

Certain piping systems are governed by separate codes or standards, which are summarized below. Generally, any failure of the piping in these systems is dangerous to the public, so local areas have adopted laws enforcing the codes.

Boiler piping. ASME *Standard* B31.1 and the ASME Boiler and Pressure Vessel Code (Section I) specify the piping inside the code required stop valves on boilers that operate above 15 psig with steam or 160 psig or 250 °F with water. These codes require fabricators and contractors to be certified for such work. The field or shop work must also be inspected while it is in progress by inspectors commissioned by the National Board of Boiler and Pressure Vessel Inspectors.

Refrigeration piping. ASME *Standard* B31.5, Refrigerant Piping, and ASHRAE *Standard* 15-78, Standard Safety Code for Mechanical Refrigeration, cover the requirements for refrigerant piping.

Plumbing systems. Local codes cover piping for plumbing systems.

Sprinkler systems. NFPA *Standard* 13, Installation of Sprinkler Systems, covers this field.

Fuel gas. ANSI Z223.1, National Fuel Gas Code, prescribes fuel gas piping in buildings.

SELECTION OF MATERIALS

Each HVAC system and, under some conditions, portions of a system require a study of the conditions of operation to determine suitable materials. For example, because the static pressure of water in a high-rise building is higher in the lower levels than in the upper levels, different materials may be required along vertical zones.

The following factors should be considered when selecting material for a piping system:

- Code requirements
- Working fluid in the pipe
- Pressure and temperature of the fluid
- External environment of the pipe
- Cost of the installed system

Table 5 lists materials currently used for heating and air-conditioning piping systems. The pressure and temperature rating of each component selected must be considered; the lowest rating establishes the operating limits of the system.

PIPE WALL THICKNESS

The primary factors determining pipe wall thickness are the hoop stresses due to internal pressure and the longitudinal stresses due to pressure, weight, and other sustained loads. Detailed stress calculations are seldom required for HVAC applications because standard pipe has ample thickness to sustain the pressure and longitudinal stresses due to weight (assuming hangers are spaced in accordance with Table 9).

STRESS CALCULATIONS

Although stress calculations are seldom required, the factors involved should be understood. The main areas of concern are (1) internal pressure stresses, (2) longitudinal stresses due to pressure and weight, and (3) stresses due to expansion and contraction.

The ASME B31 standards establish a basic allowable stress S_A equal to one-fourth of the minimum tensile strength of the material. This value is adjusted, as discussed below, because of the nature of certain stresses and manufacturing processes.

Hoop stress caused by internal pressure is the major stress on pipes. As certain forming methods form a seam that may be weaker than the base material, the standard specifies a joint efficiency factor E which, multiplied by the basic allowable stress, establishes a maximum allowable stress value in tension S_E. (Table A-1 in ASME B31.9 lists values for S_E for commonly used pipe materials.) The joint efficiency factor can be significant; for example, seamless pipe has a joint efficiency factor of 1, so it can be used to the full allowable stress (one-quarter of tensile strength). In contrast, butt-welded pipe has a joint efficiency factor of 0.60, so its maximum allowable stress value must be derated ($S_E = 0.6S$).

$$t_m = \frac{PD}{2S_E} + A \qquad (1)$$

$$P = \frac{2S_E(t_m - A)}{D} \qquad (2)$$

where

t_m = minimum required wall thickness, in.
S_E = maximum allowable stress, psi
D = outside pipe diameter, in.
A = allowance for manufacturing tolerance, threading, grooving, and corrosion, in.
P = internal pressure, psi

Equations (1) and (2) determine minimum wall thickness for a given pressure and maximum pressure allowed for a given wall thickness. Both equations incorporate an allowance factor A to compensate for manufacturing tolerances, material removed in threading or grooving, and corrosion. For the seamless, butt-welded, and electric-resistance welded (ERW) pipe most commonly used in HVAC work, the standards apply a manufacturing tolerance of 12.5%. Working pressures for steel pipe, as listed in Table 2, have been calculated using a manufacturing tolerance of 12.5%; standard allowance for depth of thread, where applicable; and a corrosion allowance of 0.065 in. for pipes 2-1/2 in. and larger and 0.025 in. for pipes 2 in. and smaller. Where corrosion is known to be greater or smaller, pressure ratings can be recalculated using Equation (2). Higher pressure ratings than shown can be obtained by using ERW or seamless pipe in lieu of CW pipe 4 in. and under and seamless pipe in lieu of ERW pipe 5 in. and over due to higher joint efficiency factor, as well as by using heavier wall pipe.

Longitudinal stresses due to pressure, weight, and other sustained forces are additive, and the sum of all such stresses must not exceed the basic allowable stress S at the highest temperature at which the system will operate. Longitudinal stress due to pressure equals approximately one-half the hoop stress caused by

Table 5   Application of Pipe, Fittings, and Valves for Heating and Air Conditioning

Application	Pipe Material	Weight	Joint type	Fitting Class	Fitting Material	Temperature, °F	System Maximum Pressure at Temperature[a], psig
Recirculating Water 2 in. and smaller	Steel (CW)	Standard	Thread	125	Cast iron	250	125
	Copper, hard	Type L	Braze or silver solder[b]	—	Wrought copper	250	150
	PVC	Sch 80	Solvent	Sch 80	PVC	75	
	CPVC	Sch 80	Solvent	Sch 80	CPVC	150	
	PB	SDR-11	Heat fusion	—	PB	160	
			Insert crimp	—	Metal	160	
2.5 to 12 in.	A53 B ERW Steel	Standard	Weld	Standard	Wrought steel	250	400
			Flange	150	Wrought steel	250	250
			Flange	125	Cast iron	250	175
			Flange	250	Cast iron	250	400
			Groove	—	MI or ductile iron	230	300
	PB	SDR-11	Heat fusion		PB	160	
Steam and Condensate 2 in. and smaller	Steel (CW)	Standard[c]	Thread	125	Cast iron		90
			Thread	150	Malleable iron		90
	A53 B ERW Steel	Standard[c]	Thread	125	Cast iron		100
			Thread	150	Malleable iron		125
	A53 B ERW Steel	XS	Thread	250	Cast iron		200
			Thread	300	Malleable iron		250
2.5 to 12 in.	Steel	Standard	Weld	Standard	Wrought steel		250
			Flange	150	Wrought steel		200
			Flange	125	Cast iron		100
	A53 B ERW Steel	XS	Weld	XS	Wrought steel		700
			Flange	300	Wrought steel		500
			Flange	250	Cast iron		200
Refrigerant	Copper, hard	Type L or K	Braze	—	Wrought copper		
	A53 B SML Steel	Standard	Weld		Wrought steel		
Underground Water Through 12 in.	Copper, hard	Type K	Braze or silver solder[b]	—	Wrought copper	75	350
Through 6 in.	Ductile iron	Class 50	MJ	MJ	Cast iron	75	250
	PB	SDR 9 & 11	Heat fusion		PB	75	
		SDR 7 & 11.5	Insert crimp		Metal	75	
Potable Water, Inside Building	Copper, hard	Type L	Braze or silver solder[b]	—	Wrought copper	75	350
	Steel, galvanized	Standard	Thread	125	Galv. cast iron	75	125
				150	Galv. mall. iron	75	125
	PB	SDR-11	Heat fusion		PB	75	
			Insert crimp		Metal	75	

[a] Maximum allowable working pressures have been derated in this table. Higher system pressures can be used for lower temperatures and smaller pipe sizes. Pipe, fittings, joints, and valves must all be considered.

[b] Lead- and antimony-based solders should not be used for potable water systems. Brazing and silver solders should be employed.

[c] Extra strong pipe is recommended for all threaded condensate piping to allow for corrosion.

internal pressure, which means at least one-half the basic allowable stress is available for weight and other sustained forces. This factor is taken into account in Table 8.

Stresses due to expansion and contraction are cyclical, and because creep allows some stress relaxation, the ASME B31 standards permit designing to an allowable stress range S_A as calculated by Equation (3). Table 1 lists allowable stresses for most commonly used piping materials.

$$S_A = 1.25\,S_c + 0.25\,S_h \qquad (3)$$

where

S_A = allowable stress range, psi
S_c = allowable cold stress at coolest temperature system will experience, psi
S_h = allowable hot stress at hottest temperature system will experience, psi

PIPE-SUPPORTING ELEMENTS

Pipe-supporting elements consist of hangers, which support from above; supports, which bear load from below; and restraints, such as anchors and guides, which limit or direct movement, as well as support loads. Pipe-supporting elements withstand all static and dynamic conditions including the following:

- Weight of pipe, valves, fittings, insulation, and fluid contents, including test fluids if heavier-than-normal flow media
- Occasional loads such as ice, wind, and seismic forces
- Forces imposed by thermal expansion and contraction of pipe bends and loops
- Frictional, spring, and pressure thrust forces imposed by expansion joints in the system

- Frictional forces of guides and supports
- Other loads as might be imposed such as water hammer, vibration, and reactive forces of relief valves
- Test loads and forces

In addition, pipe-supporting elements must be evaluated in terms of stress at the point of connection to the pipe and the building structure. Stress at the point of connection to the pipe is especially important for base elbow and trunnion supports, as the limiting and controlling parameter is usually not the strength of the structural member, but the localized stress and the point of attachment to the pipe. Loads on anchors, cast-in-place inserts, and other attachments to concrete should not be more than one-fifth the ultimate strength of attachment, as determined by manufacturers' tests. In addition, all loads on the structure should be given to and coordinated with the structural engineer.

The Code for Pressure Piping (ASME B-31) establishes criteria for the design of pipe-supporting elements and the Manufacturers Standardization Society of the Valve and Fitting Industry (MSS) has established standards for the design, fabrication, selection, and installation of pipe hangers and supports based on these codes.

The MSS *Standard* SP-69 and the catalogs of many manufacturers illustrate the various hangers and components and provide information on the types to use with various pipe systems. Table 6 shows suggested pipe support spacing and Table 7 provides maximum safe loads for threaded steel rods.

The loads on most pipe-supporting elements are moderate and can be selected safely in accordance with the information presented earlier and manufacturers' catalog data; however, some loads and forces can be very high, especially in multistory buildings and for large diameter pipe, especially where expansion joints are used at high operating pressures. Consequently, a qualified engineer should design or review the design of all anchors and pipe-supporting elements, especially for the following:

- Steam systems operating above 15 psig
- Hydronic systems operating above 160 psig or 250°F
- Risers over 10 stories or 100 ft
- Systems with expansion joints, especially for pipe diameters 3 in. and greater
- Pipe sizes over 12 in. diameter
- Anchor loads greater than 10,000 lbs (10 kips)
- Moments on pipe or structure in excess of 1000 ft · lb

PIPE EXPANSION AND FLEXIBILITY

Changes in temperature cause dimensional changes in all materials. Table 8 shows the coefficients of expansion for piping materials most commonly used in HVAC systems. For systems operating at high temperatures, such as steam and hot water, the rate of expansion is high, and significant movements can occur in short runs of piping. Even though rates of expansion may be low for systems operating in the range of 40 to 100°F, such as chilled and condenser water, they can cause large movements in long runs of piping, as commonly occur in distribution systems and high-rise buildings. Therefore, in addition to design requirements for pressure, weight, and other loadings, piping systems must accommodate thermal and other movements to prevent the following:

- Failure of pipe and supports from overstress and fatigue
- Leakage of joints
- Detrimental forces and stresses in connected equipment

An unrestrained system will operate at the lowest overall stress level. Anchors and restraints are needed to support pipe weight and to protect equipment connections. The anchor forces and bowing of pipe anchored at both ends is generally too high to be

Table 6 Suggested Hanger Spacing and Rod Size for Straight Horizontal Runs
(Adapted from MSS *Standard* SP-69)

| | Feet | | | |
| | Standard Steel Pipe[a] | | Copper Tube | Rod Size |
Inches	Water	Steam	Water	Inches
1/2	7	8	5	1/4
3/4	7	9	5	1/4
1	7	9	6	1/4
1-1/2	9	12	8	3/8
2	10	13	8	3/8
2-1/2	11	14	9	3/8
3	12	15	10	3/8
4	14	17	12	1/2
6	17	21	14	1/2
8	19	24	16	5/8
10	20	26	18	3/4
12	23	30	19	7/8
14	25	32		1
16	27	35		1
18	28	37		1-1/4
20	30	39		1-1/4

[a] Spacing does not apply where span calculations are made or where concentrated loads are placed between supports such as flanges, valves, specialties, etc.

Table 7 Capacities of ASTM A36 Steel Threaded Rods

Rod Diameter, in.	Root Area of Coarse Thread, in^2	Maximum Load[a], lb
1/4	0.027	240
3/8	0.068	610
1/2	0.126	1130
5/8	0.202	1810
3/4	0.302	2710
7/8	0.419	3770
1	0.552	4960
1-1/4	0.889	8000

[a] Based on an allowable stress of 12,000 psi reduced by 25% using the root area in accordance with ASME B31.1 and MSS SP-58.

acceptable, so general practice is to *never anchor a straight run of steel pipe at both ends*. Piping systems must be allowed to expand or contract due to thermal changes. Ample flexibility can be attained by designing pipe bends and loops or supplemental devices, such as expansion joints, into the system.

End reactions transmitted to rotating equipment, such as pumps or turbines, may deform the equipment case and cause bearing misalignment, which may ultimately cause the component to fail. Consequently, manufacturers' recommendations on the allowable forces and movements that may be placed on their equipment should be followed.

PIPE BENDS AND LOOPS

Detailed stress analysis requires involved mathematical analysis and is generally performed by computer programs. However, such involved analysis is not required for most HVAC systems because the piping arrangements and temperature ranges at which they operate lend themselves to simple analysis.

L Bends

The guided cantilever beam method of evaluating L bends can be used to design L bends, Z bends, pipe loops, branch take-off connections, and some more complicated piping configurations.

Table 8 Thermal Expansion of Metal Pipe

Saturated Steam Pressure, psig	Temperature, °F	Linear Thermal Expansion, in/100 ft Carbon Steel	Type 304 Stainless Steel	Copper
	−30	−0.19	−0.30	−0.32
	−20	−0.12	−0.20	−0.21
	−10	−0.06	−0.10	−0.11
	0	0	0	0
	10	0.08	0.11	0.12
	20	0.15	0.22	0.24
−14.6	32	0.24	0.36	0.37
−14.6	40	0.30	0.45	0.45
−14.5	50	0.38	0.56	0.57
−14.4	60	0.46	0.67	0.68
−14.3	70	0.53	0.78	0.79
−14.2	80	0.61	0.90	0.90
−14.0	90	0.68	1.01	1.02
−13.7	100	0.76	1.12	1.13
−13.0	120	0.91	1.35	1.37
−11.8	140	1.06	1.57	1.59
−10.0	160	1.22	1.79	1.80
−7.2	180	1.37	2.02	2.05
−3.2	200	1.52	2.24	2.30
0	212	1.62	2.38	2.43
2.5	220	1.69	2.48	2.52
10.3	240	1.85	2.71	2.76
20.7	260	2.02	2.94	2.99
34.6	280	2.18	3.17	3.22
52.3	300	2.35	3.40	3.46
75.0	320	2.53	3.64	3.70
103.3	340	2.70	3.88	3.94
138.3	360	2.88	4.11	4.18
181.1	380	3.05	4.35	4.42
232.6	400	3.23	4.59	4.87
294.1	420	3.41	4.83	4.91
366.9	440	3.60	5.07	5.15
452.2	460	3.78	5.32	5.41
551.4	480	3.97	5.56	5.65
666.1	500	4.15	5.80	5.91
797.7	520	4.35	6.05	6.15
947.8	540	4.54	6.29	6.41
1118	560	4.74	6.54	6.64
1311	580	4.93	6.78	6.92
1528	600	5.13	7.03	7.18
1772	620	5.34	7.28	7.43
2045	640	5.54	7.53	7.69
2351	660	5.75	7.79	7.95
2693	680	5.95	8.04	8.20
3079	700	6.16	8.29	8.47
	720	6.37	8.55	8.71
	740	6.59	8.81	9.00
	760	6.80	9.07	9.26
	780	7.02	9.33	9.53
	800	7.23	9.59	9.79
	820	7.45	9.85	10.07
	840	7.67	10.12	10.31
	860	7.90	10.38	10.61
	880	8.12	10.65	10.97
	900	8.34	10.91	11.16
	920	8.56	11.18	11.42
	940	8.77	11.45	11.71
	960	8.99	11.73	11.98
	980	9.20	11.00	12.27
	1000	9.42	12.27	12.54

(The rows from temperature 32 °F through 200 °F are bracketed under the label **Vacuum**.)

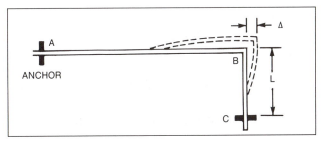

Fig. 1 Guided Cantilever Beam

Equation (4) may be used to calculate the length of leg BC needed to accommodate thermal expansion or contraction of leg AB for a guided cantilever beam (Figure 1).

$$L = \sqrt{\Delta D E / 48 S_A} \qquad (4)$$

where

L = length of leg required to accommodate thermal growth of long leg BC, ft
Δ = thermal expansion or contraction of leg AB, in.
D = actual pipe outside diameter, in.
E = modulus of elasticity, psi
S_A = allowable stress range, psi

For the commonly used A53 Grade B seamless or ERW pipe, an allowable stress S_A of 22,500 psi can be used without overstressing the pipe. However, this can result in very high end reactions and anchor forces, especially with large diameter pipe. Designing to a stress range S_A of 15,000 psi results in Equation (5), which provides reasonably low end reactions without requiring too much extra pipe. In addition, Equation (5) may be used with A53 butt-welded pipe and B88 drawn copper tubing.

For A53 continuous (butt) welded, seamless, and ERW pipe, and B88 drawn copper tubing:

$$L = 6.225 \sqrt{\Delta D} \qquad (5)$$

The guided cantilever method of designing L bends assumes no restraints; therefore, care must be taken in supporting the pipe. For horizontal L bends, it is usually necessary to place a support near point B, and any supports between points A and C must provide minimal resistance to piping movement; this is done by using slide plates or hanger rods of ample length, with hanger components selected to allow for swing of no greater than 4°.

For L bends containing both vertical and horizontal legs, any supports on the horizontal leg must be spring hangers designed to support the full weight of pipe at normal operating temperature with a maximum load variation of 25%.

The force developed in an L bend that must be sustained by anchors or connected equipment is determined by:

$$F = E_c I \Delta / 144 L^3 \qquad (6)$$

where

F = force, lb
E_c = modulus of elasticity, psi
I = moment of inertia, in⁴
L = length of offset leg, ft
Δ = deflection of offset leg, in.

In lieu of using Equation (6), for L bends designed in accordance with Equation (5) for 1 in. or more of offset, a conservative estimating value for force is 500 lb per diameter inch; *e.g.*, a 3 in. pipe would develop 1500 lb force.

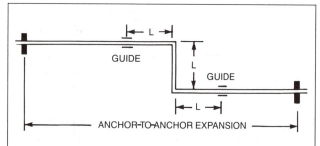

The distance from the guides, if used, to the offset should equal or exceed the length of the offset.

Offset piping must be supported with hangers, slide plates, and spring hangers similar to those for *L* bends.

Fig. 2   *Z* Bend in Pipe

Z Bends

Z bends, as shown in Figure 2, are very effective in accommodating pipe movements. A simple and conservative method of sizing *Z* bends is to design the offset leg to be 65% of the values used for an *L* bend in Equation (2), which results in Equation (7) as follows:

$$L = 4\sqrt{\Delta D} \qquad (7)$$

where

   L = length of offset leg, ft
   Δ = anchor-to-anchor expansion, in.
   D = pipe outside diameter, in.

The force developed in a *Z* bend can be calculated with acceptable accuracy by Equation (8).

$$F = 4\Delta(D/L)^2 \qquad (8)$$

where

   F = force, lb
   D = pipe outside diameter, in.
   L = length of offset leg, ft
   Δ = anchor-to-anchor expansion, in.

U Bends and Pipe Loops

Pipe loops or *U* bends are commonly used in long runs of piping. A simple method of designing pipe loops is to calculate the anchor-to-anchor expansion and, using Equation (5), determine length *L* necessary to accommodate this movement and then determine pipe loop dimensions: *W* = *L*/5 and *H* = 2*W*.

Note: Guides must be spaced no closer than twice the height of the loop, and piping between guides must be supported, as described for *L* bends, when the length of pipe between guides exceeds maximum allowable spacing for size pipe.

Table 9 provides pipe loop dimensions for pipe sizes 1 through 24 in. and anchor-to-anchor expansion (contraction) of 2 through 12 in.

No simple method has been developed to calculate pipe loop forces; however, they are generally low. A conservative estimate is 200 lb per diameter inch; *e.g.*, a 2 in. pipe will develop 400 lb force and a 12 in. pipe will develop 2400 lb force.

Cold Springing of Pipe

Cold springing or cold positioning of pipe consists of offsetting or springing the pipe in a direction opposite the expected movement. Cold springing is not recommended for most HVAC piping systems. Further, *cold springing does not permit designing a pipe bend or loop for twice the calculated movement*. For example, if a particular *L* bend can accommodate 3 in. of movement from a neutral position, cold springing does not permit the *L* bend to accommodate 6 in. of movement.

Analyzing Existing Piping Configurations

Piping systems are best analyzed by computer stress analysis programs because these provide all pertinent data including stress, movements, and loads. Services can perform such analysis if programs are not available in-house. However, many situations do not require such detailed analysis. A simple, yet satisfactory method for single and multiplane systems is to divide the system with real or hypothetical anchors into a number of single-plane units, as shown in Figure 3, which can be evaluated as *L* and *Z* bends.

EXPANSION JOINTS AND EXPANSION COMPENSATING DEVICES

Although the inherent flexibility of the piping system should be used to the maximum extent possible, expansion joints must be used where movements are too large to accommodate with pipe bends or loops or where insufficient room exists to construct a loop of adequate size. Typical situations are tunnel piping and risers in high-rise buildings, especially for steam and hot water pipes where large thermal movements are involved.

Packed and packless expansion joints and expansion compensating devices are used to accommodate movement, either axially or laterally.

In the *axial method* of accommodating movement, the expansion joint is installed between anchors in a straight line segment and accommodates axial motion only. This method has high anchor loads, primarily due to pressure thrust. It requires careful guiding, but expansion joints can be spaced conveniently to limit movement of branch connections. The axial method finds widest application for long runs without natural offsets, such as tunnel and underground piping and risers in tall buildings.

The lateral or offset method requires the device to be installed in a leg perpendicular to the expected movement and accommodates lateral movement only. This method generally has low anchor forces and minimal guide requirements. It finds widest application in lines with natural offsets, especially where there are few or no branch connections.

Packed expansion joints. These joints depend on slipping or sliding surfaces to accommodate the movement and require some type of seals or packing to seal the surfaces. Most such devices require some maintenance but are not subject to catastrophic failure. Further, with most packed expansion joint devices, any leaks that develop can be repacked under full line pressure without shutting down the system.

Packless expansion joints. These joints depend upon the flexing or distortion of the sealing element to accommodate movement. They generally do not require any maintenance, but maintenance or repair is not usually possible. If a leak occurs, the system must be shut off and drained, and the entire device must be replaced.

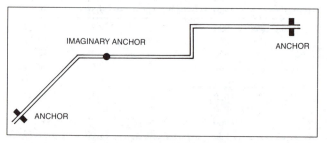

Fig. 3   Multiplane Pipe System

Table 9   Pipe Loop Design for A-53 Grade B Carbon Steel Pipe through 400°F

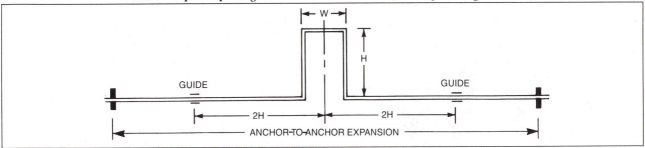

Pipe Size, in.	Anchor-to-Anchor Expansion, in.											
	2		4		6		8		10		12	
	W	H	W	H	W	H	W	H	W	H	W	H
1	2	4	3	6	3.5	7	4	8	4.5	9	5	10
2	3	6	4	8	5	10	5.5	11	6	12	7	14
3	3.5	7	5	10	6	12	6.5	13	7.5	15	8	16
4	4	8	5.5	11	6.5	13	7.5	15	8.5	17	9	18
6	5	10	6.5	13	8	16	9	18	10	20	11	22
8	5.5	11	7.5	15	9	18	10.5	21	12	24	13	26
10	6	12	8.5	17	10	20	11.5	23	13	26	14	28
12	6.5	13	9	18	11	22	12.5	25	14	28	15.5	31
14	7	14	9.5	19	11.5	23	13	26	15	30	16	32
16	7.5	15	10	20	12.5	25	14	28	16	32	17.5	35
18	8	16	11	22	13	26	15	30	17	34	18.5	37
20	8.5	17	11.5	23	14	28	16	32	18	36	19.5	39
24	9	18	12.5	25	14.5	29	17.5	35	19.5	39	21	42

W and H dimensions in feet.
  2H + W = L   L Determined from Equation (4).
  H = 2W   5W = L

Approximate force to deflect loop = 200 lb/diam. in.
Example: 8 in. pipe creates a 1600 lb force.

Further, catastrophic failure of the sealing element can occur and, although likelihood of such failure is remote, it must be considered in certain design situations.

Packed expansion joints are preferred where long-term system reliability is of prime importance (using types that can be repacked under full-line pressure) and where major leaks can be life threatening or extremely costly. Typical applications are risers, tunnels, underground pipe, and distribution piping systems. Packless expansion joints are generally used where even small leaks cannot be tolerated, such as for gas and toxic chemicals, where temperature limitations preclude use of packed expansion joints, and for very large diameter pipe where packed expansion joints cannot be constructed or cost would be excessive.

In all cases, expansion joints should be installed, anchored, and guided in accordance with expansion joint manufacturers' recommendations.

Packed Expansion Joints

There are two types of packed expansion joints—packed slip expansion joints and flexible ball joints.

Packed slip expansion joints. These are telescoping devices designed to accommodate axial movement only. Some sort of packing seals the sliding surfaces. The original packed slip expansion joint used multiple layers of braided compression packing, similar to the stuffing box commonly used with valves and pumps; this arrangement requires shutting and draining the system for maintenance and repair. Advances in design and packing technology have eliminated these problems, and most current packed slip joints use self-lubricating semiplastic packing, which can be injected under full-line pressure without shutting off the system. (Many manufacturers use asbestos-based packings, unless requested otherwise. Asbestos-free packings, such as flake graphite,

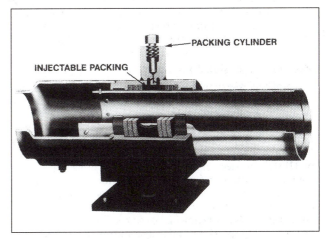

Fig. 4   Packed Slip Expansion Joint

are available and, although more expensive, should be specified in lieu of products containing asbestos.)

Standard packed slip expansion joints are constructed in sizes 1.5 to 36 in. of carbon steel with weld or flange ends for pressures to 300 psig and temperatures to 800°F. Larger sizes, higher temperature, and higher pressure designs are available. Standard single joints are generally designed for 4, 8, or 12 in. axial traverse; double joints with an intermediate anchor base can accommodate twice these movements. Special designs for greater movements are available.

Flexible ball joints. These joints are used in pairs to accommodate lateral or offset movement and must be installed in a leg perpendicular to the expected movement. The original flexible ball

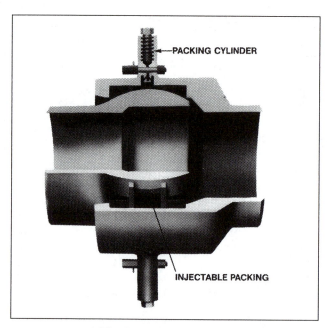

Fig. 5 Flexible Ball Joint

joint design incorporated only inner and outer containment seals that could not be serviced or replaced without removing the ball joint from the system. The packing technology of the packed slip expansion joint, as explained above, has been incorporated into the flexible ball joint design; now, packed flexible ball joints have self-lubricating semiplastic packing that can be injected under full-line pressure without shutting off the system.

Standard flexible ball joints are available in sizes 1-1/4 through 30 in. with threaded (1-1/4 to 2 in.), weld, and flange ends for pressures to 300 psig, and temperatures to 750 °F. Larger sizes, higher temperature, and higher pressure ranges are available.

Packless Expansion Joints

Metal bellows expansion joints, rubber expansion joints, and flexible hose or pipe connectors are some of the packless expansion joints available.

Metal bellows expansion joints. These expansion joints have a thin-wall convoluted section that accommodates movement by bending or flexing. The bellows material is generally Type 304, 316, or 321 stainless steel, but monel, inconel, and other materials are commonly used to satisfy service conditions. Small-diameter expansion joints in sizes 3/4 through 3 in. are generally called *expansion compensators* and are available in all bronze or steel construction. Metal bellows expansion joints can generally be designed for the pressures and temperatures commonly encountered in HVAC systems and can also be furnished in rectangular configurations for ducts and chimney connectors.

Overpressurization, improper guiding, and other forces can distort the bellows element. For low-pressure applications, such distortion can be controlled by the geometry of the convolution or thickness of the bellows material. For higher pressures, internally pressurized joints require reinforcing. Externally pressurized designs are not subject to such distortion and are not generally furnished without supplemental bellows reinforcing.

Single- and double-bellows expansion joints primarily accommodate axial movement only, similar to packed slip expansion joints. Although bellows expansion joints can accommodate some lateral movement, the *universal tied bellows expansion joint* accommodates large lateral movement. This device operates much like a pair of flexible ball joints, except bellows elements are used

instead of flexible ball elements. The tie rods on this joint contain the pressure thrust, so anchor loads are much lower than with axial-type expansion joints.

Rubber expansion joints. Similar to single-metal bellows expansion joints, rubber expansion joints incorporate a nonmetallic elastomeric bellows sealing element and generally have more stringent temperature and pressure limitations. Although rubber expansion joints can be used to accommodate expansion and contraction of the piping system, they are primarily used as flexible connectors at equipment to isolate sound and vibration and eliminate stress at equipment nozzles.

Flexible hose. This type of hose can be constructed of elastomeric material or corrugated metal with an outer braid for reinforcing and end restraint. Flexible hose is primarily used as a flexible connector at equipment to isolate sound and vibration and eliminate stress at equipment nozzles; however, flexible metal hose is well suited for use as an *offset-type expansion joint*, especially for copper tubing and branch connections off risers.

PLASTIC PIPING SYSTEMS

Nonmetallic pipe is widely used in HVAC and plumbing systems. Plastic is light in weight, generally inexpensive, and corrosion-resistant. It also has a low "C" factor, which results in lower pumping power and smaller pipe sizes. The disadvantages of plastic pipe include the rapid loss of strength at temperatures above ambient and the high coefficient of linear expansion. The modulus of elasticity of plastics is low, resulting in short support span distances. Some jurisdictions do not allow certain plastics in buildings because of toxic products emitted under fire conditions.

Plastic piping materials fall into two main categories—thermoplastic and thermosetting. Thermoplastics melt and are formed by extruding or molding. They are usually used without reinforcing filaments. Thermosets are "cured" and cannot be reformed. They are normally used with glass fiber reinforcing filaments.

For the purposes of this chapter, thermoplastic piping systems are made of the following materials:

PVC	polyvinyl chloride
CPVC	chlorinated polyvinyl chloride
PB	polybutylene
PE	polyethylene
PP	polypropylene
ABS	acrylonitrile butadiene styrene
PVDF	polyvinylidene fluoride

Thermosetting piping systems used in the HVAC industry are referred to as (1) reinforced thermosetting resin (RTR) and (2) fiberglass reinforced plastic (FRP).

The above initials are both interchangeable and refer to pipe and fittings commonly made of (1) fiberglass reinforced epoxy resin, (2) fiberglass reinforced vinyl ester, and (3) fiberglass reinforced polyester.

Since pipe and fittings made from epoxy resin are generally stronger and operate at higher temperatures than those made from polyester or vinyl ester resins, they are more likely to be used in HVAC applications.

Allowable Stresses

Both thermoplastics and thermosets have allowable stresses derived from a hydrostatic design basis stress (HDBS). The HDBS is determined by a statistical analysis of both static and cyclic stress rupture test data as set forth in ASTM *Standard* D 2837 for thermoplastics and ASTM *Standard* D 2992 for glass fiber reinforced thermosetting resins.

The allowable stress, which is called the hydrostatic design stress (HDS), is obtained by multiplying the HDBS by a service factor.

Table 10 Properties of Plastic Pipe Materials

Designation	Material Type and Grade	Cell No.	Tensile Strength, psi at 73°F	Hydrostatic[a] Design Stress, psi at 73°F Mfr.	ASME B31	Upper Temperature Limit, °F Mfr.	ASME B31	HDS[a] Upper Limit, psi	Specific Gravity	Impact Strength, ft·lb/in at 73°F	Modulus of Elasticity, psi at 73°F	Hazen-Williams C-Factor	Coefficient of Expansion per °F/10^6	Thermal Conductivity Btu·in/h·ft²·°F	Relative Pipe Cost[b]
Thermoplastics															
PVC 1120	T I,G1	1245-B	7,500	2,000	2,000	140	150	440	1.40	0.8	420,000	150	30.0	1.1	1.0
PVC 1200	T I,G2	12454-C			2,000		150				410,000		35.0		
PVC 2120	T II,G1	14333-D			2,000		150						30.0		
CPVC 4120	T IV,G1	23447-B	8,000	2,000	2,000	210	210	320	1.55	1.5	423,000	150	35.0	0.95	2.9
PB 2110	T II,G1		4,800	1,000	1,000	180	180	500	0.93		38,000	150	72.0	0.13	2.9
PE 2306	Gr. P23				630		140				90,000		80.0		
PE 3306	Gr. P34				630		160				130,000		70.0		
PE 3406	Gr. P33				630		180				150,000		60.0		
HDPE 3408	Gr. P34	355434-C	5,000	1,600	800	140	180	800	0.96	12	110,000		120.0	2.7	1.1
PP			5,000	705		212	210		0.91	1.3	120,000	150	60.0	1.3	2.9
ABS	Duraplus	6-3-3	5,500			176			1.06	8.5	240,000	150	56.0	1.7	3.4
ABS 1210	T I,G2	5-2-2		1,000	1,000		180	640			250,000		55.0		
ABS 1316	T I,G3	3-5-5		1,600	1,600		180	1,000			340,000		40.0		
ABS 2112	T II,G1	4-4-5		1,250			180	800					40.0		
PVDF			7,000	1,275		280	275	306	1.78	3.8	125,000	150	79.0	0.8	28.0
Thermosetting															
Epoxy-Glass	RTRP-11AF		44,000	8,000		210		7,000			1,000,000	150	9 to 13	2.9	
Polyester-Glass	RTRP-12EF		44,000	9,000		200		5,000			1,000,000	150	9 to 11	1.3	
For Comparison															
Steel	A 53 Grade B	ERW	60,000		12,800		800	9,200	7.80	30.0	27,500,000	100	6.31	26.2	1.3
Copper	Type L	Drawn	36,000		9,000		400	8,200	8.90		17,000,000	140	9.5		3.5

The properties listed are for the specific materials listed as each plastic has other formulations. Consult the manufacturer of the system chosen. These values are for comparative purposes.

[a] The hydrostatic design stress (HDS) is equivalent to the allowable design stress.
[b] Based on the cost of pipe only, without factoring in fittings, joints, hangers, and labor.

The HDS values recommended by some manufacturers and those allowed by the ASME/ANSI B31 Code for Pressure Piping are listed in Table 10.

The pressure design thickness for plastic pipe can be calculated using the Code stress values and the following formula:

$$t = PD/(2S + P) \qquad (9)$$

where

t = pressure design thickness, in.
P = internal design pressure, psig
D = pipe outside diameter, in.
S = design stress (HDS), psi

The minimum required wall thickness can be found by adding an allowance for mechanical strength, threading, grooving, erosion, and corrosion to the calculated pressure design thickness.

As there are many formulations of the polymers used for piping materials and different joining methods for each system, manufacturers' recommendations should be observed. Most catalogs give the pressure ratings for pipe and fittings at various temperatures up to the maximum the material will withstand.

Plastic Material Selection

The selection of a plastic for a specific purpose requires attention. All are suitable for cold water. Plastic pipe should not be used for compressed gases or compressed air if the pipe is made of a material subject to brittle failure. For other liquids and chemicals, refer to charts provided by plastic pipe manufacturers and distributors. Table 11 gives a synopsis of some applications pertinent to the HVAC industry. A brief description follows.

PVC. Since PVC has the best overall range of properties at the lowest cost, it is the most widely used plastic. It is joined by solvent cementing, threading, or flanging. Gasketed push-on joints are also used for larger sizes.

CPVC. This plastic has the same properties as PVC but can withstand higher temperatures before losing strength. It is joined by the same methods as PVC.

PB. A lightweight, flexible material, PB can be used up to 210°F. It is used for both hot and cold plumbing water piping. It is joined by heat fusion or mechanical means, can be bent to a 10 diameter radius, and is provided in coils.

PE. Low-density PE is a flexible lightweight tubing with good low-temperature properties. It is used in the food and beverage industry and for instrument tubing. It is joined by mechanical means such as compression fittings or push-on connectors and clamps.

High-density PE is a tough weather-resistant material used for large pipelines in the gas industry. Fabricated fittings are available. It is joined by heat fusion for large sizes, and flare, compression, or insert fittings can be used on small sizes.

PP. This lightweight plastic is used for chemical waste lines and also for pressure applications, as it is inert to a wide range of chemicals. A wide variety of drainage fittings are available. For pressure uses, regular fittings are made. It is joined by heat fusion.

ABS. This is a high-strength, impact- and weather-resistant material. Certain formulations can be used for compressed air, and ABS is also used in the food and beverage industry. A wide range of fittings is available. It is joined by solvent cement, threading, or flanging.

PVDF. Widely used for ultra-pure water systems and in the pharmaceutical industry, PVDF has a wide temperature range. This material is over 20 times more expensive than PVC. It is joined by heat fusion, and fittings are made for this purpose. For smaller sizes, mechanical joints can be used. Table 10 gives properties of the various plastics discussed in this section. The column labeled Cost gives the relative cost of small pipe in each category.

Table 11 Manufacturers' Recommendations for Plastic Materials

	PVC	CPVC	PB	HDPE	PP	ABS	PVDF	RTRP
Cold water service	R	R	R	R	R	R	R	R
Hot (140°F) water	N	R	R	R	R	R	R	R
Potable water service	R	R	R	R	R	R	R	R
Drain, waste, and vent	R	R	N	—	R	R	—	—
Demineralized water	R	R	—	—	R	R	R	—
Deinoized water	R	R	—	—	R	R	R	R
Salt water	R	R	R	R	R	R	—	R
Heating (200°F) hot water	N	N	N	N	N	N	—	R
Natural gas	N	N	N	R	N	N	—	—
Compressed air	N	N	—	R	N	R	—	—
Sunlight and weather resistance	N	N	N	R	—	R	R	R
Underground service	R	R	R	R	R	R	—	R
Food handling	R	R	—	—	R	R	R	R

R—Recommended
N—Not recommended
(—)—Insufficient information

Note: Before selecting a material, check the availability of a suitable range of sizes, fittings, and satisfactory joining method. Also have the manufacturer verify the best material for the purpose intended.

BIBLIOGRAPHY

ANSI. 1984. National Fuel Gas Code. ANSI *Standard* Z223.184. American National Standards Institute, New York.

ASHRAE. 1978. Safety Code for Mechanical Refrigeration. ASHRAE *Standard* 15-1978.

ASME. Boiler and Pressure Vessel Codes. American Society of Mechanical Engineers, New York.

ASME. 1989. Power piping. ASME *Standard* B31.1-89.

ASME. 1987. Refrigeration piping. ASME *Standard* B31.5-87.

ASME. 1988. Building services piping. ASME *Standard* B31.9-88.

ASME. 1985. Welded and seamless wrought steel pipe. ASME *Standard* B36.10M-85.

ASTM. 1990. Standard specification for pipe, steel, black and hot-dipped, zinc-coated welded and seamless. ASTM *Standard* A53-REV-B-90. American Society for Testing and Materials, Philadelphia, PA.

ASTM. 1990. Standard specification for seamless carbon steel pipe for high temperature service. ASTM *Standard* A106-90.

ASTM. 1989. Standard specification for seamless copper water tube. ASTM *Standard* B88-89.

ASTM. 1988. Standard specification for seamless copper tube for air conditioning and refrigeration field service. ASTM *Standard* B280-88.

ASTM. 1989. Standard specification for poly(vinyl chloride) (PVC) plastic pipe, Schedules 40, 80, and 120. ASTM *Standard* D1785-89.

ASTM. 1989. Specification for polybutylene (PB) plastic pipe (SIDR-PR) based on controlled inside diameter. ASTM *Standard* D2662-89.

ASTM. 1989. Standard specification for polybutylene (PB) plastic pipe (SIDR-PR) based on outside diameter. ASTM *Standard*.

ASTM. 1985. Specification for polybutylene (PB) plastic hot and cold-water distribution systems. ASTM *Standard* D3309-Rc.

ASTM. 1989. Standard specification for chlorinated poly (vinyl chloride) (CPVC) plastic pipe, schedules 40 and 80. ASTM *Standard* F441-89.

ASTM. 1988. Standard specification for plastic inserts fittings for polybutylene (PB) Tubing. ASTM *Standard* F845-88.

AWWA. 1986. Thickness design of ductile-iron pipe. AWWA *Standard* C150-81 (Rev. 1986). American Water Works Association, Denver, CO.

MSS. Pipe hangers and supports—materials, design and manufacturer. MSS *Standard* SP-58. Manufacturers Standardization Society of the Valve and Fitting Industry, Vienna, VA.

MSS. Pipe hangers and supports—selection and application. MSS *Standard* SP-59.

NFPA. 1987. Standard for the installation of sprinkler systems. NFPA *Standard* 13-87. National Fire Protection Association, Quincy, MA.

VALVES

FUNDAMENTALS

VALVES are the manual or automatic fluid-controlling elements in a piping system. They are constructed to withstand a specific range of temperature, pressure, corrosion, and mechanical stress. The designer selects and specifies the proper valve for the application to give the best service within the economic conditions required.

Valves have some of the following primary functions:

- Starting and stopping flow
- Regulating, controlling, or throttling flow
- Preventing backflow
- Relieving or regulating pressure

Service conditions should be considered before specifying or selecting a valve as follows:

1. Type of liquid, vapor, or gas
   - Is it a true fluid or does it contain solids?
   - Does it remain a liquid throughout its flow or vaporize as a gas?
   - Is it corrosive or erosive?
2. Pressure and temperature
   - Will these vary in the system?
   - Should worst case be considered in selecting correct valve materials?
3. Flow considerations
   - Is pressure drop critical?
   - Is the valve to be used for simple shut-off or for throttling flow?
   - Is the valve needed to prevent backflow?
4. Frequency of operation
   - Will valve be operated frequently?
   - Should valve design be chosen for maximum wear?
   - Will valve normally be open with infrequent operation?

Basic valve part nomenclature may vary from manufacturer to manufacturer and according to their applications. Figure 1 shows representative names for various valve parts.

Body Ratings

The rating of valves defines the pressure-temperature relationship within which the valve may be operated. The valve manufacturer is responsible for determining the valve rating. ANSI/ASME Standard B16.34-1988, Valves, Flanged, Threaded and Welded End, should be consulted and a valve pressure class system should be identified. Inlet pressure ratings are generally expressed in terms of the ANSI/ASME class ratings and range from ANSI Class 150 through 2500, depending on the style, size, and materials

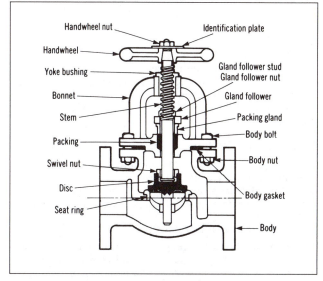

Fig. 1 Valve Components
(Courtesy Grinnell Corp.)

of construction, including seat materials. Tables in the standard show pressure ratings at various operating temperatures (Lyons 1982, Ulanski 1991, ANSI/ASME 1988).

Materials

The ANSI/ASME B16.34 Code addresses requirements for valves made from the following types of materials: forgings, castings, plate, bar stock and shapes, and tubular products. This code identifies acceptable materials from which valves can be constructed. In selecting proper valve materials, the valve body-bonnet material should be selected first and then the valve plug and seat trim.

Other factors that govern the basic materials selection include:

- Pressure-temperature ratings
- Corrosion-resistance requirements
- Thermal shock
- Piping stress
- Fire hazard

Types of materials typically available include:

- Carbon steel
- Ductile iron
- Cast iron
- Stainless steels
- Brass
- Bronze
- PVC plastic

The preparation of this chapter is assigned to TC 6.1, Hydronic and Steam Equipment and Systems.

Body materials for small valves are usually brass, bronze, or forged steel and for larger valves, cast iron, cast ductile iron, or cast steel as required for the pressure and service; typical listings are given in Lyons (1982) and Ulanski (1991).

Seats

Valve seats can be machined integrally of the body material or can be pressed or threaded (removable) in place within the valve body. Seats of different materials can be selected to suit difficult application requirements. The valve seat and the valve plug or disc are sometimes referred to as the valve trim and are usually constructed of the same material selected to meet the service requirements. The trim, however, is usually of a different material than the valve body. Removable composition disc material may be provided in the valve to provide adequate closure requirements.

Maximum permissible leakage ratings for control valve seats are defined in ANSI/Fluid Controls Institute *Standard* 70-2-1976.

Stems and Gaskets

Valve stem material should be selected to meet service conditions. Gaskets are available in different materials to seal the static surfaces of the valve flanges or the bonnet flanges. Valve manufacturers should be consulted for recommendations regarding appropriate materials available for specific fluid, temperature, and pressure.

Stem Seals and Packing

Valve stem packing and seals undergo constant wear by the movement of the valve stem exposed to the fluid and by temperature. Manufacturers can supply recommendations regarding material and lubricant for specific fluid temperature and pressure.

Flow Coefficient and Pressure Drop

Flow through any device results in some loss of pressure. Some of the factors affecting pressure loss in valves include changes in the cross section and shape of the flow path, obstructions in the flow path, and changes in direction of the flow path. For most applications, the pressure drop varies as the square of the flow when operating in the turbulent flow range. For check valves, this relationship is true only if the flow holds the valve in the full open position.

For convenience in selecting valves, particularly control valves, manufacturers express valve capacity as a function of a flow coefficient C_v. By definition in the United States, C_v is the flow of water in gallons per minute (at 60°F) that causes a pressure drop of 1 psi across a fully open valve. Manufacturers may also furnish valve coefficients at other pressure drops. Flow coefficients only apply to water. When selecting a valve to control other fluids, be sure to account for differences in viscosity.

Figure 2 shows a typical test arrangement to determine the C_v rating with the test valve wide open. HV-1 permits adjusting the supply gage reading, for example to 10 psi; HV-2 is then adjusted to 9 psi return gage to permit a test run at 1 psi pressure drop. A gravity storage tank may be used to minimize supply pressure fluctuations. The bypass valve permits fine adjustment of the supply pressure. A series of test runs are made with the weighing tank and a stopwatch to determine the flow rate. (Further capacity test detail may be found in ISA *Standard* S39.2.)

Flow coefficients only apply to water. When selecting a valve to control other fluids, be sure to account for differences in viscosity.

Cavitation

Cavitation occurs when the pressure of a flowing fluid drops below the vapor pressure of that fluid (Figure 3). In this two-step process, the pressure first drops to the critical point causing cavities of vapor to form. These are carried with the flow stream until

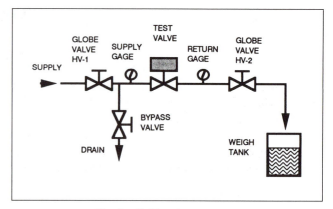

Fig. 2 C_v Test Arrangement

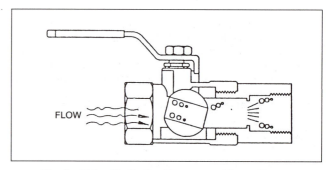

Fig. 3   Valve Cavitation Progress at Sharp Curves

reaching the second step, where an area of higher pressure is encountered. The bubbles of vapor then suddenly collapse or implode. This reduction in pressure occurs when the velocity increases as the fluid passes through a valve. After passing through the valve, the velocity decreases and the pressure increases. In many cases, cavitation manifests itself as noise. However, if the vapor bubbles are in contact with a solid surface when they collapse, the liquid rushing into the voids causes high localized pressure that can erode the surface. Premature failure of the valve and adjacent piping may occur. The noise and vibration caused by cavitation has been described as the sound of gravel flowing through the system.

Water Hammer

Water hammer is a series of pressure pulsations, of varying magnitude, above and below the normal pressure of water in the pipe. The amplitude and period of the pulsation depend on the velocity of the water as well as the size, length, and material of the pipe.

Shock loading from these pulsations occurs when any moving liquid is stopped in a short time. In general, it is important to avoid quickly closing valves in an HVAC system to minimize the occurrence of water hammer.

When flow stops, the pressure increase is independent of the working pressure of the system. For example, if water is flowing at 5 fps and a valve is instantly closed, the pressure increase is the same whether the normal pressure is 100 psig or 1000 psig.

Water hammer is often accompanied by a sound resembling a pipe being struck by a hammer—hence the name. The intensity of the sound is no measure of the magnitude of the pressure. Tests indicate that even if 15% or less of the shock pressure is removed by the addition of absorbers or arresters to eliminate the noise, adequate relief from the effect of the water hammer is not necessarily obtained.

Velocity of pressure wave and maximum water hammer pressure formulas may be found in the *Hydraulic Handbook* (1965).

Noise

Chapter 33 of the 1989 ASHRAE *Handbook—Fundamentals* points out that limitations are imposed on pipe size to control the level of pipe and valve noise, erosion, and water hammer pressure for economic reasons. One recommendation places a velocity limit of 4 ft/s for 2-in. pipe and smaller, and a pressure drop of 4 ft water/100 ft length for piping over 2-in. in diameter. Velocity-dependent noise in piping and piping systems results from any or all of four sources: turbulence, cavitation, release of entrained air, and water hammer (see Chapter 42 of the 1991 ASHRAE *Handbook—HVAC Applications*).

Some data are available for predicting hydrodynamic noise generated by control valves. ISA (1985) compiled prediction correlations in an effort to develop control valves for reduced noise levels.

Body Styles

Valve bodies are available in many configurations depending on the desired service. The usual functions are threefold—stopping the flow, allowing full flow, or controlling the flow between the two extremes. The operation of a valve can be automatic or manual.

The shapes of the bodies in both automatic and manual valves depend on the configuration of the conduit (pipe) the valve is attached to as well as the requirement to direct the streams being controlled. In both angle and straight-through valve configurations, the principle of flow is the same—the manufacturer simply provides the choice as a convenience to the installer.

Valves can be attached to pipe in one of several ways:

- Bolted to the pipe with companion flange.
- Screwed to the pipe where the pipe itself has matching threads (male) and the body of the valves with threads built into it (female).
- Welded, soldered, or sweated.
- Flared, compression, and/or various mechanical connections to the pipe where there are no threads on the pipe or the body.
- Valves of various plastic materials are fastened to the pipe if the valve body and the pipe are of compatible plastics.

Body styles can be further subdivided into types that direct flow, prevent backflow, and/or modulate flow. Generally called check valves, valves that prevent backflow are usually not operated by an external source. They are used in conjunction with screens and other filtering devices.

MANUAL VALVES

Selection

Each valve style has advantages and disadvantages for the application requirements. In some cases, inadequate information is provided in the design documents, so that the decision is based on economics and local stock availability by the installer and not on what is really required. Good submittal practice and approval by the designer is required to prevent substitutions. The questions presented in the Fundamentals section need to be carefully evaluated.

Globe Valve

In a globe valve, flow is controlled by a circular disc forced against or withdrawn from an annular ring, or seat, that surrounds an opening through which flow occurs (Figure 4). The direction of movement of the valve is parallel to the direction of the flow through the valve opening and normal to the axis of the pipe in which the valve is installed.

Globe valves are most frequently used in smaller pipes but are available in sizes of up to 12 in. They are used for throttling duty

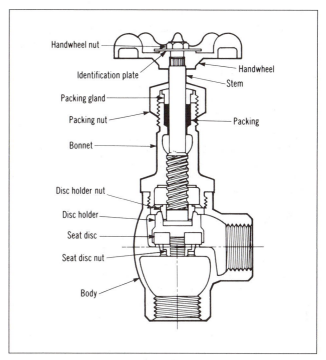

Fig. 4   Globe Valve
(Courtesy Grinnell Corp.)

where positive shut-off is required. Globe valves for controlling service should be selected by class, and whether of the straight-through or angle type, composition disc, union or gasketed bonnet, threaded, and solder or grooved ends. Manually operated flow control valves are also available with fully guided V-port throttling plugs or needle point stems for precise adjustment.

Gate Valve

A gate valve controls flow by means of a circular disc fitting against seating on machined faces (Figure 5). The straight-through opening of the valve is as large as the full bore of the pipe.

Gate valves are intended to be fully opened or completely closed. These valves are designed to permit flow or stop flow and should not be used to regulate or control flow. Various wedges available for gate valves are available for specific application. Valves located in inaccessible locations may be provided with a chain wheel or with a hammer-blow operator. More detailed information is available from valve manufacturers.

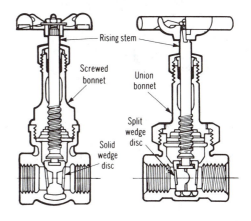

Fig. 5   Two Variations of Gate Valve

Plug Valve

The plug valve is a manual fluid flow control device (Figure 6). It operates from fully open to a complete shut-off within a 90° turn. The capacity of the valve depends on the ratio of the area of the orifice to the area of the pipe in which the valve is installed.

The cutaway view of a plug valve shows a valve with an orifice that is considerably smaller than the full size of the pipe. Lubricated plug valves are usually furnished in gas applications. A plug valve is selected as an on/off control device because (1) it is relatively inexpensive; (2) when adjusted, it holds its position definitely; and (3) its position is clearly visible to the operator. The effectiveness of this valve as a flow control device is reduced if the orifice of the valve is fully ported, *i.e.*, the same area as the pipe size.

Ball Valves

A ball valve contains a precision ball of various port sizes held between two circular seals or seats (Figure 7). A 90° turn of the handle changes operation from fully open to fully closed. Ball valves for shut-off service may be fully ported. Ball valves for throttling or controlling and/or balancing service should have a reduced port with a plated ball and valve handle memory stop. Ball valves may be of one-, two-, or three-piece body design.

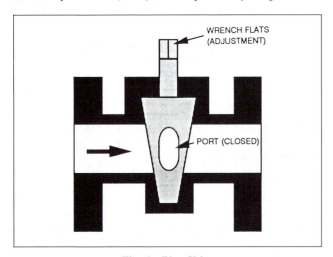

Fig. 6   Plug Valve

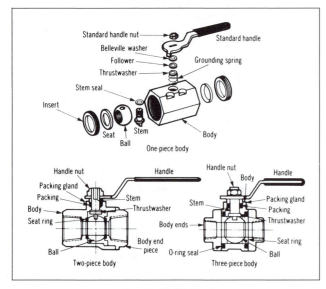

Fig. 7   Ball Valve

Butterfly Valves

Butterfly valves typically consist of a cylindrical, flanged-end body with an internal, rotatable disc serving as the fluid flow regulating device (Figure 8). Butterfly valve bodies are available in a *wafer style*, which is clamped between two companion flanges whose bolts carry the pipeline tensile stress and place the wafer body in compression, and the *lugged style*, with tapped holes in the wafer body, which may serve as a future point of disconnection. The disc's axis of rotation is the valve stem; it is perpendicular to the flow path at the center of the valve body. Only a 90° turn of the valve disc is required to change from the full open to the closed position. Butterfly valves may be operated with *hand-quadrants* (levers) for manual operation or provided with an extended shaft for automatic operation by an actuator. Special attention should be paid to manufacturers' recommendations for sizing an actuator to handle the torque requirements.

Simple and compact design, a low corresponding pressure drop, and fast operation characterize all butterfly valves. Quick operation makes them suitable for automated control, whereas the low pressure drop is suitable for high flow. Butterfly valve sizing for on/off applications should be limited to pipe sizing velocities given in Chapter 33 of the 1989 ASHRAE *Handbook—Fundamentals*; on the other hand, for throttling control applications, the valve coefficient sizing presented under automatic valves must be followed.

Pinch Valves

Two styles of pinch valve bodies are normally used—the jacket pinch and the Saunders-type bodies used for slurry control in many industries. Pressure squeezing the flexible tube jacket of a pinch valve reduces its port opening to control flow. The Saunders type employs an actuator to manually or automatically squeeze the diaphragm against a weir-type port. These valves have limited HVAC applications.

AUTOMATIC VALVES

Automatic valves are commonly considered as control valves that operate in conjunction with an automatic controller or device to control the fluid flow. Control valves can be classified as follows:

- Actuator types
- Two-way bodies (single-and double-seat)
- Three-way bodies (mixing and diverting)
- Butterfly arrangements (two- and three-way)

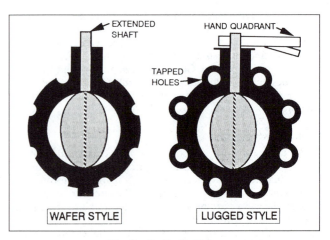

Fig. 8   Butterfly Valve

Actuators

The valve actuator converts the controller's output, such as an electric or pneumatic signal into a rotary or linear action. It changes the control variable by operating a final control element—in this case, an automatic valve. Actuators cover a wide range of sizes, types, output capabilities, and control modes.

Sizes. Actuators range in physical size from small solenoid or clock motor, self-operated radiator types no larger than a tennis ball to large pneumatic actuators.

Types. The most common types of actuators used on automatic valve applications are solenoid, thermostatic radiator, pneumatic, electric motor, electronic, and hydraulic.

Output (force) capabilities. The smallest actuators, designed for unitary commercial HVAC and residential control applications, are capable of only a very small output, while larger pneumatic or hydraulic actuators are capable of great force or thrust.

Pneumatic or diaphragm valve actuators are available with diaphragm sizes ranging from 3 to 200 in^2. The design consists of a flexible diaphragm clamped between an upper and lower housing. On direct-actuating actuators, the upper housing and diaphragm create a sealed chamber (Figure 9). A spring opposing the diaphragm force is positioned between the diaphragm and the lower housing. Increasing air pressure on the diaphragm pushes the valve stem down and overcomes the force of the load spring to close a direct-acting valve. Springs are designated by the air pressure change required to open or close the valve. A 5-lb spring requires a 5 psi control pressure change at the actuator to operate the valve . Some valves have an adjustable spring feature; others are fixed. Springs for commercial control valves usually have ± 10% tolerance, so the 5-lb spring setting is 5 psi range ± 0.5 psi.

Reverse-acting valves may use a direct-acting actuator, but they have reverse-acting valve bodies or the actuator is reverse acting and is constructed with a sealed chamber between the lower housing and the diaphragm. Simple sequencing of two valves in a control system may be accomplished with adjustable actuator springs.

The valve close-off point shifts as the supply and or the differential pressure increases across a single-seat valve due to the fixed areas of the actuator and the valve seat. The manufacturer's close-off rating tables need to be consulted to determine if the actuator is of an adequate size or if a larger actuator or a pneumatic positioner relay is required.

A pneumatic positioner relay may be added to the actuator to provide additional force to close or open an automatic control valve (Figure 9). Sometimes called positive positioners or pilot positioners, pneumatic positioners are basically high-capacity relays that add or exhaust air pressure to the actuator in relation to the stroke position of the actuator. Their application is limited to the supply air pressure available and to the actuator's spring.

Electric Actuators

Electric actuators usually consist of double-wound electric motor coupled to a gear train and an output shaft connected to the valve stem with a cam or rack-and-pinion gear linkage (Figure 10). For valve actuation, the motor shaft typically drives through 160° of rotation. Gear trains are coupled internally to the electrical actuators to provide a timed movement of valve stroke to increase operating torque, to reduce overshooting of valve movement, and can be fitted with limit switches, auxiliary potentiometers, etc. to provide position indication and feedback for additional system control functions.

In many instances, a linkage is required to convert rotary motion to the linear motion required to operate a control valve. Electric valve actuators operate with two-position, floating, proportional electric, and electronic control systems. Actuators usually operate with a 24 V(ac) low-voltage control circuit. Actuator speeds to rotate (or drive full stroke) range from 30 s to 4 min, with a 60-s actuator speed being most common.

Electric valve actuators may have a spring return, which returns the valve to a normal position in case of power failure, or it may be powered with an electric relay and auxiliary power source. Since the motor must constantly drive in one direction against the return spring, spring return electric valve actuators generally have approximately only one-third of the torque output compared to nonspring return actuators.

Electrohydraulic Actuators

Hydraulic actuators combine characteristics of electric and pneumatic actuators. In essence, hydraulic actuators consist of a sealed housing containing the hydraulic fluid, pump, and some type of metering or control apparatus to provide pressure control across a piston or piston/diaphragm. A coil controlled by a low to medium level DC voltage usually activates the pressure control apparatus.

Solenoid. A solenoid valve is an electromechanical control element that opens or closes a valve on the energization of a solenoid

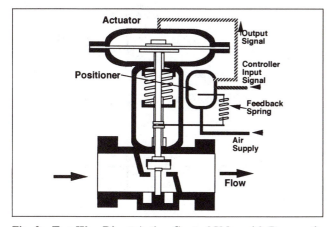

Fig. 9 Two-Way, Direct-Acting Control Valve with Pneumatic Actuator and Positioner

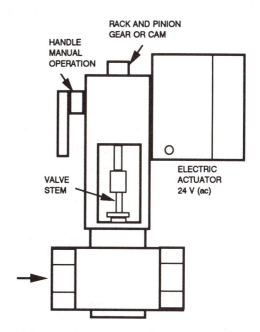

Fig. 10 Two-Way Control Valve with Electric Actuator

coil. Solenoid valves are used to control the flow of hot or chilled water and steam and range in size from 1/8 to 2 in. pipe size. Solenoid actuators themselves are two-position control devices and are available for operation from a wide range of alternating current voltages (both 50 and 60 Hz) as well as direct current. Operation of a simple two-way, direct-acting solenoid valve in an energized and deenergized state is illustrated in Figure 11.

Thermostatic Radiator

Thermostatic radiator valves are self-operated and do not require external energy. They control room or space temperature by modulating the flow of hot water or steam through free-standing radiators, convectors, or baseboard heating units. Thermostatic radiator valves are available for a variety of installation requirements with remote-mounted sensors or integral-mounted sensor and remote or an integral set point adjustment (Figure 12).

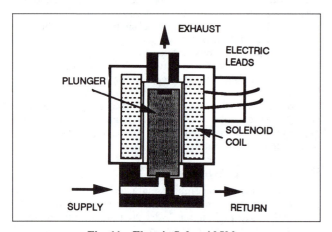

Fig. 11   Electric Solenoid Valve

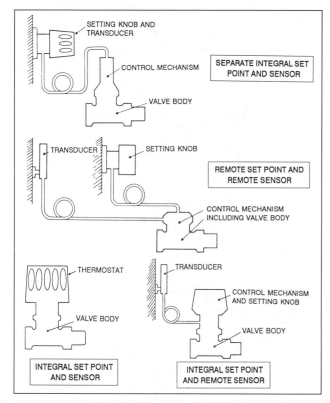

Fig. 12   Thermostatic Valves

Two-Way Valves (Single- and Double-Seat)

In two-way automatic valves, the fluid enters the inlet opening and exits the outlet port either in full volume or at a reduced volume, depending on the position of the stem and the disc as well as the seat in the valve.

Two-way valves may be single-seated or double-seated. In the case of the single-seated valve, one seat and one plug-disc close against the stream. The style of the plug-disc, however, can be quite varied depending on the requirements of the designer and the system application. For body comparison, refer to Figure 8 in Chapter 41 of the 1991 ASHRAE *Handbook—HVAC Applications*.

The double-seated valve is a special application of the two-way valve with two seats, plugs, and discs. It is generally applied to cases where the close-off pressure is too high for the single-seated valve.

Three-Way Valves

Three-way mixing valves mix two streams into one. They are usually used to mix chilled water or hot water in an HVAC system. The purpose of these valves is to control the temperature on the single stream after it leaves the mixing valve. One application would be the controlling of the air temperature in a chilled water coil downstream of the coil by mixing the chilled water going into the coil with warmer return water (Figure 13). The same could apply to a hot water coil.

Three-way valves serve different functions as they are designed to mix or divert streams of fluid. The three-way mixing valve blends two streams into one common stream based on the position of the valve plug in relation to the upper and lower seats of the valve.

On the other hand, the three-way diverting or bypass valve takes one stream of fluid and splits it into two streams for temperature control. In some limited applications, such as a cooling tower control, a diverting or bypass valve must be used in place of a mixing valve. In most cases, a mixing valve can perform the same function as a diverting or bypass valve if the piping is rearranged to suit the mixing valve. (Figure 9 in Chapter 41 of the 1991 ASHRAE *Handbook—HVAC Applications* shows typical cross sections of these valves.)

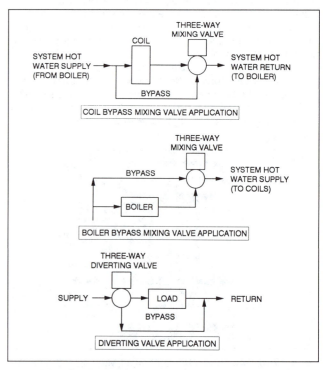

Fig. 13   Typical Three-Way Control Applications

Special Purpose Valves

Special purpose valve bodies may be used on occasion, such as the four-way valve used to allow separate circulation in the boiler loop and a heated zone. Another form of four-way valve body is used as a changeover refrigeration valve in heat pump systems to reverse the evaporator to a condenser function.

Float valves are used to supply water to a tank or reservoir or serve as a special purpose boiler feed valve to maintain an operating water level at the float level location (Figure 14).

Butterfly Valves

In some applications, it is not possible to use standard three-way mixing or standard three-way bypass valves because of size limitations or space constraints. In that case, two butterfly valves are mounted on a piping tee and cross-linked to operate as either three-way mixing or three-way bypass valves (Figure 15). Note that the flow characteristics of the butterfly valves are different from the standard seat and disc-type valves, so that their use must be limited to applications where their flow characteristics will suffice.

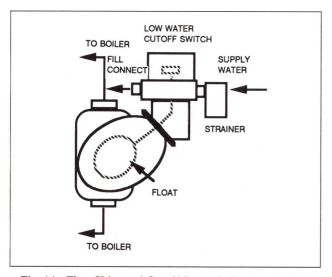

Fig. 14   Float Valve and Cutoff Steam Boiler Application

Control Valve Flow Characteristics

Generally, valves control the flow of fluids by an actuator, which moves a stem with an attached plug of various geometric shapes. The plug seats within the valve port and against the valve seat with a composition disc or metal-to-metal seating.

Based on the geometry of the plug, three distinct flow conditions can be developed as follows (Figure 16):

Quick opening. When started from the closed position, this valve allows a considerable amount of flow to pass for small stem travel. As the stem moves toward the open position, the rate at which the flow is increased per movement of the stem is reduced in a nonlinear fashion. This characteristic is used in two-position or on/off applications.

Linear. Linear valves produce equal flow increments per equal stem travel throughout the travel range of the stem. This characteristic is used on steam coil terminals and in the bypass port of three-way valves.

Equal percentage. This type of valve produces an exponential flow increase as the stem moves from the closed position to the open. The term equal percentage means that for equal increments of stem travel, the flow increases by an equal percentage for an equal stem travel. This characteristic is used on hot and chilled water terminals.

Control valves are commonly used in combination with a coil and some other valve within a circuit to be controlled. To minimize control problems and design a more efficient system, the designer should note that when these actual flow characteristics are combined with coil performance curves (heating or cooling), the resulting energy output profile of the circuit versus the stem travel alters. Taking the valve flow characteristic and the coil emission curve values, a new plot can be shown resulting in an improved percentage output emission compared to the valve position (Figure 17). For a typical hydronic heating or cooling coil, the equal percentage will result in the closest to a linear change and will provide the most efficient control (Figure 23, Chapter 12).

The three flow patterns are obtained by imposing a constant pressure drop across the modulating valve, but in actual conditions, the pressure drop across the valve is not constant and varies between a maximum (when it is controlling) and a minimum (when the valve is near full open). The ratio of these two pressure drops is known as *authority*. Figures 18 and 19 show how the valve flow characteristic for the linear and equal percentage is distorted

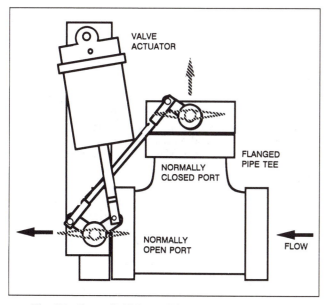

Fig. 15   Butterfly Valves—Diverting Tee Application

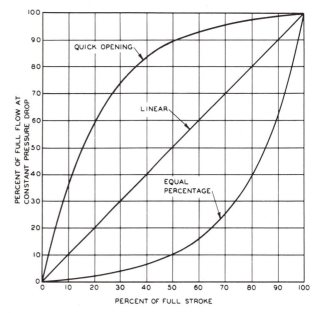

Fig. 16   Control Valve Flow Characteristics

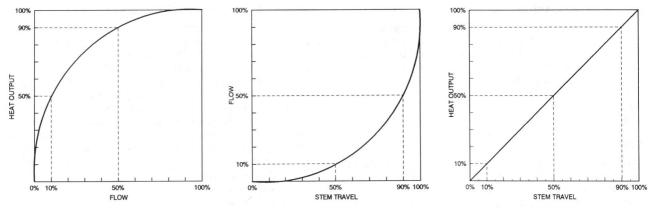

Fig. 17   Heat Output, Flow, and Stem Travel Characteristics of Equal Percentage Valve

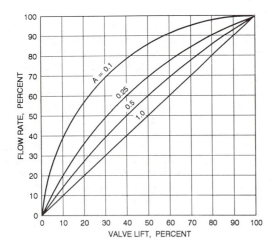

Fig. 18   Authority Distortion of Linear Flow Characteristics

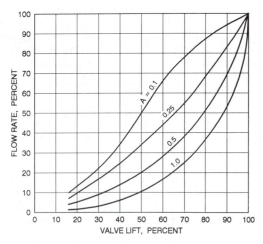

**Fig. 19   Authority Distortion of Equal Percentage
Flow Characteristic**

as the control valve authority is reduced by the choice of low valve pressure drops. The quick-opening characteristic, not shown, is distorted to the point where it approaches two-position or on/off control. The selection of the control valve pressure drop directly affects the valve authority and should be at least 25 to 50% of the system loop pressure drop, i.e., pressure drop from the pump discharge flange, supply main, supply riser, supply branch, heat transfer coil, return branch, fittings, balancing valve, and return

main to the pump suction flange. The location of the control valve in the system results in unique pressure drop selections for each control valve. The higher valve pressure drop results in a smaller valve pipe size and better control.

Control Valve Sizing

Liquids. A valve creates fluid resistance in a circuit to limit the medium flow at a calculated value. Each passive element in a circuit creates a pressure drop according to the following general equation:

$$\Delta p = RQ^n \, (\rho/\rho_w) \qquad (1)$$

where

Δp = pressure drop, psi
R = resistance
ρ = fluid density, lb/ft^3
ρ_w = density of water at 60°F, lb/ft^3
Q = flow, gpm

For turbulent flows, the coefficient n is assumed to be 2, although for steel pipes $n = 1.85$.

For a valve, $n = 2$, and solving Equation (1) for flow:

$$Q = \sqrt{(\Delta p/R) \, (\rho_w/\rho)} \qquad (2)$$

If $\sqrt{1/R}$ is replaced by the valve coefficient C_v and $\rho/\rho_w = 1$ for water for temperatures below 250°F, Equation (2) becomes:

$$Q = C_v \sqrt{\Delta p} \qquad (3)$$

or:

$$Q = 0.67 C_v \sqrt{\Delta H} \qquad (4)$$

where ΔH = pressure drop, ft of water

The control valve size should be selected by calculating the required C_v to provide the design flow at an assumed pressure drop Δp. A pressure drop of 25 to 50% of the available pressure between the supply and return riser (pump head) should be selected for the control valve. This pressure drop gives the best flow characteristic as described in the Control Valve Flow Characteristics section.

For liquids with a viscosity correction factor, Equation (2) becomes:

$$Q = (C_v/V_f)\sqrt{\Delta p \, \rho_w/\rho} \qquad (5)$$

where V_f is the viscosity correction factor.

For steam flow:

$$w_s = 2.1(C_v/K)\sqrt{\Delta p(P_1 + P_2)} \qquad (6)$$

where

w_s = steam flow, lb/h
K = 1 + 0.0007 (degrees superheat)
C_v = flow coefficient, gpm at Δp = 1 psi
P_1 = entering steam absolute pressure
P_2 = leaving steam absolute pressure
Δp = steam pressure drop across the valve ($P_1 - P_2$)

Note: Some manufacturer's list the constant in Equation (6) as high as 3.2, but most agree on a constant of 2.1. Always confirm valve sizing with the manufacturer as part of good practice.

Steam reaches critical or sonic velocity when the downstream pressure is 58% or less of the absolute inlet pressure. Increasing the pressure drop below the critical pressure produces no further increase in flow. As a result, when $P_2 \leqslant 0.58 P_1$, the following critical pressure drop formula is used:

$$C_v = w_s / 1.61 P_1 \qquad (7)$$

Applications

Automatically controlled valves are applied to control many different variables. The most common examples are temperature, humidity, flow, and pressure. However, a valve can be used directly only to control flow or pressure. When flow is controlled, a pressure drop is implied, and when pressure is controlled, some maximum flow rate is implied. These two factors must be considered in selecting control valves. For a general discussion of control theory and some typical valve applications, refer to Chapter 41 of the 1991 ASHRAE *Handbook—HVAC Applications.*

While the discussion in this chapter applies to hot water, chilled water, and steam, control valves can be used with virtually any fluid. The fluid characteristics must be considered in selecting materials for the valve. The requirements are particularly strict for use with high-temperature water and high-pressure steam.

Steam is controlled in two ways:

1. When steam pressure is too high for use in a specific application, the pressure must be reduced by a pressure-reducing valve (PRV). This is normally a globe-type valve, since modulating control is required. The valve may be externally or internally piloted and is usually self-contained, using the steam pressure to drive the actuator. Since the load may vary, it is sometimes desirable to use two or more valves in parallel, adjusted to open in sequence, for more accurate control.
2. Steam flow to a heat exchanger may be controlled in response to temperature or humidity requirements. In this case, an external control system is used with the steam valve as the controlled device. In selecting a steam valve, the maximum flow rate for the specific valve and entering steam pressure must be considered. These factors are determined from the critical pressure drop, which limits the flow.

Hot and chilled water are usually controlled in response to temperature or humidity requirements. When selecting a valve for controlling water flow, a pressure drop sufficiently large to allow the valve to control properly should be specified. The response of the heat exchanger coil to a change in flow is not linear; therefore, an equal percentage plug should be used, and the temperature of the water supply should be as high (hot water) or as low (chilled water) as required by the load conditions.

BALANCING VALVES

Two approaches are available for balancing hydronic systems—a manual valve with integral pressure taps and a calibrated port, which permits field proportional balancing to the design flow conditions; or an automatic flow-limiting valve selected to limit the circuit's maximum flow to the design flow.

Manual balancing valves can be provided with the following features:

- Manually adjustable stems for valve port opening or a combination of a venturi or orifice and an adjustable valve
- Stem indicator and/or scale to indicate relative amount of valve opening
- Pressure taps to provide a readout of the pressure difference across the valve port or the venturi/orifice
- Capability to be used as a shut-off for future service of the heat transfer terminal
- Locking device for field setting of the maximum opening of a valve
- Body capability for integral drain

Manual balancing valves may have rotary, rising, or nonrising stems for port adjustment (Figure 20).

Meters with various scale ranges, field carrying case, attachment hoses, and fittings for connecting to the manual balancing valve should be used to determine its flow by reading its differential pressure. Some meters employ analog measuring elements with direct reading mechanical dual-element bourdon tubes, and others are electronic differential pressure transducers with a digital data display.

Many manufacturers of balancing valves produce circular slide rule calculators to calculate circuit flow based on pressure difference readout across the balancing valve, its stem position, and/or the valve's flow coefficient. This calculator can also be used for determining the valve size selection and its setting when the terminal design flow conditions are known.

Automatic Flow-Limiting Valves

Differential pressure-actuated flow control valves, also called automatic flow-limiting valves (Figure 21), regulate the flow of fluid to a preset value when the differential pressure across them

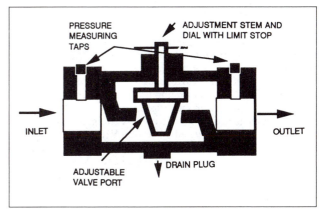

Fig. 20   Manual Balancing Valve

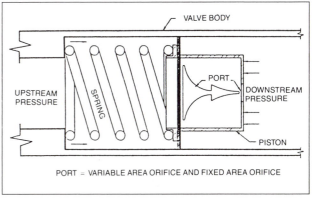

Fig. 21   Automatic Flow-Limiting Valve

is varied. This regulation helps (1) prevent an overflow condition in the circuit where it is installed, and (2) the overall system balance when other components are changing (modulating valves, pump staging, etc.).

Typically, the valve body contains a moving element containing an orifice, which, based on pressure forces, adjusts itself so that the flow passage area varies.

Change in the area of an orifice is obtained by (1) a piston or cup moving across a shear plate, and (2) increased pressure (squeeze rubber).

A typical performance curve for the valve is shown in Figure 22. The flow rate for the valve is set. The flow curve is divided in three ranges of differential pressure, which include (1) start-up range, (2) control range, and (3) above control range.

Balancing Valve Selection

The balancing valve is a flow control device that is selected for a lower pressure drop than an automatic control valve (5 to 10% of the available system pressure). Selection of any control valve is based on the pressure drop at maximum (design) flow to ensure that the valve provides control at all flow rates. A properly selected balancing valve has the capacity to proportionally balance flow to its terminal with flow to the adjacent terminal in the same distribution zone. Refer to Chapter 34 of the 1991 ASHRAE *Handbook—HVAC Applications* for balancing details.

PRESSURE RELIEF VALVES

Ordinarily, the terms safety valve and relief valve are used interchangeably. Both valves are similar in external appearance and both generally do the same job, *i.e.*, limit the pressure of a liquid, vapor, or gas by discharging some of the excess.

Safety valves are normally used for air and steam service. They are characterized by their quick opening (popping) and closing action. Relief valves, on the other hand, are generally used for liquid service and usually operate with a slow opening and slow closing action as the pressure increases and decreases.

Safety valve construction, capacities, limitations, operation, and repair are covered by ASME codes. For pressures above 15 psig, refer to ASME Section I for applicable codes. Section IV covers steam boilers for pressures not exceeding 15 psig. Unfired pressure vessels (such as heat exchange process equipment or pressure-reducing valves) are covered by Section VIII.

The capacity of a safety valve is affected by the equipment on which it is installed and the applicable code. Valves are chosen based on accumulation, which is the pressure increase above the maximum allowable working pressure of the vessel during valve discharge. Section I valves are based on 3% accumulation. Accumulation may be as high as 33.3% for Section IV valves and 10% for Section VIII. To properly size a safety valve, the required capacity and set pressure must be known. On a pressure-reducing

valve station, the safety valve must have sufficient capacity to prevent an unsafe pressure rise should the reducing valve fail in the open position.

The safety valve set pressure should be high enough to allow the valve to remain closed during normal operation, yet allow it to open and reseat tightly when cycling. A minimum differential of 5 psi or 10% of inlet pressure (whichever is greater) is recommended.

When installing a safety valve, consider the following points:

- Install the valve vertically with the drain holes open or piped to drain.
- The seat can be distorted if the valve is overtight or the weight of the discharge piping is carried by the valve body. A drip pan elbow on the discharge of the safety valve will prevent the weight of the discharge piping from resting on the valve (Figure 23).
- Use a moderate amount of pipe thread lubricant (first 2 to 3 threads) on male threads only.
- Install clean flange connections with new gaskets, properly aligned and parallel, and bolted with even torque to prevent distortion.
- Wire cable or chain pulls attached to the test levers should allow for a vertical pull and their weight should not be carried by the valve.

Testing of safety valves varies between facilities depending on operating conditions. Under normal conditions, safety valves with a working pressure under 400 psig should be tested manually once a month and pressure tested once each year. For higher pressures, the test frequency should be based on operating experience.

When steam safety valves require repair, adjustment, or set pressure change, the manufacturer or approved stations holding the ASME V, UV, and/or VR stamps must perform the work. Only the manufacturer is allowed to repair Section IV valves.

SELF-CONTAINED TEMPERATURE CONTROL VALVES

Self-contained or self-operated temperature control valves do not require an outside energy source such as compressed air or

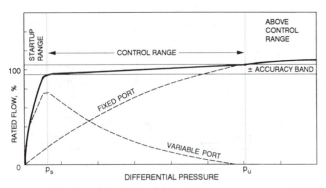

Fig. 22   Automatic Flow Valve Curve

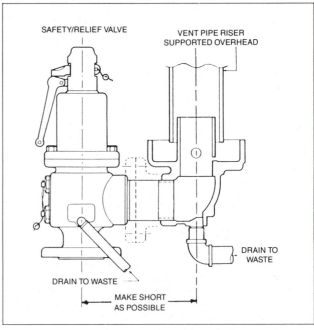

Fig. 23   Safety/Relief Valve with Drip-Pan Elbow

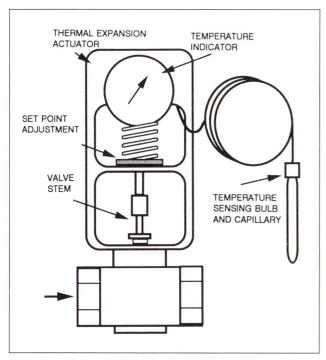

Fig. 24 Self-Operated Temperature Control Valve

electricity (Figure 24). They depend on a temperature-sensing bulb and capillary tube filled with either an oil or a volatile liquid. In an oil-filled system, the oil expands as the sensing bulb is heated. This expansion is transmitted through the capillary tube to an actuator bellows in the valve top, which causes the valve to close. The valve opens as the sensing bulb cools, and the oil contracts; a spring provides a return force on the valve stem.

A volatile liquid control system is known as a vapor pressure or vapor tension system. When the sensing bulb is warmed, some of the volatile liquid vaporizes, causing an increase in the sealed system pressure. The pressure rise is transmitted through the capillary tube to expand the bellows, which then moves the valve stem and closes the valve. Thermal systems actuate the control valve either directly or through a pilot valve. In a direct actuated design, the control directly moves the valve stem and plug to close or open the valve. These valves must compensate for the steam pressure force acting on the valve seat by generating a greater force in the bellows to close the valve. An adjustable spring adjusts the temperature set point and provides the return force to move the valve stem upward as the temperature decreases.

For self-contained temperature valves to operate as proportional controls, the bulb must sense a change in the temperature of the process fluid. The difference in temperature from a no-load to a maximum controllable load is known as the proportional band (P-Band). Since the size of this P-Band can be varied depending on valve size, the accuracy is variable. Depending on the application, P-Bands of 2 to 18°F may be selected, as shown in the following table:

Application	Proportional Band, °F
Domestic hot water heat exchanger	6 to 14
Central hot water	4 to 7
Space heating	2 to 5
Bulk storage	4 to 18

While the response time, accuracy, and ease of adjustment may not be as good as electrically or pneumatically actuated valves, self-contained steam temperature controls have been widely accepted for many applications.

A pilot-actuated valve (Figure 25) uses a much smaller intermediate pilot valve that controls a flow of steam to a large diaphragm that then acts on the valve stem. This allows the control system to work against high steam pressures due to the smaller area of the pilot valve.

PRESSURE-REDUCING VALVES

Should steam pressure be too high for a specific process, a self-contained pressure-reducing valve may be used to reduce this pressure, which will also increase the available latent heat. These valves may be direct-acting or pilot-operated (Figure 25), much like the temperature control valves previously discussed. To maintain set pressure, the downstream pressure must be sensed either through an internal port or an external line.

The amount of pressure drop below the set pressure that causes the valve to react to a load change is called *droop*. As a general rule, pilot-operated valves have less droop than direct-acting types. To properly size these valves, only the mass flow of steam, the inlet pressure, and the required outlet pressure must be known. Determine the valve line size by consulting manufacturers' capacity charts.

Due to their construction, simplicity, accuracy, and ease of installation and maintenance, these valves have been specified for most steam-reducing stations.

CHECK VALVES

Check valves prevent reversal of flow, controlling the direction of flow rather than stopping or starting flow. Some basic types include the swing check, ball check, wafer check, silent check, and stop check valves. Most check valves are available in screwed and flanged body styles.

Swing check valves have a valve body similar to gate valves or Y-pattern valves with hinge-mounted discs that open and close with flow (Figure 26). The seats are generally made of metal, while

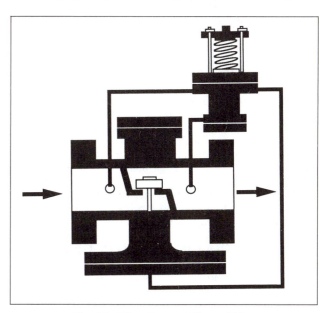

Fig. 25 Pilot-Operated Steam Valve

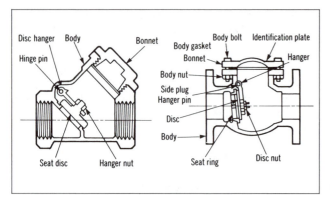

Fig. 26   Swing-Type Check Valves
(Courtesy Grinnell Corp.)

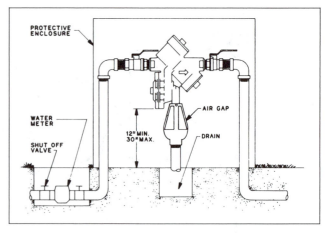

Fig. 27   Backflow Prevention Valve

the discs may be of metallic or nonmetallic composition materials. Nonmetallic discs are recommended for fluids containing dirt particles, or tighter shut-off is required. The Y-pattern check valve has an access opening to allow cleaning and regrinding in place. Pressure drop through swing check valves is lower than that through lift check valves due to the straight-through design. Weight or spring-loaded lever arm check valves are available to limit objectional slamming or chattering when pulsating flows are encountered.

Lift check valves have a body similar in design to a globe or angle valve body with a similar disc seating. The guided valve disc is forced open by the flow and closes when flow reverses. Due to the body design, the pressure drop is higher than that of a swing check valve. Lift check valves are recommended for gas or compressed air or in fluid systems not having critical pressure drops.

Ball check valves are similar to lift checks, except that the design uses a ball, rather than a disc to accomplish closure. Some ball checks are specifically designed for horizontal flow or vertical upflow installation.

Wafer design check valves are designed to fit between pipe flanges similar to butterfly valves and are used in larger sized piping (4 in. and larger). Wafer check valves have two basic designs—dual spring-loaded flapper, which operates on a hinged center post, and single flapper design similar to the swing check valve.

In silent or spring-loaded check valves, a spring positively and rapidly closes a guided, floating disc. This valve greatly reduces water hammer, which may occur with slow-closing check valves (like the swing check). Silent check valves are recommended for use in pump discharge lines.

STOP AND CHECK VALVES

Stop-check valves can operate both as a check valve and a stop valve. The valve stem does not connect to the guided seat plug, allowing the plug to operate as a conventional lift check valve when the stem is in the raised position. Screwing the stem down can limit the valve opening or close the valve. Stop-check valves are used for shut-off service on multiple steam boiler installations, in accordance with the ASME Boiler and Pressure Vessel Code, to prevent backflow of steam or condensate from an operating boiler to a shut-down boiler. They are mandatory in some jurisdictions. Local codes should be checked for compliance.

BACKFLOW-PREVENTION DEVICES

Backflow-prevention devices prevent reverse flow of the supply in a water system. A vacuum breaker prevents back siphonage in a nonpressure system, while a backflow preventer prevents backflow in a pressurized system (Figure 27).

Selection of Devices

Vacuum breakers and backflow preventers should be selected on the basis of the local plumbing codes, the water supply impurities involved and the type of cross-connection.

Impurities are classified as (1) contaminants, *i.e.*, substances that could create a health hazard if introduced into potable water, and (2) pollutants, which could create an objectionable condition but not a health hazard.

Cross-connections are classified as nonpressure or pressure connections. In a nonpressure cross-connection, a potable water pipe connects or extends below the overflow or rim of a receptacle at atmospheric pressure. When this type of connection is not protected by a minimum air gap, it should be protected by an appropriate vacuum breaker or an appropriate backflow preventer.

In a pressure cross-connection, a potable water pipe is connected to a closed vessel or a piping system that is above atmospheric pressure and contains a nonpotable fluid. This connection should be protected by an appropriate backflow preventer only. Note that a pressure vacuum breaker should not be used alone with a pressure-type cross-connection.

Vacuum breakers should be corrosion-resistant. Backflow preventers, including accessories, components, and fittings 2 in. and smaller should be made of bronze with threaded connections. Sizes larger than 2 in. should be made of bronze, galvanized iron, or fused epoxy-coated iron inside and out, with flanged connections. All backflow-prevention devices should meet applicable standards of the American National Standards Institute, the Canadian Standards Association, or the required local authorities.

Installation of Devices

Vacuum breakers and backflow preventers equipped with atmospheric vents, or with relief openings, should be installed and located to prevent any vent or any relief opening from being submerged. They should be installed in the position recommended by the manufacturer.

Backflow preventers may be double check valve (DCV) or reduced pressure zone (RPZ) types. Refer to manufacturers' information for specific application recommendations and code compliance.

STEAM TRAPS

For a description and diagram of these traps, refer to Chapter 10.

REFERENCES

ANSI/ASHRAE. Guideline 1-1989, Guideline for commissioning of HVAC systems.

ANSI/ASHRAE, *Standard* 111/88, Practices for measurement, testing, adjusting and balancing of building heating, ventilation, air conditioning and refrigeration systems.

ASME. 1986. Boiler and pressure vessel code.

ASME/ANSI. 1988. *Standard* B16.34-1988.

Fairbanks Morse Pump Company. 1965. *Hydraulic handbook*. Catalog C and #65-26313.

Heating, Piping and Air Conditioning. 1990. Piping from the beginning (October):51-70.

Heating, Piping and Air Conditioning. 1977. Valve application II: Plug valves (May):111-17.

Heating, Piping and Air Conditioning. 1977. Valve application V: Ball valves (November):97-102.

Heating, Piping and Air Conditioning. 1978. Valve application VI: Butterfly valves (January):137-42.

Heating, Piping and Air Conditioning. 1978. Valve application IX: Valve standards (October):99-107.

IEC. Industrial control valves, Part I: General considerations. IEC *Publication* 534-1. International Electrotechnical Commission, Geneva, Switzerland.

ISA. 1985. *Standard* S75.01-85. Instrument Society of America.

Lyons, J.L. 1982. *Lyon's valve designer's handbook*. ISBN 0-442-24963-2, 92-93, 209-10.

Matley, J. 1989. Valves for process control, 132-33.

McGraw Hill. 1989. *McGraw-Hill dictionary of scientific and technical terms*, 4th ed., 272.

Piping handbook. 1967. Reno C. King, ed. Mc-Graw Hill, New York, 7-9.

Ulanski, W. 1991. Valve and actuator technology, ISBN 0-07-019477-7.

Zappe, R.W. 1987. *Valve selection handbook*, 46.

BIBLIOGRAPHY

CIBSE. 1985. *Applications manual—Automatic controls and their implications for system design.*

CIBSE. 1986. Guide, Volume B, Installation and equipment data. The Chartered Institution of Building Services Engineers, London.

Crane Co. Flow of fluids through valves, fittings and pipe. Technical Paper No. 410. Crane Co., King of Prussia, PA.

Driskell, L. 1982. Control valve sizing. Instrument Society of America, ISBN 0-87664-620-8.

Driskell, L. 1982. Selection of control valves. Instrument Society of America, ISBN 0-87664-618-6.

Gupton, G. 1987. HVAC Controls, operation and maintenance

Haines, R.W. 1983. Control systems for HVAC.

Hutchinson, J.W. ISA *Handbook of control valves*, 2nd ed. Instrument Society of America, ISBN 0-87664-234-2.

Matley, J. and Chemical Engineering. 1989. Valves for process control and safety, ISBN-0-07-607010-7.

McQuiston, F.C. and J.D. Parker. 1988. *Heating, ventilating and air conditioning, Analysis and design*, ISBN 0-471-63757-2.

Merrick, R.C. 1991. *Valve selection and specification guide*, ISBN 0-442-31870-7.

Miller, R.W. 1983. *Flow measurement engineering handbook*, ISBN 0-07-042045-9.

National Board of Boiler and Pressure Vessel Inspection Code, Publication NB-23. 1989.

National Joint Steamfitter-Pipefitter Apprenticeship Committee. 1976. *Start, test and balance manual.*

Potter, P.J. 1949. Steam power plants.

Zappe, R.W. 1987. *Valve selection handbook*, 2nd ed., ISBN 0-87201-918-7.

CHAPTER 44

AIR-TO-AIR ENERGY RECOVERY

AIR-TO-AIR energy recovery systems may be categorized according to their application as (1) process-to-process, (2) process-to-comfort, and (3) comfort-to-comfort. Typical air-to-air energy recovery applications are listed in Table 1.

Table 1   Applications for Air-to-Air Energy Recovery

Method	Typical Application
Process-to-process and Process-to-comfort	Driers
	Ovens
	Flue stacks
	Burners
	Furnaces
	Incinerators
	Paint exhaust
Comfort-to-comfort	Welding
	Swimming pools
	Locker rooms
	Residential
	Smoking exhaust
	Operating rooms
	Nursing homes
	Animal ventilation
	Plant ventilation
	General exhaust

APPLICATIONS

Process-to-Process

In process-to-process applications, heat is captured from the process exhaust stream and transferred to the process supply airstream. Equipment is available to handle process exhaust temperatures as high as 1600 °F.

Process-to-process recovery devices generally recover only sensible heat and do not transfer latent heat (humidity), as moisture transfer is usually detrimental to the process. Process-to-process applications usually recover the maximum amount of energy. In cases involving condensables, less recovery may be desired to prevent condensation and possible corrosion.

Process-to-Comfort

In process-to-comfort applications, waste heat captured from a process exhaust heats the building makeup air during winter. Typical applications include foundries, strip coating plants, can plants, plating operations, pulp and paper plants, and other processing areas with heated process exhaust and large makeup air volume requirements.

Although full recovery is desired in process-to-process applications, recovery for process-to-comfort applications must be modu-

lated during warm weather to prevent overheating the makeup air. During summer, no recovery is required. Because energy is saved only in the winter and recovery is modulated during moderate weather, process-to-comfort applications save less energy over a year than do process-to-process applications.

Process-to-comfort recovery devices generally recover sensible heat only and do not transfer moisture between the airstreams.

Comfort-to-Comfort

In comfort-to-comfort applications, the heat recovery device lowers the enthalpy of the building supply air during warm weather and raises it during cold weather by transferring energy between the ventilation air supply and the exhaust airstreams.

In addition to commercial and industrial energy recovery equipment, small-scale packaged ventilators with built-in heat recovery components known as heat recovery ventilators (HRV) are available for residential and small-scale commercial applications.

Air-to-air energy recovery devices available for comfort-to-comfort applications may be sensible heat devices (i.e., transferring sensible energy only) or total heat devices (i.e., transferring both sensible energy and moisture). These devices are discussed further in the section Technical Considerations.

ECONOMIC CONSIDERATIONS

An analysis of energy recovery should consider the application over its lifetime. Neither the most efficient nor the least expensive energy recovery device may be the most economical. Many manufacturers and suppliers have computer programs to provide application-specific design and cost benefit information. Chapter 33 of the 1991 ASHRAE *Handbook—HVAC Applications* describes methods for making detailed cost/benefit analyses.

Energy costs. The absolute cost of energy and the relative costs of various energy forms are major economic factors. High energy costs favor high levels of energy recovery. In regions where electrical costs are high relative to fuel prices, heat recovery devices with low pressure drops are preferable.

Other conservation options. Energy recovery should be evaluated against other cost-saving opportunities, including reducing or eliminating the primary source of waste energy through process modification.

Amount of useable waste energy. Economies of scale favor large installations, although equipment is commercially available for air-to-air energy recovery applications from 50 cfm and more. Although using equipment with higher effectiveness results in more recovered energy, equipment costs and space requirements also increase with effectiveness.

Grade of waste energy. High-grade (i.e., high-temperature) waste energy is generally more economical to recover than low-grade energy. Large temperature differences between the waste energy source and destination are most economical.

The preparation of this chapter is assigned to TC 5.5, Air-to-Air Energy Recovery.

Coincidence and duration of waste heat supply and demand. Energy recovery is most economical when the supply is coincident with the demand and both are relatively constant throughout the year. Thermal storage may be used to store energy if supply and demand are not coincident, but this adds cost and complexity to the system.

Proximity of supply to demand. Applications with a large central energy source and a nearby waste energy use are more favorable than applications with several scattered waste energy sources and uses.

Operating environment. High operating temperatures or the presence of corrosives, condensables, and particulates in either airstream result in higher equipment and maintenance costs. Increased equipment costs result from the use of corrosion- or temperature-resistant materials, and maintenance costs are incurred by an increase in the frequency of equipment repair and washdown and additional air filtration requirements.

Effects on pollution control systems. Removing process heat may reduce the cost of pollution control systems by allowing less expensive filter bags to be used, by improving the efficiency of electronic precipitators, or by condensing out contaminant vapors, thus reducing the load on downstream pollution control systems. In some applications, recovered condensables may be returned to the process for reuse.

Effects on heating and cooling equipment. Heat recovery equipment may reduce the size requirements for primary utility equipment such as boilers, chillers, and burners, and the size of piping and electrical services to them. Larger fans and fan motors (and hence fan energy) are generally required to overcome increased static pressure losses caused by the energy recovery devices. Auxiliary heaters may be required for frost control.

Effects on humidifying or dehumidifying equipment. Selecting total energy recovery equipment results in the transfer of moisture from the airstream with the greater humidity ratio to the airstream with the lesser humidity ratio. In many situations this is desirable, since humidification costs are reduced in cold weather and dehumidification loads are reduced in warm weather (ASHRAE 1988).

TECHNICAL CONSIDERATIONS

Performance Rating

ASHRAE *Standard* 84-91, Method of Testing Air-to-Air Heat Exchangers, establishes rating and testing proceedures for commercial air-to-air heat recovery equipment. CAN/CSA-439-88, Standard Methods of Test for Rating the Performance of Heat Recovery Ventilators (HRV), is used to rate small (under 200 L/s) packaged ventilators with heat recovery.

The effectiveness of air-to-air heat exchangers is commonly measured in terms of:

- Sensible energy transfer (dry-bulb temperature)
- Latent energy transfer (humidity ratio)
- Total energy transfer (enthalpy)

ASHRAE *Standard* 84 defines effectiveness as:

$$\epsilon = \frac{\text{Actual transfer (of energy or moisture)}}{\text{Maximum possible transfer between airstreams}}$$

Referring to Figure 1,

$$\epsilon = \frac{W_s(X_1 - X_2)}{W_{min}(X_1 - X_3)} = \frac{W_e(X_4 - X_3)}{W_{min}(X_1 - X_3)} \qquad (1)$$

where

ϵ = sensible, latent, or total effectiveness

X = dry-bulb temperature, humidity ratio, or enthalpy (at the location indicated in Figure 1), respectively, and

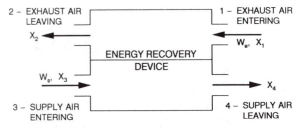

Fig. 1   Airstream Numbering Convention

W_{min} = the smaller of W_s and W_e where,

For latent and total effectiveness:
W_s = supply air mass flow
W_e = exhaust air mass flow

For sensible effectiveness:
W_s = (specific heat) × (supply air mass flow rate)
W_e = (specific heat) × (exhaust air mass flow rate)

The leaving supply air condition is:

$$X_2 = X_1 - \epsilon (W_{min}/W_s)(X_1 - X_3) \qquad (2)$$

and the leaving exhaust air condition is:

$$X_4 = X_3 + \epsilon (W_{min}/W_e)(X_1 - X_3) \qquad (3)$$

Equations (1), (2), and (3) assume that no heat transfers between the heat exchanger and its surroundings, nor are there gains from cross leakage, fans, or frost control devices. This assumption is generally true for larger commercial applications but not for HRVs. The rating term used in CAN/CSA-439-88 for HRVs is defined as the energy recovery efficiency (*i.e.*, the actual energy transfer efficiency) and the apparent sensible effectiveness (*i.e.*, a measure of the temperature rise of the supply airstream, including that resulting from external gains).

A number of variables can affect these performance factors, whether the device is designed to transfer total energy or just sensible heat. These variables include (1) humidity ratio of the warmer airstream, (2) heat transfer area, (3) air velocities through the heat exchangers, (4) airflow arrangement, (5) supply and exhaust air mass flow rates, and (6) method of frost control. The effect of some of these are shown in Figures 2, 3, and 4. The impacts of frost control method on seasonal performance are discussed in Phillips *et al.* (1989a), and sensible versus latent heat recovery for residential comfort-to-comfort applications is addressed in Barringer and McGugan (1989b).

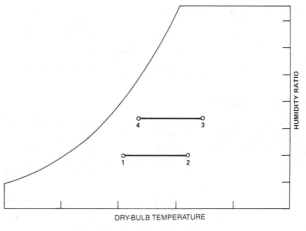

Fig. 2   Sensible Heat Recovery

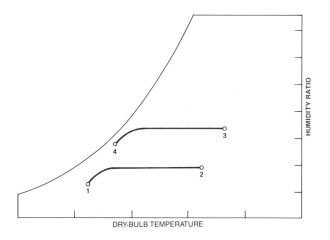

Fig. 3   Sensible Rotary Heat Exchanger Recovering Latent Heat

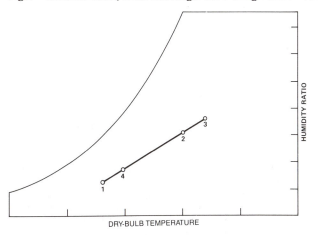

Fig. 4   Total Heat Recovery

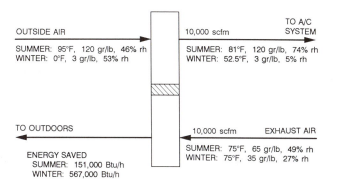

Fig. 5   Comfort-to-Comfort Sensible Heat Device

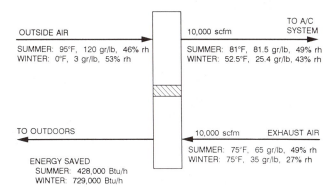

Fig. 6   Comfort-to-Comfort Total Heat Device

Sensible versus Total Recovery

Air-to-air energy recovery devices are available for sensible heat recovery and total heat recovery.

Since sensible heat devices do not transfer moisture, no latent heat is exchanged between supply and exhaust airstreams, except where the exhaust airstream is cooled below its dew point. When condensation occurs, some latent heat is transferred.

Total heat devices transfer both sensible heat and latent heat (humidity) between supply and exhaust airstreams. Unlike process-to-process and process-to-comfort applications, latent transfer is frequently desired in comfort-to-comfort applications.

A typical sensible heat recovery process between supply and exhaust airstreams is shown in Figure 2. Cold air is heated from 1 to 2, while hot air is cooled from 3 to 4. In this case, the cold air temperature is above the dew point of the hot air, and no condensation takes place.

Figure 3 illustrates a sensible heat recovery process in which condensation occurs in the hot airstream, along with evaporation in the cold one. Here latent heat transfer enhances overall effectiveness.

Figure 4 illustrates a total heat recovery process when mass flow rates and the latent and total heat effectiveness are equal. For this case, outlet states 2 and 4 lie on the straight line through cold-air inlet state 1 and hot-air inlet state 3.

Figure 5 shows a comfort-to-comfort application of a device with 70% effectiveness of sensible heat only; Figure 6 shows a device of 70% effectiveness of both sensible heat and latent heat operating under the same conditions. Under typical summer design conditions, the total heat device transfers nearly three times as much energy as the sensible heat device. Under typical winter design conditions, the total heat device recovers more than 25% more energy than the sensible heat device. Figure 7 shows the processes and a comparison of the energy transfers for these example systems for both summer and winter conditions on a psychrometric chart.

Fouling

The term fouling refers to an accumulation of dust or condensates on heat exchanger surfaces. Increasing the resistance to airflow and generally decreasing heat transfer coefficients, fouling reduces heat exchanger performance. The increased resistance increases fan power requirements and may reduce airflows.

Pressure drop across the heat exchanger core can be used as an indication of fouling and, with experience, may be used to establish cleaning schedules. Heat exchanger surfaces must be kept clean if system performance is to be maximized.

Corrosion

Process exhaust frequently contains substances requiring corrosion-resistant construction materials. If it is not known which materials are most corrosion-resistant for an application, the user and/or designer should examine on-site ductwork, review literature, and contact equipment suppliers prior to selecting materials. A corrosion study of heat exchanger construction materials in the proposed operating environment may be warranted if the installation costs are high and the environment is corrosive. Experimental procedures for such studies are described in an ASHRAE symposium (1982). Often contaminants not directly related to the process are present in the exhaust airstream (*e.g.*, welding fumes or paint carryover from adjacent processes).

Moderate corrosion generally occurs over time, roughening metal surfaces and increasing their heat transfer coefficients. Severe corrosion reduces overall heat transfer and can result in

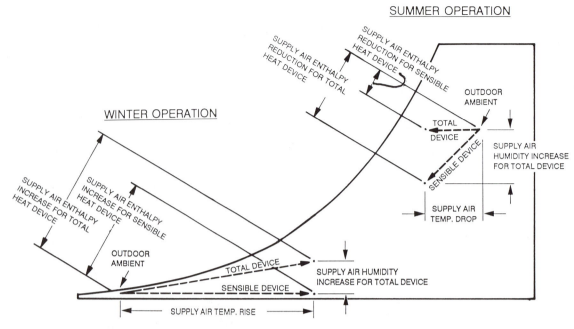

Fig. 7   Comfort-to-Comfort Sensible Heat versus Total Heat Recovery

increased cross-leakage between airstreams due to perforation or mechanical failure.

Cross-Leakage

Cross-leakage, cross-contamination, or mixing between supply and exhaust airstreams may occur in air-to-air heat exchangers. It may be a significant problem if exhaust gases are toxic or odorous. Cross-leakage varies with heat exchanger type and design, airstream static pressure differences, and the physical condition of the heat exchanger.

Condensation and Freeze-up

Condensation, ice formation, and/or frosting may occur on heat exchange surfaces. If entrance and exit effects are neglected, four distinct air/moisture regimes may occur as the warm airstream is cooled from its inlet conditions to its outlet conditions. First there is a dry region with no condensation. Once the warm airstream is cooled below its dew point, there is a condensing region which results in wetting of the heat exchange surfaces. If the heat exchange surfaces fall below freezing, the condensation will freeze. Finally, if the warm airstream temperature is reduced below 32°F, sublimation causes frost to form. The location of these regions and rates of condensation and frosting depend on the duration of frosting conditions; the airflow rates; the inlet air temperatures and humidities; the heat exchanger core temperatures; the heat exchanger effectiveness; the geometry, configuration, and orientation; and the heat transfer coefficients.

Sensible heat exchangers, which are ideally suited to applications in which heat transfer is desired but humidity transfer is not (swimming pools, kitchens, drying ovens), can benefit from the latent heat released by the exhaust gas when condensation occurs. One pound of moisture condensed transfers about 1050 Btu to the incoming air at room temperatures.

Condensation increases the heat transfer rate and thus the sensible effectiveness; it can also increase pressure drops significantly in heat exchangers with narrow airflow passage spacings. Frosting fouls the heat exchanger surfaces, which initially improves energy transfer but subsequently restricts the exhaust airflow, which in turn reduces the energy transfer rate. In extreme cases, the exhaust airflow (and supply, in the case of heat wheels) can become blocked.

Frosting and icing occur at lower temperatures in enthalpy exchangers than in sensible heat exchangers. In rotary enthalpy exchangers that use chemical absorbents, condensation may cause the absorbents to deliquesce, causing permanent damage to the exchanger. The mechanisms for condensation and frosting in air-to-air heat exchangers are discussed in ASHRAE RP543 (1989) and RP544 (1988, 1989).

An increase in the pressure drop across the heat exchanger indicates the onset of frosting or icing. Frosting and icing can be prevented by preheating the supply air; or by reducing the heat exchanger effectiveness (*e.g.*, reducing heat wheel speed, tilting heat pipes, or bypassing part of the supply airflow around the heat exchanger). Alternatively, the heat exchanger may be periodically defrosted.

Many effective defrost strategies have been developed for residential air-to-air heat exchangers. These strategies may also be applied to commercial installations. Phillips *et al.* (1989) describe frost control strategies and their impact on energy performance in various climates.

For sensible heat exchangers, system design should include drains to collect and dispose of condensation, which occurs in the warm airstream. In comfort-to-comfort applications, condensation may occur in the supply side in summer and in the exhaust side in winter.

Pressure Drop

The pressure drop for each airstream through a heat exchanger depends on many factors, including exchanger design, gas mass flow rate, temperature, moisture, and inlet and outlet air connections. The exchanger pressure drop must be overcome by fans or blowers.

An economical balance between application requirements, effectiveness, and pressure drop is required. When all other parameters are constant for a given exchanger, pressure drop increases:

- Exponentially with gas temperatures (Figure 8)
- As moisture condenses on heat exchanger surfaces

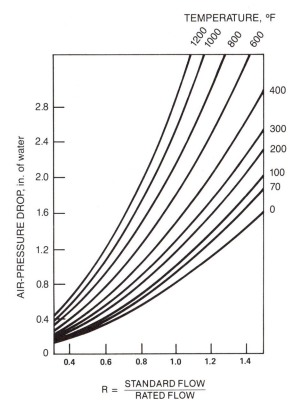

Fig. 8   Pressure Drop versus Flow at Various Temperatures
for Typical Plate Exchanger

- As plate contamination or fouling increases
- If pressure differentials between airstreams deform airflow passages
- As barometric pressures increase
- Exponentially with airflow velocities (see Figure 9)

Face Velocity

Design face velocities for heat exchangers are established on allowable pressure drop rather than recovery performance. Recovery performance (effectiveness) decreases with increasing

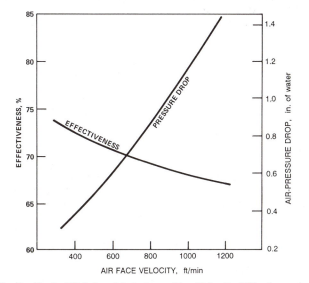

Fig. 9   Typical Relationship between Face Velocity, Effectiveness,
and Air Pressure Drop

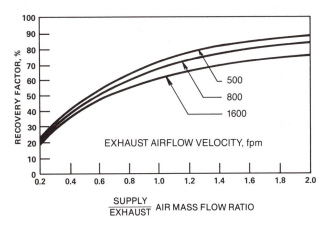

Fig. 10   Typical Relationship between Effectiveness
and Flow Ratios

velocity, but the decrease in effectiveness is not as rapid as the increase in pressure drop (Figures 9 and 10). Low face velocities give lower pressure drop, higher effectiveness, and lower operating costs but require larger units with higher capital costs and greater installation space.

Airflow Arrangements or Patterns

Airflow arrangement refers to the relative flow directions of the supply and exhaust airstreams. Counterflow is the arrangement with the highest theoretical performance, but other arrangements are common because of practical design or construction considerations. Figure 11 illustrates common airflow arrangements.

Maintenance

The method used to clean a heat exchanger depends on the transfer medium or mechanism used in the heat exchanger and the nature of the material to be removed. Grease buildup from kitchen exhaust, for example, is often removed with an automatic water-wash system. Other kinds of dirt may be removed by vacuuming, blowing compressed air through the passages, steam cleaning, manual spray cleaning, soaking of the units in soapy water or solvents, or using soot blowers. The cleaning method should be determined at the design stage so that a proper selection of heat exchanger can be made.

Cleaning frequency depends on the quality of the exhaust airstream. HVAC systems generally require only infrequent cleaning; industrial systems, usually more. Equipment suppliers should be contacted regarding the specific cleaning and maintenance requirements of the systems being considered.

Filtration

Filters should be placed in both the supply and exhaust airstreams to reduce fouling and thus the frequency of cleaning. Exhaust filters are especially important if the contaminants are sticky or greasy or if particulates can plug airflow passages in the exchanger. Supply filters eliminate insects, leaves, and other foreign materials, thus protecting both the heat exchanger and air-conditioning equipment. Snow or frost can block the air and cause severe problems. Steps to ensure a continuous flow of supply air should be incorporated into the design.

Controls

Heat exchanger controls may function to control frost formation or to regulate the amount of energy transferred between airstreams at specified operating conditions. For example, ventilation systems designed to maintain specific indoor conditions at extreme outdoor design conditions may require energy recovery modula-

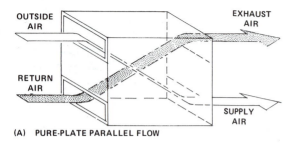

(A) PURE-PLATE PARALLEL FLOW

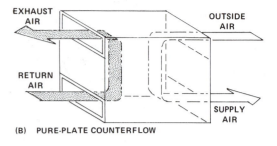

(B) PURE-PLATE COUNTERFLOW

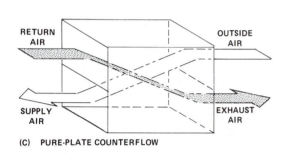

(C) PURE-PLATE COUNTERFLOW

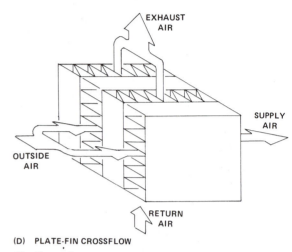

(D) PLATE-FIN CROSSFLOW

Fig. 11   Heat Exchanger Airflow Configurations

tion to prevent overheating ventilation supply air during cool to moderate weather or to prevent overhumidification. Modulation may be achieved by tilting heat pipes, changing rotational speeds of (or stopping) heat wheels, or by bypassing part of one airstream around the heat exchanger using face and bypass dampers (*i.e.*, changing the supply to exhaust mass airflow ratio).

Indirect Evaporative Air Cooling

Air passing over water absorbs water until it becomes saturated. As the water evaporates, it absorbs sensible energy from the air (about 1050 Btu per pound of water evaporated at room temperature). The sensible energy absorbed from the airstream remains in the airstream as latent energy. This process follows a constant wet-bulb line on a psychrometric chart. Thus the energy level of the airstream remains nearly constant, moisture content increases, and dry-bulb temperature decreases.

This concept may be applied as direct or as indirect evaporative air cooling. In indirect evaporative cooling applications, air-to-air heat exchangers are applied, either for year-round energy recovery or exclusively for their evaporative cooling benefits.

Indirect evaporative cooling has been applied with heat pipe, two-phase thermosiphon loops, and flat-plate heat exchangers for summer cooling (Scofield and Taylor 1986; Mathur 1990a, 1990b). Exhaust air or a scavenging airstream is cooled by passing it through a water spray, a wet filter, or other wetted media, resulting in a greater overall temperature difference between the supply and exhaust or scavenging airstreams and hence more heat transfer. Energy recovery is further enhanced by improved heat transfer coefficients due to wetted exhaust-side heat transfer surfaces. No moisture is added to the supply airstream, and there are no auxiliary energy inputs other than fan and water pumping power. The energy efficiency ratios (EER) of indirect evaporative cooling systems tend to be high, typically ranging from 30 to 70, depending on available wet-bulb depression.

Since less cooling is required, energy consumption and peak demand load are both reduced, yielding lower energy bills. Overall mechanical refrigeration system requirements are reduced, so that smaller mechanical refrigeration systems can be selected. In some cases, the mechanical system may be eliminated. Chapter 19 of this volume and Chapter 46 of the 1991 ASHRAE *Handbook—HVAC Applications* have further information on evaporative cooling.

EQUIPMENT

Table 2 provides comparative data for common types of air-to-air energy recovery devices. The following sections describe the construction, operation, and unique features of the various devices.

FIXED PLATE EXCHANGERS

Fixed surface plate exchangers have no moving parts. Alternate layers of plates, separated and sealed (referred to as the heat exchanger core) form the exhaust and supply airstream passages (Figure 12). Plate spacings range from 0.1 to 0.5 in., depending on the design and the application. Heat is transferred directly from the warm airstreams through the separating plates into the cool airstreams. Practical design and construction restrictions inevitably result in crossflow heat transfer, but additional effective heat transfer surface arranged properly into counterflow patterns can increase heat transfer effectiveness.

Normally, both latent heat of condensation [from moisture condensed as the temperature of the warm (exhaust) airstream drops below its dew point] and sensible heat are conducted through the separating plates into the cool (supply) airstream. Thus energy is transferred but moisture is not. Recovering upward of 80% of the available waste exhaust heat is not uncommon.

Table 2 Comparison of Air-to-Air Energy Recovery Devices

Airflow arrangements	Fixed Plate	Rotary Wheel	Heat Pipe	Coil Loop	Thermosiphon	Twin Towers
	Counterflow Cross Flow Parallel Flow	Counterflow Parallel Flow	Counterflow Parallel Flow	Counterflow Parallel Flow	Counterflow Parallel Flow	
Equipment size range (airflow)	50 cfm and up	50 to 70,000 cfm	100 cfm and up	100 cfm and up	100 cfm and up	
Type of heat transfer (typical effectiveness)	Sensible (50 to 80%)	Sensible (50 to 80%) Latent (45 to 55%)	Sensible (45 to 65%)	Sensible (55 to 65%)	Sensible (40 to 60%)	Sensible (40 to 60%) Latent (45 to 55%)
Face velocity (fpm) (most common design velocity)	100 to 1000 (200 to 1000)	500 to 1000	400 to 800 (450 to 550)	300 to 600	400 to 800 (450 to 550)	300 to 450
Pressure drop, in. of water (most likely pressure)	.02 to 1.8 0.1 to 1.5	0.4 to 0.7	0.4 to 2.0	0.4 to 2.0	0.4 to 2.0	0.7 to 1.2
Temperature range		−70 to +1500°F	−40 to 95°F		−40 to 104°F	−40 to 115°F
Typical method of purchase	Exchanger only Exchanger in case Exchanger and blowers Complete system	Exchanger only Exchanger in case Exchanger and blowers Complete system	Exchanger only Exchanger in case	Exchanger only Exchanger in case	Exchanger only Exchanger in case	Complete system
Unique advantages	No moving parts No cross-leakage All sizes Most materials Low pressure drop High effectiveness Easily cleaned	Latent transfer Compact large sizes Low pressure drop High effectiveness	No moving parts No cross-leakage Most sizes Fan location not critical Allowable pressure differential to 60 in. of water	Exhaust airstream can be separated from supply air	No moving parts No cross-leakage All sizes Exhaust airstream can be separated from supply air Fan location not critical	Latent transfer from remote airstreams. Multiple units in a single system. Efficient microbiological cleaning of both supply and exhaust airstreams.
Limitations	Latent available only in special units	Cold climates may increase service Cross-air contamination possible	Effectiveness limited by pressure drop and cost Few suppliers	Effectiveness may be limited by pressure drop Few suppliers	Effectiveness may be limited by pressure drop and cost Few suppliers	Few suppliers
Cross-leakage	0 to 5%	1 to 10%	0%	0%	0%	0.025%

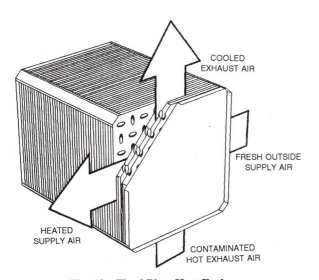

Fig. 12 Fixed Plate Heat Exchanger

Design Considerations

Plate exchangers are available in many configurations, materials, sizes, and flow patterns. Many are modular, and modules can be arranged to handle almost any airflow, effectiveness, and pres-

sure drop requirement. Plates are formed with integral separators (ribs, dimples, ovals) or with external separators (supports, braces, corrugations). Airstream separations are sealed by folding, multiple folding, gluing, cementing, and welding or any combination, depending on the application and manufacturer. Ease of access to examine and clean both heat transfer surfaces depends on the configuration.

Heat transfer resistance through the plates is small compared to the airstream boundary layer resistance on each side of the plates. Heat transfer efficiency is not substantially affected by the heat transfer coefficient of the plates. Aluminum is the most popular construction material for plates because of its corrosion resistance, ease of fabrication, heat transfer characteristics, non-flammability, durability, and cost effectiveness. Steel alloys are used for temperatures exceeding 400°F or specialized applications where cost is not a key factor. Plastic materials (and even glass) are available to satisfy low cost and corrosion resistance needs. Plate exchangers normally conduct sensible heat only, but water-permeable materials, such as treated paper, can be used to transfer latent heat (moisture), thus providing total (enthalpy) heat exchange.

Most manufacturers offer plate exchangers in modular design. Modules range in capacity from 25 to 10,000 cfm and can be arranged into configurations exceeding 100,000 cfm. Multiple sizes and configurations permit selections to meet nearly all space and performance requirements.

Performance

Fixed plate heat exchangers can economically achieve high sensible heat recovery effectiveness because they have only a primary heat transfer surface area separating the airstreams and are therefore not inhibited by the additional secondary resistance inherent in other exchanger types (*i.e.*, pumping liquid, condensing and vaporizing gases, or transporting a heat transfer media). Simplicity and no moving parts add to the reliability, longevity, low auxiliary energy consumption, and safety performance of these exchangers.

Differential Pressure/Cross-Leakage

One of the advantages of the plate exchanger is that it is a static device built so that little or no leakage occurs between airstreams.

As velocity increases, the pressure difference between the two airstreams increases exponentially. High differential pressures may deform the separating plates and, if excessive, can permanently damage the exchanger. This may reduce effectiveness, alter designed fan airflow, and cause excessive air leakage. This is not normally a problem, since the differential pressures in most applications are less than 4 in. of water. In applications requiring high air velocities, high static pressures, or both, select exchangers that are designed for these conditions.

Condensing within Exhaust Airstreams

Most plate exchangers are equipped with condensate drains, which remove both the condensate and waste water when a water-wash system is used. Heat recovered from a high humidity exhaust is better returned to a building or process by a sensible heat exchanger than an enthalpy exchanger, if humidity transfer is not desired. Figure 13 illustrates the thermodynamic process for condensation in a sensible heat exchanger on a psychrometric chart.

Table 3 illustrates the effect of moisture content on the frost threshold. Frosting can be controlled by either preheating the incoming supply air or by bypassing a portion of the incoming air. Bypassing reduces the ratio K in Table 3 and is generally more cost-effective than preheating.

ROTARY AIR-TO-AIR ENERGY EXCHANGERS

A rotary air-to-air energy exchanger, or heat wheel, has a revolving cylinder filled with an air-permeable medium with a large internal surface area. Adjacent supply and exhaust airstreams each flow through one-half the exchanger in a counterflow pattern

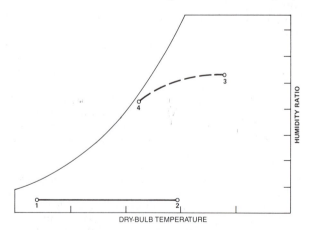

Fig. 13   Condensing within Sensible Plate Heat Exchanger for Balanced Flow

Table 3   Frost Threshold Temperature for Various Exhaust Air Conditions

| Entering Exhaust Air | | Frost Threshold Temperature (t_1), °F | | | |
| | | Ratio of Supply to Exhaust Airflow (K) | | | |
t_3, °F	rh, %	0.5	0.7	1.0	2.0
60	30	2	15	23	32
60	40	2	15	23	32
60	50	−4	9	18	32
60	60	−9	4	13	32
70	30	−13	5	17	28
70	40	−21	−3	10	21
70	50	−27	−9	3	15
70	60	−32	−13	−1	10
75	30	−25	−4	10	23
75	40	−33	−12	2	15
75	50	−40	−20	−6	7
75	60	−47	−26	−12	1
80	30	−35	−11	4	19
80	40	−44	−20	−5	10
80	50	−53	−30	−14	1
80	60	−62	−39	−23	−8
90	30	−58	−30	−11	5
90	40			−24	−8
90	50				−20

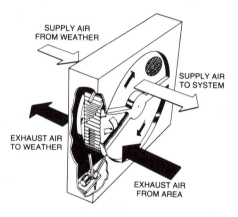

Fig. 14   Rotary Air-to-Air Energy Exchanger

(Figure 14). Heat transfer media may be selected to recover only sensible heat or total heat (sensible heat plus latent heat).

Sensible heat is transferred as the medium picks up and stores heat from the hot airstream and releases it to the cold one. Latent heat is transferred as the medium (1) condenses moisture from the airstream with the higher humidity ratio (either because the media temperature is below its dew point or by means of absorption for liquid desiccants and adsorption for solid desiccants), with a simultaneous release of heat; and (2) releases the moisture through evaporation (and heat pickup) into the airstream with the lower humidity ratio. Thus, the moist air is dried while the drier air is humidified. In total heat transfer, both sensible and latent heat transfer occur simultaneously.

Construction

Air contaminants, dew point, exhaust air temperature, and supply air properties influence the choice of structural materials for the casing, rotor structure, and medium of a rotary energy exchanger. Aluminum and steel are the usual structural, casing, and rotor materials for normal comfort ventilating systems. Exchanger media are fabricated from metal, mineral, or man-made materials and are classified as providing either random flow or directionally oriented flow through their structures.

Random flow media, consisting of knitted corrugated mesh, are made by knitting wire into an open woven cloth, which is layered to the desired configuration. Aluminum mesh, commonly used for comfort ventilation systems, is packed in pie-shaped wheel segments. Stainless steel and monel mesh are used for high-temperature and corrosive applications. These media should only be used with clean, filtered airstreams because they plug easily.

Directionally oriented media consist of small (0.0625 in.) triangular air passages parallel with the direction of airflow. The triangular shape gives the largest exposed surface for air contact per unit of face area; it is also strong and is easily produced by interleaving layers of flat and corrugated material. Aluminum foil, inorganic sheet, treated organic sheet, and synthetic materials are used for low and medium temperatures. Stainless steel and ceramics are used for high temperatures and corrosive atmospheres.

Media surface areas exposed to airflow vary from 100 to 1000 ft^2/ft^3, depending on the type of medium and physical configuration. Media may also be classified according to their ability to recover only sensible heat or total heat. Media for sensible heat recovery are made of aluminum, copper, stainless steel, and monel. Media for total heat recovery are fabricated from any of a number of materials and treated with a desiccant, typically lithium chloride or alumina, to have specific moisture recovery characteristics.

Cross-Contamination

Cross-contamination, or mixing, of air between supply and exhaust airstreams occurs in all rotation energy exchangers by two mechanisms—carryover and leakage. Carryover occurs as air is entrained within the volume of the rotation medium and is carried into the other airstream. Leakage occurs because the differential static pressure across the two airstreams drives air from a higher to a lower static pressure region. Leakage can be reduced by placing the blowers so that they promote leakage of outside air to the exhaust airstream. Carryover occurs each time a portion of the matrix passes the seals dividing the supply and exhaust airstreams. Since carryover from exhaust to supply may be undesirable, a purge section can be installed on the heat exchanger to prevent cross-contamination.

In many applications, recirculating some air is not a concern. However, critical applications, such as hospital operating rooms, laboratories, and clean rooms, require stringent control of carryover. Carryover can be reduced to below 0.1% of the exhaust airflow with a purge section (ASHRAE 1974).

The theoretical carryover of a wheel without a purge is directly proportional to the speed of the wheel and the void volume of the media (75 to 95% void, depending on type and configuration). For example, a 10-ft diameter, 8-in. deep wheel, with a 90% void volume operating at 14 rpm, has a carryover volume of:

$$(10)^2(\pi/4)(8/12)(0.9)(14) = 660 \text{ cfm}$$

If the wheel is handling a 20,000 cfm balanced flow, the percentage carryover is:

$$(660/20,000)(100) = 3.3\%$$

The exhaust fan, which is usually located at the exit of the exchanger, should be sized to include leakage, purge, and carryover airflows.

Controls

Two control methods are commonly used to regulate wheel energy recovery. In the first, supply air bypass control, the amount of supply air allowed to pass through the wheel establishes the supply air temperature. An air bypass damper, controlled by a wheel supply air discharge temperature sensor, regulates the proportion of supply air permitted to bypass the exchanger.

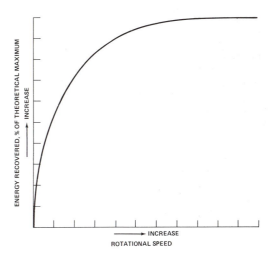

Fig. 15 Typical Wheel Energy Recovery Capacity versus Rotational Speed

The second method regulates the energy recovery rate by varying wheel rotational speed (Figure 15). The most frequently used variable-speed drives are (1) a silicon controlled rectifier (SCR) with variable speed dc motor, (2) a constant speed ac motor with hysteresis coupling, and (3) an ac frequency inverter with an ac induction motor.

A dead band control, which stops or limits the exchanger, may be necessary when no recovery is desired (*e.g.*, when outside air temperature is higher than the required supply air temperature but below the exhaust air temperature). When the outside air temperature is above the exhaust air temperature, the equipment operates at full capacity to cool the incoming air.

Maintenance

Energy exchanger wheels require little maintenance. The following maintenance procedures ensure best performance:

- Clean the medium when lint, dust, or other foreign materials build up, following the manufacturer's instructions for that medium type. Media treated with a liquid desiccant for total heat recovery must not be wetted.
- Maintain drive motor and train according to the manufacturer's recommendations. Speed control motors that have commutators and brushes require more frequent inspection and maintenance than do induction motors. Brushes should be replaced, and the commutator should be periodically turned and undercut.
- Inspect wheels regularly for proper belt or chain tension.
- Refer to the manufacturer's recommendations for spare and replacement parts.

COIL ENERGY RECOVERY (RUNAROUND) LOOPS

A typical coil energy recovery loop system (Figure 16) places extended surface, finned tube water coils in the supply and exhaust airstreams of a building or process. The coils are connected in a closed loop via counterflow piping through which an intermediate heat transfer fluid (typically water or a freeze-preventive solution) is pumped.

This system operates for sensible heat recovery only. In comfort-to-comfort applications, energy transfer is seasonally reversible—the supply air is preheated when the outdoor air is cooler than the exhaust air and precooled when the outdoor air is warmer.

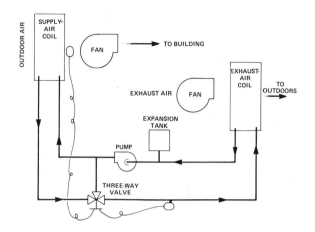

Fig. 16   Coil Energy Recovery Loop

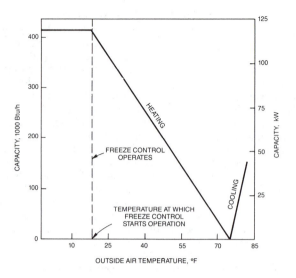

Fig. 17   Energy Recovery Capacity versus Outside Air Temperature for Typical Loop

Freeze Protection

Moisture must not freeze in the exhaust coil air passage. A dual-purpose, three-way temperature control valve prevents the exhaust coil from freezing. The valve is controlled to maintain the entering solution temperature to the exhaust coil to not less than 30 °F. This condition is maintained by bypassing some of the warmer solution around the supply air coil. The valve can also ensure that a prescribed air temperature from the supply air coil is not exceeded.

System Characteristics

Coil energy recovery loop systems are highly flexible and well-suited to renovation and industrial applications. The system accommodates remote supply and exhaust ducts and allows the simultaneous transfer of energy between multiple sources and uses. An expansion tank must be included to allow fluid expansion and contraction. A closed expansion tank minimizes oxidation when ethylene glycol is used.

Standard finned tube water coils may be used. Manufacturer's design curves and performance data should be used when selecting coils, face velocities, and pressure drops.

Effectiveness

The coil energy recovery loop cannot transfer moisture from one airstream to another. For the most cost-effective operation, with equal airflow rates and no condensation, typical effectiveness values range from 60 to 65%. Highest effectiveness does not necessarily give the greatest net cost savings.

The following example illustrates the capacity of a typical system:

Example 1. A waste heat recovery system heats 10,000 cfm of air from a 0 °F design outdoor temperature to an exhaust dry-bulb temperature of 75 °F and a wet-bulb temperature of 60 °F. The air flows through identical eight-row coils at a 400 fpm face velocity. A 30% ethylene glycol solution flows through the coils at 26 gpm.

Figure 17 shows the effect of the outside air temperature on capacity, including the effects of the three-way temperature control valve. For this example, the capacity is constant for outside air temperatures below 18.5 °F. This constant output occurs because the valve has to control the temperature of the fluid entering the exhaust coil to prevent frosting. As the exhaust coil is the source of heat and has a constant airflow rate, entering air temperature, liquid flow rate, entering fluid temperature (as set by the valve), and fixed coil parameters, energy recovered must be fixed to prevent frosting. Equation (1) may be used to calculate the sensible heat effectiveness.

When the three-way control valve operates at outside air temperatures of 18.5 °F or lower, 414,000 Btu/h is recovered. At 18.5 °F, the sensible heat effectiveness is 67.2%. At the 0 °F design temperature sensible effectiveness is 51% ($\epsilon = 414,000/810,000$), and the leaving air dry-bulb temperature of the supply coil is 38.3 °F. Above 75 °F outside air temperature, the system begins to cool the supply air.

Typically, the sensible heat effectiveness of a coil energy recovery loop is independent of the outside air temperature. However, when the capacity is controlled (as in the preceding example) the sensible heat effectiveness decreases.

Construction Materials

Coil energy recovery loops incorporate coils constructed to suit the environment and operating conditions to which they are exposed. For typical comfort-to-comfort applications, standard coil construction usually suffices. In process-to-process and process-to-comfort applications, the effect of high temperature, condensables, corrosives, and contaminants on the coil(s) must be considered.

At temperatures, above 400 °F, special construction may be required to ensure a permanent fin-to-tube bond. The effects of condensables and other adverse factors may require special coil construction and/or coatings. Chapters 21, and 24 discuss the construction and selection of coils in more detail.

Cross-Contamination

Complete separation of the airstreams eliminates cross-contamination between the supply and exhaust air.

Maintenance

Coil energy recovery loops require little maintenance. The only moving parts are the circulation pump and the three-way control valve. However, the following items must be maintained to ensure optimum operation: air filtration; cleaning of the coil surface; periodic maintenance of the pump and valve; and transfer fluid. Fluid manufacturers or their representatives should be contacted for specific recommendations.

Thermal Transfer Fluids

The thermal transfer fluid used in a closed-loop application depends on the application and temperatures of the two airstreams.

An inhibited ethylene glycol solution in water is commonly used when freeze protection is required. These solutions break down to an acidic sludge if temperatures exceed 275 °F. If freeze protection is needed and exhaust air temperatures exceed 275 °F, a nonaqueous synthetic heat transfer fluid should be used. Heat transfer fluid manufacturers and chemical suppliers should recommend appropriate fluids.

HEAT PIPE HEAT EXCHANGERS

A heat pipe heat exchanger is a passive energy recovery device. It has the outward appearance of an ordinary plate-finned water or steam coil, except that the tubes are not interconnected and the pipe heat exchanger is divided into evaporator and condenser sections by a partition plate (Figure 18). Hot air passes through the evaporator side of the exchanger and cold air passes through the condenser side. Heat pipe heat exchangers are sensible heat transfer devices, but condensation on the fins does allow latent heat transfer, resulting in improved recovery performance.

Heat pipe tubes (Figure 19) are fabricated with an integral capillary wick structure, evacuated, filled with a suitable working fluid, and permanently sealed. The working fluid is normally a Class I refrigerant, but other fluorocarbons, water, and other compounds are used for particular temperature applications.

Fin designs include continuous corrugated plate fin, continuous plain fin, and spiral fins. Fin design and tube spacing cause variations in pressure drop at a given face velocity.

Principle of Operation

Hot air flowing over the evaporator end of the heat pipe tube vaporizes the working fluid. Vapor pressure drives the vapor to the condenser end of the heat pipe tube, where the vapor condenses releasing the latent energy of vaporization. The condensed fluid is wicked or flows back to the evaporator where it is revaporized, thus completing the cycle. Thus the heat pipes operate in a closed loop evaporation/condensation cycle, which continues as long as there is a temperature difference to drive the process. Using this mechanism, heat transfer along a heat pipe is up to 1000 times faster than through copper (Ruch 1976).

Heat pipes transfer energy with a small temperature drop; consequently, the process is often considered isothermal. However, there is a small temperature drop through the tube wall, wick, and the fluid medium. Heat pipes have a finite heat transfer capacity that is affected by such factors as wick design, tube diameter, working fluid, and tube (heat pipe) orientation relative to horizontal.

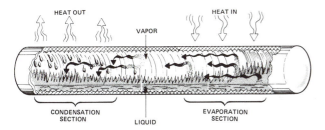

Fig. 18 Heat Pipe Heat Exchanger

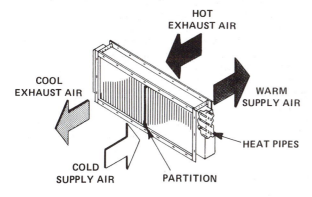

Fig. 19 Heat Pipe

Construction Material

Copper heat pipe tubes with aluminum fins are normally used in an HVAC system. Tubes and fins are usually constructed of the same material to avoid problems with different thermal expansions of materials. Heat pipe heat exchangers for exhaust temperatures below 425 °F are most often constructed with aluminum tubes and fins. Because they are more expensive than aluminum units of comparable effectiveness, copper units are generally only used where aluminum units are unsuitable because of corrosion or cleaning problems. Protective coatings designed for finned tube heat exchangers have permitted inexpensive aluminum to replace exotic metals in corrosive atmospheres. Protective coatings for corrosion protection have a minimal effect on thermal performance.

Heat pipe heat exchangers for use above 425 °F are generally constructed with steel tubes and fins. The fins are often aluminized to prevent rusting. Composite systems for special applications may be created by assembling units with different materials and/or different working fluids.

Operating Temperature Range

Selection of the proper working fluid for a heat pipe is critical to long-term operation. The working fluid should have high latent heat of vaporization, a high surface tension, and a low liquid viscosity over the operating range; it must also be thermally stable at operating temperatures. Decomposition of the thermal fluids can result in the formation of noncondensible gases that deteriorate performance.

Cross-Contamination

Heat pipe heat exchangers typically have zero cross-contamination for pressure differentials between airstreams of up to 50 in. of water. Additional protection against cross-contamination can be provided by constructing a vented double-wall partition between the airstreams. By attaching an exhaust duct to the partition space, any leakage is withdrawn and exhausted from the space between the two ducts.

Performance

Heat pipe heat transfer capacity depends on design and orientation.

Figure 20 presents a typical effectiveness curve for various face velocities and rows of tubes. As the number of rows increases, effectiveness increases at a decreasing rate. For example, doubling the number of rows of tubes in a 60% effective heat exchanger increases the effectiveness to 75%. The effectiveness of a heat

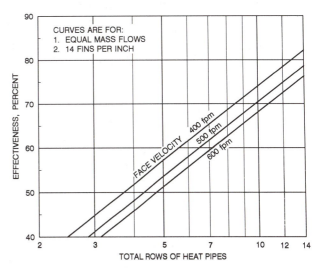

Fig. 20 Heat Pipe Exchanger Effectiveness

pipe heat exchanger depends on the total number of rows. Thus, two units in series yield the same effectiveness as a single unit of the same number of rows. Series units are often used to facilitate handling, cleaning, and maintenance.

The heat transfer capacity of a heat pipe increases roughly with the square of the inside diameter of the pipe. For example, at a given tilt angle, a 1-in. inside diameter heat pipe will transfer roughly 2.5 times as much energy as a 5/8-in. inside diameter pipe. Consequently, heat pipes with large diameters are used for larger airflow applications and where level installation is required to accommodate both summer and winter operation.

Heat transfer capacity limit is virtually independent of heat pipe length, except for very short heat pipes. For example, a 4-ft long heat pipe has approximately the same capacity as an 8-ft pipe. Since the 8-ft long heat pipe has twice the external heat-transfer surface area of the 4-ft pipe, it will reach its capacity limit sooner. Thus, in a given application, it is more difficult to meet the capacity requirements as the heat pipes become longer. A system can be reconfigured to a taller face height and more but shorter heat pipes to yield the same airflow face area while improving system performance.

The selection of fin design and spacing should be based on the dirtiness of the two airstreams and the resulting cleaning maintenance required. For HVAC applications, 14 fins/in. is common. Wider fin spacings (8 to 11 fins/in.) are usually used for industrial applications. Heat pipe heat exchangers of the plate fin type can easily be constructed with different fin spacings for the exhaust and supply airstreams, allowing a wider fin spacing on the dirty exhaust side. This increases design flexibility where pressure drop constraints exist and also prevents deterioration of performance resulting from dirt buildup on the exhaust side surface.

Controls

Changing the slope, or tilt, of a heat pipe offers a way to control the amount of heat that it transfers. Operating the heat pipe on a slope with the hot end below (above) the horizontal improves (retards) the condensate flow back to the evaporator end of the heat pipe. This feature can be used to regulate the effectiveness of the heat pipe heat exchanger.

In practice, tilt control is affected by pivoting the exchanger about the center of its base and attaching a temperature-controlled actuator to one end of the exchanger (Figure 21). Pleated flexible connectors attached to the ductwork allow freedom for the small tilting movement (6 degree maximum).

Three functions for which tilt control may be desired are:

- To change the operation from supply air heating to supply air cooling (*i.e.*, to reverse the direction of heat flow) when seasonal changeover occurs.

- To modulate effectiveness to maintain desired supply air temperature. This kind of regulation is often required for large buildings to avoid overheating the air supplied to the interior zone.
- To decrease effectiveness to prevent frost formation at low outdoor air temperatures. By reducing effectiveness, the exhaust air leaves the unit at a warmer temperature and stays above frost-forming conditions.

Other methods, such as face and bypass dampers and preheaters, can be used for individual functions.

TWIN TOWER ENTHALPY RECOVERY LOOPS

Method of Operation

In this air-to-liquid, liquid-to-air enthalpy recovery system, a sorbent liquid circulates continuously between supply and exhaust airstreams, alternately contacting both airstreams directly in contactor towers (Figure 22). This liquid transports water vapor and heat. The sorbent solution is usually a halogen salt solution, such as lithium chloride and water. Pumps circulate the solution between supply and exhaust contactor towers.

Both vertical and horizontal airflow contactor towers are manufactured. Vertical and horizontal contactor towers can be combined in a common system. Contactor towers of both configurations are supplied with airflow capacities of up to 100,000 scfm.

In the vertical configuration, the supply or exhaust air passes vertically through the contact surface counterflow to the sorbent liquid to achieve high contact efficiency. In the horizontal configuration, air passes horizontally through the contact surface crossflow to the sorbent liquid, yielding a slightly lower contact efficiency.

The contactor surface is usually made of nonmetallic materials. Air leaving the contactor tower passes through demister pads, which remove any entrained sorbent solution.

Design Considerations

Operating temperature limits. Twin tower enthalpy recovery systems operate primarily in the comfort conditioning range; they are not suitable for high-temperature applications. During summer, these systems operate with building supply air temperatures as high as 115 °F. Winter supply air temperatures as low as −40 °F can be tolerated without freezing or frosting problems, because the sorbent solution is an effective antifreeze at all useful concentrations.

Static pressure effects. Because contactor supply and exhaust air towers are independent units connected only by piping, supply and exhaust air fans can be located wherever desired; contactor towers generally operate with air inlet static pressure from −6 to 6 in. of water. The exhaust contactor tower may be operated at a higher internal static pressure than the supply contactor tower without danger of exhaust-to-supply cross-contamination.

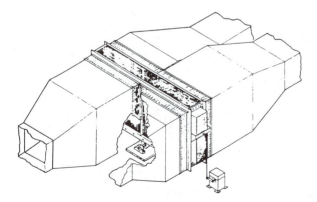

Fig. 21   Heat Pipe Heat Exchanger with Tilt Control

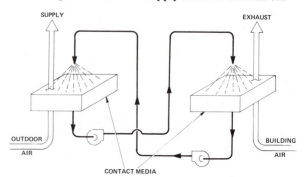

Fig. 22   Twin Tower Enthalpy Recovery Loop

Cross-contamination. Particulate cross-contamination does not occur because wetted particulates remain in the sorbent solution until they are filtered. Limited gaseous cross-contamination may occur, depending on the solubility of the gas in the sorbent solution. Tests of gaseous cross-contamination using sulphur hexafluoride tracer gas techniques have shown typical gaseous cross-contamination rates of 0.025% for the twin tower system.

Sorbent solutions—especially the halide brines—are bactericidal. Lithium chloride, as used in twin tower systems, is viricidal against some viruses. Microorganism testing of the twin tower system, using the six-plate Anderson sampling technique, has shown that microorganism transfer does not occur between supply and exhaust airstreams. Tests of actual contactor towers have shown that they effectively remove as much as 94% of the atmospheric bacteria in the supply or exhaust air.

Effect of building or process contaminants. If the building or process exhaust contains large amounts of lint, animal hair, or other solids, filters should be placed upstream of the exhaust air contactor tower.

If the building or process exhaust contains large amounts of gaseous contaminants, such as chemical fumes and hydrocarbons, the possibility of cross-contamination, as well as the possible effects of contaminants on the sorbent solution, should be investigated.

Winter operation. When using twin tower systems in controlled humidity applications in colder climates, saturation effects (which can cause condensation, frosting, and icing in other types of equipment) may overdilute the sorbent solution in the twin tower system. Heating the sorbent liquid supplied to the supply air contactor tower can prevent dilution (Figure 23). Heating raises the discharge temperature and humidity of air leaving the supply contactor tower, thus balancing the humidity of the system and preventing overdilution.

A thermostat sensing the leaving air temperature from the supply air contactor tower is commonly used to control the solution heater to deliver constant temperature air regardless of the outdoor temperature. Automatic addition of makeup water to maintain a fixed concentration of the sorbent solution enables the twin tower system to deliver supply air at fixed humidity during cold weather (Figure 23). Thus, the system provides a constant supply air temperature and humidity without preheat or reheat coils or humidifiers.

Multiple towers. Any number of supply air towers can be combined with any number of exhaust air towers, as shown in Figure 24. If a sufficient elevation difference exists between supply and exhaust towers, gravity may be used to return the sorbent solution from the upper tower(s) (Figure 24).

Maintenance

Twin tower enthalpy recovery systems operate with only routine maintenance. Complete instructions on procedures, as well as

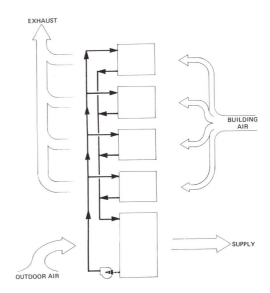

Fig. 24   Twin Tower Enthalpy Recovery Loop
(Multiple-Unit System)

spare-parts lists, are included in operating manuals relevant to each installation. Periodically the circulation pumps, spray nozzles, liquid transfer controls, and mist eliminators need checking and adjusting or maintenance.

Inhibited halide brine solutions are normally used as energy transfer media in twin tower systems. Manufacturers provide technical support, including solution sampling services to monitor and report changes in concentration, inhibitor level, and pH, thus ensuring continued maximum performance.

THERMOSIPHON HEAT EXCHANGERS

Two-phase thermosiphon heat exchangers are sealed systems that consist of an evaporator, a condenser, interconnecting piping, and an intermediate working fluid (present in both liquid and vapor phases). Two different types of thermosiphon are used—sealed tube and coil loop. In the first case, the evaporator and condenser are usually at opposite ends of a bundle of straight, individual thermosiphon tubes, and the exhaust and supply ducts are adjacent to each other (similar to the arrangement in a heat pipe system). In the second case, evaporator and condenser coils are installed independently in the ducts and are interconnected by the working fluid piping (much the same as the configuration for coil energy recovery loop systems).

In thermosiphon systems, a temperature difference and the force of gravity cause a working fluid to circulate between the evaporator and the condenser. As a result, they may be designed to transfer heat equally well in either direction (bidirectional), in one direction only (unidirectional), or anywhere between these extremes.

Although similar in form and operation to heat pipes, thermosiphon tubes are different in two ways: (1) they have no wick and hence rely only on gravity to return the condensate to the evaporator, whereas heat pipes use capillary forces; and (2) thermosiphon tubes are dependent, at least initially, on nucleate boiling, whereas heat pipes vaporize the fluid from a large, ever-present, liquid-vapor interface.

Thermosiphon loops differ from other coil energy recovery loop systems in that they require no pumps, and hence no external power supply, and the coils must be appropriate for evaporation and condensation.

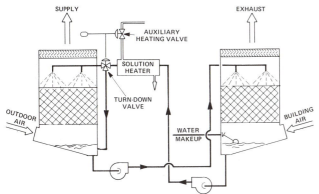

Fig. 23   Twin Tower Enthalpy Recovery Loop
(Winter Operation and Control)

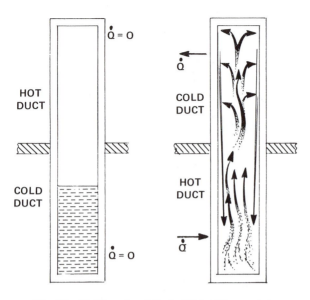

Fig. 25   Unidirectional Sealed Tube Thermosiphon

Principle of Operation

A thermosiphon is a sealed system containing a two-phase working fluid. Because part of the system contains vapor and part contains liquid, the pressure in a thermosiphon is governed by the liquid temperature at the liquid-vapor interface. If the surroundings cause a temperature difference between two regions in a thermosiphon where liquid vapor interfaces are present, the resulting vapor-pressure difference makes vapor flow from the warm to the cool region. The flow is sustained by condensation in the cool region and evaporation in the warm region. It is essential that the condenser and evaporator be oriented so that the condensate can return to the evaporator by gravity (Figure 25).

Characteristics of Two-Phase Thermosiphons

The geometric orientation of the thermosiphon, the temperature difference between the two ducts, the tube diameter(s) and length(s), and the type and quantity of working fluid govern the operating characteristics of these units. If the orientation is such that liquid is present within the tube(s) in both ducts at all times (Figures 26 and 27), the unit can transfer heat in either direction (bidirectionally). If the orientation is such that the liquid drains by gravity to one end of the heat exchanger tube(s) (Figures 25 and 28), then heat can be transferred in one direction only (unidirectionally) toward the liquid-deficient region. This may be advantageous, for example, where the evaporator operates as a solar collector absorbing energy when the sun is shining and shutting off whenever the collector (evaporator) temperature drops below the sink (condenser) temperature, without the use of auxiliary controls.

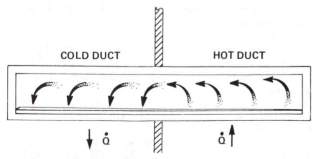

**Fig. 26   Bidirectional Sealed Tube Thermosiphon in which
Tube Conducts Equally Well in Either Direction**

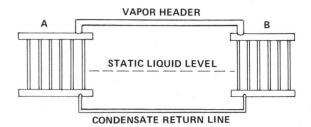

**Fig. 27   Bidirectional Thermosiphon Coil Loop—Transfers
Energy in Either Direction**

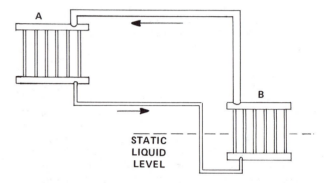

**Fig. 28   Unidirectional Thermosiphon Coil Loop—Transfers
Energy Only from B to A**

Thermosiphon systems can exhibit hysteretic behavior when operated with a temperature difference that varies between 32 °F and the temperature at which nucleate boiling starts (McDonald and Shivprasad 1989). This behavior becomes more pronounced as the system pressure decreases (relative to its critical pressure) and is highly dependent on the working fluid and the evaporator surface characteristics. Figure 29 illustrates the type of hysteretic behavior that occurs where a large temperature difference is required to initiate boiling. Any subsequent reduction in temperature difference eventually causes the number of nucleation sites to diminish until all are gone at some finite wall superheat. If the wall superheat is increased, the sites that were not cooled appreciably below the wall superheat at which the site became inactive will reactivate at a wall superheat much lower than the incipient value. The bubbles from the reactivated sites tend to seed other adjacent sites, and localized strips of nucleate boiling develop along the

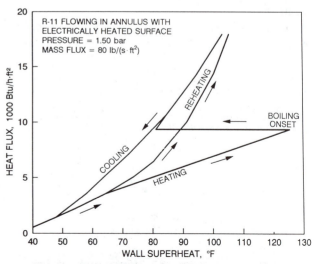

Fig. 29   Typical Boiling Hysteresis Curve for R-11

tube length yielding a third different performance curve. To minimize these hysteresis effects, a working fluid with a relatively high pressure at the operating temperature should be used when a thermosiphon is likely to be subject to small temperature difference operating conditions.

Sealed Tube Thermosiphons

Sealed tube thermosiphons are used only when the supply and exhaust air ducts are adjacent to one another. Their operating characteristics, applications, and limitations are similar to those described in the section on heat pipe heat exchangers. For a bidirectional system in which the tubes are mounted horizontally, however, thermosiphon performance is more sensitive to tube misalignment than is heat pipe performance. Capillary forces ensure that liquid returns to the evaporator end of the heat pipe, while the thermosiphon relies only on gravity. Heat transfer cannot take place if all the liquid resides at the cold end of the tube.

Coil Loop Thermosiphons

Coil loop thermosiphons may be used when the supply and exhaust air ducts are not adjacent to one another. Although similar in appearance and application to the coil energy recovery loop, the coil loop thermosiphon has several unique features. As illustrated in Figures 28 and 29, a single coil loop thermosiphon consists of two coils interconnected by a vapor and a condensate line. The loop is charged with working fluid at its saturation state, so that part of it is filled with liquid and part with vapor. The pressure within a sealed loop depends on the working fluid used and the fluid temperature at the liquid-vapor interface. The maximum loop pressure corresponds to the saturation pressure at the unit's maximum temperature. Similarly, the minimum pressure corresponds to the saturation pressure at the lowest operating temperature. As each working fluid has an optimum operating range, the proper working fluid should be selected for the intended application.

Condensation and evaporation are affected in different ways by the diameter, length, and orientation of the coil tubes, as well as by the amount of liquid charge present in each coil. As a result, bidirectional systems may be designed without external controls to provide a different effectiveness for each energy flow direction.

Unidirectional coil loop thermosiphons are more efficient than bidirectional units operating under the same conditions, because the coils and loop charge may be selected for optimum performance for only one function rather than for two functions (evaporation and condensation). In either case, to recover a higher proportion of the available energy, several coil loop thermosiphons may be mounted like wafers along the supply and exhaust ducts to create a counterflow heat exchanger with an overall effectiveness greater than that of a single loop (Figure 30). Figure 31 shows

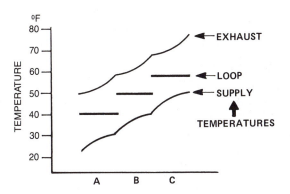

Fig. 31 **Performance of Multiple Thermosiphon Coil Loops in Series with 50% Effective Evaporator and Condenser Coils**

the temperature distribution for a system with equal exhaust and supply airflow rates and assumes that each individual coil has an effectiveness of 50%. For this case, the effectiveness of each separate thermosiphon loop is about 25%; the effectiveness of any two loops is 40%; and the effectiveness of any three loops is 50%. If the effectiveness of each individual coil was 80%, the effectiveness of three loops mounted in series, as shown, would be 67%.

The maximum possible effectiveness for these systems is limited by the minimum temperature difference required to sustain nucleate boiling. For example, if an overall temperature difference of 40°F is available between the hot and cold ducts and a 8°F temperature difference is required to sustain boiling, the maximum possible effectiveness is (40 − 8)/40 = 0.80 or 80%.

The most economical combination of coil sizes and number of loops depends on the specified design criteria, the life of the system, and economic criteria.

As with heat pipes, each loop is independent, so that different working fluids can be used in each loop if it is advantageous. In addition, thermosiphon coil loops may be used with multiple supply or exhaust ducts. Figure 32 schematically illustrates a case in which a common supply duct is being heated by three different exhaust ducts.

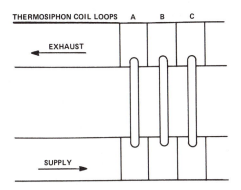

Fig. 30 **Plan View of Multiple Thermosiphon Coil Loops in Counterflow Configuration**

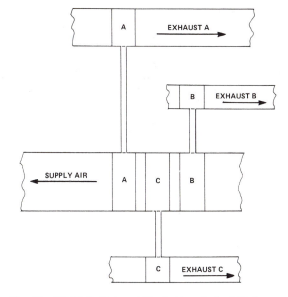

Fig. 32 **Multiple-Exhaust Single-Supply Installation**

REFERENCES

Barringer, C.G. and C.A. McGugan. 1989a. Development of a dynamic model for simulating indoor air temperature and humidity. ASHRAE *Transactions* 95(2):449-60.

Barringer, C.G. and C.A. McGugan. 1989b. Effect of residential air-to-air heat and moisture exchangers on indoor humidity. ASHRAE *Transactions* 95(2):461-74.

McDonald, T.W. and D. Shivprasad. 1989. Incipient nucleate boiling and quench study. Proceedings of Clima 2000 1:347-52. Sarajevo, Yugoslavia.

Mathur, G.D. 1990a. Indirect evaporative cooling using heat pipe heat exchangers. ASME Symposium, Thermal Hydraulics of Advanced Heat Exchangers, ASME Winter Annual Meeting, Dallas, TX, 79-85.

Mathur, G.D. 1990b. Indirect evaporative cooling using two-phase thermosiphon loop heat exchangers. ASHRAE *Transactions* 96(1):1241-49.

Phillips, E.G., R.E. Chant, B.C. Bradley, and D.R. Fisher. 1989a. A model to compare freezing control strategies for residential air-to-air heat recovery ventilators. ASHRAE *Transactions* 95(2):475-83.

Phillips, E.G., R.E. Chant, D.R. Fisher, and B.C. Bradley. 1989. Comparison of freezing control strategies for residential air-to-air heat recovery ventilators. ASHRAE *Transactions* 95(2):484-90.

Ruch, M.A. 1976. Heat pipe exchangers as energy recovery devices. ASHRAE *Transactions* 82(1):1008-14.

Scofield, M. and J.R. Taylor. 1986. A heat pipe economy cycle. ASHRAE *Journal* (October).

BIBLIOGRAPHY

ASHRAE. 1974. Symposium on air-to-air heat recovery. ASHRAE *Transactions* 80(2):302-32.

ASHRAE. 1975. Symposium on rotary air-to-air heat exchangers for energy recovery. ASHRAE *Transactions* 81(2):389-420.

ASHRAE. 1991. Method of testing air-to-air heat exchangers. ASHRAE *Standard* 84-91.

ASHRAE. 1982. Symposium on energy recovery from air pollution control. ASHRAE *Transactions* 88(1):1197-1225.

ASHRAE. 1977-1986. Research Reports for RP 140:

Ali, A.F.M. and T.W. McDonald. 1977. Thermosiphon loop performance characteristics: Part 2, Simulation program. ASHRAE *Transactions* 83(2):260-78.

Mather, G.D. and T.W. McDonald. 1986. Simulation program for a two-phase thermosiphon-loop heat exchanger. ASHRAE *Transactions* 92(2a):473-85.

McDonald, T.W. and A.F.M. Ali. 1977. Thermosiphon loop performance characteristics: Part 3, Simulated performance. ASHRAE *Transactions* 83(2):279-87.

McDonald, T.W., K.S. Hwang, and R. DiCiccio. 1977. Thermosiphon loop performance characteristics: Part 1, Experimental study. ASHRAE *Transactions* 83(2):250-59.

Stauder, F.A. and T.W. McDonald. 1986. Experimental study of a two-phase thermosiphon-loop heat exchanger. ASHRAE *Transactions* 92(2a):486-87.

ASHRAE. 1978-1986. Research Reports for RP 188:

Mather, G.D. and T.W. McDonald. 1986. Simulation program for a two-phase thermosiphon-loop heat exchanger. ASHRAE *Transactions* 92(2a):473-85.

McDonald, T.W., A.F.M. Ali, and S. Sampath. 1978. The unidirectional coil loop thermosiphon heat exchanger. ASHRAE *Transactions* 84(2):27-37.

McDonald, T.W. M. Kosnik, and G. Bertoni. 1985. Performance of a two-phase thermosiphon air-to-air heat exchanger. ASHRAE *Transactions* 91(2a):209-15.

McDonald, T.W. and S. Sampath. 1980. The bidirectional coil loop thermosiphon heat exchanger. ASHRAE *Transactions* 86(2): 37-47.

McDonald, T.W. and S. Raza. 1984. Effect of unequal evaporator heating and charge distribution on the performance of a two-phase thermosiphon loop heat exchanger. ASHRAE *Trans- actions* 90(2a):431-40.

Stauder, F.A. and T.W. McDonald. 1986. Experimental study of a two-phase thermosiphon-loop heat exchanger. ASHRAE *Transactions* 92(2a):486-87.

ASHRAE. 1988-1989. Research Reports for RP 544:

Barringer, C.G. and C.A. McGugan. 1988. Investigation of enthalpy residential air-to-air heat exchangers. Final Report for ASHRAE Research Project 544-RP (September).

Barringer, C.G. and C.A. McGugan. 1989a. Development of a dynamic model for simulating indoor air temperature and humidity. ASHRAE *Transactions* 95(2):449-60.

Barringer, C.G. and C.A. McGugan. 1989b. Effect of residential air-to-air heat and moisture exchangers on indoor humidity. ASHRAE *Transactions* 95(2):461-74.

ASHRAE. 1989. Research Reports for RP543:

Phillips, E.G., R.E. Chant, B.C. Bradley, and D.R. Fisher. 1989a. A model to compare freezing control strategies for residential air-to-air heat recovery ventilators. ASHRAE *Transactions* 95(2):475-83.

Phillips, E.G., R.E. Chant, B.C. Bradley, and D.R. Fisher. 1989b. An investigation of freezing control strategies for residential air-to-air heat exchangers. Report for ASHRAE Research Project 543 TRP (June).

Phillips, E.G., R.E. Chant, D.R. Fisher, and B.C. Bradley. 1989. Comparison of freezing control strategies for residential air-to-air heat recovery ventilators. ASHRAE *Transactions* 95(2):484-90.

Bosch, J.J., G.J. Gudac, R.H. Howell, M.A. Mueller, and H.J. Sauer. 1981. Effectiveness and pressure drop characteristics of various types of air-to-air energy recovery systems. ASHRAE *Transactions* 87(1):199-210.

Bowlen, K.L. 1974. Energy recovery from exhaust air. ASHRAE *Journal* (April):49-56.

Howell, R.H. and H.J. Sauer, Jr. 1981. Promise and potential of air-to-air energy recovery systems. ASHRAE *Transactions* 87(1):167-82.

Howell, R.H., H.J. Sauer, Jr., and J.R. Wray. 1981. Frosting and leakage testing of air-to-air energy recovery systems. ASHRAE *Transactions* 87(1):211-21.

Moyer, R.C. 1978. Energy recovery performance in the research laboratory. ASHRAE *Journal* (May).

Ruch, M.A. 1976. Heat pipe exchangers as energy recovery devices. ASHRAE *Transactions* 82(1):1008-14.

Scofield, M. and J.R. Taylor. 1986. A heat pipe economy cycle. ASHRAE *Journal*(October).

SMACNA. 1978. *Energy recovery equipment and systems*. Report.

CHAPTER 45

ROOM AIR CONDITIONERS AND DEHUMIDIFIERS

ROOM AIR CONDITIONERS

A ROOM air conditioner is an encased assembly designed as a unit primarily for mounting in a window or through a wall. These units are designed for the delivery of cool or warm conditioned air to the room, either without ducts or with very short ducts up to a maximum of about 48 in. Each unit includes a prime source of refrigeration, dehumidification, and means for circulating and cleaning air, and may also include means for ventilating and/or exhausting and heating. The ANSI/AHAM *Standard,* Room Air Conditioners (RAC 1-19-82), has further details.

The basic function of a room air conditioner is to provide comfort by cooling, dehumidifying, filtering or cleaning, and circulating the room air. It may also provide ventilation by introducing outdoor air into the room and/or exhausting room air to the outside. Also, comfort may be provided by controlling the room temperature through selection of the desired thermostat setting. The conditioner may provide heating by heat pump operation, electric resistance elements, or by a combination of both.

Figure 1 shows a typical room air conditioner. Warm room air passes over the cooling coil and, in the process, gives up its sensible and latent heat. The conditioned air is then recirculated in the room by a fan or blower.

The heat from the warm room air vaporizes the cold (low-pressure) liquid refrigerant flowing through the evaporator. The vapor then carries the heat to the compressor, which compresses the vapor and increases its temperature to a value higher than the temperature of the outdoor air. In the condenser, the hot (high-pressure) refrigerant vapor liquefies and gives up the heat from the room air to the outdoor air. The high-pressure liquid refrigerant then passes through a restrictor, which reduces its pressure and temperature. The cold (low-pressure) liquid refrigerant then reenters the evaporator to repeat this refrigeration cycle.

SIZES AND CLASSIFICATIONS

The cooling capacities of commercially available room air conditioners range from 4000 to 36,000 Btu/h.

Room air conditioners are equipped with line cords, which may be plugged into standard or special electric circuits. Most units are designed to operate at 115 V, 208 V, or 230 V, single-phase, 60-Hz power. Some units are rated at 265 V or 277 V, and the chassis or chassis assembly must provide for permanent

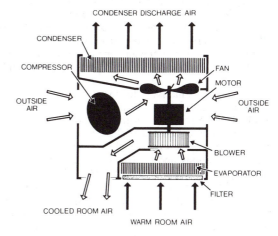

Fig. 1 Schematic View of Typical Room Air Conditioner

electrical connection, in accordance with the *National Electrical Code* (NEC). These units are typically used in multiple room installations (see Chapters 46 and 47). The maximum rating of 115-V units is generally limited to 12 A, since this is the maximum allowable current for a single-outlet, 15-A circuit permitted by the NEC.

A popular 115-V model is one rated at 7.5 A. The largest capacity 115-V units are in the 12,000 to 14,000 Btu/h range. Capacities for 230-, 208-, or 230/208-V units range from 8000 to 36,000 Btu/h. Capacities for 265- or 277-V units range from 6000 to 17,000 Btu/h. (For a more detailed discussion of wiring limitations, see the section Installation and Service.)

Models designed for countries other than the United States are generally for 50- or 60-Hz systems, with typical design voltage ranges of 100 to 120 and 200- to 240-V, single-phase. Many manufacturers apply 60-Hz units on a 50-Hz power supply, provided the supply voltage of the 50-Hz power supply is 87% of the nameplate voltage rating at 60 Hz. For example, 200 V at 50 Hz would be required for a unit rated 230 V at 60 Hz. The capacity of a unit rated at 60 Hz will be reduced by approximately the ratio 50/60 (0.833) when operated at 50 Hz.

Heat pump models are also available and are usually designed for 208- or 230-V applications. These units are generally designed for reversed refrigerant cycle operation as the normal means of supply heat, but they may incorporate electrical resistance heat to either supplement the heat pump capacity or provide the total heating capacity when outdoor temperatures drop below a predetermined value. Another type of heating model incorporates electrical heating elements in regular cooling units so that heating is provided entirely by electrical resistance heat.

The preparation of this chapter is assigned to TC 7.5, Room Air Conditioners and Dehumidifiers.

DESIGN

The design of a room air conditioner is usually based on one or more of the following criteria, any one of which automatically limits the freedom of the designer in overall system design:

- Lowest initial cost
- Lowest operating cost (highest efficiency)
- Energy efficiency ratio (EER), as legislated by the federal and/or state governments
- Low sound level
- Physical chassis size
- An unusual chassis shape (minimal depth, height, etc.)
- An amperage limitation (7.5 A, 12 A, etc.)
- Weight

Energy efficiency ratio (EER) is defined as the unit capacity in Btu/h divided by the power input to the unit in watts at the standard rating condition. The optimum design results from a carefully selected group of components consisting of an evaporator, a condenser, a compressor, one or more fan motors, air impellers for evaporator and condenser airflow, and a restrictor device.

The following combinations illustrate the effect on the various components as a result of an initial design parameter:

Low initial cost. High airflow with minimum heat exchanger surface keeps the initial cost low. These units have a low cost compressor, which is selected by analyzing various compressor and coil combinations and choosing the one that achieves optimum performance consistent with passing all tests required by UL, AHAM, etc. For example, a high capacity compressor might be selected to meet the capacity requirement with a minimum heat transfer surface, but frost tests under maximum load may not be acceptable. These tests set the upper and lower limits on what is acceptable when low initial cost is the prime consideration.

Low operating cost. Low air volumes with large heat exchanger surfaces keep operating costs low. A compressor with a low compression ratio operates at low head and high suction pressures, which results in a high EER.

Compressors

Room air-conditioner compressors range in capacity from approximately 4000 to 48,000 Btu/h. Design data are available from compressor manufacturers with rating conditions for standard and high-efficiency compressors:

	Standard, °F	High Efficiency, °F
Evaporating temperature	45	49
Evaporator suction temperature	95	51
Compressor suction temperature	95	65
Condensing temperature	130	120
Liquid temperature	115	100
Ambient temperature	95	95

Compressor manufacturers offer complete performance curves at various evaporating and condensing temperatures to aid in selection for a given design specification.

Evaporator and Condenser Coils

These coils are generally of either the tube-and-plate-fin variety or the tube-and-spine-fin variety. Information on the performance of such coils is available from suppliers, and original equipment manufacturers usually develop data for their own coils. Design parameters to be considered when selecting coils are (1) cooling rate per unit area of coil surface

($Btu/h \cdot ft^2$); (2) dry bulb and moisture content of the entering air; (3) air-side friction loss; (4) internal refrigerant pressure drop; (5) coil surface temperature; (6) air volume; and (7) air velocity.

Restrictor Application and Sizing

Essentially three types of restrictor devices are available to the designer: (1) a *thermostatic expansion valve,* which maintains a constant amount of superheat from a point near the outlet of the evaporator to a point on the suction line; (2) an *automatic expansion valve,* which maintains a constant suction pressure; and (3) a *restrictor tube* (*capillary*). The capillary is the most popular device for room air-conditioner applications because of its low cost and high reliability, even though the control of refrigerant over a wide range of ambient temperatures is not optimum. A recommended procedure for optimizing charge balance, condenser subcooling, and restrictor sizing is as follows:

1. Use an adjustable restrictor, so that a series of tests may be run with a flooded evaporator coil and various refrigerant charges to determine the optimum point of system operation.

2. Reset the adjustable restrictor to the optimum setting, remove from the unit, and measure flow pressure with a flow comparator similar to that described in ANSI/ASHRAE *Standard* 28-1988, Method of Testing Flow Capacity of Refrigerant Capillary Tubes.

3. Install a restrictor tube with the same flow rate as the adjustable restrictor. Usually, restrictor tubes are selected on the basis of cost, with shorter tubes generally being less expensive.

Fan-Motor and Air-Impeller Selections

The two types of motors generally used on room air conditioners are (1) the low-efficiency, shaded-pole type, and (2) the more efficient, permanent split-capacitor type, which requires the use of a run capacitor. Air impellers are usually of two types: (1) the forward-curved blower wheel and (2) the axial- or radial-flow fan blade. In general, *blower wheels* are used to move small to moderate amounts of air in a high-resistance system, and *fan blades* move moderate to high air volumes in low-resistance applications.

The combination of the fan motor and the air impeller is such an important part of the overall design that the designer should work closely with the manufacturers of both components. Performance curves are available for motors, blower wheels, and fans; unfortunately, however, the data are of ideal systems not usually found in practice because of physical size, motor speed, and component placement limitations.

Electronics

Microprocessors with the capability of monitoring and controlling numerous functions at essentially the same time have been incorporated into exceptionally capable control systems for room air conditioners. These microelectronic controls offer improved appearance with digital displays and touch panels that allow simple fingertip programming of desired temperature, on-off timing, modulated fan speeds, bypass capabilities, and sophisticated sensing for humidity, temperature, and airflow control.

Future capabilities include diagnostics, the monitoring of outside ambient air conditions for more precise indoor programming, and audio control, as demonstrated in other products. More effective energy saving is also possible with the inherent precision timing and temperature control.

PERFORMANCE DATA

Industry standards published for the performance of room air conditioners summarize all existing standards for rating, safety, and recommended performance levels. An industry certification program, under the sponsorship of AHAM, covers the majority of room air conditioners and certifies the cooling and heating capacities and electrical input (in amperes) of each for adherence to nameplate rating.

The following tests are specified by ANSI/AHAM *Standard RAC-1-1982*:

- Cooling capacity test
- Heating capacity test
- Maximum operating conditions test (heating and cooling)
- Enclosure sweat test
- Freezeup test
- Recirculated air quantity test
- Moisture removal test
- Ventilating air quantity and exhaust air quantity test
- Electrical input test (heating and cooling)
- Power factor test
- Condensate disposal test
- Application heating capacity test
- Outside coil deicing test

Efficiency

Efficiency for room air conditioners may be shown in either of two forms:

1. Coefficient of performance (COP—generally for heating)
   Capacity in Btu/h/Input in watts × 3.413
2. Energy efficiency ratio (EER—generally for cooling)
   Capacity in Btu/h/Input in watts

Sensible Heat Ratio

The ratio of sensible heat to total heat removal is a performance characteristic that is useful in evaluating units for specific conditions. A low ratio provides more dehumidification, and certain areas, such as New Orleans and Phoenix, might best be served with units having lower and higher ratios, respectively.

Energy Conservation and Efficiency

The rising cost of energy, the need for energy conservation, and mandated efficiency standards are key factors in the growing availability of high-efficiency units in all capacity sizes. Also, two federal energy programs have increased the demand for higher efficiency room air conditioners.

First, the *Energy Policy and Conservation Act* passed in December, 1975, Public Law 94-163, requires the Federal Trade Commission (FTC) to prescribe an energy usage label for many major appliances, including room air conditioners. The program provides consumers with operating cost data at the point of sale.

Second, the *National Energy Conservation Policy Act* (NECPA), passed in November, 1978, Public Law 95-619, directs the Department of Energy to establish minimum efficiency standards for major appliances, including room air conditioners. Subsequently, the *National Appliance Energy Conservation Act* of 1987 (NAECA) provides a single set of minimum efficiency standards for major appliances, including room air conditioners. The room air conditioner portion specifies minimum efficiencies for 12 classes, based on physical conformation. The minimums range from 8 to 9 EER and apply for all units built after January 1, 1990. These requirements will be reviewed prior to January 1995. All state and local minimum efficiency standards are automatically superseded.

Whether estimating potential energy savings associated with appliance standards or estimating consumer operating costs, the annual hours H of operation of a room air conditioner are of vital importance. These figures have been compiled from various studies commissioned by DOE and AHAM for every major city and region in the United States. The national average is estimated to be 750 h per year.

The cost of operation is as follows:

$$C = RHW/1000$$

where

C = cost of operation, dollars
R = average cost, \$/kWh
H = hours of operation
W = input, W

High-Efficiency Design

The EER can be affected by three design parameters. The first is *electrical efficiency*. Fan motors range from 25 to 65% in efficiency; compressor motors range from 60 to 85% in efficiency. The second parameter, *refrigerant cycle efficiency*, is increased by increasing the heat transfer surface to minimize the temperature differential between the refrigerant and the air. This allows the use of a smaller displacement compressor with a high-efficiency motor. The third parameter is *air circuit efficiency*, which can be increased by minimizing the airflow pressure drop across the heat transfer surface to reduce the load on the fan motor.

Table 1 shows how room air-conditioner costs and size are increased by increasing the EER from 6.5 to Table 1 values. The increases, which are approximate, range from 30 to 80% in mass and 30 to 100% in volume; small capacity units become less portable, and units larger than 20,000 Btu/h are not adaptable to window mounting.

Table 1 Effect of Increased EER over Base Value of 6.5

Capacity Range, Btu/h	EER	Weight Increment, Approx., %	Volume Increment, Approx., %	Price Increment, Approx., %
4000 to 10,000	10.0	39	48	50
12,000 to 20,000	10.5	52	60	37
24,000 to 27,000	11.0	58	80	40

Higher EERs are not the complete answer to reducing energy costs. More efficient use of energy can be accomplished by proper sizing of the unit, keeping infiltration and leakage losses to a minimum, increasing building insulation, reducing unnecessary internal loading, providing effective maintenance, and balancing the load by use of a thermostat and thermostat setback.

SPECIAL FEATURES

Some room air conditioners are designed to minimize their extension beyond the building when mounted flush with the inside wall. Low-capacity models are usually smaller and less obtrusive than higher capacity models. Units are often installed through the wall, where they do not interfere with windows. Exterior cabinet grilles are often designed to harmonize with the architecture of various buildings.

Most units have adjustable louvers or deflectors to distribute the air into the room with satisfactory throw and without drafts. The louver design should minimize recirculation of discharge air into the air inlet. Some units employ motorized deflectors for changing the air direction continuously. Discharge air velocities range from 300 to 1200 fpm, with low velocities preferred in rooms where people are at rest.

Most room air conditioners are designed for bringing in outdoor air, exhausting room air, or both. Controls usually permit these features to function independently.

Temperature is controlled by an adjustable built-in thermostat. The thermostat and unit controls may operate in one of the following modes:

1. The unit is set to the cool position and the thermostat setting is adjusted, as needed. The circulation blower runs without interruption while the thermostat cycles the compressor on and off.

2. Some air conditioners may use a two- or three-stage thermostat, which reduces the blower speed as the room temperature approaches the set temperature, cycles the compressor off on further temperature drop, and, finally, cycles the blower off on still further room temperature drop. As the room temperature rises, the sequence reverses. Any combination of these sequences is available.

3. Some air conditioners use, in addition to the control sequence above, an optional *automatic fan mode* when both the blower and the compressor are cycled simultaneously by the thermostat; this mode of operation requires close attention to proper thermostat sensitivity. One of the advantages with this arrangement is improved humidity control caused by not reevaporating moisture from the evaporator coil into the room on the off cycle. Another advantage is lower operating cost because the blower motor does not operate during the off cycle. The effective EER may be increased an average of 10% by using the automatic fan mode (Oak Ridge National Laboratory 1985).

Disadvantages to cycling fans with the compressor may be (1) changing noise level from fan cycling and (2) deterioration in room temperature control.

Room air conditioners are simple to operate. Usually, one control operates the unit electrically, while a second controls the temperature. Additional knobs or levers operate louvers, deflectors, the ventilation system, exhaust dampers, and other special features. The controls are usually arranged on the front of the unit or concealed behind a readily accessible door; however, they may also be arranged on the top or sides of the unit.

Filters on room air conditioners remove airborne dirt to provide clean air to the room and keep dirt off the cooling surfaces. Filters are made of expanded metal (with or without a viscous oil coating), glass fiber, or synthetic materials, and may be either disposable or reusable. A dirty filter reduces cooling and air circulation and frequently allows frost to accumulate on the cooling coil; therefore, filter location should allow for easy checking, cleaning, and replacement.

Some units have louvers or grilles at the rear to enhance appearance and protect the condenser fins. Sometimes, these louvers separate the airstreams to and from the condenser and reduce recirculation. Side louvers on the outside portion of the unit, when provided, are an essential part of the condenser air system. Care should be taken not to obstruct the air passages through these louvers.

The sound level of a room air conditioner (see the section Noise Standards) is an important factor, particularly when the unit is installed in a bedroom. A certain amount of sound can be expected because of the movement of air through the unit and the operation of the compressor. However, a well-designed room air conditioner is relatively quiet, and the emitted sound is relatively free from high pitched and metallic noise. Usually, fan motors with two or more speeds are used to provide a slower fan speed for quieter operation. To avoid

rattles and vibration in the building structure, units must be installed correctly (see the section Installation and Service.)

Room air conditioners require a durable finish, especially for those parts exposed to the weather. Some manufacturers use a special grade of plastic for weather-exposed parts; if metal is used, good practice calls for baked finishes over phosphatized or zinc-coated steels, and/or the use of corrosion-resistant materials such as aluminum or stainless steel.

NOISE STANDARDS

Sound reduction (both indoor and outdoor) is becoming increasingly important. Whereas indoor sound can be displeasing to the owner, excessive outdoor sound (condenser fan and compressor) can be irritating to the owner's neighbors. Many local and state outdoor noise ordinances are being considered, and some have been passed into law.

The Association of Home Appliance Manufacturers (AHAM) is developing an application standard that will allow the prediction of both outdoor sound pressure levels at distances that may be specified in a property line ordinance and the indoor sound pressure level that would exist in a typical room containing sound-absorbing furnishings.

SAFETY CODES AND STANDARDS

United States

The *National Electrical Code*, NFPA *Standard* 70; Underwriters Laboratories ANSI/UL *Standard* 484-1986, Room Air Conditioners; and ASHRAE *Standard* 15-1986, Safety Code for Mechanical Refrigeration pertain to room air conditioners. Local regulations may differ with these standards to some degree, but the basic requirements are generally accepted throughout the United States.

The *National Electrical Code*, sponsored by the National Fire Protection Association (NFPA), covers the broad area of electrical conductors and equipment installed within or on public and private buildings and other premises. Its purpose is the practical safeguarding of persons, buildings, and building contents from hazards that arise from the use of electricity for light, heat, power, radio, signaling, and other purposes. The Code contains basic minimum provisions considered necessary for safety. Proposals for modification of the Code can be made by any interested person or organization. Of primary interest to the room air-conditioner designer is Article 440, Air Conditioning and Refrigerating Equipment.

The *Safety Code for Mechanical Refrigeration* (ANSI/ASHRAE *Standard* 15-1989) covers field-erected equipment and is intended to ensure the safe design, construction, installation, operation, and inspection of every refrigerating system that vaporizes and normally liquefies a fluid in its refrigeration cycle. It provides reasonable safeguards to life, health, and property; corrects practices inconsistent with safety; and prescribes safety standards that will influence future progress and developments in refrigerating systems.

Both codes recognize equipment and materials submitted for test by an independent organization under uniform conditions; such testing agencies also have follow-up inspection services of current production runs. Underwriters Laboratories prepares standards for safety, applies such requirements to products voluntarily submitted by manufacturers, and, by means of follow-up inspections, determines that production of the equipment continues to comply with requirements. The UL Standard for Room Air Conditioners is divided into two sections — Construction and Performance; these set minimum requirements necessary to protect against casualty, fire, and electrical shock hazards.

Canada

The Canadian Standards Association has developed the Standard for Room Air Conditioners (C22.2 No. 117-70), which forms part of the *Canadian Electrical Code.*

International (Other than Canada)

Two useful documents that might assist the designer are (1) International Electrochemical Commission (IEC) *Standard* 378, Safety Requirements for the Electrical Equipment of Room Air-Conditioners, and (2) ISO *Standard* R859, Testing and Rating Room Air-Conditioners. Both standards, which are currently under revision, are available from the American National Standards Institute (ANSI), New York.

Product Standards

The CSA *Standard* C22.2 No. 117-70 and ANSI/UL *Standard* 484-1986 are similar in content. Using the UL standard as an example, safety from a casualty viewpoint involves such items as the unit enclosure (including materials), the unit's ability to protect against contact with moving and uninsulated live parts, and the means for unit installation or attachment. Attention is also given to the refrigeration system's ability to withstand operating pressures, pressure relief of the system in the event of fire, and toxicity of the refrigerant. Electrical considerations include supply connections, grounding, internal wiring and wiring methods, electrical spacings, motors and motor protection, uninsulated live parts, motor controllers and switching devices, air-heating components, and electrical insulating materials.

The performance section of the standard includes a rain test for determining the unit's ability to stand a beating rain without creating a shock hazard because of current leakage or insulation breakdown. Other tests include (1) leakage current limitations based on the ANSI *Standard* C101.1-1986 for leakage current for appliances; (2) measurement of input currents for the purpose of establishing nameplate ratings and for sizing the supply circuit for the unit; (3) temperature tests to determine whether or not components exceed their recognized temperature limits and/or electrical ratings (ANSI/AHAM *Standard* RAC-1-1982); (4) pressure tests to ensure that excessive pressures do not develop in the refrigeration system.

Abnormal conditions are also considered, such as (1) failure of the condenser fan motor, which may result in excessive pressures being developed in the system; and (2) air heater burnout, to determine whether combustibles within or adjacent to the unit may be ignited. A static load test is also conducted on window-type room air conditioners to determine whether or not the mounting hardware can support the unit adequately. As part of normal production control, tests are conducted for leakage of the refrigeration system, dielectric withstand, and grounding continuity.

Plastic materials are receiving increased consideration in the design and fabrication of room air conditioners because of their ease in forming inherent resistance to corrosion and their decorative qualities. When considering the use of plastic, the engineer should consider the tensile, flexural, and impact strength of the material; the flammability characteristics; and—from the standpoint of degradation—the resistance to water absorption, exposure to ultraviolet light, ability to operate at elevated temperatures, and thermal aging characteristics. From a product safety standpoint, some of these factors are of lesser importance, since failure of the part will not result in a hazardous condition. For other parts, such as the bulkhead, base pan, and unit enclosure, which either support components or provide structural integrity, all of the preceding factors must be considered, and a complete analysis of the material must be made to determine that it is suitable for the application.

INSTALLATION AND SERVICE

Installation procedures vary because units can be mounted in various ways. It is important to select the particular mounting for each installation that best satisfies the user and complies with existing building codes. Common mounting methods include the following:

Inside flush mounting. Interior face of conditioner is approximately flush with inside wall.

Balance mounting. Unit is approximately half inside and half outside the window.

Outside flush mounting. Outer face of unit is flush or slightly beyond outside wall.

Special mounting. Includes casement windows, horizontal sliding windows, office windows with swinging units (or swinging windows) to permit window washing, and transoms over doorways.

Through-the-wall mounts or sleeves. Used for installing window-type chassis, complete units, or consoles in walls of apartment buildings, hotels, motels, and residences.

Room air conditioners have become more compact in size because of consumer preference for minimum loss of window light and minimum projection both inside and outside the structure. Several types of expandable mounts are now available for fast, dependable installation in single- and double-hung windows, as well as windows of the horizontal sliding type. Installation kits provide all parts needed for structural mounting, such as gaskets, panels, and seals for weathertight assembly.

Adequate wiring and proper fuses must be provided for the service outlet. The necessary information is usually given on instruction sheets or stamped on the air conditioner near the service cord or on the serial plate. It is important to follow the manufacturer's recommendation for size and type of fuse. All units are equipped with grounding-type plug caps on the service cord as received from the manufacturer. Receptacles with a grounding contact correctly designed to fit the air-conditioner service cord plug cap should be used when units are installed.

Units rated 265 V or 277 V must provide for permanent electrical connection with armored cable or conduit to the chassis or chassis assembly. It is customary to provide an adequate cord and plug cap within the chassis assembly to facilitate installation and service.

One type of room air conditioner is the *integral chassis* design, with the outer cabinet fastened permanently to the chassis. Most of the electrical components can be serviced by partially dismantling the control area without removing the unit from the installation.

Another type of room air conditioner is the *slide-out chassis* design, which allows the outer cabinet to remain in place while the chassis is removed for service.

DEHUMIDIFIERS

This section deals with domestic dehumidifying units that are self-contained, electrically operated, and mechanically refrigerated. A dehumidifier consists of a motor-compressor unit, a refrigerant condenser, an air-circulating fan, a refrigerated surface (refrigerant evaporator), a means for collecting and/or disposing of the condensed moisture, and a cabinet to house these components. A typical dehumidifier unit is shown in Figure 2.

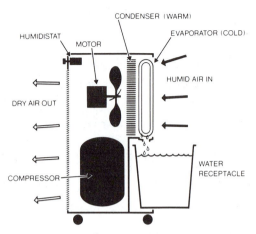

Fig. 2   Typical Dehumidifier Unit

The fan draws the moist room air through the cold coil and cools it below its dew point, removing moisture that either drains into the water receptacle or passes through a drain into the sewer. The cooled air then passes through the condenser, where it is reheated. Then, with the addition of other unit-radiated heat, the air is discharged into the room at a higher temperature but at a lower relative humidity. Continuous circulation of room air through the dehumidifier gradually reduces the relative humidity of the room.

DESIGN AND CONSTRUCTION

Dehumidifiers use hermetic-type motor-compressors of various ratings, depending on the designed output of the overall unit. The refrigerant condenser is usually a conventional finned-tube coil. Refrigerant flow metering is usually provided by a capillary tube, although some high-capacity dehumidifiers may use an expansion valve.

A propeller-type fan, direct-driven by a shaded-pole motor, moves the air through the dehumidifier. The airflow rate through typical dehumidifiers ranges from 125 to 250 cfm, depending on the unit's moisture output capacity. In general, the output capacity of a dehumidifier increases with an increase in airflow. Extremely high rates of airflow, however, may cause objectionable sound. Domestic dehumidifiers ordinarily maintain satisfactory humidity levels in an enclosed space when the airflow rate and unit placement permit the entire air volume of the space to be moved through the dehumidifier once an hour. As an example, an airflow rate of 200 cfm would provide one air change per hour in a room having a volume of 12,000 ft³.

The refrigerated surface (evaporator) is usually a bare-tube coil structure, although finned-tube coils can be used if they are spaced to permit rapid runoff of water droplets. Vertically disposed bare-tube coils tend to collect smaller drops of water, promote quicker runoff, and cause less water reevaporation than do finned-tube cooling coils or horizontally arranged bare-tube coils. Continuous bare-tube coils, wound in the form of a flat circular spiral (sometimes consisting of two coil layers) and mounted with the flat dimension of the coil in a vertical plane, are a good design compromise because they have most of the advantages of the vertical-tube coil.

Evaporators are protected against corrosion by such processes as painting, waxing, and anodizing (on aluminum). Waxing reduces the wetting effect that promotes condensate formation; tests on waxed evaporator surfaces versus non-waxed surfaces show a negligible loss in capacity. Thin paint films do not have an appreciable effect on capacity.

Removable water receptacles, provided with most dehumidifiers, hold from 16 to 24 pints and are usually made of plastic to withstand corrosion. Ease of removal and handling without spilling are important. Most dehumidifiers also provide a means of attaching a flexible hose to either the water receptacle or a fitting specially provided for that purpose. A flexible hose permits direct gravity drainage to a sewer.

Dehumidifier cabinet enclosures vary with respect to aesthetic design. The more expensive models usually have higher output capacity, are more attractively styled, and are provided with various auxiliary features. The addition of a humidity-sensing control (to cycle the unit automatically) maintains a preselected relative humidity. These humidistats are normally adjustable within a range of 30 to 80% rh. The humidistat may also provide a detent setting that causes continuous running of the unit. Most humidistats provide an on-off line switch; some models also include an additional sensing and switching device, which automatically turns the unit off when the water receptacle is full and requires emptying. This second device may or may not include an indicating light.

Dehumidifiers are designed to provide optimum performance at standard rating conditions of 80°F dry-bulb room ambient and 60% rh. When the room is less than 65°F dry bulb and 60% rh, then the refrigerant pressure and corresponding evaporator surface temperature usually decrease to the point where frost forms on the cooling coil. This effect is especially noticeable on units having a capillary tube for refrigerant flow metering.

Some dehumidifier models are equipped with special defrost controls, which cycle the compressor off under frosting conditions. This type of control is generally a bimetal thermostat that is strategically attached to the evaporator tubing, allowing the dehumidifying process to continue at a reduced rate when frosting conditions exist. The humidistat can sometimes be adjusted to a higher relative humidity setting, which reduces the number and duration of running cycles and permits reasonably satisfactory operation at low load conditions. In many instances, especially in the late fall and early spring, supplemental heat must be provided from other sources to maintain conditions within the space above those at which frosting can occur.

Dehumidifiers are usually equipped with rollers or casters. Typical sizes of dehumidifiers vary, as follows:

Width: 10 to 20 in.   Height: 12 to 24 in.
Depth: 10 to 20 in.   Weight: 35 to 75 lb

CAPACITY AND PERFORMANCE RATING

Dehumidifiers are available with moisture removal capacities ranging from 11 to 50 pints per 24-h day. The input to domestic dehumidifiers varies from 200 to 700 W, depending on the output capacity rating. Dehumidifiers are operable from ordinary household electrical outlets (115-V, single-phase, 60 Hz).

ANSI/AHAM *Standard* DH-1-1986, Dehumidifers, establishes a uniform procedure for determining the rated capacity of dehumidifiers under certain specified test conditions and establishes other recommended performance characteristics as well. An industry certification program, sponsored by AHAM, covers the great majority of dehumidifiers and certifies the dehumidification capacity (Directory of Certified Dehumidifiers). The following tests are specified by the dehumidifier standard.

Dehumidifier capacity. Units must be tested in a room maintained at the following test conditions:

Dry-bulb temperature, 80°F
Wet-bulb temperature, 69.6°F
Relative humidity, 60%

The capacity is stated in pints/24 h.

Maximum operation conditions. The test is conducted in conditions of 90°F dry bulb, 74.8°F wet bulb, and 50% rh. The unit runs continuously at this condition at 90 and 110% of rated voltage for 2 h, after which the power is cut off for 120 s. The unit should restart within 300 s and run continuously for 1 h.

Low-temperature test. The unit is operated in conditions of 65°F dry bulb, 56.6°F wet bulb, and 60% rh for a period of 8 consecutive hours. If the unit is not equipped with a defrost control, there should not be ice or frost remaining on the portion of the evaporator coil that is exposed to the entering airstream. If the unit is equipped with a defrost control, it should operate for at least 50% of the test period with no solid ice or frost remaining on the portion of the evaporator coil that is exposed to the entering airstream at the end of each cycle.

CODES

Dehumidifiers are designed to meet the safety requirements of ANSI/UL *Standard* 474-1987, Dehumidifiers; Canadian Standards Association, *Canadian Electrical Code,* Part II, Specification C22.2, No. 32, Construction and Test of Electrically Operated Refrigerating Machines; and ANSI/ASHRAE *Standard* 15-1989, Safety Code for Mechanical Refrigeration.

It is customary for UL-listed and CSA-approved equipment to have a label or data plate that indicates approval. UL also publishes the *Electric Appliance and Utilization Equipment List,* which covers this type of appliance.

ENERGY CONSERVATION AND EFFICIENCY

The *Energy Policy and Conservation Act* of 1976 (EPCA), amended in 1978 by the *National Energy Conservation Policy Act* (NECPA), also applies to dehumidifiers but on a lower priority basis. A test procedure has been issued by the DOE, including an energy factor determination to evaluate efficiency. Dehumidifiers have been exempted from the FTC Labeling Requirements (Federal Register, November 19, 1979) and the DOE Minimum Standards Program.

REFERENCES

AHAM. 1982. Room air conditioners. ANSI/AHAM *Standard* RAC-1-1982. Association of Home Appliance Manufacturers, Chicago.

AHAM. 1987. *Directory of certified dehumidifiers.* Association of Home Appliance Manufacturers, Chicago, IL.

ANSI/ASHRAE. 1988. Method of testing flow capacity of refrigerant capillary tubes. ANSI/ASHRAE *Standard* 28-1988.

ANSI/ASHRAE. 1989. Safety code for mechanical refrigeration. ANSI/ASHRAE *Standard* 15-1989.

CSA. 1982. Room air conditioners. CSA *Standard* C22.2, No. 117. Canadian Standards Association, Rexdale, Ontario, Canada.

IEC. *Safety requirements for the electrical equipment room air conditioners.* IEC378, International Electrotechnical Commission. Available from American National Standards Institute, New York.

ISO. Testing and rating room air conditioners. International Standards Organization. *ISO* R859, Available from American National Standards Institute, New York.

NFPA. 1987. *National electrical code.* ANSI/NFPA *Standard* 70-1990. National Fire Protection Association, Quincy, MA.

Oak Ridge National Laboratory. *Report* ORNL-NSF-EP-85.

UL. 1986. Room air conditioners. ANSI/UL *Standard* 484-1986. Underwriters Laboratories, Inc., Northbrook, IL.

UL. 1987. Dehumidifers. ANSI/UL *Standard* 474-1987, Underwriters Laboratories, Inc., Northbrook, IL.

CHAPTER 46

UNITARY AIR CONDITIONERS AND HEAT PUMPS

UNITARY EQUIPMENT CONCEPT

THE Air Conditioning and Refrigeration Institute (ARI) defines a unitary air conditioner as one or more factory-made assemblies that normally include an evaporator or cooling coil and a compressor and condenser combination. It may include a heating function, as well.

ARI further defines air source unitary heat pumps as consisting of one or more factory-made assemblies, which normally include an indoor conditioning coil, compressor(s), and an outdoor coil. It must provide a heating function, and possibly a cooling function as well.

Unitary equipment is divided into three general categories: residential, light commercial, and commercial equipment. The design concept is similar, but specific equipment design, application, and performance measurement methods vary. Residential equipment is single-phase unitary equipment with a cooling capacity of 65,000 Btu/h or less, and is designed specifically for residential application. Light commercial equipment is generally three phase, with cooling capacities up to 135,000 Btu/h, and is designed for small businesses and commercial properties. Commercial unitary equipment has cooling capacities higher than 135,000 Btu/h, and is designed for large commercial buildings.

In the development of unitary equipment, the following design objectives are considered: (1) user requirements, (2) application requirements, (3) installation, and (4) service.

User Requirements

The user primarily needs either comfort or a controlled environment for products or manufacturing processes. Cooling, dehumidification, filtration, and air circulation often meet those needs, although heating, humidification, and ventilation are also required in many applications.

Application Requirements

Unitary equipment is available in many secondary system configurations, such as:

Single zone, Constant volume consists of one controlled space with one thermostat that controls operations to match the sensible load.

The preparation of this chapter is assigned to TC7.6, Unitary Air Conditioners and Heat Pump.

Multizone, Constant volume has several controlled spaces, served by one unit that supplies air of different temperatures to different zones as demanded (Figure 1).

Single zone, Variable volume consists of several controlled spaces served by one unit. Supply air from the unit is at a constant temperature, with air volume to each space varied to satisfy space demands (Figure 2).

Such factors as size, shape, and use of the building, availability and cost of fuels, building aesthetics (equipment located outdoors), and space available for equipment are considered to determine the type of unitary equipment best suited to a given application. In general, roof-mounted single-package unitary equipment is limited to five or six stories because duct space and available blower power become burdensome in taller buildings. Indoor, single-zone equipment is generally less expensive to maintain and service than multizone units located outdoors.

The building load and airflow requirements determine equipment capacity, whereas the availability and cost of fuels determine the energy source. Control system requirements must be established and any unusual operating conditions must be considered early in the planning stage. In some cases, custom-designed equipment may be necessary. Of course, suitable, standard equipment is more economical.

Manufacturers' literature has detailed information about geometry, performance, electrical characteristics, application, and operating limits. The system designer then focuses on selecting suitable equipment with the capacity for the application.

Installation

Most manufacturers' installation instructions describe step-by-step procedures that allow orderly planning of labor, materials, and use of tools. A wiring diagram and installation and service manual(s) are often included to help the installer. Unitary equipment is designed to keep installation costs low. The installer must ensure that equipment is installed properly so that it functions effectively and in accordance with the criteria contained in the manufacturers' literature. Interconnecting diagrams for the low-voltage control system should be documented for proper servicing in the future. Adequate planning for the installation of large, roof-mounted equipment is important, since special rigging equipment is frequently required.

A refrigerant system must be clean, dry, and leak free. An advantage of packaged unitary equipment is that proper installation minimizes the risk of field contamination of the refrigerant system. Care must be taken to properly install split-system interconnecting tubing (including proper cleanliness, brazing, and pull-

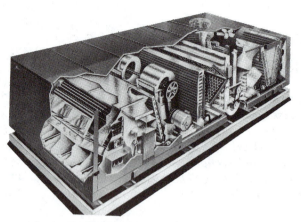

Fig. 1   Typical Rooftop Air-Cooled Single-Package Air Conditioner

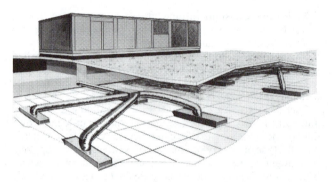

Fig. 2   Single-Package Air Equipment with Variable Air Volume

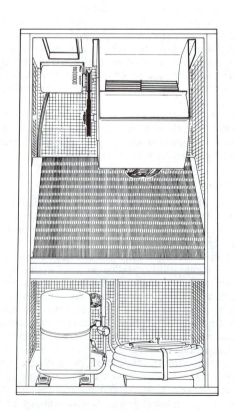

Fig. 3   Water-Cooled Single-Package Air Conditioner

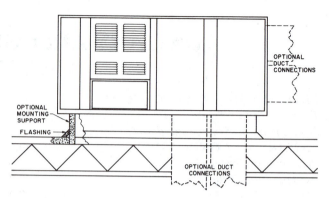

Fig. 4   Rooftop Installation of Air-Cooled Single-Package Unit

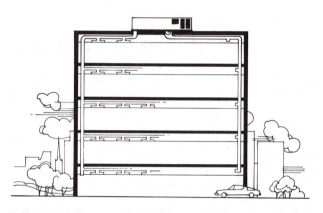

Fig. 5   Multistory Rooftop Installation of Single-Package Unit

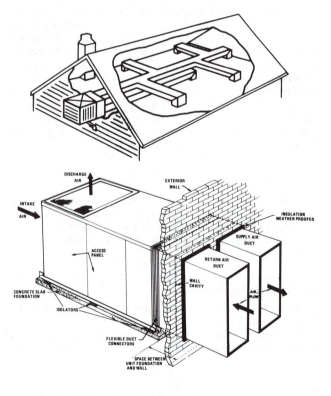

Fig. 6   Through-the-Wall Installation of Air-Cooled Single-Package Unit

ing of a vacuum on the system) to remove moisture. Some residential split systems are provided with precharged line sets and quick-connection couplings, which reduce the risk of field contamination of the refrigerant system.

In the installation of split systems, which require field installing of piping from the evaporator to the condenser, particular care must be taken to remove contaminants and moisture from the system. Lines must be properly routed and sized to assure good oil return to the compressor. Traps in the lines, particularly the suction line, also can inhibit good oil return. Chapters 6 and 7 of the 1990 ASHRAE *Handbook—Refrigeration* have more detail.

Unitary equipment must be located to avoid noise and vibration problems. Single-piece equipment of over 20 tons capacity should be mounted on concrete pads when vibration is critical. Large-capacity equipment should be roof mounted only after the structural adequacy of the roof has been evaluated. Roof-mounted units with return fans that use ceiling space for the return plenum should always have a minimum 8 ft long, lined return plenum, if they are located over occupied space. Duct silencers should be used where low sound levels are desired. Weight and sound data are available from many manufacturers. Additional installation guidelines include:

• Compressor-bearing products, in general, should be installed on solid, level surfaces.
• Avoid mounting compressor-bearing products (like remote units) on or touching the foundation of a house or building. A separate pad that does not touch the foundation is recommended to reduce any noise and vibration transmission through the slab.
• Do not box in outdoor air-cooled units with fences, walls, overhangs, or bushes. Doing so will reduce the air-moving capabilities of the unit, thus reducing the efficiency.
• When choosing an installation site for a split-system remote unit, choose a location that is close to the indoor portion of the system to minimize the line lengths and turns in the tubing.
• Contact the unitary equipment manufacturer or consult the unit's installation instructions for further information on installation locations and procedures.

Unitary equipment should be listed or certified by nationally recognized testing laboratories to ensure safe operation and compliance with government and utility regulations. The equipment should also be installed to comply with the rating and application requirements of the agency standards to ensure that it performs according to industry criteria. Larger and more specialized equipment often does not carry agency labeling. However, power and control wiring practices should comply with the National Electrical Code to facilitate acceptance by local inspection authorities. Local codes should be consulted before designing the installation; local inspectors should be consulted before installation.

Service

A clear and accurate wiring diagram and a well-written service manual are essential to the installer and service personnel. Easy and safe service access must be provided in the equipment for periodic maintenance of filters and belts, cleaning, and lubrication. In addition, access for replacement of major components must also be provided and preserved.

The availability of replacement parts aids proper service. Equitable warranty policies, covering one year of operation after installation, are offered by most manufacturers. Extended compressor warranties may be either standard or optional.

Service personnel must be qualified to repair or replace equipment components. They must also understand the importance of controlling moisture and other contaminants within the refrigerant system; they should know how to clean the hermetic system if it has been opened for service (see Chapter 7 of the 1990

ASHRAE *Handbook—Refrigeration*. Proper service procedures help ensure that the equipment will continue operating effciently for its expected life.

TYPES OF UNITARY EQUIPMENT

Table 1 shows the types of unitary air-conditioning equipment available, and Table 2 shows the types of unitary heat pumps available. The following variations apply to some types and sizes of unitary equipment:

Arrangement. Major unit components for various types of unitary air conditioners are arranged as shown in Table 1 and for unitary heat pumps as shown in Table 2.

Heat rejection. Unitary air-conditioner condensers may be air cooled, evaporatively cooled, or water cooled; the letters A, E, or W follow the ARI-type system designation.

Heat source/Sink. Unitary heat-pump outdoor coils are designated air-source or water-source by A or W following ARI practice. The same coils that act as a heat sink in the cooling mode act as the heat source in the heating mode.

Unit exterior. The unit exterior should be decorative for in-space application, functional for equipment room and ducts, and weatherproofed for outdoors.

Table 1   Classification of Unitary Air Conditioners

System Designation	ARI-Type	Heat Rejection	Arrangement
Single package	SP-A	Air	Fan | Comp ; Evap | Cond
	SP-E	Evap Cond	
	SP-W	Water	
Refrigeration chassis	RCH-A	Air	Comp ; Evap | Cond
	RCH-E	Evap Cond	
	RCH-W	Water	
Year-round single package	SPY-A	Air	Fan ; Heat | Comp ; Evap | Cond
	SPY-E	Evap Cond	
	SPY-W	Water	
Remote condenser	RC-A	Air	Fan ; Evap ; Comp ; Cond
	RC-E	Evap Cond	
	RC-W	Water	
Year-round remote condenser	RCY-A	Air	Fan ; Evap ; Heat ; Comp ; Cond
	RCE-E	Evap Cond	
	RCY-W	Water	
Condensing unit, coil alone	RCU-A-C	Air	Evap | Cond ; Comp
	RCU-E-C	Evap Cond	
	RCU-W-C	Water	
Condensing unit, coil and blower	RCU-A-CB	Air	Fan | Cond ; Evap | Comp
	RCU-E-CB	Evap Cond	
	RCU-W-CB	Air	
Year-round condensing unit coil and blower	RCUY-A-CB	Air	Fan ; Evap | Cond ; Heat | Comp
	RCUY-E-CB	Evap Cond	
	RCUY-W-CB	Water	

Note: A suffix of "O" following any of the above classifications indicates equipment not intended for use with field-installed duct systems.

Table 2 Classification of Unitary Heat Pump

Types of Unitary Heat Pumps				
	ARI-Type[a]			
System Desig-nation	Heating and Cooling	Heating Only	Arrangement	
Single package	HSP-A HSP-W	HOSP-A HOSP-W	Fan / Indoor Coil	Comp / Outdoor Coil
Remote outdoor coil	HRC-A-CB	HORC-A-CB	Fan / Indoor Coil / Comp	Outdoor Coil
Remote outdoor coil with no indoor fan	HRC-A-C	HORC-A-C	Indoor Coil / Comp	Outdoor Coil
Split system	HRCU-A-CB HRCU-W-CB	HORCU-A-CB HORCU-W-CB	Fan / Indoor Coil	Comp / Outdoor Coil
Split system, no indoor fan	HRCU-A-C	HORCU-A-C	Indoor Coil	Comp / Outdoor Coil

[a]A suffix of "O" following any of the above classifications indicates equipment not intended for use with field-installed duct systems.

Placement. Unitary equipment can be mounted on floors, walls, ceilings, roofs, or on the ground on a pad.

Indoor air. Equipment with fans may have airflow arranged for vertical upflow or downflow, horizontal flow, 90 or 180° turns, or multizone. Indoor coils without fans are intended for forced-air furnaces or blower packages. Variable-volume blowers may be incorporated with any system.

Location. Unitary equipment intended for indoor use may be placed in exposed locations with plenums or furred-in ducts, or concealed in closets, attics, crawl spaces, basements, garages, utility rooms, or equipment rooms. Wall-mounted equipment may be attached to or built into a wall or transom. Outdoor equipment may be mounted on roofs or concrete pads on the ground.

Heat. Unitary systems may incorporate gas-fired, oil-fired, electric, or hot-water or steam-coil heating sections. In unitary heat pumps, these heating sections provide supplemental heating capability.

Ventilation air. Outdoor air dampers may be built into the equipment to provide outdoor air for cooling or ventilation.

Desuperheaters. Desuperheaters may be applied to unitary air conditioners and heat pumps. These devices recover heat from the compressor discharge gas and use it to heat domestic hot water. The desuperheater usually consists of a pump, heat exchanger, and controls, and it can produce about 15 gal of heated water per hour per ton of air conditioning. Because desuperheaters improve cooling performance and reduce the degrading effect of cycling during heating, they are best applied where cooling requirements are high and in climates where a significant number of heating hours occur above the building's balance point (Counts 1988). While properly applied desuperheaters can improve cooling efficiency, they can also reduce heating capacity. A decrease in the heating capacity causes the unit to run longer, which reduces the cycling of the system.

The manufacturer of the unitary equipment should be consulted before installing any accessories or equipment not specifically approved by the manufacturer. Such installations may not only void the warranty, but could cause the unitary equipment to not function as intended.

Unitary equipment is usually designed with fan capability for ductwork, although some units may be designed to discharge directly into the conditioned space.

Figures 1 through 6 show single-package equipment. Figure 7 shows a typical installation of a split-system, air-cooled condensing unit with indoor coil, the most widely used unitary cooling system. Figures 8 and 9 also show split-system condensing units with coils and blower coil units.

Many special light commercial and commercial unitary equipment include the single-packaged air conditioner for use with variable-air-volume systems, as shown in Figure 2. These units are often equipped with a factory-installed system for controlling air volume in response to supply duct pressure.

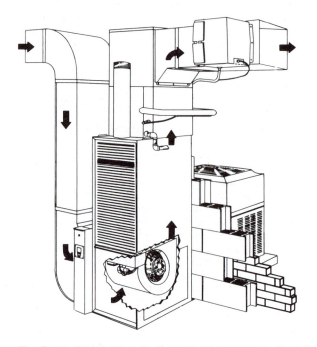

Fig. 7 Residential Installation of Split-System Air-Cooled Condensing Units with Coil and Upflow Furnace

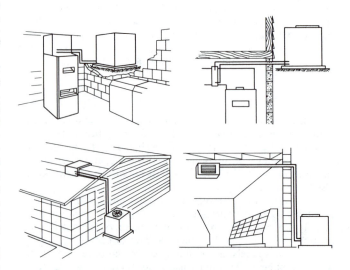

Fig. 8 Outdoor Installations of Split-System Air-Cooled Condensing Units with Coil and Upflow Furnace, or with Indoor Blower Coils

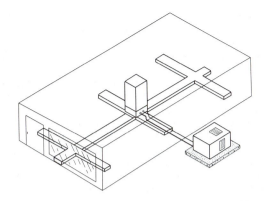

Fig. 9 Outdoor Installation of Split-System Air-Cooled Condensing Unit with Indoor Coil and Downflow Furnace

Another example of a specialized unit is the multizone unit shown in Figure 1. The manufacturer usually provides all controls, including zone dampers. The air path in these units is designed so that supply air may flow through a hot deck containing a means of heating or through a cold deck, which usually contains a direct-expansion evaporator coil.

To make multizone units more efficient, a control is commonly provided that locks out cooling by refrigeration when the heating unit is in operation and vice versa. Another variation to improve efficiency is the three-deck multizone. This unit has a hot deck, a cold deck, and a neutral deck carrying return air. Hot and/or cold deck air mix only with air in the neutral deck.

EQUIPMENT AND SYSTEM STANDARDS

Energy Conservation and Efficiency

Two U.S. programs have increased the demand for higher efficiency residential unitary air conditioners and unitary heat pumps. The Energy Policy and Conservation Act, passed in December 1977, Public Law 95-163, requires the Federal Trade Commission (FTC) to prescribe an energy label for many major appliances, including unitary air conditioners and heat pumps. The National Energy Conservation Policy Act (NECPA), passed in November 1978, Public Law 95-619, directs the Department of Energy (DOE) to establish minimum efficiency standards for major appliances, including room air conditioners. Subsequently, the National Appliance Energy Conservation Act (NAECA) passed in January 1987, Public Law 100-12, provides minimum efficiency standards for major appliances, including unitary air conditioners and heat pumps. All local minimum efficiency standards are automatically superseded.

The U.S. Department of Energy rating procedure for residential unitary equipment, updated on March 14, 1988, is in accordance with the NAECA minimum efficiency requirements for residential unitary equipment. Effective dates are the early 1990s (Public Law 100-12). The DOE testing and rating procedure is documented in Appendix M to subpart 430 of Section 10 of the Code of Federal Regulations—Uniform Test Method for Measuring the Energy Consumption of Central Air Conditioners.

This testing procedure provides a seasonal measure of operating efficiency for multiple types of residential equipment (*e.g.*, single speed, variable speed, two compressor/speed). Seasonal Energy Efficiency Ratio (SEER) is the ratio of the total seasonal cooling requirements measured in Btu to the total seasonal watt hours of input energy. This efficiency value is developed in the laboratory by conducting tests at various indoor and outdoor conditions, including a measure of performance under cyclic operation.

Seasonal heating mode efficiencies of heat pumps are similarly expressed as the ratio of the total heating requirement and the total seasonal input energy. This measure of efficiency is expressed as a Heating Seasonal Performance Factor (HSPF). In the laboratory, this efficiency measurement is determined from the results of different testing conditions, including a measure of cyclic performance. The magnitude of the HSPF measurement depends not only on the equipment performance, but also on the climatic conditions and the heating load relative to the equipment capacity.

For HSPF rating purposes, the DOE has divided the United States into six climatic regions and has defined a range of maximum and minimum design loads. This division has the effect of producing 12 different HSPF ratings for a given piece of equipment. For comparison purposes, DOE has established Region 4 (moderate northern climate) as being a typical climatic region.

SEER, HSPF, and operating costs vary appreciably with equipment design and size, and from manufacturer to manufacturer. SEER and HSPF values, ranges, and unit operating costs for DOE-covered unitary air conditioners are published semiannually in the ARI *Directory of Certified Unitary Air-Conditioners, Unitary Air-Source Heat Pumps and Sound-Rated Outdoor Unitary Equipment.*

Performance Rating—ARI Certification Programs

The Air Conditioning and Refrigeration Institute (ARI) conducts three certification programs relating to unitary equipment, which are covered in ARI *Standard* 210/240-84, Unitary Air-Conditioning and Air-Source Heat Pump Equipment, and ARI *Standard* 270-84, Sound Rating of Outdoor Unitary Equipment.

These standards include the performance requirements necessary for good equipment design. They also include the methods of testing established by ANSI/ASHRAE *Standard* 37-1988, Methods of Testing for Rating Unitary Air-Conditioning and Heat Pump Equipment.

As part of its certification program, ARI publishes the *Directory of Certified Unitary Products.* Issued twice a year, this directory identifies certified products enrolled in one or more programs and lists the certified capacity and energy efficiency for each.

Certification involves the annual audit testing of approximately 30% of the basic unitary equipment models of each participating manufacturer. These programs are limited to equipment with up to 135,000 Btu/h cooling capacity.

ARI *Standard* 210/240-84 for unitary equipment established definitions and classifications, testing and rating methods, and performance requirements. Ratings are determined at ARI standard rating conditions, rated nameplate voltage, and prescribed discharge duct static pressures with rated evaporator airflow not exceeding 37.5 cfm per 1000 Btu/h. The standard requires units to dispose of condensate properly and prohibits them from sweating under cool, humid conditions. Also, the ability to operate satisfactorily and restart at high ambient temperatures with low voltage is also tested.

For certification under ARI *Standard* 270-84, outdoor equipment is tested in accordance with ASA *Standard* 92-90. Test results obtained on a one-third octave band basis are converted to a single number for application evaluation. Application principles are covered in ARI *Standard* 275-84.

Performance Rating and Certification Program— Equipment over 135,000 Btu/h

Unitary air-conditioners exceeding 135,000 Btu/h can be tested in accordance with ARI *Standard* 360-86, Commercial Industrial Air-Conditioning Equipment.

Unitary heat pumps with capacities of 135,000 Btu/h or larger are covered by ARI *Standard* 340-86, Commercial and Industrial Heat Pump Equipment.

Unitary condensing units with capacities to 135,000 Btu/h or larger are covered by ARI *Standard* 365-87, Commercial and Industrial Unitary Air-Conditioning Condensing Units. ARI has established a certification program for large unitary air-conditioning, heat pump, and condensing units. The ARI *Applied Directory*, published twice a year, contains the certified values for such equipment.

Safety Standards and Installation Codes

Approval agencies list unitary air conditioners complying with a standard like UL 465-82, Central Cooling Air Conditioners. An evaluation of the product determines that its design complies with the construction requirements specified in the standard and that the equipment can be installed in accordance with the applicable requirements of NFPA 70, The National Electric Code; ASHRAE 15-1978, Safety Code for Mechanical Refrigeration; NFPA 90A, Installation of Air Conditioning and Ventilating Systems; and NFPA 90B, Installation of Warm Air Heating and Air Conditioning Systems.

Tests determine that the equipment and all components will operate within their recognized ratings, including electrical, temperature, and pressure, when the equipment is energized at rated voltage and operated at specified environmental conditions. Stipulated abnormal conditions are also imposed, wherein the product must perform in a safe manner. The evaluation covers all operational features (such as electric space heating) that may be used in the product.

Products complying with the applicable requirements may bear the agency listing mark. An approval agency program includes the auditing of continued production at the manufacturer's factory.

Similarly, a standard like UL 559-85, Heat Pumps, is used for agency approvals of heat pumps. Other UL standards may also apply.

UNITARY AIR CONDITIONERS

Unitary air conditioners consist of factory-matched refrigerant cycle components that are applied in the field to fulfill the requirements of the user. Often, a heating function compatible with the cooling system and a control system that requires a minimum of field wiring are incorporated by the manufacturer.

A variety of products is available to meet the objectives of nearly any system. Many different heating sections (gas- or oil-fired, electric, or condenser reheat), air filters, and heat pumps, which are a specialized form of unitary product, are available. Such matched equipment, selected with compatible accessory items, requires little field design and field installation work.

REFRIGERATION SYSTEM DESIGN

The unitary equipment concept permits optimum system design. Factory assembly of components and control devices helps ensure proper functioning and optimum reliability. Laboratory testing and continual monitoring of field performance result in reliable designs, customer satisfaction, and the attainment of cost and efficiency goals.

Chapters 21, 35, and 36 describe coil, compressor, and condenser designs. Chapters 3, 6, 7, and 8 of the 1990 ASHRAE *Handbook—Refrigeration* cover refrigerant system piping selection, chemistry, cleanliness, and lubrication. Proper coil circuiting is essential for adequate oil return to the compressor. Crankcase heaters may be used to prevent refrigerant migration to the compressor crankcase during shutdown. Oil pressure switches and pumpdown and pumpout controls are used when additional assurance of reliability is economical and required.

High-pressure limiting devices, internal pressure bypasses, and limited compressor motor torque are used to prevent excessive mechanical and electrical stresses. Low-pressure or temperature cutout controllers may be used to protect against loss of charge, coil freezeup, or loss of evaporator airflow.

Refrigerant flow is most commonly controlled by either a fixed metering device, such as an orifice or capillary tube, or by thermal expansion valves. Capillaries are simple, reliable, and economical, and they can be sized for peak performance at rating conditions. The evaporator may be overfed at high condensing temperatures and underfed at low condensing temperatures because of changing pressure differential across the capillary. When such conditions exist, a less-than-optimum cooling capacity usually results. However, the degree of loss varies with the design of the condenser, volume of the system, and the total refrigerant charge. The amount of unit charge is critical and a capillary-controlled evaporator must be matched to the specific condensing unit.

Properly sized thermal expansion valves provide constant superheat and good control over a range of operating conditions. Superheat is adjusted to ensure that only superheated gas returns to the compressor, usually 7 to 14°F superheat at the compressor inlet at normal rating conditions. This superheat setting may be higher at lower outdoor ambient temperature (cooling tower water temperature for water-cooled products) or indoor wet-bulb temperature. Compressor loading can be limited with vapor-charged expansion valves. Low discharge pressure (low ambient) operation causes a diminished pressure across valves and capillaries so that full flow is not maintained. Decreased capacity, low coil temperatures, and freezeup can result unless low ambient control is provided.

Properly designed unitary equipment allows only a minimum amount of liquid refrigerant to return through the suction line to the compressor during abnormal operation. Normally, the heat absorbed in the evaporator vaporizes all the refrigerant and adds a few degrees of superheat. However, any conditions that increase refrigerant flow beyond the heat-transfer capabilities of the evaporator can cause liquid carryover into the compressor return line. Such an increase may be caused by a poorly positioned thermal element of an expansion valve or by an increase in the condensing pressure of a capillary system, which may be caused by fouled condenser surfaces, excessive refrigerant charge, reduced flow of condenser air or water, or the higher temperature of the condenser cooling media. Heat transfer at the evaporator may be reduced by dirty surfaces, low-temperature differentials across the evaporator, or reduced airflow caused by a blockage in the air system.

Transient flow conditions are a special concern. During off periods, refrigerant migrates and condenses in the coldest part of the system. In an air conditioner, this area is typically the evaporator, if it is within a cooled space. When the compressor starts, the liquid tends to return to the compressor in slugs. The severity of slugging is affected by temperature differences, off time, component positions, and traps formed in suction lines. Various methods such as suction-line accumulators, specially designed compressors, the refrigerant pumpdown cycle, or limited refrigerant charge are used to avoid equipment problems associated with excessive liquid return. Chapter 3 of the 1990 ASHRAE *Handbook—Refrigeration* has further information on refrigerant piping.

Strainers and filter driers minimize the risk of restricting capillary tubes and expansion valves by foreign material, such as small quantities of solder, flux, and varnish. Overheated and oxidized oil may dissolve in warm refrigerant and deposit at lower temperatures in capillaries, expansion valves, and evaporators. Filter driers are recommended, particularly for split-system products, to remove any moisture that may have been introduced to the system during installation or servicing. A refrigeration system contaminated with moisture can cause oil breakdown, motor insulation failure, and freezing or other restrictions of the expansion valve.

Buildings with high internal heat loads require cooling even at low outdoor temperatures. The capacity of air-cooled condensers can be controlled by changing airflow or flooding tubes with refrigerant. Airflow can be changed by using dampers, adjusting fan speed, or by stopping some of the fan motors in a multifan system.

Suction pressures drop momentarily during start-up. Systems with a low-pressure, cutout controller may require a time-delay relay to bypass the low-pressure controller momentarily to prevent nuisance tripping.

In cool weather, air conditioners only operate for short periods. If the weather is also damp, high levels and wide variances in humidity may also occur. Properly designed capacity-controlled units operate for longer periods, which may improve humidity control and comfort. In any case, cooling equipment should not be oversized.

Units with two or more separate refrigerant systems permit independent operation of the individual systems, which reduces capacity while matching the changing load conditions better.

Larger, single-compressor systems may offer capacity reduction through the use of cylinder-unloading compressors or multispeed compressors or by the addition of hot-gas bypass controls. At full-load operation, efficiency is unimpaired. However, reduced capacity operation may increase or decrease system efficiency, depending on the type of capacity reduction method used.

Multispeed compressors can improve efficiency at part-load operation. Cylinder unloading can increase or decrease system performance, depending on the particular method used. Both multispeed systems and cylinder unloading generally produce higher comfort levels through lower cycling and better matching of the capacity to the load. Hot-gas bypass does not reduce capacity efficiently, although it generally provides a wider range of capacity reduction. Systems with capacity-reduction compressors usually have capacity-controlled evaporators, otherwise the resultant evaporator coil temperatures may be too high to provide dehumidification. Capacity-controlled evaporators are usually split, with one of the expansion valves controlled by a solenoid valve. Evaporator capacity is reduced by closing the solenoid valve. The compressor capacity-reduction controls or the hot-gas bypass system then provides maximum dehumidification, while the evaporator coil temperature is maintained above freezing to avoid coil frosting. Chapter 3 of the 1990 ASHRAE *Handbook—Refrigeration* has details on hot-gas bypass systems.

AIR-HANDLING SYSTEMS

High airflow, low static pressure performance, simplicity, economics, and compact arrangement are characteristics that make propeller fans particularly suitable for nonducted air-cooled condensers. Small diameter fans are direct-driven by four-, six-, or eight-pole motors. Low starting torque requirements allow the use of single-phase shaded pole and permanent split-capacitor (PSC) fan motors and simplify speed control for low outdoor temperature operation. Many larger units use multiple fans and three-phase motors. Larger diameter fans are belt-driven at a lower rpm to maintain low tip speeds and quiet operation.

Centrifugal fans meet the higher static pressure requirements of ducted, air-cooled condensers, forced-air furnaces, and evaporators. Indoor airflow must be adjusted to suit duct systems and plenums. Some small blowers are direct-driven with multispeed motors. Large blowers are always belt driven and may have variable-pitch motor pulleys for airflow adjustment. Vibration isolation reduces the amount of noise transmitted by bearings, motors, and fans into cabinets. (See Chapter 18 for details of fan design and Chapters 16 and 17 for information on air distribution systems.)

Disposable fiberglass filters are popular because they are available in standard sizes at low cost. Cleanable filters offer economic advantages when cabinet dimensions are not compatible with common sizes. Electronic or other high-efficiency air cleaners are used when a high degree of cleaning is desired. Larger equipment frequently may be provided with automatic roll filters or high-efficiency bag filters. (See Chapter 25 for additional details about filters.)

Provision for introducing outdoor air for cooling and/or ventilation is made in many unitary units; rooftop units are particularly adaptable for receiving outdoor air. Some units have automatically controlled dampers to permit cooling by outdoor air, which increases system efficiency.

ELECTRICAL DESIGN

Electrical controls for unitary equipment are selected and tested to perform their individual and interrelated functions properly and safely over the entire range of operating conditions. Internal linebreak thermal protectors provide overcurrent protection for most single-phase motors, smaller sizes of three-phase motors, and hermetic compressor motors. These rapidly responding temperature sensors, embedded in motor windings, can provide precise locked rotor and running overload protection.

Branch-circuit, short-circuit, and ground-fault protection is commonly provided by fused disconnect switches. Time-delay fuses allow selection of fuse ratings closer to running currents and thus provide backup motor overload protection, as well as short-circuit and ground-fault protection. Circuit breakers may be used in lieu of fuses where their use conforms to appropriate code requirements.

Some larger compressor motors have dual windings and contactors for step starting. A brief delay between contactor energization reduces the magnitude of inrush current.

The use of 24-V (NEC Class 2) control circuitry is common for room thermostats and interconnecting wiring between split systems. It offers advantages in temperature control, safety, and ease of installation.

Motor speed control is used to vary evaporator airflow of direct-drive fans, air-cooled condenser airflow for low outdoor temperature operation, and to vary compressor speed to match load demand. Multitap motors and autotransformers provide one or more speed steps. Solid-state speed control circuits can provide a continuously variable speed range. However, motor bearings, windings, overload protection, and the motor suspension system must be suitable for operation over the full speed range.

In addition to speed control, solid-state circuits provide reliable temperature control, motor protection, and expansion valve refrigerant control. Complete temperature control systems are frequently included with the unit. Control system features such as automatic night setback, economizer control sequence, and zone demand control of multizone equipment contribute to improved comfort and energy savings. Additional information on control systems may be found in Chapter 41 of the 1991 ASHRAE *Handbook—HVAC Applications*.

MECHANICAL DESIGN

Cabinet height dimensions are important for rooftop and ceiling-suspended units. Large unit design must consider the size limitations of truck bodies, freight cars, doorways, elevators, and various rigging practices. In addition, structural strength of both the unit and the crate must be adequate for handling, warehouse stacking, shipping, and rigging.

The following criteria are also important: (1) cabinet insulation must prevent excessive sweating in high-humidity ambient conditions, (2) insulated surfaces exposed to moving air should withstand air erosion, (3) air leakage around panels and at cabinet joints should be minimized, and (4) the cabinet insulation must be adequate to reduce energy transfer losses from the circulating airstream.

Also, cooling coil air velocities must be low enough to prevent blowoff of condensate. The drain pan must be sized to contain the condensate and must also be protected from high-velocity air. Service access must be provided for installation and repair. Versatility of application, such as multifan discharge direction and the ability to install piping from either side of the unit is another consideration. Weatherproofing requires careful attention and testing.

ACCESSORIES

Using standard, cataloged accessories, the system designer can apply the unitary product concept to solve special application problems. Typical examples (Figures 4, 6, 7, and 8) are plenum coil housings, return air-filter grilles, and diffuser-return grilles for single-outlet units. Air duct kits offered for rooftop units (Figure 4) permit concentric or side-by-side ducting, as well as horizontal or vertical connections. Mounting frames are available to facilitate unit support and roof flashing. Other accessories include high static pressure fan drives, controls for low outdoor temperature operation, and duct damper kits for control of outdoor air intakes and exhausts.

HEATING

For unitary air-conditioning systems it is important that cooling coils be installed downstream of furnaces so that condensation will not form inside the combustion and flue passages. Upstream cooling coil placement is permissible when the furnace has been approved for this type of application and designed to prevent corrosion. Burners, pilot flames, and controls must be protected from the condensate.

Chapter 24 describes hot water and steam coils used in unitary equipment, as well as the prevention of coil freezing from ventilation air in cold weather. Chapters 27, 29 and 32 discuss forced-air and oil- and gas-fired furnaces commonly used with, or included as part of, year-round equipment.

UNITARY HEAT PUMPS

Capacities of unitary heat pumps range from about 1.5 to 30 tons, although there is no specific limitation. This equipment is used in residential, commercial, and industrial applications. Multiunit installations, with a number of individual units, are particularly advantageous because they permit zoning, which provides the opportunity for heating or cooling in each zone on demand.

Application factors unique to unitary heat pumps include the following:

- The unitary heat pump normally fulfills a dual function—heating and cooling; therefore, only a single piece of equipment is required for year-round comfort. Some manufacturers offer heating-only heat pumps for special applications.
- A single source of energy can supply both heating and cooling requirements.
- Heat output can be as much as two to four times that of the purchased energy input.
- Vents and/or chimneys may be eliminated, thus reducing building costs.
- A moderate supply air temperature in the heating cycle provides even heat, and close adherence to air distribution principles ensures proper airflow and distribution while providing comfort.

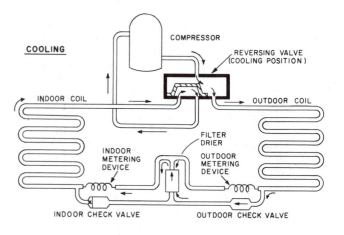

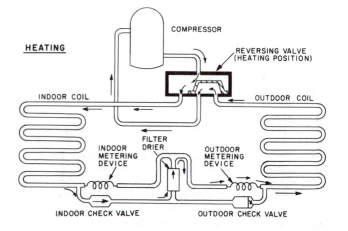

Fig. 10   Typical Schematic of Air-to-Air Heat Pump System

In an air-source heat pump, an air-to-refrigerant heat exchanger (outdoor coil in Figure 10) rejects heat to outdoor air when cooling indoor air and extracts heat from outdoor air when heating indoor air. Figure 10 shows a flow diagram of this cycle.

Most residential applications consist of an indoor fan and coil unit, either vertical or horizontal, and an outdoor fan-coil unit. The compressor is usually located in the outdoor unit. Electric heaters are commonly included within the indoor unit to provide heat during defrost cycles and during periods of high heating demand that cannot be satisfied by the heat pump alone.

Air-source heat pumps can be added to new or existing gas- or oil-fired furnaces. This unit, typically called an add-on or hybrid heat pump, normally operates as a conventional heat pump. During extremely cold weather the refrigeration system is turned off and the furnace provides the required space heating. These add-on heat pumps share the same air distribution system with the gas- or oil-fired warm air furnace. The indoor coil may be either parallel to or in series with the furnace. However, the gas or oil furnace should never be upstream of the indoor coil when both systems are operated together. This arrangement raises the condensing temperature of the refrigeration system, which could cause a compressor failure. In applications where the heat pump and furnace will be operated at the same time, the following conditions must be met: (1) the furnace and heat pump indoor coil must be arranged in parallel, or (2) the furnace combustion and flue passages must be designed to avoid condensation-induced corrosion during the cooling operation.

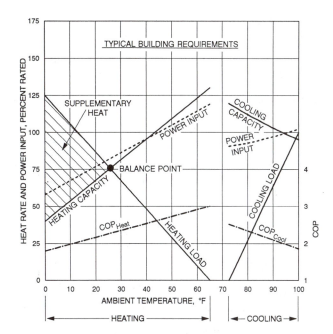

Fig. 11 Operating Characteristics of Single-Stage Unmodulated Heat Pump

HEAT PUMP SELECTION

Figure 11 shows the performance characteristics of a typical single-spaced air-source heat pump, along with the heating and cooling loads for a typical building. The heat pump heating capacity decreases as the ambient temperature decreases. This characteristic is opposite to the trend of the building load requirement. The outdoor temperature at which the heat pump capacity equals the building load is called the *balance point*. When the outdoor temperature is below the balance point, supplemental heat (usually electric resistance) must be added to make up the difference, as shown by the shaded area. The COP shown in Figure 11 below the balance point is for the refrigeration system only and does not include the supplemental heat. Water-source unitary equipment has similar operating characteristics to those shown in Figure 11, but with flatter capacity curves as a result of more constant source/sink temperatures.

In selecting the proper size heat pump, the cooling load for the building is calculated using standard practice. The balance point may be lowered to use less resistance heat by improving the thermal qualities of the structure or by choosing a heat pump larger than the cooling load requires. In practice, oversizing the cooling capacity causes excessive cycling, which results in uncomfortable temperature and humidity levels in cooling.

Use of variable-speed compressor and fans can enable matching of both heating and cooling loads over an extended range. This can reduce cycling losses and provide improved comfort levels.

HEAT PUMP REFRIGERANT SYSTEMS AND COMPONENTS

Heat pump yearly operating hours are often up to five times those of a cooling-only unit. In addition, heating extends over a greater range of system operating conditions at higher stress conditions, so the design must be thoroughly analyzed to ensure maximum reliability. Improved components and protective devices contribute to better reliability, but the equipment designer must select components that are approved for the specific application.

For a reliable and efficient heat pump system, the following factors must be considered: (1) outdoor coil circuitry, (2) defrost and water drainage, (3) refrigerant flow controls, (4) refrigerant charge management, and (5) compressor selection.

Outdoor Coil Circuitry

The outdoor coil operates as an evaporator when the heat pump is used for heating. The refrigerant in the coil is less dense than when the coil operates as a condenser. For this reason, higher flow velocities prevail, unless the number of circuits is increased to reduce the pressure drop to acceptable levels. The circuitry may be a compromise between optimum performance as an evaporator and optimum performance as a condenser.

Defrost and Water Drainage

During colder outdoor temperatures, usually below 40 to 50°F, and high relative humidities (above 50%), the outdoor coil operates below the frost point of the outdoor air. The frost that builds up on the surface of the coil is usually removed by the reverse cycle defrost method. In this method, the refrigerant flow in the system is reversed, and hot gas from the compressor flows through the outdoor coil which melts the frost. A typical defrost takes 4 to 10 min. The outdoor fan is normally off during defrost. Because the defrost is a transient process, capacity, power, and the pressures and temperatures of refrigerant in different parts of the system change throughout the defrost period (Miller 1989, O'Neal 1989a).

The performance of the heat pump during the defrost cycle can be enhanced in several ways. Improved defrost times and water removal can be achieved by ensuring that adequate refrigerant is routed to the lower refrigerant circuits in the outdoor coil. Properly sizing the defrost expansion device is critical for shorter defrost times and reducing energy use (O'Neal 1989b). If the expansion device is too small, suction pressure can be below atmospheric, defrost times become long, and energy use high. If the expansion device is too large, the compressor can be flooded with liquid refrigerant. During the conventional reverse cycle defrost, there is a significant pressure spike at defrost termination. Prestarting the fan 30 to 45 s before defrost termination can minimize the spike (Anand 1989). In cold climates, the cabinet should be raised above the ground to provide good drainage during defrost and to minimize snow and ice buildup around the cabinet. During prolonged periods of severe weather, it may be necessary to clear ice and snow from the unit.

Several methods are used to determine the need to defrost. One of the more common, simple, and reliable control methods is to initiate defrost at predetermined time intervals (usually 90 min). Other systems detect a need for defrosting by measuring changes in air pressure drop across the outdoor coil or changes in temperature difference between the outdoor coil and the outdoor air. Microprocessors are applied to control this function, as well as numerous other functions (Mueller and Bonne 1980).

Refrigerant Flow Controls

Separate refrigerant flow controls are usually used for the indoor and outdoor coils. Since the refrigerant flow reverses its direction between the heating and cooling mode of operation, a check valve bypasses in the appropriate direction around each expansion device. Either capillaries, fixed orifices, or thermostatic expansion valves may be used; however, capillaries and fixed orifices require that greater care be taken to prevent excessive flooding of refrigerant into the compressor. A check valve is not needed when an orifice-type expansion device or a biflow expansion valve is used. The reversing valve is the critical additional component required to make a heat pump air-conditioning system.

Refrigerant Charge Management

Refrigerant charge management requires extra care to control compressor flooding and the storage of refrigerant in the system during both heating and cooling. The mass flow of refrigerant during cooling is greater than during heating. Consequently, the amount of refrigerant stored may be greater in the heating mode than in the cooling mode, depending on the relative internal volumes of the indoor and outdoor coils. Usually, the internal volume of the indoor coils ranges from 110 to 70% of the outdoor coil volume. The relative volumes can be adjusted so that the coils not only transfer heat but also manage the charge.

When using capillaries or fixed orifices, the refrigerant may be stored in an accumulator in the suction line or in receivers that can remove the refrigerant charge from circulation when compressor floodback is imminent. Thermostatic expansion valves reduce the flooding problem, but storage may be required in the condenser. Use of accumulators and/or receivers is particularly important in split system products.

To maintain performance reliability, the amount of refrigerant in the system must be checked and adjusted in accordance with the manufacturer's recommendations, particularly when charging a heat pump. Manufacturer's recommendations for accumulator installation must be followed so that good oil return is assured.

Compressor Selection

Compressors are selected on the basis of performance, reliability, and probable applications of the unit. In good design practice, equipment manufacturers often consult with compressor manufacturers to verify proper application of the compressor in both design and application phases of the unitary equipment. Compressors in a heat pump operate over a wide range of suction and discharge pressures; thus, their design parameters, such as refrigerant discharge temperatures, pressure ratios, clearance volume, and motor-overload protection require special consideration. In all operating conditions, compressors should be protected against loss of lubrication, liquid floodback, and high discharge temperatures.

HEAT PUMP SYSTEM CONTROL AND INSTALLATION

The installation should always follow the manufacturer's instructions. As mentioned in the discussion under defrost, the outdoor unit should be raised from the ground in severe climate areas to permit free drainage of defrost water and to minimize the potential blocking by snow. Since the supply air from a heat pump is at a lower temperature than most heating systems, typically 90 to 100 °F, ducts and supply registers should control velocity and throw to minimize the perception of cool drafts.

Low voltage heating/cooling thermostats control heat pump operation. Models that switch automatically from heating to cooling operation and manual selection models are available. Usually, there are two stages of heating control. The first stage controls heat pump operation, and the second stage controls supplementary heat. When the heat pump cannot satisfy the first stage's call for heat, supplementary heat is added by the second stage control. The amount of supplementary heat is often controlled by an outdoor thermostat that allows additional stages of heat to be turned on only when required by the colder outdoor temperature.

Microprocessor technology has led to the use of night setback modes and intelligent recovery schemes for morning warm-up on heat-pump systems (see Chater 41 of the 1991 ASHRAE *Handbook—HVAC Applications.*

BIBLIOGRAPHY

Anand, N.K., J.S. Schliesing, D.L. O'Neal, and K.T. Peterson. 1989. Effects of outdoor coil fan pre-start on pressure transients during the reverse cycle defrost of a heat pump. ASHRAE *Transactions* 95(2).

ARI. 1984. Sound rating of outdoor unitary equipment. ARI *Standard* 270-84. Air Conditioning and Refrigeration Institute, Arlington, VA.

ARI. 1984a. Application of sound rated outdoor unitary equipment. ARI *Standard* 275-84. Air Conditioning and Refrigeration Institute, Arlington, VA.

ARI. 1986. Commercial and industrial unitary heat pump equipment. ARI *Standard* 340-86. Air Conditioning and Refrigeration Institute, Arlington, VA.

ARI. 1986a. Commercial and industrial unitary air-conditioning equipment. ARI *Standard* 360-86. Air Conditioning and Refrigeration Institute, Arlington, VA.

ARI. 1987. Commercial and industrial unitary air-conditioning condensing units. ARI *Standard* 365-87. Air Conditioning and Refrigeration Institute, Arlington, VA.

ARI. 1989. Unitary air-conditioning and air-source heat pump equipment. ARI *Standard* 210/240-89. Air Conditioning and Refrigeration Institute, Arlington, VA.

ARI. 1989a. Directory of certified applied air-conditioning products (unitary large equipment). Issued semiannually by the Air Conditioning and Refrigeration Institute, Arlington, VA.

Counts, D. 1985. Performance of heat pump/desuperheater water heating systems. ASHRAE *Transactions* 91(25):1473-87.

Miller, W.A. 1989. Laboratory study of the dynamic losses of a single speed, split system air-to-air heat pump having tube and plate fin heat exchangers. Oak Ridge National Laboratory, ORNL/CON- 253.

Mueller, D. and U. Bonne. 1980. Heat pump controls: Microelectronic technology. ASHRAE *Journal* (September).

O'Neal, D.L., N.K. Anand, K.T. Peterson, and S. Schliesing. 1989. Determination of the transient response characteristics of the air-source heat pump during the reverse cycle defrost. Final Report, ASHRAE *Project* 479-TRP, ESL/88-05F, Energy Systems Laboratory (January).

O'Neal, D.L., N.K. Anand, K.T. Peterson, and S. Schliesing. 1989a. Refrigeration system dynamics during the reverse cycle defrost. ASHRAE *Transactions* 95(2).

CHAPTER 47

APPLIED PACKAGED EQUIPMENT

AN applied packaged unit is a factory-designed, self-contained air-conditioning unit that can be integrated into an applied air-conditioning system. Products included in this definition are packaged terminal air conditioners (including the subclass, packaged terminal heat pumps); water-source heat pumps, which may be used in water-loop, groundwater, surface-water, or earth-coupled systems; and direct-expansion ground-source heat pumps.

PACKAGED TERMINAL AIR CONDITIONERS

The Air Conditioning and Refrigeration Institute (ARI) defines a packaged terminal air conditioner (PTAC) as a wall sleeve and a separate unencased combination of heating and cooling assemblies specified by the builder and intended for mounting through the wall. It includes a prime source of refrigeration, separable outdoor louvers, forced ventilation, and heating by hot water, steam, or electricity as chosen by the builder. Packaged terminal air-conditioning units with direct-fired gas heaters are also available from some manufacturers. A packaged terminal heat pump (PTHP) is a heat pump version of a PTAC that provides heat with a reverse cycle operating mode.

Packaged terminal air conditioners are designed primarily for commercial installations to provide the total heating and cooling functions for a room or zone and are specifically for through-the-wall installation. The units are mostly used in relatively small zones on the perimeter of buildings such as hotel and motel guest rooms, apartments, hospitals, nursing homes, and office buildings. In larger buildings, they may be used with nearly any system selected for environmental control of the building core.

Packaged terminal air conditioners and packaged terminal heat pumps are similar in design and construction. The most apparent difference between a PTAC and a PTHP is the addition of a refrigerant reversing valve in the PTHP. Additional components that control the heating functions of the heat pump could include an outdoor thermostat to signal the need for changes in heating operating modes, and, in the more complex designs, frost sensors, defrost termination devices, and base pan heaters.

Sizes and Classifications

Packaged terminal air conditioners are available in a wide range of capacities from 5000 to 18,000 Btu/h cooling and 2500 to 35,000 Btu/h heating.

The units are available as sectional type or integrated type. Both types include the following:

1. Heating elements available in hot water, steam, electric, or gas heat
2. Integral or remote temperature and operating controls
3. Wall sleeve or box

4. Removable (or separable) outdoor louver
5. Room cabinet
6. Means for controlled forced ventilation
7. Means for filtering air delivered to the room
8. Ductwork

The assembly is intended for free conditioned-air distribution, but a particular application may require minimal ductwork with a total external static resistance up to 0.1 in. of water.

A sectional-type unit (Figure 1) has a provision for the addition of a cooling chassis; the integrated type unit (Figure 2) has a provision for a heating option added to the chassis.

General System Design Considerations

Packaged terminal air conditioners and packaged terminal heat pumps allow the HVAC designer to integrate the exposed outdoor louver or grille with the building design. A variety of tinted grilles in various designs is available to blend with or accent most commonly used construction materials. Since the product becomes

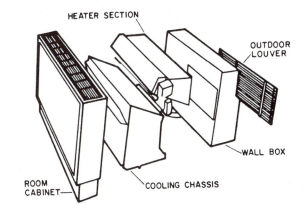

Fig. 1   Packaged Terminal Air Conditioner—Sectional Type

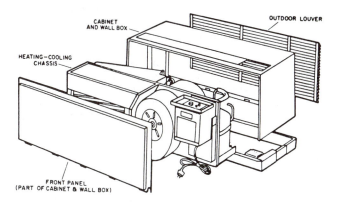

Fig. 2   Packaged Terminal Air Conditioner—Integrated Type

The preparation of this chapter is assigned to TC 7.6, Unitary Air Conditioners and Heat Pumps.

part of the building's facade, the architect must consider the product during the conception of the building. The installation of wall sleeves is more of an ironworker's, mason's, or carpenter's craft. All-electric units dominate the market. Recent market statistics indicate that 50% are PTHP, 45% are PTAC with electric heat, and 5% involve other forms of heating.

All the energy of the all-electric versions is dispersed through the building through electrical wiring, so the electric designer and the electrical contractor play a major role. The final installation is reduced to sliding in the chassis and plugging the unit into an adjacent receptacle. For these all-electric units, the traditional HVAC contractor's installation skills involving ducting, piping, and refrigeration systems are bypassed, so the conventional HVAC contractor may not be used. This results in a low-cost installation and enables deferral of installation of the actual PTAC/PTHP chassis until just before occupancy.

When comparing the selection of a gas-fired PTAC to a PTAP with electric heat or a PTHP, both operating and installation costs should be evaluated. Generally, a gas-fired PTAC is more expensive to install but less expensive to operate in the heating mode. A life cycle cost comparison is recommended.

One main advantage of the PTAC/PTHP concept is that it provides excellent zoning capability. Units can be shut down or operated in a holding condition during unoccupied periods. Present equipment efficiency-rating criteria only recognize the benefits of single units, so an efficiency comparison to other approaches may suffer.

The designer must also consider that the total capacity is the sum of the peak loads of each zone rather than the peak load of the building. Therefore, the total cooling capacity of the zonal system will exceed the total capacity of a central system. Because PTAC units are located within the conditioned space, both appearance and sound level of the equipment are important considerations. Sound attenuation in ducting is not available with the free-discharge PTAC units.

The designer must also consider the added infiltration and thermal leakage load resulting from the perimeter wall penetrations. These losses are accounted for during the on-cycle in the equipment cooling ratings and PTHP heating ratings, but during the off-cycle or with other forms of heating, they could be significant.

The widely dispersed PTAC units also present challenges relative to effective condensate disposal and the handling of extreme situations when designed condensate disposal systems are inadequate.

Most packaged terminal equipment is designed to fit into a wall aperture of approximately 42 in. wide and 16 in. high. While unitary products can increase in size with increasing cooling capacity, all PTAC/PTHP units, regardless of cooling capacity, are usually constrained to one cabinet size. The exterior of the equipment must be essentially flush with the exterior wall to meet most building codes. In addition, cabinet structural requirements and the slide-in chassis reduce the available area for outdoor air inlet and relief to less than a total of 3.5 ft^2. Manufacturers' specification sheets should be consulted for more accurate and detailed information.

Provision must be made to minimize outdoor air recirculation, and attention must be given to architectural appearance needs. This may result in increased air-side pressure drops. With a capacity range that usually spans about a 3 to 1 ratio, this causes difficulty in maintaining the efficiency of units at the higher levels of cooling capacity.

PTAC/PTHP Equipment Design

Compressor. PTAC units are designed with single-speed compressors. Both reciprocating and rotary types are used. They normally operate on a single-phase power supply and are available in 208, 230, and 265 V versions. The compressors usually are pro-

vided with electromechanical protective devices, with some of the more advanced models employing electronic protective systems.

Fan motor(s). Some PTAC units employ a single double-shafted direct-drive fan motor that provides the motive force for both the indoor and outdoor air-moving devices. This motor usually has two speeds that affect equipment sound level, throw of the conditioned air, cooling capacity, efficiency, and the sensible-to-total capacity ratio.

Full-featured models have two fan motors: one for the indoor air movement and the other for outdoor air movement. Two motors provide greater flexibility in locating components, since the indoor and outdoor fans are no longer constrained to the same rotating axis. They also allow for the possibility of different fan speeds for the indoor and outdoor systems. In this case, the outdoor fan motor is usually single speed, and the indoor fan motor has two or more speeds. Also, the designer has a broader selection of air-moving devices and can provide the user with a wider range of sound level and conditioned air throw options. Efficiency can be maintained at lower indoor fan speeds. When heating (other than with a heat pump), the outdoor fan motor can be switched off automatically to reduce electrical energy consumption, decrease infiltration and heat transmission losses through the PTAC unit, or to prevent ice from entrapping the fan blade.

Indoor air mover. The airflow quantity, air-side pressure rise, available fan motor speeds, and sound-level requirements of the indoor air system of a PTAC indicate that a centrifugal blower wheel provides reasonable indoor air performance. In some cases, proprietary mixed-flow blowers are used. Dual-fan motor units permit the use of two centrifugal blower wheels or a cross-flow blower to provide a more even discharge of the conditioned air.

Indoor air circuit. Packaged terminal air conditioners have an air filter of fiberglass, metal, or plastic foam, which removes large particulate matter from the circulating airstream. In addition to improving the quality of the indoor air, this filter also reduces fouling of the indoor heat exchanger. The PTAC also provides mechanical means of introducing outdoor air into the indoor airstream. This air, which may or may not be filtered, controls infiltration and pressurization of the conditioned space and, during portions of the year, can serve as an outdoor air economizer.

Outdoor air mover. Outdoor air movers may be either centrifugal blower wheels, mixed-flow blowers, or axial flow fans.

Heat exchangers. Packaged terminal air-conditioning units may use conventional plate-fin heat exchangers, which have either copper or aluminum smooth-bore tubes. The fins are usually aluminum, which may, in some cases, be coated to retard corrosion. Since PTACs are generally restricted in physical size, performance improvements based on increasing heat exchanger size are limited. Therefore, some manufacturers, to improve performance or reduce costs, employ heat exchangers with performance enhancements on the air side (lanced fins, spine fin, etc.) and/or the refrigerant side (internal finning or rifling).

Refrigerant expansion device. Most PTACs use a simple capillary as an expansion device. Off-rating-point performance is improved if thermal expansion valves are used.

Condensate disposal. Condensate forms on the indoor coil when cooling. Some PTACs require that a drain system be installed to convey the condensate to a disposal point. Other units spray the condensate on the outdoor coil where it is evaporated and dispersed to the outdoor ambient air. This evaporative cooling of the condenser enhances performance, but the potentially negative effects of fouling and corrosion of the outdoor heat exchanger must be considered. This problem is especially severe in a coastal installation where salt spray could mix with the condensate and, after repeated evaporation cycles, build a corrosive, salt water solution in the condensate sump.

A PTHP also produces a condensate in the heating mode. If, contrary to normal PTHP operation, the outdoor coil operates below freezing, the condensate forms frost, which is melted during defrost. This water must be disposed of in some manner. If drains are used and the heat pump operates in below-freezing weather, the drain lines must be protected from freezing. Outside drains cannot be used in this case, unless they have drain heaters. Some PTHPs introduce the condensate formed during heating onto the indoor coil, which humidifies the indoor air at the expense of heating capacity. Inadequate condensate disposal can lead to overflow at the unit and potential staining of the building facade.

Controls. Packaged terminal air-conditioning units have a built-in manual mode selector (cool, heat, fan only, and off) and a manual fan-speed selector. A thermostat adjustment is provided with set points usually identified in subjective terms such as *high*, *normal*, and *low*. Some units incorporate electronic controls, which provide room temperature limiting, evaporator freeze-up protection, compressor lockout in the case of actual or impending compressor malfunction, and service diagnostic aids. Advanced master controls at a central location are also used. This enables an operator at the master control to override the control settings registered by the occupant. These master controls may limit operation when certain room temperature limits are exceeded, adjust thermostat set points during unoccupied periods, and turn off certain units to limit peak electrical demand.

Wall sleeves. A wall sleeve is a required part of a PTAC unit. It becomes an integral part of the building structure and must be designed with sufficient strength to maintain its dimensional integrity after installation. It must withstand the potential corrosive effects of other building materials, such as mortar, and must endure long-term exposure to the outdoor elements.

Outdoor grille. The outdoor grille or louvers must be compatible with the architecture of the structure. Most manufacturers provide options in this area. A properly designed grille prevents birds, vermin, and outdoor debris from entering, impedes the entry of rain and snow and, at the same time, provides adequate free area for the outdoor airstream to enter and exit with a minimum of recirculation.

Slide-in components. The interface between the wall sleeve and the slide-in component chassis allows the components to be easily inserted and later removed for service and/or replacement. In the event of serious malfunction, the offending slide-in component can be quickly replaced with a spare, and the repair can be made off the premises. An adequate seal at the interface is essential to seal out wind, rain, snow, and insects without jeopardizing the slide-in/slide-out feature.

Indoor appearance. Since the PTAC units are located within the conditioned space, the indoor appearance must blend in with the indoor decor of the room. Manufacturers provide a variety of indoor treatments, which include variations in shape, style, and materials. In addition to metal convector-style fronts, many provide wood or simulated wood front alternatives.

Heat Pump Operation

Basic PTHP units operate in the heat pump mode down to an outdoor temperature just above the point that frosting of the outdoor heat exchanger would occur. When that outdoor temperature is reached, the heat pump mode is locked out, and other forms of heating are required. Some PTHPs use control schemes that extend the operation of the heat pump to lower temperatures. One approach permits heat pump operation down to outdoor temperatures just above the freezing point. If the outdoor coil frosts, it is defrosted by shutting down the compressor and allowing the outdoor fan to continue circulating outdoor air over the coil. Another approach permits heat pump operation to even lower outdoor temperatures by using a reverse cycle defrost sequence. In those cases, the heat pump mode is usually locked out for outdoor temperatures below 10°F.

Performance and Safety Testing

Packaged terminal air conditioners may be rated in accordance with ANSI/ARI *Standard* 310-90. Packaged terminal heat pumps are rated in accordance with ARI *Standard* 380-90, Packaged Terminal Heat Pumps.

Periodically, ARI issues a *Directory of Certified Applied Air-Conditioning Products*, which lists the cooling capacity, efficiency, heating capacity, and airflow rate for each participating manufacturer's PTAC models. The listings of PTHP models also include the heating COP.

Cooling and heating capacities, as listed in the ARI directory, must be established in accordance with ANSI/ASHRAE *Standard* 37-1988, Methods of Testing for Rating Unitary Air-Conditioning and Heat Pump Equipment, or with ASHRAE *Standard* 16-1983, Method of Testing for Rating Room Air Conditioners and Packaged Terminal Air Conditioners. All standard heating ratings should be established in accordance with ANSI/ASHRAE *Standard* 58-1986, Method of Testing Room Air-Conditioner Heating Capacity.

Additionally, PTAC units should be constructed in accordance with ANSI/ASHRAE *Standard* 15-1978, Safety Code for Mechanical Refrigeration and should comply with the safety requirements of ANSI/UL *Standard* 484-82, Room Air Conditioners.

WATER-SOURCE HEAT PUMPS

Some water-source heat pumps (WSHP) are single-package reverse-cycle heat pumps that use water as the heat source when in the heating mode and as a heat sink when in the cooling mode. The water supply may be a recirculating closed loop, a well, a lake, or a stream. Water for closed loop heat pumps is usually supplied at 2 to 3 gpm per ton of cooling capacity. Groundwater heat pumps can operate with considerably less water consumption. The main components of a WSHP refrigeration system are a compressor, a refrigerant-to-water heat exchanger, a refrigerant-to-air heat exchanger, refrigerant expansion devices, and a refrigerant-reversing valve. Figure 3 shows a typical schematic of a WSHP system.

Designs of packaged water-source heat pumps range from horizontal units located primarily above the ceiling or on the roof to vertical units usually located in basements or equipment rooms, and console units located in the conditioned space. Figures 4 and 5 illustrate typical designs.

Types of Water-Source Heat Pump Systems

Water source heat pumps are used in a variety of systems. These include the following:

1. Water loop heat pump systems (Figure 6A)
2. Groundwater heat pump systems (Figure 6B)
3. Surface water heat pump systems (Figure 6C)
4. Closed-loop surface water heat pump systems (Figure 6D)
5. Ground-coupled heat pump systems (Figure 6E)

Water-loop heat pumps (WLHP) use a circulating water loop as the heat source and the heat sink. When the loop water temperatures exceed a certain level due to heat added as the result of heat pump cooling, a cooling tower dissipates heat from the water loop into the atmosphere. When the loop water temperatures drop below a prescribed level due to heat being removed as a result of heat pump heating, heat is added to the circulating loop water, usually with a boiler. In multiple-unit installations, under conditions when some heat pumps operate in the cooling mode while others operate in the heating mode, controls are necessary so that the loop water temperatures remain between prescribed limits.

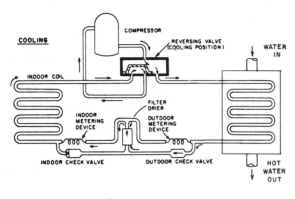

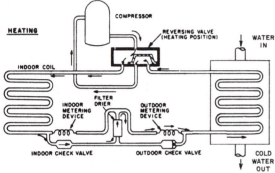

Fig. 3   Typical Schematic of Water-Source Heat Pump System

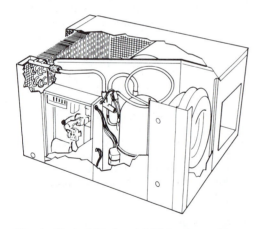

Fig. 4   Typical Horizontal Water-Source Pump

Groundwater heat pumps (GWHP) pump groundwater from a nearby well and pass it through the heat pump's water-to-refrigerant heat exchanger where it is warmed or cooled, depending on the operating mode. It is then discharged to a drain, a stream, a lake, or it is returned to the ground through a reinjection well.

Many state and local jurisdictions have enacted ordinances relating to the use and discharge of groundwater. Since aquifers, the water table, and groundwater availability vary from region to region, these regulations cover a wide spectrum.

Surface water heat pumps (SWHP) pump water from a nearby lake, stream, or canal. After passing through the heat pump heat exchanger, it is returned to the source or a drain at a temperature level several degrees warmer or cooler, depending on the operating mode of the heat pump.

Closed loop surface water heat pumps use a closed water or brine loop that includes pipes or tubing located in the surface

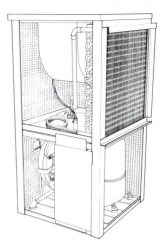

Fig. 5   Typical Vertical Water-Source Heat Pump

water (river, lake, or large pond) that serves as the heat exchanger. The adequacy of the total thermal capacity of the body of water must be considered. Usually, plastic piping is installed in either a shallow horizontal or deep vertical array to form the heat exchanger. The massive thermal capacity of the earth provides a temperature stabilizing effect on the circulating loop water. Installing this type of system requires detailed knowledge of the climate; the site; the soil thermal characteristics; and the performance, design, and installation of water-to-earth heat exchangers. The ASHRAE publication, *Design/Data Manual for Closed-Loop Ground-Coupled Heat Pump Systems* (1985), has detailed information on design and installation of GCHP systems. Additional information on these and other water-source heat pump systems is presented in Chapter 8 of this volume and Chapter 1 of the 1991 ASHRAE *Handbook—HVAC Applications*.

Entering Water Temperatures

These various applications provide a wide range of entering water temperatures to WSHPs. Not only do the entering water temperatures vary by type of application, but they also vary by climate and time of year. Due to the wide range of entering water temperatures encountered, it is not feasible to design a universal packaged product that can handle the full range of possibilities effectively. Within systems that experience a wider range of entering water temperatures (such as surface water and ground coupled), specific WSHP models optimized for the region are sometimes required to achieve optimum performance.

Performance Certification Programs

A certification program for WSHPs that is aimed primarily at heat pumps used in WLHP systems is maintained by ARI. Units are tested in accordance with ARI *Standard* 320-86, which provides standard cooling ratings at 85 °F entering water temperature and standard heating ratings at 70 °F entering water temperature. Maximum operating conditions are checked at the upper temperature level at 90 °F entering water temperature and at the lower temperature level at 65 °F leaving water temperature. This latter temperature corresponds to approximately 55 °F entering water when cooling. The range of test conditions covers the extremes of entering water temperatures typically encountered in WLHP systems.

A companion certification program for groundwater-source heat pumps is also available from ARI. In this program, units are tested in accordance with ARI *Standard* 325-85, which provides standard cooling ratings at both 50 and 70 °F entering water temperature and standard heating ratings at these same two entering

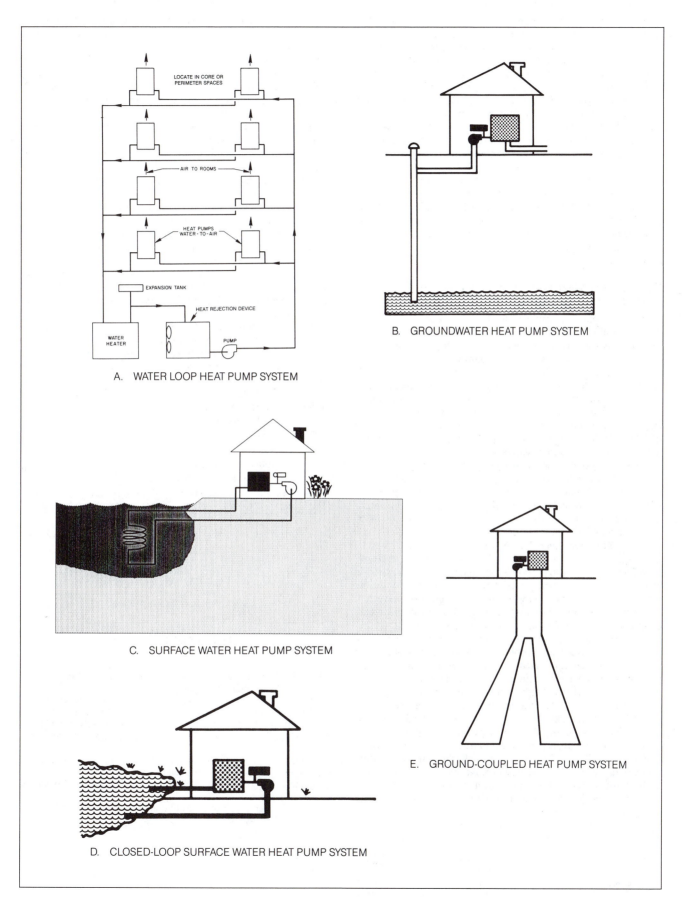

Fig. 6   Water-Source Heat Pump Systems

water temperatures. Maximum operating conditions are checked at the upper temperature level at 75 °F entering water temperature and at the lower temperature level at 45 °F entering water temperature. These entering water temperatures bracket the range of groundwater temperatures found across the United States.

The ARI *Directory of Certified Applied Air-Conditioning Products* lists the cooling capacity, cooling efficiency, heating capacity, and heating COP of units rated in accordance with ARI *Standard* 320-85. For groundwater-source heat pumps rated in accordance with ARI *Standard* 325-85, the directory also lists these same parameters for the two standard entering water temperatures and water flow rates.

ARI *Standard* 330-90 was developed to deal with ground source closed-loop heat pumps. In addition, the Canadian Standards Association (CSA) developed and published standards CSA 445 and CSA 446 for ground and water source heat pump systems.

There are no ARI certification programs for water-source heat pumps designed specifically for surface water or earth-coupled applications.

Water-Source Heat Pump Equipment Design

Water-source heat pump units are designed to match differing levels of entering water temperatures by optimizing the relative sizing of the indoor and the refrigerant-to-water heat exchangers and by matching the expansion devices to the refrigerant flow rates.

Compressors. Water source heat pumps usually have single-speed compressors. Higher capacity equipment may use multiple compressors, which provide capacity modulation. The compressors may be of the reciprocating, rotary, or scroll type. Single-phase units are available at voltages of 115, 208, 230, and 265. All larger equipment is for three-phase power supplies with voltages of 208, 230, 460, or 550. The compressors are usually provided with eletromechanical protective devices.

Indoor air system. Console WSHP models are designed for free delivery of the conditioned air and other WSHP units having ducting capability. Smaller WSHPs have multispeed, direct-drive centrifugal blower wheel fan systems. Large capacity equipment has belt-drive systems. All units have built-in air filters of fiberglass, metal, or plastic foam.

Indoor heat exchanger. The indoor heat exchanger of WSHP units is a conventional plate-fin coil of copper tubes and aluminum fins. The tubing in the coil is circuited so that it can function effectively as an evaporator with the refrigerant flow in one direction and as a condenser when the refrigerant flow is reversed.

Refrigerant-to-water heat exchanger. The heat exchanger, which couples the heat pump to source/sink water, is either of the tube-in-tube or tube-in-shell type. It must function in either the condensing or evaporating mode, so special attention is given to refrigerant-side circuitry. Heat exchanger construction is usually of copper and steel, where the source/sink water is exposed only to the copper portions. Cupronickel options to replace the copper are usually available for use with brackish or corrosive water.

Refrigerant expansion devices. Most WSHPs rated in accordance with ARI *Standard* 320-86, a single rating point standard for each operating mode, use simple capillaries as expansion devices. Units with capillaries do not perform well when rated at the dual rating points of ARI *Standard* 325-85. Normally, off-rating point performance improves if thermal expansion valves are used, since they provide peak performance over a broader range of inlet water temperatures.

Refrigerant-reversing valve. The refrigerant-reversing valves in WSHPs are identical to those used in unitary air-source heat pumps.

Condensate disposal. Condensate forms on the indoor coil when cooling is collected and conveyed to a drain system.

Controls. Console WSHP units have built-in operating mode selector and thermostatic controls. Ducted units use low-voltage remote heat/cool thermostats.

Special features. A number of special features are available on certain models of WSHPs. These features include the following:

Desuperheater. Uses discharge gas in a special water/refrigerant heat exchanger to heat water for a building.

Capacity modulation. Either multiple compressors or hot-gas bypass may be used.

Variable air volume (VAV). Reduces fan energy usage and requires some form of capacity modulation.

Automatic water valve. Closes off water flow through unit when compressor is off and permits variable water volume in the loop, which reduces pumping energy usage.

Outdoor-air economizer. Cools directly with outdoor air to reduce or eliminate the need for mechanical refrigeration during mild or cold weather when outdoor humidity levels and outdoor air quality are appropriate.

Water-side economizer. Cools with loop water to reduce or eliminate the need for mechanical refrigeration during cold weather and requires a hydronic coil in the indoor air circuit, which is valved into the circulating loop when loop temperatures are relatively low and there is a call for cooling.

Electric heaters. Used in WLHP systems that do not have a boiler as a source for loop heating.

Size. Typical space requirements and weights of WSHPs are presented in Table 1.

Table 1   Space Requirements for Typical Packaged Water-Source Heat Pumps

Water-to-Air Heat Pump	Length × Width × Height, ft	Weight, lb
1.5-ton vertical unit	2.0 ×2.0 ×3.0	180
3-ton vertical unit	2.5 ×2.5 ×4.0	250
3-ton horizontal unit	3.5 ×2.0 ×2.0	250
5-ton vertical unit	3.0 ×2.5 ×4.0	330
11-ton vertical unit	3.5 ×3.0 ×6.0	720
26-ton vertical unit	3.5 ×5.0 ×6.0	1550

See manufacturers' specification sheets for actual values.

BIBLIOGRAPHY

ARI. 1985. Ground water source heat pumps. ARI *Standard* 325-85.

ARI. 1986. Water source heat pumps. ARI *Standard* 320-86.

ARI. 1990a. Packaged terminal air conditioners. ARI *Standard* 310-90. Air-Conditioning and Refrigeration Institute, Arlington, VA.

ARI. 1990b. Packaged terminal heat pumps. ARI *Standard* 380-90.

ARI. 1990c. Ground source closed-loop heat pumps. ARI *Standard* 330-90.

ARI. 1991. *Directory of certified applied air-conditioning products.*

CSA. 1989a. Design and installation of ground ... u water source heat pumps. CAN/CSA-C445-M89. Canadian Standards Association, Ontario, Canada.

CSA. 1989b. Performance of ground and water source heat pumps. CAN/CSA-C446-M89.

CHAPTER 48

CODES AND STANDARDS

THE Codes and Standards listed in Table 1 represent practices, methods, or standards published by the organizations indicated. They are valuable guides for the practicing engineer in determining test methods, ratings, performance requirements, and limits applying to the equipment used in heating, refrigerating, ventilating, and air conditioning. *Copies can usually be obtained from the organization listed in the Publisher column.* These listings represent the most recent information available at the time of publication.

Table 1   Codes and Standards Published by Various Societies and Associations

Subject	Title	Publisher	Reference
Air Conditioners	Room Air Conditioners	AHAM	ANSI/AHAM (RA C-1)
Rooms	Method of Testing for Rating Room Air Conditioners and Packaged Terminal Air Conditioners	ASHRAE	ANSI/ASHRAE 16-1983 (RA 88)
	Method of Testing for Rating Room Air Conditioners and Packaged Terminal Air Conditioner Heating Capacity	ASHRAE	ANSI/ASHRAE 58-1986 (RA 90)
	Methods of Testing for Rating Room Fan-Coil Air Conditioners	ASHRAE	ASHRAE 79-1978 (RA 84)
	Commercial and Residential Central Air Conditioners	CSA	C22.2 No. 119-M1985
	Performance Standard for Room Air Conditioners	CSA	CAN/CSA C368.1-1990
	Room Air Conditioners	CSA	C22.2 No. 117-1970
	Room Air Conditioners (1982)	UL	ANSI/UL 484-1986
Packaged Terminal	Packaged Terminal Air Conditioners	ARI	ARI 310-90
	Packaged Terminal Heat Pumps	ARI	ARI 380-90
Transport	Air Conditioning of Aircraft Cargo (1978)	SAE	SAE AIR806A
	Nomenclature, Aircraft Air-Conditioning Equipment (1978)	SAE	SAE ARP147C
Unitary	Load Calculation for Commercial Summer and Winter Air Conditioning, 4th ed. (1988)	ACCA	ACCA Manual N
	Application of Sound Rated Outdoor Unitary Equipment	ARI	ARI 275-84
	Commercial and Industrial Unitary Air-Conditioning Equipment	ARI	ANSI/ARI 360-86
	Sound Rating of Outdoor Unitary Equipment	ARI	ARI 270-84
	Unitary Air-Conditioning and Air-Source Heat Pump Equipment	ARI	ARI 210/240-89
	Methods of Testing for Rating Heat Operated Unitary Air-Conditioning Equipment for Cooling	ASHRAE	ANSI/ASHRAE 40-1980 (RA 86)
	Methods of Testing for Rating Unitary Air-Conditioning and Heat Pump Equipment	ASHRAE	ANSI/ASHRAE 37-1988
	Methods of Testing for Seasonal Efficiency of Unitary Air Conditioners and Heat Pumps	ASHRAE	ANSI/ASHRAE 116-1983
	Air Conditioners, Central Cooling (1982)	UL	ANSI/UL 465-1984
Air Conditioning	Commercial Low Pressure, Low Velocity Duct System Design	ACCA	ACCA Manual Q
	Duct Design for Residential Buildings	ACCA	ACCA Manual D
	Load Calculation for Residential Winter and Summer Air Conditioning, 7th ed. (1986)	ACCA	ACCA Manual J
	Gas-Fired Absorption Summer Air Conditioning Appliances (with 1982 addenda)	AGA	ANSI Z21.40.1-1981
	Environmental System Technology (1984)	NEBB	NEBB
	Automotive Air-Conditioning Hose (1989)	SAE	ANSI/SAE J51 MAY89
	HVAC Systems—Applications, 1st. ed. (1986)	SMACNA	SMACNA
	HVAC Systems—Duct Design (1990)	SMACNA	SMACNA
	Installation Standards for Residential Heating and Air Conditioning Systems (1988)	SMACNA	SMACNA
Transport	Air Conditioning Equipment, General Requirements for Subsonic Airplanes (1961)	SAE	SAE ARP85D
	General Requirements for Helicopter Air Conditioning (1970)	SAE	SAE ARP292B
	Testing of Commercial Airplane Environmental Control Systems (1973)	SAE	SAE ARP217B
Unitary	Method of Rating Computer and Data Processing Room Unitary Air Conditioners	ASHRAE	ANSI/ASHRAE 127-1988
	Method of Rating Unitary Spot Air Conditioners	ASHRAE	ANSI/ASHRAE 128-1988 50.1
Air Curtains	Air Distribution Basics for Residential and Small Commercial Buildings	ACCA	ACCA Manual T
	Residential Equipment Selection	ACCA	ACCA Manual S
	Flexible Duct Performance and Installation Standards	ADC	ADC
	Laboratory Certification Manual	ADC	ADC 1062:LCM-83
	Test Code for Grilles, Registers and Diffusers	ADC	ADC 1062:GRD-84
	Metric Units and Conversion Factors	AMCA	AMCA 99-0100-76
	Test Methods for Air Curtain Units	AMCA	AMCA 220-90

Table 1   Codes and Standards Published by Various Societies and Associations (*Continued*)

Subject	Title	Publisher	Reference
Air Curtains (continued)	Air Volume Terminals	ARI	ANSI/ARI 880-89
	Method of Testing for Rating the Air Flow Performance of Outlets and Inlets	ASHRAE	ASHRAE 70-72
	Standard Methods for Laboratory Air Flow Measurement	ASHRAE	ANSI/ASHRAE 41.2-1987
	Residential Air Exhaust Equipment (1e)	CSA	C260.2-1976
	High Temperature Pneumatic Duct Systems for Aircraft (1981)	SAE	ANSI/SAE ARP699D
Air Ducts and Fittings	Commercial Low Pressure, Low Velocity Duct Systems	ACCA	ACCA Manual Q
	Duct Design for Residential Winter and Summer Air Conditioning	ACCA	ACCA Manual D
	Flexible Air Duct Test Code	ADC	ADC FD-72R1-1979
	Residential-Type Air-Conditioning Systems	CSA	B228.1-1968
	Installation of Air Conditioning and Ventilating Systems (1989)	NFPA	ANSI/NFPA 90A-1989
	Installation of Warm Air Heating and Air-Conditioning Systems (1989)	NFPA	ANSI/NFPA 90B-1989
	Ducted Electric Heat Guide for Air Handling Systems (1971)	SMACNA	SMACNA
	HVAC Air Duct Leakage Test Manual (1985)	SMACNA	SMACNA
	HVAC Duct Construction Standards—Metal and Flexible, 1st ed. (1985)	SMACNA	SMACNA
	Rectangular Industrial Duct Construction (1980)	SMACNA	SMACNA
	Round Industrial Duct Construction (1977)	SMACNA	SMACNA
	Thermoplastic Duct (PVC) Construction Manual (Rev. A, 1974)	SMACNA	SMACNA
	Factory-Made Air Ducts and Connectors (1990)	UL	UL 181
	Marine Rigid and Flexible Air Ducting (1986)	UL	ANSI/UL 1136-1986
Air Filters	Commercial and Industrial Air Filter Equipment	ARI	ARI 850-84
	Residential Air Filter Equipment	ARI	ARI 680-86
	Method of Testing Air-Cleaning Devices Used in General Ventilation for Removing Particulate Matter	ASHRAE	ASHRAE 52-1968 (RA 76)
	Method for Sodium Flame Test for Air Filters	BSI	BS 3928
	Methods of Test for Atmospheric Dust Spot Efficiency and Synthetic Dust Weight Arrestance	BSI	BS 6540 Part 1
	Electrostatic Air Cleaners (1981)	UL	ANSI/UL 867-1988
	High Efficiency, Particulate Air Filter Units (1985)	UL	ANSI/UL 586-1990
	Test Performance of Air Filter Units (1987)	UL	ANSI/UL 900-1987
Air-Handling Units	Comercial Low Pressure, Low Velocity Duct Systems	ACCA	ACCA Manual Q
	Duct Design for Residential Winter and Summer Air Conditioning	ACCA	ACCA Manual D
	Central Station Air-Handling Units	ARI	ANSI/ARI 430-89
Air Leakage	Air Leakage Performance for Detached Single-Family Residential Buildings	ASHRAE	ANSI/ASHRAE 119-1988
Boilers	A Guide to Clean and Efficient Operation of Coal Stoker-Fired Boilers	ABMA	ABMA
	Boiler Water Limits and Steam Purity Recommendations for Watertube Boilers	ABMA	ABMA
	Boiler Water Requirements and Associated Steam Purity—Commercial Boilers	ABMA	ABMA
	Fluidized Bed Combustion Guidelines	ABMA	ABMA
	Guidelines for Industrial Boiler Performance Improvement	ABMA	ABMA
	Lexicon Boiler and Auxiliary Equipment	ABMA	ABMA
	Matrix of Recommended Quality Control Requirements	ABMA	ABMA
	Operation and Maintenance Safety Manual	ABMA	ABMA
	Recommended Design Guidelines for Stoker Firing of Bituminous Coals	ABMA	ABMA
	(Selected) Summary of Codes and Standards of the Boiler Industry	ABMA	ABMA
	Thermal Shock Damage to Hot Water Boilers as a Result of Energy Conservation Measures	ABMA	ABMA
	Commercial Applications Systems and Equipment	ACCA	ACCA Manual CS
	Boiler and Pressure Vessel Code (11 sections) (1989)	ASME	ASME
	Boiler, Pressure Vessel, and Pressure Piping Code	CSA	B51-M1986
	Heating, Water Supply, and Power Boilers—Electric (1980)	UL	ANSI/UL834-1991
Cast-Iron	Ratings for Cast-Iron and Steel Boilers (1989)	HYDI	IBR/SBI
	Testing and Rating Heating Boilers (1989)	HYDI	IBR/SBI
Gas or Oil	Gas-Fired Low-Pressure Steam and Hot Water Boilers	AGA	ANSI Z21.13-1987; Z21.13a-1989
	Gas Utilization Equipment in Large Boilers (with 1972 and 1976 addenda; R-1983, 1989)	AGA	ANSI Z83.3-1971
	Control and Safety Devices for Automatically Fired Boilers	ASME	ANSI/ASME CSD.1-1988
	Oil Burning Appliance Standards (B140 Series)	CSA	B140.7.1-1976
	Oil-Fired Steam and Hot-Water Boilers for Commercial and Industrial Use (1a)	CSA	B140.7.2-1967
	Explosion Prevention of Fuel Oil and Natural Gas-Fired Single-Burner Boiler-Furnaces (1987)	NFPA	ANSI/NFPA 85A-1987
	Explosion Prevention of Natural Gas-Fired Multiple-Burner Boiler-Furnaces (1989)	NFPA	ANSI/NFPA 85B-1989

Table 1 Codes and Standards Published by Various Societies and Associations (*Continued*)

Subject	Title	Publisher	Reference
Boilers (continued)	Prevention of Furnace Explosions in Fuel Oil-Fired Multiple-Burner Boiler-Furnaces (1989)	NFPA	ANSI/NFPA 85D-1989
	Commercial-Industrial Gas Heating Equipment (1973)	UL	UL 795
	Oil Fired Boiler Assemblies (1990)	UL	ANSI/UL 726-1990
Building Codes	ASTM Standards Used in Building Codes	ASTM	ASTM
	BOCA National Building Code, 11th ed. (1990)	BOCA	BOCA
	BOCA National Property Maintenance Code, 3rd ed. (1990)	BOCA	BOCA
	CABO One- and Two-Family Dwelling Code (1989)	CABO	CABO
	Model Energy Code (1989)	CABO	CABO
	Uniform Building Code (1988)	ICBO	ICBO
	Uniform Building Code Standards (1988)	ICBO	ICBO
	Directory of Building Codes and Regulations (1991 ed.)	NCSBCS	NCSBCS
	Standard Building Code (1991)	SBCCI	SBCCI
Mechanical	Safety Code for Elevators and Escalators (plus two yearly supplements)	ASME	ANSI/ASME A 17.1-1990
	BOCA National Mechanical Code, 7th ed. (1990)	BOCA	BOCA
	Uniform Mechanical Code (1988) (with Uniform Mechanical Code Standards)	ICBO/IAPMO	ICBO/IAPMO
	Standard Gas Code (1991)	SBCCI	SBCCI
	Standard Mechanical Code (1991)	SBCCI	SBCCI
Burners	Guidelines for Burner Adjustments of Commercial Oil-Fired Boilers	ABMA	ABMA
	Domestic Gas Conversion Burners	AGA	ANSI Z21.17-1984; Z21.17a-1990
	Installation of Domestic Gas Conversion Burners	AGA	ANSI Z21.8-1984; Z21.8a-1990; Z21.176-1990
	General Requirements for Oil Burning Equipment	CAN/CSA	B140.0-M1987
	Installation Code for Oil Burning Equipment	CSA	B139-1976
	Oil Burners, Atomizing Type	CSA	B140.2.1-1973
	Pressure Atomizing Oil Burner Nozzles	CSA	B140.2.2-1971 (R 1980)
	Replacement Burners and Replacement Combustion Heads for Residential Oil Burners	CSA	B140.2.3-M1981
	Supplement No. 1 to B139-1976, Installation Code for Oil Burning Equipment	CSA	B139S1-1982
	Vaporizing-Type Oil Burners	CSA	B140.1-1966 (R 1980)
	Commercial-Industrial Gas Heating Equipment (1973)	UL	UL 795
	Oil Burners (1989)	UL	ANSI/UL 296-1989
Capillary Tubes	Capillary Tubes Method of Testing Flow Capacity of Refrigerant	ASHRAE	ANSI/ASHRAE 28-88
Chillers	Methods of Testing Liquid Chilling Packages	ASHRAE	ASHRAE 30-1978
	Absorption Water-Chilling Packages	ARI	ARI 560-82
	Centrifugal or Rotary Screw Water-Chilling Packages	ARI	ARI 550-90
	Reciprocating Water-Chilling Packages	ARI	ARI 590-86
Chimneys	Chimneys, Fireplaces, Vents, and Solid Fuel Burning Appliances	NFPA	ANSI/NFPA 211-1988
	Chimneys, Factory-Built, Medium Heat Appliance (1986)	UL	ANSI/UL 959-1986
	Chimneys, Factory-Built, Residential Type and Building Heating Appliances (1988)	UL	ANSI/UL 103-1988
Cleanrooms	Procedural Standards for Certified Testing of Cleanrooms (1988)	NEBB	NEBB-1988
Coils	Forced-Circulation Air-Cooling and Air-Heating Coils	ARI	ARI 410-87
	Methods of Testing Forced Circulation Air Cooling and Air Heating Coils	ASHRAE	ASHRAE 33-1978
Comfort Conditions	Thermal Environmental Conditions for Human Occupancy	ASHRAE	ANSI/ASHRAE 55-1981 (R 1986)
Compressors	Compressors and Exhausters (reaffirmed 1986)	ASME	ANSI/ASME PTC 10-1965
	Displacement Compressors, Vacuum Pumps and Blowers	ASME	ANSI/ASME PTC 9-1974 (R 1985)
	Safety Standard for Air Compressor Systems	ASME	ANSI/ASME B19.1-1990
	Safety Standard for Compressors for Process Industries	ASME	ASME/ANSI B19.3-1991
	Compressed Air and Gas Handbook, 5th ed. (1989)	CAGI	CAGI
Refrigeration	Ammonia Compressor Units	ARI	ARI 510-87
	Positive Displacement Refrigerant Compressors and Condensing Units	ARI	ANSI/ARI 520-90
	Methods of Testing for Rating Positive Displacement Refrigerant Compressors	ASHRAE	ASHRAE 23-1978
	Hermetic Refrigerant Motor-Compressors	CSA	CAN/CSA C22.2 No. 140.2-M89
	Hermetic Refrigerant Motor-Compressors (1990)	UL	ANSI/UL 984-1989
Computers	Protection of Electronic Computer/Data Processing Equipment	NFPA	ANSI/NFPA 75-1989
Condensers	Commercial Applications Systems and Equipment (for equipment selection only)	ACCA	ACCA Manual CS
	Remote Mechanical Draft Air-Cooled Refrigerant Condensers	ARI	ARI 460-87
	Water-Cooled Refrigerant Condensers, Remote Type	ARI	ARI 450-87

Table 1   Codes and Standards Published by Various Societies and Associations (*Continued*)

Subject	Title	Publisher	Reference
Condensers (continued)	Methods of Testing for Rating Remote Mechanical-Draft Air-Cooled Refrigerant Condensers	ASHRAE	ASHRAE 20-1970
	Methods of Testing Remote Mechanical-Draft Evaporative Refrigerant Condensers	ASHRAE	ANSI/ASHRAE 64-1989
	Methods of Testing for Rating Water-Cooled Refrigerant Condensers	ASHRAE	ASHRAE 22-1971 (RA 78)
	Addendum I, Standards for Steam Surface Condensers (1989)	HEI	HEI
	Standards for Steam Surface Condensers, 8th ed. (1984)	HEI	HEI
Condensing Units	Commercial Applications Systems and Equipment	ACCA	ACCA Manual CS
	Residential Equipment Selection	ACCA	ACCA Manual S
	Commercial and Industrial Unitary Air-Conditioning Condensing Units	ARI	ARI 365-87
	Methods of Testing for Rating Positive Displacement Condensing Units	ASHRAE	ASHRAE 14-80
	Heating and Cooling Equipment	CSA	CAN/CSA C22.2 No. 236-M90
	Refrigeration and Air-Conditioning Condensing and Compressor Units (1987)	UL	ANSI/UL303-1988
Contactors	Definite Purpose Contactors for Limited Duty	ARI	ARI 790-86
	Definite Purpose Magnetic Contactors	ARI	ARI 780-86
Controls	Quick-Disconnect Devices for Use with Gas Fuel	AGA	ANSI Z21.41-1989; Z21.41a-1990
	Energy Management Control Systems Instrumentation	ASHRAE	ANSI/ASHRAE 114-1986
	Temperature-Indicating and Regulating Equipment	CSA	C22.2 No. 24-1987
	Industrial Control Equipment (1988)	UL	ANSI/UL 508-1988
	Limit Controls (1989)	UL	ANSI/UL 353-1988
	Primary Safety Controls for Gas- and Oil-Fired Appliances (1985)	UL	ANSI/UL 372-1985
	Temperature-Indicating and Regulating Equipment (1988)	UL	ANSI/UL 873-1987
Commercial and Industrial	General Standards for Industrial Control and Systems	NEMA	NEMA ICS 1-1988
	Industrial Control Devices, Controllers and Assemblies	NEMA	NEMA ICS 2-1988
	Instructions for the Handling, Installation, Operation and Maintenance of Motor Control Centers	NEMA	NEMA ICS 2.3-1983 (R 1988)
	Maintenance of Motor Controllers after a Fault Condition	NEMA	NEMA ICS 2.2-1983 (1988)
	Preventive Maintenance of Industrial Control and Systems Equipment	NEMA	NEMA ICS 1.3-1986
Residential	Automatic Gas Ignition Systems and Components	AGA	ANSI Z21.20-1989; Z21.20a-1991
	Gas Appliance Pressure Regulators (with 1989 addenda)	AGA	ANSI Z21.18-1987
	Gas Appliance Thermostats	AGA	ANSI Z21.23-1989; Z21.23a-1991
	Manually Operated Gas Valves for Appliances, Appliance Connector Valves and Hose End Valves	AGA	ANSI Z21.15-1989; Z21.15a-1990; Z21.156-1991
	Manually-Operated Piezo Electric Spark Gas Ignition Systems and Components	AGA	ANSI Z21.77-1989
	Hot Water Immersion Controls	NEMA	NEMA DC-12-1985
	Line Voltage Integrally-Mounted Thermostats for Electric Heaters	NEMA	NEMA DC 13-1979 (R 1985)
	Quick Connect Terminals	NEMA	ANSI/NEMA DC 2-1982 (R 1988)
	Residential Controls—Class 2 Transformers	NEMA	NEMA DC 20-1986
	Residential Controls—Surface Type Controls for Electric Storage Water Heaters	NEMA	NEMA DC 5-1989
	Safety Guidelines for the Application, Installation, and Maintenance of Solid State Controls	NEMA	NEMA ICS 1.1-1984 (R 1988)
	Temperature Limit Controls for Electric Baseboard Heaters	NEMA	NEMA DC 10-1983 (R 1989)
	Wall-Mounted Room Thermostats	NEMA	NEMA DC 3-1989
	Warm Air Limit and Fan Controls	NEMA	NEMA DC 4-1986
	Electrical Quick-Connect Terminals (1985)	UL	ANSI/UL 310-1986
Coolers Air	Unit Coolers for Refrigeration	ARI	ARI 420-89
	Methods of Testing Forced Convection and Natural Convection Air Coolers for Refrigeration	ASHRAE	ANSI/ASHRAE 25-1990
	Milk Coolers	CSA	C22.2 No. 132-1973
Bottled Beverage	Methods of Testing and Rating Bottled and Canned Beverage Vendors and Coolers	ASHRAE	ANSI/ASHRAE 32-1986 (RA 90)
	Refrigerated Vending Machines (1989)	UL	ANSI/UL 541-1988
Drinking Water	Application and Installation of Drinking Water Coolers	ARI	ANSI/ARI 1020-84
	Drinking Fountains and Self-Contained, Mechanically Refrigerated Drinking Water Coolers	ARI/ANSI	ANSI/ARI 1010-84
	Methods of Testing for Rating Drinking-Water Coolers with Self-Contained Mechanical Refrigeration Systems	ASHRAE	ANSI/ASHRAE 18-1987 (RA 91)
	Drinking Water Coolers and Beverage Dispensers	CSA	C22.2 No. 91-1971 (R 1981)
	Drinking Water Coolers (1987)	UL	ANSI/UL 399-1986
Liquid	Refrigerant-Cooled Liquid Coolers, Remote Type	ARI	ARI 480-87
	Methods of Testing for Rating Liquid Coolers	ASHRAE	ANSI/ASHRAE 24-1989
Cooling Towers	Commercial Applications Systems and Equipment	ACCA	ACCA Manual CS

Table 1   Codes and Standards Published by Various Societies and Associations (*Continued*)

Subject	Title	Publisher	Reference
Cooling Towers (continued)	Atmospheric Water Cooling Equipment	ASME	ANSI/ASME PTC 23-1986
	Water-Cooling Towers	NFPA	ANSI/NFPA 214-1988
	Acceptance Test Code for Spray Cooling Systems (1985)	CTI	CTI ATC-133-1985
	Acceptance Test Code for Water Cooling Towers: Mechanical Draft, Natural Draft Fan Assisted Types, Evaluation of Results, and Thermal Testing of Wet/Dry Cooling Towers (1990)	CTI	CTI ATC-105-1990
	Certification Standard for Commercial Water Cooling Towers (1991)	CTI	CTI STD-201-1991
	Code for Measurement of Sound from Water Cooling Towers	CTI	CTI ATC-128-1981
	Fiberglass-Reinforced Plastic Panels for Application on Industrial Water Cooling Towers	CTI	CTI STD-131-1986
	Nomenclature for Industrial Water-Cooling Towers	CTI	CTI NCL-109-1983
Dehumidifiers	Commercial Applications Systems and Equipment	ACCA	ACCA Manual CS
	Dehumidifiers	AHAM	ANSI/AHAM DH 1-1986
	Dehumidifiers (3a)	CSA	C22.2 No. 92-1971
	Dehumidifiers (1987)	UL	ANSI/UL 474-1987
Desiccants	Method of Testing Desiccants for Refrigerant Drying	ASHRAE	ASHRAE 35-1976 (RA 83)
Driers	Method of Testing Liquid Line Refrigerant Driers	ASHRAE	ANSI/ASHRAE 63.1-1988
	Liquid Line Driers	ARI	ANSI/ARI 710-86
Electrical	Voltage Ratings for Electrical Power Systems and Equipment	ANSI	ANSI C84.1-1989
	Canadian Electrical Code, Part 1 (16th ed.)	CSA	C22.1-1990
	Application Guide for Ground Fault Interrupters	NEMA	NEMA 280-1984
	Application Guide for Ground Fault Protective Devices for Equipment	NEMA	NEMA PB 2.2-1988
	Enclosures for Electric Equipment	NEMA	NEMA 250-1985
	Enclosures for Industrial Control and Systems	NEMA	NEMA ICS 6-1988
	General Requirements for Wiring Devices	NEMA	NEMA WD 1-1983 (R 1989)
	Low Voltage Cartridge Fuses	NEMA	NEMA FU 1-1986
	Molded Case Circuit Breakers	NEMA	NEMA AB 1-1986
	Terminal Blocks for Industrial Use	NEMA	NEMA ICS 4-1983 (R 1988)
	National Electric Code (1990)	NFPA	ANSI/NFPA 70-1990
	Compatibility of Electrical Connectors and Wiring (1988)	SAE	SAE AIR1329 A
	Manufacturers' Identification of Electrical Connector Contacts, Terminals and Splices (1982)	SAE	SAE AIR1351 A
Energy	Air Conditioning and Refrigerating Equipment Nameplate Voltages	ARI	ARI 110-90
	Energy Conservation in Existing Buildings—Commercial	ASHRAE	ANSI/ASHRAE/IES 100.3-1985
	Energy Conservation in Existing Buildings—High Rise Residential	ASHRAE	ANSI/ASHRAE/IES 100.2-1991
	Energy Conservation in Existing Buildings—Institutional	ASHRAE	ANSI/ASHRAE/IES 100.5-1991
	Energy Conservation in Existing Buildings—Public Assembly	ASHRAE	ANSI/ASHRAE/IES 100.6-1991
	Energy Conservation in Existing Facilities—Industrial	ASHRAE	ANSI/ASHRAE/IES 100.4-1984
	Energy Conservation in New Building Design	ASHRAE	ANSI/ASHRAE/IES 90A-1980
	Energy Efficient Design of New Buildings Except Low Rise Residential Buildings	ASHRAE	ASHRAE/IES 90.1-1989
	Model Energy Code (MEC) (1989)	CABO	BOCA/ICBO/SBCCI
	Uniform Solar Energy Code	IAPMO	IAPMO
	Energy Directory	NCSBCS	NCSBCS
	Energy Management Guide for the Selection and Use of Polyphase Motors	NEMA	NEMA MG 10-1983 (R 1988)
	Energy Management Guide for the Selection and Use of Single Phase Motors	NEMA	NEMA MG 11-1977 (R 1987)
	Total Energy Management Handbook, 3rd ed.	NEMA	NEMA 05101-1986
	Energy Conservation Guidelines (1984)	SMACNA	SMACNA
	Energy Recovery Equipment and Systems, Air-to-Air (1991)	SMACNA	SMACNA
	Retrofit of Building Energy Systems and Processes (1982)	SMACNA	SMACNA
	Energy Management Equipment (1987)	UL	ANSI/UL 916-1987
Exhaust Systems	Commercial Low Pressure, Low Velocity Duct Systems	ACCA	ACCA Manual Q
	Fundamentals Governing the Design and Operation of Local Exhaust Systems	ANSI	ANSI Z9.2-1979
	Safety Code for Design, Construction, and Ventilation of Spray Finishing Operations (reaffirmed 1971)	ANSI	ANSI Z9.3-1985
	Ventilation and Safe Practices of Abrasive Blasting Operations	ANSI	ANSI Z9.4-1985
	Method of Testing Performance of Laboratory Fume Hoods	ASHRAE	ANSI/ASHRAE 110-1985
	Compressors and Exhausters	ASME	ANSI/ASME PTC 10-1974 (R 1985)
	Mechanical Flue-Gas Exhausters	CSA/CAN	3-B255-M81
	Installation of Blower and Exhaust Systems for Dust, Stock, Vapor Removal or Conveying (1990)	NFPA	ANSI/NFPA 91-1990
	Draft Equipment (1973)	UL	UL 378

Table 1   Codes and Standards Published by Various Societies and Associations (*Continued*)

Subject	Title	Publisher	Reference
Expansion Valves	Thermostatic Refrigerant Expansion Valves	ARI	ARI 750-87
	Method of Testing for Capacity Rating of Thermostatic Refrigerant Expansion Valves	ASHRAE	ANSI/ASHRAE 17-1986 (RA 90)
Fan Coil Units	Room Fan-Coil Air Conditioners	ARI	ARI 440-89
	Methods of Testing for Rating Room Fan-Coil Air Conditioners	ASHRAE	ASHRAE 79-1978 (RA 84)
	Fan Coil Units and Room Fan Heater Units (1986)	UL	ANSI/UL 883-1986
Fans	Commercial Low Pressure, Low Velocity Duct Systems	ACCA	ACCA Manual Q
	Duct Design for Residential Winter and Summer Air Conditioning	ACCA	ACCA Manual D
	Designation for Rotation and Discharge of Centrifugal Fans	AMCA	AMCA 99-2406-83
	Drive Arrangements for Centrifugal Fans	AMCA	AMCA 99-2404-78
	Drive Arrangements for Tubular Centrifugal Fans	AMCA	AMCA 99-2410-82
	Inlet Box Positions for Centrifugal Fans	AMCA	AMCA 99-2405-83
	Laboratory Methods of Testing Fans for Rating	AMCA	ANSI/AMCA 210-85
	Motor Positions for Belt or Chain Drive Centrifugal Fans	AMCA	AMCA 99-2407-66
	Site Performance Test Standard Power Plant and Industrial Fans	AMCA	AMCA 803-87
	Standards Handbook	AMCA	AMCA 99-86
	Fans and Blowers	ARI	ARI 670-90
	Laboratory Methods of Testing Fans for Rating	ASHRAE	ANSI/ASHRAE 51-1985 ANSI/AMCA 210-85
	Methods of Testing Dynamic Characteristics of Propeller Fans— Aerodynamically Excited Fan Vibrations and Critical Speeds	ASHRAE	ANSI/ASHRAE 87.1-1983
	Fans	ASME	ANSI/ASME PTC 11-1984
	Fans and Ventilators	CSA	C22.2 No. 113-M1984
	Performance of Ventilating Fans for Use in Livestock and Poultry Buildings	CSA	CAN/CSA C320-M86
	Rating the Performance of Residential Mechanical Ventilating Equipment	CSA	CAN/CSA C260-M90
	Electric Fans (1987)	UL	ANSI/UL 507-1987
Ceiling	AC Electric Fans and Regulators	ANSI	ANSI-IEC Pub. 385
Filters	Flow-Capacity Rating and Application of Suction-Line Filters and Filter Driers	ARI	ANSI/ARI 730-86
	Grease Extractors for Exhaust Ducts (1981)	UL	ANSI/UL 710-1981
	Grease Filters for Exhaust Ducts (1979)	UL	UL 1046
Fire Dampers	Fire Dampers (1990)	UL	ANSI/UL 555-1989
Fireplaces	Factory-Built Fireplaces (1988)	UL	ANSI/UL 127-1985
Fire Protection	Test Method for Surface Burning Characteristics of Building Materials	ASTM/ NFPA	ASTM E 84-89a; NFPA 255-1984
	BOCA National Fire Prevention Code, 8th ed. (1990)	BOCA	BOCA
	Uniform Fire Code (1991)	IFCI	IFCI
	Uniform Fire Code Standards (1991)	IFCI	IFCI
	Interconnection Circuitry of Non-Coded Remote Station Protective Signalling Systems	NEMA	NEMA SB 3-1969 (R 1989)
	Fire Doors and Windows	NFPA	ANSI/NFPA 80-1990
	Fire Prevention Code	NFPA	ANSI/NFPA 1-1987
	Fire Protection Handbook, 16th ed.	NFPA	NFPA
	Flammable and Combustible Liquids Code	NFPA	ANSI/NFPA 30-1987
	Life Safety	NFPA	ANSI/NFPA 101-1988
	National Fire Codes (issued annually)	NFPA	NFPA
	Smoke Control Systems	NFPA	NFPA 9ZA-1988
	Standard Method of Fire Tests of Door Assemblies	NFPA	ANSI/NFPA 252-1990
	Standard Fire Prevention Code (1991)	SBCCI	SBCCI
	Fire Tests of Building Construction and Materials (1984)	UL	ANSI/UL 263-1986
	Heat Responsive Links for Fire Protection Service (1987)	UL	ANSI/UL 33-1987
Fireplace Stoves	Fireplace Stoves (1988)	UL	ANSI/UL 737-1988
Flow Capacity	Method of Testing Flow Capacity of Suction Line Filters and Filter Driers	ASHRAE	ANSI/ASHRAE 78-1985 (RA 90)
Freezers Household	Household Refrigerators, Combination Refrigerator-Freezers, and Household Freezers	AHAM	ANSI/AHAM; HRF 1-1988
	Capacity Measurement and Energy Consumption Test Methods for Refrigerators, Combination Refrigerator-Freezers, and Household Freezers	CSA	CAN/CSA-C300-M89
	Commercial Refrigerated Equipment	CSA	C22-2 No. 120-1970 (R 1981)
	Household Refrigerators and Freezers	CSA	C22.2 No. 63-M1987
	Household Refrigerators and Freezers (1983)	UL	ANSI/UL 250-1984

Table 1 Codes and Standards Published by Various Societies and Associations (*Continued*)

Subject	Title	Publisher	Reference
Freezers (continued)			
Commercial	Dispensing Freezers	NSF	ANSI/NSF-6-1989
	Food Service Refrigerators and Storage Freezers	NSF	NSF-7
	Soda Fountain and Luncheonette Equipment	NSF	NSF-1
	Commercial Ice Cream Makers (1986)	UL	ANSI/UL 621-1985
	Commercial Refrigerators and Freezers (1985)	UL	ANSI/UL 471-1984
	Ice Makers (1958)	UL	ANSI/UL 563-1985
Furnaces	Commercial Applications Systems and Equipment	ACCA	ACCA Manual CS
	Residential Equipment Selection	ACCA	ACCA Manual S
	Direct Vent Central Furnaces	AGA	ANSI Z21.64-1988; Z21.64a-1989; Z21.64b-1989
	Gas-Fired Central Furnaces (except Direct Vent)	AGA	ANSI Z21.47-1990; Z21.47a-1990
	Gas-Fired Duct Furnaces	AGA	ANSI Z83.9-1990; Z83.9a-1989; Z83.9b-1989
	Gas-Fired Gravity and Fan Type Direct Vent Wall Furnaces (with 1989 and 1990 addenda)	AGA	ANSI Z21.44-1988
	Gas-Fired Gravity and Fan Type Floor Furnaces	AGA	ANSI Z21.48-1989; Z21.48a-1990; Z21.48b-1991
	Gas-Fired Gravity and Fan Type Vented Wall Furnaces	AGA	ANSI Z21.49-1989; Z21.49a-1990; Z21.49b-1991
	Methods of Testing for Heating Seasonal Efficiency of Central Furnaces and Boilers	ASHRAE	ANSI/ASHRAE 103-1988
	Methods of Testing for Rating Non-Residential Warm Air Heaters	ASHRAE	ANSI/ASHRAE 45-1978 (RA 86)
	Installation Code for Solid-Fuel-Burning Appliances and Equipment	CAN/CSA	B365-M87
	Solid Fuel-Fired Central Heating Appliances	CAN/CSA	B366.1-M87
	Electric Central Warm-Air Furnaces	CSA	C22.2 No. 23-1980
	Heating and Cooling Equipment	CSA	CAN/CSA C22.2 No. 236-M90
	Oil Burning Stoves and Water Heaters (2a)	CSA	B140.3-1962 (R 1980)
	Oil-Fired Warm Air Furnaces (8a)	CSA	B140.4-1974
	Installation of Oil Burning Equipment	NFPA	NFPA 31-1987
	Standard Gas Code (1991)	SBCCI	SBCCI
	Commercial-Industrial Gas Heating Equipment (1973)	UL	UL 795
	Oil-Fired Central Furnaces (1986)	UL	UL 727-1986
	Oil-Fired Floor Furnaces (1987)	UL	ANSI/UL 729-1987
	Oil-Fired Wall Furnaces (1987)	UL	ANSI/UL 730-1986
	Residential Gas Detectors (1983)	UL	ANSI/UL 1484-1983
	Solid-Fuel and Combination-Fuel Central and Supplementary Furnaces (1981)	UL	ANSI/UL 391-1983
Heat Exchangers	Remote Mechanical-Draft Evaporative Refrigerant Condensers	ARI	ARI 490-89
	Method of Testing Air-to-Air Heat Exchangers	ASHRAE	ASHRAE 84-1978
	Standard Methods of Testing for Rating the Performance of Heat Recovery Ventilators	CSA	CAN/CSA C439-88
	Standards for Power Plant Heat Exchangers, 2nd ed. (1990)	HEI	HEI
	Standards of Tubular Exchanger Manufacturers Association, 7th ed. (1988)	TEMA	TEMA
Heat Pumps	Commercial Applications Systems and Equipment	ACCA	ACCA Manual CS
	Heat Pump Systems: Principles and Applications (Commercial and Residence)	ACCA	ACCA Manual H
	Residential Equipment Selection	ACCA	ACCA Manual S
	Commercial and Industrial Unitary Heat Pump Equipment	ARI	ANSI/ARI 340-86
	Ground Water-Source Heat Pumps	ARI	ARI 325-85
	Water-Source Heat Pumps	ARI	ANSI/ARI 320-86
	Methods of Testing for Rating Unitary Air-Conditioning and Heat Pump Equipment	ASHRAE	ANSI/ASHRAE 37-1988
	Add-on Heat Pumps	CSA	C22.2 No. 186.2-M1980
	Central Forced-Air Unitary Heat Pumps with or without Electric Resistance Heat (1980)	CSA	C22.2 No. 186.1
	Heating and Cooling Equipment	CSA	CAN/CSA C22.2 No. 236-M90
	Installation Requirements for Air-to-Air Heat Pumps (2a)	CSA	C273.5-1980
	Performance Standard for Unitary Heat Pumps (1a)	CSA	C273.3-M1977
	Heat Pumps (1985)	UL	ANSI/UL 559-1985
Heat Recovery	Gas Turbine Heat Recovery Steam Generators	ASME	ANSI/ASME PTC 4.4-1981 (R 1987)
	Manual for Calculating Heat Loss and Heat Gain for Electric Comfort Conditioning	NEMA	NEMA HE 1-1980 (R 1986)
	Energy Recovery Equipment and Systems, Air-to-Air (1991)	SMACNA	SMACNA

Table 1 Codes and Standards Published by Various Societies and Associations (*Continued*)

Subject	Title	Publisher	Reference
Heaters	Direct Gas-Fired Make-Up Air Heaters	AGA	ANSI Z83.4-1989
	Direct Gas-Fired Door Heaters (with 1989 addenda)	AGA	ANSI Z83.17-1988
	Direct Gas-Fired Industrial Air Heaters	AGA	ANSI Z83.17-1988
	Gas-Fired Construction Heaters	AGA	ANSI Z83.7-1990
	Gas-Fired Infrared Heaters	AGA	ANSI Z83.6-1990
	Gas-Fired Pool Heaters	AGA	ANSI Z21.56-1989
	Gas-Fired Room Heaters, Vol. I, Vented Room Heaters (with 1989 and 1990 addenda)	AGA	ANSI Z21.11.1-1988
	Gas-Fired Room Heaters, Vol. II, Unvented Room Heaters (with 1990 addenda)	AGA	ANSI Z21.11.2-1989; Z83.6a-1989
	Gas-Fired Unvented Commercial and Industrial Heaters (with 1984 and 1989 addenda)	AGA	ANSI Z83.16-1982
	Desuperheater/Water Heaters	ARI	ARI 470-87
	Air Heaters	ASME	ANSI/ASME PTC 4.3-1968 (R 1985)
	Space Heaters for Use with Solid Fuels	CSA	B366.2 M1984
	Standards for Closed Feedwater Heaters, 4th ed.(1984)	HEI	HEI
	Fuel-Fired Heaters—Air Heating—for Construction and Industrial Machinery (1989)	SAE	SAE J1024 MAY89
	Motor Vehicle Heater Test Procedure (1982)	SAE	SAE J638 JUN82
	Electric Air Heaters (1980)	UL	ANSI/UL 1025-1988
	Electric Central Air Heating Equipment (1986)	UL	ANSI/UL 1096-1985
	Electric Heaters for Use in Hazardous (Classified) Locations (1985)	UL	ANSI/UL 823-1990
	Electric Heating Appliances (1987)	UL	ANSI/UL 499-1987
	Electric Oil Heaters (1990)	UL	ANSI/UL 574-1990
	Gas Heating Equipment, Commercial-Industrial (1973)	UL	UL 795
	Oil-Fired Air Heaters and Direct-Fired Heaters (1975)	UL	UL 733
	Oil-Fired Room Heaters (1973)	UL	UL 896
	Solid Fuel-Type Room Heaters (1988)	UL	ANSI/UL 1482-1988
	Unvented Kerosene-Fired Room Heaters and Portable Heaters (1982)	UL	UL 647
Heating	Commercial Applications Systems and Equipment	ACCA	ACCA Manual CS
	Residential Equipment Selection	ACCA	ACCA Manual S
	Determining the Required Capacity of Residential Space Heating and Cooling Appliances	CAN/CSA	CAN/CSA-F280-M86
	Automatic Flue-Pipe Dampers for Use with Oil-Fired Appliances	CSA	B140.14-M1979
	Electric Duct Heaters	CSA	C22.2 No. 155-M1986
	Heater Elements	CSA	C22.2 No. 72-M1984
	Oil-Fired Service Water Heaters and Swimming Pool Heaters (7a)	CSA	B140.12-1976
	Performance Requirements for Electric Heating Line-Voltage	CSA	C273.4-M1978
	Performance Standard for Electrical Baseboard Heaters	CSA	C273.2-1971
	Portable Industrial Oil-Fired Heaters	CSA	B140.8-1967 (R 1980)
	Portable Kerosene-Fired Heaters	CSA	CAN 3-B140.9.3 M86
	Wall Thermostats Electric Air Heaters	CSA	C22.2 No. 46-M1988
	Advanced Installation Guide for Hydronic Heating Systems, 2nd ed.	HYDI	IBR 250
	Comfort Conditioning Heat Loss Calculation Guide (1984)	HYDI	IBR H-21
	Installation Guide for Residential Hydronic Heating Systems, 6th ed. (1988)	HYDI	IBR 200
	Environmental System Technology (1984)	NEBB	NEBB
	Manual for Calculating Heat Loss and Heat Gain for Electric	NEMA	NEMA HE 1-1980 (R 1986)
	Installation and Operation of Pulverized Fuel Systems	NFPA	ANSI/NFPA 85F-1988
	Aircraft Electrical Heating Systems (1965) (reaffirmed 1983)	SAE	SAE AIR860
	HVAC Systems—Applications, 1st ed. (1986)	SMACNA	SMACNA
	Installation Standards for Residential Heating and Air Conditioning Systems (1988)	SMACNA	SMACNA
	Electric Baseboard Heating Equipment (1987)	UL	ANSI/UL 1042-1986
	Electric Central Air Heating Equipment (1986)	UL	ANSI/UL 1096-1985
Humidifiers	Appliance Humidifiers	AHAM	ANSI/AHAM HU 1-1987
	Central System Humidifiers	ARI	ANSI/ARI 610-89
	Self-Contained Humidifiers	ARI	ANSI/ARI 620-89
	Humidifiers and Evaporative Coolers	CSA	C22.2 No. 104-M1983
	Humidifiers (1987)	UL	ANSI/UL 998-1985
Ice Makers	Automatic Commercial Ice Makers	ARI	ARI 810-87
	Ice Storage Bins	ARI	ANSI/ARI 820-88
	Methods of Testing Automatic Ice Makers	ASHRAE	ANSI/ASHRAE 29-1988
	Ice-Making Machines	CSA	C22.2 No. 133-1964 (R 1981)
	Automatic Ice-Making Equipment	NSF	NSF-12
	Ice Makers (1984)	UL	ANSI/UL 563-1985
Incinerators	Incinerator Performance	CSA	Z103-1976
	Incinerators, Waste and Linen Handling Systems and Equipment	NFPA	ANSI/NFPA 82-1990
	Residential Incinerators (1973)	UL	UL 791

Table 1   Codes and Standards Published by Various Societies and Associations (*Continued*)

Subject	Title	Publisher	Reference
Induction Units	Room Air-Induction Units	ARI	ANSI/ARI 445-87
	Frame Assignments for Alternating Current Integral-Horsepower Induction Motors	NEMA	NEMA MG 13-1984
Industrial Ducts	Rectangular Industrial Duct Construction (1980)	SMACNA	SMACNA
	Round Industrial Duct Construction (1977)	SMACNA	SMACNA
Insulation	Specification for Adhesives for Duct Thermal Insulation	ASTM	ASTM C916
	Specification for Thermal and Acoustical Insulation (Mineral Fiber, Duct Lining Material)	ASTM	ASTM C1071-86
	Test Method for Steady-State Heat Flux Measurements and Thermal Transmission Properties by Means of the Guarded Hot Plate Apparatus	ASTM	ASTM C177-85
	Test Method for Steady-State Heat Flux Measurements and Thermal Transmission Properties by Means of the Heat Flow Meter Apparatus	ASTM	ASTM C518-85
	Test Method for Steady-State Heat Transfer Properties of Horizontal Pipe Insulations	ASTM	ASTM C335-89
	Test Method for Steady-State and Thermal Performance of Building Assemblies by Means of a Guarded Hot Box	ASTM	ASTM C236-89
	Thermal Insulation, Mineral Fiber, for Buildings	CSA	A101 M-1983
	National Commercial and Industrial Insulation Standards	MICA	MICA 1988
Louvers	Test Method for Louvers, Dampers, and Shutters	AMCA	AMCA 500-89
Lubricants	Method of Testing the Floc Point of Refrigeration Grade Oils	ASHRAE	ANSI/ASHRAE 86-1983
	Method for Calculating Viscosity Index from Kinematic Viscosity at 40 and 100°C	ASTM	ASTM D2270-91
	Method for Conversion of Kinematic Viscosity to Saybolt Universal Viscosity or to Saybolt Furol Viscosity	ASTM	ASTM D2161-87
	Method for Estimation of Molecular Weight of Petroleum Oils from Viscosity Measurements	ASTM	ASTM D2502-87
	Method for Separation of Representative Aromatics and Nonaromatics Fractions of High-Boiling Oils by Elution Chromatography	ASTM	ASTM D2549-91
	Recommended Practice for Viscosity System for Industrial Fluid Lubricants	ASTM	ASTM D2422-86
	Test Method for Carbon-Type Composition of Insulating Oils of Petroleum Origin	ASTM	ASTM D2140-86
	Test Method for Dielectric Breakdown Voltage of Insulating Liquids Using Disk Electrodes	ASTM	ASTM D877-87
	Test Method for Dielectric Breakdown Voltage of Insulating Oils of Petroleum Origin Using VDE Electrodes	ASTM	ASTM D1816-84a (1990)
	Test Method for Mean Molecular Weight of Mineral Insulating Oils by the Cryoscopic Method	ASTM	ASTM D2224-78 (1983)
	Test Method for Molecular Weight of Hydrocarbons by Thermoelectric Measurement of Vapor Pressure	ASTM	ASTM D2503-82 (1987)
	Test Methods for Pour Point of Petroleum Oils	ASTM	ASTM D97-87
	Semiconductor Graphite	NEMA	NEMA CB 4-1989
Measurements	A Standard Calorimeter Test Method for Flow Measurement of a Volatile Refrigerant	ASHRAE	ANSI/ASHRAE 41.9-1988
	Engineering Analysis of Experimental Data	ASHRAE	ANSI/ASHRAE Guideline 2-1986 (RA 90)
	Procedure for Bench Calibration of Tank Level Gaging Tapes and Sounding Rules	ASME	ANSI MC88.2-1974 (R 1987)
	Standard Method for Measurement of Flow of Gas	ASHRAE	ANSI/ASHRAE 41.7-1984 (RA 91)
	Standard Method for Measurement of Proportion of Oil in Liquid Refrigerant	ASHRAE	ANSI/ASHRAE 41.4-1984
	Standard Method for Temperature Measurement	ASHRAE	ANSI/ASHRAE 41.1-1986 (RA 91)
	Standard Method for Pressure Measurement	ASHRAE	ANSI/ASHRAE 41.3-1989
	Standard Methods of Measurement of Flow of Liquids in Pipes Using Orifice Flowmeters	ASHRAE	ANSI/ASHRAE 41.8-1989
	Standard Method for Measurement of Moist Air Properties	ASHRAE	ANSI/ASHRAE 41.6-1982
	Standard Methods of Measuring and Expressing Building Energy Performance	ASHRAE	ANSI/ASHRAE 105-1984 (RA 90)
	Glossary of Terms Used in the Measurement of Fluid Flow in Pipes	ASME	ANSI/ASME MFC-1M-1979 (R 1986)
	Guide for Dynamic Calibration of Pressure Transducers	ASME	ANSI MC88-1-1972 (R 1987)
	Measurement of Fluid Flow in Pipes Using Orifice, Nozzle, and Venturi	ASME	ASME MFC-3M-1989
	Measurement of Fluid Flow in Pipes Using Vortex Flow Meters	ASME	ASME/ANSI MFC-6M-1987
	Measurement of Gas Flow by Means of Critical Flow Venturi Nozzles	ASME	ASME/ANSI MFC-7M-1987
	Measurement of Gas Flow by Turbine Meters	ASME	ANSI/ASME MFC-4M-1986
	Measurement of Industrial Sound	ASME	ANSI/ASME PTC 36-1985

Table 1 Codes and Standards Published by Various Societies and Associations (*Continued*)

Subject	Title	Publisher	Reference
Measurements (continued)	Measurement of Liquid Flow in Closed Conduits Using Transit-Time Ultrasonic Flowmeters	ASME	ANSI/ASME MFC-5M-1985 (R 1989)
	Measurement of Rotary Speed	ASME	ANSI/ASME PTC 19.13-1961 (R 1986)
	Measurement Uncertainty	ASME	ANSI/ASME PTC 19.1-1985
	Measurement Uncertainty for Fluid Flow in Closed Conduits	ASME	ANSI/ASME MFC-2M-1983 (R 1988)
	Pressure Measurement	ASME	ASME/ANSI PTC 19.2-1987
	Temperature Measurement	ANSI	ANSI/ASME PTC 19.3-1974 (R 1986)
Mobile Homes and Recreational Vehicles	Load Calculation for Residential Winter and Summer Air Conditioning	ACCA	ACCA Manual J
	Recreational Vehicle Cooking Gas Appliances (with 1989 addenda)	AGA	ANSI Z21.57-1987
	Mobile Homes	CSA	CAN/CSA-Z240 MH Series-M86
	Mobile Home Parks	CSA	Z240.7.1-1972
	Oil-Fired Warm-Air Heating Appliances for Mobile Housing and Recreational Vehicles (2a)	CSA	B140.10-1974 (R 1981)
	Recreational Vehicle Parks	CSA	Z240.7.2-1972
	Recreational Vehicles	CSA	CAN/CSA-Z240 RV Series-M86
	Gas Supply Connectors for Manufactured Homes	IAPMO	IAPMO TSC 9-1989
	Manufactured Home Installations	NCSBCS	ANSI A225.1-87
	Recreational Vehicle Parks	NFPA	NFPA 501C-1990
	Plumbing System Components for Manufactured Homes and Recreational Vehicles	NSF	NSF-24
	Gas Burning Heating Appliances for Mobile Homes and Recreational Vehicles (1965)	UL	UL 307B
	Gas-Fired Cooking Appliances for Recreational Vehicles (1976)	UL	UL 1075
	Liquid-Fuel-Burning Heating Appliances for Mobile Homes and Recreational Vehicles (1990)	UL	ANSI/UL 307A-1989
	Roof Jacks for Manufactured Homes and Recreational Vehicles (1990)	UL	ANSI/UL 311-1985
	Roof Trusses for Mobile Homes (1990)	UL	UL 1298
	Shear Resistance Tests for Ceiling Boards for Mobile Homes (1980)	UL	UL 1296
Motors and Generators	Steam Generating Units	ASME	ANSI/ASME PTC 4.1-1964 (R 1985)
	Testing and Nuclear Air-Cleaning Systems	ASME	ANSI/ASME N510-1989
	Nuclear Power Plant Air Cleaning Units and Components	ASME	ANSI/ASME N509-1989
	Energy Efficiency Test Methods for Three-Phase Induction Motors/Efficiency Quoting Method and Permissible Efficiency Tolerance	CSA	C390-1985
	Motors and Generators	CSA	C22.2 No. 100-M1985
	Guide for the Development of Metric Standards for Motors and Generators	NEMA	NEMA 10407-1980
	Motion/Position Control Motors and Controls	NEMA	NEMA MG 7-1987
	Motors and Generators	NEMA	NEMA MG 1-1987
	Renewal Parts for Motors and Generators—Performance, Selection, and Maintenance	NEMA	NEMA MG 7-1987
	Electric Motors (1989)	UL	ANSI/UL 1004-1988
	Electric Motors and Generators for Use in Hazardous (Classified) Locations (1989)	UL	ANSI/UL 674-1984
	Impedance Protected Motors (1982)	UL	ANSI/UL 519-1989
Outlets and Inlets	Method of Testing for Rating the Air Flow Performance of Outlets and Inlets	ASHRAE	ANSI/ASHRAE 70-1991
Pipe, Tubing, and Fittings	Power Piping	ASME	ANSI/ASME B31.1-1989
	Refrigeration Piping	ASME	ASME/ANSI B31.5-1987
	Scheme for the Identification of Piping Systems	ASME	ANSI/ASME A13.1-1981 (R 1985)
	Specification for Acrylonitrile-Butadiene-Styrene (ABS) Plastic Pipe, Schedules 40 and 80	ASTM	ASTM D1527-89
	Specification for Polyethylene (PE) Plastic Pipe, Schedule 40	ASTM	ASTM D2104-90
	Specification for Polyvinyl Chloride (PVC) Plastic Pipe, Schedules 40, 80, and 120	ASTM	ASTM D1785-91
	Specification for Seamless Copper Pipe, Standard Sizes	ASTM	ASTM B42-89
	Specification for Seamless Copper Tube for Air Conditioning and Refrigeration Field Service	ASTM	ASTM B280-88
	Specification for Welded Copper and Copper Alloy Tube for Air Conditioning and Refrigeration Service	ASTM	ASTM B640-90
	Standards of the Expansion Joint Manufacturers Association, Inc., 5th ed. (1980 with 1985 addenda)	EJMA	EJMA
	Corrugated Polyolefin Coilable Plastic Utilities Duct	NEMA	NEMA TC 5-1990
	Corrugated Polyvinyl-Chloride (PVC) Coilable Plastic Utilities Duct	NEMA	NEMA TC 12-1990

Table 1   Codes and Standards Published by Various Societies and Associations (*Continued*)

Subject	Title	Publisher	Reference
Pipe, Tubing, and Fittings (continued)	Electrical Nonmetallic Tubing (ENT)	NEMA	NEMA TC 13-1990
	Electrical Plastic Tubing (EPT) and Conduit Schedule EPC-40 and EPC-80	NEMA	NEMA TC 2-1990
	Extra-Strength PVC Plastic Utilities Duct for Underground Installation	NEMA	NEMA TC 8-1990
	Filament Wound Reinforced Thermosetting Resin Conduit and Fittings	NEMA	NEMA TC 14-1984 (R 1986)
	Fittings, Cast Metal Boxes, and Conduit Bodies for Conduit and Cable Assemblies	NEMA	NEMA FB 1-1988
	Fittings for ABS and PVC Plastic Utilities Duct for Underground Installation	NEMA	NEMA TC 9-1990
	Polyvinyl-Chloride (PVC) Externally Coated Galvanized Rigid Steel Conduit and Intermediate Metal Conduit	NEMA	NEMA RN 1-1989
	PVC and ABS Plastic Utilities Duct for Underground Installations	NEMA	NEMA TC 6-1990
	Smooth Wall Coilable Polyethylene Electrical Plastic Duct	NEMA	NEMA TC 7-1990
	National Fuel Gas Code	NFPA/AGA/ ANSI	ANSI/NFPA 54-1988/ANSI Z223.1-1988
	Plastics Piping Components and Related Materials	NSF	NSF-14
	Refrigeration Tube Fittings (1977)	SAE	ANSI/SAE J513 OCT77
	Rubber Gasketed Fittings for Fire Protection Service (1978)	UL	UL 213
	Tube Fittings for Flammable and Combustible Fluids, Refrigeration Service and Marine Use (1978)	UL	UL 109
Plumbing	BOCA National Plumbing Code, 8th ed. (1990)	BOCA	BOCA
	Uniform Plumbing Code (1988)	IAPMO	IAPMO
	Standard Plumbing Code (1991)	SBCCI	SBCCI
Pumps	Centrifugal Pumps	ASME	ASME PTC 8.2-1965
	Displacement Compressors, Vacuum Pumps and Blowers	ASME	ANSI/ASME PTC 9-1974 (R 1985)
	Liquid Pumps	CSA	CAN/CSA C.22.2 No. 108-1989
	Performance Standard for Liquid Ring Vacuum Pumps, 1st ed. (1987)	HEI	HEI
	Hydraulic Institute Engineering Data Book, 2nd ed. (1991)	HI	HI
	Hydraulic Institute Standards, 14th ed. (1983)	HI	HI
	Circulation System Components for Swimming Pools, Spas, or Hot Tubs	NSF	NSF-50
	Electric Swimming Pool Pumps, Filters and Chlorinators (1986)	UL	ANSI/UL 1081-1985
	Motor-Operated Water Pumps (1980)	UL	ANSI/UL 778-1983
	Pumps for Oil-Burning Appliances (1986)	UL	ANSI/UL 343-1985
Radiation	Ratings for Baseboard and Fin-Tube Radiation (1989)	HYDI	IBR
	Testing and Rating Code for Baseboard Radiation, 6th ed. (1981)	HYDI	IBR
	Testing and Rating Code for Finned-Tube Commercial Radiation (1966)	HYDI	IBR
Receivers	Refrigerant Liquid Receivers	ARI	ANSI/ARI 495-85
Refrigerant-Containing Components	Refrigerant-Containing Components for Use in Electrical Equipment	CSA	C22.2 No. 140.3-M1987
	Refrigerant-Containing Components and Accessories, Non-Electrical (1986)	UL	ANSI/UL 207-1986
Refrigerants	Specifications for Fluorocarbon Refrigerants	ARI	ANSI/ARI 700-88
	Methods of Testing Discharge Line Refrigerant-Oil Separators	ASHRAE	ANSI/ASHRAE 69-1991
	Number Designation and Safety Classification of Refrigerants	ASHRAE	ANSI/ASHRAE 34-1989
	Reducing Emission of Fully Halogenated Chlorofluorocarbon (CFC) Refrigerants in Refrigeration and Air-Conditioning Equipment and Applications	ASHRAE	ASHRAE Guideline 3-1990
	Refrigeration Oil Description	ASHRAE	ANSI/ASHRAE 99-1981 (RA 87)
	Sealed Glass Tube Method to Test the Chemical Stability of Material for Use Within Refrigerant Systems	ASHRAE	ANSI/ASHRAE 97-1983 (RA 89)
	Refrigerant Recovery/Recycling Equipment (1989)	UL	UL 1963
Refrigeration	Capacity Measurement of Field Erected Compression-Type Refrigeration and Air-Conditioning Systems	ASHRAE	ANSI/ASHRAE 83-1985
	Safety Code for Mechanical Refrigeration	ASHRAE	ANSI/ASHRAE 15-1989
	Commercial Refrigerated Equipment	CSA	C22.2 No. 120-1974 (R 1981)
	Equipment, Design and Installation of Ammonia Mechanical Refrigeration Systems	IIAR	ANSI/IIAR 2-1984
	Refrigerated Medical Equipment (1978)	UL	UL 416
Refrigeration Systems Steam Jet	Ejectors	ASME	ASME PTC 24-1976 (R 1982)
	Standards for Steam Jet Vacuum Systems, 4th ed. (1988)	HEI	HEI
Transport	Mechanical Transport Refrigeration Units	ARI	ARI 1110-83
	Mechanical Refrigeration Installations on Shipboard	ASHRAE	ANSI/ASHRAE 26-1978 (RA 85)
	General Requirements for Application of Vapor Cycle Refrigeration Systems for Aircraft (1973) (reaffirmed 1983)	SAE	SAE ARP731A
	Safety Practices for Mechanical Vapor Compression Refrigeration Equipment or Systems Used to Cool Passenger Compartment of Motor Vehicles (1987)	SAE	SAE J639 JAN87

Table 1   Codes and Standards Published by Various Societies and Associations (*Continued*)

Subject	Title	Publisher	Reference
Refrigerators	Method of Testing Open Refrigerators for Food Stores	ASHRAE	ANSI/ASHRAE 72-1983
	Methods of Testing Self-Service Closed Refrigerators for Food Stores	ASHRAE	ANSI/ASHRAE 117-1986
Commercial	Food Carts	NSF	NSF 59
	Food Service Equipment	NSF	NSF-2
	Food Service Refrigerators and Storage Freezers	NSF	NSF 7
	Soda Fountain and Luncheonette Equipment	NSF	NSF 1
	Commercial Refrigerators and Freezers (1985)	UL	ANSI/UL 471-1984
	Refrigerating Units (1989)	UL	ANSI/UL 427-1989
	Refrigeration Unit Coolers (1980)	UL	ANSI/UL 412-1984
Household	Refrigerators Using Gas Fuel (with 1984 and 1989 addenda)	AGA	ANSI Z21.19-1983
	Household Refrigerators and Household Freezers	AHAM	AHAM HRF 1-1988
	Capacity Measurement and Test Methods for Household Refrigerators, Combination Refrigerator-Freezers and Household Freezers	CSA	CAN/CSA C300-M89
	Household Refrigerators and Freezers (1983)	UL	ANSI/UL 250-1984
Roof Ventilators	Commercial Low Pressure, Low Velocity Duct Systems	ACCA	ACCA Manual Q
	Power Ventilators (1984)	UL	ANSI/UL 705-1984
Solar Equipment	Method of Measuring Solar-Optical Properties of Materials	ASHRAE	ANSI/ASHRAE 74-1988
	Method of Testing to Determine the Thermal Performance of Flat-Plate Solar Collectors Containing Boiling Liquid Materials	ASHRAE	ANSI/ASHRAE 109-1986 (RA 90)
	Method of Testing to Determine the Thermal Performance of Solar Collectors	ASHRAE	ANSI/ASHRAE 93-1986 (RA 91)
	Methods of Testing to Determine the Thermal Performance of Solar Domestic Water Heating Systems	ASHRAE	ASHRAE 95-1981 (RA 87)
	Methods of Testing to Determine the Thermal Performance of Unglazed Flat-Plate Liquid-Type Solar Collectors	ASHRAE	ANSI/ASHRAE 96-1980 (RA 90)
Solenoid Valves	Solenoid Valves for Liquid Flow Use with Volatile Refrigerants and Water	ARI	ARI 760-87
Sound Measurement	Measurement of Sound from Boiler Units, Bottom-Supported Shop or Field Erected, 3rd ed.	ABMA	ABMA
	Methods for Calculating Fan Sound Ratings from Laboratory Test Data	AMCA	AMCA Standard 301-90
	Procedure for Fans	AMCA	AMCA 330-86
	Reverberant Room Method for Sound Testing of Fans	AMCA	AMCA 300-85
	Application of Sound Rated Outdoor Unitary Equipment	ARI	ARI 275-84
	Method of Measuring Machinery Sound within Equipment Rooms	ARI	ARI 575-87
	Method of Measuring Sound and Vibration of Refrigerant Compressors	ARI	ANSI/ARI 530-89
	Rating the Sound Levels and Transmission Loss of Package Terminal Equipment	ARI	ANSI/ARI 300-88
	Sound Rating of Large Outdoor Refrigerating and Air-Conditioning Equipment	ARI	ARI 370-86
	Sound Rating of Non-Ducted Indoor Air-Conditioning Equipment	ARI	ARI 350-86
	Sound Rating of Outdoor Unitary Equipment	ARI	ARI 270-84
	Guidelines for the Use of Sound Power Standards and for the Preparation of Noise Test Codes	ASA	ASA 10; ANSI S12.30-1990
	Method for the Calibration of Microphones (reaffirmed 1986)	ASA	ANSI S1.10-1966 (R 1986)
	Specification for Sound Level Meters (reaffirmed 1986)	ASA	ANSI S1.4-1983; ANSI S1.4A-1985
	Laboratory Method of Testing In-Duct Sound Power Measurement	ASHRAE	ANSI/ASHRAE 68-1986
	Measurement of Industrial Sound	ASME	ASME/ANSI PTC 36-1985
	Procedural Standards for Measuring Sound and Vibration	NEBB	NEBB-1977
	Sound and Vibration in Environmental Systems	NEBB	NEBB-1977
	Sound Level Prediction for Installed Rotating Electrical Machines	NEMA	NEMA MG 3-1974 (R 1984)
Space Heaters	Electric Air Heaters	CSA	C22.2 No. 46-M1988
	Electric Air Heaters (1980)	UL	ANSI/UL 1025-1980
Symbols	Graphic Electrical Symbols for Air-Conditioning and Refrigeration Equipment	ARI	ARI 130-88
	Graphic Symbols for Electrical and Electronic Diagrams	ASME	ANSI/IEEE 315-1975
	Graphic Symbols for Heating, Ventilating, and Air Conditioning	ASME	ANSI/ASME Y32.2.4-1949 (R 1984)
	Graphic Symbols for Pipe Fittings, Valves and Piping	ASME	ANSI/ASME Y32.2.3-1949 (R 1953)
	Graphic Symbols for Plumbing Fixtures for Diagrams used in Architecture Building Construction	ASME	ANSI Y32.4-1949 (R 1984)
	Symbols for Mechanical and Acoustical Elements as used in Schematic Diagrams	ASME	ANSI/ASME Y32.18-1972 (R 1985)
Testing and Balancing	Site, Performance Test Standard-Power Plant and Industrial Fans	AMCA	AMCA 803-87
	Procedural Standards for Certified Testing of Cleanrooms (1988)	NEBB	NEBB-1988
	Procedural Standards for Testing, Adjusting, Balancing of Environmental Systems, 4th ed. (1983)	NEBB	NEBB-1991
	HVAC Systems—Testing, Adjusting and Balancing (1983)	SMACNA	SMACNA

Table 1   Codes and Standards Published by Various Societies and Associations (*Continued*)

Subject	Title	Publisher	Reference
Terminals, Wiring	Quick Connect Terminals	NEMA	NEMA DC 2-1982 (R 1988)
	Equipment Wiring Terminals for Use with Aluminum and/or Copper Conductors (1988)	UL	ANSI/UL 486E-1987
Thermal Storage	Commissioning of HVAC Systems	ASHRAE	ASHRAE Guideline 1-1989
	Metering and Testing Active Sensible Thermal Energy Storage Devices Based on Thermal Performance	ASHRAE	ANSI/ASHRAE 94.3-1986 (RA 90)
	Method of Testing Active Latent Heat Storage Devices Based on Thermal Performance	ASHRAE	ANSI/ASHRAE 94.1-1985 (RA 91)
	Methods of Testing Thermal Storage Devices with Electrical Input and Thermal Output Based on Thermal Performance	ASHRAE	ANSI/ASHRAE 94.2-1981 (RA 89)
	Practices for Measurement, Testing and Balancing of Building Heating, Ventilation, Air-Conditioning, and Refrigeration Systems	ASHRAE	ANSI/ASHRAE 111-1988
Turbines	Land Based Steam Turbine Generator Sets	NEMA	NEMA SM 24-1985
	Steam Turbines for Mechanical Drive Service	NEMA	NEMA SM23-1985
Unit Heaters	Gas Unit Heaters	AGA	ANSI Z83.8-1989
	Oil-Fired Unit Heaters (1988)	UL	ANSI/UL 731-1987
Valves	Automatic Gas Valves for Gas Appliances (with 1989 addenda)	AGA	ANSI Z21.21-1987
	Manually Operated Gas Valves for Appliances, Appliance Connection Valves, and Hose End Valves	AGA	ANSI Z21.15-1989
	Relief Valves and Automatic Gas Shutoff Devices for Hot Water Supply Systems	AGA	ANSI Z21.22-1986
	Refrigerant Access Valves and Hose Connectors	ARI	ANSI/ARI 720-88
	Refrigerant Pressure Regulating Valves	ARI	ARI 770-84
	Solenoid Valves for Use with Volatile Refrigerants and Water	ARI	ARI 760-87
	Methods of Testing Nonelectric, Nonpneumatic Thermostatic Radiator Valves	ASHRAE	ANSI/ASHRAE 102-1983 (RA 89)
	Face-to-Face and End-to-End Dimensions of Ferrous Valves	ASME	ASME/ANSI B16.10-1986
	Large Metallic Valves for Gas Distribution (Manually Operated, NPS-2 1/2 to 12, 125 psig Maximum)	ASME	ANSI/ASME B16.38-1985
	Manually Operated Metallic Gas Valves for Use in Gas Piping Systems up to 125 psig (Sizes 1/2 through 2)	ASME	ANSI B16.33-1981
	Manually Operated Thermoplastic Gas Shutoffs and Valves in Gas Distribution Systems	ASME	ANSI/ASME B16.40-1985
	Safety and Relief Valves	ASME	ANSI/ASME PTC25.3-1988
	Valves—Flanged and Buttwelding End	ASME	ANSI/ASME B16.34-1989
	Electrically Operated Valves (1982)	UL	ANSI-UL 429-1988
	Pressure Regulating Valves for LP-Gas (1985)	UL	ANSI/UL 144-1985
	Safety Relief Valves for Anhydrous Ammonia and LP-Gas (1984)	UL	UL 132
	Valves for Anhydrous Ammonia and LP-Gas (Other than Safety Relief) (1980)	UL	UL 125
	Valves for Flammable Fluids (1980)	UL	UL 842
Vending Machines	Methods of Testing Pre-Mix and Post-Mix Soft Drink Vending and Dispensing Equipment	ASHRAE	ASHRAE 91-1976 (RA 91)
	Coin-Operated Vending Machines	CSA	C22.2 No. 128-1963 (RA 89)
	Vending Machines for Food and Beverages	NSF	NSF-25
	Refrigerated Vending Machines (1989)	UL	ANSI/UL 541-1988
Vent Dampers	Automatic Vent Damper Devices for Use with Gas-Fired Appliances	AGA	ANSI Z21.66-1988
	Vent or Chimney Connector Dampers for Oil-Fired Appliances (1988)	UL	ANSI/UL 17-1988
Venting	Draft Hoods (with 1983 addenda)	AGA	ANSI Z21.12-1981
	National Fuel Gas Code	AGA	ANSI Z223.1-1988
	Chimneys, Fireplaces, Vents and Solid Fuel Burning Appliances	NFPA	ANSI/NFPA 211-1988
	Explosion Prevention Systems	NFPA	ANSI/NFPA 69-1986
	Guide for Steel Stack Design and Construction (1983)	SMACNA	SMACNA
	Draft Equipment (1973)	UL	UL 378
	Gas Vents (1986)	UL	ANSI/UL 441-1985
	Type L Low-Temperature Venting Systems (1986)	UL	ANSI/UL 641-1985
Ventilation	Commercial Low Pressure, Low Velocity Duct Systems	ACCA	ACCA Manual Q
	Industrial Ventilation	ACGIH	ACGIH
	Method of Testing for Room Air Diffusion	ASHRAE	ANSI/ASHRAE 113-1990
	Ventilation for Acceptable Indoor Air Quality	ASHRAE	ANSI/ASHRAE 62-1989
	Residential Mechanical Ventilation Requirements	CSA	F326.1-M1989
	Residential Mechanical Ventilation System Requirements	CSA	F326.2-M1989
	Verification of the Performance of Residential Mechanical Ventilation Systems	CSA	F326.3-M1990
	Ventilation Directory	NCSBCS	NCSBCS

Table 1   Codes and Standards Published by Various Societies and Associations (*Concluded*)

Subject	Title	Publisher	Reference
Ventilation (continued)	Parking Structures; Repair Garages	NFPA	ANSI/NFPA 88A-1985; 88B-1985
	Removal of Smoke and Grease-Laden Vapors from Commercial Cooking Equipment	NFPA	ANSI/NFPA 96-1987
	Food Service Equipment	NSF	NSF-2
Water Heaters	Gas Water Heaters, Vol. I, Storage Water Heaters with Input Ratings of 75,000 Btu per Hour or Less	AGA	ANSI Z21.10.1-1990
	Gas Water Heaters, Vol. III, Storage, with Input Ratings Above 75,000 Btu per Hour, Circulating and Instantaneous Water Heaters	AGA	ANSI Z21.10.3-1990
	Methods of Testing to Determine the Thermal Performance of Solar Domestic Water Heating Systems	ASHRAE	ANSI/ASHRAE 95-1981 (RA 87)
	Construction and Test of Electric Storage Tank Water Heaters	CSA	C22.2 No. 110 M-1981
	CSA Standards on Performance of Electric Storage Tank Water Heaters	CSA	C191-Series-M90
	Oil Burning Stoves and Water Heaters	CSA	B140.3-1962 (R 1980)
	Oil-Fired Service Water Heaters and Swimming Pool Heaters	CSA	B140.12-1976
	Systems Construction and Test of Electric Storage-Tank Water Heaters	CSA	CAN/CSA C22.2 No. 110-M90
	Hot Water Generating and Heat Recovery Equipment	NSF	NSF-5
	Commercial-Industrial Gas Heating Equipment (1973)	UL	UL 795
	Electric Booster and Commercial Storage Tank Water Heaters (1988)	UL	ANSI/UL-1987
	Household Electric Storage Tank Water Heaters (1989)	UL	ANSI/UL 174-1989
	Oil-Fired Storage Tank Water Heaters (1988)	UL	ANSI/UL 732-1987
Woodburning Appliances	Method of Testing for Performance Rating of Woodburning Appliances	ASHRAE	ANSI/ASHRAE 106-1984
	Installation Code for Solid Fuel Burning Appliances and Equipment	CSA	CAN/CSA-B365-M87
	Solid-Fuel-Fired Central Heating Appliances	CSA	CAN/CSA-B366.1-M87
	Space Heaters for Use with Solid Fuels	CSA	B366.2-M1984
	Chimneys, Fireplaces, Vents and Solid Fuel Burning Appliances	NFPA	ANSI/NFPA 211-1988
	Commercial Cooking and Hot Food Storage Equipment	NSF	NSF-4
	Solid Fuel Type Room Heaters (1988)	UL	ANSI/UL 1482-1988

ABBREVIATIONS AND ADDRESSES

ABMA	American Boiler Manufacturers Association, Suite 160, 950 N. Glebe Road, Arlington, VA 22203
ACCA	Air Conditioning Contractors of America, 1513 16th Street, NW, Washington, DC 20036
ACGIH	American Conference of Governmental Industrial Hygienists, 6500 Glenway Avenue, Building D-7, Cincinnati, OH 45211
ADC	Air Diffusion Council, 230 N. Michigan Avenue, Suite 1200, Chicago, IL 60601
AGA	American Gas Association, 1515 Wilson Boulevard, Arlington, VA 22209
AHAM	Association of Home Appliance Manufacturers, 20 N. Wacker Drive, Chicago, IL 60606
AIHA	American Industrial Hygiene Association, 345 White Pond Drive, Akron, OH 44320
AMCA	Air Movement and Control Association, Inc., 30 W. University Drive, Arlington Heights, IL 60004
ANSI	American National Standards Institute, 11 West 42nd Street, New York, NY 10036
ARI	Air-Conditioning and Refrigeration Institute, 1501 Wilson Boulevard, 6th Floor, Arlington, VA 22209-2403
ASA	Acoustical Society of America, 335 E. 45 Street, New York, NY 10017-3483
ASHRAE	American Society of Heating, Refrigerating and Air-Conditioning Engineers, Inc., 1791 Tullie Circle, NE, Atlanta, GA 30329
ASME	The American Society of Mechanical Engineers, 345 E. 47 Street, New York, NY 10017
	For ordering publications: ASME Marketing Department, Box 2350, Fairfield, NJ 07007-2350
ASTM	American Society for Testing and Materials, 1916 Race Street, Philadelphia, PA 19103
BOCA	Building Officials and Code Administrators International, Inc., 4051 W. Flossmoor Road, Country Club Hills, IL 60478-5795
BSI	British Standards Institution, 2 Park Street, London, W1A 2BS, England
CABO	Council of American Building Officials, 5203 Leesburg Pike, Suite 708, Falls Church, VA 22041
CAGI	Compressed Air and Gas Institute, Suite 1230, Keith Building, 1621 Euclid Avenue, Cleveland, OH 44115
CSA	Canadian Standards Association, 178 Rexdale Boulevard, Rexdale, Ontario M9W 1R3, Canada
CTI	Cooling Tower Institute, P.O. Box 73383, Houston, TX 77273
EJMA	Expansion Joint Manufacturers Association, Inc., 25 N. Broadway, Tarrytown, NY 10591
HEI	Heat Exchange Institute, Suite 1230, Keith Building, 1621 Euclid Avenue, Cleveland, OH 44115
HI	Hydraulic Institute, 30200 Detroit Rd., Cleveland, OH 44145-1967
HYDI	Hydronics Institute, 35 Russo Place, Berkeley Heights, NJ 07922
IAPMO	International Association of Plumbing and Mechanical Officials, 20001 Walnut Drive South, Walnut, CA 91789
ICBO	International Conference of Building Officials, 5360 S. Workman Mill Road, Whittier, CA 90601
IFCI	International Fire Code Institute, 5360 S. Workman Mill Road, Whittier, CA 90601
IIAR	International Institute of Ammonia Refrigeration, 111 East Wacker Drive, Chicago, IL 60601
MICA	Midwest Insulation Contractors Association, 2017 South 139th Circle, Omaha, NE 68144
NCSBCS	National Conference of States on Building Codes and Standards, 505 Huntmar Park Dr., Suite 210, Herndon, VA 22070
NEBB	National Environmental Balancing Bureau, 1385 Piccard Drive, Rockville, MD 20850
NEMA	National Electrical Manufacturers Association, 2101 L Street, NW, Suite 300, Washington, D.C.20037
NFPA	National Fire Protection Association, 1 Batterymarch Park, Quincy, MA 02269-9101
NSF	National Sanitation Foundation, Box 1468, Ann Arbor, MI 48106
SAE	Society of Automotive Engineers, 400 Commonwealth Drive, Warrendale, PA 15096
SBCCI	Southern Building Code Congress International, Inc., 900 Montclair Road, Birmingham, AL 35213-1206
SMACNA	Sheet Metal and Air Conditioning Contractors' National Association, 4201 Lafayette Center Drive, Chantilly, VA 22021
TEMA	Tubular Exchanger Manufacturers Association, Inc., 25 N. Broadway, Tarrytown, NY 10591
UL	Underwriters Laboratories Inc., 333 Pfingsten Road, Northbrook, IL 60062-2096

ADDITIONS AND CORRECTIONS

This section supplements the current handbooks and notes technical errors found in the series. Occasional typographical errors and nonstandard symbol labels will be corrected in future volumes. The authors and editor encourage you to notify them if you find other technical errors. Please send corrections to: Handbook Editor, ASHRAE, 1791 Tullie Circle NE, Atlanta, GA 30329.

1989 Fundamentals

p. 14.12. Equation (18) should read:

$$D_{min} = M[3.16 + (0.1S/\sqrt{A_e})] \qquad 18$$

p. 15.9, Table 10. Change values for the theoretical combustion air required for liquid fuels to read:

Type of Fuel	Theoretical Air Required for Combustion
Liquid fuels	lb/gal
No. 1 fuel oil	103
No. 2 fuel oil	106
No. 5 fuel oil	112
No. 6 fuel oil	114

p. 22.3, Table 2. Footnote e should read:

e Effective emittance E of the airspace is calculated as $1/E = 1/\epsilon_1 + 1/\epsilon_2 - 1$, where ϵ_1 and ϵ_2 are emittances of the surfaces facing the air space (see Table 3).

p. 23.2. Equation (2) should read:

$$q_l = 60 \, \rho Q h_{fg} \Delta W \qquad (2)$$

p. 26.37, Table 32. The CLTD correction for SE/SW walls and roofs at 24° latitude, Jan/Nov, should read 3, not 9.

p. 26.37, Table 32. The CLTD correction for SSE/SSW walls and roofs at 24° latitude, Jan/Nov, should read 9, not 3.

p. 26.37, Table 32. The last five CLTD correction factors for walls and roofs at 32° latitude, Jan/Nov, should read as follows:

ESE WSW	SE SW	SSE SSW	S	HOR
−4	2	9	12	−15

1991 HVAC Applications

p. 3.5, 1st column, 11th line down. Delete 35 kW and insert 10 tons.

p. 3.5, 1st column, 12th line up. Replace text to read:

… air face velocity is at least 75 fpm. Face velocities of 75 to 100 fpm should be used for design, with 60 fpm as a minimum.

p. 3.5, 1st column, 7th line up. Delete 230 to 310 L/(s · m) and insert 150 to 200 cfm per linear foot.

p. 3.5, 2nd column, 4th line down. Delete 16°C and insert 60°F.

p. 3.5, 2nd column. In the paragraph Hood Exhaust System Design, delete 9 to 11 m/s and add 1800 to 2200 fpm.

p. 19.2, Figure 1. In the process block for carding, the range of humidity for cotton should read 50-55%.

p. 21.9, 1st column, 5th line up. The equation for heating infiltration air should read:

$$q_i = 0.018 \, VN(t_i - t_o)$$

p. 21.9, 2nd column, 3rd line. Change (Table 1) to read (Table 3).

p. 29.2, 1st column. In the second line below the TEMPERATURES heading, thermal energy content varies from about 40 to 680°F, not 60 to 680°F.

p. 33.6, 2nd column. Replace one i' with i_d in the equation for depreciation (for commercial systems) so that the equation reads:

$$T_{inc} \sum_{k=1}^{n} [D_k \mathrm{PWF}(i_d,k)]\mathrm{CRF}(i',n) \quad \begin{array}{l}\text{depreciation}\\ \text{(for commercial systems)}\end{array}$$

p. 33.7, 2nd column. Replace one i' with i_d in the equation below the heading Depreciation so that the equation reads:

$$T_{inc} \sum_{k=1}^{n} [D_{k,SL}\mathrm{PWF}(i_d,k)]\mathrm{CRF}(i',n) \ldots$$

p. 39.2, 1st column, 8th line up. Revise sentence to read:

Globe and plug valves are best suited for this duty because they can handle large pressure drops and they have excellent close-off characteristics.

p. 39.3, 2nd column, 17th line up. Delete term "on-hour" so that the line reads:

… used during the on-peak period (1320 ton·h) is multiplied…

p. 39.3, 2nd column, 4th line up. Change example to read:

$$85 \text{ ton} \times 1.2 \text{ kW/ton} = 102 \text{ kW}$$

p. 39.6, 1st column. Change second sentence of second paragraph to read:

To minimize mixing, diffuser dimensions should be selected to create an inlet Froude number equal to or less than 2.0…

p. 39.6, 2nd column. In the section Storage Tank Insulation, change the last sentence of the first paragraph to read:

To date, however, it has not been viable to insulate buried tanks, except on the exposed walls and within partitions.

p. 39.6, 2nd column. In the section Storage Tank Insulation, delete the last sentence of the second paragraph that begins, "Therefore, when the tank walls are made…"

p. 39.6, 2nd column. In the section Storage Tank Insulation, change the second sentence of the third paragraph to read:

Either exterior or interior insulation may be used to inhibit heat transfer to the surroundings.

p. 39.8, Figure 9. Add (Courtesy: Cryogel) to figure caption.

p. 39.8, 2nd column. In the section Ice Harvesting Systems, delete the last sentence of the first paragraph that begins, "This control prevents operation during periods..."

p. 39.8, 2nd column. In the section Ice Harvesting Systems, change the first sentence of the second paragraph to read:

When the temperature of the water entering the evaporator drops to a temperature at which ice forms on the plates, the hot gas defrost cycle is activated.

p. 39.9, 2nd column. In the first sentence of the first paragraph of the section Other Phase Change Materials (PCMs), change eutectic salts to certain salts.

p. 39.9, 2nd column. In the fourth line of the first paragraph of the section Other Phase Change Materials (PCMs), change "eutectics, which..." to read "salts that..."

p. 39.9, 2nd column. In the fifth line of the first paragraph of the section Other Phase Change Materials (PCMs), change eutectic to read PCM.

p. 39.9, 2nd column. Change the fourth and fifth line up to read:

...PCMs to be particularly appropriate for retrofit applications. Like ice storage systems, PCM systems rely on a phase change.

p. 39.10, 1st column. In the fourth line of the second paragraph, delete the word "eutectic."

p. 39.10, 1st column. In the third line of the third paragraph, change "...eutectic salts are..." to read, "...PCM is..."

p. 39.11, 1st column, line 1. Change 35 psi to 50 psi.

p. 39.11, 1st column, line 2. Change sentence to read, "The ASME Boiler Code considers such vessels unfired pressure vessels."

p. 39.14, 2nd column. In the seventh line down, delete the sentence that begins, "If energy-efficient evaporative condensing..."

p. 42.2, 1st column. In the third line of the second paragraph under the heading NC Curves, delete the word "not." The sentence should read, "...noise spectrum is assigned..."

p. 42.5, Table 2. The alternate criteria for legitimate theaters should read NC 20-25.

p. 42.11, Table 9. Change P/A ratios to read Over 0.31, 0.31 to 0.13, Under 0.13.

p. 42.11. Delete Table 10.

p. 42.11, Figure 12. The figure labels do not coordinate with Equation (5).

p. 42.15, 1st column. Replace the attenuation of duct fittings section with the following:

Attenuation of Duct Fittings

Both lined or unlined elbows attenuate sound. The amount of attenuation is a function of (1) duct size, (2) whether acoustical lining is placed before and/or after the elbow, and (3) whether or not turning vanes are provided. Tables 16 and 17 give insertion losses for square and round elbows. In Tables 16 and 17, $fw = f \times w$.

where:

f = frequency, kHz
w = duct width or depth for square duct or diameter for round duct, in.

Table 18 lists fw for various size ducts at various frequencies.

Table 16 Insertion Loss of Unlined and Lined Square Elbows with and without Turning Vanes

	Insertion Loss, dB	
	Unlined	Lined
Without Turning Vanes		
$fw < 1.9$	0	0
$1.9 < fw < 3.8$	1	1
$3.8 < fw < 7.5$	5	6
$7.5 < fw < 15$	8	11
$15 < fw < 30$	4	10
$fw > 30$	3	10
With Turning Vanes		
$fw < 1.9$	0	0
$1.9 < fw < 3.8$	1	1
$3.8 < fw < 7.5$	4	4
$7.5 < fw < 15$	6	7
$fw > 15$	4	7

Table 17 Insertion Loss of Unlined Round Elbows

	Insertion Loss, dB
$fw < 1.9$	0
$1.9 < fw < 3.8$	1
$3.8 < fw < 7.5$	2
$fw > 7.5$	3

Table 18 Values of fw

Duct Width or Dia., in.	Octave Band Frequencies, Hz						
	63	125	250	500	1000	2000	4000
5	—	—	1.25	2.5	5	10	20
10	—	1.25	2.5	5	10	20	40
20	1.25	2.5	5	10	20	40	80
40	2.5	5	10	20	40	80	160

p. 42.15, 2nd column. Equation (9) should read:

$$\Delta L = 10 \log[1 + (c/\pi f D)^{1.88}] \qquad (9)$$

p. 42.15, 2nd column. Equation (10) should read:

$$\Delta L = 10 \log[1 + (0.8c/\pi f D)^{1.88}] \qquad (10)$$

p. 42.15, 2nd column. In Equations (9) and (10), the effective diameter of a rectangular duct is $D = (4A/\pi)^{0.5}$, where A is the cross-sectional area of the duct in square feet.

p. 42.21, Table 24. Change Note 1 to read:

1. The losses listed in this table are based on laboratory tests. Actual installations have losses up to 5 dB lower.

COMPOSITE INDEX

ASHRAE HANDBOOK SERIES

This index covers the current Handbook volumes published by ASHRAE. Listings from each volume are identified as follows:

S = 1992 Systems and Equipment
A = 1991 HVAC Applications
R = 1990 Refrigeration
F = 1989 Fundamentals
E = 1988 Equipment

The index is alphabetized in a *word-by-word* format; for example, *air diffusers* is listed before *aircraft* and *heat flow* is listed before *heaters*.

Note that the code for a volume includes the chapter number followed by a decimal point and the page number(s) within the chapter. For example, F32.4 means the information may be found in the Fundamentals volume, chapter 32, page 4.